AF324098

ADVANCES IN INDUCTION AND MICROWAVE HEATING OF MINERAL AND ORGANIC MATERIALS

Edited by **Stanisław Grundas**

Advances in Induction and Microwave Heating of Mineral and Organic Materials

Edited by Stanisław Grundas

CBS, Edition **2016**

Published by InTech

Janeza Trdine 9, 51000 Rijeka, Croatia

Publishing Process Manager Ana Nikolic

Technical Editor Teodora Smiljanic

Cover Designer InTech Design Team

Image Copyright redfrisbee, 2010. Used under license from Shutterstock.com

Additional hard copies can be obtained from orders@intechopen.com

Advances in Induction and Microwave Heating of Mineral and Organic Materials,
Edited by Stanisław Grundas
p. cm.
ISBN 978-953-307-522-8

Contents

Preface

The induction and the microwave are physical phenomena that allow targeted heating of an applicable item for use in many fields of industry.

The book offers comprehensive coverage of the broad range of scientific knowledge in the fields of advances in induction and microwave heating of mineral and organic materials. Beginning with industry application in many areas of practical application to mineral materials and ending with raw materials of agriculture origin the authors, specialists in different scientific area, present their results in two parts: Part 1 – Induction and Microwave Heating of Mineral Materials, and Part 2 – Microwave Heating of Organic Materials.

The book is divided into 31 chapters, some of which are concentrated in Part 1 – devoted to mineral materials (Chapters 1-21), while those in Part 2 are focused on organic origin (Chapters 22-31).

In Part 1 the research and development achievements in induction and microwave (MW) heating in metallurgy, chemical industry, electro-technical, and biomedical applications are presented. At the beginning, recent studies on the following viewpoints at MW as a new technology are presented: MW heating of materials in separated E- and H-fields, fundamentals and smart application of thermal runway phenomena, and realization, interpretation and application of the long- debated subject of MW non-thermal effects (Chapter 1). In the next chapter (Chapter 2), a review of all kinds of application backgrounds of numerical simulation of MW heating process is given, followed with a discussion of the two most important techniques to help further the understanding of the complex MW heating process. Chapter 3 contains interesting information on the d-q modelling technique, proposed and applied in the case of a voltage inverter with LLC resonant load. The d-q model is embedded in a close-loop inverter model with voltage and frequency control. Further chapters (Chapter 4 and 5) are devoted to numerical modelling of industrial induction and describe the several involved physical phenomena corresponding with mathematical models, and numerical methods for obtaining their solution (Chapter 4), but in chapter 5 the authors present a guide for the design and control of induction heating system which has the following contents: mathematical modelling of the heating coil and load; heating coil design using finite element magnetic software, mathematical modelling of the control loop system for diverse output topologies and different control types, and control loop tuning using numerical computational tools. In Chapter 6 the authors present the way from given process parameters at the beginning to customized solution at the end at Huettinger. The chapter shows that it is advantageous for the development of customized solutions in induction heating, and there is a good cooperation between the technical know-how in the field of generator development and long-term application experience. Chapter 7 is devoted to the development of methods for setting optimal configurations for putting together inductor and work-piece while the highest electromagnetic coupling is achieved. Efforts to make the application of electricity for mass heating more efficient would make it more appealing for the industry, since nowadays environmental issues are becoming increasingly stringent, mainly those related to Global Warming, thus approaches to reduce greenhouse gas evolving are always welcome. In Chapter 8, the most popular methods for accurate power control in induction heating systems are discussed in the case of non-linearity of material properties. Some examples are shown and the classical approach to power control in the system is compared to the pulse width modulation case. The advantages and disadvantages of proposed design and process solutions are also discussed. In the next two sections (Chapters 9 and 10) the transverse flux induction heating is presented. In first one (Chapter 9) two different novel types – transverse flux induction heating (TFIH) and travelling wave induction heating (TWIH) are considered. Also, the novel crossed travelling wave induction heating (C-TWIH) system for heating thin industrial strips is reported, together with its finite element method (FEM) simulations, and the travelling wave induction heating system with distributed windings and magnetic slot wedges (SW-TWIH) is proposed to address the inhomogeneous eddy current density problem which dominates the surface thermal distribution of work strips. In the second one (Chapter 10) continual induction heating of some thin non-ferrous metal strips by means of the TFIH system are discussed. In Chapter 11, MW processing of metallic glass/polymer composite is discussed. This chapter is devoted to the development of metallic glass/polymer composites by MW processing of the blend powders in a separated magnetic field. Next, Chapter 12 is devoted to some cases of MW processing of materials, considering the conditions that could explain the "microwave effect". Discussion in this chapter is centred on ceramics where it has been demonstrated that, under certain conditions, MW can heat up these materials. In Chapter 13, historical aspects and physical principles of induction evaporation are shortly described. The further part of this chapter is devoted to devices and technology of induction evaporation for Physical Vapour Deposition of thin films and coatings. At the end of the chapter the relation to modelling of induction evaporation and mass transfer from crucible to substrate as well as to simulation of coating thickness distribution on the substrate surface is described. In Chapter 14 the potential of MW heating to recover valuable elements from industrial waste is suggested, and several practical applications are described. Chapter 15 is devoted to research processes of emolliating and fracture of rocks and other materials by their rapid MW heating. The dependence of the processes of rocks emolliating and fracture on some physical and dielectric performances is discussed, and questions linked with kimberlite fracture are considered as well. Chapter 16 describes how an induction heating system in plastic injection moulding is designed. The use of numerical simulation in order to get optimum design of the induction coil is shown. In addition to that, the need for an induction heating power supply is pointed out. In Chapter 17 the physical principles of MW irradiation in relation to coordination and organometallic compounds are discussed. A comparison of various modern physicochemical techniques with MW-irradiation method is made, with critical analysis of its advantages and disadvantages. With intention of better understanding of the effects of MW heating on the kinetics of chemical reactions and physicochemical processes, a detailed analysis of numerous chemical reactions like crystallisation of zeolite type NaA, synthesis of fullerene in the presence phase-transfer catalyst, acrylic acid polymerisation crosslinking, hydrolyses of sucrose and physicochemical processes such as ethanol adsorption from water solution onto zeolite type CMS and dehydration of poly(acrylic-co-methacrylic acid) based hydrogels, is performed in chapter 18. Chapter 19 is devoted to the use MW heating in semiconductor processing, more specifically for post ion-implantation annealing. In this chapter the relevance of MW heating in semiconductor device processing is described. In Chapter 20 the use of MW energy in dental prosthesis is described. Following themes are discussed in it: technical procedure, such as appropriated water/powder ratio, proper manipulation, and enclosing, finishing and polishing, and control of polymerisation temperature; polymerisation reaction activated by heat; MW energy used as a heat source for acrylic resin; particular measures adopted to reduce dimensional changes in resin processed by MW energy; and MW energy proposed as a method for disinfecting dentures and in the treatment of denture stomatitis. The last Chapter (21) of this group is devoted to the magnetic induction heating behaviours of ferrite nonmaterials and fluid in following aspects: basic knowledge of magnetic induction heating in ferrite nano-materials; the effect of magnetic field and frequency on heating efficiency and speed; and magnetic fluid hyperthermia behaviours.

In the Section 2 of chapters (22-31) the advances in microwave heating in agriculture and food industry are presented. This section is devoted mainly to the widely known field of research and development in agrophysics, although from physics only MW phenomenon is presented. Generally, agrophysics is focused on physical properties and processes affecting biomass production and processing. In the first Chapter of this section (22) the direct and indirect effects on the next generation crops of MW heating of wheat grain on physicochemical, protein biological activities, rheological, technological and insect resistance propertied as well as microflora contamination of grain are discussed. Chapter 23 proposes providing a general revision of the application of MW radiation energy to process cereal, tuber and root-based products. For example: the use of MW for enzyme inactivation in brown rice, drying of parboiled rice, thawing frozen bread dough, heating bread, producing hard semi-sweet cookies, drying pasta, expanding extruded pellets, producing type III resistant starch, among others, is discussed. Chapter 24 will identify the key factors, such as the applied power, exposure time, geometry of the MW applicator, and the complex dielectric constant and thermal properties of materials which determine the rate of MW energy absorption and heat distribution inside the heated material. In the next two chapters (25 and 26) the changes in the olive oil's antioxidant content during the MW heating process are discussed in comparison to those observed during conventional heating. As has been

demonstrated, the degradation of olive oil carotenoid content can be precisely monitored in–line using Raman spectroscopy (25), but in the next one (26), MW heating is
discussed as a time-saving technology and a way to accelerate oxidative reaction in
vegetable oils. In the next chapter (27), the main ideas in MW freeze drying (MWFD)
modelling are presented, and the key mechanisms of heat and mass transport governing the process are explained. In Chapter 28, the MW synthesis of porous molecular
sieve membranes is summarized, including recent development and progress in MW
synthesis of porous molecular sieve membranes: heating modes together with MW
heating and conventional heating, differences between MW synthesis and conventional hydrothermal synthesis for membrane formation, and formation mechanism and
specific MW effect in the case of MW synthesis of porous molecular sieve membranes.
Chapter 29 summarizes recent developments in MW-assisted domino reaction for the
construction of four-, five-, six-, and seven-membered small molecular skeletons and
their multicyclic derivatives. In Chapter 30 the authors introduce the applicability of
MW technology for the utilisation of recalcitrant biomass as pioneers in this field. In
this chapter the following issues are described: an introduction of MW including characteristics of MW and development of continuous flow MW irradiation apparatus spe-
cialised for biomass utilisation, characteristics of chemical components and composite
structures of woody biomass, monocotyledonous lignified biomass, agricultural and
marine biomass, a summary of MW technology of the three categories of biomass mentioned above, and the advantages of MW irradiation over other hydrothermal treatments using the conductive heating process. And, in Chapter 31 the MW pyrolysis as
an original thermochemical process of materials is presented. This chapter comprises a
general overview of the thermochemical and quantifying aspects of the pyrolysis process, including current application together with a compilation of the most frequently
used materials.

Prof. Dr. Eng. Stanisław Grundas
Bohdan Dobrzanski Institute of Agrophysics,
Polish Academy of Sciences, Lublin, Poland

e-mail: grundas@ipan.lublin.pl
www.ipan.lublin.pl

Part 1

Induction and Microwave Heating of Mineral Materials

Recent Studies on Fundamentals and Application of Microwave Processing of Materials

Noboru Yoshikawa
Graduate School of Environmental Studies,
Department of Materials Science and Engineering, Tohoku University
6-6-02, aza-Aoba, Aramaki, Aoba-ku,Sendai
Japan

1. Introduction

Microwave (MW) application for heating was discovered in 1946 and has been applied in various fields. The detailed historical tracing is presented in the next section. Among the various aplications, it is possible to classify them into two major classes (Agrawal, 2005). In the first class, MW heating applied to drying, cooking foods and to excitation of chemical reactions such as inorganic/organic synthesis. The MW activated chemical reactions has been investigated in the field categorized as "MW Chemistry". In these applications, the heating temperature is usually low, because most of the aqueous and organic liquids have the boiling point below 500ºC. Mainly, the heating is caused by a dielectric relaxation loss due to rotation of molecules, which is as the result of the interaction of the MW electric field with the electric dipole of the molecules. Electric dipoles in liquid or gas experience relatively free rotational motion, comparing with the dipoles in a solid.

On the other hand, for the MW heating application to sintering of metal/ceramics, solid state reaction, solid state phase transition (such as vitrification or devitrification) and high temperature reduction reaction, elevated temperature is needed, which often exceeds above 500ºC. In these cases, heating mechanism is not limited to dielectric loss but to the other mechanisms of ohmic loss due to eddy current, and the magnetic loss also becomes important. Most of the oxides become electric conductive above 1000ºC (Kingery, et al., 1975. From these considerations, understanding the effect of MW magnetic field interaction with materials becomes important (Roy, et al. 2002, Yoshikawa et al. 2006). As will be given in this paper, the magnetic field not only influences the magnetic heating mechanism directly, but also raises the induction current more effectively than in the electric field.

As the special characteristics of MW heating, three different heating aspects have been pointed out, namely they are the internal heating, the rapid heating and the selective heating (Bykov, et al., 2001). It is possible to consider the industrial application of MW heating from these aspects. Moreover, it has been reported that the so-called "non-thermal effect" of MW heating exists and enhances the sintering and the reaction kinetics. Although the origin of the non-thermal effect has not necessarily been clarified in the present stage, its phenomena keep providing us motivation to the reseraches on understanding and application of the MW heating.

This INTEK book chapter is intended to introduce recent researches perfomed in the authors' group especially by virture of the project supported by Japanese Ministry of Education, Sports and Culture (MEXT), to be mentioned later. This paper starts with the historical tracing of MW heating researches. And it is aimed at concentrating the following viewpoints as the new microwave technologies:

1. Fundamentals in MW heating of materials in consideration of separated heating mechanisms in E- and H- field and the static H- field imposition.
2. Fundamentals and smart application of thermal runaway phenomena
3. Realization, interpretation and application of the long debating subject of MW non-thermal effects.

Concerning the above issues, our attempts of MW heating application have been conducted for the materials processing and the environmental technologies, which are simultaneously performed with the fundamental studies on the MW heating mechanisms together with some simulation studies.

2. Historical

Microwave (MW) has been utilized for tele-communication before and during the World War II, such as radar, sensing and so on. In 1946, Percy Spencer found a candy bar in his pocket was melting during his experiments on the MW generation tube, this is the discovery of MW heating (URL, 2010). Since then, MW heating has been applied to various fields (Clark, Sutton, 1996, Katz, 1992). At first, it was used in food production areas, such as drying of potato, roasting of coffee beans. It was in '70s, when the MW oven has become popular in domestic kitchens for cooking, as mass production of magnetron became posible at this time. And it was also in '70s when the oil shock or natural gas crisis occurred, which promoted the researches on MW heating in the western countries, because of their political necessity to be dependent on electric heating methods (Katz, 1992, Oda. 1992)

Application of MW heating to the energy and the environmental fields has also been attempted and realized. Their application extends to broad areas (Oda, 1992), such as incineration of medical wastes, devulcanization of rubber tire, treatment of sewage sludges, regeneration of spent activated carbon and chemical residues of petrol industries. In the nuclear engineering, MW de-nitration for producing MOX (Mixed OXide) nuclear fuels by re-disposal of the used plutonium (Kato et al. 2005). Vitrification of nuclear wastes by MW heating had been proposed in '80 (Morita et al. 1992).

In the area of materials processing, MW application to polymer has started earlier. There are bunch of research reports, curing of thermosets (Boey et al. 1992) is one of the examples. During MW drying of Al_2O_3 castable, it was recognized that not only well dried but also they can be heated well by MW (Sutton et al., 1988). MW heating of the ceramic powders above 1400°C made it possible to sinter ceramics (Janney and Kimrey, 1988). Later, it was reported that MW sintering enhances the diffusion rate, and so-called non-thermal effect (Wroe and Rowley, 1996) has been pointed out, though its origin has been a long debate. In this field, milli-wave techniques have been developed and applied to new processing of ceramics (Clark, Sutton, 1996).

On the other hand, metal heating was a minor application area of mirowave. And thus, the studies on MW heating of metals had not been developed comparing with the other materials as mentioned above. It is known that a bulk metal reflects MW, however, the metal particles and films can be heated well. And it is also known that ferro-magnetic metals can be heated more (Walkievicz et al., 1988), indicating that the magnetism is related with the heating mechanism.

Microwave metal heating researches are reviewed next. In 1988, one of an interesting experimental result is reported by Walkievics (Walkievicz et al., 1988), who tested heating of various metal powders, and demonstrated there are differences in their heating rates. In 1991-92, some reports related with heating of metals (cermets, composites) were presented in MRS symposium (Lorenson et al., 1991, Besher 1992). In '95, Mingos attempted synthesis of metal sulfide by MW heating of metals (Whittaker and Mingos, 1995). In '99, Roy et al. (Roy et al., 1999) reported microwave full sintering of metals in Nature magazine. And later, they attempted series of studies on heating of metals in the separated Electric (E-) and magnetic (H-) MW fields (Chen et al., 2002, Roy et al. 2002). In Europe, metal heating studies also have been performed (Rodiger et al.1998, Leonelli et al. 2008). These reports and activities motivated the MW researchers to be further directed to the metal heating studies. In Japan, the authors held a special symposium on metal heating in annual meeting of Japan Institute of Metals in 2005. The author also published a review article (Yoshikawa 2009(a)) in Bulletin of Japan Institute of Metals (Materia, Japan 2009, (written in Japanese)), the application fields are classified and the authors' recent results were introduced. Some content will be also presented in this article. The separated E- and H- heating (though not only metal heating) are directly related with the basic principles of microwave interaction with materials and of the microwave heating mechanisms.

In 2006, a project called "Grant in aid for priority field area" under support of Japanese Ministry of Education and Science was adopted and the intensive research activities are being conducted under the title of "Science and Technology of Microwave Induced, Thermally Non-Equilibrium Reaction Field". The author is one of the group leaders and promotes the area of "Microwave Application to the Mateials' Processing and Environmental Technologies".

Along with these activities, Japan Institute for ElectroMagnetic Energy Application (JEMEA) was established in 2006 and held a congress in 2008 together with the other international institutes of AMPERE, IMPI, MWG.

3. Fundamentals

he energy of electromagnetic wave is expessed in terms of Poynting vector (E x H). Taking its divergence, and substituting the Maxwell's equations Eq. 1, 2 and 3. Eq. 4 is rearranged into Eq. 5, where $\mathbf{E}$, $\mathbf{H}$, $\mathbf{B}$, $\mathbf{D}$, $\mathbf{J}$ are elctric and magnetic field, magnetic flux density, electric flux density and conduction current, respectively. and $\mathbf{B}$, $\mathbf{D}$, $\mathbf{J}$ are related with $\mathbf{E}$ and $\mathbf{H}$ according to Eq. 3, where ε, μ and σ are permittivity, magnetic permeability and electric conductivity, respectively.

$$\nabla \times \mathbf{E} = -\frac{\partial \mathbf{B}}{\partial t} \tag{1}$$

$$\nabla \times \mathbf{H} = \mathbf{J} + \frac{\partial \mathbf{D}}{\partial t} \tag{2}$$

$$\mathbf{D} = \varepsilon \mathbf{E}, \quad \mathbf{B} = \mu \mathbf{H}, \quad \mathbf{J} = \sigma \mathbf{E} \tag{3}$$

$$\nabla\left(\mathbf{E}\times\mathbf{H}\right) = \mathbf{H}\cdot(\nabla\times\mathbf{E}) - \mathbf{E}\cdot(\nabla\times\mathbf{H}) \tag{4}$$

$$= -\frac{\partial}{\partial t}\left(\frac{\varepsilon\mathbf{E}^2}{2} + \frac{\mu\mathbf{H}^2}{2}\right) - \sigma\mathbf{E}^2 \tag{5}$$

Eq.5 indicates that electromagnetic energy is stored in the matters as the first (electric field energy) and second (magnetic energy) terms. And the work done by the heating due to electric conduction current is expressed as the third term. The energy loss in the matters is evaluated by the first and the second terms using the imaginary part of the permittivity and the magnetic permeability, respectively. The Joule heating (induction heating) loss is expressed by the third term. They are corresponding to the dielectric, magnetic and conduction loss of the microwave heating.

The Maxwell equations (Eq.1,2,3) are converted to the exactly same differential equations with respect to the E- and H-fields for propagation of an EM wave, as follows:

$$\nabla^2\binom{\mathbf{H}}{\mathbf{E}} - \mu\sigma\frac{\partial\binom{\mathbf{H}}{\mathbf{E}}}{\partial t} - \mu\varepsilon\frac{\partial^2\binom{\mathbf{H}}{\mathbf{E}}}{\partial t^2} = 0 \tag{6}$$

In solving the equation, the time dependence of **H** (or **E**) is assumed using exponential form, as Eq.7,

$$\binom{\mathbf{H}}{\mathbf{E}} = \binom{\mathbf{H}'}{\mathbf{E}'}e^{-i\omega t} \tag{7}$$

Then, the differential equation Eq.(6) has the same solution with respect to **H** and **E**, as shown in one dimensional form in Eq.(8).

$$\binom{\mathrm{H}'}{\mathrm{E}'} = \binom{\mathrm{H}_0'}{\mathrm{E}_0'}e^{-\alpha\cdot z}\cdot e^{-i(\omega\cdot t-\beta\cdot z)} \tag{8}$$

Here, α is a characteriastic length of H (or E) field attenuation, and $1/\alpha$ is the length of the skin depth layer (d). α is expressed in Eq.(9).

$$\alpha = \frac{\omega\sqrt{\mu\varepsilon}}{\sqrt{2}}\left[\sqrt{1+\left(\frac{\sigma}{\omega\varepsilon}\right)^2}-1\right]^{1/2} \tag{9}$$

The EM field in metal must also obey Eq.(5), so α is also evaluated as Eq.(9). This indicates not only H-field but also E-field exists within the skin layer having the same thickness. In this treatment, generally, a relation $\sigma \gg \omega\varepsilon$ is assumed for the metal case (Kraus ans Carver, 1973). Therefore ε is canceled in Eq.(9) and a can be expressed as Eq.(10).

$$\alpha = \sqrt{\frac{\omega\mu\sigma}{2}} \tag{10}$$

And for the dielectric materials, Eq.9 is expresed, using a relation $\sigma = \omega\varepsilon''$

$$\alpha = \omega\sqrt{\left(\frac{\mu_0\varepsilon'}{2}\right)\left(\sqrt{1+\left(\frac{\varepsilon'}{\varepsilon''}\right)^2}-1\right)}$$

(11)

3.1 Dielectric heating

In considering the MW heating, it is required to estimate the penetration distance of MW into materials. The penetration distance of dielectrics depends mainly on the imaginary part (ε''), so long as $\varepsilon' >> \varepsilon''$ according to Eq. 9, this inequality relation holds in the most cases. On the other hand, the real part (ε') influences the wave length within the matter. These relationships will be presented based on our recent studies.

In the case of dielectric heating, it is possible to estimate the penetration depth using Eq. 9 and the heat (P) generated per unit time and unit volume, according to Eq. 12.

$$P = \frac{1}{2}\omega\varepsilon''|E|^2$$

(12)

In order to evaluate them, permittivity values (ε' and ε'') are required to be measured at the MW frequency. In our group, the measurement was performed using cylindrical coaxial line method connected to vector network analyzer (VNA), as the schematic illustration of the apparatus shown in Fig. 1.

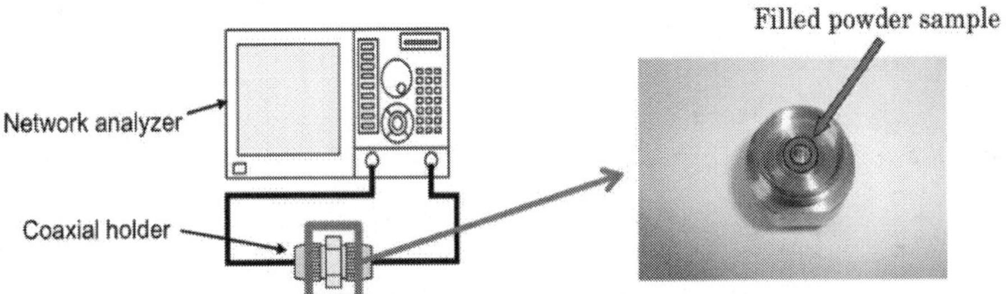

Fig. 1. Schematic illustration of apparatus for permittivity measurement.

Fist of all, the ε' effect on the wave length in a matter is demonstrated. In this case, not using the measured values by VNA but by the electromagnetic field simulation.

The simulation of the wave length of MW at 2.45GHz in a dielectric composite material with variation of ε' was performed. The composite material consists of spherical inclusion particles (perovskite) with large permittivity embedded in a matrix (spinel) with small permittivity. The inclusions were arranged in a configuration of face-centered cubic. The analyzed electric fields in the composite materials with inclusion having different particle diameters, corresponding to different average permittivity (ε') are shown in Fig. 2. The analysis demonstrates the local E-field distributions around the particles and the contours corresponding to the wave length. It can be seen that the wave length became shorter as an increase of average ε'.

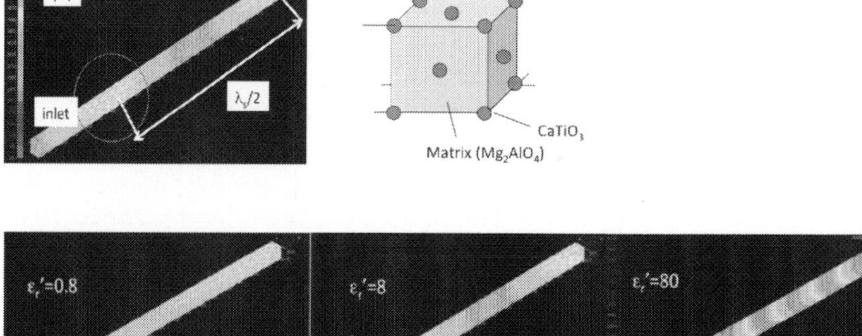

Fig. 2. Simulated electric field in a body consisting of 30 cells of ceramic composite. A cell shown upper right has dimension of 1mm in cell length. The lower three images demonstrate the electric field distributions in the bodies consisting of inclusions with various size, corresponding to different average permittivity (ε_r').

The permittivity measurement was performed for the mixture of FeOOH and graphite powders. This mixture was selected as the test example for visualization of the microwave penetration distance, because FeOOH undergoes the following reaction by heating above 200°C.

$$2FeOOH = Fe_2O_3 + H_2O$$

The color changes from dark brown (FeOOH) to brown red (Fe_2O_3) upon heating. Here, FeOOH alone is not absorbing MW and not heated well, but graphite addition makes it possible (Iwasaki et al. 2009). It was observed that penetration distance decreased as the increase of graphite fraction and the degree of compression of the powder mixture. Photographs in Fig. 3 demonstrate the color changed areas where the MW penetration occurred and heating caused the above reaction. Decrease in the MW penetration distance at the larger graphite fraction is related with the permittivity (ε''). The relationship is plotted in Fig. 4.

Fig. 3. Photo images of MW heated FeOOH/C mixture.

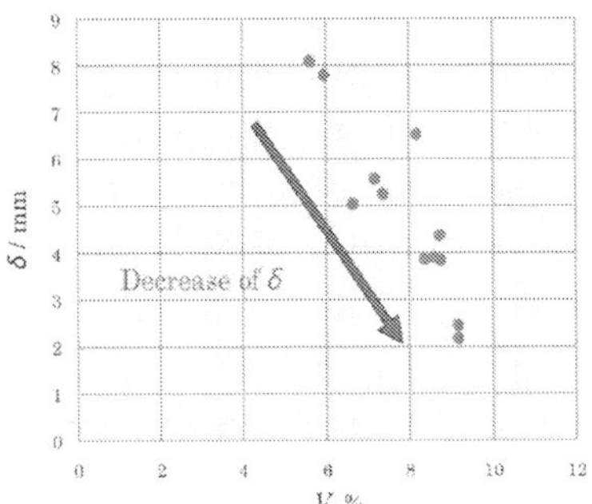

Fig. 4. Relationship between the C volume fraction and the estimated penetration distance.

3.2 Induction (Metal) heating

Because of small (order of micron or less) penetration depth of electromagnetic field into metals, temperature of a bulk metal cannot be raised very much, as mentioned above, therefore, MW heating is generally limited to metal particles or films. For example for Au case, according to Eq. 10, inputting σ =4.6x10^7 [S/m] , ω = 2πx2.45x10^9 [1/s], μ_r (relative magnetic permeability) = 1 [-], the distance is estimated to be 1.5 [μm].

Ferro-magnetic metals are well heated by MW irradiation. Magnetic mechanisms are considered to be responsible, and our recent attempts to elucidating this mechanism will be presented in the next section (3.3).

Heating of non-ferromagnetic metals in electromagnetic field is primarily understood that the induced eddy current on the metal surface is responsible for the Joule heating. Or otherwise, occurrence of arcing accompanies the discharge current coming onto the metals surface, and also the Joule heating is brought about. The energy path generating the electric current and the current intensity are different between them. In this section, the eddy current (or induction current) shall be mainly concerned.

In discussion of the metal heating which is considered to occur only on the surface, it is required to discuss the electromagnetic boundary conditions on metal surface. Especially in this section it is intended to introduce the concept of surface impedance. MW heating of metals has to be analyzed with consideration of electromagnetic (EM) field above the metal surface, which must be linked with the EM field inside the metal surface or in the skin layer. It is usually understood that the MW H- field generates the induction (eddy) current, according to the boundary conditions of the electromagnetic field defined on the metal surface. The generally utilized boundary conditions on metal surface (so-called electric wall) assume the perfect conductor (or zero resistance), and is expressed as follows:

$$\mathbf{E}_1 = 0, \quad \mathbf{n} \cdot \left(\mu \mathbf{H}_1\right) = 0 \tag{13}$$

, where $\mathbf{n}$ is a normal vector to the metal surface. and

$$\mathbf{n} \times \mathbf{H}_1 = \mathbf{J} \tag{14}$$

$\mathbf{J}$ is a surface (eddy) current, as the schematic illustration shown in Fig. 5(a). Eq. 13 and 14 are the boundary conditions with respect to E and H fields in the air side on the metal surface, required for solving the EM fields in the given space (such as a wave guide and a cavity, etc). In this treatment, however, metal is idealized and no information about the metal properties is involved.

Induction (eddy) current is generated within the skin depth of the metal surface, which, of course, is due to the generated E- and H- fields in the metal because of the finite electric resistance. In order to describe the relationship between H- and E- field within the skin layer of metals, it is required to handle the boundary conditions of electric conducting (not perfect conductor, ie. having small but finite resistivity) metals. The analysis starting from Maxwell's equation is needed, which were already given above. And the penetration distance into metal is given in Eq. 10.

There are difficulties encountered in this analytical procedure. First, dielectric constant (permittivity: ε) of metal must be evaluated for the purpose of making comparison: $\sigma \gg \omega\varepsilon$. Dielectric function of metals $\varepsilon(\omega)$ can be discussed using Drude model (Bohren and Huffman, 1983), however, this gives large negative value for the real part of ε at MW frequency, which is confirmed by measurement of the optical constants (Johnson and Christy, 1972), however, this discussion is not usually performed in the textbooks of electromagnetics. Generally, in analysis of metals, displacement current $\frac{\partial \mathbf{D}}{\partial t}$ is neglected in Eq. 2 and thus Eq.6 can be rewritten as Eq. 15.

$$\nabla^2 \begin{pmatrix} \mathbf{H} \\ \mathbf{E} \end{pmatrix} - \mu\sigma \frac{\partial \begin{pmatrix} \mathbf{H} \\ \mathbf{E} \end{pmatrix}}{\partial t} = 0 \tag{15}$$

Permittivity measurement of metal powder has been performed and the measured values of ε' and ε'' (using an equation $\sigma = \omega\varepsilon''$) are incorporated into the EM field analysis and to the heating calculations (Mishra et al., 2006, Ma et al., 2007). However, the measured values are the 'effective' permittivity, and they could contain the influence of surface oxides and the air in the powder. The permittivity of metals or dielectric function to be used in solving the Maxwell equation must be related to the electronic states of metals.

Second, as there is an eddy current $\mathbf{J}$ generated by time derivative of $\mathbf{H}$. Here, if we relate $\mathbf{J}$ to $\mathbf{E}$ by Eq.15, this indicates the $\mathbf{E}$ field also exists in the skin layer, and there are relations among $\mathbf{J}$, $\mathbf{E}$ and $\mathbf{H}$. So, the boundary condition Eq.(1) has to be modified to express them.

Landau and Lifshitz in their textbook (Landau et al., 1984) discussed this problem in terms of surface impedance of metals (originally by L.A.Leontovich, 1948). As can be derived from Eq. 9, both E and H fields exist within the surface skin layer δ as $\delta = 1/\alpha$, α is expressed by Eq. 10. On the other hand, the tangential E and H fields in metal surface is related by the following relation (t: tangential, 2: metal medium, as shown in Fig.5(b)) : $E_{2t} = \zeta_s H_{2t}$, where ζ_s is a surface impedance expressed in cgs unit (in Landau and Lifshitz) by

$$\varsigma_s = \sqrt{\frac{\omega\mu}{8\pi\sigma}} \cdot (1-i) \ , \ \text{ or } \ \ \varsigma_s = \sqrt{\frac{\mu\omega}{2\sigma}} \cdot (i-1) \ \ \text{(in MKSA unit)} \tag{16}$$

and its value has an order of $\sim \delta/\lambda$ (λ: wave length), indicating that H field is larger than E field on the metal surface at the microwave frequency ($\delta/\lambda \sim 1\times10^{-4}$). This aspect is also demonstrated in the textbook of J.D.Jackson (Jackson, 1998). Using this quantity, the boundary condition Eq. 13,14 are modified to the below form:

$$n \times E_1 = \zeta_s \, n \times (n \times H_1) \tag{17}$$

This is the boundary condition of E and H fields of the space as well, which considers the metal properties. And again, larger contribution of H-field to metal heating was confirmed.

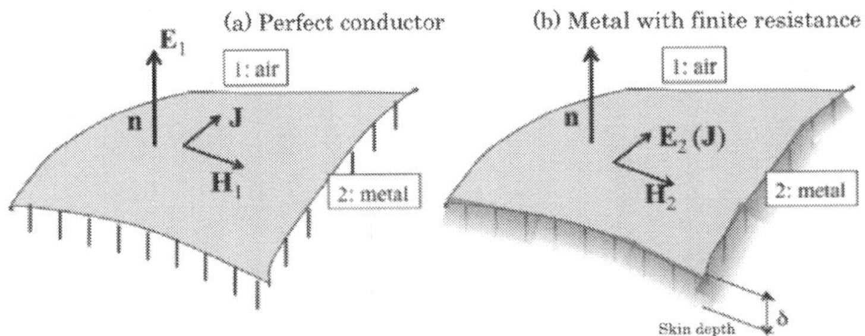

Fig. 5. Schematic illustration of EM boundary conditions on metal surface.

3.3 Magnetic heating

It is commonly recognized among the microwave processing researchers that generally the ferro (ferri)-magnetic materials are well heated by microwave, especially in H-field. And they can be heated better up to the Curie temperature (Ishizaki and Nagata, 2007). These facts indicate the magnetic mechanism plays an important role in microwave heating, however, this mechanism is not necessarily clarified enough, so far. On the other hand, in designing RF devices such as ferrite cores, the RF loss occurring in a range of $10^8 \sim 10^9$ Hz is a detrimental phenomenon and is necessary to be avoided (Pardavi-Horvath,2000). This loss is especially termed as a natural resonance (Polder and Smit, 1953). The natural resonance is related with the internal magnetic fields existing within the ferro (ferri)-magnetic materials, by which the precession movement of electron spin occurs. As the precession frequency corresponds the ranges mentioned above, the resonant absorption of microwave takes place. In general, as there are relatively large distributions of the internal magnetic field, so the resonant frequency also has broad ranges.

Natural resonance takes place without external magnetic field. Here, it is expected possible to raise the electron spin resonance (FMR) in ferro (ferri)-magnetic materials by imposition of static external magnetic field. Namely, if the natural resonance is one of the mechanism responsible for microwave magnetic heating, it might be possible to expect the extra heat generation due FMR and to observe the temperature rise.

FMR has been studied extensively from '50s, and a bunch of reports on the experimental and theories are published. However, most of the studies were performed on the spin-spin relaxation or spin wave dynamics by the physics school (Sparks, 1964). Therefore, the specimen used was mainly an yttrium iron garnet (YIG) having longer spin-spin relaxation time, and not easy to be converted into heat. Thus far, less studies was performed on the application of FMR to heating materials, except a few reports (Nikawa and Okada, 1987 and Walton et al., 1996), but no intimate examination under various FMR conditions were provided. This study (Yoshikawa and Kato, 2010) attempted to observe the temperature change of Fe_3O_4 compressed body by setting the FMR conditions. And it is intended to investigate the FMR heating behavior dependence on various conditions, and then to discuss the magnetic energy dissipation to heat.

Fe_3O_4 powder of 4N in purity, which were compressed uniaxially and formed into a rod having 7mm in diameter and 4~0.5mm in thickness, the density of which are about $1 \times 10^3 kgm^{-3}$. The specimens were placed at the bottom of silica glass rod container. The specimen was placed in a 5.8GHz microwave TE103 cavity at the H-field maximal position and the static magnetic field was imposed perpendicular to the microwave magnetic field.

The specimen temperature was measured by an optical method using a sapphire light guide. N_2 gas was flown from downward onto the specimen surface and no heat insulating materials were set above the specimens.

The specimens were heated up to the pre-determined temperature in microwave H-field without imposition of H^{ext}. Data of the temperature variation with time were digitally recorded, together with that of H^{ext}, the forward and backward reflection power of microwave (Pf and Pr, respectively).

Curves of the temperature variation with time obtained from Fe_3O_4 compressed body of 0.5mm in thickness are obtained plotted in Fig. 6, together with Pf, Pr and H^{ext}. Fe_3O_4 compact was preset at 420°C and kept for couple of ten seconds, H^{ext} was imposed. It was demonstrated that the temperature increased and had a peak (ΔT : the temperature difference from the initial preset value) was observed then decreased (a). The peak appeared at almost same H^{ext} during ascending and descending of H^{ext}, as can be seen clearly by plotting the temperature with respect to H^{ext} (b) indicating that FMR induced the temperature rise (ΔT). It can be seen from the figure that Pr decreased a little around the H^{ext} of the resonance. This is considered to be casused by the energy absorption by magnetic resonance. As mentioned above, ΔT could become larger if heat insulation condition is provided and there is no gas blow.

The resonant H^{ext} is supposed to be influenced by demagnetization field depending on the specimen shape. Namely, Kittel resonance (Kittel, 1948) condition has to be taken into consideration. Without any additional magnetic fields, such as anti-magnetic field, the resonance field is expected to be 0.21T at 5.8GHz. The resonant H^{ext} of about 0.15T is interpreted as follows: The compressed Fe_3O_4 powder is not a continuum body, and the resonant frequency is determined by each particle. The powder particles do not have complete spherical shape, and the demagnetization conditions are deviated from the sphere case (if complete sphere, H^{ext} is not affected by the demagnetization field.).

In this study, dependence of ΔT and the resonant H^{ext} on the pre-heated temperature was examined and the obtained relationships are plotted in Fig. 7(a)(b). These data provide us information how the magnetic mechanism for microwave heating varies with the temperature. It was shown that ΔT decreased as the pre-heated temperature approached to their Curie point. And the resonant H^{ext} shifted to higher. These phenomena are primarily interpreted as the decrease in the saturated magnetization (I_s). The magnetic resonance in ferro-magnetic state may be able to give rise to the observable heat generation (temperature increase) and disappears as approaching to the paramagnetic state.

The FMR is a state that the unpaired spins contributing to ferro-magnetism are precessing altogether around the (external) magnetic field direction in resonant conditions with the applied microwave magnetic field. Namely, they correspond to the spin wave number k=0 (Sparks et al., 1961). The process of the resonant precessing energy converting to heat is different from the other heating mechanisms such as Joule heating of eddy current. The differences are either magnon-phonon interaction or that of electron-phonon. it is expected that FMR heating to become a promising heat treatment method of magnetic materials giving rise of excellent properties.

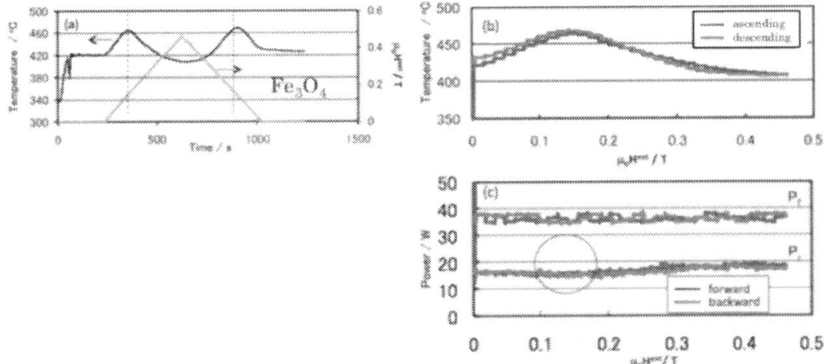

Fig. 6. Temperature of Fe_3O_4 rod and Pf,Pr variations during imposition of external magnetic field (Hext). (a) Variation of temperature and H^{ext} with time. (b) Temperature plotted with respect to H^{ext}.

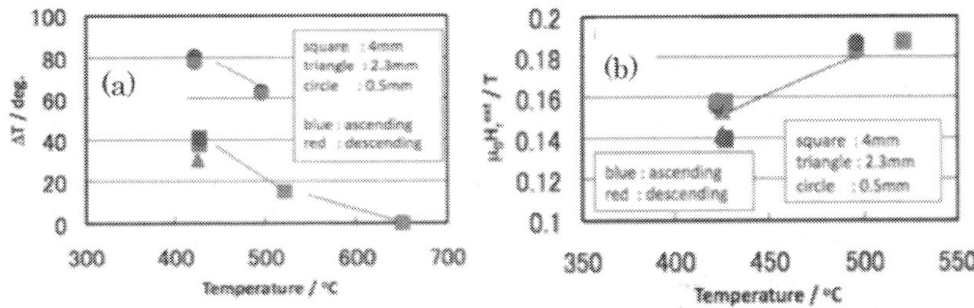

Fig. 7. (a) Temperature increase (ΔT) from the initial state plotted with respect to the pre-heated (initial) temperatures with different thickness. (b) Resonance external magnetic field (H_r^{ext}) plotted with respect to the pre-heated temperatures. (Fe_3O_4 compact) The blue and red marks represent the ΔT, H^{ext} values obtained in ascending and descending of H^{ext}, respectively. Keys in (b) are square; 4mm, triangle; 2.3mm, circle; 0.5mm.

4. Applications

4.1 E-H Separated heating of thin metal films

Recently, separated E- and H- field heating has been performed for the purpose of investigating the heating mechanisms and of discriminating the heating behaviors to fabricate new materials. In these experiments, most commonly observed tendency is that H-field is more effective for heating of metals comparing with the E- field, as first reported by Cheng et al. (Chen et al. 2001) and the authors have confirmed in series of experiments (Yoshikawa et al., 2006).

Heating behaviors of Cu particles by MW irradiation in the separated E- and H- fields was reported (Ma et al., 2007), where it is mentioned that heating rate in H-field is higher, and better heating behavior was observed in the subsequent heating. And it is to be noticed that the heating curve (time-temperature) exhibits peaks at the initial stage of heating. The authors of this report attributed this behavior to be due to the microstructural change such as the particle contacts formation and growth of average particle size, which occurs within the skin depth of MW penetration.

In our recent work on heating of Au films with nano-sized thickness, peak formation behavior was also observed, as shown in Fig. 8 (Cao et al., 2009). In this case, film thickness is less than that of the skin layer. Although there is a difference in materials, it is of interest to compare these behaviors with the Cu cases. It was also confirmed that MW H-field is effective for heating the films (same input power resulted in different temperature). The film grain structure was also altered by MW heating, as the XRD profiles indicated in Fig. 9. The changes in grain structure alter the MW absorbability and resulted in the peak formation. This must also be discussed in terms of the change in surface impedance, namely it is caused by some alteration in interaction between electromagnetic wave and metals. The reason for the better heating characteristics of metal particles and films in H-field is also interpreted as the difference in the eddy current. Further detailed studies are expected to be explored in future. Apart from the fundamentals in the metal film heating, heating of metal thin film has been successfully applied to the crystallization of PZT film in the multi layered structure (Wang et al. 2008).

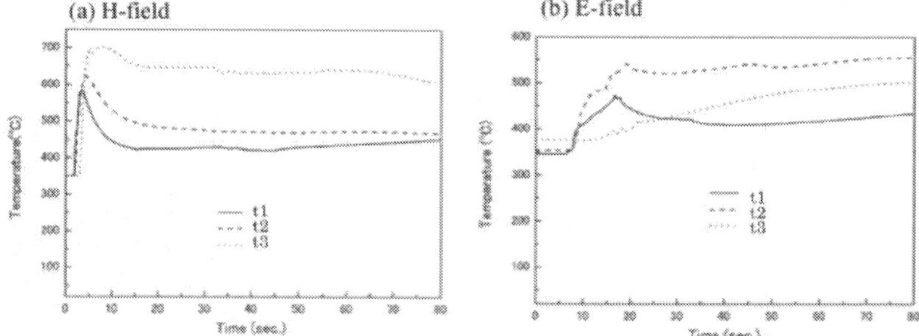

Fig. 8. Heating curves of Au films having various thickness (t1: 45nm, t2: 133nm, t3: 407nm) in H- and E-field (Cao et al., 2008).

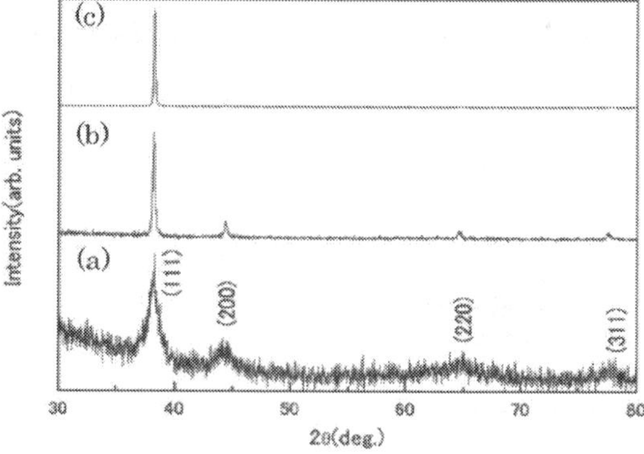

Fig. 9. XRD profiles of Au films (t2) of (a): as deposited, (b) heated in E-field and (c) heated in H-field (Cao et al., 2008).

4.2 Thermal runaway
4.2.1 MW heating behaviour of Fe/Soda-lime (Na_2O-CaO-SiO_2) glass mixture

For the purspose of glass solidification of hazardous wastes, metal pieces are mixed for promoting the MW heating. We fablicated the soda-lime (S-L) glass/metal composite materials (Yoshikawa et al., 2008(a)). It is of significance to investigate the MW heating of glass with additon of metals.

Heating behavior of spherical S-L glass beads having 0.2mm in diameter was examined, first. The heating curves by MW irradiation of constant (maximum) power indicated that the attained maximum temperature was about 200oC after 10 minutes of MW irradiation (not shown here).

Iron powder was heated up to about 700oC rather quickly and saturated within 10 min. This behavior was discussed in relation with the Curie point of iron (770oC) (Yoshikawa et al., 2006), though the maximum temperature attained is influenced by the state of thermal insulation.

Next, mixture of S-L glass beads and iron powder was heated with various volume fraction of iron powder, as shown in Fig. 10. It was observed that the mixture of small volume fraction (2.2 vol.%) of iron powder has essentially same heating behavior as the glass beads alone. On the other hand, the temperature rose quickly within one minute in the mixture of 18 vol%Fe. There is a transition state in the volume fractions of iron, in which temperature rose up suddenly after certain incubation time periods. It can be seen that the temperature rise was so rapid and this might be a phenomenon so called thermal runaway (TRW) (Comm. on Microwave Proces. Mater., 1994, and Kenkre eat al., 1991). The MW power was shut off shortly after the onset of the sudden rise in order to protect the thermocouple. The incubation time decreased as an increase of iron volume fraction. The heated specimens were observed with OM, and the microstructures are illustrated in Fig. 11. It can be seen that the iron powder fills the clearance of the beads spheres (at 800oC), and the glass beads are melted at 950oC. The iron powder was neither melted in this temperature nor reacted with the S-L glass, which was examined with SEM/EDX observation.

Pure silica glass (SiO_2) is not a good MW absorber, however the permittivity (ε'' : loss factor) increases as the temperature, because of increase in population of mobile ion or ion defects (Kenkre et al. 1991), resulting in interaction with the MW electric field. It is also known (Kingery at al., 1975) that this phenomenon is related with the large increase in electric

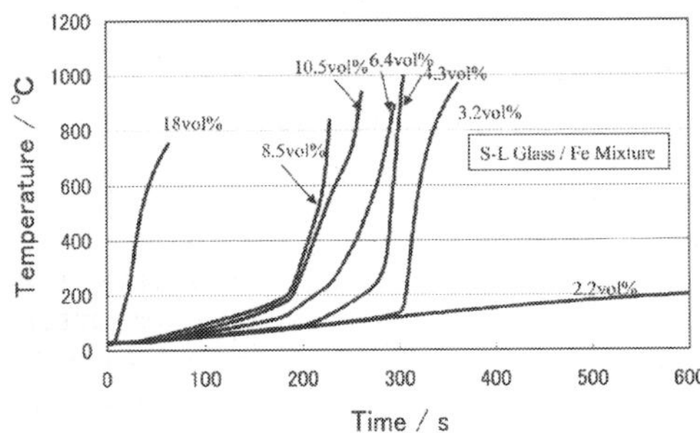

Fig. 10. Heating curves of S-L glass/iron powder mixture with different volume fraction of iron powder.

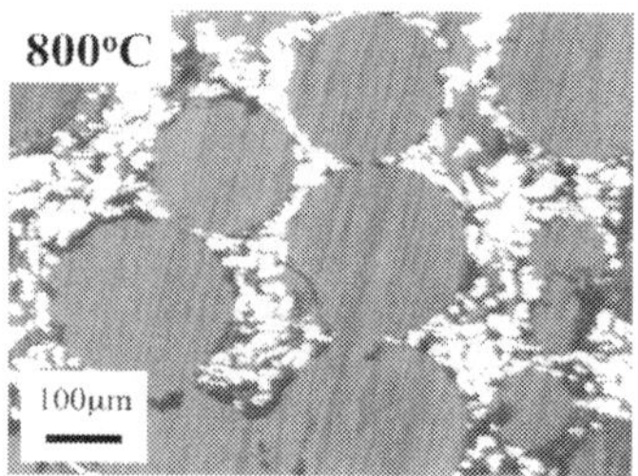

Fig. 11. Optical micrographs of MW-heated S-L glass/iron powderheated at 800°C with 18vol%Fe.

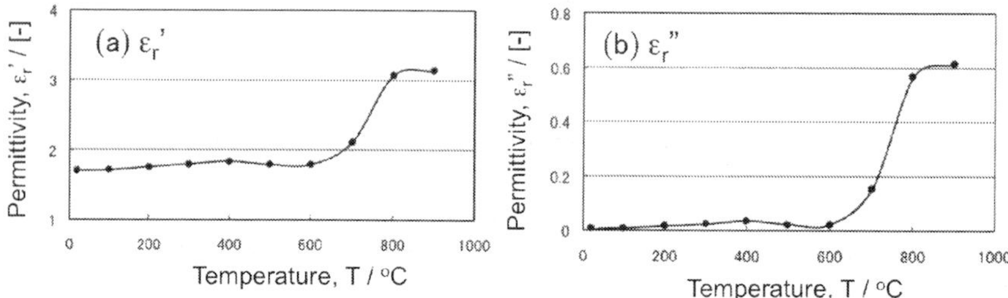

Fig. 12. Temperature dependence of (a) real (ε_r') and (b) imaginary (ε_r'') parts of S-L glass permittivity.

conductivity at high temperature as mentioned. In this study, the permittivity of S-L glass was measured up to the elevated temperature (Yoshikawa et al., 2009(b)), and plotted in Fig. 12. As can be seen from the plot, permittivity (both ε' and ε'') increases mainly above 700°C. Therefore, plausibility of the hypothesis was confirmed, and it became possible to discuss the heating behavior of the mixture of S-L glass and metal powders.

The S-L glasses were not heated well above 200°C within 10minutes in a multimode applicator without addition of iron powder, probably because the loss factor of the glass below 200°C was not large enough to cause the heating. Iron powder is heated well up to about 700°C. This makes it possible to heat the S-L glass beads in the neighbors of metal powder, and increased the loss factor of the glass, leading to the further heating of the glass beads by themselves (glass + 10vol%Fe in Fig. 10). The S-L glass might be heated by MW E-field component, mainly by dielectric heating mechanism (though interaction of the enhanced electric conductivity with the H-field is also of interest.).

Occurrence of thermal runaway (TRW) was dependent on the volume fraction of iron powder. It is of interest how the volume fraction of Fe powder determines the onset of sudden temperature rise.

Although MW penetration depth into metal is of an order of microns or less, so the heat is generated only in this surface skin layer. Heat conduction within the small metal particle is so quick, and thus the particles might be heated quickly. Therefore, it is possible to consider that the aggregates of iron powder are taken to be internally heated by MW, if we assume sufficient penetration of MW into the aggregates. In other words, the aggregates are considered to be a volumetrically heated objects by MW. The glass beads are heated by these metal aggregates. This feature is schematically illustrated in Fig. 13. As an increase of the metal powder fraction, the clearance of glass beads is filled with the larger aggregates of metals particles. If so, they can be heated more than the small aggregates, because the larger volume generates the larger heat with respect to the smaller heat transfer through the surface area. (The heat generated in the aggregate is propotional to the cube of its radius, while the heat transfered from the aggreagte is propotional to the square of ist radius. Therefore, temperature rises the higher as an aggregate has the larger size.)

Occurrence of TRW must be determined due to superposition of the following phenomena: One is the above discussed temperature dependence of the S-L glass permittivity and the other is the size dependence of the heat generated by the metal powder aggregates.

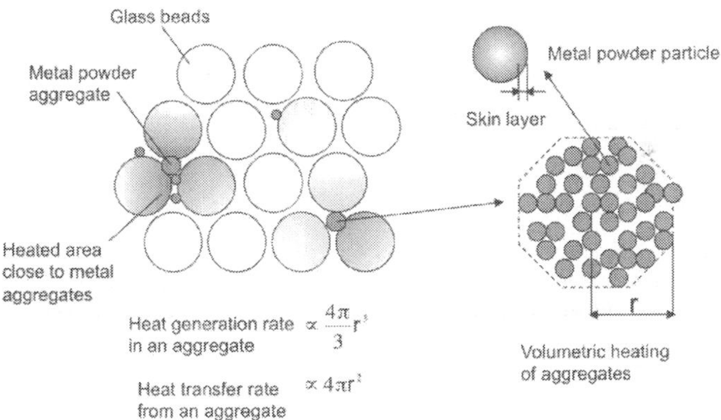

Fig. 13. Schematic illustration of S-L glass beads heating with an aid of metal aggregates.

4.2.2 MW heating for carbo-thermal reduction of Cr_2O_3 (Yoshikawa et al. 2008(b))

In order to recycle Cr metals from the wates occurring in stainless steel production process, such as the pulverized Cr-containing slags and picking sludge, they are reduced with graphite via carbo-thermal reduction reaction. Owing to the large affinity of chromium with oxygen, chromium oxide (Cr_2O_3) reqires high reduction temperature as predicted by thermodynamics. However, because of the specific dependence of MW absorbability of Cr_2O_3, it is possible to reduce the (average) reduction temperature. First of all, the reduction of real wastes are presented.

Without graphite addition, the real sludge and the Cr-containing slag powder were not heated by MW above 200°C. This is because of the fact that fractions of heavy metal oxides are lower with respect to the gangue components such as CaO, SiO_2, and the silicates. Mixture of either slag or sludge with sufficient amount of graphite was successfully heated by MW and the reduced Cr-containing alloys were obtained.

SEM/EDX photographs of the reduced specimens are shown in Fig. 14. The Fig. 14(a) illustrates the Cr-containing slag specimen heated at 1000°C for 10min., with graphite addition of 5wt%. The bright area corresponds to the reduced metal (alloy). Composition of the reduced alloy was determined. The obtained Cr concentration values differed from the different areas, probably because of the original segregation of Cr in the oxide states.

MW heating behavior of Cr_2O_3 was investigated by measuring its heating curves (time-temperature relationship) in a multi-mode applicator and compared with that of graphite and NiO. They were measured under the same input power of 670W (maximum), and the plots are shown in Fig. 15.

Graphite can be heated very well, and temperature rose up to 1000ºC within 2min. On the other hand, NiO and Cr_2O_3 had incubation time before onset of the rapid temperature increase. Although the incubation time periods of Cr_2O_3 is similar to NiO, temperature of NiO saturated at 900ºC, but temperature of Cr_2O_3 still increased. In this case, power was switched off in order to avoid damaging the thermo-couple.

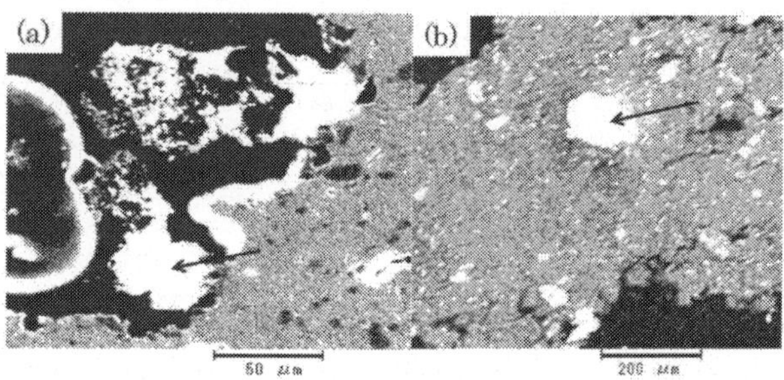

Fig. 14. SEM images of MW-reduced alloys from (a) Cr-containing steel making slag and (b) stainless steel pickling sludge. The bright areas indicated by arrows are the reduced alloys, the EDX-analyzed compositions (at%) are (a) Fe 88.5(bal.), Cr 11.5 and (b) Fe 58.3(bal.), Cr 7.3, Ni 33.9

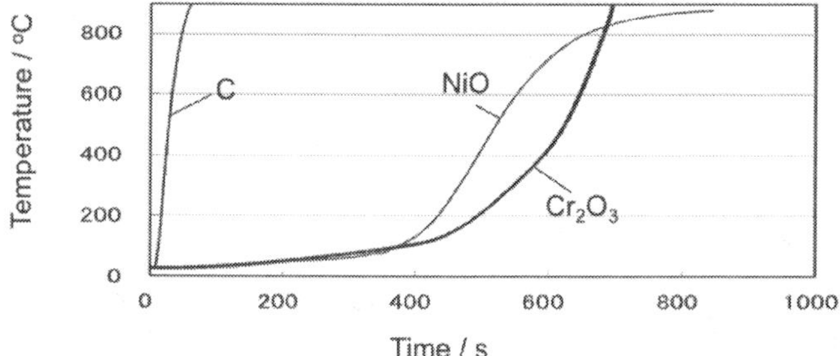

Fig. 15. Heating curves of graphite (C), NiO and Cr_2O_3 powders in a multi-mode applicator at maximum power (670W).

Next, Cr_2O_3 powders were heated in a single-mode MW applicator. Separated E- and H-field heating provides information to discuss the MW heating mechanisms. The heating curves are plotted in Fig. 16 (NiO heating curve in E- and H-field (Yoshikawa, 2007) Although there was an incubation time before onset of the temperature rise, E- field MW irradiation enabled high temperature heating, while H- field application was not successful to heat it above the detection limit of the pyrometer (saphire light pipe). These facts show that the dielectric heating mechanism is important in heating of Cr_2O_3, other than the magnetic and the ohmic loss mechanisms.

Not only the Cr-containing oxides' waste but also pure Cr_2O_3 were reduced at lower temperature than that expected from thermodynamics. In these experimental settings, the temperature is measured from the localized area in the specimen. Namely, some finite area on the surface temperature is measurable by an optical method (2mm in the optical rod diameter), and a thermocouple sheath in the inside of the powder (having diameter of 3mm). This means there could be some areas having much higher temperatures locally, as schematically illustrated in Fig. 17. In these considerations, it is expected that the reduction occurred at the high temperature regions where the thermodynamic conditions are satisfied. It is plausible to assume existence of the local temperature fluctuation at the initial stage of the MW heating, for example, because of the Cr_2O_3 and graphite grain size distribution or their aggregates' size in the mixture according to the mixing conditions, although the degree of their difference is not clear. If there are regions having differnces in local temperature, it is inferred that the high temperature region can be heated preferentially, because of the permittivity increase, as the case of S-L glass shown in 4-2.1, and thus the temperature fluctuations are expected to be amplified. However, other possibility of arcing effects cannot be still discarded.

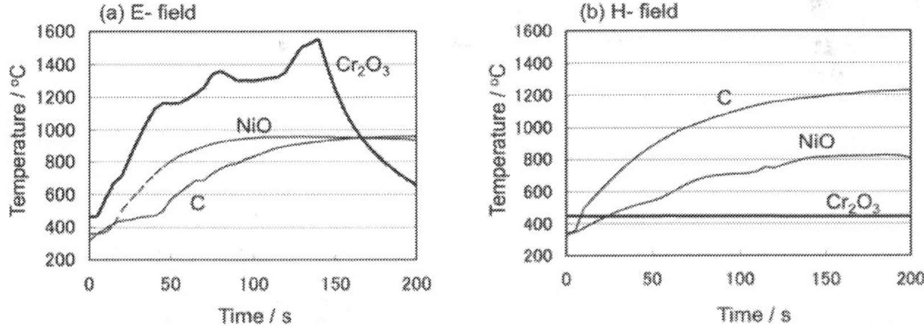

Fig. 16. Heating curves of graphite (C), NiO and Cr_2O_3 powders in a single-mode applicator at input power of 200W, (a) in E-field and (b) in H-field.

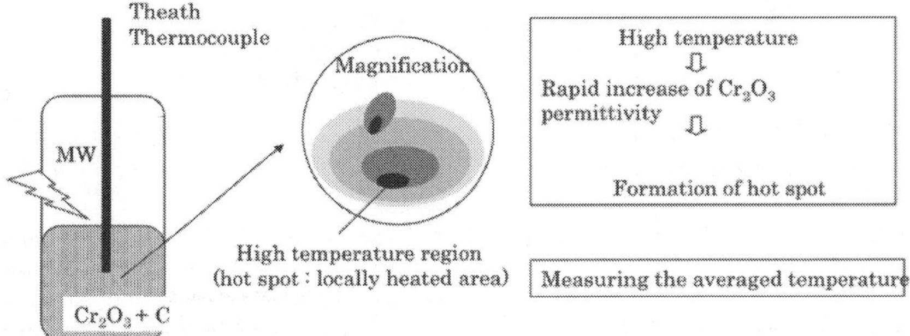

Fig. 17. Schematic illustration of models of the enhanced reaction kinetics and formation of metal pieces.

Accordingly, it is possible to cause reactions using TRW at the average temperature lower than that expected from thermodynamics. And the reaction process could be finished within the shorter time period. Therefore, it is of significance to take advantage of this phenomenon for the practical application. However, it is important to note that the controllability of the TRW process is not necessarily easy.

5. Discussion on the non-thermal effects

Classification of the phenomena

There have been many reports concerning on the so-called non-thermal effect in MW heating. It would be of significance to describe the interpretation and possibilities for realization of the non-thermal effects in discussion of the our results presented in this paper. The arguments are to be divided into three classes.

5.1 Case 1: The thermal effects appear to be non-thermal

Hot spot is sometimes brought about because of selective heating. Starting with the fluctuation of the temperature, rapid increase in temperature occurs locally due to the large temperature dependence of permittivity (loss factor) at high temperature. The examples are presented in 4.2 for the cases of S-L glass and Cr_2O_3. It was mentioned however, that taking advantage of the rapid increase of temperature or TRW is one of the new points of view for the application of MW heating.

On the other hand, it has to be noted that the temperature measurement performed in the presence of hot spot or temperature distribution provides the erroneous information. For examle, thermo couple measurement gives the temperature value at one spot. Optical method provides the average temperture which contains spectrum from different areas of high and low temperature. It is likely to misunderstand the observed phenomena to be the non-thermal effects.

For consideration of selective heating, instructive results on the temperature difference evolution were obtained by the numerical simulation of the heat conduction (Yoshikawa and Tokyama, 2009(c)), though the simulation was peformed without coupling the electromagnetic analysis. The ceramic spherical inclusions having large loss factor are embedded in a ceramic matrix having low loss factor. The inclusion particles are placed in the configulation of face centered cubic (fcc) cells, using symmetric boundary, 8 cells are made to be connected, as the schematic illustartion shown in Fig. 18. The same fcc configuration was set as the analysis presented in 3.1. Fig. 19 indicates the temperature evolution for the cases of different particle diameters (This calculation was conducte without using the radiation heat flux boundary condition. All the faces are assumed to be adiabatic.). The larger temperature difference can be evolved in the larger diameter case. This is because the heat generated is propotional to the particle volume but the heat transfered to the surrounding matrix is propotional to the surface area of the particle. This was discussed in 4.2.1. The heat conduction distance (d) can be simply evaluated by

$$d = \sqrt{2D \cdot t} \tag{17}$$

,where t is time and D is the thermal diffusivity D = $\lambda/(\rho Cp)$, λ: thermal conductivity, ρ: density and Cp: specific heat at constant pressure. Using the material properties of spinel for the matrix and perovskite for the inclusions, d is estimated to be 100μm for 19ms. Large temperature difference can be evolved only within several ten milli seconds, if the particles are separated by each other in an order of hundred microns. The simulated temperature distributions of Cr_2O_3/graphite mixture at 300ms is shown in Fig. 20.

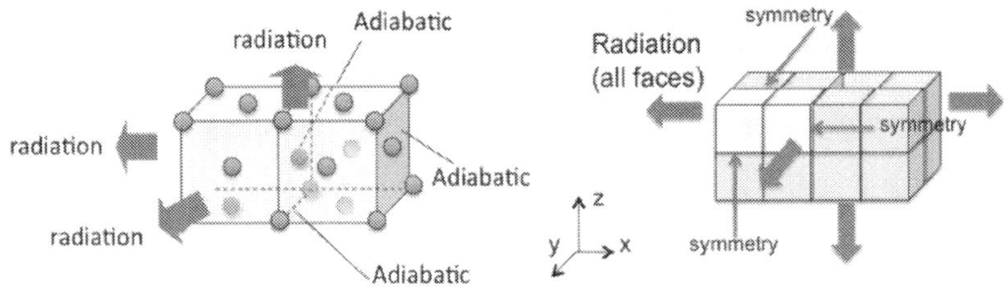

Fig. 18. Calculation region and the boundary conditions.

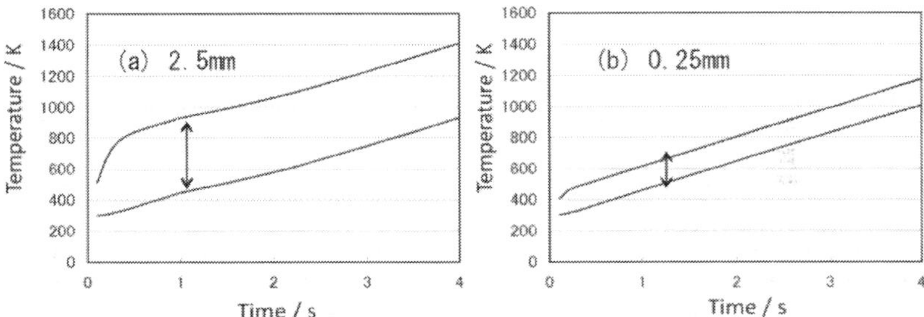

Fig. 19. Temperature difference between the particle center (upper curve) and the matrix minimum (lower curve), for different diameter cases (a): 2.5mm and (b) 0.25mm. Simulations were performed under adiabatic boundary conditions.

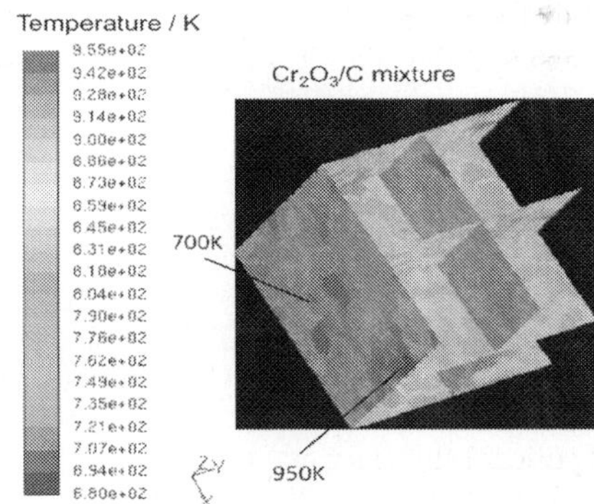

Fig. 20. Temperature distribution in a model cell of powder mixture at 300ms.

Here, in regards of the ceramic thermal conductivity, important conclusions are that the large temperature difference in ceramics composite or ceramic powder mixture can be evolved only several tens or hundreds seconds. And the temperature difference cannot be evolved in the small scale. Inclusions of millimiter size is required for several hundred degree in ceramics. It is considered that the inhomogeneities in micron scale embedded in ceramics matrix or some crystal imperfectional areas such as the grain boundaries could be different in MW absorbing behavior, however, the temperature difference, if generated, is too small to be detected.

5.2 Case 2: The heating accompanied with other phenomena

Two examples are to be pointed out. One is occurrence of arcing. The arcing or discharge could be generated at large electric field gradient, which is dependent on the material shape. Arcing readily occur at high temperature, because of emission of thermal electrons. Arcing or local plasma could be generated within the small voids, but usually it is difficult to detect them. Moreover, the local arcing or plasma could alter the surface energy and activate the sintering kinetics and some reactions, which are likely to be regarded as the non-thermal effect. The arcing phenomena if happened in the small area, it is not possible to be detected. In our studies, the arcing effect might also be related with the reduction kinetics in Cr_2O_3/graphite mixture wherever the large electric field gradients exist.

The other effect is the generation of high pressure. In heating of liquids, it is reported that super heating occurs. This is simply interpreted that MW internal heating of liquid brings about high temperature area in the internal region of the liquids, however, the difficulty in (homogeneous) nucleation of vapor in the internal region (nucleation boiling) resulted in the temperature rise above the boiling point. This is accompanied by areas having high pressure, comparing with the ambient. So, the observed phenomena, such as increasing the reaction kinetics might be caused by the conditions of high temperature/pressure. The MW digestion processes usually used for dissolving solid particles in the solvents for preparation of ICP analysis specimens are conducted in a closed container and the high pressure/temperature conditions could accelerate the dissolution kinetics. It is still difficult to discuss which effect is dominant to influence the kinetics, either of hydrothermal state or MW irradiation.

5.3 Case 3: Inherent phenomena raised by MW irradiation

Atoms or molecules in a gas or liquids, ions in the solid crystals or glass are thermally vibrated. The frequency of thermal vibration is an order of several tens to hundred THz and is much (10^4~10^5 times) higher than that of MW. These thermal vibration frequencies correspond to the „temperature" of the matter. MW is an electromagentic wave, so it is expected the electric or magnetic dipoles in the matter are directly responding to the alternating field, however, it is not necessarily clear how these dipole vibration or rotation excite the thermal vibration of 10^4 times higher frequency. Generally, it is understood that the delay of the rotational motion corresponds to the loss factor and is believed to dissipate and convert to heat, however there no description on these specific processes.

It is of interest to know whether the atomic vibration spectrum excited by MW and the direct excitation by infra-red light (general heating) are different from each other, and whether MW excitation generates the non-Maxwellian distribution of the spectrum. It is expected that the phenomena related with the low frequency vibration spectrum could occur and detected as the non-thermal effects. The magnetic heating mechanism in 3.3 is brought about by resonant electron spin precession motion, inducing the wave number k=0 spin wave is generated and interacts with thermal phonons or creates another phonons. It is possible to interact with the elastic waves which could be observed as the coherent phonon and could give rise the specific effect on the materials' nano structures (Turner, 1960).

As mentioned above, the diference in the local vibration states could be raised in some crystal imperfectional areas such as the grain boundaries could be different in MW absorbing behavior, and effect of nonlinear coupling with the elastic wave was proposed before (Booske, 1992). These phenomena, though not detected as the temperature difference, could induce the non-thermal effect in MW heating.

6. Conclusion

Based on the recent studies by authors' group on the new aspects in MW heating, the essence of the MW heating was discussed. The heating mechanisms were divided into three classes. The magnetic mechanism of which has been recognized but not fully understood with the experimental evidences. This paper provided arguments on contribution of the FMR and discussed its possibility to give rise the non-thermal effect of MW heating.
Separated E- and H- field heating of metal thin film was reported and the application of its heat treatment was introduced. The rapid thermal treatment was proposed taking advantages of the thermal runaway phenomena for the puropse of oxide reduction and glass melting. The non-thermal effect was discussed with possible interpretation of the origins. Some effects to be expected based on the non-Maxwellian distribution of the atomic vibration were suggested important for the future prospects of MW heating applications.

7. Acknowledgement

This study was supported by the Grant-in-Aid of Ministry of Education, Sports,Culture, Science and Technology, Japan, Priority Area on Science and Technology of Microwave-Induced, Thermally Non-Equilibrium Reaction Field (2006-10). Author is grateful to Prof. S. Taniguchi at Tohoku Univ, and Prof. M.Sato at NIFS. The efforts done by the following students and fellows are greatly acknowledged ; Ms.Ishizuka.E, Mr.Mashiko.K, Mr.Sasaki.Y, Mr.Saito.Y, Mr.Kawahira,K, Mr.Kato,T, Dr.Cao,ZP.

8. References

Agrawal,D. (2005). in his talk at Japan Institute of Metals (JIM) Symp., Fall Annual Meeting, Hiroshima, Japan.

Bescher, E.P. Sakkar, U. and Mackenzie, J. (1992) „Microwave Processing of Aluminum Silicon Carbide Cermets", *Mat.Res.Symp, Proc.*, 269 371-378.

Boey,F. Gosling,I. Lye,S.W. (1992). „High-Pressure Microwave Curing Process for an Epoxy-Matrix/Glass-Fibre Composite", *J.Mater.Process.Technol.*, 29, 311-319.

Bohren,C and Huffman,D. (1983). "Absorption and Scattering of light by small particles", Wiley and Sons Inc. New York

Booske,J.H. Cooper,R.F and Dobson,I. "Mechanisms for Nonthermal Effects on Ionic Mobility duringMicrowave Processing of Crystalline Solids"(1992) *J.Mater. Res.*, 7 [2] 495-501.

Bykov,Y.V. Rybakov,K.I. and Semenov,V.E. (2001). "High-temperature microwave processing of materials", *J.Phys. D: Appl. Phys.* 34, R55-R57.

Cao,Z.P. Yoshikawa,N and Taniguchi,S. (2009) "Microwave heating behavior of nanocrystalline Au thin films in single-mode cavity", *J.Mater.Res.*, 24 [1] 268-273.

Cheng,J Roy,R and Grawal,D.A. (2001) "Experimental proof of major role of magnetic field losses in microwave heating of metallic composites" *J.Mater.Sci.Lett.* 20 1561-1563.

Cheng,J. Roy,R, and Agrawal,D. (2002). "Radically different effects on materials by separated microwave electric and magnetic fields.", *Mater. Res. Innovat.*, 5 170-177.

Clark,D.E. and Sutton,W.H. (1996). "Microwave processing of materials", Annu. Rev. Mater. Sci., 26 299-331.

Committee on Microwave Proc. of Mater. (Ed.), (1994). "Microwave processing of materials", An Emerging Industrial. Technology. National Academy Press, Washington D.C. pp.36

Ishizaki,K Nagata,K. (2007) "Selectivity of Microwave Energy Consumption in the Reduction of Fe_3O_4 with Carbon Black in Mixed Powder", ISIJ Int. 47 [6] 811-816.

Iwasaki,K. Mashiko,K. Saito,Y. Todoroki,H. Yoshikawa,N. Taniguchi, S. (2009). *ISIJ Int.*, 49 [4] 596-601.

Janney.M.A. and Kimrey,H. (1988). "Microwave sintering of alumina at 20 GHz", Ceramic Powder Science II, Ceramic Trans. Am. Ceram. Soc., 1 919-924.

Jackson,J.D. (1998), "Classical Electrodynamics", 3rd ed., Wiley and Sons Inc., new York.

Johnson,P.B. and Christy,R.W., (1972) "Optical Constants of the Noble Metals", *Phys. Rev.*, B6, 4370-4379.

Kato,Y, Shibata,T. and Anbe,T. (2005). "Reaction Mechanism of De-nitration of $UO_2(NO_3)_2$ by the Microwave Heating", *Jpn. J.Nucl. Sci. Technol.* 4 [1] 77-83.

Katz, J.D. (1992). "Microwave Sintering of Ceramics", *Annu. Rev. Mater. Sci.*, 22 153-170.

Kenkre,V.M. Skala,L. and Weiser,M.W. (1991). "Theory of microwave interactions in ceramic materials: the phenomenon of thermal runaway", J.Mater. Sci., 26 (1991) 2483-2489.

Kingery,W.D.Bowen,H.K. and Uhlmann, D.R., (1975). Introduction to Ceramics, 2nd Ed., John Wiley and Sons, New York, pp. 240.

Kittel,C. (1948). "On the theory of ferromagnetic resonance", *Phys. Rev.*, 73 155-161.

Kraus,J.D. and Carver,K.R. (1973), "Electromagnetics",2nd ed. Int. Student Ed. pp.402. McGraw-Hill, New York, USA (For example)

Landau,L.D. and E.M.Lifshitz,E.M. and Pitaevskii,L.P., (1984) "Electrodynamics of Continuous Media",Landau and Lifshitz Course of Theoretical Physis, 2nd ed. Vol. 8 Pergamon Press, Oxford UK.

Leonelli,C. Veronesi,P. Denti,L. Gato.A and Iuliano,L. (2008). "Microwave assisted sintering of green metal parts", J.Mater.Process. Technol. 205 489-496.

Leontovich,L.A. (1948), Investigation of propagation of radiowaves, Pt II.

Lorenson,C. Ball,M.D. Herzig,R and Shaw,H. (1991) „The Microwave Heating Behavior of Metallic-Insulator Composite SYstems", Mat.Res. Symp. Proc., 189, 279-282.

Ma,J. Diehl, J.F. Johnson,E.J. Martin,K.R. Miskovsky,N.M. Smith,C.T. Weisel.G.J. Weiss,B.L. and Zimmerman,D.T. (2007) "Systematic study of microwave absorption, heating, and microstructure evolution of porous copper powder metal compacts.", *J.Appl.Phys.*, 101 074906,1-8.

Mishra,P. Sethi.G and Upadhyaya,A. "Modeling of Microwave. Heating of Particulate Metals", *Met. Trans.* 37B (2006) 839-845.

Morita,K. Nguyen, V.Q. Nakaoka, R. and Mackenzie, D. (1992), "Immobilization of ash by microwave melting", *Mater. Res. Symp. Proc.* 269 pp. 471-476.

Oda,S.J. (1992). "Dielectric processing of hazardous materials – prrsent and future opportunities", *Mater. Res. Symp. Proc.* 269 (1992) pp. 453-464.

Pardavi-Horvath,M., (2000). "Microwave applications of soft ferrites", *J. Magn. Magn. Mater.*, 215-216 171-183.

Polder,D. and Smit,J., (1953). "Resonance Phenomena in Ferrites", *Rev. Modern Phys.*, 25 [1] 89-90.

Rodiger,K. Dreyer,K. Gerdes,T. and Willert-Porada,M. (1998). "Microwave Sintering of Hardmetals.", *Int.J.Refractoy Met and hard Mater.* 16 4-6, 409-416.

Roy,R. Agrawal,D. Cheng.J and Gedevanishvili,S. (1999) "Full sintering of powdered-metal bodies in a microwave field", *Nature*, 399 668-670.

Roy.R. Peelamedu,R. Hurtt,L. Cheng.J and Agrawal,D. (2002) "Definitive experimental evidence for Microwave Effects: radically new effects of separated E and H fields, such as decrystallization of oxides in seconds", *Mater. Res. Innovat.*, 6 128-140.

Nikawa,Y and Okada,F. (1987). "Selective Heating for Microwave Hyperthermia Using Ferrimagnetic Resonance ", *IEEE Trans. Magn.* vol. Mag-23,[5] 2431-2433.

Sparks,M. Loudon,R. and Kittel,C. (1961) "Ferromagnetic Relaxation I. Theory of the Relaxation of the Uniform Precession and the Degenerate Spectrum in Insulatord at Low Temperatures", *Phys. Rev.*, 122 [3] 791-803.

Sparks,M. (1964). "Ferromagnetic Relaxation Theory", McGraw-Hill, New York, USA

Turner.E (1960). "Interaction of Phonons and Spin Waves in Yttrium Iron Garnet", *Phys. Rev. Lett.*, 5 [3] 100-101.

URL: http://www.gallawa.com/microtech/history.html (browsed in Aug. 2010)

Walkievicz,J. Kazonich,G. and MacGill,S.L. (1988). "Microwave Heating Characteristics of Minerals and Compounds", Min. Metall. Processing 5 39-42.

Walton,D. Snape,S. Rolph,T.C. Shaw.J and Share,J. (1996). "Application of ferrimagnetic resonance heating to palaeointensity determinations", Phys. Earth Planet.Inter., 94 (1996) 183-186.

Wang, Z.J. Cao,Z.P Ohtsuka,Y. Yoshikawa,N Kokawa,H and Taniguchi,S. (2008) "Low temperature growth of ferroelectric lead zirconium titanate thin films using the magnetic field of low power 2.45GHz microwave irradiation" *Appl. Phys. Lett.* 92 222905-

Whittaker,A.G. and Mingos, D.P. (1995). "Microwave-assisted solid-state reactions involving metal powders", J. Chem. Soc., Dalton Trans. 2073-2079.

Wroe,R and Rowley,A.T. (1996) "Evidence for a Non-Thermal. Microwave. Effect in the Sintering of Partially. Stabilized Zirconia", J.Mater.Sci., 31 2019-2026.

Yoshikawa,N. Ishizuka,E and Taniguchi,S. (2006). "Heating of Metal Particles in a Single-Mode Microwave Applicator." *Mater. Trans.*, 47 [3] 898-902.

Yoshikawa,N Ishizuka,E Mashiko,K. Taniguchi,S. (2007). "Carbon Reduction Kinetics of NiO by Microwave Heating of the Separated Electric and Magnetic Fields.", *Metal. Mater. Trans.* B,38 [6],863-868.

Yoshikawa,N. Wang,H.C. Mashiko,K and Taniguchi,S. (2008a). "Microwave heating of Soda-Lime Glass by Addition of Iron Powder." *J. Mater. Res.,* 23 [6] 1564-1569.

Yoshikawa,N. Mashiko,K. Sasaki,Y. Taniguchi,S. and Todoroki,H. (2008b). "Microwave Carbo-thermal Reduction for Recycling of Cr from Cr-containing Steel Making Wastes." *ISIJ Int.* 48 [5] 697-702.

Yoshikawa,N. (2009a). "", Fundamentals and Application of Mcrowave heating of Metals", *Materia, Japan (Bulletin of Japan Inst.Metals)*, 48 [1] 3-10.

Yoshikawa,N· Wang,H.C. and Taniguchi,S. (2009b) "Application of Microwave Heating to Reaction between Soda-Lime Glass and Liquid Al for Fabrication of Composite Materials" *Mater. Trans.* 50 [5] 1174-1178.

Yoshikawa,N. Tokuyama,Y. (2009b). "Numerical Simulation of Temperature Distribution in Multi-Phase Materials as a Result of Selective Heating by Microwave Energy", *J. Microwave Power and Electromagnetic Energy,* 43 [1], 27-33.

Yoshikawa,N. and Kato,T. (2010) "Ferromagnetic Resonance Heating of Fe and Fe_3O_4 by 5.8 GHz Microwave Irradiation", J.Phys.D: Appl. Phys. 43 425403.

2

Review of Numerical Simulation of Microwave Heating Process

Xiang Zhao, Liping Yan and Kama Huang
College of Electronics and Information Engineering, Sichuan University, Chengdu 610064,
China

1. Introduction

Since the great experiments were carried out by Dr. Percy Spencer in 1946, microwave heating has been used widely in food industry, medicine, chemical engineering and so on [1][2][3][4][5], for processing many materials ranging from foodstuffs [1][6][7][8] to tobaccos [9][10][11], from wood [12][13][14] to ceramics [15][16][17][18], and from biological objects [19][20] to chemical reactions [21][22][23]. Typical applications include heating and thawing of foods, drying of wood and tobaccos, sintering of ceramics, killing of cancer cells and accelerating of chemical reactions.

Microwaves are electromagnetic (EM) waves with frequencies between 300MHz and 300GHz. Under microwave radiation, dipole rotation occurs in dielectric materials containing polar molecules having an electrical dipole moment. Interactions between the dipole and the EM field result in energy in the form of EM radiation being converted to heat energy in the materials. This is the principle of microwave heating. For each material, permittivity, or dielectric constant, ε is the most essential property relative to the absorption of microwave energy. Permittivity is often treated as a complex number. The real part is a measure of the stored microwave energy within the material, and the imaginary part is a measure of the dissipation (or loss) of microwave energy within the material. The complex permittivity is usually a complicated function of microwave frequency and temperature.

Compared with conventional heating, microwave heating has many advantages such as simultaneous heating of a material in its whole volume, higher temperature homogeneity, and shorter processing time. However, the nonlinear process of interaction between microwave and the heated materials has not been sufficiently and deeply understood. Some phenomena arising from this process, such as hotspots and thermal runaway, decreases the security and efficiency of microwave energy application and prevents its further development.

Hotspot means that where temperature non-uniformity in heated material of local small areas having high temperature increase occurs. For example, some zones in material are overheated, and even burned, while some others have not reached the required minimum temperature. This phenomenon has become one of the major drawbacks for domestic or industrial applications.

"Thermal runaway" refers to temperature varies dramatically with small changes of geometrical sizes of the heated material or microwave power. It may lead to a positive feedback occurring in the material — the warmer areas are better able to accept more energy than the colder areas. It is potentially dangerous, especially when the heat exchange between the hot spots and the rest of the material is slow. This phenomenon occurs in some ceramics.

These existing problems related to temperature distribution inside materials must be considered. Although experimental studies and qualitative theoretical analysis are important, numerical simulations are indispensable for helping us further the understanding of the complex microwave heating process, predict and control its behaviors. Thanks to rapid hardware and software development of numerical calculation techniques, it has been possible to simulate the microwave heating process on a personal computer. As a result, fully empirical design of heating system and expensive prototype equipment for testing can be avoided.

Numerical simulation of microwave heating process is basically an analysis of multi-physical coupling, so at least two kinds of partial differential equations (PDE) are associated together, the EM field equations (i.e. Maxwell's equation) and the heat transport equation. To describe their interaction relationship, the inhomogeneous term of the heat transport equation representing the heating source is provided by microwave dissipated power; meanwhile, the temperature variation during the heating process can cause the change of complex permittivity, which directly affects the space and time variation of the EM field. Sometimes, more kinds of differential equations may need to be joined, for example, the mass transfer equation may be required when drying the porous materials.

To obtain accurate solution of the above coupled equations, analytical methods are hardly to be used. There are many numerical methods being used to simulate microwave heating. The popular methods include Finite Difference (FD) methods (e.g. Finite Differential Time Domain / FDTD method), Finite Element Method (FEM), the Moment of Method (MoM), Transmission Line Matrix (TLM) method, Finite-Volume Time-Domain (FVTD) method and so on.

In this chapter, we first review all kinds of application backgrounds of numerical simulation of microwave heating process, and then discuss the two most important techniques - numerical modeling and numerical methods. Finally we conclude and outline some desirable future developments.

2. Application backgrounds

According to the statistics of the open publications related to numerical simulation of microwave heating, we find that so many kinds of objects, from common substances such as water [24][25[26], food [27][28][29][30] and wood [31][32][33][34][35][36], to chemical engineering materials such as ceramic material [37][38][39][40][41][42][43][44], minerals[45], SOFC materials[46] and zinc oxide [47], and even human tissues [48][49] and some chemical reactions [50][51][52][53][25], have been investigated.

Water strongly absorbs microwaves, so usually microwave heating works by heating the water in foods in food industry. Furthermore, water is such an important substance whose permittivity shows clear relationship with temperature, that it is natural to regard it as the most typical object in microwave heating and a benchmark example when comparing the results of different numerical methods. At the frequency of 2.45 GHz, the curves plotted on Fig. 1 show the complex permittivity of water varying with temperature.

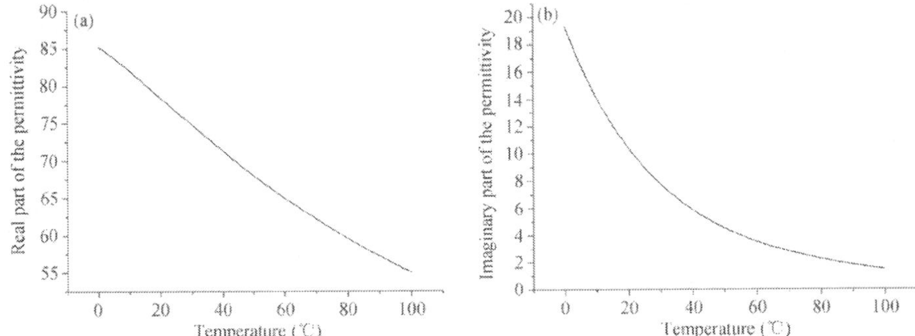

Fig. 1. Variation of the complex permittivity of water with temperature at 2.45 GHz.
(a) Real part; (b) Imaginary part.

When microwave energy being used to dry wood from the green state, some benefits can be offered, such as moisture leveling and increased vapor migration from the interior to the surface of the wood. Perré, et al.,[31] combined a two-dimensional transfer code with a three-dimensional EM computational scheme, obtained the overall behavior of combined microwave and convective drying of heartwood spruce remarkably well and predicted the occurrence of thermal runaway within the material. To provide a detailed algorithm for simulating the nonlinear process of microwave heating, Zhao, et al.,[32] presented a three-dimensional computational model for the microwave heating of wood with low moisture contents. The model coupled a FVTD algorithm for resolving Maxwell's equations, together with an algorithm for determining the thermal distribution within the wood sample. Hansson, et al.,[33] used FEM as a tool to analyze microwave scattering in wood. They used a medical computed tomography scanner to measure distribution of density and moisture content in a piece of Scots pine, calculated dielectric properties from measured values for cross sections from the piece and used in the model. Hansson and his colleagues also studied moisture redistribution in wood during microwave heating process [34]. Rattanadecho [35] used a three dimensional FD scheme to determine EM fields and obtained temperature profiles of wood in a rectangular wave guide. Temperature dependence of wood dielectric properties was simulated by updating dielectric properties in each time step of temperature variation. The influence of irradiation times, working frequencies and sample size were illustrated. To take into consideration of the influence of moisture diffusion during microwave heating of moist wood, Brodie [36] derived of the PDE, which described simultaneous heat and moisture diffusion, yielded two independent values for the combined heat and moisture diffusion coefficient.

Microwave sintering of ceramic material has also been studied widely. It is a promising technique for the densification of ceramics. Compared to conventional heating, microwave sintering can obtain higher densities with a finer grain structure. Thermal runaway is a common phenomenon in microwave sintering. Numerical simulation may help to understand the mechanism of the thermal runaway, and then avoid it. Iskander, et al., [39] developed a FDTD code that was used to model some of the many factors that influence a realistic microwave sintering process in multimode cavities. The factors included the conductivity of the insulation surrounding the sample and the role of the Sic rods in modifying the EM field distribution pattern in the sample. Chatterjee, et al., [40] analyzed the microwave heating of ceramic materials by solving the equations for grain growth and porosity[54] during the late stages of sintering, coupled with the heat conduction equation and electric field equations for 1-D slabs. Gupta, et al., [41] analyzed the steady-state behavior of a ceramic slab under microwave heating by transverse magnetic illumination. They observed that for a certain set of parameters, there are periodically recurring ranges of slab thickness for which thermal runaway may be avoided. The runaway dependence on other parameters critical to the operation of the process was also studied. Riedel, et al., [42]modeled microwave sintering by the Maxwell's equation, the heat conduction equation and a sintering model describing the evolution of density and the grain size, so as to predict runaway instabilities and to find process parameters leading to a homogeneously sintered product with a uniform, fine grain structure. Liu, et al., [43] solved Maxwell's equations (by using FDTD) coupled with a heat transfer equation (by using FD), and obtained the temperature variation in a ceramic slab during microwave heating. Various ceramic parameters and applied microwave powers were simulated so as to analyze the condition under which thermal runaway is introduced. Furthermore, they presented a microwave power control method based on a single temperature threshold and dual applied microwave powers, which can improve microwave heating efficiency and controls thermal runaway.

Although reports of the potential of microwave heating in chemical modification can be traced back to the 1950s, microwave heating began to gain wide acceptance with the publication of several papers in 1986 [55]. Presently, microwave has been widely used in various chemical domains from inorganic reaction to organic reaction, from medicine chemical industry to food chemical industry and from simple molecules to complex life process. However, most of studies on the mechanism of interaction between microwave and chemical reaction are only related with the experimental studies of some concrete reactions and qualitative theoretical analysis. The main reason is that the complex effective permittivities of many chemical reactions are still unknown. This fact brings difficulty to implement numerical simulation. Wang, et al., [50] presented a numerical model of microwave heating of a chemical reaction, used FDTD technology to solve Maxwell's equation, and the explicit FD scheme to solve heat equation. Huang, et al., [51] presented a numerical model to study the microwave heating on saponification reaction in test tube, where the reactant was considered as a mixture of dilute solution. They solved the coupled EM field equations, reaction equation (RE) and heat transport equation by using FDTD method, and employed dual-stage leapfrog scheme and method of time scaling factor to overcome the difficulty of long time calculation with FDTD. Yamada, et al., [52] simulated microwave plasmas inside a reactor for thick diamond syntheses. In a model reactor used in the simulation, a diamond substrate with finite thickness and area is taken into account. Distributions of electric field, density of microwave power absorbed by the plasma, temperatures and flow field of gas were studied not only in a bulk region inside a reactor but also a local region around the substrate surface. Numerical results implied that the adopted arrangement of the substrate is not desirable for continuous growth of large diamond crystals. Huang, et al., [53] studied the precise condition of thermal runaway in microwave heating on chemical reaction with a feedback control system. They derived the condition of thermal runaway according to the principle of the feedback control system, simulated the temperature rising rates for 4 different kinds of reaction systems placed in the waveguide and irradiated by microwave with different input powers, and discussed the relationship between the thermal runaway and the sensor's response time. Zhao, et al., [25] carried out preliminary study on numerical simulation of microwave heating process for chemical reaction. They considered calculation of 2D multi-physical coupling, solved the integral equation of EM field by using MoM, and the heat transport equation by using a semi-analysis method. Moreover, a method to determine the equivalent complex permittivity of reactant under the heating was presented in order to perform the calculation. The numerical results for water and a dummy chemical reaction showed some features of the hotspot and the thermal runaway phenomenon.

Besides the above-mentioned typical applications of numerical simulation, there are more application cases used in other materials [56][57][58]. Furthermore, numerical simulation was used to analyze and design a variety of microwave heating applicators [59][60][61].

3. Numerical simulation technology

In essence, numerical simulation of microwave heating process is an analysis of multi-physical coupling. The mathematic model consists of at least EM field equation (i.e. Maxwell's equation) and the heat transport equation. There is an interaction relationship between them (Fig. 1). The inhomogeneous term of heat transport equation, i.e., the heating source is provided by microwave dissipated power. And meanwhile, the temperature variation during the heating process can cause the change of complex permittivity, which also affects the space and time variation of the EM field. Consequently, to perform the simulation, one must model the two equations together with their coupling relation at first, and then solve them by using some numerical methods.

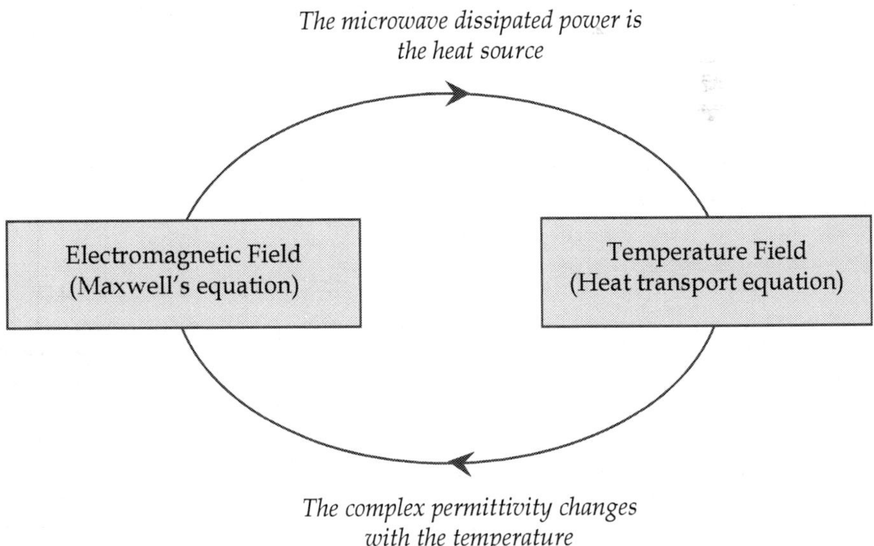

Fig. 1. Multi-physical coupling in microwave heating process

In some cases of more complex application scenarios, more kinds of differential equations may need to be joined, for example, the mass transfer equation may be required when drying the porous materials [62][63][64].

As the key techniques, the numerical modeling and the numerical method will be discussed below.

A. Numerical modeling

- **Maxwell's equation**

The governing equation of the EM field vectors is based on the well-known Maxwell's equation. The differential and integral form of Maxwell's equation can be expressed as

$$\nabla \times \vec{E} = -\frac{\partial \vec{B}}{\partial t} , \quad \oint_C \vec{E} \cdot d\vec{l} = -\int_S \frac{\partial \vec{B}}{\partial t} \cdot d\vec{S} , \tag{1}$$

$$\nabla \times \vec{H} = \vec{J} + \frac{\partial \vec{D}}{\partial t} , \quad \oint_C \vec{H} \cdot d\vec{l} = \int_S \vec{J} \cdot d\vec{S} + \int_S \frac{\partial \vec{D}}{\partial t} \cdot d\vec{S} , \tag{2}$$

$$\nabla \cdot \vec{D} = \rho , \quad \oint_S \vec{D} \cdot d\vec{S} = \int_V \rho dV , \tag{3}$$

$$\nabla \cdot \vec{B} = 0 , \quad \oint_S \vec{B} \cdot d\vec{S} = 0 , \tag{4}$$

where $\vec{E} = \vec{E}(x,y,z,t)$ is the fields electric, $\vec{H} = \vec{H}(x,y,z,t)$ the magnetic fields, $\vec{D} = \vec{D}(x,y,z,t)$ the flux density, $\vec{B} = \vec{B}(x,y,z,t)$ the magnetic flux density, $\vec{J} = \vec{J}(x,y,z,t)$ the current density and $\rho = \rho(x,y,z,t)$ the charge density. The constraint relations of them are

$$\vec{D} = \varepsilon\vec{E} , \quad \vec{B} = \mu\vec{H} , \quad \vec{J} = \sigma\vec{E} , \tag{5}$$

and

$$\nabla \cdot \vec{J} + \frac{\partial \rho}{\partial t} = 0 , \quad \oint_S \vec{J} \cdot d\vec{S} = -\frac{d}{dt}\int_V \rho dV , \text{ (Continuity of electric current)} \tag{6}$$

where ε is the dielectric constant (or electrical permittivity), μ the magnetic permeability and σ the electric conductivity.

When there exists discontinuity of the parameters of medium (ε, μ and σ), e.g., at the surface of the medium, the boundary conditions are required to specify the behavior of EM field at the boundary. For example, at the boundary of a perfect conductor, boundary conditions are given as

$$E_t = 0 , \quad H_n = 0 , \tag{7}$$

where E_t is the tangential component of $\vec{E}$, H_n the normal component of $\vec{H}$. And the boundary conditions along the interface between two different dielectric materials are given as

$$E_{t1} = E_{t2} , \quad H_{t1} = H_{t2} , \quad D_{n1} = D_{n2} , \quad B_{n1} = B_{n2} . \tag{8}$$

There are some equivalent or derivative forms of Maxwell's equation, such as wave equation, Helmholtz equation (the time-harmonic form of wave equation), scalar potential equation, vector potential equation, integral equation based on Green's theorem and so on. All of these equations can be used to model the behavior of EM field.

- **Heat transport equation**

Heat transport equation is also called heat transfer equation or thermal conduction equation, which describes the space and time behavior of the temperature field, that is

$$\rho_m C_m \frac{\partial T}{\partial t} - \nabla \bullet (k_t \nabla T) = P ,\tag{9}$$

where ρ_m, C_m and k_t denote material density, specific heat capacity and thermal conductivity, respectively. $T = T(x,y,z,t)$ is the absolute temperature , $P = P(x,y,z,t)$ the EM power dissipated per unit volume. For simplicity in solving this equation, the parameters ρ_m, C_m and k_t are usually taken as constants that are independent of position, time and temperature.

According to Newton's law of cooling, the convective boundary condition at the material surfaces is used such as

$$h(T_a - T) = k_t \frac{\partial T}{\partial n} ,\tag{10}$$

and the adiabatic boundary condition as

$$\frac{\partial T}{\partial n} = 0 ,\tag{11}$$

where h is the convective heat transfer coefficient, T_a the temperature of the surrounding air. Also, the initial condition is needed to determine the unique solution of (9), which can be given as

$$T(x,y,z,t)\big|_{t=0} = T_{ini}(x,y,z) .\tag{12}$$

In some cases, the heat radiation of the material can be considered by subtracting it from the right-hand side of (9)[65][43]. Its ideal case is a blackbody radiation, which is described by the Stefan-Boltzmann law.

- **Coupling relationship**

The inhomogeneous item P in (9) is determined by the EM power density that is dissipated in the nonmagnetic and nonconductive material due to its dielectric losses, and can be expressed by [24]

$$P = \frac{1}{2}\left(\vec{E} \bullet \frac{\partial \vec{D}}{\partial t} - \vec{D} \bullet \frac{\partial \vec{E}}{\partial t} \right) .\tag{13}$$

For the case of steady-state time harmonic EM fields, it yields

$$\overline{P} = \frac{1}{2}\omega \varepsilon_0 \varepsilon'' E^2\tag{14}$$

where ω is the angular frequency, ε_0 the vacuum permittivity, ε'' the imaginary part of the permittivity of the material, E the amplitudes of $\vec{E}$.

On the other hand, the space and time variation of the temperature can change the ε, μ and σ of the material. For nonmagnetic and nonconductive material, we can only discuss ε.

It is well known that the response delay of the materials to external EM fields generally depends on the frequency of the field. For this reason permittivity is often treated as a complex function of the (angular) frequency ω. The complex permittivity $\varepsilon^* = \varepsilon^*(\omega)$ is defined by

$$ \mathbf{D} = \varepsilon^* \mathbf{E} = (\varepsilon' - j\varepsilon'')\mathbf{E}, \tag{15} $$

where $\mathbf{D} = \mathbf{D}(x,y,z)$ and $\mathbf{E} = \mathbf{E}(x,y,z)$ are the phasor representation of time harmonic field $\vec{D}$ and $\vec{E}$ respectively, which satisfy the following

$$ \vec{D}(x,y,z,t) = \mathrm{Re}\left[\mathbf{D}(x,y,z)e^{j\omega t} \right] \text{ and } \vec{E}(x,y,z,t) = \mathrm{Re}\left[\mathbf{E}(x,y,z)e^{j\omega t} \right]. \tag{16} $$

And j is the imaginary unit, $j^2 = -1$. Note that the sign of the imaginary part of ε^* is constantly negative when the time harmonic factor is $e^{j\omega t}$.

For conductive material, one can consider ohmic loss by taking σ into the imaginary part of ε^* as follows

$$ \varepsilon^* = \varepsilon' - j\left(\varepsilon'' + \frac{\sigma}{\omega} \right). \tag{17} $$

Debye (1926) deduced the well known equation for the complex relative permittivity ε_r^* ($\varepsilon_r^* = \varepsilon^*/\varepsilon_0$) as

$$ \varepsilon_r^* = \varepsilon_\infty + \frac{\varepsilon_s - \varepsilon_\infty}{1 + j\omega\tau}, \tag{18} $$

where ε_∞ is the permittivity at the high frequency limit, ε_s is the static, low frequency permittivity, and τ is the relaxation time. The variation of temperature can affect the parameters of ε_∞, ε_s and τ, and consequently affect ε_r^*. So ε_r^* is a function of the temperature, that is

$$ \varepsilon_r^* = \varepsilon_r^*(\omega, T). \tag{19} $$

Particularly, for a chemical reaction, it is also the function of the reaction duration time.

Although in some cases, thermal and dielectric properties may be assumed to be temperature-independent [78], the knowledge of the function $\varepsilon_r^*(\omega, T)$ of a material is a necessary and key condition to implement the numerical simulation of interaction between microwave and the material. There is a lot of basic work that has been done for this area [66][67][68][69][70][26][71]. Besides, the complex permittivities of many specific materials are studied [72][73][74][75][76][77][78]. Obviously, it is more difficult for the chemical reaction to determine the complex permittivity of the reactant, because it varies not only with temperature, but also with reaction duration time.

- **Geometric model**

As far as the geometric model is concerned, one-dimensional (1D)[43][40][79][80] and two-dimensional (2D)[94][81][25] models are more generally used for theoretical studies, and three-dimensional (3D)[24][82][83][90][84][85] model (e.g. the heated object is placed in a microwave oven) is more close to practical applications.

Besides, Plaza-Gonzalez, et al,[86] studied the case of mode stirrers which are widely used by industrial applicators.

B. Numerical methods

Typically, four iterative steps are used to solve the coupling between the EM and thermal processes [87][88][89][90][91][92][93][94][95] (See Fig. 2).

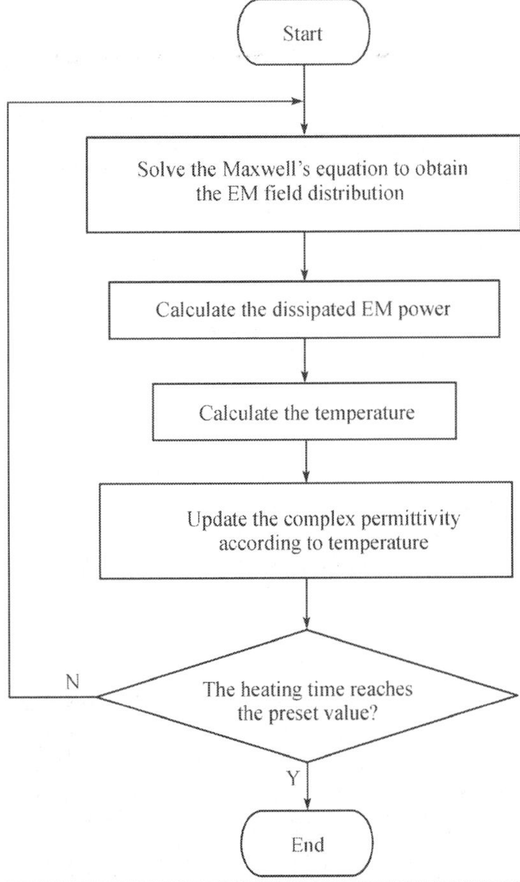

Fig. 2. The general simulation procedure of microwave heating

From the point of view of solving the PDE, the EM field equations are more difficult to solve comparing with the heat transport equation, because the former in fact include two vector equations and the latter is a scalar equation. For this reason, we only discuss below the solving of the EM field equations. There are many numerical methods being used: the Finite Differential Time Domain (FDTD) method, the Finite Element Method (FEM), the Moment of Method (MoM), the Transmission Line Matrix (TLM) method, the Finite Integral Method (FIM)[96] and so on. In some 1D/2D cases, some analytical methods or semi-analytical methods can be applied.

- **FDTD**

The FDTD method [97][98][99] is an application of the FD method for solving EM field equation. It is based on Yee grid as shown in Fig. 3, which is proposed by Kane Yee in 1966 [100].

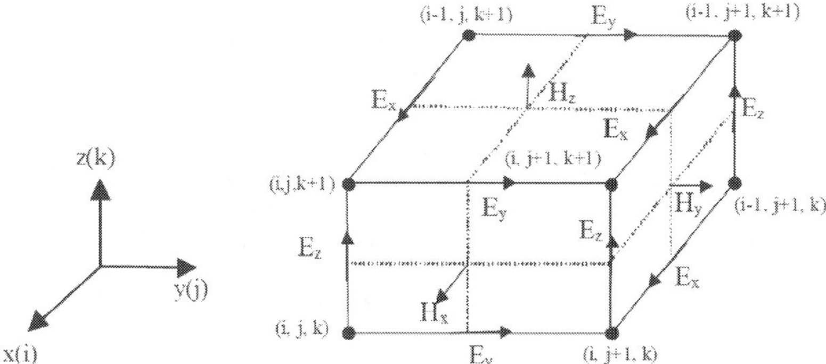

Fig. 3. The Yee grid

The main idea of the FDTD method is that the time-dependent Maxwell's equations (in partial differential form) are discretized using difference approximations to the space and time partial derivatives. It contains of three key elements: differential scheme, stability of solutions and absorbing boundary conditions.

In FDTD method, the differential grid is first set up in space. And at the instant time $n\Delta t$, $F(x,y,z)$ can be written as $F^n(i,j,k) = F(i\Delta x, j\Delta y, k\Delta z)$. Using central difference, the space and time are discretized as follows, respectively

$$\frac{\partial F^n(i,j,k)}{\partial x} = \frac{F^n\left(i+\frac{1}{2},j,k\right) - F^n\left(i-\frac{1}{2},j,k\right)}{\Delta x} + O\left((\Delta x)^2\right) \tag{20}$$

$$\frac{\partial F^n(i,j,k)}{\partial t} = \frac{F^{n+\frac{1}{2}}(i,j,k) - F^{n-\frac{1}{2}}(i,j,k)}{\Delta t} + O\left((\Delta t)^2\right) \tag{21}$$

Therefore, the two curl equations in Maxwell's equation (1) and (2) can be converted to six difference equations, for example

$$
E_x^{n+1}\left(i+\frac{1}{2},j,k\right) = \frac{1-\dfrac{\sigma\left(i+\frac{1}{2},j,k\right)\Delta t}{2\varepsilon\left(i+\frac{1}{2},j,k\right)}}{1+\dfrac{\sigma\left(i+\frac{1}{2},j,k\right)\Delta t}{2\varepsilon\left(i+\frac{1}{2},j,k\right)}}\cdot E_x^{n}\left(i+\frac{1}{2},j,k\right)+
$$

$$
\frac{\Delta t}{\varepsilon\left(i+\frac{1}{2},j,k\right)}\cdot\frac{1}{1+\dfrac{\sigma\left(i+\frac{1}{2},j,k\right)\Delta t}{2\varepsilon\left(i+\frac{1}{2},j,k\right)}}\cdot
$$

$$
\left[\frac{H_z^{n+\frac{1}{2}}\left(i+\frac{1}{2},j+\frac{1}{2},k\right)-H_z^{n+\frac{1}{2}}\left(i+\frac{1}{2},j-\frac{1}{2},k\right)}{\Delta y}+\frac{H_y^{n+\frac{1}{2}}\left(i+\frac{1}{2},j,k-\frac{1}{2}\right)-H_y^{n+\frac{1}{2}}\left(i+\frac{1}{2},j,k+\frac{1}{2}\right)}{\Delta z}\right]
\tag{22}
$$

By this way, the electric field vector components in a volume of space are solved at a given instant in time, and then the magnetic field vector components are solved at the next instant in time. The process is repeated until the desired transient or steady-state EM field behavior is fully evolved.

While, in order to ensure the numerical stability condition, the time increment Δt must satisfy with

$$
\Delta t \le \frac{\sqrt{\mu\varepsilon}}{\sqrt{\dfrac{1}{(\Delta x)^2}+\dfrac{1}{(\Delta y)^2}+\dfrac{1}{(\Delta z)^2}}}.
\tag{23}
$$

Moreover, in order to solve the contradiction between the computing space and the computer memory, the absorbing boundary conditions are introduced, such as PML, Mur and so on.

Nowadays, the FDTD method is the most popular EM simulation method due to its simplicity. Most studies on numerical simulation of microwave heating were carried out by using FDTD method [83][101][102][51][103].

- **FEM**

The Finite Element Method (FEM)[104][105][106][107] is a technique to solve the PDE numerically. According to the variational principle, the PDE is transformed to an equivalent variational. As a result, the solving of the PDE is transformed to the minimization of a certain functional as follows

$$
L(u) = f \Leftrightarrow \min_{u} I(u),
\tag{24}
$$

where $L(u) = f$ is the PDE, L denotes the differential operator, f a known forcing function, u the unknown solution and I the functional to be minimized with respect to u.

After variational formulation is deduced, the discretization needs to be performed to concrete formulae for a large but finite dimensional linear algebraic equation. One can expand the function u in a series of orthogonal basic functions $\{v_i(\cdot)\}_{i=1,\cdots,n}$ as

$$u(\cdot) = \sum_{i=1}^{n} c_i v_i(\cdot) , \tag{25}$$

and then, use the minimization condition

$$\partial I / \partial c_i = 0, (i = 1, \cdots, n) , \tag{26}$$

to concrete a n-dimensional linear algebraic equation with respect to the unknown coefficients $c_i(i = 1, \cdots, n)$. As soon as $c_i(i = 1, \cdots, n)$ are solved, one can obtain u by substituting them into (25).

One of the main advantages of the FEM, in comparison to the FDTD method, lies in the flexibility concerning the geometry of the solving domain.

Some numerical simulations of microwave heating were studied by using FEM[90][81][85][108] or using a FEM software COMSOL[109][45].

- **MoM**

The Method of Moments (MoM)[110] is an efficient numerical computational method for solving integral or differential equations for the analyses of electromagnetic characteristics of complex objects. Conceptually, it works by constructing a "mesh" over the modeled surface and so can be used to solve open boundary problems without needing to truncate the domain. The MoM solves EM problems mostly in frequency domain.

For an EM problem, u in the operator function (24) represents the unknown function (e.g. the induced charge or current) and f the known excitation source (e.g. the incident field).

Just like FEM, let us expand the function u in a series of orthogonal basic functions $\{v_i(\cdot)\}_{i=1,\cdots,n}$ as (25), and use the so-called weight functions $\{w_j(\cdot)\}_{j=1,\cdots,n}$ to take the inner product at both sides of (24), which yields

$$\sum_{i=1}^{n} c_i \langle L(v_i), w_j \rangle = \langle f, w_j \rangle, \ j = 1, \cdots, n , \tag{27}$$

or in matrix form,

$$\left[L_{ij} \right]\left[c_i \right] = \left[F_j \right], \tag{28}$$

where $L_{ij} = \langle L(v_i), w_j \rangle$ and $F_j = \langle f, w_j \rangle$.

So we can get $c_i(i = 1, \cdots, n)$ by solving the n-dimensional linear algebraic equation (27), and then get u by substituting $c_i(i = 1, \cdots, n)$ into (25).

Usually, the weight functions are taken same as the basic functions. The basis functions (including *local* and *global* basis functions) are chosen to model the expected behavior of the unknown function throughout its domain.

The MoM is also used to the numerical simulations of microwave heating [25].

- **TLM**

The transmission-line matrix (TLM) method[111][112][113] is a time-domain, differential method for solving EM field problems using circuit equivalent. Based on the analogy between the wave propagation in space and the voltage and current propagation in a transmission line, it can be regarded as a discrete version of Huygens's continuous wave model.

When using this method, the computational domain is modeled by a transmission line network organized in nodes of a grid (or mesh). The value of electric and magnetic fields at each node is related to the voltages and the currents at corresponding node of the mesh.

Take a 2D case as an example. As can be seen in Fig. 4, a voltage pulse of amplitude 1V is launched on the central node in the space modeled by a Cartesian rectangular mesh. This pulse is partially reflected and transmitted according to the transmission line theory. Assume that each line has a characteristic impedance Z, then the reflection coefficient and the transmission coefficient are given as follows

$$R = \frac{Z/3 - Z}{Z/3 + Z} = -\frac{1}{2}, \; T = 1 + R = \frac{1}{2}. \tag{29}$$

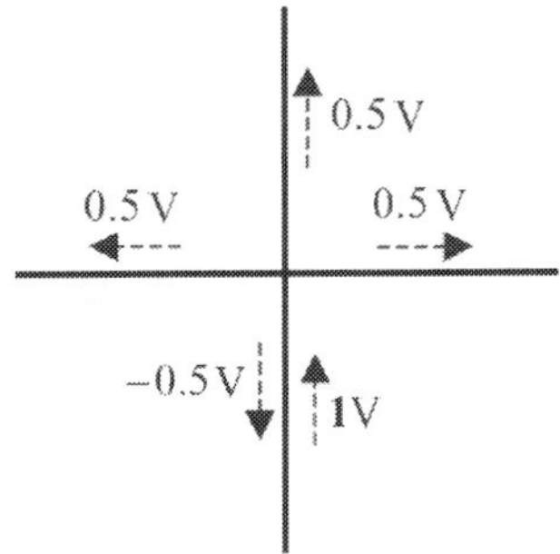

Fig. 4. The scattering of incident pulse of 1V in TLM mesh

Fig. 5. The typical 3D structure of TLM model

Therefore, we can get values of reflected and transmitted voltage shown in Fig. 4. Obviously, the energy conservation law is fulfilled by the model. And then, the reflected and transmitted voltages excite the neighboring nodes again. Fig. 5 shows the typical 3D structure of TLM model – Symmetrical Condensed Node. One can see [114] for the relative formulae and more details of TLM method.

Some numerical simulations of microwave heating were studied by using the TLM method[115][116].

When using the time domain methods such as the FDTD method, one has to consider the compatibility of the two time steps in the EM equation and the heat transport equation. Because the variation speed of the EM field (e.g. time step is 10^{-12} s) is much faster than the temperature field (e.g. time step is 1s). There are two techniques – the leapfrog scheme [51] and the time scaling factor method[24] which have been used to deal with this difficulty.

4. Conclusions

Remarkable progress has been achieved in recent years in the numerical simulation of microwave heating, which range from numerical modeling and computing to many applications in processing kinds of materials. We believe that the development will continue based on more refined models, faster numerical methods and more complete description of the complex permittivity of more materials.

With the helping of the numerical simulation, the complex process of interaction between microwave and materials will be further understood, the critical parameters influencing the process will be identified, and a rapid design of optimized and controllable applicators will be provided.

5. Acknowledgement

The authors would like to thank Juan Liu, Huabin Zhang, Lei Li, Kai Hu, Changsong Wang and Guanghua Li for their helping and thank National Natural Science Foundation of China for its financial support of this work (No. 60801035).

6. References

[1] A. J. H. Sale, "A review of microwaves for food processing," *International Journal of Food Science & Technology*, vol. 11, no. 4, pp.319–329, 1976.

[2] A. C. Metaxas, and R. J. Meredith, *Industry Microwave Heating*. London: Peter Peregrinus Ltd, 1983.

[3] D.A. Copson, *Microwave heating, Second Edition*. Westport: AVI Publishing Co. Inc., 1975.

[4] J. M. Osepchuk, "A History of Microwave Heating Applications", *IEEE Transactions on Microwave Theory and Techniques*, vol. 32, no. 9, pp.1200 - 1224, 1984.

[5] Materials Research Society (MRS) Proceedings on *"Microwave Processing of Materials"*, vol. 269, 1992; vol. 347, 1994, and vol. 430, 1996.

[6] M. E. C. Oliveira, A. S. Franca, "Microwave heating of foodstuffs," *Journal of Food Engineering*, vol. 53, no. 4, pp. 347-359, 2002

[7] Y. Wang, T. D. Wig, and J. Tang, etc., "Sterilization of foodstuffs using radio frequency heating," *Journal of Food Science*, vol. 68, no. 2, pp. 539-544, 2003.

[8] F. Marra, J. Lyng, and V. Romano, etc., "Radio-frequency heating of foodstuff: Solution and validation of a mathematical model," *Journal of Food Engineering*, vol. 79, no. 3, pp. 998-1006, 2007.

[9] G. S. Michael, and T. P. Gaines, "Microwave drying as a rapid means of sample preparation," *Tobacco Science*, vol. 6, pp. 67-68, 1975.

[10] X. Zhu, Q. Su, and J. Cai, etc., "Optimization of microwave-assisted solvent extraction for volatile organic acids in tobacco and its comparison with conventional extraction methods," *Analytica Chimica Acta*, vol. 579, no.1, pp. 88-94, 2006.

[11] H. Y. Zhou, and C. Z. Liu, "Microwave-assisted extraction of solanesol from tobacco leaves," *Journal of Chromatography A*, vol. 1129, no. 1, pp. 135-139, 2006.

[12] A. Oloyede, and P. Groombridge, "The influence of microwave heating on the mechanical properties of wood," *Journal of Materials Processing Technology*, vol. 100, no.1-3, pp. 67-73, 2000.

[13] M. Miura, H. Kaga, and A. Sakurai, etc., "Rapid pyrolysis of wood block by microwave heating", *Journal of Analytical and Applied Pyrolysis*, vol. 71, no. 1, pp. 187-199, 2004.

[14] P. Rattanadecho, "The simulation of microwave heating of wood using a rectangular wave guide: Influence of frequency and sample size", *Chemical Engineering Science*, vol. 61, no. 14, pp. 4798-4811, 2006.

[15] M. L. Li, G. O. Beale, Y. L. Tian, etc., "Modeling and control for microwave heating of ceramics," in *Proc.1995 American Control Conference*, Jun 1995, vol. 2, pp. 1235-1239.

[16] A.W. Fliflet, R.W. Bruce, and A.K. Kinkead, etc., "Application of microwave heating to ceramic processing: Design and initial operation of a 2.45-GHz single-mode furnace", *IEEE Transactions on Plasma Science*, vol. 24, no. 3, pp. 1041-1049, 1996.

[17] S. Aravindan, and R. Krishnamurthy, "Joining of ceramic composites by microwave heating," *Materials Letters*, vol. 38, no. 4, pp. 245-249, 1999.

[18] A.W. Fliflet, S.H. Gold, R.W. Bruce, etc. , "Temperature Profiles in Ceramic Cylinders Produced by Microwave and Millimeter-wave Heating," in *Vacuum Electronics Conference, 2006 held Jointly with 2006 IEEE International Vacuum Electron Sources., IEEE International*, 2006, pp. 385-386.

[19] A. Sekkak, V. N. Kanellopoulos, and L. Pichon, etc., "A thermal and electromagnetic analysis in biological objects using 3D finite elements and absorbing boundary conditions", *IEEE Transactions on Magnetics*, vol. 31, no. 3, pp. 1865-1868, 1995.

[20] R. I. Kovach , Calculation of the temperature distribution in a biological object heated by microwaves, Biomedical Engineering Volume 5, Number 6 , 336-339.

[21] J. Berlan, "Microwaves in chemistry: Another way of heating reaction mixtures," *Radiation Physics and Chemistry*, vol. 45, no. 4, pp. 581-589, 1995.

[22] Y. Wang, Y. Jing, A. Wei, and W. Song, "Numerical model of microwave heating to organic reaction," in *Proc. 1999 International Conference on Computational Electromagnetics and Its Applications*, 1999, pp. 505-508.

[23] K. Huang, and L. Bo, "The precise condition of thermal runaway in microwave heating on chemical reaction," *Science in China Series E: Technological Sciences*, vol. 52, no. 2, pp. 491-496, 2009.

[24] F. Torres, and B. Jecko, "Complete FDTD analysis of microwave heating processes in frequency-dependent and temperature dependent media," *IEEE Trans. on Microwave Theory Tech.*, vol. 45, pp. 108-116, 1997.

[25] X. Zhao, K. Huang, L. Yan, and Y. Yao, "A preliminary study on numerical simulation of microwave heating process for chemical reaction and discussion of hotspot and thermal runaway phenomenon," *Science in China Series G-Physics Mechanics Astron*, vol. 52, no. 4, pp. 551-562, 2009.

[26] R. W. P. King, and G. S. Smith, *Antennas in Matter: Fundamentals, theory, and application*. Cambridge, NASA: MIT Press, 1981.

[27] S. Ryynänen, "Microwave heating uniformity of multicomponent prepared foods," *Academic Dissertation*. University of Helsinki, 2002.

[28] R. Malheiro, I. Oliveira, M. Vilas-Boas, S. Falcão, A. Bento, and J. A. Pereira, "Effect of microwave heating with different exposure times on physical and chemical parameters of olive oil," *Food and Chemical Toxicology*, vol. 47 pp. 92–97, 2009.

[29] L. A. Campañone, C.A. Paola, and R.H. Mascheroni, "Modeling and Simulation of Microwave Heating of Foods Under Different Process Schedules," *Food and Bioprocess Technology*, DOI: 10.1007/s11947-010-0378-5, 2010.

[30] H. Zhang, and A. K. Datta, "Coupled electromagnetic and thermal modeling of microwave oven heating of foods," *Journal of Microwave Power & Electromagnetic Energy*, vol. 35, no. 2, pp. 71-86, 2000.

[31] P. Perré, and I. W. Turner, "The use of numerical simulation as a cognitive tool for studying the microwave drying of softwood in an over-sized waveguide," *Wood Science and Technology*, vol. 33, pp. 445-464, 1999.

[32] H. Zhao, and I. W. Turner, "The use of a coupled computational model for studying the microwave heating of wood," *Applied Mathematical Modelling*, vol. 24, no. 3, pp. 183-197, Mar. 2000.

[33] L. Hansson, N. Lundgren, A.L. Antti, and O. Hagman, "Finite element modeling (FEM) simulation of interactions between wood and microwaves," *Journal of Wood Science*, vol. 52, no. 5, pp. 406-410, 2006.

[34] L. Hansson, and L. Antti, "Modeling Microwave Heating and Moisture Redistribution in Wood," *Drying Technology*, vol. 26, no. 5, pp. 552-559, May 2008.

[35] P. Rattanadecho, "The simulation of microwave heating of wood using a rectangular wave guide: Influence of frequency and sample size," *Chemical Engineering Science*, vol. 61, pp. 4798- 4811, 2006.

[36] G. Brodie, "Simultaneous heat and moisture diffusion during microwave heating of moist wood," *Applied Engineering in Agriculture*, vol. 23, no. 2, pp. 179-187, 2007.

[37] A. Birnboim, and Y. Carmel, "Simulation of microwave sintering of ceramic bodies with complex geometry," *J. Am. Ceram. Soc.*, vol. 82, pp. 3024-3030, 1999.

[38] D. Bouvard, S. Charmond, and C.P. Carry, "Finite element modeling of microwave sintering," *Advances in Sintering Science and Technology, Ceram. Trans.*, vol. 209, pp. 173-180, 2010.

[39] M. D. Iskander, and A. O. N. M. Andrade, "FDTD simulation of microwave sintering of ceramics in multimode cavities", *IEEE Trans. Microwave Theory and Techniques*, vol. 42, pp. 793-799, 1994.

[40] A. Chatterjee, T. Basak, and K.G. Ayappa, "Analysis of microwave sintering of ceramics", *AIChE Journal*, vol. 44, no. 10, pp. 2302–2311, Oct. 1998.

[41] N. Gupta, V. Midha, and V. Balakotaiah, etc, "Bifurcation Analysis of Thermal Runaway in Microwave Heating of Ceramics", *Journal of the Electrochemical Society*, vol.146, no.12, pp.4659-4665, 1999.

[42] R. Riedel, and J. Svoboda, "Simulation of microwave sintering with advances sintering models", *Advances in Microwave and Radio Frequency Processing*. Springer Press, 2006.

[43] C. Liu, and D. Sheen, "Analysis and control of the thermal runaway of ceramic slab under microwave heating", *Science in China Series E-Technological Sciences*, vol.51, no.12, pp. 2233-2241, 2008.

[44] P. V. Kozlov, I. R. Rafatov, and E. B. Kulumbaev, et al, "On modeling of microwave heating of a ceramic material", *J Phys D: Appl Phys*, vol.40, pp. 2927-2935, 2007.

[45] M. Lovás, M. Kováčová, G. Dimitrakis, and S. Čuvanová, "Modeling of microwave heating of andesite and minerals", *International Journal of Heat and Mass Transfer*, vol. 53, no. 17-18, pp. 3387-3393, 2010.

[46] K. Darcovich, P. S. Whitfield, G. Amow, K. Shinagawa, and R.Y. Miyahara, "A microstructure based numerical simulation of microwave sintering of specialized SOFC materials", *Journal of the European Ceramic Society*, vol. 25, no. 12, pp. 2235-2240, 2005.

[47] P. Fischer, "Numerical Simulation of Microwave Sintering of Zinc Oxide," Master's Thesis, Virginia Technologic Institute, 1997.

[48] H. S. Ho, A. W. Guy, R. A. Sigelmann, and J. F. Lehmann, "Microwave heating of simulated human limbs by aperture sources," *IEEE Transactions on Microwave Theory and Techniques*, vol. 19, no. 2, pp. 224-226, 1971.

[49] A. Sekkak, V. N. Kanellopoulos, L. Pichon, and A. Razek, "A thermal and electromagnetic analysis in biological objects using 3D finite elements and absorbing boundary conditions," *IEEE Transactions on Magnetics*, vol.31, no.3, pp.1865-1868, 1995.

[50] Y. Wang, Y. Jing, A. Wei, and W. Song, "Numerical model of microwave heating to organic reaction," in Proc. 1999 International Conference on Computational Electromagnetics and Its Applications, 1999, pp. 505-508.

[51] K. Huang, Z. Lin, and X. Yang, "Numerical simulation of microwave heating on chemical reaction in dilute solution", *Progress In Electromagnetics Research*, vol. 49, pp. 273-289, 2004.

[52] H. Yamada, A. Chayahara, Y. Mokuno, Y. Horino, and S. Shikata, "Numerical analyses of a microwave plasma chemical vapor deposition reactor for thick diamond syntheses", *Diamond and Related Materials*, vol. 15, no. 9, pp. 1389-1394, 2006.

[53] K. Huang, and B. Lu, "The precise condition of thermal runaway in microwave heating on chemical reaction", *Science in China Series E: Technological Sciences*, vol. 52, no. 2, pp. 491-496, 2009.

[54] J. Svoboda, and H. Riedel, "Pore-Boundary Interactions and Evolution Equations for the Porosity and the Grain Size during Sintering", *Acta Metallurgica et Materialia*, vol. 40, no. 11, pp. 2829-2840, 1992.

[55] R. Gedye, F. Smith, K. Westaway, H. Ali, L. Baldisera, L. Laberge, and J. Rousell, "The use of microwave ovens for rapid organic synthesis," *Tetrahedron Letters*, vol. 27, no. 3, pp. 279-282, 1986.

[56] P. Chen, and P. S. Schmidt, "An integral model for drying hygroscopic and nonhygroscopic materials with dielectric heating", *Drying Technology*, vol. 8, pp. 907-930, 1990.

[57] W. G. Engelhart, "Microwave hydrolysis of peptides and proteins for amino acid analysis", *Am. Biotechnol. Lab*, vol. 8, pp. 30-34, 1990.

[58] T. Basak, and S. Durairaj, "Numerical Studies on Microwave Heating of Thermoplastic-Ceramic Composites Supported on Ceramic Plates", *J. Heat Transfer*, vol. 132, no. 7, 2010.

[59] P. H. Harms, Y. Chen, R. Mittra, and Y. Shimony, "Numerical modeling of microwave heating systems", *Journal of Microwave Power & Electromagnetic Energy*, vol. 31, no. 2, pp. 114-122, 1996.

[60] A. W. Filfet, "Application of microwave heating to ceramic processing: Design and initial operation of a 2.45-GHz single-mode furnace", *IEEE Trans. Plasma Sci.* vol. 24, no. 3, pp. 1041-1049, 1996.

[61] J. M. Tranquilla, M. Feng, and H. M. Al-Rizzo, "A Cartesian-Cylindrical Hybrid FD-TD Analysis of Composite Microwave Applicator Structures," *Journal of Microwave Power and Electromagnetic Energy*, vol.34, no.2, pp.97-105, 1999.

[62] P. PerreÂ, and I.W. Turner, "The use of macroscopic equations to stimulate heat and mass transfer in porous media: some possibilities illustrated by a wide range of configurations that emphasis the role of internal pressure," *Mathematical modeling and Numerical Techniques in Drying Technology*, pp.83-156,(edited by I.W. Turner and A. Mujumdar), Marcel Dekker, Inc., 1996.

[63] H. Feng, J. Tang, R. P. Cavalieri, and O. A. Plumb, "Heat and Mass Transport in Microwave Dying of Hygroscopic Porous Materials in a Spouted Bed," *AIChE J.*, vol.47, no.7, pp.1499-1512, 2001.

[64] F. Marra, M. V. De Bonis, and G. Ruocco, "Combined microwaves and convection heating: A conjugate approach," *Journal of Food Engineering*, vol. 97, pp. 31-39, 2010.

[65] F. Ying, T. Wang, and L. Johan, "Microwave-transmission, heat and temperature properties of electrically conductive adhesive," *IEEE Transactions on CPMT, Part A*, vol.26, no.1, pp.193-198 March, 2003.

[66] J. G. Kirkwood, "The dielectric polarization of polar liquids," *J. Chem. Phys.*, vol. 7, pp. 911-919, Oct. 1939.

[67] K. S. Cole, and R. H. Cole, "Dispersion and absorption in dielectrics," *J. Chem. Phys.*, vol. 9, pp. 341-351, Apr. 1941.

[68] P. Debye, *Polar Molecules*, New York: Dover, 1945.

[69] H. Fröhlich, *Theory of Dielectrics*, Oxford: Clarendon, 1949.

[70] R. H. Cole, "Theories of dielectric polarization and relaxation," in *Progress in Dielectrics*, J. Birks and J. Hart, Eds. London, U.K: Heywood, vol. 3, 1961.

[71] C. Gabriel, S. Gabriel, E. H. Grant, B. S. J. Halstead, and D. M. P. Mingos, "Dielectric parameters relevant to microwave dielectric heating," *Chemical Society Reviews*, vol.27, pp.213-223, 1998.

[72] E. Peyskens, M. De Pourcq, and M. stevens, etc., "Dielectric properties of softwood species at microwave frequencies," *Wood Science and Technology*, vol.18, no.4, pp.267-280, 1984.

[73] B. Zhou, and S. Avramidis, "On the loss factor of wood during radio frequency heating," *Wood Science and Technology*, vol.33, pp.299-310, 1999.

[74] *Ceramic Source*, compilted by American Ceramic Society, Vol. 7, 1991-1992.

[75] R. M. Hutcheon, M. S. De Jong, F. P. Adams, P. G. Lucuta, J. E. McGregor, and L. Bahen, "RF and Microwave Dielectric Measurements to 1400°C and Dielectric Loss Mechanisms," *in: Materials Research Society Symposium Proceedings*, vol.269, pp.541-551, 1992.

[76] G. W. Padua, "Microwave heating of agar gels containing sucrose," *Journal of Food Science*, vol. 58, no. 60, pp. 1426-1428, 1993.

[77] N. Sakai, W. Mao, Y. Koshima, and M. Watanabe, "A method for developing model food system in microwave heating studies", *Journal of Food Engineering*, vol. 66, no.4, pp. 525-531, 2005.

[78] K. G. Ayappa, H. T. Davis, G. Crapiste, E. A. Davis, and J. Gordon , "Microwave heating: an evaluation of power formulations", *Chemical Engineering Science*, vol. 46, no. 4, pp. 1005-1016, 1991.

[79] P. Kopyt, "A one-dimensional semi-analytical model of the microwave heating effect in verification of numerical hybrid modeling software", *EUROCON 2007 The Intl.Conf. on "Computer as a Tool"*, Warsaw, Sept. 2007, pp.65-72.

[80] Y. Alpert, and E. Jerby, "Coupled Thermal-Electromagnetic Model for Microwave Heating of Temperature-Dependent Dielectric Media", *IEEE Trans. on Plasm Science*, vol. 27, no. 2, pp. 555- 562, Apr. 1999.

[81] E. C. M. Sanga, A. S. Mujumdar, and G. S. V. Raghavan, "Simulation of convection-microwave drying for a shrinking material", *Chem. Eng. Process*, vol. 41, pp. 487-499, 2002.

[82] J. Haala, and W. Wiesbeck, "Modeling microwave and hybrid heating processes including heat radiation effects," *IEEE Trans. Microwave Theory Tech.*, vol. 50, no. 5, pp. 1346-1354, 2002.

[83] L. Ma, D.-L. Paul, N. Pothecary, C. Railton, J. Bows, L. Barratt, J. Mullin, and D. Simons, "Experimental validation of a combined electromagnetic and thermal FDTD model of a microwave heating process," *IEEE Trans. Microwave Theory Tech.*, vol. 43, no.11, pp. 2565-2572, 1995.

[84] B. Milovanović, N.Dončov, and J.Joković, "Microwave Heating Cavities: Modelling and Analysis", *Microwave Review*, vol. 10, no. 2, pp. 26-35, Nov. 2004.

[85] A. Sekkak, V. N. Kanellopoulos, L. Pichon, and A. Razek, "A thermal and electromagnetic analysis in biological objects using 3D finite elements and

absorbing boundary conditions", *IEEE Trans. on Magnetics*, vol. 31,no. 3, pp. 1865-1868, May 1995.

[86] P. Plaza-Gonzalez, et al., "New approach for the prediction of the electric field distribution in multimode microwave-heating applicators with mode stirrers", *IEEE Trans. on Magnetics*, vol. 40, no. 3, pp. 1672-1678, May 2004.

[87] K. D. Paulsen, D. R. Lynch, and J. W. Strohbehn, "Three dimensional finite boundary and hybrid element solution of the Maxwell equation for lossy dielectric media," *IEEE Trans. Microwave Theory Tech.*, vol. 36, pp. 682-693, Apr. 1988.

[88] C. T. Choi, and A. Konrad, "Finite element modeling of the RF heating process," *IEEE Trans. Magn.*, vol. 27, pp. 4227-4230, Sept. 1991.

[89] W. Li, M. A. Ebadian, T. L. White, R. G. Grubb, and D. Foster, "Heat and mass transfer—In a contaminated porous concrete slab with variable dielectric properties," *Int. J. Heat Mass Transfer*, vol. 37, no. 6, pp. 1013-1027, Apr. 1994.

[90] A. Sekkak, L. Pichon, and A. Razek, "3-D FEM magneto-thermal analysis in microwave ovens," *IEEE Trans. Magn.*, vol. 30, pp. 3347–3350, Sept. 1994.

[91] L. E. Lagos, W. Li, M. A. Ebadian, T. L. White, R. G. Grubb, and D. Foster, "Heat transfer within a concrete slab with a finite microwave heating source," *Int. J. Heat Mass Transfer*, vol. 38, no. 5, pp. 887–897, Mar. 1995.

[92] L. Ma, D. Paul, N. Pothecary, C. Railton, J. Bows, L. Barratt, J. Mullin, and D. Simons, "Experimental validation of a combined electromagnetic and thermal FDTD model of a microwave heating process," *IEEE Trans.Microwave Theory Tech.*, vol. 43, pp. 2565–2572, Nov. 1995.

[93] A. Sekkak, V. N. Kanellopoulos, L. Pichon, and A. Razek, "A thermal and electromagnetic analysis in biological objects using 3D finite elements and absorbing boundary," *IEEE Trans. Magn.*, vol. 31, pp. 1865-1868, May 1995.

[94] J. Clemens, and C. Saltiel, "Numerical modeling of materials processing in microwave furnaces," *Int. J. Heat Mass Transfer*, vol. 39, no. 8, pp. 1665-1675, 1996.

[95] J. Braunstein, K. Connor, S. Salon, and L. Libelo, "Investigation of microwave heating with time varying material properties," *IEEE Trans. Magnetics* , vol. 35, no. 3, pp. 1813-1816, 1999.

[96] M. Clemens, E. G. Jonaj, P. Pinder, and T. Weiland, "Numerical simulation of coupled transient thermal and electromagnetic fields with the finite integration method," *IEEE Trans. Magnetics,* vol. 36, no. 4, pp. 1448-1452, 2000.

[97] A. Taflove and S. C. Hagness, *Computational Electrodynamics: The Finite-Difference Time-Domain Method, 3rd ed.* Artech House Publishers, 2005.

[98] A. Taflove, and M. E. Brodwin, "Numerical solution of steady-state electromagnetic scattering problems using the time-dependent Maxwell's equations", *IEEE Transactions on Microwave Theory and Techniques*, vol. 23, pp. 623-630, 1975.

[99] J. Berenger, "A perfectly matched layer for the absorption of electromagnetic waves", *Journal of Computational Physics*, vol. 114, pp. 185–200, 1994.

[100] K. Yee, "Numerical solution of initial boundary value problems involving Maxwell's equations in isotropic media", *IEEE Transactions on Antennas and Propagation*, vol. 14, pp. 302-307, 1966.

[101] H. M. Al-Rizzo, M. Feng, and J. M. Tranquilla, "Incorporation of waveguide feed and cavity wall losses in a Cartesian/cylindrical hybrid finite-difference time domain (FD-TD) analysis of microwave applicator", *J. Microwave Power & Electromag. Energy*, vol. 35, no. 2, pp. 110-118, 2000.

[102] H. M. Al-Rizzo, J. M. Tranquilla, and M. Feng, "A finite difference thermal model of a cylindrical microwave heating applicator using locally conformal overlapping grids: Part I - Theoretical formulation", *J Microwave Power & Electromag. Energy*, vol. 40, no. 1, pp. 17-29, 2005.

[103] H. Chen, J. Tang, and F. Liu, "Simulation model for moving food packages in microwave heating processes using conformal FDTD method", *Journal of Food Engineering*, vol. 88, no. 3, pp. 294-305, 2008.

[104] G. Strang, and G. Fix, *An Analysis of the Finite Element Method 2nd Edition*, Wellesley-Cambridge, 2008.

[105] I. Babuska, U. Banerjee, and J. E. Osborn, "Generalized Finite Element Methods: Main Ideas, Results, and Perspective," *International Journal of Computational Methods*, vol. 1, no. 1, pp. 67-103, 2004.

[106] G. Pelosi, "The finite-element method, Part I: R. L. Courant [Historical Corner]," *IEEE Antennas and Propagation Magazine*, vol. 49, no. 2, pp.180 – 182, 2007.

[107] R. L. Ferrari, "The Finite-Element Method, Part 2: P. P. Silvester, an Innovator in Electromagnetic Numerical Modeling ," *IEEE Antennas and Propagation Magazine*, vol. 49, no. 3, pp.216-234, 2007.

[108] M. Pauli, T. Kayser, and G. Adamiuk, etc., "Modeling of mutual coupling between electromagnetic and thermal fields in microwave heating," in *Proc. IEEE MTT-S International Microwave Symposium 2007*, Jun. 2007, pp. 1983-1986.

[109] D. Bouvard, S. Charmond, and C. P. Carry, "Multiphysics Simulation of Microwave Sintering in a Monomode Cavity", in *Proc. 12th Seminar Computer Modeling in Microwave Engineering & Applications (Advances in Modeling of Microwave Sintering)*, Grenoble, France, March 8-9, 2010, pp. 21-26.

[110] W. C. Gibson, *The Method of Moments in Electromagnetics*, London: Chapman & Hall/CRC, 2008.

[111] W. J. R. Hoefer, "The Transmission-Line Matrix Method—Theory and Applications," *IEEE Transactions on Microwave Theory and Techniques*, vol. 33, no. 10, pp.882-893, 1985.

[112] P. B. Johns, and R. L. Beurle, "Numerical solution of two-dimensional scattering problems using a transmission-line matrix," *Proc. IEE*, vol. 118, no. 12, pp.1203-1208, 1971.

[113] P. B. Johns, "A symmetrical condensed node for the TLM method," *IEEE Trans. MTT*, vol. 35, no. 4, pp.370-377, 1987.

[114] C. Christopoulos. *The Transmission-Line Modeling (TLM) Method in Electromagnetics*, First Edition, Morgan & Claypool, 2006.

[115] C. Flockhart, V. Trenkic, and C. Christopoulos, "The simulation of coupled electromagnetic and thermal problems in microwave heating", in *Proc. Second International Conference on Computation in Electromagnetics*, London, Apr. 1994, pp. 267-270.

[116] C. Christopoulos, "The application of time-domain numerical simulation methods to the microwave heating of foods," *IMA Journal of Management Mathematics*, vol. 5, no. 1, pp.385-397, 1993.

3

Modelling and Analysis of the Induction-Heating Converters

András Kelemen and Nimród Kutasi
Sapientia Hungarian University of Transylvania
Romania

1. Introduction

The aim of this chapter is the presentation of some aspects related to the modelling and analysis of the induction-heating load-resonant converters as components of the induction-heating equipments.

The induction-heating converters are used to produce a high-intensity alternative magnetic field in an inductor. The workpiece placed in the inductor is heated up by eddy-current losses produced by the magnetic field at a frequency defined by technological requirements (Rudnev et al., 2003).

The basic principles of power electronic converter modelling are presented with emphasis on the specific solutions that can be applied in case of load-resonant converters. Detailed description and modelling of the electromagnetic, thermal and metallurgical phenomena involved in the induction-heating process is behind the scope of this chapter.

Mathematical modelling of the induction-heating equipments means the representation of the physical behaviour of these systems by mathematical means.

Generally, modelling should be concentrated on the phenomena which are of interest for the solution of the given engineering problem. These phenomena can be labelled as "dominant behaviour". All the other phenomena which have no influence from the point of view of the given problem are regarded as "insignificant phenomena". For example, the parasitic circuit components like the parasitic capacitances of power transistors and the stray inductances of busbars have negligible influence on the power control characteristics of a low-frequency or medium-frequency induction-heating inverter, while their influence is very important from the point of view of the transistor switching losses.

The model should be as simple as possible, because a simple model yields better physical insight. For instance, in the model of the resonant load circuit used for converter analysis, an inductor for induction heating can almost always be modelled by a serial LR circuit with constant inductance and resistance. Variation of these parameters with the temperature of the workpiece and with the process (for example with the position of the workpiece in a scanning inductor) can be taken into account by defining a possible range of inductor parameters. The behaviour of the converter is normally verified for several sets of parameters from this range, but these are kept constant during the converter analysis.

The refinement of the model can be made if necessary after the basic phenomena have been clarified. Depending on the problem to be analyzed, one can decide to refine the existing model or to go for other simplified models. For instance, the analysis of the DC-link voltage of an induction-heating converter can be performed using a detailed rectifier and filter model, with the inverter input represented by equivalent resistive loads corresponding to different operating states of the inverter. With the variation range of the DC-link voltage determined this way, the operation of the inverter can be studied using an ideal voltage source. While this approach is useful for defining the component stresses and clarifying basic phenomena concerning the operation of these two sides of the converter, it can be naive in case of the controller design. Due to the dynamic influence between the filter and the inverter and due to the presence of AC components in the rectified voltage, the complete model is needed for controller design.

In the following, several modelling procedures are presented, which form a set of useful tools for analysis of the induction-heating converters when they are used according to the concept of hierarchical (multilevel) modelling.

2. The hierarchical approach to resonant converter modelling

One of the basic ideas emerging from the above principles is that modelling of the induction-heating load-resonant converters has to be made in a *hierarchical* way (Mohan et al., 2002). This means that different methods and tools have to be used depending on the details of the phenomena into which we want to get insight. Each modelling method or simulation program has its advantages and limitations, which make it suitable for the analysis of some of the aspects of a converter's behaviour.

It should be clarified that modelling is always an iterative process, even if some modelling steps are usually substituted by a-priori knowledge about the converter. For instance, a certain operation mode of a converter is normally the result of a closed-loop (controlled) operation, which can be assessed by higher-level modelling, but knowledge about the existence of this mode and about the circumstances it creates can be used for lower-level analysis (for example calculation of the switching device losses in a certain mode of operation).

The hierarchical approach is closely related to the main results expected from the modelling and simulation of the induction-heating converters, which are:
- calculation of the stress on the power devices in different operating conditions;
- reliable understanding of the converter operation based on the waveforms of different circuit variables;
- determination of the steady-state and dynamic performance of the system.

Modelling methods and analysis tools suitable for obtaining these results are presented in the following, with detailed discussion of some methods that make possible the analysis of the induction-heating load-resonant converters at higher hierarchical levels.

In case of a low-level analysis the electrical devices and circuits are modelled in detail, while in case of higher-level analysis simplifying assumptions are introduced, which ignore some details, but which make it possible to concentrate on the behaviour of the system as a whole.

The computing effort is kept almost the same at different levels of the hierarchy at the expense of simplifying assumptions. At higher levels the detailed circuit models are replaced by behavioural ones.

Most knowledge about the converter operation can be obtained by dynamic modelling, which makes also possible the synthesis of the control system.

In case of the induction-heating converters the main control tasks are the control of the switching conditions and the control of the heating power.

The switching conditions are usually controlled by means of some PLL loop version. The task of this controller is either to assure some essential operating conditions (for example a capacitive phase shift between the output current and voltage of a current-source thyristor-based inverter) or to create optimal switching conditions (for example zero voltage switching in case of a transistor-based voltage inverter).

The power control is performed either by means of an auxiliary converter, or by means of the resonant inverter. The use of an auxiliary converter makes possible the decoupling of the two main control tasks (Kelemen & Kutasi, 2007a).

Examples for the use of auxiliary converters are the application of controlled rectifiers in case of medium frequency induction-heating converters with thyristor-based current inverter and the application of step-down DC-DC converters in case of the high-frequency voltage inverters with power control by pulse amplitude modulation.

The power control by means of the inverter can be performed in several ways including different pulse width modulation (PWM) methods, pulse density modulation (PDM) (Fujita & Akagi, 1996), frequency shift control (FS) and their different combinations (Kelemen, 2007).

Small-signal (linearized) models can be used successfully to design control loops relying on classical control theory, while large-signal (usually nonlinear) models can be used for controller design based on the nonlinear control theory.

Sometimes the delay introduced by the processing time of the controller should also be modelled. Thus, the presence of the controller itself becomes an input data for the controller design.

Detailed modelling is generally used to determine the stress on the power devices and for detailed analysis of the high-speed processes, like transistor switching. Such high-speed processes are strongly influenced by circuit components like snubbers, which do not influence the main behaviour of the converter.

This level of the detailed analysis can be performed using general-purpose simulation programs, which solve in each step of the simulation a complicated nonlinear system of equations. The detailed modelling approach can be problematic due to the large scale of the time constants of the processes involved. Due to the fast variation of the circuit variables during switching, it can happen that the initial guess used in the corresponding time step is not close enough to the solution to assure the convergence of the solver.

Besides, for an acceptable computing time it is important to set the initial conditions of the whole simulation close to the steady-state values of the state variables.

These values can be obtained using a less detailed analysis based for instance on a **switched-circuit model**, which contains only the main circuit components and which can be used to determine the dominant circuit waveforms. This kind of model can offer insight into the principles of the converter operation, but handling of the switches needs special techniques.

Switched-circuit models can be built based either on ideal or on non-ideal switches. Non-ideal power electronic switches are often represented as piece-wise linear networks.

The main problems to be solved are:

- the switching instants of "internally controlled" switches have to be determined accurately;
- switching can cause discontinuity of the state variables of the circuit (Dirac pulses);
- the circuit topology subsequent to the switching has to be determined correctly;
- the initial conditions of a switching state have to be determined in a "consistent" way, i.e. given the solution of the circuit immediately before the switching instant, the solution in the moment immediately following the switching instant has to be determined.

Not surprisingly, the switched-circuit modelling with ideal switches is similar to the approach used in case of the **hybrid systems**, a class of systems which nowadays represents the subject of intense research activity.

The load-resonant induction-heating inverters are inherently hybrid systems consisting of circuits with continuous dynamics and of switching devices. Hybrid modelling and analysis is a powerful tool for describing the behaviour of the system.

The key problem is to model the discrete transitions (when and how) and to associate to each discrete state a corresponding continuous dynamics. To address this problem one can use the hybrid automaton theory (Torrisi & Bemporad, 2004), (Borelli, 2003).

In the following the idea of hybrid modelling is outlined in case of the induction-heating voltage inverter with LLC resonant load (Fig. 1).

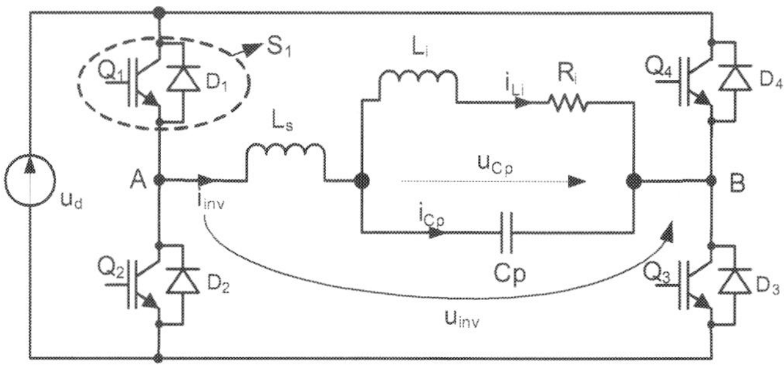

Fig. 1. Voltage inverter with LLC resonant load

Let $X \in \Re^n$ be the continuous space of the inverter states and let $Q \in \{q_1, q_2, ..., q_m\}$ be a finite set of discrete states. In case of the inverter from Fig. 1 and the control strategy presented below, n=3 and m=3. The continuous state space specifies the possible values of the continuous states for all switching configurations $q \in Q$ of the converter. For each $q \in Q$ the continuous dynamics can be modelled by differential equations of the form:

$$\underline{\dot{x}}(t) = f_q(\underline{x}(t)) = A_q \cdot \underline{x}(t) + b_q ;$$
$$A_q \in \Re^{n \times n}, b_q \in \Re^{n \times 1}, \underline{x} \in X .$$

(1)

The states of the circuit from Fig. 1 are: $\underline{x}(t) = \begin{bmatrix} i_{inv} & i_{Li} & u_{Cp} \end{bmatrix}^T$. The discrete states of the system are defined by the states of the two-quadrant switches S1, S2, S3 and S4. Each of the switches consists of a transistor Q_i and its anti-parallel diode D_i. The transistors are turned on and off by gate control, while the diodes are controlled by their currents and AK voltages. There are several possible strategies for the control of the inverter from Fig. 1 (Kelemen, 2007). A possible sequence of the gate control signals is shown in Fig. 2 along with characteristic waveforms of the i_{inv} inverter current and of the u_{Cp} capacitor voltage. This control mode is characterized by three discrete states (2) and is called *discontinuous current mode* due to the existence of the state q_3.

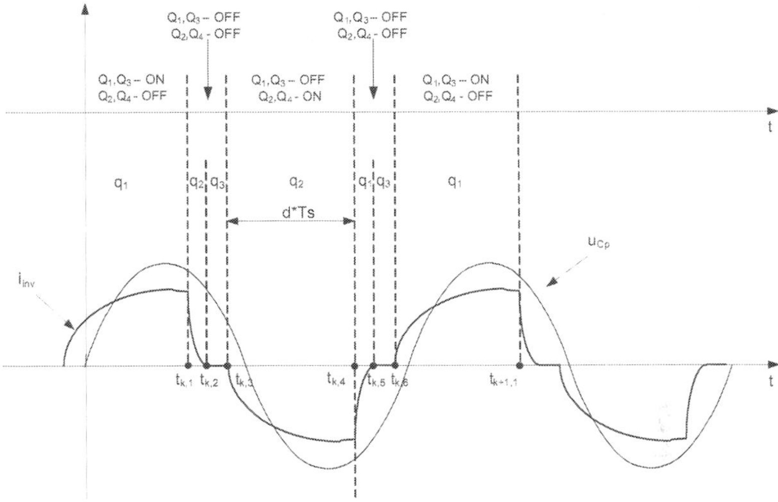

Fig. 2. The resonant load capacitor voltage u_{Cp} and the inverter output current i_{inv} in discontinuous current operation mode

$$q_1 = \left(S_1 - ON, S_3 - ON, S_2 - OFF, S_4 - OFF\right)$$
$$q_2 = \left(S_1 - OFF, S_3 - OFF, S_2 - ON, S_4 - ON\right) \quad \cdot \tag{2}$$
$$q_3 = \left(S_1 - OFF, S_3 - OFF, S_2 - OFF, S_4 - OFF\right)$$

The transitions between the different discrete states are related to the time instances denoted by $t_{k,i}$. For instance the first discrete state q_1, characterized by the conduction of Q_1 and Q_3, is followed from the moment $t_{k,1}$ by the state q_2, characterized by the conduction of the diodes D_2 and D_4. The third discrete state means that all switches are open, no transistor or diode is in conduction, and consequently the inverter output current is interrupted.

The hybrid model of the system is depicted in Fig. 3. The hybrid automaton H has been decomposed in two hybrid automata H_1 and H_2. H_1 is a finite state machine governing the discrete transitions which depend on the continuous state vector $\underline{x}$ delivered by H_2. The continuous states $\underline{x}$ of H_2 evolve according to the discrete variable $\delta \in \Delta$ ($\Delta = \{\delta_1, \delta_2, \delta_3\}$) received from H_1. The discrete transitions are governed by the time instances determined by the controllers and by the internal switching conditions. The linear circuits with continuous dynamics corresponding to the discrete states are also shown in Fig. 3.

Fig. 4 presents the structure of the inverter controller, indicating the influence of its components on the different switching instances. Thus, $t_{k,1}$ and $t_{k,4}$ are determined by the frequency controller and the voltage-controlled oscillator (VCO), while $t_{k,2}$ and $t_{k,5}$ are determined by the voltage (power) controller.

The uncontrolled ("internally controlled", "indirectly controlled") switching instances $t_{k,3}$ and $t_{k,6}$ depend on the parameters of the resonant load and on the DC supply voltage.

The hybrid automaton H can be transformed into different models (PWA – **Piece-Wise Affine**, DHA – **Discrete Hybrid Automaton**) for analysis and controller design using high-level hybrid-system describing languages like HYSDEL (Torrisi & Bemporad, 2004).

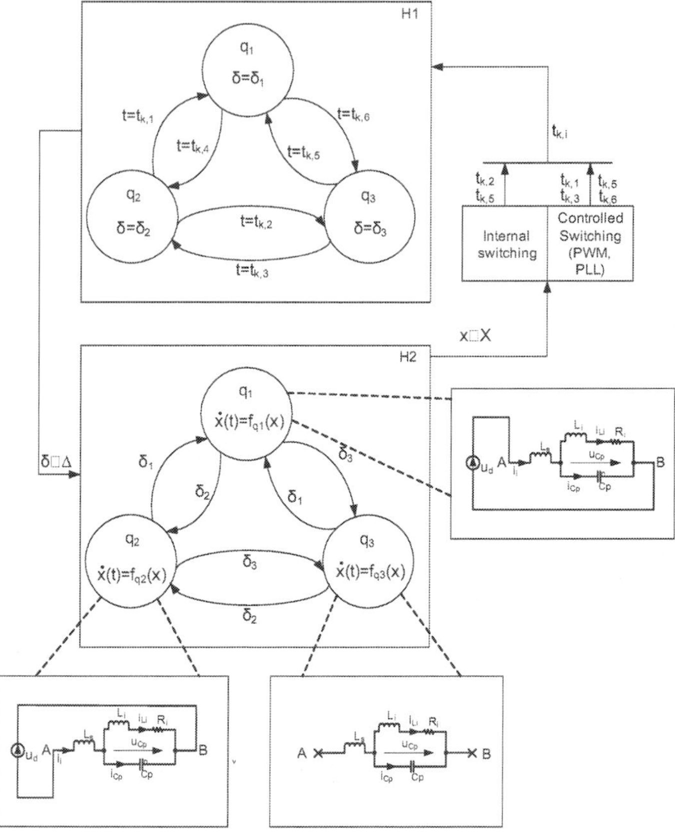

Fig. 3. The hybrid model of the inverter with LLC resonant load in discontinuous current mode of operation

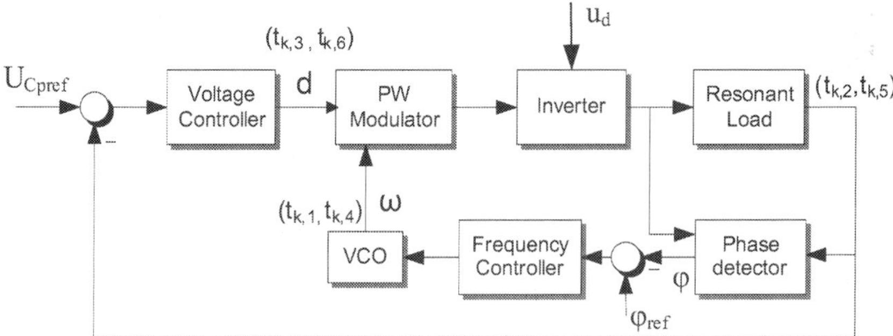

Fig. 4. The principle of closed-loop control of the inverter with LLC resonant load in discontinuous current operation mode

Taking into account the hybrid nature of the power electronic converters, we can say that the hybrid modelling and analysing approach is the natural way in this field.

A further step towards the simplification of the converter model can be made by **averaging** performed either over the state-space equations or directly over the converter circuit. In case of the resonant converters, the simulation of the closed-loop, large-signal behaviour is a challenge due to the high-frequency large variation of the state variables, which makes the state-space averaging method useless. Due to large variation of the state variables within a switching cycle, the state-space averaging method would pass over the essential phenomena.

The averaging method has been generalized in order to be applicable also in case of resonant converters. Application examples of the classical and generalized averaging methods are given in the next subsections, with detailed presentation of the *d-q* modelling technique.

In order to avoid the need to model the high-frequency variations of the state variables, the **sampled-data modelling technique** has been proposed for description of both large-signal and small-signal dynamics of resonant converters (Elbuluk et al., 1998). In this case, the states of the converter are redefined to be equal with their values at the ends of the switching periods.

3. Modelling of the induction-heating resonant load circuit

The induction-heating resonant load is composed of usually linear and time-invariable reactive components and the induction-heating inductor. The structure of the resonant load is strongly dependent on the induction-heating converter topology and the load either can be viewed as power-factor compensation scheme, or can be considered a component of the load-resonant converter.

The description of the inductor's behaviour is a coupled-field problem, with electromagnetic, thermal, metallurgical and hydrodynamic phenomena with more or less strong coupling.

For instance, the temperature of the workpiece has a strong influence on the electromagnetic phenomena, like the penetration depth and the power density. This influence becomes strongly non-linear when the metallurgical phase transformation leads to a sudden change of the magnetic properties of the workpiece at the Curie temperature.

The induction heating represents a transient process both from thermal and electromagnetic point of view, but the behaviour of the electric circuit is usually analysed assuming a permanent thermal state.

For a given operating frequency and temperature distribution, the inductor and the workpiece can be represented by an equivalent RL circuit.

Usually, the quality factor of the inductor is 3...40 ($\cos\varphi < 0.4$), with lower values in case of inductors with field concentrators and higher values in case of weak coupling between the inductor and the load (ex. vacuum melting). The equivalent parameters are valid only for a given intensity of the magnetic field. In the simple case of a cylindrical workpiece in a long inductor, the equivalent impedance has the form (3) (Sluhotkii & Rîskin, 1982).

$$\underline{Z}_e = a\sqrt{f} + j\left(b\sqrt{f} + cf\right).$$

(3)

In case of strong coupling and large variation of the operating frequency, a "large bandwidth" model can be used (Forest et al., 2000) in order to take into account the frequency-dependence of the equivalent parameters (Fig. 5).

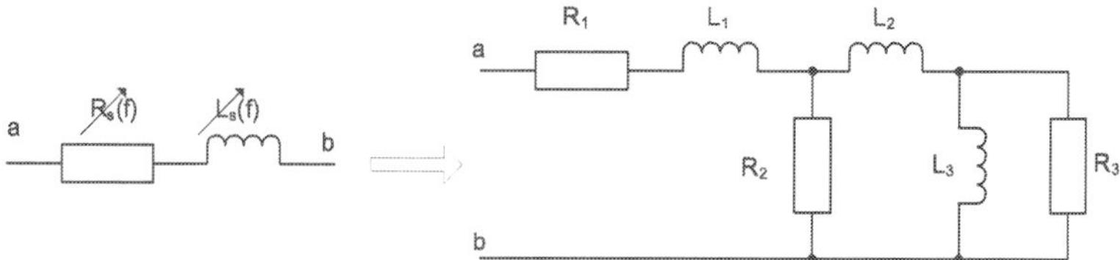

Fig. 5. Serial RL inductor model with frequency-dependent parameters and its large-bandwidth equivalent circuit with constant parameters

4. Converter modelling by averaging techniques

A popular class of analysis methods for power electronic converters is based on the **averaging technique**. Different versions of this technique have been developed based on assumptions fulfilled by different classes of converters and these methods can provide large-signal (nonlinear) and small-signal (linear) converter models.

In fact, averaging represents a kind of transformation, which replaces the original "fast" variables with "slow" variables. For instance, the evolution in time of the inductor current from a step-down DC-DC converter can be seen as the evolution of the DC component of this current, with a superimposed ripple. The DC component means in fact the moving average, with slow variation.

The question can be raised whether the trajectory of this moving average is represented well enough by the trajectory of the corresponding quantity from the averaged model, and which are the conditions that must be satisfied in order to have a good approximation of the converter operation? (Krein et al., 1990). There are also questions, whether the averaged model is valid for large variations of the signals and whether stability properties of the system are preserved by the transformation, or not.

Periodically driven systems, like the power-electronic converters can be modelled using the **generalized averaging** method, presented in (Sanders et al., 1991).

The generalized averaging concept is based on the calculation of the time-dependent complex Fourier coefficients of the circuit variables, using a sliding frame equal with a switching period.

For a system-level analysis the circuit variables can be represented with acceptable accuracy by the first one or two terms of their Fourier series expansion. For instance, in case of a switch-mode DC-DC converter, the evolution of the average output voltage (i.e. of the first Fourier term) is important from the point of view of the controller design, while in case of a load-resonant induction-heating inverter, the amplitude of the resonant capacitor voltage (i.e. the second term) is of interest.

For a circuit variable x, the coefficients of the series (4) are defined by (5) (Sanders et al., 1991).

$$x(t-T+\tau) = \sum_k \langle x \rangle_k (t) e^{jk\omega(t-T+\tau)} \tag{4}$$

$$\langle x \rangle_k (t) = \frac{1}{T} \int_0^T x(t - T + \tau) e^{-jk\omega(t-T+\tau)} d\tau \tag{5}$$

T is the switching period, $\omega = 2\pi / T$ and τ is the time from the beginning of the sliding frame.

The question is how do the Fourier coefficients vary with time if the circuit variables satisfy a given set of differential equations (6).

$$\frac{d}{dt} x = f(\underline{x}, \underline{u}) \tag{6}$$

In (5) $\underline{x}$ is the state vector and $\underline{u}$ is the periodic input of the system.

After expansion of (6), imposing the equality of the corresponding coefficients, it results:

$$\left\langle \frac{d}{dt} x \right\rangle_k = \langle f(\underline{x}, \underline{u}) \rangle_k \tag{7}$$

Using the differentiation rule (8)

$$\frac{d}{dt} \langle \underline{x} \rangle_k = \left\langle \frac{d}{dt} \underline{x} \right\rangle_k - jk\omega \langle x \rangle_k , \tag{8}$$

the relation (7) becomes:

$$\frac{d}{dt} \langle \underline{x} \rangle_k = -jk\omega \langle x \rangle_k + \langle f(\underline{x}, \underline{u}) \rangle_k . \tag{9}$$

This differential equation doesn't contain the k-th order high-frequency terms. Instead, by means of the Fourier coefficients, it describes the "slow" variation of the amplitude of the state variables.

This method is similar to the time-varying phasor analysis presented in (Rim & Cho, 1990), which introduces a modified phasor $X(t)$, defined by:

$$x(t) \equiv \mathrm{Re}\left\{ \sqrt{2} \, \underline{X(t)} e^{j\omega t} \right\} . \tag{10}$$

In (10), $x(t)$ represents a sinusoidal function of time with slowly varying amplitude.

Substituting $k = 0$ into (9) the result is the well-known **state-space averaging** method introduced by Cuk and Middlebrook in 1977, which can be used to describe the operation of hard-switching PWM DC-DC converters, but it is not suitable for analysis of quasi-resonant or resonant converters.

For steady-state analysis of the resonant converters, the so-called "cyclic-averaging" technique is useful (Foster et al., 2003).

In subsections 4.1 and 4.2 it is demonstrated, how the averaging techniques can be applied for modelling of the induction-heating converters. First the d-q modelling method (Zhang & Sen, 2004) is applied as a version of generalized averaging in order to build an inverter model which can be used for the design of the PLL and power controllers.

In the case presented in subsection 4.1 the power is controlled by means of the inverter. In the case of the converter presented in subsection 4.2, the power control is solved by an auxiliary converter, which is modelled using the state-space averaging method.

4.1. The *d-q* model of a voltage inverter with LLC resonant load

The resonant inverter with hybrid LLC load and its characteristic waveforms are presented in Fig. 6 (Kelemen & Kutasi, 2007a). This load matching solution is also called "resonance transformation" (Fischer & Doht, 1994).

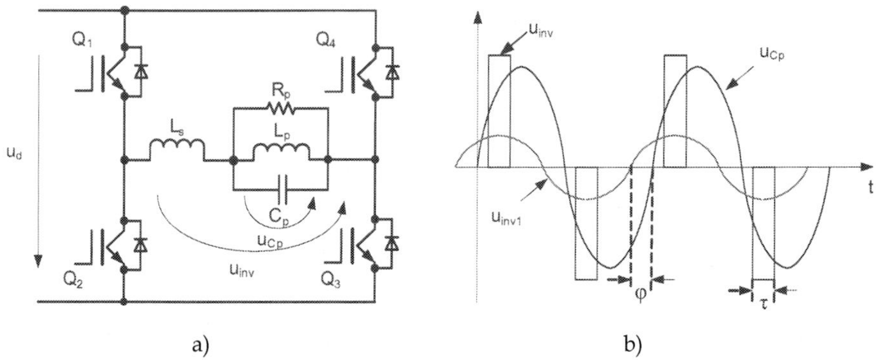

a)   b)

Fig. 6. Voltage inverter, a) with LLC resonant load and parallel inductor model and b) its characteristic waveforms

The complex circuit of the resonant inverter is composed from the original and its orthogonal circuit (Fig. 7). Fig. 8 shows the high-frequency complex circuit variables expressed in d-q form.

Substituting the complex current of the L_s inductor into the inductor voltage equation, it results:

$$\underline{u}_{Ls}(t) = \left(u_{Lsd}(t) + ju_{Lsq}(t) \right) e^{j\omega t} = L_s \left(\frac{d}{dt} \left(i_{Lsd}(t) + ji_{Lsq}(t) \right) + j\omega \left(i_{Lsd}(t) + ji_{Lsq}(t) \right) \right) e^{j\omega t} . \quad (11)$$

The complex current of capacitor C_p results from the complex capacitor voltage as

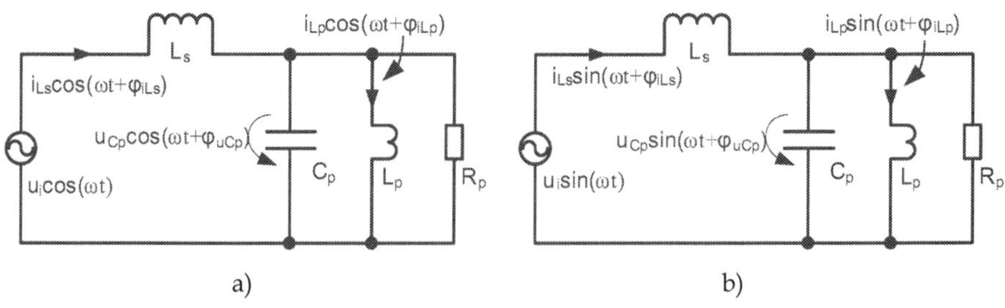

a)   b)

Fig. 7. The inverter with resonant load and its orthogonal circuit

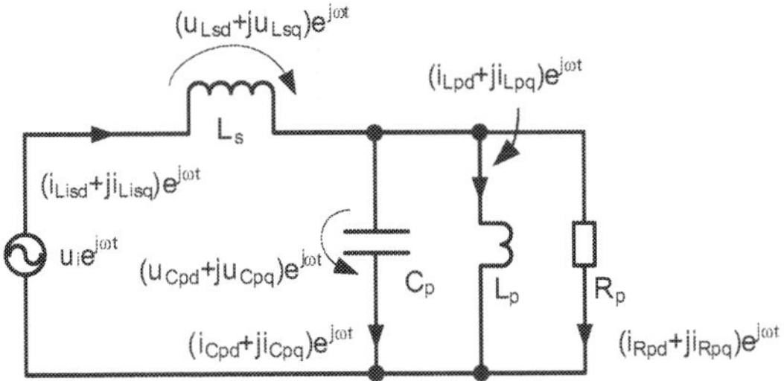

Fig. 8. The complex inverter circuit

$$\underline{i}_{Cp}(t) = \left(i_{Cpd}(t) + j i_{Cpq}(t)\right) e^{j\omega t} = C_p\left(\frac{d}{dt} + j\omega\right)\left(u_{Cpd}(t) + j u_{Cpq}(t)\right) e^{j\omega t}. \tag{12}$$

Assuming low-frequency variation, thus negligible second derivative of $u_{Cpd}(t)$ and $u_{Cpq}(t)$, the current of the inductor L_p can be expressed as follows:

$$\underline{i}_{Lp}(t) = \frac{1}{L_p}\int\left(u_{Cpd} + j u_{Cpq}\right) e^{j\omega t} = \frac{1}{\omega L_p}\left(\frac{1}{\omega}\frac{d}{dt} - j\right)\left(u_{Cpd} + j u_{Cpq}\right) e^{j\omega t}. \tag{13}$$

After removal of the high-frequency terms from (11), (12) and (13), the corresponding relations can be modelled with passive circuit components and controlled voltage or current sources. The low-frequency complex d-q model is obtained (Fig. 9), with low-frequency complex circuit variables, the time-dependence of which has not been shown for simplicity of notation. Decomposition of the complex d-q model leads to the low-frequency d-q model from Fig. 10.

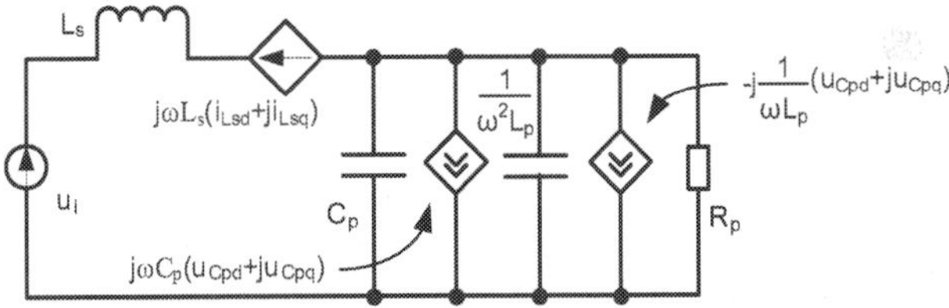

Fig. 9. Low-frequency complex d-q model of the resonant inverter

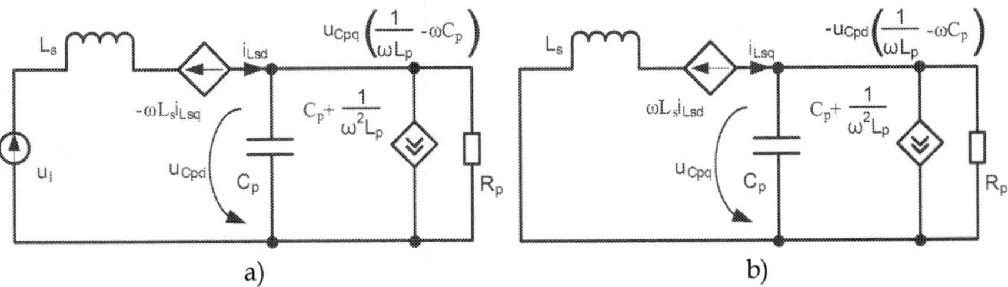

Fig. 10. Low-frequency d (a) and q (b) circuits

Start-up evolution of the resonant capacitor tank voltage under various load conditions is shown in Fig. 11.a. The circuit parameters of the resonant tank are $C_p = 63\ \mu F$, $L_s = 20\ \mu H$ and $L_p = 4\ \mu H$. The resonant capacitor tank voltage is controlled by means of pulse width modulation of the inverter output voltage, while the phase shift φ is controlled by a phase locked loop. Start-up evolution of the phase shift φ is shown in Fig. 11.b. The initial duty factor, the start-up frequency and the controller coefficients are $D_{start} = 0.005$, $f_{start} = 15\ \text{kHz}$, $K_1 = 0.005$, $K_2 = 0.3$, $K_3 = 1000$, $K_4 = 50K_3$, while the voltage and phase shift set-values are $U_{Cpref} = 300\ V$ and $\varphi_{ref} = 50°$. Unstable operation can be observed in case of a high quality factor.

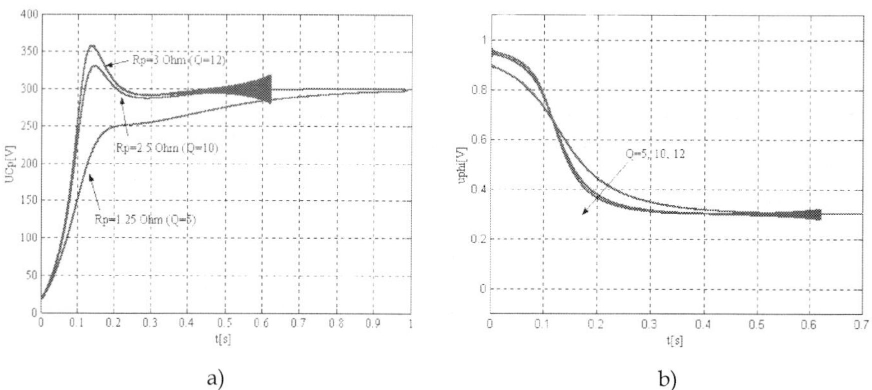

a)   b)

Fig. 11. Start-up evolution of the resonant capacitor tank voltage u_{Cp} (a), and of the phase shift φ (u_{phi} is the output voltage of the phase detector, 0.5 V corresponding to $\varphi = 90°$), for different values of the quality factor ($Q = 5, 10, 12$)

The small-signal model resulting from the perturbation of the above d-q model is shown in Fig. 12. The steady-state d-q components of the state variables and of the angular frequency are denoted with I_{Lsd}, I_{Lsq}, U_{Cpd}, U_{Cpq} and Ω.

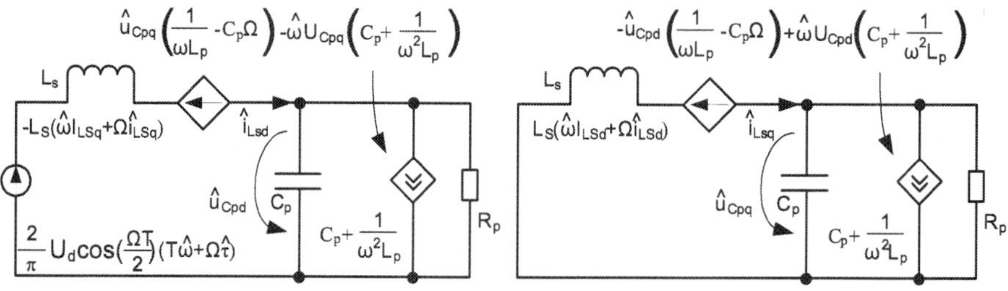

Fig. 12. Small-signal low-frequency d-q model of the inverter with LLC resonant load and parallel equivalent inductor circuit

4.2. State space averaging in case of a converter with pulse amplitude modulation

State-space averaging is an efficient method for analysis of converters characterized by the fulfilment of certain simplifying conditions. These are generally formulated as the small ripple condition and the linear ripple approximation (Erickson & Maksimović, 2004), (Sanders et al., 1991).

Small ripple means that the DC component of the signal for which this condition is applied is much larger than the other terms of its Fourier expansion. Of course the meaning of "much larger" is not easy to be defined and only the accuracy of the final result can tell whether its interpretation had been correct or not.

An application example of the state-space averaging can be given for the induction-heating converter with voltage inverter fed from a two-quadrant DC-DC converter (Fig. 13). In this case this intermediary converter solves the modulation of the DC voltage, providing the power control by pulse amplitude modulation.

The output voltage of this step-down DC-DC converter can always be handled from the point of view of the converter analysis as a small-ripple signal. This is simply the requirement from the load side and the converter must be designed to have a small output voltage ripple. The result is that the output voltage can be considered a DC voltage from the point of view of its influence on other circuit variables. Knowledge of the circuit variables determined by the consequently simplified analysis further gives the possibility to determine the small ripple itself.

The linear ripple approximation means that the given signal is assumed to have a linear variation during a switching state of the converter. This is the case of the inductor current from the two-quadrant DC-DC converter used to supply the induction-heating inverter with pulse amplitude modulation from Fig. 13.

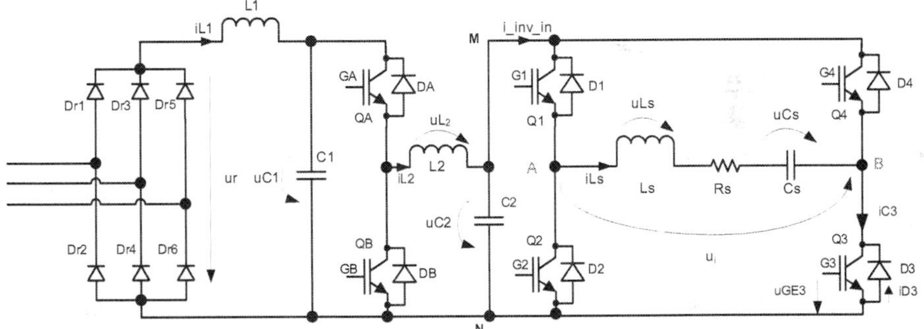

Fig. 13. Induction-heating converter with power control by pulse amplitude modulation

Let us consider the variation of the inductor current over a switching period. The switching states are:

$$q_1 = \left(Q_A - ON, Q_3 - OFF\right) \ and \ q_2 = \left(Q_A - OFF, Q_3 - ON\right). \tag{14}$$

The u_{L2} inductor voltage is assumed to be approximately constant during a switching period i.e. $u_1(t) = u_{L2\_q1} = u_{C1}(t) - u_{C2}(t) \cong ct.$ and $u_1(t) = u_{L2\_q1} = u_{C1}(t) - u_{C2}(t) \cong ct..$
Thus, the inductor current has linear variation in each switching state i.e. it satisfies the linear ripple condition (Fig. 14).

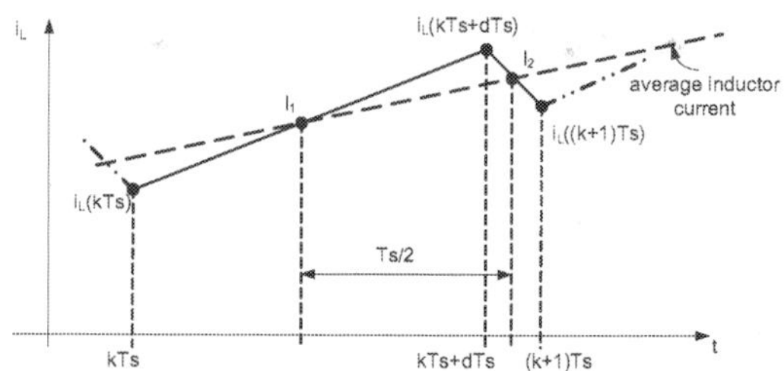

Fig. 14. Variation of the instantaneous and average inductor current in a step-down converter

Consequently:

$$\left(i_L((k+1)\cdot T_s) - i_L(k\cdot T_s)\right)/T_s = d\cdot\langle u_1(t)\rangle_{dT_s} + (1-d)\cdot\langle u_2(t)\rangle_{(1-d)T_s} = d\cdot u_1(t) + (1-d)\cdot u_2(t) \qquad (15)$$

It results, that the average variation of the current is equal to the weighted sum of the inductor voltages (the weights are defined by the duty factor, which has a slow variation with time). The variation of the average current can be found as follows (Fig. 14):

$$\frac{d\langle i_L(t)\rangle_{T_s}}{dt} = \frac{I_2 - I_1}{T_s/2} = \frac{i_L((k+1)\cdot T_s) - i_L(k\cdot T_s)}{T_s}. \qquad (16)$$

From (15) and (16) it results, that

$$\frac{d\langle i_L(t)\rangle_{T_s}}{dt} = d\cdot u_1(t) + (1-d)\cdot u_2(t), \qquad (17)$$

which is a differential equation defined by the "slow" variables $\langle i_L(t)\rangle_{T_s}$, $u_1(t)$, $u_2(t)$ and called "averaged state-space equation".

The small-signal model is obtained by perturbing the steady-state values of the state variables and of the duty factor d according to (18).

$$x = X + \hat{x}, \; d = D + \hat{d}. \qquad (18)$$

In case if the inverter from Fig. 13 is replaced by a resistive load R between points "M" and "N", the averaged state-space equations and the corresponding small-signal equations of the step-down converter are:

$$
\begin{aligned}
\frac{d\langle i_{L1}\rangle_{T_s}}{dt} &= -\frac{1}{L_1}\langle u_{C1}\rangle_{T_s} + \frac{1}{L_1}\langle u_r\rangle_{T_s} &\Rightarrow&\quad \frac{d\hat{i}_{L1}}{dt} = -\frac{1}{L_1}\hat{u}_{C1} + \frac{1}{L_1}\hat{u}_r \\[2mm]
\frac{d\langle u_{C1}\rangle_{T_s}}{dt} &= \frac{1}{C_1}\langle i_{L1}\rangle_{T_s} - \frac{d}{C_1}\langle i_{L2}\rangle_{T_s} &\Rightarrow&\quad \frac{d\hat{u}_{C1}}{dt} = \frac{1}{C_1}\hat{i}_{L1} - \frac{1}{C_1}D\hat{i}_{L2} - \frac{1}{C_1}\hat{d}I_{L2} \\[2mm]
\frac{d\langle i_{L2}\rangle_{T_s}}{dt} &= \frac{d}{L_2}\langle u_{C1}\rangle_{T_s} - \frac{1}{L_2}\langle u_{C2}\rangle_{T_s} &\Rightarrow&\quad \frac{d\hat{i}_{L2}}{dt} = \frac{1}{L_2}D\hat{u}_{C1} + \frac{1}{L_2}\hat{d}U_{C1} - \frac{1}{L_2}\hat{u}_{C2} \\[2mm]
\frac{d\langle u_{C2}\rangle_{T_s}}{dt} &= \frac{1}{C_2}\langle i_{L2}\rangle_{T_s} - \frac{1}{RC_2}\langle u_{C2}\rangle_{T_s} &\Rightarrow&\quad \frac{d\hat{u}_{C2}}{dt} = \frac{1}{C_2}\hat{i}_{L2} - \frac{1}{RC_2}\hat{u}_{C2}
\end{aligned}
\qquad (19)
$$

4.3. The model of the converter with pulse amplitude modulation

The small-signal model of the induction-heating converter from Fig. 13 with power control by means of pulse amplitude modulation (PAM) is built using the model of the step-down converter obtained by state-space averaging and the d-q model of the voltage inverter with serial resonant load. The small-signal inverter model for serial load is built using the procedures described in 4.1 for the case of the voltage inverter with hybrid LLC load.

This model is meant for controller design using the classical control theory. The controller design can also be made starting from a model with the DC-DC converter supplied from an ideal voltage source. Later the influence of the L_1C_1 (Fig. 13) input filter is taken into account using the Middlebrook's extra element theorem (Erickson & Maksimović, 2004), i.e. the condition can be established to have similar performances to those of the system with ideal voltage source.

The load of the step-down converter is modelled by a current source, controlled by the inverter output current. The inverter operates with rectangular output voltage and pulse amplitude modulation control is being considered.

For a given operating point of the converter, the duty factor D, the capacitor voltage components and amplitude U_{Csd}, U_{Csq} and U_{Cs}, which are parameters of the small-signal model, are known, respectively can be determined from the steady-state circuit equations.

The small-signal model of the system composed of filter, step-down converter, inverter and resonant load is shown in Fig. 15.

The block diagram of the controlled converter, built using the principle of superposition, is shown in Fig. 16 (Kelemen, 2007).

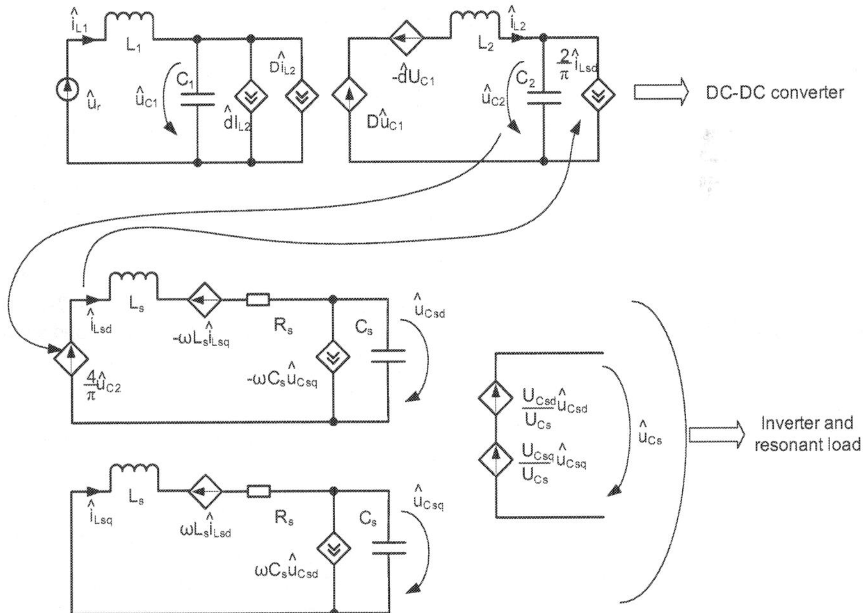

Fig. 15. The small-signal low-frequency model of the converter composed of filter, DC-DC converter, inverter and resonant load

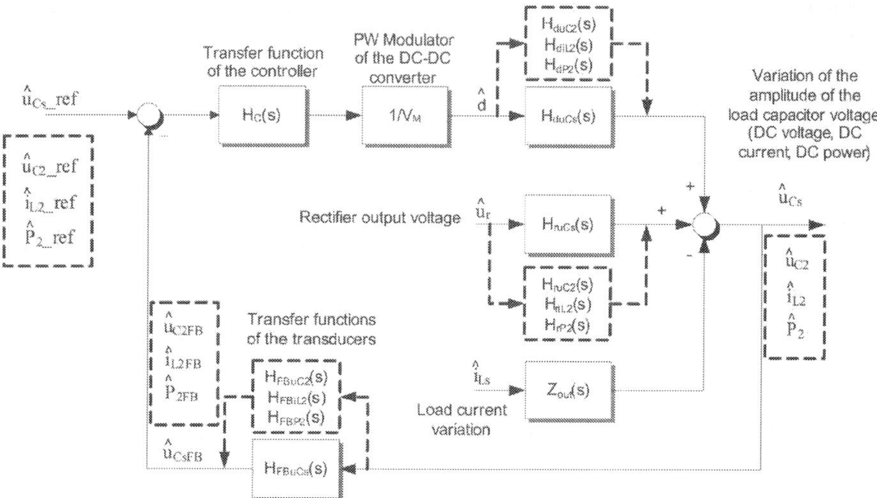

Fig. 16. Block diagram of the controlled converter

In Fig. 16 $H_{duCs}(s)$, $H_{duC2}(s)$, $H_{diL2}(s)$, $H_{dP2}(s)$ represent the transfer functions between the small variation of the duty factor of the DC-DC converter control around its steady-state value D and the small variation of the load capacitor voltage, output voltage, current and power of the step-down converter around the steady-state values U_{Csmax}, U_{C2}, I_{L2}, and P_2.

In a similar way, $H_{ruCs}(s)$, $H_{ruC2}(s)$, $H_{riL2}(s)$, $H_{rP2}(s)$ represent the small-signal transfer functions between the rectified voltage and the load capacitor voltage, output voltage current and power of the step-down converter. $H_{FBuCs}(s)$, $H_{FBuC2}(s)$, $H_{FBiL2}(s)$, $H_{FBP2}(s)$ represent the transfer functions of the transducers used to measure the controlled variables. $Z_{out}(s)$ is the output impedance of the converter.

$$\hat{u}_{Cs} = \hat{u}_{Cs\_ref} \frac{1}{H_{FBuCs}(s)} \frac{T_{uCs}}{1+T_{uCs}} + \hat{u}_r(s) \frac{H_{ruCs}(s)}{1+T_{uCs}} - \hat{i}_{Ls} \frac{Z_{out}}{1+T_{uCs}}, \qquad (20)$$

where the open-loop gain is:

$$T_{uCs} = H_{FBuCs}(s) H_c(s) H_{duCs}(s) / V_M. \qquad (21)$$

In (21), $1/V_M$ denotes the gain of the PWM modulator, which is inversely proportional with the V_M amplitude of the carrier.

The model introduced above offers the possibility to analyze the behaviour of the induction-heating converter in case of "slow" variation of different quantities like the inverter operation frequency and the duty factor of the step-down converter. The envelope of different signals has such a slow variation and the controllers are designed to control these slow variations. In other words, the control of the resonant load voltage means the control of its RMS value or the control of its peak value, without caring for its oscillatory character during a switching cycle.

The "slow" or "low-frequency" model of the high-frequency converter is used to derive the frequency characteristics applied in classical controller design. For example, Fig. 17 and Fig. 18 present the Bode plots of the open voltage control loop in case of different operation frequencies of the inverter, respectively in case of different duty factors of the DC chopper control.

The circuit parameters and the operation conditions are:

$U_r = 500V$, $L_1 = 1.6mH$, $C_1 = 1850\mu F$, $L_2 = 800\mu H$, $C_2 = 200\mu F$, $L_s = 64\mu H$, $C_s = 39nF$, $R_s = 4\Omega$.
The transfer function of the voltage transducer is represented by a $1:5000$ gain. The gain of the PWM modulator is $V_M = 1$, while the parameters of the PI voltage controller are $K_p = 1$, $T_i = 1s$.

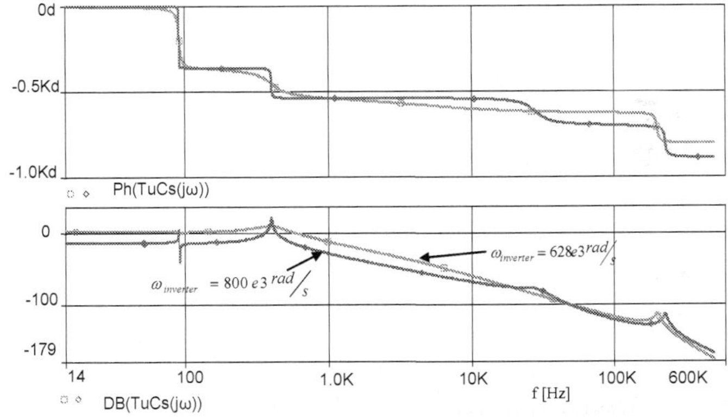

Fig. 17. Bode diagrams of the u_{Cs} voltage control loop for different inverter operation frequencies (ω = 800 krad/s and ω = 628 krad/s) in case of D=0.5

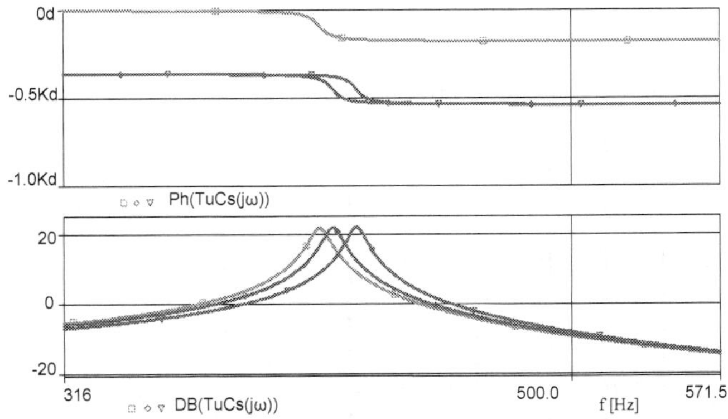

Fig. 18. Bode diagrams of the u_{Cs} control loop for different values of the step-down converter duty cycle: D = 0.2, D=0.5 and D=0.8, for ω = 800 krad/s

Keeping in mind that the control of the high frequency voltage is performed by means of the step-down DC-DC converter, i.e. by means of the variation of its control duty factor, it can be observed how the value of this duty factor and the value of the inverter operating frequency influence the frequency characteristics of the control loop.

5. Sampled-data modelling

The sampled-data models can be used for large-signal and small-signal analysis of the resonant converters.

The digital control of the converters is inherently based on sampled measurement data and discrete-time controller output.

Let's consider the classical thyristor-based, medium-frequency induction-heating converter from Fig. 19. This converter performs the power control by means of the controlled rectifier. The current inverter can operate only with a capacitive phase shift between the inverter output current and the resonant load voltage, as shown in Fig. 20, necessary for the recovery of the thyristors. The capacitive phase shift is controlled by a dedicated controller in a manner that positive AK voltage is applied to the thyristors only after the elapse of the circuit-commutated recovery time.

The presentation of the controllers which are able to guarantee this time delay between the end of the commutation period and the zero travel of the resonant capacitor voltage (see Fig. 20) is behind the scope of this chapter. Nevertheless, in order to develop a model which can be used for the power controller design, it is enough to assume that the problem is solved and this time delay is being controlled accurately by a control loop which is much faster than the power control loop. The result is the inverter operation at a frequency that assures the phase shift corresponding to the above condition. For a given resonant load the switching frequency can be determined using a fundamental-frequency approach. For instance, in case of the converter from Fig. 19, with the circuit parameters $L_d = 10\ mH$, $L_i = 10.14\ \mu H$, $R_i = 63.7\ m\Omega$, $C_p = 25\ \mu F$, the resonance frequency of the parallel resonant load is $f_0 = 10\ kHz$, and operation at $f_0 = 10.2\ kHz$ assures a $T_s = 7.5\ \mu s$ thyristor recovery time.

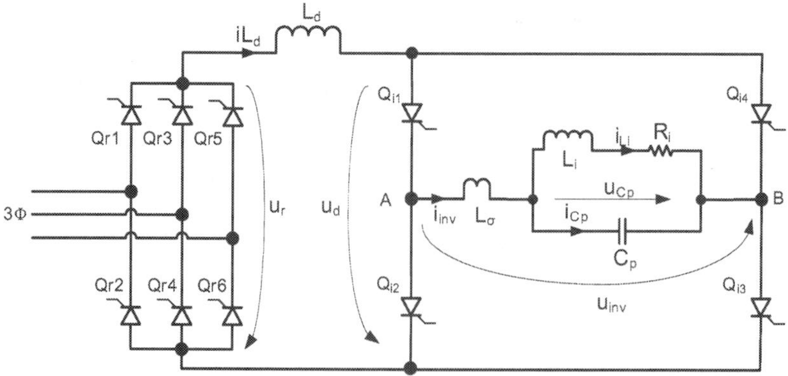

Fig. 19. Thyristor-based induction-heating converter with parallel resonant load

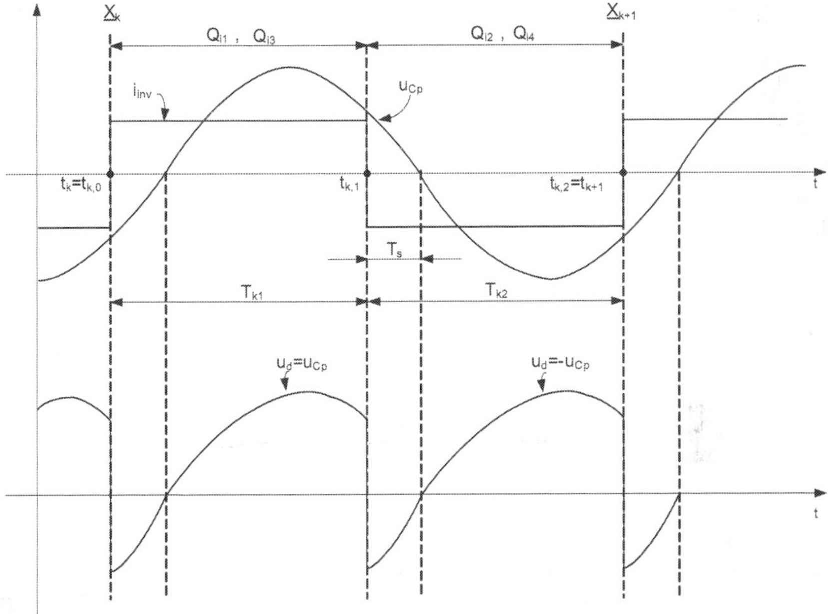

Fig. 20. Characteristic waveforms of the converter from Fig. 19

The above assumptions make possible the division of the switching period into time intervals with known lengths, corresponding to known switching states. This approach assumes the a priori knowledge of the converter operation and can be used only at a certain level of the hierarchical modelling.

As a simplifying assumption, the stray inductance L_σ is omitted. In consequence, the switching time is reduced to zero and only two states have to be considered corresponding to the conduction of the diagonals Q_{i1}, Q_{i3} and Q_{i2}, Q_{i4}.

The switching states of the converter are described by the equations (22). These equations form the state-space description of the linear time-invariant systems corresponding to the two switching states.

T_{ki} represents the i-th switching state of the k-th switching period.

A_i and B_i (23) represent the system matrices corresponding to the i-th switching state.

$$\text{switching state } "T_{k1}" \qquad \text{switching state } "T_{k2}"$$

$$\left\{ \begin{aligned} \frac{di_{Ld}}{dt} &= -\frac{1}{L_d} \cdot u_{Cp} + \frac{1}{L_d} \cdot u_r \\ \frac{di_{Li}}{dt} &= -\frac{R_i}{L_i} \cdot i_{Li} + \frac{1}{L_i} \cdot u_{Cp} \\ \frac{du_{Cp}}{dt} &= \frac{1}{C_p} \cdot i_{Ld} - \frac{1}{C_p} \cdot i_{Li} \end{aligned} \right. \quad \left\{ \begin{aligned} \frac{di_{Ld}}{dt} &= \frac{1}{L_d} \cdot u_{Cp} + \frac{1}{L_d} \cdot u_r \\ \frac{di_{Li}}{dt} &= -\frac{R_i}{L_i} \cdot i_{Li} + \frac{1}{L_i} \cdot u_{Cp} \\ \frac{du_{Cp}}{dt} &= -\frac{1}{C_p} \cdot i_{Ld} - \frac{1}{C_p} \cdot i_{Li} \end{aligned} \right. \qquad (22)$$

$$A_1 = \begin{bmatrix} 0 & 0 & -\dfrac{1}{L_d} \\[2mm] 0 & -\dfrac{R_i}{L_i} & \dfrac{1}{L_i} \\[2mm] \dfrac{1}{C_p} & -\dfrac{1}{C_p} & 0 \end{bmatrix}, \quad B_1 = \begin{bmatrix} \dfrac{1}{L_d} \\[2mm] 0 \\[2mm] 0 \end{bmatrix} \quad A_2 = \begin{bmatrix} 0 & 0 & \dfrac{1}{L_d} \\[2mm] 0 & -\dfrac{R_i}{L_i} & \dfrac{1}{L_i} \\[2mm] -\dfrac{1}{C_p} & -\dfrac{1}{C_p} & 0 \end{bmatrix}, \quad B_2 = \begin{bmatrix} \dfrac{1}{L_d} \\[2mm] 0 \\[2mm] 0 \end{bmatrix} \tag{23}$$

$$\begin{cases} \dfrac{d\underline{x}(t)}{dt} = A_i \cdot \underline{x}(t) + B_i \cdot \underline{u}(t) & t \in (t_{k,i-1}, t_{k,i}), \quad i = 1,2 \\[3mm] \underline{x}(t) = \begin{bmatrix} i_{Ld}(t) & i_{Li}(t) & u_{Cp}(t) \end{bmatrix}^T \end{cases} \tag{24}$$

The state vector at the end of the k-th switching period denoted by $\underline{x}_{k+1}$ can be determined based on the state vector at the beginning of this period, denoted by $\underline{x}_k$.
According to the formulation from (Elbuluk et al., 1998),

$$\underline{x}(t_{k+1}) = f\left(\underline{x}(t_k), P_k\right). \tag{25}$$

$P_k = \begin{bmatrix} u_r & L_d & L_i & R_i & C_p & t_{k,1} & t_{k,2} \end{bmatrix}^T$ represents the vector of controlling parameters, including the controlled transition times defined by the thyristor firing instants.
A more accurate model should take into account also the commutation period characterized by the simultaneous conduction of the inverter thyristors. The end of this switching state is controlled "internally", i.e. it is not directly determined by the control system, but by the evolution of some circuit variables (i.e. by the zero transition of the thyristor currents). The internally controlled switching instants become parameters of (25), which forms a system of nonlinear equations together with the equations which formulate the conditions of these internally controlled transitions.
The state vector is continuous across the transitions between the switching states. In order to determine $\underline{x}_{k+1}$, the state transition matrix is built using the system matrices (23).
The state transition from $t_k = t_{k,0}$ to $t_{k,1}$ i.e. during the time interval T_{k1} is determined by the equation (26), where $u_r(t_k)$ represents the sampled (and held) value of the rectified voltage. In case of a 10 kHz medium-frequency induction-heating converter, this voltage has a slow variation (a 300 Hz component) in comparison with the switching frequency. Thus, the assumption that it is constant during a switching period is acceptable. In case of low-frequency converters this assumption is not valid any more and the steps must be refined.

$$\underline{x}(t_{k,1}) = e^{A_1 T_{k1}} \underline{x}(t_k) + \left(\int_0^{T_{k1}} e^{A_1 \alpha} d\alpha \right) B_1 u_r(t_k) = \Phi_1 \underline{x}(t_k) + \Gamma_1 u_r(t_k) = \Phi_1 \underline{x}(t_k) + A_1^{-1}\left(\Phi_1 - I\right) B_1 u_r(t_k) \tag{26}$$

The state transition from $t_{k,1}$ to $t_{k+1} = t_{k,2}$ i.e. during the time interval T_{k2} can be described according to the relation (27).

$$\begin{aligned} \underline{x}(t_{k+1}) &= e^{A_2 T_{k2}} \underline{x}(t_{k,1}) + \left(\int_0^{T_{k2}} e^{A_2 \alpha} d\alpha \right) B_2 u_r(t_k) = \Phi_2 \underline{x}(t_{k,1}) + \Gamma_2 u_r(t_k) = \\ &= \Phi_2 \underline{x}(t_{k,1}) + A_2^{-1}\left(\Phi_2 - I\right) B_2 u_r(t_k) \end{aligned} \tag{27}$$

By substitution of (26) into (27), it results:

$$\underline{x}(t_{k+1}) = \Phi_2\left(\Phi_1\underline{x}(t_k) + A_1^{-1}(\Phi_1 - I)B_1 u_r(t_k)\right) + A_2^{-1}(\Phi_2 - I)B_2 u_r(t_k) =$$
$$= \Phi_2\Phi_1\underline{x}(t_k) + \left(\Phi_2 A_1^{-1}(\Phi_1 - I)B_1 + A_2^{-1}(\Phi_2 - I)B_2\right)u_r(t_k) = \Phi\underline{x}(t_k) + \Gamma u_r(t_k)$$

$$(28)$$

Fig. 21, Fig. 22 and Fig. 23 show the results of sampled-data analysis superimposed with the results of a detailed circuit analysis performed using the Matlab- Simulink environment.
Fig. 21 demonstrates the transition of the state variable u_{Cp} between the starting instants of the switching periods.
Its sampled values are shown to be equal with the values of the continuous-time u_{Cp} in the moments of the $Q_{i2}, Q_{i3} \rightarrow Q_{i1}, Q_{i4}$ commutation, obtained by detailed analysis.

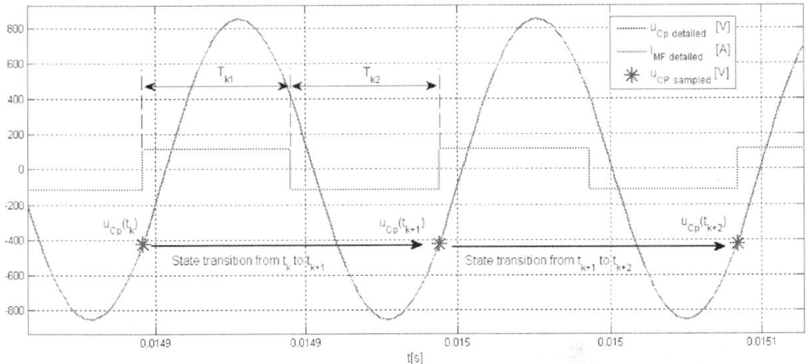

Fig. 21. The steady-state waveform of the resonant capacitor tank voltage obtained using the detailed model and the corresponding sampled-data result

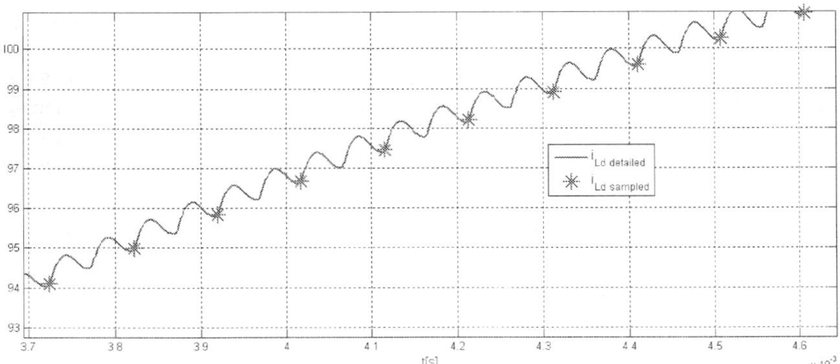

Fig. 22. The steady-state waveform of the DC-link current and the corresponding sampled-data result

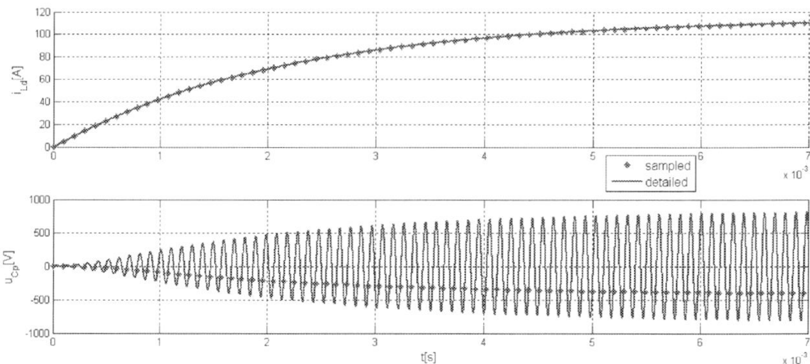

Fig. 23. Evolution of the DC-link current and of the capacitor voltage during the converter start-up

Fig. 22 and Fig. 23 demonstrate that i_{Ld} is a "slow" variable and that its evolution is described well enough by the sampled-data model.

Thus, the sampled-data modelling can be used for power controller design, resulting in a much shorter processing time, than the application of a detailed model.

6. Inverter analysis using the describing function method

The sinusoidal describing function method characterizes the output of the nonlinear part of a system by its fundamental component. If the output of the nonlinear part is fed to the input of a linear part of the system with low pass filter characteristics, the higher harmonic components are filtered out and the fundamental component approach would introduce only small errors.

The induction-heating inverters usually have a quasi-rectangular output voltage or current, which feeds a resonant load with good band-pass filter characteristics. This is the reason why most of the electrical quantities from the load can be considered sinusoidal.

That's why the describing function method can be introduced in a natural way for the analysis of the resonant converters (Sewell et al., 2004), (Kelemen & Kutasi, 2007b).

Let us consider the inverter circuit from Fig. 24., with lossless snubber capacitors and series resonant load modelled by a sinusoidal current source.

The describing function of the inverter output voltage is derived in order to represent its non-linear dependence on the inverter output current. The waveforms of the inverter output current and of the half-bridge voltage are shown in Fig. 25 in case of an optimal switching strategy characterized by the minimum value of the turn-off current that still assures zero voltage turn-on of the transistors.

The investigation is started from the moment when $Q3$ is turned off. In the $(0...\gamma)$ interval, the iLs current flows through C_3 and C_4.

Assuming that $C_i = C$ and $u_{C3} + u_{C4} = U_d = ct.$, it results:

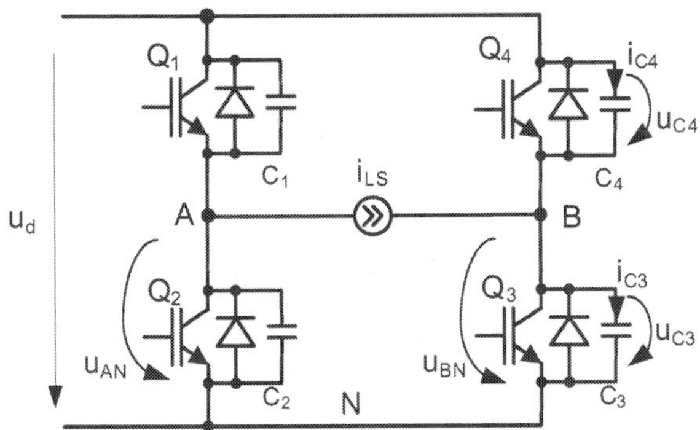

Fig. 24. Voltage-source inverter with snubber capacitors and the resonant load modelled by a current source

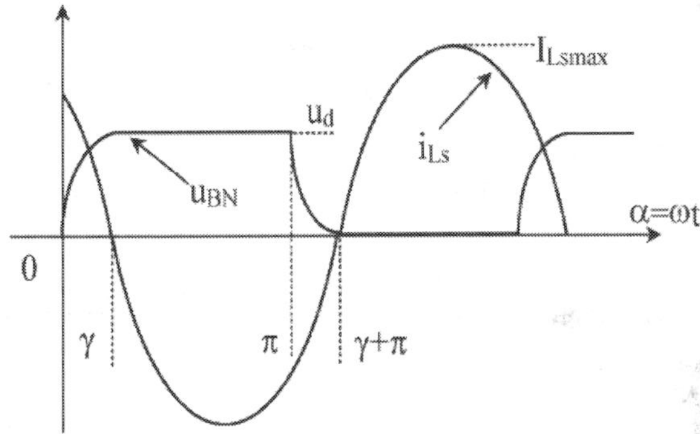

Fig. 25. The resonant load current and half-bridge voltage waveforms in case of optimal ZVS control of the inverter from Fig. 24

$$u_{BN}(\alpha) = \begin{cases} \dfrac{1}{2\omega C}\displaystyle\int_0^{\alpha}(-I_{Ls\,max})\sin(\varphi-\gamma)d\varphi = \dfrac{I_{Ls\,max}}{2\omega C}\left(\cos(\alpha-\gamma)-\cos\gamma\right) & for\quad \alpha\in(0,\gamma) \\[2mm] U_d + \left(\dfrac{I_{Ls\,max}}{2\omega C}\left(\cos(\alpha-\gamma)+\cos\gamma\right)\right) & for\quad \alpha\in(\pi,\pi+\gamma) \\[2mm] 0 & for\quad \alpha\in(\pi+\gamma,2\pi) \end{cases} \quad (29)$$

From the $u_{BN}(\gamma)=U_d$ condition,

$$\cos\gamma = 1 - \frac{2\omega C U_d}{I_{Ls\,max}}. \tag{30}$$

The inverter output voltage is defined by the equations:

$$u_{AB} = \begin{cases} U_d - \dfrac{I_{Ls\max}}{\omega C}\big(\cos(\alpha-\gamma)-\cos\gamma\big) & \alpha \in (0,\gamma) \\[2mm] -U_d & \alpha \in (\gamma,\pi) \\[2mm] -U_d - \dfrac{I_{Ls\max}}{\omega C}\big(\cos(\alpha-\gamma)+\cos\gamma\big) & \\[2mm] & \alpha \in (\pi,\pi+\gamma) \\[2mm] +U_d & \alpha \in (\pi+\gamma,2\pi) \end{cases} \tag{31}$$

The complex representation of the fundamental term of the inverter output voltage is:

$$\underline{u}_{AB1} = A_1 e^{j\omega t} e^{j\left(-\frac{\pi}{2}+\varphi\right)} = \big(u_{AB1d}+ju_{AB1q}\big)e^{j\omega t}, \text{ where}$$

$$A_1 = \sqrt{a_1^2+b_1^2}, \quad a_1 = \frac{2U_d}{\pi}\frac{\sin\gamma-\gamma\cos\gamma}{1-\cos\gamma}, \quad b_1 = \frac{2U_d}{\pi}\frac{(-\gamma\sin\gamma)}{1-\cos\gamma} \tag{32}$$

The relationship between the inverter output voltage and the load current can be represented by the describing function defined as the ratio of the complex representations of their fundamental components:

$$N\big(I_{Ls\max},\omega\big) = \frac{\underline{u}_{AB1}}{\underline{i}_{Ls}} = \frac{2}{\pi}\frac{U_d}{I_{Ls\max}}\left(1+\cos\gamma+j\frac{\gamma-\sin\gamma\cos\gamma}{1-\cos\gamma}\right). \tag{33}$$

The phase shift between the fundamentals of the inverter output voltage and current is:

$$\varphi\big(\underline{u}_{AB1},\underline{i}_{Ls}\big) = \arg\big(N\big(I_{Ls\max},\omega\big)\big) \quad\Rightarrow\quad \varphi = arctg\frac{\gamma-\sin\gamma\cos\gamma}{\sin^2\gamma}. \tag{34}$$

Thus, the inverter can be seen as a system with the structure from Fig. 26.

One should bear in mind that the nonlinear relationship between the inverter output voltage and current results both from the existence of an intrinsic feedback loop due to the presence of the lossless snubber capacitors, and from the presence of the fast control loop of the switching angle γ.

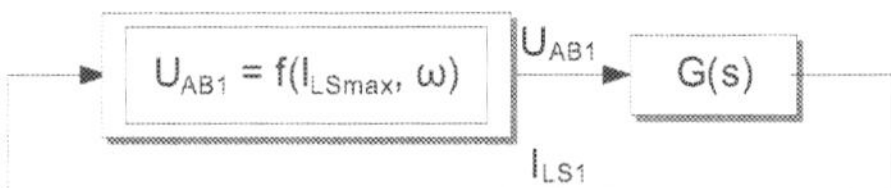

Fig. 26. System structure with nonlinear part represented by the describing function (31), and linear part represented by the linear resonant load

The existence of limit cycles in the optimal ZVS control mode can be verified using the describing function of the nonlinear part of the system and the transfer function of the serial load (35)

$$G(j\omega) = \frac{I_{Ls1\max}(j\omega)}{U_{AB1\max}(j\omega)} = \frac{j\omega C_s}{1-\omega^2 C_s L_s + j\omega R_s C_s}, \tag{35}$$

where R_s, L_s, C_s denote the parameters of the serial resonant induction heating load.
The structure of the model of the PAM converter from Fig. 13, defined using the describing function of the inverter is shown in Fig. 27.

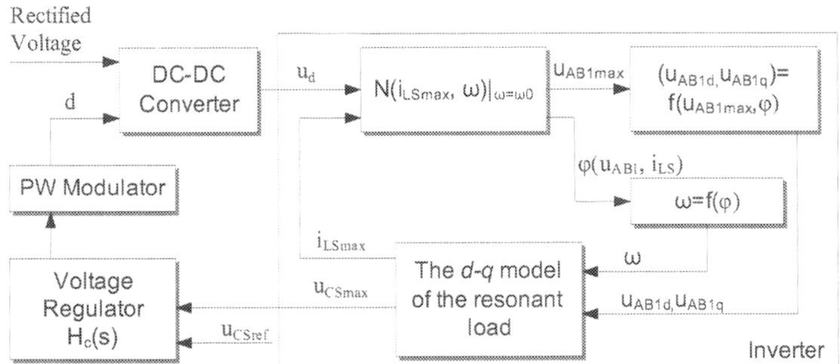

Fig. 27 Induction heating converter model with optimal (ZVS and close to ZCS) switching mode control of the inverter

The model from Fig. 26 encloses the *d-q* model of the resonant load (Kelemen & Kutasi, 2007a) and the model of the DC-DC converter obtained by the classical state-space averaging method.

7. Conclusion

Specific procedures have been presented that can be applied for modelling and analysis of the induction-heating power electronic converters. The basic idea behind this brief overview is that different tools have to be identified in order to handle the tasks arising at different hierarchical levels of the converter analysis.

Details have been presented in case of the methods that prove to be useful for system-level modelling. Thus, averaging and sampled-data modelling techniques have been applied in case of load-resonant transistor-based and thyristor-based inverters, taking into account the operating conditions imposed by the specific control tasks of these converters.

It has been shown that the describing function method is suitable for modelling of the nonlinearities in case of the inverters that operate close to the resonance frequency of the load.

The presented procedures have been used for development of several induction-heating equipments in the frequency range of $50\ Hz \div 500\ kHz$ and in power range of $4 \div 1000\ kW$.

8. References

Borelli, F. (2003). *Constrained Optimal Control of Linear and hybrid Systems*, Springer Verlag, ISBN: 3-540-00257-X, Berlin, Heidelberg

Elbuluk, M. E., Verghese G. C. & Kassakian, J.G. (1998). Sampled-data modeling and digital control of resonant converters, *IEEE Transactions on Power Electronics*, Vol. 3, No. 3, July 1988, 344–354, ISSN: 0885-8993

Erickson, R. W. & Maksimović, D. (2004). *Fundamentals of power electronics- second edition*, Kluwer Academic Publishers, Norwell, Massachusetts, ISBN: 0-7923-7270-0

Fischer, G. L. & Doht, H. C. (1994). An inverter system for inductive tube welding utilizing resonance transformation, *Proceedings of the Industry Applications Society Annual Meeting*, pp. 833–840 Vol. 2, ISBN: 0-7803-1993-1, Denver, CO, October 1994

Forest, F., Labouré, E., Costa, F. & Gaspard, J. Y. (2000). Principle of a multi-load/single converter system for low power induction heating, *IEEE Transactions on Power Electronics*, Vol. 15, No. 2, March 2000, 223-230, ISSN: 0885-8993

Foster, P. F., Sewell H. I., Bingham C. M., Stone D. A., Hente, D. & Howe D. (2003). Cyclic-averaging for high-speed analysis of resonant converters, *IEEE Transactions on Power Electronics*, Vol. 18, No. 4, July 2003, 985–993, ISSN: 0885-8993

Fujita, H. & Akagi, H. (1996). Pulse-density-modulated power control of a 4 kW, 450 kHz voltage-source inverter for induction melting applications, *IEEE Transactions on Industry Applications*, Vol. 32, No. 2, March/April 1996, 279-286, ISSN: 0093-9994

Kelemen, A. (2007). *Power control of the induction-heating power-electronic converters- PhD thesis*, Transylvania University of Brasov

Kelemen, A. & Kutasi, N. (2007a). Induction-heating voltage inverter with hybrid LLC resonant load, the D-Q model, *Pollack Periodica*, Vol. 2, No. 1, 2007, 27–37, ISSN: 1788-1994

Kelemen, A. & Kutasi, N. (2007b). Describing function analysis of a voltage-source induction-heating inverter with pulse amplitude modulation, *Acta Electrotehnica*, Vol. 48, No. 3, 2007, 223-229, ISSN 1841-3323

Krein, P. T., Bentsman, J., Bass, R. M. & Lesieutre, B. C. (1990) On the use of averaging for the analysis of power electronic systems, *IEEE Transactions on Power Electronics*, Vol. 5, No. 2, April 1990, 182-190, ISSN: 0885-8993

Mohan, N., Undeland, T. M. & Robbins, W. P. (2002). *Power Electronics - Converters, Applications and Design, 3rd edition*, John Wiley & Sons INC., New York, ISBN: 978-0-471-22693-2

Rim, C.T. & Cho, G. H. (1990) Phasor transformation and its application to the DC/AC analysis of frequency-phase-controlled series resonant converters (SRC), *IEEE Transactions on Power Electronics*, Vol. 5, No. 2, April 1990, 201–211, ISSN: 0885-8993

Rudnev, V., Loveless, D, Cook, R. & Black, M. (2003). *Handbook of induction heating*, Marcel Dekker, Inc., New York, Basel, ISBN: 0-8247-0848-2

Sanders, S. R., Noworolski, J. M., Liu, X. Z. & Verghese, G. C. (1991). Generalized averaging method for power conversion circuits, *IEEE Transactions on Power Electronics*, Vol. 6, No. 2, April 1991, 251–259, ISSN: 0885-8993

Sewell, H. I., Stone, D. A. & Bingham, C. M. (2004). A describing function for resonantly commutated H-bridge inverters, *IEEE Transactions on Power Electronics* ,Vol. 19, No. 4, July 2004, 1010-1021, ISSN: 0885-8993

Sluhotkii, A. E. & Rîskin, S. E. (1982) *Inductoare pentru încălzirea electrică* , Editura Tehnică, Bucureşti

Torrisi, F. D. & Bemporad, A. (2004). HYSDEL — A Tool for Generating Computational Hybrid Models for Analysis and Synthesis Problems, *IEEE Transactions on Control Systems Technology*, Vol. 12, No. 2, March 2004, 235–249, ISSN: 1063-6536

Zhang, Y. & Sen, P. C. (2004). D-Q models for resonant converters, *Proceedings of the 35th Annual IEEE Power Electronics Specialists Conference PESC 04*, pp. 1749–1753, Aachen, Germany, November 2004, ISBN: 0-7803-8399-0

4

Numerical Modelling of Industrial Induction

A. Bermúdez[1], D. Gómez[1], M.C. Muñiz[1], P. Salgado[1] and R. Vázquez[2]
[1]*Departamento de Matemática Aplicada*
Universidade de Santiago de Compostela, 15782 Santiago de Compostela
[2]*Istituto di Matematica Applicata del CNR, via Ferrata 1, 27100, Pavia*
[1]*Spain*
[2]*Italy*

1. Introduction

Induction heating is a physical process extensively used in the metallurgical industry for different applications involving metal melting. The main components of an induction heating system are an induction coil connected to a power-supply providing an alternating electric current and a conductive workpiece to be heated, placed inside the coil. The alternating current traversing the coil generates eddy currents in the workpiece and by means of ohmic losses the workpiece is heated (see Fig. 1).

The design of an induction heating system mainly depends on its application. In this chapter we are interested in modelling the behavior of a coreless induction furnace like those used for melting and stirring. A simple sketch of this furnace is presented in Fig. 2(a). It consists of a helical copper coil and a workpiece formed by the crucible and the load within, which is the material to melt. The crucible is a cylindrical vessel made of a refractory material with higher temperature resistance than the substances it is designed to hold in. The coil, which is water-cooled to avoid overheating, is usually enclosed into a refractory material for safety reasons. Alternating current passing through the coil induces a rapidly oscillating magnetic field which generates eddy currents in the workpiece. These induced currents heat the load. When the load melts, the electromagnetic field produced by the coil interacts with

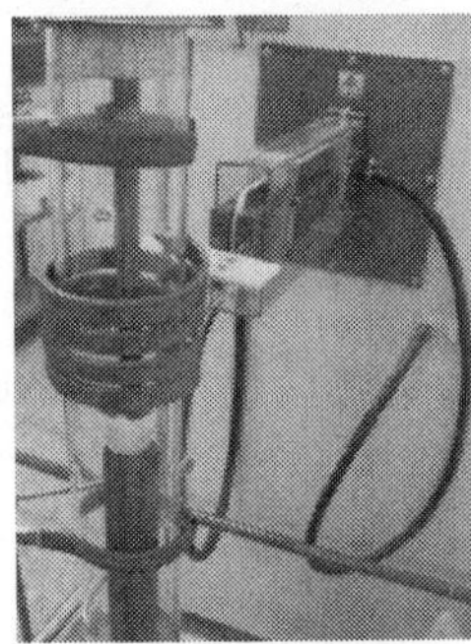

Fig. 1. Induction heating furnace off (left) and on (right). Photographs courtesy of Mr. Víctor Valcárcel, Instituto de Cerámica, Universidade de Santiago de Compostela.

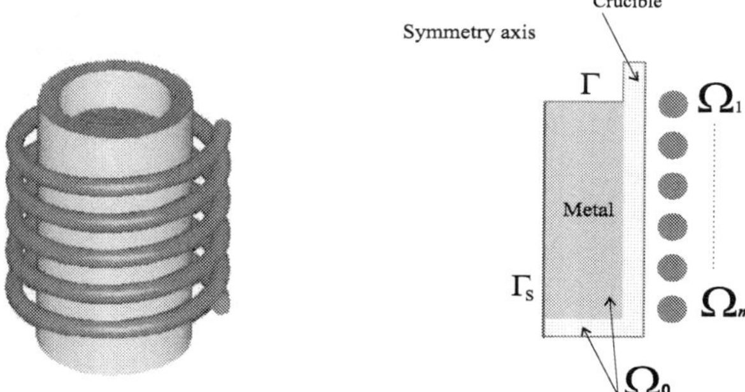

Fig. 2. Sketch of the induction furnace (a) and radial section of inductors and workpiece (b).

the electromagnetic field produced by the induced current. The resulted force causes a stirring effect helping to homogenize the melt composition and its temperature.

It is well known that the high frequencies used in induction heating applications gives rise to a phenomenon called skin effect. This skin effect forces the alternating current to flow in a thin layer close to the surface of the workpiece, at an average depth called the skin depth. Thus, the ohmic losses are concentrated in the external part of the workpiece in such a way that the higher the frequency the thinner the skin depth. It is crucial to control the distribution of these ohmic losses, since they could cause very high temperatures in the crucible thus reducing its lifetime. Moreover, the frequency and intensity of the alternating current also affect the stirring of the molten bath. Since stirring mainly determines some of the properties of the final product, it is convenient to accurately know the influence of these parameters to achieve the desired result.

Our goal is to understand the influence on the furnace performance of certain geometrical parameters such as the crucible thickness, its distance to the coil, the number of turns of the coil, or physical parameters such as the thermal and electrical conductivity of the refractory materials. In the last years, numerical simulation reveals as an important tool for this purpose, since it allows to introduce changes in the above parameters *in silico*, thus avoiding long and costly trial-error procedures in plant.

The overall process is very complex and involves different physical phenomena: electromagnetic, thermal with phase change and hydrodynamic in the liquid region; all of them are coupled and it is essential to consider a suitable mathematical model to achieve a realistic simulation.

Many papers have been published concerning the numerical simulation of induction heating devices, from some pioneering articles published in the early eighties (see, for instance, (Lavers, 1983) and references therein) to more recent works dealing with different coupled problems, such as the thermoelectrical problem appearing in induction heating (Chaboudez et al., 1997; Rappaz & Swierkosz, 1996), the magnetohydrodynamic problem related to induction stirring (Natarajan & El-Kaddah, 1999) and also a thermal-magneto-hydrodynamic problem (Henneberger & Obrecht, 1994; Katsumura et al., 1996) but not fully coupled because material properties are not supposed to be dependent on temperature. Some other related works include mechanical effects in the workpiece (Bay et al., 2003; Hömberg, 2004). A more extensive bibliographic review can be found in (Lavers, 2008).

The aim of this chapter is to deal with the coupled thermo-magneto-hydrodynamic simulation of an induction heating furnace like the one described above. It is a survey of previous works by the authors (Bermúdez et al., 2007;b; 2009; Vázquez, 2008). Section 2 is devoted to describe the mathematical model to compute the electromagnetic field and the temperature in the furnace, and the velocity field in the molten region; the coupled mathematical model is a system of partial differential equations with several non-linearities. In Section 3 a weak formulation of the electromagnetic model, defined in an axisymmetric setting, is derived. In Section 4 we present the time-discretization of the thermal and hydrodynamic models and the corresponding weak formulations. In Section 5 we describe the spatial discretization and the algorithms used to deal with the coupling formulation and the non-linearities. Finally, in Section 6, several numerical results are given for an industrial heating furnace designed for melting.

2. Mathematical model for the induction furnace

In this section we introduce the mathematical model to study the thermo-magneto-hydro dynamic behavior of an induction furnace.

2.1 The electromagnetic model

In order to model the electric current traversing the coil and the eddy currents induced in the workpiece, we introduce an electromagnetic model which is obtained from Maxwell equations. We state below a three-dimensional model defined in the whole space $\mathbb{R}^3$ and deduce an axisymmetric formulation in Section 3. Since the full description from the 3D model to the axisymmetric one involves a lot of technical steps and it has been done in (Bermúdez et al., 2009), we only give here a brief description and refer the reader to the quoted paper for further details.

We begin by introducing some assumptions and notations related with the geometry that will be used in the sequel. Since the material enclosing the coil is not only a good refractory but also a good electrical insulator, the induction process in the workpiece is almost unaffected by the presence of this material. Thus, in the electromagnetic model, it will be treated as air. For simplicity, we do not consider the water refrigerating the coil.

In order to state the problem in an axisymmetric setting, the induction coil has to be replaced by m rings having toroidal geometry. Let Ω_0 be the radial section of the workpiece and $\Omega_1, \Omega_2, \ldots, \Omega_m$ be the radial sections of the turns of the coil, which are assumed to be simply connected (see Fig. 2(b)). Moreover, let us denote by Ω the radial section of the set of conductors, i.e., inductors and conducting materials in the workpiece, given by,

$$\Omega = \bigcup_{k=0}^{m} \Omega_k,$$

and $\Omega^c = \mathbb{R}^2 \setminus \bar{\Omega}$. In particular, the refractory layer enclosing the coil is part of Ω^c.

Let $\Delta \subset \mathbb{R}^3$ be the bounded open set generated by the rotation about the $z-$axis of Ω and Δ^c the complementary set of $\bar{\Delta}$ in $\mathbb{R}^3$. Notice that Δ^c is an unbounded set corresponding to the air surrounding the whole device. Analogously, we denote by Δ_k, $k = 0, \ldots, m$ the subset of $\mathbb{R}^3$ generated by the rotation of Ω_k, $k = 0, \ldots, m$, respectively, around the $z-$axis (see Fig. 3). For simplicity, we will assume that Δ_0 is simply connected and that Δ_k, $k = 1, \ldots, m$ is a solid torus, in the sense that one single cutting surface Ω_k is enough for the set $\Delta_k \setminus \Omega_k$ to be simply–connected (see Fig. 3).

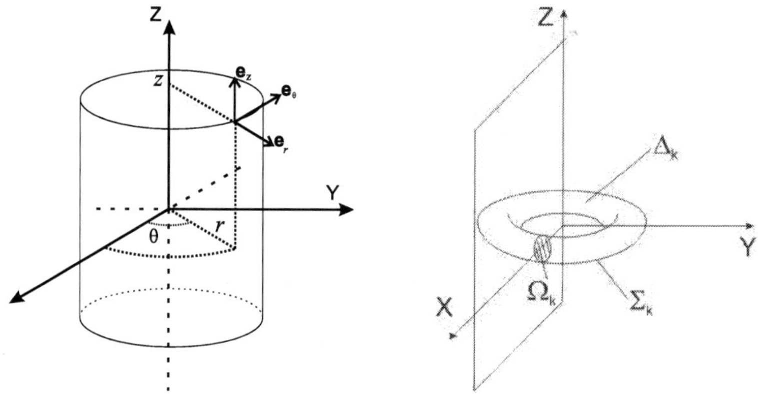

Fig. 3. Cylindrical coordinate system (a) and sketch of a toroidal turn Δ_k (b).

We denote by Σ the boundary of Δ and by Γ its intersection with the half-plane $\{(r,z) \in \mathbb{R}^2; r > 0\}$. We notice that $\Sigma = \cup_{k=0}^m \Sigma_k$, where Σ_k denotes the boundary of Δ_k. Moreover, we assume that the boundary of Ω is the union of Γ and Γ_s, the latter being a subset of the symmetry axis (see Fig. 2(b)).

The induction furnace works with alternating current so we can consider that all fields vary harmonically with time in the form:

$$\mathcal{F}(\mathbf{x},t) = \mathrm{Re}\left[e^{i\omega t}\mathbf{F}(\mathbf{x})\right], \tag{1}$$

where t is time, $\mathbf{x} \in \mathbb{R}^3$ is the spatial variable, i is the imaginary unit, $\mathbf{F}(\mathbf{x})$ is the complex amplitude of field $\mathcal{F}$ and ω is the angular frequency, $\omega = 2\pi f$, f being the frequency of the alternating current.

For low and moderate frequencies, displacement currents can be neglected. Then, Maxwell's equations can be reduced to the so-called *eddy current* model (see (Bossavit, 1998)):

$$\mathbf{curl\,H} \;=\; \mathbf{J} \quad \text{in } \mathbb{R}^3, \tag{2}$$

$$i\omega\mathbf{B} + \mathbf{curl\,E} \;=\; \mathbf{0} \quad \text{in } \mathbb{R}^3, \tag{3}$$

$$\mathrm{div}\,\mathbf{B} \;=\; 0 \quad \text{in } \mathbb{R}^3, \tag{4}$$

where $\mathbf{H}$, $\mathbf{J}$, $\mathbf{B}$ and $\mathbf{E}$ are the complex amplitudes associated with the magnetic field, the current density, the magnetic induction and the electric field, respectively.

System (2)-(4) needs to be completed with the following constitutive relations:

$$\mathbf{B} \;=\; \mu\mathbf{H} \quad \text{in } \mathbb{R}^3, \tag{5}$$

$$\mathbf{J} \;=\; \begin{cases} \sigma\mathbf{E} & \text{in } \Delta, \\ 0 & \text{in } \Delta^c, \end{cases} \tag{6}$$

where μ is the magnetic permeability –which is supposed to be independent of $\mathbf{H}$– and σ is the electric conductivity. Both are assumed to be bounded from below by a positive constant and dependent on both the spatial variable and the temperature T, i.e., $\mu = \mu(\mathbf{x},T)$ and $\sigma = \sigma(\mathbf{x},T)$. Nevertheless, for the sake of simplicity, we drop this dependency in the notation until forthcoming sections.

Notice that we have neglected the fluid motion in the Ohm's law (6), because a dimensionless analysis shows that it is negligible in comparison with the other terms for this kind of applications.

Since the previous equations hold in $\mathbb{R}^3$, we also require for the fields a certain behavior at infinity. More specifically, we have,

$$\mathbf{E}(\mathbf{x}) = O(|\mathbf{x}|^{-1}), \qquad \text{uniformly for } |\mathbf{x}| \to \infty, \tag{7}$$

$$\mathbf{H}(\mathbf{x}) = O(|\mathbf{x}|^{-1}), \qquad \text{uniformly for } |\mathbf{x}| \to \infty. \tag{8}$$

Model (2)-(8) must be completed with some *source data* related to the energizing device. In particular, we would like to impose the current intensities $\mathbf{I} = (I_1, \ldots, I_m)$ crossing each transversal section of the inductor, i.e.,

$$\int_{\Omega_k} \mathbf{J} \cdot \nu = I_k, \ k = 1, \ldots, m, \tag{9}$$

where ν denotes a unit normal vector to the section Ω_k.

Remark 2.1 *Imposing these constraints is not trivial at all and requires to relax in some sense equation (3). We refer the reader to Remarks 2.1 and 2.2 of (Bermúdez et al., 2009). To this respect, we also cite the paper (Alonso et al., 2008) where the authors give a systematic analysis to solve eddy current problems driven by voltage or current intensity in the harmonic regime and in bounded domains.*

To solve the eddy current problem in an axisymmetric setting, we are going to propose a formulation based on the magnetic vector potential. To do that, we also need to introduce a suitable scalar potential. Firstly, equation (4) allows us to affirm that there exists a magnetic vector potential $\mathbf{A}$ defined in $\mathbb{R}^3$ such that,

$$\mathbf{B} = \mathbf{curl}\, \mathbf{A}, \tag{10}$$

and from equation (3), we obtain

$$i\omega\, \mathbf{curl}\, \mathbf{A} + \mathbf{curl}\, \mathbf{E} = 0 \quad \text{in } \Delta.$$

Taking into account the form of the kernel of the **curl** operator in each connected component of the conductor, we can say that (see, for instance (Amrouche et al., 1998))

$$i\omega \mathbf{A} + \mathbf{E} = -\mathbf{v} \quad \text{in } \Delta, \tag{11}$$

where

$$\mathbf{v} = \widetilde{\mathbf{grad}}\, \widetilde{U},$$

and $\widetilde{U}$ is a scalar potential having a constant jump through each Ω_k, for $k = 1, \cdots, m$. We have denoted by $\widetilde{\mathbf{grad}}$ the gradient operator in the space $\mathrm{H}^1(\Delta \setminus \cup_{k=1}^{m} \Omega_k)$. As we will show below, this representation allows us to impose the sources in the closed circuits Δ_k (see also Section 5.2 of (Hiptmair & Sterz, 2005)).

For $k = 1, \ldots, m$, let us denote by η_k the solution in $\mathrm{H}^1(\Delta_k \setminus \Omega_k)$, unique up to a constant, of the following weak problem:

$$\int_{\Delta_k \setminus \Omega_k} \sigma\, \widetilde{\mathbf{grad}}\, \eta_k \cdot \mathbf{grad}\, \xi = 0, \quad \forall \xi \in \mathrm{H}^1(\Delta_k), \tag{12}$$

$$[\eta_k]_{\Omega_k} = 1, \tag{13}$$

where $[\eta_k]_{\Omega_k}$ denotes the jump of η_k through Ω_k along ν.

By using functions η_k, $k = 1, \cdots, m$, the scalar potential $\widetilde{U}$ can be written as (see again (Amrouche et al., 1998)),

$$\widetilde{U} = \Phi + \sum_{k=1}^{m} V_k \eta_k, \tag{14}$$

with $\Phi \in H^1(\Delta)$ and $V_k, k = 1, \ldots, m$ being some complex numbers. From the definition of η_k we deduce that V_k is the constant jump of $\widetilde{U}$ through each surface Ω_k, $k = 1, \cdots, m$. From a physical point of view, these complex numbers V_k can be interpreted as voltage drops (Hiptmair & Sterz, 2005).

Then, taking into account that $\mathbf{H} = \mu^{-1} \mathbf{curl}\, \mathbf{A}$ and equation (2) we obtain

$$i\omega\sigma\mathbf{A} + \mathbf{curl}(\frac{1}{\mu}\mathbf{curl}\,\mathbf{A}) = -\sigma\mathbf{v} = -\sigma(\mathbf{grad}\,\Phi + \sum_{k=1}^{m} V_k \widetilde{\mathbf{grad}}\,\eta_k). \tag{15}$$

Notice, however, that the vector potential $\mathbf{A}$ is not unique because it can be altered by any gradient. Thus, in order to get uniqueness we need to impose some *gauge conditions*. In the conductor region Δ, we set $\mathrm{div}(\sigma\mathbf{A}) = 0$ and the boundary condition $\sigma\mathbf{A} \cdot \mathbf{n} = 0$ on Σ, where $\mathbf{n}$ is a unit normal vector to Σ outward from Δ.

From these gauge conditions, we can easily deduce that $\mathbf{grad}\,\Phi = \mathbf{0}$. Consequently,

$$\mathbf{v} = \sum_{k=1}^{m} V_k \widetilde{\mathbf{grad}}\,\eta_k$$

and equation (15) can be written as

$$i\omega\sigma\mathbf{A} + \mathbf{curl}(\frac{1}{\mu}\mathbf{curl}\,\mathbf{A}) = -\sigma\mathbf{v} = -\sigma\sum_{k=1}^{m} V_k \widetilde{\mathbf{grad}}\,\eta_k \quad \text{in } \Delta. \tag{16}$$

Notice that, in particular, the electric field in Δ is given by

$$\mathbf{E} = -i\omega\mathbf{A} - \sum_{k=1}^{m} V_k \widetilde{\mathbf{grad}}\,\eta_k. \tag{17}$$

On the other hand, in the air region, we impose the gauge conditions

$$\mathrm{div}\,\mathbf{A} = 0 \text{ in } \Delta^c \quad \text{and} \quad \int_{\Sigma_j} \mathbf{A} \cdot \mathbf{n} = 0, \quad j = 0, \ldots, m. \tag{18}$$

In Section 3 we will obtain a weak formulation from equations (16), (18) and the imposition of the current intensities (9) in the axisymmetric setting.

2.2 The thermal model

The electromagnetic model must be coupled with the heat equation to study the thermal effects of the electromagnetic fields in the workpiece. We will describe the equations of the model in an axisymmetric setting, paying attention to the terms coupling the thermal problem with the electromagnetic one.

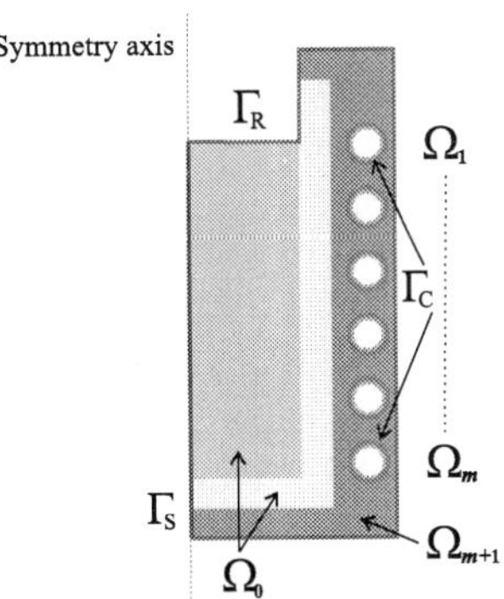

Fig. 4. Computational domain for the thermal problem

The computational domain for the thermal model is the radial section Ω_T of the whole furnace, that is,

$$\Omega_T := \Omega_0 \cup \Omega_1 \cup \cdots \cup \Omega_m \cup \Omega_{m+1},$$

where Ω_{m+1} denotes the radial section of the dielectric parts of the furnace (see Fig. 4). We notice that we now take into account that the coil is water-cooled and replace each turn by a hollow torus. An appropriated boundary condition will be considered on the inside boundary.

Remark 22 *For simplicity, in Section 2.1 the coil was replaced by rings with toroidal geometry, i.e., by solid tori. Since we are now considering a refrigeration tube along the coil, it will be replaced by hollow tori. We refer the reader to Remark 4.4 of (Vázquez, 2008) to see that, under axisymmetric assumptions, the electromagnetic model does not change in this topology.*

Since the metal is introduced in solid state and then melted, we use the transient heat transfer equation with change of phase, written in terms of the enthalpy. Furthermore, since the molten metal is subject to electromagnetic and buoyancy forces, we also need to consider convective heat transfer. Let us suppose that we already know the velocity field $\mathbf{u}$ which is null in the solid part of the workpiece, and the current density $\mathbf{J}$. Then, the equation for energy conservation is

$$\dot{e} - \operatorname{div}(k(\mathbf{x},T)\,\mathbf{grad}\,T) = \frac{|\mathbf{J}|^2}{2\sigma} \quad \text{in } \Omega_T, \tag{19}$$

where T is the temperature, k is the temperature-dependent thermal conductivity and $\dot{e}$ is the material derivative of the enthalpy density. It is given by

$$\dot{e} = \frac{\partial e}{\partial t} + \mathbf{u} \cdot \mathbf{grad}\,e.$$

The source term on the right-hand side of (19) represents the heat released due to the Joule effect; it is obtained by solving the electromagnetic model introduced in the previous section.

Remark 23 *Notice that the time scale for the variation of the electromagnetic field is much smaller than the one for the variation of the temperature. Indeed, the physical parameters used in a typical industrial situation give a time scale for temperature of the order of 60 seconds while for the* *electromagnetic problem it is of the order of 10^{-5} seconds. Thus, we may compute the electromagnetic field in the frequency domain using the eddy current model and then the heat source due to Joule effect is determined by taking the mean value on a cycle, namely,*

$$\frac{\omega}{2\pi} \int_0^{2\pi/\omega} \mathcal{J}(\mathbf{x},t) \cdot \mathcal{E}(\mathbf{x},t)\, dt, \tag{20}$$

$\mathcal{J}$ and $\mathcal{E}$ being the current density and the electric field, respectively, written in the form of (1). It is not difficult to prove that expression (20) coincides with the right-hand side of (19).

In general, the metal in the crucible undergoes a phase change during the heating process. For this reason, the enthalpy in (19) must take into account the latent heat involved in the phase change. This can be done by introducing an enthalpy function similar to that used for Stefan problems (see (Elliot & Ockendon, 1985)). More precisely, the enthalpy density is expressed as a multi-valued function of temperature by

$$e(\mathbf{x},t) \in \mathcal{H}(\mathbf{x},T(\mathbf{x},t)), \tag{21}$$

with

$$\mathcal{H}(\mathbf{x},T) = \begin{cases} \Psi(\mathbf{x},T) & T < T_s(\mathbf{x}), \\ [\Psi(\mathbf{x},T_s), \Psi(\mathbf{x},T_s) + L\,\rho(\mathbf{x},T_s)] & T = T_s(\mathbf{x}), \\ \Psi(\mathbf{x},T) + L\,\rho(\mathbf{x},T_s) & T > T_s(\mathbf{x}), \end{cases} \tag{22}$$

and

$$\Psi(\mathbf{x},T) = \int_0^T \rho(\mathbf{x},\zeta)c(\mathbf{x},\zeta)\, d\zeta, \tag{23}$$

ρ being the density, c the specific heat and L the latent heat per unit mass, i.e., the heat per unit mass necessary to achieve the change of state at temperature T_S.

Remark 24 *It is to be noted that $\mathcal{H}$ is a multivalued function rather than discontinuous. Indeed, an element of volume at the solidification temperature may have any enthalpy value in the interval $[\Psi(\mathbf{x},T_s), \Psi(\mathbf{x},T_s) + \rho(\mathbf{x},T_s)\,L]$.*

2.2.1 Thermal boundary conditions

Equation (19) must be completed with suitable initial and boundary conditions. Let Γ_T be the boundary of Ω_T; we denote by Γ_S its intersection with the symmetry axis, by Γ_R the part of the boundary in contact with air and by Γ_C the internal boundary of the coil which is in contact with the cooling water (see Fig. 4). On Γ_C we consider a convection condition

$$k(\mathbf{x},T)\frac{\partial T}{\partial \mathbf{n}} = \alpha(T_w - T), \tag{24}$$

α being the coefficient of convective heat transfer and T_w the temperature of the cooling water. On boundary Γ_R we impose the radiation-convection condition

$$k(\mathbf{x},T)\frac{\partial T}{\partial \mathbf{n}} = \alpha(T_c - T) + \gamma(T_r^4 - T^4), \tag{25}$$

where T_c and T_r are the external convective and radiative temperature, respectively, and γ is the product of the emissivity by the Stefan-Boltzmann constant (5.669e-8 W/m^2K^4). Finally, on Γ_S we set the symmetry condition

$$k(\mathbf{x},T)\frac{\partial T}{\partial \mathbf{n}} = 0. \tag{26}$$

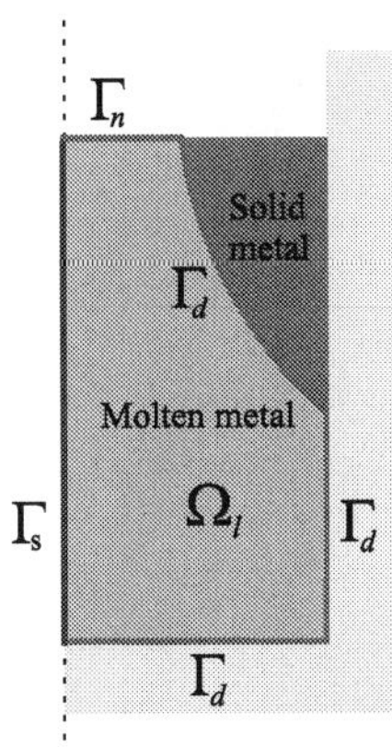

Fig. 5. Computational domain $\Omega_l(t)$ for the hydrodynamic problem.

2.3 The hydrodynamic model: Boussinesq approximation

As mentioned before, in order to achieve a realistic simulation of the overall process occurring in the furnace, convective heat transfer must be taken into account. The hydrodynamic domain is the molten region of the metal, which varies as the metal melts or solidifies, making our hydrodynamic domain time-dependent. This molten metal is subjected to both electromagnetic and buoyancy forces, the latter due to the variation of mass density with temperature.

Since the range of temperatures in the molten region is not very large, we use the Boussinesq approximation to model the fluid motion. Roughly speaking, this approximation consists on assuming that the mass density is constant in the inertial term, and that it depends linearly on the temperature in the right-hand side. Denoting by $\Omega_l(t)$ the radial section of the molten metal, and by $\Gamma_s(t)$, $\Gamma_d(t)$ and $\Gamma_n(t)$ the different parts of the boundary at time t (see Fig. 5), the fluid motion is modeled by the following equations

$$\rho_0 \left(\frac{\partial \mathbf{u}}{\partial t} + (\mathbf{grad}\,\mathbf{u})\,\mathbf{u} \right) - \mathrm{div}(2\eta_0 D(\mathbf{u})) + \mathbf{grad}\,p \quad = \quad -\rho_0\beta_0(T - T_0)\mathbf{g} + \mathbf{f}_l, \qquad (27)$$

$$\mathrm{div}\,\mathbf{u} \quad = \quad 0, \qquad (28)$$

where $\mathbf{u}$ is the velocity field, p is the (modified) pressure, and $D(\mathbf{u})$ is the strain rate tensor, namely $D(\mathbf{u}) = (\mathbf{grad}\,\mathbf{u} + \mathbf{grad}\,\mathbf{u}^t)/2$. Constants ρ_0, η_0 and β_0 denote the mass density, the dynamic viscosity and the thermal expansion coefficient, respectively, at the reference temperature T_0. The first term in the right-hand side takes into account the buoyancy forces ($\mathbf{g}$ denotes the gravity acceleration), whereas $\mathbf{f}_l$ represents the Lorentz force which is computed as

$$\mathbf{f}_l = \frac{\omega}{2\pi} \int_0^{2\pi/\omega} \mathcal{J}(\mathbf{x},t) \times \mathcal{B}(\mathbf{x},t)\,dt, \qquad (29)$$

where $\mathcal{B}$ is the magnetic induction field written in the form (1).

Equations (27)-(28) are completed with the following boundary and initial conditions:

$$\mathbf{u} = \mathbf{0} \quad \text{on } \Gamma_d(t), \tag{30}$$

$$S\mathbf{n} = \mathbf{0} \quad \text{on } \Gamma_n(t), \tag{31}$$

$$S\mathbf{n} = \mathbf{0} \quad \text{on } \Gamma_s(t), \tag{32}$$

$$\mathbf{u} = \mathbf{0} \quad \text{in } \Omega_l(0), \tag{33}$$

where S denotes the Cauchy stress tensor, $S = -pI + 2\eta_0 D(\mathbf{u})$ and I is the identity tensor. When using the Boussinesq approximation, the thermal model is modified in such a way that the material properties in the molten region are considered at the reference temperature, that is to say,

$$\rho_0 c_0 \left(\frac{\partial T}{\partial t} + \mathbf{u} \cdot \mathbf{grad}\, T \right) - \operatorname{div}(k_0 \,\mathbf{grad}\, T) = \frac{|\mathbf{J}|^2}{2\sigma}, \tag{34}$$

where c_0 and k_0 represent the specific heat and the thermal conductivity at the reference temperature, respectively. We remark that this approximation is only used in the molten region of the metal. In the rest of the domain the heat equation remains non-linear.

2.3.1 An algebraic turbulence model: Smagorinsky's model

To take into account the possibility of turbulent flows, it is necessary to introduce a turbulence model. The basic methodology consists in modifying equations (27), (28) and (34), in the form

$$\rho_0 \left(\frac{\partial \hat{\mathbf{u}}}{\partial t} + (\mathbf{grad}\, \hat{\mathbf{u}})\, \hat{\mathbf{u}} \right) - \operatorname{div}(2\eta_{eff} D(\hat{\mathbf{u}})) + \mathbf{grad}\, \hat{p} = -\rho_0 \beta_0 (\hat{T} - T_0)\mathbf{g} + \mathbf{f}_l, \tag{35}$$

$$\operatorname{div} \hat{\mathbf{u}} = 0 \tag{36}$$

$$\rho_0 c_0 \left(\frac{\partial \hat{T}}{\partial t} + \hat{\mathbf{u}} \cdot \mathbf{grad}\, \hat{T} \right) - \operatorname{div}(k_{eff} \,\mathbf{grad}\, \hat{T}) = \frac{|\mathbf{J}|^2}{2\sigma}, \tag{37}$$

where η_{eff} and k_{eff} denote the effective viscosity and thermal conductivity, respectively; namely, $\eta_{eff} = \eta_0 + \eta_t$ and $k_{eff} = k_0 + k_t$, η_t being the turbulent viscosity and k_t the turbulent thermal conductivity.

Notice that fields $\mathbf{u}$, p and T in (27), (28) and (34) have been replaced by their corresponding filtered fields $\hat{\mathbf{u}}$, $\hat{p}$ and $\hat{T}$ (see (Wilcox, 2008)). Hereafter, and for the sake of simplicity, we will eliminate the symbol $\hat{\ }$ when referring to these variables.

For a rigorous derivation of turbulence models we refer the reader to, for instance, (Mohammadi & Pironneau, 1994; Wilcox, 2008). The turbulence models differ essentially on the way the turbulent viscosity η_t and the turbulent conductivity k_t are computed. An efficient and easy to implement model is the one proposed by Smagorinsky, which has been considered in the present work. It belongs to the family of Large Eddy Simulation (LES) models. In this case, η_t and k_t are given by

$$\eta_t = \rho_0 C h^2 |D(\mathbf{u})|, \quad C \cong 0.01, \quad k_t = c_0 \frac{\eta_t}{Pr_t}, \tag{38}$$

where $h(\mathbf{x})$ is the mesh size of the numerical method around point $\mathbf{x}$, and Pr_t is the turbulent Prandtl number, which is taken equal to 0.9.

It is worth noting that more accurate turbulent models, such as the well-known k-ϵ model, can also be used. The drawback is that the computational effort to apply those models is much higher, since they require the solution of some additional partial differential equations in order to compute the turbulent viscosity η_t. However, if an accurate simulation of the fluid motion is necessary -as for instance in electromagnetic stirring- the use of one of these models might become unavoidable.

3. Weak formulation of the electromagnetic problem in an axisymmetric setting

In this section we present the weak formulation of the electromagnetic problem. We start by introducing some functional spaces and sets. Let $\mathcal{X}$ be the Beppo-Levi space (see (Nédélec, 2001), Section 2.5.4)

$$\mathcal{X} = \{\mathbf{G} : \frac{\mathbf{G}(\mathbf{x})}{\sqrt{1+|\mathbf{x}|^2}} \in \mathbf{L}^2(\mathbb{R}^3), \mathbf{curl}\,\mathbf{G} \in \mathbf{L}^2(\mathbb{R}^3)\},$$

and its subset

$$\mathcal{Y} = \{\mathbf{G} \in \mathcal{X} : \operatorname{div}\mathbf{G} = 0 \text{ in } \Delta^c, \int_{\Sigma_j} \mathbf{G} \cdot \mathbf{n} = 0, j = 0,\ldots,m\}.$$

We recall that we are interested in finding a solution of the eddy current problem satisfying the intensity conditions (9). To attain this goal, we start by writing these constraints in a weak sense.

Since current density $\mathbf{J} = \sigma\mathbf{E}$ satisfies $\operatorname{div}\mathbf{J} = 0$ in Δ and $\mathbf{J} \cdot \mathbf{n} = 0$ on Σ, we have for $k = 1,\ldots,m$,

$$\int_{\Delta_k\backslash\Omega_k} \mathbf{J} \cdot \widetilde{\mathbf{grad}}\,\eta_k = -\int_{\Delta_k\backslash\Omega_k} \operatorname{div}\mathbf{J}\,\eta_k + \int_{\partial(\Delta_k\backslash\Omega_k)} \mathbf{J} \cdot \mathbf{n}\,\eta_k = \int_{\Omega_k}[\eta_k]\mathbf{J} \cdot \nu_k = I_k, \quad (39)$$

where ν_k is the unit vector normal to the radial section Ω_k. Hence, we can impose the current intensities as follows:

$$\sum_{k=1}^{m} \bar{W}_k \int_{\Delta_k\backslash\Omega_k} \sigma\mathbf{E} \cdot \widetilde{\mathbf{grad}}\,\eta_k = \sum_{k=1}^{m} I_k\bar{W}_k, \quad \forall\mathbf{W} = (W_1,\ldots,W_m) \in \mathbb{C}^m.$$

Then, taking into account (17), we obtain the following weak form of constraint (9) which is well defined for any vector function $\mathbf{A} \in \mathcal{Y}$ and vector $\mathbf{W} \in \mathbb{C}^m$:

$$-\sum_{k=1}^{m} \bar{W}_k \int_{\Delta_k\backslash\Omega_k} i\omega\sigma\,\widetilde{\mathbf{grad}}\,\eta_k \cdot \mathbf{A} - \sum_{k=1}^{m} \bar{W}_k \int_{\Delta_k\backslash\Omega_k} \sigma V_k |\widetilde{\mathbf{grad}}\,\eta_k|^2 = \sum_{k=1}^{m} I_k\bar{W}_k. \quad (40)$$

On the other hand, multiplying equation (16) by the complex conjugate of a test function $\mathbf{G}$, denoted by $\bar{\mathbf{G}}$, integrating in $\mathbb{R}^3$ and using a Green's formula we can easily obtain,

$$i\omega\int_{\mathbb{R}^3} \sigma\mathbf{A} \cdot \bar{\mathbf{G}} + \int_{\mathbb{R}^3} \frac{1}{\mu} \mathbf{curl}\,\mathbf{A} \cdot \mathbf{curl}\,\bar{\mathbf{G}} + \sum_{k=1}^{m} V_k \int_{\Delta_k\backslash\Omega_k} \sigma\,\widetilde{\mathbf{grad}}\,\eta_k \cdot \bar{\mathbf{G}}, \quad \forall\mathbf{G} \in \mathcal{Y}. \quad (41)$$

Thus, we are led to solve the following mixed problem:

Problem MPI.- *Given* $\mathbf{I} = (I_1,\ldots,I_m) \in \mathbb{C}^m$, *find* $\mathbf{A} \in \mathcal{Y}$ *and* $\mathbf{V} \in \mathbb{C}^m$, *such that*

$$i\omega\int_{\mathbb{R}^3} \sigma\mathbf{A} \cdot \bar{\mathbf{G}} + \int_{\mathbb{R}^3} \frac{1}{\mu} \mathbf{curl}\,\mathbf{A} \cdot \mathbf{curl}\,\bar{\mathbf{G}} + \sum_{k=1}^{m} V_k \int_{\Delta_k\backslash\Omega_k} \sigma\,\widetilde{\mathbf{grad}}\,\eta_k \cdot \bar{\mathbf{G}} = 0, \forall\mathbf{G} \in \mathcal{Y}.$$

$$\sum_{k=1}^{m} \bar{W}_k \int_{\Delta_k\backslash\Omega_k} \sigma\,\widetilde{\mathbf{grad}}\,\eta_k \cdot \mathbf{A} + \frac{1}{i\omega}\sum_{k=1}^{m} \bar{W}_k \int_{\Delta_k\backslash\Omega_k} \sigma V_k |\widetilde{\mathbf{grad}}\,\eta_k|^2 = -\frac{1}{i\omega}\sum_{k=1}^{m} I_k\bar{W}_k, \forall\mathbf{W} \in \mathbb{C}^m.$$

Notice that the vector field $\mathbf{V}$ of voltage drops, can be interpreted as a Lagrange multiplier introduced to impose the current intensities in a weak sense. The mathematical analysis of this mixed formulation can be found in (Bermúdez et al., 2009), where the simpler formulation with the voltage drops as data is also studied.

3.1 An axisymmetric BEM/FEM formulation of problem MPI

Notice that the above formulation is written in the whole space $\mathbb{R}^3$; in (Bermúdez et al., 2007) we have approximated the problem in a bounded domain by defining approximated boundary conditions far from the conducting region and using a finite element technique. However, in this work, we consider a hybrid boundary element/finite element method (in the sequel BEM/FEM) to solve the problem in the whole space. To attain this goal, we are going to write the problem $\mathbf{MPI}$ in another form involving only the values of the magnetic vector potential $\mathbf{A}$ in Δ and on its boundary Σ. Notice first that the field $\mu^{-1}\,\mathbf{curl}\,\mathbf{A}$, which is the intensity of the magnetic field, belongs to $\mathcal{X}$, and then its tangential trace $\mu^{-1}\,\mathbf{curl}\,\mathbf{A} \times \mathbf{n}$ is continuous across Σ. Besides

$$\mathbf{curl}\left(\frac{1}{\mu_0}\,\mathbf{curl}\,\mathbf{A}\right) = \mathbf{curl}\,\mathbf{H} = 0 \quad \text{in } \Delta^{\mathrm{c}}, \tag{42}$$

where μ_0 denotes the vacuum magnetic permeability. Then, by using a Green's formula in Δ^{c}, we have,

$$\int_{\mathbb{R}^3}\frac{1}{\mu}\,\mathbf{curl}\,\mathbf{A}\cdot\mathbf{curl}\,\bar{\mathbf{G}} = \int_{\Delta}\frac{1}{\mu}\,\mathbf{curl}\,\mathbf{A}\cdot\mathbf{curl}\,\bar{\mathbf{G}} + \int_{\Delta^{\mathrm{c}}}\frac{1}{\mu_0}\,\mathbf{curl}\,\mathbf{A}\cdot\mathbf{curl}\,\bar{\mathbf{G}}$$

$$= \int_{\Delta}\frac{1}{\mu}\,\mathbf{curl}\,\mathbf{A}\cdot\mathbf{curl}\,\bar{\mathbf{G}} - \int_{\Sigma}\frac{1}{\mu_0}\,\mathbf{curl}\,\mathbf{A}\times\mathbf{n}\cdot\bar{\mathbf{G}}, \;\forall\mathbf{G}\in\mathcal{Y}.$$

Thus, the first equation of problem $\mathbf{MPI}$ can be formally written as:

$$i\omega\int_{\Delta}\sigma\mathbf{A}\cdot\bar{\mathbf{G}} + \int_{\Delta}\frac{1}{\mu}\,\mathbf{curl}\,\mathbf{A}\cdot\mathbf{curl}\,\bar{\mathbf{G}} - \int_{\Sigma}\frac{1}{\mu_0}\,\mathbf{curl}\,\mathbf{A}\times\mathbf{n}\cdot\bar{\mathbf{G}}$$

$$+ \sum_{k=1}^{m}V_k\int_{\Delta_k\setminus\Omega_k}\sigma\,\widetilde{\mathbf{grad}}\,\eta_k\cdot\bar{\mathbf{G}} = 0, \;\forall\mathbf{G}\in\mathcal{Y}. \tag{43}$$

We notice that the value of $\mu_0^{-1}\,\mathbf{curl}\,\mathbf{A}\times\mathbf{n}$ on Σ can be determined by solving an exterior problem in Δ^{c}. Since we are interested in the numerical solution of the problem in an axisymmetric domain, we directly transform this term in the axisymmetric case.

We consider a cylindrical coordinate system (r,θ,z) with the $z-$axis coinciding with the symmetry axis of the device, (see Figure 3). Hereafter we denote by $\mathbf{e}_r$, $\mathbf{e}_\theta$ and $\mathbf{e}_z$ the local unit vectors in the corresponding coordinate directions. Now, cylindrical symmetry leads us to consider that no field depends on the angular variable θ. We further assume that the current density field has non-zero component only in the tangential direction $\mathbf{e}_\theta$, namely

$$\mathbf{J} = J_\theta(r,z)\,\mathbf{e}_\theta.$$

We remark that, due to the assumed conditions on $\mathbf{J}$, (3), (6) and (10), only the θ-component of the magnetic vector potential, hereafter denoted by A_θ, does not vanish, i.e.,

$$\mathbf{A} = A_\theta(r,z)\mathbf{e}_\theta. \tag{44}$$

Thus $\mathbf{A}$ automatically satisfies (18). Let $\mathbf{G} = \psi(r,z)\mathbf{e}_\theta$ be a test function and $\mathbf{n} = n_r\,\mathbf{e}_r + n_z\,\mathbf{e}_z$. By taking into account the expression for **curl** in cylindrical coordinates and the fact that

$$\widetilde{\mathbf{grad}}\,\eta_k = \frac{1}{2\pi r}\mathbf{e}_\theta, \quad \text{in } \Delta_k,\ k = 1,\ldots,m, \tag{45}$$

the axisymmetric version of the problem formally writes as follows:

Given $\mathbf{I} = (I_1,\ldots,I_m) \in \mathbb{C}^m$, *find* A_θ *and* $\mathbf{V} \in \mathbb{C}^m$, *satisfying,*

$$i\omega \int_\Omega \sigma A_\theta \cdot \bar{\psi} r\,drdz + \int_\Omega \frac{1}{\mu r}\frac{\partial(rA_\theta)}{\partial r}\frac{\partial(r\bar{\psi})}{\partial r}\,drdz + \int_\Omega \frac{1}{\mu}\frac{\partial A_\theta}{\partial z}\frac{\partial\bar{\psi}}{\partial z}r\,drdz$$

$$- \int_\Gamma \frac{1}{\mu_0}\frac{\partial(rA_\theta)}{\partial\mathbf{n}}\bar{\psi}\,d\gamma + \frac{1}{2\pi}\sum_{k=1}^m V_k\int_{\Omega_k}\sigma\bar{\psi}\,drdz = 0, \tag{46}$$

$$\frac{1}{2\pi}\sum_{k=1}^m\Big(\int_{\Omega_k}\sigma A_\theta\,drdz\Big)\bar{W}_k + \frac{1}{4\pi^2 i\omega}\sum_{k=1}^m\Big(\int_{\Omega_k}\sigma\frac{V_k}{r}\,drdz\Big)\bar{W}_k = -\frac{1}{2\pi i\omega}\sum_{k=1}^m I_k\bar{W}_k, \tag{47}$$

for all test function ψ *and* $W \in \mathbb{C}^m$.

On the other hand, the term $\int_\Gamma \mu_0^{-1}\partial(rA_\theta)/\partial\mathbf{n}\,\bar{\psi}\,d\gamma$ can be transformed by using the single-double layer potentials. We refer the reader to (Bermúdez et al., 2007b) for further details concerning this transformation and introduce the same notation of that paper, namely

$$\begin{aligned} A_\theta' &= rA_\theta, \\ \lambda'(r,z) &= \frac{\partial A_\theta'}{\partial r}n_r + \frac{\partial A_\theta'}{\partial z}n_z. \end{aligned}$$

We are led to solve the following weak problem

Problem WEPI.- *Given* $\mathbf{I} = (I_1,\ldots,I_m) \in \mathbb{C}^m$, *find* $A_\theta' : \Omega \longrightarrow \mathbb{C}$, $\mathbf{V} \in \mathbb{C}^m$ *and* $\lambda' : \Gamma \longrightarrow \mathbb{C}$ *such that*

$$i\omega\int_\Omega \frac{\sigma}{r}A_\theta'\bar{\psi}'\,drdz + \int_\Omega \frac{1}{\mu r}\mathbf{grad}\,A_\theta'\cdot\mathbf{grad}\,\bar{\psi}'\,drdz - \int_\Gamma \frac{1}{\mu r}\lambda'\bar{\psi}'\,d\gamma + \frac{1}{2\pi}\sum_{k=1}^m V_k\int_{\Omega_k}\frac{\sigma}{r}\bar{\psi}'\,drdz = 0,$$

$$\frac{1}{2\pi}\sum_{k=1}^m\bar{W}_k\int_{\Omega_k}\frac{\sigma}{r}A_\theta'\,drdz + \frac{1}{4\pi^2 i\omega}\sum_{k=1}^m\bar{W}_k\int_{\Omega_k}\sigma\frac{V_k}{r}\,drdz = -\frac{1}{2\pi i\omega}\sum_{k=1}^m I_k\bar{W}_k,$$

$$\int_\Gamma \frac{1}{\mu r}A_\theta'\bar{\zeta}\,d\gamma - \int_\Gamma \frac{1}{\mu}(\mathcal{G}_n A_\theta')\bar{\zeta}\,d\gamma + \int_\Gamma \frac{1}{\mu}(\mathcal{G}\lambda')\bar{\zeta}\,d\gamma = 0$$

for all test functions ψ', ζ *and* $W \in \mathbb{C}^m$. In the previous equations, $\mathcal{G}$ and $\mathcal{G}_n$ denote the fundamental solution of Laplace's equation and its normal derivative, respectively, in cylindrical coordinates (see again (Bermúdez et al., 2007b)).

4. Time discretization and weak formulation of the thermal and hydrodynamic problems

In this section we introduce a time discretization and a weak formulation of the thermal and hydrodynamic models. Again, we exploit the cylindrical symmetry and we assume that $\mathbf{u}$

does not depend on θ and it has zero component in the tangential direction $\mathbf{e}_\theta$. In order to simplify the notation, in what follows we shall drop index t for $\Omega_l(t)$.

To obtain a suitable discretization of the material time derivative in equations (19) and (27) we shall use the characteristics method (see, for instance, (Pironneau, 1982)).

Given a velocity field $\mathbf{u}$ we define the characteristic curve passing through point $\mathbf{x}$ at time t as the solution of the following Cauchy problem:

$$\begin{cases} \dfrac{\mathrm{d}}{\mathrm{d}\tau}\mathbf{X}(\mathbf{x},t;\tau) = \mathbf{u}(\mathbf{X}(\mathbf{x},t;\tau),\tau), \\ \mathbf{X}(\mathbf{x},t;t) = \mathbf{x}. \end{cases} \tag{48}$$

Thus $\mathbf{X}(\mathbf{x},t;\tau)$ is the trajectory of the material point being at position $\mathbf{x}$ at time t. In terms of $\mathbf{X}$, the material time derivative of e is defined by

$$\dot{e}(\mathbf{x},t) = \frac{d}{d\tau}[e(\mathbf{X}(\mathbf{x},t;\tau),\tau)]_{|\tau=t}. \tag{49}$$

Let us consider a time interval $[0,t_f]$ and a discretization time step $\Delta t = t_f/N$ to obtain a uniform partition $\Pi = \{t^n = n\Delta t, 0 \leq n \leq N\}$. Let e^n and $\mathbf{u}^n$ be the approximations of e and $\mathbf{u}$ at time t^n, respectively. We approximate the material time derivative of e at time t^{n+1} by

$$\dot{e}(\mathbf{x},t^{n+1}) \simeq \frac{e^{n+1}(\mathbf{x}) - e^n(\chi^n(\mathbf{x}))}{\Delta t}, \tag{50}$$

where $\chi^n(\mathbf{x}) = \mathbf{X}^n(\mathbf{x},t^{n+1};t^n)$ is obtained as the solution of the following Cauchy problem

$$\begin{cases} \dfrac{\mathrm{d}}{\mathrm{d}\tau}\mathbf{X}^n(\mathbf{x},t^{n+1};\tau) = \mathbf{u}^n(\mathbf{X}^n(\mathbf{x},t^{n+1};\tau),\tau), \\ \mathbf{X}^n(\mathbf{x},t^{n+1};t^{n+1}) = \mathbf{x}, \end{cases} \tag{51}$$

at time t^n. Notice that, since $\mathbf{u} = \mathbf{0}$ in the solid region, the solution of this Cauchy problem is $\mathbf{X}^n(\mathbf{x},t^{n+1};\tau) = \mathbf{x}$ for any τ. Hence equation (50) in the solid part is equivalent to a standard time discretization without using the method of characteristics.

Analogous to (50), we consider in (35) the above two-point discretization for the material time derivative of the velocity. Thus, taking into account the cylindrical symmetry, multiplying equations (35) and (36) by suitable test functions, and integrating in the liquid domain Ω_l we obtain, after using the Green's formula, the following weak formulation of the semi-discretized hydrodynamic problem (the : representing the scalar product of two tensors)

Problem WHP.- *For each $n = 0,1,\dots,N-1$, find functions $\mathbf{u}^{n+1}$ and p^{n+1} such that $\mathbf{u}^{n+1} = \mathbf{0}$ on Γ_d and furthermore*

$$\frac{1}{\Delta t}\int_{\Omega_l}\rho_0\mathbf{u}^{n+1}\cdot\mathbf{w}\,rdrdz + \int_{\Omega_l}\eta_{eff}\left(\mathbf{grad}\,\mathbf{u}^{n+1}:\mathbf{grad}\,\mathbf{w}\right)rdrdz +$$

$$\int_{\Omega_l}\eta_{eff}\left((\mathbf{grad}\,\mathbf{u}^{n+1})^t:\mathbf{grad}\,\mathbf{w}\right)rdrdz - \int_{\Omega_l}p^{n+1}\operatorname{div}\mathbf{w}\,rdrdz =$$

$$-\int_{\Omega_l}\rho_0\beta_0(T^n - T_0)\mathbf{g}\cdot\mathbf{w}\,rdrdz + \int_{\Omega_l}\mathbf{f}_l^{n+1}\cdot\mathbf{w}\,rdrdz + \frac{1}{\Delta t}\int_{\Omega_l}\rho_0(\mathbf{u}^n\circ\chi^n)\cdot\mathbf{w}\,rdrdz,$$

$$\int_{\Omega_l}\operatorname{div}\mathbf{u}^{n+1}\,qrdrdz = 0,$$

for all test functions **w** *null on* Γ_d, *and q.*

On the other hand, assuming cylindrical symmetry and the fact that the temperature does not depend on the angular coordinate θ, we can write equation (19) in cylindrical coordinates. Then, applying the time discretization (50), multiplying by a suitable test function and using a Green's formula we obtain the following weak formulation of the semi-discretized thermal problem:

Problem WTP.- *For each* $n = 0, 1, \ldots, N - 1$, *find a function* T^{n+1} *such that*

$$\int_{\Omega_T} \frac{1}{\Delta t} e^{n+1} Z \, r \, drdz + \int_{\Omega_T} k_{eff}(r, z, T^{n+1}) \, \mathbf{grad} \, T^{n+1} \cdot \mathbf{grad} \, Z \, r \, drdz$$

$$= \int_{\Gamma_C} \alpha(T_w - T^{n+1}) Z \, r \, d\Gamma + \int_{\Gamma_R} (\alpha(T_c - T^{n+1}) + \gamma(T_r^4 - (T^{n+1})^4)) Z \, r \, d\Gamma$$

$$+ \int_{\Omega_T} \frac{1}{\Delta t} (e^n \circ \chi^n) Z \, r \, drdz + \int_{\Omega_T} \frac{1}{2\sigma(r, z, T^{n+1})} |J_\theta^{n+1}|^2 Z \, r \, drdz,$$

for all test function Z.

Remark 4.1 *Note that, from equations (6), (17) and the expression for functions* η_k, *we can infer that*

$$J_\theta = -i\omega\sigma A_\theta \qquad in \ \Omega_0,$$

$$J_\theta = -i\omega\sigma A_\theta - \frac{V_k}{2\pi r} \qquad in \ \Omega_k, \quad k = 1, \ldots, m,$$

$$J_\theta = 0 \qquad in \ \Omega_{m+1}.$$

Hence, to determine the heat source we need to compute an approximation of field A_θ *at time* t^{n+1}. *This is determined as the solution of the weak formulation* **WEPI** *with the physical parameters* μ *and* σ *being evaluated at temperature* T^{n+1}.

5. Space discretization and iterative algorithm

Problem **WTP** has been spatially discretized by a piecewise linear finite element method defined in a triangular mesh of the domain Ω_T. For the spatial discretization of problem **WEPI** we have used a BEM/FEM method (see (Bermúdez et al., 2007b) for further details). Finally, problem **WHP** has been spatially discretized by the finite element couple P_1-*bubble*/P_1, which is known to satisfy the *inf-sup* condition (Brezzi, 1991); this last problem only takes place in the time-dependent liquid domain $\Omega_l(t)$, which must be computed at each time step.

It is important to notice that, at each time step, there are several terms coupling the three problems. However, since we are neglecting the velocity in Ohm's law, and the method of characteristics is used with the velocity at the previous time step, the hydrodynamic problem can be solved uncoupled from the two others. Nevertheless, the coupling between the thermal and the electromagnetic models cannot be avoided: the heat source in the thermal equation is the Joule effect and the solution of the electromagnetic problem depends on the electrical conductivity and the magnetic permeability, which may vary with temperature.

Moreover, the thermal problem **WTP** contains several nonlinearities:

– The thermal conductivity k depends on temperature.

– The external convective temperature T_w also varies with the heat flux of the inner boundaries of the coil.

- The enthalpy e depends on temperature and it is a multivalued function.
- The radiation boundary condition depends on T^4.

To treat the nonlinear terms we are going to introduce several iterative algorithms. The dependence of the thermal conductivity k can be easily treated, just by taking in the heat equation the thermal conductivity evaluated at the temperature of the previous time step (or of the previous iteration of an outer loop). This can be done because the thermal conductivity is a smooth function. Nevertheless, as we will show in the next section, the other nonlinearities need other more sophisticated iterative algorithms to be introduced.

5.1 Iterative algorithm for the temperature of cooling water

As we said before, the induction coil of the furnace is water-cooled to avoid overheating; this fact is modeled by means of boundary condition (24), where the temperature of the cooling water, T_w, depends on the solution of our problem which introduces a nonlinearity in the equations.

To deal with this nonlinearity we will seek the convergence of the heat flux from the coil to the cooling water. From the solution of the thermal problem this heat flux is computed as

$$H = 2\pi \int_{\Gamma_C} k \frac{\partial T}{\partial \mathbf{n}} r \, d\Gamma, \tag{52}$$

where the integral is multiplied by 2π because Γ_C is only a radial section of the boundary. Using (52) the outlet water temperature can be computed as

$$T_o = T_i + \frac{H}{\rho_w c_w Q}, \tag{53}$$

with ρ_w and c_w the density and specific heat of water, respectively, T_i the inlet temperature of the cooling water, and Q the water flow rate.

To solve the problem, we assume that water temperature T_w is constant along the coil which is actually a reasonable assumption, since the difference between the inlet and the outlet temperature is seldom higher than $10\,^\circ\mathrm{C}$.

The temperature of the cooling water is then computed using an iterative algorithm. Let us suppose $T_{w,j}$ is known. Then, at iteration $j+1$, we compute T_{j+1} as the solution of thermal problem **WTP** with $T_w = T_{w,j}$. Then, we compute the heat flux H from equation (52) and set

$$T_{w,j+1} = T_i + \frac{H}{2\rho_w c_w Q},$$

i.e., $T_{w,j+1}$ is the mean value of the given inlet temperature and the computed outlet temperature.

It is worth noting that an implicit algorithm is needed to avoid instabilities. The iterations of the implicit method can be merged with the iterations of the thermoelectrical coupling leading to a low computational cost.

5.2 Iterative algorithms for the enthalpy and the radiation boundary condition

In order to solve the non-linearities due to the multi-valued character of the enthalpy and to the exterior radiation boundary condition, we propose a fixed point algorithm which is described in detail in references (Bermúdez et al., 2003; 2007; Vázquez, 2008). It is summarized here for the reader's convenience.

At time step $(n+1)$ and for a positive β we introduce the new function

$$q^{n+1} = e^{n+1} - \beta T^{n+1}.$$

Using (21) and the fact that $\mathcal{H}$ is a maximal monotone operator, we get

$$q^{n+1}(r,z) = \mathcal{H}_\lambda^\beta((r,z), T^{n+1}(r,z) + \lambda q^{n+1}(r,z)) \text{ for } 0 < \lambda \leq \frac{1}{2\beta}, \qquad (54)$$

$\mathcal{H}_\lambda^\beta$ being the Yosida approximation of the operator $\mathcal{H}^\beta = \mathcal{H} - \beta\mathcal{I}$ (Bermúdez-Moreno, 1994). The same idea is introduced to deal with the nonlinearity associated to the fourth power of the boundary temperature in (25): we consider the maximal monotone operator $\mathcal{G}(T) = |T|T^3$, which coincides with T^4 for $T > 0$, and we define the new function

$$s^{n+1} = \mathcal{G}(T^{n+1}) - \kappa T^{n+1}.$$

Then using $\mathcal{G}_\delta^\kappa$, the Yosida approximation of operator $\mathcal{G}^\kappa = \mathcal{G} - \kappa\mathcal{I}$, we get

$$s^{n+1}(r,z) = \mathcal{G}_\delta^\kappa(T^{n+1}(r,z) + \delta s^{n+1}(r,z)) \text{ for } 0 < \delta \leq \frac{1}{2\kappa}. \qquad (55)$$

Now, in order to solve problem **WTP**, the idea is to replace e^{n+1} by $q^{n+1} + \beta T^{n+1}$ and $\mathcal{G}(T^{n+1})$ by $s^{n+1} + \kappa T^{n+1}$. Finally, to determine T^{n+1}, q^{n+1} and s^{n+1} we introduce an iterative process using (54) and (55).

We notice that the performance of the proposed algorithm is known to depend strongly on the choice of parameters β, λ, δ and κ. In (Vázquez, 2008) an automatic procedure is proposed to compute these parameters as functions of (r,z), which accelerates the convergence of the method.

5.3 Iterative algorithm for the whole problem

Now we present the iterative algorithm to solve the three coupled models, along with the nonlinearities. Basically, it consists of three nested loops: the first one for the time discretization, the second one for the thermoelectrical coupling, and the third one for the Bermúdez-Moreno algorithms for the enthalpy and the radiation boundary condition presented above. As we have said before, the hydrodynamic problem is solved at each time step, uncoupled from the two others. Moreover, the nonlinearities of the thermal conductivity k and the cooling water temperature T_w are also treated by iterative algorithms and their corresponding loops are in fact merged with that of the thermoelectrical coupling. For the sake of simplicity we present in Fig. 6 a sketch of the algorithm with the three nested loops.

5.3.1 The algorithm

Let us suppose that the initial temperature T^0 and velocity $\mathbf{u}^0$ are known. From T^0 we determine the initial enthalpy e^0 and set the temperature of cooling water $T_w^0 = T_i$, being T_i the inlet temperature. Then, at time step n, with $n = 1,\ldots,N$ we compute A_θ^n, T^n and $\mathbf{u}^n$ by doing the following steps:

1. If $\mathbf{u}^{n-1} \neq 0$, compute $\chi^{n-1}(\mathbf{x})$, the solution of (51).

2. Calculate the turbulent viscosity $\eta_t^n = \rho_0 C h^2 |D(\mathbf{u}^{n-1})|$ and the turbulent thermal conductivity $k_t^n = c_0 \eta_t^n / Pr_t$.

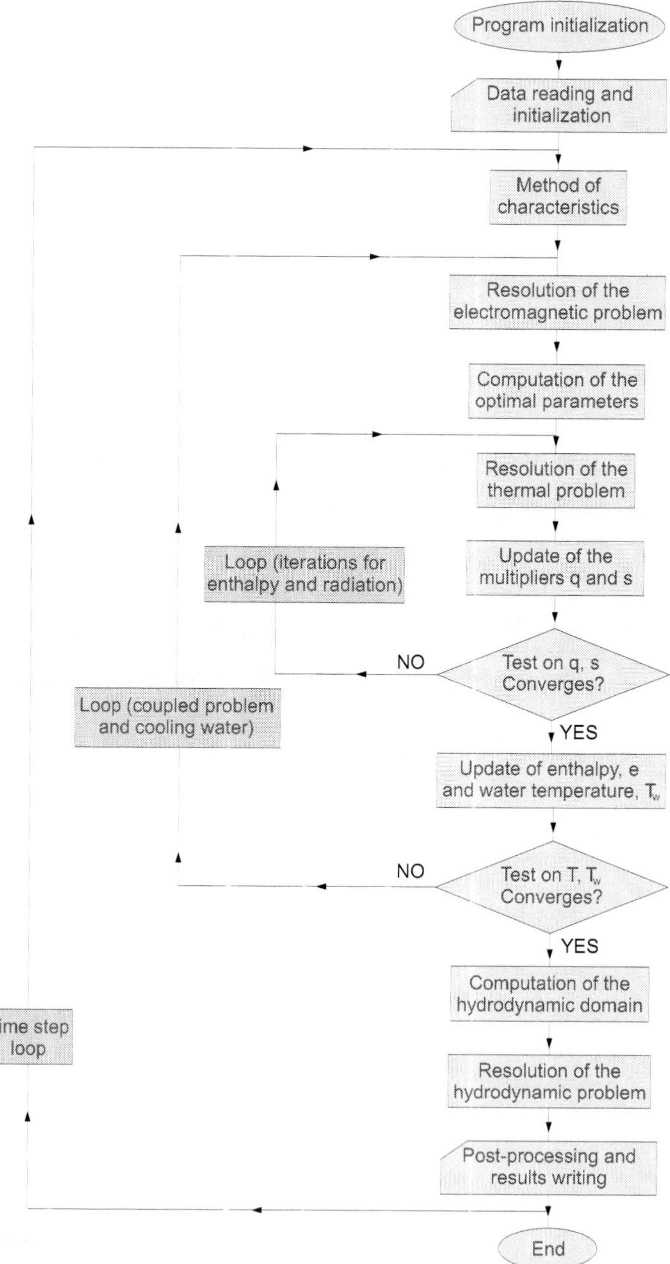

Fig. 6. Scheme of the numerical algorithm considering all the nested loops.

3. Set $T_0^n = T^{n-1}$, $e_0^n = e^{n-1}$ and $T_{w,0}^n = T_w^{n-1}$. Then compute A_θ^n, T^n, e^n and T_w^n as the limit of $A_{\theta,j}^n$, T_j^n and $T_{w,j}^n$, obtained from the following iterative procedure:

   1. For $j \geq 1$ let us suppose that T_{j-1}^n and $T_{w,j-1}^n$ are known. Then set $A_{\theta,j}^n = A_\theta' / r$, with A_θ', λ' and $\mathbf{V} \in C^m$ the solution of **WEPI** by taking $\sigma = \sigma(r,z,T_{j-1}^n)$ and $\mu = \mu(r,z,T_{j-1}^n)$.

   (b) Set $J_{\theta,j}^n = -i\omega\sigma(r,z,T_{j-1}^n)A_{\theta,j}^n - V_k/(2\pi r)$ in Ω_k, $k = 0,\ldots,m$ (with $V_0 = 0$).

   (c) Determine the optimal parameters $\kappa_j^n(r,z)$ and $\beta_j^n(r,z)$ with the method described in (Vázquez, 2008).

   (d) Set $T_{j,0}^n = T_{j-1}^n$ and also

$$
\begin{aligned}
q_{j,0}^n &= e_{j-1}^n - \beta_j^n T_{j-1}^n, \\
s_{j,0}^n &= |T_{j-1}^n|(T_{j-1}^n)^3 - \kappa_j^n T_{j-1}^n.
\end{aligned}
$$

Then compute T_j^n, q_j^n and s_j^n as the limit of the following iterative procedure:

   1. For $k \geq 1$ let us suppose that $q_{j,k-1}^n$ and $s_{j,k-1}^n$ are known. Then compute $T_{j,k}^n$ as the solution of the linear system

$$
\int_{\Omega_T} \frac{1}{\Delta t} \beta_j^n T_{j,k}^n Z r \, drdz + \int_{\Omega_T} (k(T_{j-1}^n) + k_t^n) \, \mathbf{grad}\, T_{j,k}^n \cdot \mathbf{grad}\, Z r \, drdz + \int_{\Gamma_C} \alpha T_{j,k}^n Z r \, d\Gamma
$$

$$
+ \int_{\Gamma_R} (\alpha + \gamma\kappa_j^n) T_{j,k}^n Z r \, d\Gamma = \int_{\Gamma_C} \alpha T_{w,j-1}^n Z r \, d\Gamma + \int_{\Gamma_R} (\alpha T_c + \gamma T_r^4 - \gamma s_{j,k-1}^n) Z r \, d\Gamma
$$

$$
+ \int_{\Omega_T} \frac{1}{\Delta t}(e^{n-1} \circ \chi^{n-1} - q_{j,k-1}^n) Z r \, drdz + \int_{\Omega_T} \frac{1}{2\sigma(r,z,T_{j-1}^n)} |J_{\theta,j}^n|^2 Z r \, drdz \quad \forall Z \text{ test function.}
$$

   ii. Update multipliers $q_{j,k}^n$ and $s_{j,k}^n$ by means of the formulas

$$
\begin{aligned}
q_{j,k}^n &= \mathcal{H}_\lambda^\beta \left(T_{j,k}^n + \lambda q_{j,k-1}^n \right), \\
s_{j,k}^n &= \mathcal{G}_\delta^\kappa (T_{j,k}^n + \delta s_{j,k-1}^n).
\end{aligned}
$$

   (e) Update the value of the enthalpy and the value of the cooling water temperature by computing

$$
e_j^n = q_j^n + \beta_j^n T_j^n,
$$

$$
H = 2\pi \int_{\Gamma_C} k \frac{\partial T_j^n}{\partial \mathbf{n}} r \, d\Gamma, \qquad \text{and} \qquad T_{w,j}^n = T_i + \frac{H}{2\rho_w c_w Q}.
$$

4. Determine the hydrodynamic domain from the value of enthalpy e^n.

5. Find $\mathbf{u}^n$ and p^n solution of

$$\frac{1}{\Delta t}\int_{\Omega_l} \rho_0 \mathbf{u}^n \cdot \mathbf{w}\, r\, drdz + \int_{\Omega_l}(\eta_0 + \eta_t^n)\,(\mathbf{grad}\,\mathbf{u}^n : \mathbf{grad}\,\mathbf{w})\, r\, drdz$$

$$+ \int_{\Omega_l}(\eta_0 + \eta_t^n)\,((\mathbf{grad}\,\mathbf{u}^n)^t : \mathbf{grad}\,\mathbf{w})\, r\, drdz - \int_{\Omega_l} p^n \operatorname{div}\mathbf{w}\, r\, drdz =$$

$$- \int_{\Omega_l}\rho_0\beta_0(T^n - T_0)\mathbf{g}\cdot\mathbf{w}\, r\, drdz + \int_{\Omega_l}\mathbf{f}_l(A_\theta^n)\cdot\mathbf{w}\, r\, drdz +$$

$$\frac{1}{\Delta t}\int_{\Omega_l}\rho_0(\mathbf{u}^{n-1}\circ\chi^{n-1})\cdot\mathbf{w}\, r\, drdz, \quad \forall \text{ test function } \mathbf{w},$$

$$\int_{\Omega_l}\operatorname{div}\mathbf{u}^n\, q = 0, \quad \forall \text{ test function } q.$$

6. Numerical simulation of an industrial furnace

In this section we present some numerical results obtained in the simulation of an industrial furnace used for melting and stirring. The results have been performed with the computer Fortran code THESIF (http://www.usc.es/~thesif/) which implements the algorithm described above.

6.1 Description of the furnace

We briefly describe the geometry and working conditions of the furnace and refer the reader to Chapter 4 of (Vázquez, 2008) for further details.

The inductor of the furnace is a copper helical coil with 12 turns which contains a pipe inside carrying cool water for refrigeration. Inside the coil a crucible is placed, containing the metal to be melted. The crucible is surrounded by an alumina layer to avoid heat losses. The reason to take alumina for this layer is that it is not only a good refractory but also a good electrical insulator. Thus the induction process in the crucible and in the metal is almost unaffected by the presence of alumina. For safety reasons, the induction coil is also embedded in the alumina layer. Above this alumina layer there is a layer of another refractory material, called Plibrico. Moreover, for safety reasons there is a metal sheet surrounding the furnace, which prevents from high magnetic fields outside the furnace. The furnace rests on a base of concrete, which will be also considered in the computational domain. For the same purpose of thermal insulation, there is also placed a lid over the load, but it will not be considered it in the numerical simulation. Otherwise, we should solve an internal radiation problem, instead of imposing the convection-radiation boundary condition given in (25). In (Bermúdez et al., 2006; 2011) an improved thermal model including an internal radiation condition is introduced.

The computational domain for the electromagnetic problem consists of the furnace (load, crucible and coil) as it is shown in Fig. 7. The computational domain for the thermal problem is composed by the metal, the crucible, the refractory layers and the coil. The computational domain for the hydrodynamic problem is the molten metal and it is computed at each time step.

Concerning the values of the physical properties of the different materials, we show only the electrical conductivity of the crucible and metal to be melted because it plays a major role in the results. We notice that the electrical conductivity depends on temperature, being this dependency really important in the second case, because the load is assumed to be an insulator when solid and a good conductor when it melts, as can be seen in Fig. 8.

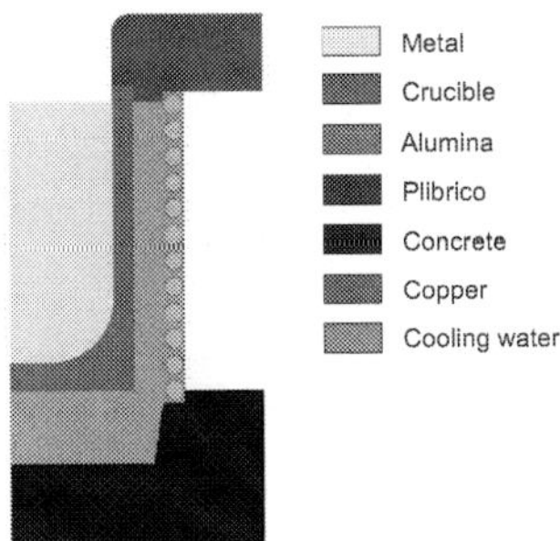

Fig. 7. Distribution of materials in the radial section of the furnace.

In the electromagnetic model that we have presented two working parameters have to be chosen: the frequency of the alternating current and the corresponding current intensity. However, when trying to simulate the real furnace, the given data is the power, and the current intensity is adjusted to obtain that power in the furnace. Moreover, since the electrical conductivity of the materials varies with temperature, the relationship between power and intensity also changes, and the intensity is dynamically adjusted during the process. To deal with this difficulty, the algorithm was slightly modified (see (Bermúdez et al., 2011) for further details) to provide the power as the known data and then to compute the intensity to attain the given power. Thus, for the numerical simulation, we have used the total power of the system as data.

6.2 Numerical results

We have performed two numerical simulations of the furnace with the same value of the power and two different values of the frequency, 500 Hz and 2650 Hz, to see how the frequency affects the heating and stirring of the metal.

In Fig. 9 we represent the temperature in the furnace for each simulation. As it can be seen the temperatures obtained in the furnace are very similar, but a little higher in the case of the highest frequency. We notice the strong influence of the refrigeration tubes in the temperature: the temperature in the copper coil and the surrounding refractory is about 50 °C, causing a very large temperature gradient within the refractory layer. In Fig. 10 we show a detail of the temperatures in the crucible and in the load. As it can be seen, higher temperatures are

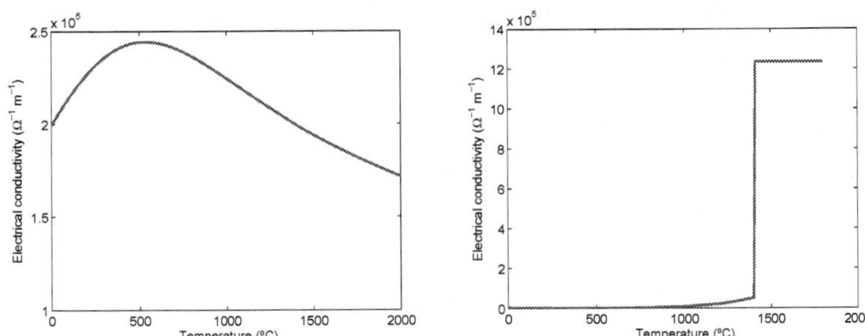

Fig. 8. Electrical conductivity for the crucible (left) and the load (right).

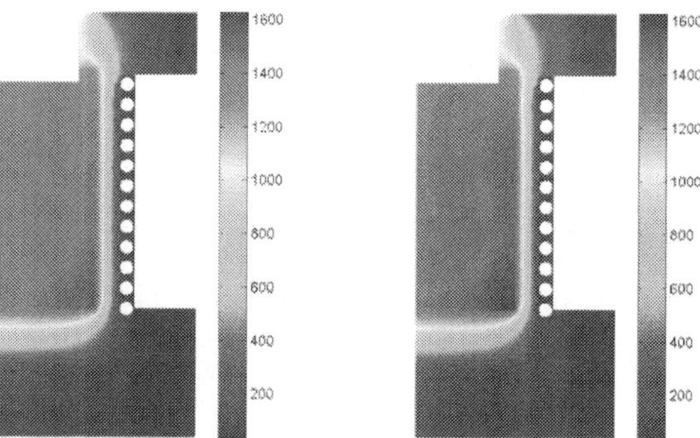

Fig. 9. Temperature after three hours for 500 Hz (left) and 2650 Hz (right).

reached when working with high frequency. This fact is explained due to the distribution of ohmic losses, as we will explain below.

In Fig. 11 we show the Joule effect in the load and the crucible after five minutes, when the metal is still solid. In Fig. 12, the same field is represented after three hours, when the metal has been melted. Comparing the results for different frequencies, we can see that the higher the frequency the higher the maximum values of the Joule effect, but due to the skin effect they are concentrated on the external wall of the crucible. Decreasing the frequency allows a better power distribution, at the cost of using higher intensities, thus causing larger power losses in the coil (see Fig. 14).

Moreover, comparing the results for solid and molten load, we see how the high conductivity of the molten metal affects the performance of the furnace. At low frequency the ohmic losses in the load become higher when the material melts, thus heating the load directly and reducing the crucible temperature (see Fig. 14 for the ohmic losses and Fig. 10 for the crucible temperature). Moreover, the presence of molten material increases the skin effect on

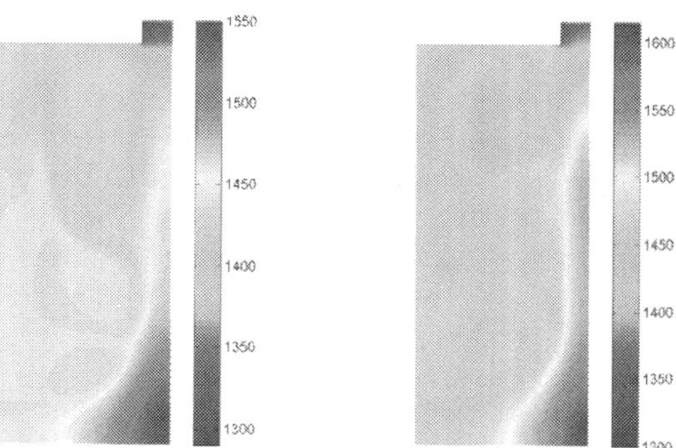

Fig. 10. Metal temperature after three hours for 500 Hz (left) and 2650 Hz (right).

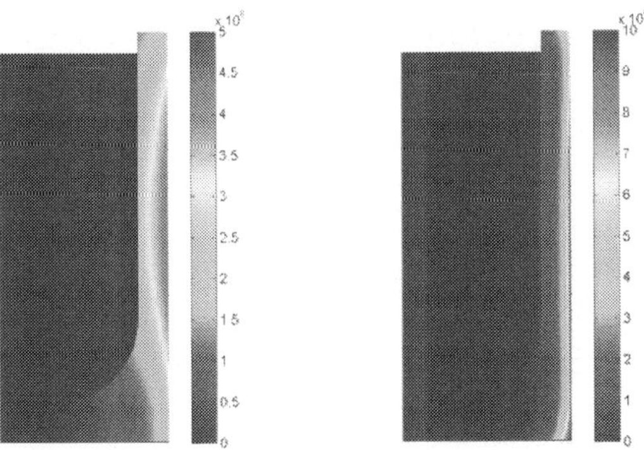

Fig. 11. Joule effect after five minutes for 500 Hz (left) and 2650 Hz (right).

the crucible wall. On the contrary, when working with high frequency the power distribution in the furnace remains almost unaffected in the presence of molten material.

We also present in Fig. 13 the velocity field for both frequencies. When working with low frequency the depth of penetration is higher, and Lorentz's force becomes stronger than buoyancy forces. At high frequency, instead, the low skin depth makes Lorentz's force almost negligible and buoyancy forces become dominant. This can be seen in this figure: at high frequency the molten metal is moving by natural convection, thus it tends to go up near the hot crucible, except in the upper part, probably due to the boundary condition we are imposing. At low frequency the electromagnetic stirring enforces the metal to go down close to the crucible, and a new eddy comes up in the bottom of the furnace.

Finally, in Fig. 14 we represent the variation in time of the Joule effect in each material, along with the heat losses through the refrigeration tubes. In order to attain the desired power, higher intensities are needed working at low frequency, which causes stronger ohmic losses in the copper coil. Moreover, at low frequency the load begins to melt after 60 minutes, affecting

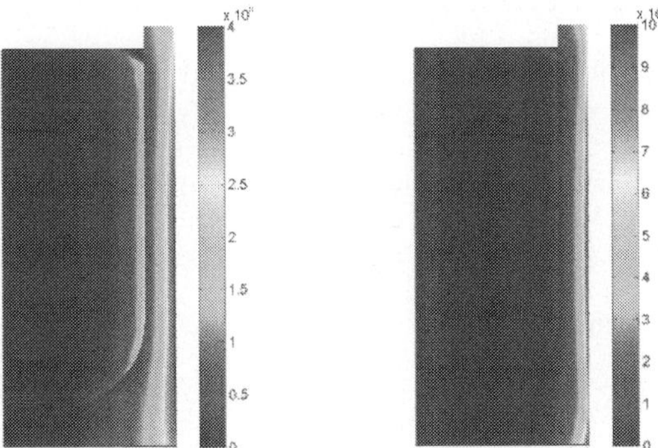

Fig. 12. Joule effect after three hours for 500 Hz (left) and 2650 Hz (right).

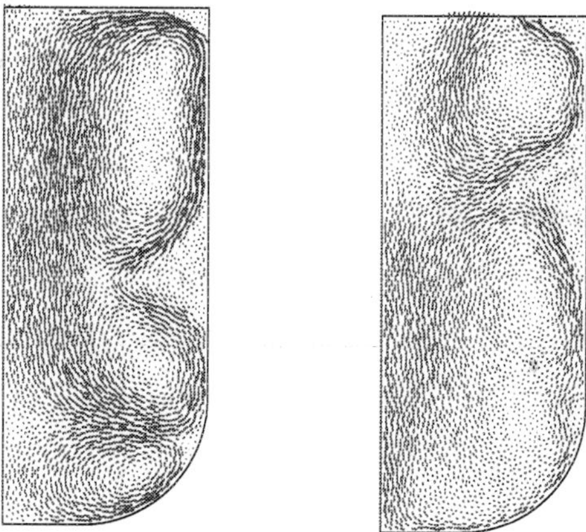

Fig. 13. Velocity fields after three hours for 500 Hz (left) and 2650 Hz (right).

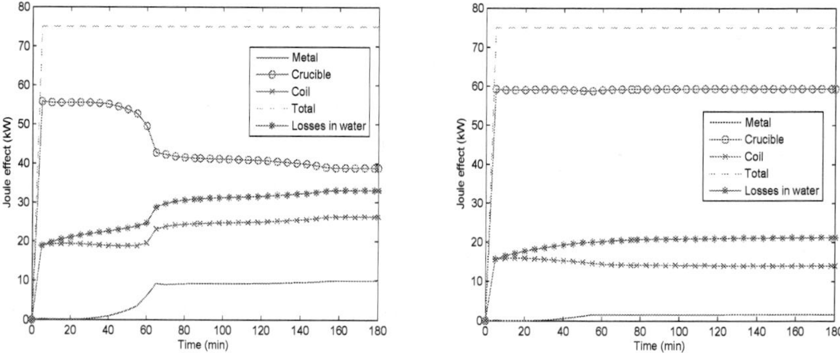

Fig. 14. Joule effect and heat losses through the tubes, for 500 Hz (left) and 2650 Hz (right).

the performance of the furnace and enforcing to increase the intensity, consequently increasing the power losses in the coil. When working at high frequency the performance of the furnace is almost unaffected by the presence of molten material. It is also remarkable that at the first time steps the power losses in the coil match the heat losses through the water tubes. When time increases the heat losses through the tubes become higher due to the heat conduction from the crucible across the refractory layer.

7. Acknowledgements

This work was supported by Ministerio de Ciencia e Innovación (MICINN), Spain, under research projects MTM2008-02483 and Consolider MATHEMATICA CSD2006-00032, by Xunta de Galicia, Spain, under grant number INCITE09-207047-PR (A. Bermúdez, D. Gómez and P. Salgado) and research project 10PXIB291088PR (M.C. Muñiz) and by the European Research Council through the FP7 Ideas Starting Grant 205004 (R. Vázquez).

8. References

Alonso Rodríguez, A. M. & Valli, A. (2008). Voltage and current excitation for time-harmonic eddy-current problems. *SIAM J. Appl. Math.*, Vol. 68, 1477-1494.

Amrouche, C.; Bernardi, C.; Dauge, M. & Girault, V. (1998). Vector potentials in three-dimensional non-smooth domains, *Math. Meth. Appl. Sci.*, Vol. 21, 823-864.

Ammari, H.; Buffa, A. & Nédélec, J.C. (2000). A justification of eddy currents model for the Maxwell equations, *SIAM J. Appl. Math.*, Vol. 60, No. 5, 1805-1823.

Bay, F.; Labbe, V.; Favennec, Y. & Chenot, J.L. (2003). A numerical model for induction heating processes coupling electromagnetism and thermomechanics, *Int. J. Numer. Meth. Engng.*, Vol. 5, 839-867.

Bermúdez, A.; Moreno, C.; (1994). Duality methods for solving variational inequalities, *Comput. Math. Appl.*, Vol. 7, 43-58.

Bermúdez, A.; Bullón, J.; Pena, F. & Salgado, P. (2003). A numerical method for transient simulation of metallurgical compound electrodes, *Finite Elem. Anal. Des.*, Vol. 39, 283-299.

Bermúdez, A.; Gómez, D.; Muñiz, M.C. & Salgado, P. (2007). Transient numerical simulation of a thermoelectrical problem in cylindrical induction heating furnaces, *Adv. Comput. Math.*,Vol. 26, 39-62.

Bermúdez, A.; Gómez, D.; Muñiz, M.C. & Salgado, P. (2007). A FEM/BEM for axisymmetric electromagnetic and thermal modelling of induction furnaces, *Internat. J. Numer. Methods Engrg.*,Vol. 71, No. 7, 856-882.

Bermúdez, A.; Gómez, D.; Muñiz, M.C. & Salgado, P. & Vázquez, R. (2009). Numerical simulation of a thermo-electromagneto-hydrodynamic problem in an induction heating furnace, *Applied Numerical Mathematics*, Vol. 59, No. 9, 2082-2104.

Bermúdez, A.; Gómez, D.; Muñiz, M.C. & Vázquez, R. (2011). A thermoelectrical problem with a nonlocal radiation boundary condition, *Mathematical and Computer Modelling*, Vol. 53, 63-80.

Bermúdez, A.; Leira, R.; Muñiz, M.C. & Pena, F. (2006). Numerical modelling of a transient conductive-radiactive thermal problem arising from silicon purification, *Finite Elem. Anal. Des.*, Vol. 42, 809-820.

Bossavit, A. (1998). *Computational electromagnetism*, Academic Press Inc., San Diego, CA.

Brezzi, F. & Fortin, M. (1991). *Mixed and hybrid finite element methods*, Springer Verlag, New York.

Chaboudez, C.; Clain, S.; Glardon, R.; Mari, D.; Rappaz, J. & Swierkosz, M. (1997). Numerical Modeling in Induction Heating for Axisymmetric Geometries, *IEEE Trans. Magn.*,Vol. 33, 739-745.

Henneberger, G. & Obrecht, R. (1994). Numerical calculation of the temperature distribution in the melt of industrial crucible furnaces, *Second International Conference on Computation in Electromagnetics.*

Hiptmair, R. & Sterz, O. (2005). Current and Voltage Excitation for the Eddy Current Model, *Int. J. Numer. Modelling*, Vol. 18, No. 1, 1-21.

Hömberg, D. (2004). A mathematical model for induction hardening including mechanical effects, *Nonlinear Anal. Real World Appl.*, Vol. 5, 55-90.

Katsumura, Y.; Hashizume, H. & Toda, S. (1996). Numerical Analysis of fluid flow with free surface and phase change under electromagnetic force, *IEEE Trans. Magn.*, Vol. 32, 1002-1005.

Lavers, J. (1983). Numerical solution methods for electroheat problems, *IEEE Trans. on Magn.*,

Vol. 19, 2566-2572.

Lavers, J. (2008). State of the art of numerical modelling for induction processes, *COMPEL*, Vol. 27, 335-349.

Mohammadi, B. & Pironneau, O. (1994). *Analysis of the k-epsilon turbulence model*, Wiley/Masson, New York.

Natarajan, T.T. & El-Kaddah, N. (1999). A methodology for two-dimensional finite element analysis of electromagnetically driven flow in induction stirring systems, *IEEE Trans. Magn.*,Vol. 35, No. 3, 1773-1776.

Nédélec, J.C. (2001). *Acoustic and electromagnetic equations. Integral representations for harmonic problems*, Springer-Verlag, New York.

Elliot, C. & Ockendon, J.R. (1985). *Weak and Variational Methods for Free Boundary Problems*, Pitman, London.

Pironneau, O. (1982). On the transport-diffusion algorithm and its applications to the Navier-Stokes equations, *Numer. Math.*, Vol. 38, No. 3, 309-332.

Rappaz, J. & Swierkosz, M. (1996). Mathematical modelling and numerical simulation of induction heating processes, *Appl. Math. Comput. Sci.*, Vol. 6, No. 2, 207-221.

Vázquez, R. (2008). *Contributions to the mathematical study of some problems in magnetohydrodynamics and induction heating*, PhD thesis, Universidade de Santiago de Compostela, Spain.

Wilcox, D.C. (1998). *Turbulence modeling for CFD*, DCW Industries, Inc.

5

Using Numerical Methods to Design and Control Heating Induction Systems

Julio Walter and Gerardo Ceglia
Universidad Simón Bolívar
Venezuela

1. Introduction

Although induction heating systems exists since 1906 (Curtis, 1950) its design always has been an art, with a great emphasis in the designer's experience. This state of things has not really changed much until the advent of computers and analysis software powerful enough. Now it is possible to use relatively common programs, even free software in some cases, to design and simulate the heating system before any physical device needs to be mounted. This chapter is devoted to show some of these techniques and to present examples that illustrates it.

2. Modeling of the heating coil and load.

2.1 The transformer effect

The principle involved in transferring energy from a heating inductor to a load is equivalent to a transformer which primary and secondary are composed by the induction coil and the load, respectively. This can be better seen in Fig. 1, where an alternating current applied to the heating coil induces another current into the load which may be modeled as a single loop inductance with a series resistance formed by the outside ring (dotted region) of the load. This ring is forced by the Skin Effect (Marion, 1980) that coerces the current to flow in a shell which geometry is dependant from various parameters (frequency, conductivity, dimensions, topology, etc).

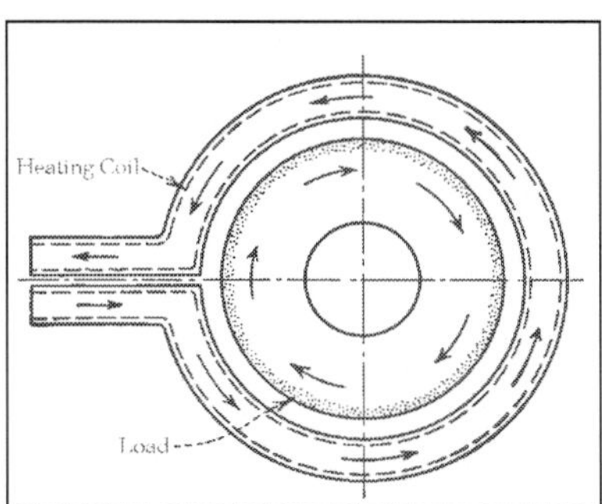

Fig. 1. Current distribution in an induction heating system

The shell formed by the Skin Effect is the main influence that determine load's inductance and resistance and they should be calculated using a Magnetic Finite Element program since they are too complex (or perhaps impossible) to evaluate in an analytical way, except for the most simplest cases which has academic interest only.

The coil-load system represented in the figure 2A shows a standard heating system viewed in the sense of the current flow. Its electrical equivalent can be seen in Fig. 2B, where L_p and L_s are, respectively, the coil and the load self inductances; M represents the mutual inductance between L_p and L_s while R_s is the real part of the load resistance and V_1, V_2, I_1 and I_2 are the system voltages and currents.

2.2 Mathematical representation of the heating system

Using the Kirchoff Voltage Law it is possible to set the equations (1) and (2) which describes mathematically the model derived from the circuit shown in Fig. 2B.

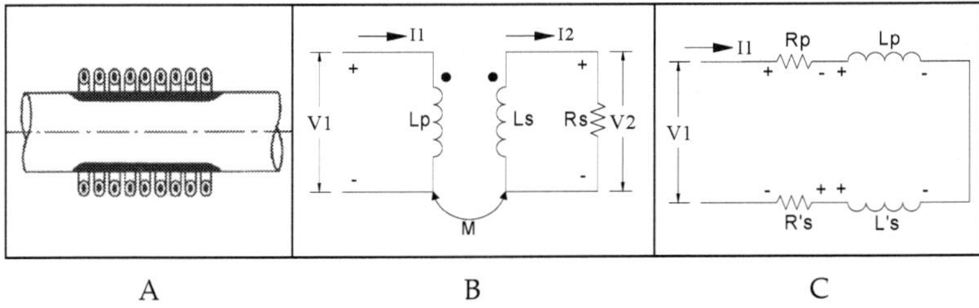

Fig. 2. A: Coil-Load system, B: Electrical Equivalent, C: Electrical Reduced Equivalent

$$V_1 = j\omega L_p I_1 - j\omega M I_2 \tag{1}$$

$$-V_2 = -j\omega M I_1 + j\omega L_s I_2 \tag{2}$$

Taking in consideration the dot standard, it is possible to write I_2 in function of I_1, as follows

$$I_2 = \frac{j\omega M}{R_s + j\omega L_s} I_1 \quad where \quad R_s = \frac{V_2}{I_2} \tag{3}$$

Using (3) to eliminate I_2 from (1),

$$V_1 = \left[j\omega L_p + \frac{\omega^2 M^2}{R_s + j\omega L_s} \right] I_1 \tag{4}$$

Equation (4) has an interesting implication. It suggests that the impedance $Z=V_1/I_1$ seen from the primary side is equivalent to a single non coupled inductor of L_P Henrys serially connected with an additional impedance. Expanding this second term:

$$Z_R(j\omega) = \frac{\omega^2 M^2}{R_s + j\omega L_s} = \frac{\omega^2 M^2}{R_s + j\omega L_s} \cdot \frac{R_s - j\omega L_s}{R_s - j\omega L_s} = \frac{R_s \omega^2 M^2}{R_s^2 + \omega^2 L_s^2} - \frac{j\omega^3 L_s M^2}{R_s^2 + \omega^2 L_s^2} \tag{5}$$

Rewriting equation (5)

$$Z_R(j\omega) = \frac{R_s\,\omega^2 M^2}{R_s^2 + \omega^2 L_s^2} - j\omega\left(\frac{\omega L_s M^2}{R_s^2 + \omega^2 L_s^2}\right) = R_s' + j\omega\left(-L_s'\right) \tag{6}$$

The first term of equation (6) is the impedance's real part and it corresponds to the ohmic resistance of the load reflected to the primary side of the induction heating coil. The second term reflects as negative impedance that reduces the total system's inductance.

The circuit that represents the total heating system is shown in Fig. 2C, where R_p and L_p again represents the primary's coil resistance and self inductance, R_s' and L_s' are the values of R_s and L_s reflected to the primary side by equation (5).

3. Heating coil design using finite element magnetic software

Having deducted the mathematical relationship between the heating coil and the load, it is necessary now to obtain the parameters needed to feed this model. These parameters are induction coil inductance (L_p) and resistance (R_p), load inductance (L_s) and resistance (R_s) and its mutual inductance. These parameters plus the working frequency, which is an external variable, permit to evaluate completely the impedance of the heating coil-load system.

3.1 Finite Elements Methods Magnetics software (FEMM)

Although there are several Finite Elements Methods Magnetics programs in the market (www.comsol.com), Visimag (www.vizimag.com), etc., we chose FEMM (www.femm.info) because is powerful enough and is free software. It limits the simulations to only 2 dimensions, but most of the heating coils have some degree of symmetry that permits accurate results using 2D analysis. An in deep examination of the FEMM software is beyond the purpose of this book but we encourage to the reader that he or she review extensively the tutorials and documentation to gain know-how in the area, there is also some terminology which is not defined here but it is in the FEMM user's manual (Meeker, 2009).

Although FEMM is capable of giving the total inductance and resistance for the associated system, it is necessary to obtain independently the primary resistance (R_p) and the load resistance (R_s'), because the first is associated with losses in the induction coil and it is quite recommendable to minimize it. On the contrary, the second is associated with the energy transfer to the load and should be maximized.

Having both R_p and R_s' it is possible to define an important parameter, the heating efficiency of the system. Since the flowing current is the same for both (see Fig. 2C):

$$\eta_{heat} = \frac{W_{LOAD}}{W_{Total}}\cdot 100\% = \frac{I^2\cdot R_s'}{I^2\cdot R_s' + I^2\cdot R_p}\cdot 100\% = \frac{R_s'}{R_s' + R_p}\cdot 100\% \tag{7}$$

The associated resistances are evaluated within the FEMM postprocessor as we shall study in the next paragraph.

3.2 Defining the structure of the induction coil and load system

To obtain the needed parameters it is necessary first to define the geometry of the heating system and the material to be heated; the material used to make the heating coil is always copper, so it is already predetermined. Once the geometry is defined, it is necessary to draw it on the FEMM user's interface or, better, in a CAD (Computer Aided Design) program capable of export the draw in DXF (Drawing Exchange File) format which will be then imported in FEMM. This last approach is very useful because the FEMM user's interface lack some drawing facilities that are included in other CAD programs, like AutoCAD®.

The next step, once in FEMM, is defining the materials, boundaries and other properties required to run the finite element algorithm (see the FEMM tutorials and manual). The postprocessor has the facilities to extract the values needed to calculate the parameters mentioned before. To clarify the entire simulation process, it will be shown a working example.

3.3 Working example: Design of a small aluminum melting furnace
3.3.1 Initials definitions

It is required to design a furnace to melt approximately 0,5 kg of aluminum. The crucible is given and is shown in the figure 3A. Since the crucible has symmetry on its axis, the magnetic problem reduces to an axisymmetric 2D system; figure 3B shows the transversal cutting for the internal side of the crucible, where the metal will be melted. In this case the crucible is formed by a nonconductive material, so it will not be taking into account for the finite element calculation.

For the axisymmetric problem is necessary only to define half of the transversal cutting, as the FEMM program determine the contribution of the 2D slice and calculate then the total produced by the corresponding solid of revolution. Figure 4A shows this draw made in AutoCAD® which includes also the heating coil; figure 4B shows the system already imported in FEMM. The measurements units in the CAD should be defined either in millimeters, centimeters, meters or inches since these dimensions will be exported to FEMM via DXF and is information relevant to the program. Now it is required to define all the regions with its corresponding materials. No region could be left undefined; in case the user needs to leave out some regions, it must be defined as "no mesh" as showed in Figure 4C where these definitions supports the working problem.

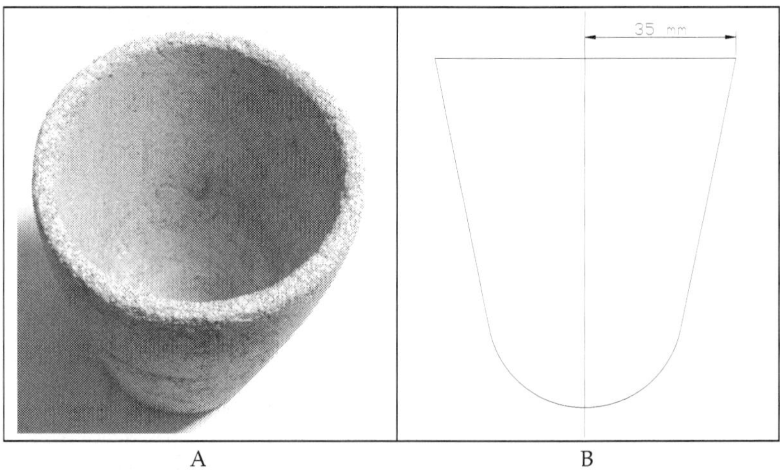

A   B

Fig. 3. A: Crucible, B: Internal crucible transversal cutting

It is necessary to point out that all the turns of the heating coil has to be defined as a "circuit" in FEMM to accomplish correctly the simulation (see "circuits" in FEMM manual). In the working example this circuit was named "HeatingCoil" and it is necessary also to apply to the circuit a current different from zero, which, for this example, was chosen as 100 amperes.

3.3.2 Simulation

Simulation requires having some knowledge of the final working frequency. There are various ways to estimate it but one of the easiest method is using tables or graphics of the penetrations deep (δ) against frequency. Figure 5 (Martínez & Esteve, 2003) shown one example of this, where is possible to obtain δ for several common materials and frequencies. To prevent current cancelation inside the load and thus maximize the power transfer within it, a simple rule of thumb state that the radius of the piece to be heated has to be greater than 10 times the penetration deep, so if the load has a radius of 35 mm, δ must be less than 3,5 mm. Locating this region in fig. 5 and searching the intersection with the material used (aluminum at 500 °C) it is possible to obtain an adequate frequency range to operate the system. A frequency of 40 kHz and a 0,5 mm deep are good estimates for this study case.

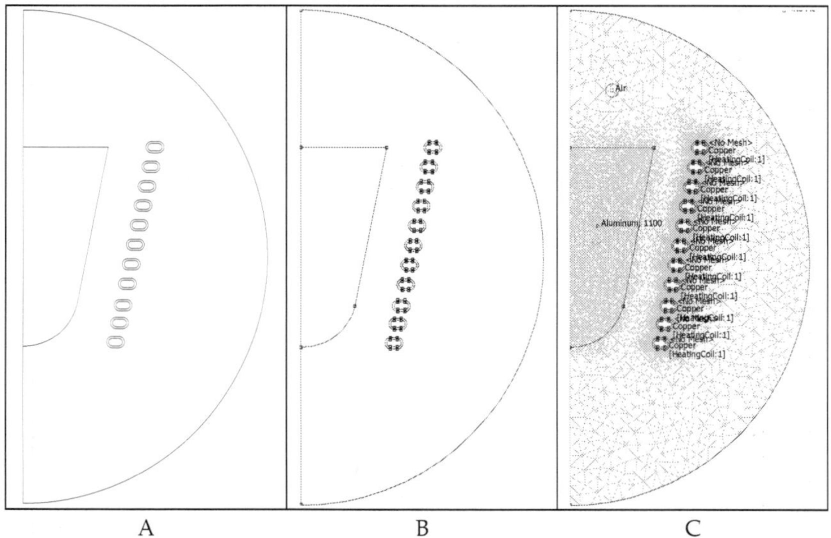

Fig. 4. A: CAD's drawing of the heating system, B: FEMM imported, C: Materials Definitions

Since all the relevant parameters are already defined, it is possible to make a simulation of the heating system. Fig. 6A illustrates the results, showing the current density both in the coil and the load. Fig. 6B shows the same in a detailed view. It can be seen graphically the Skin Effect since the current tends to go to the conductor's surface and Proximity Effect (Martínez & Esteve, 2003) where currents tend to face each other, incrementing then the effective resistance and changing the system's topology, which means a change in the inductance.

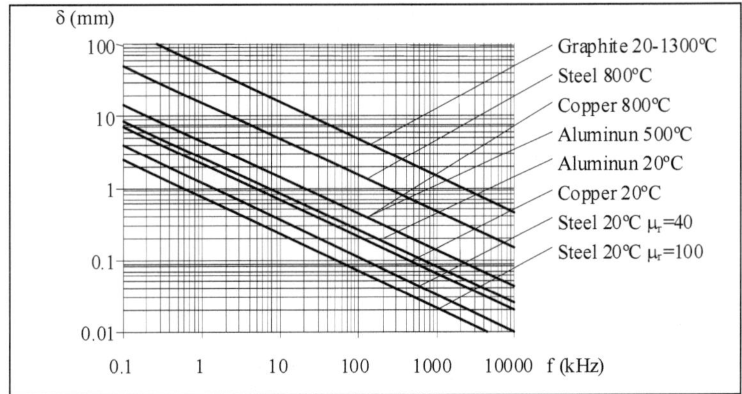

Fig. 5. Penetration deep against frequency for different materials

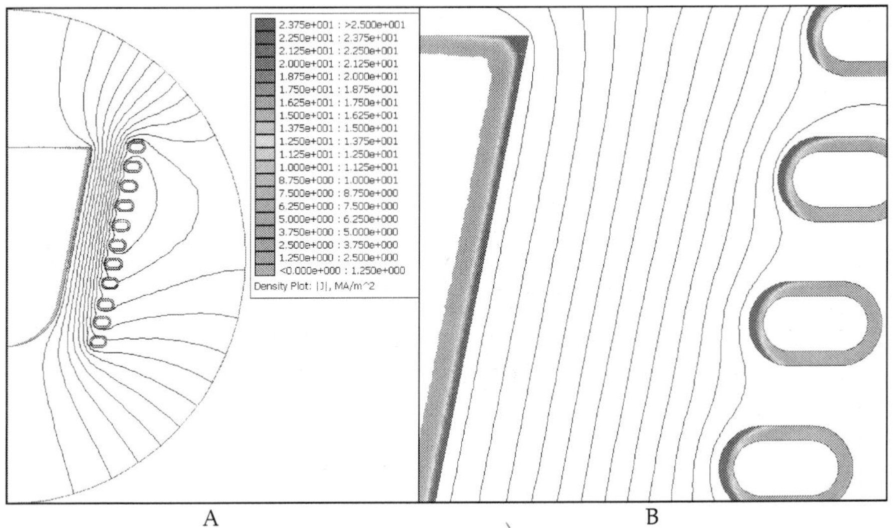

A   B

Fig. 6. A: Current density simulation at 40 kHz, B: Load and coil details

Using FEMM postprocessor facility is possible to obtain the parameters needed. With the command ***View->Circuits Props*** a window pops up with the information shown in Fig. 7. The line "Voltage/Current = 0.0307749+I*1.27647 Ohms" directly gives the values or $R_p+R'_s$ and L_p-L'_s, according equations (8) and (9).

$$R_p + R'_s = 0,0307749\Omega \cong 30,7m\Omega \tag{8}$$

$$L_p - L'_s = \frac{1,27647j}{j\omega} = \frac{1,27647j}{2\pi \cdot 40000j} = 5,0789 \cdot 10^{-6} H \cong 5,08\mu H \tag{9}$$

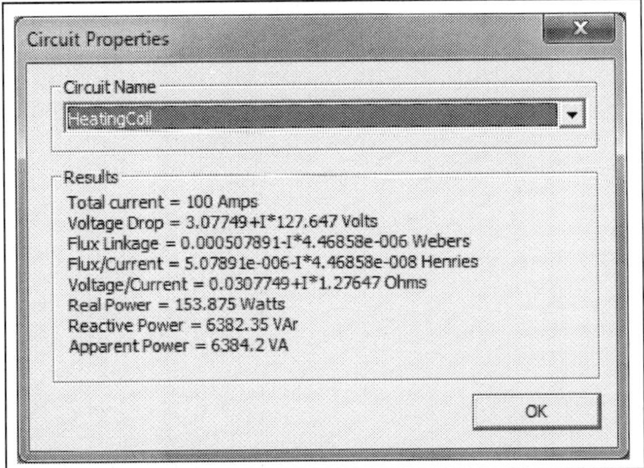

Fig. 7. Heating coil-load circuit properties

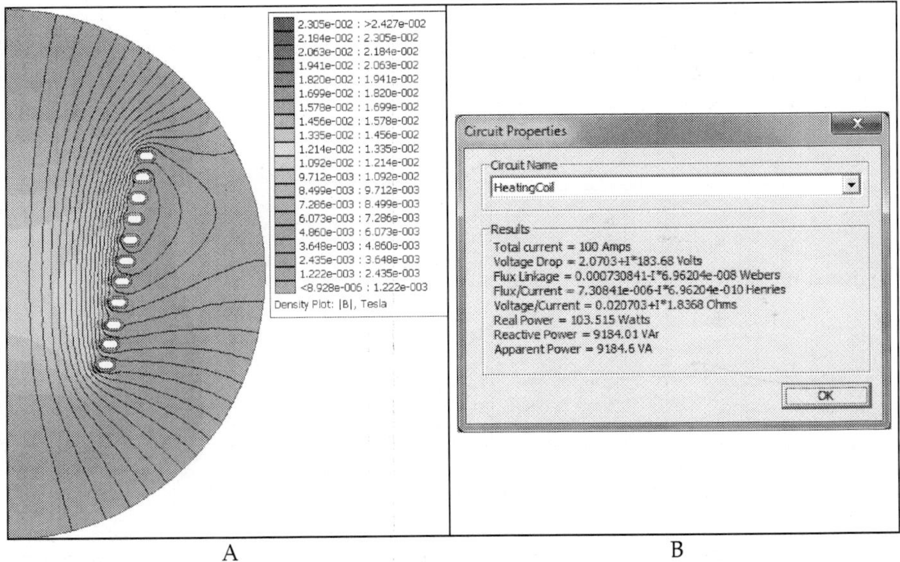

A

B

Fig. 8. A: Heating coil only simulation, B: Heating coil circuit properties

Making another simulation eliminating the aluminum load from the FEMM entry data (Fig. 8A), either by deleting or redefining it as "no mesh" will provide the numerical data shown in Fig. 8B.

Repeating the calculation done before

$$R_p = 0,020703\,\Omega \cong 20,7\,m\Omega \tag{10}$$

$$L_p = \frac{1,8368j}{2\pi \cdot 40000j} = 7,30839 \cdot 10^{-6} H \cong 7,31 \mu H \tag{11}$$

Now it is possible to obtain the values for R's and L's. Both are given in equation (12) and (13).

$$R_s' \cong 30,7 m\Omega - 20,7 m\Omega = 10 m\Omega \tag{12}$$

$$L_s' \cong 7,31 \mu H - 5,08 \mu H = 2,23 \mu H \tag{13}$$

All these values are referred to the primary side of the circuit showed in Fig. 2B, therefore these are directly applicable to the circuit of the Fig. 2C. With this information is possible to check the heating efficiency to have an overall idea about the performance of the heating system. Accordingly to equation (7)

$$\eta_{heat} = \frac{R_s'}{R_s' + R_p} \cdot 100\% = \frac{10}{30,7} \cdot 100\% = 32\% \tag{14}$$

This is a rather low figure. It means that 68% of the heating power is dissipated as losses in the induction coil and only 32% contributes to heating the load. Since the last one cannot be modified, it is possible to design and test some other topologies for the induction coil to improve the efficiency. It is left to the reader, as an exercise, to experiment other coil's configurations.

A prototype for the furnace was made with the dimensions used in the FEMM simulations and its partial assembly is shown in Fig. 9.

Fig. 9. Melting furnace partially assembled

4. Mathematical model for the control loop system

Using the terminology adopted in control engineering, it is necessary to define the "plant". We partially did that by creating the model of the heating system showed in Fig. 2C. To complete it requires defining the feeding circuit topology and control type.

All induction heating systems presents to the power generator a large reactive input which must be compensated either partially or totally. Since the heating coil has an inductive behavior, the compensation is made using specially design capacitors which may be connected to the coil either serially, parallel or in a mixed way.

4.1 Power generator review

There are four basic power topologies to energize an induction heating system and they are closely related to the reactive power compensation. Basically these are LC, both serial and parallel and a combination of them (LLC or LCC). The following paragraphs will show a review of these topologies.

4.1.1 Serial compensation

Fig. 10 shows a serially compensated system, often used in relative small systems (Kamli, M. et al. 1998), (Okuno, M. et al. 1998), (Praveen, M. et al. 1998)

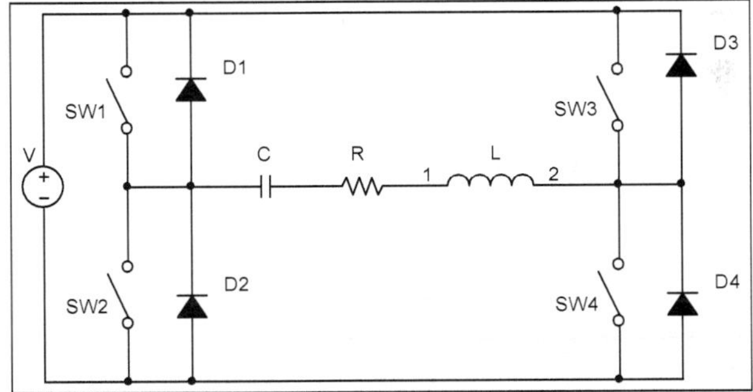

Fig. 10. Serially compensated induction heating system

This topology is fed by a voltage source V which could be fixed or variable. The switches SW1 to SW4 provide the high frequency AC to energize the induction coil L. Typically the coil includes a transformer which is not showed in this circuit, to maintain simplicity. C is the reactive compensation capacitor and D1 to D4 are necessary to maintain a current path when the associated switch is open. It is important to note that the heating coil L represents the total inductance L_p-L'_s and the resistor R is formed by R_p+R'_s.

4.1.2 Parallel compensation

Figure 11 shows a system with a parallel type reactive compensator which is usually used in high power systems (Dede et al., 1993a), (Dede et al., 1993b), (Dawson & Jain 1993). This compensator has complementary characteristics to the serial type and must have a current source to feed it instead a voltage source. Again, the heating coil L represents the total inductance L_p-L'_s and the resistor R is formed by R_p+R'_s.

4.1.3 Mixed compensation

It is feasible to use a voltage source to power a parallel compensation network (Bonert & Lavers 1999) (Kamli et al.1996) or a current source to supply a serial network (Espí et al. 2000) (Dieckerhoff et al. 1999), but it is necessary then to add an extra inductor or capacitor, correspondingly. These configurations are shown in Fig. 12A and 12B. R and L have the same meaning than in previous paragraphs.

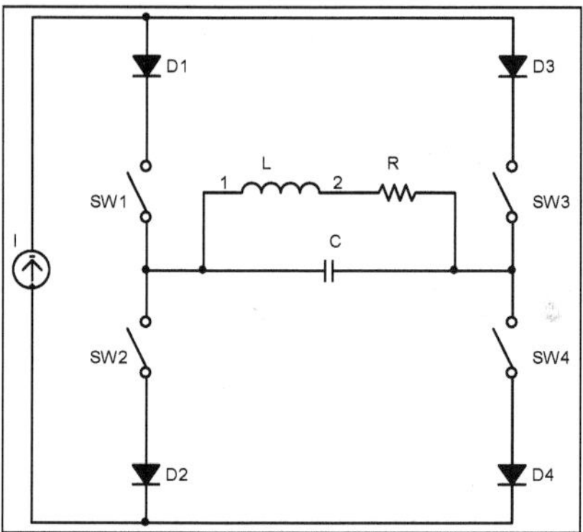

Fig. 11. Parallel compensated induction heating system

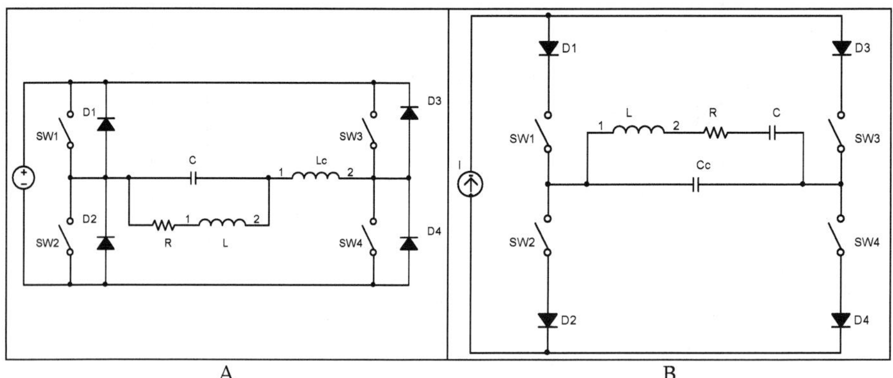

A   B

Fig. 12. Mixed compensated induction heating system

4.2 Plant modeling of the compensated induction system

Continuing with the melting oven working example and analyzing the four power topologies, the serial compensation is the best choose for a small system, since it has the simplest topology and the power is supplied by a voltage source which is the most common one.

4.2.1 Response of the serially compensated heating system

The serially compensated circuit shown in Fig. 10 could be simplified to calculate the voltage transfer function, which is one of the easiest variables to control and is directly related to the power injected to the heating system. It is possible also to obtain the current over R but the problems are equivalent. Using Kirchoff Voltage Law in the simplified circuit shown in Fig. 13 and taking the direct Laplace Transform, we have:

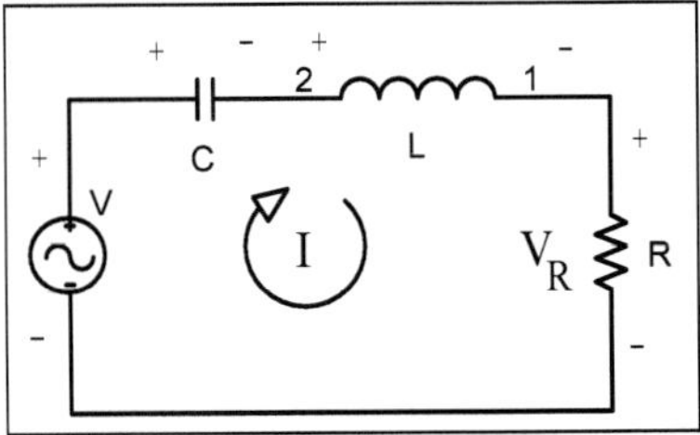

Fig. 13. Simplified serial compensation circuit

$$V(s) - \frac{I(s)}{sC} - sLI(s) - RI(s) = 0 \tag{15}$$

$$V_R(s) = RI(s)$$

Eliminating **I** from (15)

$$V(s) - \frac{V_R(s)}{sRC} - \frac{sLV_R(s)}{R} - V_R(s) = 0 \tag{16}$$

Placing V_R in function of **V**

$$V_R(s) = \frac{V(s)}{\left(\dfrac{1}{sRC} + \dfrac{sL}{R} + 1\right)} \tag{17}$$

Arranging the terms and normalizing equation (17)

$$V_R(s) = \frac{s\,R/L}{s^2 + s\,R/L + 1/LC}V(s) \tag{18}$$

Finding the denominator's roots of equation (18)

$$s = -\frac{R}{2L} \pm \sqrt{\left(\frac{R}{2L}\right)^2 - \frac{1}{LC}} \tag{19}$$

Normally is desired that the system has an under-damped (oscillatory) response because in most cases the control system locks via a PLL (Phase locked Loop) to this natural frequency, so the roots must be complex and must satisfy the condition:

$$\left(\frac{R}{2L}\right)^2 < \frac{1}{LC} \tag{20}$$

Under this condition, equation (18) could be written as

$$V_R(s) = \frac{s\,R/L}{(s + (a + j\omega))(s + (a - j\omega))}V(s) = \frac{s\,R/L}{((s + a)^2 + \omega^2)}V(s) \tag{21}$$

Where

$$a = \frac{R}{2L} \quad y \quad j\omega = \sqrt{\left(\frac{R}{2L}\right)^2 - \frac{1}{LC}} \tag{22}$$

The normalized response of this RLC system to an AC voltage source of magnitude 1 is showed in Fig. 14. The graph was obtained dividing the real part of the network impedance and its modulus.

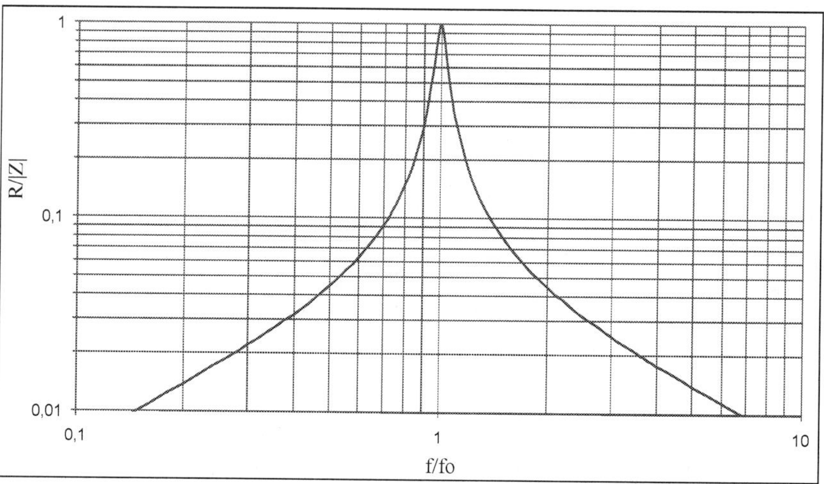

Fig. 14. Normalized response of an RLC system

Figure 14 has also information about which control type could be designed. It is possible to vary the frequency in the region above $f/f_0>1$ to have a continuous control of the voltage over **R** or use a discontinuous control (Hee et al. 1993) or PWM type control (Ogiwara et al. 2001) at exactly the resonance frequency ($f/f_0=1$). The zone $f/f_0<1$ cannot be used because represent a capacitive-type switching that easily could destroy the switches. The RLC system acts as a filter that eliminates the influence of all the harmonics excluding the first one and for this reason, although the commutation of the switches presents a square wave-type signal to the heating system, only the first harmonic matters. For this reason, it is possible to define the source voltage as

$$V(t) = V_i \sin(\omega' t)$$

$$V(s) = V_i \frac{\omega'}{(s^2 + \omega'^2)} \tag{23}$$

Equation (21) could be written now as

$$V_R(s) = \frac{s\,R/L}{((s+a)^2 + \omega^2)} \cdot V_i \frac{\omega'}{(s^2 + \omega'^2)} = V_i \frac{R}{L} \frac{s\omega'}{((s+a)^2 + \omega^2)(s^2 + \omega'^2)} \tag{24}$$

Where V_i corresponds to the peak voltage of the source, $R= R_p+R'_s$ and $L= L_p-L'_s$. Equation (24) represents the Laplace Transform of the voltage over resistance R and it will be used as the system variable to control the heat over the load. The same procedure could be realized over the other three compensation topologies explained before and are left to the readers as exercises.

4.2.2 Defining the control system

The heat needed to treat any metal by induction has a linear relation with the power injected to the load. In our example it is necessary to melt aluminum and until now it has been developed the dependence of the voltage over the aluminum load in function of parameters extracted from the physical system and the excitation voltage V_i. Now it is necessary to obtain the relation between voltage and power over the resistance R, which is given by

$$W = \frac{V_{RMS}^2}{R} = \frac{V_{Pk}^2}{2R} \tag{25}$$

The chosen control for the working example is discontinuous delta at the resonance frequency ($f/f_0=1$ therefore $\omega=\omega'$) for the reason that it has the capacity of zero switching loses because they are made exactly when the current trough the switches is zero. Since the power is directly related to the capacity of heating the load, the control loop will be a linear one if the control variable is power. Under these conditions, equation (24) could be written as

$$V_R(s) = V_i \frac{R}{L} \frac{s\omega}{((s+a)^2 + \omega^2)(s^2 + \omega^2)} \tag{26}$$

Doing the partial fractions expansion of 26 and taking its Laplace inverse transform

$$V_R(t) = V_i \frac{R}{L} \frac{1}{(a^2 + 4\omega^2)} \left[(1 - \omega e^{-at})\cos(\omega t) + \frac{2\omega^2 - (a^2 + 2\omega^2)e^{-at}}{a} sen(\omega t) \right] \tag{27}$$

Since $2a=R/L$, (27) may be written as

$$V_R(t) = V_i \frac{2}{(a^2 + 4\omega^2)} \left[a(1 - \omega e^{-at})\cos(\omega t) + sen(\omega t)(2\omega^2 - (a^2 + 2\omega^2)e^{-at}) \right] \tag{28}$$

According (25) the power is proportional to the square of the peak voltage, which could be obtained dropping the oscillatory terms of (28) and taking its value squared. This is equivalent to obtain the envelope of (28). Equation (29) shows this envelope function.

$$V_R(t) = V_i \frac{2}{(a^2 + 4\omega^2)} \left[a(1 - \omega e^{-at}) + 2\omega^2 - (a^2 + 2\omega^2)e^{-at} \right] \tag{29}$$

Arranging the terms of (29) and factorizing it

$$V_R(t) = V_i \frac{2(a + 2\omega^2)}{(a^2 + 4\omega^2)} \left[1 - \frac{(a^2 + a\omega + 2\omega^2)}{(a + 2\omega^2)} e^{-at} \right] = V_i K_1(1 - K_2 e^{-at}) \tag{30}$$

With

$$K_1 = \frac{2(a + 2\omega^2)}{(a^2 + 4\omega^2)} \quad y \quad K_2 = \frac{(a^2 + a\omega + 2\omega^2)}{(a + 2\omega^2)} \tag{31}$$

It could be conclude from (30) that if the heating system, which natural frequency is ω, is excited with a function $sin(\omega t)u(t)$, its envelope evolves as a first order system. Figure 15 presents the evaluation of (28) and (30) and shows effectively the relation between the peak voltage and the envelope function as well as its variation in time.

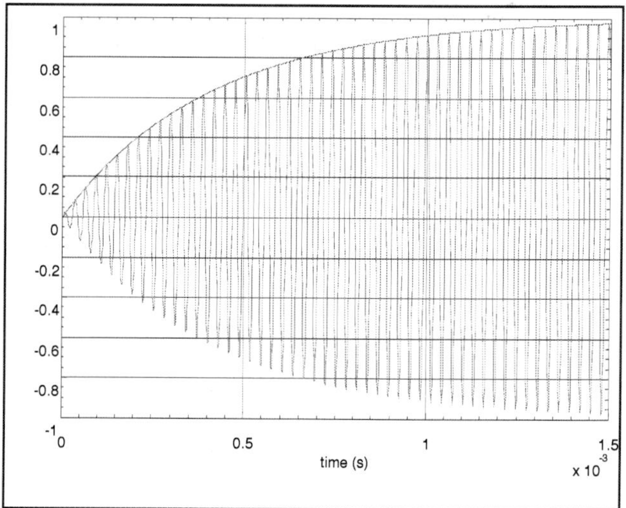

Fig. 15. Normalized response of an RLC system to a resonant excitation

Our interest is obtaining the power over the heating system, so, replacing (30) in (29)

$$W(t) = \frac{V_R^2(t)}{2R_T} = \frac{V_i^2}{2R_T}K_1^2(1 - K_2 e^{-at})^2 = \frac{V_i^2}{2R_T}K_1^2(1 - 2K_2 e^{-at} + K_2 e^{-2at}) \tag{32}$$

The control system theory normally requires that the control variable should be represented in the Laplace field, therefore taking the Laplace transform of (32)

$$W(s) = \frac{V_i^2}{2R_T}K_1^2(\frac{1}{s} - \frac{2K_2}{s+a} + \frac{K_2}{s+2a}) \tag{33}$$

Expanding (33)

$$W(s) = \frac{V_i^2}{2R_T}K_1^2(\frac{(K_2^2 - 2K_2 + 1)s^2 + (3 - 4K_2 + K_2^2)as + 2a^2}{s^3 + 3as^2 + 2a^2 s}) \tag{34}$$

Now (34) represents the power over the load as a response to a step $\dfrac{V_i^2}{2R_T}u(s)$, or

$$W(s) = H(s)u(s) \tag{35}$$

Where $H(s)$ is the plant's transfer function and $u(s)=1/s$ is the transform of the step function $u(t)$, consequently

$$sW(s) = H(s) = \frac{V_i^2}{2R_T}K_1^2\left(\frac{(K_2^2 - 2K_2 + 1)s^2 + (3 - 4K_2 + K_2^2)as + 2a^2}{s^2 + 3as + 2a^2}\right) \tag{36}$$

Figure 16 shows the response of the plant's transfer function $H(s)$ to a value 1 step. This transfer function contains the system information needed to design the control strategy. The procedure to tune this plant will be developed in the next paragraph.

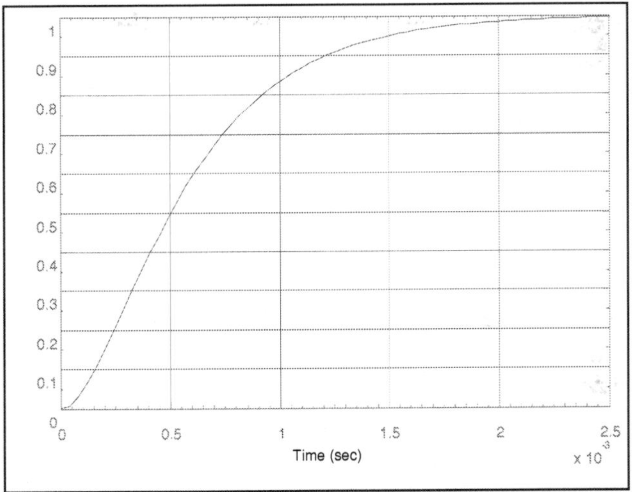

Fig. 16. Step response of the transfer function to a unitary step

It has to be mentioned that all the evaluations were made using MATLAB® (www.matlab.com), but it is possible to do the same using SciLAB (www.scilab.org) or Octave (http://octave.sourceforge.net/), which are free software.

5. Control loop tuning using numerical computational tools

Feedback is the standard control topology systematically used in hundred of thousand control systems, from a simple DC power supply to the navigational system of a jet. It is based in taking a sample of the system's output; compare it with a set point and correct the output according this difference. Although an in deep review of the control theory is beyond the scope of this book, there are several books (Ogata 1990) that treat very well this topic. Henceforth, all the calculations are made in MATLAB® using the Control System Toolbox.

It has been already decided that the control variable for the working example is power, but it is necessary to reiterate that is possible to control other variables like material's temperature. The procedure is the same than the one shown in the example.

5.1 System description

Figure 17 shows the building blocks of the DELTA control loop for the working example induction furnace. Its meaning is explained below.

- Wref: Power set-point
- We: Power error
- Wf: Feedback signal
- Wo: Load Power
- Comparator and FlipFlop: Synchronous Delta generator
- Power Stage: Power Switches, Load Network and Power Supply
- PID: Compensation network
- Sensor: Real Power measurement system

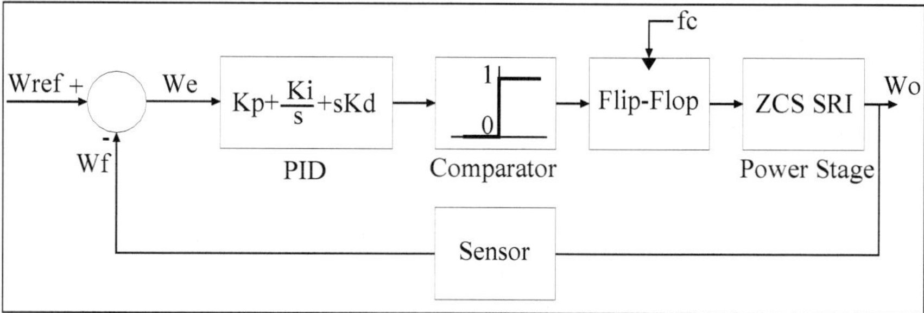

Fig. 17. Block diagram of the closed loop induction system.

Each block has its own properties that contribute to the system response, however, the most important are the Coil-Load Network, the Sensor and the PID because the others blocks only contributes with a scaling factor that can be included in the total response as constants. The coil-load transfer function is already defined, so it remains to obtain the sensor and the PID (Proportional-Integral-Differential compensation network).

5.1.1 Power stage model

As shown in Fig. 17, the PID compensation network output feed a comparator which has a binary output given by (37)

$$Output = \begin{cases} 0 & if \;\; PID < 0 \\ 1 & if \;\; PID > 0 \end{cases} \tag{37}$$

This state is stored in a flip-flop with the zero-crossing current of the load as its synchronous clock, maintaining then the permanent resonant state of the system. The output of the flip-flop manages a driver that activates the high power switches. This conform the Delta Modulator and function in its average behavior as a method to regulate the output power, so it do not contribute to the shape of the transfer function.

The terms a and ω of equation (36) are evaluated using the FEMM obtained parameters shown in equations (10) to (13) and using the formulas for K_1 and K_2 given in (31). The capacitor C used in the heating network to place the system in resonance has a nominal value of 2,7 µF. Equation (38) present the final numerical evaluation of the power transfer function and includes all the constants given by the power supply and coupling transformer.

$$H_{Load}(s) = \frac{0,3177s^2 + 2,695 \cdot 10^5 s + 1,150 \cdot 10^{11}}{s^2 + 7361s + 1,204 \cdot 10^7} \tag{38}$$

5.1.2 Sensor model

Figure 18 shows the heating coil simplified topology. It could be stated that the power is obtained multiplying the instantaneous voltage and current. This multiplication gives a value that differs from the instantaneous power value only by a scale factor.

As was affirmed, the heating coil-load system could be visualized as a serially connected inductor and resistor as is shown in Fig. 18. The current trough L has a sinusoidal form.

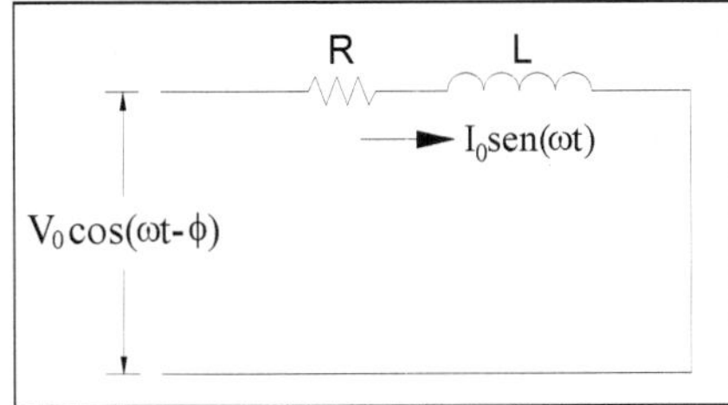

Fig. 18. Induction coil and load model.

The phasor relationships came from the fact

$$V_L = L\frac{di}{dt} = LI_0 \frac{d}{dt} sen(\omega t) = LI_0 \cos(\omega t) \tag{39}$$

But accordingly the Kirchoff Voltage Law

$$V = V_L + V_R = LI_0 \cos(\omega t) + RI_0 sen(\omega t) = V_0 \cos(\omega t - \phi) \tag{40}$$

Now, the power over the heating system is **VI**, then

$$W = VI = V_0 I_0 sen(\omega t)\cos(\omega t - \phi) \tag{41}$$

By trigonometric identities

$$W = \frac{V_0 I_0}{2} sen(2\omega t - \varphi) + \frac{V_0 I_0}{2} sen(\phi) \tag{42}$$

Equation (42) shows that the power signal has to components; one is oscillatory with a frequency double than the original while the other is a constant component that depends only from the phase difference ϕ between voltage and current. For the inductive behavior of the system ϕ is small so $sin(\phi)$ could be approximated to ϕ. After passing W through a low-pass filter (42) gives

$$\overline{W} = \frac{V_0 I_0}{2} sen(\phi) \approx \frac{V_0 I_0}{2} \phi \tag{43}$$

Figure 19 shows the block diagram needed to obtain a signal proportional to the power injected into the load

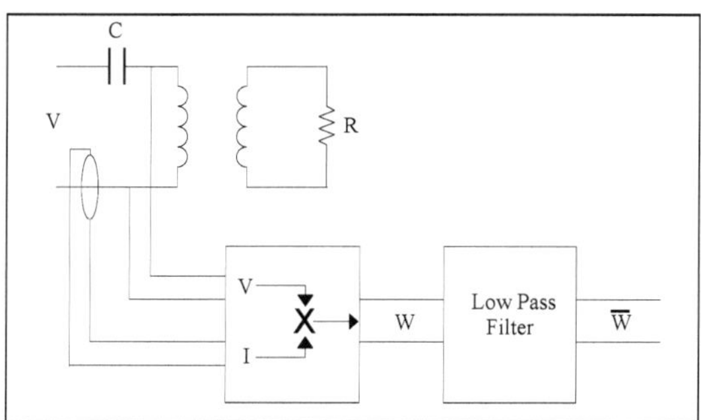

Fig. 19. Block diagram of the Power Sensor.

The Low Pass Filter is somewhat critical because it must have simultaneously a low line delay, a high attenuation and a fast response. As a compromise, it was chosen a fourth order Bessel low pass filter with a cut-off frequency of 5kHz, that gives a maximum response time of 200 µs, which is sufficient to maintain the parameters of the semiconductor well within its maximums for this particular circuit. The filter is showed in Fig. 20 and was designed using a free program supplied by Texas Instruments (Texas Instrument 2004)

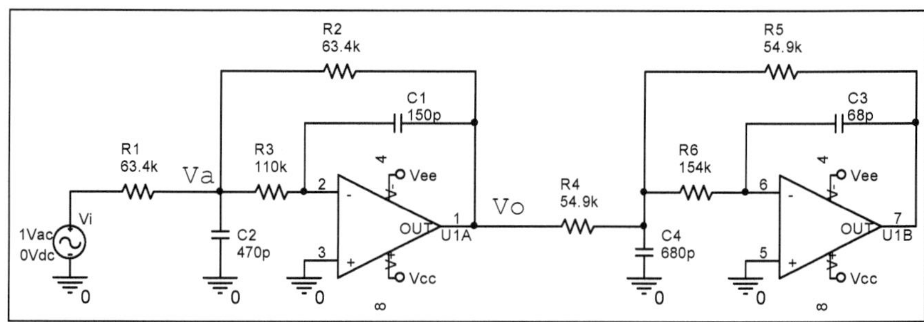

Fig. 20. Fourth order Bessel low pass filter.

It is necessary to obtain now the filter's transfer function to include it in the control loop. Using operational amplifier circuit theory it could be stated

$$\frac{V_i - V_a}{R_1} - \frac{V_a}{\dfrac{1}{sC_2}} - \frac{V_a}{R_3} - \frac{V_a - V_O}{R_2} = 0 \tag{44}$$

$$\frac{V_a}{R_3} - \frac{V_a - V_O}{\dfrac{1}{sC_1}} \tag{45}$$

V_a is obtained from 45

$$V_a = -\frac{V_O sR_3 C_1}{1 - sR_3 C_1} \tag{46}$$

And from 44 is obtained Vi

$$V_i = V_a R_1 \left(\frac{R_1 R_2 + R_1 R_3 + R_2 R_3 + sR_1 R_2 R_3 C_2}{R_1 R_2 R_3} \right) - \frac{R_1 V_O}{R_2} \tag{47}$$

Substituting 46 in 47

$$V_i = -\frac{V_O sR_3 C_1}{1 - sR_3 C_1} \left(\frac{R_1 R_2 + R_1 R_3 + R_2 R_3 + sR_1 R_2 R_3 C_2}{R_2 R_3} \right) - \frac{R_1 V_O}{R_2} \tag{48}$$

Simplifying and obtaining common factor V_O

$$\frac{V_i}{V_O} = -\left[\frac{(R_1 R_2 + R_1 R_3 + R_2 R_3 + sR_1 R_2 R_3 C_2)sC_1}{(1 - sR_3 C_1)R_2} + \frac{R_1}{R_2} \right] \tag{49}$$

Adding the fractions and inverting

$$\left(\frac{V_O}{V_i} \right)_1 = -\left[\frac{(1 - sR_3 C_1)R_2}{(R_1 R_2 + R_1 R_3 + R_2 R_3 + sR_1 R_2 R_3 C_2)sC_1 + (1 - sR_3 C_1)R_1} \right] \tag{50}$$

The form of the gain of the second stage is identical to the first one, so it could be inferred directly its transfer function

$$\left(\frac{V_O}{V_i} \right)_2 = -\left[\frac{(1 - sR_6 C_3)R_5}{(R_4 R_5 + R_4 R_6 + R_5 R_6 + sR_4 R_5 R_6 C_4)sC_3 + (1 - sR_6 C_3)R_4} \right] \tag{51}$$

The subscripts of (50) and (51) are referred to the particular filter stage. The complete filter's transfer function was evaluated as

$$\left(\frac{V_O}{V_i} \right)_T = \left[\frac{0{,}6014 s^2 - 9{,}388 \cdot 10^4 s + 3{,}481 \cdot 10^9}{1{,}020 \cdot 10^{-9} s^4 + 1{,}292 \cdot 10^{-4} s^3 + 7{,}766 s^2 + 2{,}436 \cdot 10^5 s + 3{,}481 \cdot 10^9} \right] \tag{52}$$

The final transfer function of the power sensor is obtained by multiplying the sensor's transfer function by the filter's transfer functions and was evaluated using the MATLAB symbolic facility. It is represented by equation (55)

$$\overline{W}(s) = H_{Sens}(s) = \frac{3{,}868 \cdot 10^{-4} s^2 - 135{,}1 s + 6{,}766 \cdot 10^6}{5{,}736 \cdot 10^{-10} s^4 + 1{,}046 \cdot 10^{-4} s^3 + 5{,}801 s^2 + 1{,}884 \cdot 10^5 s + 1{,}433 \cdot 10^9} \tag{53}$$

5.1.3 PID network model

The PID is a classical analog topology (Ogata 1990) showed in figure 21 and represented mathematically by (54).

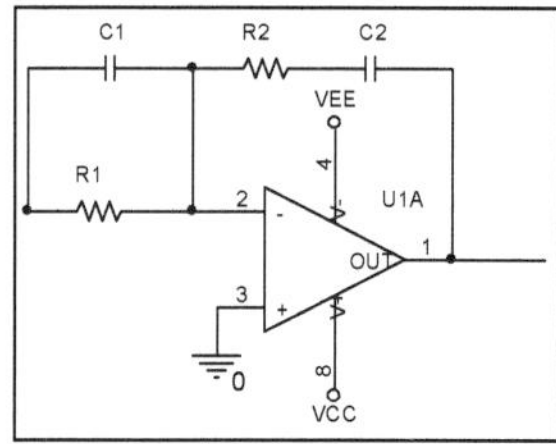

Fig. 21. PID Compensation Network.

$$\frac{V_O(s)}{V_i(s)} = \frac{(sR_2C_2 + 1)(sR_1C_1 + 1)}{sR_1C_2} \tag{54}$$

For the moment, these values are left free. These will be fixed when the system will be tuned using a closed-loop simulator in MATLAB®.

5.2 Tuning the system

Once were determined all the components that shapes the system's transfer function, the remaining task is closing the loop. This will be made using the SISOTOOL (Control 2006) from MATLAB, which permits to visualize the open loop behavior and change the poles and zeroes of the PID compensation network, until a stable configuration could be reach.

Figure 22 shows the system's open loop pole-zero diagrams, as well as the Bode Plot, showing that the system is unstable.

After manipulate the pole-zeroes scheme of the PID network, which is an interactive process, a good result was obtained using the values shown in (55)

$$H_{PID}(s) = 100\frac{(0,00039s + 1)(0,00022s + 1)}{s} = \frac{(sR_2C_2 + 1)(sR_1C_1 + 1)}{sR_1C_2} \tag{55}$$

Stating one component value, for example for R_1, the others automatically are fixed, so let $R_1 = 100k\Omega$, then

$$R_1 \cdot C_2 = 0,01 = 1 \cdot 10^5 \cdot C_2 \Rightarrow C_2 = 0,1\mu F$$

$$R_2 \cdot C_2 = 0,00039 = 0,1 \cdot 10^{-6} \cdot R_2 \Rightarrow R_2 = 3,9k\Omega \tag{56}$$

$$R_1 \cdot C_1 = 0,00022 = \cdot 10^5 \cdot C_1 \Rightarrow C_1 = 2,2nF$$

Placing these values on the PID compensating network and then simulates it on SISOTOOL, gives the result shown in Fig. 23

This behavior fulfills the Nyquist Stability Theorem (Ogata 1990), and gives enough phase and gain margin to maintain the loop stable against variations outside the calculated operational point.

With the values shown in equation (56), it is possible now to close physically the control loop showed in Fig 17 and have the possibility of changing the total power in the system simple adjusting the set point (Wref)

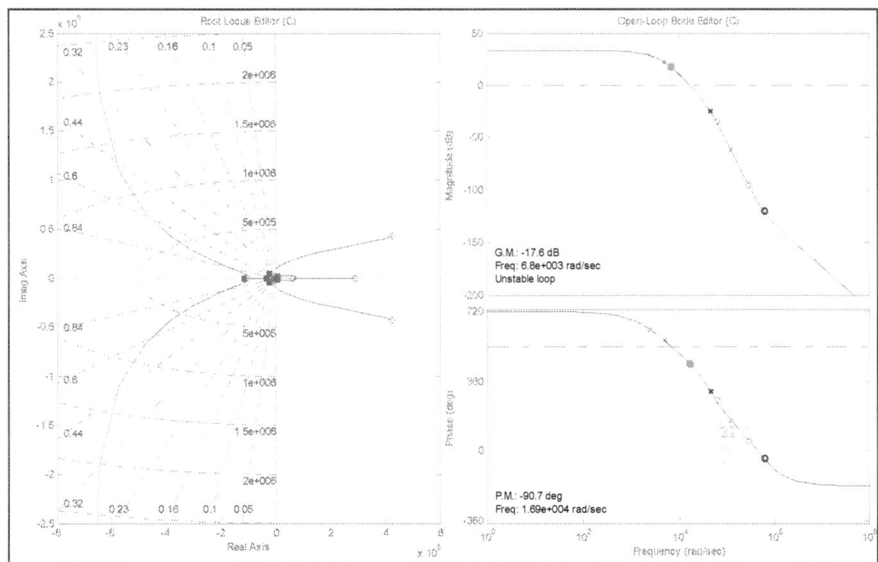

Fig. 22. System's open loop behaviour: Pole-Zeroes Diagram (left); Bode-Plot (right).

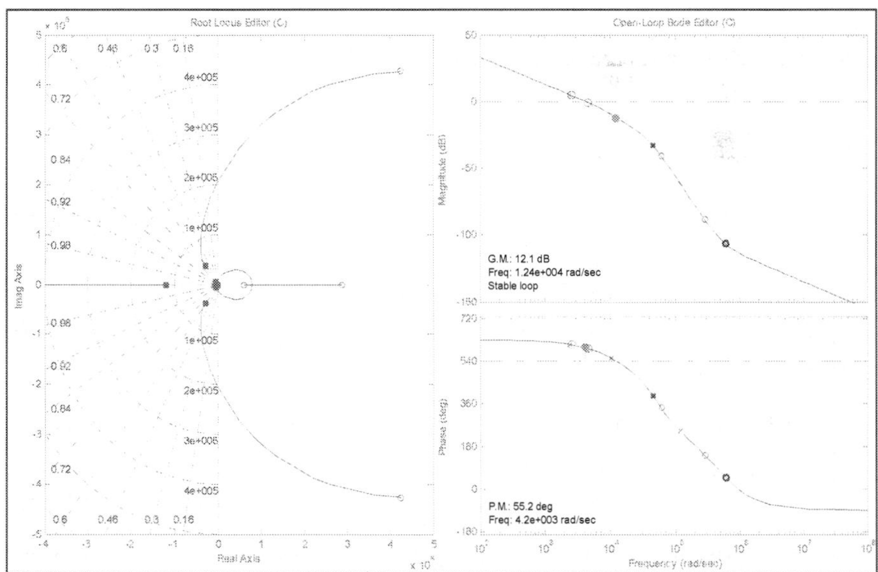

Fig. 23. System's Closed Loop Behavior: Pole-Zeroes Diagram (left); Bode-Plot (right).

6. Results

The results of the working example could be seen in figures 24 and 25. The first correspond to the power output against the voltage on the DC bus for various set points. While the power output is less than the set point is clear the V^2 (parabolic) behavior, but in the moment that the output reaches it, the system maintains a constant power output independently of variations in the bus voltage.

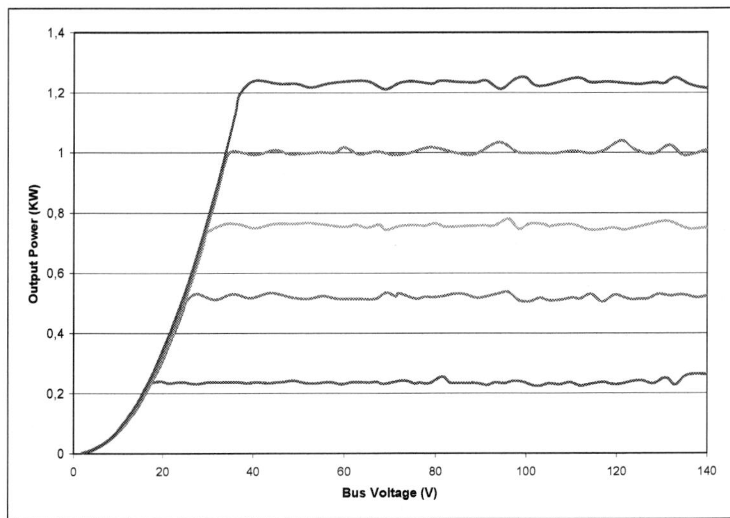

Fig. 24. Output Power response Family vs. Bus voltage

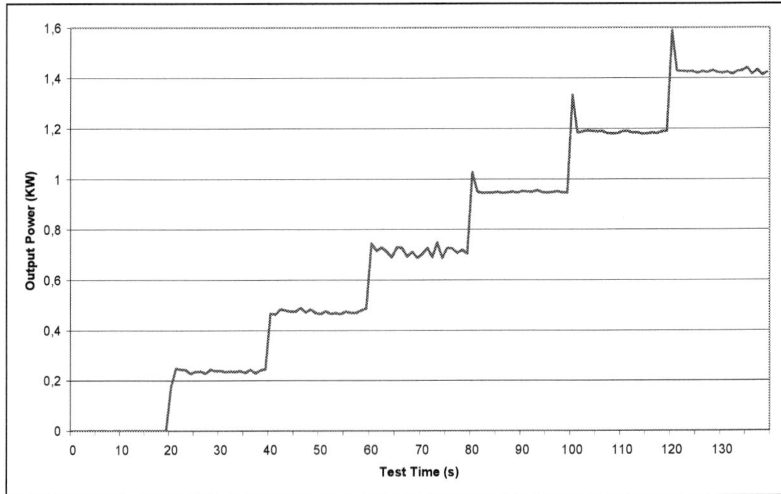

Fig. 25. Output power against set-point

Fig. 26. Melted Aluminum load

Figure 25 shows the output power variation against changes in the set point, for a constant DC bus. It could be seen that the system remains stable when the set point abruptly changes. In the high scale of power, there are smalls overshoots (aprox. 10% of the scale), but it go back fast to the predetermined level.

Figure 26 shows the operation of the constructed furnace using the techniques developed in this book. 0,5 kg of aluminum load is melted in the crucible at a temperature of 600 C.

7. Conclusion

It was developed a methodology that permit to obtain a precise mathematical model of the coil-load system for an induction heating device, useful for any heating topology. Since the heating coil-load usually is complicated enough to be calculated in a closed form, a Finite Element Magnetic Method is used to extract the relevant parameters that feed the developed model, so the designer may test easily several configurations of the induction coil in a quick way without the need of building any physical device.

Once the heating device is defined and its parameters are extracted, using the techniques developed it is possible to obtain the transfer functions of the compounds blocks for the control system and using ready available simulation tools, adjust the overall system's behavior without the need of use error-and-trial method that could easily destroy the power semiconductors in a resonant system.

It was designed a small prototype as a working example applying the techniques illustrated in the book. This prototype was built and tested showing the correctness of the principles developed.

8. References

Bonert, R.; Lavers, J.D. (1994), *Simple starting scheme for a parallel resonance inverter for induction heating*, Power Electronics, IEEE Transactions on, Volume: 9 Issue: 3 , May 1994, pp: 281-287

Control (2006) *Control System Toolbox*, The Mathworks, 18Feb2006
    http://www.mathworks.com/products/control/

Curtis F., W. (1950). *High-Frequency Induction Heating*, McGraw-Hill, New York, USA

Dawson, F.P.; Jain, P. (1993), *A comparison of load commutated inverter systems for induction heating and melting applications*, Power Electronics, IEEE Transactions on, Volume: 6 Issue: 3, July 1991, pp: 430-441

Dede, E.J.; Esteve, V.; Garcia, J.; Navarro, A.E.; Maset, E.; Sanchis, E. (1993b), *Analysis of losses and thermal design of high-power, high-frequency resonant current fed inverters for induction heating*, Industrial Electronics, Control, and Instrumentation, 1993. Proceedings of the IECON'93., International Conference on, 1993, pp: 1046-1051 vol.2

Dede, E.J.; Jordan, J.; Esteve, V.; Gonzalez, J.V.; Ramirez, D. (1993a) *Design considerations for induction heating current-fed inverters with IGBT's working at 100 kHz*, Applied Power Electronics Conference and Exposition, 1993. APEC '93, Conference Proceedings 1993, Eighth Annual, 1993, pp: 679 -685

Dieckerhoff, S.; Ruan, M.J.; De Doncker, R.W. (1999) *Design of an IGBT-based LCL-resonant inverter for high-frequency induction heating*, Industry Applications Conference, 1999. Thirty-Fourth IAS Annual Meeting Conference Record of the 1999 IEEE, Volume: 3, 1999, pp: 2039-2045 vol.3

Espi, J.M.; Navarro, A.E.; Maicas, J.; Ejea, J.; Casans, S. (2000) *Control circuit design of the L-LC resonant inverter for induction heating*, Power Electronics Specialists Conference, 2000. PESC 00. 2000 IEEE 31st Annual, Volume: 3, 2000, pp: 1430 -1435 vol.3

Hee Wook Ahn; Byeong Rim Jo; Myung Joong Youn (1993) *Improved performance current regulated delta modulator in series resonant inverter for induction heating application*, Industrial Electronics, Control, and Instrumentation, 1993. Proceedings of the IECON'93., International Conference on, 1993, pp: 857-862 vol.2

Kamli, M.; Yamamoto, S.; Abe, M. (1996) *"A 50-150 KHz Half Bridge Inverter for induction Heating Applications"*, IEEE Transactions on Industrial Electronics, Vol. 43, No. 1, February 1996, pp: 163-172

Kamli, M.; Yamamoto, S.; Abe, M. (1996) *A 50-150 KHz Half Bridge Inverter for induction Heating Applications, IEEE Transactions on Industrial Electronics*, Vol. 43, No. 1, February 1996, pp: 163-172.

Marion, J.; Heald, M. (1980). *Classical Electromagnetic Radiation, Second Edition*, pp 150-156, Academic Press, New York, USA.

Martínez, J; Esteve, V. (2003) *Inversores Resonantes Serie para Calentamiento por Inducción, Primera Edición*, Pág. 9, Editorial Moliner-40, Valencia, 2003, España

Meeker D. C. (2009), *Finite Element Method Magnetics*, Version 4.2 (02Nov2009 Build), http://www.femm.info/wiki/Documentation/

Ogata, K. (1990) *Modern Control Engineering Second Edition*, Prentice-Hall International, Englewood Cliffs, N.J., USA, pp 536-606, 1990

Ogiwara, H.; Gamage, L.; Nakaoka, M. (2001) *Quasiresonant soft switching PWM voltage-fed high frequency inverter using SIT for induction heating applications*, IEE proceedings for electronic power applications, Vol. 148 . 5, Septiembre 2001, pp: 385-392.

Okuno, A.; Kawano, H.; Sun, J.; Kurokawa, M., Kojina, A.; Nakaoka, M. (1998) *Feasible development of soft-switched SIT inverter with load-adaptive frequency-tracking control scheme for induction heating*, IEEE Transactions on Industry Applications, Vol. 34, No. 4, July/August 1998, pp: 713-718.

Praveen, K.; Espinoza, J.; Dewan, S. (1998) *Self-started voltage-source series-resonant converter for high-power induction heating and melting applications*, IEEE Transactions on Industry Applications, Vol. 34, No. 3, May/June 1998, pp: 518-525.

Texas Instruments (2004) *FilterPro, Filter Design Application* (03May2004 Build) http://focus.ti.com/docs/toolsw/folders/print/filterpro.html.

6

Development of Customized Solutions – an Interesting Challenge of Modern Induction Heating

Jens-Uwe Mohring and Elmar Wrona
HÜTTINGER Elektronik GmbH + Co. KG, Freiburg
Germany

1. Introduction

Induction heating is a reliable and innovative technology which conquered a fix place in the most different markets. Today the fields of application run from the classical heat treatment in the metallurgical industry to modern crystal growing processes in the semiconductor industry. It is characterized by the fact that the required energy is non-contacting transmitted into the work piece where the transformation into warmth energy finally takes place. The most important advantages of induction heating against heat transmitting applications are reachable high rates of heat up speed, high efficiency, excellent reproducibility, and the possibility of automation of the process. For an extensive use of these advantages, it is necessary to have both the possibility to develop a suitable induction coil design and an innovative technique for the used power supply. These two aspects of the modern application development shall be highlighted in this report.

2. Effective development of customized solutions for industrial application

2.1 Definition of efficiency rates

The solution of a heating task for a given application starts with a calculation of the required usage power P_u which depends on the weight of material which has to be heated up in a certain time. In processes where the work piece is moved the time unit can be assigned to the mass of material instead of the temperature increase per time. An additional power which is necessary for the thermal heat losses by radiation and convection at the surface has to be considered for the whole required thermal power P_{th}. Up to 100°C the convection losses are dominant. Since the radiation losses depend on the temperature to the 4th over 900°C the convection can be neglected. The ratio between the usage and the thermal power/energy is the thermal efficiency rate η_{th} which is a time dependant value (see Figure 1). To reduce the energy costs the heating time should be as low as possible. On the other hand, the installation costs rise with a shorter heating time. At last, the customer has to take into account the usage ratio of the system. For defining the optimal size he has to balance the three aspects.

For an induction heating application the required generator power P_G is the summation of the thermal power P_{th} and the ohmic loss power within the induction coil P_I which has to be removed by the cooling water. The ratio between P_{th} and P_G is the electrical efficiency rate ηI of the induction coil. The value of η_I is strongly influenced by the material properties of the work piece and the design of the induction coil. Typical values can vary between 0.2 for materials like copper and 0.9 for magnetic steel below the Curie point. The generator itself has to deliver the power P_G and is supplied by the main with the power P_N. The ratio between P_G and P_N is the efficiency rate of the generator η_G. A modern electronic power supply can achieve a value of $\eta_G > 0.85$.¶

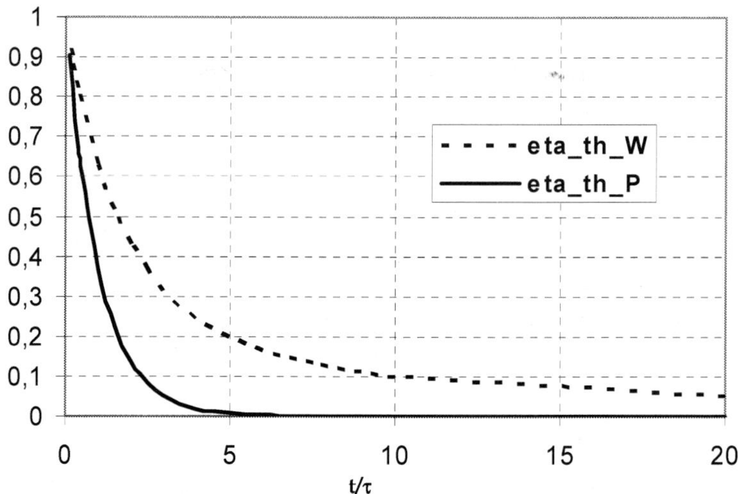

Fig. 1. Thermal efficiency rate for power (η_{th_P}) and energy (η_{th_W}) vs. t/τ. t is the heating time and τ is a constant time characterized by the work piece. Radiation losses are neglected.

2.2 Specifying the coil design

The most important and difficult part during the development of an induction heating application is to specify the optimal coil design. Solving this task the application engineer has to consider the aspects of reaching the heating goal, of maximising the efficiency rate η_L, and of optimising the load impedance of induction coil and work piece for the generator output. A classical cylindrical induction coil is defined by its number of turns, its length and its inner diameter. There is no possibility to calculate these parameters directly. The calculation can only be performed in one direction: from a certain coil design to the resulting heating of the work piece. Therefore, an iteration loop has to be passed to find suitable parameters of the induction coil. There are three different methods: the analytical calculation, the numerical calculation and the performance of experiments. Analytical methods for the calculation of cylindrical arrangements of coil and work piece are described by Nemkov [1] and Sluchotsky [2].

Meanwhile, for more sophisticated geometries the numerical calculation has developed to an essential tool for the application engineer. There are a couple of commercial software packages on the market using the method of Finite Elements (FEM), and since the personal computers have become very powerful in the last few years numerical calculations can be performed quickly. Nevertheless, for an efficient and short-time development of induction heating application the engineer has to do some simplifications. For industrial needs in the most cases it is advantageous to calculate as accurate as necessary and not as accurate as possible. Typical simplifications are the use of geometrical symmetry planes to reduce the number of elements and the negligence of dependencies of material properties on temperature and/or electromagnetic field strength in case of magnetic materials. The last point makes sense since these material properties are very often unknown. For a lot of applications a precise knowledge of the local temperature distribution is not necessary too. Thus the resulting temperature field can be estimated by a calculation of the specific power losses distribution. So the computing time can be rapidly reduced. Hence, there are heating tasks which have to be computed by considering the temperature dependent material properties, e.g. hardening and soldering. Because of the greater complexity of such problems the performing of these calculations should be done in scientific laboratories.

The third method to develop customized solutions is to perform experimental tests. This way offers three important benefits. At first, the heating results can be directly controlled and measured at the real work piece. At second, after performing successful experiments a suitable induction coil design has been obtained. And the last not least, the heating effect can be impressively presented in combination with the generator to the customer. Recently, we can observe the trend that more and more customers tend to use this opportunity. Therefore, to have the facility of a well-equipped application laboratory with experienced engineers is a big advantage for the supplier of induction heating systems.

3. Investigation of a System for Crystal Growing by Epitaxy

3.1 Crystal Growing by Epitaxy

The crystal growing by epitaxy is used in the semiconductor industry, e.g. for the production of light emitting diodes (LEDs). Mostly planetary reactors (see Figure 2) are used, which basically consist of a water cooled stainless steel chamber and a replaceable graphite susceptor. The susceptor is the support plate for rotating carrier discs - so called satellites - with the substrates on it.

Fig. 2. Typical reactor arrangement for epitaxy processes

The customer has three basic demands. At first he wants to have a reliable heating system which delivers the necessary process power over many years. Furthermore, to reach good process results, the susceptor has to be very homogeneously heated up to temperatures of more than 1000°C in the region where the satellites are located. And nevertheless, the third demand of the operator is preferably low energy consumption and respectively a high efficiency factor. Therefore, the two main components of an induction heating system – the generator and the induction coil – must perfectly play together. The generator has to deliver very reliably his full power at chosen frequency. Moreover, it should be able to work over a wide frequency range so that the customer has a couple of opportunities to match different arrangements of coils and work pieces. The internal power losses should not exceed 15% of the nominal output power. The product family TruHeat MF 3010 – 7040 fulfils these requirements.

3.2 Dimensioning of a typical system

We want to present the different opportunities to develop an induction heating system for a typical arrangement for an epitaxy process. The susceptor to be heated is a slice of graphite with a thickness of 8 mm and a diameter of 320 mm. The induction coil has to be located at its lower side. At first we have to calculate the required power. For a goal temperature of 1100°C the convective heat losses can be neglected against the losses by radiation. With an emission rate of 0.92 we find a loss power of around 30 kW. This power has to be induced at least in the graphite to hold the temperature during the process. A suitable induction coil for a flat work piece is a so-called pancake coil (see Figure 3). It is a helical wound copper profile (γ = 58·106 S/m) in one layer. We chose a coil with 6 turns made of a 15x10 mm tube with a wall thickness of 1 mm. The connection end of the inner winding is located at the lower side of the turns. The graphite slice (γ = 70000 S/m [3]) has a coupling gap of 10 mm to the coil surface.

Fig. 3. Pancake coil

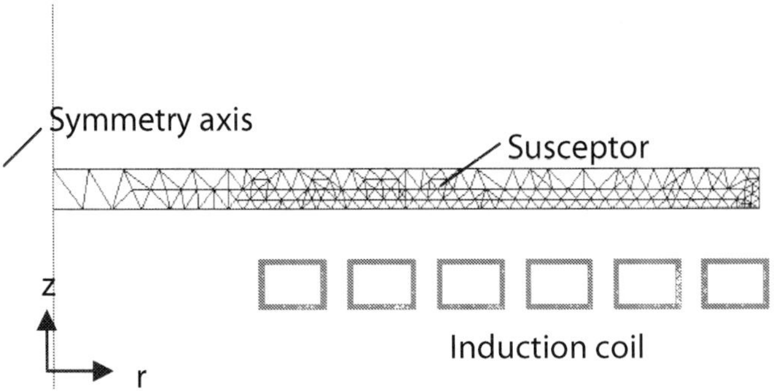

Fig. 4. Numerical 2d model with finite element mesh

To find out a suitable power supply we have to calculate the ohmic resistance R and the inductivity L of the coil and the work piece at the coil ends. The most effective and accurate way to obtain these values is by performing a numerical simulation of the electromagnetic field distribution. Since the turns of the coil are not circular there is no exact symmetry in azimuthally direction. But we neglect this small disturbance and only have to solve an axisymmetric problem in the (r,z) plane. The geometry mesh is represented in Figure 4. The numerical simulation is performed with the commercial software package MAXWELL. At the beginning we chose a certain current I of a certain frequency f in the coil, e.g. I=100 A and f=50 kHz.

From the simulation we obtain the integral parameters of the total electrical power P and the magnetic energy W_m. So we find the total ohmic resistance R and the inductivity L by:

$$R = P/I^2, \; L = \frac{2W_m}{I^2} \tag{1a,b}$$

The resulting load Z is a complex serial impedance of R and L:

$$\underline{Z} = R + j\omega L \;\; \text{with} \;\; \omega = 2\pi f \tag{2a,b}$$

Now we consider that our generator works with a parallel oscillating circuit, i.e. a capacity C is switched parallel to the induction coil in the tank circuit. We have to calculate C with the used frequency f by:

$$C = \frac{1}{\omega^2 L + \dfrac{R^2}{L}} \tag{3}$$

Since the system works at the resonance frequency the resulting load R' for the generator is given by:

$$R' = R(1 + Q^2) \;\; \text{with} \;\; Q = \omega L/R \tag{4}$$

The factor Q is called quality of the oscillating circuit. The suitable nominal power P_G of the generator is obtained from the required power in the work piece P_{th} and the electrical efficiency rate of the coil η_I which can be calculated from the power values. The nominal load R_G of the generator depends on the nominal output voltage U_G and P_G and should be equal to R'. In case of R'>R_G we decrease f, if R'<R_G we increase f and repeat the simulation until we find R'~R_G and C = n·C_p where C_p is an available size of capacitors, e.g. 0.66 µF. At a frequency of 85 kHz we find a load which is good matched to a TruHeat MF 3040 generator which is presented in the next chapter. At a voltage of 82% (492 V) we have a generator current of 93% (68 A) and an induced power in the susceptor of more than 31 kW. The efficiency rate of the coil ηI is more than 90%. Figure 5 shows the distribution of the density of Joule heat sources p_V in the susceptor which is defined by:

$$p_V = \frac{J^2}{\gamma} \qquad (5)$$

In Fig. (5) J is the current density. We can make out three areas of different values of p_V. In the centre of the susceptor a region of low intensity exists which comes from the type of the used coil. The current flows circularly in the slice and is equal zero in the axis. The second zone is the most important for the process because the satellites are located in this region. For the success of the process this zone should have a homogeneous temperature distribution. In the third zone at the outer perimeter of the susceptor the highest power density is induced. This is advantageous because of the higher radiation losses at the edge.

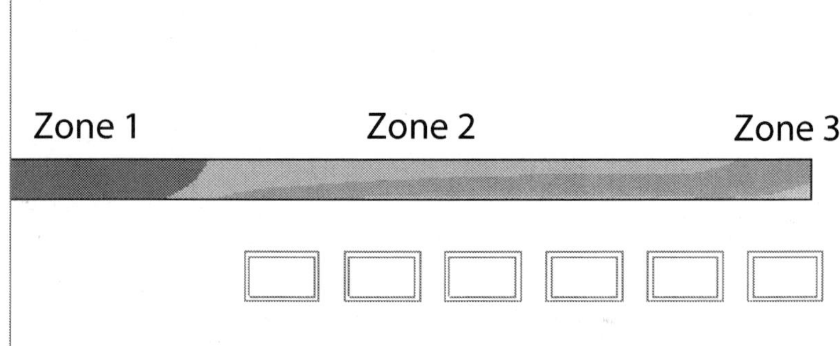

Fig. 5. Distribution of the density of Joule heat sources

3.3 Validation of the Numerical Model

To validate the results of the numerical simulation we performed measurements of the temperature distribution at the surface of the susceptor using an infrared camera system. The 2d-solver of the used software package MAXWELL only offers the possibility to compute steady state temperature fields. Therefore, we performed an experiment at a reduced generator voltage (20% of the nominal output voltage U_G). At this voltage (120 V) the susceptor surface was heated up to steady state temperature below 500°C. These values can also be reliably measured with our infrared camera system. At first we had to determine the emission rate of the graphite surface by comparing the temperatures measured with both a pyrometer and a thermocouple. We found a value of 0.92. For the numerical simulation we had to recalculate the distribution of the heat sources with the real current through the coil I_I at the chosen generator voltage U. The current can be calculated by (6) from the results of the former calculation since the values of R and L are independent on the voltage.

$$I_I = U / \sqrt{(R^2 + \omega^2 L^2)}$$

(6)

The 2d modeller of MAXWELL allows a direct coupling between an eddy current and a thermal simulation. In this feature the distribution of the Joule heat sources are directly imported by the temperature field solver. Due to the temperature dependence of the conductivity of graphite we calculated the value at 400°C (γ = 99200 S/m [3]). At the surface of the susceptor the boundary condition was set as radiation and convection while at the surface of the water-cooled coil a constant temperature of 25°C was defined. The convection coefficient was set to 7.1 W/m2K [4]. Figure 6 shows the calculated and the measured temperature distribution along the diameter of the susceptor. The measured curve was obtained after heating up the susceptor over a time of 12 min with a constant generator voltage of 120 V. We see a good qualitative agreement between both curves but a reasonable difference along the diameter of the slice. The measured temperatures are lower than the computed. We see two reasons for that. During the measurement the convection coefficient could be increased in comparison to the theoretical value because of air flows in the laboratory.

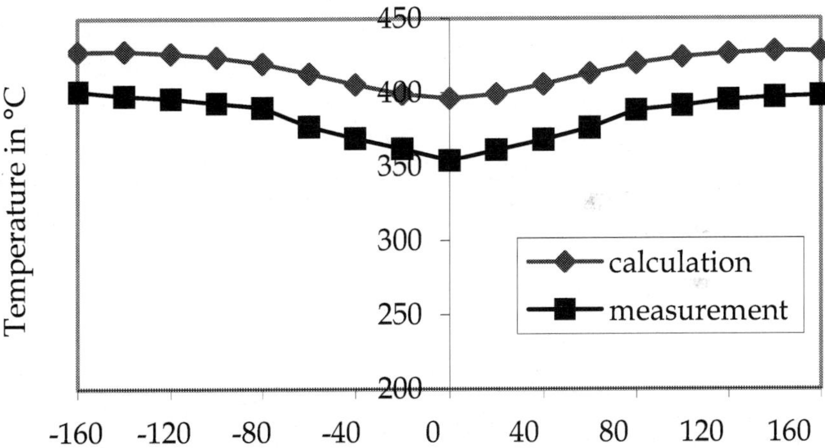

Fig. 6. Temperature distribution along diameter

Furthermore, we can assume that the steady state temperature distribution had not been reached after the heating time of 12 min. Considering this facts the numerical model seems to have a good accuracy. For the process of epitaxy the distribution is not homogeneous enough due to the temperature drop between the highest and the lowest value of nearly 10%. It is possible to decrease this value by changing the coil geometry. The application engineer has the opportunity of a local modification of the coupling gap and/or the distance between the turns of the coil. The effects of these modifications can be conveniently investigated with numerical simulations. In this manner an optimal solution for the customer can be developed.

4. Product Family TruHeat MF 3010 - 7040

4.1 General Information

The experimental investigations were done with the innovative induction heating generator TruHeat MF 3010 - 7040. This product family is available with 10, 20, 30 or 40 kW output power. The frequency range reaches from 5 to 100 kHz. The user has the possibility to operate at any frequency from 5 to 30 kHz or from 20 to 100 kHz. The TruHeat MF 3010 - 7040 was designed for mains voltages from 400 to 480 V. The generator works at any frequency in this range.

Three different variants of the TruHeat MF 3010 - 7040 are available. The basic version is the 19-inch plug-in module, designed for integration in a 19-inch rack (see Figure 7). This very compact variant offers the possibility to integrate the power supply very easy into the facility, e.g. in machines for modern crystal growing by epitaxy. This variant already offers the full functionality of the product family. The second version is the tabletop unit, which is the ideal heating equipment for stand alone operation. This variant is the best choice for laboratories, universities or companies with low amount of space. The third version is the cabinet. This is the traditional variant for industrial environment with a high protection class IP54. This enables the use even in rough conditions, e.g. in the field of induction hardening.

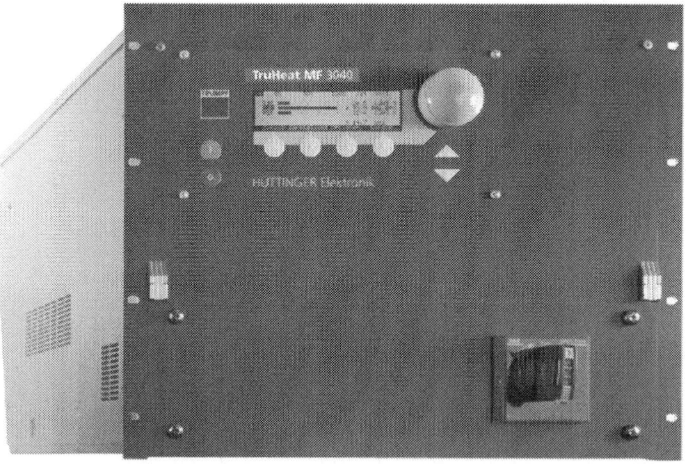

Fig. 7. TruHeat MF 3040, 19-inch plug-in

4.2 Technical specification

The 6-puls rectifier is connected directly – without using a relatively space consuming mains transformer – to the mains. This saves space and money.

The DC output voltage of the rectifier is controlled by a chopper unit, which enables to vary the output power (voltage or current) from almost 0 to 100 %. Therefore, the TruHeat MF 3010 – 7040 is ideal for high power as well as for low power applications, or for processes which need the full power range, e.g. to reach special temperature profiles.

The inverter is operated by an advanced control unit. It recognizes the frequency of the oscillating circuit and chooses the control parameters for the different frequency ranges by itself. As a result, the whole TruHeat MF 3010 – 7040 family needs only one inverter control board.

The output transformer fulfils two functions. It separates the mains potential from the application area and adapts the load to the power supply. The output voltage of the transformer can be switched from 600 to 300 V or vice versa very easy. Using this feature, the user has the full flexibility to solve different application tasks.

The described components are integrated in the power supply. It is connected with power cables to a parallel oscillating circuit, which has two main advantages in comparison to a serial oscillating circuit. The current in the induction coil – thus the electromagnetic field – can be increased by incrementing the factor Q. In addition to that, the matching of the generator and the load can easily be done by changing the capacity. An overview of the features and benefits of TruHeat MF 3010 – 7040 is shown in Figure 8.

Features	Benefits
▩ Compact design, high power density	▩ Low space requirement for easy system integration
▩ Mains voltage of any value between 400 V -10% and 480 V +10%	▩ Application in almost every country possible
▩ One MF-transformer for 300 V and 600 V output voltage	▩ Full flexibility to solve different application tasks
▩ Parallel oscillating circuit	▩ Quick process adaptation by changing the capacity ➜ low set up time and high flexibility in use
▩ Control of the process power from almost 0 % to 100 %	▩ Ideal for temperature controlled processes

Fig. 8. Features and Benefits of the TruHeat MF 3010 - 7040

5. Conclusions

The induction heating is an innovative technology which offers important advantages against heat transmitting application. For an extensive use of these advantages it is necessary to have both the possibility to develop a suitable induction coil design and an innovative technique for the used power supply. We have highlighted these two aspects in the report. We have discussed three methods for analyzing induction heating systems, and have shown that the numerical simulation of electromagnetic and thermal fields has been developed to an essential tool for the application engineer. Together with the possibility to perform measurements it allows the effective and quick development of customized solutions. Our generator family TruHeat MF 3010 – 7040 can deliver its nominal output power over a wide frequency range. Because of its low space requirement and its high flexibility in use it is the ideal power supply for modern induction heating application.

6. References

[1] Nemkov, W.S., Demidovich, W.B. (1988). Theory and Computation of Set-ups for Induction Heating. Energoatomisdat, Agency of Leningrad

[2] Sluchotsky, A.E., Ruiskin, S.E. (1974). Induction Coils for Induction Heating. Energia, Agency of Leningrad

[3] Philippow, E. (1988). Taschenbuch Elektrotechnik. Band 6. VEB Verlag Technik Berlin, p. 440.

[4] Gröber, Erk, Grigull (1955). Grundgesetze der Wärmeübertragung. Springer-Verlag Berlin/Göttingen/Heidelberg, p.274.

7

Configuration Proposals for an Optimal Electromagnetic Coupling in Induction Heating Systems

Carrillo, E.
RHI/ BU Glass-Refmex
Mexico

1. Introduction

Induction Heating (IH) is a mature technique for heating metals due it occurs mainly as a consequence of their high electric conductivity. That condition, no necessarily magnetic permeability, enables a suitable electromagnetic coupling that in turns generates intense heat throughput at very high rates at well-defined locations that reduce process cycle time with repeatable quality.

When it comes to induction heating, one must keep in mind that ferromagnetic materials loose their magnetic properties above Curie temperature, thus good electric conductor condition prevails as to be taken into account for a proper work-piece candidate in a IH system, whereas low conducting materials will require high frequency for the excitation coil in order to get the above advantages. Other way to overcome the poor electromagnetic coupling in low conductive work-pieces is by means of assembling a high melting temperature susceptor, which due to its high conductivity heats indirectly the low conductive insert so as to avoid high frequency ancillary equipment thus lowering capital costs.

Despite IH is a well-established heating method; it realizes on empirical rules relating size, frequency and the skin depth of the work-piece to fit a convenient IH scheme.

This chapter is devoted to the development of analytic parameters, which are intended to be taken as guidance to set optimal IH configurations for putting together Inductor and work-piece while the highest electromagnetic coupling is achieved. They should match those optimal conditions that are already known for HI practitioners. The simplest and more ubiquitous geometries dealt in massive induction heating systems happen to be the cylinder and the slab, which are subject matter in this chapter.

Since an IH system can be depicted as a step-down electrical transformer in which the secondary winding is short-circuited and performing as a workpiece, the more intense the influence on the idle excitation coil impedance when the secondary loop is assembled, the greater the occurrence of eddy currents on it. Thus, a proposed reflection coefficient is developed, which is the common factor that precedes both secondary resistance and secondary inductance as appearing in the effective impedance formula for an ideal electrical transformer. The larger the reflection coefficient, the better is the electromagnetic coupling of the workpiece. To undertake that task, a dimensionless treatment of simplified governing differential equation of the eddy current process is given, featuring the shielding parameter

$R\omega$. The above approach has been presented for cylindrical systems (Carrillo, 2008), by considering as boundary values: either a longitudinal or a transverse magnetic flux. In this chapter same treatment also will be given to rectangular workpieces, namely slabs being excited by a transverse flux.

For confronting the proposed model, practical scenarios are taking into account involving the abovementioned workpieces to feed them as an input for computer runs using the commercial FEM software package Quick Field from TERA ANALISYS LTD. Here, the electrical efficiency, or coil efficiency as being the ratio of the power dissipated as heat in the workpiece to the total input power, is referred as a measure of electromagnetic coupling. Results on coil efficiencies confirm the proposed analytical approach.

Furthermore regarding the use of composite loads for improving weakly coupled IH systems as found in susceptor practices, a treatment on a susceptor-insert induction heated device, carried out in Germany (Reichert, 1965) is taken as a reference, for comparing the performance of alternative HI schemes featuring the highest coil efficiency.

Exploring the feasibility of the heating processes other than combustion for both; non-electrical or poor electrical conductors would be an interesting task. Efforts to make more efficient the application of electricity for mass heating would turn it more appealing for Industrial activities, since nowadays environmental issues are becoming constantly stringent, mainly on those related to the Global Warming, hence approaches to reduce green house gas evolving are always welcome.

2. Impedance parameters for an HI system

Eddy currents drive the IH process. The electromagnetic diffusion equations are the governing equations for eddy currents, which are to be solved so as to mathematically model the HI process for a given HI device.

A lumped parameter model is a representation in a mathematical model of a physical system where field variables are simplified down to single scalars. The mathematical analysis of an electrical circuit diagram is much simpler than solving the electromagnetic diffusion equations for the actual HI physical system. Thus impedances, as lumped parameters, are electrically equivalent to those electromagnetic fields distributed in an existing inductor or an element of a circuit. Thereby, magnetic strength and current density distributions for a given HI configuration are to be featured from which basic impedance parameters such as resistances and inductances are stemmed from and then treated in such a way that a reflection coefficient is obtained as a result. In the following a description of how electromagnetic diffusion equations are simplified and set up for being solved, are presented just for cylindrical frames. The same procedure has been carried out using Cartesian coordinates for characterizing the slab geometry.

2.1 Governing equations for featuring eddy currents:
Eddy currents (Krawczyk 1993) occurring in a solid workpiece can be derived by applying Ampere's circuit law for low frequencies, which is:

$$\nabla \times \mathbf{H} = \mathbf{J} \qquad (1)$$

The magnetic strength distributions in the workpiece are the solutions of the electromagnetic diffusion equation (Krawczyk 1993), which for rigid bodies is:

$$\nabla^2 \mathbf{H} = \sigma\mu_0 \frac{\partial \mathbf{H}}{\partial t} \tag{2}$$

where, σ (S m^{-1}) is the electric conductivity, μ (V s A^{-1} m^{-1}) is the magnetic permeability of the conductor, t is time (s), $\mathbf{J}$ the current density (A m^{-2}), and $\mathbf{H}$ (A m^{-1}), the magnetic strength field. Components of Eq. (2) are expressed according to the coordinate system suitable to the geometry of the HI configuration. Depending on certain simplifications owing to the symmetry of the excitation source configuration, the components of the magnetic strength vector in Eq. (2) can be uncoupled for the following general cases, namely: an infinitely long conducting cylinder and semi-infinite slab: This consideration usually takes place for neglecting the end and edge effects.

$$\begin{aligned} &H_\varphi(z) = cons\tan t, \quad H_z(\varphi) = constant, \quad H_r = 0 \quad Cylinder \\ &H_y = H_z = 0 \qquad\qquad\qquad\qquad\qquad\qquad\qquad Slab \end{aligned} \tag{3}$$

Excitation fluxes are imposed in cylinders by: A solenoidal winding, which impresses an axial magnetic strength on the workpiece and axial coils, which set up an angular magnetic strength inside the workpiece. And for the slab, straight coils, which impose a transverse flux inside the slab, being the coordinate Z for the length, X for the depth and Y for the thickness respectively. Taking into account all the pertinent considerations, the eddy current process will be varying only along the radius for the cylinder and along the thickness "y" for the slab. Furthermore, in the frequency domain, solutions to Eq. (2), are expressed in a general expression as:

$$H_g(\vec{\mathbf{x}},t) = \left(\mathrm{Re}\,\hat{H}_g(\ell) + i\,\mathrm{Im}\,\hat{H}_g(\ell) \right) \mathbf{expi}(\omega t - p\varphi) \tag{4}$$

The position vector $\mathbf{x}$ simplifies to ℓ, which is the coordinate either radius r or thickness y depending on what coordinate system is being considered, ω (s^{-1}) is the electric angular frequency and p is the pair of magnetic poles. Equation (4) is a generic function to be used for solution of Eq. (2), depending on the corresponding subscript g (i.e., Cylindrical φ, z, or Cartesian x), Thus simplified electromagnetic diffusion equations for a given coordinate system then become:

$$\frac{d^2\hat{H}_\varphi}{dr^2} + \frac{1}{r}\frac{d\hat{H}_\varphi}{dr} - \frac{1+p^2}{r^2}\sigma\mu_0\hat{H}_\varphi = 0 \tag{5}$$

when the axial coils are performing as an excitation source.

$$\frac{d^2\hat{H}_z}{dr^2} + \frac{1}{r}\frac{d\hat{H}_z}{dr} - \sigma\mu_0\hat{H}_z = 0 \tag{6}$$

and Eq. (6) for a solenoidal winding. Finally, for a slab being exposed to the magnetic field from a rectangular cage inductor type, the simplified governing equation becomes:

$$\frac{d^2\hat{H}_x}{dy^2} + \frac{d\hat{H}_x}{dy} - \sigma\mu_0\hat{H}_x = 0 \tag{7}$$

2.2 Solution procedure for eddy current equations

The following normalizations for the primitive variables convert Eqs. (5), (6) and (7) into a pair of dimensionless coupled equations.

i) The dimensionless length χ, defined as either the ratio of the outer radius R to any inner radius r for the cylinder, and the ratio of slab thickness Y to any distance along the "y" coordinate:

$$\chi = \frac{r}{R}, \quad \cdot \quad \chi = \frac{y}{Y} \tag{8}$$

R or Y might be referred as the work-piece characteristic length as used in the principle of similitude technique.

ii) The electromagnetic variables are normalized to the difference between the real parts of the magnetic strength at the surface and at the axis. The magnetic strength subscript g takes labels of φ, z, or y, according to the configuration being analyzed. Since there is no magnetic field at the workpiece axis owing to the magnetic shielding process, then Re H_g (ℓ =0) and Im H_g (ℓ =0) are equal to zero. Thus, news dimensionless magnetic variables turn out as:

$$\Theta_g = \frac{\mathrm{Re}\,\hat{H}_g(\ell)}{\mathrm{Re}\,\hat{H}_g(surface)} \cdot \tag{9}$$

$$\Psi_g = \frac{\mathrm{Im}\,\hat{H}_g(\ell)}{\mathrm{Re}\,\hat{H}_g(surface)} \cdot \tag{10}$$

The electromagnetic properties of the material are gathered in the dimensionless number $R\omega$ termed the screen or shielding parameter, which relates the ability of a conducting medium to exclude high-frequency magnetic fields from its core:

$$R\omega_{cylinder} = \sigma\mu\omega R^2, \qquad R\omega_{slab} = \sigma\mu\omega Y^2. \tag{11}$$

$R\omega$ also can be expressed in term of the electromagnetic skin depth δ, being $\delta = \sqrt{(2/\omega\sigma\mu)}$

$$R\omega_{cylinder} = 2\left(\frac{R}{\delta}\right)^2, \qquad R\omega_{slab} = 2\left(\frac{Y}{\delta}\right)^2. \tag{12}$$

in terms of the workpiece thickness,

$$\frac{D}{\delta} = \sqrt{2R\omega_{cylinder}}, \qquad \frac{Y}{\delta} = \sqrt{2R\omega_{slab}}. \tag{13}$$

Following the procedure for solving the simplified governing equation for the dimensionless angular magnetic fields is given. Same treatment is to be given for the remaining cases dealt so far. By substituting Eqs. (9), (10), and (11) into Eq. (5), a set of simultaneous ordinary equations for both the real and imaginary parts of the dimensionless angular magnetic is built:

$$\frac{d^2\Theta_\varphi}{d\chi^2} + \frac{1}{\chi}\frac{d\Theta_\varphi}{d\chi} - \left(\frac{1+p^2}{\chi^2}\right)\Theta_\varphi + (R\omega)\Psi_\varphi = 0 \tag{14}$$

$$\frac{d^2\Psi_\varphi}{d\chi^2} + \frac{1}{\chi}\frac{d\Psi_\varphi}{d\chi} - \left(\frac{1+p^2}{\chi^2}\right)\Psi_\varphi - (R\omega)\Theta_\varphi = 0 \tag{15}$$

With regard to the complex plane, real zero is set equal to the right-hand terms of Eqs. (14) and (15). The systems of coupled equations are solved by means of the Newton shooting method (Constantinides, 1987) with the following labels: The boundary values for the magnetic fields at the axis are fixed to zero to resemble the cancellation of fields at that point:

$$\left.\frac{d\Theta_g}{d\chi}\right|_{\chi=0} = y_1(y_4 = 0) = 0 \tag{16}$$

$$\left.\frac{d\Psi_g}{d\chi}\right|_{\chi=0} = y_2(y_4 = 0) = 0 \tag{17}$$

$$\left.\Psi_g\right|_{\chi=0} = y_3(y_4 = 0) = 0 \tag{18}$$

The value of y_4 at the left boundary is assumed to be nearly zero, e.g. 10^{-8}, to avoid singularities. The real magnetic strength is set to unity at the cylinder surface:

$$\left.\Theta_g\right|_{\chi=1} = y_5(y_4 = 1) = 1 = \frac{H_g}{H_g} \tag{19}$$

The missing initial condition is the left-boundary value y_5, which cannot be set because the right-boundary value has already been granted. Therefore, this missing initial condition is estimated by means of an initial shoot γ_5, keeping the notion of magnetic shielding, as required by Lenz's Law:

$$\left.\Theta_g\right|_{\chi=0} = y_5(y_4 = 0) \approx \gamma_5 \approx 0 \tag{20}$$

The field should fade toward the axis. The Newton shooting procedure is balanced by a variational equation set, which weights the certainty of the guessed shoot γ.

$$y_6 = \frac{d}{d\chi}\left(\frac{\partial\Theta_g}{\partial\gamma_5}\right) = \frac{\partial y_1}{\partial\gamma_5}, \quad y_7 = \frac{d}{d\chi}\left(\frac{\partial\Psi_g}{\partial\gamma_5}\right) = \frac{\partial y_2}{\partial\gamma_5},$$

$$y_8 = \frac{\partial\Psi_g}{\partial\gamma_5} = \frac{\partial y_3}{\partial\gamma_5}, \quad y_9 = \frac{\partial\chi}{\partial\gamma_5} = \frac{\partial y_4}{\partial\gamma_5}, \tag{21}$$

$$y_{10} = \frac{\partial\Theta_g}{\partial\gamma_5} = \frac{\partial y_5}{\partial\gamma_5}$$

The error to converge the solution is less than 10^{-8}.

2.3 Approaching of an HI system as a short-circuited electric transformer

An induction heating system can be depicted as a step-down electrical transformer in which the secondary winding is short-circuited. The secondary winding performs as a workpiece in which ohmic heat is dissipated due to an induced electric field.

2.3.1 Circuit equations

The initial approach to feature an electric transformer is carried out by analyzing the following coupled differential equations

$$V_0 = R_c \frac{\mathrm{d}I_c}{\mathrm{d}t} + L_c \frac{\mathrm{d}I_c}{\mathrm{d}t} + M \frac{\mathrm{d}I_w}{\mathrm{d}t}$$
$$0 = R_w \frac{\mathrm{d}I_w}{\mathrm{d}t} + L_w \frac{\mathrm{d}I_w}{\mathrm{d}t} + M \frac{\mathrm{d}I_c}{\mathrm{d}t} \tag{22}$$

The variables are the electric current I (A), electrical resistance R (Ω), and self-inductance L (H). The mutual inductance M (H) is common to both loops. The primary circuit with excitation voltage V_0 is denoted using subscript c and the corresponding secondary circuit, as featured by the eddy current pattern inside the workpiece, is labeled with subscript w. We assume that the voltages and currents are harmonic variables:

$$V_0 = \hat{V}_0\, e^{i\omega t}$$
$$I_c = \hat{I}_c\, e^{i\omega t} \tag{23}$$

where a hat ($^\wedge$) on top of variables denotes a complex phasor and i represents the complex number $\sqrt{(-1)}$. Then, by substituting Eqs (23) into Eq. (22) and some rearrangement, the effective impedance Z can be obtained:

$$\hat{Z} = \left(R_c + \frac{\omega^2 M^2}{R_w^2 + \omega^2 L_w^2} R_w \right) + i \cdot \left(\omega L_c - \frac{\omega^2 M^2}{R_w^2 + \omega^2 L_w^2} \omega L_w \right) \tag{24}$$

Equation (24) shows that the load is electrically equivalent to a resistance and a negative inductance placed in the primary circuit. The common factor that precedes both secondary resistance and secondary inductance might be envisaged as a reflection coefficient for the excitation circuit:

$$\text{reflection coefficient} \equiv \frac{\omega^2 M^2}{R_w^2 + \omega^2 L_w^2} \tag{25}$$

The larger the reflection coefficient, the better is the electromagnetic coupling of the workpiece. Hence, the optimization process consists of identifying conditions under which the reflection coefficient is enhanced.

2.3.2 Impedance parameters

The resistive and inductive load of a HI system can be derived from the balance of energy inside an electromagnetic field, as described by Poynting (Krawczyk 1993) as follows:

$$\oint_A \left(\hat{\mathbf{E}} \times \hat{\mathbf{H}} \right) \bullet \mathbf{n} \, \mathrm{d}A = -\mathrm{i} \cdot 2\omega\mu_0 \int_V \frac{1}{4} \left| \hat{\mathbf{H}} \right|^2 \mathrm{d}V - \int_V \frac{1}{2} \left| \hat{\mathbf{J}} \right|^2 \mathrm{d}V \tag{26}$$

Equation (26) comprises the vectors $\mathbf{B}$, the induction flux (V s m^{-2}), $\mathbf{E}$, the electric field (V m^{-1}), $\mathbf{J}$, the current density (A m^{-2}), and $\mathbf{H}$ (A m^{-1}), the magnetic strength field. The terms within parentheses are the complex modules of the corresponding variables.

The workpiece resistance R_w is taken from the second term on the right-hand side of Eq. (26), the real part of the influx of the Poynting vector:

$$R_w = \frac{1}{\sigma_w \left(I_w \right)^2} \iiint \frac{1}{2} \left| \hat{J}_w \right|^2 \cdot \mathrm{d}z \cdot \mathrm{d}\varphi \cdot r \mathrm{d}r \tag{27}$$

The dummy variables for the integral in Eq. (27) are expressed in cylindrical coordinates. Let us assume that

$$I_w = N_c I_c \tag{28}$$

since the secondary loop is equivalent to a one-turn coil as an idealized transformer. Here N_c is the number of coil turns in the excitation winding. By substituting Eq. (28) into Eq. (27) and integrating for angular and axial coordinates, this yields:

$$R_w = \frac{\pi}{\sigma_w} \frac{Z_\infty}{N_c^2 I_c^2} \cdot \int_0^R \left| \hat{J}_w \right|^2 \cdot r \mathrm{d}r \tag{29}$$

The upper limit R in the integral is the outer radius of the workpiece, and Z_∞ represents the length of an infinitely long cylinder. The workpiece inductance L_w stems from the first term on the right-hand side of Eq. (26), which represents the magnetic energy stored in that volume.

$$L_w = \frac{\mu_0}{\left(I_w \right)^2} \iiint \frac{1}{2} \left| \hat{H}_w \right|^2 \cdot \mathrm{d}z \cdot \mathrm{d}\varphi \cdot r \mathrm{d}r \tag{30}$$

Similarly, by substituting Eq. (28) into Eq. (30), the workpiece inductance can be expressed as:

$$L_w = \omega\mu_0 \pi \frac{Z_\infty}{N_c^2 I_c^2} \cdot \int_0^R \left| \hat{H}_w \right|^2 \cdot r \mathrm{d}r \tag{31}$$

In addition, the mutual inductance M can be derived from the expression for the magnetic energy stored in the space shared by the excitation coil and the workpiece, as follows:

$$M = \frac{1}{I_c I_w} \iiint \left(\mathbf{H}_c \bullet \mu_0 \mathbf{H}_w \right) \mathrm{d}z \cdot \mathrm{d}\varphi \cdot r \mathrm{d}r \tag{32}$$

Assuming harmonic behavior for the electromagnetic variables, Eq. (32) can be expressed as:

$$M = \frac{\mu_0}{I_c I_w} \iiint \frac{1}{2} \mathrm{Re}\left(\hat{H}_c \cdot \hat{H}_w^* \right) \cdot \mathrm{d}z \cdot \mathrm{d}\varphi \cdot r \mathrm{d}r \tag{33}$$

The superscript (*) for H_w represents its complex conjugate. Thus, by substituting Eq. (28) into Eq. (33) and integrating for axial and angular coordinates yields:

$$M = \frac{Z_\infty \mu_0 \pi}{N_c I_c^2} \int_0^R \left(\operatorname{Re} \hat{H}_c \operatorname{Re} \hat{H}_w + \operatorname{Im} \hat{H}_c \operatorname{Im} \hat{H}_w \right) \cdot r dr \tag{34}$$

Thus, by substituting Eqs. (29), (31) and (34) into Eq. (25), the reflection coefficient can be expressed as:

$$\frac{\omega^2 M^2}{\left(R_w^2 + \omega^2 L_w^2 \right)} = N_c^2 \frac{\omega^2 \mu_0^2 \left[\int_0^R \left(\operatorname{Re} \hat{H}_c \operatorname{Re} \hat{H}_w + \operatorname{Im} \hat{H}_c \operatorname{Im} \hat{H}_w \right) \cdot r dr \right]^2}{\left[\frac{1}{\sigma_w^2} \left(\int_0^R |\hat{J}_w|^2 \cdot r dr \right)^2 + \omega^2 \mu_0^2 \left(\int_0^R |\hat{H}_w|^2 \cdot r dr \right)^2 \right]} \tag{35}$$

Multiplying both the numerator and denominator in Eq. (35) by $R^4 \sigma_w^2$ yields:

$$\frac{1}{N_c^2} \frac{\omega^2 M^2}{\left(R_w^2 + \omega^2 L_w^2 \right)} = \frac{\sigma_w^2 \omega^2 \mu_0^2 R^4 \left[\int_0^R \left(\operatorname{Re} \hat{H}_c \operatorname{Re} \hat{H}_w + \operatorname{Im} \hat{H}_c \operatorname{Im} \hat{H}_w \right) \cdot r dr \right]^2}{R^4 \left(\int_0^R |\hat{J}_w|^2 \cdot r dr \right)^2 + \sigma_w^2 \omega^2 \mu_0^2 R^4 \left(\int_0^R |\hat{H}_w|^2 \cdot r dr \right)^2} \tag{36}$$

This expression describes the reflection coefficient in terms of fundamental electromagnetic variables such as the magnetic strength and current density.

2.3.3 Dimensionless reflection coefficient

The reflection coefficient represents a measure of the coupling of the induced electromagnetic field to the excitation source. Therefore, the electromagnetic parameters that result in a maximum reflection coefficient while keeping minimal either the operational frequency or workpiece size for a given induction heating design need to be identified. For generalization, the task involves searching for the lowest shielding parameter $R\omega$ that matches the highest reflection coefficient.

Variables in the reflection coefficient, such as the dimensions, the electromagnetic properties of the workpiece and the excitation frequency, can be expressed in terms of shielding parameter $R\omega$ by substituting Eq. (11) into Eq. (36), resulting in:

$$\frac{1}{N_c^2} \frac{\omega^2 M^2}{\left(R_w^2 + \omega^2 L_w^2 \right)} = \frac{(R\omega)^2 \left[\int_0^R \left(\operatorname{Re} \hat{H}_{c,g} \operatorname{Re} \hat{H}_{w,g} + \operatorname{Im} \hat{H}_{c,g} \operatorname{Im} \hat{H}_{w,g} \right) \cdot r dr \right]^2}{R^4 \left[\int_0^R |\hat{J}_{w,g}|^2 \cdot r dr \right]^2 + (R\omega)^2 \left[\int_0^R |\hat{H}_{w,g}|^2 \cdot r dr \right]^2} \tag{37}$$

Therefore, the procedure to be followed comprises finding both the magnetic strength and the current density distributions inside the workpiece that match the minimum shielding parameter $R\omega$, which maximizes the reflection coefficient, as shown in Eq. (37).

As required by the similarity criteria for scaled process analysis, the terms inside the square brackets in Eq. (37) are dimensionless variables, and hence they are defined as follows. The dimensionless complex magnetic strength is

$$\Xi_g = \Theta_g + i\Psi_g \tag{38}$$

and the relationship between the time-averaged value of Ξ with its complex module is

$$\overline{\Xi_g^2} = \tfrac{1}{2}\mathrm{Re}\left(\Xi_g \cdot \Xi_g^*\right) = \frac{\left|\Xi_g\right|^2}{2} \tag{39}$$

Similarly expressions for the dimensionless current densities Γ_g, are obtained by developing Ξ with components Θ_g and Ψ, following the Ampere Law while holding the corresponding dimensionless conditions. Then time-averaged dimensionless axial current density is obtained:

$$\frac{\left|\Gamma_z\right|^2}{2} = \frac{1}{2}\left(\left(\frac{d\Theta_\varphi}{d\chi} + \frac{\Theta_\varphi}{\chi}\right)^2 + \left(\frac{d\Psi_\varphi}{d\chi} + \frac{\Psi_\varphi}{\chi}\right)^2\right) \tag{40}$$

The scalar component of Γ_r has been neglected for simplicity. The expressions in Eq. (39), and (40) are then substituted in Eq. (37). The outer radius assigned as the upper limit in the integrals is set to unity, in accordance with Eq. (8)

$$\frac{1}{N_c^2}\left(\frac{\omega^2 M^2}{R_w^2 + \omega^2 L_w^2}\right)_{\text{dimensionless}} = \frac{\left[\int_0^1\left(\Theta_{c,g}\Theta_{w,g} + \Psi_{c,g}\Psi_{w,g}\right)\cdot\chi d\chi\right]^2}{\dfrac{1}{(R\omega)^2}\left(\int_0^1\left|\Gamma_{w,g}\right|^2\cdot\chi d\chi r\right)^2 + \left(\int_0^1\left|\Xi_{w,g}\right|^2\cdot\chi d\chi\right)^2} \tag{41}$$

The dimensionless excitation magnetic strength $\Xi_{c,g}$ is taken as a constant through the linked flux and its complex module is assumed to equal 1. Under ideal conditions, since the coil is non-ferrous, there will be a $\pi/4$ phase shift inside the imposed field $\Xi_{c,g}$, and hence it dissembles into its real and imaginary parts as follows:

$$\hat{\Xi}_{c,g} = \Theta_{c,g}\frac{\sqrt{2}}{2} + i\Psi_{c,g}\frac{\sqrt{2}}{2} \tag{42}$$

In addition, the gap between the workpiece and the excitation coil is assumed to be negligible, so the value of the magnetic strength in the excitation coil is considered to be the same as the real magnetic strength at the workpiece surface, which was taken as a normalizing value:

$$\Theta_{w,g} = 1 = \Theta_{c,g} = \Psi_{c,g} \tag{43}$$

Thus, Eq. (42) is transformed to:

$$\hat{\Xi}_{c,g} = \frac{\sqrt{2}}{2}(1+i) \tag{44}$$

By substituting Eq. (44) into Eq. (41), the expression for the dimensionless reflection coefficient becomes for the cylindrical workpiece:

$$\frac{1}{N_c^2}\left(\frac{\omega^2 M^2}{R_w^2+\omega^2 L_w^2}\right)_{dimensionless}=\frac{\dfrac{\sqrt{2}}{2}\left[\int_0^1\left(\Theta_{w,g}+\Psi_{w,g}\right)\cdot\chi d\chi\right]^2}{\dfrac{1}{(R\omega)^2}\left(\int_0^1\left|\Gamma_{w,g}\right|^2\cdot\chi d\chi\right)^2+\left(\int_0^1\left|\Xi_{w,g}\right|^2\cdot\chi d\chi\right)^2} \tag{45}$$

whereas for the slab, taking into account the Cartesian variable is: $v = y/Y$ then the expression for its reflection coefficient is:

$$\frac{1}{N_c^2}\left(\frac{\omega^2 M^2}{R_w^2+\omega^2 L_w^2}\right)_{dimensionless}=\frac{\dfrac{\sqrt{2}}{2}\left[\int_0^1\left(\Theta_{w,cage}+\Psi_{w,cage}\right)\cdot dv\right]^2}{\dfrac{1}{(R\omega)^2}\left(\int_0^1\left|\Gamma_{w,cage}\right|^2\cdot dv\right)^2+\left(\int_0^1\left|\Xi_{w,cage}\right|^2\cdot dv\right)^2} \tag{46}$$

3. Analysis on electromagnetic coupling

In this section some numerical results stemmed from the previous analytical assessment on optimal conditions where electromagnetic coupling occurs in several HI systems are presented and some of them confronted to those already known by IH practitioners.

Thereby, for obtaining reflection coefficients for cylindrical and rectangular workpieces when exposed to either longitudinal or transverse excitation fluxes, whatever the case is selected, values of the shielding parameter $R\omega$ are to be varied and fed as an input for the dimensionless electromagnetic diffusion equations, thus dimensionless magnetic strength and current density distributions are calculated and integrated and then treated as in Eq (43) or Eq. (44) whichever is proper.

Fig.1 gathers various reflection coefficients as dependant of the shielding parameter.

Fig. 2 shows same coefficients but now as a function of the ratio of the workpiece thickness to skin depth, which is more familiar to HI practitioners.

The relationship between shielding parameter and skin depth is regarded in Eq. (36). Reflection coefficent, hence electromagnetic coupling, seems to enhance in workpieces subjected to excitation transverse flux as the size decreases being the slab the shape that dissipates more power while either minimum frequency is applied or the smallest size is employed.

Nevertheless, the reflection coefficient for slabs is of the order of 10^{-4} so it is displayed in the right-hand side ordinate of Figs. 1 and 2 as multiplied by the square of 100 coil turns, which might mean when heating a slab, a multilayer inductor should be used in order to have proper electromagnetic coupling.

For longitudinal flux applications, the reflection coefficient is the highest for the workpiece diameter / skin depth at 5.7. This ratio together with the range of the width of the "bell" of the corresponding curve, which goes from 4.7 to 6.25, matches what is stated by Rudnev (Rudnev et al., 2003) There, a rule of thumb is referred which stipulates current cancelation, hence no induction heating, will take place if the ratio of workpiece diameter to current

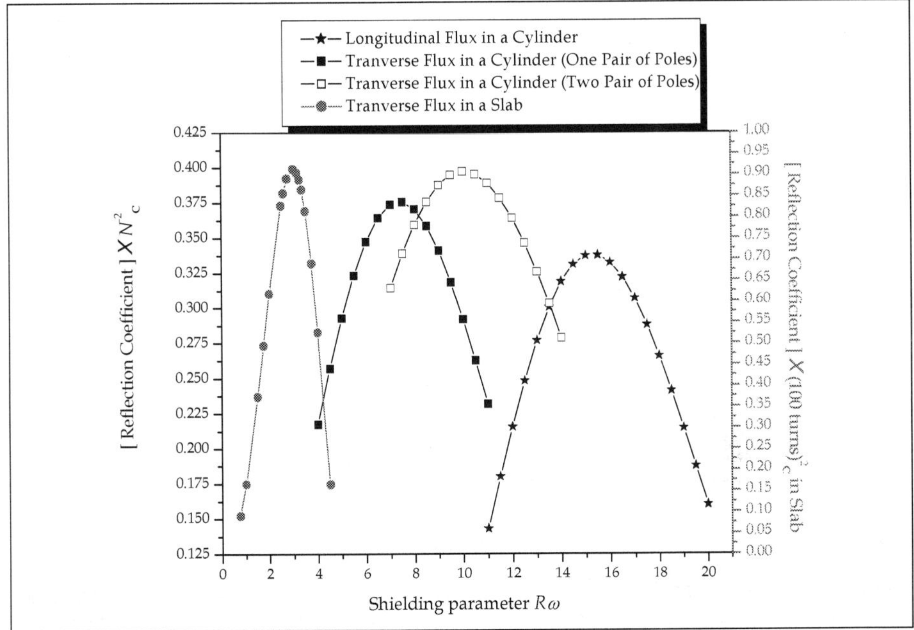

Fig. 1. Influence of the Excitation Coil Configuration on Electromagnetic Coupling in HI Systems, as a Function of Shielding Parameter.

penetration depth is less than four. Beyond of ratio 6 it will only slightly increase the heating induction efficiency.

Simpson (Davies & Simpson,1979) tried to find analytically those optimal conditions by invoking the Bessel functions ber and bei from which a maximum p parameter should have been associated to above mentioned workpiece diameter to skin depth ratios, actually resulting in ratios lower than those found in practice. Nevertheless, it is close to the range where the cylindrical devices excited by one-pair of poles transverse flux achieve their maximum reflection coefficients as it can be seen in Fig.2.

For a single phase device, the one-pair-of-poles windings would need the full 180 ° arc in each hemisfere of the cylinder to accomodate the excitation coils, thus the HI process would be affected by the so called proximity effect due to the closeness of the ends of the coils carrying opposite currents. A resource to maintain far enough the coils could be placing the excitation coils half way the full arc of the hemisphere in such a manner that the whole circumference is divided in four parts having the same length: two parts opposing each other as the locus for the windings and the other section 90° apart to give room and avoid the proximity effect. See Figs. 6 b) and 11 a). This layout is equivalent to two-pair-of-poles windings whose configuration also is analysed and reflection coefficient plotted in Figs. 1 and 2. This arrangement seems to be more practical hence it is considered for further Finite Element Software simulations. The range where higher shielding parameters occur lies between those corresponding to one-pair of poles and solenoidal configurations. It shows the highest reflection coefficient among the analysed configurations.

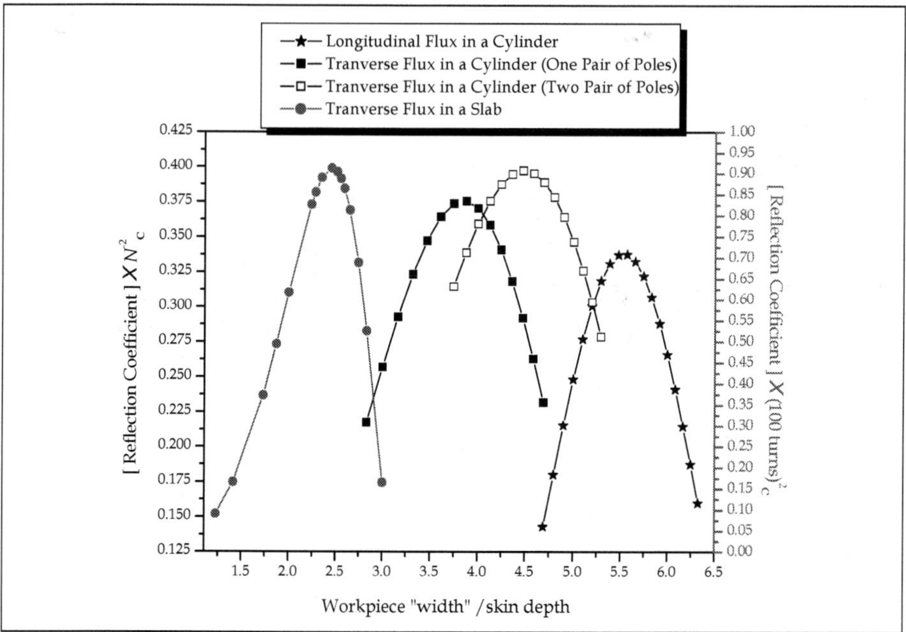

Fig. 2. Influence of the Excitation Coil Configuration on Electromagnetic Coupling in HI Systems, as a Function of the Ratio of Workpiece Width to Skin Depth.

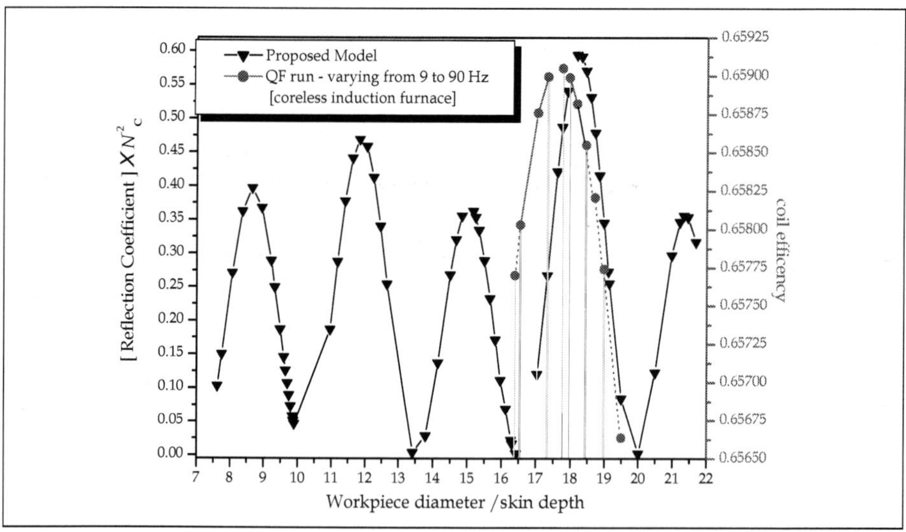

Fig. 3. Electromagnetic Coupling Performance at Higher Shielding Parameters for a Solenoid Coil Configuration

On the other hand, Fig. 2 shows the lowest workpiece diameter to skin depth ratio range for solenoidal coil configurations, e.g. longitudinal flux applications, where a mininum either frequency or HI device size can be fixed while keeping high electromagnetic coupling, thus high efficiency in the energy conversion. Nevertheless, there are also industrial applications such as metal melting when using mains frequency coreless induction furnaces whose sizes are above that optimal range and performig with relatively high efficiencies. That problem was tackled in Germany at the end of the 1950's (Davies & Simpson, 1979). The proposed reflection coefficient also could explain that break through, since those obtained at higher shielding parameters seems to have cyclic behaviour resulting in waves of maximums and minimums depending on the value of a given workpiece to skin depth ratio, see Fig. 3.

The range shielding parameters corresponding to the highest reflection coefficients has been chosen and compared to certain computed coil efficiencies, which will be discussed in further section.

3.1 Coil efficiency as a means of measuring of electromagnetic coupling in HI systems

Since being the reflection coefficient an analytical concept difficult to be measured, if not impossible, indirect parameters that are enhanced when eddy currents occur in the workpiece have to be found for comparison. There exists a practical index that relates the electromagnetic coupling of a HI assemblage, namely the electric efficiency or coil efficiency defined as the fraction of the total energy that takes into account only the ohmic losses released in the workpiece. The later is more commonly known and it will be referred from now on. Therefore parameters of interest would be: the overall Ohmic losses for the global HI system, e.g. those occurring in the inductor coil and workpiece, and the heat released solely in the work piece as well. Then, the expression for the coil efficiency, η_c, becomes:

$$\text{Coil efficiency} \equiv \frac{\text{Ohmic losses in workpiece}}{\text{Total Losses}} = \eta_c \tag{47}$$

4. Simulating of coil efficiencies for single HI assemblages

In this section the calculation of coil efficiencies is undertaken for mains frequency HI schemes considering several inductor configurations for massive slabs and billets, these as a solid workpieces made out of an aluminium alloy having 2.32 X 10^7 [S m^{-1}] of electric conductivity and a thermal insulation gap of 0.001 [m] . Those scenarios have been modelled using the 2D Finite Element software QUICK FIELD 5.7 from TERA ANALYSIS LTD, which will be referred as QF from now on. All of the exercises were carried out in the high precision mode.

Table 1 shows parameters whose reflection coefficients at 60 Hz are the highest on each of the three schemes dealt so far. From there, corresponding dimensions as in 6th and 7th rows have been taken as an input for the QF software for obtaining the joule heat developed in inductor and workpiece hence the coil efficiency for a short range of frequencies around 60 Hz.

4.1 Multilayer inductor configurations

When dealing with mains frequency in HI systems, the increase of number of coil turns has to be taken into account so as to avoid high amperage in the excitation coil due to the low impedance driven by very low frequencies. The total current then is uniformly distributed

Excitation Magnetic Flux Application	Longitudinal	Transverse	Transverse
Inductor Configuration	Solenoidal	One Pair of Poles	Rectangular Cage
Shielding Parameter $R\omega$	15.5	7.75	3
Workpiece "Thickness"/ skin depth	5.55	3.75	2.45
Workpiece Geometry	Cylinder	Cylinder	Slab
Workpiece „Thickness"	0.0751 m	0.0531 m	0.0313 X 0.358 width
Workpiece length	0.36 m	0.2555 m	0.306 m
Inductor-workpiece Insulation gap	0.0121 m	0.0121 m	0.0121 m

Table 1. Parameters for an Optimal Electromagnetic Coupling at 60 Hz for Solid Aluminium

over the whole radial depth of the winding. Too many turns imply small conductors, hence, small water passages, giving the risk of clogging, when windings are made from hollow water-cooled copper. In addition, by using multiple layers of tubular coils, outer layers, those further away from the workpiece, heat inner layers as well as the workpiece, worsening that way coil efficiency. Therefore, multilayer does not allow the use of internally cooled conductors. Furthermore, multi-layers are suitable if the conductors are thin compared with the skin depth.

To overcome the above shortcomings, recourse is made for applying external cooling to the winding, (Harvey, 1977) where coolant is forced to circulate trough spacing between bunchs of individual conductors as follows:

a.   Axial cooling.- Radial gaps separate turns of each layer, whereby coolant flows over the interlayer room as it is directed axially through the winding. See Fig. 4 a).

b.   Edge cooling.- The coil comprises several stacks of tight layers forming a disc wrapped around and twisting itself into a spring in such a manner that a gap is allowed between each stack. Coolant is caused to flow in the gap, and radially guided in a sinuously fashion. That disc becomes a plate for the Slab /Cage-shaped inductor configuration. See Fig 4 b).

The two of the multilayer coil arrangements are simulated for the longitudinal flux case, e.g. solenoidal inductor configuration whereas the edge cooling arrangement is simulated just for the Transverse Flux cases, namely: Two-pair of poles and rectangular cage configurations. In all cases the pitch coil is neglected for simplification.

4.3 HI design data for quick field software input

Table 2 shows the dimension and number of turns that were fed to the QF runs for the configurations that were dealt so far. Current source was fixed to 150 [A] for all the cases. The section of the coil is 4.1 mm X 7 mm. The Inductor-Workpiece gap was considered the same for all the exercises. It is assumed from a practical point of view, a given constant thickness for covering thermal insulation issues is needed to withstand same levels of high

Inductor Type	Edge Cooling	Axial Cooling	One Pair of Poles	Rectangular Cage
Excitation Flux	Longitudinal	Longitudinal	Transverse	Transverse
Layers	30	16	30	30
Stack of Coils	26	49	11 each side	26 each side
Turns	780	784	330	780

Table 2. HI layout input parameters for QF simulations

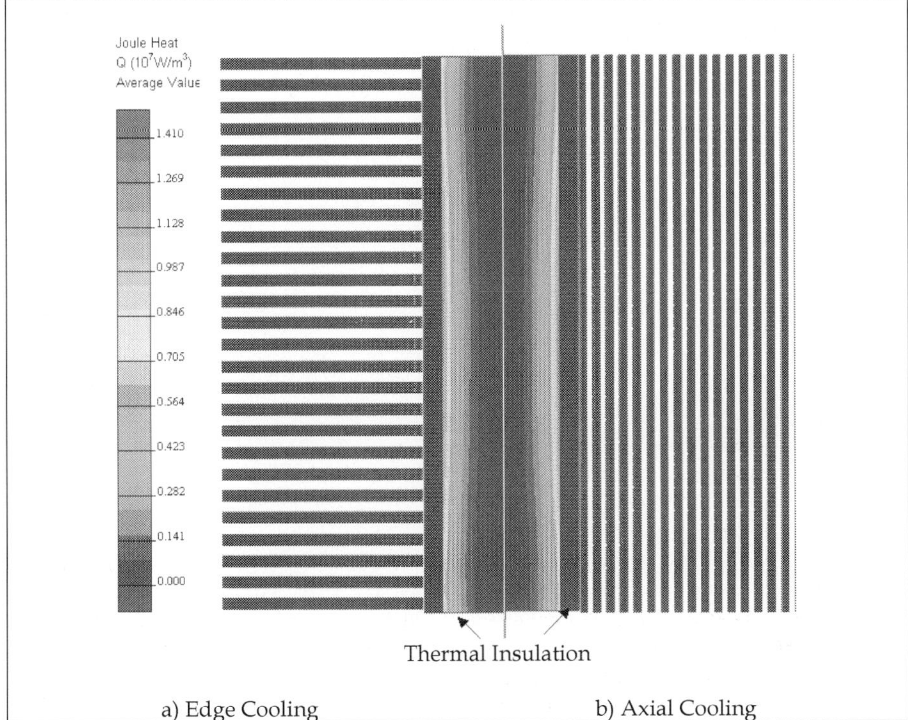

Fig. 4. Layout of Multilayered Coil excited by Longitudal Flux. Left hand Side: Half Section of an Edge Cooled Inductor. Right Hand Side: Half Section of an Axial Cooled Inductor

power density dissipation no matter the workpiece size, despite the smaller sizes as for the slab and the pole paired transverse flux cases, would call for a gap size proportional to corresponding workpiece sizes so as to have a fair comparison. The layout for each case can in Figs: 4 a), 4 b), 6 b) and 6 a) respectively.

4.4 Multilayered coil schemes powered by a longitudinal excitation flux
In Fig. 4 the Joule heat distribution in each half of the longitudinal section for the above cases have been put together as QF displayed results. The left-hand-side shows the edge cooled multilayered coil whereas the right hand side does axial cooled multilayered coil. Coolant impingement aims 90° when it is compared each other: Towards the gap between wounded discs for the former whereas towards the interlayer gap for the later. Both coils are externally cooled.

Fig. 5 shows the electromagnetic coupling parameter for two solenoidal multilayered schemes both with almost same number of coils, eg, 784 for axial cooling configuration and 780 for edge cooling configuration. The former shows slightly better coil efficiency than later and both of them matches the locus where the maximum reflection coefficient is predicted to occur.

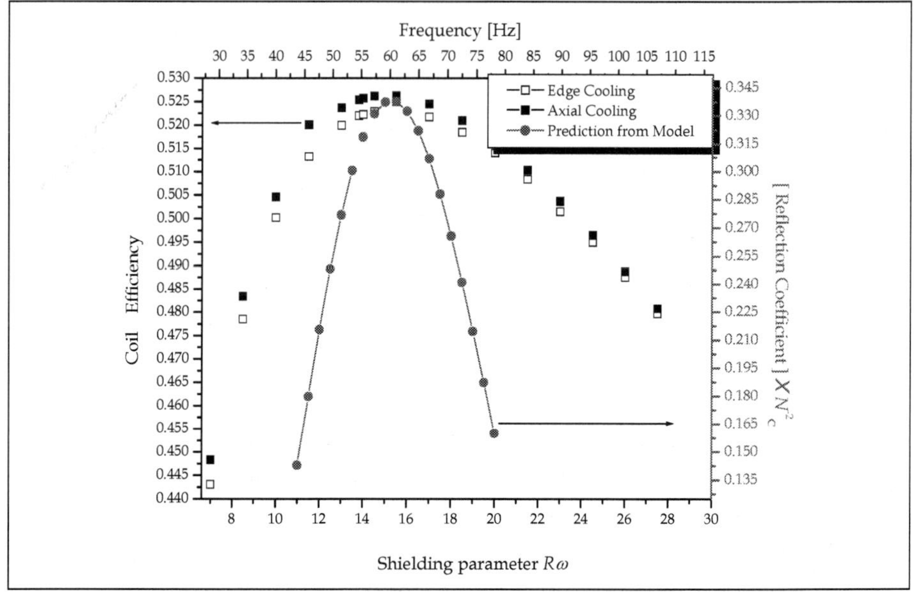

Fig. 5. Electromagnetic Coupling Parameter Performance in Longitudinal Flux Schemes

4.5 Multilayered coil scheme powered by one one–pair-of poles excitation transverse flux

In Fig. 6 b) shows the transverse section of a cylinder being heated by means of an inductor wounded in such a manner that it looks shaped as a star from a top view. Current is directed towards inside the plane in the top half whereas current is going outside the plane in the bottom half and it goes alternatively according to frequency. The coolant is caused to flow along the edges of the branches of the star. Broad spaces between branches make easier the cooling that is carried out externally on the inductor. Straight coils of each half of the inductor might be connected each other at the ends with radial coils in a progressive "rotating X" pattern for the blind side and for the opposite side, where the workpiece might be inserted, circular coils as in windings at the ends of an induction motor stator.

The cylinder radius was chosen 0.052 m as the minimum size for maximum reflection coefficient for a one –pair- of poles transverse excitation flux. Both coil efficiency and one pole pair flux configuration reflection coefficient curves develop the same slope at lower shielding parameters. Nevertheless the maximum coil efficiency and reflection coefficient occur at shielding parameter $R\omega = 10$, but now for the two-pair of poles configuration. By looking the regions at both sides, it can be seen that room for an extra pair of poles can be accommodated in that place. Thereby, the reflection coefficient modeled for two-pair of poles configuration can be approached to the presented example by appealing geometry considerations. As a drawback of this configuration, concentrated Joule heat dissipation may be located near to the centres of each winding, leaving a "cold spot" at right angles from them, which are located at the left-hand side and at the right-hand side of the Fig 6 b). In order to spread the released power, a rotating mechanism fixed to one end of the billet would enable to expose all the surface of a solid workpiece to a uniform excitation transverse flux. Another alternative is to fix an inductor similar to a stator of a one-pair-of poles tri-phase induction motor for setting a travelling transverse excitation flux, so as to avoid the formation of uneven-heated regions in the workpiece.

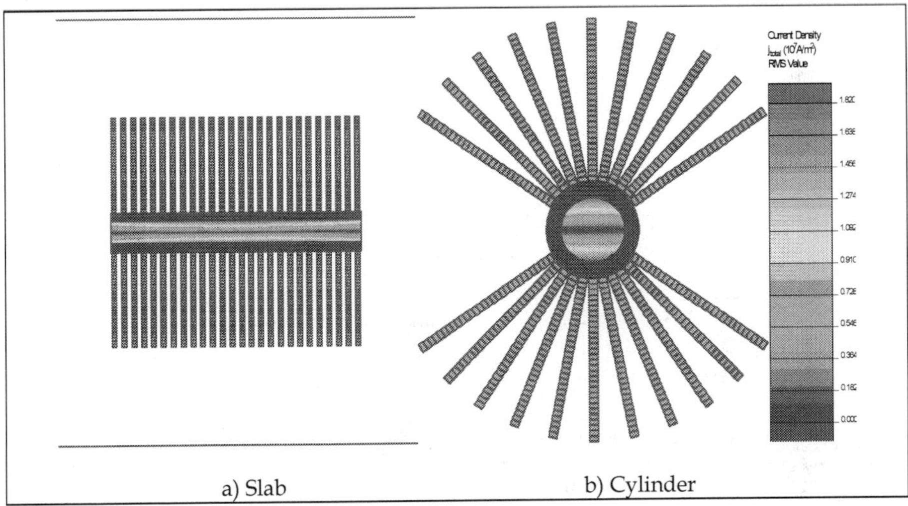

Fig. 6. Top View of Layouts of Multilayered Coil excited by Transverse Flux. Left hand Side: Rectangular Cage Inductor. Right Hand Side: Star Inductor.

4.6 Multilayered rectangular coil scheme powered by a transverse excitation fFlux

Fig. 6 a) shows the transverse section of a solid slab being heated by means of an inductor where plates of piled coils are separated to allow room for the coolant. Since this configuration is being excited by a transverse flux, current circulates aiming inside the plane in the top half whereas current leaves outside the plane in the bottom half in an alternated fashion according to frequency. Fig. 8 shows that the maximum coil efficiency is achieved at $R\omega = 3$ which corresponds to a slab thickness to skin depth ratio of 2.45, see Eq. (12), which matches the well-known practical rule for heating inductively slabs. Both coil efficiency and reflection coefficient start to decay beyond $R\omega = 3$. The former decays slowly whereas the later decays abruptly.

5. Simulating of coil efficiencies for HI susceptor assemblies

The suceptor is generally a relatively good electrically conductive insert possessing high melting point, which is assembled together with the material to be heated, so as to the later is heated indirectly by the following heat transfer modes: conduction and radiation. This is the main use for susceptors. It is required due to the poor electromagnetic coupling of electrically low conductors, unless high frequency is used. The aim here is to improve energy conversion efficiency. With the same purpose, susceptors are also used for the heating of very-good conductors such as aluminium and copper, since it is well-known that the efficiency in an induction heating device is improved as the overall electric conductivity of the load decreases between certain limits. In this section, molten aluminium at 700°C, around 7.88 times more resistive than the previous solid aluminium, is taken for showing the second application involving susceptors. The worked example and results of K. Reichert (Reichert, 1965) are referred. There, a coreless induction furnace using a conductive crucible made out of graphite as a susceptor, is analysed for melting aluminium. Despite graphite is not that good conductor as molten aluminium is, the more resistive the load the more is enhanced the electromagnetic coupling. In Reichert's paper an analytical approach using Bessel functions is presented and confronted to field data measured in a real device

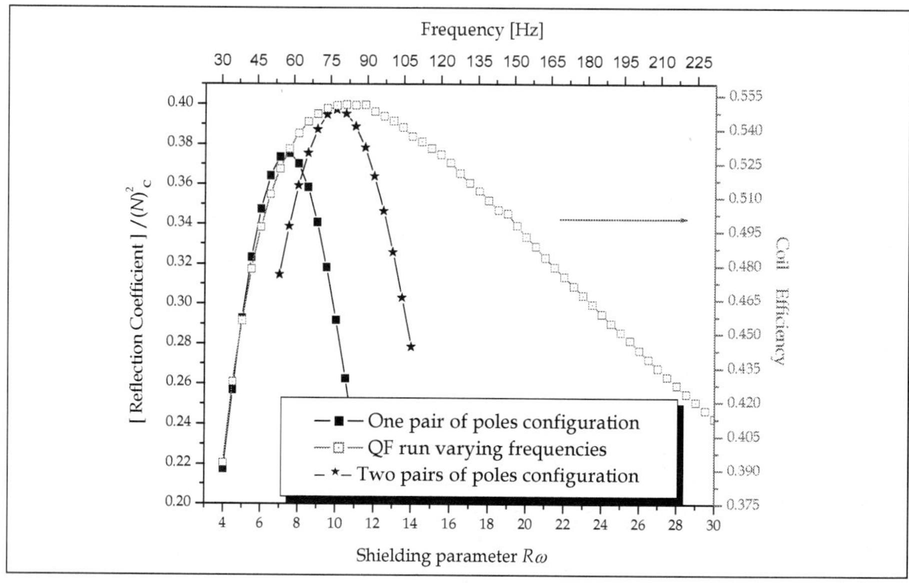

Fig. 7. Electromagnetic Coupling Parameter Performance in a Pole Paired Transverse Flux Scheme

In the following, besides of comparing his results on the coreless furnace features as far coil efficiency is concerned, further QF exercises are set up according to the various inductor configurations previously analyzed, now as a monolayer winding, and then corresponding coil efficiencies compared.

5.2 Reichert's assembling layout.

Table 3 shows the lay out and the electric conductivities of each component of the Reichert's assembling. 80 % of the total radius of the induced load corresponds to the molten pool and the remainder to the crucible thickness. Magnetic properties for magnetic yokes are not displayed in the worked example, therefore a typical non-linear magnetic permeability chart as shown in Fig. 9, is considered as corresponding QF input data. Also their electric conductivity was taken as null, in order to consider the insert/molten pool as only constituents of the induced resistive load in the QF simulation. Figs.10 and 11 show sketches for the layout of the Table 3. Details on coil cooling requirements are supposed to be already covered in Reichert's design, thus they are not discussed in this section.

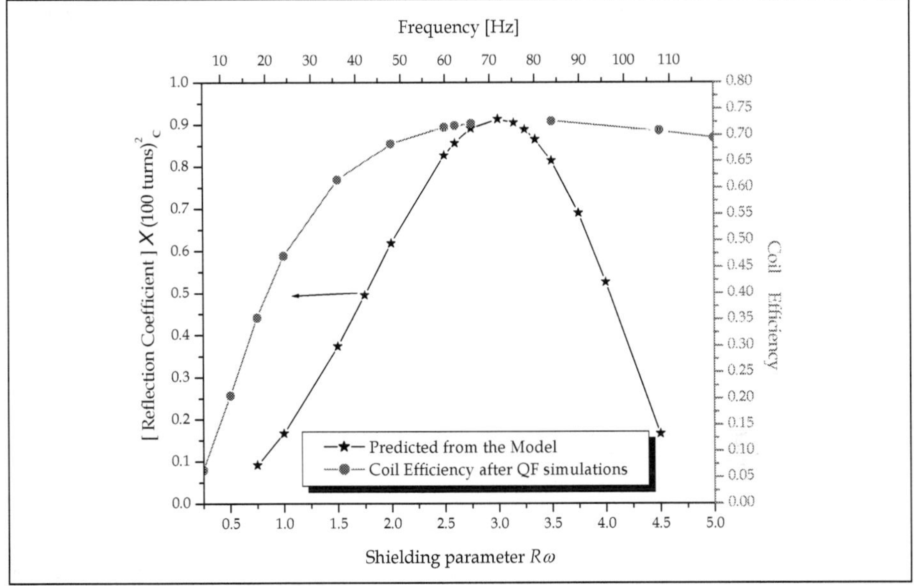

Fig. 8. Electromagnetic Coupling Parameter Performance in a Slab being excited by a Magnetic Transverse Flux.

Heigth = 0.51 m	Outer Radius [m]	Thickness [m]	Electric Conductivity [S m⁻¹]
Molten Aluminium	0.34	-	2 941 176
Graphite Crucible	0.425	0.085	166 666
Insulation Layer	0.475	0.05	0
Inductor Copper	0.492	0.017	55 556 000
Magnetic Shunts	0.52	0.028	0

Table 3. Lay out and electric conductivities for Reichert's assembling

5.3 Reichert's assembling coil efficiency featuring

When it comes to melting of metals at mains frequency, a crucible made out of a dense refractory has to contain the molten metal to stand mechanical erosion due to the strong stirring, which is driven by the induced Lorenz force. As a shortcoming, dense refractory has high thermal conductivity, which transfers heat to the excitation coil. Then a high porosity refractory is placed as thermal insulation layer in between to protect the coil copper and save energy. The addition of non-conductive regions to the gap between inductor and workpiece pays a penalty to the coil efficiency.

Table 4 shows the frequency and voltage supply, as being the input for both the worked example and QF exercises. As a result, coil efficiencies and fractions of the heat released to the total power in each component are also shown in the same Table. In the last row of Table 4, the basic features of a HI system where molten aluminium is performing only as a

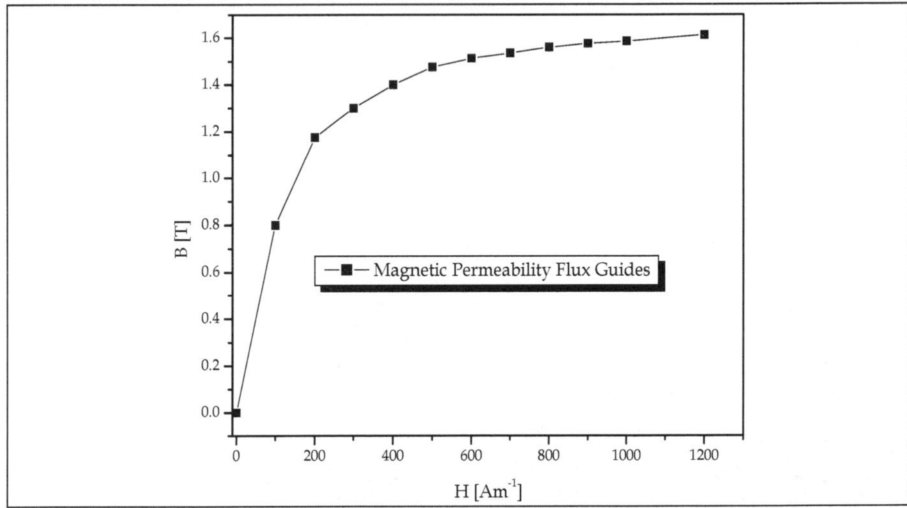

Fig. 9. B vs H behaviour for the Magnetic Yokes in the Graphite-Aluminium assemblages

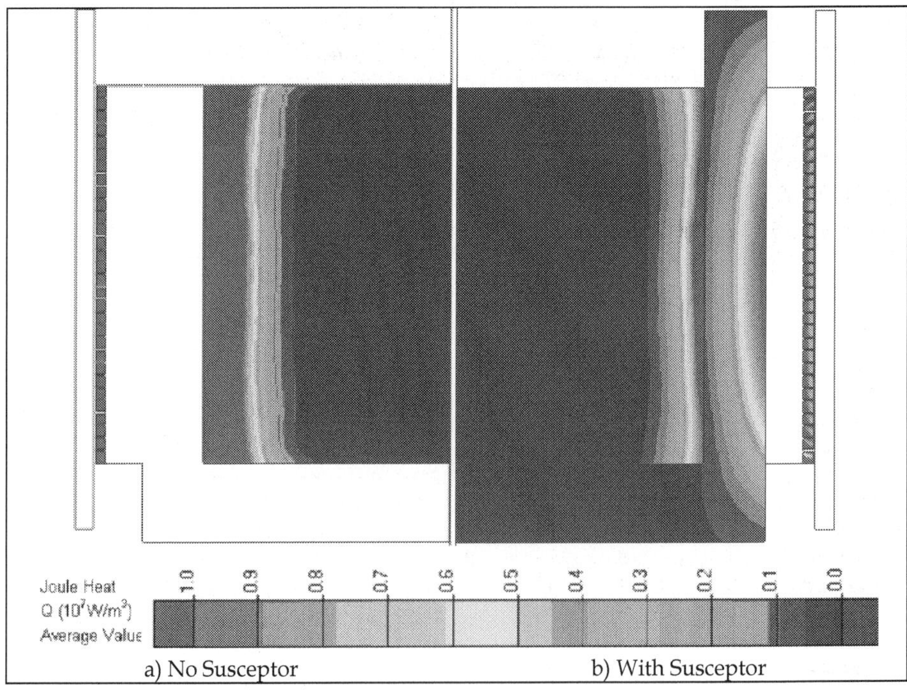

Fig. 10. Layout for a coreless induction furnace: a) Only Molten Aluminium as the workpiece b) Graphite Crucible and Molten Aluminium.

Supply: 380 V, 50 Hz Coil turns: 30	Current [A]	Rated Power [kW]	Susceptor Power Fraction	Molten Pool Power Fraction	Coil Efficiency
Field Data	1 640	134	-	-	-
Reichert's Model	1 690	125	0.39	0.33	0.72
QF run	1 670	137.73	0.4442	0.3406	0.7848
QF run*(Input: 391 V)	1 640	84.07	-	0.657	0.657

(*)Non-conductive Crucible

Table 4. Power Distribution in Reichert's Assemblage

workpiece is displayed. Here it is supposed that a dense crucible that contains molten aluminium is a non-electric conductor, thus the thickness of both the crucible and the insulation layer becomes an enlarged inductor-workpiece gap, which lowers coil efficiency. This layout is depicted in the left-hand-side of the Fig. 10. The corresponding Shielding parameter $R\omega$ is 134.23.

Coil efficiencies for $R\omega$ between 125 and 190 are obtained by means of QF runs, by taking into account same Inductor-Workpiece arrangement but varying frequency between 50 and 70 Hz. Those efficiencies are shown at the right-hand-side ordinate in Fig 3. Although the maximum difference in coil efficiencies displayed is 0.003, they resemble the rise and fall of the corresponding reflection coefficient curve. Both curves are bell-shaped and share approximately the same width. On the other hand, the peak of the reflection coefficient for the Fig.3 is 0.59427 whereas in Fig 2 is 0.337, hence the $R\omega$ region for the former might lead to a given HI layouts seemingly more robust as far as coil efficiency is concerned.

By fixing the same current as measured in the real device, 1640 [A] for the QF runs, the results show that both throughput and coil efficiency are the lowest for the molten aluminium by itself, as can be seen in last row of the Table 4. Once a crucible made out of dense graphite is placed instead of a non-conductive one, the efficiency (+19.45%) and the throughput of the assembling as well increase, according to rows 3 and 4 of Table 4. Deviations of QF results with respect to the field data of the Reichert´s assembling are: 2.78 % on rated power and 1.83 % on electric current which are close fair enough.

6. Proposals for new schemes of HI susceptor assemblies for improving coil efficiency

So far, this chapter has dealt with the finding of optimal configurations of HI systems regarding: size and shapes of inductors, application of either longitudinal or transverse excitation flux when appropriate and the use of susceptors in coreless furnaces (solenoidal scheme). All of these while keeping the lowest operational frequency, so as to avoid frequency converters or at least lowering eddy currents losses in the ancillary equipment.

In this section, combinations of the above schemes are analyzed, taking into account same materials, same frequency, similar both size and operational features, e.g. voltage and current as in the previous susceptor analysis.

6.1 Rectangular susceptor assembly excited by a transverse flux

In order to emulate the heating process that undergoes a slab under the effect of an excitation transverse flux on the previous HI susceptor schemes, a case is simulated below with the QF software, considering molten aluminium contained in a box made out of dense graphite which in turn is surrounded by straight coils connected in series, resembling a rectangular cage. This particular coil design is known as "Ross "coil. The graphite box might be shifted out from one side of the monolayer coil cage for being either fed with raw materials or tilted for pouring out the melt. The thicknesses for the graphite walls and the back up insulation layer are the same as for the Reichert's assembly. The top view of the Ohmic losses spreading for this device is included in Fig 11 b). Coil efficiency and power distribution are shown in Table 5. Its performance looks better than the solenoidal assembly. Generally speaking, straight coils are easier to shape than circular ones. As a matter of fact, the real graphite "cylindrical" crucible as referred in Reichert's paper was actually an octagon 0.08 m thick, 0.51 m tall.

6.2 Cylindrical susceptor assembly excited by one-pair-of- poles transverse flux

Here the induced cylindrical load, which is comprised the dense graphite crucible, the inner molten pool, together with the outer insulation layer, have same dimensions and conductivities as in the Reichert's layout. The only thing that changes is the winding. Now this time straight coils run outside insulation layer parallel to the cylinder axis in each half of the cylinder, starting approximately at 90° arc apart each other so as to minimize the

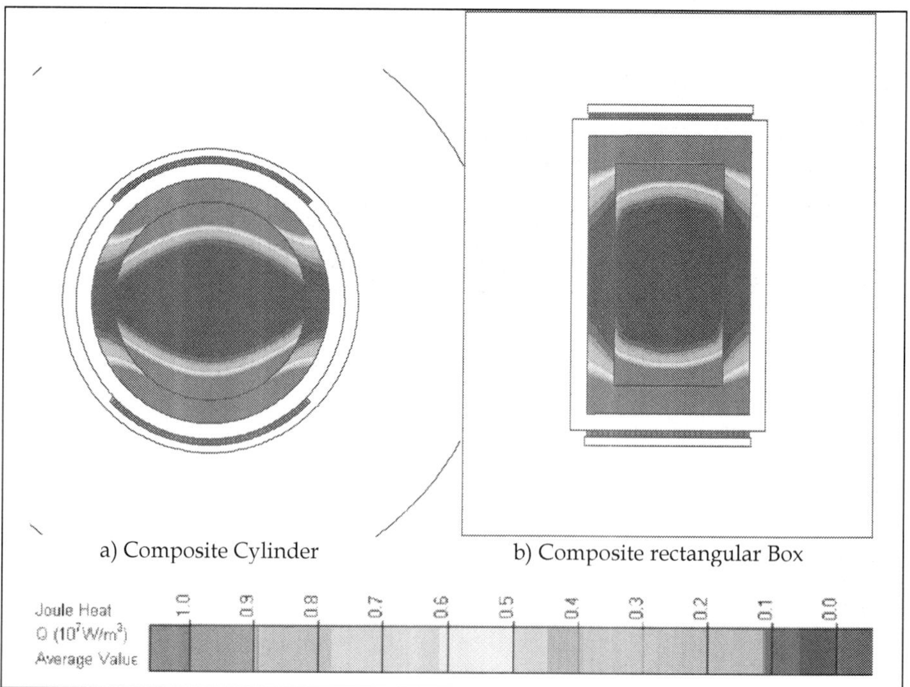

Fig. 11. Top view of the QF simulations for the Joule Heat Distribution in a Composite Load being excited by a Transverse Flux

proximity effect between opposite windings. See the sketch in Fig 11 a). The ends of this single winding would connect as depicted in Fig. 12 where just one half of the straight winding is presented for purposes of clarity. Coil efficiency and power distribution are shown in Table 5. It seems this configuration is the most promising among those analysed in section 6. The heavy Joule heat dissipation located near to centres of each winding, as shown in Fig 11 a) would fade due to the vigorous stirring that occur in melting of metals at mains frequency.

Frequency: 50 Hz Height: 0.51 m	Current [A]	Rated Power [kW]	Susceptor Fraction	Molten Pool Fraction	Coil Efficiency
QF run (Reichert's Scheme] Supply: 380 V, Coil turns: 30	1 670	137.7	0.4442	0.3406	0.7848
Susceptor: Box Supply: 292 V Straight Coils: Top 30 /30 Bottom	1 652	66.96	0.537	0.3222	0.8592
One pair of poles Susceptor: Cylinder Supply: 274 V Straight Coils: Top 45 /45 Bottom	1 642	60.83	0.5935	0.3960	0.8842

Table 5. Power Distributions in New Schemes of HI Susceptor Assemblies.

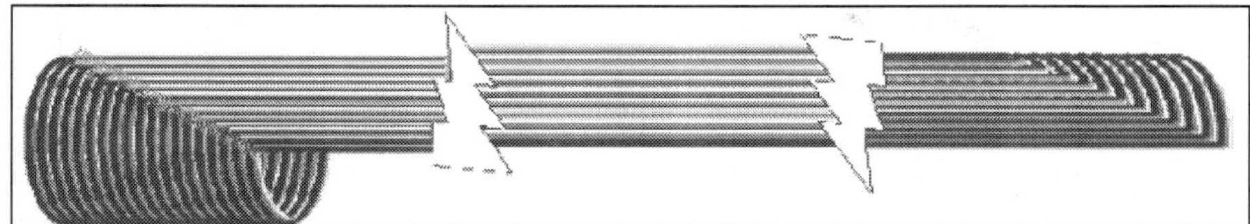

Fig. 12. Proposal for Connecting Coils at the Ends of the Device depicted in Fig. 11 a)

7. Conclusion

A mathematical model for obtaining a reflection coefficient for the workpiece contribution to the impedance of a given HI system was given. This is intended to asses how workpiece is linked to the excitation loop thus in order to find out conditions where the eddy current occurrence is enhanced, thus proving empirical rules already know for HI practitioners and to extrapolate those findings to design new possibilities. Quick Field Element Finite Sofwtare runs featuring coil efficiency of several HI systems were carried out so as to demonstrate the validity of the proposed approach. Calculated reflection coefficient shown rapid decay as long as the workpiece diameter to skin depth ratio goes beyond the optimum when comparing with coil efficiency. For simplification purposes, eddy currents process occurring inside the inductor itself have been neglected in this analytical approach. In the future, taking into account eddy currents process occurring inside the inductor might predict the above slow decay in coil efficiency among others refinements to take. Nevertheless, this simplification enabled explanations of current practices in HI systems as well to set a generic guide for exploring new alternatives to those already known

8. References

Carrillo, E. (2008). Optimization of Electromagnetic Coupling in Induction Heating Systems Excited by Solenoidal and Axial Windings, *International Journal of Applied Electromagnetics and Mechanics*, 28, (3), 2008, pp. 395-412.

Krawczyk, A & Tegopoulos J. (1993). *Numerical Modeling of Eddy Currents*, Clarendon Press, Oxford, pp. 5, 6, 27.

Constantinides, A., (1987). *Applied Numerical Methods with Personal Computers*, McGraw-Hill, pp. 406–412, 418–439.

Black M., Cook R.., Loveless D., & Rudnev V. (2003) *Handbook of Induction Heating*, Marcel Dekker, New York, pp. 467, 527–528, 536–538, 599.

Davies, J. & Simpson, P. (1979). *Induction Heating Handbook*, McGraw-hill Book Company (UK) Limited, pp. 146,321,327.

Harvey, I.G. (1977). A method of improving the energy transfer in induction heating processes and its application in a 1 MW billet heater, In: *IEE Conference Publication, 149*, Electricity for Materials Processing and Conservation, 1977, pp. 16–20.

Reichert, K. (1965). Die berechnung von kernelosen inductiontiegelöfen mit elektrisch eitenden Tiegel (zweischttiegelofen), *Archiv für Elektrotechnik*, Vol. 6, (1965) pp. 376-397.

8

Criterions for Selection of Volume Induction Heating Parameters

Niedbała Ryszard and Wesołowski Marcin
Warsaw Technical University
Warsaw, Poland

1. Introduction

Induction heating, with regard to a great number of applications in material processing, can be divided into two domains: surface and volumetric heating. In the first case, the criteria for parameters selection in induction heating installations are determined for skin depth and process time. In the case of volumetric heating, additional parameters, such as electrothermal efficiency and power control ability, must be taken into account. These requirements in connection with large variances of major electrical and thermal parameters of workpieces make the method multi-parameter and have a strong influence on high-frequency power sources. Without the knowledge of physical phenomenon and quantitative influence on temperature distribution it is unfulfillable to prepare a set of input data for effective modeling of the technological processes and directions for optimal selection of power sources.

Therefore the volumetric induction heating issues are a very complicated discipline that requires using of specialized calculating procedures for optimal selection of power sources for realizing an established technological processes. In this chapter some methods for modeling, designing and power controlling in volumetric induction heating systems were discussed.

In the chapter, the most popular methods for accurate power control in induction heating systems were analyzed in the case of nonlinearity material properties. Some examples were shown and classical approach for power control in the systems was compared to pulse width modulation case. The advantages and disadvantages of proposed construction and process solutions were discussed.

2. Volumetric heating – the basics

Volumetric induction heating is a very popular non-contact technology that has very wide applications in material processing. Operating frequencies are depended on the technology used. The spectrum ranges from low frequencies (50 Hz for heating a massive details), through the middle range (50 kHz for rafination and recasting processes) to high values (250 kHz and more for levitation melting etc.). Describing of all technologies as an universal problem is not possible, so in this chapter only selected topics were presented.

Induction heating modeling and simulation is a very popular domain, earlier by plurality of mathematical field and circuit models. Nowadays, there are numerous of numerical systems that enable the simulation of complex coupled-physics electromagnetic and thermal problems. However, calculating systems usage often does not lead to practical applications solving, but to working out computational exercises. Analytical descriptions which used to be helpful in getting to know and making appraisals of electrothermal effects are being forgotten nowadays. Moreover, in the latest specialist literature concerning the issue of radio frequency induction heating, the electromagnetic part seems to be preferable. But it is worth to remember that basic knowledge on heating processes is essential to create input data that enable us to reaching our established targets quickly. An attempt to indicate the possibility of acquiring the first numerical approximation to the planned field decomposition has been undertaken on the example of inductive volumetrically heated charges. Not only material and geometrical properties of the charge but also extortive values- continually modified in the process of heating have the major impact on the way it is shaped. Therefore, cooperation of the source of power and inductive heating system is extremely important for the final result. The influence of dense electrical circuit parameters on the temperature field in the charge and vice versa is high enough to be taken under consideration. Moreover, simple relations arising from analytical models may determine the basis for an estimation of the quality of numerical calculations. Validated solutions to analytical equations may not only constitute the first approximation, but also explain physical basis of electrothermal effects in inductive heating much better.

When heated material is ferrous, material parameters are significantly important. Its main characteristics such as resistance and inductivity change during heating. A knowledge of the character of load changes enables to choose the right way of control. Defining the range of changes, including power (magnetic field strength on the surface of the charge) limits resonance frequency adjustment of the charge in magnetic state compared to the other states, including the final- nonmagnetic. It allows working out some more safe power adjustment algorithms in the frequency ranges which are being adapted.

Articles, in which electrothermal devices powered by real sources are being simulated, come out in the specialist literature exceptionally rarely. Attempts to verify results received are made even less often. It seems to be significantly important, when electric energy is being converted into heat in strongly nonlinear materials. Many comparisons show that simplifications assumed by many authors cannot provide useful results. Supplying of induction heating systems from high frequency energy sources is not optimally exploited. Generally energy sources are produced as current inverters. However, transistors frequency converters with voltage inverters can be recognized as very efficiency sources because of little self-losses. Unfortunately, nonlinearity of workpiece parameters can reduce efficiency of voltage sources by introducing additional inductances in high-frequency circuit.

2.1 Temperature distribution in heated bodies

In many cases the main goal of induction heating is to realize the heating processes with uniform temperature distribution within the heated body. This requirement is often hard to satisfy because of necessity to providing the maximum electrical efficiency of induction heating systems. In practice, the efficiency is increased by reducing process time and uniform energy consumption. Other important factors include maximum production rate, environmental friendliness and providing compact systems.

To display the temperature distribution within inductively heated body, a several analysis of long cylindrical bars were realized. The results were classified and carried to the average temperature of charge at the end of heating process. This average temperature was calculated according to equation:

$$t_{\acute{s}r} = \frac{\sum\limits_{n} V_n \cdot (t_n - t_0)}{V_C} + t_0 \tag{1}$$

Where: t_n - temperature of elementary volume V_n, V_c- total volume of charge, starting temperature, elementary thermal capacitance of charge assumed as constant.

All of the temperature deviations varying from the average value $\Delta t_n = t_n - t_{\acute{s}r}$ are the same for charges of the same thermal efficiency $\eta_c = const$ surrounded by a time-variable electromagnetic field that has the same relative frequencies. Relative frequencies were described as a proportion of a charge specific dimension (r_2) to the skin depth (δ_2): $\xi = r_2/\delta_2$. Another important factor, the elementary surface power p_F, which has a strong influence on the temperature uniformity, has been temporary neglected. According to the assumption which has been made earlier that the relation with the same radius of a charge is constant $(p_F \cdot r_2 = const)$ (Rapport E. at al, 2006; Sajdak C. at al, 1985) proves that the omission of this parameter from designing has not influenced the final results. To show the results calculated for temperature distribution in inductively heated steel charge $p_F \cdot r_2 = 10\,\text{W/m}$, some deviations from the average temperature $t_{\acute{s}r}=1100°C$ carried to the relative radius $rw=r/r_2$ were shown in figures 1 and 2. It has been proved that regardless of the heat loses (fig. 1) or of the heat sources distribution (fig. 2), all temperature curves were intersected into a common point located in the center of the charge volume (mass).

Results presented provides to modify the criterial temperature equation that applies in the case of heat conduction with arbitrary located heat sources (Zgraja J. at al, 2003).

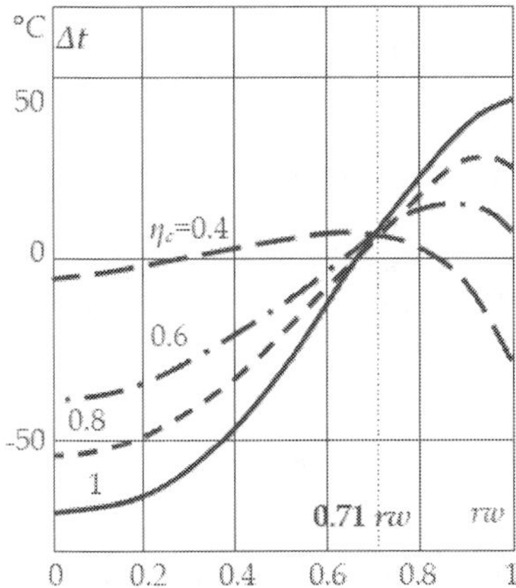

Fig. 1. Temperature divergences within inductively heated charges by frequencies of ξ=3, for different heat efficiencies

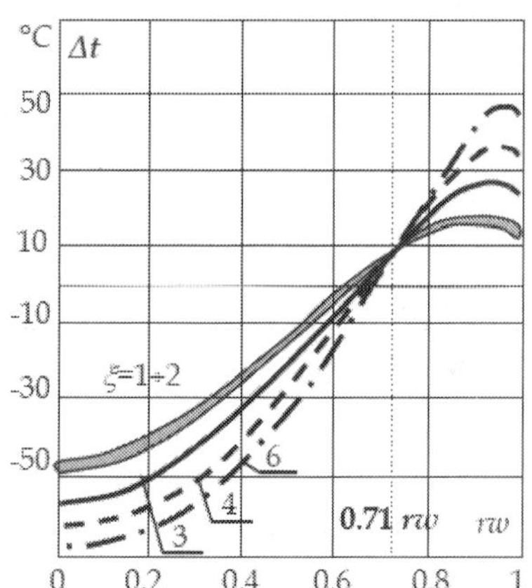

Fig. 2. Temperature divergences within inductively heated charges by different frequencies for constant heat efficiency η_c=0,8

$$K(\xi,\eta_c) = \frac{2 \cdot \lambda \cdot \Delta t}{p_F \cdot r_2} = \eta_c \cdot \left[1 - \frac{2.2}{\xi \cdot \sqrt{2} \cdot (\eta_c + 1)} \cdot \frac{B(\xi) - 1}{A(\xi)} \right] \tag{2}$$

Where:

λ - thermal conductivity, $B(\xi) = ber^2(\xi) + bei^2(\xi)$, $A(\xi) = ber(\xi) \cdot ber'(\xi) + bei(\xi) \cdot bei'(\xi)$.

In the equation presented above momentary heat efficiency (as the proportion of stored power and dissipate power of charge) and total heat efficiency were determined as a linear approximation in time domain, calculated as $0.5 \cdot (\eta_c + 1)$.

On the other hand, maximal temperature divergences Δt_{max} in a cross-section of the charge were calculated. The divergences refer to $\Delta t = (p_F \cdot r_2)/(2 \cdot \lambda)$ that enable us to calculate the criterial temperature $K_{\xi\eta c}$ based on field analysis. The comparison of numerical and analytical results has been presented in the figure 3. The convergence between calculating results was very high for both cases of all frequency ranges (fig. 3a) and heat efficiency (fig 3b) practical spectrum. All results were compared within the same, constant material properties. This simplification does not result in significant errors at the end of volumetric heating process.

Additional verification of K coefficient has been accomplished by determining maximal temperature differences (calculated by using equation no 2) and using forward calculations shown in figures 1 and 2.

The results has been presented in figure 4. The convergence between analytical and numerical calculation results have reached a high value. Main difference was that the extreme value of the temperature was not located in the external surface of the charge.

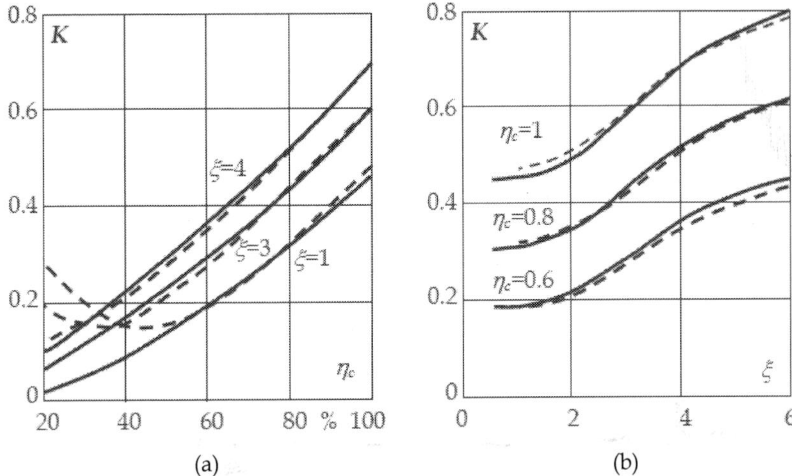

(a)   (b)

Fig. 3. Criterial temperature values from analytical (solid lines) and numerical (dash lines) calculations as a function of heat efficiency η_c (a) and relative coordinate ξ (b)

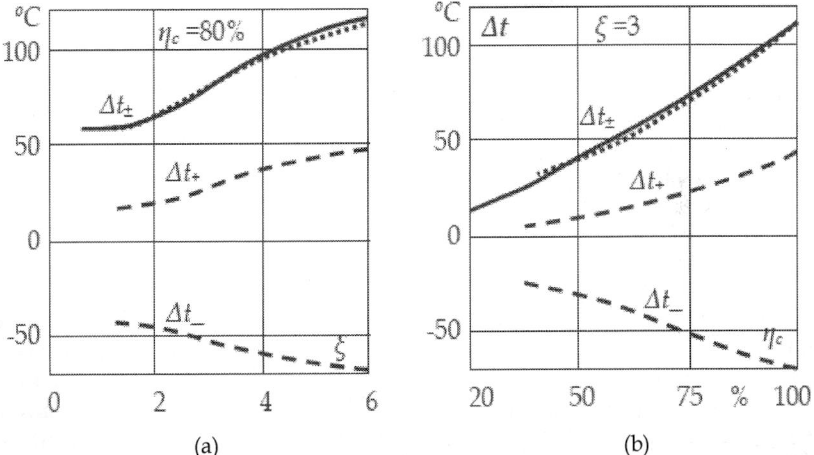

(a)   (b)

Fig. 4. Positive Δt_+, negative Δt_- and summary $\Delta t_\pm$ temperature differences in cross-section from numerical (dash lines) and analytical (solid lines) analysis as a functions of heat efficiency η_c (a) and relative coordinate ξ (b)

2.2 Time and energy in induction heating process

The design criteria for induction mass heating systems which seems to be the requirement for temperature uniformity of the charges is only one of the goals. Another important factor includes minimum process time and energy consumption. Heating time (fig. 5) is determined by the temperature differences in charges cross-section, frequency and heat

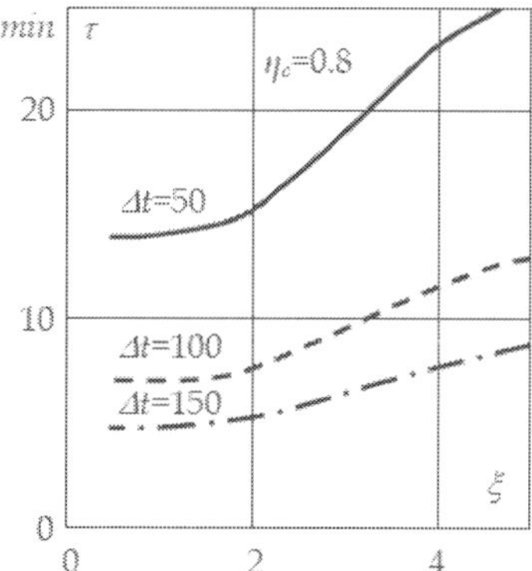

Fig. 5. Exemplary heating times calculated for workpiece of outer radius r_2=0,05 as a function of ξ, for different temperature uniformities

efficiency. Additional, this parameter is a function of radius of workpieces and can be determined by the following equation:

$$\tau(r_2,\Delta t,\xi,\eta_c) = \frac{q_a \cdot r_2^2 \cdot K(\xi,\eta_c)}{2 \cdot \Delta t \cdot \lambda \cdot (\eta_c +1)} \tag{3}$$

Where: $q_a = \gamma_{śr} \cdot c_{śr} \cdot (t_{śr} - t_0)$ - elementary heat capacity of charge
The elementary electrical energy consumption is determined only by the efficiency of heating system, without taking into consideration the temperature uniformity requirement:

$$e(\xi,\eta_c,m,r) = \frac{2 \cdot c_{śr}}{(\eta_c +1) \cdot \eta_e(\xi,m,r)} \tag{4}$$

Where: $c_{śr}$ - average specific heat, $m = \sqrt{\dfrac{\rho_2 \cdot \mu_2}{\rho_1 \cdot \mu_1}}$, $r = \dfrac{r_1}{r_2}$ - proportion of material properties
and radiuses of inductor and workpiece.
The energy consumption rate for steel heating has been presented in figure 6. The model of long cylindrical bar of the radius r=1,4 was used.
The criteria for heating time and energy consumption minimizing cannot be satisfied by optimization of process frequency. Optimal frequency range for exact process has to be described using criterion criteria of minimal time which is limited by the temperature uniformity and type of power source.

2.3 Circuit parameters of induction heating system
Due to the assumption that value of relative coordinate $\xi=r_2/\delta_2$ is dependent on the frequency, the only independent property from the workpiece geometry is absolute heating power. The maximum value of the power can be determined by equation 5.

$$Pc(\Delta t,\xi,\eta_c,m,r) = \frac{4 \cdot \pi \cdot \lambda \cdot \Delta t}{K(\xi,\eta_c) \cdot \eta e(\xi_2,m,r)} \tag{5}$$

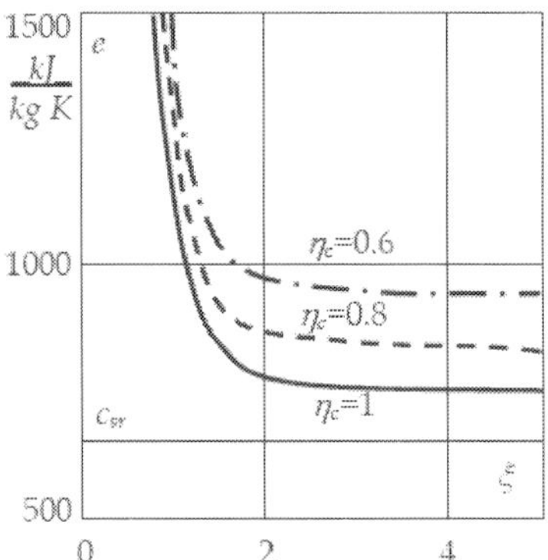

Fig. 6. Elementary electric energy consumption as a function of ξ, for different heat efficiencies

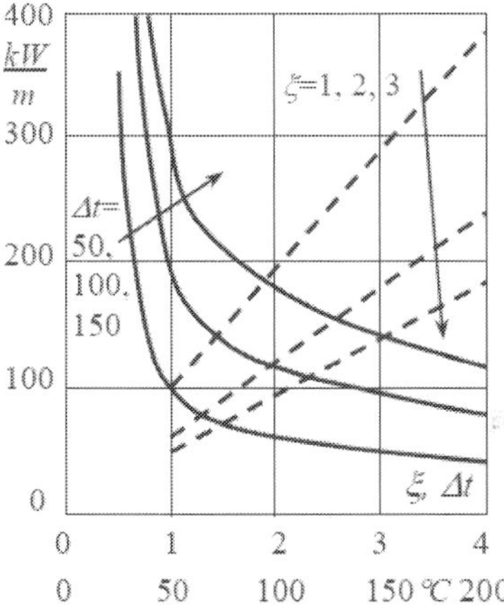

Fig. 7. Maximal power values of induction heating system for the case of heating the steel cylindrical bar (r=1.4) as a function of ξ i Δt.

Power values have been shown in the figure 7. The values can be used to determine the total source power or current linkage $I \cdot n = H_0$.

$$H_0(\Delta t, \xi, \eta_c) = \sqrt{\frac{2 \cdot \lambda \cdot \Delta t}{\rho_2 \cdot \xi \cdot \Phi r(\xi) \cdot K(\xi, \eta_c)}} \qquad (6)$$

Current linkage or magnetic field intensity, depending on the power source type, are highly variable by the heating time. These variables should be corrected basing on momentary value of system impedance, especially in induction mass heating systems, where the variable of system parameters can reach a high values. The range of impedance in steady state can be determined by using the method of equivalent resistances, which are independent from workpiece geometry:

$$R_2(t,H) = 2 \cdot \pi \cdot \rho_2(t) \cdot \xi(t,H) \cdot \varPhi r(t,H)$$

$$X_2(t,H) = R_2(t,H) \cdot \frac{\varPhi x(t,H)}{\varPhi r(t,H)} \tag{7}$$

Equivalent resistances, supplied by workpiece resistance and inductive leakage reactance (8) can be used for determination of impedance variance from material properties (9)

$$R_1 = \frac{R_2(t,H) \cdot r}{\varPhi r(t,H)} \sqrt{\frac{\rho_1}{\rho_2(t) \cdot \mu_2(H)}} \qquad X_0 = R_2(t,H) \cdot \frac{\xi(t,H)}{\varPhi r(t,H)} \cdot (r^2 - 1) \tag{8}$$

$$Z(t,H) = R_2(t,H) \sqrt{\left[(1 + \frac{r}{\varPhi r(t,H)} \sqrt{\frac{\rho_1}{\rho_2(t) \cdot \mu_2(H)}} \right]^2 + \left[\frac{\varPhi x(t,H) + \xi(t,H)(r^2 - 1)}{\varPhi r(t,H)} \right]^2} \tag{9}$$

Impedance shown above or its resistance and reactance enable for determinate the induction system parameters. Power values in specific time intervals of heating process for the cases of current source (10a), power source (10b) and voltage source (10c)

$$Pc(t,H) = (I \cdot n)^2 \cdot \frac{R_2(t,H)}{\eta_e(t,H)} \tag{10a}$$

$$Pc(t,H) = Pc \tag{10b}$$

$$Pc(t,H) = \left(\frac{U}{n} \right)^2 \cdot \frac{\cos \varphi(t,H)}{Z(t,H)} \tag{10c}$$

3. Ferrous bodies heating

Material properties have a strong influence on induction heating process, especially in cases of ferrous materials. Resistance and reactance of charges are strongly variable in heating time. The parameters range should be taken into consideration for proper selection of power source and control systems. Determination of the range of resistance and reactance variances and also minimal value of momentary power (magnetic field intensity in the workpiece surface) reduce the range of resonant frequency at the beginning of process (in magnetic state) from other states (non- magnetic at the end of heating process). Determination of the power (from equation 10) values for different states, for example for low temperature range, for Curie point and at the end of the process, can be used for planning the heating process with extremely electromagnetic and thermal values. Such calculations enables users to design safety algorithms for precisely power controlling in frequency domain.

Analysis of energetic relations in induction heating system was based on example of the system supplied from voltage source. Let us assume the workpiece of outer radius r_2=0,05 m, placed in high-intensity electromagnetic field. The air gap between workpiece and inductor was extremely short (r=1,1) to underline variable of workpiece parameters influence on the instantaneous absorbed power. The heating process was very dynamic, such in the figure 8. The marked area in this figure was used to illustrate the time range of two states (magnetic and non- magnetic).

In the figure 9 the results of static characteristics comparison was shown. Except from low temperatures and Curie temperature, even simply linear characteristic (parameters for high temperatures) is characterized by quite convergence to real process.

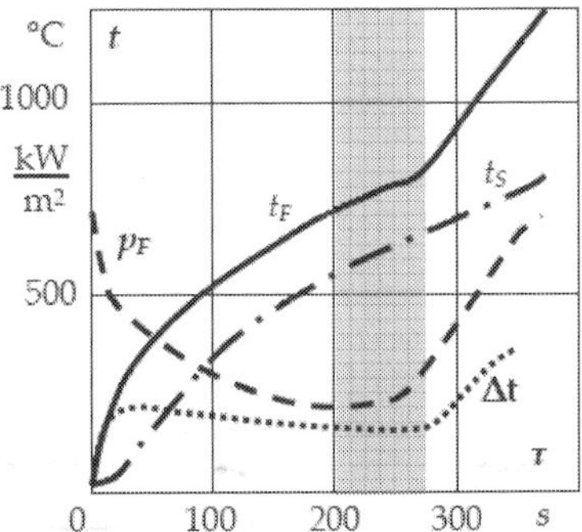

Fig. 8. Time characteristics of temperature and surface power for the case of voltage source: t_F - temperature of external surface of workpiece, t_S - temperature in the centre of workpiece, p_F - surface power

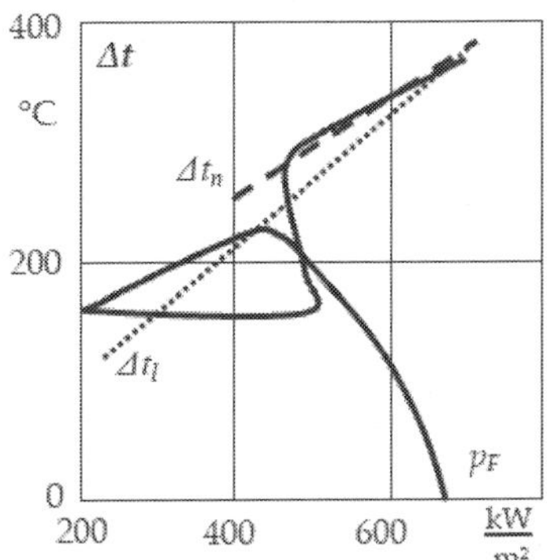

Fig. 9. Numerical and analytical static characteristics comparison $\Delta t = f(p_F)$ for the case of voltage source. Solid line- field calculations; dash and dot lines- analytical calculations Δt_l - linear, Δt_n - nonlinear

4. Accuracy of numerical modeling of induction heating systems

Modeling of induction heating systems seems to be the well-known and often used discipline. There is a lot of commercial and non- commercial calculating systems (for example FEM systems) that enable users to solve such problems automatically. The numerical system used different types of techniques for solving coupled – field problems. In present time, many authors (Galunin S. at al, 2006; Paya B. et al., 2002) describes a very advanced numerical models and take into consideration many of physical phenomenon like multiply reflection effect in radiation heat transfer or phase changes. This approach complicates models and prevent them from having the ability of calculations verification. There is a less information about accuracy of numerical modeling of coupled problems. This fact essentially limits the utility of numerical calculations in practical applications. In this chapter the numerical modeling and simulation accuracy problems were taken into account. Some popular MES calculating systems were used and simulation results were compared to the analytical ones and measurement results of physical model. In the first case, the induction heating system shown in figure 10 was used. The dimensions and properties of the model were the same like in the previous chapters. Nonferrous long cylindrical bar was used as the workpiece during simulations. Outer diameter of the charge was 24 mm. Air gap between worpiece and inductor were minimized (1 mm). Some of the basic material properties of charge – inductor system were shown in table 1.

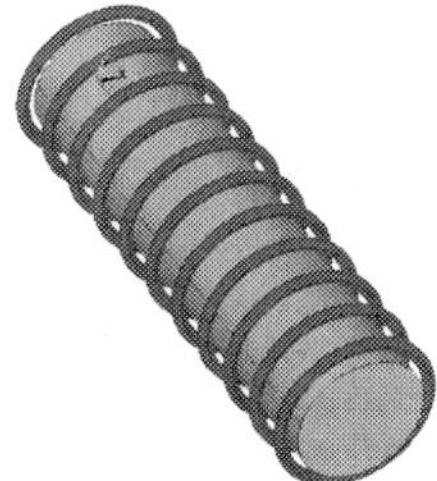

Fig. 10. Geometry of model

	μ_w	ρ	λ	c_w	γ
	-	$\Omega \cdot m$	W/(mK)	J/(kgK)	kg/m^3
Workpiece	1	$3{,}8 \cdot 10^{-7}$	12	654	7800
Charge	1	$1{,}78 \cdot 10^{-8}$	200	380	8933

Table 1. Material properties of the model

During modeling process the two - dimensional axisymmetric numerical models were used. The task has been solved analytically at first. The power consumption rate was used for the energy equation determination as well as for determination of basic parameters of induction heating system. Calculated parameters were used as an input data for numerical calculations. The final results of such calculation were temperature characteristics of thermal analysis of the induction heating problem. Numerical calculations were made by using the coupled – field analysis of electromagnetic field (harmonic problem) and thermal field (transient heat transfer problem). Such approach was necessary in view of different time constants for analyzed fields.

At first, some results of analytically and numerically calculated magnetic field intensity were compared. Two cases were compared for different sources of energy: the impressed current source (500 A) and equivalent current density source. The basic goal of the comparison was to calculate the errors rate by assuming simplification of using current density source instead of the general voltage or current source. The numerical results were compared to analytical ones and radial distribution of magnetic field intensities was shown in figure 11.

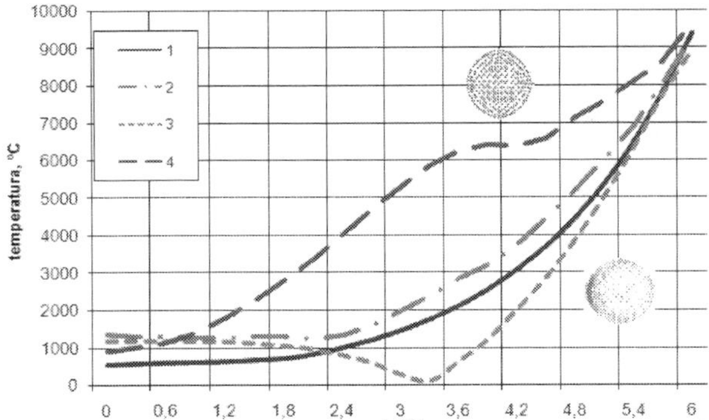

Fig. 11. Radial distribution of magnetic field intensity: 1 - ANSYS results for current source; 2 - ANSYS results for current density source; 3 - Quick Field results for current source; 4 - Analytical results

The results shown in figure 11 proves some huge differences, as quantitative as qualitative in magnetic field distribution within the workpiece calculated by using both analytical and numerical methods. Relative insignificant differences from analytical solution were obtained by using the Quick Field program. Simulation results from ANSYS in the case of current source were characterized by errors level reached 50% from analytical results. However, maximal errors were received in ANSYS in the case of current density source. For numerical results errors values were calculated for skin depth and shown in the figure 12.

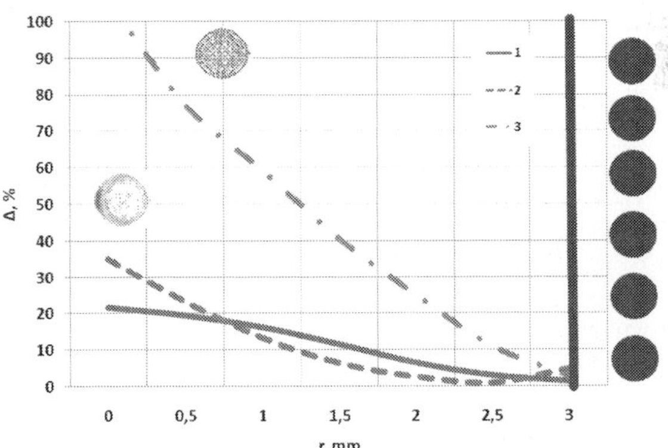

Fig. 12. Magnetic field intensity differences between analytical and numerical simulation results for the cases of current source (1) and equivalent current density source (2) in ANSYS system and current source in QF system(3)

The results presented above shows that in such heating systems the proximity effect determinates current conduction conditions in the workpiece. This rule is especially important in high efficiency induction heating systems, where dimmensions of air gap between workpiece and the inductor should be as minor as possible. In systems of large dimmensions of the air gap, the proximity effect is less important and establishment of refered simplification can affor much less calculation errors.

Basing on simulation results of magnetic field energy, the heat source distribution (Joule heat generation) in the workpiece was calculated and used in heat transfer analysis. In view of differences between results, both as a funcion of radius and heigh of workpiece, thermal analysis was solved by using of most accuracy results. Transient heat transfer phenomenon was solved by heating time 600 s. Boundary conditions were the same like in the analytical solution. The problem was solved in two different FEM systems (QF and ANSYS). Three algorithms were used for solving the coupled field problems. The simulation results were used to compare both methods and calculate a global accuracy of numerical simulation of induction heating phenomenon. Heating characteristics were shown in the figure 13.

It is noteworthy that simulation results of induction heating process as well as in ANSYS and QF systems are significantly different and slightly compared to conventional methods. Maximal differences between the results has reached the value of 170 K. Additionally numerical results are not compared to analytical ones. The differences from assumed temperature has reached 80 K for both ANSYS and QF results. Even simulation results from equivalent (Physical environment and MFS) solvers were significantly different in the evaluation of dynamics temperature characteristics and temperature chart after 600 s of heating.

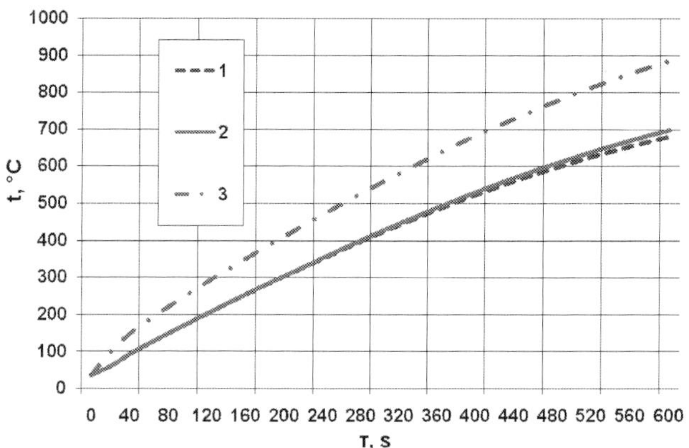

Fig. 13. Calculated temperature evolution curves at the longitudal center of bar: 1 - ANSYS physical environment results; 2 - ANSYS MFS results; 3 - QF results

The simulation results of magnetic field distribution show a significant differences between analyzed case depended of many factors, such as the manner of assuming power source in inductor, discretization rate or solver used during simulation process. This study has shown the disadvantages of the coupled field modeling with analytical approach. The calculated results of magnetic field intensity and current densities were two times different. The differences reached during the simulation process enable us to affirm the low usefulness of numerical simulation in practical problems and quantitative simulation of induction heating process.

There is a small number of articles and references, where the problems of induction heating systems were analyzed, and where the real power sources have been taken into account as well. Additionally, it is not many reliable verifications of numerical simulations with real processes. These results are significantly important when electrical properties are slightly dependent on the temperature. Many of verifications of experimental validations with the computed ones provides the influence of numerical models simplification onto results. The simplifications can afford the low practical usefulness results. As an example the authors of the article (Paya B. et al., 2002) presents the results of a series of tests made in order to validate the magneto – thermal module of the FLUX3D system. The physical model used for studying the induction heating was a cylindrical tube of magnetic steel, external diameter 100 mm, thickness 5 mm, length 300 mm. The inductor has been supplied by thyristor inverter working at 2,75 kHz. Figure 14 presents the comparison of the model and the experiment of the evolution of temperature at certain point at outer surface. It shows a huge differences between the model and experiment, especially in the case of ferrous materials heating.

This study has shown the advantages and the drawbacks of the coupling way compared to the linking way. The linking way may be profitable when the electrical properties are slightly dependent on the temperature because the computation time is strongly reduced. Otherwise, for example, when reached Curie temperature, the coupling way is the only one which can provide acceptable results. According to the experiment, it is also necessary to put more accurate physical properties at different temperatures to describe the behavior of the material correctly.

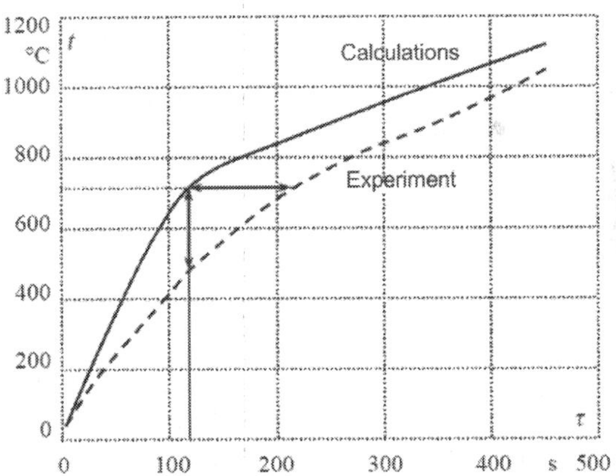

Fig. 14. Comparison of the experimental results and the computed ones (Paya B. et al., 2002)

5. Frequency inverters in induction heating systems

Basic calculations of induction heating systems in most cases takes into consideration only the workpiece – inductor system. There is a small number of articles, where electro-thermal converters were simulated and fed from the real power sources. Supplying the induction heating systems by energy sources based on the topology of frequency converters cannot be fully exploited. In many cases current source inverters have been used. In spite of minimal power losses of transistor frequency inverters, the variable workpiece properties make the cooperation with heating system difficulty. Properties of such sources can afford some complications in control system and reduce the system efficiency (generator and heating system) by inserting an excessive inductances to the system.

Such power sources are commonly used for melting of precious metals, continuous heating of sheets (ribbons, wires), remelting, refinement, tempering and other processes when the properties are slightly dependent on temperature. On other hand, in processes like steel heating, hardering etc (when impedance are highly dependent on temperature), utility of such sources is highly limited and application of complex control algorithms is required. In many cases, to avoid this requirement, maximal internal inductances and multi-layer inductors are being exploited. Increasing inductance stabilizes the charge variables but decreasing the system efficiency. In many articles the advantages of frequency converters are emphasized. But there is not many analysis about using converters in specific technologies.

Disregarding from power and high frequency ranges, the number of thermal application of heating materials in multi geometry inductors is wide enough. Induction heating can be divided into two domains: surface and volumetric heating, but based on conditions for selection of power source, into huge and low scattering systems. The scattering rate depends on air gap reactance and additional reactance of system chokes. The scattering effect do not have any influence on induction heating system efficiency (11) (Rudnev V., 2003)

$$\eta e(f,t,H,Dd) = \cfrac{1}{1 + \sqrt{\cfrac{\rho_w}{\rho(t)\cdot\mu(H)} \cdot \cfrac{Dd}{\phi[f,\rho(t),\mu(H)]}}} \tag{11}$$

Where: f - frequency, t - temperature, $\rho(t)$ - workpiece resistivity, H - magnetic field intensity, $\mu(H)$ - magnetic permeability, ρ_w - inductor resistivity (based on workpiece resistivity), Dd - geometry of heating system, ϕ - shape coefficient.

Scattering effect have a strongly influence on heating efficiency of the workpiece (12).

$$\eta c(f,t,H,Dd,t_F) = 1 - \cfrac{P_{SC}(t_F,Dd)}{P2[f,\rho(t),\mu(H),\rho_w,Dd]} \tag{12}$$

Together with increasing frequency, the voltage on air gap increases and power rate dissipates in the charge ($P2$) decreases. Basing on references, maximal value of heat loses (Psc) depends only on geometry and on the type of insulating material, without taking into consideration the kind of power supply system (Haimbaugh R. E., 2001; Rudnev V., 2003; Rapoport E. at al, 2006) . The power system influence is not practically determined in respect of necessity of coupling the field (inductor – workpiece) and the circuit (power supply – load) analysis. Such analysis is a very time-consuming process. Additionally a special calculating procedures should be used and a variance of dynamic impedance during heating process must be taken into account.

The indicator of energy conversion quality is electro – thermal efficiency of induction heating system. In all time steps, for the workpiece of known properties can be calculated from equation (13).

$$\eta et(f,t,H,Dd,t_F) = \eta e(f,t,H,Dd)\cdot\eta c(f,t,H,Dd,t_F) \tag{13}$$

Circuit parameters of induction heating system were calculated basing on approximate method of equivalent resistances (Rudnev V., 2003). Load parameters and voltage source model were used for induction heater parameters determination. The physical model used for studying a induction heating was a ferrous steel cylindrical bar of external diameter d = 12 mm , surrounded by a 7 turns coil, internal diameter 22 mm, length 31 mm. The series capacitor was selected in manner that frequency in high temperatures (at the end of the process) satisfied the volumetric heating requirements (Rudnev V., 2003). The operating frequency was at f_g=30 kHz. The system was supplied by a step-down transformer (transfer ratio 40:1) by the voltage inverter. In the figure 15 some efficiency characteristics in frequency domain has been shown. The electric efficiency for analyzed case was constant in practice (dash line in fig. 15). So that, only thermal efficiency determinates the rate of total electro – thermal efficiency. If the value is smaller, the maximal value of efficiency curve has reached a lower value of frequency.

Characteristics presented in the figure 15 were determined for nonferrous charge for one time step. The momentary values of maximal efficiencies during heating the steel charge has been shown in figure 16. Additionally, the changes of power coefficient were shown.

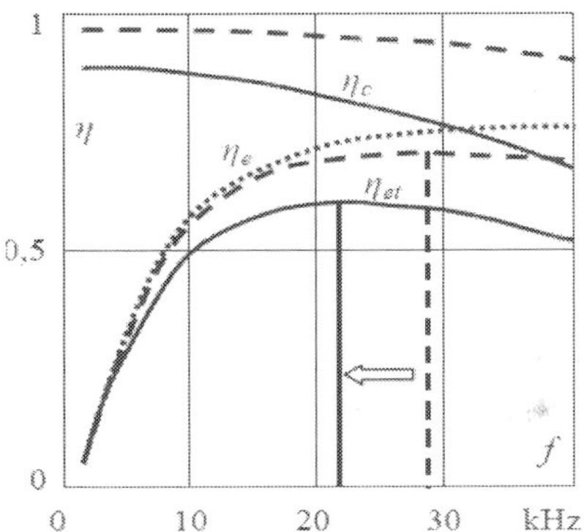

Fig. 15. Efficiencies of induction heating system as a function of frequency

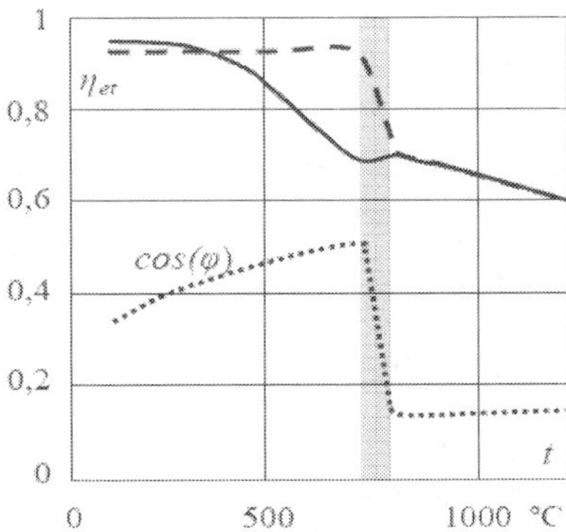

Fig. 16. Efficiency of heating process and the power coefficient in function of workpiece temperature

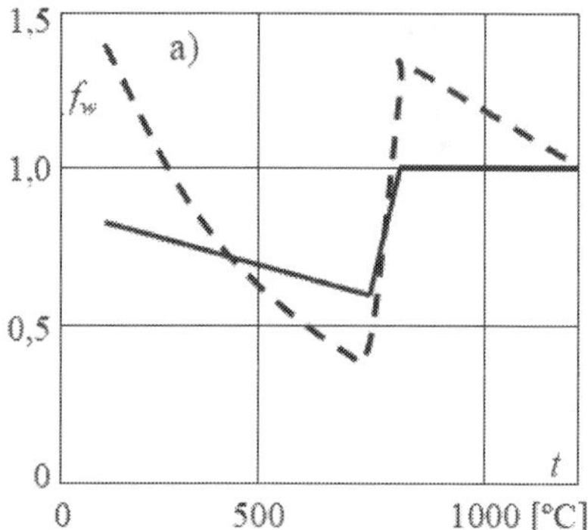

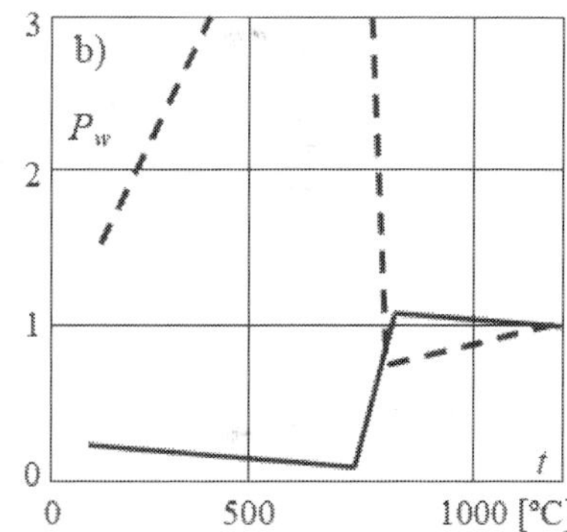

Fig. 17. Relative variances of chosen parameters of induction heater for maximal value of efficiency during all process time: a) – frequency with reference to the end process value (30 kHz), b) - active power with reference to the end process power (2600 W). Before (dash line) and after (solid line) reactive power compensation

In the next step, the heating process in two cases was analyzed: with (solid lines in figure 17) and without (dash lines in figure 17) the series capacitor. The value of capacitor was determined by satisfying the requirement for reactive power compensation at the end of the heating process. Comparison of such heating processes have been made for the case with and without the capacitor. In second case, the source voltage on inductor was pulling up to the specific value, when active power in the charge has reached the same value as in the compensation process. Non-compensating inductor voltage is larger: $Q = \omega \cdot L/R$. To provide the maximal efficiency values during the whole process, the frequency range should be changing. The frequency changes, carried to the end of the process has been shown in figure 17.a. The active power curves during the heating process were shown in figure 17.b.

The series capacitor variant, the resonant voltage inverter, minimizes the range of characteristics values variances. The case was characterized by better, corresponding to heat losses, power profile, that produces isothermal heating process.

6. Voltage frequency inverters

In induction heating systems of medium frequency range, both thyristors and transistors inverters are used as an power suppliers (Haimbaugh R. E., 2001). Voltage supply systems are commonly configured as the bridge systems. Voltage source inverters are not often used for direct supplying of the induction heating systems. The reason of such practice is complicated control ability, especially in the cases of ferrous materials. In such systems, the current source inverters are commonly used. The inverters are characterized by simplicity of process realizing but their energetic efficiencies are significantly smaller.

Thyristors and transistors inverters, except from operating frequency ranges, are characterized by the different operating frequency changes during the heating processes. The thyristors should work with frequencies smaller then resonant ones. Different characteristics describes transistor circuits, which are characterized by large operating frequency range. The power sources based on transistors are better for feeding an universal induction heating systems.

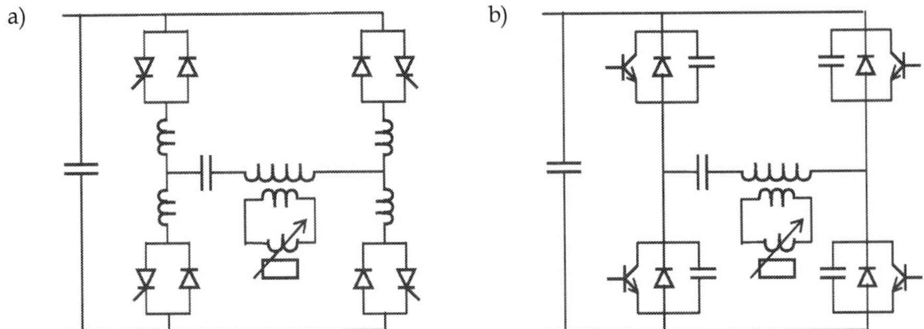

Fig. 18. Frequency inverters for induction heating systems: a) thyristors inverters $f < f_0$; b) transistor inverters $f > f_0$

Some of the new innovative manners for inverters system control (fig. 18) can provide more useful and better operating characteristics in induction heating systems. The datasheets focusing on developed transistor generators are limited in most of the cases just to revealing basic information (Rudnev V., 2003; Rapoport E. at al, 2006) about frequency range, maximal output power and range of power regulation, efficiency, etc. Information about power and frequency adjustment ability to realized processes or inductors parameters are not frequently published. Lack of the basic data makes it difficult to precisely design the energetic save processes.

Transistor inverters are characterized by the ability to control frequency, voltage and pulse width modulation. In the model described previously, some calculations have been done using different types of commonly used power control systems of induction heaters:

- voltage;
- frequency;
- pulse width modulation.

7. Power control systems

Power control during the induction heating process seems to be basic factor for precise realization of the temperature uniformity requirement. Maximal value of the power in analyzed heating system was determined by following parameters of power source:

constant voltage U_0=300 V, resonant frequency f_g=30kHz and the constant duty factor w=1.

Three different analysis have been solved for momentary power ranges as a functions of feeding voltage, frequency and a duty factor. The curves of relative variances of power p_g, p_z were shown in the figure 19. The results shown in the following figure are presented as the functions of the relative variables (14) described as the proportions of momentary values of voltage, frequency and duty factor to the characteristic values described above.

$$x = \frac{U_i}{U_0}, \frac{f_i}{f_0}, \frac{w_i}{1} \tag{14}$$

Presented results shows the phenomenon that for significant frequency increase, the dissipated Power decreases rapidly. Changes of the duty factor enables to linear power controlling. In this kind of power control system, the insignificant increase (for the high temperatures range) or decrease (for the low temperatures range) of operating frequency is necessary. This phenomenon was shown in figure 19 (dash line fw). Constant value of frequency can be supported only in the case of voltage control for nonferrous material. In the case of ferrous materials, the heating efficiency increases strongly for less power values. In the figure 20 the phenomenon of efficiency increase for ferrous and constant values for nonferrous have been shown for three analyzed control systems.

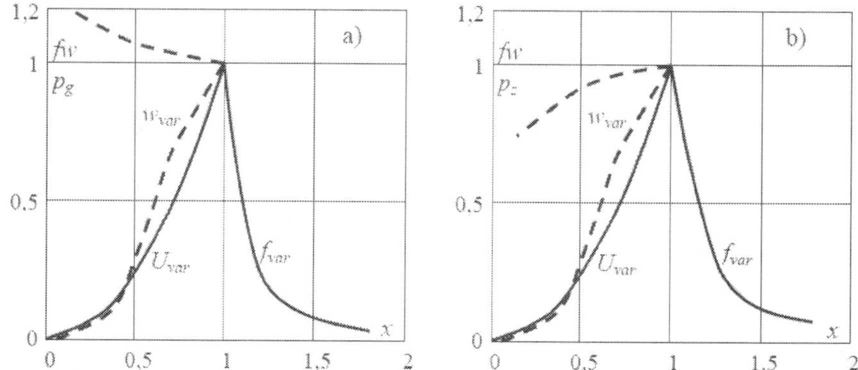

Fig. 19. Power control of the heater: voltage U_{var}, frequency f_{var}, duty factor w_{var} with variables fw, a) nonferrous charge, b) ferrous charge.

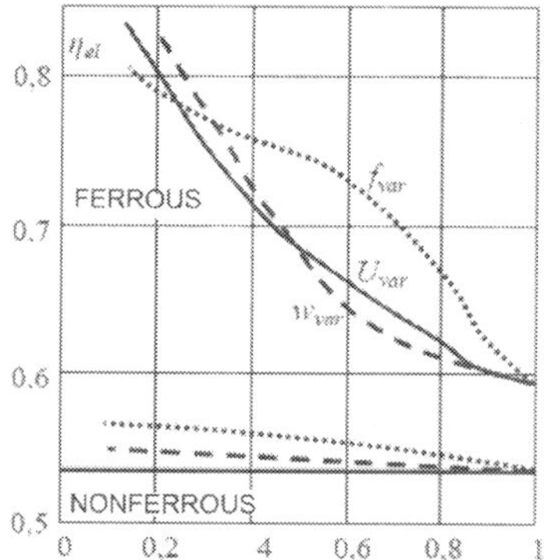

Fig. 20. System efficiency in function of variable set values for the case of ferrous and nonferrous charge

7.1 PWM controls system

Classical approach for power control of induction heating systems by frequency or intermediate voltage regulation is commonly used, well-known domain. For the case of voltage duty factors controllers, operating characteristics should be additionally discussed. Without instantaneous phase control (moment of transistors firing), the switches triggers at the moment of voltage wave synchronizing with maximal value of inductor current. The phenomenon has been shown in the figure 21. In this case, for frequencies near resonant, the current waveform is quite similar to sinusoidal one. The current wave deformation can be observed only for low pulse duty factors. In the figure 21 the inductors current (dot line) with its first harmonic (solid line) were compared for the pulse duty factor $w=0,1$.

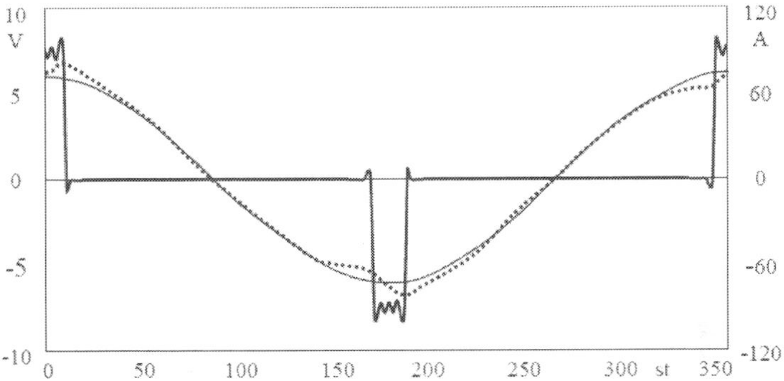

Fig. 21. Voltage and current waveforms for basic pulse duty factor controlling.

This manner for PWM control have some faults. Basic hazard is ability to decrease adjustment frequency below the resonance. Transistors can be destroyed in this condition. Other ability is forced controlling of the moment of transistors firing. This manner is quite better and can provide a lower current wave deformation. The case of transistors firing at the moment of current wave polarity change have been shown in figure 22.

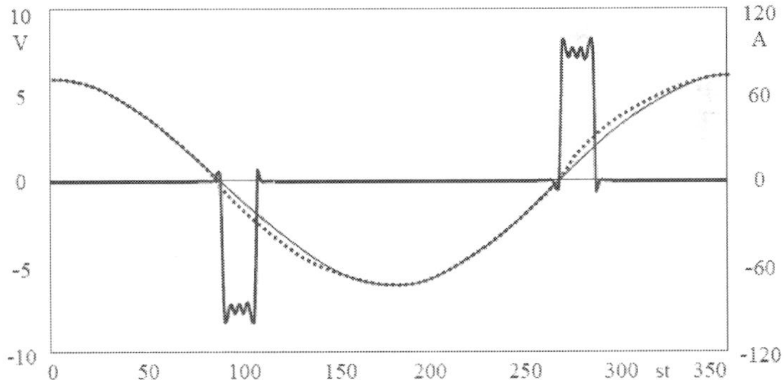

Fig. 22. Voltage and current waveforms for forced pulse duty factor controlling

Controlling of voltage impulse generating can provide the ability of keeping load characteristic of induction heater generator as an resistive or inductive-resistive during the process. Advanced transistors firing (from current wave) can be used for saturating power control in the system. The phenomenon of such approach is from that in this case two power control manners were used: the pulse width modulation of voltage and frequency modulation. In optimal operating conditions of generator (near resonance frequency) the changes of active power values as a function of inductors current can be observed. The characteristics presented in figure 23 shows the phenomenon. The shape of the curves is quite similar to the case of resistance furnace controlled by thyristors.

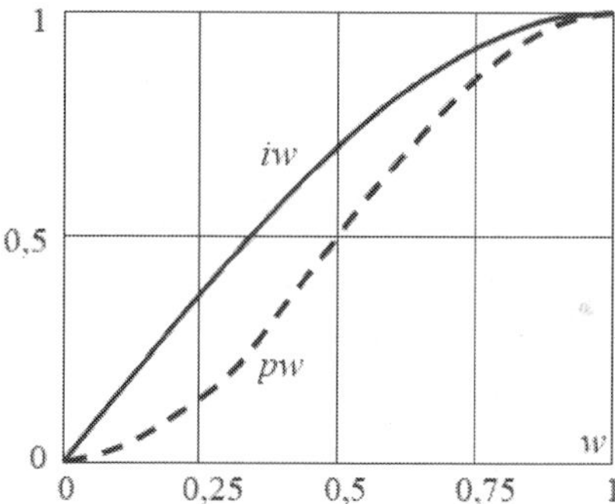

Fig. 23. Inductors current and active power of generator as a function of PWM value

7.2 Simulation results

Simulation of voltage and current waveforms and power control methods as a result were solved by using the harmonics method. The equation of voltage waveform was written as:

$$U_\tau = \sum_k u_{k,\tau} \tag{15}$$

Where all of harmonics depends from amplitude U_0, time (frequency) $\tau\,(\omega)$ and duty factor w:

$$u_{k,\tau} = \frac{4 \cdot U_0}{\pi \cdot (2 \cdot k - 1)} \cdot sin[(2 \cdot k - 1) \cdot \pi \cdot w] \cdot cos[(2 \cdot k - 1) \cdot \omega \cdot \tau] \tag{16}$$

Based on previous equation, the ability of power control range can be estimated. For studying a control systems $k=1...40$ of odd harmonics were used. The impedance of heating generator was calculated by using the equivalent resistances method and particular current harmonics were calculated for different temperature values:

$$i_{k,\tau,s} = \frac{4 \cdot U_0}{\pi \cdot (2 \cdot k - 1) \cdot Z_{k,s}} \cdot sin[(2 \cdot k - 1) \cdot \pi \cdot w] \cdot cos[(2 \cdot k - 1) \cdot \omega \cdot \tau - \varphi_{k,s}] \tag{17}$$

Current in the series branch of generator was determined as the consequence.

$$I_{k,s} = \sum_k i_{k,\tau,s} \tag{18}$$

The simulation results for chosen steady state has been show in the figure 24. Voltage waveforms of PWM control system and the equal current waveform were depicted. In this kind of power control system, increasing of the operating frequency is necessary as described in previous chapter. Frequency increasing additionally provides to decreasing of momentary power value. So that it has been proved that controlled power value decreases faster than pulse width.

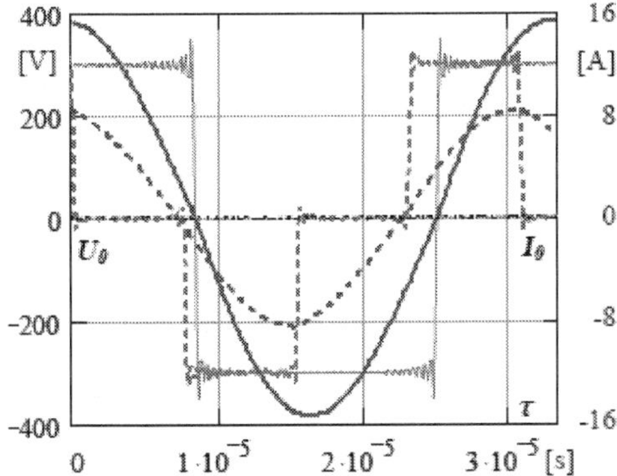

Fig. 24. Voltage and current waveforms in voltage inverter without the power control (solid lines) and for the case of PWM (dot lines) system

8. Conclusions

This chapter describes very widely the issues of the volume induction heating phenomenon. The criterions for selection optimal parameters of the induction heaters were discussed. It has been proved that in the case of volumetric heating, additional parameters, like electrothermal efficiency and power control ability must be taken into account to satisfy all requirements for optimal process implementation. Without the knowledge of physical phenomenons and quantitative influence on temperature distribution it is unfulfillable to prepare a set of input data for effective modeling of the technological processes and directions for optimally selection of power sources.

Basics rules of induction mass heating were described at first. Classical and modern calculating methods of such electrothermal devices were discussed very widely. Some solutions and analytical methods for accurate modeling of such systems were described.

In the chapter some basic sources of numerical analysis errors were discussed and simulation results were compared to analytical description of the problem and experimental data. In spite of many advantages of calculating systems, the accuracy of results should be always taken into account as a basic requirement. Numerical simulation results of magnetic field distribution show a significant differences between analyzed case depended of many factors, such as the manner of assuming power source in inductor, discretization rate or solver used during simulation process. This study has shown the disadvantages of the coupled field modeling with analytical approach. The calculated results of magnetic field intensity and current densities were two times different. The differences reached during the simulation process enable us to affirm the low usefulness of numerical simulation in practical problems and quantitative simulation of induction heating process. According to the experiment, it is also necessary to put more accurate physical properties at different temperatures to describe the behavior of the material correctly.

Most popular methods for accurate power control in induction heating systems were discussed in the case of nonlinearity material properties. Some examples were shown and classical approach for power control in the systems was compared to pulse width modulation case. The advantages and disadvantages of proposed construction and process solutions were discussed. It has been proved that the power system influence is not practically determined in respect of necessity of coupling the field (inductor – workpiece) and the circuit (power supply – load) analysis. Omission of such coupling can provide a very unserviceable results.

9. References

Davies E. J.: Conduction and induction heating, *Peter Peregrinus Ltd.*, London, United Kingdom 1990.

Haimbaugh R., E.: Practical induction heat treating, *ASM International*, 2001.

Rudnev V.: Handbook of induction heating, *Marcel Dekker*, 2003.

Rapoport E., Pleshivtseva Y.: Optimal control of induction heating processes, *CRC/Taylor & Francis*, 2006.

Sajdak C., Samek E.: Nagrzewanie indukcyjne. *Wyd. Śląsk*, Katowice 1985.

UIE: Electromagnetic Induction and Electric Conduction in Industry. Centre Français l'Électricite 1997.

Zgraja J., Bereza J., (2003) "Computer simulation of induction heating system with series inverter", *COMPEL: The International Journal for Computation and Mathematics in Electrical and Electronic Engineering*, Vol. 22 Iss: 1, pp.48 – 57.

Paya B., Fireteanu V., Spahiu A., Guerin C.: 3D Magneto-Thermal Computations in Induction Heating. Models and Experimental Validations, *10th International IGTE Symposium on Numerical Field Calculation in Electrical Engineering*, September 16-18, 2002, Graz, Austria.

Niedbała R.: Charakterystyka energetyczna indukcyjnego nagrzewania objętościowego. *Zeszyty Naukowe Politechniki Łódzkiej „Elektryka"* nr 101, s 107-114, Łódź 2003.

Kainuma K., Sakuma M.: Application of Numerical Analysis to Industrial Electrical Heating Products. Fuji Electric Journal 1998, Vol. 71, No.5.

Chaboudez C., Clain S., Glardon R., Rappaz J., Swierkosz M., Touzani R.: Numerical Modelling of Induction Heating of Long Workpieces, *IEEE Treansaction on magnetics*, vol.30, no.6, 1994.

Biro O., Preis K., Richter K., R.: Various FEM formulations for the calculation of transient 3d eddy currents in nonlinear media, *IEEE Treansaction on magnetics*, vol.31, no.3, 1995.

Matthes H., Jürgens R.: 1.6 MW 150 kHz Series Resonant Circuit Converter incorporating IGBT Devices for welding applications. *International Induction Heating Seminar*, 1998 Padova, pp 25-31.

Niedbała R.: Tranzystorowy falownik napięcia zasilający indukcyjny układ grzejny. *Przegląd Elektrotechniczny* 2004, nr 11, s.1149...1152.

Dede J., Jordan J., Esteve V., Ferreres A., Espi J.: On the Behaviour of Series and Parallel Resonant Inverters for Induction Heating under Short-Circuit Conditions. *PCIM Europe 1998 Power Conversion*, pp 301-307.

Nerg J., Partanen J.: Numerical Solution of 2D and 3D Induction Heating Problems with Non Linear Material Properties Taken into Account. *IEEE Transactions on Magnetics*, vol. 36 no.

Galunin S., Zlobina M., Blinov Yu., Nikanorov A., Zedler T., Nacke B., Electrothermal Modeling and Numerical Optimization of Induction System for Disk Heating, *International Scientific Colloquium Modeling for Material Processing*. Riga, 2006.

Niedbała R.: Sterowanie energią elektryczną w procesie indukcyjnego nagrzewania stali. *Przegląd Elektrotechniczny* 2009, nr 6, s.58...62.

9

Two Novel Induction Heating Technologies: Transverse Flux Induction Heating and Travelling Wave Induction Heating

Youhua Wang[1], Junhua Wang[2], S. L. Ho[2], Xiaoguang Yang [1], and W. N. Fu[2]
[1]Province-Ministry Joint Key Laboratory of Electromagnetic Field and Electrical Apparatus Reliability, Hebei University of Technology, Tianjin
[2]Department of Electrical Engineering, The Hong Kong Polytechnic University, Kowloon
[1]China
[2]Hong Kong

1. Introduction

The increasing demand for industrial furnace heaters has attracted a lot of attentions from researchers and practitioners. Traditional longitudinal method is inadequate to meet the ever increasing demand on quality performance in regard to temperature uniformity on the work strips, especially on thin and long strips. Among the new systems being developed recently, transverse flux induction heating (TFIH) and the traveling wave induction heating (TWIH) processes are potentially very promising and attractive.

Longitudinal induction heating was the most often used method in the past and it is still used today. However, there are many disadvantages of the traditional induction heating method. It is well known that we can increase the operational frequency to decrease the skin depth in order to increase the heating efficiency. For the metal strip with certain depth, 10 kHz or above frequency has to be used if using longitudinal induction heating method. The traditional induction heating method can only applied to metal strips with large thickness.

TFIH is prior to the conventional axial flux induction heating in many aspects, e.g. lower frequency for thin strips. But the temperature distribution is more complex such as may higher at one place but lower at the other across the entire strip width. The fundamental principles of the TWIH are known for many years [1-3]. TWIH, as one of the multiphase induction heating systems, has particular features which make them attractive for application to some heating and melting processes in industry. Among the advantages we can mention the possibility to heat quite uniformly thin strips or regions of a body without moving the inductor above its surface, to reduce the vibrations of inductor and load due to the electro-dynamic forces and also the noise provoked by them, to obtain nearly balanced distributions of power and temperature [4].

In this chapter these two different novel types – TFIH and TWIH technologies are considered. Some standard and specially developed FEM codes based on numerical methods were successfully used for the study of TFIH and TWIH heaters. The main advantages and the particular characteristics of TFIH and TWIH systems such as the effect of slots on induced power distributions were studied. The FEM codes and the results of numerical simulation may be therefore used easily and reliably, for the design of these two systems. While TFIH systems have been studied for many years, TWIH systems are not yet fully appreciated with respect to their main advantages and possible industrial applications [1, 2].

2. Transverse Flux Induction Heating (TFIH)

Transverse flux induction heating (TFIH) is prior to the conventional axial flux induction heating in many aspects, e.g. lower frequency for thin strips. But the temperature distribution is more complex such as may higher at one place but lower at the other across the entire strip width. It is influenced by the inductor structure, which must be optimized. Considering the long computation time, Optimization is done through orthogonal experiment. The design of the TFIH equipment necessitates accurate prediction for the thermal characteristic. So it is essential to obtain a clear understanding of the eddy current in the workpiece and further more the power losses in temperature field [5, 6].

The mathematical model for this sinusoidal quasi-static eddy current problem results from the Maxwell equations and is described by the complex magnetic vector potential $\underline{\vec{A}}$ and an electrical complex scalar potential $\underline{\phi}$.

$$rot\frac{1}{\mu}rot\ \vec{\underline{A}} - grad(\frac{1}{\mu}rot\underline{\vec{A}}) + j\omega\kappa\left(\underline{\vec{A}} + grad\ \underline{\phi}\right) = \kappa\vec{\underline{E}}_S \tag{1}$$

where $\vec{\underline{E}}_s$ is the electric field strength impressed by the power source. The requirement of a zero divergence condition of current density must be fulfilled:

$$div\left(\kappa\underline{\vec{A}} + \kappa grad\ \underline{\phi}\right) = 0 \tag{2}$$

The eddy current density is computed with

$$\vec{\underline{J}} = j\omega\kappa\left(\underline{\vec{A}} + grad\ \underline{\phi}\right) + \kappa\vec{\underline{E}}_S \tag{3}$$

And it determines the heat source distribution:

$$p_v = \left|\vec{\underline{J}}^2\right| / k \tag{4}$$

The temperature field $\vartheta(x,y,z)$ is computed based on the Fourier's thermal conduction equation,

$$\frac{\partial(c\rho\vartheta)}{\partial t} = div\left(\lambda\ grad\ \vartheta\right) + p_v - \vec{v}\ grad(c\rho\vartheta) \tag{5}$$

where c is the specific heat, λ is the thermal conductivity coefficient, ρ is the mass density and $\vec{v}$ is the strip velocity.

The computation of the electromagnetic field by approximation is performed on the basis of the 3D FEM with Galerkin method applied to equation (1) and (2). Fig. 1 shows the mesh of the TFIH equipment and we can easily find two additional planes of symmetry, situated in the middle of the strip (x-y plane) and in the middle of the inductor-yoke combination (y-z plane). So the solution area is reduced to 1/8 of the total volume. In the x-z and the y-z planes of symmetry, the electric current densities are oriented perpendicular to them respectively. Hence on the x-z plane the components of the magnetic vector potential $\vec{\underline{A}}$ and the scalar potential $\underline{\phi}$ are

$$\underline{\vec{A}}_X = \underline{\vec{A}}_Z = 0 \tag{6}$$

$$\frac{\partial \underline{\vec{A}}_Y}{\partial y} = 0 \tag{7}$$

$$\underline{\phi} = 0 \tag{8}$$

And on the y-z plane

$$\underline{\vec{A}}_y = \underline{\vec{A}}_Z = 0 \tag{9}$$

$$\frac{\partial \underline{\vec{A}}_x}{\partial x} = 0 \tag{10}$$

$$\underline{\phi} = 0 \tag{11}$$

On the x-y plane of symmetry, the current density has no perpendicular component and

$$\underline{\vec{A}}_z = 0 \tag{12}$$

$$\frac{\partial \underline{\vec{A}}_x}{\partial x} = \frac{\partial \underline{\vec{A}}_y}{\partial y} = 0 \tag{13}$$

$$\frac{\partial \underline{\phi}}{\partial z} = 0 \tag{14}$$

In this problem, the scalar potential in the electrical non-conductive field areas is practically insignificant, because in the equations (1) and (2) it always appears in conjunction with the electrical conductivity. The perpendicular component of the current density must be zero on the surface of electrical conducting areas, except for the forced source current density. Hence the following equation is valid:

$$j\omega k(\underline{\vec{A}} + grad\underline{\phi}) \cdot \vec{n} = 0 \tag{15}$$

The present problem has an open boundary condition because the vector potential disappears only in the infinity. With the aids of comparative computations, an enveloping surface can, however, be determined for which the condition is valid within a known upper limit of error.

$$\underline{A} = 0 \tag{16}$$

For computation of temperature field, there is no y-z symmetric plane. The heat source density in the newly added area can be obtained by reflection at the y-z plane or by separated computation for a corresponding temperature distribution.

Power losses on the surfaces were taken into account by using non-homogenous Neuman boundary conditions for convection and radiation (17). Where ϑ_s is the surface temperature, ϑ_α is the ambient temperature, α_c is the heat transfer coefficient for convection and α_R is the equivalent heat transfer coefficient for radiation. The calculation of α_R based on the Stefan-Boltzman law and is given by equation (18), where ε is the emissivity coefficient of the surface material and c_R is the radiation number.

$$-\lambda \left.\frac{\partial \vartheta}{\partial n}\right|_{\Gamma_3} = \alpha_c (\vartheta_s - \vartheta_\alpha) + \alpha_R (\vartheta_s - \vartheta_\alpha)\big|_{\Gamma_3} \tag{17}$$

$$\alpha_R = c_R \varepsilon \frac{\left(\dfrac{\vartheta_s + 273}{100}\right)^4 - \left(\dfrac{\vartheta_\alpha + 273}{100}\right)^4}{\vartheta_s - \vartheta_\alpha} \tag{18}$$

Furthermore, at an adequate distance from the inductor, boundary conditions must be determined over the cross section of the workpiece. On the side from which the strip enters the solution area, temperature (e.g. ambient temperature) is given.

$$\vartheta(x,y,z)\big|_{\Gamma_1} = \vartheta_0(x,y,z)\big|_{\Gamma_1} \tag{19}$$

On the exit side, however

$$\text{grad } \vartheta \cdot \vec{n} = 0 \tag{20}$$

is indicated, where $\vec{n}$ corresponds to the velocity direction.

The electromagnetic field and the temperature field are coupled via the temperature dependence of the electrical conductivity and the magnetic permeability. The relationship between eddy current distribution and coil geometry in TFIH equipment was analyzed. The non-linear coupled electromagnetic-thermal problem was solved through finite element method (FEM). Based on that, new coil geometry was designed to get a homogeneous temperature distribution on the surface of the strip, shown in Fig.1 [7].

Fig. 1. New structure of the TFIH equipment

Through the orthogonal experiment above, the more suitable structure can be determined by comparing the distribution of the eddy current and heat source, furthermore that of the temperature on workpiece surface. Fig. 2 gives the distribution of heat sources density computed with equation (4). Fig. 3 shows the temperature distribution resulted from the computation with induced heat sources density shown in Fig. 2.

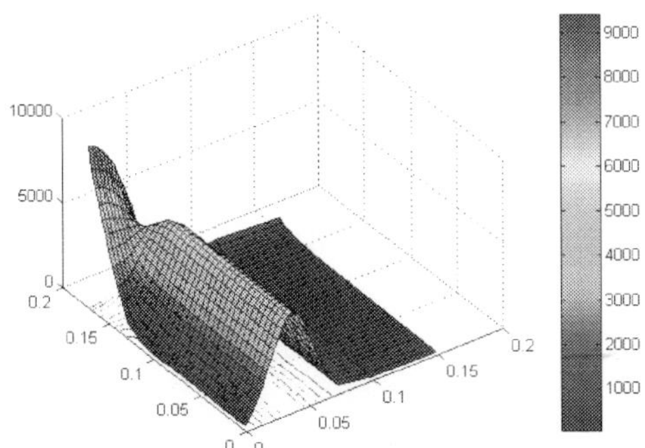

Fig. 2. Distribution of heat sources density on the1/4 surface of the strip

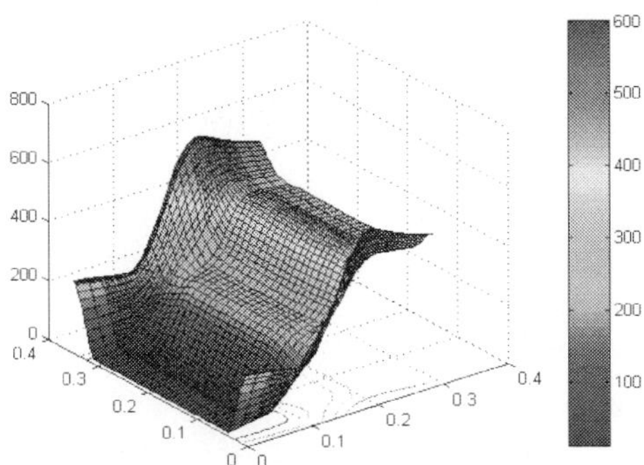

Fig. 3. Temperature distribution on the surface of the strip

In the design of TFIH equipment with orthogonal experiment the computation of eddy current and heat source distributions are beneficial for the optimum chosing of the design parameters in order to obtain a high efficient and homoginous heating across the strip width, the exactly controll of the working process varies from different workpieces by means of the numerical predication model under the industry conditions.

In the case of TFIH, the flux is perpendicular to the surface of the work piece; therefore the eddy current is across the thickness of the work piece. So compared to longitudinal flux induction heating equipments the TFIH equipments have considerable advantages such as a much lower frequency resulting in a higher electrical efficiency, lower reactive power, more suitable for continuously heating treatment and local heating because of its inductor not encircling the work piece.

Neural Networks with FEM was also introduced to predict eddy current and temperature distribution on the continuously moving thin conducting strips in TFIH equipments. With the temperature prediction results, a good guessed value of the temperature dependent parameters for each finite element and initial values for temperature field solution are provided. The predicted results are presented in Fig. 4. Network prediction results speed up the iteratively solution process for the non-linear coupled electromagnetic-thermal problems.

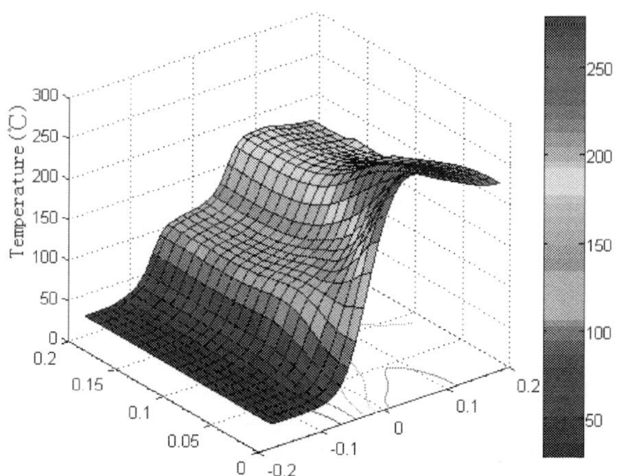

Fig. 4. The predicted result of temperature field distribution on the surface of one half of the thin metal work piece

3. Travelling Wave Induction Heating systems (TWIHs)

Unlike the longitudinal or the single-phase transverse flux heaters, TWIHs is energized with three-phase windings. Parametric analysis is used to assess the key parameters. TWIHs is intrinsically more suitable than its single phase counterpart in the provision of uniform thermal density on the work strips [18]. Indeed, three-phase induction heaters not only have the same advantages as their single-phase counterparts, they can produce more uniform temperature distributions, operate with less industrial noise and vibration. For operations in which the electromagnetic force may increase significantly and the heating parameters could change as the temperature rises, the advantage of TWIHs is even more significant.

3.1 Typical Travelling Wave Induction Heating system (TWIH)

Typical travelling wave induction heating system (TWIH), as one of the multiphase induction heating systems, has particular features which make them attractive for application to some heating and melting processes in industry. Among the advantages we can mention the possibility to heat quite uniformly thin strips or regions of a body without moving the inductor above its surface, to reduce the vibrations of inductor and load due to the electro-dynamic forces and also the noise provoked by them, to obtain nearly balanced distributions of power and temperature [1].

TWIH introduces three-phase induction heating heater, using the typical three-phase windings and parametric analysis to assess the key parameters (transmission of electricity to the work-piece, efficiency, power factor, et al) and the distribution of electricity. Compared with the single-phase induction heating heater of Transverse Flux Induction Heating inductors, three-phase induction heating heater not only has the same advantages, but owns the ability to produce more uniform temperature distributions, and reduces industrial noise with low vibration. Especially in the conditions that electromagnetic force increases significantly and heating parameters change with the rise of temperature, the advantage is very important.

TWIH is of particular interest for theoretical analysis and some industrial applications of induction heating. Since the analysis of bibliography shows that TWIH systems have not been studied so deeply as the single phase ones, the present paper has been devoted to the simulation and study of some peculiar characteristics of this type of inductors. Particular attention has been paid to the simulation of TWIH systems for heating thin strips and plain bodies including TFIH.

Fig. 5 (a) shows the schematic of a typical configuration of the heater with two linear inductors on the opposite sides of the strip, slots perpendicular to the direction of the movement and a relatively large air-gap due to the thickness of the refractory material interposed between the inductor and the strip.

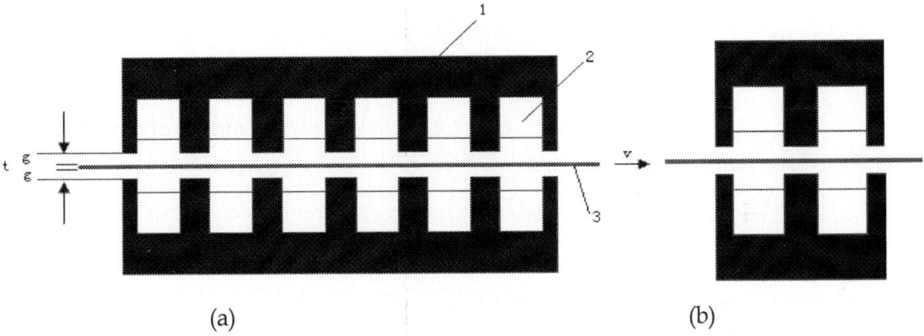

Fig. 5. (a) Schematic of a double-side TWIH [1-magnetic yoke; 2-exciting windings; 3-load metal sheet; t-strip thickness; g-airgap between inductor and load; v-strip movement directions]; (b) Schematic of a double-side TFIH.

A schematic of a TFIH system is displayed in Fig. 5 (b). Besides the phases, other parameters are similar to those in TWIH system in Fig. 5. The TFIH inductor configuration analyzed in this chapter has been chosen to be as close as possible to an existing system for assuring the possibility of checking the results by tests.

The TFIH system, shown in Fig. 5 (a), consists of two inductors with current carrying conductors located inside the slots of two magnetic yokes and perpendicular to the length of the strip and to the direction of its movement. The inductors are situated on both sides of the strip [2].

On each inductor there are three coils and six conductors in every coil. The aluminum work-piece has the following dimensions: thickness 2.5 mm, width from 1200 to 1600 mm, while the air gap between magnetic yokes is 60 mm.

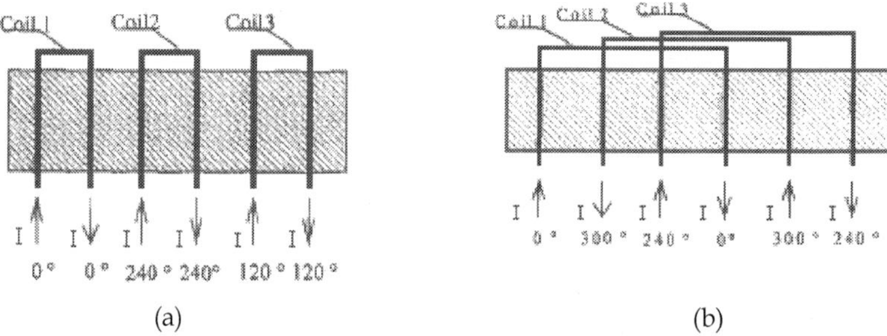

Fig. 6. (a) Coil connections for TFIH; (b) Coil connections for TWIH.

For the purpose of comparison, the same configuration has been used for the TWIH system, but with different schemes of coils connections in order to obtain a travelling wave field, shown in Fig. 6(b). The reference to a similar design for the two systems, which have approximately the same efficiency, gives the possibility of understanding more clearly some basic differences between TFIH and TWIH inductors. As stated before one of the problems. Maxwell equations are the theoretical basis of all macro- electromagnetic phenomena, and its differential equations are as following:

$$\nabla \times \vec{H} = \vec{J} + \frac{\partial \vec{D}}{\partial t} \tag{21}$$

$$\nabla \cdot \vec{B} = 0 \tag{22}$$

$$\nabla \times \vec{E} = -\frac{\partial \vec{B}}{\partial t} \tag{23}$$

$$\nabla \cdot \vec{D} = \rho \tag{24}$$

This project overlooks the functions of the accumulative charges, that is $\rho = 0$. From (21) to (24), $\vec{H}$ -magnetic field intensity, $\vec{J}$ -current density, $\vec{E}$ -electric field intensity, $\vec{B}$ -magnetic field intensity, $\vec{D}$ -electric displacement vector. According to the relationship between $\vec{H}$, $\vec{J}$, $\vec{E}$, $\vec{B}$ and $\vec{D}$, we can get

$$\nabla \times \vec{H} = \vec{J} + \frac{\partial \vec{D}}{\partial t} = \sigma \vec{E} + \frac{\partial}{\partial t}(\varepsilon \vec{E}) \tag{25}$$

The corresponding vector form

$$\nabla \times \tilde{\vec{H}} = (\sigma + j\varepsilon\omega)\tilde{\vec{E}} \tag{26}$$

When $\sigma \gg \varepsilon\omega$, $\frac{\partial \vec{D}}{\partial t}$ can be neglected. And when the frequency is up to thousands of hertz, it

satisfies $\sigma \gg \varepsilon\omega$. Therefore, in this project we can neglect $\frac{\partial \vec{D}}{\partial t}$. Equation (21) is simplified as

$$\nabla \times \vec{H} = \vec{J} \tag{27}$$

For the eddy current field, basic equations can be established with $\vec{A} - \phi$ method. Then we introduce magnetic vector potential $\vec{A}$, define it as $\vec{B} = \nabla \times \vec{A}$, and eventually attain the current density formula

$$\vec{J} = -\sigma(\frac{\partial \vec{A}}{\partial t} + \nabla \frac{\partial \phi}{\partial t}) + \vec{J}_s \tag{28}$$

TFIH and TWIH comply with the Maxwell equations strictly in theory. From the equations above we can see that the current density in the strip is decided by the magnetic flux and the field intensity, so magnetic field inducted by TFIH and TWIH can create different vortex effect in the strip.

Both analytical and numerical techniques can be used for the study of these heating systems. Analytical methods are more convenient for the integral parameters determination and analysis, while the numerical techniques are more universal and particularly useful for investigating the induced current and power distributions, taking into account the inductor edge-effects and the slots effects which are usually well pronounced in TWIH systems. The analytical methods which make use of Fourier integral transformation or series are effective for the simulation of lD, 2D and even 3D multiphase devices, but some simplifications and assumptions must be made.

Recently some dedicated analytical codes, described in previous papers, have been developed by the authors and applied to the study of such systems [3, 6, 8, 9]. However, they introduce in the simulation some rough simplifying assumptions (e.g. current sheet inductors, yoke without slots) which, for some purposes, are too far from reality. On the other hand, numerical methods are more universal and able to solve problems for more complex geometries. Their use is practically mandatory when the tooth pulsation, the influence of the slot-tooth ratio and slot depth or the inductor and load edge effects have to be taken into account.

Since in this chapter the influence of the relative slot-tooth position of the upper and lower inductor has to be considered, the analysis has been performed by 2D FEM code, overcoming many of the above difficulties by the use of a specially built computer programmer in which the FEM code is called as an external subroutine [10, 11]. Given the system geometry, the programmer automatically performs with the changes of the geometrical and electrical data, without the direct intervenes of the FEM code users. In order to study the efficiency of TWIH system, the magnetic flux density distribution is calculated. It is displayed in Fig. 7. (Current density is 3 A/mm^2, and frequency is 600 Hz).

TFIH system magnetic induction intensity distributes symmetrically along both sides of axis and reach the top above magnetic yokes, and the intensity is relatively weak above the coils. Near the center of axis the intensity decreases to the minimum. The curve of TWIH system's magnetic flux density in Fig. 7 (a) is relatively more homogeneous than TFIH heater's which shows low degree only between 130 mm and 175 mm region.

Fig. 8 show results about eddy current density distributions in TFIH and TWIH systems. As TWIH using three-phase induction heaters, upper and lower induction heaters can be installed properly to make up better for the low-fields, so as to achieve uniform heating results.

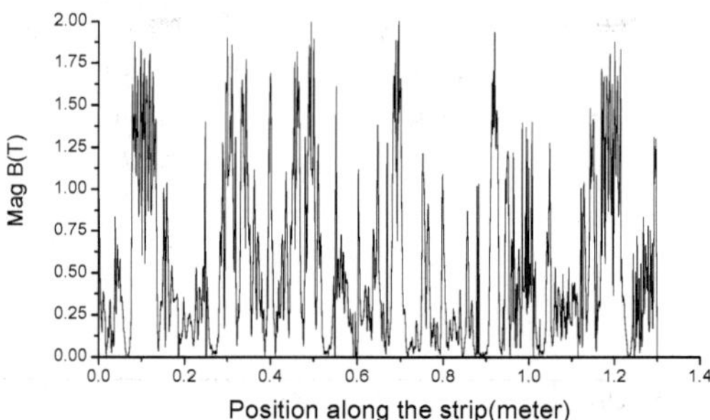

Fig. 7. The magnetic flux density distribution in the strip of TWIH system at 600Hz.

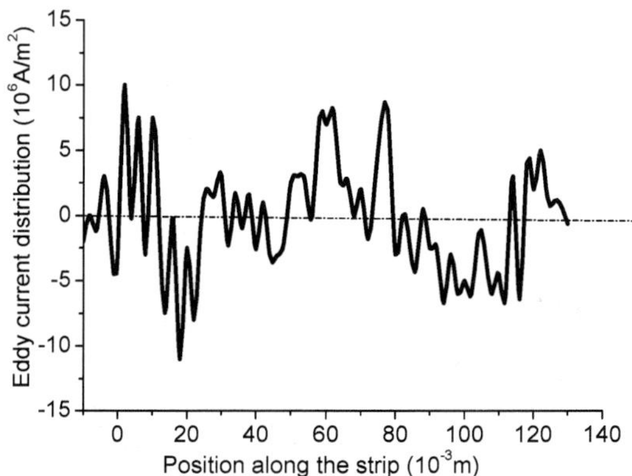

Fig. 8. The eddy current density distribution along the strip perpendicularly of TWIH system at 600Hz

Another characteristic of the TFIH system here considered is the presence of many high eddy current peaks along the direction of the movement along the edge of the sheet, which can give rise to, in some cases, dangerous strip deformations. In the corresponding TWIH system this problem is reduced since few and less sharp peaks with their highest value much less than in TFIH system are present. The same situation arises in centre of the sheet, but due to the very similar configuration of TWIH and TFIH inductors, in both systems eddy current peaks are present.

3.2 Crossed traveling wave induction heating (C-TWIH)

A novel design for a crossed traveling wave induction heating (C-TWIH) system is proposed to address the inhomogeneous eddy current density which dominates the temperature distribution on the surface of work strips. FEM is used to calculate the eddy current and temperature distributions. Simulation results of the novel C-TWIH system are presented and compared with those of typical TWIH and TFIH systems to showcase its better performance.

The schematic of a typical heater configuration is shown in Fig. 9 (a). There are two linear inductors on the opposite sides of the strip and twelve slots along the direction of the movement. There is a relatively large airgap between the inductor and the strip due to the thickness of the interposing refractory materials.

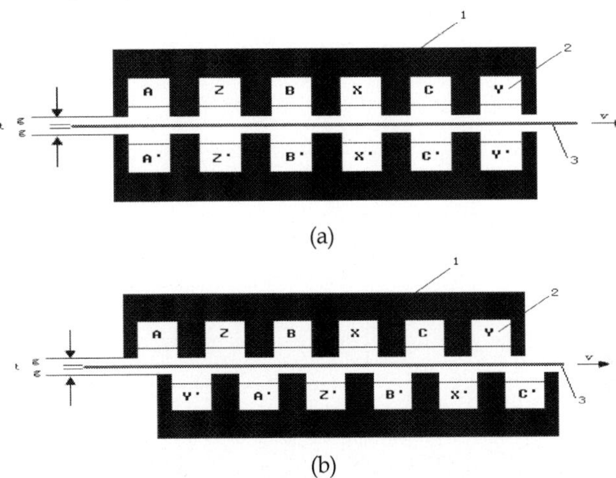

(a)

(b)

Fig. 9. (a) Schematic of a typical double-sided TWIH system; (b) Schematic of the double-sided C-TWIH system (1-magnetic yoke; 2-exciting windings; 3-load metal sheet; t-strip thickness; g-airgap between inductor and load; v-strip movement velocity. Phase degree: A-X and A'-X'=0°, B-Y and B'-Y' =-120°, C-Z and C'-Z' =-240°).

The proposed C-TWIH system is shown in Fig. 9 (b). With this structure the lower inductor is shifted 60 degrees along the movement direction of the strip and the winding starts from phase Y'. On each inductor there are three phases and six windings. Alternating current through every two in-phase sets of windings sets up the magnetic field, which is perpendicular to the surface of the sheet, and eddy currents are induced by the alternating magnetic flux on the work strip.

Number of phases	3
Number of windings	6
Number of slots	12
Airgap distance (g)	1.0 mm
Magnetic yoke length	1300mm
Magnetic yoke height	200mm
Aluminum strip width	130 mm
Aluminum strip thickness (t)	2.0 mm
Strip movement velocity (v)	1.0 m/min

Table I. Design Data of Typical TWIH and C-TWIH Heaters of Fig. 1

In the following FEM simulation, the amplitude of the exciting current is 1200A at 600Hz. The phase current waveforms are as shown in Fig. 10. It can be seen from Figs. 10 and 11 that at any instant, there are always four inductor phase windings acting as the field windings to generate the flux in the C-TWIH device; while in typical TWIH device, there are two phase windings only.

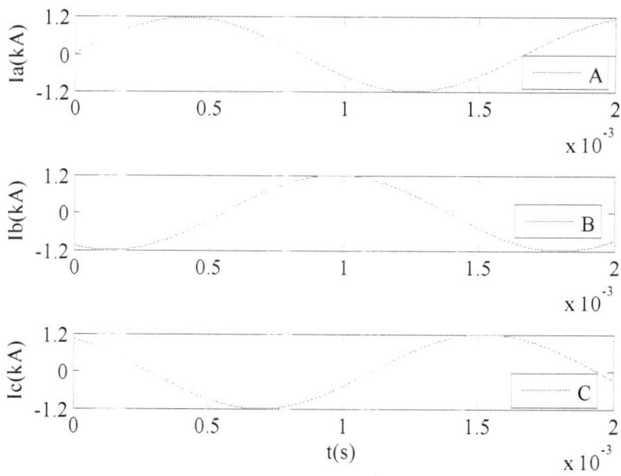

Fig. 10. Three-phase exciting current waveforms for typical TWIH and C-TWIH systems

In this section, the operation of the two three-phase induction heating devices, namely the typical TWIH and C-TWIH, with the phase currents of Fig. 10 are investigated using FEM analysis. Particular attention is paid to the analysis of the magnetic flux density, eddy current distribution, power density and the temperature field.

Because AC current through every winding generates a magnetic field, which induces eddy currents to produce the thermal field, the total magnetic field is the additive contribution of six upper-and-lower windings.

$$\oint Hdl = \sum i = N_c i \qquad (29)$$

where N_c is the number of turns per phase per winding, and

$$i = \sqrt{2}I \sin(\omega t + \theta) \qquad (30)$$

where I is the r.m.s. current and θ is the phase angle.

The result of the airgap flux density calculation of the C-TWIH system is shown at different positions in Fig. 11. The average trend of the airgap flux density B has a symmetrical waveform having similarly triangle-topped amplitude. Flux densities in the upper and lower inductors make up for the weak areas of each other. This leads to a more uniform eddy current density in the work strips, as discussed below.

The magnetic flux densities in the work strip of the C-TWIH system are depicted in Fig. 12 (b). In order to study the efficiency of the C-TWIH system, the flux density distribution along the work strips of a typical TWIH system is also calculated, which is shown in Fig. 12 (a).

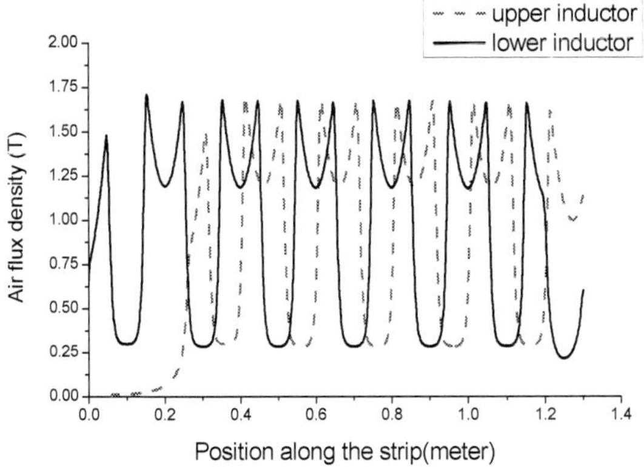

Fig. 11. The airgap flux density distribution in the work strip from the upper and lower inductors

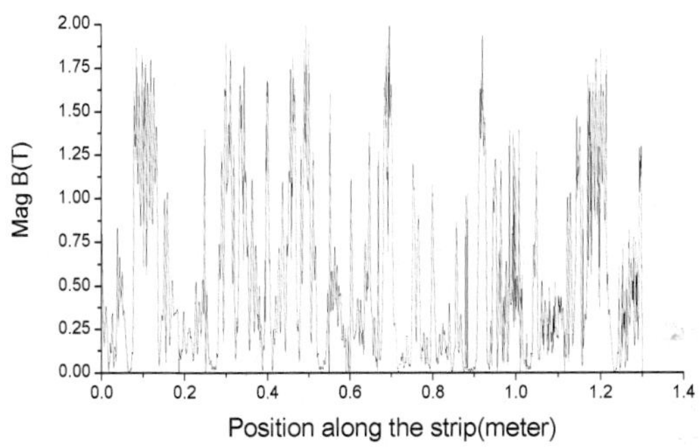

(a) In typical TWIH system

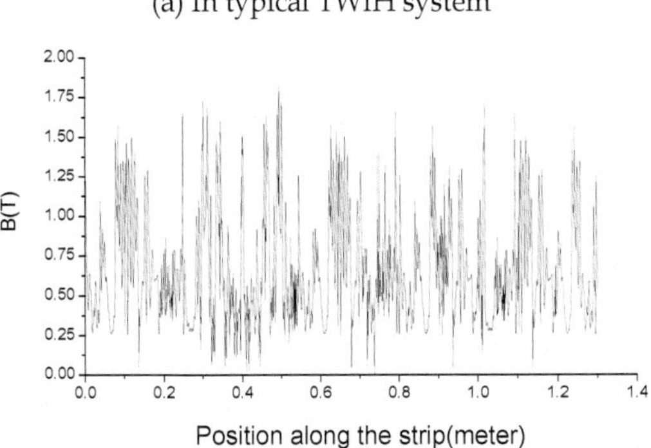

(b) In C-TWIH system

Fig. 12. The magnetic flux density distribution in the work strip.

It can be seen that typical TWIH system's magnetic flux is distributed symmetrically along both sides of the axis and reaches the maximum above the magnetic yokes, while the magnetic flux intensity is relatively weak above the windings. Near the center of axis the intensity decreases to the minimum. The curve of the C-TWIH system's magnetic flux density is relatively more homogeneous than that of typical TWIH heater, which shows low density only in several limited regions. Thus the eddy current field and the temperature field distributions on the work strip should be more uniform in the proposed C-TWIH system than that in typical TWIH systems as confirmed below.

For the induced eddy current field, the basic equations are established using $\bar{A} - \phi$ formulation. The magnetic vector potential $\bar{A}$ is defined as $\bar{B} = \nabla \times \bar{A}$ and the current density is

$$\bar{J} = -\sigma(\frac{\partial \bar{A}}{\partial t} + \nabla \varphi) + \bar{J}_s \tag{31}$$

where σ is the electric conductivity; $\bar{J}_s$ is the impressed exciting current density; $\bar{E}$ is the electric field intensity; and

$$\nabla \varphi = -\bar{E} \tag{32}$$

The induced current density in the strip is generated by variations in the magnetic flux. The vortex effects in the strip are attributed to the magnetic fields produced by different structures of typical TWIHs or other types of heaters like TFIH. By using FEM, the magnetic flux distributions and induced eddy current density in the strip of typical TWIH and C-TWIH systems are shown, respectively, in Figs. 13 (a) and (b) for electrical phase angles of 90°and 270°.

It is shown in Fig. 13 that the novel C-TWIH system has a relatively more uniform power density distribution. This is accomplished mainly because of the application of crossed three-phase AC excitations to create uniform magnetic fields that govern the eddy current density distribution. On the other hand, the combination of six induction heaters serves to widen the directions of the induced magnetic field which is distributed along the magnetic yokes. These inductors interact with each other and compensate for the weak magnetic areas existing between the gap districts.

Fig. 14 shows the eddy current density distributions in typical TWIH and C-TWIH systems. As C-TWIH uses crossed three-phase induction heaters, the upper and lower induction heaters are installed properly to compensate for the low-fields so as to realize uniform heating results. It can be seen that the eddy current distributions are more uniform in C-TWIH when compared to those in typical TWIH.

Another characteristic of TWIH systems that needs to be considered is the presence of many high eddy current peaks along the movement directions on the edges of the sheet, which can give rise to, in some cases, dangerous strip deformations. In the proposed C-TWIH system this problem is reduced since sharp peaks of eddy currents are much fewer than those in typical TWIH systems.

The best cases among both typical TWIH and C-TWIH system [12], which has the similar geometry, are compared in Fig. 15, where the irregularities due to the use of a relatively small number of mesh elements have been smoothed. It can be seen that from 3 mm to 130 mm, the C-TWIH system induces higher power than any other systems being studied.

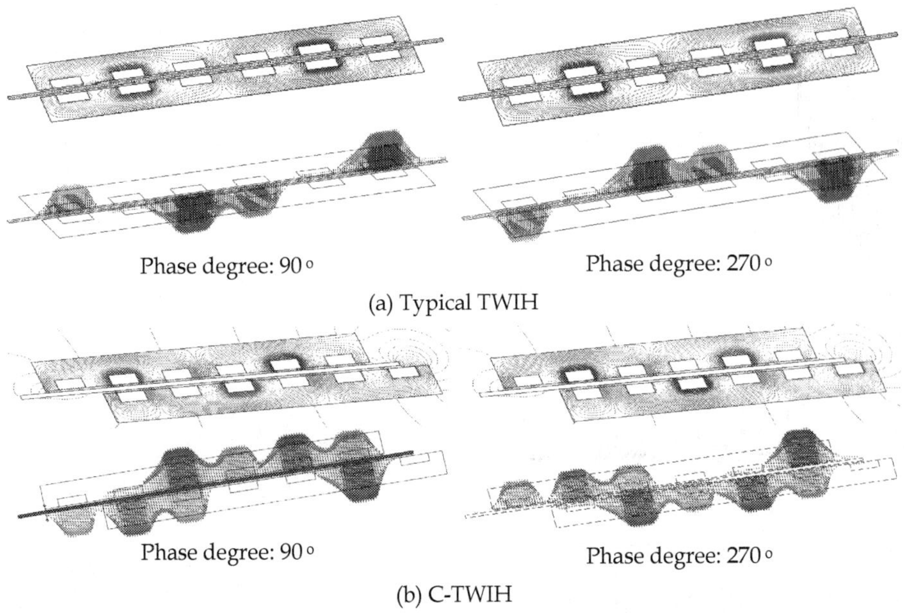

Fig. 13. Magnetic flux distributions in the strip and induced power density distributions along the direction of the strip movement (yoke size - 1300×30mm; slot size - 150 ×50 mm; airgap - 1 mm; strip thickness - 2 mm; currents supply I –1200A; f - 600 Hz).

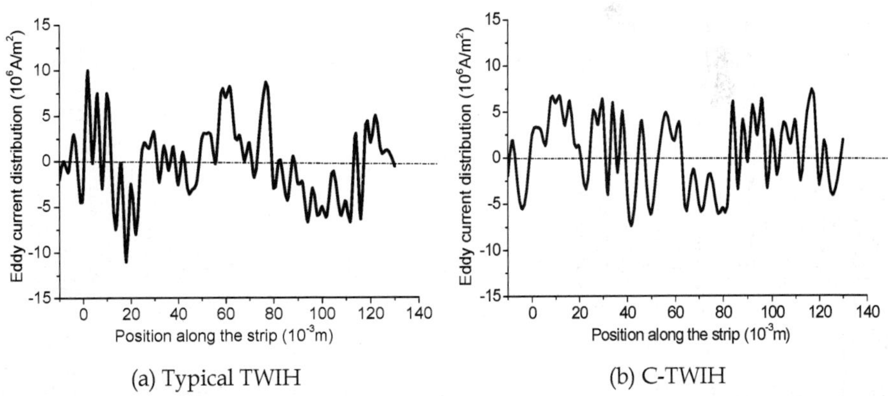

Fig. 14. The eddy current density distribution along the strip perpendicularly of typical TWIH and C-TWIH systems

The thermal properties of the materials are functions of temperature. Around the surface of the heating regions, there are dramatic changes in the high temperature zones. In order to improve the accuracy of the simulation, 2D nonlinear transient heat conduction equations are introduced to describe the thermal fields.

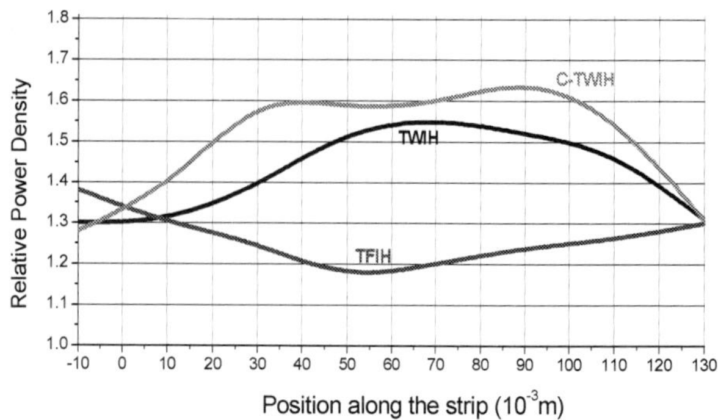

Fig. 15. Relative Power density (P$_2$/P$_1$) distribution of C-TWIH, TWIH and TFIH

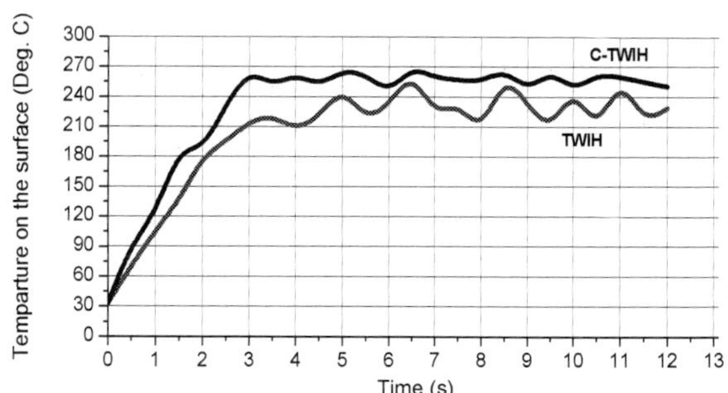

Fig. 16. Temperature distribution along the surface of the strip

Applying eddy current density to be the heat sources during the heating process, the thermal source is

$$P_V = \frac{\left|\bar{J}\right|^2}{\sigma} \tag{33}$$

where $\bar{J}$ is the eddy current density and σ is the electric conductivity.

The temperature field is computed based on Fourier's thermal conduction equation

$$\frac{\partial(c\rho\vartheta)}{\partial t} = \nabla \cdot (\lambda \nabla \vartheta) + P_V + \bar{v} \cdot (c\rho\vartheta) \tag{34}$$

where c is the specific heat; ρ is the mass density; λ is the thermal conductivity coefficient; P_V is the thermal source; and $\bar{v}$ is the strip velocity.

Fig. 16 shows that the C-TWIH reaches a higher temperature faster and with fewer variations when compared to typical TWIH. After three seconds, the temperature of typical TWIH is observed to fluctuate between 212°C and 256°C, while C-TWIH has a temperature range from 251°C to 270°C and its unexpected temperature difference is reduced by about 43% compared to that in typical TWIH device.

In addition, the proposed C-TWIH system can overcome problems of metal deformation and reduces the noise from the system, where the uniformity of power and eddy currents in the work strips plays an important role.

3.3 Traveling wave induction heating system with distributed windings and magnetic slot wedges (SW-TWIH)

A traveling wave induction heating system with distributed windings and magnetic slot wedges (SW-TWIH) and its improved system are proposed to address the inhomogeneous eddy-current density that dominates the temperature distribution on the surface of work strip. Magnetic flux distributions and eddy current distributions are computed using FEM. Performance of the novel SW-TWIH and an improved version is also presented and compared with a typical TWIH [6, 13, 14] using FEM to showcase the merits of the proposed system.

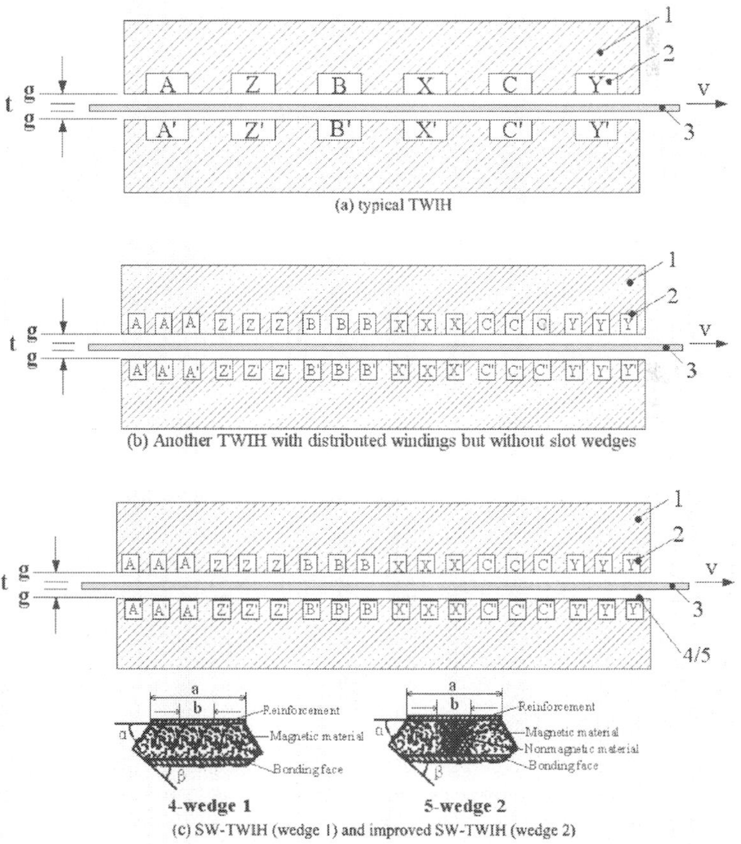

Fig. 17. Schematics of the systems.

Similar to the study of traditional longitudinal heaters, two-axis equivalent circuit method is usually employed in the steady-state performance study of TWIH [3]. For transient performance, such as eddy current and temperature field distributions, the results from traditional methods are not as accurate as expected, due to skin effects on the solid pole surface and serious magnetic nonlinearities. With the advent of powerful computing workstations, 2D and 3D finite-element analyses (FEAs) have now become feasible in practical applications, not only for steady-state field analysis, but also for transient performance study of induction heaters [1, 6, 11, 13]. For complicated TWIHs, transient 3D FEA study to include the eddy current and temperature field distributions is however computationally expensive and are hence not feasible for industrial applications due to its complex 3D meshing process and excessively long solution time required.

In order to study the performance of the proposed SW-TWIH system, an interpolative FEA modeling method is introduced [15]. The salient performance of the proposed SW-TWIH device can then be obtained accurately using a 2D transient FEA simulation.

Fig. 17 (a) is a typical TWIH system and Fig. 17 (b) represents the TWIH system with distributed windings but without slot wedges. The proposed SW-TWIH system is shown in Fig. 17 (c), which has the same structure but with magnetic slot wedges (SW) in each coil. Two linear inductors and thirty three coils are equipped on the opposite sides of the strip along the direction of movement. Due to the thickness of the interposing refractory materials, there is a relatively large airgap between the inductor and the strip. For the improved system, a nonmagnetic strip is equipped in the middle of SW. Table II gives the parameters of the proposed system.

For the transient analysis, the time step needs to be sufficiently fine, such as 0.01 ms in this study, so as to simulate the eddy current distribution correctly. Based on the observations, it is estimated that it takes more than a few days in order to obtain the complete solution with 3D FEA. Such a long computation period is impractical for engineering applications.

	Number of phases	3
	Number of windings	6
	Number of slots	36
The proposed heater	Airgap distance (g)	1.0 mm
	Magnetic yoke length	1400 mm
	Magnetic yoke height	120 mm
	Slot width	33.3 mm
	Slot height	60 mm
Width of aluminum strip		130 mm
Thickness of aluminum strip (t)		3.0 mm
Movement velocity of strip (v)		1.0 m/min
The width of nonmagnetic material in improved SW-TWIH system		2 mm

Table II. Design Data of The Proposed Heaters

Two main causes that contribute to inaccurate simulation results when using 2D models are the end-turn leakage inductance and the equivalent effective core length as 3D simulations are needed. A 2D transient FEA model with the above parameters from 3-D FEA is proposed in this study and the end-turn leakage inductance and effective core length are computed according to the method mentioned in [15].

The operation of the novel 3-phase induction heating devices, namely the SW-TWIH system, and its improved system are investigated using FEM with phase currents as given in Table II. The amplitude of the exciting current is 1200A at 500Hz. The winding current phases of W1, W2, and W3 are 0°, -120°, and -240°, respectively, for the equivalent 2-D transient FEA model. Particular attention is paid to the analysis of the magnetic flux density in the airgap, the eddy current distribution, and the power density.

For the magnetic flux density field, the basic equations are established using $\bar{T} - \Omega$ formulation. $\bar{T}$ is the magnetic vector potential and Ω is the magnetic scalar potential. Applying Ampere's law, Faraday's law, and Gauss's law for the solenoidality of the flux density yields the differential equations in conducting region as

$$\nabla \times \left(\frac{1}{\sigma} \nabla \times \bar{T} \right) + \frac{d}{dt} \left(\mu \bar{T} + \mu \nabla \Omega \right) = -\frac{d}{dt} \left(\mu \bar{H}_S \right) \tag{35}$$

$$\nabla \cdot \left(\mu \bar{T} + \mu \nabla \Omega \right) = -\nabla \cdot \left(\mu \bar{H}_S \right) \tag{36}$$

where $\bar{H}_S$ is a source field; σ is the electric conductivity; and μ is the relative permeability.

Because AC current through every winding generates a magnetic field, which induces eddy currents to produce the thermal field, the total magnetic field is the additive contribution of six pairs of upper-and-lower windings.

Results of the airgap flux densities of the typical TWIH, the TWIH system with distributed windings but without slot wedges, the proposed SW-TWIH and the improved SW-TWIH system are shown in Fig. 18. In the typical TWIH system, the average trend of the airgap flux density B has a symmetrical waveform having similarly triangular-topped amplitudes and large peaks. After applying the distributed windings without slot wedges, the airgap flux density becomes larger and more uniform but also has large differences between the flux peaks. For SW-TWIH system and its improved partner, the weak areas of the flux densities between two neighboring coil slots in the inductors are compensated due to the magnetic SW, which makes the whole flux density distribution more uniform. And compared to SW-TWIH, the improved SW-TWIH system also provides higher flux density because of the nonmagnetic strip in the SW. In fact, the curve of the improved SW-TWIH system's magnetic flux density is relatively more homogeneous than those of others, which shows low density only in several limited regions. Thus the eddy current field and then the temperature field distributions on the work strip should be more uniform in the improved SW-TWIH system than those in other systems mentioned earlier as discussed and confirmed below.

The relationship between eddy current density $\bar{J}_S$ and the source field $\bar{H}_S$ is deduced by Maxwell electromagnetic equations as

$$\nabla \left(\nabla \cdot \bar{H}_S \right) - \mu \sigma \frac{\partial \bar{H}_S}{\partial t} = \nabla \times \bar{J}_S \tag{37}$$

The induced current density in the strip is generated by variations in the magnetic flux. The vortex effects in the strips are attributed to the magnetic fields produced by different structures of TWIHs or other types of heaters like TFIH devices.

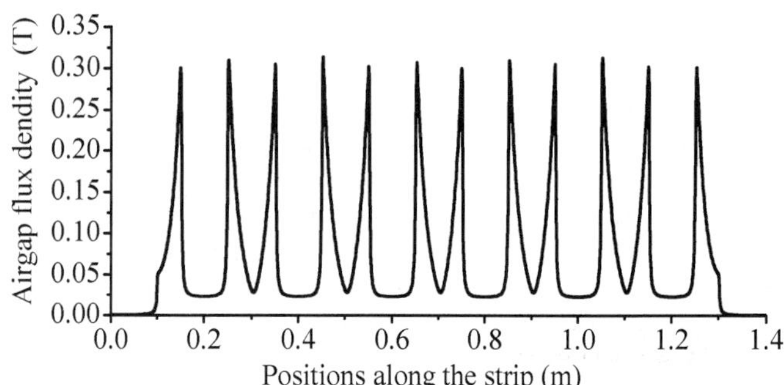

(a) Typical TWIH system

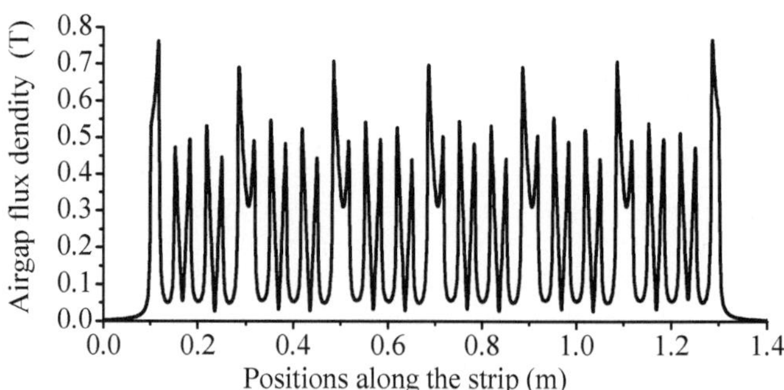

(b) TWIH system with distributed windings but without slot wedges

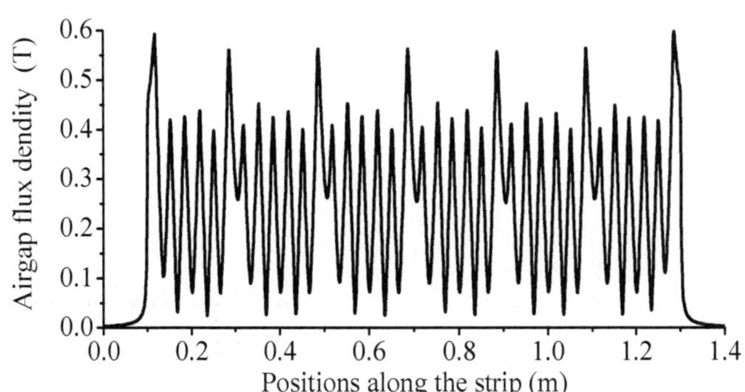

(c) Proposed SW-TWIH

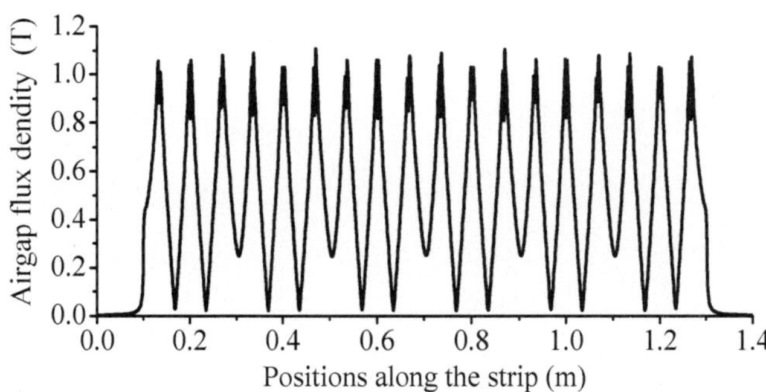

(d) Improved SW-TWIH system

Fig. 18. The airgap flux density distribution along the direction of the strip movement.

By using FEM, the magnetic flux distributions of the improved SW-TWIH system are shown in Figs. 29 at the phase of 0º, 90º, 180º, 270º and 360º. It can be seen in Fig. 20 that the novel improved SW-TWIH system has a relatively more uniform magnetic flux density distribution. This is accomplished mainly because of the application of three-phase AC excitations to create uniform magnetic fields that largely govern the eddy-current density distribution. On the other hand, the combination of distributed windings and SWs serves to widen the directions of the induced magnetic field which is distributed along the magnetic yokes. These factors interact with each other and compensate for the weak magnetic areas that exist between the gap districts.

Fig. 20 shows the eddy-current density distributions in typical TWIH, the proposed SW-TWIH and the improved SW-TWIH system. It can be seen that the eddy-current distributions are more uniform in the two proposed systems, especially in the improved SW-TWIH, when compared to those in typical TWIH system because of the application of the distributed windings and the magnetic SWs. Also, the combination of six induction heaters serves to widen the directions of the induced magnetic fields along the magnetic yokes. For the improved SW-TWIH system, the nonmagnetic material strip decreases the magnetic flux leakage to result in a relatively higher magnetic flux density.

Another characteristic of typical TWIH system that needs to be considered is the presence of many high eddy-current peaks along the movement directions on the edges of the sheet, which can give rise to, in some cases, dangerous strip deformations. In the proposed SW-TWIH system and its improved partner, this problem is reduced since sharp peaks of eddy currents are much fewer. The eddy current mean square deviations are 18.7303 for the typical TWIH being studied, 8.606 for the SW-TWIH and 11.731 for the improved SW-TWIH, respectively.

The best cases among the typical TWIH, the proposed SW-TWIH and the improved SW-TWIH system are compared in Fig. 21, where the irregularities due to the use of a relatively

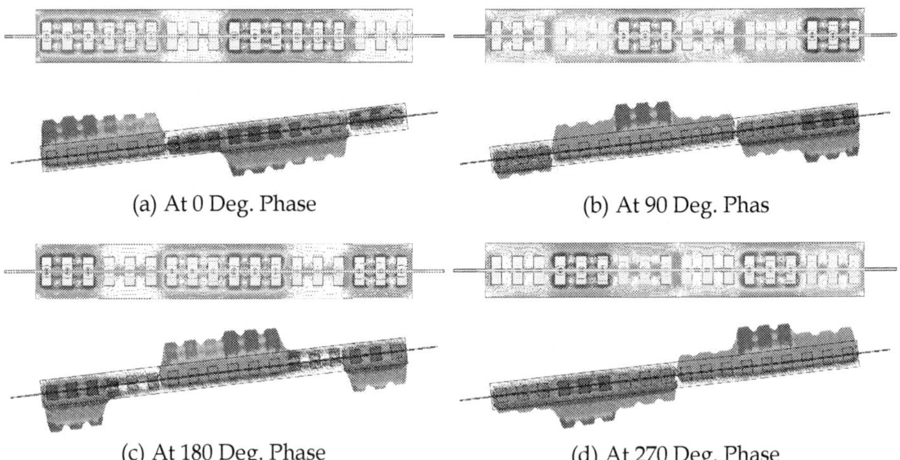

(a) At 0 Deg. Phase
(b) At 90 Deg. Phas

(c) At 180 Deg. Phase
(d) At 270 Deg. Phase

Fig. 19. Magnetic flux distributions of the improved SW-TWIH system along the direction of the strip movement.

small number of mesh elements have been smoothened. It can be seen that from 0.16 m to 1.32 m, the SW-TWIH and improved SW-TWIH systems, particularly the latter, produce more uniform power density than that in the typical TWIH system being studied. The improved SW-TWIH system also attains nearly double power density distributions comparing to the other two systems. Therefore, the improved SW-TWIH system provides relatively more homogeneous power distributions and much larger power supply.

For the improved SW-TWIH, the power factor changes with the airgap width of the nonmagnetic material in the magnetic slot wedges, and Fig. 22 shows the trend. The power factor becomes higher when the width is increased, but the distribution uniformity becomes worse according to the analysis. Hence it is essential to select a compromise value based on practical requirements in order to balance uniformity and power factor.

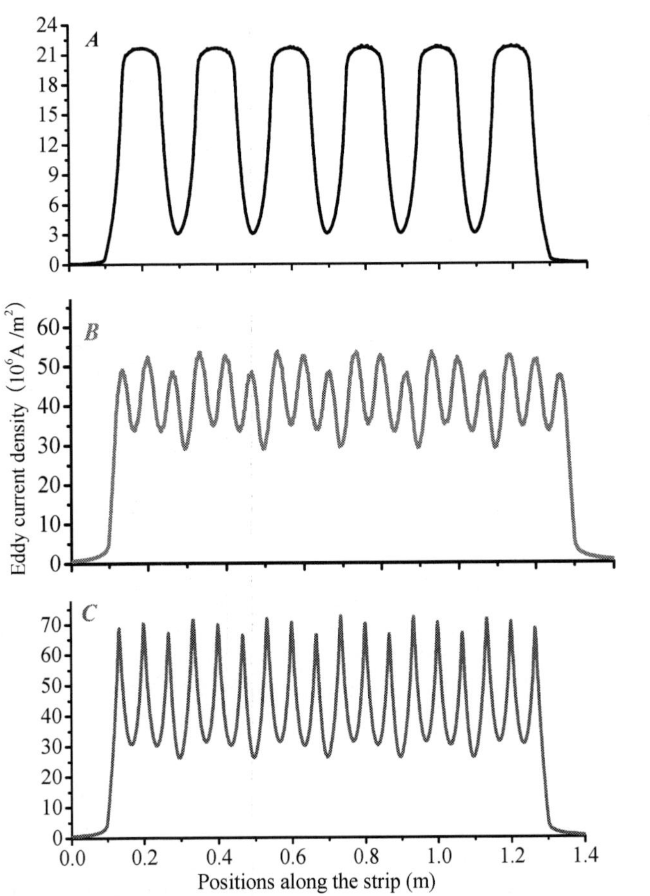

Fig. 20. The eddy current density distribution along the strip parallelly: (A) typical TWIH, (B) SW-TWIH and (C) improved SW-TWIH.

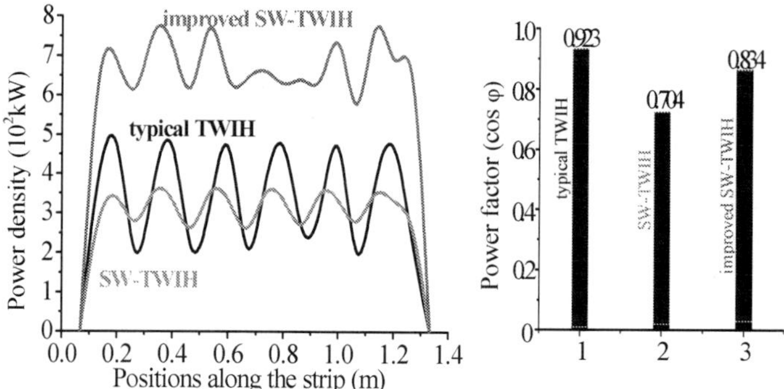

Fig. 21. Relative power density distributions and the power factor.

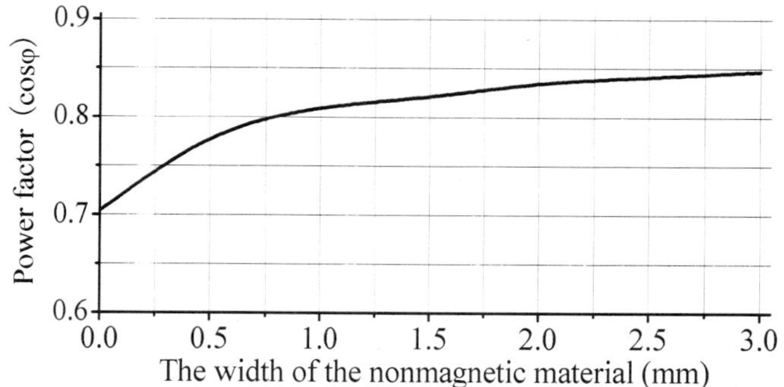

Fig. 22. The power factor with different width of the nonmagnetic material.

4. Conclusion

The simulations have shown that a relative good power and eddy currents distribution across the strip can be obtained for both systems using the method of spatial control of the electromagnetic field. The theoretical values of efficiency (with uniform current density distribution in the inductor coil cross-section) are quite high for all the configurations. The TWIH systems, especially for the improved C-TWIH and the SW-TWIH systems, look to be more promising for further study because it might solve some problems related to metal deformation and noise from the system, which play important roles to the uniform of power and eddy currents. Moreover, the values of power density in the load are able to produce good heating of the strip.

Also, it is demonstrated that a convenient choice of the systems and their parameters (inductor geometry, poly-phase winding arrangement and supply, air gap values and control of the relative position of the inductors) gives the opportunity of obtaining very good energetic performances and a certain degree of control of the value and distribution of the power transferred to the load. Some standard and specially developed FEM codes based on numerical methods were successfully used for the study of TWIH heaters. The main advantages and the particular characteristics of TWIH systems such as the effect of slots on induced power distributions were studied. The FEM codes and the results of numerical simulation may be therefore used easily and reliably, for the design of different TFIH systems.

It should be added that eddy current distributions and power intensity are not the only problems for the use of these systems and for this reason the comparison must be studied in depth. First of all, the electromagnetic problem solution must be coupled with the thermal and the mechanical ones in order to have a better understanding of the different temperature distributions, mechanical deformations and noise. Moreover an economic evaluation of the different power supply systems must be carried out.

5. Referring

We have referred many materials from the following papers we have published.

[1] Ho S. L., Wang Junhua, Fu W. N., Wang Y. H. "A novel crossed traveling wave induction heating system and finite element analysis of eddy current and temperature distributions," *IEEE Trans. Magn.*, vol: 45(10), pp: 4777-4780, Oct. 2009.

[2] J. H. Wang, S. L. Ho, W. N. Fu, Y. H. Wang "Design and analysis of a novel traveling wave induction heating system with magnetic slot wedges for heating moving thin strips ," *IEEE Trans. Magn.*, vol.20, no.6, pp. 2175 – 2178, 2010.

[3] Wang Junhua, Wang Youhua, "The study of two novel induction heating technology," *ICEMS 2008*, Oct. 17-20 2008, Wuhan China, pp: 572-574. .

[4] Wang Youhua, Wang Junhua, Li Jiangui, Li Haohua, "Analysis of induction heating eddy current distribution based on 3D FEM," *2008 IEEE Region 8 International Conference on Computational Technologies in Electrical and Electronics Engineering*, July 21-25 2008, Novosibirsk, RUSSIA, pp: 238-241.

[5] Yang Xiaoguang, Wang Youhua, "New Method for Coupled Field Analysis in Transverse Flux Induction Heating of Continuously Moving Sheet," *Heat Treatment of Metals*, vol.29, no.4, pp.53-57, 2004.

6. References

[1] A.L.Bowden, E.J.Davies, "Travelling Wave induction Heaters Design Considerations," *BNCE-UIE Electroheat for Metals Conference*, 11.5.2, Cambridge (England), 21-23 Sept. 1982

[2] S.Lupi, M.Forzan, F.Dughiero, et al. "In the corresponding TWIH system this problem is reduced since less and not sharp peaks are present with their highest," *IEEE Transactions on Magnetics*, 1999, vol35, no.5, pp.3556-3558.

[3] F.Dughiero, S.Lupi, P.Siega, "Analytical Calculation of Traveling Wave Induction Heating Systems," *International Symposium on Electromagnetic Fields in Electrical Engineering* 1993, 16-18 September 1993, Warsaw-Poland, 207-210.

[4] F.Dughiero, S.Lupi, V.Nemkov, et al. "Travelling wave inductors for the continuous induction heating of metal strips," *Proceedings of the Mediterranean Electrotechnical Conference-MELECON*.1994, vol.3 , no.3, pp.1154-1157.

[5] W. Andree, D. Schulze, Z. Wang , "3D FEM Eddy current computation in transverse flux induction heating equipment," *IEEE Transaction on Magnetic*, vol. 30, No.4, July 1994.

[6] Zanming Wang, Xiaoguang Yang, Youhua Wang, et al. "Eddy current and temperature field computation in transverse flux induction heating equipment," *IEEE Trans. Magn.*, vol. 37, no. 5, pp. 3437-3439, 2001.

[7] Z. Wang, X. Yang et al, " FEM Simulation of Transverse Flux Induction Heating for Galvanizing Line" Proceedings of the Fourth International Conference on Electromagnetic Field Problem and Applications.

[8] S. Lupi, M. Forzan, F. Dughiero, et al. "Comparison of edge-effects of transverse flux and travelling wave induction heating inductors," *IEEE Trans. Magn.*, vol. 35, no. 5, pp. 3556 -3558, 1999.

[9] F.Dughiero, S.Lupi, V.Nemkov, et al. "Travelling wave inductors for the continuous induction heating of metal strips," *Proceedings of the Mediterranean Electrotechnical Conference-MELECON*.1994, vol.3 , no.3, pp.1154-1157.

[10] A.Ali, V.Bukanin, F.Dughiero, et al. "Simulation of multiphase induction heating systems," *IEE Conference Publication*, 1994, vol.38, no.4, pp.211-214.

[11] V.V.Vadher, I.R.Smith. "Travelling Wave Induction Heaters with Compensating Windings", *ISEF"93, Warsaw (Poland)*. 16-18 Sept. 1993, pp. 211-217.

[12] X.G. Yang, Y.H. Wang, "The effect of coil geometry on the distributions of eddy current and temperature in transverse flux induction heating equipment," *Heat Treatment of Metals*, 2003, vol.28, no.7, pp.49-54.

[13] S. L. Ho, J. H. Wang, W. N. Fu, Y. H. Wang , "A novel crossed traveling wave induction heating system and finite element analysis of eddy current and temperature distributions," *IEEE Trans. Magn.*, vol. 45, no. 10, pp. 4777-4780, 2009.

[14] Nicola Bianchi, Fabrizio Dughiero, "Optimal design techniques applied to transverse-flux induction heating systems," *IEEE Trans. Magn.*, vol. 31, no. 3, pp. 1992-1995, 1995.

[15] Y. B. Li, S. L. Ho, W. N. Fu, W. Y. Liu, "An interpolative finite-element modeling and the starting process simulation of a large solid pole synchronous machine," *IEEE Trans. Magn.*, vol. 45, no. 10, pp. 4605-4608, 2009.

10

Induction Heating of Thin Strips in Transverse Flux Magnetic Field

Jerzy Barglik
Silesian University of Technology
Poland

1. Introduction

An idea of transverse flux induction heating has been well known for more than sixty years. Some of its main principles were presented for instance by Russian physicist Volovgin (Mühlbauer, 2008). Many papers on the topic considering mainly theoretical aspects of the task were published in the last sixty years (Baker, 1950), (Jackson, 1972), (Barglik, 1992), (Mühlbauer et al, 1995), (Tudorache & Fireteanu, 1998), (Nacke et al, 2001), (Dughiero et al, 2003). Usage of transverse flux induction heating system have been effective in case of thin strips of the thickness comparable with the depth of electromagnetic field penetration δ. It makes it possible to obtain required parameters of the process like: uniformity of temperature distribution within the workpiece and big total energy efficiency at rather low frequencies of the field current (in many cases also for mains frequency f = 50 or 60 Hz) in comparison with more often used classical induction heaters working with longitudinal magnetic field. The chapter deals with continual transverse flux induction heating of thin non-ferrous metal strips. The three-dimensional model used for the analysis was based on a system of non-linear differential equations for coupled electromagnetic and temperature fields that were solved by numerical methods. Development of the mathematical modelling of transverse flux induction heating as well as the computations by means of professional software and supplementary user codes reach quite a high level. But there is still a problem with the accuracy of calculations that was associated mainly with reliability of existing mathematical and numerical models with respect to the physical reality. The reason consists in various simplifications accepted for shortening the time of calculations and level of knowledge of temperature dependent material properties as well as convection and radiation heat transfer coefficients for the particular task. So the best way for obtaining data necessary for an optimal design and construction of induction heaters seems to be a numerical analysis supported by well planned experiments. Illustrative examples of continual transverse flux induction heating were analyzed. Results of numerical simulation were presented and compared with measurement data taken from laboratory stands. Quite good accordance between calculations and measurements was achieved. An illustrative example demonstrates that one of the ways making possible to obtain an uniform temperature distribution along the width of the workpiece could be a proper selection of suitable frequency of the field current.

2. Idea of transverse flux induction heating

Induction heating of thin flat workpieces, similar like levitation or cold crucible melting, induction heating for semi-liquid state before plastic working and surface induction hardening belongs to the quickest developed areas of modern electroheat (Barglik, 2010), (Baake & Nacke, 2010). Usage of typical induction heaters with longitudinal magnetic field do not allow to obtain high enough electrical efficiency of inductor-workpiece system. In order to achieve high enough value of electrical efficiency for induction heating system with longitudinal magnetic field the ratio of workpiece thickness g to electromagnetic field penetration δ should be equal or bigger than 3.5

$$g \geq 3.5\delta \tag{1}$$

However it requires to use a field current with properly increased frequency and consequently necessity of an expensive power source. Usage of induction heater with transverse flux field makes it possible to obtain big enough electrical efficiency of inductor-workpiece system at distinctly lower frequency of field current. So it is possible to use in that case cheaper power source. Transverse flux magnetic field penetrates the workpiece and has practically only normal component of magnetic flux density perpendicular to external plane of the flat workpiece. The tangent component of magnetic flux density is more and more smaller and often could be neglected. Figure 1 presents comparison of electrical efficiency of inductor-workpiece systems with transverse flux and longitudinal magnetic fields. It shows dependencies of electrical efficiency for both systems in case of stationary heating of brass strip with thickness of 3.2 mm.

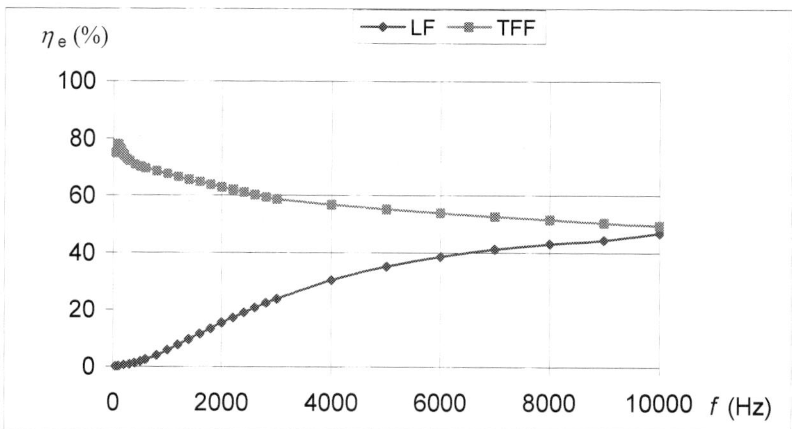

Fig. 1. Comparison of electrical efficiency of induction heater with longitudinal (LF) and transverse flux (TFF) fields on field current frequency (Barglik, 2005)

Two-dimensional and three dimensional models making possible to determine a distribution of electromagnetic and temperature fields in conductive workpiece were elaborated. Some kind of summary was presented for instance in (Barglik, 2002). It contains a broad overview of earlier presented papers on application of transfer flux induction heating systems. A special emphasis was put on flexible transverse flux induction heating systems for non-ferrous workpieces, making possible to heat strips with different widths by means of universal type of inductor. In order to solve such a problem the inductor could be divided into a few single inductors with different widths and also various positions to the axis of the strip. Usage of transverse flux induction heating systems is especially effective in case of heating of strips of thickness comparable with electromagnetic field penetration. It makes it possible to achieve expected process parameters like uniformity of temperature distribution in a whole volume of the workpiece and big total efficiency of the heater at relatively low field current frequency (sometimes even at mains frequency f = 50 or 60 Hz) in comparison with much bigger values typically obtained for classical longitudinal induction heating systems. Three-dimensional calculation model based upon partial differential equations for electromagnetic and temperature fields was elaborated and solved by proper numerical methods. Development of mathematical methods of tranverse flux induction heating systems and used professional software, numerical methods have reached a satisfied big level. A kind of problem seems to be an accuracy of calculations connected mainly with necessary simplifications of mathematical models making possible shortening of time of calculations and uncertainty of material properties of their temperature dependencies (Barglik, 2000). The best way for obtaining of data for calculation and designing of any transverse flux induction heating systems seems to be numerical analysis supplemented by well planned experimental part. Continual induction heating of thin non-ferrous metals strips (copper, brass, aluminum) was presented in the paper. Illustrative examples describing continual induction heating were described and analyzed. Results of simulation were compared with measurements done at laboratory stands.

3. Induction annealing

Term induction annealing means softening of treated material with usage induction heating (Davies, 1990). The process consists of rapid induction heating of treated material to a particular temperature, keeping it through some period of time and then cooling with a selected velocity. As a result requested microstructure and suspected softness of the material was obtained. Scheme of the industrial technologicial line for continual annealing of copper and its alloys strips in a classical horizontal system and a modified annealing line applied initial induction heating were presented in Fig. 2. Upper figure 2 presents a classical line. Strip (2) moves from uncoiler drum (1) through two zones of the technological line: heating chamber (3) and cooling chamber (5) to the final drum (6). In order to protect the strip against oxidation any protective atmosphere was used. In order to achieve requested velocity of strip movement of about 0.5 m/s it is necessary to use a long heating zone. It causes that the total length of the device often could be longer than 30 m. Lower figure 2 shows modified annealing line. Strip (2) moves from uncoiler drum (1) to a zone of induction heating. The induction heater has created a chamber jointly with resistance device (the induction-resistance device (7)). At the end of the induction heater zone the strip has reached the requested final temperature with uniform distribution within the annealed material. Resistance part of the device (7) makes it possible keeping of constant temperature for requested period of time. It is possible to use instead of indirect resistance device another electroheat device like for instance infrared heater (Hering, 1992).

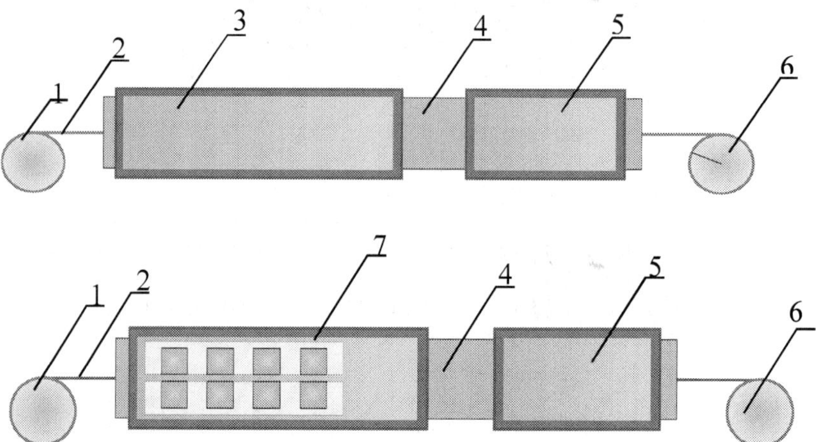

Fig. 2. Block scheme of technological annealing line in a horizontal system: classical line (upper figure), modified line with induction heating (lower figure). 1 – uncoiler drum, 2 – non-ferrous strip, 3 – heating chamber, 4 – container with inert atmosphere, 5 – cooling chamber, 6 – final drum, 7 – induction- resistance device.

Based upon exploitation experiences of the technological lines for annealing of copper and brass strips with initial transverse flux induction heating in Poland and abroad it has been possible to confirm that due to the modification total energy consumption has decreased for about 30 %. A different solution could be a vertical technological line. Besides of clear advantages of such a solution there was also some disadvantages (Barglik, 2000).

4. Basic of transverse flux induction heating

Let us consider transverse flux induction heating system used for instance for low temperature heating of thin non-ferrous metals strips (copper, brass, aluminum, lead and their alloys) (Fig.3). The moving strip (1) was placed symetrically in a narrow air-gap of two-side multipolar inductor. The strip moves with a constant velocity in x direction. Winding (2) was placed in grooves of the laminated magnetic core (3). It was constructed from copper profile conductors cooled by water. Coils were connected in series. The transverse flux induction heating system was supplied by one-phase harmonic current of frequency f and effective value of current I_z. As a power source directly power network or more frequently thyristor convertor were used. Usage of thyristor power source makes possible easy regulation of both quantities frequency of field current and power. Sometimes transistor power sources were used. A way of inductor connection and consequent direction of field current was indicated in Fig. 3 by red arrows. For such a system of connections in the moving strip time-varying transverse flux magnetic field was generated. Of course, if we modify a way of winding connection we could arrange a classical induction heating system with the longitudinal magnetic field (Barglik, 2002). A view of such an universal inductor was shown in Fig. 4.

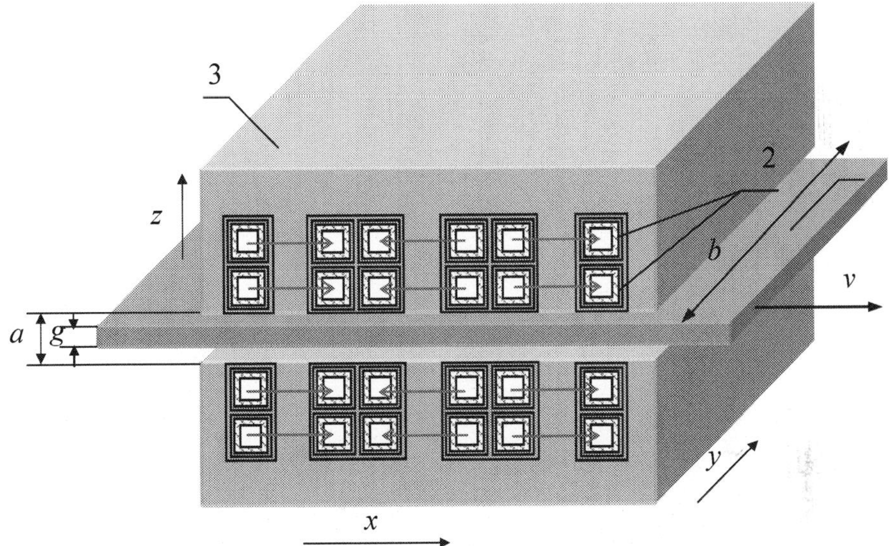

Fig. 3. Scheme of TFIH system. 1 – strip, 2 – winding, 3 – magnetic core, a – air-gap, b – width of the strip, g – thickness of the strip. Red arrows show current direction in winding

Fig. 4. A view of inductor from Fig.3

5. Mathematical model

Mathematical model of such multiphysic problem is quite complicated because of interaction between electromagnetic and temperature fields. In many cases it is also necessary to take into consideration also other physical fields like for instance heat stress field. The most important task is to recognize such parameters of the system like: frequency of field current, velocity of strip movement and its dimensions in order to obtain required exploitation parameters like: average value of workpiece temperature, uniformity of its distribution and big value of electrical efficiency of the system. Block scheme of calculation algorithm was shown in Fig. 5.

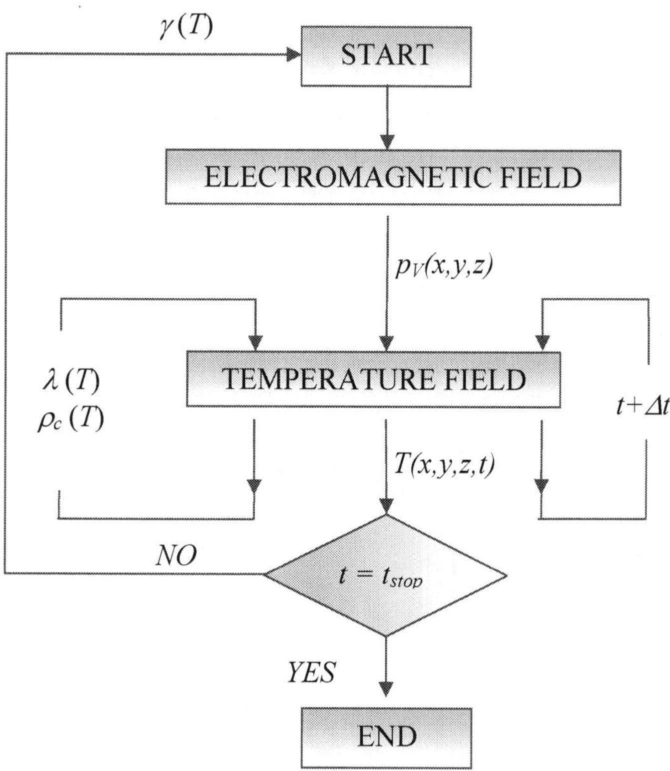

Fig. 5. Algorithm for calculation of coupled electromagnetic and temperature field

Calculations begin with the determination of electromagnetic field distribution in a whole area. Based upon that specific Joule losses released in the workpiece were calculated. Then temperature field was achieved. During these calculations non-linear dependencies of material properties on temperature were taken into consideration. If changes of the strip temperature were bigger than ΔT proper correction of electric conductivity were done and we start again with the electromagnetic calculations. The algorithm stops, if the proper selected criterion was satisfied. It was assumed that the criterion was satisfied for $t = t_{stop}$

Three-dimensional model for calculation of quasi-stationary electromagnetic field was shown in Fig. 6.

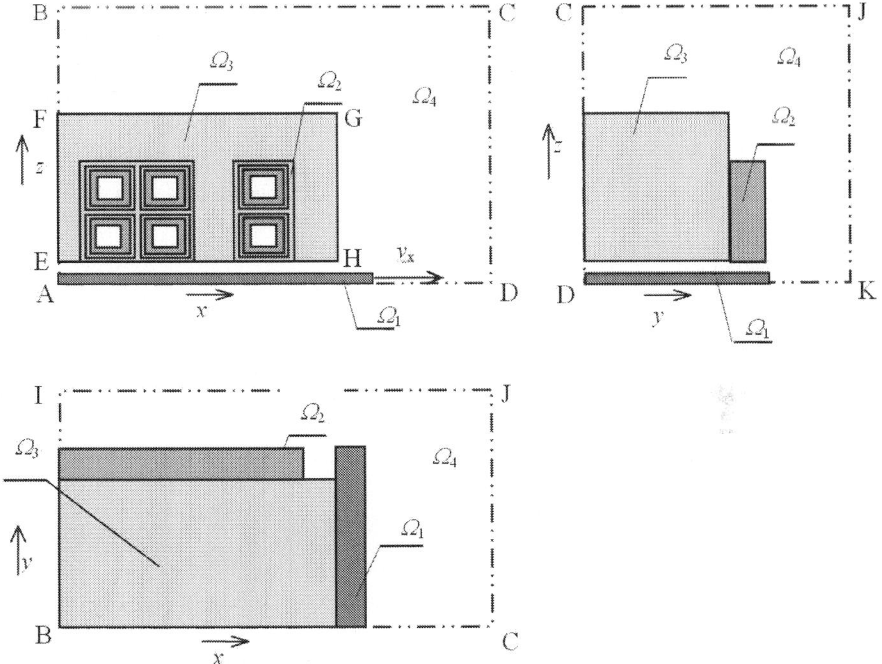

Fig. 6. Three-dimensional calculation model of electromagnetic field distribution.
Subdomains: Ω_1- strip, Ω_2- winding, Ω_3- magnetic core, Ω_4- surroundings (air),

Calculations were provided only for 1/8 part of the total calculation domain taking into account symmetry conditions at planes $x = 0$, $y = 0$, $z = 0$ (see Fig. 6) (Barglik, 2002). The task being in fact a kind of the open problem could be treated as a close one. It means that all the system was put to an interior part of rectangular prism which walls being external borders of the system were taken far enough to the inductor-workpiece arrangement. The borders were depicted in Fig. 6 by broken lines. Based upon testing calculations dimensions of the rectangular prism were determined. It was assumed that AB = 7 AF and AD = 5.5 AH (see Fig. 6). Electromagnetic field was described in a classical way by means of magnetic vector potential A.

$$\operatorname{curl}\frac{1}{\mu}\operatorname{curl}A + \gamma\frac{\partial A}{\partial t} - \gamma\left(v \times \operatorname{curl}A\right) = J_z \tag{2}$$

where μ denotes magnetic permeability, γ - electric conductivity, v – velocity, J_z - vector of field current density.

In case when magnetic permeability of magnetic core (subdomain Ω_3) could be considered as constant equation (2) transforms to a following form for phasor of magnetic vector potential $\underline{A}$

$$\operatorname{curl}\operatorname{curl}\underline{A} + j\omega\mu\gamma\underline{A} - \mu\gamma\left(v \times \operatorname{curl}\underline{A}\right) = \mu\underline{J}_z \tag{3}$$

where j denotes imaginary unit, ω - angular frequency.

For each subdomain equation (3) was modified by putting suitable values of angular frequency ω, magnetic permeability μ, electric conductivity γ, velocity of strip movement v and field current density $\underline{J}_z$ For instance for the subdomain Ω_1 (strip moving in direction of x axis with velocity v_x) equation (3) was transformed to a following form:

$$\operatorname{curlcurl}\underline{A} + j\omega\mu\gamma\underline{A} - \mu\gamma\left(v_x,0,0\right) \times \operatorname{curl}\underline{A} = 0 \tag{4}$$

Equation (3) could be additionally simplified by neglecting of the third component of the left side. It could be done for not big values of strip velocity. In order to check what kind of unaccuracy could be introduce as a result of such a simplification additional calculations were provided making possible to compare specific Joule losses caused by movement of the strip p_{vm} with specific Joule losses caused by eddy currents induced in the strip by time-varying electromagnetic field p_{vi}. Let us consider a ratio of these two values of specific Joulle losses as velocity coefficient k_r.

$$k_r = \frac{\displaystyle\int_V p_{vr}\mathrm{d}V}{\displaystyle\int_V p_{vi}\mathrm{d}V} \tag{5}$$

Calculations were done for two different frequencies of field current (f = 50, 2000 Hz) and for several velocities of strip movement. Exemplary results showing dependence of velocity coefficient on velocity of strip movement were presented in Fig. 7. For mains frequency (f = 50 Hz) and big velocity of strip movement (v_x = 1 m/s) velocity coefficient k_r reaches value of 3.12 %. If strip velocity two times smaller, the coefficient k_r decreases to a value of 0.78 %. When frequency of field current increases velocity coefficient k_r rapidly decreases. For frequency of field current f = 2000 Hz and velocity of strip movement v_x = 1 m/s velocity coefficient k_r= 0.0186 % only. For the same frequency and two times smaler velocity v_x = 0.5 m/s velocity coefficient k_r= 0.0047. More detailed analysis of the subject was presented in (Barglik, 2005).

Based upon the presented above analysis equation (3) for magnetic vector potential could be transformed into the form (6)

$$\operatorname{curl}\operatorname{curl}\underline{A} + j\omega\mu\gamma\underline{A} = \underline{J}_z \tag{6}$$

Equation (6) should be completed by adequate boundary conditions for magnetic vector potential. For planes of external borders of the domain (BIJC, DCJK, I′KJI – see Fig.6) Dirichlet condition for magnetic vector potential should be satisfied:

$$\underline{A} = 0 \tag{7}$$

The same condition was applied for symmetry plane (I′ABI – see Fig.6) characterized by antisymmetry of field currents. For other symmetry planes of the calculation domain: z = 0, ADKI′ and y = 0, ABCD (see Fig.6) condition (8) was applied:

$$\underline{A} \times n = 0 \tag{8}$$

a)

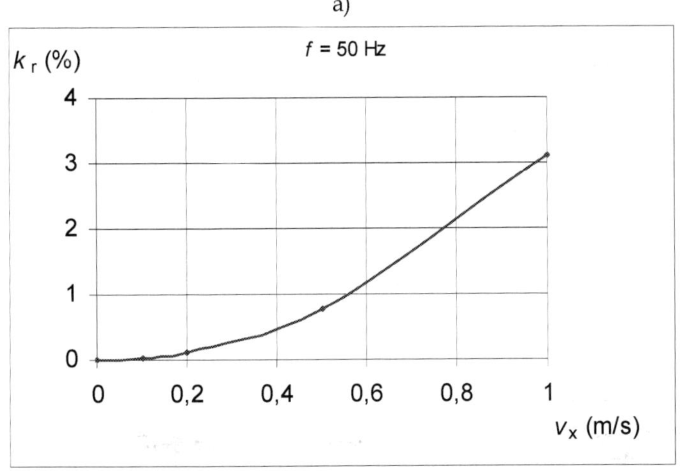

b)

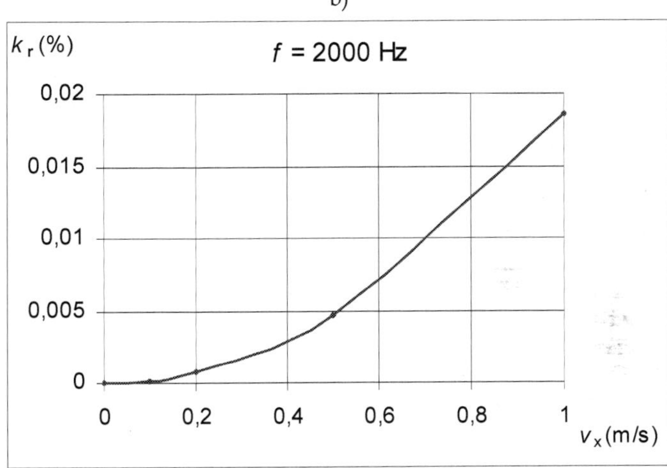

Fig. 7. Dependence of velocity coefficient on velocity of strip movement frequency of field current a) f = 50 Hz, b) f =2 kHz

Eddy current density induced in the strip $\underline{J}$ and specific Joule losses p_v were expressed by equations (9-10)

$$\underline{J} = j\omega\gamma\underline{A} \tag{9}$$

$$p_v = \frac{\underline{J}\cdot\underline{J}^*}{\gamma} \tag{10}$$

where $\underline{J}^*$ denotes the complex conjugate to $\underline{J}$.

Model for temperature field calculations was presented in Fig.8

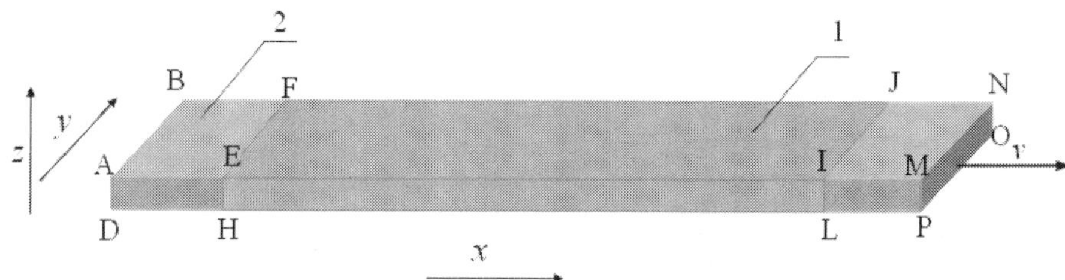

Fig. 8. Model for calculation of temperature field in the workpiece: 1 – part of the strip located inside air-gap of the inductor 2 – remaining part of the strip

As long as possible part of the moving strip was taken in order to correctly formulate boundary conditions on planes ABCD and MNOP (see Fig.8). Calculations were provided for ¼ part of the strip taking into account symmetry of temperature field to planes AMPD and DCOP (see Fig.8). Temperature distribution within the strip was determined by means of Kirchhoff-Fourier equation completed by specific Joule losses taken from electromagnetic calculations (Holman, 2009):

$$\mathrm{div}\left(\lambda\,\mathrm{grad}T\right) - \rho c\left(v\,\mathrm{grad}T\right) = -p_v \tag{11}$$

where λ denotes specific thermal conductivity, ρ - density and c – specific heat
Differential equation (11) was completed by a set of boundary conditions. At the artificial boundary plane ABCD (see Fig. 8) located sufficiently far from the inductor the condition (12) should be satisfied:

$$T\left(y,z,t\right) = T_p \tag{12}$$

where T_p denotes temperature of strip before heating.
At the second artificial boundary plane MNOP (see Fig.8) located also sufficiently far from the inductor the condition (13) should be satisfied:

$$\frac{\partial T}{\partial x} = 0 \tag{13}$$

At part 1 of upper plane of the strip located in the narrow air-gap of the inductor FJIE (see Fig. 8) classical boundary condition for convection was satisfied. Radiation was neglected

$$-\lambda\frac{\partial T}{\partial y} = \alpha_g\left(T - T_o\right) \tag{14}$$

where α_g denotes convection heat transfer coefficient in a narrow air-gap, T_o - ambient temperature.
Coefficient α_g in a narrow air-gap of the inductor was several times smaller than outside the inductor (Barglik, 2005). Neglecting of radiation seems to be reasonable hovewer it could a source of additional errors (Barglik et al, 2008). So for other external planes of the strip

ABFE, IJNM, BNOC (see Fig. 8) boundary conditions were formulated in a simplified way both for convection and radiation by introduction of total heat transfer coefficient α_z

$$-\lambda\frac{\partial T}{\partial n} = \alpha_z\left(T - T_o'\right) \tag{15}$$

where T_o' denotes average ambient temperature representing both radiation and convection. Total coefficient α_z was given by relation (16)

$$\alpha_z = \alpha_k + \sigma_o\cdot\left(T^2 + T_r^2\right)\cdot\left(T + T_r\right) \tag{16}$$

where α_k denotes convection heat transfer coefficient, σ_o - Stefan-Boltzman constant, T_r – temperature of surrounding surfaces involved in radiation heat transfer phenomena.
At both symmetry planes (AMPD, DCOP – see Fig.8) condition (17) should be satisfied:

$$\frac{\partial T}{\partial n} = 0 \tag{17}$$

In order to calculate electrical efficiency of the inductor-workpiece system was enough to take into consideration simplified two-dimensional calculation model shown in Fig.9. Phenomena in width of the strip could be neglected.

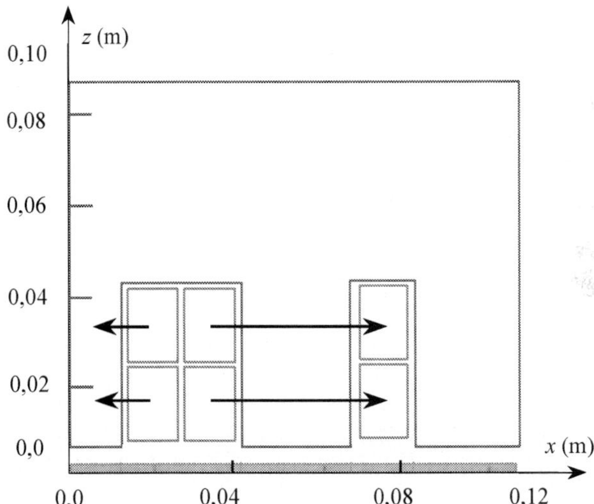

Fig. 9. Model for determination of electrical efficiency of inductor-workpiece system

Electrical efficiency of the inductor-workpiece system η_e were calculated by definition based upon equation (18)

$$\eta_e = \frac{\displaystyle\int_{V_s} p_v\mathrm{d}V_s}{\mathrm{Re}\{\boldsymbol{E}\times\boldsymbol{H}\}\mathrm{d}V_t} \tag{18}$$

where V_s - volume of the strip, Re – real part of phasor, $\boldsymbol{E}$ – vector of electric field intensity, $\boldsymbol{H}$ – vector of magnetic field intensity, V_t - total volume of the system

6. Illustrative example

Based upon the presented above model:
- temperature distribution within the moving strip,
- electrical efficiency of the inductor-workpiece system

were calculated. For the temperature calculations 3D model was used in weak-coupled formulation. First, the distribution of electromagnetic field was calculated by means of FEM-based software OPERA 3D. Eddy current, specific Joule losses and finally temperature distribution were determined. Non-linear dependencies on temperature of material properties and parameters characterized heat transfer were taken into account. For calculations of electrical efficiency of the transverse flux induction heating system 2D model in quasi-coupled formulation was used. Calculations were done for many variants. Only small part of them was presented in the paper. Some basic input parameters and main dimensions of the system were presented below:

- frequency of field current f = (50-3000 Hz),
- inductor current I = 1000 A,
- length of inductor l_i = 0.217 m,
- total width of inductor b_i = 0.35 m,
- width of magnetic core of the inductor b_c = 0.25 m,
- height of inductor (half of two-side inductor) h_i = 0.08 m,
- relative magnetic permeability of the magnetic core μ_r = 1000,
- thickness of the air-gap of the inductor a = 0.01 m or a = 0.02 m,
- material: copper, brass (Cu60), aluminium,
- length of the strip taken to the calculations l = 0.34 m,
- strip thickness g = 0.0032 m or g = 0.002 m or g = 0.001 m or g = 0.0005 m,
- strip velocity v = 0.01 or v = 0.02 or v = 0.04 m/s,
- ambient temperature T_0 = 20°C,
- temperature on inlet of the inductor T_p = 20°C,
- material properties in ambient temperature,
    - electric conductivity: copper γ_{Cu} = 5.7·10⁷ S/m, brass γ_b = 1.43·10⁷ S/m, aluminum γ_{Al}= 3.7·10⁷ S/m,
    - relative magnetic permeability μ_r = 1,
    - density: copper ρ_{Cu} = 8900 kg/m³, brass ρ_b = 8600 kg/m³, aluminum ρ_{Al} = 2700 kg/m³,
    - thermal conductivity: copper λ_{Cu} = 395 W/mK, brass λ_b = 144 W/mK, aluminum λ_{Al} = 211 W/mK,
    - specific heat: copper c_{Cu} = 381 J/kgK, brass c_b = 410 J/kgK, aluminum c_{Al} 933 J/kgK,
- convection heat transfer coefficient in a narrow gap a_g = 5 W/m²K
- convection heat transfer coefficient outside the inductor a_k = 20 W/m²K,
- Stefan –Boltzmann constant σ_o = 5.6676 ·10⁻⁸ W/(m² ·K⁴)

As it was mentioned earlier non-linear characteristics on temperature of all material properties were taken into account.

6.1 Mains frequency induction heating

Distribution of eddy-current density for two different widths of the brass strip was shown in Fig. 10.

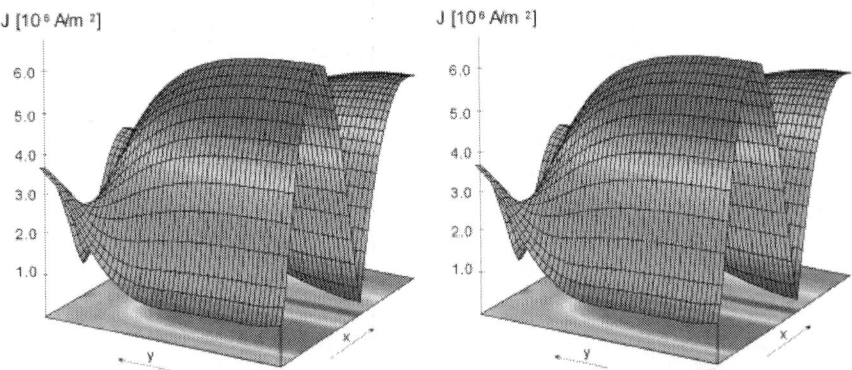

Fig. 10. Distribution of absolute value of eddy current density at internal plane of flat brass workpiece for two different widths of the brass strip $b = 0.35$ m (left histogram), $b = 0.25$ m (right histogram).

In the first case shown in left histogram of Fig. 10 the width of the strip was equal to the total width of the inductor. In the second case shown in right histogram of Fig. 10 the width of the strip is equal to the width of the magnetic core. Absolute value of eddy current density reaches its maximum exactly in an axis of the workpiece. In both cases the distribution of eddy current was strongly non-uniform. Let us calculate a coefficient of non-uniformity of eddy current distribution defined as an absolute value of eddy currents in the axis of the strip to absolute value of eddy currents near its edges:

$$k_J = \left. \frac{J_{y=0}}{J_{y=b}} \right|_{x=x_{max}} \tag{19}$$

where x_{max} denotes such a value of coordinate x for which abslolute value of eddy current density reaches its maximum. Results were selected in Tab. 1

Based upon comparison of two cases characterized by different width of the strip it was assumed that the distribution of eddy current density was more uniform in case of narrow strip. Distribution of specific Joule losses released in the brass strip was presented in Fig. 11. Parameters describing specific Joule losses distribution were collected in Tab. 2. Coefficient of non-uniformity of specific Joule losses k_p was defined as a ratio of value of specific Joule losses in an axis if the strip to the value of specific Joule losses near the edge of the strip.

$$k_p = \left. \frac{p_{v_{y=0}}}{p_{v_{y=b}}} \right|_{x=x_{max}} \tag{20}$$

In both presented cases average values of specific Joule losses seem to be practically the same. But non-uniformity of distribution of specific Joule losses was relatively bigger.

Absolute value of eddy current density (10^6 A/m²)	Value	Width b (m)	
		0.35	0.25
	average J_a	3.14	3.53
	in an axis $J_{y=0}$ for x_{max} = 0.0275 m	7.132	6.736
	near the edge $J_{y=b}$ for x_{max} = 0.0275 m	1.96	3.7
Coefficent of non-uniformity k_J (-)		3.63	1.819

Table 1. Parameters describing non-uniformity of eddy current distribution (f = 50 Hz)

specific Joule losses (10^7 W/m³)	Value	width b (m)	
		0.35	0.25
	average $p_{vśr}$	1.067	1.264
	in the axis $p_{v(y=0)}$ for x_{max} = 0.0275 m	4.168	4.096
	Near the edge $p_{v(y=b)}$ For x_{max} = 0,0275 m	0.32	2.02
Non-uniformity coefficient k_p (-)		13.1	2.03

Table 2. Parameters describing non-uniformity of specific Joule losses distribution

For the first case of induction heating of the broader strip (b = 0.35 m) coefficient of non-uniformity of specific Joule losses k_p= 13.1, however fo the second case of transverse flux induction heating of the narrower strip the coefficient seems to be several times smaller (k_p = 2.03).

Some examplary results of specific Joule losses distribution in the moving brass strip were shown in Fig. 11. Non-uniform specific Joule losses distribution in the strip width was observed. In both cases the specific Joule losses reaches its maximum in the axis of the strip. Shape of the curie strongly depends on the width of the strip. More precisely speaking it

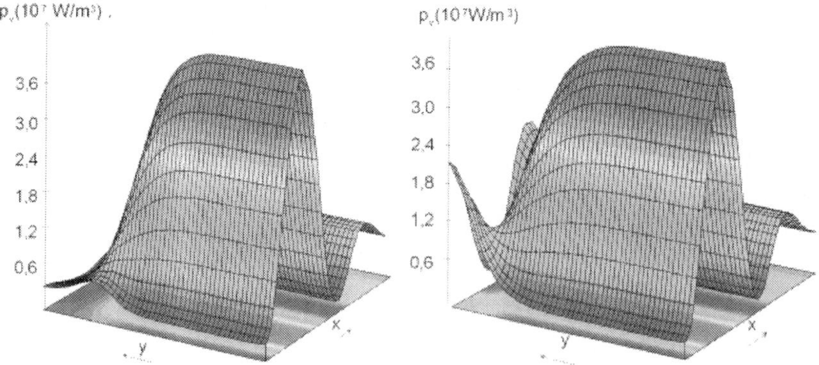

Fig. 11. Specific Joule losses for two different strip widths b = 0.35 m (left histogram), b = 0.25 m (right histogram)

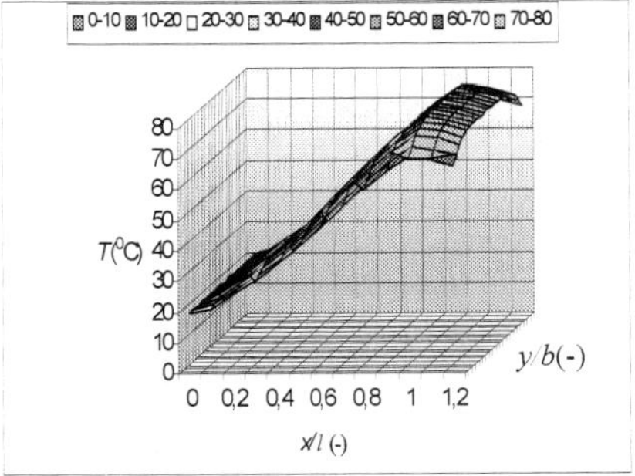
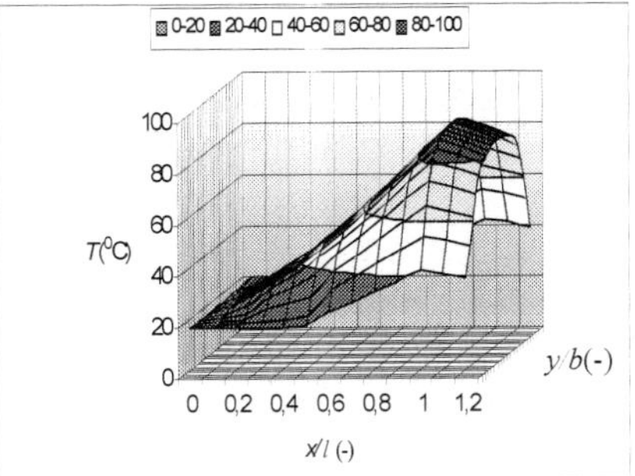

Fig. 12. Temperature distribution on upper plane of brass strip heated by induction moving with velocity $v = 0.01$ m/s for two different widths $b = 0.35$ m (left diagram) and $b = 0.25$ m (right diagram)

depends on the ratio between the width of the strip and the total width of the inductor. In case of broader strip ($b = b_i$) decrease of temperature seems to be more clear (about 53 % at the outlet of the inductor). In case of narrower strip decrease of temperature was distincly smaller ($b = b_c$) reaching of about 11 %. Applied supply system with mains frequency makes it possible to heat up the strip to relatively low temperature (see Fig.12), even for relatively small velocity of strip movement. In order to obtain big temperatures it is necessary to increase the frequency of field current.

6.2 Medium frequency transverse flux induction heating

The most important parameter strongly influenced on uniformity of eddy current distribution and consequently on specific Joule losses was the frequency of field current. Distribution of specific Joule losses at the internal plane of brass strip for two selected frequencies of field current ($f = 200$ Hz and $f = 2000$ Hz) was shown in Fig. 13.

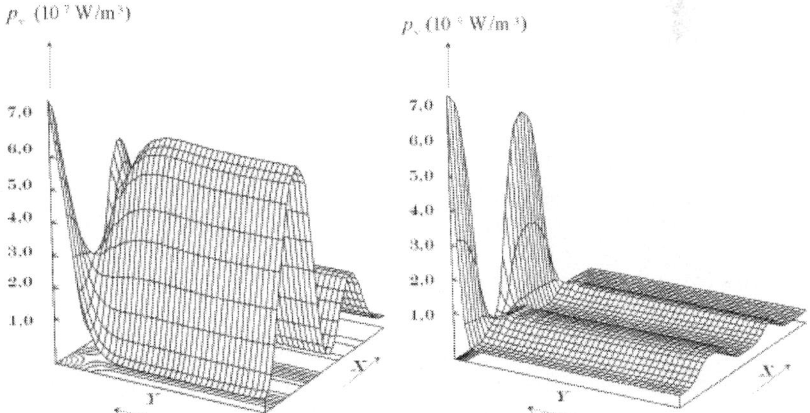

Fig. 13. Distribution of specific Joule losses at internal plane of brass strip ($b = 0.25$ m) for $f = 200$ Hz (left histogram) and $f = 2000$ Hz (right histogram)

Parameters describing distribution of specific Joule losses in the workpiece for three different frequencies were presented in tab. 3

Specific Joule losses(10^7 W/m³)	Frequency f (Hz)					
	Brass			Aluminum		
	50	200	2000	50	200	2000
Average value $p_{v,av}$	1.264	5.4	15.12	2.11	6.4	15.82
Value at the axis $p_v\|_{y=0}$ for x_{max} = 0.0275 m	4.096	6.3	10.18	3.1	6.8	10.5
Value at the axis $p_v\|_{y=b}$ for x_{max} = 0.0275 m	2.02	7.05	72.45	2.5	9.5	91.45
Coefficient of non-uniformity k_p (-)	2.03	0.888	0.14	1.24	0.674	0.1148

Table 3. Parameters describing specific Joule losses in the workpiece

If frequency of field current increases average value of specific Joule losses also increase. If we try to compare shapes of specific Joule losses histograms for different field current frequencies we learn about big differences between them. For mains frequency (f = 50 Hz) (see Fig. 11 right histogram) specific Joule losses reach their maximum at the axis of the strip. For bigger frequency f = 200 Hz distribution of specific Joule losses (see Fig. 13 left histogram) seems to be more uniform. Difference between maximum value reached near edge of the strip and value at the axis is rather small. Coefficient of non-uniformity is equal to 0.888. At bigger frequency (Fig. 13 – right drawing) distribution of the specific Joule losses become again strongly non-uniform with a distinct maximum near the edge of the strip. Coefficient of non-uniformity was equal to 0.138 for that case. Exemplary results of temperature calculations in moving strip for frequency of field current f = 2000 Hz was shown in Fig. 14.

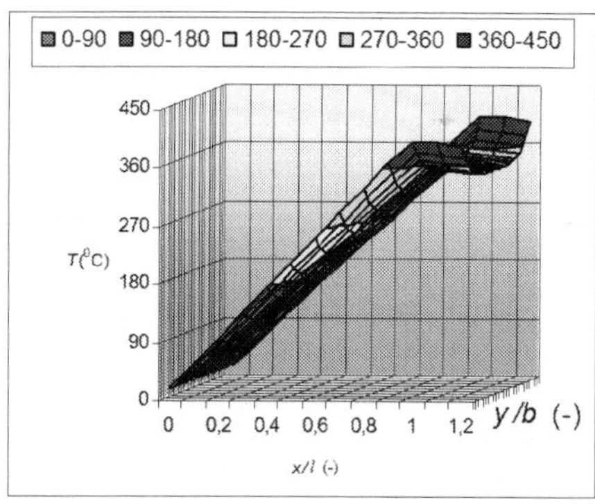

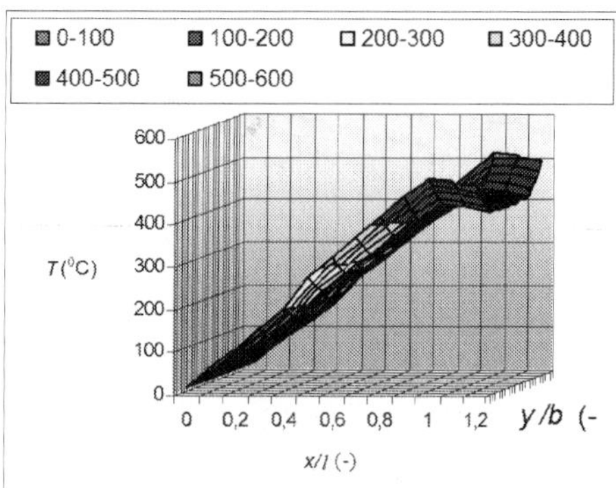

Fig. 14. Temperature distribution at upper plane of brass (left diagram) and aluminum (right diagram) strip

Non-uniform temperature distribution in width of the strip was observed with maximum near the edges. In we take into consideration the criterion of temperature distribution uniformity we could assume that it was only one frequency of field current being the optimal one from that point of view. For presented cases of induction transverse flux heating the optimal frequency of field current was about 150 – 300 Hz. For lower frequencies shape of temperature curve was again non-uniform with the maximum near the axis of the strip. The optimal frequency for transverse flux induction heating of brass strip of thickness g = 3.2 mm was 275 Hz (Barglik, 2002) an for aluminium strips of same thickness was slightly smaller (f_{opt} = 175 Hz) (Barglik, 2005).

6.3 Electrical efficiency

As it was mentioned earlier for calculation of electrical efficiency of the transverse flux inductor-workpiece system a simplified two-dimensional calculation model for used. Below were presented some results of the calculations. Dependence of electrical efficiency of the inductor-brass strip system on frequency for two different values of relative air-gaps of the inductor defined as a ratio of thickness of the inductor air-gap a and the thickness of the strip g was shown in Fig. 15.

$$a' = \frac{a}{g} \tag{19}$$

For induction heating of the strip with thickness g = 5 mm electrical efficiency reaches its maximum for frequency of field current $f \approx 100$ Hz. In case of strip with thickness g = 3.2 mm the maximal value of the electrical efficiency of the system was three times bigger ($f \approx 300$ Hz).

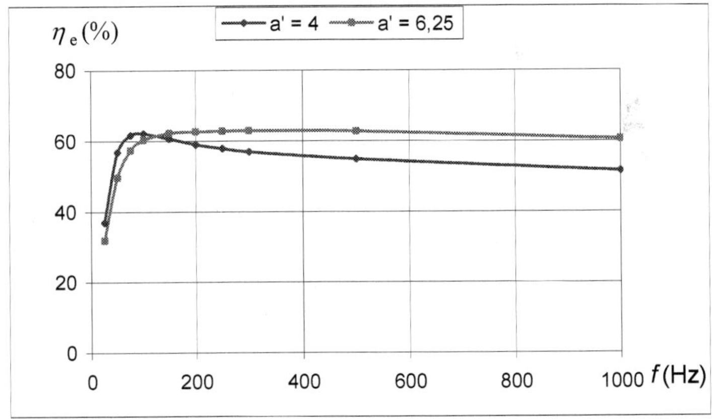

Fig. 15. Dependence of electrical efficiency of inductor-brass strip for two relative air-gap of the inductor-strip system (a = 20 mm, g = 3.2 mm, 5 mm).

Dependence of electrical efficiency of inductor-brass strip system for heating of strips with thicknesses: g = 1 mm, 2 mm was presented in Fig. 16

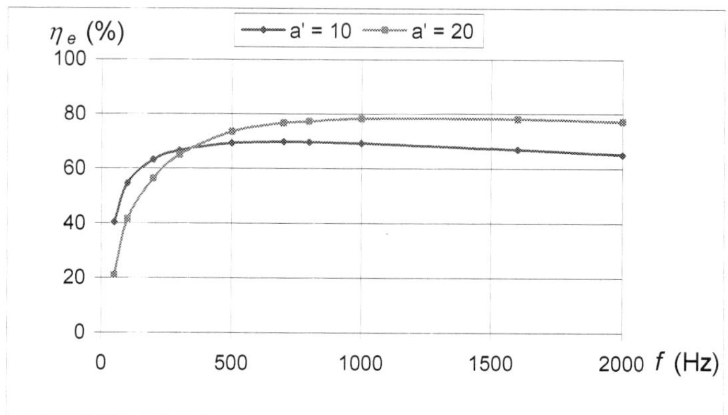

Fig. 16. Dependence of electrical efficiency of inductor-brass strip system on field current frequency for g = 2 mm, a' = 10 and g = 1 mm, a' =20

The air-gap of the transverse flux inductor was again a = 20 mm. Electrical efficiency of the transverse flux inductor-strip system η_e reaches its maximum at bigger frequencies of field current. For thickness of the strip g = 2mm and the relative air-gap of the inductor a' = 10, the electrical efficiency of the system reaches its maximum ($\eta_e = \eta_{max}$) at frequency of field current $f \approx 700$ Hz. For smaller strip thickness g = 1 mm and the relative air-gap of the inductor a' = 20 the electrical efficiency of the system reaches its maximal value ($\eta_e = \eta_{max}$) for almost two times bigger frequency of field current $f \approx 1300$ Hz.

Dependence of electrical efficiency of the transverse flux inductor-brass strip system for two kinds of the heated material: brass and copper was presented in Fig. 17. Comparison was provided for very thin strips (thickness of the strip g = 0.5 mm) heated in the basic configuration of the inductor (air-gap of the inductor a = 20 mm). So the relative air-gap of the inductor strip-system was big (a' = 40).

During the transverse flux induction heating of very thin brass strips maximum of electrical efficiency of the system η_e was reached at the frequency of field current $f \approx 2000$ Hz. For bigger frequencies of the field current the electrical efficiency of the transverse flux inductor – brass strip system slowly decreses. Similarly for smaller frequencies the electrical efficiency of the transverse flux inductor – brass strip system quickly decreases reaching value of only η_e =0.25 for frequency of field current f = 50 Hz. Completely different situation was in case of the transverse flux induction heating of similar copper strips. Maximal value of the electrical efficiency of the transverse flux inductor – copper brass system ($\eta_e = \eta_{max}$ = 0.79) was obtained for frequency of field current $f \approx 200$ Hz. Then, for bigger frquencies η_e the electrical efficiency of the system slowly decreases up to the value of η_e = 0.71 for the frequency of field current $f \approx 200$ Hz. For the mains frequency of field current f = 50 Hz the electrical efficiency of the system was equal to η_e = 0.7.

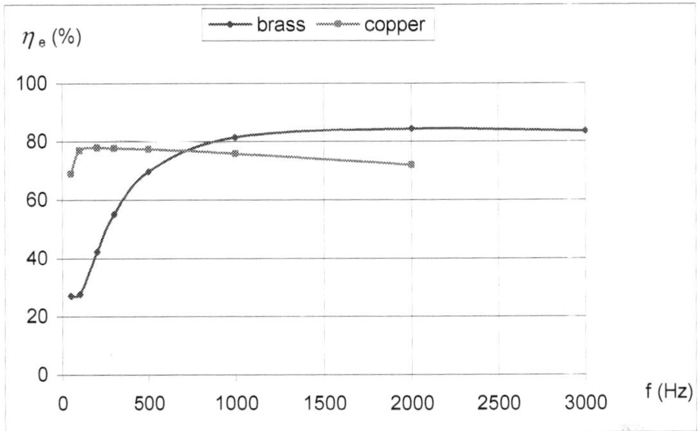

Fig. 17. Dependence of electrical efficiency of inductor – strip system for two kinds of material: brass and copper

Dependence of the electrical efficiency of the transverse flux inductor-strip system η_e on frequency of field current for brass strip (thickness g = 3.2 mm and air-gap of the inductor a =10 mm) was shown in Fig. 18

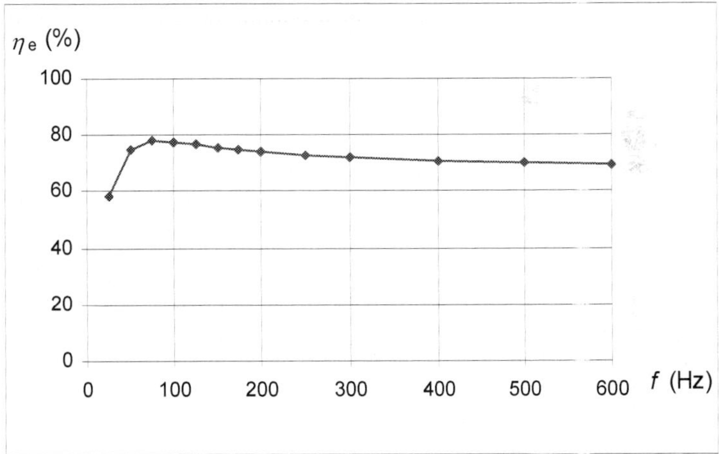

Fig. 18. Dependence of electrical efficiency of the inductor – brass strip system on field current frequency (g = 3.2 mm, a =10 mm)

For such a configuration of the system the maximum value of the electrical efficiency of the inductor-strip system was reached for frequency of field current of about $f \approx 75$ Hz.

7. Experimental part

7.1 Laboratory stand

A block scheme of the laboratory stand equipped with a thyristor power source of medium frequency used for measurements necessary for experimental validation of computations was shown in Fig. 19. The similar laboratory stand not presented in the paper, making possible to supply the transverse flux induction heater by current of mains frequency was described in details in (Barglik, 2004). Brass strip (6) moves with the constant, but regulated velocity from an uncoiler drum (5) to a final drum (7) through the narrow air-gap of two-side, multipolar transverse flux inductor (4) (see Fig. 3 and Fig. 4). Velocity of strip movement was regulated fluently by means of stearing panel (9) containing two separate thyristor power sources: one of them for each direction of strip movement. In order to measure the final temperature of the strip T_k at the outlet of the inductor, pyrometer (8) was used. A specialized measured system (10) contaning converters making possible to register of measurement data in the proper database computer system (12) was applied. The thyristor power source (1) has its nominal active power of about 200 kW and frequency of harmonic field current regulated in a range of (1 - 2.5 kHz) by means of block of capacitors (2). The transverse flux inductor was connected to the thyristor power source through the water cooled matching transformer (3).

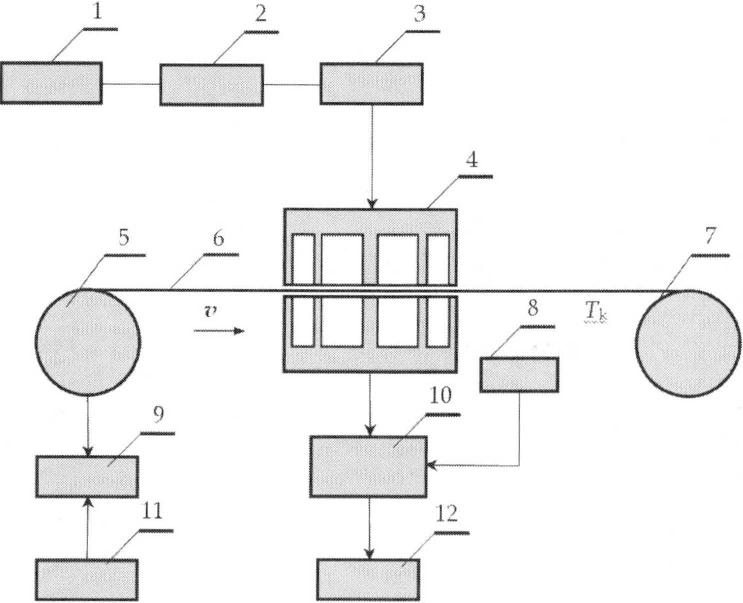

Fig. 19. Block scheme of the laboratory stand. 1- thyristor power source of medium frequency, 2 – block of capacitors, 3 – matching transformer 4 - transverse flux field inductor, 5 – uncoiler drum, 6 –brass strip, 7 – final drum, 8 –pyrometer, 9 – stearing panel, 10 – block of measurement converters, 11 – control desk, 12 – database computer system

The overall view of the laboratory stand (block scheme in Fig. 19) was shown in Fig. 20. The stand was built in the laboratory of Electroheat of the Silesian University of Technology in Katowice, Poland. At the first plane of the photograph the control desk with the computer for registration of measurement have been seen. Then we could see the transverse flux inductor. It was placed at the special basis making possible a fluent regulation of the thickness of the inductor air-gap. At the last plane of the photograph we can see the thyristor power source and other elements of the supply system: the cubicle with capacitors (on left) and the matching transformer of medium frequency.

7.2 Measurements

The measurements were done at two separate laboratory stands. The transverse flux field inductor could be supplied in two ways: directly from 50 Hz regulated transformer (construction of such the laboratory stand was described in (Barglik, 2004)) or by field current of medium frequency through the thyristor power source, block of capacitors and the matching transformer (see Figs. 19, 20). The temperature distribution at the outlet of the inductor ($x = 1.25\ l$) supplied by field current of mains frequency in the width of the moving brass strip for two different velocities of strip movement was shown in Fig. 21. Computations (blue line in Fig. 21) were compared with the measurements done at the laboratory stand (red points in Fig. 21). Characteristic was a shape of temperature curves

Fig. 20. View of the laboratory stand

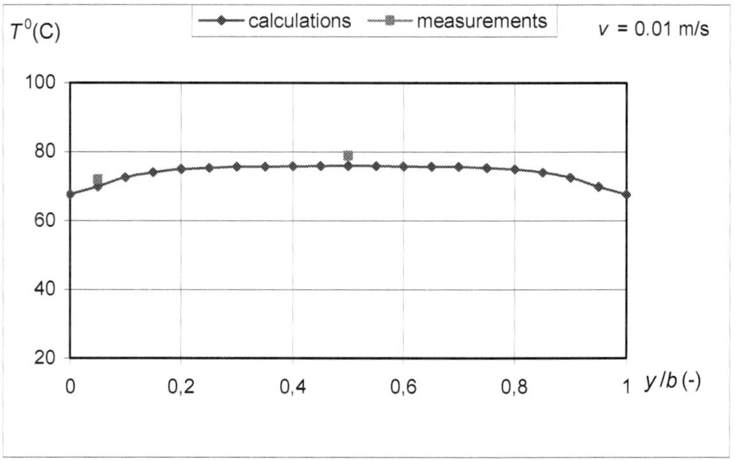

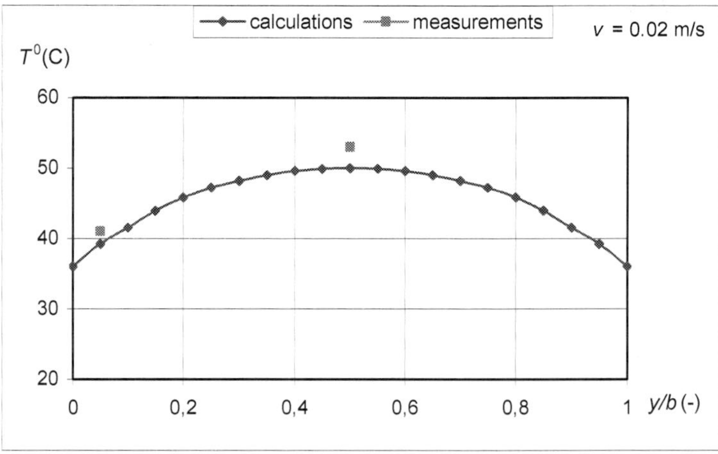

Fig. 21. Dependence of temperature on relative width of the brass strip at the outlet of the inductor ($x = 1.25\ l$) supplied by current of mains frequency ($f = 50$ Hz) for two different strip movement ($v = 0, 01$ m/s – upper diagram, $v = 0{,}02$ m/s – lower diagram)

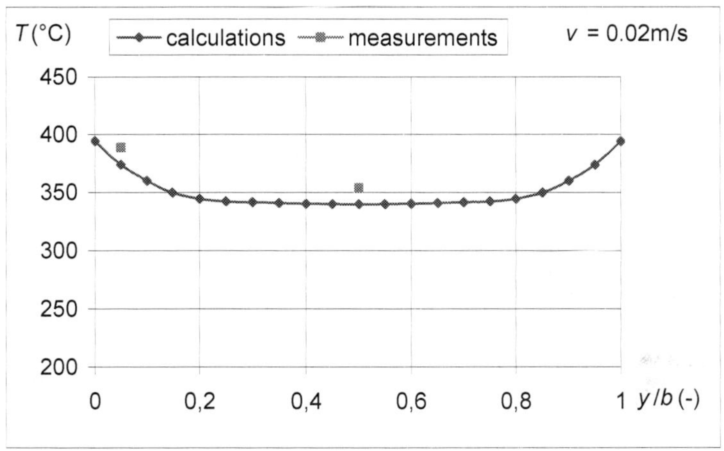

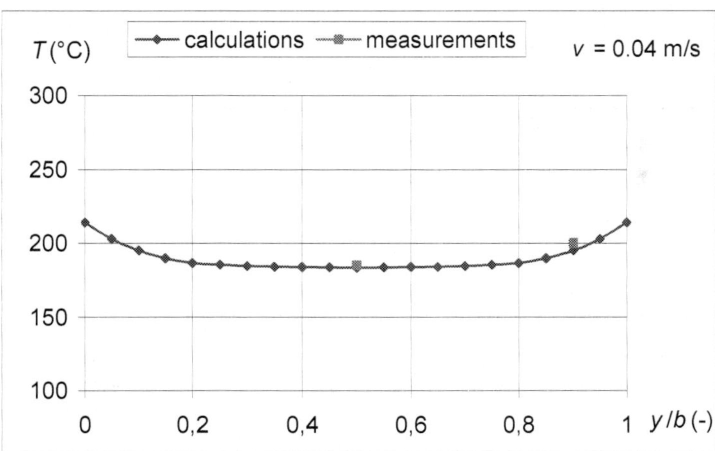

Fig. 22. Dependence of temperature on relative width of the bras strip. Frequency of field current f = 2000 Hz. Velocity of strip movement v = 0,02 m/s – upper diagram, v =0,04 m/s – lower diagram

showing temperature distribution in width of the strip with the distinct maximum in the axis of the strip.

In both cases quite good accordance of about not more than 3.5 % between calculations and measurements was achieved. The temperature distribution in width of the strip for medium frequency (f = 2000 Hz) was shown in Fig. 22. Characteristic was a shape of temperature curves showing temperature distribution in width of the strip with the distinct maximum near the axis of the strip. In both presented cases a quite good accordance between calculations and measurements data of about 4.8 % was achieved. In order to satisfy the criterion of uniform temperature distribution along the width of the strip it was necessary to find the optimal frequency of field current. As a result of computations such a frequency was obtained for a case of transverse flux induction heating of brass strip with thickness g = 3.2 mm. It was equal to f_{opt} = 275 Hz (Barglik, 2002). The uniform temperature distribution in width of the brass strip for optimal frequency of field current was obtained. The selection of optimal frequency based upon the criterion of temperature uniformity was valid for the particular configuration of the inductor-strip system. In case of usage of one inductor to heat up a wide range of the strips from materials of different properties and dimensions more complicated criteria should be applied (Barglik, 2002). Besides a classical division of the transverse flux inductor in a few single segments of different configuration to the axis of the strip it was also a concept of the flexible VABID inductor (Schülbe et al, 2004).

8. Conclusions

Three-dimensional analysis of electromagnetic and temperature fields in transverse flux induction heater for thin strips by means of FEM-based professional software supplemented by user codes was provided. Eddy current density and specific Joule losses released in the workpiece, temperature distribution in the width of the moving strip and the electrical efficiency of the transverse flux inductor-workpiece system were determined. The shapes of the temperature curves for different frequencies of field current were compared. The results were discussed in order to analyze possibilities of obtaining the requested parameters of the induction heating system. The computations were compared the measurement data. Quite good accordance between computations and measurements was noticed. Next steps in the investigations should be aimed at increasing the accuracy of calculations and shortening time of computations.

9. Acknowledgement

The financial support of the Polish Ministry of Science and Higher Education in several research projects realized in years 1995 - 2010 including ongoing projects N N510 256 338 and N N508 388 637 are highly grateful. Some research activities were done also within the framework of the multigovernment Polish-Czech collaboration project on mathematical modelling of induction heating processes.

10. References

Bakke, E. & Nacke B. (2010). Efficient heating by electromagnetic sources in metallurgical processes; recent applications and development trends. *Electrotechnic Review*. No 7, July 2010, pp. 11-14 ISSN 0033-2097

Baker, R. (1950). Transverse Flux Induction Heating, *AIEE Transactions*, vol. 69, pp. 711–719.

Barglik, J. (1992). Induction heating of thin non-ferrous strips, *Proceedings of 37. Wissenschaftliche Kolloqium* pp. 94-99, Ilmenau

Barglik, J. (2000) Influence of input material data on accuracy of thermal calculations at induction heating, *Acta Technica CSAV*, Vol. 45, pp. 323–336. ISSN 0001-7043

Barglik, J. (2002) Induction heating of flat charges in transverse flux field. Computer simulation and investigation verification. *Scientific Issues of the Silesian University of Technology* Vol.1575, No 65 (November 2002), pp. 1-120, PL ISSN 0324-802-X

Barglik, J. (2004). 3D analysis of coupled electromagnetic and temperature fields in transverse flux induction heating system supplied by current of mains frequency. *Acta Technica*, Prague, pp. 393-410, ISSN 0001-7043

Barglik, J. (2005). Computer modelling of induction heating of thin workpieces from non-ferrous metals, In: *Development in materials science and metallurgy*. Publisher of the Silesian University of Technology. pp. 7-32 ISBN 83-7335-214-7. Gliwice

Barglik, J., Czerwiński, M., Hering, M. and Wesołowski,M. (2008). Radiation In Modelling of Induction Heating Systems. In: *Advanced Computer Techniques in Applied Electromagnetics*. IOS Press. pp. 202-211. ISSN 1383- 7281. Amsterdam, Berlin, Oxford, Tokyo, Washington, DC.

Barglik,J. (2010). Induction heating in technological processes – selected examples of applications *Electrical Review*, No 5 May 2010 pp.294 – 297 ISSN 0033-2097

Davies, E. (1990) *Conduction and Induction Heating*. Peter Peregrinus Ltd. on behalf of IEEE ISBN 0 86341 1746 Exeter

Dughiero, F., Lupi S., Mühlbauer A. and Nikanorov A., (2003). TFH – Transverse Flux Induction Heating of Non-Ferrous and Precise Strips. Results of a EU research project,. *Compel International Journal for Computation and Mathematics in Electrical Engineering.*, pp. 134 – 148 ISSN: 0332-1649

Hering, M. (1992). *Basic of Electroheat*. WNT 1998 ISBN 83-204-2319-8 Warsaw

Holman, J. (2009). *Heat transfer*, McGraw-Hill, ISBN 13: 9780073529363 New York

Jackson,W. (1972). Transverse Flux Induction Heating Flat Metal Products, *Proceedings of the seventh UIE congress*, Paper No 206. Warsaw, July 1972.

Mühlbauer. A, Leßmann H., Mohring, J., Demidowitch. V. Nikanorov (1995). Modelling of 3D Electromagnetic Processes in Transverse Flux Induction Heaters, *Proceedings of Compumag Conference on Computation of Electromagnetic Fields*, , pp. 444 - 445, Berlin

Mühlbauer, A., (2008). *History of Induction Heating & Melting*, Vulkan-Verlag GmbH, ISBN 978-3-8027-2946-1, Essen

Nacke, B. Mühlbauer, A., Nikanorov A., Nauvertat, G. and Schülbe, H., (2001). Transverse Flux Heating of Metals Rolling and Treatment with reduced energy demand, *Proceedings of the conference "Energy Savings in Electrical Engineering"*, pp. 123-126, July 2001, Publisher of Warsaw Univ. of Technology, ISBN 83-913405-3-8 Warsaw

Schülbe, H.,Nikanorov,A., Nacke B. (2004) *Flexible Transverse Flux Induction Heaters of Metal Strip*, Proceedings of the International Symposium on Heating by Electromagnetic

Sources.   pp. 293–300, ISBN 88-86281-92-7, Padua, June 2004, Servizi Grafici Editoriali, Padua

Tudorache ,T. & Fireteanu, V (1998). 3D Numerical Modelling of New Structures for Transverse Flux Heating of Metallic Sheets,, *Proceedings of the International Heating Seminar HIS 1998* pp. 117-123, 88-86281, Padua, May 1998, Servizi Grafici Editoriali, Padua

11

Microwave Processing of Metallic Glass/polymer Composite Material in A Separated H-Field

Song Li[1], Dmitri V. Louzguine-Luzgin[1], Guoqiang Xie[2],
Motoyasu Sato[3] and Akihisa Inoue[4]
[1]*WPI Advanced Institute for Materials Research, Tohoku University*
[2]*Institute for Materials Research, Tohoku University*
[3]*National Institute for Fusion Science, Toki*
[4]*Tohoku University*
Japan

1. Introduction

As a rapid developing technique, microwave heating has shown its potential in the field of material processing. Compared with conventional heating, it can offer some interesting advantages including significant processing time and energy savings, high heating rate, lower environmental hazards and provision of high product quality and reliability since it is a volumetric heating in that thermal energy is internally and instantaneously generated within the material by the interaction between the material and electromagnetic field, instead of originating from external radiant or resistance heating [Clark & Sutton, 1996; Katz, 1992; Roy et al., 1999, 2002; Clark et al., 2000]. Furthermore, some excess effects, which are referred as nonthermal, or athermal effects, may also occur during microwave heating by the interaction of the material with the electromagnetic field [Bergese, 2006; Saitou, 2006; Roy et al., 2002]. It is generally known that the way microwaves interact with matter depends mainly on the electrical conductivity, complex permittivity, and complex permeability of the materials. Currently, microwave heating is applied widely to food, rubber, semiconductors, polymers, ceramics and metallic materials. For dielectric materials, the heating mechanisms in microwave field are mainly the polarization dissipation of electric dipoles under the alternating electric (E-) field and direct conduction effects. However, microwave heating of metals is different from that observed in dielectric materials. As good electrical conductors, bulk metals can not be well heated up because microwaves interaction with metals is restricted to its surface, but fine metal powders may be couple with microwaves at room temperature by Joule heating and magnetic-induced effects. Now a variety of metallic materials have been successfully sintered and processed in a pure microwave system or a hybrid heating system in which an external microwave susceptor is used [Anklekar, et al., 2005; Gedevanishvili, et al., 1999]. Although the mechanisms of microwave heating have not been fully understood until now, there is still an increasing interest in microwave processing of various materials, especially new materials with potential applications.

Metallic glasses are MGs known as a new class of metallic materials with the unique combination of physical, chemical and mechanical properties due to their disordered atomic arrangement different from that in crystalline metallic materials [Inoue, 2000; Johnson, 1999; Wang et al., 2004]. Compared with its crystalline counterparts, MGs exhibit significant viscous flow if they are heated to supercooled liquid region (temperature range between glass transition (T_g) and crystallization onset temperature (T_x)), which can be utilized for heating and sintering process. This makes it a possibly good candidate for microwave heating. In fact, microwave heating and sintering of metallic glassy powders and their composites has been performed in a single-mode cavity[Yoshikawa, et al., 2007; Louzguine-luzgin, et al., 2009a], and nanocrystallization of some metallic glassy powders and ribbons has been induced in the E-field[Nicula, et al., 2009]. Otherwise, microwave processing of metallic glass matrix composites has also been reported because the composite materials may exhibit the desired properties that can not be produced by conventional monolithic materials [Li, et al., 2008; Xie, et al., 2009a].

Based on the above consideration, we tried to heat of the powdered mixture MG of and polymer in microwave field and intended to develop a composite material based on MG and polymer using microwave heating technique because MGs and typically polymers show complementary properties, such as electrical, optical and mechanical characteristics and the liquidus or glass transition temperature of the commonly-used thermoplastics are relatively low (not more than 450°C) and very close to the Tg of some MGs. If two components can be combined in a single material, some interesting properties may be expected in the proposed composites.

The present paper describes microwave processing of the powdered mixtures of metallic glasses and polymers using a single-mode 915 MHz applicator in a separated magnetic (H-) field. A single-mode applicator is chosen because it has high field strength, the most efficient by far, can be operated continuously and used to focus the microwave field at a given location. We investigated the heat response of the powdered mixtures of the $Cu_{50}Zr_{45}Al_5$ metallic glass and polyphenylene sulfide (PPS) $[C_6H_4S]_n$ in a separated microwave H-field. PPS was chosen because of its good thermal stability, excellent chemical resistance, and inherent flame resistance. The effect of the PPS content in the mixtures on the heating behaviors of the composites was also investigated.

2. Experimental procedure

The $Cu_{50}Zr_{45}Al_5$ metallic glassy powders were fabricated by a high-pressure gas-atomized method. The preparation procedure has been presented elsewhere [Xie, et al., 2007]. Size distribution of the gas-atomized alloy powders was measured by a conventional sieving method. PPS used in this work was in the form of off-white neat powder and its average molecular weight was about 10,000. The blended powders were obtained by mixing $Cu_{50}Zr_{45}Al_5$ metallic glassy alloy powders with the size under 90 µm and PPS powders according to various volumetric ratios in a mixer. To obtain larger-size samples and improve sintering, we have designed and built a single-mode microwave applicator (915MHZ, 5kW) with the pressure loading system. The illustration schematic of a single-mode 915 MHz microwave system is shown in Fig. 1. It consists of a single-mode 915 MHz generator with a power output adjustable from 1 to 5 kW in steps of 40 W, a waveguide, three-stub tuner, separator, a single-mode tunable applicator, a pressure loading system, an infrared pyrometer and a computer control system. Microwave-induced heating was performed in a separated H-field maximum. The samples were placed in an alumina (Al_2O_3) mould because Al_2O_3 has not obvious heat response in the H-field. The applied pressure was kept at about 5 MPa for avoiding the fracture of the alumina mould during heating process. To order to initiate microwave absorption, the resonating cavity was tuned by a remotely controlled three-stub tuner and the slide plunger at the end of the waveguide. The temperature was controlled by adjusting the input power (P_i). The temperature measurement was performed in situ using an infrared pyrometer through a quartz window on the side of the press. The pyrometer was calibrated by comparison of the reading with thermocouple in a resistance furnace. It is not possible to measure the sample temperature under about 450 K. All experiments were carried out under a flowing nitrogen gas or low vacuum.

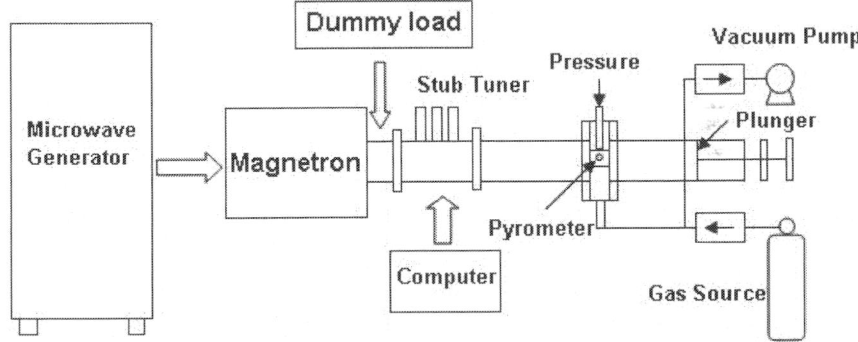

Fig. 1. Illustration schematic of a single-mode 915 MHz microwave system

The structures of the initial powders and the sintered composites were determined by X-ray diffractometry (XRD) in reflection with a monochromatic Cu-Kα radiation. The thermal stability of metallic glassy powder was examined using differential scanning calorimetry (DSC) at a heating rate of 40 *K/min*. The density of the bulk sintered specimen was calculated by measuring the mass and dimension of the sample (mass per volume). The microstructures of the sintered specimens and initial metallic glassy alloy powders were characterized by scanning electron microscopy (SEM).

3. Results and discussion

3.1 Characterization of $Cu_{50}Zr_{45}Al_5$ gas-atomized powders

Here the characteristics of the gas-atomized $Cu_{50}Zr_{45}Al_5$ are determined. Figure 2 shows the XRD pattern and DSC curve of $Cu_{50}Zr_{45}Al_5$ alloy powders with particle size under 125 µm. It can be seen in Fig. 2(a) that only typical broad diffraction maxima without any observable sharp diffraction peaks corresponding to crystalline phases occur, indicating that a fully amorphous phase is formed in the particle size range below 125 µm. As shown in Fig. 2(b), a distinct glass transition characteristic followed by two crystallization events occurs, and the values of T_g, T_x, and supercooled liquid region width (ΔT) were determined to be 708 K, 776 K and 68 K, respectively, which are similar to the ones reported for the as-cast $Cu_{50}Zr_{45}Al_5$ rod and ribbon[Inoue & Zhang, 2002]. Figure 3 shows a typical SEM micrograph of the gas-atomized $Cu_{50}Zr_{45}Al_5$ alloy powders with particle sizes under 125 µm. Most of the gas-atomized powders have a spherical morphology with clean surface. On the outer surface of the particles no observable contrast revealing the formation of crystalline phase is seen. These results show that the gas-atomized method is flexible for fabricating metallic glassy powders.

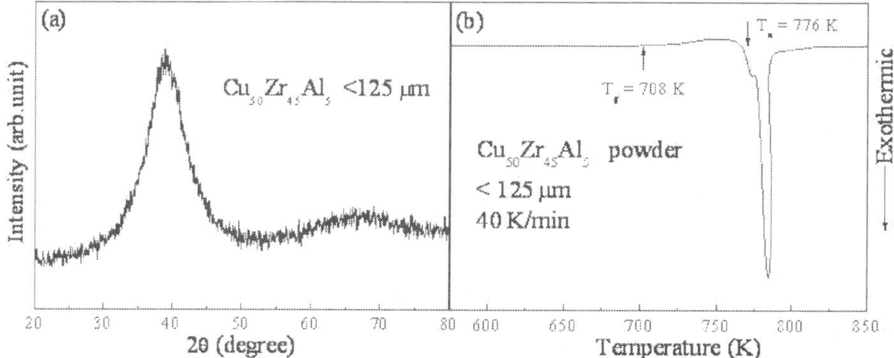

Fig. 2. XRD pattern and DSC curve of Cu$_{50}$Zr$_{45}$Al$_5$ alloy powders with particle size below 125 µm.

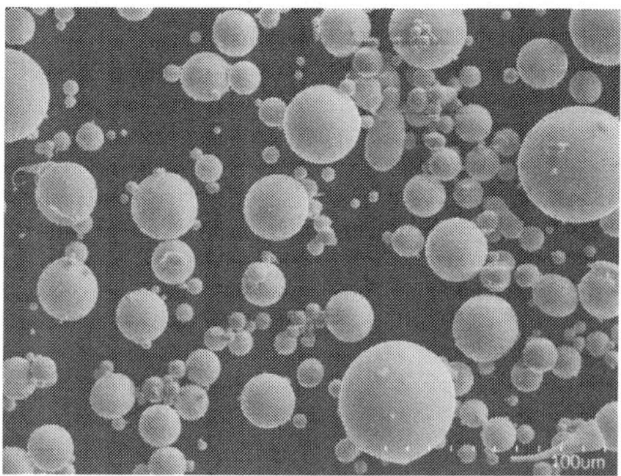

Fig. 3. SEM micrograph of the gas-atomized Cu$_{50}$Zr$_{45}$Al$_5$ alloy powders with particles sizes under 125 µm

3.2 Microwave processing of the mixtures of Cu$_{50}$Zr$_{45}$Al$_5$ metallic glassy powder and PPS powder

We firstly simulated the electromagnetic field distribution in the waveguide based on the finite element method and by using a JAMG-Studio program. The results have indicated that the separated E- and H-field maxima can be obtained by adjusting the position of the alumina loading pressure system in the waveguide [Xie, et al., 2009[b]]. Using this applicator, a two-phase Cu$_{50}$Zr$_{45}$Al$_5$/Fe sintered sample was successfully heating up in H-field maximum [Louzguine-luzgin, et al., 2009[b]]. The combination of two materials better offers temperature control on heating compared to heating of metallic glass separately. We also processed the $Cu_{50}Zr_{45}Al_5$ metallic glass matrix composites containing Sn powders in H-field maximum with an applied pressure of about 5 MPa[Li, et al., 2010]. Now we tried to heat a two-phase material which consists of $Cu_{50}Zr_{45}Al_5$ and PPS in H-field maximum.

We investigated the heat response of the powdered mixture of the $Cu_{50}Zr_{45}Al_5$ metallic glass and PPS in H-field maximum. Figure 4 shows the temperature profiles and energy consumption of the PPS powder and blended powders heated by microwaves in a separated H-field. It can be seen that all curves of the temperature and energy consumption are not smooth because microwave absorption capacity of the material is very sensitive to the tuner adjustment. It has been reported that fine metallic glassy powders can be heated up like most of pure metals in H-field likely by the eddy current loss effect[Buchelnikov, et al., 2008]. Compared with metallic glassy powders, PPS did not show obvious heat response under a flowing nitrogen gas with the increase of P_i because most of polymers are loss dielectric constant materials at room temperature. In H-field magnetic effect did not contribute to microwave absorption of PPS powders. Only under low vacuum the obvious increase of the temperature was detected by microwave-induced plasma at the surface of the sample which is not a normal heating process. The addition of $Cu_{50}Zr_{45}Al_5$ metallic glassy powders strongly affected the heating behaviors of the composites in the H-field. In the case of 75 vol.% PPS, although the temperature was not detected like pure PPS because its surface temperature was too low, its inner temperature was far higher than its surface temperature due to the reverse thermal gradient inside the composite, temperature decreasing rapidly from inside to outside. This is attributed to conductive and radiant heat loss from the surface and selecting heating achieved by the differential coupling of metallic glassy powder and PPS with microwaves. With the further increase of the content of $Cu_{50}Zr_{45}Al_5$ metallic glassy powders, the relatively uniform temperature distribution could be realized in the cases of 25 and 50 vol.% PPS. This also means that the increase of the temperature in the blended powders at room temperature depends mainly on $Cu_{50}Zr_{45}Al_5$ metallic glass and obvious heat response only occurred at a sufficient content of $Cu_{50}Zr_{45}Al_5$ metallic glass.

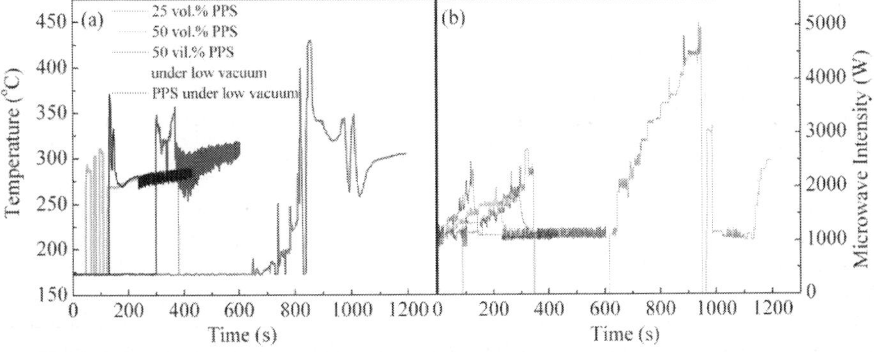

Fig. 4. Temperature profiles and energy consumption of the PPS powder and blended powders heated by microwaves in a separated H-field

Figure 5 shows the XRD patterns of the initial blended powders and composites with 25, 50 and 75 vol.% PPS powder heated by microwaves under a flowing nitrogen gas. As shown in Fig. 5(a), in the case of 50 vol.% PPS, except for the phase corresponding to PPS, only typical broad diffraction maxima was seen without any detectable sharp diffraction peak corresponding to a crystallization peak, indicating that an amorphous phase was retained. Partial crystallization was induced in the case of 25 vol.% PPS because of its relatively high sintering temperature. Different from the composites with 25 and 75 vol.% PPS, a gradient structure in the composite with 75 vol.% PPS was induced by microwave heating in a separated H-field. To confirm its thermal gradient, the structural changes from the center to surface were investigated. The composite have had three different zones. It is shown that in Fig. 5(b) that zone A, close to the surface of the sample, did not show obvious heat response and had a similar structure as the initial blended powders. Zone C was located in the central region of the sample and exhibited a porous structure. Due to the decomposition of PPS, the formation of cubic ZrO_2, $ZrS_{0.67}$ and $ZrH_{0.25}$ phase was detected except for $Cu_{10}Zr_7$ and $AlZr_2$ phase during microwave processing. In the zone B, PPS powders were cured by microwaves and a high fraction of $Cu_{50}Zr_{45}Al_5$ metallic glassy phase was also retained. Such a gradient structure formed by microwave heating is closely linked with different dielectric properties of $Cu_{50}Zr_{45}Al_5$ metallic glass and PPS and their contents in the composite. Due to the low content of $Cu_{50}Zr_{45}Al_5$ metallic glass, the composite has a poorer heat conductivity compared to those with a higher fraction of $Cu_{50}Zr_{45}Al_5$ metallic glass, so the reverse thermal gradient was more easily induced in a short time.

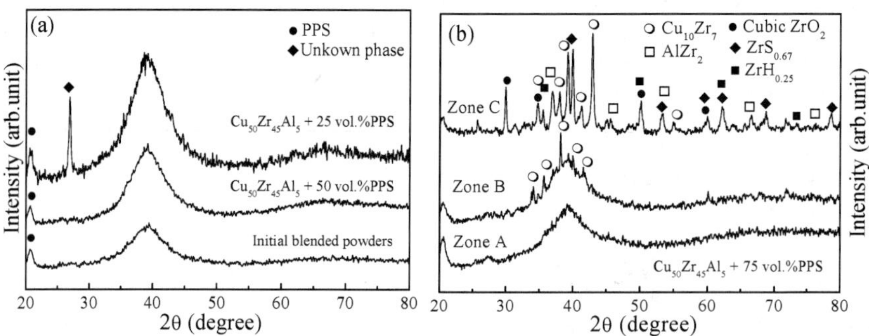

Fig. 5. XRD patterns of the initial blended powders and composites with 25, 50 and 75 vol.% PPS powder heated by microwaves under a flowing nitrogen gas.

Table 1 lists the common features and relative densities of Pure PPS powder and the composites with 25, 50, 75 and 90 vol.% PPS powder heated by microwaves in a separated H-field. Pure PPS sample obtained under low vacuum shows very low relative density (57%) because the plasma induced by the interaction between microwaves and atmosphere interrupted the normal heating process and the actual sintering temperature is lower than the reading temperature by pyrometer. Under a flowing nitrogen gas, the introduction of 10 vol.% $Cu_{50}Zr_{45}Al_5$ metallic glassy powders only makes the mixture be weakly sintered. In the case of 75 vol.% PPS, the powdered mixture was heterogeneous heated and sintered to lead to the formation of a gradient structure. The composite with 50 vol.% PPS obtained

Composition	Atmosphere	Relative Density	Common Feature
$Cu_{50}Zr_{45}Al_5$+25 vol.% PPS	N_2	84%	Sintered and partial crystallization
$Cu_{50}Zr_{45}Al_5$+50 vol.% PPS	N_2	72%	Sintered and amorphous
$Cu_{50}Zr_{45}Al_5$+50 vol.% PPS	Vacuum	59%	Weakly sintered and amorphous
$Cu_{50}Zr_{45}Al_5$+75 vol.% PPS	N_2		Sintered and gradient structure
$Cu_{50}Zr_{45}Al_5$+90 vol.% PPS	N_2		Weakly sintered
PPS	N_2		Not sintered
PPS	Vacuum	58%	Weakly sintered

Table 1. Common features and relative densities of pure PPS powders and the composites with 25, 50, 75 and 90 vol.% PPS powder heated by microwaves.

under a flowing nitrogen gas also has a higher relative density than that under low vacuum. Bulk $Cu_{50}Zr_{45}Al_5$ metallic glass/PPS composite with a high relative density (84%) was obtained by microwave heating of the blended powders containing 25 vol.% PPS. This shows that the higher sintering temperature leads to the melting of PPS particles and then the resulted composites are much denser with the significantly reduced number of voids. The compressive strength of these composites was also measured and no more than 100 MPa, similar to that of common polymers. These results further manifested that a better sintering quality could be obtained in the composite with a high fraction of $Cu_{50}Zr_{45}Al_5$ metallic glassy phase.

Figure 6 shows the appearance and microstructure of the composite with 25 vol.% PPS powder heated by microwaves in H-field under a flowing nitrogen gas. As shown in Fig. 6(a), a bulk sintered composite with a relative density of about 84% was obtained under an applied pressure of about 5 MPa. It can be seen in Fig. 6 (b) that PPS powders were melted by microwave heating and permeated into the voids and interstices between the $Cu_{50}Zr_{45}Al_5$ particles. Figure 7 presents the SEM micrograph of the polished cross section of the zone B of the composite with 75 vol.% PPS powder heated by microwaves under a flowing nitrogen gas. In the zone B, PPS powders were rapidly and well consolidated by microwaves heating and the wetting seemed good with no obvious cracking observed on the interface between metallic glassy particle and PPS particle. A good bonding state between $Cu_{50}Zr_{45}Al_5$ metallic glassy particles and PPS was seen. Recently it has been reported that some MGs with low glass transition temperature have similar viscosities at the same temperature as some polymers [Kündig, et al., 2007]. In addition, $Zr_{55}Cu_{30}Al_{10}Ni_5$ metallic glass was also successfully deposited on various engineering polymer substrates by magnetron sputtering as thin homogeneous layers of about 400 nm and a good adhesion was achieved[Soinila, et al., 2009]. The above results show that microwave-induced heating process can offer a possibility to fabricate a single composite material that consists of metallic glass and polymer.

(a)   (b)

Fig. 6. Appearance and microstructure of the composite with 25 vol.% PPS powder heated by microwaves in H-field

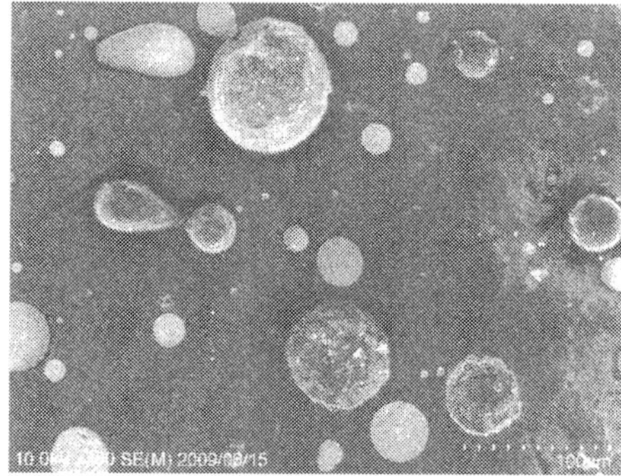

Fig. 7. SEM micrograph of the polished cross section of the zone B of the composite with 75 vol.% PPS powder heated by microwaves under a flowing nitrogen gas

4. Conclusion

In the present work, using a single-mode 915 MHz microwave applicator, we investigated the heating behaviors of the mixtures of $Cu_{50}Zr_{45}Al_5$ metallic glassy powder and PPS powder in a separated H-field. The $Cu_{50}Zr_{45}Al_5$ metallic glass/PPS composites were fabricated with an applied pressure of about 5 MPa under a flowing nitrogen gas. The mixtures with a high fraction of $Cu_{50}Zr_{45}Al_5$ metallic glassy phase could be heated well and exhibited a relatively uniform structure. With the increase in PPS content above 90 vol.%, the mixture was hardly heated up under a nitrogen-flow atmosphere, but was heated well in low vacuum by microwave-induced plasma heating. A bulk sintered composite with retention of an amorphous structure were obtained. The bonding state between metallic glassy and PPS particles was found. The gradient structure could be formed by microwave processing of the composites with a high fraction of PPS phase. It is evident that microwave induced heating process in the mixture of $Cu_{50}Zr_{45}Al_5$ metallic glassy powder and PPS powder in a separated H-field offers a possibility to fabricate metallic glass/polymer composites with large dimensions.

5. References

Anklekar, R.M.; Bauer, K.; Agrawal, D.K. & Roy, R. (2005). Improved mechanical properties and microstructural development of microwave sintered copper and nickel steel PM parts, *Powder Metallurgy*, Vol. 48, 39-46

Bergese, P. (2006). Specific heat, polarization and heat conduction in microwave heating system: A nonequilibrium thermodynamic point of view, *Acta Materialia*, Vol. 54, 1843-1849

Buchelnikov, V.D.; Louzguine-luzgin, D.V.; Xie, G.Q.; Li, S.; Yoshikawa, N.; Sato, M. & Inoue, A. (2008). Heating of metallic powders by microwaves: Experiment and theory, *Journal of Applied Physics*, Vol. 104, 113505-1-6

Clark, D.E. & Sutton, W.H. (1996). Microwave processing of materials, *Annual Review of Materials Science*, Vol. 26, 299-331

Clark, D.E.; Folz, D.C. & West, J.K. (2000). Processing materials with microwave energy, *Materials Science and Engineering: A*, Vol. 287, 153-158

Gedevanishvili, S.; Agrawal, D.K. & Roy, R. (1999). Microwave combustion synthesis and sintering of intermetallics and alloys, *Journal of Materials Science Letters*, Vol. 18, 665-668

Inoue, A. (2000). Stabilization of metallic supercooled and bulk amorphous alloys, *Acta Materialia*, Vol. 48, 279-306

Inoue, A. & Zhang, W. (2002). Formation, thermal stability and mechanical properties of Cu-Zr-Al bulk glassy alloys, Materials Transition, Vol. 43, 2921-2925

Johnson, W.L. (1999). Bulk glass-forming metallic alloys: science and technology, *Materials Research Bulletin*, Vol. 24, 42-56

Katz, J.D. (1992). Microwave sintering of ceramics, *Annual Review of Materials Science*, Vol. 22, 153-170

Kündig, A.; Schweizer, T.; Schafler, E. & Löffler, J. (2007). Metallic glass/polymer composites by co-processing at similar viscosities, *Scripta Materialia*, Vol. 56, 289-292

[a]Louzguine-luzgin, D.V.; Xie, G.Q.; Li, S.; Inoue, A.; Yoshikawa, N.; Mashiko, S.; Taniguchi, S. & Sato, M. (2009). Microwave-induced heating and sintering of metallic glasses, *Journal of Alloys and Compounds*, Vol. 483, 78-81

[b]Louzguine-luzgin, D.V.; Xie, G.Q.; Li, S.; Inoue, A.; Yoshikawa, N. & Sato, M. (2009). Microwave-induced heating of a single glassy phase and a two-phase material consisting of a metallic glass and Fe powder, *Philosophical Magazine Letters*, Vol. 89, 86-94

Li, S.; Xie, G.Q.; Louzguine-lozgin, D.V.; Cao, Z.P.; Yoshikawa, N.; Sato, M. & Inoue, A. (2008). Microwave sintering of Ni-based bulk metallic glass matrix composite in a single-mode applicator, *Materials Transaction*, Vol. 49, 2850-2853

Li, S.; Xie, G.Q.; Louzguine-lozgin, D.V.; Sato, M. & Inoue, A. (2010). Microwave-induced sintering of Cu-based metallic glass matrix composites in a single-mode 915-MHz applicator, *Metallurgical and Materials Transactions A*, DOI: 10.1007/s11661-010-0346-8

Nicula, R.; Stir, M.; Ishizaki, K.; Catalá-Civera, J.M. & Vaucher, S. (2009). Rapid nanocrystallization of soft-magnetic amorphous alloys using microwave induction heating, *Scripta Materialia*, Vol. 60, 120-123Roy, R.; Agrawal, D.K.; Cheng, J.P. &

Gedevanishvili, S. (1999). Full sintering of powdered metals using microwaves, *Nature*, Vol. 399, 668-670

Roy, R.; Peelamedu, R.D.; Cheng, J.P.; Grimes, C.A. & Agrawal, D.K. (2002). Major phase transformations and magnetic property changes caused by electromagnetic fields at microwave frequencies, *Journal of Materials Research*, Vol. 17, 3008-3011

Roy, R.; Peelamedu, R.; Hurtt, L.; Cheng, J.P. & Agrawal, D.K. (2002). Definitive experimental evidence for microwave effects: radically new effects of separated E and H fields, such as decrystallization of oxides in seconds, *Journal of Materials Research Innovation*, Vol. 6, 128-140

Saitou, K. (2006). Microwave sintering of iron, cobalt, copper and stainless steel powders, *Scripta Materials*, Vol. 54, 875-879

Soinila, E.; Sharma, P.; Heino, M.; Pischow, K. & Inoue, A. (2009). Bulk metallic glass coating of polymer substrates, *Journal of Physics: Conference Series*, Vol. 144, 012051

Wang, W.H.; Dong, C. & Shek, C.H. Bulk metallic glasses, *Materials Science and Engineering: R: Reports*, Vol. 44, 45-89

Xie, G.Q.; Louzguine-luzgin, D.V.; Kimura, H. & Inoue, A. (2007). Nearly full density Ni52.5Nb10Zr15Ti15Pt7.5 bulk metallic glass obtained by spark plasma sintering of gas atomized powders, *Applied Physics Letters*, Vol. 90, 241902-1-3

[a]Xie, G.Q.; Li, S.; Louzguine-lozgin, D.V.; Cao, Z.P.; Yoshikawa, N.; Sato, M. & Inoue, A. (2009). Effect of Sn on microwave-induced heating and sintering of Ni-based metallic glassy alloy powders, *intermetallics*, Vol. 17, 274-277

[b]Xie, G.Q.; Suzuki, M.; Louzguine-luzgin, D.V.; Li, S.; Inoue, A; Yoshikawa, N. & Sato, (2009). M. Analysis of electromagnetic field distributions in a 915 MHz single-mode microwave applicator, *Progress in Electromagnetics Research*, Vol. 89, 135-148

Yoshikawa, N.; Louzguine-luzgin, D.V.; Mashiko, K.; Xie, G.Q.; Sato, M.; Inoue, A. & Taniguchi, S. (2007). Microstructural changes during microwave heating of $Ni_{52.5}Zr_{15}Nb_{10}Ti_{15}Pt_{7.5}$ metallic glasses, *Materials Transaction*, Vol. 48, 632-634

12

Thermal Microwave Processing of Materials

Juan A. Aguilar-Garib

Universidad Autónoma de Nuevo León, Facultad de Ingeniería Mecánica y Eléctrica
México

1. Introduction

Practically all kind of materials; namely polymers, metals and ceramics have been heated up with microwaves. There are many reasons for conducting research in this area, microwaves can be applied in vacuum and can be turned on and off in the same way than an electric resistance, but without thermal inertia. Other attributes have been investigated, such as higher efficiency and uniformity in heat transfer and high reaction rates. Researches have reported that processing of materials using microwaves for supplying energy is improved, decreasing the sintering temperature, increasing reaction rate and reducing the activation energy.

The term "microwave effect" has emerged from the many serious reports describing a higher reaction rate due to microwave application available in literature; however a satisfactory explanation has not been given because the results that are presented are often extremely specific and can not be applied to other conditions.

Some results are controversial because they consider that the energy is not being transferred to the system through conversion of microwave energy into thermal energy. It is assumed that microwaves interact directly with the molecules or atoms in the lattice considering non-thermal activation of processes such as improved atomic diffusion.

There are other explanations based on thermodynamic and kinetic effects, given by selective heating of specific reaction components, rapid heating rates and high temperature gradients due to self-generated heat within the sample rather than convection and conduction heat transfer.

Good microwave absorbers, meaning materials where microwaves are clearly converted into heat, will not show non-thermal effects. Heating mechanism due to the presence of an electric field in the frequency of 1 up to 300 GHz includes dielectric: dipolar losses, ion jump relaxation, and resistive heating. In ceramics, and especially in semiconductors, it is possible to have dielectric and resistive contributions simultaneously. These materials can be heated with microwaves, but the amount of energy that they absorb limits their application to specific processes, such as synthesis, sintering or some thermal treatment. Many examples of materials showing non-thermal effects of microwaves are taken from these materials.

The aim of this book chapter is to show some cases of microwave processing of materials considering the conditions that could explain the "microwave effect" often reported. It is necessary to consider that always that the conditions are changed over a system there will be always an effect over the process; effects where a thermodynamic or kinetic explanation is possible will not be considered as "microwave effect" just as heating water with fire is not considered a "fire effect".

Discussion in this document is centered on ceramics where it has been demonstrated that under certain conditions, microwaves heat these materials up. It is considered processing at 2.45 GHz of, calcium zirconate, spinel alumina–magnesia, silicon carbide and manganites. In each of the cases the processes are described in terms of providing an explanation for the speed up of the reaction based on thermodynamic and kinetic approaches, considering that a reaction could be speed up based on activating a given mechanism with a different activation energy, or simply having thermal gradients that are not considered when average temperatures are taken as the actual ones, rather than non-thermal diffusion which is described by the same Arrhenius equation only with the activation energy changed. Enhance in diffusion comes then from a claimed reduction of this activation energy with no further explanation or change of the phenomena or thermal considerations, such as thermal uniformity due to a combination of having small samples compared to the penetration depth of the microwaves, and high thermal diffusivity that results in smaller thermal gradients.

"Microwave effect" is described in terms of thermodynamic and kinetic aspects for the cases presented here, looking forward to extrapolate this explanation to other similar cases where this effect has been accepted without further argumentation. Although there are still many non-thermal cases that are not easily explained invocating known kinetic arguments, the purpose of this research is to reduce the amount of those cases whenever is possible.

2. Background

It has been reported that processing of materials using microwaves for supplying energy is improved, decreasing the sintering temperature, increasing the reaction rate (Katz et al., 1991) and reducing the activation energy (Janney et al., 1991). These kinds of reports induce to consider that the microwaves have an especial effect over the reactions. There has been outlined several possibilities for explaining this phenomenon, ranging from an improved atomic diffusion in this kind of materials (Binner et al., 1995), to iron providing local warming that causes the reactions to be more rapid (Roy et al., 2000).

There are several examples where a "microwave effect" has been proven (Binner et al., 2008). The amount of serious reports is impressive and respectfully although a satisfactory explanation has not been given. However the results that are presented are often extremely specific and can not be applied to other conditions. The mechanisms that have been proposed to explain the warming through microwaves do not seem to sustain the "microwave effect" affirmation, indeed, there are results where such effect is not observed (Roy et al., 2000), then this issue is controversial (Garbacia et al., 2003; Wittaker, 2005; Hoogenboom et al., 2005) and deserves more study.

Polymers (Popescu et al., 2008), metals (Roy et al., 1999) and ceramic (Xiang et al., 2005) have been heated up with microwaves. Ceramics can be heated with microwaves; however the amount of energy that they absorb limits their application to specific processes, such as synthesis (Rao & Ramesh, 1995), sintering (Brandon et al., 1992) or some thermal treatment (Sorescu et al., 2004).

Microwave heating is very sensitive to the kind of material that is being heated, following are some examples of processing where the potentiality of this method is presented with spinel alumina–magnesia, calcium zirconate, silicon carbide and manganites, and how some observations regarding that could be taken as non-thermal effect or special effect of microwave can be justified with classical thermodynamics and kinetics arguments.

3. Microwave processing

3.1 Microwave synthesis: calcium zirconate

Perovskite-structured oxides such as the zirconates are of considerable interest because their conductive properties at high temperature (Islam et al., 2001). One that has been often studied is the $CaZrO_3$ (Jacob, 1997; Yamaguchi et al., 2000) and in this case it is proposed that synthesis of this material using microwaves, as an energy source, is possible.

Mechanisms such as polarization increase of conductivity improving Joule effect and ion vacancy jump are explained in several references and results from other research works show that ZrO_2 absorbs energy from microwaves (Aguilar et al., 1996). Hence, it seems viable that this form of energy is an alternative for producing these kinds of materials. Formation of $CaZrO_3$ by heating $CaCO_3$ and ZrO_2 powders conducted in a microwave oven has been reported earlier (Aguilar et al., 1996). It was expected that absorption improves with temperature because it has been reported (Willert-Porada & Gerdes, 1991) that the loss factor increases as function of temperature. $CaZrO_3$ is a perovskite that is a ferroelectric material; the characteristics of these materials include high dielectric permeability, ferroelectric hysteresis and polarization saturation (Hippel, 1995). The use of a graphite susceptor has been reported (Aguilar et al., 1996), the maximum temperature that graphite alone under air atmosphere can reach is below 1200 K, but temperature in the tests would be higher because the mixture also absorbs microwaves. The purpose of this study is to process mixtures of CaO-ZrO_2 with microwaves for producing zirconate considering the pressing level, mass and applied power.

The experimental design was based on three parameters: pressing level, mass and applied power. The mixtures were sat over a bed of graphite as susceptor. It was found that 10 minutes are enough to reach melting point temperatures and therefore the nominal testing time was that long. The selected powers were 1000 W and 2000 W. It must be considered that not all of the applied power was actually being deposited in the sample, there are heat losses and the material itself is not a perfect absorber.

The mixtures were prepared in a 1:1 molar ratio of ZrO_2: CaO. The conditions of the tests are shown in table 1. Average particle sizes were 12μm for CaO and 8μm for ZrO_2. The pressing was carried out in just one direction thus the final shape of the sample in the compacted case was a tablet of 2.9 cm diameter and either 0.45 cm or 0.89 cm height, depending if the mixture was 8 g or 16 g.

The size of the non-compacted samples was 3.5 cm diameter, which is the inner diameter of the crucible, while height was according to the mass, 0.80 cm for the 8 g mixture or 1.5 cm for the 16 g one. The crucible was 5.0 cm outer diameter, 3.5 cm high and it was made of high purity alumina and was isolated with ceramic fibber.

The crucible with the mixture was introduced into a multimode cavity. Forward and reflected power was monitored while tuning was performed in order to keep this reflected power to a minimum value.

The power source was a magnetron working at 2.45 GHz and variable power up to 3 KW, the microwaves were conducted through a WR284 (7.2 cm x 3.6 cm) waveguide attached to the cavity, and the temperature measurement was performed by optical pyrometry.

The results and the conditions of experiments are also presented in table 1. Regarding X-rays analysis, all the samples exhibited $CaZrO_3$, although the only ones that did not present free CaO or ZrO_2 are those that were processed at 2000 Watts. Presence of $CaZrO_3$ at lower temperatures demonstrates that there were hot spots within the samples. In all of the cases there was formation of $(CaO)_{0.15}(ZrO_2)_{0.85}$.

Exp.	Pressure (MPa)	Applied Power (W)	Mass (g)	Maximum temperature (K)	Time for achieving maximum temperature (sec)
1	0	1000	8	705	553
2	147	1000	8	678	503
3	0	2000	8	2488	266
4	147	2000	8	2291	352
5	0	1000	16	884	546
6	147	1000	16	498	525
7	0	2000	16	1240	272
8	147	2000	16	1561	378

Table 1. Experimental design and results.

For the samples that did not show residual CaO, or at least was not detected, the amount of $(CaO)_{0.15}(ZrO_2)_{0.85}$ must be small because the mixtures were prepared in 1:1 molar ratio and a large removal of CaO is necessary, which is rather difficult, for having an appreciable amount of this solid solution. One interesting aspect is that the compacted mixtures got molten while the non-compacted ones exhibited sintering only, even when the achieved temperatures were similar in both cases. This behavior was expected considering that pressing supplies energy to the mixture. Other aspect can be appreciated by looking at the thermal evolution (Figures 1 and 2), compacted samples took longer to achieve the maximum temperature than non-compacted, which is more notorious in the 16 g mixtures.

It is interesting to observe that in the non-compacted case the temperature remains around 1000 K until heating is activated. It is possible that this behavior is present in all of the cases, but temperatures lower than 900 K were not registered by the optical pyrometer, and

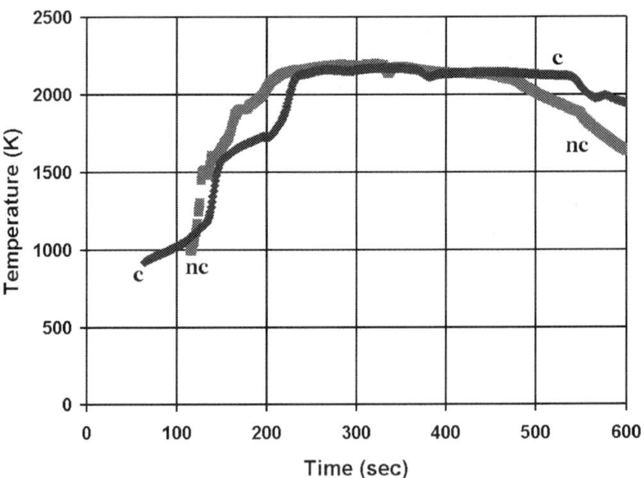

Fig. 1. Comparison between the average thermal evolution of compacted (c) and non-compacted (nc) mixtures of 8 g exposed to 2000 Watts. Temperatures below 900 K where not registered by the optical pyrometer.

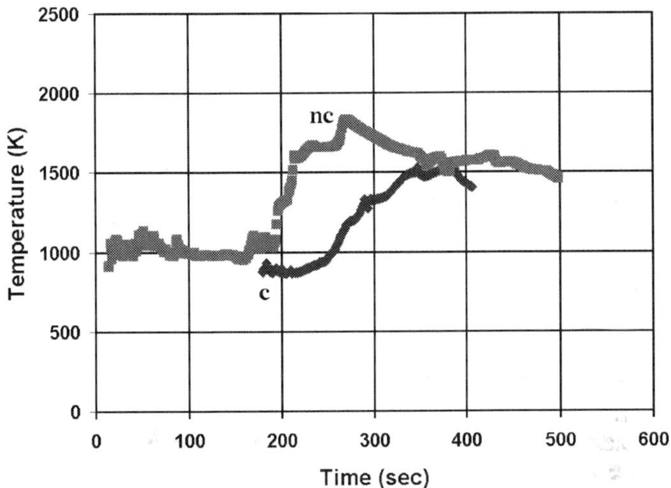

Fig. 2. Comparison between the average thermal evolution of compacted (c) and non-compacted (nc) mixtures of 16 g exposed to 2000 Watts.

therefore this "activation" temperature is not evidenced. In any case, times for activation are shorter for non-compacted mixtures than compacted and for high power/mass ratios than low (2000/8 against 2000/16).

Notice how a time for activation of the non-compacted mixture is about double when energy supply is the half (75 sec against 190 sec). The same behavior is found in the compacted case (115 sec against 250 sec), meaning that the necessary energy for having an excitable (microwave absorbent) mixture is about constant. For interpreting the differences between compacted and non-compacted mixtures, regardless the energy input, it is necessary to describe the CaO-ZrO_2 system and propose a reaction path.

There are three defined compounds in the CaO-ZrO_2: $CaZrO_3$, $CaZr_4O_9$ and $Ca_6Zr_{19}O_{44}$ (Tanabe & Nagata, 1996). They also reported the activity of CaO in a CaO-ZrO_2 system, which is low compared to its (Table 2). Hence, it is possible to assume that Ca gets into the ZrO_2 lattice to form a solid solution in the proportions already showed. $(CaO)_{0.15}(ZrO_2)_{0.85}$ was found in all the cases. According to Tien (Tien, 1964) the electrical conductivity of this system increases with temperature and with the amount of CaO up to 13 mole%, although there are other reports for electrolytes made of $CaO+ZrO_2$ that are higher, but in the same order of magnitude (Kingery et al., 1975).

Temperature (K)	Cubic solid solution	
	CaO	ZrO_2
1673	0.02051	0.9295
1773	0.02222	0.9295
1873	0.02430	0.9274

Table 2. Activities of CaO and ZrO_2 in a solid solution of 15 mole% of CaO.

The differences in the thermal behavior of the non-compacted and compacted mixtures can be explained by the more intimate contact among the particles in the compacted case and the energy that was already stored due to the press process. Therefore, as the solid solution is being formed, the conductivity increases and Joule effect becomes a heat contributor of the mixture and promotes even more reaction thermally. Further melting and heat losses determine the maximum achievable temperature.

Conductive material could form a shield effect, but before this happens there is dielectric heating without melting and then Joule heating as described above. The supplied power was kept to the nominal value during the test, the reflected power varied as the properties of the mixture change with composition and temperature. The absorbed power against temperature was estimated by monitoring these changes with the directional coupler (Roussy & Pearce, 1995). Notice in figure 3 how in each case, compacted and non-compacted exhibit different energy absorption related to temperature, which means a different absorption mechanism or at least different dielectric properties.

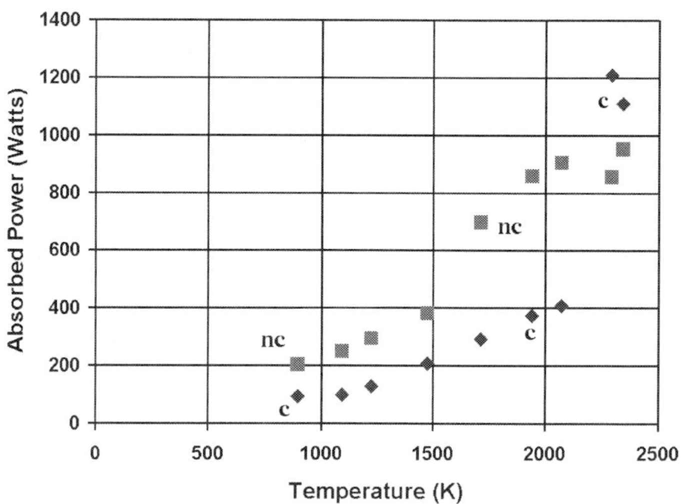

Fig. 3. Power taken by the system (cavity, crucible and mixture) estimated from the forward and reflected power ratio during the test of mixtures of 8 g Compacted (c) and non-compacted (nc) mixtures exposed to 2000 Watts.

Then, it can be suggested that heating during the reaction is aided by the formation of $(CaO)_{0.15}(ZrO_2)_{0.85}$, which was found in all the samples and has a composition close to those where the electric conductivity of a CaO-ZrO_2 solution is maximum in the experimental temperature range. CaO-ZrO_2 solid solutions are considered as a high conduction electrolyte and because any other composition was not found, it is speculated that among the ZrO_2, the phase, $(CaO)_{0.15}(ZrO_2)_{0.85}$, is thermally responsible of the heating with microwaves.

3.2 Microwave synthesis: silicon carbide

The main applications of silicon carbide are in composite materials, abrasives, high hard tools and machining equipment. It is also used as a coating material due to wear resistance.

Another application in the refractory industry is related to the manufacture of crucibles. These are just structural applications, although there are other electric ones when other elements are added (Lee et al., 1994).

Traditionally, this material is produced by means of a procedure named Acheson process, which consists in placing a mixture of silica and carbon in an electric furnace. Heating is accomplished by a core of graphite and coke placed at the center of the furnace. The mixture is set around this core where temperature is 2700 °C approximately. The graphite remains at the core, while Si reacts with carbon to form different SiC polytypes in colder parts of the oven. Among the several polytypes there are more than 200 non cubic polytypes known as α-SiC and only one that is cubic, known as β-SiC or SiC-3C, that is interesting because of its applications. Being one among many, often β-SiC is produced in mixtures with large amounts of α-SiC.

Since the development of the microwave devices there has been a growing interest, within the scientific community, on the possible applications of microwave radiation as an energy source for the processing of materials. Microwave synthesis of silicon carbide from silicon and activated charcoal has been reported at temperatures as low as 727 °C (Ramesh et al., 1994), which is far below the temperatures achieved in the Acheson process. From previous experiences (González et al., Aguilar, 1996) is known that temperatures around 2000°C are achievable with microwaves when the materials are good microwave absorbers.

Silicon carbide production can be described as a reaction where silicon oxide is reduced by graphite resulting in substitution of the oxygen by carbon. As a process, this reaction can be described thermodynamically as:

$$SiO_2 + 3C = SiC + 2CO \tag{1}$$

This endothermic reaction requires about 528 KJ/mole and industrially it is carried out at temperatures ranging between 1600 °C and 2500 °C.

Then, regarding microwave processing, the idea is to producing β-SiC from a mixture of silica and graphite by means of microwaves as an energy source knowing that it is possible to keep the temperature within the range of β-SiC growth.

Silicon carbide production tests were carried out using a magnetron working at 2.45 GHz and a power supply of up to 2000 Watts. Powders of SiO_2 and C of average size of 80 µm and 50 µm respectively, were mixed thoroughly to get intimate contact between the particles. The used ratio was 1:3 molar. SiO_2 and C are good microwave absorbers even at low temperatures, and then the mixture is able to achieve the required temperature for the process to occur. The mixture was placed inside a thermally insulated crucible made of high purity alumina which was in turn placed into the cavity in a specific location that was found to be the best according to the heating rate. The mixtures were exposed to the conditions given in table 3; temperature was taken optically.

The obtained samples exhibited two separated phases; first one was conformed by a black shape, while the second one consisted in a whitish material in drops. Both of phases were analyzed by X-ray diffraction of powders (λ=1.54Å), and the only compound found in the black material was SiC-3C, with un-reacted graphite, while the drops were SiO_2 only. One example of the diffraction pattern is shown in figure 4, which corresponds to test 8.

The X-rays diffraction patterns took from different zones in the sample show homogeneity. In all of the cases the found silicon carbide corresponded to SiC-3C, according to the diffraction center (ICSD) it is cataloged as (29-1129) which is the cubic β-SiC.

Test number	Time (sec)	Power (Watts)	Mass (g)
1	600	1000	10
2	600	2000	20
3	900	2000	10
4	1500	2000	10
5	900	1000	10
6	1200	1000	20
7	600	1000	20
8	600	2000	20
9	1200	2000	20
10	1200	1000	20

Table 3. Experimental conditions for SiC production.

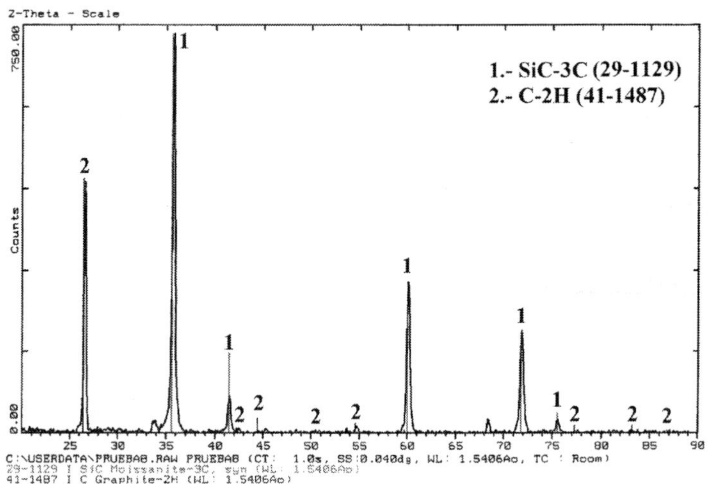

Fig. 4. X-ray diffraction pattern from the black phase in test number 8. The only species found are silicon carbide and graphite. Small unidentified peaks do not form a set for other species, although they might be other silicon carbide polytypes or silicon as element.

Table 4 shows the SiC/C ratio and the maximum achieved temperatures were between 1800°C and 2000°C, which are adequate for having growth of this particular structure. Cavity was overheated in several occasions and for that reason some tests were terminated earlier as the actual exposition time column shows. Figure 5 corresponds to the pattern of a sample of commercial SiC where β-SiC is not present.

This commercial sample, taken a resistance electrode that according to the producer was made of a single kind of SiC is made of α-SiC, which is valid considering that all of the present phases are specified that way. The microwave method gave the temperature conditions for producing β-SiC (SiC-3C) as confirmed by X-rays diffraction. SEM images show wiskers of SiC at the center of the obtained mass (Figure 6), while grains were formed at the edge (Figure 7). Achieved temperatures are in the adequate range for producing β-SiC, no other polytype was produced, which is a difference against other methods. Then, mixtures of SiO_2 and C exposed to microwaves at temperatures between 1700°C and 2100°C actually produced β-SiC. The difference between the center and the edge is an evidence of thermal profile and that microwaves acted thermally.

Test number	Achieved max temp. (°C)	SiC/C peak ratio	Actual exposition time (sec)
1	1702	0.92	589
2	2063	0.17	605
3	2025	0.32	969
4	1968	1.05	1468
5	1873	2.76	830
6	1930	0.54	1243
7	1876	0.38	600
8	2126	1.55	605
9	1957	1.3	1050
10	2054	0.30	1100

Table 4. Results of the tests described in table 3.

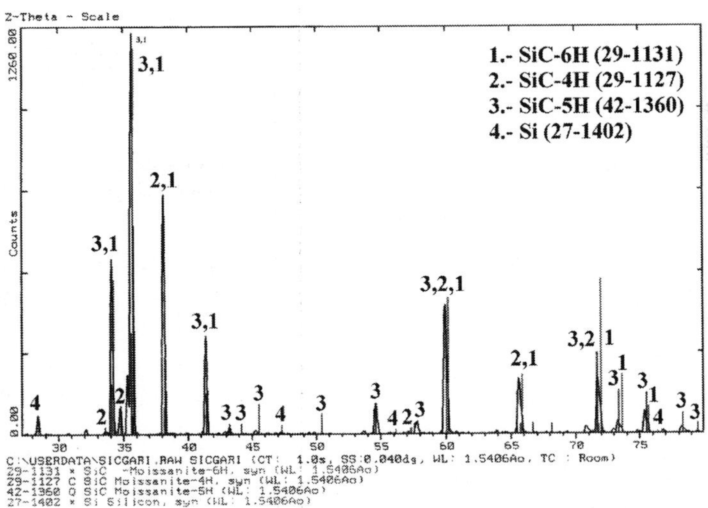

Fig. 5. X-ray diffraction pattern from commercial silicon carbide, notice the mixture of present structures, including silicon. β-SiC is not present in this case.

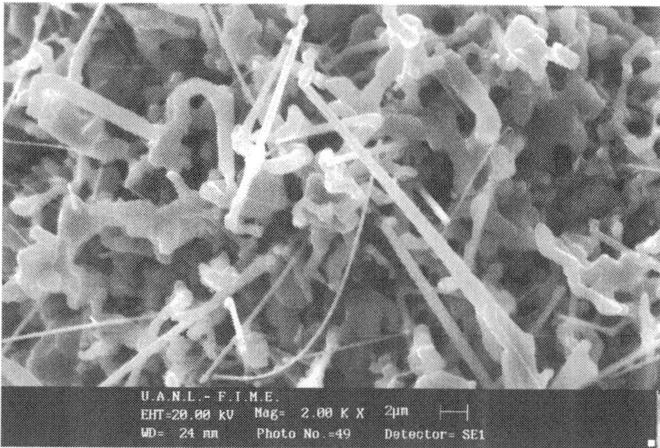

Fig. 6. Whiskers of silicon carbide (test number 5). This is one of the tests where more silicon carbide was observed; just few zones in the sample (not shown) exhibited free graphite.

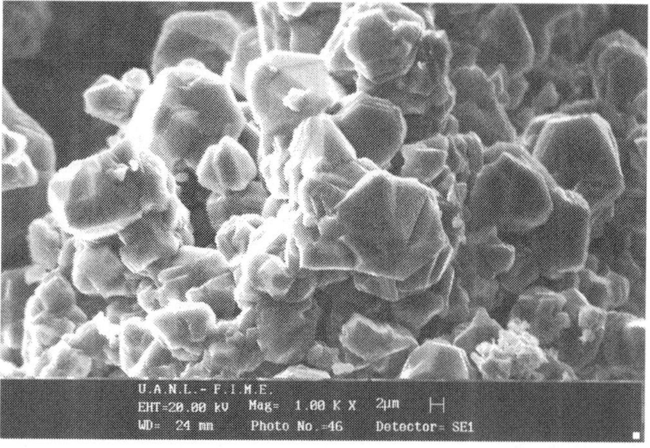

Fig. 7. Grains of silicon carbide (test number 5). This area was far from the center, closer to the crucible walls where thermal losses are higher.

4. Microwave effect

4.1 Temperature measurement

Discussions about the improvement of the processing rates when microwaves are used can not be solved while temperature is not well known for conducting kinetic comparisons. That is the reason for emphasizing the measurement techniques, optical pyrometry is a good option because it does not interact with the sample, but it is only surface measurement and does not work at low temperatures. Optical fiber is an option for low temperatures and it is claimed that they do not affect the system. Thermocouples work in a wide range of temperature and are quite simple to use, but it is common to have arguments against them for taking temperatures in processes that are taking place under microwave fields. Thermocouples placed perpendicularly to the electric field are reliable, however this configuration is possible in waveguides only, and the test was made with two thermocouples, one in front of the other, tip to tip, with the tip at different distance from the waveguide wall and perpendicular to the electric field. The sample was an alumina crucible filled with alumina (total weight 145 g) in a WR284 waveguide. Reflected and forward power were monitored as well as temperature, power was switched on and off as points 1 to 5 (Figure 8) are switching of power and it can be noticed that there are not discontinuities in temperature. This discussion (Aguilar & Pearce, 2003) is taken in more detail below.

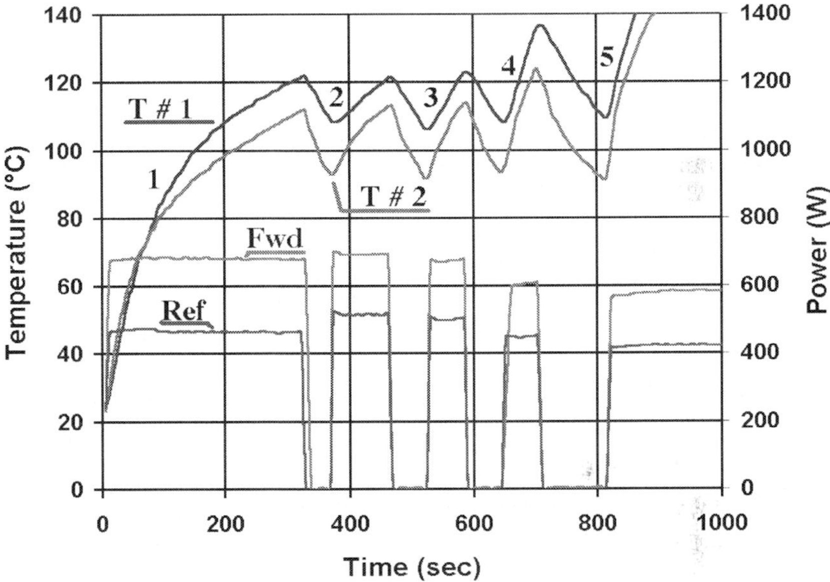

Fig. 8. Temperature measurement of 145 g of alumina, T#1 is one thermocouple located at 3 mm from the centre of the waveguide and T#2 is another one located at 1 mm. Switching of the power supply (points 1-5) shows reliability of thermocouples. Forward and reflected power is also shown.

However, its simplicity makes it attractive for being considered in multimode cavities where it is not possible to locate them in a particular position referred to the electric field. One question that arises when thermocouples are employed is if the electric field perturbs the measurement, and if the thermocouple affects the processing. Therefore, the first issue is to evaluate the reliability of the thermocouple and its possible influence on the processing in a multimode cavity.

One way for confirming reliability of thermocouples is by conducting experiments in a process that is suitable to be perform while other evidences related to temperature and reaction are considered. The chosen process was microwave synthesis of spinel ($MgAl_2O_4$) with hematite (Fe_2O_3). This system is interesting because alumina-based systems are very common in refractory industry. Analysis of the obtained samples was carried out by X-ray diffraction of powders.

There are several reports of thermocouples employed for temperature measurements in a microwave field (Binner, et al., 1995; Roussy & Pearce, 1995; Gómez, et al., 1996).

The sample or powder mixture was placed into a high purity alumina crucible of 115 g (Figure 9), placed in an appropriate location in a cavity, and isolated with Kwool. The thermocouple was a shielded (3.17 mm diameter) ungrounded type K was placed at the center of the crucible.

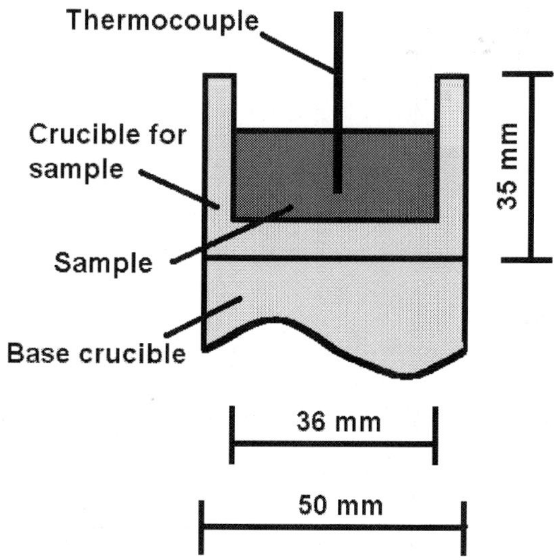

Fig. 9. Scheme of the crucible and the thermocouple inserted in the mixture.

The tests consisted in having a microwave-heating device with an on/off control. The first tests were conducted on hematite, which is know to be a good microwave absorber alone with the temperature setting to 800 °C and 1200 °C An on/off cycle produce ripples in temperature with the frequency given by the controller (Figure 10). Temperature sampling time was 5 readings per second. Notice that the line forming the ripples due to the control action is continuous even during power switching. This test proves that temperature measurement is not affected by the microwaves, although that given process would not be affected by them. Reliability of thermocouples consists on compliance of these two conditions.

Proving the second condition requires analysis of the products of a reaction. The chosen process was synthesis of $MgAl_2O_4$ spinel processed with microwaves from magnesia (MgO) and alumina (Al_2O_3) and hematite (Fe_2O_3) as a susceptor given than neither MgO nor Al_2O_3 are microwave absorbers (Ortiz et al., 2001). Tests were performed on the same arrangement of 7 g incompact, approximately 10 cm³, and mixtures with different compositions of the system $MgO-Al_2O_3-Fe_2O_3$, according to table 5.

Magnesia is not a good microwave absorber, and alumina absorbs above 500 °C (Aguilar & Pearce, 2003), then Fe_2O_3 in the mixture acts as a susceptor that absorbs energy at room

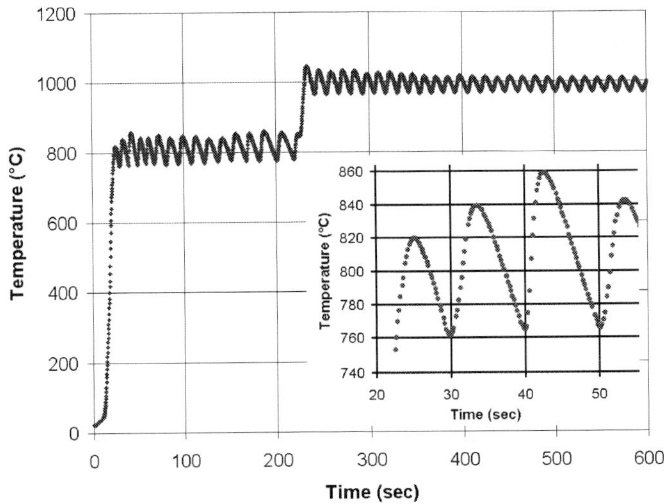

Fig. 10. Test of 20 g of hematite heated with microwaves at 2.45 GHz and 800 Watts in a multimode cavity to 800 °C and 1200°C. Notice the continuity of the temperature in the insert showing early stage of heating.

Compound/Mixture	1	2	3
MgO	45	35	30
Al_2O_3	50	50	50
Fe_2O_3	5	15	20

Table 5. Compositions of the mixtures tested in this work (mole percentage).

temperature and heats up the mixture to 500 °C where Al_2O_3 as well as the just formed spinel, absorb energy from the microwaves. Following the example of previous experience with this kind of material (Aguilar et al., 1997), mixtures were exposed to 800 W microwaves at 2.45 GHz for 30 minutes.

By chemical analysis of the products, it is possible to determine if the thermocouple affected the process. With this idea, a comparison against tests without a thermocouple was also performed in order to detect possible differences. In the thermocouple case, when temperature was about to exceed 1200 °C, which is the maximum for a K type thermocouple, it was removed from the system, and process was allowed to continue.

In practice, temperatures above 1200 °C are necessary for producing $MgAl_2O_4$ from MgO and Al_2O_3 in reasonable time scales. There are two hematite phases, α Fe_2O_3 and the γ Fe_2O_3. The latter has a lattice structure similar to Al_2O_3. It has been reported (Bogdandy & Engell, 1971) that at temperatures between 500 °C and 700 °C, the electron diffraction of the hematite crystal is different from those of α or γ, meaning that there is another phase that has been designated as β Fe_2O_3. It could be a superstructure effect or an order-disorder reaction. In any case, this process has been reported endothermic, absorbing 160 cal/mole at 677 °C. This property is used in this work for confirming reliability in the temperature measurement.

The samples were removed from the crucible and analyzed by means of X-Ray diffraction of powders; relative peak intensity was used for estimating amounts of the species after calibration with specimens of known composition. The maximum temperature reached was estimated to be around 2000 °C based on two aspects: the temperature of the insulator taken with an optical pyrometer at the end of the test was 2000 °C, and the melting point of the compounds found in this system range between 2000 °C and 2135 °C.

Simultaneous pyrometer – thermocouple measurements were not considered because the thermocouple was inside the mixture and the insulation was outside, hence the temperatures would not be the same. Besides, the thermocouple could not work above 1200°C and the pyrometer could not measure temperature below 600 °C. Results of tests with and without the thermocouple inserted are presented in table 6.

	With	Without	With	Without	With	Without
Added Fe_2O_3 (Mole %)	5	5	15	15	20	20
MgO	22	23	4	2	2	0
Al_2O_3	14	15	2	0	0	0
$MgAl_2O_4$	28	37	47	49	0	0
$Mg(Al,Fe)_2O_4$	30	18	36	38	98	93
$FeAl_2O_4$	6	6	11	11	0	7

Table 6. Analysis of the obtained samples from the tests with and without thermocouples.

Analysis shows approximately the same composition for tests with and without thermocouple, the difference that is observed at 5% hematite can be explained considering that the melting point of the spinel decreases as hematite is added, therefore the mixture becomes molten and it is difficult to identify each zone into the sample. At this low hematite content the sample was highly heterogeneous; it was even possible to identify different portions of the sample because it was incipiently molten.

Despite this difference at that concentration, from the rest of the results it can be expressed that the thermocouple did not affect the process. Heterogeneity of the sample is an evidence of the importance of the thermal gradient that indeed is more important than the possible error introduced by the thermocouple. Sensitivity of the thermocouples can be evaluated from the thermal evolution of the three chosen hematite compositions (Figure 11).

Plotting those data as heating slopes against temperature (Figure 12) it can be noticed a drop around 600 °C in all cases, including the sample of pure hematite. The thermocouple was sensitive enough to detect that change. Magnitude of the change in the heating slope is related to hematite content, and all of them coincided at a temperature around 650 °C.

This can be related to the hematite transformation (β Fe_2O_3) at 677 °C. This value is above the coincidence point and although endothermic curve begins around 400 °C, a wide range of temperature for this transformation has been reported in literature (Bogdandy & Engell, 1971). However, the most important issue here is that the thermocouple is sensitive enough to follow such a thermal change.

The configuration was reliable while similitude in the chemical analyses with and without a thermocouple confirms that at least by the time that the thermocouple was inserted the mixture reaction profile was not altered. Thermocouple is a very simple device that with certain considerations can be used as a measurement probe in microwave processing and helps to have one of the most important parameters for kinetics studies.

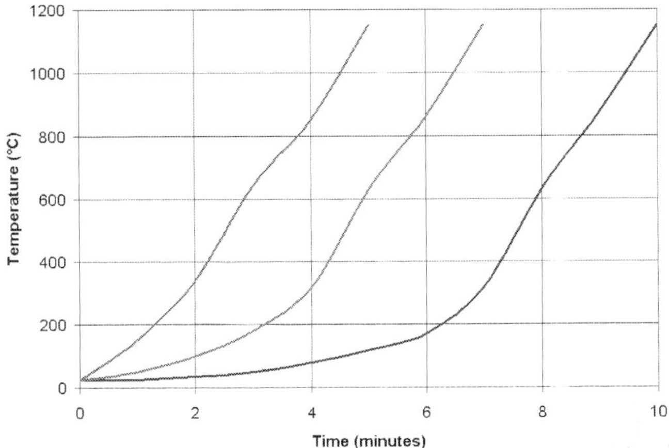

Fig. 11. Heating evolution of mixtures with different hematite content: 20 mole% (violet), 15 mole% (orange) and 5 mole% (blue).

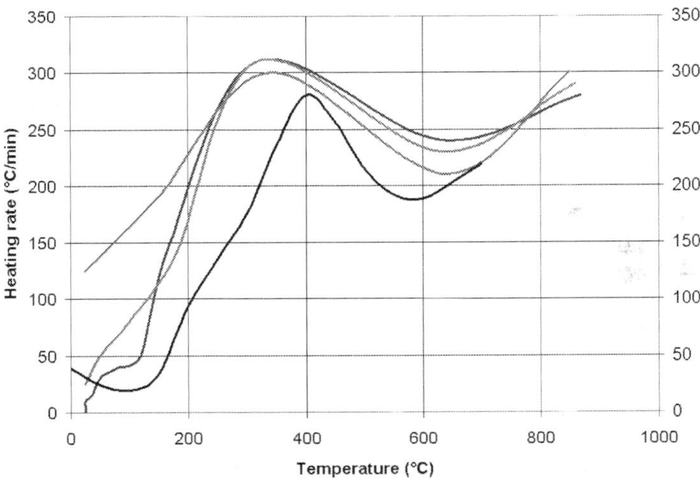

Fig. 12. Heating rates for the tests of the figure 5, as function of temperature. The hematite content are: 20 mole% (violet), 15 mole% (orange) and 5 mole% (blue). The right scale is only for the hematite (black).

4.2 Catalytic effect: synthesis of $MgAl_2O_4$

There are many papers referring to "microwave effects" of the microwaves during materials processing that consists in improvement of sintered materials as well as sintering temperatures, reaction rate increase and activation energy reduction. For instance researchers at Oak Ridge National Laboratory (Janney & Kimrey, 1991) found that activation energy for sintering was reduced to 160KJmole^{-1} for the microwave case, instead of 575 KJmole^{-1} commonly reported. For crystal growth they report a reduction from 590 KJ mole^{-1} to 480 KJmole^{-1}. In other example, Boch (Boch et al., 1992) proposes that microwaves reduce the temperature for mullite and aluminum titanate processing up to 100 °C. Therefore, one of the "microwave effects", described often as non thermal effects, concerns to "catalysis effect" which is tested with synthesis of $MgAl_2O_4$ from MgO and Al_2O_3 as the following lines explain.

The test were conducted over 20 g mixtures of $MgO{:}Al_2O_3$ in a 1:1 proportion at 1200 °C for producing $MgAl_2O_4$ by:

$$MgO + Al_2O_3 \rightarrow MgAl_2O_4 \tag{2}$$

A real comparison requires two kinds of tests, the first one consisted on heating the mixture in an electric resistance furnace, and taking the temperature with a thermocouple inserted in the sample. In the second case the mixture and graphite as a susceptor, were heated in a multimode microwave cavity at 800 W and 2.45GHz with the thermocouple inserted as explained above. A very important aspect in a comparison is maintaining conditions that can be compared; in this case temperature in the microwave process was matched to conventional one at least in the range where reactions are taken place, sudden heating was not possible to control below 600 °C, but initially it was considered that the process is not being carried out under those conditions.

More tests were conducted at 2000 °C without external control; the temperature was maintained due to the heat balance between the mixtures getting energy and dispreading by radiation when at high temperature providing a self control.

Different exposition times where tried and X-rays diffraction was used for estimating the composition of the obtained sample for confirming spinel formation. X-rays patterns of known samples were used for calibration purposes and then a sample of 1 cm^3 around the thermocouple tip was analyzed.

Reliability and sensibility of the thermocouples has been already shown, figure 13 shows the matching temperature in the two kinds of tests. Testing time is 1800 seconds, but matching was achieved since temperature was 600 °C. The insert shows how close the actual temperature was to the setting point of 1200 °C without discontinuities during switching. Thermal runaway of alumina (Aguilar & Pearce, 2003) make difficult to control temperature, therefore the most appropriate condition in this case is to control above 600 °C after the increase of permittivity of alumina has taken place.

Once that it is confirmed that temperature measurement is reliable, the kinetic aspect can be taken in the search for a microwave effect. Accepting that the process is controlled by diffusion the following expression can be taken:

$$R^2 = 2Kt \tag{3}$$

Where R is the molar fraction of formed spinel $MgAl_2O_4$, K is the reaction constant ant t is the time.

Then it is possible to plot R^2 against nominal testing time as it is shown in figure 14 and 15. Testing time and temperature are not appropriate for having full transformation, nut that is the essence of a kinetic study, to have different transformation degrees for comparing against time. Error are more noticeable at low transformation because it is squared, however trends and differences are clear and they are evident despite the dispersion of the points.

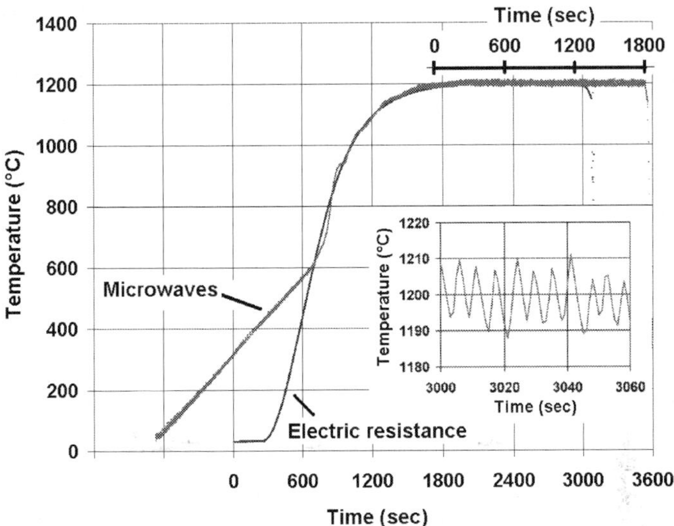

Fig. 13. Matching of the temperatures for the electric resistance and the microwave cavity. The test at 1200 °C lasted 1800 seconds. The insert shows the action of the on/off controller.

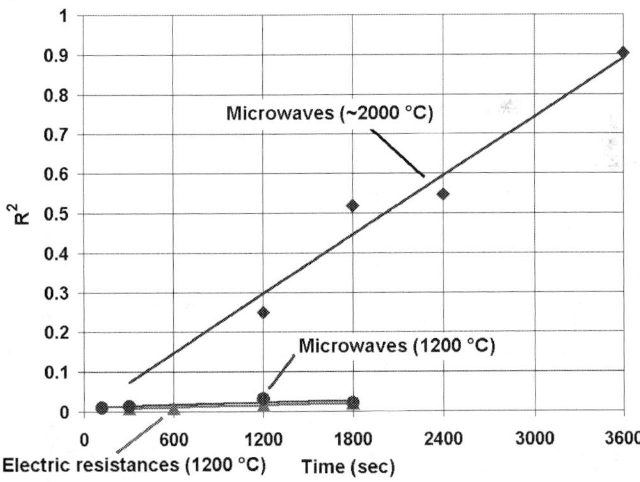

Fig. 14. Experimental results of the test at 1200 °C and 2000 °C. Parabolic regressions are also shown.

It looks like indeed the process was faster with microwaves, the slope in figures 14 and 15 represents 2K and value of K is presented in table 7.

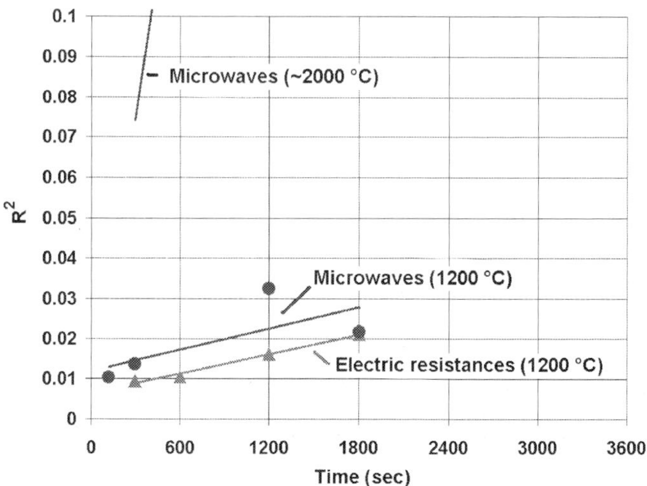

Fig. 15. Amplification of figure 13 showing the experimental results and the parabolic regressions for calculating K.

Test	K (sec^{-1})	Ordinate interception
Electric resistance (1200°C)	4.07×10^{-6}	0.0064
Microwaves (1200°C)	4.41×10^{-6}	0.0119
Microwaves (2000°C)	1.23×10^{-4}	0

Table 7. Reaction constants

Interception at ordinate can not be null because the testing time begins when the temperature is 1200 °C and, although a very low rate, reaction is taking place at least since temperature exceeded 600 °C. In tests conducted previously at temperatures below 600 °C spinel was not detected by X-rays diffraction after 20 hours.

Differences in K values at 1200 °C are not extraordinary for supporting an argumentation of "microwave catalysis". Difference is in the order of 8% when catalytic effects are claimed to be around 400%.

The differences in the origin interception can be justified because the test with microwaves was at higher temperature larger than the electric resistance case before reaching 600 °C, which means higher energy transfer at that stage. This can be also justified by observing the test at 2000 °C, conducted under free heating without any control. Interception is practically null because temperature was achieved in about 600 seconds, instead of the 2400 that took to the 1200 °C tests. Reaction rate is higher at 2000 °C and hence the thermal history is less important.

Activation energy can be also estimated using the tests conducted at 2000 °C with:

$$K = Ae^{-Q/RT} \tag{4}$$

Where T is the absolute temperature, Q is the activation energy and R is the universal constant of the gases. From the values in table 7, Q/R is -13949.64 and A is 0.0573 s^{-1}. These values are useful for calculating the fraction of spinel formed in 20 hours at 600 °C, which was 0.03, too low for practical detection with X-rays diffraction of powders, being in agreement with the experimentation the assumption of no appreciable reaction below 600 °C. However, reaction took place and it is accumulative to the found spinel after the 1800 seconds of testing time.

These values are also useful for estimating the reaction degree taking into account the all the tests, including the whole thermal history, not only the portion after the mixture reached 1200 °C. This calculation is presented in figure 15, the time scale corresponds to the testing time only. Looking in first instance the curve 1 that was calculated with the temperatures recorded with the tests conducted at 1200 °C with electric resistances. The points obtained from these tests are following the kinetic equation with the calculated parameters. The curve calculated with the thermal history of the tests conducted with microwaves is practically the same, at the scale of the plot the difference can not be appreciated which corroborates that matching the tests above 600 °C is a good experimental condition and that regardless the differences in temperatures below 600 °C the samples are the same at the moment that they reached that point. It can be also noticed that at the time that the tests begins there is already some formed spinel.

However, all the microwave points, but the last one, are above curve 1 in the figure 16. The difference looks large, but it is R^2 and not R, hence it is not that large that it could be call special kinetic effect of microwaves. However it still is a difference that must be explained. Separated tests confirmed that the samples obtained from the experiments conducted in the electric resistance furnace are more homogenous than the samples from the microwave testing. It was random and although always the less reaction occurred on the surface, a profile around the thermocouple could not be determined.

An idea was that given that permittivity of alumina increases with temperature, and there is even thermal run away about 600 °C, it was possible that some parts of the mixture could

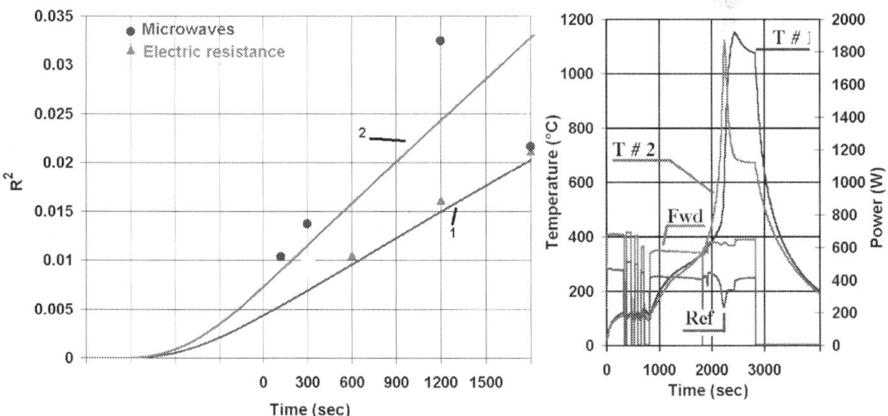

Fig. 16. The curve 1 was calculated with the parameters found in this work considering the thermal history, while the curve 2 was calculated with the same parameters at a temperature 6.4% above the registered one. Right chart shows thermal run away of alumina, T#1 and T#2 as the conditions of figure 8.

reach temperatures higher than those sensed by the thermocouple, this phenomena can not happen with the electric resistance due to the heat transfer mechanism. Thermal run away of alumina is shown in figure 16 right, alumina increase its temperature rapidly after reaching about 400 °C (T#1 and T#2) as the reflected power decrease due to the increase of the dielectric heating.With this idea it was calculated which must be the temperature in order to have the points showed by the microwave tests. It was decided to estimate an overheating error along the whole process which was 6.4%. This argument seems to be against reliability of the thermocouples, however in 1200 °C (1473 K) this error means that the average temperature was 1295 °C, which compared with test where temperature has been taken optically to the surface, could provide similar accuracy.

However, although this results could bring the idea of a microwave "kinetic effect" what is better to say according to the results is that there is a microwave effect, because the heating mechanism is different, that there is heterogeneity, which suppose selective heating rather than uniform, but all of these observations are explained thermally, described by classical thermodynamics and kinetics. There are many variables involved in the process, from the powder distribution, location of the thermocouple, and change in permittivity as the reaction evolves and heat transfer. However a non thermal kinetic effect would be much more evident.

5. Microwave sintering: manganites

Applications of NTC (negative temperature coefficient) thermistors ceramics such as $Fe_xNi_{0.7}Mn_{2.3-x}O_4$ are of high interest. Sintering of these materials is a crucial part in their fabrication; because electrical properties dependent of the microstructure and cation distribution. Sintering affects microstructure, including decomposition of NiO, that affects overall quality of the final product. The first approach for solving this situation is to reduce the sintering temperature below 900°C, which is the decomposition one (Töpfer et al., 1994; Csete de Györgyfalva et al., 2001) but sintering process takes too long that becomes economically impractical: Increase the sintering temperature for reducing time of processing is against product performance because the structure could be fully decomposed. There is a competition between required time for achieving full dense materials and kinetics of decomposition of the structure. Current research is dedicated to study such competition, where new and different methods that can contribute reduction of time processing are being designed. Microwaves have been used as a power source for synthesize and sinter different materials. The main benefits that many researchers claim (Bykov et al., 2001) are enhanced diffusion, selective heating, increased reaction rates, and sometimes decreased reaction temperatures. No data were reported of the application of microwave energy in the production of $Fe_xNi_{0.7}Mn_{2.3-x}O_4$ ceramics for application as NTC thermistors.

Different manganite compositions of the Mn-Ni-Fe-O system were tested in preliminary studies (Aguilar et al., 2002), the manganite $Fe_xNi_{0.7}Mn_{2.3-x}O_4$ was produced by coo-precipitation in order to have homogeneous powders to be sintered. The powders were axially compacted at 250 MPa in tablets of 6 mm diameter and 1.5 mm thickness. Conventional sintering was performed in an electrical resistance oven at 1170 °C, while the microwave sintering was conducted in a microwave oven operating at 1100 W and 2.45 GHz, modified with a temperature control within 5% error (Valdez, 2001). The samples were placed in a graphite bed, with a thermocouple located 1 mm above it. From dilatometry tests and a relationship from literature that considers the heating and cooling slopes for determining the appropriate sintering time (Binner, 1995); temperature schedules are given in figure 17. It was confirmed by X-rays diffraction that the powders obtained to be sintered were formed of spinel type oxides (figure 18).

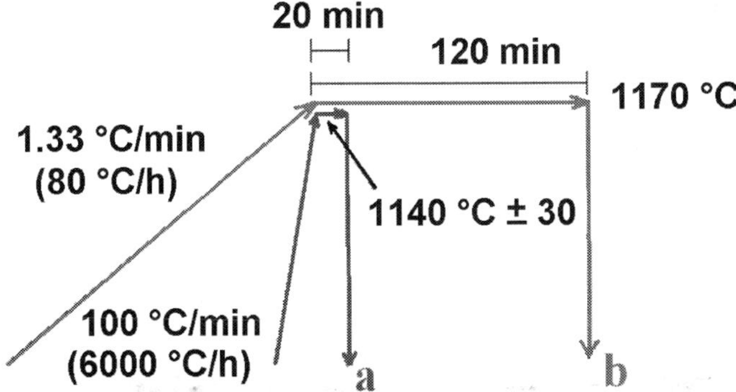

Fig. 17. Temperature schedules for the sintering, a) microwave and b) conventional processing.

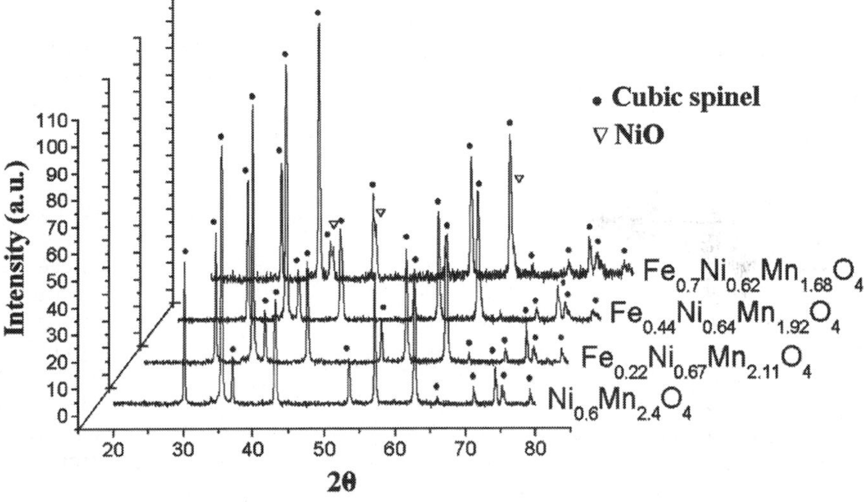

Fig. 18. X-ray diffraction pattern of the powders prior to sintering.

Apparent density of the sintered powders is compared in figure 19, final density was higher in the case of microwave sintered sample, and then it is easy to suppose that a "microwave effect" actually exists for the manganites with high iron content.

However, it is very important to evaluate if the obtained materials are the same in both processes. X-rays diffraction patterns of the samples obtained showed that the material is different for each case, free nickel was found in the sample processed with microwaves while in the conventional sintering there is no free nickel.

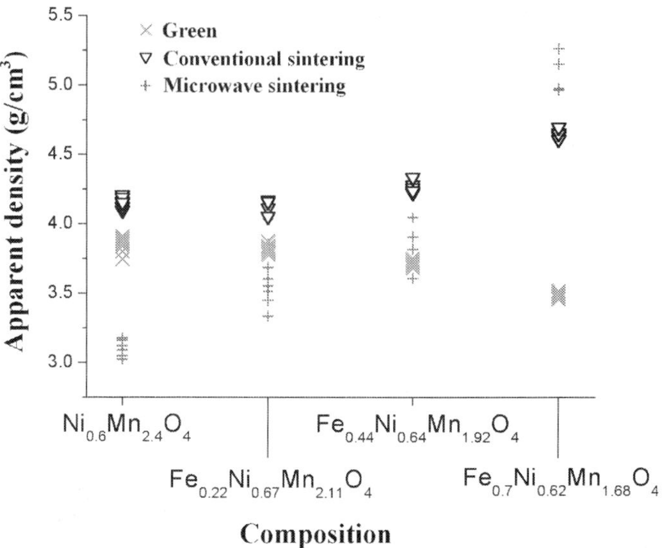

Fig. 19. Final densities of the processed powders.

An explanation is that nickel was reduced due to the presence of graphite as heating susceptor as has been suggested by Csete (Csete de Györgyfalva & Reaney, 2001). Hence, if the products are different that means that a direct comparison between microwave and conventional sintering was not actually performed and that the first idea from figure 19 is not precisely right. In studies of diffusion of isotope ^{59}Fe in a structure type ferrite (Fe, Mn) was found that the diffusion coefficient was around 10^{-7} cm^2/s at a partial oxygen pressure of 10^{-12} atm and 1000 °C (Franke & Dieckmann, 1988).

Peterson (Peterson et al., 1980) found that the diffusion coefficients of ^{59}Fe in a magnetite (Fe_3O_4) were in the order of 10^{-9} cm^2/s at an oxygen pressure of 10^{-12} and 1000 °C. This means that the mobility of the iron in the magnetite is different than in the ferrite, furthermore the diffusion of Mn is two orders of magnitude smaller in MnO (D^{*}_{Mn}= 2.01x10^{-10} cm^2/s, T=997 °C, PO_2=10^{-12} atm) (Peterson & Chen, 1982) and Fe diffusion in FeO (D^{*}_{Fe}=7.29x10^{-8} cm^2/s, T=1007 °C, PO_2=10^{-12} atm) (Chen & Peterson, 1975). Then, diffusion of iron is greater in a wustite structure (Fe, Mn)O than in a spinel, and the diffusion coefficient of manganese is smaller than the iron one. Consequently, while in the conventional heating the powders are being sintered only, in the microwaves case there is a reduction of the oxides, and what is being sintered does not have the spinel structure, but the wustite, and following the iron/manganese diffusion coefficients argument, sintering of $Fe_{0.7}Ni_{0.62}Mn_{1.68}O_4$ is easier than sintering of $Ni_{0.6}Mn_{2.4}O_4$. Hence, for this specific case, the "microwave effect" is actually a different processing path.

6. Conclusions

The tests that were presented in this document show that microwaves are a possible method for supplying energy for heating and therefore drive processes. There are several very serious research works that present evidence of non-thermal effect that improve efficiency of the processes that are taken place.

Indeed from a strictly perspective of the final product with microwaves against any other process, there are differences due to selective heating, which is a microwave effect but it is thermal; higher temperatures than actually measured, but that is also thermal. In the case of the manganites, the found product was different; therefore the special "microwave effect" was not present.

In classic thermodynamics there is no difference in how the energy is supplied to a given system for conducting a process; it is only a matter of system states. However, kinetics depends on the path that is actually taken for the system to change its state; it could be in different stages. In endothermic processes the reaction is limited by the energy supply, changing the way the energy is being supplied could change global kinetics, without affecting the essence of the process or changing activation energies. In this case the microwave effect consists in the possibility of supplying all the energy that the system is demanding when the system is demanding it.

Each mechanism has its own activation energy; activation of a different mechanism will give as a result a change in this value, activating another process. In kinetic classes there are experiences of identifying a mechanism from evaluating its activation energy.

Assigning these effects, which are real, to non thermal phenomena deviate attention from the real challenge of the non thermal observations that are hard to explain with classical thermodynamics and kinetics. The amount of non-thermal observations available by serious researchers can not be denied. Therefore it is important to look for classical explanations always that is possible in order to identify and leave room for giving attention to those special cases.

7. Acknowledgements

This document contains part of different works developed during the thesis of students of the Programa de Doctorado en Ingeniería de Materiales en la Facultad de Ingeniería Mecánica y Eléctrica de la Universidad Autónoma de Nuevo León, México, under the direction of the author and sponsored by the Mexican Council for Science and Technology (CONACYT), Evaluation Cooperation and Scientific Orientation (ECOS-France-México) and the university itself through its Aid Program to Scientific and Technological Research (PAICYT). The students were: Dra. Idalia Gómez, Ing. Javier Rodríguez, Dra. Oxana Kharissova and Dr. Zarel Valdez. Scientific contributions, among many others, of Prof. Bernard Durand, Prof. John Pearce, Dra. Sophie Guillemet and Dr. Ubaldo Ortiz, are also highly appreciated and recognized.

8. References

Aguilar, J. & Pearce, J. (2003). Measurement of Dielectric Properties of Aluminum Oxide While Exposed to Microwaves. *British Ceramics Transactions*, 102, 2 (52-56) ISSN 0967-9782

Aguilar, J.; Gómez, I. & González, M. (1996). Fabrication of Calcium Zirconium Oxide Ceramics by Microwave Processing, *Proceedings of the 31st Microwave Power Symposium*, pp. 100-103, Boston MA, July 1996, International Microwave Power Institute

Aguilar, J.; Guillemet, S.; Valdez, Z. &, Garza, F. (2002) Processing of $Ni_{0.60}Fe_{0.82}Mn_{1.57}O_4$ with microwaves. *XI International Materials Research Conference*, Cancún Qr, México

Aguilar, J.; González M. & Gómez I. (1997). Microwaves as an Energy Source for Producing Magnesia-Alumina Spinel, *Journal of Microwave and Electromagnetic Energy*, 32, 2 (74-79) ISSN 0832-7832

Binner J.; Hassine N. & Cross T. (1995). The Possible Role Of The Pre-Exponential Factor In Explaining The Increased Reaction Rates Observed During Microwave Synthesis Of Titanium Carbide, *Journal of Materials Science* 30, 21, (5389-5393), ISSN 0022-2461

Binner, J.; Vaidhyanathan, B.; Wang J. & Price, D. (2008). Evidence for Non-Thermal Microwave Effects Using Single and Multimode Hybrid Conventional/Microwave Systems, *Journal of Microwave and Electromagnetic Energy*, 42, 2 (47-63) ISSN 0832-7832

Boch, P.; Lequeoux N. & Piluso, P. (1992). Reaction sintering of ceramics materials by microwave heating, *Microwave Processing of Materials III, Materials Research Society Proceedings*, 269, p. 211-216, San Francisco CA, MRS

Bogdandy, L. & Engell H. (1971). *The Reduction of Iron Ores*, Springer Verlag

Brandon, J.; Samuels, J. & Hodgkins, W. (1992). Microwave Sintering of Oxides Ceramics, *Symposium Proceedings: Microwave Processing of Materials III*, pp. 237-243 San Francisco CA, MRS, October 1992 MRS

Bykov, Y.; Rybakov, K. & Semenov, V. (2001). High Temperature Microwave Processing of Materials, *Journal of Applied Physics D* 34, (55-R75) ISSN 0022-3727

Chen, W. & Peterson, N. (1975). Effect of the Deviation from Stoichiometry on Cation Self-diffusion and Isotope Effect in Wustite, $Fe_{1-x}O$, *Journal of Physics and Chemistry of Solids*, 36, 10 (1097-1103) ISSN 0022-3697

Csete De Györgyfalva, G. & Reaney, I. (2001), Decomposition of $NiMn_2O_4$ Spinel: An NTC Thermistor Material, Journal of the European Ceramics Society, 21 (2145-2148) ISSN 0955-2219

Franke, P. & Dieckmann, R. (1988). Defect Structure and Transport Properties of Mixed Iron-Manganese Oxides, *Solid State Ionics*, 26, 2 (817-23) ISSN 0167-2738

Garbacia, S.; Bimibisar, D.; Lavastre, O. & Kappe, O. (2003). Microwave Assisted Ring Closing. Metathesis Revisited. On the Question of Non Thermal Microwave Effect. *Journal of Organic Chemistry*, 68, 23, (9136-9139) ISSN 0022-3263

González, M.; Gómez, I. & Aguilar, J. (1996). Microwave Processing Applied to Ceramic Reactions. *Microwave Processing of Materials V, Materials Research Society*, 430, pp. 107-112, San Francisco CA, April 1994, MRS

Hippel, A. (1995). *Dielectric Materials and applications*, Artech House Microwave Library, ISBN 1580531237

Hoogenboom, R.; Leenen, M.; Wiesbrock, F. &, Schubert, U. (2005). Microwave Accelerated Polymerization of 2 Pheniyl-2 Oxizoline. Microwave or Temperature Effects? *Macromolecular Rapid Communications*, 26, (1773-1778) ISSN 1521-3927

Islam, M.; Davies, R.; Fisher, C. & Chadwick, A. (2001). Defects and Protons in the $CaZrO_3$ Perovskite and $Ba_2In_2O_5$ Brownmillerite: Computer Modelling and EXAFS Studies, *Solid State Ionics*, 145, 1-4 (333-338) ISSN 0167-2738

Jacob, K. (1997). Discussion of Use of Solid Electrolyte Galvanic Cells to Determine the Activity of CaO in the $CaO-ZrO_2$ System and Standard Gibbs Free Energies of

Formation and CaZrO$_3$ From CaO and ZrO$_2$, *Metallurgical and Materials Transactions B*, 28B (723-725)

Janney, M. & Kimrey, H. (1991). Diffusion-Controlled Processes in Microwave–Fired OxideCeramic, *Microwave Processing of Materials II, Materials Research Society Proceedings*, 189, pp. 215-227, San Francisco CA, MRS

Janney, M.; Calhoun, C. & Kimrey, H. (1991). Microwaves: Theory and Applications in Materials Processing, *Ceramics Transactions* 21, (311-317), American Ceramics Society, Westerville OH

Katz J.; Blake R.; Kimrey H. & Kenkre V. (1991). Microwaves: Theory and Applications in Materials Processing, *Ceramics Transactions* 21, (95-105), American Ceramics Society, Westerville OH

Kingery, W.; Bowen, H. & Uhlmann, D. (1975). *Introduction to Ceramics*, John Wiley and Sons, ISBN 0471478601

Lee, W.; Phil, D. & Rainford W. (1994). *Ceramic Microstructures, Property Control by Processing*. Department of Materials Engineering and Sorby Center for Electron Microscopy, University of Sheffield, UK. First Edition 1994.

Ortiz U.; Aguilar J. & Kharissova O., (2001). Effect of Iron Over the Magnesia-Alumina Spinel Lattice, *Journal of Advances in Materials and Materials Processing*, 2, 2 (107-116)

Peterson, N. & Chen, W. (1982). Cation Self-diffusion and the Isotope Effect in Mn$_{1-\delta}$O, *Journal of Physics and Chemistry of Solids*, 43, 1 (29-38) ISSN 0022-3697

Peterson, N.; Chen, W. & Wolf, D. (1980). Correlation and Isotope Effects for Cation Diffusion in Magnetite, Journal of Physics and Chemistry of Solids, 41, 7 (709-719) ISSN 0022-3697

Popescu L.; Stanca A; Branzoi, I. & Branzoi, F. (2008). Antiscale Efficiency of New Polymers, Synthesized in Microwave Field *World Academy of Science, Engineering and Technology* 47 (11-15)

Ramesh, P.; Vaidhyanathan, B.; Ghandi, M. & Rhao, K. (1994). Synthesis of β-SiC powder by use of microwave radiation. Journal of Materials Research 9, (3025-3027) ISSN 0884-2914

Rao, K. & Ramesh, P. (1995). Use of Microwaves for the Synthesis and Processing of Materials, *Bulletin of Materials Science* 18 (447-465)

Roussy, G. & Pearce, J. (1995). *Foundations and Industrial Applications of Microwaves and Radio Frequency Field*, John Wiley and Sons, ISBN 978-0-471-93849-1

Roy, R.; Agrawal, D. & Cheng, J. (2000). New First Principles of Microwave-Material Interaction. Discovering the Role of the H Field, and Anisothermal Reactions. Microwaves: Theory and Applications in Materials Processing V. *Proceedings of the Second World Congress on Microwave and Radio Frequency Processing* pp. 471-485, Orlando, FL, April, 2000, American Ceramics Society, Westerville OH

Roy, R.; Agrawal, D.; Cheng, J. & Gedevanishvili, S. (1999). Full Sintering of Powdered-Metal Bodies in A Microwave Field, *Nature* 399 (668-670)

Sorescu, M.; Diamandescu, L.; Peelamedu, R.; Roy, R. & Yadoji, P. (2004). Structural and Magnetic Properties of NiZn Ferrites Prepared By Microwave Sintering, *Journal of Magnetism and Magnetic Materials*, 279, 2-3 (195-201) ISSN 0304-8853

Tanabe, J. & Nagata, K., (1996). Use of Solid Electrolyte Galvanic Cells to Determine the Activity of CaO in the CaO-ZrO$_2$ System and Standard Gibbs Free Energies of

Formation and $CaZrO_3$ From CaO and ZrO_2, *Metallurgical Transactions B*, 27B (658-662)

Tien, T. (1964). Electrical Conductivity in the System ZrO_2-$CaZrO_3$, *Journal of the American Ceramic Society*, 47 (430-436)

Töpfer, J.; Neupert, L.; Schütze, D. & Wartewig, P. (1994). Structure, Properties and Cation Distribution of Spinels of the Series $Fe_2Ni_{1-z}Mn_2O_4$ (0<=Z<=2/3). Journal of Alloys and Compounds, 215 (97-103) ISSN 0925-8388

Valdez-Nava, Z. (2001). Thesis Influencia de la alúmina como absorbedor de microondas en la reacción de formación de espinela alúmina-magnesia". Universidad Autónoma de Nuevo León, México

Willert-Porada, M. & Gerdes, T. (1991). Metalorganic and Microwave Processing of Eutectic Al_2O_3-ZrO_2 Ceramics, Materials *Research Symposium Proceedings, Microwave Processing of Materials IV*, pp. 563-569, San Francisco CA, April 1994, MRS

Wittaker, A. (2005). Diffusion in Microwave Heated Ceramics, *Chemistry of Materials*, 17, (3426-3432) ISSN 0897-4756

Xiang, S.; Jiann-Yang H.; Xiaodi H.; Bowen L. &, Shangzhao S. (2005) Effects Of Microwave On Molten Metals With Low Melting Temperatures Journal of Minerals & Materials Characterization & Engineering, 4, 2, (107-112) ISSN 1539-2511

Yamaguchi, S.; Kobashi, K.; Higuchi, T.; Shin, S. & Iguchi, Y., (2000). Electronic Transport Properties and Electronic Structure of $InO_{1.5}$-Doped $CaZrO_3$, *Solid State Ionics*, 136-137 (305-311) ISSN 0167-2738

13

Evaporators with Induction Heating and Their Applications

Anatoly Kuzmichev and Leonid Tsybulsky
National Technical University "Kiev Polytechnical Institute"
Ukraine

1. Introduction

Induction heating is widely used in different fields due to high process productivity and universality, cleanness and possibility of full automation. This process is friendly for environment. Since the induction allows reaching temperatures of 2000 °C and more, this method is very suitable for evaporation. This process is known to be a process of emitting atoms or molecules by hot matter surface. The atoms or molecules form "the vapor". Since vapor pressure depends on temperature, the latter is conditionally accepted as "the evaporation temperature" when pressure of saturated vapor reaches 1 Pa. Data on evaporation temperatures are collected in numerous handbooks, *e.g.* (Banshah, 1994).

The evaporation is a very important process for industrial production and scientific researches. Its main application is deposition of thin film layers and coatings on semiconductors, glass and polymers in electronics and optics as well as on other industrial products including food packaging films, metal strips, *etc.* The evaporation is mostly related with technology of Physical Vapor Deposition (shortly, PVD) that means matter deposition from vapor or atomic/molecular medium. In PVD, the vapor medium, as a source of depositing matter, is generated physically (not chemically!) due to thermal evaporation of some primary source materials; deposition, in turn, must happen as atom-by-atom process, at least as multi-atom cluster process but without any drops or even smallest droplets. There are various evaporation methods for PVD except induction: evaporation from boats and crucibles with resistive (Joule) or radiation heating, evaporation with electron and laser beam heating, cathode arc evaporation, *etc.* The induction technology competes with the mentioned methods and occupies own niche in industrial technology and for preparation of scientific objects.

The evaporation also is a very useful process for refining materials to obtain high purity, for producing micro- and nanoparticles and, thus, it is a base of so-called evaporation metallurgy.

The main features and fundamental possibilities of the evaporating induction technology are the followings:

-   electromagnetic induction may provide not only heating evaporated materials but also vapor ionization and high density vapor plasma generation for Ion-Plasma Enhanced (Activated, Assisted) PVD processes, often called as the Ion Plating, accompanied by self-ion bombardment for obtaining high quality coatings;

- the inductor may be disposed aside of a heated object and inside or outside of a technological chamber since electromagnetic induction acts through non-conducting medium (vacuum, gas, liquid, solid) and dielectric envelopes; this allows, in particular, to evaporate chemically active, radioactive and toxic substances in closed envelopes, for example, for electron device or special material manufacturing;
- the inductor may operate at voltages and in conditions avoiding electrical discharges (arcing) around the inductor placed into gas medium under pressure from vacuum to atmospheric one; therefore, it is possible to evaporate not only in vacuum but in gas medium for synthesis of compound coatings and production of micro- and nanoparticle materials by "gas evaporation/condensation" method;
- on the other hand, generation of plasma by electromagnetic field can provide production of micro- and nanoparticle materials by "plasma assisted evaporation/condensation" method;
- usually induction evaporation occurs from crucible, therefore the maximal evaporation temperature is limited by the maximal operation temperature of crucible material;
- the direct contact between the inductor and evaporated material is absent, the operation temperature of the water-cooled inductor is less than 80 °C; hence, the method ensures introducing minimal impurities into coatings; however, the problem of reaction of crucible material with evaporated substance remains;
- electromagnetic field of the inductor-heater may provide mechanical confining of molten metal in a crucible or levitation effect on solid and molten metal and evaporation in the levitation mode, not from crucible, for obtaining very pure thin films and coatings.

Due to advantages and performances of evaporators with induction heating they are interesting for thin film and coating technology and special metallurgy.

Historically, we are not aware of the first researchers and engineers employing induction heating for evaporation but one may find patents, published as far back as at the 1930's, that dealt with induction evaporation. This method has been periodically used for commercial purposes (to produce coatings, thin films and various materials) since. Mainly as simple thermal evaporation for the 1940's-1960's, and then also as with vapor ionization.

The problem of induction evaporation and its applications has been discussed in literature from time to time but very shortly, see, for example, handbooks (Banshah, 1994; Bishop, 2007) and survey (Spalvins, 1980). The more special and full review papers (Anderson, 1975; Tada I.. et al., 1989), devoted to this theme, were published many years ago. So, there is a lack of fresh information on contemporary solutions in this field and potential profit of the induction evaporation for industry and science. The authors of the present paper try to fill up this gap but they do not describe all known evaporators with induction heating. Only principle solutions and new those developed in Kiev Polytechnical Institute are presented in the paper.

In this chapter we consider base principles of induction evaporation design, then
- evaporation without vapor ionization, in particular, for microelectronic industry;
- evaporation with vapor ionization for Ion-Plasma Enhanced PVD (Ion Plating); herein, different approaches and electromagnetic structures, including additional electron thermoemitters and ionizing inductor with pulsed power supplying for effective vapor ionization will be presented;
- design of special induction evaporators, in particular, for deposition of coatings onto the internal surfaces of hollow substrates and for levitation evaporation.

2. Principles of induction evaporator design and operation

The base design of the typical induction evaporator (vapor source) is shown in Fig. 1. A crucible 1 charged by evaporated material 2 is concentrically surrounded by an inductor 3 that is a circular coil made from water-cooled copper pipe and powered by an alternative current (AC) generator 4. The arrows 5 depict the water flow through the coil tube. The crucible is needed as a container for the melted and evaporated material 2. A medium frequency (MF: f = 1-66 kHz) or radio frequency (RF: f = 0.44-1.76 MHz) current is ordinarily used for power supply of inductors in the evaporators. The inductor generates an alternative magnetic field that, in turn, induces an eddy current (also called as Foucault current) in the crucible body, if it is made from conducting material, or directly in the conducting evaporated material, if the crucible is from non-conducting material. The eddy current provides Joule heating of the conducting crucible or, accordingly, the conducting evaporated material up to the desired temperature. In the case of conductive crucible, the evaporated material is heated from the hot crucible due to heat transfer. The physical laws and main features of induction heating of the crucible and the charge are the same as at the other induction heating processes.

The crucible is disposed in a technological chamber for isolation from the atmosphere air. Vacuum is the ordinary operation environment in the chamber and the lower the pressure of the residual gas in the chamber after pumping out the better the conditions for stable work of the evaporator and for obtaining the high quality products without gas impurities are. The residual gas pressure is order of 10^{-2}-10^{-3} Pa in big web coaters but the pressure of 10^{-5} Pa and less is desired for very high vacuum processes. But in some cases, a working gas is introduced into the chamber.

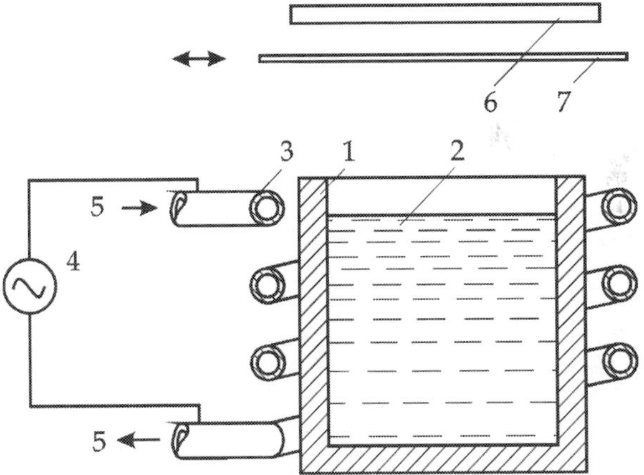

Fig. 1. Simplified diagram of typical induction evaporator. See position names in the text

In PVD system, a substrate 6 is disposed above the crucible to collect and condensate vapor species, and to provide formation of thin film or coating from the evaporated source material on the substrate. A movable shutter 7 is disposed at the front of the substrate 6 to open or close its surface from the vapor flow. There may be additional parts in induction evaporators, for example, heat shields between the crucible and the inductor, and electrical or protective shield for the inductor. The crucibles for induction evaporation may be of different shape (cylindrical, conical, rectangular) and dimensions (from 1-2 cm in diameter to tens centimetres in length) and with different mass of charge (from grams to tens and hundreds kilograms). Fig. 2 presents photographs of practical design of some water-cooled evaporating inductors.

Fig. 2. Photographs of water-cooled evaporating inductors for use in electronics, two of them are shown with crucibles. The left inductor has the top turn with opposite winding as compared with the rest turns

The diagram in Fig. 3 depicts the structure of PVD process with induction evaporation and the aggregation (phase) state transformation of the primary source materials, charged into crucible, during the process: solid state $\rightarrow$ liquid $\rightarrow$ vapor $\rightarrow$ solid state. Note that some metals (*e.g.* Cr) evaporate by sublimation without melting and passing into the liquid phase at the evaporation temperatures.

If evaporation is carried out in high vacuum, the evaporated species do not collide with residual gas molecules and go by straight-lines without scattering from the crucible to the substrate surface. As a rule, the angular distribution of vapor flow density is non-uniform that leads to a non-uniform coating thickness distribution across the immovable substrates. Very often, they are moved with help of various mechanisms relatively the crucible either to reach high uniformity of coating thickness, or to provide deposition on large substrate surfaces. Substrates are often heated (up to 200-600 °C) in order to improve coating properties.

In systems for vacuum distillation of the primary materials, cooled crystallizers are used instead the substrates. Work of the distillation systems is analogous to the PVD system operation (see Fig. 3).

Powder collectors are used instead of the substrates in systems for manufacturing of micro- and nano-particles (powders) by means of condensation of vapor species directly in the dense working gas introduced into the chambers. In this case the aggregation state

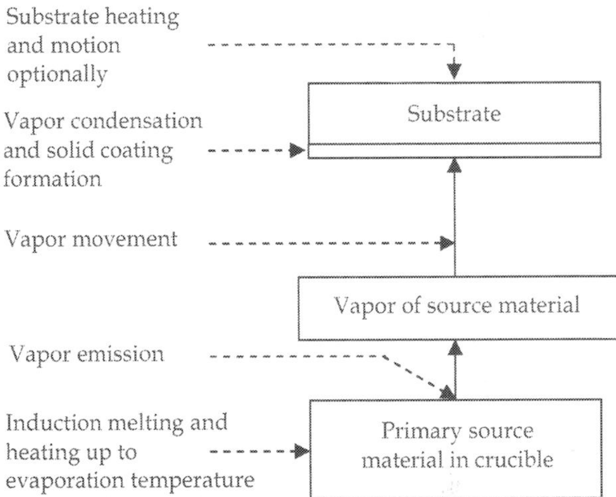

Fig. 3. Diagram of the aggregation state transformation of the primary source material during PVD process

transformation diagram, shown in Fig. 3, changes: vapor condensation occurs on the stage of vapor movement in the gas medium.

The important moment is choice of crucible material. Firstly, the latter must be refractory material. Secondly, it must be chemically inert relatively evaporated materials. Thirdly, the reason for the crucible failure is that some molten metals have a tendency to creep and migrate into pores of the crucible material. When the crucibles are cooled after finishing deposition process, there would be a differential contraction between the bulk crucible wall material and the metal-containing crucible wall surface. The wall would then be under stress and would tend to crack or to spall a part of its surface. One method used to extend the life of the crucibles for evaporating Al is the deposition of alumina into the pores in the crucibles, thus preventing the access of Al to the pores. Zr carbide is also used for coating the internal crucible surface for this purpose. The crucible material, often used in induction evaporators, are practically the same as for resistive-heated boat and crucible (Banshah, 1994) and those are, for example, graphite, pyrolytic carbon, pyrolytic BN, sintered composites TiB_2+BN and $TiB_2+TiC+AN$ as well as different other ceramics. The best crucibles material has a dense structure and ensures hundreds hours of operation life or tens of deposition cycles.

The next question, what crucible is to be: conducting or non-conducting? When thickness of conducting crucible walls is greater than the skin-layer thickness, the crucible fully absorbs electromagnetic energy, electromagnetically shields the molten metal and prevents interaction of the inductor field with the metal. As the result, quantity and volume of the melted charge do not affect work of AC generator, and monitoring of the crucible temperature is sufficient for monitoring the molten metal temperature as they are correspond each other. Besides, in this case mechanical effects of the inductor field on the molten metal are absent that ensures the quiet splashless evaporation process without turbulence of the melt. On the contrary, when the crucible is non-conducting, the magnetic field of the inductor directly interacts with the molten metal. The work of AC generators and transfer of electromagnetic energy into the evaporated material will depend on quantity and volume of the melted charge within the crucible. Also, the inductor field will interact with the eddy current in the molten metal and generate mechanical force acting on the melt. The force can compress the melt, and considerable turbulence with ejection of metal droplets towards a substrate may occur during induction evaporation. See the more detailed explanation of this effect in Section 6. The mechanical effect of the inductor field on the melt has its own advantages and disadvantages but it is often desired to avoid it during deposition of precise thin films for electronics and optics.

Many induction evaporators comprise both a non-conducting crucible and a specially introduced conducting element that serves as an absorber of electromagnetic power generated by the inductor. This element, called as "susceptor", heats the non-conducting crucible due to heat radiation and heat conductivity. When the susceptor surrounds the crucible, it also shields the crucible from the inductor field.

The crucible volume must allow charging a sufficient quantity of evaporating material to complete, at least, a single technological run without the need for replenishing the crucible. When the evaporator is to employ continuously or for long runs, wire feeders are used to add the new portions of evaporated material in the wire form. Sometimes it is possible to introduce an ingot into the crucible through a bottom port.

Consider the problem addressed to wettability of the hot crucible wall surface by the evaporating metal. Such active molten metals as Al, keeping at the high evaporation temperature, are able to wet the wall surface, flow in the form of liquid film upwards to the top due to the effect of surface tension and migrate through the edge to the outer crucible surface, and then evaporate onto the inductor coil. This leads to shorting the coil turns or to electrical breakdown and firing of some kind of parasitic electrical discharge. To avoid appearance of the evaporating metal films on the outer crucible surface, several approaches for crucibles from conducting materials may be proposed. The one is to provide overheating the upper part of the conducting crucible walls or the top edge due to stronger coupling with the inductor electromagnetic field. The overheating must cause enhancing metal atom evaporation from the mentioned crucible surface parts before the atoms can reach the outer crucible surface. The first way to increase the electromagnetic coupling is to provide the crucible with an outwardly directed lip or projection that will be closer to the inductor coil than the cylindrical crucible wall, and the induced current in the lip and, hence, its temperature will be higher. The second way to increase the electromagnetic coupling and to overheat the liquid metal film on the top edge is to make the thickness of upper part of the conducting crucible walls less than the skin-layer thickness at the operation frequency, while the lower part of the crucible, where the melt is, has the wall thickness larger than the skin-layer (Ames, 1966). On the top part, the induced current will heat not only the crucible body but also the metal film on the internal surface of the wall as RF field is able to penetrate through the relatively thin crucible wall. The lower part of the crucible wall is relatively thick and the wall will shield the melt from RF field to avoid turbulence of the liquid metal and undesirable possibility that droplets of the molten metal may be ejected towards the substrate being coated.

We propose to solve the problem of crucible wall wettability by attaching a shield element with low heat-conductivity on the top edge of the crucible (Kuzmichev & Tsybulsky, 2008). Fig. 4 shows the induction evaporator with such element on the crucible. The crucible 1, surrounded by the inductor 2, contains the molten evaporating material 3. Due to wettability of the crucible wall surface and the high temperature (> 1000 °C), the molten metal wets the wall surface and migrates in the form of liquid film upwards to top edge 4, and then overflows to the outside of the crucible. Vapor back condensation also contributes to deposition of liquid metal on the top of the crucible. The crucible top edge outside is tightly bonded with the ring shield element 5 with low heat conductivity. The outer edge of the shield element 5 is disposed on the water-cooled element 6. The low heat conductivity of the shield element 5 prevents excessive cooling of the upper edge of the crucible 4 and formation of solid metal growths on it. At the same time, due to the lower temperature of the outer edge of the shield element 5 the liquid metal migration is excluded here. The heat-removing element 6 is located outside the lines of sight from the crucible 1, and its dimensions (radial length and cross-sectional area in azimuth) are chosen from the following condition for its thermal resistance R_t (1):

$$\frac{0{,}1T_{\mathrm{m}} - T_{\mathrm{col}}}{Q} \leq R_t < \frac{T_{\mathrm{m}} - T_{\mathrm{col}}}{Q}, \tag{1}$$

where T_{m} is the melting temperature of the evaporated metal 3, T_{col} is temperature of the cooling system 7, Q is the heat flow to the heat-removing element 6. When selecting the dimensions and material of the heat-removing element 6 are in accordance with condition (1), the temperature at the points of its contact with the shield element 5 does not exceed the melting temperature. This completely stops the flow of liquid metal towards the outside of the crucible below the shield 5, if it is tightly bonded with the crucible wall or forms with the crucible a one body.

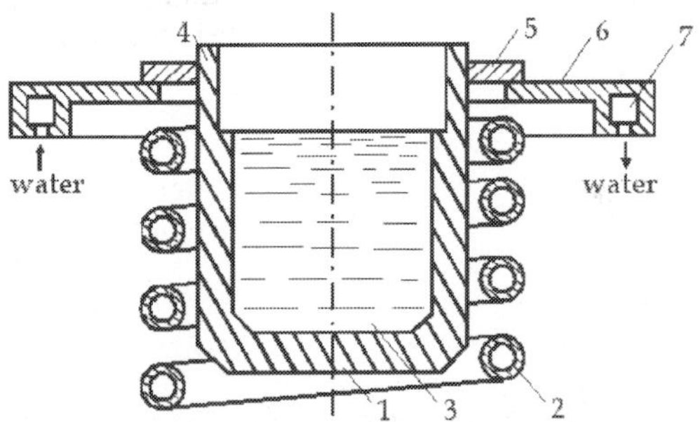

Fig. 4. Evaporator with the elimination of overflow of the melt to the outer crucible side (see position names in the text)

The exemplary crucible with the top-edge shield for evaporation active metals such as Al may be made from sintered conducting ceramics (*e.g.* of TiB_2+BN composition that was proposed (Ames, 1966) earlier) or from pyrolytic BN. The former case with conducting crucible is illustrated by Fig. 4. The latter case is presented by photograph in Fig. 5, where the top-edge shield of the BN crucibles is, in fact, an outwardly directed funnel-shaped lip. The crucible is produced by plasma chemical deposition from the gas phase; herein, the crucible and its top-edge shield are made as one body. Since boron nitride is dielectric, an external hollow cylindrical susceptor is needed for coupling with RF field of the inductor. Then the susceptor will absorb RF power and heat Al charge within the crucible by heat-transfer. The susceptor must be from conducting material, for example, from high-pure graphite. The susceptor plays the important role of RF shielding the molten Al and preventing the melt from mechanical interaction with the magnetic field of the inductor. By this reason the wall thickness of the conducting crucible and the susceptor must be larger than the skin-layer thickness at operation frequency.

Fig. 5. Pyrolytic BN crucibles with outwardly directed funnel-shaped lips to avoid migration of liquid metal towards the outer crucible surface

3. Induction evaporation without ionization for electronic component manufacturing

Metal deposition on semiconductor wafers and other substrates is the essential part of technological processes for manufacturing electronic components including very large integrated circuits, processors and memory for computers. Metal thin films and coatings fulfil different roles, in particular, they serve as electrodes and electrical wiring in integrated circuits. The commonly used vacuum evaporation methods for this purpose are evaporation from resistance-heated boats and crucibles, induction and electron-beam evaporation. Like the resistance-heated evaporation, the induction-heated one may be free from the ionizing radiation that is very important for many microelectronic structures, for example, metal-dielectric/oxide-semiconductor devices, which are very sensitive to the ionizing radiation. In this relation, the electron-beam evaporation is very dangerous because of accompanying by emission of X-rays from the crucible. By the way note, many plasma deposition processes (*e.g.* ion sputtering or chemical plasma deposition) also may be dangerous due to ultraviolet and high-energy particle generation. On the other hand, the induction evaporation is able to provide deposition rate that is much higher than at the resistance-heated evaporation and comparable with rate at the electron-beam evaporation. The vapor-emitting surface of induction-heated crucibles is larger than the resistance-heated surface; therefore the former generates more uniform and wide vapor flow than the latter. This allows simplifying the substrate fixture/holder design for obtaining uniform deposited layers. The cost of induction evaporators is higher than the cost of resistance-heated evaporators but comparable with the cost of the electron-beam evaporators. The common feature of the all evaporation methods is generation and employing of low-energy species for formation of metal thin films. The energy of the species is thermal by nature and is order of 0.1 eV. For comparison, the average energy of species, forming thin films in ion sputtering processes, is about 5-10 eV. The low energy species are required to create delicate micro- and nanoelectronic structures. Thus, the induction-heated evaporation may compete with other evaporation methods in PVD technologies in electronics and close fields.

In order to ensure no ionizing radiation, one must prevent the induction evaporator from electrical breakdown, sparking, ignition and firing of any kind of electrical discharges as well as from electron and ion emission by surfaces of evaporator parts. These electrical processes may generate charge particles (electrons and ions) with energy up to 1 keV and higher, ultraviolet radiation and soft X-rays, so, the factors promoting electrical and radiation damage of the sensitive electronic structures. Obviously, the first condition for preventing from any discharges is providing the high vacuum in the deposition chamber that is the lowest residual gas pressure (less than 10^{-4} Pa) and cleanliness of evaporator parts. However in the vacuum evaporating systems, other factors may cause the discharge processes, too.

Discuss from this point of view the simple induction system shown in Fig. 1. The system is quite able to deposit thin films in some operation regimes but does not guarantee no ionization processes by the following reasons: i) high inductor voltage (hundreds volts and even up to 1 kV), then high difference of potentials between the inductor and the crucible or other chamber parts and presence of the strong electromagnetic field promote ignition and firing of an electrical discharge in vapor flow going towards the substrate; ii) the migration and overflow of liquid metal to the outside of the crucible and then firing an electrical discharge in vapor medium between the lateral crucible surface and the inductor coil are possible; iii) high temperature heating of the crucible and, in particular, overheating of the crucible top edge may cause thermoelectron and thermion emission from the crucible surface; thermoelectrons and ions, in turn, are accelerated by the inductor voltage and may affect the structures on the substrate or ignite the vapor discharge. Proceeding from the said, one may conclude the evaporator with the shielded inductor area and stopped liquid metal migration to the outer crucible side, presented in Fig. 4, as well as the crucibles, presented in Fig. 5, are more suitable for thin film deposition without ionization effects than the shown in Fig. 1.

In order to illustrate another variant of induction evaporator quite acceptable for metal deposition on semiconductors, consider the evaporator proposed by (Phinney & Strippe, 1988) and presented in Fig. 6.

The evaporator 50 with an inductor 22 is disposed in a vacuum chamber on a base 11. The evaporator comprises a crucible 58, a RF susceptor 57 for coupling with the electromagnetic field of the inductor, an inner heat shield 55, surrounding the crucible, and an outer heat shield 53. A thermally isolating hollow column or spacer 52 supports the outer heat shield 53. The contained within the outer heat shield 53 is a second thermally isolating spacer 54, which supports and thermally isolates the inner heat shield 55 from the outer heat shield 53. Within the inner heat shield 55 there is disposed a third thermally isolating spacer 56, upon which there is disposed the RF susceptor 57, which in turn supports the crucible 58. This crucible is provided with a flared lip 59, which extends outwardly from the inner edge of the crucible above the upper edges of the two heat shields 53 and 55. The lip 59 protects the top exposed edges of the shields 53 and 55 and prevents any of the species, being evaporated from within the crucible, from depositing on the inner edges of the heat shield 55. The crucible 58 contains a charge 19 of material (Cu in the given case) to be evaporated.

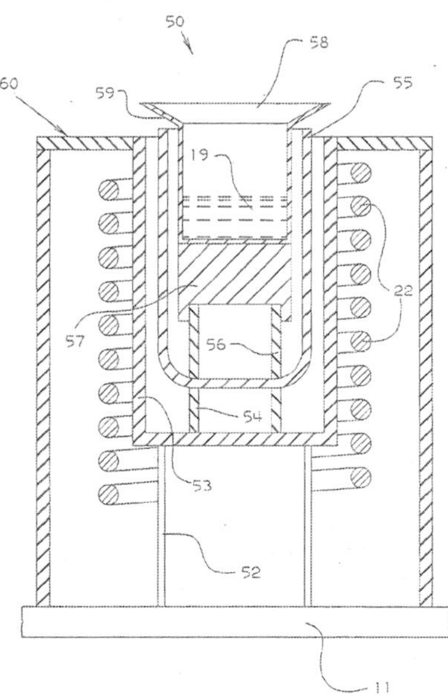

Fig. 6. Induction-heated evaporator by patent (Phinney & Strippe, 1988). Ssee position names in the text

The multi-turn inductor 22 surrounds the above-described structure. There is also an outer casing 60 coupled to the outer heat shield 53 to surround and enclose the inductor 22. The casing 60 prevents any of the evaporated species from depositing on the inductor 22, the outer heat shield 53, and the thermal-isolating column 52. This casing may not be used but because it aids in extending the life of the inductor and eliminates discharge phenomena in the vicinity of the inductor as well as the need of cleaning the column 52 and outer heat shield 53 after every evaporation run, it is very desirable.

The RF susceptor 57 is made of a solid, columnar block of graphite carbon 3.5 cm in diameter and 1.3 cm height. The crucible 58 is 2.1 cm in diameter and 1.7 cm deep, and is made of a material (*e.g.* Mo) that has a lower vapor pressure than the material to be evaporated (*e.g.* Cu). The crucible is filled with approximately 30 g of material, such that it is partially filled (*i.e.* the material within the crucible forms a pool approximately 0.8 cm height). The susceptor 57, being a relatively large mass and being very receptive to inductively heating by RF frequencies, assures that the charge 19 contained within the crucible 58 is quickly heated in controllable manner to the desired levels necessary to melt and evaporate the charge 19 as well as to provide it self-fractionating after its replenishment. Such susceptors are especially required when the charge is a low vapor pressure material (*e.g.* Cu or Au) that is not very absorbent of RF energies. Thus, it is necessary to select both the configurations and compositions of the susceptor 57 and crucible 58 so as to absorb the maximum amount of RF energy and convert it into heat, and to transfer the heat by radiation and conduction to the charge 19 within the crucible 58. The other components of the evaporator 50 are selected to minimize this RF absorption so that the level to which they become heated is minimized.

One can see from the above-described crucible dimensions the depth of the crucible is less than its diameter that is the evaporator generates the sufficiently wide vapor flow to the substrate and does not create the problem with uniformity of deposited films. The problem arises if the depth/diameter ratio is increased to the point where the depth is greater than twice the diameter.

Since it is necessary to heat the contents of the crucible rapidly and effectively (*i.e.* with low heat losses), one uses the thermally isolating system for the crucible and the RF susceptor. The system consists from two concentrically disposed cup-shaped heat shield 53 and 55 as well as three thermal insulator/supports/spacers 52, 54 and 56 in the form of hollow columns. The materials of the thermally isolating system parts must be refractory, transparent to radio frequencies and have low heat conductivity. Practically, these materials are found among refractory dielectrics.

Pyrolytic BN is the exemplary material for the inner heat shield 55 and the thermal insulating spacers 56. Pyrolytic BN is used as the inner heat shield because it prevents infrared radiation from being emitted from either the susceptor or the crucible. In this way the inner shield acts to retain heat within itself, *i.e.* in the crucible area and susceptor area, while shielding the outer shield 53 from infrared radiation. Pyrolytic BN is used for the support spacer 56 not only because it is a good thermal isolator but also because it is formed with the highest purity and does not interact with the carbon of the susceptor 57 at the high temperatures at which evaporator operates. Silica and quartz are the exemplary materials for the other parts of the thermally isolating system. Other materials having the similar properties could be used. Of cause, the common demand to the all materials used for manufacturing the evaporator parts is high purity and no interaction each with other and with evaporated substances.

A special mechanism is provided within the evaporation chamber for automatically feeding a wire, of the material in the crucible being evaporated, into the crucible to automatically replenish the charge 19 as it becomes diminished by evaporation.

It is important to note, the choice of above-described dimensions of the evaporator parts, and especially the commensurable volumes of the graphite carbon susceptor and molten material prior to the evaporation provides the process of self-fractionation of the molten material before the evaporation and thus enhances the properties of the film as-deposited, in particular, eliminates conductivity-detracting impurities from the deposited film. Since this moment is critical for technology of precision thin films for microelectronics and optics, discuss it more detail.

It is known that during the initial heating and melting phase any impurities contained within the charge would, during the evaporation thereof, emit species of impurity material (*e.g.* oxide, slag, *etc.*) from the melt. This phenomenon (known as "spitting") results in large quantity of species of the source material and impurities in solid or liquid state being transported to substrates. The spits or impurity species in any form cause irregularities, such as spheres, lumps, or voids, in the deposited films. The irregularities reduce the conductivity of the films, causing electrical and structural defects and the like that result in a lower production yield. Patent (Phinney & Strippe, 1988) proposes the following procedure of self-fractionation during the film deposition cycle with periodical replenishment of the evaporated charge.

The Cu charge 19 of 30 g is disposed within the crucible 58 (see Fig. 6). At the beginning of the cycle, the substrates (semiconductor wafers) are located in the vacuum chamber and the one is evacuated. 450 W of RF power is then applied to the inductor. The susceptor 57 begins to heat and transfer this heat by a radiation and conduction process into the charge in the crucible, and the charge begins to melt with little or no evaporation occurring. This power level is maintained until the temperature of the charge stabilizes at 1200 °C. During this time, to protect the substrates from any undesired evaporation or contamination, a shutter closes the substrates. When they are to be coated, the RF power is ramped to about 2 kW, and is maintained there for approximately 135 seconds. During this time the temperature of the charge 19 is increased and self-fractionation, i.e. fractional distillation, of the charge occurs. As this fractional distillation begins, any contaminants in the charge (such as oxides, or the like) adhere to the surface of the melt to form a slag-like surface layer. As the RF power is increased, the temperature of the charge also increases and the fractional distillation continues until the contaminants collected on the liquid surface of the melted charge become either sublimated or vaporised due to limits set by respective vapor pressures of the material used in the charge and its oxides and other contaminants. The impurities migrate to the wall of the crucible where they are drawn up. In copper, this occurs because a high vapor pressure of copper evaporated species is created, which pushes the surface contaminants and the like to the side walls of the crucible that are at a temperature slightly higher than that of the melted charge. Once this fractionation process is completed the temperature of the crucible has peaked at 1900 °C.

The RF power is then decreased to a power level of 780 W. This power level corresponds to the desired rate of deposition of 4 Å/sec of the charged material onto the substrate surface (for the crucible-substrate distance of 71 cm). 45 seconds after this decrease in RF power, when the temperature of the charge 19 stabilizes at about 1500 °C, the shutter is withdrawn to open the substrates and to permit the evaporated species of the source material to coat the substrate surface. This power level is maintained for an additional 150 seconds at which time the shutter is then positioned in front of the substrates to prevent further deposition. Simultaneously with the shutter closing, the power level is raised to 1000 W and is maintained at this level for 90 seconds. This causes a slight rise in temperature of the charge. 24 seconds after the shutter closes and the power level rises, the wire of the source material is fed by the wire feed mechanism into the crucible 58 to replenish the charge 19 therein. As the wire melts and replenishes the charge the temperature of the charge decreases. At the end of 90 seconds the power is again decreased to the standby power level of 450 W and the temperature of the charge continues to decline to 1200 °C at which temperature it again stabilizes.

By carrying out the technological cycle, as described above, the crucible 58 is heated in a thermally controlled manner to provide self-fractionation. This self-fractionation eliminates any spitting of Cu or Cu oxides. Self-fractionation of the charge is especially important when the material of the charge has impurities that are introduced during its replenishment. The employing of several shields for the outer surface of the crucible 58 and the total shielding of the inductor 22 together with evaporation of copper, not inclined to wet the crucible wall surface and to migrate upwards to the top edge of the crucible, ensure no parasitic electrical discharges and no ionizing radiation emission from the evaporator.

Thus, the right constructive design and employing various heat and electromagnetic shields are very important in order to create the induction evaporator being without electrical discharges and ionizing radiation emission. However the electrical design of the induction evaporator is important too, because electrical discharges are, in the first instance, electrical processes caused by electromagnetic fields. The often-used approaches, well known from the induction heating technique to avoid electrical discharges, are as follows:

- lowering the operation frequency, *e.g.* from 440 kHz down to 66 kHz and less;
- grounding the upper inductor turn with a copper strap well connected to the base plate of the chamber near the inductor, it provides the ground potential for the upper turn which would be most subjected to the vapor;
- eliminating a stray magnetic field above the inductor in the space of vapor movement by several ways: i) to introduce an electromagnetic shield above the inductor, the shield must have an orifice for the vapor flow and separate the inductor area and the space where vapor flies, ii) to wind one or two top turns of the inductor in the opposite direction relatively the rest lower turns in order to compensate the magnetic field of the lower turns in the space where vapor flies (see the photography of such inductor in Fig. 2, that is the left inductor), iii) to introduce a shorten turn (*i.e.* a ring) over the upper inductor turn, the current induced in the ring will generate the opposite magnetic field relatively the inductor field, iv) to decrease as much as possible the inductor voltage that is to use a single-turn inductor instead the multi-turn inductor.

The most of the listed approaches are realized in the induction evaporator shown in simplified form in Fig. 7 (Kuzmichev & Tsybulsky, 2008). The real sample of the evaporator is presented by its photograph in Fig. 2. The induction system of the evaporator belongs to the concentrator-type inductors known from the induction heating technique. The induction system factually is a step-down RF transformer comprising the concentrator body 1 and the multi-turn coil 2 from copper pipe connected to a RF power source. The concentrator 1 is manufactured from copper and has the narrow radial gap (split) 3 coupling the outer surface 4 of the concentrator with its internal surface 5. The latter surrounds the space where the crucible 6 and other parts may be disposed.

The coil 2 serves as a primary winding of the transformer. The outer surface 4 of the concentrator 1 under the coil 2 serves as a second one-turn winding that is as a load winding of the transformer. The internal surface 5 of the concentrator 1 serves as a one-turn inductor for heating the crucible 6 (or RF susceptor if it introduced into the evaporator). The facing each to other surfaces 7 of the radial gap 3 in the concentrator body 1 serve as plate buses (conductors) connecting the second one-turn transformer winding with the one-turn heating inductor. It is clear that all the mentioned surfaces 4, 5, and 7 pass RF current induced by the coil 2 within the skin-layer. Such arrangement of the secondary circuit of the RF transformer allows to minimize the working voltage of the heating inductor and the stray magnetic field above the concentrator in the space where vapor flies as well as to simplify grounding the top turn of the coil 2 together with the concentrator body 1. All these factors ensure no electrical discharge and no ionizing radiation emission from the evaporator. It is worth to note the RF transformer play a role of a matching element providing effective transportation of RF power from the generator to the heated crucible. Our use of such an inductor-concentrator reduces the power of the RF generator working at frequency of 440 kHz by more than 2 times.

The described induction evaporator is combined with the pyrolytic BN crucible preventing from liquid metal migration towards the outer crucible side (see Fig. 5 and Fig. 8). The crucible is produced by plasma chemical deposition from the gas phase onto the graphite base, removed after deposition. The crucible and its top-edge shield in the form of the outwardly directed funnel-shaped lip are made as one body. The conical shape of the shield surface provides the alignment of the crucible and the concentrator and drain back of liquid metal into the crucible from the upper surface of the shield. The pyrolytic BN crucible is disposed in a cup-shaped graphite susceptor and they together are disposed within a cup-shaped pyrolytic BN heat shield and then within the central hole of the concentrator. The evaporator has been designed for deposition of Al and Al alloy thin films for electronics, the power of RF generator is 5 kW, frequency is 440 kHz, the deposition rate is about 0.1 μm / min.

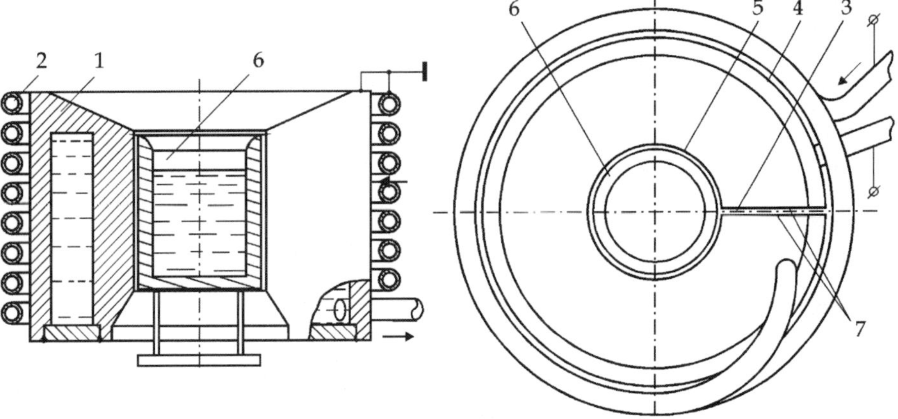

Fig. 7. Evaporator with inductor-concentrator (see position names in the text)

Fig. 8. Evaporator with inductor-concentrator equipped with crucible, which allows avoiding migration of liquid metal towards the outer crucible surface

4. Induction evaporation with vapor ionization for ion plating

4.1 What means the ion plating

Ion bombardment of substrate surfaces during deposition of thin films or coatings is well known to strongly modify the properties of the deposited material. The modification may be very useful as it may enhance adhesion of the films and coatings, make denser their microstructure, increase anticorrosion properties, *etc.* A source of bombarding ions is usually the plasma of electrical discharge maintained in a gas medium or, that is more right for evaporation process, in the gas/vapor medium. Argon is the often-used gas for ion generation since it is inert and well ionizable gas, hence, the ions Ar^+ and few of metal ions are considered to bombard the substrate. In reactive processes a proper reactive gas is added, and reactive ions also bombard the substrate. A negative potential is applied to the substrate for acceleration of the bombarding ions. D. Mattox has described such technology in the 1960's and named it as "the ion plating" by analogy with the galvanic electrodeposition process. In his review book (Mattox, 2003), he gives the history and foundations of this technology as well as cites the other names for similar deposition processes, in particular, "ion assisted deposition" – IAD or "ionized physical vapor deposition" – iPVD.

As a prior art for the ion plating technology, it may be pointed out the similar process patented by B. Berghaus in 1938, which includes placing coated articles under a negative bias potential into gas medium, firing gas glow discharge with the articles as a cathode and use of a crucible with molten material as a vapor source (Berghaus, 1939). However the proposal by B. Berghaus is not known, at least, by literature data as practically realized. So, the ion plating technology has been developed since the 1960's due to very attractive results obtained by D. Mattox.

Initially, the ion plating employed a thermal vapor source, mainly resistance-heated boats, and a plasma gas discharge. The resistance-heated vapor sources are employed, as they are simplest ones. The gas discharge is employed since it is a simple way to obtain ions. Besides, the ion plating in gas environment provides good surface coverage over three-dimension substrates due to scattering evaporated species and randomization of their trajectory in the gas medium when gas pressure is about 1 Pa and higher. The application of electron beam evaporators to ion plating with the higher deposition rates than the provided by resistance-heated sources meets with difficulties of electron gun exploitation in the gas medium. So, the induction evaporators that are able to work in the gas environment and ensure the high deposition rate of wide range of materials are better candidates for the ion plating processes. May be by this reason, B. Berghaus proposed to use just the induction evaporator in his ion plating process (Berghaus, 1939).

A simple evaporation system with a bare inductor coil, without shielding, similar to the presented in Fig. 1 has been used for ion plating directly in a gas containing chamber without producing any sparks and arcs in the inductor area (Spalvins, 1976). The system had a ceramic crucible 2.5 cm in diameter and a four-turn inductor connected to a 5 kW generator through a 20:1-ratio step-down transformer. The voltage across the inductor was 70 V with the operation frequency of 75 kHz (the conventional frequency for induction generators of 440 kHz could not be utilized in the gas ion plating process because severe arcing in the inductor occurred in gas glow discharge conditions). The operation conditions were Ar pressure: 2-2.5 Pa; the negative substrate bias potential: 3-5 kV; substrate ion current density: 0.3-0.8 mA/cm². Very good results of metal and alloy (Au, Pt, Ni, Fe, Cr, Inconel) deposition on various metal surfaces with excellent adhesion of coatings with thicknesses from 0.15 to 50 μm have been obtained (Spalvins, 1976).

4.2 Induction evaporation with vapor ionization in gasless environment for ion plating

Obligatory introducing of the inert gas into the evaporation chamber in the "classic" gas ion plating leads to some shortcomings, in particular, to gas capture by the coatings and limitations of operation parameters. The much better approach is to use the metal vapor plasma as an ion source and to provide the bombardment by own metal ions that is the self-ion bombardment. Such gasless approach was named as "the pure ion plating" (Hale et al., 1975). And in this case the induction evaporation has advantages over the resistive-heated or electron beam-heated evaporation since the electromagnetic field of the inductor may provide effective vapor ionization unlike the mentioned processes. To confirm this, consider some results of our experiments with an evaporation system similar to the presented in Fig. 1.

A bare three-turn inductor (its inductance was 0.3 μH) surrounded a molybdenum crucible 3 cm in diameter (see the photograph of the middle inductor in Fig. 2). An ion collector was placed 20 cm above the crucible. A 5 kW 440 kHz generator powered the inductor through a step-down transformer. The maximum working temperature of the crucible was 1400 °C. The residual gas pressure after gas evacuation from a chamber was about 10^{-3}-10^{-4} Pa. No special gas was introduced into the chamber.

With the crucible grounded and the top inductor turn under a high potential (the lower turn was grounded) a plasma discharge appeared at the inductor voltage of about 300 V. Fig. 9 shows the voltage-current characteristics of the ion collector, when it was under negative voltage ($-U_c$), for the discharges in Cu and Cr vapor. As can be seen from the figure, the curves resemble the voltage-current characteristics of the negatively biased plasma probe with the trend toward saturation of the ion current I_c. The collector ion current was order of hundreds of milliamperes, with the average current density being 2-13 mA/cm² for Cu evaporation and 2-8 mA/cm² for Cr evaporation. The ratio of the metal ion flow to that of neutral metal atoms at the collector was 0.05-0.1. The latter number may be considered as a vapor ionization degree.

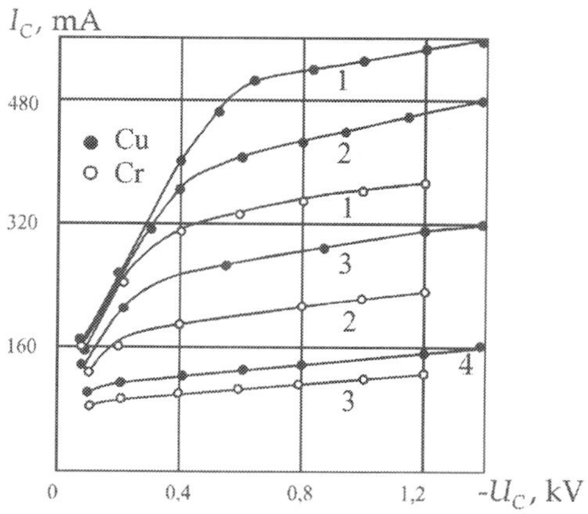

Fig. 9. Voltage-current characteristics of the ion collector for inductor voltage of 500 V (1), 400 V (2) and 350 V (3)

Analysis of the above-mentioned and other related data testifies to the fact that in the described unshielded induction system the appearance and maintenance of the metal plasma discharge essentially depends on shape and disposition of the crucible and the top inductor turns, what parts are grounded as well as the secondary ion-electron emission from the crucible walls and the top turns. The RF magnetic field above the crucible affects electron trajectories, making them longer, and raises, due this, the efficiency of vapor ionization. Therefore this kind of vapor discharge may be classified as a version of magnetic field supported glow discharge with RF power supply. By increasing the height of crucible walls, one can easily produce a discharge with a hollow-cathode/crucible as an electron-emitting electrode; in this case the upper inductor turn serves as an anode for the hollow-cathode discharge at positive half-waves of the inductor voltage. Thus, the tested induction evaporation system showed that the inductor could simultaneously function as an evaporator and as an ionizer of metal vapor in gasless environment.

It is well known from gas discharge electronics that introduction of a thermionic cathode into a discharge system strongly facilitates discharge ignition due to free electron injection by the cathode. This approach was realized with the above-described induction evaporator. A ring cathode, made from tungsten filament 0.3 mm in diameter, was placed above the crucible and was grounded together with the lower turn of the inductor; so, the filament and the upper turn created a diode with crossed electric and magnetic fields. Passing heating current through the filament from an external power source to cause thermoelectron emission gave effect in some decreasing of the ignition voltage for the vapor plasma discharge, but after discharge ignition the filament heating current had no visible effect on the collector ion current. The same data were obtained with the simpler system with a closed filament cathode at the ground potential. In the latter case the filament was a closed turn heated by the induction current. These results may be explained by the lower electron emission of the filament as compared with the secondary ion-electron emission of the wall surface of the grounded crucible because of small filament surface area and high work of electron exit from tungsten.

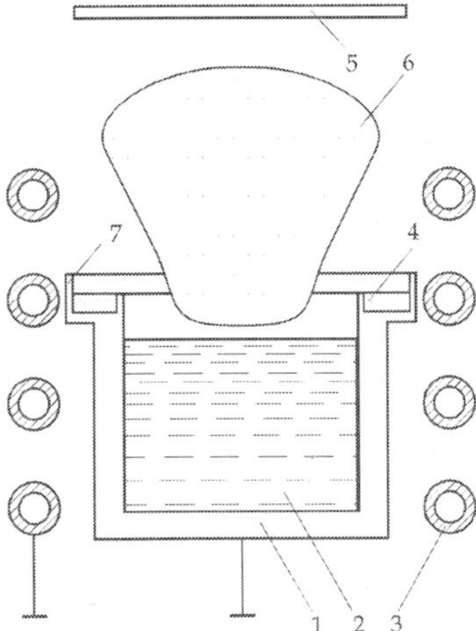

Fig. 10. Schematic diagram of evaporator with crucible containing electron emitting insert (see position names in the text)

In order to successfully realize the approach based on employing thermoelectron emission, an induction system with a special crucible containing an electron emitting insert is proposed (Kuzmichev & Tsybulsky, 2006a). Fig. 10 schematically depicts the system. A cylindrical cup-shaped conducting crucible 1 with a molten metal 2 is surrounded by an inductor 3; its upper turn located above the top edge of the crucible 1. The top crucible edge is a lip flaring towards the inductor coil, and it contains the insert 4. The latter is made from refractory material with low work of thermoelectron exit, for example, from sintered LaB_6 or composition $W(Mo)+LaB_6$. Due to location of the insert 4 close to the inductor coil 3 the insert temperature is sufficiently high to cause thermoelectron emission. The high temperature also provides cleaning of the insert surface from any contaminants by evaporation. An ion collector 5 is disposed above the crucible 1. When the inductor 3 is powered from a 440 kHz generator, the metal 2 evaporates from the crucible and the thermoelectron emission from the insert 4 takes place, a brightly shining plasma cloud 6 appears above the crucible that means ignition of metal vapor discharge. Providing the top edge of the crucible with an additional wall/shield 7 increases vapor ionization due to longer trajectories of the thermoelectrons emitted by the insert. Testing the induction evaporation system with crucible containing the thermoelectron emission insert showed that its use gave increase of ion collector current about 1.5-2 times as compared with the system without the insert.

Lifting the upper inductor turn towards a substrate or the ion collector in our case (see Fig. 10) means, in fact, dividing the inductor coil 3 by two parts: the lower part, surrounding the crucible, is the heating inductor; and the upper part above the crucible is the ionizing inductor; herein, both the parts are connected in series. There is a proposal on induction evaporating systems (Japanese patent application, 1978) provided with two inductors separately for evaporation (the first heating inductor surrounds a crucible) and for vapor ionization (the second inductor is disposed above a crucible and surrounds the vapor flow), see Fig. 11. The first system (Fig. 11a) is provided with the inductors connected in series. This approach needs one RF generator that simplifies RF circuits and allows to automatically stabilize evaporation and ionization processes. Indeed, when evaporation rate increases, the vapor density grows up and the power, consumed by the ionization inductor, increases. Hence, the power for the evaporation inductor decreases and evaporation rate

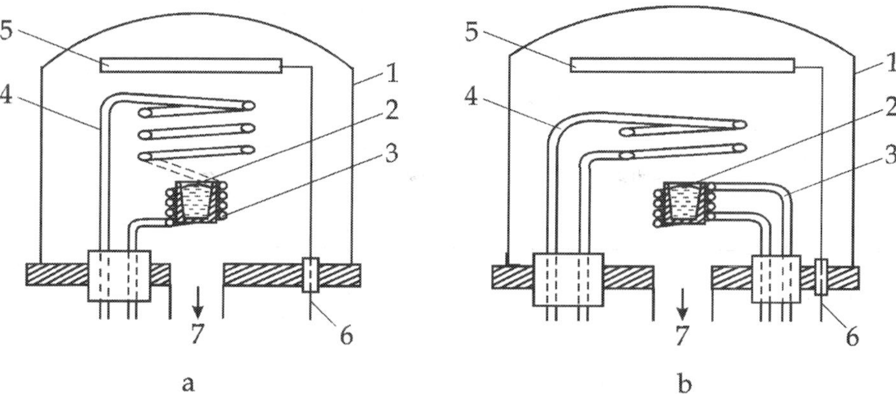

a   b

Fig. 11. Schematic diagrams of evaporators with two inductors for evaporation and vapor ionization. (a): the inductors are connected in series; (b): the inductors are connected to different power sources (Japanese patent application, 1978). 1 is a chamber, 2 is a crucible with molten evaporating material, 3 is the lower heating inductor for evaporation, 4 is the upper inductor for vapor ionization, 5 is a substrate, 6 is a conductor for applying a bias voltage to the substrate, 7 is pumping out of the chamber

decreases, too; that is such two-inductor system works in some equilibrium regime. If one desires to rise the evaporation rate, it needs to increase the RF generator power. The second system with separate power supply of the inductors (Fig. 11b) has advantage in separate control of evaporation and ionization processes; also one can use two different frequencies, which are optimal for evaporation and vapor ionization, but such system is more complex.

The considered-above induction evaporators with vapor ionization employ bared (unshielded) inductors. This may lead to limitations in use of high power generators at radio frequencies (*e.g.* 440 kHz and higher) because of arcing in the crucible/inductor units. High RF power is needed for enhancing vapor ionization degree. Besides, ion sputtering of induction system parts, which are under negative potential relatively the vapor plasma and contact with the plasma, leads to introduction of undesired impurities into the deposited coatings. So, the high power stable pure ion plating needs the specially designed shielded induction evaporator/ionizer systems. Consider two examples of realization of shielded ion plating sources.

The first source, called "the i-Gun Ion Source", is described in (Hale et al., 1975). It is similar to the induction system presented in Fig 1, but in this case the bare inductor is surrounded by a grounded cylindrical cup-shaped metal shield with a cut on its wall along the axis; a heat shield is placed between the crucible and the inductor. The grounded metal shield serves to inhibit coupling of the RF into the surrounding plasma, which would otherwise result in side discharge appearance and significant power losses. The plasma discharge may appear only above the crucible. The heat shield, made from alumina and being transparent to RF, allows the RF energy to couple to the conducting carbon crucible while to inhibit crucible heat losses. The tube conductors for connecting the inductor with a RF generator are also covered by grounded metal shields. In result, the electrical design of the i-Gun Ion Source allows the bare inductor to continuously operate without arcing at 440 kHz power up to 5 kW. The deposition rates of about 25 μm/min at distance of 25 cm are possible for Al or Cu. The efficient use of power offers approximately three-fold operation energy saving over electron beam evaporation of such metals as Al.

The second high rate ion plating source is described in (White, 1977) and schematically depicted in Fig. 12. A conducting carbon crucible 1, containing evaporated material, is supported by a nonmagnetic dielectric support 2 and surrounded by a heat shield 3, made from alumina. All these parts are surrounded by a multi-turn working inductor 4, which is externally covered by a cup-shaped metal shield 5. The conductors 6 for connecting the inductor 4 with a RF generator are also covered by metal shields 7. A conductor 8 is to apply the ground potential to the middle turn of the inductor 4. The inductor 4 and the conductors 6 are made from copper pipe to provide water-cooling. The shields 5 and 7 have cuts along their walls in order to inhibit RF power losses. The copper pipe forming the turns of the inductor 4 is generally covered completely by fiber insulation 9 (such as aluminium oxide fiber batting) that is required to maintain proper spacing and to prevent shorting between turns. The fiber insulation batting 9 supports the tubular heat shield 3 and the dielectric cylindrical block 2, supporting, in turn, the crucible 1. The conductors 6 are covered by dielectric material such as Teflon to prevent shorting between conductors 6 and their shields 7. The metal shields 5 and 7 as well as the crucible 1 are at floating potential that serves to inhibit coupling of the RF power into the plasma. The shield 5 has no metal part on its top to provide coupling RF field of the inductor with a plasma discharge that can appear only above the crucible.

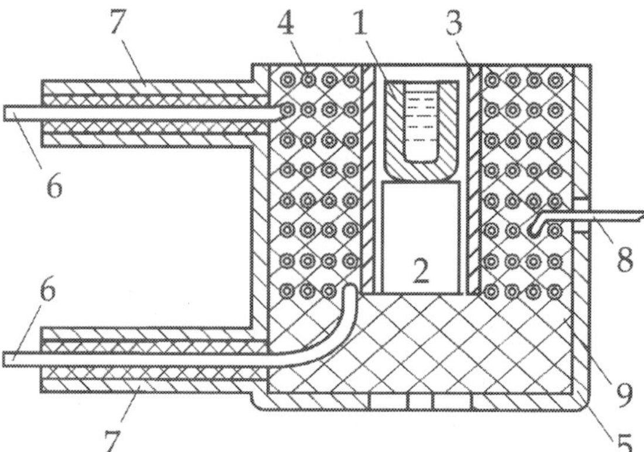

Fig. 12. Induction evaporator for ion plating source (White, 1977) . See position names in the text

Vapor species of the material, being heated and evaporated in the crucible 1, pass along the source axis through the RF electromagnetic field, generated by the inductor 4, in the area of the strongest field just above the crucible. When strength of the RF field is higher than the threshold value, the plasma discharge arises due to vapor species ionization. This plasma discharge relates to so called "Inductively Coupled Plasma". The main feature of the discharge is its firing in metal vapor gasless environment. Electrical design of the ion plating source prevents from arcing and parasitic discharges and allows the source to operate throughout a pressure range from ultra high vacuum to greater than atmospheric pressures. So, the considered shielded ion plating sources can operate in the reactive mode of deposition with introducing some quantity of proper reactive gas in the chamber. The sources may provide ion plating of metals and semiconductors such as Al, Cr, Cu, Si, Ti, *etc.* and of oxides, nitrides, sulphides Al_2O_3, HfO_2, HfN, SiO_2, Si_3N_4, TiO_2, ZnS, CdS, *etc.* on different substrates (metal articles and foil, semiconductor wafers, glass, polymer films).

4.3 Evaporator with pulse modulation of inductor power supply

The evaporator, in which the same inductor generates the electromagnetic field for heating of the evaporated material and for ionization of the vapor, has the disadvantage associated with the inability to separately adjust the evaporation rate (*i.e.* vapor density) and the degree of vapor ionization because of simultaneous influence of the inductor electromagnetic field on the processes of evaporation and ionization. This disadvantage can be overcome by introduction of the amplitude pulse modulation of the RF power, at which the average value of the inductor power will determine the intensities of the crucible (or susceptor) heating and the metal evaporation, while the amplitude of the inductor power will determine the degree of vapor ionization. The evaporation will be practically continuous because of the great thermal inertia of the crucible (susceptor) and the molten metal, but the vapor ionization will be pulse modulated. Indeed, when the inductor power amplitude is above a threshold value, the plasma discharge occurs in the vapor medium, and when the inductor power amplitude is below the threshold value, the discharge breaks off at once. In such approach the strong non-linear dependence of the degree of vapor ionization upon the instantaneous values of the RF power that is the inductor electromagnetic field strength is used.

There are many ways to provide amplitude pulse modulation of RF power well known for radio engineers but, in induction heating technique, one uses mainly the self-exciting triode tube generators, power control of which is commonly made by regulation of either the filament current (to control the cathode emission current), or the anode voltage, or the grid resistor. The first method has great inertia and the constraints on the allowable minimum and maximum filament currents, as well as a decrease of triode tube life; so this option does not fit for pulse modulation. The control of the anode rectifier voltage with thyristors or thyratrons is advisable only to adjust the average power of RF heating. The fast amplitude pulse modulation of the anode voltage with vacuum tubes is quite complex and cumbersome. Also, the modulation of RF power by the regulation of the grid resistor is of little interest.

One may suppose, the quite effective is the modulation by periodical applying of negative voltage pulses to the grid of the generating triode in order to periodically close and open the triode. Earlier, the applying of high negative voltage to the grid was used for emergency shutdown of the generator. The regulation of the ratio "vapor ionization degree/vapor density" at the fixed amplitude pulse modulation may be carried out on the principles of pulse-width or pulse-frequency modulation.

The method of the pulsed grid modulation was proved with the self-exciting triode generator working at 440 kHz, the triode was of GU-89A type (made in Russia). The rectangular modulation pulses of negative polarity with amplitude up to 800 V and duration of 0.03-5.0 ms (the pulse front duration was about 10 µsec) were generated by a transistor submodulator with independent control of the pulse duration and the frequency of pulse repetition. Three-turn inductor, made from water-cooled copper pipe, surrounded the crucible 32 mm in diameter, made from a conductive material (Mo or TiB_2+TiC+AlN ceramics). See design of such induction system in Fig. 1. The charge of Cu or Cr was loaded into the crucible. The upper turn of the inductor was connected to the high terminal of the secondary winding of the RF step-down matching transformer and was disposed above the crucible, that provided establishing the RF electromagnetic field over the crucible and the metal vapor ionization in a glow-like plasma discharge, when the voltage applied to the inductor was above 300 V.

Fig. 13 shows the simplified time diagrams of the voltages on the triode grid u_g and the inductor u_i. Here T is the repetition period of the modulating grid pulses. When the negative modulation pulse opened the generator triode and stopped the RF generation, the damping of oscillations in the inductor did not happen instantly, but over time τ_d about 40 µsec, due to the high Q-factor of the triode anode circuit. When the pulse duration $\tau < \tau_d$, some inductor voltage was maintained throughout the negative modulation pulse (it did not down to zero) as shown in Fig. 13a. Then after the end of the negative pulse, the oscillations in the anode circuit self-excited and the inductor voltage u_i recovered nominal value during the time interval shown as Θ. The recovery process duration Θ depended on the degree of damping of oscillations in the anode circuit and the inductor and the parameters of the heated matter.

If the pulse duration $\tau > \tau_d$ and the RF oscillations were completely damped during the action of the negative modulation pulse, then the process of their recovery in the anode circuit and the inductor strongly delayed (see Fig. 13b). The recovery time Θ reached 500 µsec. There was even a "dead" period Θ_1 in the generation of the inductor current when the voltage on the inductor was practically absent, although the negative modulation pulse already passed. The reported features of transients were caused by that the triode generator operated in the self-excitation mode.

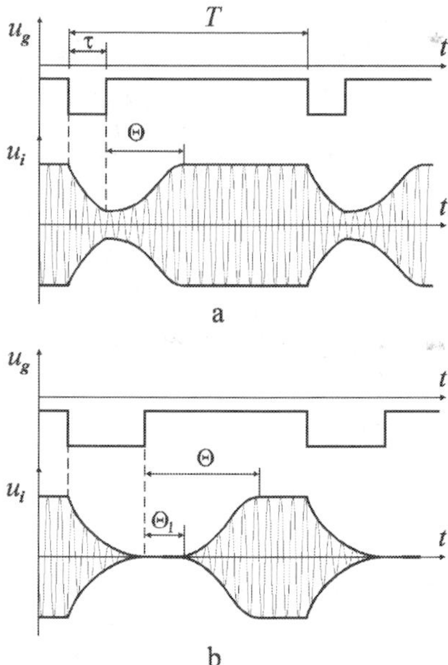

Fig. 13. Diagrams of voltages on the generator triode grid and the inductor

Thus, there are the restrictions on the minimum duration of the modulation pulses (in our case, it must choose at least 50-60 μsec) if one wants to get 100 % of the amplitude modulation. With a less pulse duration, the modulation depth decreases, *i..e.* RF heating continues during the action of the negative modulation pulses but the ionization stops. It is also worth-while to choose the duration of the pause between pulses $(T-\tau)$ more Θ, and, of course, much more Θ_1. In order to exclude the regime with the "dead" period Θ_1 and provide more fast regulation, one should use the partial amplitude modulation or external excitation of the triode generator during the "dead" period.

By varying the duration of the modulation pulses (τ) and the frequency of their repetition $(1/T)$, one could choose different modes of the evaporation and the vapor ionization. This allowed to independently regulate the temperature of the crucible within 1000-1600 °C and the ion current to the substrate, located above the crucible, within 100-500 mA. Since the pulse modulation enables continuous control of the average power of heating, the employing of the stabilized anode voltage source for power supply of the RF generator may be excluded. The same approach may be applied to semiconductor RF generators for induction heating.

Pulse modulation of the inductor power supply for the separate regulation of the evaporation and the vapor ionization may be also accomplished with the thyratron (or thyristor) pulser, which provides impact exciting of a ringing oscillating circuit. The inductor serves as an inductance in this oscillating circuit. Similar pulsers were used for induction heating in the electronic industry (Mastyaev, 1978). The pulser generates megawatt pulses, the average power of which is order of some kilowatts and supports the evaporation, but the instantaneous megawatt power provides the effective plasma discharge firing and the vapor ionization. Herein, the frequency and the amplitude of the inductor current oscillation as well as the frequency of repetition of the impact inductor excitations determine the ration "vapor ionization degree/vapor density". The work of such pulser is also considered in Section 4.4

4.4 Two-inductor evaporator with impact excitation of vapor ionizing inductor

The induction evaporation with vapor ionization may be realized in two variants, namely: with use of one inductor, which simultaneously evaporates and ionizes, and with two inductors, one of which evaporates and the other ionizes vapor. The former case was considered in Section 4.2 and Section 4.3. Now consider the case with two inductors. The schematic design of such system is shown in Fig. 11b. This variant gives possibility to use two different frequencies for separate power supply of the evaporating inductor and the ionizing inductor. As a rule, the higher the frequency, the stronger the vapor ionization is, therefore the ionization needs frequencies of about 1 MHz and more (the often used frequency is 13.56 MHz). The optimal evaporation frequency is less than 1 MHz, mainly 66-440 kHz for relatively small crucibles. The ordinary way to power supply the ionizing inductor is use of the continuos non-modulated RF current generated with electronic tube or transistor generators. But there is another way to power supply the ionizing inductor that is employing the thyratron (or thyristor) pulser with impact exciting of a ringing oscillating circuit (see the end of Section 4.3). The pulser is able to generate megawatt pulses, providing the effective vapor ionization at frequencies less than 1 MHz. Employing such frequencies lower than 13.56 MHz gives a technical profit in simplification of RF schemes and impedance matching systems. This approach has been successfully proved with the thyratron pulser, the simplified electrical scheme of which is depicted in Fig. 14.

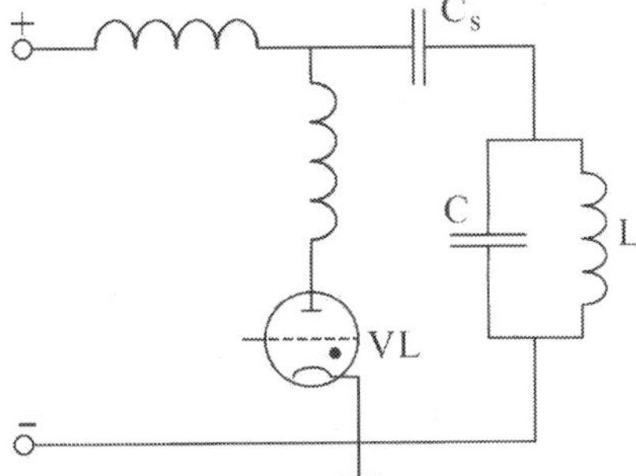

Fig. 14. Schematic diagram of the thyratron pulser for power supply of the inductor for vapor ionization

The five-turn ionization inductor L together with the capacitor C forms a self-oscillating circuit or, in other words, the oscillating contour with resonance frequency f of 0.88 MHz. Also in Fig. 14, VL is the hydrogen thyratron of TGI-1000/25 type (made in Russia) and C_s is an energy storage capacitor charged from a high voltage anode power source. The repetition frequency f_r of thyratron ignition is set of 1-20 kHz. As the thyratron being ignited, the storage capacitor C_s quickly discharges through the thyratron and the LC-circuit, exciting in the latter the self-maintaining free sine-like oscillations but these oscillations are dying ones. When the oscillations cease, the storage capacitor C_s again charges from a high voltage power source and the process repeats itself after the next act of thyratron ignition. So, the pulser provides periodical current oscillations in the ionizing inductor and generation of the electromagnetic field by him. This field is able to ionize the vapor and maintain a plasma discharge in the vapor medium due to the very high field strength. It is clear the plasma discharge is a pulse-like one and repeats itself with the frequency f_r.

The ionization inductor was disposed above the crucible with Al charge as an evaporated material. The power used for evaporation was up to 3 kW, so the metal vapor pressure in the inductor area was order of 0.1 Pa. Electrostatic probes were placed in several points in the vicinity of the inductor, and an ion collector under negative bias voltage of 100-1000 V was disposed above the inductor. The collector was used as a substrate holder, too. The vapor plasma discharge occurred after energizing the inductor L above some critical level (the level of energizing was determined by the voltage of charging C_s from the anode power supply source). Yet, the ionizing process had a delay that was several half-waves of the inductor current oscillation. Then the plasma discharge continued until oscillations in the inductor died down.

The pulse mode of the discharge led to the pulse modulation of the ion current in the collector (or substrate) circuit. There was a delay (some tens microseconds) in appearance of the ion collector current relatively the inductor excitation pulse as seen in Fig.15, where t is time, j – collector ion current density and u - envelope of free dying sine-like oscillations of the inductor voltage. It is a result of the finite time of plasma expansion outside the inductor. This delay as well as the ion current pulse duration are greater for larger distances between the inductor and collector. Within the 15 cm distance from the crucible with the inductor current oscillation duration < 10 μsec, the ion current density on the negative probe was 0.1-12 mA/cm^2 and the ion current pulse duration ~ 50 μsec. The measurements made with help of the probe above the inductor have shown that the charge particles density of Al-vapor pulsed plasma was the order of 10^{10} cm^{-3}, the electron temperature ~ 15 eV, plasma potential was 50-90 V with the 2-3 kV inductor voltage, f = 0.88 MHz and 2.5 kW evaporation power. The detail study of this vapor plasma discharge allows concluding that was a glow-like discharge in crossed fields and the inductor served as a magnetic field generator, on the one hand, and as a spreaded electrical electrode, on the other hand. The efficiency of energy transfer from the inductor to the vapor plasma discharge depended on the charge particle concentration or the plasma density inside the inductor. That is why it increased when the following inductor current oscillations were laid on the tail of the previous ones. In this case there was some residual concentration of charge particles within the inductor area already at the moment of applying the following pulse to the inductor, and the discharge began to develop again but without the delay in the first half-wave of the inductor current.

Therefore, the higher repetition frequencies of excitation pulses or the pulse packet regime is more suitable for vapor ionization and vapor plasma generation. However in the former case the average power of the discharge rises substantially and, hence, the pulse packet regime is more preferable. The latter regime presents itself as the succession of groups of excitation pulses and each group or packet in turn contains a few pulses. The pauses between pulses inside the packet are to be smaller than the deionization time of pulsed discharge plasma. The packet repetition frequency is chosen to obtain the required average power of vapor plasma discharge.

Fig. 15 shows the time diagrams of collector ion current density and envelope of the free dying sine oscillations of the inductor voltage for packets containing one (a), two (b) and three (c) excitation pulses. Inside packets, the following pulses were laid on the tail of

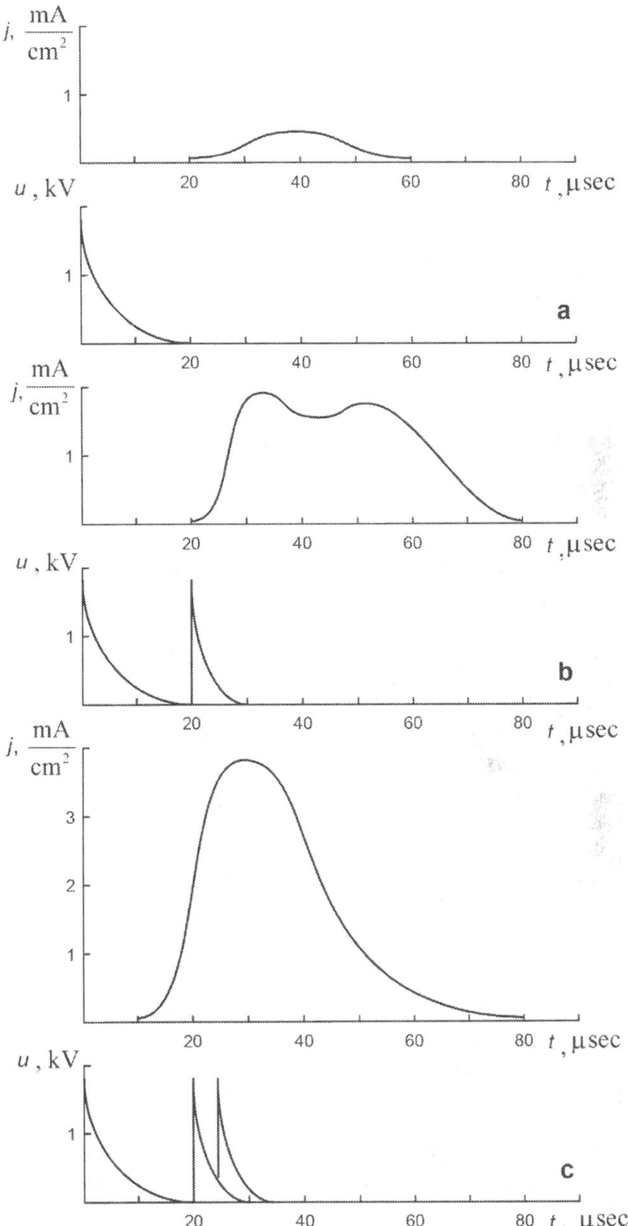

Fig. 15. Time diagrams of ion collector current density j and envelope of sine-like inductor voltage u at impact excitation of vapor ionizing inductor. Al-vapor, $f = 0.88$ MHz

previous ones. As it can be seen, the ion charge given to the collector during one pulse packet and divided by quantity of pulses inside packets is greater for two- and three-pulse packets. Increasing the quantity of pulses inside packets and the delays between them led to varying the shape and amplitude of the ion current pulses. It is seen also, inductor current oscillations after the second and the following excitation pulses died more quickly that indicates more effective energy absorption by the plasma discharge during these pulses.

This system is used for deposition of metal thin films (Al, Ni, Cr, Cu, *etc.*) on a substrate surface with self-ion bombardment. It is suitable especially for coating the temperature sensitive substrates. Other advantages consist in the low plasma potential and weak interaction of the discharge plasma with chamber walls because the duration of the active discharge stage is smaller than the time of flying of vapor plasma species through the space between the ionizing inductor and the substrate. Due to high pulse power of inductor current oscillations, their frequency may be lower than for usual continuos low pressure RF discharges.

5. Evaporators with induction heating through dielectric walls of closed envelopes

The production of tube ozonizers and some electronic devices (often called as electronic tubes) deals with metallization of inner surface of dielectric tubes. Also, manufacturing of some special materials and electron components needs evaporation of chemically active, toxic and radioactive substances within closed envelopes. Since induction heating is carried out by the electromagnetic field, which is able to penetrate without losses through dielectric walls, it may be used for these purposes with employing relatively simple equipment. For instance, consider the typical induction system for metallization of the inner surface of tubes, schematically depicted in Fig. 16. A water-cooled inductor 1, made of copper pipe, surrounds a crucible 2 with a charge of evaporated material. The crucible 2 together with a

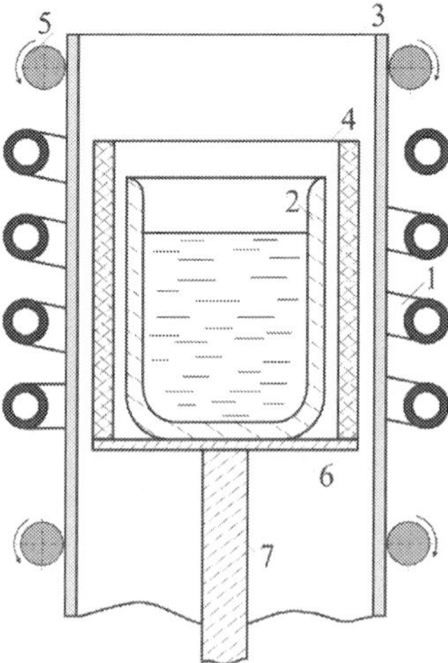

Fig. 16. Schematic diagram of induction evaporator for metallization of the inner surface of dielectric tubes (see position names in the text)

Fig. 17. Photograph of the induction evaporator for metallization of the inner surface of dielectric tubes. 1 is the secondary winding of the RF auto-type transformer, 2 is the working inductor for crucible heating, 3 is the heat shield for protection of substrate from overheating, 4 is the glass tube (substrate)

cylindrical heat shield 4 is disposed within a tube 3 coated inside by metal. The heat shield 4 reduces heat losses of the crucible and protects the tube 3 from overheating. The crucible material is sintered conductive ceramics, for example, of $TiB_2+TiC+AlN$ composition, graphite or Mo. The shield material is BN or alumina. The tube 3 is centred with help of guides 5. The crucible is mounted on a table 6, attached to a rod rack 7. Fig. 17 presents the photograph of the induction evaporator for metallization of the inner surface of dielectric tubes developed in Kiev Polytechnical Institute. The feature of this evaporator is combination of the one-turn working inductor with the step-down (11:1) RF auto-type transformer. Both windings of the transformer and the inductor are water-cooled. This evaporator is intended for deposition of metal coating (Al, Cr, Cu, Zn) with thickness of 0.1-2.0 µm on the inner surface of glass tubes 28-34 mm in diameter. After the air evacuation from a chamber, where the evaporator is set up, the glass tube is heated with an incandescent infrared lamp, then a generator of 440 kHz is switched on, and the metal charge in the crucible is heated to a temperature of evaporation. During the coating process the tube slowly rises up to get the thin film coating even on its length.

There may be various modifications of evaporation systems with the induction heating through dielectric walls, and all of them have such an advantage as possibility of the induction system disposition outside the vacuum chamber, in the atmosphere side, that makes service of them easier in comparison with disposition of the systems in vacuum. The evaporator presented in Fig. 17 is able to work in such manner.

6. Evaporation with electromagnetic confinement of melt. Levitation evaporation

Common technique of inductive evaporation employs different types of crucibles as containers for melted materials. However, crucibles are often the weak and even undesirable element of evaporating devices because of interaction and chemical reaction of the molten materials, especially, reactive metals such as Al, Ni, Ti, Zr, etc., with the crucible. Great difficulty is usually encountered in selection of crucible materials for evaporating high reactive metals. This problem is more serious than in the case of inductive remelting technology for refining of materials since the evaporation temperatures are much higher than the remelting temperatures. Interaction of molten materials with crucible leads to introducing undesirable impurities into coatings. For example, impurities of carbon in Ti-based coatings may reach some percents and more when Ti is evaporated from graphite crucibles. Moreover, crucibles may be simply destroyed due to solving their materials or reaction products in molten metals and cracking or crumbling crucible walls. For example, when Al is evaporated from ordinary graphite crucible, the latter becomes embrittled because of forming Al carbide.

One of the ways to solve this problem is to realize evaporation process without a crucible for containing evaporated material or, at least, to remove the molten material from crucible walls towards the central axis of the crucible. But, how to do this? The answer may be in applying the mechanical forces, generated due to electromagnetic induction, to molten conducting non-magnetic matter. Such approach is based on the electromagnetic confinement effect well known in magnetohydrodynamics and plasma physics. Herein, the inductor not only generates the electromagnetic force for keeping liquid metal but also provides heating the metal.

As well known from textbooks, magnetic field with induction vector $\mathbf{B}$ affects unit element of conducting matter, through which current of density $\mathbf{j}$ passes, with the force $\mathbf{F}$ directed accordingly to "the left-hand rule". Mathematically, it is expressed as $\mathbf{F} = [\mathbf{j}\,\mathbf{B}]$. The electromagnetic force compresses, squeezes molten non-magnetic metal and removes it into the space with weaker magnetic field. When the strong skin-effect occurs, the value of electromagnetic pressure p_{com} (in pascals) may be calculated by the following well known formula (2)

$$p_{com} = 3.16 \cdot 10^{-4} P_0 (\rho f)^{-1/2}, \tag{2}$$

where P_0 is density of induction power on the molten metal surface (W/m^2), ρ is specific resistivity of the molten (liquid) metal (Ohm·m), f is frequency (Hz). The pressure p_{com} in the steady state is balanced by hydrostatic pressure (by liquid metal weight). One can see increase of f leads to decrease of the electromagnetic pressure.

Fig. 18 schematically illustrates the effect of liquid non-magnetic metal electromagnetic confinement. Dots and crosses show the current direction in the inductor coil 1 and in the molten metal 2. In Fig. 18, 3 is a crucible made from non-conducting material, 4 is a substrate.

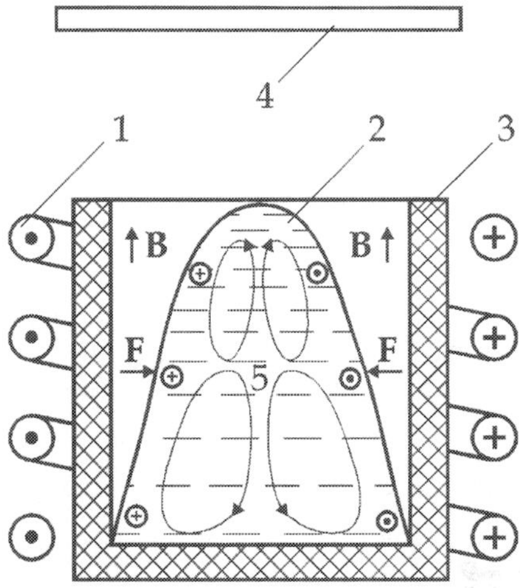

Fig. 18. Schematic illustration of electromagnetic confinement of liquid non-magnetic metal (see position names in the text)

Fig. 18 shows situation when the electromagnetic force **F**, generated by the inductor 1, compresses the molten metal 2 and removes it from the vertical crucible walls towards the central axis so, that the liquid metal forms a meniscus at the top. The lower part of the liquid metal lies on the crucible bottom. Accordingly, such system is usually named as providing "the electromagnetic confinement of melt on a support". If the molten metal has sufficient temperature, its surface emits vapor species going towards the substrate 4 and the crucible walls. The metal, condensing on the walls, flows down to the bottom and then rise up. The balance of electromagnetic and hydrostatic forces as well as surface tension forces defines the shape and position of the meniscus.

The hot metal, condensing on the wall, cannot avoid reaction with the crucible material, therefore such an induction system is suitable for evaporation of non-reactive materials. If induction heating is not sufficient for obtaining the evaporation temperature, one may use an additional heating of the meniscus by electron beam, for instance. The advantage of this system is internal motion of the liquid metal under the external electromagnetic pressure force that provide remixing of chemical elements during metal alloy evaporation and saving coating composition. Fig. 18 depicts simplified trajectories of liquid metal motion (closed lines with arrows): from points with higher p_{com} (near the mid-plane of the inductor) to points with lower values of the inductor magnetic field. However motion of conducting liquid leads to its interaction with both own magnetic field and the external field of the inductor, and the real trajectories of liquid metal motion are more complicated than the shown in Fig. 18. The considered case is met in induction coaters with dielectric crucibles and direct heating of evaporated metallic material by eddy current at relatively low frequency (~1-10 kHz); mass of material charged into crucibles may reach 10 kg and more, RF power is up to 100 kW. It is worth to note the following, forming the meniscus at the centre of the molten metal (*e.g.* Al) allows impurities of metal oxide or carbide to flow down from the top metal surface. Thus, the metal evaporating surface is maintained fresh for a long time to prevent splash generation, and a high quality coatings without pin-holes and other defects are obtained.

The induction system with the melt removed from crucible walls, presented in Fig. 18, is very preferable for remelting reactive materials to refine and obtain high-purity ones since this technology needs temperatures much below the evaporation temperatures, and reactions with the crucible material do not take place. In this technology the crucible bottom is cooled by water, so the lowest part of metal is solid and it serves, in fact, as a crucible bottom and a support for the melt.

Consider what will be if we strongly increase the inductor power. In this case, the electromagnetic force in the inductor mid-plane increases so, that a big metal drop or ball can tear off from the rest liquid metal and soar above the inductor. This is the effect of electromagnetic levitation. To avoid a random non-controlled trajectory of the metal ball, an additional inductor with opposite electrical current is needed above the lower inductor 1 as shown in Fig. 19. The upper inductor 2 will generate electromagnetic force pushing the metal "ball" down and, in result, the "ball" will levitate on a height between the lower and the upper inductors. Fig. 20 depicts the levitation system with the lower inductor of conical shape.

Magnetic fields of both inductors are to create a three-dimensional potential well or electromagnetic trap for the metal ball. For this, the space with weak magnetic field, where the ball will levitate, is to be surrounded by more strong field. The dashed lines in Fig. 20 illustrate the magnetic force line configuration for obtaining such potential well. The liquid metal levitates in a stable space position if the well is symmetric and the balance between gravitation and electromagnetism forces takes place at each point of the levitating liquid metal surface. Accordingly, the primary shape of the metal ball transforms into "top" form

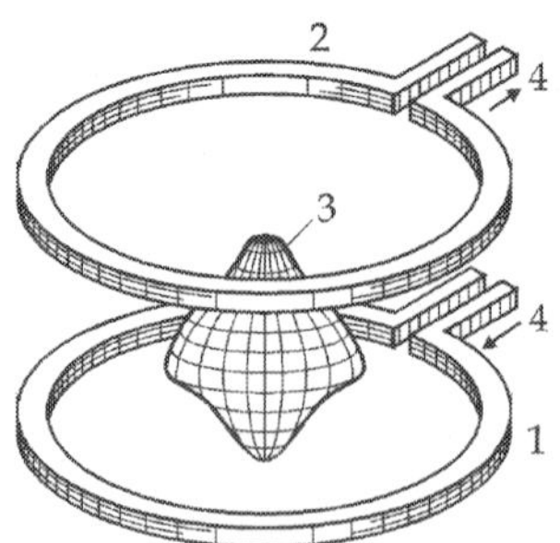

Fig. 19. Two-inductor levitation system. 1 and 2 are inductors, 3 is levitating "ball" of liquid metal, 4 is current direction in the inductors

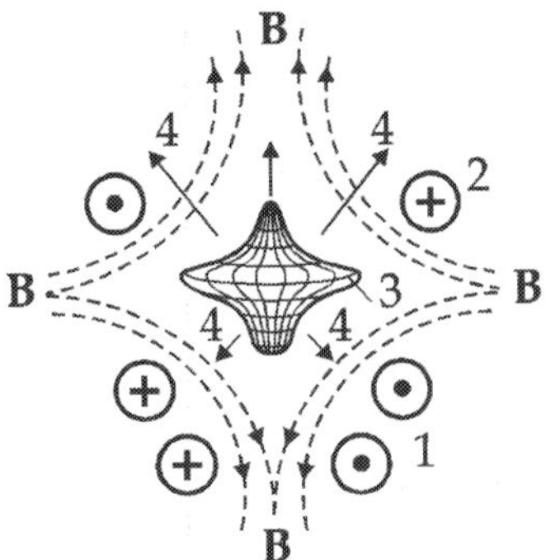

Fig. 20. Two-inductor levitation system. 1 is the lower inductor of conical shape, 2 is the upper inductor with opposite winding, 3 is levitating "top" of liquid metal, 4 is directions of vapor flow; the dashed lines are magnetic force lines

shown in both Fig. 19 and Fig. 20. It is interesting to note, the eddy currents is not induced at "the equator" and "the poles" of the liquid top and the electromagnetic pressure, confining the melt, is absent here, but liquid metal can levitate, not spilling, due to action of surface tension forces. Unfortunately, because of weakness of the surface tension forces, only small quantity of liquid metal (up to tens grams) can levitate.

The feature of evaporating system of such kind is change of vapor flow density during the evaporation process caused by two factors: decrease of the evaporating surface area and continuous movement of the levitating top in the space between the inductors towards a new position with the less magnetic field. Both the factors arise from decreasing the levitating top mass. Nevertheless it is possible to stabilize the vapor flow density if one introduces control of RF power supplying the inductors by the flow density sensor signal.

The evaporating levitation system usually contains the lower conical inductor (as in Fig. 20), the one-turn upper inductor, connected in series with the lower inductor but with opposite current direction, and one RF generator for supplying both the inductors. The evaporation system also contains means to locate a piece of evaporated material (*e.g.*, granule) between the indictors in the space of future levitation before the evaporation process and a crucible (or box) for reception of the residual material after evaporation. Commonly, the evaporated granule is placed in a dielectric holder, which can move along the axis of inductors and lift up the granule to the space of levitation. When the RF generator is switched-on, the electromagnetic force begins to support the granule and the holder is removed down. After heating the granule by the induced eddy current up to the evaporation temperature a substrate shutter (not shown in Fig. 20) is opened and the process of coating the substrate (not shown in Fig. 20) by evaporated material begins. For finishing the process, the RF generator is switched-off and the reception crucible picks up the rest of evaporated granule. The deposition process may be repeated by charging a new portion of evaporated material in the inductor system. The details of evaporating levitation system design and deposition procedure may be find elsewhere (Audenhove, 1965; Vyrelkin et al., 1985).

Since the evaporated material deposits not only on the substrate but also on all surfaces, surrounding the levitating liquid top, and, in the first turn, on the inductor coil, the total evaporated material mass or the time of continuos work of the evaporation system is limited by shorting of the coil. It is possible to enlarge the total evaporated mass at the same time of

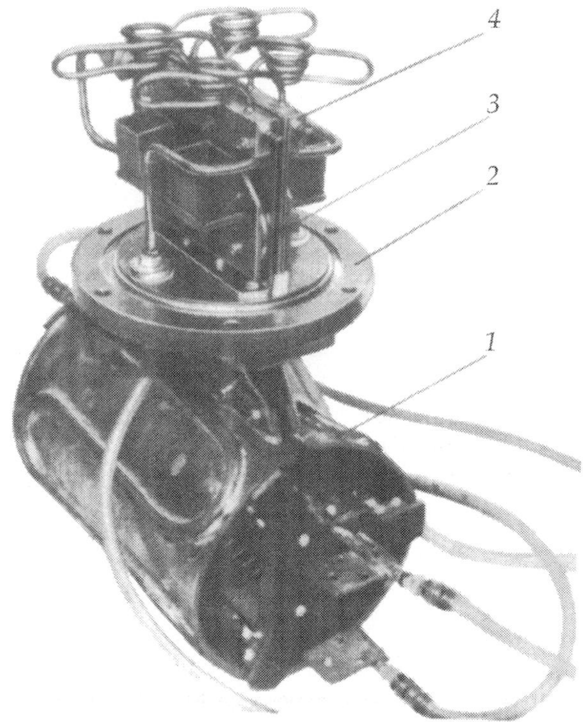

Fig. 21. Photograph of levitation evaporation system with 4-fold inductor unit. 1 is step-down RF transformer, 2 is flange, 3 is crucible unit, 4 is 4-fold inductor unit (Vyrelkin et al., 1985)

continuos work by increase of inductor number. Fig. 21 shows the example of 4-fold inductor unit employed as a part of a compact evaporation system designed in Russia for application in the standard industrial high-vacuum deposition machine instead of resistance-heated evaporators (Vyrelkin et al., 1985). Increase of the inductor number provides also more uniform coating thickness distribution on substrates. The induction system, presented in Fig. 21, has the following parameters given in Tabl. 1.

Average rate of Al deposition during 20 min process with evaporator-substrate distance of 220 mm	0.1 μm/min
Non-uniformity of coating thickness across substrate of 76 mm diameter	5 %
RF power for supply of inductors	10 kW
RF frequency	440 kHz
Evaporated granule mass	up to 2 g
Base gas pressure in technological chamber	less 10^{-4} Pa

Table 1. Parameters of evaporating levitation systems with 4 inductors (Vyrelkin et al., 1985)

The main field of application of evaporating levitation system is obtaining high-purity coatings and thin films in laboratory conditions.

7. Conclusion

Thus, the principle approaches to design of evaporators with induction heating, determining the main fields of their applications, have been considered in the given paper. A small volume of the chapter does not permit to discuss in details such important branch of application as web coaters with induction evaporation but information on this theme may be find elsewhere (Bishop, 2007). Also, this relates to induction evaporation technology for obtaining and treatment of micro- and nanodispersed materials (Kuzmichev & Tsybulsky, 2006b).

The induction evaporators are able to provide different characteristics of vapor flows including their ionization for very perspective processes of ion plating and, although other evaporation technologies are strong competitors for the induction evaporation, the unusual and interesting possibilities of the induction methods allow them to occupy own niche in industrial technology and for preparation of scientific objects. Moreover, the solutions based on employing pulse technique are to incite engineers towards creation of new types of induction technological and electrophysical equipment. So, we can say the future belongs to the induction evaporation.

8. References

Ames, I. et al. (1966). Crucible type evaporation source for aluminum. *Rev. Sci. Instr.*, Vol. 37, No. 12, pp.1737-1738.

Anderson, E.R. (1975). Induction heated vapor sources, *Proc. 18th Ann. Tech. Conf. SVC*, pp. 114-118, USA.

Audenhove, V. (1965). Vacuum evaporation of metals by high frequency levitation heating. *Rev. Sci. Instr.*, Vol. 36, No. 3, pp.133-135.

Banshah, R.F. (1994). Evaporation: processes, bulk microstructures and mechanical properties, In: *Handbook of deposition technologies for films and coatings*, R.F. Banshah, (Ed.), pp. 131-248, Noyes Publications, ISBN 0-8155-1337-2, New Jersey, U.S.A.

Berghaus, B. (1939). Improvements in and relating to the coating of articles by means of thermally vaporised material. *Patent of Great Britain*, No. 510993.

Bishop, C.A. (2007). *Vacuum Deposition onto Webs, Films, and Foils*, William Andrew Inc., ISBN 978-0-8155-1535-7

Hale, G.J.; White, G.W. & Meyer, D.E. (1975). Ion plating using a pure ion source: an answer looking for problems. *Electronic Packaging and Production*, May, pp. 39-45.

Japanese patent application (1978). No. 53-9191.

Kuzmichev, A. & Tsybulsky, L. (2006a). Ionized vapor flow source. *Patent of Ukraine*, No. 13134.

Kuzmichev, A. & Tsybulsky, L. (2006b). Obtaining and treatment of micro- and nanodispersed materials with help of inductive heating facilities, *Proc. Kharkov Nanotechnological Assembly*, Vol. 2, pp. 50-59, ISBN 966-8855-21-3, Kharkov, Ukraine (in Russian).

Kuzmichev, A. & Tsybulsky, L. (2008). Evaporators with induction heating and their applications. *Przegląd elektrotechniczny*, Vol. 84, No. 3, pp. 32-35. ISSN 0033-2097.

Mastyaev, V.Ya. (1978). *Pulse thyratron based generators for induction heating*, Energiya, Moscow (in Russian).

Mattox, D.M. (2003). *The foundations of vacuum coating technology*, Noyes Publications/William Andrew Publishing, ISBN 0-8155-1495-6, N.Y., U.S.A.

Phinney, R.R. & Strippe, D.C. (1988). Crucible for evaporation of metallic film. *Patent of USA*, No. 4791261.

Spalvins, T. (1976). Industrialization of the ion plating process. *Research/Development*, October, pp. 45-48.

Spalvins, T. (1980). Survey of ion plating sources. *J. Vac. Sci. Technol.*, Vol. 17, No. 1, pp. 315-321.

Tada, I.; Yamamori, T.; Matsumori, K. & Yoneda, Y. (1989). Comparison of evaporation sources for vacuum web coaters, *Proc.* 32nd *Ann. Tech. Conf. SVC*, pp. 131-149, USA.

Vyrelkin, V.P.; Denisov, A.G.; Kozlitin, V.P.; Muzlov, D.P. & Fedotov, S.M. (1985). Film deposition from electromagnetic crucible in set-ups of UVN-type. *Electronic industry*, No. 2 (140), pp. 22-24 (in Russian).

White, G.W. (1977). High rate ion plating source. *Patent of USA*, No. 4016389.

14

Application of Microwave Heating to Recover Metallic Elements from Industrial Waste

Joonho Lee and Taeyoung Kim
Korea University
Republic of Korea

1. Introduction

Metals have been used in human society for thousands of years, and the amount of metal production continuously increased. In modern society, the production of metals has explosively increased, and the use of metals becomes inevitable in our daily life throughout society. However, the source of metals seems to be changed from high concentrations of ores to low concentrations or industrial wastes.

Among the global primary metals production, crude steel shares approximately 90% (Worrell, 2004), and major metallurgical industrial waste comes from iron- and steel-making industries. In 2007, the total production of crude steel reaches 1,351 million tons (Worldsteel Association, 2009), and it is expected to increase continuously. Along with the continuous growth in steel production, the generation of iron- and steel-making industrial wastes such as mill scale, slag, dust, sludge, *etc.* have increased considerably. Considering that these industrial wastes contain lots of useful resources such as Fe, Zn, etc., many researchers have suggested various processes recycling the valuable metallic elements.

Pyrometallurgical process can be a general solution to recover valuable elements from industrial wastes. However, traditional pyrometallurgical recycling processes (e.g. Waelz process) have some problems inherently (Lee *et al.*, 2008). (1) Industrial wastes are difficult to handle due to their physical characteristics as fine particles: apparent low density and flying. Therefore, recycling processes require pre-treatment such as pelletizing with binders. (2) During the process, erosion of refractory materials easily occurs, because the heating zone temperature is much higher than the reaction zone. Therefore, maintenance cost is a critical barrier, when the process is extended from a pilot plant to a commercial one. (3) After processing, the metallic components mainly iron-based alloys are difficult to be separated from residues (originated from the waste as well as the binders).

Recently, many researchers have paid attentions to the microwave heating process as an alternative recycling method due to its unique characteristics such as fast heating and direct internal heating (Nishioka *et al.*, 2002; Morita *et al.*, 2001; Morita *et al.*, 2002; Saidi & Azari, 2005; Cho & Lee, 2008; Lee *et al.*, 2008; Kim *et al.*, 2009; Lee *et al.*, 2010; Kim *et al.*, 2010; Lee *et al.*, 2010). It is also believed that the microwave heating may provide savings in both time and energy (Kelly & Rowson, 1995; Ishizaki *et al.*, 2006).

2. Mirowave heating fundamentals

Details of the microwave heating are precisely described in a reference (Gupta, 2007). Here, a brief explanation of the microwave heating is suggested. Microwaves are electromagnetic waves with frequencies ranging from 300 MHz to 300 GHz. Accordingly, the wavelength of microwaves ranges from 1 mm to 1 m, which is much larger than molecular (nm) or crystalline grain size (μm). Therefore, microwave may supply energy on valence electrons, yielding electron fluctuation. Microwave can be transmitted, absorbed, or reflected, depending on the sorts of materials (Fig. 1). When microwave is absorbed in a material, the electron fluctuation is eventually transferred to lattice ions and generates vibrations in matrix, which can be transferred to heat energy. The heating rate depends on the electromagnetic properties of a material (complex permittivity and permeability), and the average power absorbed by a material is the sum of electric loss and magnetic loss.

$$P_{av} = \omega\varepsilon_0\varepsilon''_{eff}E^2_{rms} + \omega\mu_0\mu''_{eff}H^2_{rms} \qquad (1)$$

where ω is the angular frequency ($=2\pi f$), ε_0 the permittivity of free space ($= 8.854 \times 10^{-12}$ F/m), ε''_{eff} the effective relative dielectric loss factor, E_{rms} the root mean square of the electric field, μ_0 the permeability of free space ($= 2\pi \times 10^{-7}$ H/m), μ''_{eff} the effective relative magnetic loss factor, H_{rms} the root mean square of the magnetic field.

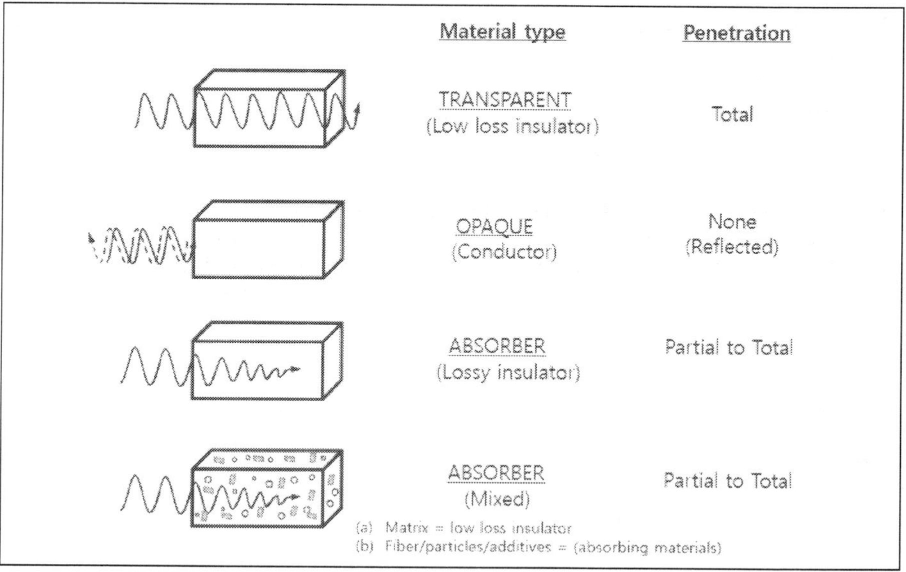

Fig. 1. Microwave penetration, reflection and absorption (Sutton, 1989)

Yoshikawa *et al.* investigated the heating behaviour of NiO and carbon under electric field and magnetic field respectively using a single mode microwave applicator (Yoshikawa, 2007). They found that NiO was heated only in the electric field, whereas carbon was heated both in the electric and magnetic fields. Typical microwave heating furnaces make use of multi mode applicators. Therefore, it is inherently difficult to distinguish the effect of the electric field from that of the magnetic field, and vice versa. This unique feature of microwave heating results in so-called "thermal runaway" behaviour. For several ceramic materials, the dielectric constant shows strong temperature dependence, namely, above a critical temperature the dielectric constant rapidly increases with temperature (Birnboim *et al.*, 1998). When the microwave is concentrated on a certain spot, the local temperature becomes higher than neighbours. As the local temperature exceeds a critical value, the heating rate becomes much faster, and results in the acceleration of the heating rate of the neighbours. This thermal runaway behaviour would be very useful to accelerate high temperature reactions and reducing energy consumption.

Reduction of metallic component from oxide mixtures was firstly investigated by Standish and Worner (Standish & Worner, 1990). Standish and Huang recovered metallic component from iron ore (magnetite and hematite) using the microwave heating and reported that the microwave heating was 5 times faster than the conventional heating (Standish & Huang, 1991). Mourao *et al.* investigated the effects of carbonaceous materials on the reduction behaviour of hematite iron ore (Mourao *et al.*, 2001). Ishizaki *et al.* also obtained pig iron from magnetite ore-coal composite pellets by microwave heating (Ishizaki *et al.*, 2006). Chen *et al.* investigated the effect of the microwave power on the heating rate of self-fluxing pellets containing coal, and found the heating rate was increased 8 times by increasing the microwave power from 5 kW to 15 kW (Chen *et al.*, 2003). Most of these works have focused on the reduction of iron ore as an alternative iron-making process. Iron- and steel-making industrial wastes generally contain lots of iron oxide, so that similar reduction behaviour can be expected.

3. Recovery of metallic elements from industrial waste

3.1 Recovery of metallic elements from slag

Table 1 shows typical industrial slag compositions (Sano, 1977). Not only copper and lead blast furnace slag but also converter slag of iron shows high Fe content. Morita *et al.* firstly invetigated the recovery of metallic iron from the converter steelmaking slag (Morita, 2001, 2002, 2009). Fig. 2 shows a typical experimental setup of the lab scale microwave heating experiments, and reduced metal droplets (Kim *et al.*, 2010). Morita *et al.* examined the effect of the ionic status of Fe in synthetic slags on the heating rate, and found that the heating rate strongly depends on Fe^{3+}/Fe^{total} fraction. The maxium heating rate was obtained when the Fe^{3+}/Fe^{total} fraction equals approximately 0.15~0.16 (Morita *et al.*, 2001). This behavior could be explained by the existance of $CaFe_3O_5$ phase, whose dielectric loss was as high as that of Fe_3O_4 (Morita *et al.*, 2001). They also found that the heating rate increased with increasing the amount of carbon, which implied that cabon was used as the reductant as well as the microwave absorber.

Based on the experimental results with synthetic slags, Morita *et al.* investigated the reduction behavior of industrial steel-making slag (Morita *et al.*, 2002). Industrial steel-making slag contained relatively high concentration of phosphorus (approximately 4.0 wt%). Accordingly, the reduced iron also contained high content of phosphorus, which should be removed before recycling the reduced iron. Morita *et al.* suggested a novel steel-making slag recycling process shown in Fig. 3 (Morita *et al.*, 2002). They suggested a process through which Fe-P-C alloy was obtained by using the microwave heating, and phosphorus was removed further with K_2CO_3 flux. Reduced K_3PO_4 was supplied as a fertilizer, and Fe-C alloy was recharged to the steel-making process stream. In the reduction of industrial

	Chemical composition (wt%)					Others
	SiO_2	CaO	FeO	Al_2O_3	MgO	
Blast furnace slag (Cu)	30-40	5-15	35-50	5-10	1-3	Zn,S,Cu
Blast furnace slag (Pb)	25-40	10-25	30-40	5-10	-	Zn, Pb, S
Blast furnace slag (Fe)	30-40	35-45	-	5-10	-	MnO
Converter steelmaking slag (Fe)	10-20	40-50	10-25	-	4-10	MnO, P_2O_5, CaF_2
Electric furnace slag (Fe) (reduced period)	15-20	65-65	<1.0	<3.0	5-10	CaF_2

Table 1. Typical composition of industrial slag (Sano, 1997)

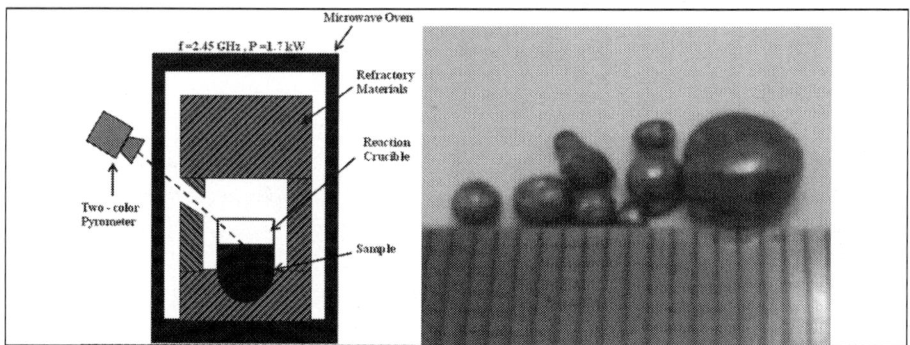

Fig. 2. Typical example of a setup of the microwave heating experiments and reduced iron (Kim *et al.*, 2010)

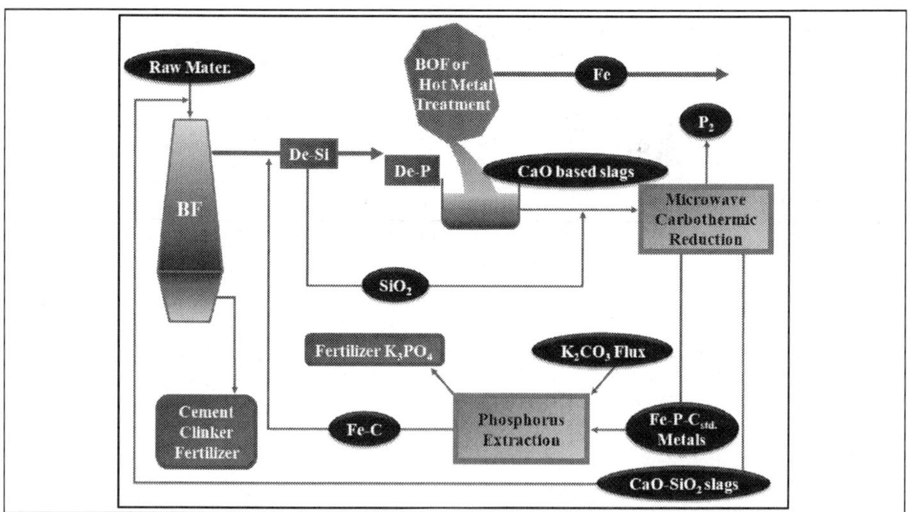

Fig. 3. Steel-making slag recycling process suggested by Morita *et al.* (Morita *et al.*, 2002)

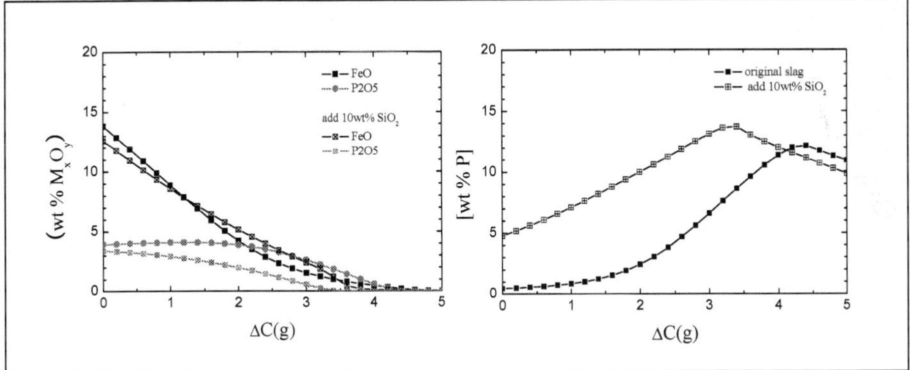

Fig. 4. Effect of SiO$_2$ addition on the change in the composition of (a) FeO and P$_2$O$_5$ in slag and (b) P in metal at 2073K by the carbothermic reduction. Initial slag weight is assumed to be 100g, and ΔC is the amount of carbon consumption. (Lee *et al.*, 2010)

steel-making slag, the formation of 2CaO.SiO$_2$ would decrease the reduction rate by decreasing the fluidity of slag. They reported that the reduction rates of Fe and P could be increased by adding extra 10 wt% SiO$_2$ (Morita *et al.*, 2002). Although the recovery rate of Fe slightly increased, that of P considerabley increased in their experiments. Therefore, there seems to be another reason except the increase of fluidity. Morita *et al.* supposed that the change of activity coefficient of P$_2$O$_5$ would increase the reduction rate (Morita *et al.*, 2002). Recently, Lee *et al.* reproduced the Morita's experimental results using thermodynamic calculations based on recently developed thermodynamic database with FactSage (Fig. 4), and confirmed that the addition of 10 wt% SiO$_2$ considerably increased the reduction rate of P due to the increase of the activity coefficient of P$_2$O$_5$ (Lee *et al.*, 2010).

Morita and Guo also examined the recovery of Fe and Cr from stainless steelmaking slag (Morita & Guo, 2009). They investigated the heating rate of slag before and after Fe-Si reduction. Although the slag after the Fe-Si reduction contained very small amount of Fe and Cr, the heating rate was not so much different from that before reduction. Accordingly, carbon was considered dominant microwave absorber in their stainless steel-making slag reduction.

3.2 Recovery of metallic elements from mill scale

Mill scale is very attractive industrial waste due to high contents of iron (more than 70 wt%). However, generation of mill scale is relatively small, and economical benefit is hard to be expected when conventional pyrometallurgical methods are applied. In the whole world, 13.5 M tons of mill scales are generated annually, and among them only 500 K tons from Korea (Cho & Lee, 2008). Moreover, when a conventional heating method is applied, 180 min is required to obtain 100% metallization at 1373-1473 K (Kim *et al.*, 1986). Therefore, most of them are currently used as coolants in steel-making process or additives in sintering process.

Microwave process seems to be very suitable for the recovery of metallic elements from mill scale, because the reaction rate is much faster and the process facilities are much smaller than the conventional ones. In addition, pre- and post-treatments such as pelletizing and crushing are not required. Moreover, gas burner and related facilities can be eliminated from the recycling process. Cho and Lee suggested a novel process recovering metallic elements from mill scale (Cho & Lee, 2008). The unique feature of this process was the minimization of the secondary waste emission. Since the reduced metallic particles were coagulated due to the microwave-induced vibration, self-assembled particles (1-5 mm in diameter) were easily separated from the residues (Fig. 5). Most of the residues were carbon particles, which could be re-charged in the process stream. Moreover, this process was successfully applied to stainless steel mill scale. Stainless steel mill scale contains environmentally harmful elements such as chromium and nickel, which could be recovered as metallic particles. Therefore, microwave processing is very useful not only to recover valuable metallic elements from mill scale but to prevent the effluence of environmentally harmful elements from stainless steel mill scale.

Fig. 5. Reaction products and a flowsheet for the recovery of metallic components from mill scales using the microwave heating. (1 : carboneous materials, 2 : mill scales, 3 : mixer, 4 : supplier, 5 : microwave furnace, 6 : gravitational separator, 7 : recharge of remaining particles, 8 : cooling tower, 9 : chimney) (Cho & Lee, 2008)

3.3 Recovery of metallic elements from dust

Several commercialized processes for recovery of metallic elements from dust from iron-making and steel-making processes have been developed based on Wealz process. Most of these processes pay attention to the recovery of Zn from the electric arc furnace dust (EAF dust), which generally contains 25 wt% Fe and 20 wt% Zn (Kim *et al.*, 2009). However, EAF dust is categorized as hazardous waste in many countries, and the conventional process yields elution of Pb and Cr from clinker. Moreover, the total amount of waste generation cannot be reduced by using additives such as binder and flux. Therefore, environmentally-kindly alternative processes like plasma process, flame reactor process, top submerged lance process have been developed. However, these methods still have some economical, environmental or technical problems.

The recovery of metallic elements from EAF dust using the microwave heating was firstly attempted by Nishioka *et al.* (Nishioka *et al.*, 2002). In their experiments, synthesized dust-carbon composites (38.5 wt% Fe_2O_3, 38.5 wt% C, 19.2 wt% ZnO and 3.8 wt% PbO) were used. They reported that the reduction of ZnO was slower than that of Fe_2O_3. Lee *et al.* examined the reduction behavior of EAF dust and found that the reduction of ZnO occurred after flanklinte was fully reduced from XRD analysis of the reacted samples: Fe was recovered as fine metallic particles from 1100K and Zn was separated as gas phase from

1200K (Lee *et al.*, 2008) (Fig. 6). Sun *et al.* carried out similar experiments and successfully recovered metallic iron and metallic zinc (Sun *et al.*, 2009). Saidi and Azari investigated the reduction behavior of zinc oxide concentrate by carbon under microwave irradiation (Saidi & Azari, 2005). They found that the reduction started from the center on the sample with microwave, whereas without microwave the reduction started from the surface. Recently, Kim and Lee studied the reduction kinetics of ZnO with solid carbon at constant temperatures under microwave irradiation (Kim & Lee, 2009). They found that the microwave heating enhanced the reaction rate. When the reaction rates were compared with and without microwave, the reaction rates with microwave showed much higher values. Moreover, when additional microwave power was applied to increase the reaction temperature, the reaction rate became much faster than a expected value (stronger temperature dependence than each elementary reaction).

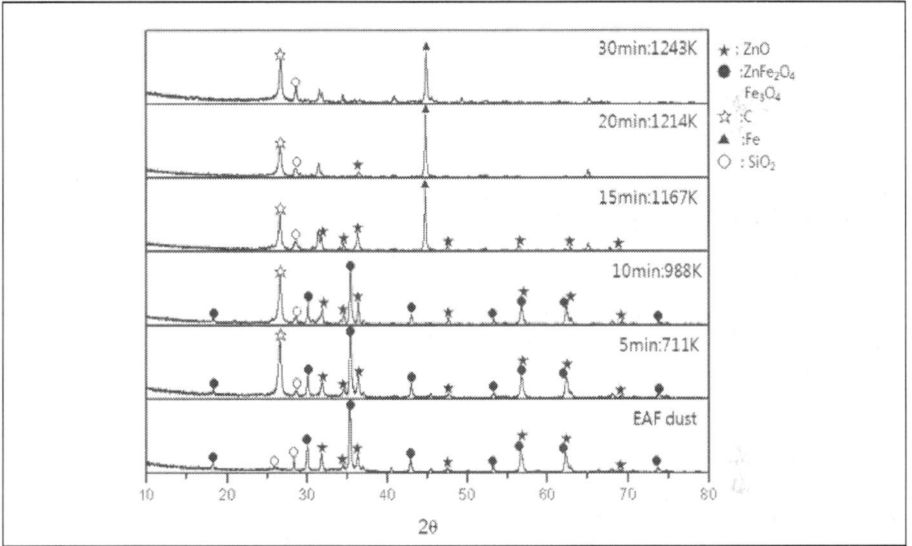

Fig. 6. XRD analysis results of EAF dust and carbon mixture after microwave irradiation (Lee *et al.*, 2008)

3.4 Recovery of metallic elements from other industrial wastes

Morita *et al.* applied the microwave heating to fly ash treatments (Morita *et al.*, 1992; Nakaoka *et al.*, 1993). Recently, Chou *et al.* also used the microwave heating and successfully fixed the hazardous elements (Chou *et al.*, 2009). Ma *et al.* recovered metallic elements from sludge (Ma *et al.*, 2005). In the microwave process with sludge, removal of water is very important (Standish *et al*, 1988). It is expected that the removal of water can be enhanced by combining vacuum and microwave heating methods. There still have challenges to recover metallic elements form industrial wastes not only from iron- and steel-making plants but also other sectors in many industries. One example is the recovery of Ni from used Ni ion batteries (Yoshikawa *et al.*, 2007). Due to its simplicity and freedom from process limitations, the microwave process is expected to be extensively used in many recycling processes.

4. Prospective

The microwave heating process has gained many attentions from many pyrometallurgy industries, who are searching for an innovative process to reduce carbon dioxide emission. In principle, the microwave heating process generates lower carbon dioxide emission, when the electricity is supplied by a nuclear power plant. Therefore, the use of microwave heating process can be determined based on the social electric power generation system of each country.

In order to commercialize the microwave heating process for the industrial waste treatment, several points should be improved. First, the energy efficiency should be increased from the current status (approximately 10-20%) to a much higher level (at least 30-50%). Second, the capacity and powder of the microwave furnace should be increased. Most of the lab scale experiments have been carried out with 1-2 kW microwave furnaces. Currently, Michigan Institute of Technology and Tokyo Institute of Technology are developing pilot plant type furnaces, Rotary Hearth and Rotary Kiln based furnaces, respectively. Furnace design and process optimization are very time consuming, but should be improved further. Third, more fundamental studies should be carried out. We have very limited knowledge on the electromagnetic properties of materials and related chemical reactions at high temperatures. Extensive works on the microwave treatment of industrial waste have been carried out in the University of Tokyo and Korea University. Nevertheless, there remain many theoretical and technical problems solved.

5. Conclusion

Microwave can penetrate to the core of a target material, and directly heat the inside. Therefore, high energy efficiency can be obtained when heat transfer is a rate-determining step. Microwave heating also reduces the erosion of refractory materials due to internal heating characteristic. Microwave also accelerates agglomeration of the particles, yielding direct contact between oxides and carbonaceous materials. Therefore, fast reaction rates can be expected. Moreover, microwave heating system does not need any gas injection for combustion or reduction, so that fine particles can be directly used without any pre-treatment. After the microwave process, recovered metallic components can be easily separated from the remaining materials. Therefore, reduction in process costs can be expected. From the lab scale experiments, it was confirmed that metallic elements were successfully recovered in a very short period of time. In a near future, mass production of recycled metals can be achieved with the microwave heating method.

6. Acknowledgement

The authors thank Prof. Kazuki Morita for his kind discussion and valuable information. This work was supported by Korea Environmental Industry & Technology Institute.

7. References

Birnboim, A.; Gershon, D.; Calame, J.; Birman, A.; Carmel, Y.; Rodgers, J.; Levush, B.; Bykov, Y. V.; Eremeev, A. G.; Holoptsev, V. V.; Semenov, V. E.; Dadon, D.; Martin, P. L. & Rosen, M. (1998). Comparative Study of Microwave Sintering of Zinc Oxide at 2.45,

30, and 83 GHz, *Journal of American Ceramic Society*, Vol. 81, No. 6, (June 1998), 1493-1501, 0002-7820

Chen, J.; Liu, L.; Zeng, L.; Zeng, J.Q.; Liu, J.Y. & Ren, R.G. (2003). Experiment of Microwave Heating on Self-fluxing Pellet Containing Coal. *Journal of Iron and Steel Research International*, Vol. 10, No. 2, (May 2003) 1-4, 1006-706X

Cho, S. & Lee, J. (2008). Metal Recovery from Stainless Steel Mill Scale by Microwave Heating. Metals and Materials International, Vol. 14, No. 2, (April 2008) 193-196, 2005-4149

Chou, S.; Lo, S.; Hsieh, C. & Chen, C. (2009). Sintering of MSWI Fly Ash by Microwave Energy. *Journal of Hazardous Materials*, Vol. 163, No. 1, (April 2009) 357-362, 0304-3894

Gupta, M. & Wong E. (2007). Microwave Heating, In: *Microwaves and Metals*, 43-64, John & Wiley & Sons (Asia) Pte Ltd, 978-0-470-82272-2, Singapore

Ishizaki, K., Nagata, K. & Hayashi, T. (2006). Production of Pig Iron from Magnetite Ore-Coal Composite Pellets by Microwave Heating. *ISIJ International*, Vol. 46, No. 10, (October 2006) 1403-1409, 0915-1559

Kelly, R.M. & Rowson, N.A., (1995). Microwave Reduction of Oxidized Ilmenite Concentrates. *Minerals Engineering*, Vol. 8, No. 11, (November 1995) 1427-1438, 0892-6875

Kim, E.; Cho, S. & Lee, J. (2009). Kinetics of the Reactions of Carbon Containing Zinc Oxide Composites under Microwave Irradiation. *Metals and Materials International*, Vol. 15, No. 6 (December 2009), 1033-1037, 2005-4149

Kim, K. H.; Shin, H. K.; Chung, Y. C. & Kim, T. D. Preparation of Iron Powders from Mill Scale, *POSCO Technical Report*, Vol. 8, No. 3, (1986) 387-394

Kim, T.; Kim, E.; Shin, M. & Lee, J. (2010). An Experimental Study on Iron Recovery from Steelmaking Slag by Microwave Heating. *Journal of the Korean Institute of Resources Recycling*, Vol. 19, No. 1, (February 2010) 21-26, 1225-8326

Lee, J.; Cho, S. & Kim, E. (2008). Metal Recovery from Industrial Wastes by Microwave Heating, *Proceeding of 3rd International Conference on Process Development in Iron and Steelmaking*, pp. 547-550, 978-91-633-2269-3, Lulea, June 2008, MEFOS, Lulea

Lee, J.; Kim, E.; Kim, T. & Kang Y. (2010). Thermodynamic Study for P Reduction from Slag to Molten Steel by using the Microwave Heating. *Korean Journal of Materials Research*, Vol. 20, No. 1, (January 2010), 42-46, 1225-0562

Ma, P.; Lindblom, B. & Bjorkman, B. (2005). Experimental Studies on Solid-State Reduction of Pickling Sludge generated in the Stainless Steel Production. *Scandinavian Journal of Metallurgy*, Vol. 34, No. 1, (February 2005) 31-40, 0371-0459

Morita, K.; Nguyen, V.Q.; Mackenzie, J.D. & Nakaoka, R. (1992). Immobilization of Ash by Microwave Heating, *Proceedings of the 11th Incineration Conference*, pp. 507-514, IDS BY93L, Albuquerque, May 1992, Univ. California Irvine, Irvine

Morita, K.; Guo, M.; Miyazaki, Y. & Sano, N. (2001). The Heating Characteristics of CaO-SiO₂-FeₜO System Slags under Microwave Irradiation. *ISIJ International*, Vol. 41, No. 7, (July 2001) 716-721, 0915-1559

Morita, K.; Guo, M.; Oka, N. & Sano, N. (2002). Resurrection of the Iron and Phosphorus Resource in Steel-making Slag. *Journal of Material Cycles Waste Management*, Vol. 4, No. 2, (October 2002), 93-101, 1438-4957

Morita, K & Guo, M. (2009). Metal Recovery from Slags, *Proceedings of 1st International Slag Valorisation Symposium*, pp. 151-162, 978-94-6018-049-1, Leuven, April 2009, ACCO, Leuven

Mourao, M.B., Parreiras de Carvalho, Jr., I. & Takano C. (2001). Carbothermic Reduction by Microwave Heating. *ISIJ International*, Vol. 41, (February 2001) S27-S30, 0915-1559

Nakaoka, R.; Morita, K. & Mckenzie, J.D. (1993). Immobilization Process for Transuranic Incinerator Ash, Fusion and Processing of Glass, *Proceedings of the 3rd International Conference on Advances in Fusion and Processing of Glass*, pp. 553-560, 0-944904-56-4, New Orleans, June 1992, American Ceramic Society, Westerville

Nishioka, K.; Maeda, T. & Shimizu, M. (2002). Dezincing Behavior from Iron and Steelmaking Dusts by Microwave Heating. *ISIJ International,* Vol. 42, (February 2002) S19-S22, 0915-1559

Saidi, A. & Azari, K. (2005). Carbothermic Reduction of Zinc Oxide Concentrate by Microwave. *Journal of Materials Science & Technology*, Vol. 21, No. 5, (September 2005) 724-728, 1005-0302

Sano, N. (1997). Thermodynamics of Slag, In: *Advanced Physical Chemistry for Process Metallurgy*, Sano, N. *et al.* (Ed.), 45-86, Academic press, 0-12-618920-X, San Diego

Standish, N., Worner, H. & Kaul, H. (1988). Microwave Drying of Brown Coal Agglomerates. Journal of Microwave Power and Electromagnetic Energy, Vol. 23, No. 3, (1988) 171-175, 0832-7823.

Standish, N. & Worner, H. (1990). Microwave Application in the Reduction of Metal-Oxides with Carbon. *Journal of Microwave Power and Electromagnetic Energy*, Vol. 25, No. 3, (1990) 177-180, 0832-7823

Standish, N. & Huang, W. (1991). Microwave Application in Carbothermic Reduction of Iron Ores. *ISIJ International*, Vol. 31, No. 3, (March 1991) 241-245, 0915-1559

Sun, X.; Hwang, J.-Y.; Huang, X. & Li, B. (2009). Petroleum Coke Particle Size Effects on the Treatment of EAF Dust through Microwave Heating. *Journal of Minerals and Materials Characterization & Engineering*, Vol. 8, No. 4, (December 2009) 249-259, 1539-2511

Sutton, W.H. (1989). Microwave Processing of Ceramic Materials, *Ceramic Bulletin*, Vol. 68, No. 2, (February 2989) 376-386, 0002-7812

Worldsteel Committee on Economic Studies (2009). Total Production of Crude Steel, In: *Steel Statistical Yearbook 2008*, 3-5, Worldsteel Association, Brussels

Worrell, E. (2004). Recycling of Metals, In: *Encyclopedia of Energy*, Cleveland, C., (Ed.), 245-252, Elsevier Inc., 0-12-176485-0, San Diego

Yoshikawa, N.; Ishizuka, E.; Mashiko, K. & Taniguchi, S. (2007). Carbon Reduction Kinetics of NiO by Microwave Heating of the Separated Electric and Magnetic Fields, *Metallurgical and Materials Transactions B*, Vol. 38B, No. 6, (December 2007) 863-868, 1073-5615

15

Microwave Heating for Emolliating and Fracture of Rocks

Aleksander Prokopenko
National Research Nuclear University MEPhI
Russia,
Moscow

1. Introduction

Fracture and emolliating of rocks has been traditionally used in the process of ore dressing. This process has always been given great attention. At the present stage of development of mining and processing industries except energy the efficiency of crushing rock and the increase the degree of extraction the problems arise preservation of individual inclusions crushed material or separation. Therefore, along with traditional methods of mechanical fracture investigates new ways of emolliating (loss of strength) and fracture of rocks based on different physical phenomena. These methods include: electrical (sputter-ion, electrostrictive and piezoelectric), magnetic (magnetostrictive), electromagnetic (laser), sound (impact plastic, ultrasonic), beam (electrons, protons and plasma) and thermal methods of destruction. Common to all methods of destruction is that as a result of such exposure in a material creates mechanical stresses. These stresses exceed fracture stress for most rocks. The stresses can lead to the formation of microcracks in the rock, which leads to a significant emolliating of rocks. Research on the electromagnetic effects on the rocks actively carried out in Russia. There have been many monographs devoted to new methods of destruction (Emelin, 1990), (Novik & Zilbershmitt 1996) and (Gridin & Goncharov, 2009).

Many of these methods can be used only for fracture of certain class of materials or have little capacity and selectivity. Thermal methods of destruction are rather promising. This method is limited to difficulties rapid heating of rocks at great depth because of the small coefficient of thermal conductivity. Conventional heating can not be uniformly and promptly heat the rock. From this point of view only using microwave heating can be achieved volumetric heating of the material at high speed. The majority of rocks are dielectrics with losses, which are well absorbed microwave energy. Microwave discharge on dissemination elements of rock caused by the high electric fields can also lead to its destruction.

Work on the use of microwave energy for the destruction and emolliating of rocks began in the USSR in the late 80's of the last century. It dealt mainly thermomechanical destruction of frozen soils (Nekrasov, 1979) and (Riabec, 1991). Revitalization of the work took place at the end of the 90's of last century, which is associated with the end of the Cold War, when many developments of powerful microwave electronics were unclaimed. In the Soviet Union have been created the most powerful and advanced generators of microwave frequencies, which power and efficiency is significantly faster than foreign. There were many recent trends of microwave energy use: light sources based on microwave discharge, prompt microwave heating for fine destruction of coal, high-temperature microwave heating in the nuclear power, the use of pulsed microwave generators for cleaning and modification of metal surfaces, the use of microwave energy to fracture the kimberlites (Didenko, 2003), (Arkhangelsiy, 1998), (Petrov a, 2002) (Petrov b, 2002). (Didenko, en al., 2005).

Investigations on the emolliating and fracture of rocks by the microwave field are actively carried out abroad. In the last decade in the world issued more than 20 patents on microwave softening ore before grinding and removing the reservoir. Researches are made at universities and commercial organizations in Australia, USA, Canada, UK and Slovenia (Walkiewicz en al., 1991), (Tranquilla, (1998), (Lindroth en al., 1991), (Badhurst et al., 1990) and (Michal et al., 2010). Research is being carried in the direction of emolliating the iron and other ores before extraction or milling in order to ore-dressing. In (Salsman et al., 1996) the selectivity of the effects of microwave energy on the individual components of the rock and the experimentally growth of microcracks due to the thermomechanical stresses are displayed. A significant decrease in power consumption for milling and further separation of ferruginous ore after microwave treatment is shown in (McGill en al. 1988), (Walkiewicz en al., 1991). All researches note the perspectives of using microwave energy for emolliating and fracture of the rock.

The research is needed to find new methods of rocks destruction, which will allow better and more efficiently produce fracture and dressing of the rocks. The advantage of microwave processing of rocks before grinding may be increasing the selectivity of the disclosure of rocks. Preliminary microwave emolliating of rocks is the most effective for fine grinding rocks to improve the efficiency of separation of individual elements or dressing. Reduction of energy consumption and cost of equipment depreciation for fine grinding will contribute to increased formation of microcracks in the microwave processing of rock.

Paper purpose is to describe and analyze existing knowledge's about the processes of fracture and emolliating of rocks at microwave heating. Methods for measuring the dielectric characteristics of rocks, analysis of physical principles loss of strength and fracture of the rock, possible equipment for use in these processes, review of work done and the results of work on the destruction of the kimberlite at prompt microwave heating are presented in the paper. The author hopes that this paper will help to find new application of microwave energy.

2. Dielectric characteristics of rocks in microwave range

When working at microwave frequencies (300 - 30000 MHz) is necessary to know the electrodynamics characteristics of substances with which to work. They define the processes of absorption, reflection and the penetration of electromagnetic waves in a substance. For dielectric properties of minerals are divided into dielectrics (diamond, sulfur, fluorite, feldspar, etc.) and conductors or semiconductors (gold, graphite, pyrite, nickel, etc.). Dielectric loss in the first group is characterized by dissipation factor (dielectric loss tangent tgδ). In the second group the power dissipation parameter is conductivity σ. Power absorbed in the dielectrics and conductors is proportional to an electric field intensity quadrate in them, as well as the permittivity, dissipation factor and conductivity, respectively. The inhomogeneity factor of the microwave heating of minerals within the rock depends on the dielectric properties of minerals. Nonuniform heating of minerals in the rock causing thermomechanical stresses, which lead to emolliating and destruction of the rock.

For rocks it is necessary to consider that the dielectric ε' and the magnetic μ' permeabilities of substance are complex. Typically the magnetic properties of materials in microwave range are feebly expressed $\mu'=1$. It is necessary to consider only the complex permittivity $\varepsilon'=\varepsilon(1-i\cdot tg\delta)$, where ε – permittivity (dielectric constant); $tg\delta$ - dissipation factor. The magnetic properties should be considered in ores containing ferromagnetic materials. Dissipation factor characterizes the losses in the dielectric due to polarization effects of molecules under the influence of electric field and the electronic and ionic conductivity. The majority of rocks belong to the class of imperfect dielectrics, for which the depth δ at which the value of the electric field is reduced to e=2.71828 times makes

$$\delta=\lambda/(\pi\cdot\varepsilon^{0.5}\cdot tg\delta),\qquad\qquad(1)$$

where λ - wavelength of electromagnetic radiation. Thus, knowing the dielectric constant can be judged on uniformity of the rock heating by volume and effectiveness of this heating. Knowledge of the dielectric properties of the rock is necessary in the design and setting up installation for superprompt microwave heating.

Substance	Composition	ε (on axes a, b, c)	$tg\delta$
Not polar minerals			
Diamond	C	5.7	$2\cdot10^{-4}$
Quartz	SiO_2	a 4.5, c 4.65	$2\cdot10^{-4}$
Chalcedony	SiO_2	4	10^{-4}-10^{-3}
Corundum	Al_2O_3	a 11, c 9	$5\cdot10^{-5}$
Rutile	TiO_2	a 88, c 170	$2\cdot10^{-4}$
Periclase	MgO	9.8	$4\cdot10^{-4}$
Halite	NaCl	5.9	$3\cdot10^{-4}$
Sylvite	KCl	4.4	10^{-4}
Fluorite	CaF_2	6.6	10^{-4}
Zincite	ZnO	a 8.2, c 11	10^{-4}-10^{-3}
Mullite	$3Al_2O_3 2SiO_2$	7	10^{-4}-10^{-3}
Olivine	$(Mg, Fe)_2 [SiO_4]$	a 7.2, b 7.6, c 7.0	10^{-3}
Arizona ruby	$Mg_3Al_2 [SiO_4]_3$	11	10^{-3}
Polar minerals			
Topaz	$Al_2 [SiO_4] (F, OH)_2$	a 6.8, b 6.8, c 6.5	10^{-2}-10^{-3}
Tourmaline	B-aluminosilicate	a 6.3, c 7.1	10^{-3}
Talcum	$Mg_3 [Si_4O_{10}] (OH)_2$	5	10^{-2}
Asbestos	$Mg_6 [Si_4O_{10}] (OH)_8$	3.1	10^{-2}
Serpentine	$Mg_6 [Si_4O_{10}] (OH)_8$	6.1	10^{-2}
Muscovite	$KAl_2 [AlSi_3O_{10}] OH$	a 13, c 7	a 0.2, c 10^{-4}
Phlogopite	$KMg_3 [AlSi_3O_{10}] OH$	a 30, c 6	a 0.3, c 10^{-3}
Orthoclase	$K [AlSi_3O_8]$	7	10^{-3} - 0.05
Hornblende	Na, Ca, Mg, Fe-aljumosilikat	5	10^{-2}

Table 1. The dielectric characteristics of minerals with small dissipation factor.

The values of ε and $tg\delta$ rocks are determined by existing mechanisms of polarization. Most crystals have electronic polarization. The crystals, with only e-polarization (with covalent bonds), have little ε (2.0-3.0, and the biggest - 5.7 - diamond) and very low $tg\delta$ order 10^{-4}. In crystals with ionic bonds an additional contribution to ε is made the ionic polarization. Therefore, the dielectric constant is higher: halite - 5.9, fluorite - 6.7, etc. Dielectric loss of these nonpolar mineral remain small until IR range, where the resonance begins variance.

In polar dielectrics there are permanent dipoles. They are capable of reorientation, giving an additional contribution to ε' and losses. The average time of reorientation is usually in the range of 10^{-3}-10^{-9} s. At frequencies of 10^3 - 10^9 Hz are observed relaxation dispersion and maximum of dielectric losses, because the dipoles do not have time to shift between the electric field. Polar minerals have higher losses and their $tg\delta$ usually in the range of 10^{-3}-10^{-2}. The dielectric properties of some materials are presented in (Petrov a, 2002). In Table 1. presents data on the dielectric parameters of polar and nonpolar minerals. Due to lack of data for some polar dielectric characteristics of minerals are presented in the radio frequency range. Power dissipation of the electromagnetic microwave field in dielectrics is directly proportional to the dielectric constant and dielectric loss tangent. Measuring of dielectric characteristics is attracting the considerable interest for specialists.

Mineral	Composition	γ, $1/(Ohm \cdot m)$
Gold	Au	$4.4 \cdot 10^7$
Graphite crystalline	C	a 10^4, c $2.5 \cdot 10^6$
Cuprite	Cu_2O	1-10
Calcocite	Cu_2S	10-30
Covellite	CuS	10^2-10^6
Calcopirite	$CuFeS_2$	10^2-10^5
Haematite	Fe_2O_3	10-10^2
Magnetite	Fe_3O_4	10-10^2
Titaniferous magnetite	$(Fe, Ti)_2O_3$	0.1-1
Chromite	$FeCr_2O_4$	> 1
Cassiterite	SnO_2	10-10^3
Uraninite	UO_2	> 1
Pyrite	FeS_2	10-10^5
Marcasite	FeS_2	10-10^2
Pentlandite	$(Fe, Ni)_9S_8$	> 1
Molibdenite	MoS_2	10-10^3
Blende	ZnS	10^{-5}-10
Galenite	PbS	10^3-10^5
Arsenopirite	FeAsS	0.1-1
Tennantite	$Cu_3 (Sh, As) S_2$	> 1
Nickelite	NiAs	10^5-10^6
Ilmenite	$FeTiO_3$	1-10
Polianite	MgO_2	0,1-1
Manganite	MgO (OH)	> 1
Tantalite	$(Fe, Mn) (Nb, Ta)_2O_6$	> 1

Table 2. Conductance of ore minerals.

In conductive substances significant contribution to the dielectric loss makes their electrical conductivity. The depth δ in such substances is less than a millimetre. Therefore, power dissipation of the electromagnetic field in such minerals is directly proportional to the conductivity. The conductivities of some minerals are given in Table 2.

In reality, the rocks have a complex structure and dielectric parameters are determined by different mechanisms of losses - relaxation and conduction. The majority of rocks are minerals with high losses. It is well known that the loss of conductivity in semiconducting materials greatly increase with increasing temperature, as well as with increasing humidity. In anisotropic crystals the values of ε depend on crystallographic direction. In tables permittivities indicate along different axes. Dielectric characteristics also depend on impurities.

In porous rocks the conductivity, dielectric constant and dissipation factor increase with increasing humidity. This dependence is because of increase of water content in the pores. Water have great values of $\varepsilon=80$ and $tg\delta=0.21$ on 3.0 GHz. In addition, water is a good solvent salt, which leads to an increase in dielectric loss. This relationship is particularly well observed in the study of the dielectric characteristics of kimberlites (Didenko en al., 2008).

Many rocks are heterogeneous materials containing conducting inclusions. Conductivity of inclusion leads to the migration of the dielectric losses. The dissipation factor becomes dependent on frequency. Typically, the maximum migration losses lie in the low frequencies. In the ores with strongly conductive minerals maximum of dissipation factor is in meter or decimetres range. Contribution migration mechanism is proportional to the content of conductive inclusions.

Substance	Composition	ε (on axes a, b, c)	$tg\delta$
Calcite	$CaCO_3$	8	10^{-2}
Magnesite	$MgCO_3$	10.6	$10^{-2} - 10^{-3}$
Dolomite	$CaMg(CO_3)_2$	7.7	10^{-2}
Siderite	$FeCO_3$	7	10^{-2}
Cerussite	$PbCO_3$	23	$10^{-1}-10^{-2}$
Strontianite	$SrCO_3$	7	$10^{-2} - 10^{-3}$
Anglesite	$PbSO_4$	14	10^{-2}
Gypsum	$CaSO_42H_2O$	7.9	10^{-2}
Diaspore	$HAlO_2$	6.2	10^{-2}
Goethite	$HFeO_2$	12	0.1
Limonite	$Fe_2O_3nH_2O$	32	0.1
Jacobsite	$MnFe_2O_4$	15	0.1
Scheelite	$CaWO_4$	a 11, c 9.5	0.1
Powellite	$CaMoO_4$	a 24, c 20	10^{-2}
Wulfenite	$PbMoO_4$	a 30, c 40	0.1
Crocoite	$PbCrO_4$	9.6	0.1
Proustite	Ag_3AsS_3	30	0.1 - 1
Pyrargyrite	Ag_3SbS_3	20	1

Table 3. The dielectric characteristics of minerals with high dissipation factor.

It should be noted that the majority of rocks in the microwave range are dielectrics with losses. The dielectric characteristics of minerals with high losses are presented in Table 3. Characteristics of minerals depend on exterior requirements and deposits. In each specific case it is necessary to measure dielectric characteristics. All data for the tables have been taken from (Petrov b, 2002). These data can be used only for preliminary assessments of the process.

To study the emolliating and fracture of rocks under the influence of electromagnetic microwave field, to determine the effectiveness and to describe processes it is necessary to measure the dielectric characteristics of rock in ranges of frequency and temperature. The information obtained will allow to optimize a processing of ores and maximize energy efficiency, when most of microwave energy is absorbed by the ores.

There are waveguide and resonator methods of measurement of dielectric characteristics, which can be used to measure substances with high dissipation factor. All these methods have their advantages and disadvantages. Which one to choose depends on your hardware and tools, experience and preferences of the researcher. In (Petrov, 1979) and (Petrov a, 2002) for the dielectric characteristics measurement of ore with very high dielectric losses (tg$\delta \approx$ 0.2-1.0) is proposed to use the coaxial waveguide method for measuring line. The advantage of the method is the ability to measure the characteristics in the frequency range 1 GHz on the same equipment.

In the study of fracture processes of kimberlites by superprompt microwave heating (Didenko en al., 2008) preference to resonance method presented in (Esaulkov en al., 2008) was given. Resonance method for measuring the complex dielectric constant, despite the different character of its technical realization, is to study the resonance curves before and after addition of a dielectric in an oscillating circuit. Method allows to determine the real and imaginary parts of permittivity of the dielectric by measuring Q-factor of the circuit and its resonant frequency. The value of ε is determined experimentally from the difference frequency $(f_0 - f_\varepsilon)$. Dependence tgδ from (Q_0, Q_ε) - much more difficult. The real part of complex permittivity is measured more precisely than imaginary. As one of the resonance methods for measuring permittivity of arbitrary shape samples is designed probe method of comparison. The method is based on comparison of resonance frequency bias in the standard resonator with investigated dielectric sample of permittivity ε and identical shape sample of precisely known permittivity ε_p. The method is useful for measuring substance permittivity with different dissipation factor. The modern numerical methods of resonator calculation can also be applied in this method.

Dielectric behaviours of kimberlites were measured by this method using cylindrical cavity with oscillations such as TM010 at resonator frequency 2.8 GHz. Cylindrical discs of plexiglas (ε=2.57; tgδ=0.001), teflon (ε=2.1; tgδ=0.0001) and ebony (ε=2.8 и tgδ=0.007) were used as reference with known characteristics. The kimberlites (ksenotufobrekchy, porphyritic kimberlite, kimberlite breccia autolithic) were investigated at relative humidity of muck from 1 to 8 %. Cylindrical samples (volume 0.2 cm^3) axisymmetric placed in the cavity and electrodynamics characteristics of it were measured. The same were made with standard samples having the same shape and size. Processing the results of measurements were performed using the analytical expressions given in (Didenko en al., 2008) and using numerical methods implemented in the existing software package (Esaulkov en al., 2008). The measurement accuracy of permittivity was 5%, and dissipation factor - 10%. Dielectric characteristics of kimberlites of different types differ in one and half times and they are highly dependent on relative humidity breeds. Averaging of observed data is executed. Kimberlites with a relative humidity of less than 1% is ε=6.5 and tgδ=0.15. Kimberlites with relative humidity of more than 4% - ε=12 and tgδ=0.25.

Thus, knowledge of the dielectric characteristics of rocks will allow to identify the basic mechanisms of its emolliating and destruction, to assess the efficiency of this process and to implement development and optimization of equipment for the processing of rock.

3. Physical principles of rocks distraction

For today there is not comprehensive description of processes of emolliating and fracture of rocks under the influence of microwave electromagnetic fields. Description of the processes of emolliating and fracture of rocks lying at the intersection of sciences as microwave electronics, thermal engineering, geophysics, solid state physics, plasma physics, chemistry. In (Emelin, 1990), (Nekrasov, 1979), (Didenko, 2003), (Petrov a, 2002), (Walkiewicz en al., 1991) and (Michal et al., 2010) are attempted to describe the fracture processes at non-uniform microwave heating. It is assumed that there are two mechanisms of crack formation and fracture of rocks - heat and the discharge. The thermal mechanism is predominant in the processes of destruction and emolliating rocks, as this requires less strength of electromagnetic fields. In experimental investigations discharge mechanism of failure is more difficult to isolate and describe. Only in some cases, it is possible to assume its presence. Thermal mechanism of rocks destruction at non-uniform microwave heating is more brightly allocated. It depends on the type of rock, structure and chemical composition. Cracking and fracture of rocks due to heat of the microwave energy can occur for two reasons:

- Due to the linear expansion of solids, when the stresses K exceed the breaking stress: $\alpha\Delta T = K/E$, where α - coefficient of linear expansion, ΔT - temperature change during heating and E - Young's modulus. In fact, fracture stress depends more complicated from Young's modulus. The rocks are the brecciated structure consisting of clusters. The clusters have different dielectric characteristics and different coefficients of linear expansion. The non-homogeneous microwave heating of the clusters leads to non-uniform tension and thermomechanical stresses arise. The destruction of such a mechanism occurs at the boundaries of clusters.
- Due to the rapid evaporation of water contained in rocks pores and microcracks when the vapor pressures inside the water-filled cavities exceed the breaking stress. Some rocks have micropores that could contain water. The relative humidity of rocks can reach 15%.

At normal thermal heating of rocks also occur thermomechanical stresses. There are a process of evaporation of moisture, decrepitation, a relaxation of residual stresses, polymorphous and phase transitions burning off of organic compounds (Emelin, (1990). The reducing of rocks strength with increasing temperature shows in Fig. 1. Conventional heating is slow heating by thermal conductivity from surface of rock.

Let us consider the mechanism of thermomechanical stresses in rock under microwave exposure. At the expense of dielectric losses there is a volume heating of rock. The increment of rocks temperature is

$$\Delta T = (P \cdot t)/(c \cdot \rho), \tag{2}$$

where P - microwave power absorbed per unit volume of rock; t - exposure time of the field, c - heat capacity, ρ - soil density.

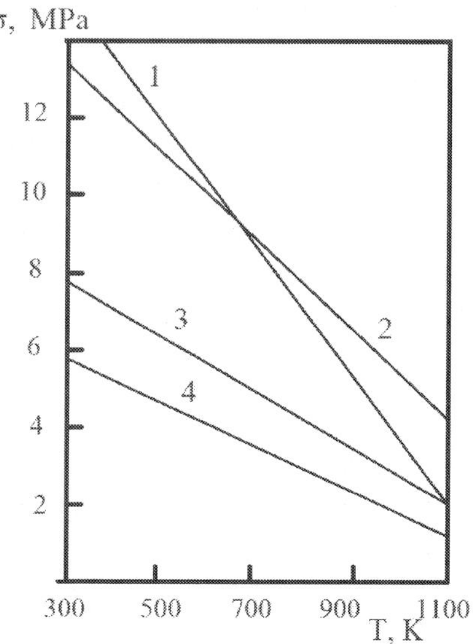

Fig. 1. Dependence of ultimate strength on extension from temperature for samples of ferruginous quartzite (1), gneiss (2), harburgite (3) and dolerite (4).

In (Petrov a, 2002) proposed to place minerals at the time 100 microseconds in the microwave field with frequency 2.45 GHz and strength of 1 MV/m (that is close to brake down limit for air). The heating of different types minerals are following:

quartz, halite, etc. non-polar dielectrics - $\Delta T = 0.005$ K;

felspars, asbestos, etc. polar dielectrics - $\Delta T = 0.5$ K;

goethite, limonite, etc. semiconducting soils - ΔT more than 5 K;

pyrite, hematite, magnetite, etc. conductive soils - ΔT more than 500 K.

Consequently, conducting and semiconducting minerals considerably heat up in the microwave electric field, whereas the dielectric minerals are feebly heated. To heat insulator dielectrics can use higher frequencies, but this prevents decrease in the penetration depth δ of the microwave field. Increasing the radiating time or microwave power leads to significant volumetrical heating of rocks. A significant increasing of radiating time t leads to significant increase in power consumption for rocks emolliating. In the case of loss strength and destruction of the microwave heating tend to prompt heating (temperature excursion) of rocks. The thermomechanical stresses at the temperature excursion exceed the fracture stress. Only in this case, the required energy consumption for heating rocks is decreased.

Let's estimate the energy efficiency of emolliating. To increase the temperature of rock mass m at ΔT needs the energy $W = m \cdot c \cdot \Delta T$. The typical heat capacity of minerals is 700-800 J/(kg·K). To heat soil ton on 100 degrees 75 MJ (20 kW·hour) energy is required. Total energy efficiency (efficiency of microwave generator and absorption factor) is amount to about 30 kWh per ton of rock, which is comparable to energy consumption at purely mechanical crushing.

Microwave technology of emolliating is energetically favorable for rocks containing small amount (up to 10%) of ore (semiconducting or conducting) minerals. Microwave energy heats only the mineral without heating the dielectric mineral tailings. In this case, lose of emolliating is spent no more than 3 kW·h/ton and subsequent operations of crushing and dressing saved - 10-30 kW·h/ton. Consider this process in more detail. Treated rock contains minerals of metals (pyrite, nickel, manganite, etc.) and waste rock (quartz, calcite, granite, etc.). The first minerals have major losses over microwave range, and second - rather small. Therefore, in the microwave field ore minerals are heated, and the waste remains in the beginning cold. The temperature is equalized only after a while. To determine the temperature distribution near the phase boundary it is necessary to solve the dynamic heat equation for two-phase mediums media.   The solution displays that the temperature distribution is described by an error integral with a diffusion length:

$$L = 2 \cdot (\lambda_{cond} \cdot t / (c \cdot \rho))^{0.5}, \tag{3}$$

where λ_{cond} - coefficient of thermal conductivity. If phase 1 absorbs microwave energy, and phase 2 is not, then the diffusion length L characterizes the distance to which the heat from the phase 1 is distributed in the phase 2 at time t. The qualitative picture of temperature distribution in the two-phase medium with flat phase boundaries is displayed in Fig. 2.

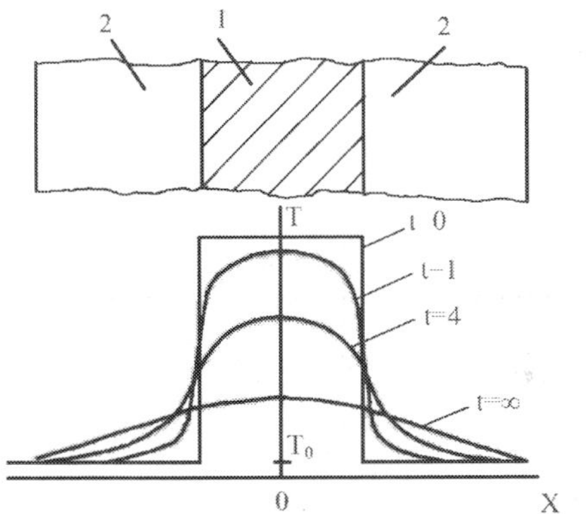

Fig. 2. Temperature allocation in a plate consisting from interior absorbent phase 1 and an outdoor non absorbent phase 2, at instantaneous (t = 0 ms), pulsing (t = 1 ms and 4 ms) and static (t =∞) microwave heating.

For instant heating of phase 1 of the pulse duration t = 0 high temperature is only created in this phase, in phase 2 it is equal to the ambient temperature T_0. In the static case the temperature is aligned on both phases. The distribution of temperature near the border is sharply inhomogeneous at pulsed heating. The temperature gradient increases with decreasing pulse duration.

Substituting (3) the typical numerical values of minerals, we get $L = (10^{-5} \cdot t)^{0.5}$. Consequently, during the 100 ms heat has time to be spread for 1 mm from phase boundary, for 1 ms at 0.1 mm and for 10 microseconds at 10 microns. To reduce heat losses in the waste minerals it is necessary to do the heating time of ore mineral whenever possible small - no more than 1 ms. It is possible only when using a microwave generators with major pulsing power.

The temperature drop at phase boundary under pulsed microwave heating leads to significant thermomechanical stresses caused by differential thermal expansion. Numerical calculation of mechanical stresses in calcite with pyrite inclusions (balls of diameter 0.15 mm) heated by pulses of microwave fields of different duration is presented in (Gridin & Goncharov, 2009). The calculation was made by finite element method and its results are presented in Fig. 3. Mechanical stresses in pyrite are compressive and it has a negative sign. Mechanical stresses in calcite are stretching and it has a positive sign. The temperature have time to equalize the volume when the heating time 1 s (curves 1; Fig. 3.). Emerging thermomechanical stresses are small (in calcite, a maximum of 20 MPa). After irradiation for 40 ms heat is also greatly distributed over both phases. The mechanical stress have maximum of 30 MPa (curves 2; Fig. 3.). In the case of prompt microwave heating (40 microseconds, curves 3; Fig. 3.) the temperature gradient between the phases increase up to 1200 K, and mechanical stress - 360 MPa. At the same time energy consumption at microsecond pulses is of an order of magnitude smaller than the millisecond pulses. This confirms the advantages of prompt heating by high microwave power.

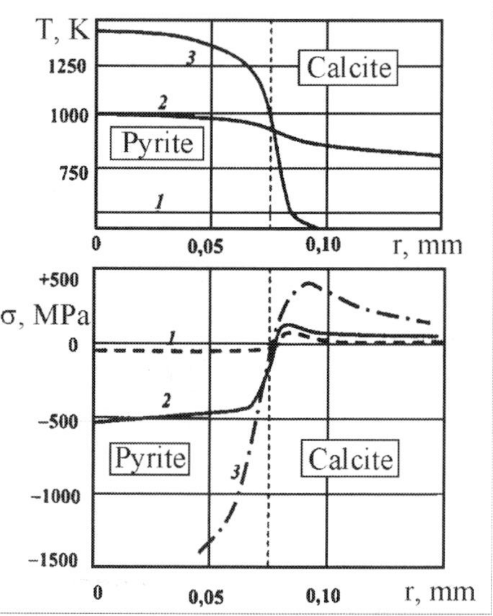

Fig. 3. The radial allocation of temperature T and thermomechanical stress σ the sample of calcite with focus globe turning on of pyrite at microwave heating: **1** - power density of 10^{10} W/m³, duration t=1 s; **2** - power of 10^{12} W/m³, t=40 ms; **3** - 10^{14} W/m³, t=0.04 ms.

Obtained in the calculation of thermomechanical stresses in calcite (20, 30 and 360 MPa) is certainly more than the ultimate strength of rock, (5-15 MPa for granite; 3.5-5.5 MPa for limestone; 2.1-6.9 MPa for sandstone). Therefore, calcite will be cracking around pyrite inclusions. This effect should facilitate and reduce the price of the subsequent grinding and separation.

Consider other mechanism of emolliating and fracture of porous rocks. In such rocks the mechanism of failure due to saturated vapor pressure of water in the rocks pores is predominant. The porosity of rocks ranges from 1 to 15% by volume. The dependences of the dielectric characteristics of porous rocks from its moisture are presented in Table 4. The moisture increase leads to permittivity increase and losses increase in rocks.

Rock	Porosity, %	Permittivity		Resistivity, Ohm	
		Dry	Wet	Dry	Wet
Sandstone	14	5	10	10^4	10^3
Clay slate	20	5	20	10^4	10^3
Granite	2.8	6	7	10^6	10^4
Marble	3	7	8	$3 \cdot 10^8$	10^6
Basalt	4	18		$3 \cdot 10^4$	$2 \cdot 10^3$
Porphyry	3.2			$2 \cdot 10^7$	$3 \cdot 10^3$
Diabase	1.4	8.5		10^3	$6 \cdot 10^4$
Chalkstone	To 20	8	12	10^2	10

Table 4. Change of the dielectric characteristics of porous mucks at moistening.

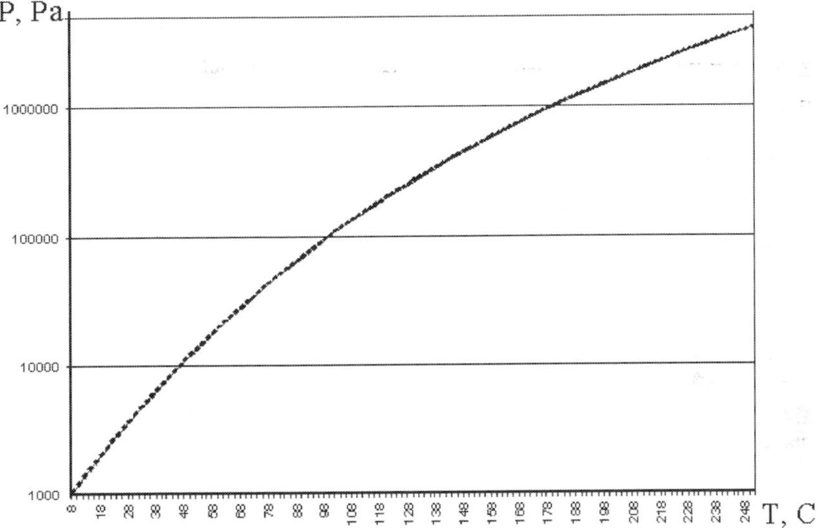

Fig. 4. The pressure of water saturated steams on temperature.

To develop a rigorous theoretical description of rocks destruction due to water pressure in the micropores it is represented the difficult physical and mathematical problem. Therefore, based on empirical data it is possible only to estimate rock fracture processes. Dependence of saturated steams pressure of water on its temperature is given on Fig. 4. Thus, at rock temperature of 100 °C vapour pressure in pores will be 10^5 Pa, and at 200 °C - $2 \cdot 10^6$ Pa. In fact, water temperature and pressure in rocks pores will be much greater because water in pores will better absorb microwave energy than dielectric rocks minerals. To assess the energy efficiency of the rock destruction is required to determine the temperature of beginning of rock destruction. It is necessary to mean for many rocks that the tension fracture stress is less than compression. Thus, vapour pressure 2 MPa may be sufficient to destroy rock. In the experiments (Didenko en al., 2005) and (Didenko en al., 2008) noted that with increasing heating rate the efficiency of destruction increases also. This means that stream has not time to percolate from pores on rocks surface and pressure in pores does not collapse. For some rocks, for example kimberlite, such mechanism of emolliating and fracture will be predominating.

Short pulses of microwave field duration of 10 ns - 1 microsecond with power of 10 MW - 10 GW can also be used to rocks destruction. Energy of these pulses reduces to breakdowns of air gaps between electrodes and rocks pieces. Microwave energy is spent plasma heating of the microwave discharge. This causes formation of shock waves, which may lead to the destruction of portions of rocks. Such experiments on destruction of surface coatings of metals are described in (Novik & Zilbershmitt, 1996). This indicates the possibility of using microwave discharges for destruction of rocks. In some cases the discharge mechanism of fracture can appear more favourable energy, than thermal.

Thus, processes of emolliating and fracture of rocks in microwave electromagnetic field are determined.

4. The equipment for rocks destruction

Establish of effective equipment for emolliating and fracture of rocks is complicated, but completely realized the problem. The technologies of microwave processing of various materials have already been worked out. They are applied in industry (vulcanization of rubber, sintering of ceramics, wood drying, the synthesis of new materials, etc.), agriculture (disinfection of grain) and just at home (microwave ovens). The methods and the approaches for creating energy-efficient systems that meet the requirements for electromagnetic compatibility are already developed (Didenko, 2003), (Arkhangelsiy, 1998). In mining there are the features associated with large tonnage, severe conditions of equipment requirements, minimum energy consumption and major life equipment expectancy. The cyclic duty is not comprehensible to the industrial equipment. Loading and unloading operations of processed material reduce productivity. Industrial equipment requires a continuous operating mode. At designing the installations it is required to consider matching condition for the microwave generator working at chamber with rock. The mismatch of the generator with working chamber leads to significant energy losses, failure of operation and even failure of high-power microwave generators.

Dielectric parameters of rocks are changed at heating up. In some minerals the changes of phase or chemical transformations are observed at heating that also modifies their dielectric properties. In the dump rocks at microwave heating there are moisture reduction and decrease of permittivity and losses. Any changes at the rock dielectric parameters at microwave heating change the working camera input impedance that leads to mismatch of microwave line and microwave generator. To ensure constant coordination of microwave generator is necessary to use continuous mode of rocks loading into the chamber at constant speed. Such mode regime can be achieved by continuously spilling out rock through the working chamber with constant speed. In this case, the input impedance of working camera will be determined by average parameters of cold soil on its inlet and hot on an exit. Average input impedance of working camera will not change during processing and the generator will always work mode of maximum power.

At emolliating and fracture of rock choose appropriate types of installations: radiation type for processing of rock massif before the mechanical destruction and spilling type for irradiating of rock before mechanical grinding. The system of antennas creating high microwave power flux density in the rock is applied at radiation type. Rocks is been transporting across the active zone of working chamber in installation of spilling type. Rock is heated in active zone.

From the point of view of efficiency it is desirable that microwave generator operated in continuous mode, but created an impulse heating on piece of rock. It can be achieved in the working chamber spilling type, when pieces of pre-crushing rocks wake up in the active zone of the working chamber. If the pieces of rock fall from height h and holding path in active zone is l, the action time is determined by the relation:

$$t = l / (2 \cdot g \cdot h)^{0.5}, \tag{4}$$

where g - a free fall acceleration, $l \ll h$. So, at $l = 15$ mm and $h = 3$ m processing time is only 2 milliseconds. Approximate parameters of such working cabinets of spilling type are presented in the patent (Kingman, 2002). To increase the action time of the rock in working chamber in the active zone set a dielectric pipe with small losses at certain angle to the horizon. The moving pieces of rock are transported in pipe. Transporting speed of rock is determined by angle of inclination of pipe and rock friction coefficient about pipe wall.

In developing the working chambers of emolliating and fracture of rocks it is necessary to have maximum transfer efficiency of microwave power to heating rock. The loss power in dielectric P_d fractionally or completely filling the working chamber is recorded

$$P_d = \frac{\sigma}{2} \int_{V_d} \left| \vec{E}_d \right|^2 dV = \omega_0 \varepsilon' tg\delta \int_{V_d} \left| \vec{E}_d \right|^2 dV , \tag{5}$$

where - V_d volume of dielectric; $\varepsilon' = \varepsilon\varepsilon_0$ - permittivity of dielectric; $tg\delta$ - dissipation factor and E_d electric intensity in a dielectric. Therefore for magnification of transfer efficiency of microwave power the heated rock it is necessary to increase an electric intensity in rock or to make maximum filling of working chamber.

Technological installation for emolliating and fracture of rocks (Fig. 5.) consists of a transmission line, a powerful generator and a working chamber with spilling system of continuous supply and shipment of rocks. Size of rock fragments is determined by the following criteria: necessity to maintain the ecological safety of microwave installation at acting spilling system, field penetration depth into rock should be comparable to size of rocks pieces, homogeneity of electric intensity in piece with its location at the maximum electric field. On frequency of 2.45 GHz the maximum size of pieces should not exceed 30 mm. Working chamber for destruction of rocks can be irradiating (Fig. 5.a), waveguide (Fig. 5.b) and resonator (Fig. 5.c). Design of these cameras should ensure maximum efficiency of microwave power transmission processed rock.

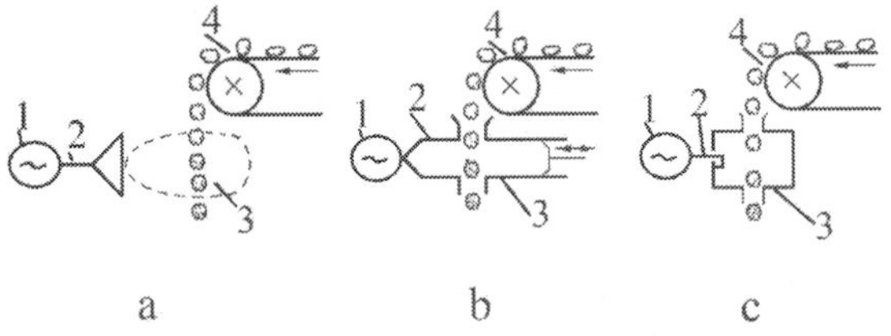

Fig. 5. Installations for processing rocks by microwave field: 1 - microwave generator; 2 - feeder; 3 - working chamber; 4 - spilling system.

Irradiating working chambers represent a system of antenna radiators matching with microwave line. Irradiating working chambers can be used both for processing of rock before its extraction and for loss of strength and fracture in spilling mode. In this chamber the stream of rock impinging in free space is irradiated by an antenna, such as horn (Fig. 5.a). Such a simple design of the installation requires the difficult protection against microwave radiation. In such working chamber it is possible to irradiate uniformly materials. It is impossible to get big efficiency of microwave power transmission to rock because of power losses in the space.

Waveguide working chambers operate in traveling wave mode. Waveguide working chambers is sufficiently broadband. They may be in the form of waveguide inserts and matched loads. The matched loads efficiency of microwave power transmission to rocks can reach 80% at wide range. Matched loads (Fig. 5.b) can use different waveguides working on dominant mode and active zone is placed in the maximum electric field. Matched loads work with large loss dielectrics and can have a system of adjustment of waveguide pistons.

For effective heating of samples of small volumes with low dielectric losses it is preferable to use resonator working chamber.

Resonator chambers provide high electric intensity for installed microwave power (Fig. 5.c). They should be used to influence the small pieces of rock with low dielectric losses. In the resonator chambers transfer coefficient of microwave energy to rock may reach 85-90%. Resonators are singlemode and multimode. Multimode cavities with introduction of processed material begin to work on new oscillations mode with self-resonant frequencies close to the frequency of microwave generator. Unfortunately, in these cavities can be observed unevenness of microwave heating of rocks and transfer coefficient of microwave energy to rock does not exceed 70%. However, resonator working chambers are the most promising for microwave heating for emolliating and fracture of rocks.

Waveguide and resonator working chambers can be made on the basis of rectangular and circular waveguides, strip line and coaxial transmission lines, unclosed and enclosed slowing structures. Using modern software packages numerical simulation of microwave devices, it is possible to develop working chamber of any construction. Installations for microwave heating for emolliating and fracture of rocks also include matching and trimmer microwave units, security features microwave components from the destructive and polluting effects of the rock and protection from microwave radiation. Microwave field shielding in places loading and unloading rocks is carried out with the help of below-cutoff waveguides and throttle elements.

Thus, at the present stage of development of microwave electronics it is possible to create effective systems for rocks destruction.

5. Application of microwave heating for emolliating and fracture

The application of microwave energy for emolliating and fracture of rocks is considered. The major cycle of research on microwave energy use has been executed in Russia. In other countries also carried out work on the loss of rocks strength. Great attention was always paid to the questions of mineral extraction and raw dressing. In this chapter will describe the experimental research results of microwave action for thawing of frozen soils, emolliating of various ores, extraction of hard gold, a dispersion of hard coals and fracture of kimberlites.

Great volume of research on microwave energy application to the emolliating of frozen soils is executed in Russia. The essence of a problem consists that strength of frozen soils is very high and increases with decreasing temperature. At dynamic loads and temperature of minus 30°C the compressive strength of frozen sand attains 15 MPa, and frozen clay - 75 MPa which is higher than that of many rocks and minerals. Therefore mining of the frozen soils in a winter time and the permafrost usual excavation technics is possible only with prior thawing. The microwave heating for loss of frozen soils strength is actual.

Works in this direction were intensively until 1990 (Nekrasov, 1979) and (Riabec, 1991). The complex for emolliating of the frozen soil during the digging of trenches and the installation for the wells expansion in the rocks have been created at the Leningrad Mining Institute (Russia). The complex consists of microwave installation and the earth-moving machine. The depth of mining for one pass is 0.3-0.5 m at power inputs of 5-6 kW·hour/m³. Microwave equipment has microwave power of 50 kW at a frequency of 915 MHz. The installation for the wells expansion in the rocks emolliates by well directed stream of the microwave field and breaks rocks by cutting tool. Performances of installation as follows: the frequency of 2.45 MHz, microwave power 5 kW; diameter hole 800 mm; productivity of 4-10 m/hour depending on the a ground.

The study of dielectric characteristics of frozen soils depending on microwave frequency and the development of installations for emolliating were occupied in Yakutsk Scientific Centre (Russia). It was showed advantage of switching to the frequency of 430 MHz for frozen soils. The microwave installation for layering sinking pits in the frozen soil with depth to 1 m, power consumption of 30 kW·hour/m³ and productivity 0.85 m³/hour have been created. Comparison of microwave thawing and emolliating of frozen soils indicates advantages in operating velocity.

Let's execute survey of research on the microwave destruction of rock. In the patent (Maksimenko et al., 1977) is proposed an original method of rock destruction by irradiation of two microwave generators of Fig. 6. At first a rock massif 3 is irradiated with small microwave density (150-300 W/cm²) from the generator 1 to the formation of the heat trace. Then in a perpendicular direction the massif is irradiated waves of higher density (300-5000 W/cm²) from the generator 2. The first generator creates in the rock heated zone 4 with

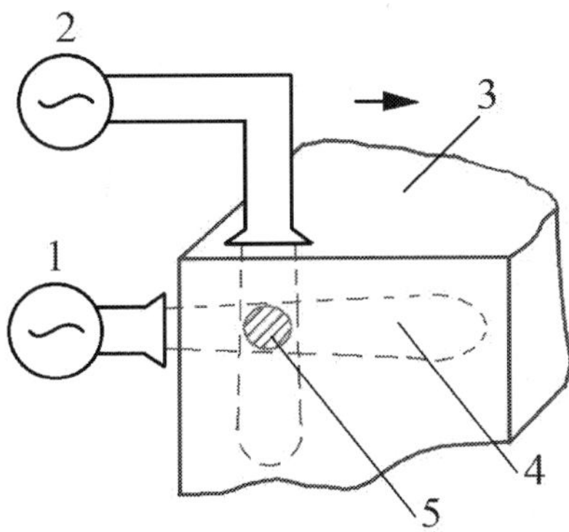

Fig. 6. The advantageous process of making emolliating channels in rocks with two microwave generators.

increased dissipation factor. Microwave power from generator 2 is absorbed mainly in the area of intersection of irradiation 5. The rapid heating zones 5 leads to thermal expansion, phase transitions and the formation of gas phase in this area, which leads to the destruction of rocks. Translocating the antennas it is possible to create the channel of broken rocks. Crystalline shale, granites, sandstones and other rooks were destroyed with use of generators at frequency of 2.45 GHz.

Similar theoretical and experimental researches for sandstone fracture were studied in (Obrazcov, 1976) and (Krasnovskij & Uvarov, 1991). At irradiation density of 300 W/cm² the destruction occurs in the form of periodic separation husks thickness of 2-4 cm. The husks thickness is decrease and the fracture frequency is increase then the density of power flux was increased. Explosive fracture of strong sandstones with damp from 0.1 to 2 % volume was observed at densities of 1000-3000 W/cm². Damp essentially influences both the energy intensity of explosive destruction and the critical density of microwave power. Explosive fracture of various sandstones by a microwave field is studied at the radiant density of 500 W/cm² from the generator by power of 50 kW. The slot by depth 5-20 cm is generated at beam translocating travel in block sandstone. Power consumption amounted to 80-200 kW·hour/m³. Fracture of strong sandstones at radiation by millimeter waves (3.5 mm) with densities of pulsing energy to 40 kW/cm² and pulse duration of 30 ms is featured in (Krasnovskij & Uvarov, 1991). Power consumption of fracture process at major densities is 240-250 J/cm³, whereas at smaller (a few kW/cm²) - 360-700 J/cm³. Hence, the adiabatic regime of a heating a high power allows to reduce power consumption of microwave destruction essentially. The density of microwave field in 20-40 kW/cm² produces the surface electric breakdown of the irradiated rock. In (Krasnovskij & Uvarov, 1991) the possibility of practical use combined plasma-wave destruction is viewed.

Comprehensive theoretical and experimental investigations of emolliating by electromagnetic fields conducted in Moscow Mining Institute (Russia) (Gridin & Goncharov, 2009). Samples of iron ore (about 45% magnetite, 10% hematite and 45% quartz) with dimensions 15×15×30 mm had been irradiated from the generator with power 0.5 kW at frequency 2.45 GHz. Their flour milling and dressing were studied. Because of small microwave power heating up of the sample was slow - to 200°C for 50 seconds with, 450 °C for 3 mines and to 900 °C for 10 minutes. At a heating to a redness explosive fracture was observed. The scraps which boundary transited on mining stratums separated from samples. Such fracture is caused by presence in iron ore of abound water and its evaporation. At the cooling of the heated samples in water the cracking was faster and often leads to their destruction. According to the made calculations the temperature increase at one degree leads to thermomechanical stresses to the magnetite grains - tensile 0.17-0.23 MPa, the hematite grains - tensile 0.35-0.40 MPa and the quartz grains - compressive 0.21-0.27 MPa. At temperature spring on 100 degrees thermomechanical stresses should exceed ultimate strength of minerals. The appearance of cracks coming mainly along the grain boundaries is confirmed by structural studies of processed ore. The flour milling and mesh analysis of the samples processed by the microwave heating testifies to loss of ore strength and reducing of power consumption. Dressing of samples by magnetic separation in a strong magnetic field is found to improve the quality concentrate in comparison with raw ore. The yield of concentrate increases at 11.6%. The greatest effect is achieved on the samples heated to 200 °C, instead of to 900 °C. Researches (Kondrashov & Moskalev, 1971) and (Chelyshkina & Korobsky, 1988) have convincingly proved necessity of selection of an optimum regime of microwave processing for each rocks type.

Researches on the emolliating and destruction of rock are carried out in countries where there is a developed mineral industry. In (Lindroth en al., 1991) was patented a method for removing the formation of solid rock with strength of 170 MPa by microwave irradiating rocks in front of the cutting tool. As a result of thermomechanical stresses in the rock formation of microcracks and discontinuities. That leads to the decrease of fracture energy consumption.

In the patent (Tranquilla, 1998) company EMR Microwave Technology Corporation is proposed to process the ore or concentrate by spilling through a rectangular microwave cavity in the maximum of electric intensity. Preferable parameters of generator are power 10-50 kW and frequency of 915 MHz or 2.45 GHz. The resonator Q-factor should be within 1000-25000. With these parameters the processing time is no more than 6 seconds.

In (Walkiewicz en. al 1991). the flour milling of iron ore after processing in microwave field was explored. Samples of ore sizes of 3 cm contained hematite, magnetite and goethite. The generator with power of 3 kW and frequency of 2.45 GHz heated them up to temperature 940 °C. By means of a scanning electronic microscope the change of microscopic structure of samples at their processing was studied. The flour milling and screen analysis was made also. Microcracks at the grain boundaries are visible on a photomicrography of the processing samples. The research noted that energy consumption for heating is reduced with increasing power of the magnetron to 6 kW.

In research (Salsman et al., 1996) correctness of calculation of microwave heating of ores was mustered by experiment with the artificial composites containing a quartz and 1-7% chalcopyrite absorber. The composite heated up in microwave oven (1 kW) during 120 seconds. The energy analysis microwave loss of strength was presented. The thermomechanical stresses created in ore are 50 MPa (temperature jump at 200 degrees) if the processed ore contains 2.5% by weight of the absorbing minerals. Power consumption in pulsed mode of processing was only 0.8 kW·hour/t. Thus, processing of rocks for the purpose of a loss of strength by powerful microwave impulses allows to reduce power consumption.

One of problems that can be solved in the processing of the rock it is more effective dressing and extraction of minerals. In recent years, the world scientific community became interested in the problem of improving the gold extraction from the tailings of old minings.

However, the extraction of gold from the waste pile is problematic. In addition to the low percentage of gold content - (1-3 g/ton), it is difficult to extract. Gold corpuscles particles in this material have a size from hundredths to tenths of a micrometer and are concluded in other minerals (mainly pyrite and arsenopyrite). Therefore they are not extracted by a usual method at use of cyanides solutions. For extraction of this gold it is necessary to uncover or destroy shell of sulfides before cyanation. Current production technology is very energy intensive and not environmentally friendly. Using the thermomechanical stress at the prompt microwave heating and the microwave discharges it is possible to release dust grains of gold from sulfide shells. The sulfide shell which is absorbed microwave energy is rapid heated and cracked by the energy of microwave pulse. As a result metal becomes accessible to extraction by solutions. Microwave processing allows to raise degree of disclosure to increment extraction of precious metals, to use for their extraction poor rock. Besides, fine grinding of a concentrate to tens micrometers is not needed at microwave processing, there is enough sizes 0.5-2 mm.

Work to extraction of gold using microwave field has been begun in Canada (Haque, 1987). At a usual heating the concentrate starts to decompose at temperature 550 °C. At microwave heating that is at 420 °C. Extraction of gold and silver by cyanidation grows depending on microwave power. In fact, the complete extraction of gold (98%) occurs at powers of 5-6 kW. EMR Microwave Technology Corp. (Murray, 1998) was working on manufacturing equipment for microwave processing and optimization technology. It had been shown that microwave heating allows to reduce costs of preparing base ore to cyanation. Thus, the processing of 200 tons of concentrate per day by the microwave heating costs was $ 3.84 million, whereas the normal firing - $ 6.9 million.

In Russia has also carried out active research on the microwave energy application in the process of gold extraction (Lunin et al., 1997) and (Kolesnik et al., 2000). In the studies note that the microwave heating of raw auric concentrate causes the appearance of thermal stresses and the formation of intergranular microcracks. Extraction of gold from the concentrate by the scheme "microwave firing up to 360 °C - cyanation" was 94.8%, that on 8.3 % above in comparison with normal firing scheme.

In research (Chanturija et al., 1999) it is offered to process raw auric concentrate by the short electromagnetic impulses which spectrum covers the microwave range. In the same study display the prospects of nano-pulse radiation of raw auric concentrate. It is offered to explore agency of an electric intensity on formations of pin-holes of a disruption and on fractures of sulphidic shells.

In (Boriskov et al., 2000) presents experimental data on the irradiation of the tail concentrate by nanosecond pulses. It is shown that with increase in the number of pulses the degree of gold extraction grows at the subsequent cyanation. Thus, pulsed radiation also may be used for a loss of strength of rocks.

Research of applications microwave energy in mining industry carried out in National Research Nuclear University MEPhI (NRNU MEPhI) (Russia) (Didenko, 2003), (Didenko en al., 2005), (Didenko et al., 2008) and (Prokopenko & Zverev, 2008) Superprompt microwave heating was used to the fine destruction of coals and the fracture of kimberlites. It was showed that during rapid heating of coals up to 120 °C is emolliating and distraction of coal (Didenko, 2003). The mechanism of dispersion is based on the extremely prompt heating of water located in the pores of rocks. Heating occurs in single-mode resonator working chambers at a frequency of 2.45 GHz. At such process it is possible to gain coil corpuscles with a size 10 microns. Fracture goes on boundaries of clusters that give the chance to clean up ash-forming impurities from coil dust. Estimates show that by using the microwave generator 5 kW it is possible to disperse 100 kg of coal. Stripped of ashes and impurities of sulphur the coal dust is supposed to be used for making of solar oil analogues without chemical methods.

In NRNU MEPhI together with JST ALROSA (Russia) in 2006-2008 is executed extensive amount of studies the processes of kimberlites fracture and development of the equipment for analysis of these processes. Studies have shown the promise of this destruction method. Some research results are presented in (Didenko en al., 2005), (Didenko en al., 2008). and (Prokopenko & Zverev, 2008). In the next chapter the research results of kimberlite fracture by microwave energy will be presented in more detail.

6. Microwave heating for fracture of kimberlite

One of the main problems in the diamond-mining industry is preservation of rough diamonds extraction. Up to 20% of extracted natural diamonds have faults of facets gained as a result of extraction and dressing of kimberlite rocks. The faults obtained on the wet self-grinding mill significantly reduce the cost of rough diamonds. New methods of safety processing are demanded.

In 2006-2008 in NRNU MEPI supported by Joint Stock Company ALROSA made series of researches (Didenko en al., 2005) on the processes of kimberlites fracture. The purposes of this study were to determine the technological parameters of kimberlites fracture, assessing the applicability and effectiveness of the new method of fracture, development of theoretical bases for creating spilling type installations, experimental study of operation of manufactured installation and make recommendations.

Kimberlite is the magmatic ultrabasic brecciate rock representing a carbonate-serpentine rock with insignificant amounts olivine, pirosseno, granite, ilmenite, phlogopite, apatite, magnetite and other minerals. Kimberlite represents assemblage of various dielectrics: a carbonate-calcium, silicates and oxides of iron, aluminium, magnesium, chrome, titanium, etc. Bound water contains only in a phlogopite (about 10 %) and consequently does not influence fracture process as the phlogopite content is insignificant. The brecciate structure of kimberlites means presence in it of pores with free water.

Studies have displayed that the mechanism of kimberlites fracture mixed in which will dominate destruction due to overheating of water in the pores of the rock. The study of dielectric and thermal characteristics kimberlite rocks is executed. At examination the kimberlite of following types was used: ksenotufobrekchy, porphyritic kimberlite, kimberlite breccia autolithic. The question of water content of kimberlites should pay special attention. The separation procedure of kimberlite on conditionally "dry" and artificial "wet" is produced. The results of density change indicate rocks porosity. The "wet" kimberlite contains from 4 to 8 percent water. For a more complete investigation it is interesting to study the natural moisture of kimberlite extracted immediately from opencast.

The rock fracture mechanism due to the saturated steam in the rocks pores is predominant. The advantage of this destruction method for the "wet" rocks has been shown on numerous experiments. To develop the strict theoretical exposition of these processes it is represented the difficult physical and mathematical problem. Therefore, based on empirical data, it is possible to execute only qualitative assessment of the destruction processes. Knowing temperature of an initiation of a fracture (Fig. 4.) it is possible to determine the stresses of rocks destruction. To describe the processes kimberlites fracture it is required to determine the destruction temperature and the desired heating rate for effective destroy.

Dielectric characteristics of kimberlite were determined by the probe method of comparison presented above. The dielectric for average kimberlite characteristics are gained: the "dry" kimberlite (with a relative humidity of less than 1%) is $\varepsilon = 6.5$ and $tg\delta = 0.15$; "wet" kimberlite (with a relative humidity of more than 4%) - $\varepsilon = 12$ and $tg\delta = 0.25$. The measurements were made at a frequency of 2.8 GHz. Dielectric characteristics of kimberlite slightly dependent on frequency therefore on frequency of 2.45 MHz they remain practically without changes. To ensure uniformity of heating of kimberlite pieces size should be comparable to the penetration depth δ. At frequency of 2.45 GHz, according to (1), the penetration depth will be 10.2 cm in the "dry" piece and 4.5 cm in the "wet" kimberlite. Thus, fragments with an average size of 3 cm heat up almost microwave field and a large dissipation factor promotes significant losses in the rock.

To determine the energy characteristics of kimberlites fracture processes it is necessary to execute estimates of thermal rate of a prompt microwave heating. The heat exchange processes are important in determining the efficiency of kimberlites destruction. Microwave heating power of mass $\dot{m}_{kimb}$ per second from the initial temperature T_0 to a temperature of destruction T_d is defined by well-known relation

$$\eta_p P_g = P_{kimb} = \dot{m}_{kimb} \cdot C_k \cdot (T_d - T_0), \tag{6}$$

where C_k - specific heat of kimberlite; $P_{kimb} = \eta_p \cdot P_g$; $\dot{m}_{kimb}$ - mass of kimberlite destroyed in unit of time; η_p - transfer coefficient of microwave energy to kimberlites; P_g - power microwave generator. Expression (6) does not consider a heat transfer from heated kimberlite in free space. Since heating rate is significant these thermal losses can be neglected. The most important question to determine the energy efficiency of the fracture process is the temperature of kimberlite fracture.

For power estimations of microwave heating to temperature of kimberlite destruction and of heating rate of the kimberlite samples it is required to know the specific heat of kimberlite. Conducted search of the specific heat of kimberlites in the specialized literature has not yielded any results. We executed measuring of specific heat of kimberlites by standard laboratory calorimetric methods. Measuring had estimate character with an accuracy of about 15%. For the evaluation of destruction processes the average value of specific heat is $C_k=1.1$ kJ/(kg·K). Averaging was carried out by type of kimberlite and its moisture content. The measured specific heat of kimberlites in different samples differs no more than on 30%. The measurements of the specific heat for "wet" kimberlite differ much by not more than 5%. Accepted value $C_k=1.1$ kJ/(kg·K) is not strongly different from specific heat of similar materials (basalt - 0.85 kJ/(kg·K); granite - 0.65 kJ/(kg·K); volcanic lava - 0.84 kJ/(kg·K) at 20 °C).

Magnetron generator with a frequency of 2.45 GHz and microwave power 600 W was used to research fracture of kimberlites. The high-voltage power supply of magnetron allowed to adjust anode current and was powered by a single-phase AC 220 V, 50 Hz. Control of the microwave power was carried by the anode current of magnetron. The tunable cylindrical resonator with TE111 mode was used as working chamber for destruction of kimberlite. This cavity has a large value of the electric field E_r, transfer efficiency of microwave power to the heated rock, acceptable size and convenient system of resonant frequency adjustment. Cavity had a magnetic coupling element and the flange.

Kimberlite was introduced into the cavity through the sliding cover and placed in the maximum electric intensity E_r in the centre on the wall cavity. In a resonator upper there was the hole with below-cutoff waveguide for temperature performances measuring of kimberlites fracture processes. On lateral cylindrical wall of the resonator there was the below-cutoff grid for visual fixation of destruction.

The experimental installation consists of the following elements: power supply, a magnetron, a rectangular waveguide adapter 90×45 mm, the phase shifter and tunable cylindrical cavity connected. This installation allows to heat up the rocks sample (2-4.5 cm^3) to high temperatures for a few seconds and to investigate impact prompt heating processes of kimberlites destruction.

Measuring of the temperature and a fracture time was executed, fractional composition of the destroyed material was spotted and searching of an optimum regime of fracture was executed on the installation featured above. The general scheme of the experiments is presented at Fig. 7. Pyrometric methods were applied at temperature measuring of kimberlites fracture. The pyrometer placed so that its optical system has been interfaced to the resonator hole and the measuring square of spot has made ~ 5 mm^2 on the heated sample.

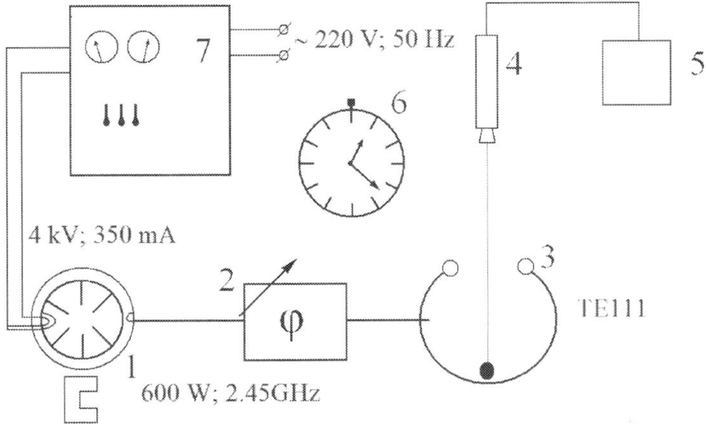

Fig. 7. The experimental setup to determine temperature characteristics of kimberlite destruction: 1 – magnetron; 2 – phase changer; 3 – cavity with kimberlite; 4 – pyrometer; 5 – display block of pyrometer; 6 – stopwatch; 7 – power supply.

It was produced more than 50 samples with a typical size 2.4 cm³ of the samples kimberlite three types. Their mass was measured. In all experiments sample was heated up during the 2-5 seconds to partial or total destruction accompanied by a strong bang. Partial destruction occurs because of a scatter sample fragments at the beginning of the destruction from zone with intense heating. Temperature and time of the destruction was fixed at the beginning of the destruction process. The microwave generator was switched off on termination of the noise effects. Processed kimberlite extracted from the working chamber. Thus "dry" kimberlite were crushed to pieces of larger size.

The experiments have displayed that efficient processes for the kimberlites destruction begin at microwave heating to 150 °C and at rate of temperature rise of at least 40 deg/sec. Pressure of saturated steams, according Fig. 4, in rocks pores attained 5×10^5 Pa. Such fracturing temperature at times is less than the temperature of diamond thermal destruction. Experiments on fractional analysis of kimberlite after fracture were executed to describe the technological parameters of the kimberlites fracture process. Kimberlite samples were weighed before fracture with accuracy 5 mg. Fracture is executed under the standard scheme with the control for temperature and time responses. The destroyed samples were removed from the working chamber and carefully separated on three fractions. Each fraction was carefully weighed. Fractions were parted on the characteristic particle size: first (I) - equally or less than 1 mm; second (II) - more than 1 mm and less 2.5 mm; the third (III) - equally or more than 2.5 mm. As a result of examinations the relative fractional composition on weight was the following: fraction I - from 15 to 20 %; fraction II - from 40 to 60 %; fraction III - to 40 %. The fractional composition depend from following factors: the growth rate of the temperature, humidity and structural characteristics of kimberlite, as well as the characteristics and design features of the installation for kimberlites destruction.

Let's execute an energy and technological estimate of kimberlites fracture processes. Assuming the initial temperature equal to T_0 = 20 °C and the final T_d = 150 °C, specific heat of kimberlite C_k=1.1 kJ/(kg·K) and microwave power of 600 W consumed by kimberlites we will gain productivity 4 g/s (240 g/min or 14.4 kg/hour). This theoretical result is close to the experimental value of productivity 2.5 g/sec, which is obtained by research installation. For definition of the complete productivity it is necessary to consider a transfer efficiency of the microwave power to rock in the cavity, which is at 85%, the frequency instability of the generator and a definition of temperature and heat destruction of kimberlites. All these factors will directly affect the performance.

Productivity of installation is of the order 1 t/hour if to use a generator of microwave power of 50 kW and transfer efficiency of microwave energy to kimberlite 80%. For an intrusion of new fracture method in commercial scale it is required installations with productivity ~ 20 t/hour. Productivity growth is limited by the power of existing microwave generators. Besides, the thermal method of fracture was more power intensity than fracture in the wet self-grinding mills. At microwave method the energy consumption per ton of kimberlite fracture increases several times.

However, this method may be more promising and enhance the safety of extracted diamonds. This assumption is based on the fact that linear expansion coefficient of diamond is about an order less than the linear expansion coefficient of kimberlite. The face of diamond will not be pressured the heating and fracturing kimberlite. Diamond is highly nonuniform impregnation. At such method fracture it can appear that the cracking of rock will occur on diamond facets. It is possible that there will be no need for further destruction of the remaining rocks to the size of 1-2 mm. In this case efficiency of installation is considerably risen. Such method of destruction may be able to get the diamonds without any changes that will significantly increase their cost.

Small productivity of designed installation does not allow to investigate the applicability of this method in industrial scale. Investigation of fracture should be performed on a more productive installation, where the kimberlite heating will be carried out during its spilling through the cavity (Prokopenko & Zverev, 2008).

Experimental installation of resonator type with microwave power P=5 kW and a frequency f=2.45 GHz is developed in NRNU MEPhI. Its productivity is not less than 80 kg/hour. The working camera of installation is used cylindrical cavity with TM020 mode. The resonance wavelength of the cavity does not depend on the resonator length and is given $\lambda=2\pi R/5.521$, where R - radius of the resonator; $\lambda=122.3$ mm at a frequency f = 2.45 GHz. On the resonator axis in the first maximum of an electric field the ceramic pipe by caliber of 28 mm is disposed. The resonator had a length of 270 mm and its radius was chosen taking into account its frequency displacement by pipe and kimberlite. Electrodynamics characteristics of cavity are counted and optimized. Designed cavity has transfer efficiency of microwave energy to kimberlite from 75 to 80 % at volume filling of cavity by kimberlite from 0.04 to 0.08 %. The resonator axis is canted on an angle 45°±5° to horizon for sliding kimberlite on the dielectric pipe by gravity. For greater uniformity of kimberlite spilling pipe can rotate around its own axis with changeable velocity and to be scavenged by a stream of warm air. The below-cutoff waveguides on the outputs of ceramic pipe ensure the environmental safety of microwave radiation.

Developed installation based on the magnetron microwave generator 5 kW consists of the following elements: power supply of magnetron, magnetron, plate shifters, ferrite circulator, mismatched segment and the working chamber. Installation is executed under the plan of a frequency control of a magnetron by resonator working chamber. Productivity of this installation would be around 90 kg/hour.

Experimental studies of this installation were performed. Fife or six rubbles of kimberlite with a total volume 16 cm^3 was loaded into dielectric pipe. The coefficient of volume filling of cavity by kimberlite was 0.2%. The effective process of crushing kimberlite rubbles accompanied by a loud bang was started on the microwave power 4.4 kW. The portion of rubble was fractured no more than 4 seconds. With initiation of fracture the new portion of kimberlite rubbles was loaded. The grain size of crushed kimberlite were from (0.1 - 0.5) mm to 5.0 mm. The relative weight of small fractions was not less than 45 % from the total mass of the crushed material. Unfortunately, experiments were conducted on kimberlite without diamonds. There was no possibility to estimate effect preservation of diamonds facets at microwave heating.

Designed installation on spilling fracture of kimberlite was served as a prototype for making of the resonator installation working in geological laboratory JSC ALROSA. Now in geological laboratory the microwave heating processes for fracture of kimberlites are explored in more details.

7. Conclusion

Recent trend of power microwave engineering is presented. Microwave heating for emolliating and fracture of rocks allows to reduce energy consumption in the mining industry, help to convert collected waste and dumps of mining and processing enterprise and protect environment. The survey works by microwave heating of rocks confirmed the prospects of using this method. Installations for rocks destruction already find the application of and processing industry.

Presented in the paper model of the microcracks growth and rocks fracture by prompt and uniform microwave heating explains the influence mechanism and allows to predict the nature of emolliating and fracture of rocks depending on the degree of microwave impact. The effectiveness and mechanism of the emolliating are determined by the dielectric properties and structure of rocks. Therefore, examination of the dielectric characteristics over the microwave range is very important. The dielectric characteristics of many minerals have not been studied at all. Important temperature dependences of the dielectric characteristics were not explored. Modern methods and equipment allow to execute these examinations.

Possibility of making of effective installations of irradiating, waveguide and resonator types for emolliating and failure of rock is displayed. Problems of filling and retrieval of processed rock and safety on an electromagnetic radiation are solved in these installations. Development of working chambers of installations is facilitated at using software packages for modelling microwave devices. Transmission efficiency of microwave energy to rocks in the working chamber can reach 90%.

Studies on emolliating and fracture of rocks by the microwave field are actively carried out in Russia, the USA, Canada, UK and Slovenia. Microwave power action for an emolliating of frozen soils, a loss of strength of various ores containing black and nonferrous metals, a destruction of hard coals and a fracture of kimberlites are explored. All studies indicated to increase extraction of minerals after microwave processing of rocks. To enhance efficiency of rocks destruction it is offered to process rocks by powerful microwave pulses of micro and nanosecond duration. Using pulse microwave generators will allow to cut the cost of rocks destruction essentially.

The importance of processes examinations of emolliating and fracture of rocks is displayed on example of kimberlites destruction by superprompt microwave heating. The paper presents experimental results to determine the technological parameters of the kimberlite destruction and the effectiveness of new method of fracture. The recommendations about making of spilling type installations are produced. It is shown that the predominant fracture mechanism of kimberlite is the vapor pressure of water in pores of rock. Despite the large power-consuming of microwave heating processes, studies show the applicability and usefulness of this fracture method.

Thus, the accumulated knowledge about the processes of destruction and emolliating of rocks by microwave heating is performed in the present study. It is shown that microwave energy can be effectively used in the mining and processing industry.

8. References

Arkhangelsiy G.S. (1998). *Microwave electrothermics.* Saratov:SSTU. 2000 408 p. Russia. *(in Russian).* ISBN 5-7433-0451-3.

Badhurst D.H. et al. (1990). The application of microwave energy in mineral processing and pyrometallurgy in Australia. *SPRECHSAAL,* Vol. 123, No. 2, 1990, pp. 194-197.

Boriskov F.F., Filatov A.L., Korukin B.M., Kotov J.A. & Mesiac G.A. (2000). Complex processing pyrite waste of mining and processing enterprise by nanosecond impulse excitations. Reports of RAS, 2000, Vol. 372, No. 5, pp. 654-656 Russia. *(in Russian).*

Chanturija V.A., Guliaev Ju.V., Lunin V.D., Bunin I.G., Cherepenin V.A. & Korgenevskiy A.V. (1999). Opening persistent auriferous ores at affecting of thick electromagnetic pulses. *Reports of RAS,* 1999, Vol. 366, No. 5, pp. 680-683. Russia. *(in Russian).*

Chelyshkina V.V. & Korobsky V.K. (1988). Influence of ore processing in electromagnetic field on decomposing effects. *Gorniy Journal* No 3, 1988, pp. 115-117 Russia. *(in Russian).*

Didenko A.N. (2003). *Microwave power engineering: The theory and practice.* Moscow:Nauka, 2003. 448p. Russia. *(in Russian).* ISBN 5-02-002869-X.

Didenko A.N., Prokopenko A.V. & Zverev B.V. (2005). Microwave fracture and refinement of rocks on the kimberlite instance. *Reports of RAS,* 2005, Vol. 403, No. 2, pp. 1-2. Russia. *(in Russian).*

Didenko A.N., Prokopenko A.V. & Zelberg A.S. (2008). Research of processes of kimberlite fracture by a microwave field. *News of RAS: series Power engineering,* No. 2, 2008, pp. 40-54. Russia. *(in Russian).*

Emelin M.A. (1990). *New methods of rocks fracture.* Moscow:Nedra, 1990, Russia. *(in Russian).*

Esaulkov D.O., Lukashev A.A. & Prokopenko A.V. (2008). Measuring of dielectric substance characteristics by probe method of comparison. *Proceedings of 18 Int. Crimean conf. "Microwave and Telecomunication technology"* Vol. 2, pp. 766-768 . ISBN 978-966-335-171-1 Sevastopol, Ukraine, September, 2008.

Gridin O.M. & Goncharov S.A. (2009). *Electromagnetic processes.* Moscow:Publishing house "Mountain book", 2009. 498 p. Russia. *(in Russian).* ISBN 978-5-98672-109-5.

Haque K.E. (1987). Microwave irradiation pretreatment of a refractory gold concentrate. *Proc. of the Internet. Symposium on gold metallurgy,* pp. 327-339, Winnipeg, Canada, 1987.

Kingman S.U. (2002). Method used for formation of the electromagnertic radiation with the high intensity of the field and treatments of the materials with its help, for example, softening of the multiphase materials GB Pat. No 0207530.7, 02.04.2002.

Kolesnik V.G, et al. (2000). Agency of microwave processing on extraction of gold from mineral raw materials. *Non-ferrous metals,* No. 8, 2000, pp. 72-75. Russia. *(in Russian).*

Kondrashov V.A. & Moskalev A.N. (1971). Examination of strength of strong rocks at irradiation by microwave energy. In. "Physics of rocks and processes". pp. 179-180. Moscow: Nedra 1971, Russia. *(in Russian).*

Krasnovskij S.S. & Uvarov A.P. (1991). Examination of possibility of lowering of power consumption of rocks fracture at affecting of thick streams of microwave energy of millimeter wave. *Proceedings of 6 Russian scientific conf. "Microwave energy application in processes and scientific examinations",* pp. 139-140, Saratov, 1991. Russia. *(in Russian).*

Lindroth D.P., Morrell R.J. & Blair J.R. (1991). *Microwave assisted hard rock cutting.* US Pat. No 5003144, 26.03.1991.

Lunin V.D. et al. (1997). Model of process of microwave affecting on persistent auriferous concentrate. *Physicotechnical problems of minerals mining,* No. 4, 1997 pp. 89-94. Russia. *(in Russian).*

Maksimenko A.G., et al. (1977). Expedient of fracture of rocks electromagnetic waves. USSR Pat. No 724731, 1977. Russia. *(in Russian).*

McGill S.L., Walkiewicz J.W. & Smyres G.A. (1988). The effects of power level on the microwave heating of selected chemicals and minerals. *Proc. Materials Res. Soc. Symp. Microwave Processing Materials,* Vol. 124, pp. 247-253. Reno, 1988.

Michal L., Milota K., et. al. (2010). Modeling of microwave heating of andesite and minerals. *International Journal of Heat and Mass Transfer* Vol. 53, Issues 17-18, 2010, pp. 3387-3393.

Murray G. (1998). Microwave to slash refractory gold costs? *Mining Magazine,* Vol. 178, No. 4, 1998, pp. 276-278.

Nekrasov L.B. (1979). The base of electrothermomechanical fracture of frozen mucks. Novosibirsk:Nauka, 1979. Russia. *(in Russian).*

Novik G.J. & Zilbershmitt M.G. (1996). *Control of rocks properties in mining processes.* Moscow:Nedra, 1994, Russia. *(in Russian).*

Obrazcov A.P., Moskalev A.N. & Uvarov A.P. (1976). Examination of effect of rocks volume fracture in the strong microwave fields In: *Thermomechanics methods of fracture of rocks,* Kiev:Naukova dumka, pp. 149-150, 1976, Ukraine. *(in Russian).*

Petrov V.M. (1979). Method and resorts of microwave dielectric measurings of materials with high inductivity. *Proceedings of Methods and measuring apparatuses of electromagnetic performances of radio materials on microwave frequency.* Novosibirsk, 1979, pp. 8-12. Russia. *(in Russian).*

Petrov V.M. (2002). a) Loss of rocks strength by microwave power. *Radio electronics and telecommunications,* No2, 2002, pp. 35-41. Russia. *(in Russian).*

Petrov V.M. (2002). b) Rocks destruction by microwave power. *Radio electronics and telecommunications,* No4, 2002, pp. 34-42. Russia. *(in Russian).*

Prokopenko A.V. & Zverev B.V. (2008). Spilling type installation for thermal fracture of rocks by microwave power of 5 kW. *News of RAS: series Power engineering*, No. 2, 2008, pp. 22-39. Russia. *(in Russian)*.

Riabec N.I. (1991). Fundamental of emolliating and thawing of mucks by microwave energy. Yakutsk: NSC Siberian Branch of the RAS. 1991. Russia. *(in Russian)*.

Salsman J.B., Williamson R.L. et al. (1996) Short-pulse microwave treatment of disseminated sulfide ores. *Minerals Eng.*, Vol. 9, No. 1, 1996, pp. 43-54.

Tranquilla J.M. (1998). *Microwave treatment of metal bearing ores and concentrates*. US Pat. No 5824133, 20.10.1998.

Walkiewicz J.W., Clark A.E. & McGill S.L. (1991). Microwave assisted grinding. *IEEE Trans. on Industry Appl.*, 1991, Vol. 27, No. 2, pp. 239-243.

16

Use of Induction Heating in Plastic Injection Molding

Udo Hinzpeter and Elmar Wrona
*Kunststoff-Institut für die mittelständische
Wirtschaft NRW GmbH (K.I.M.W.), Lüdenscheid
HÜTTINGER Elektronik GmbH + Co. KG, Freiburg
Germany*

1. Introduction

Variothermal processes are needed in the field of plastic injection molding. They avoid surface faults, enable the modeling of micro structures and the production of micro molded parts. Conventional variothermal processes, which heat with water or oil, have the disadvantage of high cycle times and low productivity. The density of heat flux is very low, and one has to cool down large areas before starting a new injection cycle.

Induction heating enables low cycle times due to excellent heat flux density and the possibility of partial heating. Moreover, induction heating systems save costs. Surface qualities, that are absolutely comparable with varnished parts, are possible with up to 50 times lower cost.

2. Design of the heating system

Heat and cool processes become more and more important in the area of injection molding. For economical production it is imperative to use a technology which guarantees high heating and cooling rates. Comparative to all the other possibilities for heat and cool injection molding processes the induction heating of injection molds should be the first election. Figure 1 gives an overview of different induction heating technologies.

This paper shows the technology INDUMOLD®. Some design criteria have to be considered for the design of internal inductors. This is important to reach the required qualitative and economical targets. The major design criteria are:

- sufficient uniform temperature on the mold wall surface
- no increase of the cycle time influenced by the heating and cooling phase
- determination of the generator performance

The arrangement of the inductors is an important step during the design phase of an injection mold since it isn't possible to make a post correction on the existing mold. Because of the complex thermal processes during the injection cycle it is a must to use a suitable simulation program for calculating the temperature distribution (heating and cooling phase). Based on the results of the calculation it is during the mold design phase possible to optimize the inductor arrangement and the cooling channel layout and also to avoid other failure.

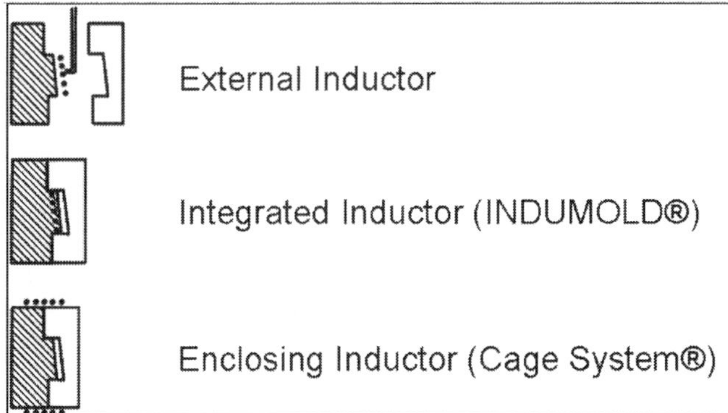

Fig. 1. Overview induction heating techniques for plastic injection molding

The thermal simulation gives information about the reached uniform temperature on the mold wall surface at the end of the heating phase. Furthermore hot and cold spots can be safely detected. Depending on the cooling channel layout it is also feasible to see how the temperature profile during the cooling phase courses and how efficient the mold cooling works. All these results and improvements guarantee that cycle times do not unneeded increase because the design and layout of the inductors and the mold cooling are optimal adjusted of the heat and cool phase.

Figure 2 shows the temperature distribution on the mold wall surface at the end of the heating time. The coloured picture shows the temperature distribution at the end of the heating time. The curves present the temperature course over the time.

It can be identified that the temperature in the middle area is higher as in comparison to the lower handle area. This uneven temperature distribution leads in practice to negative part properties like gloss differences, undue warpage or silver streaks which are shown in Figure 3.

3. Numerical simulation

To avoid those failures in practise it is imperative to use convenient simulation software for the induction heating of injection molds. The shown results have been calculated with a simplified thermal simulation program which doesn't consider any electromagnetic effects. This simplification is sufficient for most of the calculations for the induction heating of injection molds based on the INDUMOLD® system.

The software Moldex3D is used for the described calculations which are offered by the Kunststoff-Institut Lüdenscheid. Moldex3D is 3D CAE software for plastic injection molding. The modular structure of the software package makes it possible to adjust it to the individual requirements of the product range of plastic injection molding.

This way it is possible to avoid aberrations in product and mold design before they occur and to realize short developing times at low development costs. It can be used to analyze and optimize any type of complex sprue and cooling geometries. Variothermal and close-to-contour cooling is possible as well as coupled warping for multi-component systems. Based on a 3D hybrid grid and a powerful and robust finite volume algorithm, Moldex3D allows a safe analysis of any type of complex component geometries, multi-component parts with extremely varying wall thicknesses. The scientifically proven models and the extensive material data base help to analyze and interpret the processes inside the mold and to tune process parameters. A sufficiently exact analysis of the problems occurring in connection with the injection molding process requires three-dimensional calculation models on the level of the component as well as on the sprue, temperature control and mold level.

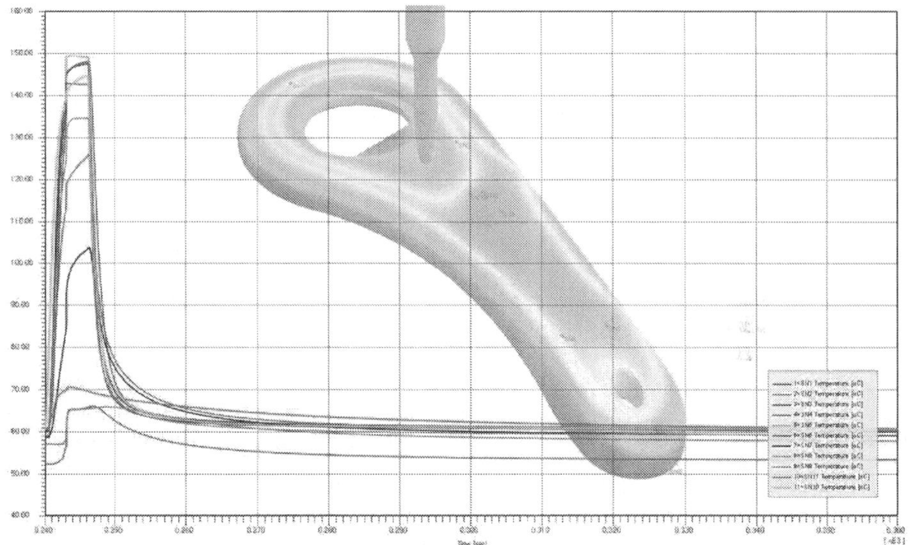

Fig. 2. Temperature distribution on the mold wall surface at the end of the heating time, bottle opener

Fig. 3. Warpage and silver streaks at bottle Opener

The needed boundary conditions for a complete simulation of a variothermal injection molding process are:
- part design
- raw material
- processing parameters such as

- melt and mold wall temperature
- estimated holding pressure and holding pressure time
- injection time
- cycle time
- estimated temperature of the inductor
- heating time

Based on the boundary conditions it is possible to optimize the inductor design and also the whole injection molding process. Depending on the target and the part complexity the duration of the necessary calculations is app. 3-5 days.

A typical CAE service for an injection molding process starts with the generation of a FEM model based on the 3D geometry data. Therefore the general CAD data formats like igs, stl or step can be used. This step could be very labour intensive if the part design is complex or the CAD data quality is bad. If the FEM model is generated the aforementioned boundary conditions have to be assigned. According to the calculation aim a cooling system and also an inductor arrangement have to be added. This could be done by a customer proposal or on a first draft developed by Kunststoff-Institut Lüdenscheid. The first calculation results are evaluated and discussed with the customer for necessary optimization.

4. Installation concept

The induction coil consists of a flexible stranded wire, which can be placed very easy in the milled mold. The stranded wire is connected to the external oscillating circuit of the induction heating generator. The generater is controlled by the control system of the plastic injection molding machine via the A/D-interface or Profibus. The control system gives a start signal to the generator. It delivers power for the necessary time by using the internal timer functionality.

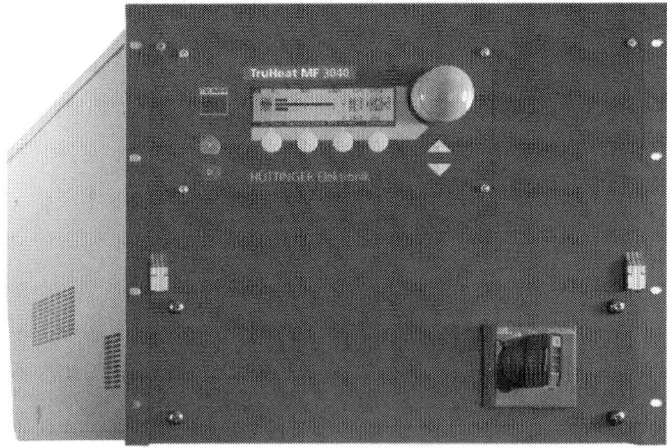

Fig. 4. TruHeat MF 3040, 19-inch plug-in

The required output power depends on the length of the stranded wire and the corresponding mass of the material to be heated. The typical output power is in the range of

20 to 40 kW. In order to heat only thin layers, frequencies in the range of 70 to 100 kHz are used. If you heat only thin layers, a fast cooling after the injection molding process is possible. Therefore, the cycle times are short.

Typical heating temperatures are between 120 and 180 °C. This temperature range must be realized between 2 and 7 seconds.

5. Product family TruHeat MF 3010 -7040

5.1 General information

The induction heating is done with the innovative induction heating generator TruHeat MF 3010 - 7040. This product family is available with 10, 20, 30 or 40 kW output power. The frequency range reaches from 5 to 100 kHz. The user has the possibility to operate at any frequency from 5 to 30 kHz or from 20 to 100 kHz. The TruHeat MF 3010 - 7040 was designed for mains voltages from 400 to 480 V. The generator works at any frequency in this range.

Three different variants of the TruHeat MF 3010 - 7040 are available. The basic version is the 19-inch plug-in module (so called TruHeat MF Series 3000), designed for integration in a 19-inch rack (see Figure 4). This very compact unit offers the possibility to integrate the power supply very easy into the facility, e.g. in machines for modern plastic injection molding. This variant already offers the full functionality of the product family. The second version is the tabletop unit (TruHeat MF Series 5000), which is the ideal heating equipment for stand alone operation. This variant is the best choice for laboratories, universities or companies with low amount of space. The third version is the cabinet (TruHeat MF Series 7000). This is the traditional variant for industrial environment with a high protection class IP54. This enables the use even in rough conditions, e.g. in the field of induction hardening.

5.2 Technical specification

The 6-puls rectifier is connected directly – without using a relatively space consuming mains transformer – to the mains. This saves space and money.

The DC output voltage of the rectifier is controlled by a chopper unit, which enables to vary the output power (voltage or current) from almost 0 to 100 %. Therefore, the TruHeat MF 3010 - 7040 is ideal for high power applications as well as for low power applications, or for processes which need the full power range, e.g. to reach special temperature profiles.

The inverter is operated by an advanced control unit. It recognizes the frequency of the oscillating circuit and chooses the control parameters for the different frequency ranges by itself. As a result, the whole TruHeat MF 3010 - 7040 family needs only one inverter control board.

The output transformer fulfils two functions. It separates the mains potential from the application area and adapts the load to the power supply. The output voltage of the transformer can be switched from 600 to 300 V or vice versa very easy. Using this feature, the user has the full flexibility to solve different application tasks.

The described components are integrated in the power supply. It is connected with power cables to a parallel oscillating circuit, which has two main advantages in comparison to a serial oscillating circuit. The current in the induction coil – thus the electromagnetic field – can be increased by incrementing the factor Q. In addition to that, the matching of the generator and the load can easily be done by changing the capacity.

6. Summary

Induction heating is an innovative technology in the field of plastic injection molding. It enables high quality surfaces at low cycle times and saves costs. To tap the full potential of this technology, numerical simulations are necessary in order to get the best inductor shape. Moreover, the induction heating system and the plastic injection molding facility have to work together perfectly.

17

Microwave-assisted Synthesis of Coordination and Organometallic Compounds

Oxana V. Kharissova, Boris I. Kharisov and Ubaldo Ortiz Méndez
Universidad Autónoma de Nuevo León, Monterrey,
México

1. Introduction

Microwave irradiation (MW) as a "non-conventional reaction condition" (Giguere, 1989) has been applied in various areas of chemistry and technology to produce or destroy diverse materials and chemical compounds, as well as to accelerate chemical processes. The advantages of its use are the following (Roussy & Pearce, 1995):

1. Rapid heating is frequently achieved,
2. Energy is accumulated within a material without surface limits,
3. Economy of energy due to the absence of a necessity to heat environment,
4. Electromagnetic heating does not produce pollution,
5. There is no a direct contact between the energy source and the material,
6. Suitability of heating and possibility to be automated.
7. Enhanced yields, substantial elimination of reaction solvents, and facilitation of purification relative to conventional synthesis techniques.
8. This method is appropriate for green chemistry and energy-saving processes.

The substances or materials have different capacity to be heated by microwave irradiation, which depends on the substance nature and its temperature. Generally, chemical reactions are accelerated in microwave fields, as well as those by ultrasonic treatment, although the nature of these two techniques is completely distinct.

Microwave heating (MWH) is widely used to prepare various refractory inorganic compounds and materials (double oxides, nitrides, carbides, semiconductors, glasses, ceramics, etc.) (Ahluwulia, 2007), as well as in organic processes (Oliver Kappe et al, 2009; Leadbeater, 2010): pyrolisis, esterification, and condensation reactions. Recent excellent reviews have described distinct aspects of microwave-assisted synthesis of various types of compounds and materials, in particular organic (Martínez-Palou, 2007; Oliver Kappe et al, 2009; Besson et al, 2006) and organometallic (Shangzhao Shi and Jiann-Yang Hwang, 2003) compounds, polymers, applications in analytical chemistry (Kubrakova, I.V., 2000), among others. Microwave syntheses of coordination and organometallic compounds, discussed in this chapter, are presented by relatively a small number of papers in the available literature in comparison with inorganic and organic synthesis. The use of microwaves in coordination chemistry began not long ago and, due to the highly limited number of results, these works can be considered as a *careful pioneer experimentation*, in order to establish the suitability of this technique for synthetic coordination chemistry. Classic ligands, whose numerous derivatives have been used as precursors for obtaining their metal complexes, are shown in Table 1.

2. Physical principles of microwave irradiation and laboratory equipment

Microwave heating is a physical process where the energy is transferred to the material through electromagnetic waves. Frequencies of microwaves are higher of 500 MHz. It is known that a non-conductive substance can be heated by an electric field, which polarizes its charges without rapid reversion of the electric field. For some given frequencies, the current component, resulting in the phase with electric field, produces a dissipation of the potency within the dielectric material. Due to this effect, a dielectric can be heated through the redistribution of charges under the influence of external electric fields. The potency dissipated within the material depends on the established electric field within the material. This potency is diminished as the electromagnetic field penetrates to the dielectric.

The most common microwave application is that of multimode type which accepts broad range thermal charges with problems of microwave uniformity. The application of multimode type is given in a closed metallic box with dimensions of various wave lengths and which supports a large number of resonance modes in a given range of frequencies. A resonance cavity or heater consists on a metallic compartment that contains a microwave signal with polarization of the electromagnetic field; it has many reflections in preferential directions. The superposition of the incident and reflected waves gives place to a combination of stationary waves. If the configuration of the electric field is precisely known, the material to be treated can be put to a position of electric field maximum for an optimal transference of electromagnetic energy.

Typical microwave equipment consists of a magnetron tube (Fig. 1) (Roussy & Pearce, 1995). Just as other vacuum tubes, the anode has a higher potential with respect to the cathode (source of electrons). So, the electrons are accelerated to the anode in the electric field. The cathode is heated till the high temperature expulse electrons. Generally, the anode is close to earth potential and the cathode has a high negative potential. The difference between the magnetron and other vacuum tubes is that the electron flow passes along a spiral; this route is created by external magnetic field B (Fig. 1). The electron cloud produces resonance cavities several times in its trip to the anode. These cavities work as *Helmholtz* resonators and produce oscillations of fixed frequency, which is determined by the cavity dimensions: small cavities produce higher frequencies, large cavities give smaller frequencies. The antenna in the right zone collects the oscillations.

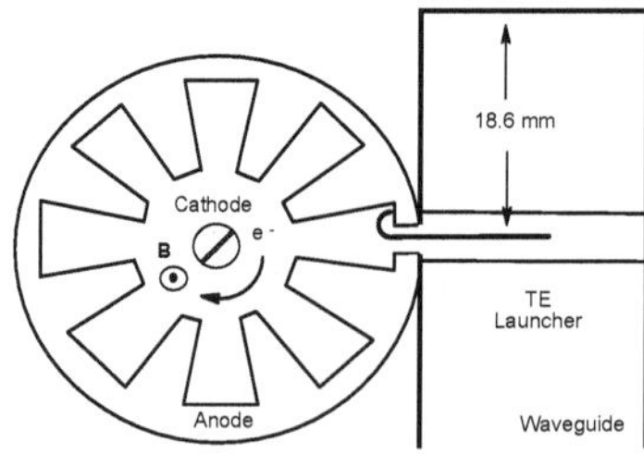

Fig. 1. Scheme of microwave equipment.

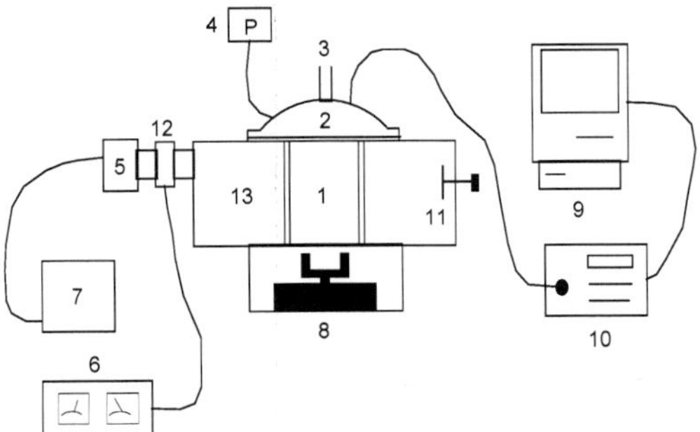

Fig. 2. Reactor for batchwise organic synthesis (with permission): 1, reaction vessel; 2, top flange; 3, cold finger; 4, pressure meter; 5, magnetron; 6, forward/reverse power meters; 7, magnetron power supply; 8, magnetic stirrer; 9, computer; 10, optic fiber thermometer; 11, load matching device; 12, waveguide; 13, multimodal cavity (applicator).

The use of a microwave reactor for batchwise organic synthesis (Raner et al, 1995), described in Fig. 2), permits to carry out synthetic works or kinetic studies on the 20-100 mL scale, with upper operating limits of 260°C and 10 MPa (100 atm). Microwave-assisted organic reactions can be conducted safely and conveniently, for lengthy periods when required, and in volatile organic solvents. The use of water as a solvent is also explored.

A typical reactor used for organic and/or organometallic syntheses (Matsumura-Inoue et al, 1994) is presented in Fig. 3, which can be easily implemented using a domestic microwave

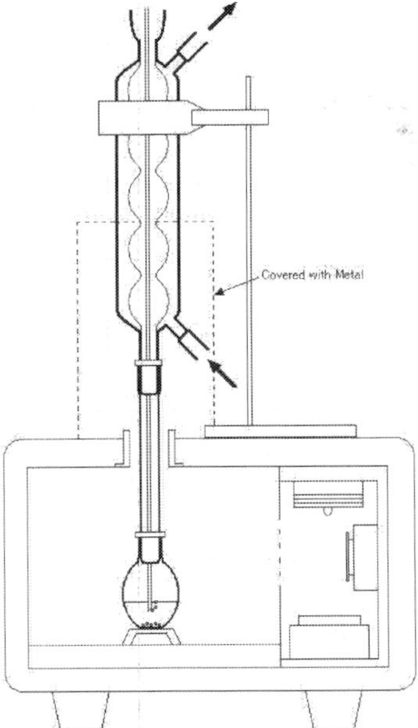

Fig. 3. Typical MW-reactor for organic and/or organometallic synthesis. With permission.

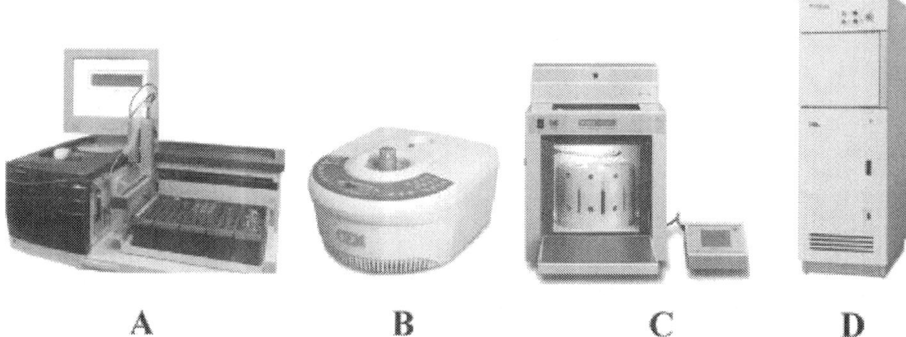

Fig. 4. Microwave reactors for chemical syntheses. **A:** Emrys Liberator (Biotage, Sweden, www.biotage.com); **B:** CEM Discover BenchMate (CEM, USA, www.cem.com) Copyright CEM Corporation; **C:** Milestone Ethos TouchControl (Milestone, Italy, www.milestonesci.com); **D:** Lambda MicroCure2100 BatchSystem (Lambda, USA, www.microcure.com). With permission.

oven. Due to some problems occurring during microwave treatment, for example, related with the use of volatile liquids (they need of an external cooling system *via* copper ports), original solutions to these problems are frequently found in the reported literature. More modern laboratory MW-reactors (Wiesbrock et al, 2004) are shown in Fig. 4.

A combination of different techniques can frequently improve yields of final compounds or synthetic conditions. Reunion of microwave and ultrasonic treatment was an aim to construct an original microwave-ultrasound reactor (Chemat et al, 1996) suitable for organic synthesis (pyrolysis and esterification) (Fig. 5). The ultrasound (US) system is a cup horn type; the emission of ultrasound waves is made at the bottom of the reactor. The US probe is not in direct contact with the reactive mixture. It is placed a distance from the electromagnetic field in order to avoid interactions and short circuits. The propagation of the US waves into the reactor is made by means of decalin introduced into the double jacket. This liquid was chosen because of its low viscosity that induces good propagation of US and its inertia towards MW.

Some years ago, an alternative method for performing microwave-assisted organic reactions, termed "Enhanced Microwave Synthesis" (EMS), has been examined in an excellent review (Hayes, 2004). By externally cooling the reaction vessel with compressed air, while simultaneously administering microwave irradiation, more energy can be directly applied to the reaction mixture. In "Conventional Microwave Synthesis" (CMS), the initial microwave power is high, increasing the bulk temperature (TB) to the desired set point very quickly. However, upon reaching this temperature, the microwave power decreases or shuts off completely in order to maintain the desired bulk temperature without exceeding it. When microwave irradiation is off, classical thermal chemistry takes over, losing the full advantage of microwave-accelerated synthesis. With CMS, microwave irradiation is predominantly used to reach TB faster. Microwave enhancement of chemical reactions will only take place during application of microwave energy. This source of energy will directly activate the molecules in a chemical reaction. EMS ensures that a high, constant level of microwave energy is applied.

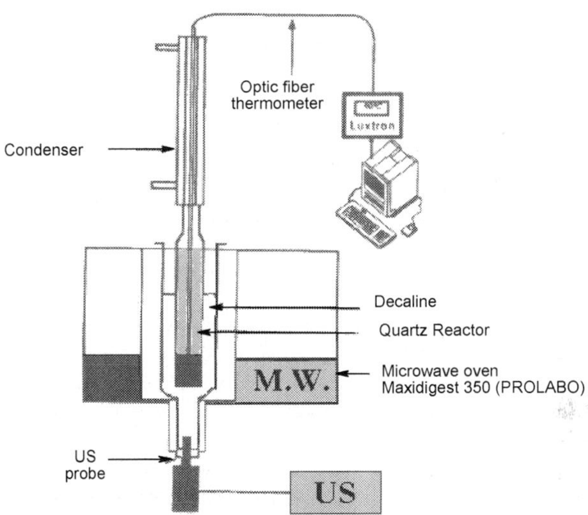

Fig. 5. Combined MW-US reactor. With permission.

3. Complexes with O-containing ligands

3.1 β-Diketonates, alkoxides and alcohol adducts

The MWH of metal β-diketonates or their precursors, represented by acetylacetonates, has been used both for their synthesis (rarely) and destruction (mainly), leading, in the last case, to various inorganic materials, nanostructures and nanocomposites. *Synthesis route* is represented by only a few examples. Thus, a rapid and environmentally benign method for the coupling of 2-naphthols is described using copper(II) acetylacetonate under microwave irradiation in dry media (Meshram et al, 2003). The procedure was found to be very convenient and avoids the use of excess solvent for reaction. Microwave synthesis method was developed for the synthesis of a series of cyclometalated platinum complexes with long chain β-diketone ancillary ligands, with which reaction time was greatly reduced from 32 h to several minutes (Luo et al, 2007; Luo et al, 2007). The formed compounds were used for fabrication of organic light-emitting diodes. A protected ethynyl group was introduced into a γ-position of a (β-diketonato)*bis*(bipyridine)ruthenium(II) complex through the reaction of the bromo complex and (tri*iso*propylsilyl)-acetylene with very good yield under MWH (Munery et al, 2008). Two mononuclear mixed-ligand ruthenium(II) complexes with bipyridine (bpy) and functionalized acetylacetonate ion (acac⁻), [Ru(bpd)(bpy)₂](PF₆) (bpy = 2,2'-bipyridine, bpd = 3-Bromo-2,4-pentanedionate ion) and [Ru(tipsepd)(bpy)₂](PF₆) {tipsepd = 3-((triisopropylsilyl)ethynil)-2,4-pentanedionate ion} were then prepared as candidates for building blocks. Also, microwave-assisted synthesis method enabled the preparation of the (*tris*-acetylacetonate)(2,9-dimethyl-4,7-diphenyl-1,10-phenanthrolinate) terbium(III) {Tb(acac)₃(dmdpphen)} complex with outstanding high green luminescence and good thermal stability (Nakashima et al, 2008). This complex was expected to be used in functional materials for electronic products. Zirconium acetylacetonate, $Zr(acac)_4$, was prepared from its hydrate $Zr(acac)_4 \cdot 10H_2O$ by microwave dehydration of the latter (Berdonosov et al, 1992). Additionally, a convenient method for ^{68}Ga-labeling under anhydrous conditions using solid-phase derived gallium-68-acetylacetonate $\{[^{68}Ga]Ga(acac)_3\}$ in a microwave-enhanced radiosynthesis was offered (Zoller et al, 2010). [68]Ga was absorbed quantitatively in a cation exchange resin; more than 95% of the generator-eluted ^{68}Ga was obtained from the cation exchange resin with a 98% acetone/2% acetylacetone mixture providing $[^{68}Ga]Ga(acac)_3$ as labeling agent for further use in labeling porphyrin derivatives (^{68}Ga-labeled porphyrins may facilitate the medical application for molecular imaging via positron emission tomography).

MW-decomposition of metal acetylacetonates is represented much more frequently. Thus, silicalite (Si-MFI) zeolite crystals with incorporated tetravalent metal ions were used to MW-synthesize metallosilicalite (M-MFI; M = Sn, Zr, Sn/Zr, Ti/Zr) zeolites crystals (Hwang et al, 2006). Acetylacetonates were applied as chelating ligands of the metal precursors, to reduce their hydrolysis rates and, therefore, to enhance framework incorporation of each metal in the syntheses of M-MFI zeolites. The resulting zeolite crystals formed showed puck-like morphology and were stacked to form fibers with the degree of self-assembly varied depending on the nature of the tetravalent metal ion used. Chromium-substituted β-diketonate complexes of aluminum were synthesized and employed as precursors for a "soft chemical" process, wherein MWH of a solution of the complex yielded, within minutes, well-crystallized needles of $\alpha\text{-}(Al_{1-x}Cr_x)_2O_3$ measuring 20-30 nm in diameter and 50 nm long (Gairola et al, 2009). By varying the microwave irradiation parameters and using a surfactant such as polyvinyl pyrrolidone, the crystallite size and shape can be controlled and their agglomeration prevented. Mg-Al hydrotalcite-like compounds $\{HT, Mg_6Al_2(CO_3)(OH)_{16} \bullet 4(H_2O)\}$ were prepared by the microwave method with ethoxide-acetylacetonate or acetylacetonate as precursors (Paredes et al, 2006; Paredes et al, 2006). Hydrotalcites prepared with ethoxide-acetylacetonate were found to be better sorbents for [131]I- than those with acetylacetonate. Also, it was established that organic residues presented in the samples prepared by the microwave method favored the sorption of radioactive anions, in particular [131]I- if compared with nitrate and/or carbonate interlayered hydrotalcites. Ferric acetylacetonate, among other iron salts, was used as a precursor to obtain black magnetic Fe_3O_4 nanoparticles in polyhydric alcohols in presence of surfactants (polyethylene glycol, cetyltrimethylammonium bromide, sodium dodecyl benzene sulfonate, etc.) and cosolvents (ethylenediamine, formamide, 1,4-butanediamine and/or butanolamine) (Gao et al, 2009). The product can be used in biomedical, mechanic or electronic fields with strong magnetism, controllable size, and good dispersibility. Additionally, as described in a related work (Bilecka et al, 2008),highly crystalline metal oxide nanoparticles such as CoO, ZnO, Fe_3O_4, MnO, Mn_3O_4, and $BaTiO_3$ were synthesized in just a few minutes by reacting metal alkoxides, acetates or acetylacetonates with benzyl alcohol under microwave heating. At last, organically dispersible nanoalloys were prepared from mixture of salts and metal acetates/acetylacetonates in oleyamine (OAm) and oleic acid (OA), for instance $Pd(acac)_2\text{-}Ni(HCO_2)_2\text{-}OAm\text{-}OA$ (nanoalloy PdNi) or $Ag(ac)\text{-}Cu(ac)_2\text{-}OAm\text{-}OAc$ (AgCu) (Abdelsayed et al, 2009). High activity and thermal stability have been observed for the nanoalloys according to the order CuPd>CuRh>AuPd>AuRh>PtRh>PdRh>AuPt.

CVD techniques have been successfully applied to decompose metal complexes, in particular microwave plasma aerosol-assisted chemical vapor deposition (MWAACVD), which was used, among other varieties of AACVD, to prepare Y_2O_3 stabilized ZrO_2, Y_2O_3 doped CeO_2, Gd_2O_3 doped CeO_2 and $La_{0.8}Sr_{0.2}MnO_3$ thin films on various ceramic substrates starting froml β-diketonate chelates as the source materials (Meng et al, 2004). Amorphous GaF_3 and GaF_3-BaF_2 thin films were synthesized by electron cyclotron resonance microwave plasma-enhanced CVD (MWPECVD) using metal β-diketonates and a NF_3 gas as starting materials and a fluorinating reagent, respectively (Takahashi et al, 2003). A thin zirconia electrolyte film for a solid oxide fuel cell was prepared on a porous Al_2O_3 substrate by MPE CVD using two zirconia sources: zirconium acetylacetonate and zirconium tetra-n-butylate (Okamura et al, 2003). As-deposited electrolyte film grown indicated the columnar structure, but this was deformed to a crystal structure with a large crack or pore occurred at grain boundary in film by annealing at 400°C. Additionally, MWPECVD was shown to be a promising method for the solvent free preparation of catalytic materials (Dittmar & Herein, 2009), such as, for example catalytic active chromia species on zirconia and lanthanum doped zirconia supports. During this process, the adsorption of $Cr(acac)_3$ probably took place by cleavage of one ligand on both supports. Furthermore, the utilization of the PECVD method can inhibit the formation of large CrO_x agglomerates or α-Cr_2O_3 on both supports and, after upscaling, this method can be used for the preparation of catalysts for fine chemicals in larger scale. In a related work (Dittmar et al, 2004), where cobalt oxide supported on titania, CoO_x/TiO_2, was obtained starting from cobalt(III) acetylacetonate, $Co(acac)_3$, (precursor) and TiO_2 (support), the $Co(acac)_3$ was evaporated and adsorbed on carrier surface in a first step and afterwards decomposed during the microwave-plasma treatment in oxygen atmosphere. Volatile copper(II) acetylacetonate was used for preparation of copper thin films in Ar-H_2 atmosphere at ambient temperature by MWPECVD (Pelletier et al, 1991). The formed pure copper films with a resistance of 2-3 $\mu\Omega\cdot$cm were deposited on Si substrates. It was noted that oxygen atoms were never detected in the deposited material since Cu-O intramolecular bonds were totally broken by microwave plasma-assistant decomposition of the copper complex.

Additionally to the examples described above on the use of β-diketonate-alkoxide mixtures, *alkoxides* themselves were also reported as precursors for MW-obtaining of inorganic films and structures. Thus, synthesis of TiO_2 and V-doped TiO_2 thin layers was significantly improved and extended under application of microwave energy during the drying and/or calcination step (Zabova et al, 2009). Thin nanoparticulate titania layers were prepared via the sol-gel method using titanium n-butoxide as a precursor. The photocatalytic activities of prepared layers were quantified by the decoloring rate of Rhodamine B.

Another type of coordination compounds, molecular *adducts of alcohols* of the composition $VOPO_4\cdot C_nH_{2n+1}OH$ (1-alkanols, n=1-18) and $VOPO_4\cdot C_nH_{2n}(OH)_2$ (1,ω-alkanediols, n=2-10) were prepared long ago (Beneš et al, 1997) by the direct reaction of various liquid alcohols with solid and finely ground $VOPO_4\cdot 2H_2O$ in a MW field. According to X-ray diffraction data, the structures of all these polycrystalline complexes retained the original layers of $(VOPO_4)_\infty$. Alcohol molecules were placed between the host layers in a bimolecular way, being anchored to them by donor-acceptor bonds between the oxygen atom of an OH group and a vanadium atom as well as by hydrogen bonds. Other adducts, $[(n\text{-}Bu)_4N][TlMS_4]$ (M=Mo, W), were also prepared in the conditions of microwave treatment and their nonlinear optical properties were studied (Lang et al, 1996).

3.2 Carboxylates

MW-synthesized carboxylates are represented mainly by aromatic derivatives possessing multiple carboxylic groups. These complexes are sometimes isolated as adducts with stabilizing ligands as 2,2'-bipy or 1,10-phen, as well as solvent molecules. Thus, by treating $Cu(NO_3)_2 \cdot 3H_2O$ with a V-shaped ligand 4,4'-oxydibenzoic acid (H_2oba), a dynamic metal-carboxylate framework $[Cu_2(oba)_2(DMF)_2] \cdot 5.25DMF$ (MCF-23; DMF = N,N-dimethylformamide) was synthesized, which features a wavelike layer with rhombic grids based on the paddle-wheel secondary building units (Wang et al, 2008). These layers stack *via* strong offset π-π stacking of the Ph groups of oba ligands to give 3D porosity. A MWAS solvothermal method was proven to be a faster and greener approach to synthesize phase-pure MCF-23 in high yield without impurities, typical for conventional synthesis. In contrast, the product obtained by the conventional solvothermal method was not phase-pure. Two isostructural coordination polymers, $M_3(NDC)_3(DMF)_4$ (M = Co, Mn; H_2NDC = 2,6-naphthalenedicarboxylic acid), crystallizing in the monoclinic system with space group C2/c, were prepared through conventional and MWAS solvothermal methods (Liu et al, 2008). These microporous cobalt(II) and manganese(II) coordination polymers underwent reversible structural change upon desolvating, giving stable microporous frameworks containing unsaturated metal sites.

Trimesic acid **1** and its analogue, containing four carboxylate units, have been reported in a series of publications related to MWAS of metal complexes. Thus, two isostructural coordination polymers $(EMim)_2[M_3(TMA)_2(OAc)_2]$ (M = Ni or Co, EMim = 1-ethyl-3-methylimidazolium, H_3TMA = trimesic acid) with anionic metal-organic frameworks were synthesized under microwave conditions using an ionic liquid EMIm-Br as solvent and template (Lin et al, 2006). In a related report, the microwave solvothermal reaction of nickel nitrate with trimesic acid provided the $[Ni_3(BTC)_2(H_2O)_{12}]_n$ (BTC = benzene-1,3,5-tricarboxylate anion of trimesic acid), which is a metal coordination polymer composed of 1D zigzag chains (Hsu et al, 2009). In the asymmetric unit, two types of Ni atoms were found: one of the NiO_6 groups was coordinated to only one carboxylate group and thus terminal, the other is bridging, forming the coordination polymer. Magnesium coordination polymers, $[Mg_2(BTEC)(H_2O)_4] \cdot 2H_2O$, $[Mg_2(BTEC)(H_2O)_6]$, and $[Mg_2(BTEC)(H_2O)_8]$ (BTEC = 1,2,4,5-benzenetetracarboxylate anion), were synthesized from magnesium nitrate and 1,2,4,5-benzenetetracarboxylic acid with variable ratios of organic base under MW solvothermal reactions at 150-180ºC (Liu et al, 2009). Structure of MW-synthesized complex $\{[Co_2(C_2O_4)(C_6H_2(COO)_4)(H_2O)]_4 \cdot 4H_2O \cdot (NH_2CH_2COOH)\}_n$ (crystallized in the monoclinic system and the space group Cc), had a flattened octahedral configuration (Xu & Fan, 2007). Three mixed-ligand cobalt(II) complexes $[Na_2Co(\mu_4-btec)(H_2O)_8]_n$, $[Co_2(\mu_2-btec)(bipy)_2(H_2O)_6] \cdot 2H_2O$, and $[Co_2(\mu_2-btec)(phen)_2(H_2O)_6] \cdot 2H_2O$ (H_4btec = 1,2,4,5-benzenetetracarboxylic acid, bipy = 2,2'-bipyridine, phen = 1,10-phenanthroline) were synthesized using hydrothermal and microwave methods (Shi et al, 2009). All three complexes were found to be bridged by the ligands to form 3D (first complex) and binuclear (other complexes) structures.

Three isostructural 2D metal-organic frameworks, $[M(bpydc)(H_2O) \cdot H_2O]_n$ (where M = Zn; Co; Ni and bpydc is 2,2'-bipyridine-5,5'-dicarboxylate), were prepared by hydrothermal, ultrasonic and MWAS methods (Huh et al, 2010).The coordination environment of the metal ions was found to be a distorted octahedral geometry. The metal ions were found to be coordinated by two nitrogen atoms from the bipyridyl moiety, two oxygen atoms from one

$$1$$

carboxylate in a bidentate manner, one oxygen atom from another carboxylate in a monodentate manner, and one oxygen atom from the aqua ligand. $[Zn(bpydc)(H_2O) \cdot H_2O]_n$ displayed strong solid state blue luminescence. Additionally, the green synthesis of a variety of 3,4-disubstituted-1-H-pyrrole-2-carboxylates was described (Dickhoff et al, 2006).

As an example of MW-decomposition of metal carboxylates leading to nanostructures, Ni nanoparticles with average sizes of 43, 71, and 106 nm were obtained by the intramolecular reduction of Ni^{2+} ion contained in a formate complex having long-chain amine ligands {oleylamine (=(Z)-9-octadecenylamine), myristylamine (=tetradecylamine), and laurylamine (=dodecylamine)} within an extremely short time under MW conditions (Yamauchi et al, 2009). Formate ion coordinated to Ni^{2+} ion acted as a reducing agent for Ni^{2+} in this reaction and finally decomposed to hydrogen and carbon dioxide. Also, microwave synthesis of metal oxide nanoparticles, γ-Fe_2O_3, NiO, ZnO, CuO and Co-γ-Fe_2O_3 were carried out by microwave-assisted route through the thermal decomposition of their respective metal oxalate precursors employing polyvinyl alcohol as a fuel (Lagashettya et al, 2007).

3.3 Nitrogen-containing ligands

N-Containing ligands are widely represented (although lesser in comparison with N,O-ligands) by substituted derivatives of classic heterocycles with 1÷3 nitrogen atoms (Table 1), such as azoles, azines (in particular polypyridines), frequently together with carboxylate anions (see also the section on N,O-containing ligands) or CO-groups. Thus, among *azole complexes*, MWAS of the neutral complex *fac*-[ReL(CO)$_3$Cl] and isomers thereof were carried out by reacting the chelating ligand 4-[4,6-*bis*(3,5-dimethyl-1H-pyrazol-1-yl)-1,3,5-triazin-2-yl]-N,N-diethyl-benzenamine, L, with pentacarbonylchlororhenium in toluene (Salazar et al, 2009). Further substitution of the carbonyl and/or the chloride attained multiple products with remarkable luminescence properties that included thermochromism, rigidochromism, solvatochromism, and/or vapochromism. Cobalt(II) pyrazolate metal-organic frameworks comprising bridging *bis*-pyrazolyl ligands $(3,5$-$R_{1,2}C_3HN_2)$-$(1,4$-$C_6H_3R_2)_n$-$(3,5$-$R_{1,2}C_3HN_2)$ ($C_3H_3N_2$ = 1H-pyrazol-4-yl; n = 0-3, R^1, R^2 = H, halo, CF_3, OH, NH_2, CHO, C_{1-6} alkyl, alkenyl, alkynyl, alkoxy), tetrahedrally coordinated to Co(II) ions, useful as redox-active materials, oxidation catalysts, adsorbents and storage materials for H_2 and methane, gas sensors, were prepared by conventional or MWH of solutions containing Co(II) salts with F-, Cl-, Br-, I-, NO_3-, SO_4^{2-}, AcO- anions, and the bis-pyrazolyl ligands above in water, MeOH, EtOH, DMF, N,N-diethylformamide, PhCl, N-methylpyrrolidone at 80-140°C for 1-150 h (Bahnmueller et al, 2009). The complex $[ReO_3\{\mu_3$-$SO_3C(pyz)_3\}]$ **2** was prepared in 42% yield by reacting lithium *tris*(1-pyrazolyl)methanesulfonate with rhenium(VII) oxide in water at ambient temperature during 5 h (or 30 min under microwave irradiation at 20°C) (Pombeiro et al,

2007). These complexes were used as catalysts in the following reactions: a) partial oxidation of ethane into acetic acid or its carboxylation into propionic acid in the atmosphere of CO; b) partial oxidation of ethane into acetaldehyde; c) partial peroxidative oxidation of cyclohexane into cyclohexanol and cyclohexanone.

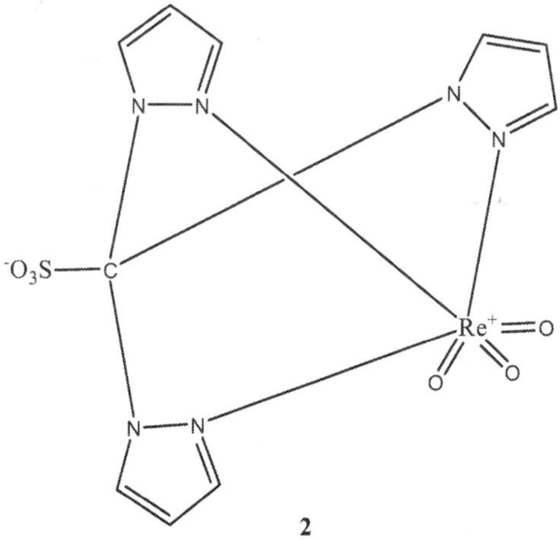

2

Benzimidazole complexes have also been MW-prepared. Thus, a cobalt(II) complex, [Co(H$_2$bzimpy)$_2$](ClO$_4$)$_2$, with tridentate ligand 2,6-*bis*(benzimidazol-2-yl)pyridine (H$_2$bzimpy) was synthesized by microwave irradiation method (Tan et al, 2004). The *bis*(2-benzimidazolylmethyl)amine was synthesized under the microwave irradiation, and the complex ([DyL$_2$(NO$_3$)$_2$]NO$_3$) {where L is *bis*(2-benzimidazolylmethyl)amine} was synthesized (Ouyang et al, 2009). The dysprosium (III) complex was found to bind to DNA base pairs by partial intercalation and electrostatic binding. Additionally, pincer-type, pyridine-bridged *bis*(benzimidazolylidene)-palladium complexes **3** (R = *n*-C$_{16}$H$_{33}$, X = Br; R = *n*- C$_{16}$H$_{33}$, X = I; R = *n*-C$_8$H$_{17}$, X = I; R = *n*-C$_4$H$_9$, X = I) were synthesized from cheap commercial precursors under microwave assistance

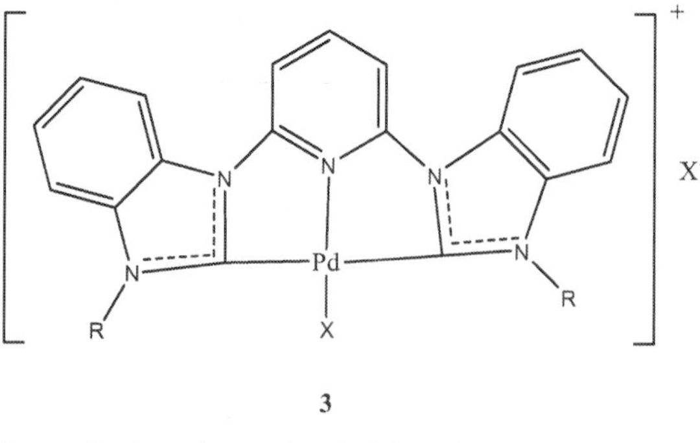

3

Among *azine metallocomplexes*, cyclometalated chloroplatinum complexes containing neutral monodentate ligands such as 2-phenylpyridine, 2-(2'-thienyl)pyridine or 4-

methoxypyridine, as well as the cyclometalated benzo[h]quinoline chloride complex with 4-methoxypyridine, were synthesized in a few minutes in 63-99% yields by irradiating the reaction mixture with microwaves (Godbert et al, 2007). The availability of this class of complexes in a few minutes offers the possibility of a combinatorial approach for the preparation of libraries of homologous compounds of potential interest for large-scale screening studies. MWAS of $[Cu_2(pz)_2(SO_4)(H_2O)_2]_n$ (pz = pyrazine) produced monocrystal suitable for X-ray diffraction studies, reducing reaction time and with higher yield than the classical hydrothermal procedures (Amo-Ochoa et al, 2007). The iridium complexes, obtained by cyclometalation of 5-(3-cyanophenyl)-2,3-diphenylpyrazine by 0.22 g of $IrCl_3 \cdot H_2O$ in 2-ethoxyethanol under a 100 W MW for 30 min, featured emission at 632 nm in chloroform solution, by reaction with sodium acetylacetonate (Inoue & Seo, 2010). The products were found to be useful as a phosphorescent compound for use in organic light-emitting devices. MWAS of the ligands *bis*(2-pyridylmethyl)amine (BMPA), Me 3-[*bis*(2-pyridylmethyl)amino]propanoate (MPBMPA), 3-[*bis*(2-pyridylmethyl)amino]propanamide (PABMPA), 3-[*bis*(2-pyridylmethyl)amino]propionitrile (PNBMPA), (3-aminopropyl)*bis*(2-pyridylmethyl)amine (APBMPA), and lithium 3-[*bis*(2-pyridylmethyl)amino]propanoate (LiPBMPA) were reported (Pimentel et al, 2007). A series of 2-(1-alkyl/aryl-1H-1,2,3-triazol-4-yl)pyridine (pytrz) ligands were synthesized using microwave-assisted Huisgen-Meldal-Fokin 1,3-dipolar cycloaddition and were used to prepare homoleptic and heteroleptic ruthenium(II) complexes with 4,4'-dimethyl-2,2'-bipyridine as second ligand (Happ, 2009). The iridium-quinoxaline complex **4** (where X is H or F) was prepared from iridium trichloride hydrate as metal source precursor in ethylene glycol by MWH for 4-5 min (Zhang et al, 2008). The complex had high solubility in common organic solvents, and can be used as electrophosphorescent material with high luminescence efficiency.

4

Complexes of such classic *polypyridine ligands* as 2,2'- or 4,4'-bipyridine (bipy) {or closely related 1,10-phenantroline (phen), which in terms of its coordination properties is similar to 2,2'-bipyridine} and their derivatives were also prepared by MWH, for instance $[Co(phen)_2Cl_2]ClO_4$ (Jin et al, 2009). A metal organic-inorganic coordination framework formulated as $\{[Cu(4,4'-bipy)(H_2O)_3(SO_4)] \cdot 2H_2O\}_n$ were similarly synthesized (Phetmung et al, 2009). The resulting compound was an 1D polymer in which 4,4'-bipy acted as a bridging ligand supporting the formation of infinite $[Cu(4,4'-bipy)(H_2O)_3(SO_4)]$ chains. Several reports are dedicated to noble metals, in particular ruthenium complexes; thus, microwave mediated reaction of [Ru(COD)Cl$_2$]$_n$ with 4,4'-di-*tert*-butyl-2,2'-bipyridine (tbbpy) in DMF gave 97.5% Ru(tbbpy)$_2$Cl$_2$ which on treatment with 5,5',6,6'-tetramethyl-2,2'-bibenzimidazole (tmbibim)/NH$_4$PF$_6$ gave 63% [Ru(tbbpy)$_2$(tmbibim)](PF$_6$)$_2$ (Walther et al, 2005). Ruthenium(II) polypyridine complex [Ru(Hdpa)$_3$](ClO$_4$)$_2$ {Hdpa = *bis*(2-pyridyl)amine} was prepared from RuCl$_3$·3H$_2$O in a few minutes in 91% yield (Xiao et al, 2002). Poly(bipyridine)ruthenium complexes [RuCl$_2$(dcmb)$_2$] (dcmb = 4,4'-dimethoxycarbonyl-2,2'-bipyridine) and [Ru(dcmb)$_{3-n}$(tbbpy)$_n$](PF$_6$)$_2$ (*n* = 0-3 and tbbpy = 4,4'-di-*tert*-butyl-2,2'-bipyridine) were also synthesized (Schwalbe et al., 2008). With the same tbbpy ligand, the oxodiperoxo complex MoO(O$_2$)$_2$(tbbpy) was isolated from the reaction of MoO$_2$Cl$_2$(tbbpy) in water under MWH at 120°C for 4 h (Amarante, 2009). It was established that the Mo^{VI} centre is seven-coordinated with a geometry which strongly resembles a highly distorted bipyramid. The crystal structure is formed by the close packing of the columnar-stacked complexes. Interactions between neighbouring columns are essentially of van der Waals type mediated by the need to effectively fill the available space. The authors noted that their synthesis route was surprising, since all known standard procedures for the synthesis of this type of complex involved a peroxide source such as H$_2$O$_2$ or *tert*-butyl hydroperoxide (TBHP). Rh^{III} complexes [Ru(L)$_2$][PF$_6$]$_2$ (L = L^1 or L^2) of enantiomerically pure, chiral terpyridyl-type ligands ligands L^1 ('dipineno'-[5,6:5'',6'']-fused 2,2':6',2''-terpyridine, 2,6-*bis*(6,6-dimethyl-5,6,7,8-tetrahydro-5,7-methanoquinolin-2-yl)pyridine) and L^2 ('dipineno'-[4,5:4'',5'']-fused 2,2':6',2''-terpyridine, 2,6-*bis*(7,7-dimethyl-5,6,7,8-tetrahydro-6,8-methanoisoquinolin-3-yl)pyridine), synthesized in high yields starting from 2,6-diacetylpyridine and enantiopure α-pinene, with Rh^{III} and Ru^{II} were prepared (Ziegler et al, 1999) and studied spectroscopically. These complexes had a helically distorted terpyridyl moiety, as shown by the considerable optical activity in the ligand centered and metal to ligand charge transfer transitions. Additionally, the MWAS (from Bpy$_2$OsCl$_2$·6H$_2$O as Os source) and photophysical properties of heterometallic dinuclear complex based on ruthenium and osmium *tris*-bipyridine units, Ru-mPh$_3$-Os (**5**), in which the metal complexes were linked via an oligophenylene bridge centrally connected in the *meta* position, were described (Aléo et al, 2005).

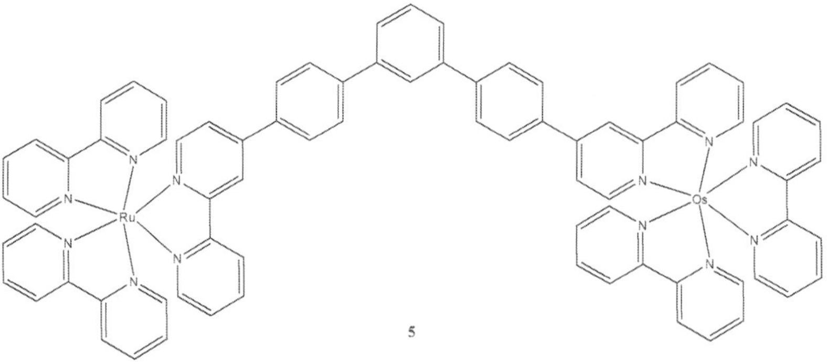

5

In a difference with acetylacetonates, the N-containing complexes are more rarely applied as precursors using MW-treatment for obtaining metal alloys and nanostructures. Thus, cyanogel coordination polymers (amorphous Prussian blue analogs formed in a hydrogel state by the reaction of a chlorometalate with a cyanometalate in aqueous solution) can be thermally auto-reduced to form transition-metal alloys {binary and ternary transition-metal alloys (Pd/Co, Pt/Co, Ru/Co, Ir/Co, Pd/Ni, Pt/Ni, Pt/Ru, Pd/Fe, Pd/Fe/Co) and intermetallics (Pt$_3$Fe, Pt$_3$Co, PtCo)}, in particular by MWH (Vondrova et al, 2007). The authors showed that the cyanogel polymers are susceptible to microwave dielectric heating, which leads to a sufficient temperature increase in the sample to cause the reduction of the metal centers, thus allowing for the conversion of cyanogels to metal alloys in a few minutes instead of hours needed in the traditional furnace heating.

3.4 Porphyrins

MWAS techniques have been developed for the synthesis and/or rearrangements of both metal-free porphyrins (Yaseen et al, 2009; Hou et al, 2007) and their metal complexes. Thus, several substituted 5,10,15,20-tetraarylporphyrins {5,10,15,20-tetraphenylporphyrin, 5,10,15,20-*tetrakis*(4-chlorophenyl)porphyrin, and 5,10,15,20-*tetrakis*(3-hydroxyphenyl)porphyrin} and insertion of five different transition metals into the porphyrin core were achieved with high yields using MW (Nascimento et al, 2007). Experimental protocols were characterized by extremely short reaction times and quite small quantities of solvents employed. 5,10,15,20-*Tetrakis*(2-pyridyl)porphyrin (H$_2$TPyP) and its complex, Mn(III)TPyP, were synthesized under MWH in the presence of propionic acid (Zhang et al, 2006). A *tetrakis*(terpyridinyl)porphyrin derivative and its Ru^{II} complexes, obtained through microwave-enhanced synthesis, were found to have photovoltaic properties and nanowire self-assembly (Jeong et al, 2007). Condensation adducts of the Ni(II) and Cu(II) complexes of β-amino-*meso*-tetraphenylporphyrin with di-Me acetylenedicarboxylate (DMAD) and di-ethylethoxymethylenemalonate were converted into the corresponding esters of pyridinone-fused porphyrins by using different cyclization protocols, including MW (Silva et al, 2009), resulting high yields in a short period of time under closed-vessel conditions. Soluble 5,10,15,20-*tetrakis*(4-*tert*-butylphenyl)metalloporphyrins [M(TBP), M = Mg, Cu, Tb(OAc), Lu(OAc), La(OAc)] were rapidly synthesized by microwave irradiation from 5,10,15,20-*tetrakis*(4-*tert*-butylphenyl)porphyrins [H$_2$(TBP)] or from pyrrole and 4-*tert*-butylbenzaldehyde with appropriate metal salts (Liu et al, 2005). The observed fluorescent properties of metalloporphyrins depend on their central metals due to heavy-atom effect. In a related work, soluble 5,10,15,20-*tetrakis*(4-*tert*-butylphenyl)magnesium porphyrins {Mg(TBP)}, perylene tetracarboxylic derivative [N,N'-bis(1,5-dimethylhexyl)-3,4:9,10-perylenebis(dicarboximide), PDHEP], and porphyrin-perylene tetracarboxylic complex were quickly prepared under MWH (Liu et al, 2004). It was revealed that porphyrin-perylene tetracarboxylic complex exhibited better fluorescent quantum yield and photo-electricity conversion effect than Mg(TBP) and PDHEP, respectively. MWAS methods were also developed to cleanly produce the tetra(2',6'-dimethoxyphenyl)porphyrin and its Fe, Zn, and Ni complexes (Wolfel et al, 2009). Additionally, Cu(II) complexes with asymmetrical (5-(3-hydroxyphenyl)-10,15,20-*tris*-(4-carboxymethylphenyl)-21,23-Cu(II)-porphine) **6** and symmetrical (5,10,15,20-*meso-tetrakis*-(4-carboxymethylphenyl)-21,23-Cu(II) porphine) **7** porphyrinic ligands were synthesized with superior yields using MW to be used in unconventional treatment of various diseases by means of photodynamic therapy (PDT) (Boscencu et al, 2010). The results of the biological *in vitro* tests indicated a low cytotoxicity of the compounds for the studied cells. At last, the one-step 15 min (instead of 24 h by classic preparation) synthesis of metalloporphyrazines with enhanced yields directly from substituted maleonitriles, involving tetramerization using hexamethyldisilazane and *p*-toluenesulfonic acid and DMF in a sealed tube under MW (Chandrasekharam et al, 2007).

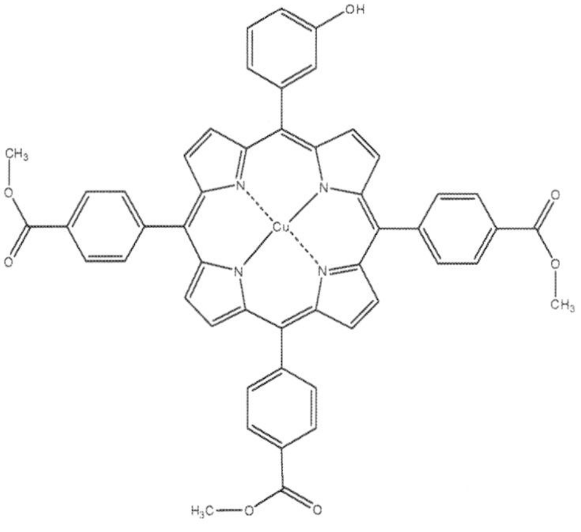

6

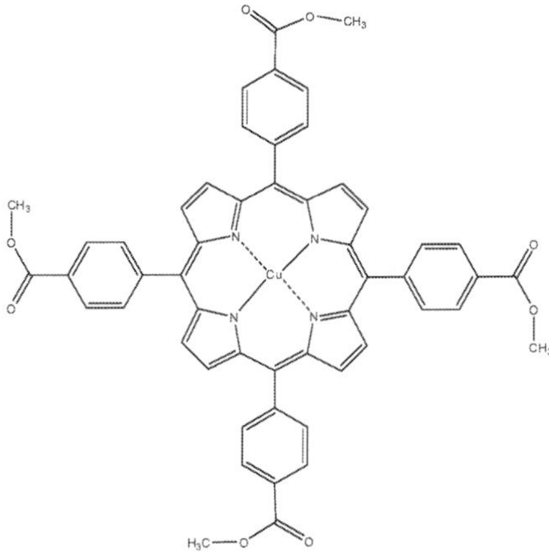

7

3.5 Phthalocyanines

Phthalocyanine (Pc) area is industrially important, in a difference with major part of N-containing ligands having an academic interest only, since both metal free phthalocyanines and their several metal complexes (Cu, Zn, Ni, Fe, etc.) are produced during several decades in large quantities and used as pigments, in compact disk production, and catalysis, among many other applications. So, novel techniques for their production are permanently in search, as for classic Pcs as for substituted (generally R_4Pc for symmetrical Pcs; R = alkyl, aryl, Cl, NO_2, ethers, crowns, etc.). In particular, a variety of metal phthalocyanine complexes has been fabricated via MWH allowing absence of solvents (we note that solvent nature is very important for tetramerization of phthalonitrile and other Pc precursors). Thus, metal substituted octachlorophthalocyanines (M = Fe, Co, Ni, Cu, Zn), hexadecachlorophthalocyanines (M = Fe, Co, Ni, Cu) and tetranitrophthalocyanines (M = Fe, Co, Ni, Cu, Pd) were synthesized by exposure to MW under solvent free and reflux conditions (Safari et al, 2004; Shaabani et al, 2003). The synthesis of various axially substituted Ti phthalocyanines in high yield using MW without solvent was reported (Maree, 2005). The times of reaction, as expected, were short (generally <10 min). Substituted Fe and Co octachloro-, tetranitro-, tetracarboxy- or polyphthalocyanines were easily prepared by MWH of the starting materials under solvent free condition, which reduced reaction time considerably and used as epoxidation catalysts of cyclooctene in homogeneous and heterogeneous conditions by iodosylbenzene as an oxidant (Bahadoran & Dialameh, 2005). Their catalytic activities showed that the electron withdrawing groups on the phthalocyanine ring have a very small effect on stability of the catalyst during the reactions. The tetrasubstituted metal-free phthalocyanine 8 (R = SO_2NH-*p*-C_6H_4Me) and its nickel and zinc metallophthalocyanines bearing four 14-membered tetraaza macrocycles moieties on peripheral positions were synthesized by cyclotetramerization reaction of phthalonitrile derivative **9** in a multi-step reaction sequence (Biyiklioglu et al., 2007). Additionally, a reaction mixture containing perfluoro-phthalonitrile reacted in a vessel with application of microwave energy for a reaction period sufficient to yield a fluorinated phthalocyanine (Fraunhofer-Gesellschaft et al, 2009), having wide ranging applications, e.g., corrosion-related applications, coating-related applications, catalysis, and the production of optical and electronic materials.

8

9

Thermal and microwave reactions between [PcSn^{IV}Cl$_2$] and the potassium salts of eight fatty acids led to *cis*-[(RCO$_2$)$_2$Sn^{IV}Pc] compounds {R = (CH$_2$)$_n$Me (*n* = 4, 6, 8, 10, 12, 14, 16) and (CH)$_7$-*cis*-CH:CH(CH$_2$)$_7$Me} in yields ranging from 54 to 90% (Beltran et al, 2005). Some products revealed anticorrosion properties. Triazol-5-one substituted phthalocyanines were prepared quickly by the reactions (*1*) of 4-nitrophthalonitrile with anhydrous metal (M = Co, Cu, Zn, Ni) salts in DBU (1,8-diazabicyclo[5,4,0]undec-7-ene) and DMAE (dimethylaminoethanol) by MW. Microwave yields were found to be higher than those of the conventional synthesis methods (Kahveci et al, 2006). We note that some metal-free substituted phthalocyanines {2,9(10),16(17),23(24)-tetra(3,5-dimethylphenoxy) phthalocyanine, 2,9(10),16(17), 23(24)-tetra(4-*tert*-butylphenoxy) phthalocyanine, and 2,9(10),16(17),23(24)-tetra(3,5-di-*tert*-butyl-4-hydroxyphenyl) phthalocyanine} were also obtained by similar routes with higher yields in comparison with conventional methods (Seven et al, 2009). These Pc-compounds had high thermal stability, which was determined at 520ºC (midpoint), 549ºC, and 400ºC, respectively, as a maximum weight loss temperature.

Bis- and *sub*-phthalocyanines, as well as mixed phthalocyanine-porphyrin complexes, were also reported as MW-fabricated. Thus, starting with phthalic and 4-*tert*-butylphthalic acid derivatives, the *bis*phthalocyanines of rare earth elements and Hf and Zr were MW-prepared (Kogan et al, 2002). *Sub*-phthalocyanine (SubPc) derivatives with different kind of substituent groups were synthesized from various phthalonitriles using conventional and microwave heating sources (Kim et al, 2009). Compared to the conventional synthesis, it was found that SubPc derivatives were synthesized in a shorter reaction time with a higher synthetic yield by MW. A soluble phthalocyanine-porphyrin complex {Lu(TBPor)Pc} was quickly obtained by MWH; Lu(TBPor)Pc was shown to have better photoelectric conversion properties than porphyrin {Lu(TBPor)OAc}, phthalocyanine {H_2(TBPc)}, and Lu(TBPor)OAc/H_2(TBPc) blend (Liu et al, 2004). More information on MW-synthesis of phthalocyanines was reported: Ga (Masilela & Nyokong, 2010), and other metals (Co, Ni, Cu, Mg, Al, Pd, Sn, Tb, Lu, Ce, La, Zn) (Hu et al, 2002; Park et al, 2001).

4. Complexes with N,O-containing ligands

These coordination compounds are widely represented by a series of oximes, amines, imines, Schiff bases, as well as such cyclic N,O-ligands as oxadiazoline. Cluster complexes have been also reported, in particular those that cannot be obtained by standard non-microwave techniques. Thus, tetradentate N_2O_2 ligand [HO(Ar)CH:N-$(CH_2)_2$-N:CH(Ar)OH] (Ar = o-C_6H_4) and manganese(II), cobalt(II), nickel(II), and zinc(II) diimine complexes ML were synthesized by classical and MW techniques (Pagadala et al., 2009). It was proposed that, probably, the metal is bonded to the ligand through the phenolic oxygen and the imino nitrogen. The reaction of Ni(ClO_4)$_2$·6H_2O with 2-hydroxybenzaldehyde and an aqueous solution of methylamine in acetonitrile/MeOH under MWH and controlled temperature/pressure gave trinuclear cluster [Ni_3(mimp)$_5$-(MeCN)]ClO_4 (mimp = 2-methyliminomethylphenolate anion) in only 29 min and also resulted in higher yields in contrast to other synthesis methods (Zhang et al, 2009). This complex displayed dominant ferromagnetic interactions through μ_3-O (oxidophenyl) and μ_2-O (oxidophenyl) binding modes. Another cluster, unusual for a specific group of complexes, was found for an oxime complex. Thus, the microwave-assisted reaction of Fe(O_2CMe)$_2$ with salicylaldoxime (saoH_2) in pyridine produced an octametallic cluster [Fe_8O_4(sao)$_8$(py)$_4$] in crystalline form in 2 min (Gass et al, 2006). The core of the complex contained a cube encapsulated in a tetrahedron while sao^{2-} exhibited an unique coordination mode $\eta^2:\eta^1:\eta^1:\mu_3$ among the structurally characterized metal complexes containing the sao^{2-} ligand. The authors noted that [Fe_8O_4]$^{4+}$ core is uncommon, observed earlier only in one other complex: [Fe_8O_4(pz)$_{12}$$Cl_4$] (pz = pyrazolate anion). The MW-heating had not only led to the isolation of a beautiful and unusual {Fe^{III}_8} cluster, impossible to produce under ambient reaction conditions, but has also greatly improved the reaction rate and enhanced the yield in comparison to solvothermal methods.

Among other oxime complexes, the metal-mediated iminoacylation of *ketoximes* R^1R^2C:NOH (R^1 = R^2 = Me; R^1 = Me, R^2 = Et; R^1R^2 = C_4H_8; R^1R^2 = C_5H_{10}) upon treatment with the platinum(II) complex *trans*-[$PtCl_2$(NCCH$_2$$CO_2$Me)$_2$] with an organonitrile bearing an acceptor group proceeded under mild conditions in dry CH_2Cl_2 or in microwave field to give the *trans*-[$PtCl_2$\{NH:C(CH$_2$$CO_2$Me)ON:C$R^1R^2$\}$_2$] isomers in moderate yield (Lasri et al, 2006). Nine cobaloximes of the type *trans*-[Co(dmgH)$_2$(B)X], where dmgH$^-$ = dimethylglyoximate anion, X$^-$ = Cl$^-$, Br$^-$ or I$^-$ and B = pyrazine, Pz (1 to 3), pyrazine carboxylic acid, PzCA (4 to 6), pyrazine carboxamide, PzAM (7 to 9), imidazole (Imi) or histidine (His), were prepared (an example of the complex, N,N'-dihydrogenpiperazonium dichloridobis(dimethylglyoximato-k^{2}N,N')cobaltate(III) dihydrate, PpH$_2$[Co(dmgH)$_2$$Cl_2$]$_2$·2$H_2$O, is shown by formula 10) (Martin et al, 2008; Dayalan et al,

2009). The free ligands Pz, PzCA and PzAM showed antibacterial activity in the order: Pz > PzCA > PzAM whereas, the free equatorial ligand $dmgH_2$ was inactive against all the bacteria tested. The cobaltoximes were more active than the corresponding pyrazine and its derivatives as axial ligand in the complexes. It was revealed that the bromo complexes dissociated at higher temperatures compared to the chloro complexes, the iodo cobaloximes being unstable even at low temperature decomposing without any sharp change in mass. Iodocobaloximes were found to be more active than the corresponding chloro- and bromo-cobaloximes with the antibacterial activity order for the axial halides as I^- > Cl^- > Br^- and that of the axial nitrogen heterocycles as histidine > imidazole. Additionally, a 3D coordination polymer, $[Cd(\mu_3\text{-HIDC})(bbi)_{0.5}]_n$ {H_3IDC = 4,5-imidazoledicarboxylic acid, bbi = 1,1'-(1,4-butanediyl)*bis*(imidazole)}, was synthesized under MWH solvothermal conditions (Liu et al, 2008). Its crystal structure consisted of 2-D brickwall-like networks of $[Cd(\mu_3\text{-HIDC})]_n$, which are further linked through μ_2-bbi to generate a 3D structure.

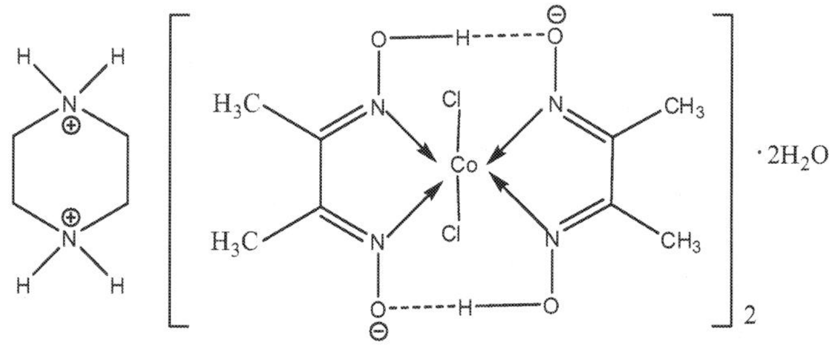

10

Microwave-assisted [2+3] cycloaddition of nitrones to the nitrile ligands in *cis*- or *trans*-[PtCl$_2$(PhCN)$_2$] occurred under ligand differentiation and allowed for selective synthesis of *cis*- or *trans*-[PtCl$_2$(oxadiazoline)(PhCN)] (Desai et al, 2004). Reaction of the *trans*-substituted mono-oxadiazoline complexes with a nitrone different from the one used for the first cycloadditionj step gave access to mixed *bis*-oxadiazoline compounds *trans*-[PtCl$_2$(oxadiazoline-a)(oxadiazoline-b)]. The corresponding *cis*-configured complexes, however, did not undergo further cycloaddition. In case of palladium complexes, the reaction between the nitrone p-MeC$_6$H$_4$CH:N(Me)O and *trans*-[PdCl$_2$(RCN)$_2$] (R = Ph, Me) in the corresponding RCN (or of the nitrone in neat RCN in the presence of PdCl$_2$) proceeded at 45°C (R = Ph) or reflux (R = Me) for 1 day and gave the Δ4-1,2,4-oxadiazoline complexes [PdCl$_2$\{Na:C(R)ON(Me)CbH(C$_6$H$_4$Me-p)\}$_2$(Na-Cb)] (R = Ph, Me) in ~50 and ~15% yields, respectively (Bokach et al, 2005). The reaction time can be drastically reduced by focused MW of the reaction mixture.

Phenylantimony chloride and Sb chloride complexes with Schiff base ligands having N-S and N-O donor systems were synthesized under MW using a domestic microwave oven from hours to a few seconds with improved yield as compared with conventional heating (Mahajan et al, 2008). The treatment with the ligands and their phenylantimony derivatives at dose levels of 20 mg per rat per day did not cause any significant change in body weight, but a significant reduction in the weights of reproductive organs was observed. Transition metal complexes of Cu(II), Ni(II), Co(II), Mn(II), Zn(II), Hg(II), and Sn(II) were synthesized from the Schiff base (L) derived from 4-aminoantipyrine and 4-fluoro-benzaldehyde using traditional synthetic methodology and microwave-induced organic reaction enhancement (MORE) technique (Ali et al, 2010). Neat reactants were subjected to microwave irradiation giving the required products more quickly and in better yield compared to the classical methodology. As an example of use of Schiff base complexes for catalytic purposes, we note an octahedral titanium binaphthyl-bridged Schiff base complex **11**, investigated in respect of catalytic behavior toward epoxidation of allylic alcohols (Soriente et al, 2005). It was established that a mixture of monoterpene alcohol **12**, *tert*-Bu hydroperoxide, the complex **11**, and CH_2Cl_2, being irradiated with microwave for 15 min, gave 87% terpene epoxy alcohol **13**.

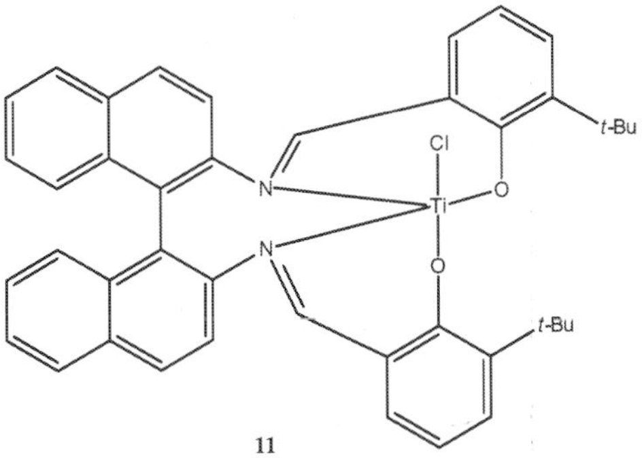

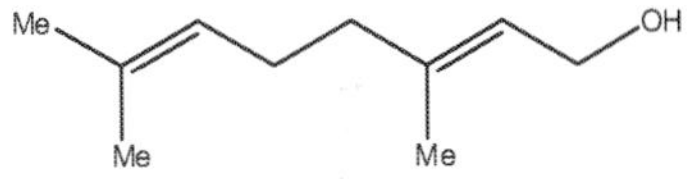

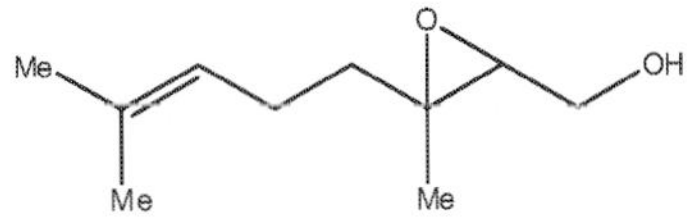

Four ligands *i.e.* N,N'-*bis*(3-carboxy-1-oxopropanyl)-1,2-dimethylethylenediamine (CDMPE), N,N'-*bis*(3-carboxy-1-oxoprop-2-enyl)-1,2-dimethylethylenediamine (CDMPE-2) N,N'-*bis*(3-carboxy-1-oxopropanyl)-1,2-diethylethylenediamine (CDEPE), N,N'-*bis*(3-carboxy-1-oxoprop-2-enyl)-1,2-diethylethylenediamine (CDEPE-2) and their manganese complexes were prepared by microwave method (Bhojak et al, 2008). Antibacterial activity of the ligands and complexes were also reported on *S. aureus* and *E. coli*. Complexes of Mn(II) with 4 amide group containing ligands (Bhojak et al, 2007) {N, N'-*bis*-(3-carboxy-1-oxopropanyl)-1,2-ethylenediamine (CPE), N,N'-*bis*-(3-carboxy-1-oxo-propanyl)-1,2-phenylenediamine (CPP), N,N'-*bis*-(2-carboxy-1-oxophenelenyl)-1,2-phenylenediamine (CPPP), N,N'-*bis*-(3-carboxy-1-oxoprop-2-enyl)-1,2-phenylenediamine (CPP-2), obtained by MW-heating of amine and carboxylic acid} were MW-synthesized. Typical preparation of these complexes included simple steps: a slurry of ligand (i.e. CPE, CPP, CPPP or CPP-2) was prepared in water or in water-ethanol mixture; in this a solution of $Mn(CH_3COO)_2 \cdot 4H_2O$ was added, and the resulting mixture was irradiated in a microwave oven for 2 to 6 minutes at medium power level (600 W) maintaining the occasional shaking. Proposal structures of complexes are shown by formulae **14-17**. The antibacterial activity of the ligands and complexes was studied. Additionally, the Chinese-lantern-type $Co_2(O_2CBut)_4\{2,6\text{-}(NH_2)_2C_5H_3N\}_2$ complex reacted with RCN (R = Me or Pr) under microwave irradiation to give the mononuclear *amidine complexes* $Co(O_2CBut)_2\{H_2N(C_5H_3N)NHC(R):NH\}$ (R = Me or Pr) (Bokach et al, 2006).

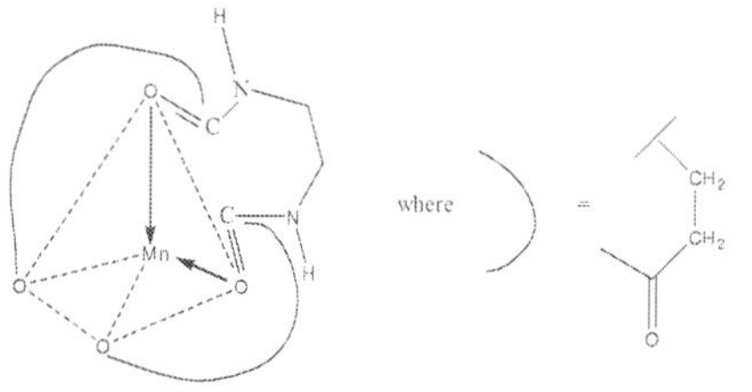

Al-containing mesoporous silicates (Al-MCM-41 and Al-HMS) supported Mn(salen) catalysts were prepared by three different methods: impregnation of salen ligand and support with dichloromethane and then irradiated by microwave (method A), direct solid-state interaction between salen complex and support under microwave irradiation (method B), as well as the conventional ion exchange (method C) (Yin et al, 2005; Zhang et al, 2003). The effect of catalyst preparation methods on the catalytic activity and selectivity in the styrene epoxidation indicate that the catalyst of Mn(Salen)/Al-HMS-IP prepared by method A showed similar activity to the neat complex and the best selectivity for styrene epoxide. In comparison with the traditional adsorption method, the MW-assisted approach was efficient and environmentally friendly, and improved the loadings of Mn(III)-salen complexes on HMS via a strengthening axial coordination of the surface NH_2 groups of HMS toward the Mn(III)-salen complexes (Fu et al, 2007). The effects of several extrinsic physical fields, such as the magnetic field, the ultrasonic wave and the MW, on the rate and yield of chitosan-Fe(II) complexing reaction were investigated (Jiang et al, 2008), showing that ultrasound had the greatest effect on the reaction rate and complexing capacity, followed by the magnetic field and the MW. A mechanism for the enhancement of the complexing reaction by the three physical fields was proposed.

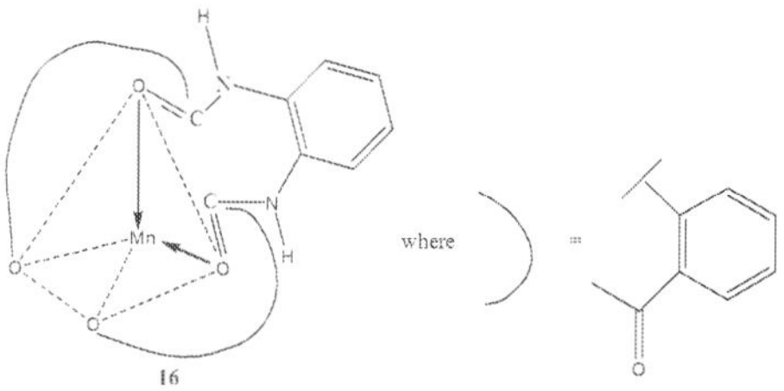

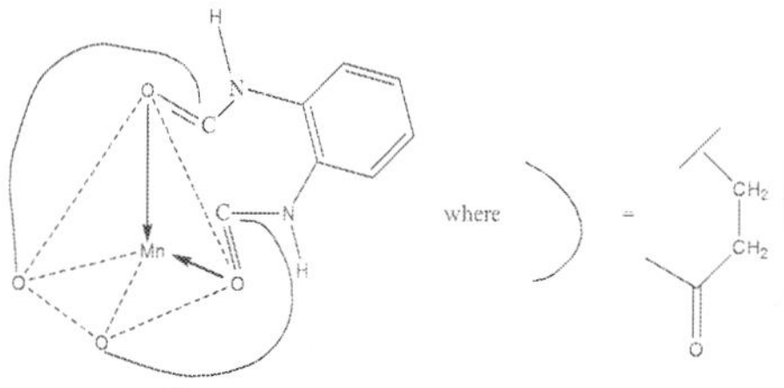

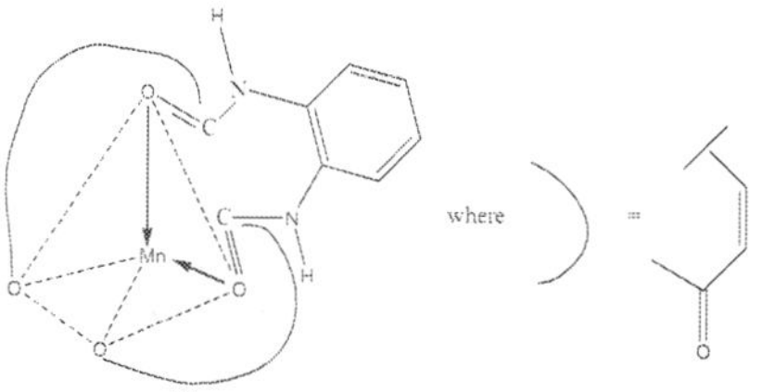

5. Complexes with S- and N,S-containing ligands

According to the available literature, microwave-synthesized complexes of S-containing ligands without other donor heteroatoms are represented by coordination compounds of dithiolene. Thus, dithiolene-transition metal complexes **18** were obtained by a series of steps (Kim et al, 2009) including microwave heating in the first steps of the mixture of benzaldehyde and KCN in EtOH.

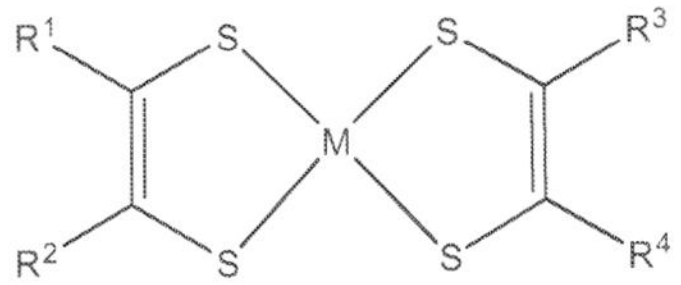

18

N,S-Complexes are represented by a series of different types of compounds: thiolates, thioimidazoles, thiazoles, thiosemicarbazones, among others. Thus, the solid phase reaction of 1-alkyl-2-{(o-thioalkyl)phenylazo}imidazoles (SRaaiNR) and RuCl$_3$ on silica gel surface upon MW yielded [Ru(SRaaiNR)(SaaiNR)](PF$_6$) (Mondal et al, 2009). BiPh$_3$ was treated with thiols of varying pKa and functionality (2-mercaptobenzothiazole, 2-mercaptobenzoxazole, 2-mercaptopyrimidine, 2-mercapto-1-methylimidazole and 2-mercaptobenzoic acid) in a 1:3 ratio under a variety of reaction conditions: with toluene or mesitylene under standard reflux conditions and under microwave irradiation, and solvent free with conventional and microwave heating (Andrews et al, 2007). As a result, several reactions yielded the *tris*-substitution product in good yield and high purity; 2-mercaptobenzoic acid gave the complex Bi$_2$L$_3$ in all reactions carried out in solvent and PhBiL when solvent free, both complexes containing the doubly deprotonated dianion (L = -O$_2$C-C$_6$H$_4$-S-). The authors noted that reactions carried out in the microwave reactor generally gave comparable yields to the conventional methods but in significantly shorter times; however, the solvent free microwave reactions of 2-mercaptobenzoxazole and 2-mercaptopyrimidine caused partial decomposition to give microcrystalline Bi$_2$S$_3$. MWH of racemic *cis*-[Ru(bpy)$_2$(Cl)$_2$] (bpy = 2,2'-bipyridine) or racemic *cis*-[Ru(phen)$_2$(Cl)$_2$] (phen = phenanthroline) with either (R)-(+)-or (S)-(-)-Me *p*-tolyl sulfoxide yielded the ruthenium *bis*(diimine) sulfoxide complexes, for example **19** (Pezet et al, 2000). This source of energy improved both yields and reaction rates with a very good diastereoselectivity (73-76%) and represented a significant advance in the asymmetric synthesis of octahedral ruthenium complexes.

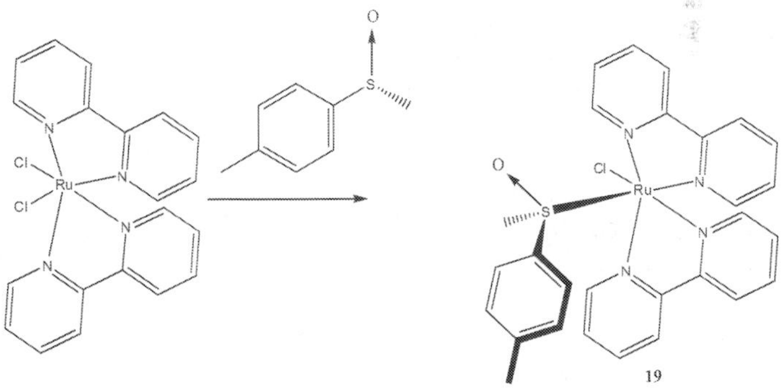

19

A lot of complexes of thiosemicarbazone and its derivatives have been MW-obtained. Thus, molybdenum(VI) complexes MoO$_2$(L)$_2$ of the ligands HL {3,4,5-trimethoxybenzaldehydethiosemicarbazone (TBTSCZH), 3,4,5-trimethoxybenzaldehydesemicarbazone (TBSCZH), 3,4,5-trimethoxybenzaldehydebenzothiazoline (TBBZTH) and 3,4,5-trimethoxybenzaldehyde-S-benzyldithiocarbazate (TBDTCZH)} were MW-fabricated(Maanju et al, 2007) by the reactions between dioxo*bis*(2,4-pentanedionato-O,O')molybdenum(VI) and the ligands TBTSCZH, TBSCZH, TBBZTH and TBDTCZH by MW-assisted and conventional thermal methods. All four ligands and their complexes were screened for their biological activity on several pathogenic fungi and bacteria and the data show good activity of these complexes and ligands. The synthesis of some Mn(II), oxovanadium(V) and dioxomolybdenum(VI) complexes with 5-chloro-1,3-dihydro-3-[2-(phenyl)ethylidene]-2H-indol-2-one thiosemicarbazone (L^{1}H) and 5-chloro-1,3-dihydro-3-[2-(phenyl)ethylidene]-2H-indol-2-one semicarbazone (L^{2}H) were carried out in unimolar and bimolar ratios in an open vessel under MW using a domestic microwave oven. In the case of the oxovanadium complexes, the metal was found to be in the penta- and hexa-coordinated environments. The ligands and complexes possessed antimicrobial properties. Trigonal bipyramidal and octahedral complexes of Sn(IV) were synthesized by the reaction of dimethyltin(IV) dichloride with 4-nitrobenzanilidethiosemicarbazone (L^{1}H), 4-chlorobenzanilidethiosemicarbazone (L^{2}H), 4-nitrobenzanilidesemicarbazone (L^{3}H) and 4-chlorobenzanilidesemicarbazone (L^{4}H) from dimethyltin(IV) dichloride and monobasic bidentate ligands using MW as the thermal energy source (Singh et al, 2008). The antifungal, antibacterial and antifertility activities were examined and the results were indeed very encouraging. A series of mixed ligand ruthenium(II) containing diimines and thiosemicarbazones with general formula [Ru(N-N)$_2$(N-S)](PF$_6$)$_2$ where N-N = bipyridine or 1,10-phenanthroline and N-S = 9-anthraldehyde thiosemicarbazone and the 4-alkyl substituted (R = Me, Et and phenyl) analogs were synthesized using microwave energy (Beckford et al, 2009; Beckford et al, 2010). The compounds quenched the fluorescence of the complex between ethidium bromide and calf-thymus DNA with the Stern-Volmer quenching consisted in the range 1.18-2.71·10^4 M^{-1}. Additionally, the Pd(II) and Pt(II) complexes were synthesized using microwave heating by mixing metal salts in 1:2 molar ratios with heterocyclic ketimines, 3-acetyl-2,5-dimethylthiophene thiosemicarbazone (C$_9$H$_{13}$N$_3$OS$_2$) and 3-acetyl-2,5-dimethylthiophene semicarbazone (C$_9$H$_{13}$N$_3$OS), obtained by reactions of 3-acetyl-2,5-dimethylthiophene with thiosemicarbazide and semicarbazide hydrochloride (Sharma et al, 2010). The authors proposed that the ligands coordinate to the metal atom in a monobasic bidentate manner and square planar environment around the metal atoms. The antiamoebic activity of both the ligands and their palladium compounds against the protozoan parasite *Entamoeba histolytica* was tested. Other data on MW-obtaining thiosemicarbazone complexes were discussed in (Chaudhary et al, 2009; Shen et al, 2008).

In case of thiophene derivatives, MW-assisted condensation of salicylaldehyde with 2-amino-3-carboxyethyl-4,5-dimethylthiophene in the absence of solvent was efficiently performed to form a potentially tridentate Schiff base, 2-(N-salicylideneamino)-3-carboxyethyl-4,5-dimethylthiophene (HSAT), which acted as neutral tridentate with ONO donor sequence towards the lanthanide(III) ions, forming 1:2 metal-ligand complexes of the type [Ln(HSAT)$_2$Cl$_3$] where Ln = La(III), Ce(III), Pr(III), Nd(III), Sm(III), Eu(III) and Gd(III) (Kumasi et al, 2009). Additionally, it is known that thiophene can react with elemental iron in the form of metal atoms in cryosynthesis conditions or its carbonyls carrying out the desulfurization of the ligand. In reactions with iron carbonyls, the use of MWH evidently led (Singh et al, 1996) to acceleration of reported reactions of thiophene and its tellurium analogue and its derivatives with [Fe$_3$(CO)$_{12}$]. The following dechalcogenation reactions take place, forming binuclear complexes **20-21** (reactions 2). Among other organometallic compounds, prepared this way, it is necessary to mention chromium, molybdenum, and tungsten carbonyls (Van Atta et al, 2000).

$$[Fe_3(CO)_{12}] + \quad \longrightarrow \quad \longrightarrow \quad + FeS \qquad (2)$$

X = S, Te

20 21

Cyanobipyridine-derived zinc(II) *bis*(thiolate) complexes [Zn(L)(SAr)$_2$] (L = 2-methyl-4-(4-biphenyl)-6-(2-pyridyl)nicotinonitrile and 2-methyl-4-(4-(diphenylamino)phenyl)-6-(2-pyridyl)nicotinonitrile, Ar = Ph, 4-MeOC$_6$H$_4$, 2-naphthalenyl) were prepared rapidly and efficiently by a microwave-assisted cross-coupling/complexation sequence and display luminescence that can be modulated using intrinsic functionality and ancillary ligands (Bagley et al, 2010). Organotin complexes of thiol- and thione-containing Schiff bases were MW-prepared and tested for antifungal activity, using Ph$_3$SnCl, Ph$_3$SiCl, Ph$_2$SnCl$_2$ as metal source and the sodium salts of ligands, 2-HSC$_6$H$_4$N:CMeCH$_2$(*i*-Pr) (L^{1}H) and (*i*-PrCH$_2$C(Me):NNHC(S)NH$_2$ (L^{2}H), synthesized by condensation of 4-methyl-2-pentanone with 2-aminobenzenethiol and thiosemicarbazide, respectively (Gaur et al, 2005). Pentacoordinated complexes Ph$_3$SnL^1, Ph$_3$SiL^1, Ph$_2$SnClL^1, Me$_2$SnClL^1, Ph$_3$SnL^2, Ph$_3$SiL^2 and hexacoordinated complexes Ph$_2$SnL$^1_2$ and Me$_2$SnL$^1_2$ were isolated and tested against a number of microorganisms exhibiting inhibition of growth of *Aspergillus niger*, *Fusarium oxysporum* and *Alternaria alternata*. MWAS and spectroscopic studies of dimethyl-, diphenyl- and triphenyl- Si(IV) chelates derived from the reactions of organochlorosilanes with the sodium salt of a biologically active N-donor ligand [1-(furan-2-yl)ethylidene][4-[(pyridin-2-yl)sulfamoyl]phenyl]amine was reported (Singh et al, 2005). The biological activity of the ligand and its corresponding complexes with regard to antifungal and antibacterial activity against pathogenic fungi and bacteria was revealed; all the compounds also acted as nematicides and insecticides, by reducing the number of nematodes (*Meloidogyne incognita*) and insects (*Trogoderma granarium*).

Antimony complexes with substituted thioimines (**22**) were prepared by reaction of Ph$_3$Sb and [1-(2-naphthyl)ethylidene]hydrazinecarbodithionic acid phenyl ester, [1-(2-thienyl)ethylidene]hydrazinecarbodithionic acid phenyl methyl ester, [1-(2-pyridine)ethylidene]hydrazinecarbodithionic acid phenyl ester, and and [1-(2-furanyl)ethylidene]hydrazinecarbodithionic acid phenyl ester by MWH (Mahajan et al, 2007). Reactions of Ph$_3$Sb and monobasic bidentate ligands having N⌢S donor set in 1:1 and 1:2 molar ratios proceeded with the cleavage of the antimony carbon bond of Ph$_3$Sb and yielded monosubstituted derivatives (reactions 3-4).

$$Ph_3Sb + N{\frown}SH \rightarrow Ph_2Sb(N{\frown}S) + C_6H_6 \qquad (3)$$

$$Ph_3Sb + 2N{\frown}SH \rightarrow Ph_2Sb(N{\frown}S)_2 + 2C_6H_6 \qquad (4)$$

22a   22b

Thione form   Thiol form

R = CH,   R' =   and

The reaction product of 2-hydroxy-N-phenylbenzamide with 2-aminobenzenethiol, 2-(2-hydroxyphenyl)-2-(phenylamino)benzothiazoline (H$_2$-Saly·BTZ), reacted with PhSbCl$_2$, SbCl$_3$, and BiCl$_3$ under varied reaction conditions (microwave, as well as conventional method) leading to corresponding antimony(III) and bismuth(III) compounds (an example is **23**) (Mahajan et al, 2009). The ligand was found to bifunctional tridentate, as well as monodentate for different starting materials of metal (Sb/Bi). The complexes were more toxic than the corresponding ligand.

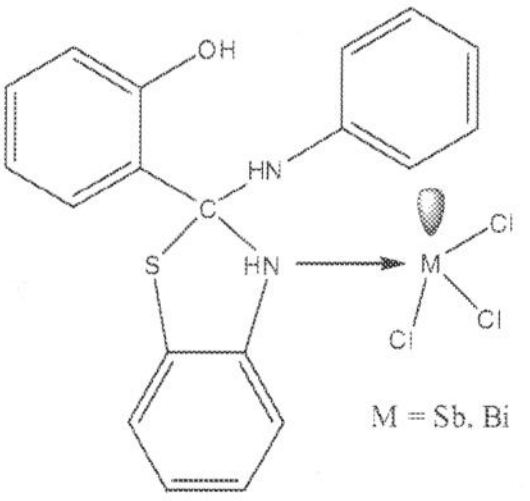

23

A highly semiconducting 1D coordination polymer architecture was obtained by the reaction of a Cu^{II} salt with 2,2'-dipyridyldisulfide under microwave solvothermal conditions, proceeding with an unusual C-S and S-S bond cleavage of the 2,2'-dipyridyldisulfide ligand to give $\{[Cu^I{}_9(L)_8(SH)_8](BF_4)\}_n$ (SH = 2-pyridylthiolate) and $\{[Cu^{II}(2\text{-}dps)_2]_2(\mu\text{-}S)\}(BF_4)_2\cdot 4CH_2Cl_2$ (2-dps = 2,2'-dipyridylsulfide) (reaction 5) (Delgado et al, 2008). The unprecedented architecture of the first compound consisted of a 1D polymeric chain formed by the assembling of Cu$_9$ cluster cages. In a related report of the same research group, an unprecedented microwave C(sp^2)-S and S-S bond activation of 2,2'-dipyridyldisulfide (2-dpds) and the formation of an architecture of coordination networks obtained by reaction of Cu(HCO$_2$)$_2\cdot x$H$_2$O with 2-dpds in the same conditions were described (Delgado et al, 2010). Partial oxidation of 2,2'-dipyridyldisulfide to sulfate was found to take place, resulting a Cu(I) dimetallic complex [Cu$_2$(μ-Hpyt)$_2$(Hpyt)$_4$](SO$_4$)·~5EtOH (Hpyt = pyridine-2(1H)-thione), a Cu(I,II) polycationic coordination polymer [Cu(H$_2$O)$_6$][Cu$_6$(μ-Hpyt)$_{12}$](SO$_4$)$_4\cdot$4H$_2$O, and a dimetallic Cu(II) complex [Cu(2-dps)(μ-SO$_4$)(H$_2$O)]$_2\cdot$3H$_2$O. The strong red and yellow-orange luminescence was shown for the first two complexes.

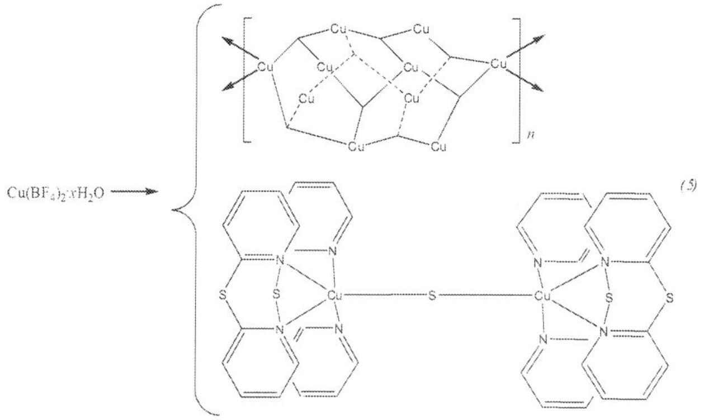

6. σ- and π-organometallic compounds

Microwave heating has been applied to obtain a series of metal complexes with classic ligands forming σ- and π-organometallic compounds: carbonyls, cyclopentadienyls, dienes, and arenes, among others. Generally, as well as for the case of the coordination compounds above, main advantages of MW-application are frequently higher yields and almost always considerably shorter reaction times.

6.1 Carbonyls

Among fundamental generalizing publications on MW-fabricated metal carbonyls, we note a review on Group 6 metals, describing, in particular, metal carbonyls synthesis in a conventional MW-oven (Holder, 2005), and a report (Ardon et al, 2004) dedicated to the preparation of a series of mixed Group 6 metal carbonyl complexes with other ligands {cis-[Mo(CO)$_4$(dppe)], [CpMo(CO)$_3$]$_2$, [Cp$_2$Mo$_2$(CO)$_4$(μ-RC$_2$R)], [CpMo(CO)$_2$]$_2$, cis-[W(CO)$_4$(pip)$_2$], [Cr(CO)$_5$Cl][NEt$_4$], where dppe = 1,2-bis(diphenylphosphino)ethane, pip = piperidine}. Also, mixed carbonyl-arene complexes are known; thus, the microwave-assisted synthesis of (η^6-arene)tricarbonylchromium complexes from hexacarbonylchromium and arenes gave high yields of (η^6-arene)chromium tricarbonyl complexes (Lee et al, 2006). In case of noble metals, by using a gas-loading accessory, microwave-assisted synthesis of Ru$_3$(CO)$_{12}$, Ru$_3$(CO)$_9$(PPh$_3$)$_3$, HRu$_3$(CO)$_9$(C≡CPh) and H$_4$Ru$_4$(CO)$_{12}$ was performed (Leadbeater et al, 2008). Ligand substitution reactions of Ru$_3$(CO)$_{12}$ with triphenylphosphine were also studied in real time by means of a digital camera interfaced with the microwave unit. Microwave-assisted ligand substitution reactions of Os$_3$(CO)$_{12}$ in a remarkably short period of time led to the labile complex Os$_3$(CO)$_{11}$(NCMe) in high yield without the need for a decarbonylation reagent such as trimethylamine oxide (Jung et al, 2009). Additionally, MWH of Os$_3$(CO)$_{12}$ in a relatively small amount of acetonitrile was shown to be a useful first step in two-step, one-pot syntheses of the cluster complexes Os$_3$(CO)$_{11}$(py) and Os$_3$(CO)$_{11}$(PPh$_3$). Microwave-assisted reactions of 3,3,3-$tris$(3'-substituted pyrazolyl)propanol ligands [(3-Rpz)$_3$CCH$_2$CH$_2$OH, R = H] and [Re(CO)$_5$Br] yielded [Re(CO)$_3$]Br and degradation products [(HpzR)$_2$Re(CO)$_3$Br] [R = t-Bu (7b), Ph] (Kunz et al, 2009). These complexes were also prepared directly from [Re(CO)$_5$Br] and the corresponding pyrazoles by microwave-assisted synthesis. Beginning with MO$_4^-$ (M = ^{99m}Tc, $^{186/188}$Re), the carbonyl precursor [M(CO)$_3$(H$_2$O)$_3$]$^+$ was synthesized in 3 min in quantitative yield in a microwave reactor (Causey et al, 2008). When di-picolyl ligand (HL = 5-[bis(2-pyridinylmethyl)amino]pentanoic acid) was added to the reaction mixture, the chelate complex [M(CO)$_3$(L)]$^+$ was formed in high yield in 2 min using MWH at 150ºC. These and further syntheses under MW-heating represented a move away from traditional instant kits toward more versatile platform synthesis and purification technologies that are better suited for producing modern molecular imaging and therapy agents.

6.2 Cyclopentadienyls

As metal-Cp complexes, MW-obtained ferrocene derivatives are the most common. Among relatively old and already classic achievements in this area, we emphasize the following condensation reactions. Thus, according to the conventional techniques, *Claisen-Schmidt* template reactions of acetylferrocene **24** and ferrocene carboxaldehyde **26** are usually performed under classical homogeneous conditions in ethanol. Using MWH of the reaction system, it became possible (Villemin et al, 1994) to prepare (reactions *6* and *7*) ferrocenylenones **25** and **27** without solvent in presence of solid KOH with higher yields in comparison with those reported earlier. It is noted that the reactions may be accelerated efficiently by microwave irradiation.

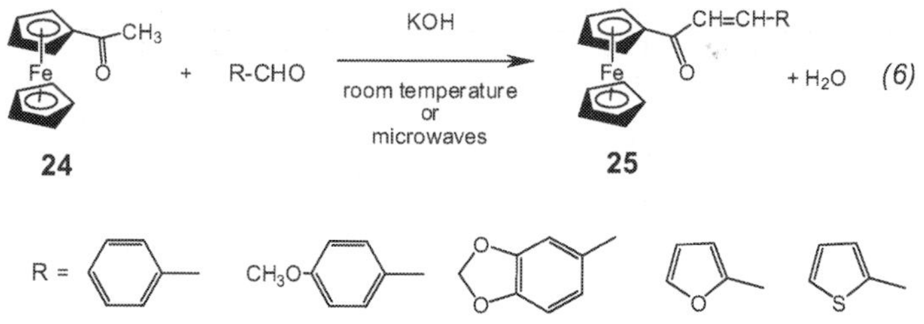

Condensation of acetylferrocene

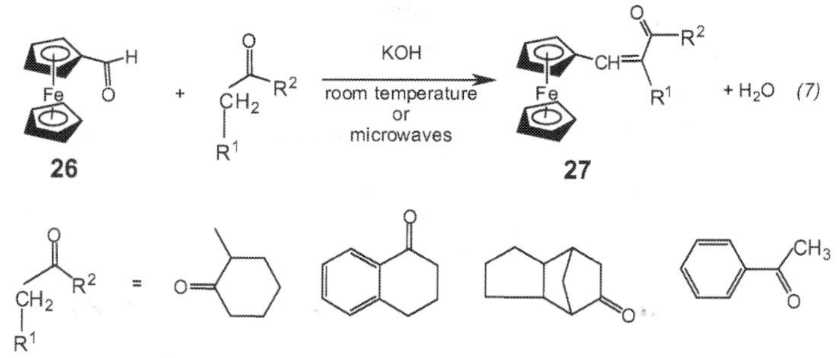

Condensation of ferrocene carboxaldehyde

A significant accelerating effect by MWH for the ligand exchange reaction of ferrocenes was observed; this effect was due to the absorption of microwave energy by the adduct between the ferrocene and the Lewis acid (Okada et al, 2009). Six ferrocenyl α,β-unsaturated ketones, FcCOCH:CHAr (Fc = ferrocenyl, Ar = Ph, 4-MeOC$_6$H$_4$, 2-furyl, 4-Me$_2$NC$_6$H$_4$, ferrocenyl, 4-O$_2$NC$_6$H$_4$) were prepared by MW-assisted reaction of ArCHO with FcCOMe in the presence of KF-Al$_2$O$_3$ as catalyst (Lu et al, 2003). 1,5-dioxo-3-(*p*-methylbenzyloxyphenyl)[5]ferrocenophane (**28**) was MW-prepared (50 W for 30 min with 80°C) in 2-step reaction from 4-hydroxybenzaldehyde in acetone, 4-methylbenzylbromide, CsCO$_3$, and further addition of diacetylferrocene in 90% yield (Patti et al, 2009).

Additionally, the species MW-synthesized include ferrocene and acetylferrocene, piano stool complexes such as $CpFe(CO)_2I$, $CpFe(PPh_3)(CO)I$, and $CpFe(PPh_3)(CO)(COMe)$, and bisphosphine iron complexes. The use of microwave-assisted reactions decreased reaction times while maintaining or improving yields as compared to traditional methods (Garringer et al, 2009). Mixed (η^6-arene)(η^5-cyclopentadienyl)iron(II) complexes are also known (Roberts, 2006; Roberts, 2006).

Group 4 is represented by all three transition metals (Ti, Zr, and Hf). Thus, the reactions of *bis*(cyclopentadienyl)titanium(IV) chloride with Schiff bases (LH$_2$), derived by condensing 3-(phenyl/2-chlorophenyl/4-nitrophenyl)-4-amino-5-hydrazino-1,2,4-triazoles with salicylaldehyde or 2-hydroxyacetophenone, were studied both by conventional stirring method and also by using microwave heating, isolating [(η^5-C$_5$H$_5$)$_2$Ti(L)] in both cases (Banerjee et al, 2008). The ligands behaved as dibasic, tetradentate chelating agents and a six-coordinated structure were assigned to these derivatives. The same precursor was applied in reactions with *bis*(thiosemicarbazones) (H$_2$L), derived by condensing isatin with different N(4)-substituted thiosemicarbazides, were studied both by a conventional stirring method and also using MW technology isolating binuclear [{(η^5-C$_5$H$_5$)$_2$TiCl}$_2$(L)] compounds (Banerjee et al, 2009). The ligands and complexes possessed inhibiting potential against various fungal, viral and bacterial strains. Similarly, reactions of (η^5-C$_5$H$_5$)$_2$HfCl$_2$ with benzil *bis*(hydrazones) (LH$_2$), derived from benzil and aromatic acid hydrazides (benzoic, 2-chlorobenzoic, 4-chlorobenzoic, 2-methylbenzoic or 4-methoxybenzoic) were studied in anhydrous THF in the presence of *n*-butylamine by both conventional methods and by MWH, isolating binuclear complexes of type [{(η^5-C$_5$H$_5$)$_2$HfCl}$_2$(L)] (Sinha et al, 2008). The stoichiometric reactions of titanocenedichloride or zirconocenedichloride with monofunctional bidentate ligands **29** and **30**, derived from heterocyclic ketones and semicarbazide hydrochloride and 2-hydroxy-N-phenyl benzamide, resulted in the formation of unsymmetrical complexes **31** (reactions *8-11*) (Poonia et al, 2007; Poonia et al, 2008). A comparison of conventional and microwave route revealed that the second way was 100 times faster.

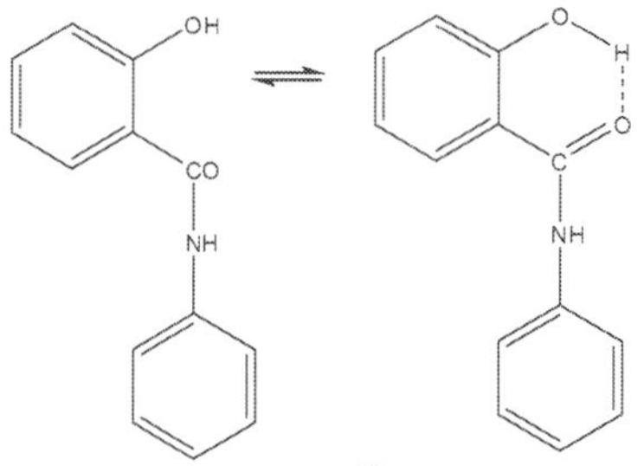

29 (O OH)

30 (N OH)

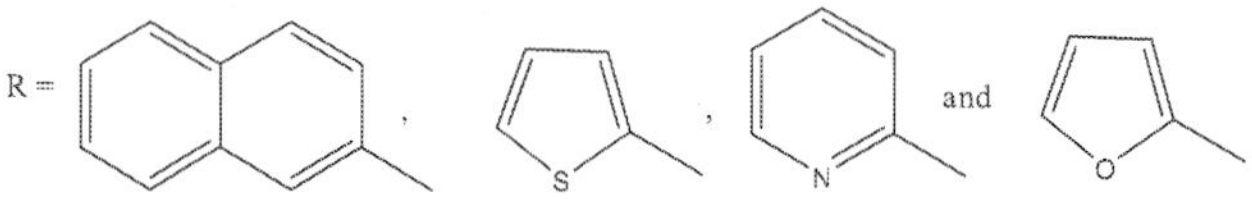

$$R = \text{, , and}$$

$$Cp_2MCl_2 + O\!\!\frown\!\!OH + N\!\!\frown\!\!OH \xrightarrow[\text{THF}]{Et_3N} Cp_2M(O\!\!\frown\!\!O)(N\!\!\frown\!\!O) + 2\,Et_3N\!\cdot\!HCl \quad (8)$$

$$Cp_2MCl_2 + 2HPOH \xrightarrow[\text{THF}]{Et_3N} Cp_2M(HPO)_2 + 2\,Et_3N\!\cdot\!HCl \quad (9)$$

$$Cp_2MCl_2 + HO\!\!\frown\!\!N\!\!\frown\!\!OH \xrightarrow[\text{THF}]{Et_3N} Cp_2M(O\!\!\frown\!\!N\!\!\frown\!\!O) + 2\,Et_3N\!\cdot\!HCl \quad (10)$$

$$Cp_2MCl_2 + HO\!\!\frown\!\!N\!\!\frown\!\!SH \xrightarrow[\text{THF}]{Et_3N} Cp_2M(O\!\!\frown\!\!N\!\!\frown\!\!S) + 2\,Et_3N\!\cdot\!HCl \quad (11)$$

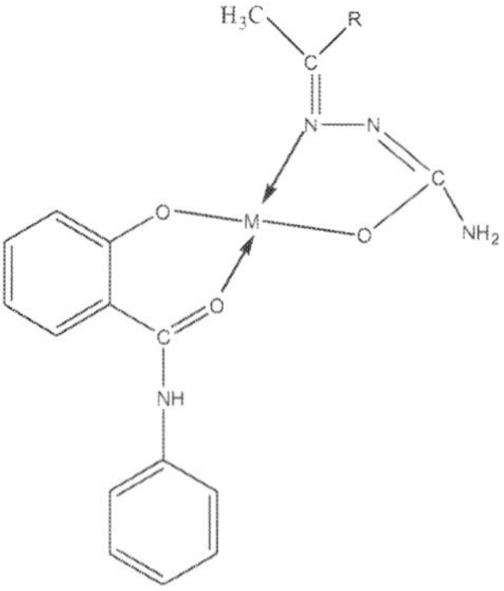

31

A few of other metal cyclopentadienyls have been also reported, for example a very efficient MWAS of [RuCp(η^6-naphthalene)][PF$_6$] (Mercier et al, 2009). The synthesis of cyclopentadienyl *bis*-phosphine ruthenium thiolato complexes of the type [RuCp(dppm)SR] (R = Ph, CH$_2$CH$_2$Ph, CH$_2$(2-furyl), CH$_2$CO$_2$Et, CH$_2$CH(NHAc)CO$_2$H) from [RuCp(PPh$_3$)$_2$Cl] using conventional heating and MW using a focused monomode reactor was described (Kuhnert & Danks, 2002). Sealed tube microwave dielectric heating of diaryl acetylenes with cyclopentadienyl Co dicarbonyl at elevated temperature in *p*-xylene provided access to metallocenes in both the cyclobutadiene (Ar$_4$C$_4$CoCp, 3-52% yields) and cyclopentadienone (Ar$_4$C$_4$(CO)CoCp, 14-85% yields) families (Harcourt et al, 2008). In the case of an especially bulky diarylacetylene (1,1'-dinaphthylacetylene), the microwave approach allowed access to a complex that cannot be readily obtained under traditional thermal conditions.

6.3 Arene complexes

Microwave-mediated syntheses of [(η-arene)(CO)$_3$Mn](PF$_6$) complexes (Dabirmanesh et al, 1997), [Fe(η-C$_5$H$_5$)(η-arene)][PF$_6$] salts from reactions of Fe(C$_5$H$_5$)$_2$ with arenes and chloroarenes, as well as [Fe(η-arene)$_2$][PF$_6$]$_2$ salts from reaction of arenes and FeCl$_3$ (Dabirmanesh et al, 1993) were reported relatively long ago. Reaction times were reduced from several hours by conventional methods to a few minutes using an unmodified domestic microwave oven. Microwave heating was employed to promote arene displacement in reactions of [(*p*-cymene)RuCl$_2$]$_2$ or [{(1,3,5-C$_6$H$_3$(*i*-Pr)$_3$)RuCl$_2$}$_2$] with neutral chelate ligands L-L' [L-L': 1,1'-*bis*(diphenylphosphanyl)methane, 1,1'-*bis*(diphenylphosphanyl)ferrocene, (S)-BINAP, (S,S)-DIOP, N,N'-*bis*(2,4,6-trimethylphenyl)-1,2-ethanediylidenediamine], (R)-Ph-PHOX, and 3-(phenylsulfanylpropyl)diphenylphosphine giving [(arene)Ru(μ-Cl)$_3$-RuCl(L-L')] in good yield (Albrecht et al, 2009). Also, the MWAS of (η^6-arene)tricarbonylchromium complexes from hexacarbonylchromium with arenes gave high yields of products **32** (reaction *12*) (Lee et al, 2006).

$$R\text{–arene} + Cr(CO)_6 \xrightarrow{MW} (\eta^6\text{-arene})Cr(CO)_3 \quad (12)$$

32

6.4 Other organometallics

The effect of the microwave irradiation on the reaction of alkynyl alkoxy *carbene complexes* with urea derivatives was studied (Spinella et al, 2003), showing that in these conditions (CO)$_5$W:C(OEt)C≡CPh reacted with ureas, (RNH)C(O)(NHR') (e.g., R , R' = H, Me, allyl, Et), with reduced reaction times to give uracil derivatives **33**. It is noteworthy that the use of large amounts of solvents could be drastically reduced or even avoided and, in any case, reaction times were dramatically shortened. The MWAS of two different types of N-heterocyclic carbene-palladium(II) complexes, (NHC)Pd(acac)Cl (NHC = N-heterocyclic carbene; acac = acetylacetonate) and (NHC)PdCl$_2$(3-chloropyridine), led to drastic reduction in reaction times (20 to 88 times faster, depending on the complex) (Winkelmann & Navarro,

2010). The complex (IPr)Pd(acac)Cl [IPr=1,3-*bis*(2,6-diisopropylphenyl)imidazol-2-ylidene] was similarly obtained. Bridged and unbridged N-heterocyclic carbene (NHC) ligands were metalated under MW-conditions with [Ir(COD)Cl]$_2$ to give Ir(I) mono- and biscarbene substituted catalysts [Ir(COD)NHC(Cl)] and [Ir(COD)(NHC)$_2$][X] (X = I, PF$_6$, BF$_4$, CF$_3$COO, OTf) (Rentzsch et al, 2009). Palladium(II) carbene complexes were also reported in (Scarborough et al, 2009).

33

Diene derivatives are represented by the first example of microwave-promoted solid-state synthesis of Na complex [Na(L)(µ-EP)·H$_2$O]$_2$ derived from a heteromacrocyclic compound (Tusek-Bozic et al, 2007). This alkali complex, as a diphosphonate-bridged dinuclear species, was prepared from the 15-membered mixed dioxa-diaza macrocycle 5,6,14,15-dibenzo-1,4-dioxa-8,12-diazacyclopentadeca-5,14-diene (L) by reaction with Na Et [4-α-(benzeneazoanilino)-N-benzyl]phosphonate (NaEP·1.5H$_2$O). MW-heating of *metal-allyl complexes* can result organic products. Thus, nucleophilic attack of 3-hydroxycoumarin on η^3-π-allylpalladium complex, formed from substituted cinnamyl alcohols **34** (R^1 = R^2 = H, MeO, R^3 = H; R^1 = OCH$_2$Ph, R^2 = MeO, R^3 = H) and acetyls **34** (R^3 = COMe) in the presence of palladium acetylacetonate and triphenylphosphine, resulted in normal addition products like 4-(3'-phenylallyl)-3-hydroxycoumarin, except for cinnamyl acetate, which provided an unusual product, 4-(1'-phenylallyl)-3-hydroxycoumarin, by conventional and MWH (Mitra et al, 2003).

34

Table 1. Main containing ligands' units present in the studied MW-synthesized complexes.

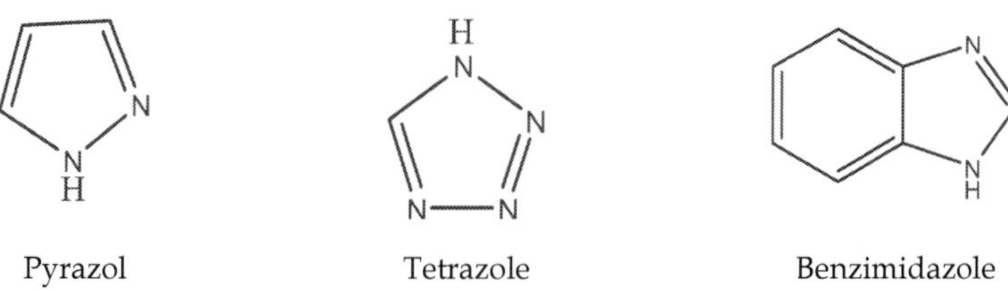

Pyrazol   Tetrazole   Benzimidazole

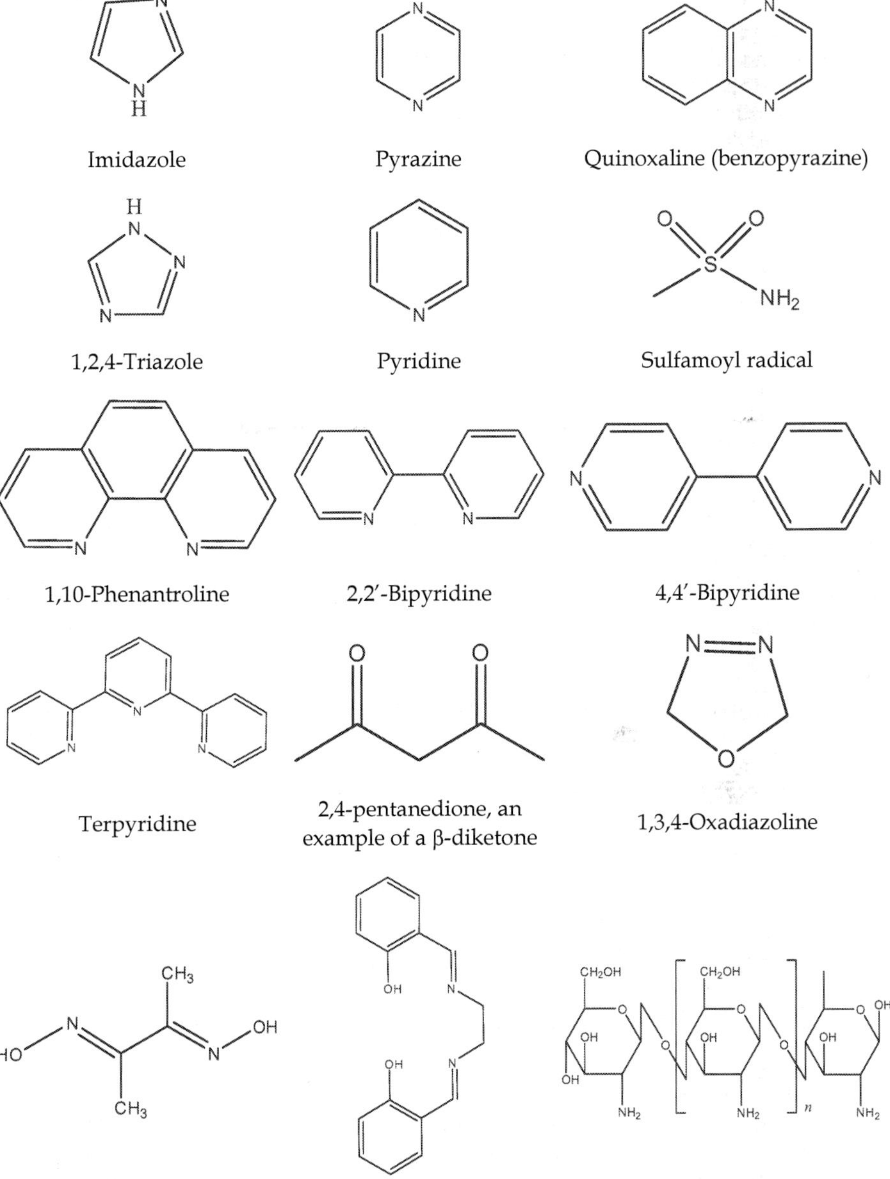

Imidazole

Pyrazine

Quinoxaline (benzopyrazine)

1,2,4-Triazole

Pyridine

Sulfamoyl radical

1,10-Phenantroline

2,2'-Bipyridine

4,4'-Bipyridine

Terpyridine

2,4-pentanedione, an example of a β-diketone

1,3,4-Oxadiazoline

Dimethylglyoxime

Salen

Chitosan

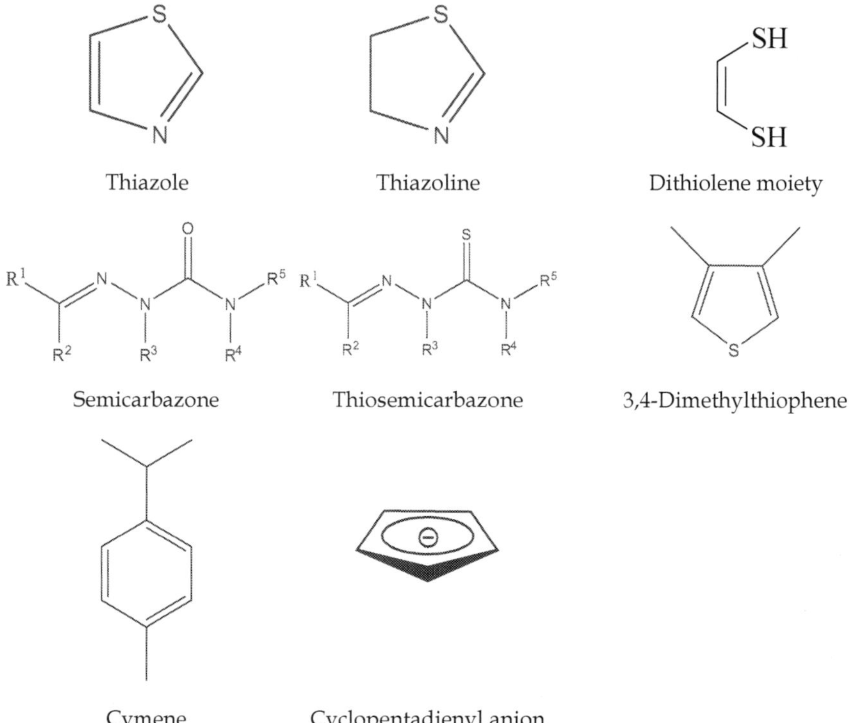

7. Microwave-assisted catalysis using metal complexes

Several reports are dedicated to the use of metal (mainly noble metals, such as Rh, Pd, Os, which in free form are used in catalytic processes) complexes in MWAS or rearrangements of organic compounds. Thus, a highly efficient C-C bond cleavage of unstrained aliphatic ketones bearing β-hydrogens with olefins was achieved using a chelation-assisted catalytic system consisting of $(Ph_3P)_3RhCl$ and 2-amino-3-picoline by MW under solvent-free conditions (Ahn et al, 2006). The addition of cyclohexylamine catalyst accelerated the reaction rate dramatically under microwave irradiation compared with the classical heating method. Microwave-assisted Rh-diphosphane-complex-catalyzed dual catalysis, providing [2+2+1] cycloadducts by sequential decarbonylation of aldehyde or formate and carbonylation of enynes within a short period of time, was reported (Lee et al, 2008). Various O-, N-, and C-tethered enynes were transformed into the corresponding products in good yields. The first enantioselective version of this microwave-accelerated cascade cyclization was realized. In the presence of chiral Rh-(S)-bisbenzodioxanPhos complex, the cyclopentenone products were achieved with ee values up to 90%. Osmium complex (μ-H)Os_3(μ-O:CPh)(CO)$_{10}$ was an active catalyst for the allylic rearrangement N-allylacetamide under MW-radiation (Afonin et al, 2008).

An efficient method for intermolecular hydroarylation of aryl and aliphatic alkenes with indoles using a combination of [(PR$_3$)AuCl]/AgOTf as catalyst under thermal and microwave-assisted conditions was developed (Wang ert al, 2008), achieving the gold(I)-catalyzed reactions of indoles with aryl alkenes in toluene at 85°C over a reaction time of 1-3 h with 2 mol% of [(PR$_3$)AuCl]/AgOTf as catalyst (yields 60-95%). Under microwave irradiation, coupling of unactivated aliphatic alkenes with indoles gave the corresponding adducts in up to 90% yield. Additionally, metal acetates were found to be effective catalysts; thus, a rapid and efficient method for the synthesis of β-arylalkenyl nitriles by a one-pot three component coupling reaction of diphenylacetylene, K$_4$Fe(CN)$_6$, and aryl halides using Pd(OAc)$_2$ as a catalyst and water as a solvent under MW (Velmathi et al, 2010). The method employed a cyanide source which is safe and inexpensive. Copper-catalyzed cyanation of aryl halides was improved to be more economical and environmentally friendly by using water as the solvent and ligand-free Cu(OAc)$_2$·H$_2$O as the catalyst under MW (Ren et al, 2009). The suggested methodology was applicable to a wide range of substrates including aryl iodides and activated aryl bromides.

8. Conclusions

In the coordination and organometallic chemistry, the microwave-assisted synthesis is not developed such sufficiently as for the preparation of inorganic compounds, composites and materials or in the organic synthesis, where microwave heating can be considered as a common preparative tool. However, during the last decade a considerable growth of related reports has been registered. The most number of reports corresponds to MW-reactions of the N-, N,O-, and N,S-containing ligands with sources of metal ions. Some MW-fabricated classic π- and σ-organometallic compounds are also presented.

Practically in all reports, main attention of researchers is paid to extreme fastness of MW-assisted reactions in comparison with classic protocols. The same reactions in the MW-field take place in 10-100 times more rapidly. Moreover, higher or comparable yields are frequently reported. Sometimes, the MW-route leads to products, which it is impossible to get *via* traditional routes, for instance preparation of several metal cluster complexes.

Despite of the development of novel synthesis techniques in chemistry and especially nanotechnology (for example, laser-, sputtering-, CVD-, electron- and ion-beam-, radiation-, or combustion-assisted methods, among many others, the microwave heating remains very attractive for chemists due to its obvious advantages, noted at the beginning of this chapter.

9. Abbreviations

2-dpds = 2,2'-dipyridyldisulfide
acac = acetylacetonate
APBMPA = (3-aminopropyl)*bis*(2-pyridylmethyl)amine
BMPA = *bis*(2-pyridylmethyl)amine
bbi = 1,1'-(1,4-butanediyl)*bis*(imidazole)
bpd = 3-bromo-2,4-pentanedionate ion
bpy, bipy = 2,2'-bipyridine
bpydc = 2,2'-bipyridine-5,5'-dicarboxylate
BTEC = 1,2,4,5-benzenetetracarboxylate anion
CDEPE = N,N'-*bis*(3-carboxy-1-oxopropanyl)-1,2-diethylethylenediamine

CDEPE-2 = N,N'-*bis*(3-carboxy-1-oxoprop-2-enyl)-1,2-diethylethylenediamine
CDMPE = N,N'-*bis*(3-carboxy-1-oxopropanyl)-1,2-dimethylethylenediamine
CDMPE-2 = *bis*(3-carboxy-1-oxoprop-2-enyl)-1,2-dimethylethylenediamine
CPE = N, N'-*bis*-(3-carboxy-1-oxopropanyl)-1,2-ethylenediamine
CPP = N,N'-*bis*-(3-carboxy-1-oxo-propanyl)-1,2-phenylenediamine
CPP-2 = N,N'-*bis*-(3-carboxy-1-oxoprop-2-enyl)-1,2-phenylenediamine
CPPP = N,N'-*bis*-(2-carboxy-1-oxophenelenyl)-1,2-phenylenediamine
DBU = 1,8-diazabicyclo[5,4,0]undec-7-ene
dcmb = 4,4'-dimethoxycarbonyl-2,2'-bipyridine
DMAD = di-methylacetylenedicarboxylate
DMAE = dimethylaminoethanol
DMF = N,N-dimethylformamide
dppe = 1,2-*bis*(diphenylphosphino)ethane
2-dps = 2,2'-dipyridylsulfide
EMim = 1-ethyl-3-methylimidazolium
H_2bzimpy = 2,6-*bis*(benzimidazol-2-yl)pyridine
Hdpa = *bis*(2-pyridyl)amine
H_2NDC = 2,6-naphthalenedicarboxylic acid
H_2oba = 4,4'-oxydibenzoic acid
H_2-Saly-BTZ = 2-(2-hydroxyphenyl)-2-(phenylamino)benzothiazoline
HSAT = 2-(N-salicylideneamino)-3-carboxyethyl-4,5-dimethylthiophene
H_2TPyP = 5,10,15,20-*tetrakis*(2-pyridyl)porphyrin
H_3IDC = 4,5-imidazoledicarboxylic acid
Hpyt = pyridine-2(1H)-thione
H_3TMA = trimesic acid
H_4btec = 1,2,4,5-benzenetetracarboxylic acid
LiPBMPA = 3-[*bis*(2-pyridylmethyl)amino]propanoate
MW = microwave irradiation
MWAS = microwave-assisted synthesis
MWH = microwave heating
Mg(TBPor) = 5,10,15,20-*tetrakis*(4-tert-butylphenyl)magnesium porphyrins
mimp = 2-methyliminomethylphenolate anion
MORE = microwave-induced organic reaction enhancement
MPBMPA = Me 3-[*bis*(2-pyridylmethyl)amino]propanoate
MWAACVD = microwave plasma aerosol-assisted chemical vapor deposition
MWPECVD = microwave plasma enhanced chemical vapor deposition
OA = oleic acid
OAm = oleylamine
PABMPA = 3-[*bis*(2-pyridylmethyl)amino]propanamide
PDHEP = [N,N'-*bis*(1,5-dimethylhexyl)-3,4:9,10-perylenebis(dicarboximide)
phen = 1,10-phenanthroline
pip = piperidine
PNBMPA = 3-[*bis*(2-pyridylmethyl)amino]propionitrile
pytrz = 2-(1-alkyl/aryl-1H-1,2,3-triazol-4-yl)pyridine
pyz = pyrazolyl ligand
Pz = pyrazine

PzAM = pyrazine carboxamide
PzCA = pyrazine carboxylic acid
saoH$_2$ = salicylaldoxime
SRaaiNR = 1-alkyl-2-{(o-thioalkyl)phenylazo}imidazoles
tbbpy = 4,4'-di-*tert*-butyl-2,2'-bipyridine
TBHP = *tert*-butyl hydroperoxide
TBBZTH = 3,4,5-trimethoxybenzaldehydebenzothiazoline
TBDTCZH = 3,4,5-trimethoxybenzaldehyde-S-benzyldithiocarbazate
TBSCZH = 3,4,5-trimethoxybenzaldehydesemicarbazone
TBTSCZH = 3,4,5-trimethoxybenzaldehydethiosemicarbazone
tipsepd = 3-((tri*iso*propylsilyl)ethynil)-2,4-pentanedionate ion
TRISPHAT-N = 2,3-pyridinyldioxy anionic phosphate

10. Acknowledgements

The authors are very grateful to Professors *Yurii E. Alexeev* and *Alexander D. Garnovskii* (Southern Federal University, Rostov-na-Donu, Russia) for critical revision of the final manuscript.

11. References

Abdelsayed, V.; Aljarash, A.; El-Shall, M.S.; Al Othman, Z.A.; Alghamdi, A.H. (2009). Microwave synthesis of bimetallic nanoalloys and CO oxidation on ceria-supported nanoalloys. *Chemistry of Materials*, 21(13), 2825-2834.

Afonin, M. Yu.; Maksakov, V. A. (2008). Effect of microwave radiation on the allylic rearrangement catalyzed by cluster complexes with chemically labile ligands. *Russ. J. Coord. Chem.*, 34(11), 869-870.

Ahluwulia, V.K. (2007). *Alternate Energy Processes in Chemical Synthesis: Microwave, Ultrasonic, and Photo Activation*. Alpha Science Int'l Ltd, 280 pp.

Ahn, J.-A.; Chang, D.-H.; Park, Y.J.; Yon, Y.R.; Loupy, A.; Jun, C.-H. (2006). Solvent-free chelation-assisted catalytic C-C bond cleavage of unstrained ketone by rhodium(I) complexes under microwave irradiation. *Advanced Synthesis & Catalysis*, 348(1+2), 55-58.

Albrecht, C.; Gauthier, S.; Wolf, J.; Scopelliti, R.; Severin, K. (2009). Microwave-assisted organometallic syntheses: formation of dinuclear [(arene)Ru(μ-Cl)$_3$RuCl(L-L')] complexes (L-L': chelate ligands with P-, N-, or S-donor atoms) by displacement of arene ligands. *Eur. J. Inorg. Chem.*, (8), 1003-1010.

Aléo, A. D'; Welter, S.; Cecchetto, E.; De Cola, L. (2005). Electronic energy transfer in dinuclear metal complexes containing meta-substituted phenylene units. *Pure Appl. Chem.*, 77(6), 1035-1050.

Ali, P.; Ramakanth, P.; Meshram, J. (2010). Exploring microwave synthesis for co-ordination: synthesis, spectral characterization and comparative study of transition metal complexes with binuclear core derived from 4-amino-2,3-dimethyl-1-phenyl-3-pyrazolin-5-one. *J. Coord. Chem.*, 63(2), 323-329.

Amarante, T.R.; Almeida Paz, F.A.; Gago, S.; Gonçalves, I.S.; Pillinger, M.; Rodrigues, A.E.; Abrantes, M. (2009). Microwave-Assisted Synthesis and Crystal Structure of

Oxo(diperoxo)(4,4'-di-tert-butyl-2,2'-bipyridine)-molybdenum(VI). *Molecules*, 14, 3610-3620.

Amo-Ochoa, P.; Givaja, G.; Miguel, P.J.S.; Castillo, O.; Zamora, F. (2007). Microwave assisted hydrothermal synthesis of a novel CuI-sulfate-pyrazine MOF. *Inorg. Chem. Commun.*, 10(8), 921-924.

Andrews, P.C.; Deacon, G.B.; Junk, P.C.; Spiccia, N.F. (2007). Exploration of solvent free and/or microwave assisted syntheses of bismuth(III) thiolates. *Green Chemistry*, 9(12), 1319-1327.

Ardon, M.; Hogarth, G.; Oscroft, D.T.W. (2004). Organometallic chemistry in a conventional microwave oven: the facile synthesis of group 6 carbonyl complexes. *J. Organomet. Chem.*, 689(15), 2429-2435.

Bagley, M.C.; Lin, Z.; Pope, S.J.A. (2010). Microwave-assisted synthesis and complexation of luminescent cyanobipyridyl-zinc(II) bis(thiolate) complexes with intrinsic and ancillary photophysical tunability. *Dalton Trans.*, 39(13), 3163-3166.

Bahadoran, F.; Dialameh, S. (2005). Microwave assisted synthesis of substituted metallophthalocyanines and their catalytic activity in epoxidation reaction. *J. Porphyr. Phthalocyan.*, 9(3), 163-169.

Bahnmueller, S.; Langstein, G.; Hitzbleck, J.; Volkmer, D.; Lu, Y.; Tonigold, M. (2009). Cobalt(II)-based redox- and catalytic active metal-organic framework (MOF) compounds. German patent DE 102008027218, 11 pp.

Banerjee, P.; Pandey, O.P.; Sengupta, S.K. (2008). Microwave assisted synthesis, spectroscopic and antibacterial studies of titanocene chelates of Schiff bases derived from 3-substituted-4-amino-5-hydrazino-1,2,4-triazoles. *Transition Metal Chemistry*, 33(8), 1047-1052.

Banerjee, P.; Pandey, O.P.; Sengupta, S.K. (2009). Microwave-assisted synthesis, spectroscopy and biological aspects of binuclear titanocene chelates of isatin-2,3-bis(thiosemicarbazones). *Applied Organometallic Chemistry*, 23(1), 19-23.

Beckford, F.A.; Shaloski, M., Jr.; Leblanc, G.; Thessing, J.; Seeram, N.P.; Li, L. (2009). Mixed ligand diimine-thiosemicarbazone complexes of ruthenium(II): Synthesis, biophysical reactivity and cytotoxicity. *Abstracts of Papers, 237th ACS National Meeting, Salt Lake City, UT, United States, March 22-26, 2009*, CHED-585.

Beckford, F.; Shaloski, M.; Thessing, J.; Morgan, C. (2010). Microwave synthesis of mixed ligand diimine-thiosemicarbazone complexes of ruthenium(II): Reactivity with DNA and human serum albumin. *Abstracts of Papers, 239th ACS National Meeting, San Francisco, CA, United States, March 21-25, 2010*, INOR-225.

Beltran, H.I.; Esquivel, R.; Lozada-Cassou, M.; Dominguez-Aguilar, M.A.; Sosa-Sanchez, A.; Sosa-Sanchez, J.L.; Hoepfl, H.; Barba, V.; Luna-Garcia, R.; Farfan, N.; Zamudio-Rivera, L.S. (2005). Nanocap-shaped tin phthalocyanines: Synthesis, characterization, and corrosion inhibition activity. *Chemistry--A European Journal*, 11(9), 2705-2715.

Beneš, L.; Melánova, K.; Zima, V.; Kalousová, J.; Votinský, J. (1997). Preparation and Probable Structure of Layered Complexes of Vanadyl Phosphate with 1-Alkanols and 1,ω-Alkanediols. *Inorg. Chem.* 36, 2850-2854.

Berdonosov, S.S.; Kopilova, I.A.; Lebedev, V.Ya.; Chesnokov, D.E. (1992). Use of microwave irradiation for preparation of non-aqueous zirconium acetylacetonate. *Neorg. Mater.* 28(5), 1022.

Besson, T.; Thiery, V.; Dubac, J. (2006). Microwave-assisted reactions on graphite. *Microwaves in Organic Synthesis (2nd Edition)*, 1, 416-455.

Bhojak, N.; Gudasaria, D.D.; Khiwani, N.; Jain, R. (2007). Microwave Assisted Synthesis Spectral and Antibacterial Investigations on Complexes of Mn(II) With Amide Containing Ligands. *E-Journal of Chemistry*, 4(2), 232-237.

Bhojak, N.; Singh, B. (2008). Microwave assisted synthesis, spectral and antibacterial investigations on complexes of Mn(II) with amide containing ligands. *Rasayan Journal of Chemistry*, 1(1), 105-109.

Bilecka, I.; Djerdj, I.; Niederberger, M. (2008). One-minute synthesis of crystalline binary and ternary metal oxide nanoparticles. *Chem. Communs*, (7), 886-888.

Biyiklioglu, Z.; Kantekin, H.; Oezil, M. (2007). Microwave-assisted synthesis and characterization of novel metal-free and metallophthalocyanines containing four 14-membered tetraaza macrocycles. *J. Organomert. Chem.*, 692(12), 2436-2440.

Bokach, N.A.; Krokhin, A.A.; Nazarov, A.A.; Kukushkin, V.Yu.; Haukka, M.; Frausto da Silva, J.J.R.; Pombeiro, A.J.L. (2005). Interplay between nitrones and (nitrile)Pd[II] complexes: Cycloaddition vs. Complexation followed by cyclopalladation and deoxygenation reactions. *Eur. J. Inorg. Chem.*, (15), 3042-3048.

Bokach, N.A.; Kukushkin, V.Yu.; Haukka, M.; Mikhailova, T.B.; Sidorov, A.A.; Eremenko, I.L. (2006). Co(II)-Mediated and microwave assisted coupling between 2,6-diaminopyridine and nitriles. A new synthetic route to N-(6-aminopyridin-2-yl)carboximidamides. *Russ. Chem. Bull.*, 55(1), 36-43.

Boscencu, R.; Ilie, M.; Socoteanu, R.; Sousa Oliveira, A.; Constantin, C.; Neagu, M.; Manda, G.; Vieira Ferreira, L.F. (2010). Microwave Synthesis, Basic Spectral and Biological Evaluation of Some Copper (II) Mesoporphyrinic Complexes. *Molecules*, 15, 3731-3743.

Causey, P.W.; Besanger, T.R.; Schaffer, P.; Valliant, J.F. (2008). Expedient Multi-Step Synthesis of Organometallic Complexes of Tc and Re in High Effective Specific Activity. A New Platform for the Production of Molecular Imaging and Therapy Agents. *Inorg. Chem.*, 47(18), 8213-8221.

Chandrasekharam, M.; Rao, C.S.; Singh, S.P.; Kantam, M.L.; Reddy, M.R.; Reddy, P.Y.; Toru, T. (2007). Microwave-assisted synthesis of metalloporphyrazines. *Tetrahedron Lett.*, 48(14), 2627-2630.

Chaudhary, P.; Swami, M.; Sharma, D.K.; Singh, R. V. (2009). Ecofriendly synthesis, antimicrobial and antispermatogenic activity of triorganotin(IV) complexes with 4'-nitrobenzanilide semicarbazone and 4'-nitrobenzanilide thiosemicarbazone. *Applied Organometallic Chemistry*, 23(4), 140-149.

Chemat, F.; Poux, M.; Berlan, J. J. (1996). An original microwave-ultrasound combined reactor suitable for organic synthesis: application to pyrolysis and esterification. *Microwave Power & Electromag. Energy*. 31(1), 19.

Cho, T.J.; Shreiner, C.D.; Hwang, S.-H.; Moorefield, C.N.; Courneya, B.; Godinez, L.A.; Manriquez, J.; Jeong, K.-U.; Cheng, S.Z.D.; Newkome, G.R. (2007). 5,10,15,20-Tetrakis[4'-(terpyridinyl)phenyl]porphyrin and its Ru[II] complexes: Synthesis, photovoltaic properties, and self-assembled morphology. *Chem. Commun.*, (43), 4456-4458.

Dabirmanesh, Q.; Roberts, R.M.G. (1993). The synthesis of iron sandwich complexes by microwave dielectric heating using a simple solid CO_2-cooled apparatus in an unmodified commercial microwave oven. *Journal of Organometallic Chemistry*, 460(2), C28-C29.

Dabirmanesh, Q.; Roberts, R.M.G. (1997). The application of microwave dielectric heating to the synthesis of arene-metal complexes. Synthesis of [(η-arene)(CO)$_3$Mn]PF$_6$ complexes and [(η-arene)(η-cyclopentadienyl)Fe][PF$_6$] complexes with triphenylphosphine, *tert*-butylbenzenes and a sterically hindered phenol as arene ligands. *Journal of Organometallic Chemistry*, 542(1), 99-103.

Dayalan, A.; Meera, P.; Balaraju, K.; Agastian, P.; Ignasimuthu, S. (2009). Halocobaloximes containing axially coordinated imidazole or histidine: microwave assisted synthesis, characterization and antibacterial activity. Journal of the Indian Chemical Society, 86(6), 628-632.

Delgado, S.; Sanz Miguel, P.J.; Priego, J.L.; Jimenez-Aparicio, R.; Gomez-Garcia, C.J.; Zamora, F. (2008). A Conducting Coordination Polymer Based on Assembled Cu$_9$ Cages. *Inorg. Chem.* 47(20), 9128-9130.

Delgado, S.; Santana, A.; Castillo, O.; Zamora, F. (2010). Dynamic combinatorial chemistry in a solvothermal process of Cu(I,II) and organosulfur ligands. Dalton Trans., 39(9), 2280-2287.

Desai, B.; Danks, T.N.; Wagner, G. (2004). Ligand discrimination in the reaction of nitrones with [PtCl$_2$(PhCN)$_2$]. Selective formation of mono-oxadiazoline and mixed bis-oxadiazoline complexes under thermal and microwave conditions. *Dalton Trans.*, (1), 166-171.

Dickhoff, J.N.; Hoeltje, N.; Kellen-Yuen, C. (2006). Microwave synthesis of 3,4-disubstituted-1-H-pyrrole-2-carboxylates. *Abstracts of Papers, 231st ACS National Meeting, Atlanta, GA, United States, March 26-30, 2006*, CHED-432.

Dittmar, A.; Kosslick, H.; Muller, J.-P.; Pohl, M.-M. (2004). Characterization of cobalt oxide supported on titania prepared by microwave plasma enhanced chemical vapor deposition. *Surface and Coatings Technology*, 182(1), 35-42.

Dittmar, A.; Herein, D. (2009). Microwave plasma assisted preparation of disperse chromium oxide supported catalysts. *Surface and Coatings Technology*, 203(8), 992-997.

Fraunhofer-Gesellschaft; Gorun, Sergiu M.; Schnurpfeil, Guenter; Hild, Olaf; Woehrle, Dieter; Tsaryova, Olga; Gerdes, Robert. (2009). Microwave-assisted synthesis of perfluorophthalocyanine molecules. PCT Int. Appl. WO 2009139973, 20 pp.

Fu, Z.; Liao, H.; Xiong, D.; Zhang, Z.; Jiang, Y.; Yin, D. (2007). A highly-efficient and environmental-friendly method for the preparation of Mn(III)-salen complexes encapsulated HMS by using microwave irradiation. *Microporous and Mesoporous Materials*, 106(1-3), 298-303.

Gairola, A.; Umarji, A.M.; Shivashankar, S.A. (2009). Microwave irradiation-assisted method for the rapid synthesis of fine particles of α-Al$_2$O$_3$ and α-(Al1-XCr$_x$)$_2$O$_3$ and their coatings on Si(100). *Ceramic Transactions, 203(Processing and Properties of Advanced Ceramics and Composites)*, 15-21.

Gao, F.; Yang, D.; Cui, D.; He, R. (2009). Preparation of particle-size-controllable magnetic ferroferric oxide nanoparticle microspheres by microwave heating in nonaqueous solvent. *Patent of China* CN 101337695, 9 pp.

Garringer, S.M.; Hesse, A.J.; Magers, J.R.; Pugh, K.R.; O'Reilly, S.A.; Wilson, A.M.C. (2009). Microwave Synthesis of Benchmark Organo-Iron Complexes. *Organometallics*, 28(23), 6841-6844.

Gass, I.A.; Milios, C.J.; Whittaker, A.G.; Fabiani, F.P.A.; Parsons, S.; Murrie, M.; Perlepes, S.P.; Brechin, E.K. (2006). A Cube in a Tetrahedron: Microwave-Assisted Synthesis of an Octametallic Fe^{III} Cluster. *Inorg. Chem.*, 45(14), 5281-5283.

Gaur, S.; Maanju, S.; Fahmi, N.; Singh, R. V. (2005). Synthesis and biological properties of organotin and organosilicon derivatives using microwave irradiations. *Main Group Metal Chemistry*, 28(5), 293-300.

Giguere, R.A. (1989). In: *Organic Synthesis: Theory and Application.* (Hudlicky, T., Edit.). Vol. 1, JAI Press Inc., p.103-172.

Godbert, N.; Pugliese, T.; Aiello, I.; Bellusci, A.; Crispini, A.; Ghedini, M. (2007). Efficient, ultrafast, microwave-assisted syntheses of cycloplatinated complexes. *European Journal of Inorganic Chemistry*, (32), 5105-5111.

Happ, B.; Winter, A.; Hager, M.D.; Friebe, C.; Schubert, U.S. (2009). Alternative bipyridines analogs via click chemistry. *Polymer Preprints*, 50(2), 254-255.

Harcourt, E.M.; Yonis, S.R.; Lynch, D.E.; Hamilton, D.G. (2008). Microwave-Assisted Synthesis of Cyclopentadienyl-Cobalt Sandwich Complexes from Diaryl Acetylenes. *Organometallics*, 27(7), 1653-1656.

Hayes, B.L. (2004). Recent Advances in Microwave-Assisted Synthesis. *AldrichChimica Acta*, 37(2), 66-76.

Hou, J.; Zhou, Z.-Y.; Wu, X.; Zhu, D.-X.; Li, P. (2007). Synthesis and properties of novel porphyrin compounds. *Gaodeng Xuexiao Huaxue Xuebao*, 28(8), 1424-1427.

Holder, A.A. (2005). Chromium, molybdenum and tungsten. *Annual Reports on the Progress of Chemistry, Section A: Inorganic Chemistry*, 101, 161-193.

Hsu, S.-C.; Chiang, P.-H.; Chang, C.-H.; Lin, C.-H. (2009). Catena-Poly[[tetraaquanickel(II)]-μ_3-benzene-1,3,5-tricarboxylato-3':1:2-$\kappa^4 O^1$:O^3,O^3':O^5-[tetraaquanickel(II)]- μ_2-benzene-1,3,5-tricarboxylato-2:3$\kappa^2 O^1$:O^3-[tetraaquanickel(II)]]. *Acta Crystallographica, Section E: Structure Reports Online*, E65(6), m625-m626.

Hu, A.T.; Tseng, T.-W.; Hwu, H.-D.; Liu, L.-c.; Lee, C.-C.; Lee, M.-C.; Chen, J.-R. (2002). Synthesis of phthalocyanines by microwave irradiation. US Patent 6491796.

Huh, S.; Jung, S.; Kim, Y.; Kim, S.-J.; Park, S. (2010). Two-dimensional metal-organic frameworks with blue luminescence. *Dalton Transactions*, 39(5), 1261-1265.

Hwang, Y.K.; Jin, T.; Kim, J.M.; Kwon, Y.-Uk; Park, S.-E.; Chang, J.-S. (2006). Microwave synthesis of metallosilicate zeolites with fibrous morphology. *Journal of Nanoscience and Nanotechnology*, 6(6), 1786-1791.

Inoue, H.; Seo, S. (2010). Cyclometalated organometallic complexes as phosphorescent compounds for use in light-emitting elements, light-emitting devices, and electronic devices. US Patent 010105902, 35 pp.

Jiang, Y.; Wang, Z.-m.; Ye, S.-q.; Guo, H.; Guo, S.-y. (2008). Effects of extrinsic physical fields on coordination reaction for chitosan-Fe(II). *Huanan Ligong Daxue Xuebao, Ziran Kexueban*, 36(7), 62-66, 71.

Jin, J.; Meng, Q.; Liu, J.; Niu, S. (2009). Synthesis, crystal structure and Photo-physical property of a series of Co^{n+}(n=2, 3) coordination supramolecules. *Liaoning Shifan Daxue Xuebao, Ziran Kexueban*, 32(3), 336-340.

Jung, J.Y.; Newton, B.S.; Tonkin, M.L.; Powell, C.B.; Powell, G.L. (2009). Efficient microwave syntheses of the compounds Os$_3$(CO)$_{11}$L, L = NCMe, py, PPh$_3$. *J. Organomet. Chem.*, 694(21), 3526-3528.

Kahveci, B.; Sasmaz, S.; Ozil, M.; Kantar, C.; Kosar, B.; Buyukgungor, O. (2006). Synthesis and Properties of Triazol-5-one Substituted Phthalocyanines by Microwave Irradiation. *Turk. J. Chem.*, 30, 681-689.

Kim, J.H.; Heo, J.; Kang, B.M.; Son, D.-H.; Lee, G.-D.; Hong, S.-S.; Park, S.S. (2009). Microwave syntheses of subphthalocyanine derivatives and their properties. *Kongop Hwahak*, 20(2), 154-158.

Kim, C.H.; Lee, H.D.; Park, S. S.; Son, D.H.; Hwang, T.G.; Jeon, S.Y. (2009). Method for preparing dithiolene-metal complex. Patent KR 2009034208, 16 pp.

Kogan, E.G.; Ivanov, A.V.; Tomilova, L.G.; Zefirov, N.S. (2002). Synthesis of mono- and bisphthalocyanine complexes using microwave irradiation. *Mendeleev Commun.*, (2), 54-55.

Kubrakova, I.V. (2000). Effect of Microwave Radiation on Physicochemical Processes in Solutions and Heterogeneous Systems: Applications in Analytical Chemistry. *Journal of Analytical Chemistry*, 55(12), 1113-1122. Translated from *Zhurnal Analiticheskoi Khimii*, 55(12), 2000, 1239-1249.

Kuhnert, N.; Danks, T.N. (2002). Microwave accelerated synthesis of cyclopentadienyl bis-phosphine ruthenium (II) thiolato complexes using focused microwave irradiation. *J. Chem. Res., Synopses*, (2), 66-68.

Kumari, B.S.; Rijulal, G.; Mohanan, K. (2009). Microwave assisted synthesis, spectroscopic, thermal and biological studies of some lanthanide(III) chloride complexes with a heterocyclic Schiff base. *Synthesis and Reactivity in Inorganic, Metal-Organic, and Nano-Metal Chemistry*, 39(1), 24-30.

Kunz, P.; Berghahn, M.; Brueckmann, N.E.; Dickmeis, M.; Kettel, M.; Spingler, B. (2009). Functionalized tris(pyrazolyl)methane ligands and $Re(CO)_3$ complexes thereof. *Z. Anorg. Allg. Chem.* 635(3), 471-478.

Lagashettya, A.; Havanoorb, V.; Basavarajab, S.; Balajib, S.D.; Venkataraman, A. (2007). Microwave-assisted route for synthesis of nanosized metal oxides. *Science and Technology of Advanced Materials*, 8, 484-493.

Lang, J.; Tatsumi, K.; Kawaguchi, H.; Lu, J.; Ge, P.; Ji, W.; Shi, S. (1996). Microwave Irradiation Synthesis of Mo(W)/Tl/S Linear Chains and Their Nonlinear Optical Properties in Solution. *Inorg. Chem.* 35, 7924-7927.

Leadbeater, N.E. (2010). *Microwave Heating as a Tool for Sustainable Chemistry*. CRC Press, 282 pp.

Lasri, J.; Charmier, M.A.J.; Guedes da Silva, M.F.C.; Pombeiro, A.J.L. (2006). Direct synthesis of (imine)platinum(II) complexes by iminoacylation of ketoximes with activated organonitrile ligands. *Dalton Trans.*, (42), 5062-5067.

Leadbeater, N.E.; Shoemaker, K.M. (2008). Preparation of ruthenium and osmium carbonyl complexes using microwave heating: demonstrating the use of a gas-loading accessory and real-time reaction monitoring by means of a digital camera. *Organometallics*, 27(6), 1254-1258.

Lee, Y.T.; Choi, S.Y.; Lee, S.I.; Chung, Y.K.; Kang, T.J. (2006). Microwave-assisted synthesis of (η^6-arene)tricarbonylchromium complexes. *Tetrahedron Letters*, 47(37), 6569-6572.

Lee, H.W.; Lee, L.N.; Chan, A.S.C.; Kwong, F.Y. (2008). Microwave-assisted rhodium-complex-catalyzed cascade decarbonylation and asymmetric Pauson-Khand-type cyclizations. *Eur. J. Org. Chem.*, (19), 3403-3406.

Lin, Z.; Wragg, D.S.; Morris, R.E. (2006). Microwave-assisted synthesis of anionic metal-organic frameworks under ionothermal conditions. *Chem. Commun.*, (19), 2021-2023.

Liu, B.; Zou, R.-Q.; Zhong, R.-Q.; Han, S.; Shioyama, H.; Yamada, T.; Maruta, G.; Takeda, S.; Xu, Q. (2008). Microporous coordination polymers of cobalt(II) and manganese(II) 2,6-naphthalenedicarboxylate: preparations, structures and gas sorptive and magnetic properties. Microporous and Mesoporous Materials, 111(1-3), 470-477.

Liu, H.-K.; Tsao, T.-H.; Zhang, Y.-T.; Lin, C.-H. (2009). Microwave synthesis and single-crystal-to-single-crystal transformation of magnesium coordination polymers exhibiting selective gas adsorption and luminescence properties. *Cryst. Eng. Comm.*, 11(7), 1462-1468.

Liu, M.O.; Tai, C.-H.; Chen, C.-W.; Chang, W.-C.; Hu, A.T. (2004). The fluorescent and photoelectric conversion properties of porphyrin-perylene tetracarboxylic complex. *Journal of Photochemistry and Photobiology, A: Chemistry*, 163(1-2), 259-266.

Liu, M.O.; Tai, C.-H.; Hu, A.T. (2005). Synthesis of metalloporphyrins by microwave irradiation and their fluorescent properties. *Materials Chemistry and Physics*, 92(2-3), 322-326.

Liu, M.O.; Hu, A.T. (2004). Microwave-assisted synthesis of phthalocyanine-porphyrin complex and its photoelectric conversion properties. *J. Organomet. Chem.*, 689(15), 2450-2455.

Liu, W.; Ye, L.; Liu, X.; Yuan, L.; Lu, X.; Jiang, J. (2008). Rapid synthesis of a novel cadmium imidazole-4,5-dicarboxylate metal-organic framework under microwave-assisted solvothermal condition. Inorg. Chem. Commun., 11(10), 1250-1252.

Lu, J.; Chen, W. (2003). Microwave-assisted synthesis of ferrocenyl unsaturated ketones in solid phase. *Huaxue Yanjiu Yu Yingyong*, 15(2), 265-266, c3.

Luo, K.-j.; Xu, L.-l.; Peng, J.-b.; Wei, X.Q.; Xie, M.-k.; Jiang, Q.; Zhang, Li-f. (2007). Microwave synthesis and electroluminescent characteristic of phosphorescent cyclometalated platinum complexes with long chain β-diketonate ancillary ligands. *Sichuan Daxue Xuebao, Gongcheng Kexueban*, 39(5), 53-58.

Luo, K.-j.; Xu, L.-l.; Wei, X.-q.; Xie, M.-g.; Jiang, Q. (2007). Rapid synthesis of phosphorescent cyclometalated platinum complexes by microwave irradiation. *Huaxue Yanjiu Yu Yingyong*, 19(9), 991-993.

Maanju, S.; Poonia, K.; Chaturvedi, S.; Singh, R.V. (2007). Synthesis, characterization and biological activity of dioxomolybdenum(VI) complexes with nitrogen-sulfur and nitrogen-oxygen donor ligands. *International Journal of Chemical Sciences*, 5(3), 1247-1257.

Mahajan, K.; Fahmi, N.; Singh, R.V. (2007). Synthesis, characterization and antimicrobial studies of Sb(III) complexes of substituted thioimines. *Indian Journal of Chemistry*, 46A, 1221-1225.

Mahajan, K.; Swami, M.; Joshi, S.C.; Singh, R.V. (2008). Microwave-assisted synthesis, characterization, biotoxicity and antispermatogenic activity of some antimony(III) complexes with N-O and N-S donor ligands. *Appl. Organomet. Chem.*, 22(7), 359-368.

Mahajan, K.; Swami, M.; Singh, R.V. (2009). Microwave synthesis, spectral studies, antimicrobial approach, and coordination behavior of antimony(III) and bismuth(III) compounds with benzothiazoline. *Russian Journal of Coordination Chemistry*, 35(3), 179-185.

Maree, S.E. (2005). Microwave assisted axial ligand substitution of titanium phthalocyanines. *J. Porphyr. Phthalocyan.*, 9(12), 880-883.

Martin, S.; Revathi, C.; Dayalan, A.; Mathivanan, N.; Shanmugaiya, V. (2008). Halocobaloximes containing coordinated pyrazine, pyrazine carboxylic acid and pyrazine carboxamide: microwave assisted synthesis, characterization and antibacterial activity. *Rasayan Journal of Chemistry*, 1(2), 378-389.

Martínez-Palou, R. (2007). *Ionic Liquid and Microwave-Assisted Organic Synthesis: A "Green" and Synergic Couple. J. Mex. Chem. Soc.* 2007, 51(4), 252-264.

Masilela, N.; Nyokong, T. The synthesis and photophysical properties of water soluble tetrasulfonated, octacarboxylated and quaternised 2,(3)-tetra-(2 pyridiloxy) Ga

phthalocyanines.
http://eprints.ru.ac.za/1525/1/nyokong_synthesis_and_photophysical3.pdf.

Matsumura-Inoue, T.; Tanabe, M.; Minami, T.; Ohashi, T. (1994). Aremarkably rapid synthesis of ruthenium(II) polypyridine complexes by microwave irradiation. *Chem. Lett.* 2443.

Meng, G.; Song, H.; Dong, Q.; Peng, D. (2004). Application of novel aerosol-assisted chemical vapor deposition techniques for SOFC thin films. *Solid State Ionics*, 175(1-4), 29-34.

Mercier, A.; Yeo, W.C.; Chou, J.; Chaudhuri, P.D.; Bernardinelli, G.; Kundig, E.P. (2009). Synthesis of highly enantiomerically enriched planar chiral ruthenium complexes via Pd-catalysed asymmetric hydrogenolysis. Chem. Commun., (35), 5227-5229.

Meshram, H. M.; Reddy, G. S. (2003). Microwave thermolysis. Part III. A rapid and convenient coupling of 2-naphthols in solvent-free condition. *Indian Journal of Chemistry, Section B: Organic Chemistry Including Medicinal Chemistry*, 42B(10), 2615-2617.

Mitra, A.K.; Karchaudhuri, N.; De, Aparna; M., Jayati; Mahapatra, T. (2003). Unusual regioselectivity in nucleophilic addition to allylpalladium complex in conventional heating and under microwave irradiation. *ARKIVOC*, (9), 96-103.

Mondal, T. K.; Wu, J.-S.; Lu, T.-H.; Jasimuddin, Sk.; Sinha, C. (2009). Syntheses, structures, spectroscopic, electrochemical properties and DFT calculation of Ru(II)-thioarylazoimidazole complexes. *J. Organomet. Chem.*, 694(21), 3518-3525.

Munery, S.; Jaud, J.; Bonvoisin, J. (2008). Synthesis and characterization of bis(bipyridine)-ruthenium(II) complexes with bromo or protected ethynyl β-diketonato ligands. *Inorg. Chem. Commun.*, 11(9), 975-977.

Nakashima, K.; Masuda, Y.; Matsumura-Inoue, T.; Kakihana, M. (2008, Volume Date 2009). Microwave-assisted synthesis of (tris-acetylacetonato)(2,9-dimethyl-4,7-diphenyl-1,10-phenanthroline)terbium(III) complex with outstanding high green luminescence. *Journal of Luminescence*, 129(3), 243-245.

Nascimento, B.F.O.; Pineiro, M.; Rocha Gonsalves, A.M.d'A.; Silva, M.R.; Beja, A.M.; Paixao, J.A. (2007). Microwave-assisted synthesis of porphyrins and metalloporphyrins: a rapid and efficient synthetic method. *J. Porphyr. Phthalocyan.*, 11(2), 77-84.

Okada, Y.; Nakano, S. (2009). Studies on ferrocene derivatives. Part XVIII. Microwave irradiation effect on the ligand exchange reaction between ferrocene derivatives and aromatic compound. *Inorg. Chim. Acta*, 362(13), 4853-4856.

Okamura, I.; Nagata, A. (2003). Property of a thin zirconia electrolyte film using solid oxide fuel cell grown by ECR plasma CVD method. *Shinku*, 46(3), 310-313.

Oliver Kappe, C.; Dallinger, D.; Murphree, S. (2009). *Practical Microwave Synthesis for Organic Chemists: Strategies, Instruments, and Protocols*. Wiley-VCH, 310 pp.

Oliver Kappe, C. (2009). Controlled Microwave Heating in Modern Organic Synthesis. *Angew. Chem. Int. Ed.* 2004, 43, 6250 –6284.

Ouyang, Z.-Y.; Xiao, X.-M.; Zhang, J.-Y.; Xu, S.-S. (2009). Synthesis and spectrum study of compound dysprosium (III)-bis(2-benzimidazolylmethyl)amine and its interaction with DNA. *Guangpu Shiyanshi*, 26(1), 67-70.

Pagadala, R.; Ali, P.; Meshram, J.S. (2009). Microwave assisted synthesis and characterization of N,N'-bis(salicylaldehydo)ethylenediimine complexes of Mn(II), Co(II), Ni(II), and Zn(II). *J. Coord. Chem.*, 62(24), 4009-4017.

Paredes, S. P.; Fetter, G.; Bosch, P.; Bulbulian, S. (2006). Iodine sorption by microwave irradiated hydrotalcites. *Journal of Nuclear Materials*, 359(3), 155-161.

Paredes, S. P.; Fetter, G.; Bosch, P.; Bulbulian, S. Sol-gel synthesis of hydrotalcite - like compounds. *Journal of Materials Science*, 2006, 41(11), 3377-3382.

Park, S.S.; Jung, K.S.; Lee, J.Y.; Park, J.H.; Lee, G.D.; Suh, C.S. (2001). Microwave synthesis of metal phthalocyanine. *Kongop Hwahak* , 12(7), 750-754.

Patti, A.; Pedotti, S.; Paolo Ballistreri, F.; Trusso Sfrazzetto, G. (2009). Synthesis and Characterization of Some Chiral Metal-Salen Complexes Bearing a Ferrocenophane Substituent. *Molecules*, 14, 4312-4325.

Pelletier, J.; Pantel, R.; Oberlin, J. C.; Pauleau, Y.; Gouy-Pailler, P. (1991). Preparation of copper thin films at ambient temperature by microwave plasma-enhanced chemical vapor deposition from the copper (II) acetylacetonate-argon-hydrogen system. *J. Appl. Phys.* 70(7), 3862-3866.

Pezet, F.; Daran, J.-C.; Sasaki, I.; Aiet-Haddou, H.; Balavoine, G.G.A. (2000). Highly Diastereoselective Preparation of Ruthenium Bis(diimine) Sulfoxide Complexes: New Concept in the Preparation of Optically Active Octahedral Ruthenium Complexes. *Organometallics*, 19(20), 4008-4015.

Phetmung, H.; Wongsawat, S.; Pakawatchai, C.; Harding, D.J. (2009). Microwave synthesis, spectroscopy, thermal analysis and crystal structure of an one-dimensional polymeric {[Cu(4,4'-bipy)(H$_2$O)$_3$(SO$_4$)]·2H$_2$O}$_n$ complex. *Inorg. Chim. Acta*, 362(7), 2435-2439.

Pimentel, L.C.F.; de Souza, A.L.F.; Fernandez, T.Lopez; Wardell, J.L.; Antunes, O.A.C. (2007). Microwave-assisted synthesis of N,N-bis(2-pyridylmethyl)amine derivatives. Useful ligands in coordination chemistry. *Tetrahedron Lett.*, 48(5), 831-833.

Pombeiro, A.J.L.; Martins, L.M.D.R.S.; Alegria, E.C.B.A.; Kirilova, M.V. (2007). Complexes of rhenium with pyrazole or tris(1-pyrazolyl)methanes and their preparation and use as catalysts in partial oxidation, peroxidation or carboxylation of ethane and cyclohexane in mild conditions. Port. Pat. Appl. PT 103735, 25 pp.

Poonia, K.; Maanju, S.; Chaudhary, P.; Singh, R.V. (2007). Use of microwave technology for the synthesis of organotitanium(IV) and organozirconium(IV) complexes with HPOH, HO\N\OH and HO\N\SH type of ligands. *Transition Metal Chemistry*, 32, 204–208.

Poonia, K.; Swami, M.; Chaudhary, A.; Singh, R.V. (2008). Microwave assisted synthesis, characterization and antimicrobial studies of unsymmetric imine complexes of organotitanium(IV) and organozirconium(IV). *Indian Journal of Chemistry*, 47A, 996-1003.

Raner, K.D.; Strauss, C.R.; Trainor, R.W. J. (1995). A new microwave reactor for batchwise organic synthesis. *J. Org. Chem.* 60, 2456.

Ren, Y.; Wang, W.; Zhao, S.; Tian, X.; Wang, J.; Yin, W.; Cheng, L. (2009). Microwave-enhanced and ligand-free copper-catalyzed cyanation of aryl halides with K$_4$[Fe(CN)$_6$] in water. *Tetrahedron Lett.*, 50(32), 4595-4597.

Rentzsch, C.F.; Tosh, E.; Herrmann, W.A.; Kuehn, F.E. (2009). Iridium complexes of N-heterocyclic carbenes in C-H borylation using energy efficient microwave technology: influence of structure, ligand donor strength and counter ion on catalytic activity. *Green Chemistry*, 11(10), 1610-1617.

Roberts, R.M.G. (2006). Synthesis of (η^6-arene)(η^5-cyclopentadienyl) iron (II) complexes with heteroatom and carbonyl substituents. Part I: Oxygen and carbonyl substituents. *J. Organomet. Chem.*, 691(12), 2641-2647.

Roberts, R.M.G. (2006). Synthesis of (η^6-arene)(η^5-cyclopentadienyl)iron(II) complexes with heteroatom and carbonyl substituents. Part II. Amino substituents. *J. Organomet. Chem.*, 691(23), 4926-4930.

Roussy, G.; Pearce, J.A. (1995). *Foundations and Industrial Applications of Microwave and Radio Frecuency Fields.* John Wiley & Sons. Chichester-New York-Brisbane-Toronto-Singapore.

Safari, N.; Jamaat, P.R.; Pirouzmand, M.; Shaabani, A. (2004). Synthesis of metallophthalocyanines using microwave irradiation under solvent free and reflux conditions. J. Porphyr. Phthalocyan., 8(10), 1209-1213.

Salazar, G.A.; Determan, J.J.; Yang, C.; Omary, M.A. (2009). Synthesis and photophysics of novel rhenium(I) pyrazolyl-triazine complexes. *Abstracts of Papers, 238th ACS National Meeting, Washington, DC, United States, August 16-20, 2009*, INOR-183.

Scarborough, C.C.; Bergant, A.; Sazama, G.T.; Guzei, I.A.; Spencer, L.C.; Stahl, S.S. (2009). Synthesis of PdII complexes bearing an enantiomerically resolved seven-membered N-heterocyclic carbene ligand and initial studies of their use in asymmetric Wacker-type oxidative cyclization reactions. *Tetrahedron*, 65(26), 5084-5092.

Schwalbe, M.; Schaefer, B.; Goerls, H.; Rau, S.; Tschierlei, S.; Schmitt, M.; Popp, J.; Vaughan, G.; Henry, W.; Vos, J.G. (2008). Synthesis and characterisation of poly(bipyridine)ruthenium complexes as building blocks for heterosupramolecular arrays. *European Journal of Inorganic Chemistry*, (21), 3310-3319.

Seven, O.; Dindar, B.; Gultekin, B. (2009). Microwave-Assisted Synthesis of Some Metal-Free Phthalocyanine Derivatives and a Comparison with Conventional Methods of their Synthesis. *Turk. J. Chem.*, 33, 123-134.

Shaabani, A.; Safari, N.; Bazgir, A.; Bahadoran, F.; Sharifi, N.; Jamaat, P.R. (2003). Synthesis of the tetrasulfo- and tetranitrophthalocyanine complexes under solvent-free and reflux conditions using microwave irradiation. *Synth. Commun.*, 33(10), 1717-1725.

Shangzhao Shi and Jiann-Yang Hwang (2003). Microwave-assisted wet chemical synthesis: advantages, significance, and steps to industrialization. *Journal of Minerals & Materials Characterization & Engineering*, 2 (2), 101-110.

Sharma, K.; Singh, R.; Fahmi, N.; Singh, R.V. (2010). Microwave assisted synthesis, characterization and biological evaluation of palladium and platinum complexes with azomethines. *Spectrochimica Acta, Part A: Molecular and Biomolecular Spectroscopy*, 75A(1), 422-427.

Shen, Y.; He, B.-j.; Li, B.-j.; Lin, J.-j. (2008). Synthesis and bioactivity of salicylaldehyde thiosemicarbazone and its metal complexes under microwave irradiation. *Huaxue Gongchengshi*, 22(7), 68-70.

Shi, Z.-F.; Jin, J.; Li, L.; Xing, Y.-H.; Niu, S.-Y. (2009). Syntheses, structures, and surface photoelectric properties of Co-btec complexes. Wuli Huaxue Xuebao, 25(10), 2011-2019.

Silva, A.M.G.; Castro, B.; Rangel, M.; Silva, A.M.S.; Brandao, P.; Felix, V.; Cavaleiro, J.A.S. (2009). Microwave-enhanced synthesis of novel pyridinone-fused porphyrins. *Synlett*, (6), 1009-1013.

Singh, K.; McWhinnie, W.R.; Li Chen, H.; Sun, M.; Hamor, T.A. (1996). Reactions of heterocyclic organotellurium compounds with triiron dodecacarbonyl: reactions of thiophenes revisited. *J. Chem. Soc., Dalton Trans.* 1545-1549.

Singh, R.V.; Jain, M; Deshmukh, C.N. (2005). Microwave-assisted synthesis and insecticidal properties of biologically potent organosilicon(IV) compounds of a sulfonamide imine. *Appl. Organomet. Chem.,* 19(7), 879-886.

Singh, R.V.; Chaudhary, P.; Poonia, K.; Chauhan, S. (2008). Microwave-assisted synthesis, characterization and biological screening of nitrogen-sulphur and nitrogen-oxygen donor ligands and their organotin(IV) complexes. *Spectrochimica Acta, Part A: Molecular and Biomolecular Spectroscopy,* 70A(3), 587-594.

Sinha, S.; Srivastava, A.K.; Sengupta, S.K.; Pandey, O.P. (2008). Microwave assisted synthesis, spectroscopic and antibacterial studies of bis(cyclopentadienyl)hafnium(IV) derivatives with benzil bis(aroyl hydrazones). Transition Metal Chemistry, 33(5), 563-567.

Soriente, A.; De Rosa, M.; Lamberti, M.; Tedesco, C.; Scettri, A.; Pellecchia, C. (2005). Synthesis, crystal structure and application in regio- and stereoselective epoxidation of allylic alcohols of a titanium binaphthyl-bridged Schiff base complex. *J. Mol. Catal. A: Chemical,* 235(1-2), 253-259.

Spinella, A.; Caruso, T.; Pastore, U.; Ricart, S. (2003). Improving methodology for the preparation of uracil derivatives from Fischer carbene complexes. Microwave activation. *J. Organomet. Chem.,* 684(1-2), 266-268.

Takahashi, S.; Shojiya, M.; Kawamoto, Y.; Konishi, A. (2003). Preparation and characterization of amorphous GaF_3 and GaF_3-BaF_2 thin films by ECR microwave plasma-enhanced CVD. Thin Solid Films, 429(1-2), 28-33.

Tan, N.Y.; Xiao, X.M.; Li, Z.L.; Matsumura-Inoue, T. (2004). Microwave synthesis, characterization and DNA-binding properties of a new cobalt(II) complex with 2,6-bis(benzimidazol-2-yl)pyridine. *Chinese Chemical Letters,* 15(6), 687-690.

Tusek-Bozic, L.; Marotta, E.; Traldi, P. (2007). Efficient solid-state microwave-promoted complexation of a mixed dioxa-diaza macrocycle with an alkali salt. Synthesis of a sodium ethyl 4-benzeneazophosphonate complex. *Polyhedron,* 26(8), 1663-1668.

Van Atta, S.L.; Duclos, B.A.; Green, D.B. (2000). Microwave-Assisted Synthesis of Group 6 (Cr, Mo, W) Zerovalent Organometallic Carbonyl Compounds. *Organometallics.* 19(12), 2397 (2000).

Velmathi, S.; Vijayaraghavan, R.; Pal, R.P.; Vinu, A. (2010). Microwave Assisted Ligand Free Palladium Catalyzed Synthesis of β-Arylalkenyl Nitriles Using Water as Solvent. *Catalysis Lett.,* 135(1-2), 148-151.

Villemin, D.; Martin, B.; Puciova, M.; Toma, S. (1994). Dry synthesis under microwave irradiation: Synthesis of ferrocenylenones. *J. Organomet. Chem.* 484, 27.

Vondrova, M.; Burgess, C.M.; Bocarsly, A.B. (2007). Cyanogel Coordination Polymers as Precursors to Transition Metal Alloys and Intermetallics - from Traditional Heating to Microwave Processing. *Chem. Mater.,* 19(9), 2203-2212.

Walther, D.; Schebesta, S.; Rau, S.; Schaefer, B.; Gruessing, A. (2005). Procedure for the production of light-emitting ruthenium(II) compounds. Patent DE 102004009551, 6 pp.

Wang, X.-F.; Zhang, Y.-B.; Huang, H.; Zhang, J.-P.; Chen, X.-M. (2008). Microwave-Assisted Solvothermal Synthesis of a Dynamic Porous Metal-Carboxylate Framework. *Crystal Growth & Design,* 8(12), 4559-4563.

Wiesbrock, F.; Hoogenboom, R.; Schubert, U.S. (2004). Microwave-Assisted Polymer Synthesis:
State-of-the-Art and Future Perspectives. *Macromol. Rapid Commun.* 25, 1739–1764.

Wang, M.-Z.; Wong, M.-K.; Che, C.-M. (2008). Gold(I)-catalyzed intermolecular hydroarylation of alkenes with indoles under thermal and microwave-assisted conditions. *Chemistry--A European Journal*, 14(27), 8353-8364.

Winkelmann, O.H.; Navarro, O. (2010). Microwave-Assisted Synthesis of N-Heterocyclic Carbene-Palladium(II) Complexes. *Advanced Synthesis & Catalysis*, 352(1), 212-214.

Wolfel, A.M.; Walters, B.M.; Zovinka, E.P. (2009). Microwave metallation of porphyrins. *Abstracts of Papers, 238th ACS National Meeting, Washington, DC, United States, August 16-20, 2009*, CHED-214.

Xiao, X.; Sakamoto, J.; Tanabe, M.; Yamazaki, S.; Yamabe, S.; Matsumura-Inoue, T. (2002). Microwave synthesis and spectroelectrochemical study on ruthenium(II) polypyridine complexes. *Journal of Electroanalytical Chemistry*, 527(1-2), 33-40.

Xu, X.-x.; Fan, H.-t. (2007). Microwave synthesis of Co(III) complex and its crystal structure. *Hefei Gongye Daxue Xuebao, Ziran Kexueban*, 30(4), 458-460.

Yamauchi, T.; Tsukahara, Y.; Sakamoto, T.; Kono, T.; Yasuda, M.; Baba, A.; Wada, Y. (2009). Microwave-assisted synthesis of monodisperse nickel nanoparticles using a complex of nickel formate with long-chain amine ligands. *Bulletin of the Chemical Society of Japan*, 82(8), 1044-1051.

Yaseen, M.; Ali, M.; NajeebUllah, M.; Munawar, M.A.; Khokhar, I. (2009). Microwave-assisted synthesis, metalation, and Duff formylation of porphyrins. *J. Heterocyclic Chem.*, 46(2), 251-255.

Yin, D.; Liu, J.; Zhang, Y.; Gao, Q.; Yin, D. (2005). Immobilization of Mn(salen) complex on aluminum-containing mesoporous materials by microwave heating for epoxidation of styrene. *Studies in Surface Science and Catalysis*, 156(Nanoporous Materials IV), 851-858.

Zabova, H.; Sobek, J.; Cirkva, V.; Solcova, O.; Kment, S.; Hajek, M. (2009). Efficient preparation of nanocrystalline anatase TiO_2 and V/TiO_2 thin layers using microwave drying and/or microwave calcination technique. *Journal of Solid State Chemistry*, 182(12), 3387-3392.

Zhang, G.; Wang, C.; Wu, Q.; Song, X.; Ran, F. (2008). Method for synthesizing quinoxaline-series ligand-iridium complexes. Patent CN 101215299, 15 pp.

Zhang, Z.-y.; Hao, W.; Hu, W.-x. (2006). Synthesis of 5,10,15,20-tetrakis(2-pyridyl)porphyrin and its Mn(III) complex. *Hecheng Huaxue*, 14(5), 497-499.

Zhang, S.-H.; Tang, M.-F.; Ge, C.-M. (2009). Microwave synthesis, crystal structure and magnetic behavior of a Schiff base trinuclear nickel cluster. *Z. Anorg. Allg. Chem.*, 635(9-10), 1442-1446.

Zhang, Y.; Yin, D.; Fu, Z.; Li, C.; Yin, D. (2003). Preparation of Mn(salen)/Al-MCM-41 catalyst under microwave irradiation and its catalytic performance in epoxidation of styrene. *Cuihua Xuebao*, 24(12), 942-946.

Ziegler, M.; Monney, V.; Stoeckli-Evans, H.; Von Zelewsky, A.; Sasaki, I.; Dupic, G.; Daran, J.-C.; Balavoine, G.G.A. (1999). Complexes of new chiral terpyridyl ligands. Synthesis and characterization of their ruthenium(II) and rhodium(III) complexes. *J. Chem. Soc., Dalton Trans.: Inorg. Chem.*, (5), 667-676.

Zoller, F.; Riss, P. J.; Montforts, F.-P.; Roesch, F. (2010). Efficient post-processing of aqueous generator eluates facilitates ^{68}Ga-labelling under anhydrous conditions. *Radiochimica Acta*, 98(3), 157-160.

The Effect of Microwave Heating on the Isothermal Kinetics of Chemicals Reaction and Physicochemical Processes

Borivoj Adnadjevic and Jelena Jovanovic
Faculty of Physical Chemistry, University of Belgrade
R. Serbia

1. Introduction

Microwave heating is a widely accepted, non-conventional energy source which is now applied in numerous fields. In many of the examples published in the literature, microwave heating has been shown to lead to dramatically reduce reaction times, increased product yields and enhanced product purities by reducing unwanted side reactions compared to the conventional method of heating.

Several reviews have been published on the application of microwaves in organic synthesis (Kingston & Haswall, 1997; Loupy, 2002), polymer synthesis (Bogdal & Prociak, 2007), material science (Perelaer et al., 2006), nanotechnology (Tsuji et. al., 2005), homogeneous and heterogeneous catalyses (Larhed et. al., 2002; Will, et. al., 2004), medicinal chemistry (Shipe et. al., 2005), drug delivery (Kappe & Dallinger, 2006), biochemical processes (Collins & Leadbeather, 2007), synthesis of radioisotopes (Elander et. al., 2000), combination chemistry (Kimini et. al., 2000), green chemistry (Nücther et. al., 2004), sintering (Saberi et. al., 2009), ceramics (Vanderach, 2002), extraction (Luchesi, et. al., 2004), nucleation and crystallization (Sung et. al., 2007; Cundy 1998), annealing (Bhaskar, et. al., 2007), solid-state reactions (Li et. al., 2007), adsorption (Turner, 2000) and combustion (Y-Pee, et. al., 2005).

The region of microwave radiation in the electromagnetic spectrum is located between infrared radiation and radio waves. Microwaves have wavelengths of 1 mm to 1 m, corresponding to frequencies between 0.3 and 300 GHz. In industrial and household microwave equipment, a frequency of 2.45 GHz with a wave number 12.2 cm is used.

The process by which matter absorbs microwave energy is known as dielectric heating. Three mechanism of dielectric heating of materials under the influence of microwave irradiation are recognized: the dipolar polarization mechanism, the conduction mechanism and the interphase polarization mechanism (Gabriel, et. al., 1998). Microwave heating of materials which occurs due to the dipolar polarization mechanism is provoked by the molecular friction caused by the re-orientation of the dipole moments of the material under the influence of a microwave field, *i.e.*, by the breaking and reforming of intermolecular bonds, due to which irreversible dissipation of the kinetic energy and thermal energy of the sample occurs. In the case when a substance contains mobile ions, the solvated ion due to attending the change in the oscillation of electrical component of the microwave field, increase the energy of collision and provokes the breakage and re-establishment of intermolecular bonds. This means that the substance will be heated in the same manner as in the case of the dipolar polarization mechanism, *i.e.*, by the conduction mechanism. When metals are used, the third mechanism, of heating materials under the influence of microwave field, interphase polarization, is possible.

The coupling of microwave energy in the medium depends on the dielectric properties of the substance to be heated, *i.e.*, it depends on the quantity of microwave radiation that fails to penetrate the substance. The dielectric constant or relative permittivity (ε) is a property which describes the ability of a substance to store charge irrespective of the dimensions of the sample.

In an electromagnetic field, ε_r is defined by the equation:

$$\varepsilon_r = \varepsilon' + i\varepsilon''$$ (1)

where ε' is the real part of the dielectric constant and ε'' is the dielectric loss factor (dynamic dielectric coefficient) and i is imaginary number. The dielectric loss factor is obtained by comparing the irradiated microwave energy to the energy that has coupled with the sample. The degree of energy coupling in a reaction system depends on both parameters ε' and ε'' and is called the dissipation factor D:

$$D = \frac{\varepsilon''}{\varepsilon'}$$ (2)

The dissipation factor defines the ability of a medium at a given frequency and temperature to convert electromagnetic energy into heat. The efficiency on conversion of microwave energy into thermal energy depends on both the dielectric and thermal properties of a material. The fundamental relationship is:

$$P = \sigma|E|^2 = \left(\omega\varepsilon_0\varepsilon''\right)|E|^2 = \left(\omega\varepsilon_0\varepsilon'D\right)|E|^2$$ (3)

where ε_0 is permittivity of free space, P is the power dissipation per unit volume, σ is the conductivity of the material, E is the strength of the electric field in the sample and ω is the angular frequency.

Assuming negligible heat loss and diffusion, the rate of heating or temperature rise (ΔT) in a time interval can be expressed as:

$$\frac{\Delta T}{t} = \frac{\sigma|E|^2}{\rho \cdot c}$$ (4)

where ρ is the density and c is the specific heat capacity.

The effect of microwave irradiation in chemical reactions and physico-chemical processes is a combination of thermal effects (overheating (Baghurst & Mingos, 1992), hot-spots (Zhang et. al., 1999) and selected heating (Raner et. al., 1993), as well as specific microwave effects (Hoz et. al., 2005). The thermal effects are connected with the different characteristics of microwave dielectric heating and conventional heating. Table 1 presents basic characteristics of microwave and conventional heating.

Microwave heating is based on the ability of some components of a material to transform electromagnetic energy into heat, whereby the magnitude of the heating depends on the

MICROWAVE HEATING	CONVENTIONAL HEATING
Energetic coupling	Conduction/ convection
Coupling at the molecular level	Superficial heating
Rapid	Slow
Volumetric	Super ficial
Selective	Non-selective
Dependant on the properties of the materials	Less dependant on the properties of the materials

Table 1. The basic characteristics of microwave and conventional heating

dielectric properties of the molecules. Energy transmission is produced by dielectric losses, which is in contrast to conduction and convection processes observed in conventional heating. Microwave heating is rapid and volumetric, with the whole material being heated simultaneously. In contrast, conventional heating is slow and is introduced into the sample from the surface.

The thermal effects observed under microwave heating are a consequence of the inverted heat transfer, the inhomogeneties of the microwave field within the sample and the selective absorption of the radiation by polar compounds. These effects can be used efficiently to improve processes, modify selectivities or even to perform reactions that do not occur under conventional heating.

The issue of specific microwave effects is still a controversial matter. Several theories have been postulated and also some predicted models have been published.

Berlan et al., 1998, observed the acceleration of a cycloaddition reaction under isothermal microwave reaction conditions and explained the change in entropy of the reaction systems.

Microwave enhanced the imidization reaction of a sulfone and ketone group-containing polyamic acid, which Lewis et al., 1992, elucidated with a mechanism based on "excess dipole energy", in which it was proposed that the localized energy (temperature) of the dipole groups was higher compared to the non-polar bonds within these systems.

Rybakov and Semenov, 1994, explained the effect of microwave reaction conditions on the kinetics of reactions in the solid state with the formation of a ponderomotive force, which influences the time-average motion of charged particles and enhances ionic transport in the solid state.

The ability of microwave radiation to excite rotational transitions and thus enhance the internal energy of a system was used by Strauss and Trainor, 1995, to explain the effects of microwave fields on the kinetics of chemical reactions. Binner et al., 1995, investigating the effect of a microwave field on the kinetics of titanium carbide formation, concluded that in the presence of a microwave field, the molecular mobility increases which leads to the increasing value of Arrhenius pre-exponential factor which further causes an acceleration of the synthesis of titanium carbide. Stuerga and Gaillard, 1996, explained the acceleration of the reaction rate in condensed states under microwave reaction conditions in comparison to conventional reacting conditions by the enhanced rate of collisions in condensed phases, which induces transfer between rotational and vibrational energy levels and reaction acceleration. Booske et al., 1997, used the existence of non-thermal energy distributions to explain microwave-enhanced solid-state transport.

Based on the experimentally confirmed decrease in the values of the activation energy of sodium bicarbonate decomposition under microwave reaction conditions, Shibata et al., 1996, concluded that the effect of a microwave field on dielectric materials is to induce rapid rotation of the polarized dipoles in the molecules. This generates heat due to friction while simultaneously increasing the probability of contact between the molecules, thus enhancing the reaction rate and reducing the activation energy.

Perreux and Loupy, 2000, studied and classified the nature of the microwave effects considering the reaction medium and reaction mechanism, *i.e.*, the polarity of the transition state and the position of the transition state along with the reaction coordinate. Blanco and Auerbach, 2003, theoretically proved that the energy of a microwave field is initially transferred to select molecular modes (transition, vibration and/or rotation) and used this information to explain the inverse desorption of benzene and methanol from zeolite, compared to thermal desorption.

Conner and Tumpsett, 2008, explained specific microwave effects with the capability of microwaves to change the relative energies of rotation of intermediates in a given sequences. Bearing the preceding discussion in mind, this Chapter provides results of investigations of the kinetics of a) chemical reactions: isothermal PAA hydrogel formation, fullerole formation, and sucrose hydrolyses and b) physicochemical process: ethanol adsorption and PAM hydrogel dehydration;   under conventional and microwave conditions with the aim to present the effects of microwave reaction conditions on kinetics model, values of kinetic parameters, the complexity of the kinetics of the investigated processes and to explain the effects of a microwave field on the kinetics of chemical reaction and physico-chemical processes.

2. Results

2.1 Isothermal kinetics of acrylic acid polymerization and crosslinking

Due to their extraordinary properties, such as swellability in water, hydrophilicity, biocompatibility, low toxicity and to their abilities to respond to a variety of changes in the surrounding environment, superabsorbing hydrogels of crosslinked poly(acrylic acid) (PAA) and their copolymers have been utilized in a wide range of biological, medical and pharmaceutical applications. Despite the importance of PAA hydrogels and the numerous publications concerning investigations of the absorptive properties of PAA hydrogels, information regarding the kinetics of polymerization is sparse.

Presently, crosslinked PAA hydrogels are usually synthesized following a procedure based on the radical polymerization of acrylic acid and crosslinking of the formed PAA, which could be realized either by a solution polymerization technique (Zhou et al. 1997a; Zhou et al. 1997b; Takeda, 1985) or by inverse suspension polymerization (Mayoux et.al., 2000; Chen & Zhao, 1999).

In a previous paper, the kinetics of the overall process of poly(acrylic acid) hydrogel formation, in both the case of conventional and microwave reacting conditions, were investigated (Jovanovic & Adnadjevic, 2010). It was found that the reaction rate was accelerated by about 50 times under isothermal microwave reaction conditions (MWIRC) when compared to conventional isothermal reaction conditions (CIRC), while the kinetics model is the same for both processes. Herein, a comparison of the kinetics of PAA hydrogel synthesis by aqueous solution polymerization and crosslinking will be presented. The procedure of PAA hydrogel formation, which occurred through the polymerization of acrylic acid and the crosslinking of the formed poly(acrylic acid), is thoroughly described in previous papers (Adnadjevic & Jovanovic, 2008; Jovanovic & Adnadjevic, 2010). The microwave-assisted reactions were conducted using a commercially available monomode microwave unit. (Discover, CEM Corporation, Matthews, North Carolina, USA) presented in Figure 1.

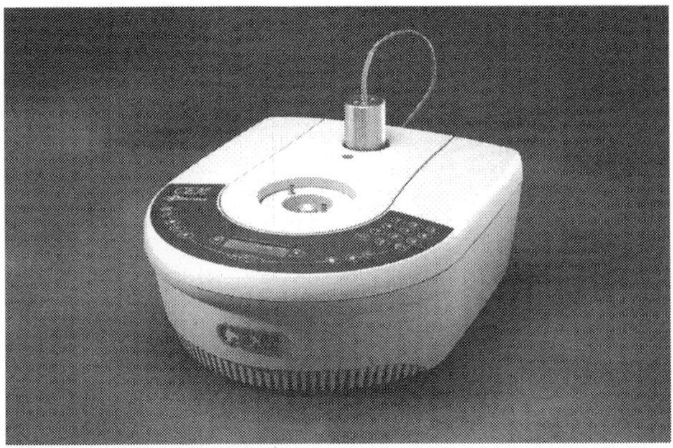

Fig. 1. Microwave unit

The machine consists of a continuously focused microwave power delivery system with an operator selectable power output from 0–300 W. All the reactions were realized in a microwave field of 2.45 GHz. The reactions were performed in glass-tubes (working volume 10 mL) which were supplied by the manufacturer of the microwave reactor. The temperature of the content of the reactor was monitoring using a calibrated fiber-optic probe inserted into the reactor. The temperature, pressure, and profiles were monitored using commercially available software provided by the manufacturer of the microwave reactor.

The temperature of the reaction mixture was controlled by the simultaneously varying of the input power of the microwave field and cooling of the reaction mixture by dry nitrogen. The temperature ranged from 303 to 323 K (±1 K) during 0.5 min to 5 min, which was sufficient to approximately reach 100 % yield of the PAA hydrogel.

The yield (%) of the obtained cross-linked PAA product was determined gravimetrically by measuring the weight of the washed-out hydrogel, dried to constant weight (W_t). The yield, Y, was calculated as the ratio of W_t and the weight of the monomers (monomer and crosslinker) in the reaction mixture (W_o) (Eq. 5):

$$Y = \frac{W_t}{W_o} \tag{5}$$

The degree of PAA hydrogel formation (α) was calculated using Eq. 6:

$$\alpha = \frac{Y}{Y_{max}} \tag{6}$$

where Y is the yield of the obtained PAA hydrogel at time t, and Y_{max} is the maximum value of the hydrogel yield obtained at a certain temperature. Figure 2 shows the isothermal kinetics of PAA hydrogel formation under conventional conditions (a) and under microwave conditions (b) at different temperatures.

All the conversion curves for the formation of PAA hydrogels for both CIRC and MWIRC, are similar in shape. For all the curves, three specific shape of change of PAA hydrogel formation with time can be clearly observed: a linear, non-linear and a saturation stage.

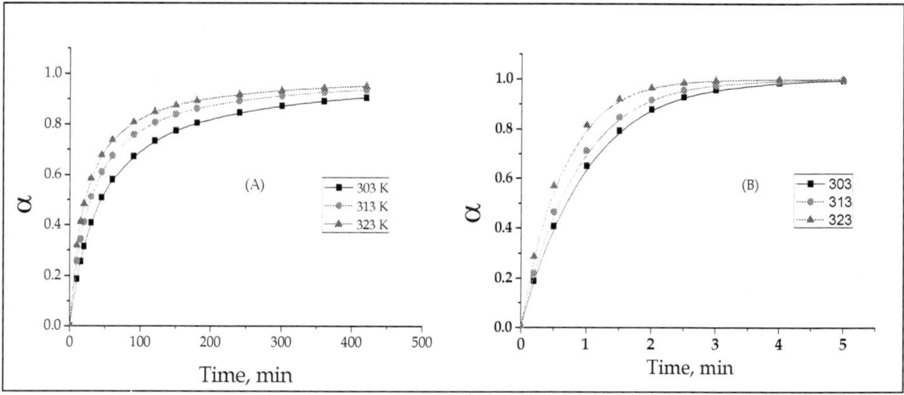

Fig. 2. Conversion curves for the formation of PAA hydrogel under the (A) CIRC conditions and (B) MWIRC conditions

With increasing temperature, for both conventional and microwave heating, the duration of the linear change in the degree of PAA hydrogel formation and the time required to achieve the saturation stage decreased.

With the intention of initially finding a kinetic model of the formation of PAA hydrogel formation both under conventional and microwave conditions, in agreement with the method of Khawan, 2006, the dependence of $d\alpha/dt$ on the degree of PAA hydrogel formation (α) was analyzed (presented in Figure 3 for conventional conditions and under microwave conditions).

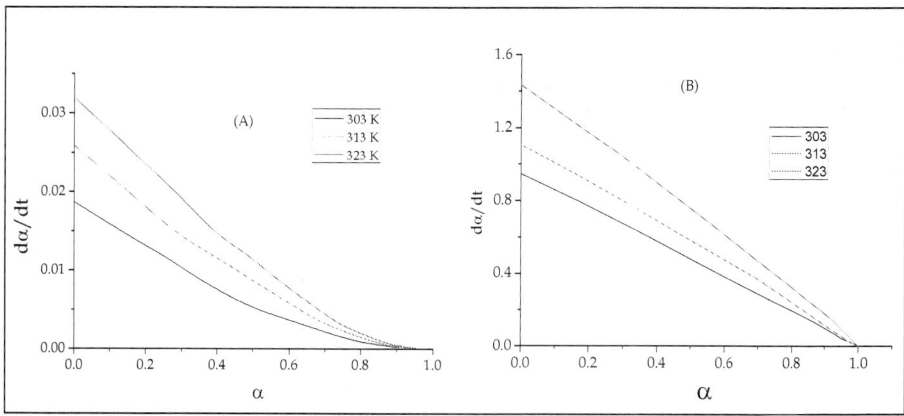

Fig. 3. Dependence of $d\alpha/dt$ on the degree of PAA hydrogel formation under (A) conventional conditions and (B) microwave conditions

As depicted, for the both CIRC and MWIRC, the maximal rate was achieved at the beginning of the process and it decreased with increasing degree of PAA hydrogel formation but in a different manner. Under microwave conditions, the rate of PAA hydrogel formation decreased linearly with increasing degree of PAA hydrogel formation, while under conventional conditions, the decrease was concave shaped. The established maximal values of the rate of PAA hydrogel formation and the changes in the dependence of da/dt on a imply different kinetic models of PAA hydrogel formation under conventional and microwave conditions. In order to prove this hypothesis, the values of kinetic parameters, activation energy and pre-exponential factor were determined by application of the "stationary point" method of Klaric, 1995, and the "model fitting" method Brown et.al., 1980; Adnadjevic et al., 2008) was used with the aim of determining the kinetic model of PAA hydrogel formation. The changes of $(da/dt)_{max}$ of PAA hydrogel formation and the kinetic parameters for both processes are presented in Table 2.

T, K	Conventional process		Microwave process	
	$(da/dt)_{max}$, [min⁻¹]	Kinetic parameters	$(da/dt)_{max}$, [min⁻¹]	Kinetic parameters
303	0.019		0.947	
313	0.023	E_a=22±0.2 kJ/mol	1.105	E_a=17±0.3 kJ/mol
323	0.032	ln A = 5±0.1	1.433	ln A=7±0.1

Table 2. The changes of maximal reaction rate with temperature and kinetic parameters for PAA hydrogel formation for the conventional and the microwave process

The isothermal maximal rates for PAA hydrogel formation are approximately 50 times higher for the microwave heated process than for the conventionally heated process. With increasing temperature, the values of maximal rates for both processes exponentially increased, which enabled the kinetic parameters of PAA hydrogel formation (E_a and ln A) to be determined using the Arrhenius Equation (given in Table 1). The activation energy for PAA hydrogel formation under MWIRC was E_a=17 kJ/mol, which is 1.3 times lower than the E_a of the process under CIRC, while the value of the pre-exponential factor for microwave conditions is 7.4 times higher than the value for the conventional conditions.

By application of the "model fitting method", it can be proved that the isothermal kinetics curves of PAA hydrogel formation during the conventional and microwave conditions can be described by equations (7) and (8), respectively:

$$\alpha = \frac{k_M \cdot t}{1 + k_M \cdot t} \tag{7}$$

$$\alpha = 1 - \exp\left(-k_M \cdot t\right) \tag{8}$$

which are characteristic for the so-called kinetic model of a second-order chemical reaction and a first-order chemical reaction, respectively. If Equations (7) and (8) describe the isothermal kinetics of PAA hydrogel formation, this means that the dependences of $\left[\frac{1}{1-\alpha} - 1\right]$ and of $[-\ln(1-a)]$ on reaction time should give straight lines for the conventional and microwave conditions, respectively, based on the slopes of which, the isothermal model constants of PAA hydrogel formation could be determined. Isothermal plots of $\left[\frac{1}{1-\alpha} - 1\right]$ and $[-\ln (1-a)]$ *versus* reaction time at different temperatures are shown in Figure 4 for the conventional and microwave conditions, respectively.

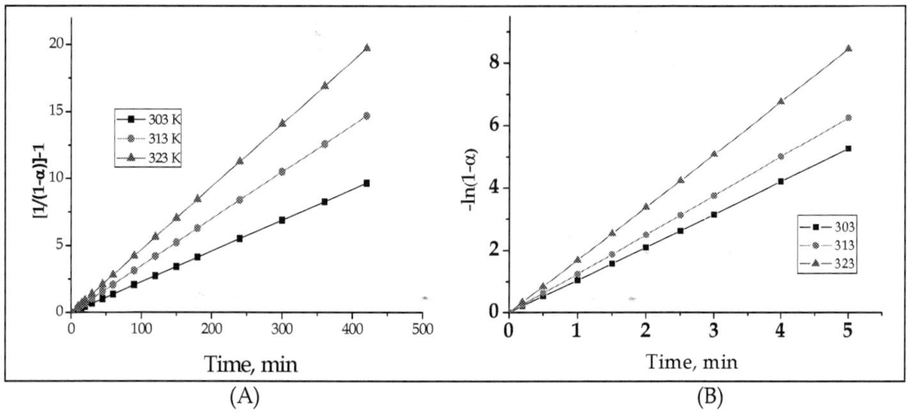

(A)   (B)

Fig. 4. A. Isothermal dependence of $[[1/(1-a)]-1]$ on reaction time for PAA hydrogel formation under the conventional conditions; B. Isothermal dependence of $[-\ln(1-a)]$ on reaction time for PAA hydrogel formation under microwave conditions.

As depicted in Figure 4 A and B, both the dependence of $\left[\frac{1}{1-\alpha}-1\right]$ and $[-\ln (1-a)]$ on reaction time gave straight lines, which confirms the postulated kinetic model and enabled to calculate the models constant of PAA hydrogel formation to be calculated, the values of which are given in Table 3 together with the kinetic parameters determined using the Arrhenius Equation.

T, K	Conventional process		Microwave process	
	k_M^{conv}, min⁻¹	Kinetic parameters	k_M^{mw}, min⁻¹	Kinetic parameters
303	0.023	$E_{a,M}$=21±1 kJ/mol	1.008	$E_{a,M}$=17±1 kJ/mol
313	0.035	ln $(A_M$/min⁻¹)=	1.180	ln $(A_M$/min⁻¹)=8.6±0.2
323	0.047	9.4±0.2	1.590	

Table 3. The model's rate constants and kinetic parameters for PAA hydrogel formation for the conventional and the microwave process

The values of the rate constant of the model for the microwave process are temperature dependant and 34 to 44 times higher than the constants for the conventional process at the same temperature.

As the values of k_M exponentially increase with increasing temperature for both the conventional and the microwave process, the kinetic parameters of PAA hydrogel formation (E_a and ln A) were determined using the Arrhenius Equation and the obtained values are presented in Table 2.

The energy of activation of PAA hydrogel formation under MWIRC is E_a=17kJ/mol, which is 1.2 times lower then the energy of activation under CIRC, while the ln A value is 10 % lower for the microwave process. The values of the kinetic parameters calculated using the rate constants of the model agree well with those based on the stationary point method.

The increase in the rate of some chemical reactions and physico-chemical processes are most frequently explained by the existence of local overheating, *i.e.*, by the presence of "hot-spot" points (Baghurst, 1992; Zhang, 1999; Raner, 1993; Hoz, 2005).

The established values of the rate constants of the model and their changes with temperature under conventional and microwave reaction conditions enabled the effects of so-called "hot-spots" on the kinetics of the investigated process to be objectively evaluated. In fact, if it is assumed that the values of the activation energy and the pre-exponential factor obtained for microwave reaction conditions are identical to the values for conventional reaction conditions, then, based on the experimentally determined values of the rate constants of the model under the microwave reacting conditions (k_M^{mw}), it is possible to calculate the temperature of the reacting system under microwave reacting conditions – the calculated microwave temperature (T^*) using the following equation:

$$T^* = \frac{E_{a,M}^c}{R\left(\ln A_M^c - \ln k_{mw}^T\right)} \tag{9}$$

where $E_{a,M}^c$ is the activation energy under conventional conditions, R is the universal gas constant, $\ln A_M^c$ is the pre-exponential factor for CIRC and k_{mw}^T is the reaction rate constant for MWIRC at a defined temperate. Table 4 presents values of the calculated microwave temperature T*.

T, K	303	313	323
T*, K	269	274	283

Table 4. Values of calculated microwave temperature

As the calculated values of the temperatures of the reaction system are significantly lower than the real macroscopic values, it can be stated that the acceleration of PAA hydrogel formation under microwave condition is not the consequence of the existence of "hot spots" in the reaction system.

Bearing in mind the established changes of the kinetic model of PAA hydrogel formation under microwave reaction conditions compared to conventional conditions, the Vyazovkin method was applied to examine the complexity of the kinetics of this process (Vyazovkin & Lesnikovich, 1990). In order to apply the Vyazovkin method, the dependences of $E_{a,\alpha}$ on the degree of conversion were established by the Friedman differential isoconversional method (Friedman, 1964) for the investigated process under conventional and microwave conditions and the results are presented in Figure 5.

The results shown in Figure 5 enabled it to be said with great certainty that the E_a of PAA hydrogel formation is independent of the degree of formation in both the conventional and microwave-heated process. This confirms that the investigated processes occurred through the elementary, overall single-stage formation the PAA hydrogel in both processes.

The established decrease in the values of the kinetic parameters of PAA hydrogel formation under microwave isothermal reaction conditions in comparison to conventional isothermal reacting conditions permit the possibility of an entirely new explanation of the action of a microwave field on the kinetics of chemical reactions and physico-chemical processes. The essence of this novel approach is the existence of a functional relationship between the values of the kinetic parameters of PAA hydrogel formation under the conventional and microwave reacting conditions, which is expressed as:

$$\ln A_F = 5.23 + 0.195\, E_{a,F} \tag{10}$$

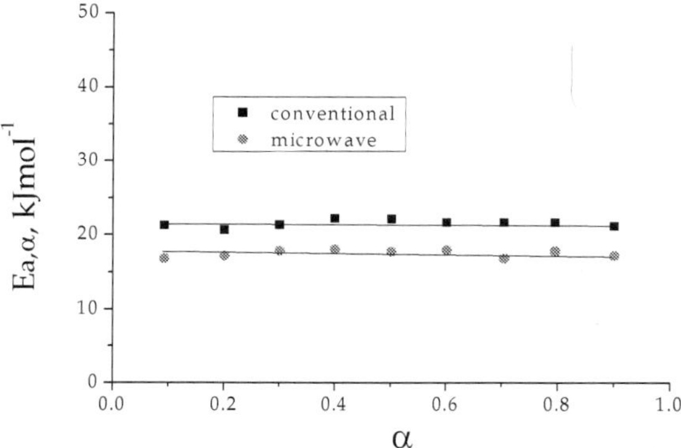

Fig. 5. The dependence of the $E_{a,a}$ on the degree of conversion for PAA hydrogel formation

where $E_{a,,F}$ and $\ln A_F$ are the activation energy and pre-exponential factor in a defined physical field (thermal or microwave). The form of equation (10) is recognized as the equation of the compensation effect, which is in relation to distinct changes in the conditions of a reaction or process (Vyazovkin & Wight, 1997). According to the Larsson Model of selective energy transfer during the interaction of a catalyst with reacting molecules, the existence of a compensation effect and the formation of an "active complex" (Larson, 1989) is explained as the consequence of transfer of the necessary amount of vibrational energy from an energetic reservoir onto the reacting molecule, which is aimed at "active complex" formation. The established changes in the kinetics parameters (decrease) can be explained with the decrease in: required amount of vibration energy necessary for "active complex" formation and in resonance frequency of the energetic transfer the between the oscillators under MWIRC in comparison to CIRC.

It can be assumed that a reacting molecule can be modeled as a sum of normal oscillators and that that molecule convert into an "active complex" anytime when it accepts the necessary amount of energy (energy of activation) and that during the formation of such an "active complex", vibrational changes occur which are in relation with changes connected with the localization of the energy on a defined bond (normal oscillator). Then, in accordance with the Larsson Model (Larsson, 1989) the wave number of the resonant oscillation (v) and the energy of activation (E_a) of the reaction can be calculated. The wave number of the resonant oscillation is given by the Equation (11):

$$v = 0.715 / R \cdot b \tag{11}$$

where b is the energetic parameter: slope of the compensation effect equation, while the values of the activation energy is quantized and given as:

$$E_a = (nv + n^2\, vx) + RT \tag{12}$$

where n is the number of quantum of vibrational energy transferred from one to another oscillator or from an energy reservoir („heat bath") to the resonant oscillator, which are necessary to overcome the energetic barrier (activation energy) and x is the anharmonicity constant of the oscillator. The calculated values for v, n, x for the formation of a PAA hydrogel are presented in Table 5.

Variable	CIRC	MWIRC
v (cm^{-1})	429	429
n	4	3
x	−0.022	−0.0017

Table 5. The values of v, n and x under CIRC and MWIRC

During the formation of the "active complex" in the process of PAA hydrogel formation, resonance between the oscillator of the energetic reservoir and a molecules of acrylic acid under both CIRC and MWIRC is achieved at the same frequency of v = 429 m^{-1}. That frequency corresponds to the so-called C–C–C skeleton bending vibration along the acrylic acid (Rufino & Monteiro, 2003) and therefore the formation of the "active complex" commences with the focusing and increasing of the energy of this oscillator. The number of quanta of vibrational energy that must be transferred from the energetic reservoir to the activated oscillator of a molecule of acrylic acid in order to overcome the energetic barrier for active complex formation under CIRC is higher than under MWIRC. Therefore, the established decrease in activation energy of the reaction under the MWIRC is a consequence of the increased ground vibration energetic level of the activated oscillator due to energy absorption from the microwave field. In addition, the recognized decrease in the pre-exponential factor of a reaction is a result of the decreasing number of energetic transfers per time unit (frequency of transfer), which is caused by an increase in the energy of an oscillator and by an increase in anharmonic oscillators.

2.2 Isothermal kinetics of fullerol formation under conventional and microwave reacting conditions

Water-soluble C_{60} fullerols are one of the most interesting fullerene derivatives extensively investigated nowadays. They have potential applications in aqueous solution chemistry, electrochemistry, material chemistry and biochemistry. Fullerols have been used as preparatory material for syntheses of dendritic star-shaped polymers (Dai, 1997) and hypercrosslinked networks, as probes for investigating the surface properties of biomaterials (Zhao et. al., 2004) and as coatings for solid-phase micro extraction (Lu et. al., 2004). They exhibited excellent free-radical scavenging abilities against reactive oxygen species and radicals under physiological conditions (Lai et. al., 2000; Dugan et. al., 1996). Also, they show, *in vivo* and *in vitro*, potentially high biological activity in many important biological processes such as inhibition of HIV protease, lipid peroxidation and neuronal degeneration in Alzheimer's disease, antiviral and antioxidant effects, specific cleavage of DNA (Bosi et al., 2003).

Different methods have been developed to synthesize water-soluble C_{60}. C_{60} - fullerols can be obtained either directly from C_{60} (Li et al., 1993) or from their derivatives (Taylor, 1995). Some of the most published methods involve the hydrolysis of a fullerene intermediate made by nitronium chemistry (Chiang et al., 1992a), aqueous acid chemistry (Chiang et al., 1993), mechanochemical synthesis (Chen et al., 1992b), nitrogen dioxide radicals (Chiang et al., 1996), hydroboration (Schneier et al., 1994) or by hydrolysis of polycyclosulfated precursors (Chiang et al., 1994). A much simpler method of polyhydroxylation involves the reaction of an aqueous NaOH solution in contact with a toluene or benzene solution of C_{60}

in the presence of a phase transfer catalyst. Phase transfer is usually accomplished with tetrabutyl-amonium hydroxide (Li et al., 1993), cetyl trimethyl ammonium bromide (CTAB) (Lu et. al., 2004) or polyethilene-glycol (Alves et. al., 2006).

Fullerol was synthesized by the procedure described in detail previously (Adnadjevic et al, 2008). The degree of transformation of C_{60} *to* $C_{60}(OH)_{24}$ (*a*) at time (*t*) was calculated using Equation (13):

$$\alpha = \frac{C_o - C_i}{C_o} \tag{13}$$

where C_o is the starting concentration of C_{60} in the benzene solution and C_i is the concentration of C_{60} in the organic layer of the reaction mixture at time (*t*). The concentration of C_{60} was determined by measuring the absorption at $\lambda = 538$ nm and using Beer's law. The VIS spectra of both the C_{60} solution in benzene and the organic layer of the reaction mixture were obtained using a UV-Visible Spectrophotometer Cintra 10e, UK. The C_{60} benzene solution was deep violet in color and had two characteristic absorption maxima at $\lambda_1 = 538$ nm and $\lambda_2 = 592$ nm. During the reaction of fullerol formation, the intense color of the C_{60} benzene solution decreased, due to the decrease of fullerene concentration in the reaction mixture.

Figure 6 shows the isothermal changes of the degrees of transformation of fullerene (C_{60}) to fullerol ($C_{60}(OH)_{24}$) with reaction time (conversion curves) at different temperatures for conventional and microwave reacting conditions.

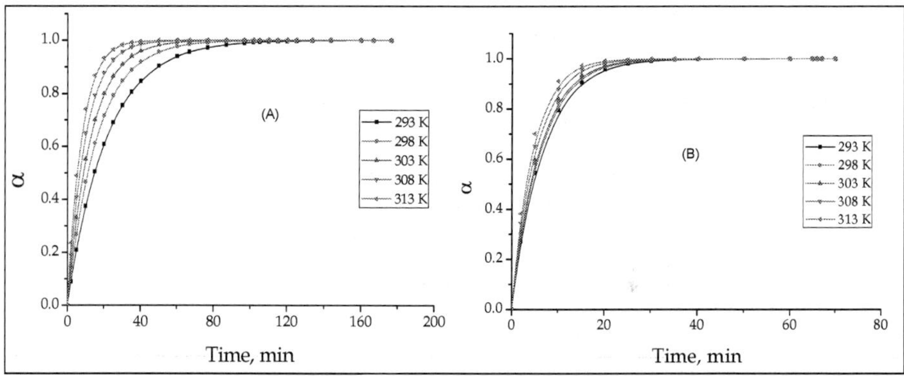

Fig. 6. The isothermal conversion curves for fullerol formation under (A) conventional reacting conditions and (B) microwave reacting conditions

Conversion curves of fullerol formation both for conventional and microwave reacting conditions are of same shape which points out to independence of kinetic model of fullerol formation on reacting conditions. By applying the "model-fitting method" it is easy to conclude that the kinetic's model for the formation of fullerol both under the conventional and microwave reacting conditions can be described with the Equation which is characteristic for first-order chemical reactions:

$$\alpha = 1 - \exp\left(-k_M \cdot t\right) \tag{14}$$

where k_M is a model constant for the first-order chemical reaction rate. The dependences of $[-\ln(1-a)]$ on reaction time for fullerol formation at different temperatures, for the microwave and conventional reacting conditions are shown in Figure 7.

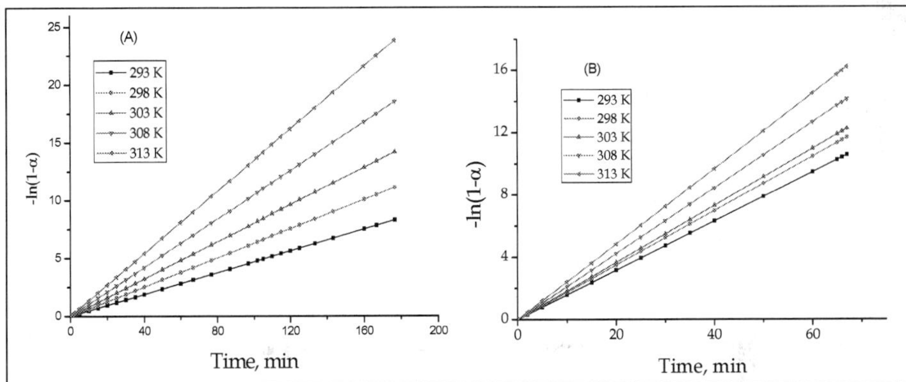

Fig. 7. The dependence of $[-\ln(1-a)]$ on reaction time for fullerol formation for (A) conventional reacting conditions and (B) microwave reacting conditions

The dependences of $[-\ln(1-a)]$ on time give straight lines for both conventional and microwave processes, which implies on properly selected kinetic's model and make possible to calculate the model's reaction rate constants from the slopes of these linear dependences. The calculated values of the k_M for the microwave and conventional reacting conditions at investigated temperatures are shown in Table 6.

T, K	Conventional process		Microwave process	
	k_M^{conv} , min⁻¹	Kinetics parameters	k_M^{mw} , min⁻¹	Kinetics parameters
293	0.047		0.101	
298	0.066	$E_{a,M}$=38±2 kJ/mol $\ln(A_M/\text{min}^{-1})=$ 12.6±0.5	0.175	$E_{a,M}$=10.5±0.5 kJ/mol $\ln(A_M/\text{min}^{-1})$=2.44±0.04
303	0.080		0.186	
308	0.106		0.192	
313	0.135		0.211	

Table 6. The temperature influence on the model's reaction rate constants and the kinetics parameters for fullerol formation, for the conventional and the microwave reacting conditions

The k_M values for microwave reacting conditions are 3.4 to 1.5 times higher in comparison to the values at same temperatures for conventional reacting conditions. As reaction temperature increase the k_M values increase exponentially for the each reacting conditions. By applying the Arrhenius equation, the kinetic parameters of fullerol formation, activation energy and pre-exponential factor, were calculated and are also given in Table 6.

The activation energy for the fullerol formation process under the microwave reacting conditions is about 4 times lower than the activation energy for the conventionally reacting conditions and the pre-exponential factor for the microwave process is 25000 times lower than the pre-exponential factor for the conventional process.

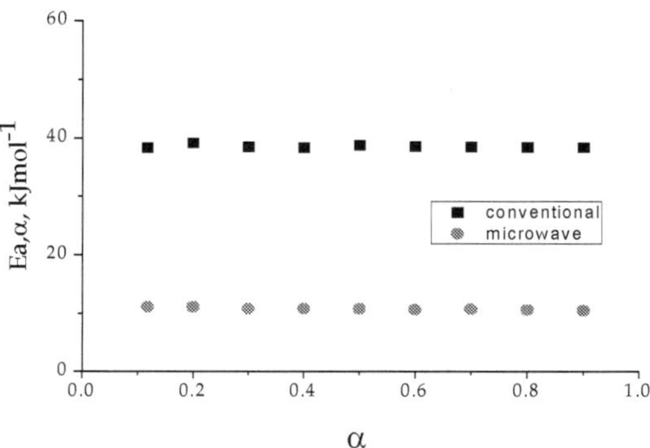

Fig. 8. The dependence of the $E_{a,a}$ on the degree of fullerol formation

The dependence of $E_{a,a}$ as a function of the degree of fullerol formation, for microwave and conventional reacting conditions is presented in Figure 8.

The values of $E_{a,a}$ for each of the reacting conditions are independent on the degree of fullerol formation, which verify elementary character of the formerly presented kinetic model of fullerol formation. As in previously given examples, the values of kinetic parameters of fullerol formation for conventional and microwave reacting conditions are in mutually relationship by the compensation effect equation:

$$\ln A_F = -1.440 + 0.368 E_{a,F} \tag{15}$$

Based on the compensation effect's parameters, values of the wave number of the resonant oscillator, the number of quantum of vibrational energy and the anharmonicity constant, under the CIRC and MWIRC were calculated and given in Table 7.

Parameter	CIRC	MWIRC
ν (cm^{-1})	233	233
n	14	3
x	-0.004	-0.0013

Table 7. The values of the ν, n and, x under the CIRC and MWIRC

The microwave reacting conditions do not lead to the change in the value of wave number of resonant vibrator. That, once again proves that reaction of fullerole formation is kinetically elementary process. The wave number of resonant oscillations corresponds to the so called H_g normal C-C vibrations in molecule of fullarene (Schettino et al., 2001). Therefore, activation of fullerene molecule for reaction of fullerol formation starts with the intensification of the so called Hg normal C-C vibrations and with new bond formations. Due to enhanced ground vibration level of the fullerene's molecule resonant oscillator, the number of vibrational quanta which are necessary for „active complex " formation is significantly lower for MWIRC then for the CIRC. As, with the increasing energy of vibrational level frequency of energy transport from energetically reservoir on resonant oscillator decreases as well as the lnA value while the x value increases under the microwave reacting conditions when compared to the conventional reacting conditions.

2.3 Isothermal kinetics of acid catalysed sucrose hydrolyses under the conventional and microwave conditions

Recently, biomass attracts much attention as a renewable feed stock due to its CO_2 neutral impact on the environment. Sugars are key intermediates from biomass to chemicals and conversion of sugar also requires a green process. However, the present enzyme-catalyzed process for hydrolyses of starch to glucose need two steps with several drawbacks such as generation of waste, low thermal stability of enzymes and difficulty in separations of products and enzymes (Reilly,1999). Heterogeneous catalysts may find an opportunity for replacing the enzyme catalyst if glucose is obtained in one-step reaction with high catalytic activity and selectivity.

Kinetics of acid catalyzed sucrose hydrolyze on ion-exchange resins is widely investigated in the papers of (Gilliland, 1971; Siggers & Martinola, 1985; Dhepe et al., 2005). Plaze presented results for sucrose hydrolyses, under the conventional and microwave conditions in stirred tank reactor (Plaze et al., 1995).

Herein, strongly acid cation–exchanged resin, IRA-120 (Amberlit), was used as catalyst for hydrolyses of sucrose (pro analysis, Merck, Germany) to glucose and fructose. Water solution of sucrose at the initial concentration Cs=57 g/L and 15 g acid cation–exchange was used in all experiments. During the conventional reacting conditions (CRC) reactor (V=100mL) was placed in thermostated controlled water bath. The contents were agitated at a stirring rate of 400 rpm to suspend the solid catalysts, to eliminate the mass transport resistance and to assure a thermally homogeneous suspension. The temperature was maintained within ±0.5 K and monitored with a thermometer. Microwave reactions were conducted using a commercially available monomode microwave unit (CEM Discover) as it is thoroughly described in previous parts of this Chapter. Reactions were performed in100mL glass rector. In all cases, the contents of reactors were stirred by means of rotating magnetic plate located below the floor of the microwave cavity and a Teflon –coated magnetic stirrer bar in the reactor.

Glucose concentration in a reaction mixture was determined by spectrophotometry method (Miller, 1959). Degree of sucrose conversion was determined by the Eq 16:

$$\alpha = \frac{C_i}{C_{max}} \tag{16}$$

where C_i is glucose concentration in reacting time and a C_{max} is theoretically maximal value of glucose in reacting system. Figure 9 presents conversion curves of acid catalyzed sucrose hydrolyses, under the conventional and microwave conditions.

Conversion curves of sucrose hydrolyses are of same shape which point out on identical kinetic's model for both reacting conditions, conventional and microwave.

By applying the "model fitting method" it was revealed that kinetics of acid catalyzed sucrose hydrolyses under each reacting conditions can be modelled with the equation characteristic for kinetics of first order chemical reaction:

$$\alpha = 1 - \exp\left(-k_M \cdot t\right) \tag{17}$$

where the k_M is a model constant for the first-order chemical reaction rate. If kinetic of hydrolyses occurs in agreement with equation (17), then the dependences of $[-\ln(1-a)]$ on reaction time should give straight line. Figure 10 (A and B) show the dependences of $[-\ln(1-$

a)] on reaction time for sucrose hydrolyses at different temperatures, for microwave and conventional reacting conditions, respectively.

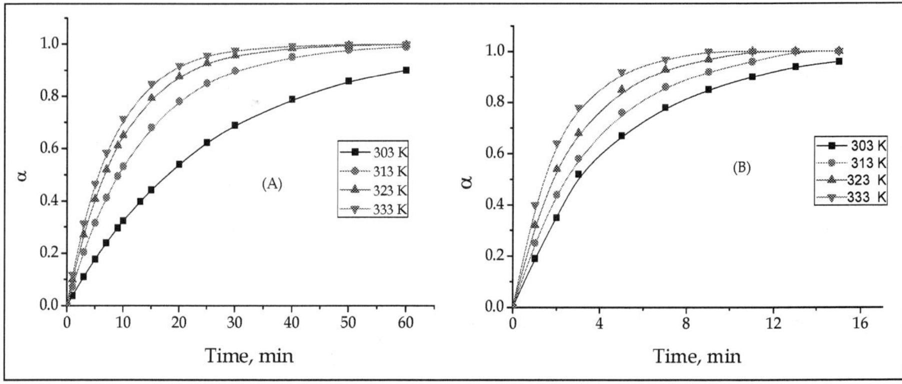

Fig. 9. The isothermal conversion curves for sucrose hydrolyses under (A) conventional and (B) microwave reacting conditions

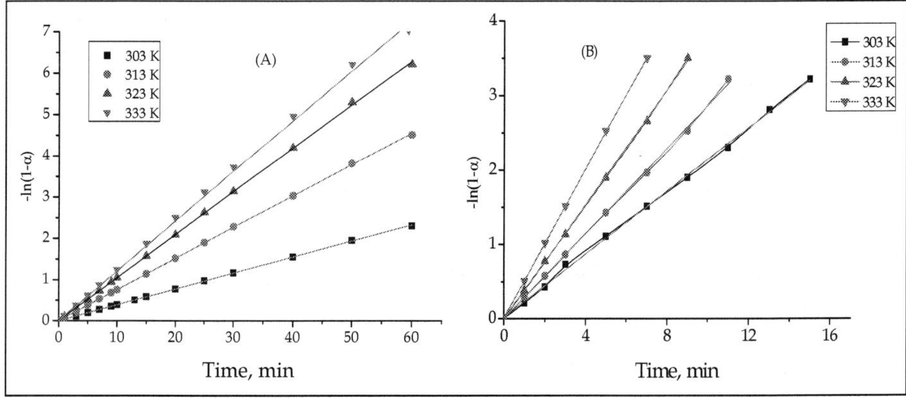

Fig. 10. Dependence of [–ln(1-a)] on reaction time for (A) conventional and (B) microwave reacting conditions of sucrose hydrolyses

The dependences of [-ln(1-a)] on time give straight lines for both conventional and microwave sucrose hydrolyses, which enables to calculate the reaction rate constants of the model. Table 8 presents dependence of the calculated values for k_M on temperature for both microwave and conventional reacting conditions.

The k_M values for microwave reacting conditions are ~5 to 10 times higher than the comparable values for conventional reacting conditions. The increasing reaction temperature leads to the exponentially increase of the k_M values for the both reacting conditions. The kinetic parameters of sucrose hydrolyses, activation energy and pre-exponential factor, were calculated by applying the Arrhenius Equation and they are shown in Table 8.

T, K	Conventional process		Microwave process	
	k_M^{conv}, min⁻¹	Kinetics parameters	k_M^{mw}, min⁻¹	Kinetics parameters
303	0.024		0.204	
313	0.039	$E_{a,M}$=32±2 kJ/mol	0.287	$E_{a,M}$=25.15±1 kJ/mol
323	0.076	$\ln(A_M/\mathrm{min}^{-1})$= 9±1	0.386	$\ln(A_M/\mathrm{min}^{-1})$=8.4±0.1
333	0.105		0.510	

Table 8. The temperature influence on the k_M and the kinetic parameters for sucrose hydrolyses under the conventional and the microwave reacting conditions

The activation energy of sucrose hydrolyses under the conventionally reacting conditions is ~1.3 times higher than the value of E_a for the microwave reacting conditions, while the value of the pre-exponential factor for the conventional hydrolyses process is for ~7 % higher than the corresponding value for the microwave process.

The dependence of $E_{a,a}$ as a function of the degree of sucrose hydrolyses, for microwave and conventional reacting conditions is presented in Figure 11.

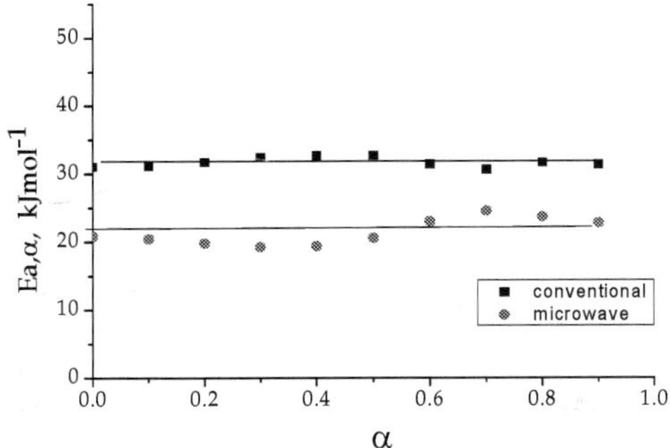

Fig. 11. The dependence of activation energy on the degree of sucrose hydrolyses

As in previously given examples of PAA hydrogel formation and fullerol formation, the values of $E_{a,a}$ are independent on the degree of sucrose hydrolyses, for conventional and microwave reacting conditions. Therefore, acid catalyzed sucrose hydrolyses on ion-exchange resin, presents an elementary chemical reaction with unique kinetic model and mechanism, both under the conventional and microwave reacting conditions. The kinetic's parameters of sucrose hydrolyses, for conventional and microwave reacting conditions are interconnected by the compensation effect Equation as in former examples:

$$\ln A_F = 5.407 + 0.119 E_{a,F} \tag{18}$$

Based on the parameters of the compensation effect, the values of the v,n, x for microwave and conventional reacting conditions were calculated and presented in Table 9.

During the formation of the "activated complex" for the reaction of sucrose hydrolyses, for conventional and microwave reacting conditions, resonance between the oscillators of energetically reservoir takes place at the same frequency. As wave number of resonant frequency corresponds to the so called C-H out of plane deformation vibration (Ring, 2010)

Variable	CIRC	MWIRC
ν (cm⁻¹)	723	723
n	4	3
x	-0.035	-0.042

Table 9. The values of the ν, n, x for sucrose hydrolyses, under the CIRC and MWIRC

of sucrose molecule, it may be concluded that activation of sucrose molecule for hydrolyses begins with the intensification of that vibration, which is in the following stage accompanied with the breaking of glycoside bond. The presence of microwave field leads to the decreasing number of vibrational quanta which should be accepted from the sucrose molecule in order to form "activated complex". That decreases, as in previously given examples, is a consequence of the increased energy of the ground vibration level of resonant oscillator due to the absorption of microwave energy. The decreased value of the pre-exponential factor and the enhanced values of anharmonicity factor under the microwave reacting conditions in comparison to the conventional conditions, is caused with the decreasing frequency of energy transfer from one to another oscillators which is provoked with the increase in vibrational energetically level of oscillators.

2.4 Isothermal kinetics of ethanol adsorption from aqueous solutions onto carbon molecular sieve under the conventional and microwave conditions

The bioethanol which is produced from renewable resources (waste lignocelluloses materials, algae, is now main alternatively source for promising economic production of novel fuels and chemicals (Kosaric et al, 2001). Selective ethanol adsorption from low concentrated aqueous solutions, which are usually formed after fermentation process, on hydrophobic materials of the zeolite type presents an effective technological process for development of profitable production of bioethenol (Adnadjevic at. al., 2007).

Kinetics adsorption curves of ethanol from aqueous ethanol solution onto carbon molecular sieve (CMS-3A) were performed by batch adsorption experiments. A measured quantity (5 g) of thermally activated CMS-3A granules was added to a predetermined quantity (50 g) of 10 wt.% aqueous ethanol solution thermostated at the required temperature (±0.2 °C). The adsorption system was stirred at a rotation rate of 500 rpm at the given temperature. In the case of microwave conditions, the adsorption system was placed in a focused microwave reactor (Discover, CEM). Samples were taken from the adsorption system at regular time intervals. After their centrifugation, the zeolite was removed from solution and the ethanol concentration remaining in the solution was determined by measuring the refraction index of the solution using a Reichart-Jung Auto Abbe refractometer at 298K.

The amount of adsorbed ethanol (x) at a certain reaction time of the adsorption process can be calculated from Eq. 19:

$$x = \frac{C_o - C}{m} \cdot m_s \tag{19}$$

where: C_o is ethanol concentration in solution before adsorption (wt.%); C is ethanol concentration after a definite adsorption time (wt.%); m_s is the mass of solution (g) and m is mass of the zeolite (g). The degree of ethanol adsorption (α) was calculated from Eq. 20:

$$\alpha = \frac{x}{x_{max}} \tag{20}$$

where: x_{max} is the experimentally determined maximal value of specific CMS-3A adsorption capacity at a given temperature. The isothermal dependences of the amount of adsorbed ethanol *versus* adsorption time at different temperatures under conventional and microwave conditions are shown in Figure 12.

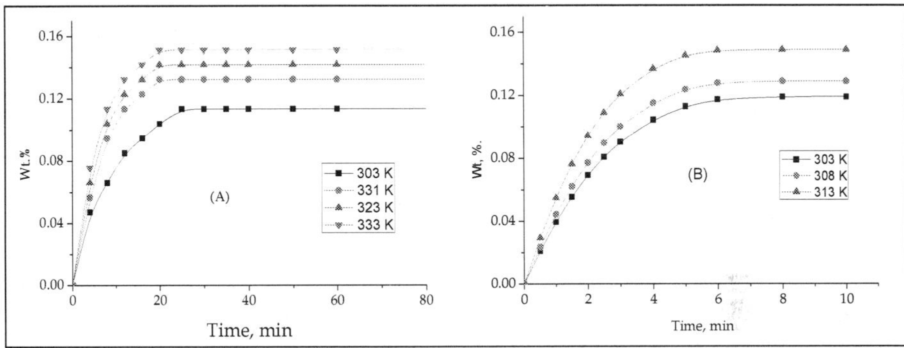

Fig. 12. The kinetic curves of ethanol adsorption on CMS-3A for (A) conventional and (B) microwave conditions

The kinetic curves of ethanol adsorption on CMS-3A for both conventional and microwave conditions have same shape at all the investigated temperatures. Three specific shapes of change of the degree of ethanol adsorption with time can be clearly seen at all kinetic curves, a linear, nonlinear and saturation region. With increasing temperature for both conventional and microwave conditions the period of linear change of the adsorption capacity with time became shorter, while the maximal value of adsorption capacity increased.

In order to preliminary find kinetic model of the isothermal adsorption of ethanol on CMS-3, for both conventional and microwave conditions, the dependence of da/dt on the degree of adsorption of ethanol was analyzed and presented in Figure 13.

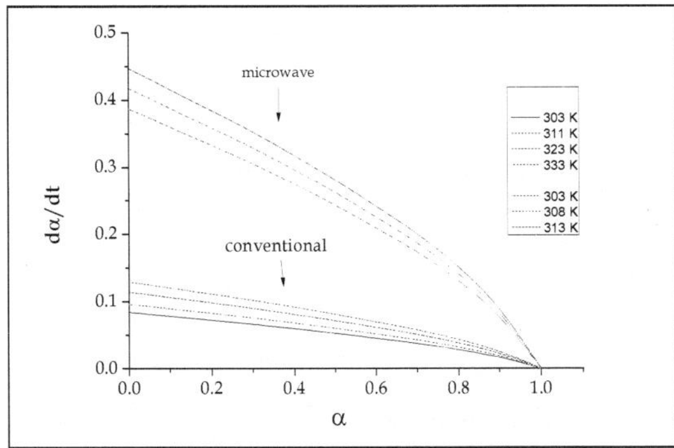

Fig. 13. Dependence of da/dt on the degree of ethanol adsorption

The changes in the (da/dt) on the degree of adsorption of ethanol are of identical shape; the so called concavely decreasing shape, for both conventional and microwave reacting conditions. The identical shapes of the changes in the (da/dt) on the degree of ethanol adsorption, imply on unique kinetic model of adsorption of ethanol on CMS under the conventional and microwave conditions. For the each reacting conditions the maximal rate is achieved at the beginning of the process. Table 10 presents the temperature influence on the $(da/dt)_{max}$ for the adsorption of ethanol and kinetic parameters, for the conventional and microwave reacting conditions.

The maximal rates of the isothermal adsorption of ethanol on CMS-3A are approximately 4.6 times higher for the microwave process then for the conventional process. With the increasing temperature of adsorption system the values of maximal rates exponentially increase for both conventional and microwave conditions. Based on that, by using the Arrhenius Equation the kinetic parameters of adsorption of ethanol (E_a and lnA) were determined by applying the method of stationary point and are given in Table 10.

Conventional process			Microwave process		
T, K	$(da/dt)_{max}$, [min⁻¹]	Kinetics parameters	T, K	$(da/dt)_{ma}$ [min⁻¹]	Kinetics parameters
303	0.084		303	0.387	
311	0.094	E_a=11.9±0.2 kJ/mol	308	0.417	E_a=10.9±0.2 kJ/mol
323	0.114	lnA = 2.2±0.1	313	0.447	lnA=2.6±0.1
333	0.129				

Table 10. Temperature influence on $(da/dt)_{max}$ for adsorption of ethanol and kinetic parameters, for the CIRC and MWIRC processes

The activation energy for adsorption of ethanol under the microwave reacting conditions is lower for 8.4 % than under the conventional conditions. The value of the pre-exponential factor for microwave conditions is higher, for 3 times than the value for the conventional process.

By application of the "model fitting method", it was established that isothermal kinetics of adsorption of ethanol can be described by the Equation (21) both for the conventionally and microwave conditions. The Equation (21) is characteristic for the phase-boundary controlled reaction (tridimensional contracting volume):

$$\alpha = 1 - \left(1 - k_M \cdot t\right)^3 \tag{21}$$

where k_M is a model constant of the adsorption rate. The isothermal dependences of $[1-(1-\alpha)^{1/3}]$ on adsorption time for the conventionally and microwave conditions are shown in Figure 14.

The dependences $[1-(1-a)^{1/3}]$ on time, over the whole range of the adsorptions degrees and at all investigated temperatures for both CIRC and MWIRC are linear, which confirms that the correct model of kinetic was selected for describing ethanol adsorption on CMS-3A and it enable to calculate the model's constant of the adsorption rate. The changes of the model's constant with temperature for conventional and microwave conditions are given in Table 11.

The values of the k_M for the microwave conditions are approximately 4.5 times higher than the k_M values for the conventional conditions at the same temperatures. As the temperature increases, the values of the model constant of the ethanol adsorption rate exponentially increase, so the model parameters ($E_{a,M}$ and $\ln A_M$) of ethanol adsorption can be determined by applying the Arrhenius Equation. The obtained results are given in Table 11.

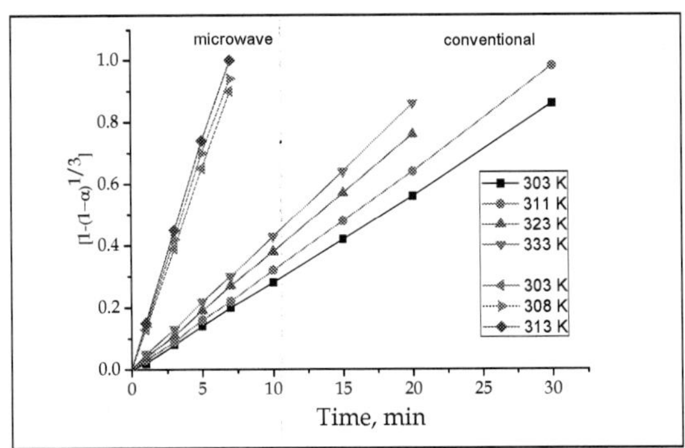

Fig. 14. Dependence of $[1-(1-a)^{1/3}]$ on adsorption time for the (A) conventional and (B) microwave conditions

Conventional process			Microwave process		
T, K	k_M^{conv} [min⁻¹]	Kinetics parameters	T, K	k_M^{mw} [min⁻¹]	Kinetics parameters
303	0.028		303	0.129	
311	0.032	E_a=12±0.5 kJ/mol	308	0.139	E_a=11.2±0.5 kJ/mol
323	0.038	lnA =1.15±0.05	313	0.149	lnA=2.4±0.1
333	0.043				

Table 11. The influence of temperature on model's kinetic constants of the ethanol adsorption rate and kinetic parameters, for the CIRC and MWIRC processes

The activation energy for the adsorption of ethanol under the MWIRC is lower (~6%) while the pre-exponential factor is 3.5 times higher than for the CIRC.

The dependence of the activation energy on the degree of adsorption for conventional and microwave heating are presented in Figure 15.

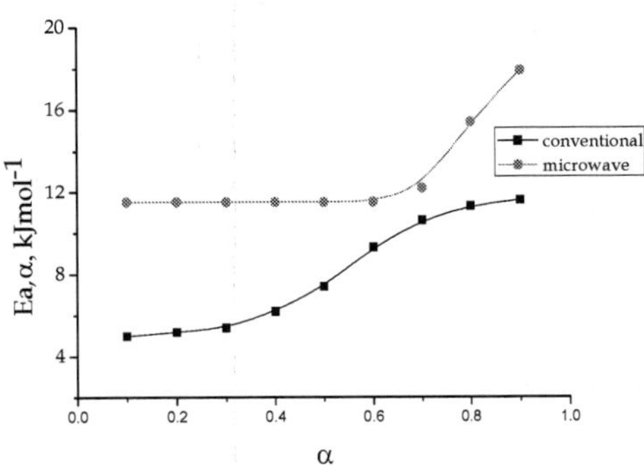

Fig. 15. The dependence of the $E_{a,a}$ on the degree of conversion

As can be seen in Figure 15, the change in activation energy with degree of adsorption using conventional heating is completely different from the same change applying microwave heating. This difference indicates a different mechanism (such as the order of the elemental steps) of adsorption using microwave heating in comparison to conventional heating. With conventional heating, three specific regions of change of the activation energy with degree of adsorption can be easily seen. At $\alpha \leq 0.3$, the value of activation energy does not change with increasing degree of adsorption, which indicates the dominant influence of an elemental step with $E_{a,\alpha} \approx 5$ kJ/mol on the mechanism and kinetics of adsorption. Likewise, at $\alpha \geq 0.8$, an increase of value of the degree of adsorption does not lead to a change in the activation energy. Therefore, another elemental step with $E_{a,\alpha} \approx 11$ kJ/mol has a dominant influence on the adsorption. At values $0.3 \leq \alpha \leq 0.8$, with increasing degree of adsorption, the value of the activation energy rapidly increased, indicating an increase in the influence of other elemental steps on the adsorption.

Contrary to this, with microwave heating, only two specific regions of change in the activation energy with degree of adsorption can be seen. In the region $\alpha \leq 0.8$, the value of the activation energy is independent of changes in the degree of adsorption, *i.e.*, an elemental step with $E_{a,\alpha} \approx 11.5$ kJ/mol has a dominant influence on the adsorption kinetics. At $\alpha \geq 0.8$, activation energy value increase rapidly with increasing degree of adsorption.

The established changes in the activation energy with degree of adsorption by applying microwave and conventional heating indicate that microwave heating leads, in comparison to conventional heating, to changes in the complexity of the reaction mechanism of the adsorption process.

The change of lnA_a for both conventional and microwave process of adsorption of ethanol are in functional relationship with the change in $E_{a,a}$ (Eqs 22 and 23) and confirms the existence of a compensation effects against the degree of adsorption:

$$\ln A_\alpha^{conv} = -1.907 + 0.295 E_{a,\alpha} \tag{22}$$

$$\ln A_\alpha^{mw} = -1.937 + 0.223 E_{a,\alpha} \tag{23}$$

In accordance with the previously exposed model of influence of microwave field on the kinetic of the process, the values of the wave number of the resonant oscillator, the number of quantum of vibrational energy and the anharmonicity constant, under the CIRC and MWIRC were calculated and given in Table 12.

Variable	CIRC	MWIRC
ν (cm⁻¹)	292	386
n	3	2
x	-0.025	-0.0016

Table 12. The values of the ν, n and , x under the CIRC and MWIRC of ethanol adsorption process

The wave number of resonant frequency is higher for the MWIRC then the value for the CIRC of ethanol adsorption process. The wave number of resonant oscillations corresponds to the OH twisting vibrations in molecule of ethanol (Plyler, 1952) while the wave number of resonant frequency under the CIRC corresponds to the first overtone of rotation of ethanol molecule. Thus, in the presence of microwave field, molecules of ethanol absorb microwave energy and distribute to the rotation and vibration levels. Consequently, ground vibrational level of resonant oscillator increases and also the wave number of resonant frequency. As in the case of all previous examples, number of transferred vibrational quanta is lower for MWIRC then for the CIRC. The decreased numbers of transferred quanta is a result of the increased energy of the ground level of vibrational level of the resonant oscillator which is formed due to the absorption of microwave energy. Also, established decreases of the pre-exponential factor and increases of value of anharmonicity factor is consequence of the increasing energy of the ground rotational level of the resonant oscillator.

2.5 Isothermal dehydration of poly(acrylic -co-methacrylic acid) hydrogels

Despite the enormous practical importance of the dehydration of hydrogels, especially concerning their possible application in agrochemistry and ecology, and the evaluation of the kinetics of dehydration process of hydrogels, which could lead to significant advancement in the theoretical consideration of the molecular mechanisms of the dehydration process, which is furthermore interconnected with the state of water in hydrogels, investigations of the dehydration of hydrogels are sparse.

The kinetics of isothermal dehydration under the conventional conditions of equilibrium swollen poly(acrylic acid) hydrogels (PAA) was investigated (Jankovic et. al., 2008; Adnadjevic et al., 2009). Applying the model-fitting method, Jankovic et al., 2008, established that a change in the dehydration temperature caused a change in the dehydration kinetic model. Adnadjevic et al., 2009, established that the isothermal dehydration of a PAA hydrogel under conventional conditions could be mathematically described by a Weibull distribution function (WDF) of reaction times.

The poly(acrylic acid-co-methacrylic acid) hydrogel (PAM) was synthesized by a procedure based on the radical polymerization of acrylic acid and methacrylic acid (1:1 mol ratio) and cross-linking of the formed polymers, using a previously described procedure (Adnadjevic & Jovanovic, 2009).

Dehydration under the conventional isothermal process conditions (CIPC) was performed by the usual thermogravimetric measurements. The isothermal thermogravimetric curves were recorded on a simultaneous DSC-TGA thermal analyzer, Model 2960, TA Instruments, USA. The analyses were performed with (20 ± 2 mg) samples of equilibrium swollen hydrogel in platinum pans under a nitrogen atmosphere at a gas flow rate of 10 ml min^{-1}. Isothermal runs were performed at nominal temperatures of 293, 313, and 333 K. The samples were heated from the ambient to the selected dehydration temperature at a heating rate of 300 K min^{-1} and then held at that temperature for a given reaction time.

Dehydration under the microwave isothermal process conditions (MWIPC) was performed in a focused microwave reactor, as described in a former subchapter. At the beginning of each experiment, the equilibrium swollen hydrogel was placed in a glass device specially designed for these experiments. The analyses were performed with (500 ± 10 mg) samples of equilibrium swollen hydrogel at temperatures of 313 K, 323 K and 333 K.

The degree of dehydration under both conventional and microwave conditions was calculated as:

$$\alpha = \frac{m_0 - m}{m_0 - m_f} \tag{24}$$

where m_0, m, m_f refer to the initial, actual and final mass of the sample, respectively. The isothermal conversion curve represents the dependence of the degree of conversion (α) on the reaction time (t) at a constant value of the experimental temperature (T).

The experimentally obtained isothermal conversion curves of dehydration at different operating temperatures for the dehydration of the PAM hydrogel under the conventional and microwave conditions are shown in Figure 16.

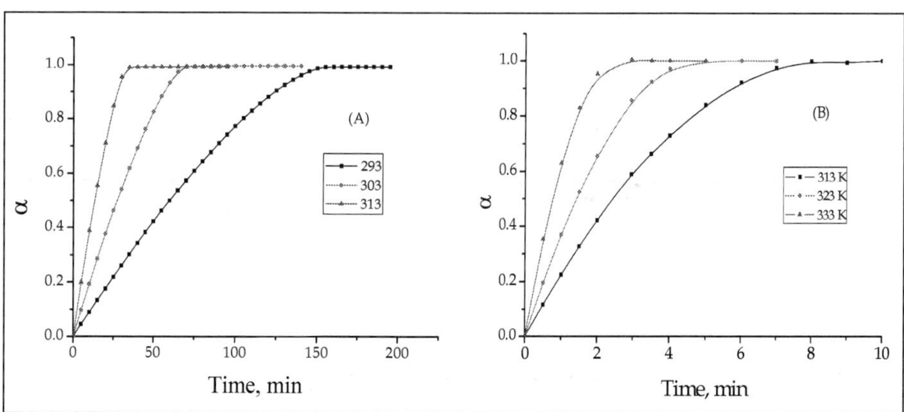

Fig. 16. The conversion curves for (A) conventional and (B) microwave dehydration process of the hydrogel

The conversion curves of PAM dehydration are of characteristic shape and exhibit three specific shapes of changes of the degree dehydration on dehydration time (linear, non-linear and plateau).

By means of the "model fitting method", it was established that the kinetics of isothermal PAM hydrogel dehydration, for both conventional and microwave conditions, can be modeled by an equation which is specific for a phase-boundary controlled reaction model (contracting area), as follows:

$$\left[1 - \left(1 - \alpha \right)^{1/2} \right] = k_{M} \cdot t \tag{25}$$

where k_{M} is the model rate constant for the dehydration the PAM hydrogel. The dependence of $[1 - (1 - \alpha)^{1/2}]$ on time for the dehydration of PAM at different temperatures is presented in Figure 17 for conventional and microwave conditions.

Since, the dependence of $[1 - (1 - \alpha)^{1/2}]$ on time for the dehydration of PAM gives a straight line over the entire range of degrees of dehydration at all of the investigated temperatures for both the processes, the validity of the selected kinetics model was confirmed, enabled the calculation of the rate constant of the dehydration model. The effects of temperature of isothermal dehydration on the k_{M} values and the kinetic parameters of the process of isothermal dehydration the PAM hydrogel are given in Table 13.

Similar to the process of ethanol adsorption (described formerly), the values of the model rate constant for PAM hydrogel dehydration are approximately 5.0 times higher under microwave conditions than for the conventional heating process. As the k_{M} values for both processes exponentially increase with the increasing temperature, it is possible to determine the kinetic parameters of PAM hydrogel dehydration (E_{a} and ln A) by means of the Arrhenius Equation. The obtained results are also presented in Table 13.

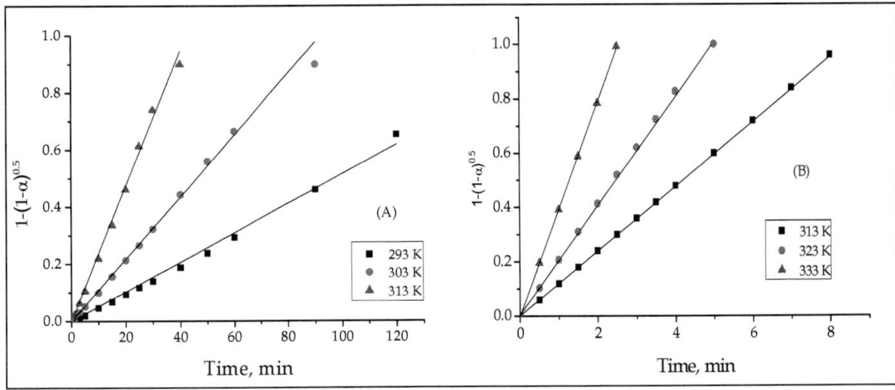

Fig. 17. The dependence of $[1-(1-\alpha)^{1/2}]$ on time for (A) conventional and (B) microwave dehydration process of the hydrogel

| T, K | Conventional process | | T, K | Microwave process | |
	k_M [min⁻¹]	Kinetics parameters		k_M [min⁻¹]	Kinetics parameters
293	0.0052	E_a=58.5±0.5	313	0.120	E_a=51.6±0.5 kJ/mol
303	0.0109	kJ/mol	323	0.207	ln A=17.7±0.2
313	0.0236	ln A = 18.5±0.2	333	0.393	

Table 13. The effect of temperature on the k_M values and kinetic parameters of the isothermal dehydration of the PAM hydrogel by the conventional and the microwave process

The activation energy for PAA hydrogel dehydration under microwave conditions is E_a=51.6±0.5 kJ/mol, which is 12 % lower than the value of activation energy for the conventional process. The value of the pre-exponential factor for the microwave process is also lower than for the conventional process.

The plots of $E_{a,\alpha}$ *versus* the degree of PAA hydrogel dehydration for the investigated processes under conventional and microwave conditions are presented in Figure 18.

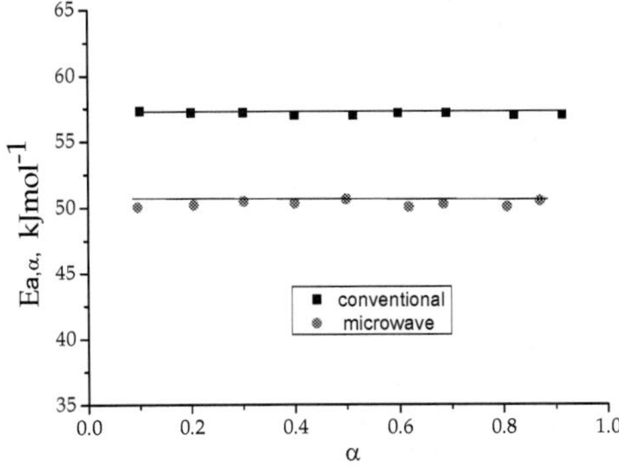

Fig. 18. The dependence of the $E_{a,a}$ on the dehydration degree of the hydrogel

In the whole range of degrees of dehydration 0.1 a ≤0.9, the activation energy is practically independent of the degree of dehydration, for both the investigated dehydration processes. As the independence of $E_{a,\alpha}$ on the degree of dehydration is distinctive for an elementary (single-stage) processes, it may conclude that the investigated process of dehydration is an elementary one and that the microwave field does not lead to changes in the mechanism of the process. A functional relationship exists between the values of the kinetic parameters for conventional and microwave process, which is defined by the compensation effect in relation to an interactive field, which can be presented as:

$$\ln A_F = 11.71 + 0.116 \cdot E_{a,F} \qquad (26)$$

where $\ln A_F$ and $E_{a,F}$ are the values of the pre-exponential factor and the activation energy, respectively, in a specific field. The existence of a compensation effect against the interactive field, thermal or microwave, enables to calculate the values of the wave number of the vibrational energy of the vibrators which are in resonant interaction (v), the vibrational quantum number (n) and the anharmonicity constant (χ) under conventional and microwave conditions, as shown in Table 14.

Variable	CIPC	MWIPC
v	726	830
n	7	6
χ	−0.011	−0.015

Table 14. The values of v, n, and χ under conventional and microwave conditions

Microwave reacting conditions does not lead to the change in wave number of resonant oscillation, which proves that dehydration process is an elementary kinetic process, both under the conventional and microwave conditions. Wave number of the resonant vibrations corresponds to the bond of intermolecular oscillations of molecules of water in cluster (Eisenberg, 1975). Therefore, activation of the absorbed water molecules for dehydration begins with the intensification of intermolecular oscillation of absorbed water in the clusters and with the breakage the certain intermolecular bonds. In the presence of microwave field, due to absorption of energy, energy of the ground vibrational level increases. For that reason, number of vibrational quanta required to overcome potential barrier under the microwave reacting conditions decreases compared to the conventional conditions. Due to increases in the vibrational energy of resonant oscillator in the molecules of absorbed water decreases energy transfer and for that reason pre-exponential factor has lower value for the microwave reacting conditions while the anharmonicity constant has higher value than the corresponding values for conventional conditions.

3. Conclusions

Under the influence of microwave reacting conditions, depending on dielectric properties of the reacting material, kinetics model and kinetics complexity of chemical reaction and/ or physical-chemical process could be altered or not altered.

Constant of chemical reaction or physical-chemical process are always higher for process under the influence of microwave field than when compared to their values obtained for the chemical reaction or process performed under the conventional reacting conditions.

The increase in the temperature of reaction and/or process leads to the changes in the values of rate constant of the reaction /process in accordance with the Arrhenius dependence.

The increase in the rate of reaction/ process under the microwave reacting /process conditions is a consequence of the decreased values of activation energy and pre-exponential factor of the investigated reacting /process.

The values of kinetics parameters of the reaction / process under the microwave reacting and conventionally reacting conditions are mutely interconnected with the Equation of compensation effect.

The decrease in the value of activation energy of the reaction/ process under the microwave reacting and conventionally reacting conditions is consequence of the increased energy of the ground level of the resonant oscillator which happened due to the absorption and distribution of microwave energy.

The decreased value of pre-exponential factor and the increased value of the anharmonicity under the influence of microwave reacting conditions is caused with the decreased value of energy transfer frequency and with the change in oscillator due to the change in energetically state.

In the investigated reaction of PAA hydrogel formation through acrylic acid polymerization and crosslinking, microwave reacting condition leads to change in the kinetics model of reaction when compared to the same process under the conventionally reacting conditions.

Activation energy of chemical reaction is quantized value which is predetermined with the number of transferred quant of resonant oscillator from energetically reservoir onto the active oscillator of reactants.

4. References

Adnadjevic, B; Gigov, M.; Sindjic, M. & Jovanovic, J. (2008). Comparative study on isothermal kinetics of fullerol formation under conventional and microwave heating. *Chemical Engineering J.*, Vol 140, p.p. 570-577.

Adnadjevic, B; Janković, B.; Kolar-Anić, Lj. & Jovanović, J. (2009). Application of the Weibull distribution function for modeling the kinetics of iso thermal dehydration of equilibrium swollen poly(acrylic acid) hydrogel. *React. & Funct.Polym.* Vol. 69, p.p.151- 158.

Adnadjevic, B. & Jovanovic, J., (2009). A comparative kinetics study of isothermal drug release from poly(acrilic acid) and poly (acrylic-co-methacrylic acid) hydrogels. *Coll. Surf. B. Biointerfaces*, Vol. 69, p.p. 31-42.

Adnadjevic, B. & Jovanovic, J. (2008). Novel approach in investigation of the poly (acrylic acid) hydrogel swelling kinetics in water. *J. Appl. Polym. Sci.*, Vol 107 (No 7), 3579-3587.

Adnadjevic, B. & Jovanovic, J. & Krkljus I. (2008). Isotermal kinetics of(E)-4-(4 metoxyphenyl)-4-oxo-2-butenoic acid release from poly(acryliccid-comethacrylic acid)hydrogel. *J. Appl. Polym. Sci.*, Vol 107 (No 5), p.p. 2768-2775.

Adnadjevic, B; Mojovic, Z.; Rabi, A. & Jovanovic, J. (2007). Isoconversional kinetic of isothermal selective ethanol adsorption on zeolite type Na-ZSM-5. *Chemical Engineering and Technology* Vol. 30 (No 9), p.p. 1228-1234.

Alves, G.; Ladeira, L.; Righi, A.; Krambrock, K.; Calado, H..; Gil, R. & Pinheiro, M. (2006). Synthesis of $C_{60}(OH)_{18-20}$ in aqueos alkaline solution under O_2-atmosphere. *J. Braz. Chem. Soc.*, 17 p.p. 1186-1190.

Baghurst, D. & Mingos, D. (1992). Superheating effects associated with microwave dielectric heating. *J. Chem. Soc. Chem. Commun.*, No 9, p.p. 674-677.

Berlan, J.; Giboran, P.; Lefeuvre, S. & Marcheno, C. (1998). Synthese organique suos champ microndos premier exemple d'activation specifiquen phase homogene. *Tetrahedron Letters*, Vol. 32, p.p. 2363-2366.

Bhaskar, A.; Chang, H.; Chang, T. & Cheng, S. (2007). Effect of microwave annealing temperatures on lead zirconate titanate thin films. *Nanotechnology*, Vol.18 (No 1-4), p.p. 395-404.

Blanco, C. & Auerbach, S. (2003). Nonequilibrium molecular dynamics of microwave driven zeolite – guest systems. Loading dependence of athermal effects. *J. Phys.Chem. B.*, Vol. 107, p.p. 2490-2499.

Binner, G.; Hassine, A. & Cross, T. (1995). The possible role of the pre-exponential factor in explaining the increased   reaction rates observed during the microwave synthesis of titanium carbide. *J. Mater. Sci.*, Vol 30, p.p. 5389- 5393.

Bogdal, D. & Prociak A. (2007). *Microwave-Enhanced of polymer Chemistry and Technology*, Blackwell Publ., Oxford, UK, p.p.1-296.

Booske, J.; Cooper, R. & Freeman; S.(1997). Microwave enhanced reaction kinetics in ceramics. *Mat. Res. Innovat.* Vol.1, p.p. 77-84.

Bosi, S.; DaRos, T.; Spalluto, G. & Prato, M. (2003). Fullerene derivatives an attractive tool for biological applications. *Eur. J. Med. Chem.*, Vol. 38, p.p. 913-923.

Brown M.E.; Dollimore, D. & Galway, A.K. (1980). *Reaction in the Solid State*. In *Comprehensive Chemical Kinetics*. Ed. Elsevier, p.p. 87.

Chen J. & Zhao Y., (1999). An efficient prepartion method for superabsorbent polymers. *J. Appl. Polym. Sci.*, Vol 74 (No 1), p.p. 119-234.

Chiang, L.; Upasani, R. & Swirczewski, J.(1992). Versatile nitronium chemistry for C60 fullerene functionalization. *J. Am. Chem. Soc.*, Vol. 114, p.p. 10154-10157.

Chiang, L.Y.; Upasani, R.B.; Swirczewski, J.W. & Soled, S. (1993). Evidence of hemiketals incorporated in the structure of fullereols derived from aqueous acid chemistry. *J. Am. Chem. Soc.*, Vol. 115, p.p. 5453-5457.

Chiang, L.Y. Bhonsle, J.; Wang, L.; Shu, S.; Chang, T. & Hwu, J.(1996). Efficient one-flask synthesis of water soluble [60] fullerenols. *Tetrahedron* Vol. 52, p.p. 4963-4972.

Chiang, L.Y.; Wang, L.Y. ; Swirczewski, J.W.; Soled, S. & Cameron, S. (1994). Efficient synthesis   of   polyhydroxylated   fullerene   derivatives   via   hydrolysis   of polycyclosulfoted precursors. *J. Org. Chem.*, Vol. 59, p.p. 3960-3968.

Collins, J. & Leadbeater, N. (2007). Microwave energy: a versatile tool for the biosciences. *Org. Biomol. Chem.*, Vol. 5, p.p. 1141-1150.

Conner, V. & Tumpsett, G. (2008). How could and do microwaves influence chemistry at interfaces. *J. Phys.Chem.B*, Vol. 111, p.p. 2110-2119.

Cundy, C. (1998). Microwave techniques in the synthesis and modification of zeolite catalysts, A Review. *Collect Czech. Chem. Commun.*, Vol. 63, pp. 1699-1723.

Dai, L. & Mou, A. (1997). A facile route to fullerol-containing polymers. *Synthetic Metals*, Vol. 86, p.p. 2277-2278.

Dhepe, P.L.; Ohashi, M.; Inagaki, S.; Ishikawa, M. & Fukuoka, A. (2005). Hydrolysis of sugars catalyzed by water-tolerant sulfonated mesoporous silicas. *Catalysis Letters*, Vol.102 (No. 3-4), p.p. 163-169.

Eisenberg, D. & Kauzman, W. (1975). *The structure and properties of water*, Oxford, U.P.N.Y., p.p. 246-248.

Elander, N.; Jones, J.; Lu, S. & Stone-Elander, S. (2000). Microwave-enhanced Radiochemistry. *Chem. Soc. Rev.*, Vol. 29, p.p. 239-249.

Friedman, H. (1964). Kinetics of thermal degradation of char-forming plastic from thermogravimetry ; Application to a phenolic plastic. *J. Polym. Sci.*, C6, pp. 183-195.

Gabriel, C.; Gabriel, S.; Grant, E.; Halstead, B. & Mingos, D. (1998). Dielectric parameters relevant to microwave dielectric heating. *Chem. Soc. Rev.*, Vol.27, p.p 213-223.

Gilliland, E.; Bixler, H.& O'Convell, J. (1971). Catalysis of sucrose inversion in ion-exchange resins. *Indust. Eng. Chem. Fundam.* Vol 10 (No 2), p.p.185-191.

Hoz, A.; Diaz-Ortiz, A. & Moreno, A. (2005). Microwaves in Organic Synthesis, Thermal and non-thermal microwave effects. *Chem.Soc. Rev.*, Vol. 34, p.p.164-178.

Janković, B.; Adnadjević, B. & Jovanović, J. (2008). Isothermal kinetics of dehydration of equilibrium swollen poly(acrylic acid) hydrogel. *J. Therm. Anal. Cal.* Vol. 92 (No 3), 821-827.

Jovanovic, J. & Adnadjevic, B. (2010). Influence of microwave heating on the kinetic of acrylic acid polymerisation and crosslinking, *J. Appl. Polym. Sci.*, Vol. 116 (No 1), p.p. 55-63.

Kappe, C. & Dallinger, D. (2006). The impact of microwave synthesis on drug discovery, *Drug Discovery*, Vol. 5, p.p 51-68.

Khawam, A. & Flanagen, D. (2006). Solid-state kinetic models: basic and mathematical fundamentals, *J. Phys. B*, Vol 110, p.p. 17315-17328.

Kingston, H.M & Haswall, S.J. (1997). *Microwave Enhanced Chemistry*. American Chemical Society, Washington, DC., p.p. 1-749.

Kiminami, R.; Morell, M.; Folz, D. & Clark, D. (2000). Microwave synthesis of alumina powders. *American Ceramic Society Bulletin*, Vol.79 (No3), p.p. 63-67.

Klaric, I.; Roje, U. & Kovacic, T. (1995). Kinetics of isothermal thermogravimetrical degradation of PVC/ABS blends. *J. Therm. Anal. Calorim*, Vol. 45, p.p. 1373-1380.

Kosaric, N.; Vardar-Sukan, F. & Pieper, H.J. (2001.) *The biotechnology of ethanol classical and future application*. Wiley-VCH, ed. M.Roeher, Verlag Gmbh, pp.1-226.

Lai, H.S.; Chen, W.J. & Chiang, L.Y. (2000). Free radical activity of fullerenol on the ischemia –reperfusion intestine in dogs. *World J. Surgery* Vol. 24, p.p. 450–454.

Larsson, R. (1989). Model of selective energy transfer at the active site of the catalyst, *J. of Molecular Catalysis*, Vol(55), p.p. 70-83.

Larhed, M.; Moberg, C. & Halleberg, A. (2002). Microwave-accelerated homogeneous catalysis in organic chemistry. *Acc. Chem. Res.*, Vol. 35, p.p. 717-727.

Lewis, D.; Summers, J.; Ward, T. & McCrath, J. (1992). Accelerated imidization reactions using microwave radiations, *J.Polym.Sci., Part A.; Polym.Chem.*, Vol. 30, p.p. 1647-1653.

Li, J.; Y-Li, J.; X-Gang, Z. & Yang, H. (2007). Microwave solid state synthesis of spinel $Li_4Ti_5O_{12}$ nanocrystallites as anode material for lithium-ion batteries. *Solid state ionies*, Vol. 178 (No 29-30), p.p. 1590-1594.

Li, J.; Takeuchi, A.; Ozawa, M.; Li, X.; Saigo, K. & Kitazawa, K. (1993). C_{60} fullerol formation catalyzed by Quaternary Ammonium hydroxides. *J. Chem. Soc. Chem. Commun.,* p.p. 1784-1792.

Lin, T. & Choi, D. (1996). Buckminster fullerenol free radical scavengers reduce excitotoxic and a poptotic death of cultured cortied neurons. *Neurobiology of Disease,* Vol. 3, p.p. 129–135.

Lucchesi, M.; Chemat, F. & Smadja, J. (2004). Solvent free microwave extraction of essential oil from aromatic herbs: comparison with conventional hydro-destillation. *J. Chromatography A.,* Vol. 1043 (No 2), p.p 323-327.

Loupy, A. (2002). *Microwave in Organic Synthesis,* Wiley –VCH, Weinheim, p.p. 1-500.

Mayoux C.; Dandurand, J.; Ricard, A.& Lacabanne C. (2000). Inverse suspension polymerization of sodium acrylate: Synthesis and characterization. *J. Appl. Polym. Sci.,* Vol 77 (No 12), 2621-2630.

Miller, L. (1959). Use of dinitrosalicylic acid reagent for determination of reducing sugar. *Analytical Chemistry,* Vol 31 (3), p.p. 426-428.

Nüchten, M.; Ondruschka, B.; Bonrath, W. & Gum, A. (2004). Microwave-Assisted Synthesis – a critical technology overview. *Green Chem.,* Vol. 6, p.p. 128-141.

Perelaer, J.; deGans, B. & Schubert, U. (2006). Ink –gel printing and microwave sintering of conductive silver tracks, *Adv. Mater.,* Vol. 18 (No16), p.p. 2101-2104.

Perreux, L. & Loupy, A. (2000). A tentative rationalization of microwave effects in organic synthesis according to the reaction medium, and mechanistic considerations, *Tetrahedron,* Vol. 57, p.p. 9199-9223.

Plaze, I.; Leskovšek, S. & Koloni, T. (1995). Hydrolysis of sucrose by conventional and microwave heating in stirred tank reactor. *Chem. Eng. J.,* Vol. 59, p.p. 250-257.

Plyler, E. (1952). Infrared spectra of methanol, ethanol and n-propanol. *Journal of Research of Burean of Standards* Vol. 48 (No 4), p.p. 281-286.

Raner, K.D.; Strauss, C.R. Vyskoc, F. & Mokbel, L. (1993). A comparison of reaction kinetics observed under microwave irradiation and conventional heating. *J. Org. Chem.,* Vol. 58 (No 4), p.p. 950-953.

Reilly, P. (2000). Protein engineering of glycoamylase to improve industrial performance-A Review. *Starch,* Vol.51 (No 8-9), p.p. 269-274.

Ring, T.A. (2010). Comparation of Raman and ATR-FTIR spectroscopy of aqueous sugar solutions. URL:www.chem.utah.edu

Rufino, E. & Monteiro, E. (2003). Infrared study on methyl mathacrylite-methacrylic acid copolymers and their sodium salts, *Polymer,* Vol 44, p.p. 7189-7190.

Rybakov, K. & Semenov, V. (1994). Possibility of plastic deformation of an ionic crystals due to the nonthermal influences of a high frequency electric field. *Phys. Rev. B,* Vol. 49, p.p. 64-68.

Saberi, A.; Golestani-Fara, F.; Willert-Porada, M.; Negahdam, Z.; Liebscher, C. & Gossler, B. (2009). A novel approach to synthesis of nanosize $MgAl_2SO_4$ spinel powder through sol-gel citrate technique and subsequent heat treatment. *Ceramic International,* Vol.35 (No 3), p.p. 933-937.

Schettino, V.; Paglioi, M.; Ciabini, L. & Cardini, G. (2001). The vibrational spectrum of fullerene C_{60}. *J. Phys. Chem. A,* Vol. (105), p.p. 11192-11196.

Schneider, S.; Darwish, A.; Kroto, H.; Taylor, R. & Walton, D. (1994). Formation of fullerols via hydroboration of fullerene - C60. *J. Chem. Soc., Chem. Commun.*, p.p. 463-464.

Shibata, C.: Kashima, T. & Ohuchi, K. (1996). Nonthermal influence of Microwave Power on Chemical Reactions. *Jpn. J. Appl. Phys.*, Vol. 35, p.p. 316-319.

Shipe, W.; Wolkenberg, S. & Linfsley, C. (2005). Accelerating lead development by microwave - enhanced medicinal chemistry, *Nature Rev. Drug Discov. Today Technol.*, Vol. 2, p.p. 155-161.

Siggers, G. & Martinola, F. (1985). Sucrose inversion with cation exchange resins. *Int. Sugar Journal*, Vol. 87, pp. 23-26.

Strauss, C. & Trainor, R. (1995). Developments in Microwave-Assisted Organic Chemistry. *Aust. J. Chem.*, Vol. 48 (No 10), p.p. 1665-1692.

Stuerga, D. & Gaillard, P. (1996). Microwave athermal effects in chemistry: a myth' anthopsy –Part II: orienting effect and thermodynamic consequences of electric field. *J. Microwave Power Electromagn. Energy*, Vol. 31 (No 2), p.p. 101-113.

Sung, H., J.; Taihuan, J.; Young, K., & Jong-San, C. (2007). Microwave effect in the fast synthesis of microporous materials: which stage between nucleation and crystal growth is accelerated by microwave irradiation, *Chemistry A., European J.*, Vol. 13 (No 6), p.p. 4410-4417.

Takeda H, Tanaguchi Y, (1985). Production process for highly water absorbable polymers. *US. Patent* 4, 525, 527.

Taylor, R. (1995). *The chemistry of fullerene*, World scientific publishing Co. Pte. Ltd. Singapore, pp. 67 – 109.

Tsuji, M.; Hashimoto, Y.; Nishizawa, Y.Kubokawa, M. & Tsuji, T. (2005). Microwave-assisted synthesis of metalic nanostructures in solution. *Chem. Eur. J.* Vol.1 1, p.p440-452.

Turner, M.; Laurence, R.; Conner, W. & Yugvessor, K. (2000). Microwave Radiations influence on sorption and competitive sorption on zeolites. *J. Aiche*, Vol. 46, p.p.758-768.

Vanderah, T. (2002). Talking ceramic. *Science* , Vol. 298 (No 5596), p.p. 1182-1184.

Vyazovkin, S. & Lesnikovich, A. (1990). An approach to the solution of the inverse kinetic problem in the case of complex processes. 1. Methods employing a series of thermoanalytical curves. *Thermochimica Acta*, Vol. 165, p.p. 273-228.

Vyazovkin, S. & Wight, C. (1997). Kinetics in solids. *Annu.Rev.Phys.Chem.*, Vol. 48, p.p. 125-149.

Will, H.; Scholz, P. & Ondruschka, B. (2004). Heterogeneous gas-phase catalysis under microwave irradiation- a new multy-mode microwave. *Topics in Catalysis*, Vol. 29 (No 3-4) p.p. 175-182.

Y-Pei, F.; Ch-Hsiung, L. & Ch-Sang, H. (2005). Preparation of ultrafine CeO_2 powders by microwave –induced combunation and precipitation. *J. of Alloys and Compounds*, Vol. 391 (No 1-2), p.p. 110-114.

Yu, J.; Dong, L.; Wu, C.; Wu, L. & Xing, J. (2002). Hydroxyfullerene as a novel coating for solid-phase microextraction fiber with sol-gol technology. *J. of Chromatography A*, Vol. 978, p.p. 37–48.

Zhang, X.; Hayward, D. & Mingos, D. (1999). Apparent equilibrium shifts hot-spot formation for catalytic reactions induced by microwave dielectric heating. *Chem. Commun.*, No 9, p.p. 975-976.

Zhou W., J. Yao K.J. & Kurth J, (1997)a. Studies of crosslinked poly(AM-MSAS-AA) gels, I. Studies and characterization. *J. Appl. Polym. Sci.,* Vol. 64 (No 5), 1001-1008.

Zhou W., J. Yao K.J. & Kurth J, (1997)b. Studies of crosslinked poly(AM-MSAS-AA) gels, II. Effects of polymerization conditions on the absorbency. *J. Appl. Polym. Sci.,* Vol. 64 (No 5), p.p. 1009-1016.

Zhao, G.; Zhang, P.; Wei, X. & Yang, Z. (2004). Determination of proteins with fullerol by a quaternary ammonium hydroxides. *Analytical Biochemistry,* Vol. 334, p.p. 297-302.

Zhang, P.; Pan, H.; Liu, D.; Zhang, F. & Zhu, (2003). Effective mechanochemical synthesis of [60] fullerenols. *Sinthesis Communications,* Vol. 33 (No 14), p.p. 2469-2474.

19

The Use of Microwave Energy in Dental Prosthesis

Célia M. Rizzatti-Barbosa, Altair A. Del Bel Cury
and Renata C. M. Rodrigues Garcia
State University of Campinas (Unicamp)
Brazil

1. Introduction

Acrylic resins are used as denture base materials since 1937 (Craig, 1993) and heat-cured Poly(methyl-methacrylate) (PMMA) (Lay et al., 2004) composed by methyl methacrylate monomers chains is the most popular of them. Virtually, dentures are constructed from conventional polymer/monomer (Lay et al., 2004) processed by heat water-bath system and compression molding technique (Ganzarolli et al., 2002). It is prepared by mixing the monomer and polymer in a specific ratio, resulting in a dough mass that is packed and pressed in a flask for polymerization (Ganzarolli et al., 2007). Addition polymerization of PMMA requires the activation of the initiator (benzoyl peroxide) to provide free radicals. Polymerization takes place as the free radicals open the double bonds of the methyl methacrylate, creating a chain reaction where the monomer attaches to polymer free radicals (Bartoloni et al., 2000). When the polymerization reaction is activated by heat, the monomers form polymeric chains joined by high energy linkings (crossed-links) and this reaction would finish when all monomers have supposal been reacted (Machado et al., 2004).

Although water-bath polymerization is extensively used to process PMMA, new resins and processing methods have been proposed in order to obtain better physical properties and to simplify the technique (Souza Jr. et al., 2006). Heating curing, chemical curing by pouring technique, light curing and microwave curing resin have been extensively studied for denture base processing (Harman, 1949; Nishii, 1968; Jerolimov et al., 1989; Urabe et al., 1999).

Polymerization of heat-cured PMMA is usually carried out in a temperature-controlled water bath for at least 9 hours. However, the use of microwave energy to polymerize PMMA decreases the time to three minutes only, producing acrylic resin bases with the same quality as those polymerized by water bath technique (Nishii, 1968; Kimura et al, 1983; Hayden, 1986; Jerolimov et al., 1989; Del Bel Cury et al., 1994; Rizzatti-Barbosa et al., 1995; Braun et al., 1988; Ganzarolli et al., 2002; Yunus et al., 2005; Rizzatti-Barbosa et al., 2009). The microwave polymerization technique has advantages such us less equipment is required (Schnieder et al., 2002), the method is fast and clean, the final product has the same quality of physical properties, and it presents better accuracy of fit, resulting in improvement adaptation of denture base (Rizzatti-Barbosa et al., 1995; Rodrigues-Garcia et al., 1996; Compagnoni et al., 2004; Shibayama et al., 2009). The advantages of microwave heating over conventional heating are: (1) the inside and outside of substance are almost equally heated, and (2) temperature rises rapidly (Lai et al., 2004). Microwave energy acts on the monomer promoting an uniform and immediate heating of the polymer mass (Machado et al., 2004), that activates the decomposition of benzoyl peroxide, the reaction initiator, and quickly yields free radicals for the polymerization process (Sarac et al., 2005), which decreases in the same proportion as polymerization increases (Compagnoni et al., 2004).

Polymerization of PMMA by microwave energy is possible because methyl methacrylate (MMA) is a polar liquid at room temperature, and microwaves create an electronmagnetic field inside microwave oven that allows the MMA molecules to orient themselves at a frequency of 2450 MHz (Sanders et al., 1991). Numerous polarized molecules are flipped over rapidly and generate heat due to molecular friction. Because of this, microwave heating is independent of the thermal conductivity; therefore, curing cycles involving the application of rapid heat may be used without the development of a high exothermic temperature.

The use of microwave energy was first reported in 1968 by Nishii as an alternative PMMA processing method and has become increasingly popular as to conventional water-bath processing (Schnieder et al., 2002). Several studies showed that physical properties of denture base resins polymerized with microwave energy or by heat water bath (conventional method) exhibited similar behavior being both of methods adequate to be used in denture acrylic resins fabrication (Kimura et al., 1983; Kimura et al., 1984; Kimura et al., 1987). Some studies reported that it is possible to polymerize acrylic resin in a very short time using microwave energy, however the metal flask must be substituted by the fiber-reinforced plastic (FRP) (Kimura et al., 1983, 1984 and 1985; Reitz et al., 1985; Rizzatti-Barbosa et al., 1995).

Since that, several studies have been investigated different properties of acrylic resin when processed by microwave energy. In Brazil, the first microwave research was reported in 1994 by Del Bel Cury et al., who investigated physical properties of acrylic resin processed by microwave energy compared to water bath. Their results showed differences among the resins that were attributed to the composition of acrylic resins.

2. Solubility, residual monomer and water sorption

The solubility is a property of acrylic resin, representing the not reacted substances releasing (residual monomer, plasticizers and initializers). It is characterized as an undesirable property of resins, since they should be insoluble in oral fluids. Residues releasing from a polymerized resin base can promote tissular reactions in users of prosthesis (Machado et al., 2009). From this point of view, the degree of conversion is one of the most important characteristic of resin, on account of the high residual monomer levels that could be unreacted. Its presence has an adverse effect on physical and mechanical properties (Yunus et al., 1994) as well as on the biocompatibility (Ilbay et al., 1994). Acrylic resin also presents water sorption that is directly related to the polar properties of resin molecules, the physical process of water diffusion through intermolecular space (Takahashi et al., 1998), and the amount of residual monomer in the polymerized mass (Wong et al., 1999). Thus, polymerization degree is directly related to resin's ability of absorbing water (Meloto et al., 2006). Polymerization takes place as the free radicals open the double bonds of the methyl-methacrylate, creating a chain reaction where the monomer attaches to polymer free radicals. The degree of monomer conversion to polymer structure of PMMA is a measure of the carbon double bonds (C=C) converted into carbon single bonds (C–C) (Bartoloni et al., 2000).

According to Azzarri et al. (2003), from the appropriate selection of power and the resin's curing time, it is possible to optimise the level of residual monomer and a low cytotoxicity keeping at the same time better mechanical properties. Because of the quick and instantaneous heating, the polymerization of monomers could be damaged by the microwave energy. However, classical studies have been shown that microwave-polymerized acrylic resin presents lower or the same residual monomer levels relative to conventionally polymerized resins (McCabe & Basker, 1976; Austin & Basker, 1980; Austin & Basker, 1982; Huggett et al., 1984; DeClerck, 1987; Truong & Thomasz, 1988; Al Doori et al., 1988; Sadamori et al., 1994; lbay et al., 1994; Yunus et al., 1994; Shlosberg et al., 1989; Jacob et al., 1997). Other researches showed resin polymerized by microwave energy presented similar solubily, residual monomer or water sorption levels than conventional or fast polymerization process.

Authors and date	Materials and groups	Methods and measurements	Comments and conclusions
Urban et al. (2007)	Four hard chair-side reline resins (Duraliner II-D, Kooliner-K, Tokuso Rebase Fast-TRF and Ufi Gel hard-UGH) (short and long polymerization cycles) One heat-polymerized denture base resin (Lucitone 550-L) (short and long polymerization cycles) Immersion in hot water	High performance liquid chromatography, expressed as a percentage of residual monomer.	Statistical differences in residual monomer percentage were found among all materials (K >D >UGH > LLong >TRF > LShort). The reduction in residual monomer promoted by water bath and microwave post-polymerization treatments could improve the mechanical properties and biocompatibility of the relining and denture base materials.
Meloto et al. (2006)	Vipi Cril heat-polymerized acrylic resin (73°C for 9 h) and Vipi Wave microwave resin (20 min/90 W and 5 min/450 W)	Specimens (m1) were stored in distilled water at $37 \pm 1°C$ during 30 days (m2), and weighed. Water sorption (g/cm3) was calculated using the formula: WS = (m2 - m1)/ V.	No difference was found between the group fabricated using water bath and microwave polymerization.
Machado et al. (2004)	Vipi Cril heat-cured acrylic resin (water bath - $74 \pm 1 °C$ for 9 h; Microwave oven – 500 W/ 3 min.)	Solubility test and percentile solubility	The association between polymerization by microwave irradiation and mechanical polishing showed lower percentile solubility, indicating lower residual substances releasing.

Authors and date	Materials and groups	Methods and measurements	Comments and conclusions
Lai et al. (2004)	PMMA denture base polymer (Optilon-399) – (water-bath at 70 8C for 9 h; resin blocks was processed at 80, 160, 240, and 560 W for 15, 10, 7, and 2 min, separately + additional 2 min at 560 W.	Internal temperature – Thermocouples. Solubility – samples weigh percentage of after specific treatment.	Large difference in the curing temperature was observed. The results of the temperature measurement and the solubility analysis indicated that a microwave oven for curing resin was much faster than a conventional water-bath, and the degree of curing also increased a little. Microwave energy can efficiently polymerize and cure denture base polymer from the results of insoluble weight percent and the curing temperature.
Phoenix et al. (2004)	6 commonly used polymethyl methacrylate denture base resins. ADA Specification No. 12. Thermal assessments (differential scanning calorimetry and dynamic mechanical analysis).	Sorption and solubility	The microwaveable resins displayed better results than other denture base resins included in the investigation.
Azzarri et al. (2003)	Four different conditions of power and curing time	Release of residual monomer in water was evaluated by spectrophotometric method up to 24 h.	It is possible to optimise the level of residual monomer and a low cytotoxicity keeping from the appropriate selection of power and time of curing of the resin.
Oliveira et al. (2003)	Acron MC – 500W/3m or 4.5m. One simple flask centrally placed on the turning plate; two flasks, one in the centre and the other peripherally placed in the plate; two flasks centrally, one above and the other below.	Monomer levels - spectrophotometry at 206 nm.	Monomer release was affected by the position of the flask.

Authors and date	Materials and groups	Methods and measurements	Comments and conclusions
Bartoloni et al. (2000)	Lucitone 199® (conventional long-cure), Ac-celar 20® (rapid cure), and Acron MC® (microwave cure)	Fourier transform infrared spectrometry (FTIR)	All curing methods obtained similar degrees of conversion determined by FTIR.
Blagojevic et al. (1999)	TS1195 ; Acron MC; Biocryl NR. Microwave oven for 3 min at 600 W and hot-water bath for 14 h at 70 °C followed by a 3h terminal boil.	Gas liquid chromatography.	General trend suggested that the water-bath method enhanced the degree of polymerization resulting in a lower level of residual monomer. Residual monomer was considerably lower for TS1195, Acron MC and Biocryl NR following water both polymerization.
Del Bel Cury et al. (1994)	Acrylic resins: Acron MC (microwave – 500W/3m); Lucitone 550 (water bath – 73oC/90m); Ortho-Class (self polymerized).	Water sorption and solubility – load of samples	There were differences among tested materials that can be resulted from composition and polymerization methods.

Table 1. Studies on solubility, residual monomer and water sorption of microwaved resin

3. Porosity and color stability

Porosity has been attributed to a variety of factors such as air entrapped during mixing, monomer contraction during polymerization, monomer vaporization associated with the exothermic reaction, and the presence of residual monomer (Keller & Lautenschlager, 1985; Wolfaardt et al., 1986). According to Tager (Tager, 1978) porosity is a property of solids that relates to their structure and is expressed in the presence of voids (pores) between separate grains, layers, crystals, and other elements of a coarse structure of a solid. This definition emphasizes the fact that the concept of porosity can be applied to solids, and that pores are spaces not between molecules, but between super molecule structures. Other authors (Taylor, 1941; Sweeney et al., 1942;) demonstrated that insufficient mixing of monomer and polymer, processing temperatures higher than 74°C, packing of the mold, and inadequate compression on the flask may cause porosity in denture base resin. Depending on polymerization conditions, up to 11% porosity has been observed associated with decreased mechanical properties and poor esthetics (Keller & Lautenschlager, 1985) and the potential harboring of organisms and retention of fluids (Davenport, 1970; Compagnoni et al., 2004).

Irregularities and porosities present on denture surfaces offer a favorable niche to retain stain and bacterial biofilm, and because of this denture base polymers are susceptible to color shifting. It depends not only of the polymerization method but also on the chemical characteristics of the material (Polyzois et al., 1999). Considering the processing methods for acrylic resins, some authors (Austin et al., 1982; May et al.,1992) have stated that polymerization for short period of time promotes higher color instability.

Porosity and its consequences have been studied to resin processed by microwave energy (Reitz et al., 1985; Wolfaardt et al., 1986; De Clerck, 1987; Al Doori et al., 1988; Truong & Thomasz, 1988; Levin et al., 1989; Shlosberg et al., 1989; Alkhatib et al., 1990; Bafile et al.,1991; Sadamori et al., 1994; Ilbay et al.,1994;) and some authors verified that it depends on the base thickness (Sanders et al., 1987) or the selection of the material, despite of the microwave processing. (Yannikakis et al., 2002). Prevalent studies on porosity and its consequences are listed on Table 2 and show similar results when comparing microwave and water bath or light cure processing.

Authors and date	Materials and groups	Methods and measurements	Comments and conclusions
Rizzatti-Barbosa et al. (2009)	Water bath-cured resin (Vipi-Cryl) – (74° C for 9 hours) Microwave-cured resin (Vipi-Wave) – (90 W for 13 minutes + 450 W for 5 minutes)	Solution of permanent black ink for 12 hours. Number of pores per area.	Microwave curing is not a factor that alters surface superficial porosity.
Assunção et al. (2009)	Heat cured resin (Microwave polymerization - 500 W for 3 minutes; water bath - 74°C for 9 hours)	Spectrophotometer that measured visible ultraviolet reflection. Color changes calculatedby the Commission Internationale de l'Eclairage system - D65 standard illumination.	The color of denture teeth was not affected by the polymerization methods, as there was no significant difference between microwave and conventional polymerization.
Paranhos et al. (2009)	Microwave-polymerized acrylic resin (after immersion in 0.5% NaOCl, 1% NaOCl, and Clorox/Calgon)	Portable colorimeter (Color-guide 45/0). Color changes (ΔE) were calculated with the use of CIELab color space	No statistically significant differences (p>0.05) among the solutions for color teeth stability.
Pero et al. (2007)	Onda-Cryl (microwave: 500 W for 3 min; 90 W for 13 min + 500 W for 90 s; 320 W for 3 min + 0W for 4min + 720 W for 3 min); Classico (water bath: 74° C for 9h). Sample thickness: 2.0 mm; 3.5 mm; 5.0 mm.	Polymers and water (classical sorption method and mercury porosimetry - porosity of a sorbet is estimated quantitatively by total pore volume (W0)).	The influence of microwave polymerization cycle on porosity of acrylic resin appears only on the thinnest specimens (2.0 mm). Water bath polymerization group presented similar porosity results for specimens of all tested thicknesses.

Authors and date	Materials and groups	Methods and measurements	Comments and conclusions
Lai et al. (2004)	PMMA denture base polymer (Optilon-399) – (water-bath at 70 °C for 9 h; resin blocks was processed at 80, 160,240, and 560 W for 15,10, 7, and 2 min, separately + additional 2 min at 560 W. Thickness samples was less than 10 mm.	Pulse echo C-scans - Testech ultrasonic tester (model VI-100) – frequency of 10 MHz / 25.4 mm	The amount of porosity increased with an increase in microwave energy level. It is important to control temperature accurately and ensure correct timing to minimize porosity when microwave polymerization is used.
Compagnoni et al. (2004)	Acrylic resin samples: Onda-Cryl, microwave polymerized (500 W for 3 minutes, 90 W for 13 minutes+500 W for 90 seconds, 320 W for 3 minutes+0 W for 4 minutes+720 W for 3 minutes); Clássico, heat-polymerized (water 74 o C/9h).	Porosity – measurement of the specimen volume before and after its immersion in water. Specimen was weighed in air and in water to calculate percent mean porosity. The absolute density of acrylic resin was used to calculate the percent mean porosity.	No differences in mean porosity were found among resin specimens polymerized by microwave energy. The porosity of microwave-polymerized resin was similar in porosity to the heat-polymerized resin.
Oliveira et al. (2003)	Acron MC – 500W/3m or 4.5m. One simple flask centrally placed on the turning plate; two flasks, one in the centre and the other peripherally placed in the plate; two flasks centrally, one above and the other below.	Porosity - permanent ink and counting the porous in a stereo light microscope.	There is no difference among the groups.
Yannikakis et al. (2002)	Two heat-activated denture base resins (short and long microwave cycle) One conventional (Paladon 65) (water bath cured) One designed for microwave polymerization (Acron MC) 3 mm and 6 mm thickness.	Photographed under a microscope at ×100 magnification. Pore were measured with a digital planimeter. Total area of pores per surface was calculated in percentage form	There are no pores in the group polymerized in water bath. Minor porosity was identified in thin areas and more severe porosity in thicker areas of conventional resin specimens that underwent microwave polymerization. No significant porosity was observed in the resin designed specifically for microwave polymerization. Severe porosity in thicker

Authors and date	Materials and groups	Methods and measurements	Comments and conclusions
			areas of conventional resin specimens that underwent microwave polymerization, and no significant porosity in the resin designed specifically for this technique.
Rodrigues Garcia et al. (1996)	Denture bases of conventional acrylic resin (water bath 73oC/ 9 hours) and specific resin for microwave polymerization (500 W/3m). Dentures were relined by addition method.	Porosity (immersion of samples in a solution black - pores were counted / stereo light microscope under magnification of 6.3 x)	Conventional resin cured by water bath or microwaves energy showed the highest number of pores after relining
May et al. (1996)	Seven conventional denture base materials One microwave heat cured denture base material processed with the microwave method. Conditions of accelerated aging to test for color stability.	Color measurements were made before weathering and at 300, 600, and 900 hours.	Color changes occurred after accelerated aging in heat-cured denture base resins and Acron GC microwave acrylic resins processed with the microwave.

Table 2. Studies on porosity and color stability of microwaved resin

4. Hardness, transverse strength, flexural strength, shear bond strength, tensile bond strength, impact strength, roughness, and modulus of elasticity

Curing processes have been modified in order to improve the physical and mechanical properties of those materials, and also to afford the technical work of the professionals. The relationship between the physicochemical characteristics and the final properties of a material are of fundamental importance to obtain a resin with the desired properties. The constituent polymer of the powder exhibits a high average molecular weight and a broad of molecular weight distribution. The powder – liquid ratio determines the time dependence of monomer conversion and the rate of polymerization for the formation of the cross-linked network that grows throughout the chains of the base polymer (Wallace et al., 1994). Both characteristics of the polymerization rate and conversion, are proportional to the concentration of the reactive species as well as to the instantaneous temperature (Urabe et al., 1999).

For the microwave-cured acrylic resins, it has been demonstrated that the temperature developed during the reaction is not constant: it increases quickly at the beginning, goes through a maximum and then decays, being able to reach peaks of the order of 150–200 °C, depending on the working conditions (Gourdinne et al., 1979; Jacob et al., 1997). Hence, both the power of the microwave and the time of exposition can be regulated to control in these systems the rate of polymerization and the conversion degree. The long time of microwaves exposition could enhance the rate of secondary reactions of bond breaking on the pending chains breaking bonds by free radical mechanisms, which would be competitive with the main curing reaction. Properties like hardness, transverse strength, flexural strength, tensile bond strength, and modulus of elasticity, in some studies, are not modified by the longer sample exposition time or by the microwave power, probably because these secondary reactions do not change the cross-link density of the material. Only the impact strength would feel their effect because of the shortening of the pendant chains. The reduction of the impact strength for longer molecules was clearly stated, as well as the influence of the length of the cross-linking agent on the resin mechanical properties (Caycik & Jagger, 1992).

Researchers investigated mechanical properties of microwave polymerized resins and showed that acrylic resin processed by microwave energy presented the same characteristics of conventional procedures of processing (Nishii, 1968; Stafford & Handley, 1975; Stafford & Huggett, 1978; Faraj & Ellis, 1979; Gourdinne et al., 1979; Kimuta et al., 1983; Reitz et al., 1985; Hayden, 1986; De Clerck, 1987; Kimura et al., 1987; Truong & Thomasz, 1988; Al-Doori et al., 1988; Al-Mulla et al., 1988; Shlosberg et al., 1989; Levin et al., 1989; Alkhatib et al., 1990; Hogan & Mori, 1990; Al-Hanbali et al., 1991; Smith et al., 1992; Caycik & Jagger, 1992;. Chen et al., 1993; Frangou et al., 1993; Jacob et al., 1997). Studies on mechanical properties also showed similar or better results of conventional or light cured resins (Table 3).

Authors and date	Materials and groups	Methods and measurements	Comments and conclusions
Faot et al. (2008)	Onda Cryl resin. Microwave processing: (3 min at 360 W, 4-min pause, and 3 min at 810 W; and 6 min at 630 W).	Impact strength: Charpy method. Types and morphology of fractures: all fragments analyzed in morphology, crack propagation angles and microstructure.	Both polymerization cycles are adequate to polymerize the denture resin studied.
Ganzarolli et al. (2007)	Conventional water bath polymerized resin. Microwave polymerized resin. Injection-molded resins.	Transverse strength test - universal machine under axial load, at a crosshead speed of 5 mm/min. Impact strength test - Charpy's test performed with a 40 kJ/cm load.	There was no relevant improvement of transverse and impact strength. Microwaveable resin showed similar transverse and impact strengths.
Souza Jr. et al. (2006)	Acrylic resins samples: microwave- (Onda Cryl); visible light- (Triad); water bath polymerized (Clássico). Cobalt-chromium metal bar included in resin samples.	Roughness - profilometer (Surfcorder SE 1700) Knoop hardness - (Kg/mm²) - microhardness tester (Shimadzu HMV 2000)	The presence of metal did not influence roughness and hardness values of any of the tested acrylic resins

Authors and date	Materials and groups	Methods and measurements	Comments and conclusions
Tacir et al. (2006)	PMMA conventional acrylic resin. Reinforcing effect of glass fibres.	Fracture resistance and Flexural strength. 3- points of the samples. Universal testing machine	Flexural strength of heat-polymerized PMMA denture resin was improved after reinforcement with glass fibres.
Sarac et al. (2005)	Denture base material (a conventionally molded, heat polymerized resin [Meliodent, M]; an injection-molded, Heat polymerized resin[SRIvocap, I], and a microwave polymerized resin [Acron MC, A]). Repaired with an auto polymerizing acrylic resin (Meliodent). Surfaces treated with chemical etchants: acetone (30 s), methylene chloride (30 sec), MMA (180 sec).	Shear bond strength (MPa): universal testing machine.	Chemical treatment showed improvement on the bond strength of the base materials. Microwave-polymerized acrylic resin showed the lowest shear bond strength compared to the control groups.
Vergani et al. (2005)	Heat-polymerized resin (Lucitone 550) Autopolymerizing reline resins: (Duraliner II, Kooliner, Ufi Gel Hard, and Tokuso Rebase Fast). Postpolymerization by microwave energy (500, 550, or 650 W) for (3, 4, or 5 minutes)	Flexural strength – load measurements (Newtons) / crosshead speed of 5 mm/min using a 3-point bending and span of 50 mm.	Microwave postpolymerization irradiation can be an effective method for increasing the flexural strength of Duraliner II (at 650 W) and Kooliner (at 550 W and 650 W for 5 minutes).
Yunus et al. (2005)	Nylon denture base material (Lucitone FRS); conventional compression-moulded heat-polymerized (Meliodent); compression-moulded microwave-	Flexural modulus; flexural strength: three point bending test.	Flexural modulus of nylon was significantly lower than the three PMMA polymers Flexural strength of nylon was significantly lower than those of Acron MC (microwaved) and

Authors and date	Materials and groups	Methods and measurements	Comments and conclusions
	polymerized (Acron MC); injection-moulded microwave polymerized (Lucitone 199) PMMA polymers. Water stored at 37°C/30 day Disinfectant stored for 24 h		comparable with Lucitone 199 (microwaved).
Phoenix et al. (2004)	6 commonly used polymethyl methacrylate denture base resins. ADA Specification No. 12. Thermal assessments (differential scanning calorimetry and dynamic mechanical analysis).	Color stability, flexural stiffness, and hardness.	The microwaveable resins displayed greater stiffness, and greater surface hardness than other denture base resins. Elastomeric toughening agents yielded decreased stiffness, decreased surface hardness, and decreased glass transition temperatures.
Lai et al. (2004)	PMMA denture base polymer (Optilon-399) – (water-bath at 70°C for 9 h; resin blocks was processed at 80, 160, 240, and 560 W for 15, 10, 7, and 2 min, separately + additional 2 min at 560 W. Thickness samples were less than 10 mm.	Hardness - Shimadzu hardness tester (HMV-2000 - 300g/30s - 15 areas along uniformly selected points of surfaces). Flexural strength - MTS dynamic tensile testing machine (Model 810) at a crosshead speed of 1.25 mm/min. Three-point-bending test.	Highly statistical differences in flexural properties were evident in a comparison of processing methods. The size and the volume fraction of the rubber phase are in favor of the water-bath method. Water-bath cured specimens showed better flexural strength and fexural modulus than the microwave-cured specimens. There were no significant differences in the surface hardness and the domain size distribution of the effective rubber phase.
Oliveira et al. (2003)	Acron MC – 500W/3m or 4.5m One simple flask centrally placed on the turning plate; two flasks, one in the centre and the other peripherally placed in the plate; two flasks centrally, one above and the other below.	Hardness test – 12 indentations in the surface of specimen.	There is no difference among the groups.

Authors and date	Materials and groups	Methods and measurements	Comments and conclusions
Azzarri et al. (2003)	Acrylic denture base resin microwave polymerized (samples were prepared in 200, 500, and 800, for 5 and 10 min each side up).	Hardness (Rockwell method) Strength (Charpy method). Young's modulus of elasticity – technique described by Stafford & Handley (1975) according to ISO 1567.	The mechanical properties of the acrylic denture base resin microwave polymerized depend both on the exposition time and microwave power. From the appropriate selection of power and time of curing of the resin it is possible to obtain the best mechanical properties.
Schneider et al. (2002)	Acrylic resins samples: heat polymerized (Lucitone 199); and Microwave polymerized (Acron MC). Acrylic resin denture teeth: (IPN, SLM, Vitapan, and SROrthotyp-PE). ADA specification n° 15.	Strength until sample fracture – custom alignment device debonded specimens. Cohesive failures - scanning electron microscope.	Type of denture base material and denture tooth selected for use may influence the tensile bond strength of the tooth to the base.
Memon et al. (2001)	Microwave polymerized polyurethane-based denture material processed by an injection-molding technique, a conventional microwave polymerized denture material, and a heat polymerized compression-molded poly(methyl methacrylate) (PMMA) denture material.	Impact strength - Charpy-type impact tester. Transverse strength and the flexural modulus - three-point bending test.	Impact and flexural strengths – microwave polymerized injection molded, offered no advantage over the existing heat- and microwave-polymerized PMMA-based denture base polymer. It has rigidity comparable to that of the microwave-polymerized PMMA polymer.
Polyzois et al. (2001)	Heat-polymerized denture base material repaired with heat polymerized resin. Auto polymerized resin alone. Auto polymerized resin with glass fiber	Fracture force, deflection at fracture, toughness: 3-point bending test.	The most effective was microwave-irradiated, auto polymerized resin reinforced with round wire or monolayer glass fiber ribbon.

Authors and date	Materials and groups	Methods and measurements	Comments and conclusions
	or wire reinforcement. Auto polymerized resin repairs; heat polymerized resin repairs (with no reinforcement, treated with microwave irradiation after polymerization, with monolayer or multilayer glass fiber reinforcement, and with round or braided wire reinforcement).		
Rached et al. (2001)	Heat-cured acrylic resin (Lucitone 550) – repaired with a microwave acrylic resin (Acron MC) – 500W/3m. Chemical treatments (AC monomer dipping/30 s; acetone dipping/30 s; acetone dipping/15 s + blast of air + AC monomer dipping/15 s) in the cut ends.	Surface texture - scanning electron microscopy. Flexural strengh.	Surface treatments affected the bond strength between the two acrylic resins. There are no differences in strength between intact heat-cured denture base material and the same material repaired with microwave acrylic resin.
Blabojevic et al. (1999)	A without crosslinking agents homopolymer (TS1195). A cross-linked microwave resin (Acron MC) (Microwave Panasonic oven with rotating turntable). A cross-linked for water-bath polymerization (Biocryl NR) (Conventional water bath). An autopolymerizing reline:repair material (CroForm).	Indentation hardness – Wallace hardness tester Model H6B:SA:C Impact strength – Zwick pendulum impact model 5102 with 0·5 J pendulum.	Microwaving of autopolymerizing resin improved analyzed mechanical properties.

Authors and date	Materials and groups	Methods and measurements	Comments and conclusions
Polyzois et al. (1995)	Denture base resins repaired (standard heat activated polymerising resin, a denture base resin especially formulated for microwave activated polymerisation, auto polymerizing resin). Conventional water bath; microwave curing, and auto polymerized resin repairs.	Transverse Bend impact tests	The transverse strength, and impact resistence of the resin specimens repaired with microwave irradiation were generally superior to specimens repaired by using a water bath curing cycle or the use of an autopolymerising resin.
Yunus et al. (1994)	Acrylic resin repair material bench-cure, hydroflask-cure and microwave irradiation cure.	Repair strength.	On the three methods, polymerization using microwave energy resulted in the strongest repair.
Ilbay et al. (1994)	Twenty-one different polymerization methods were used by varying radiation power and curing time. (3 min at 550 W)	Vickers hardness test transverse load transverse deflection	Resin cured by microwave energy is more resistant to mechanical failure than conventionally cured acrylic.
Del bel Cury et al. (1994)	Acrylic resins: Acron MC (microwave – 500W/3m); Lucitone 550 (water bath – 73°C/90m); Ortho-Class (self polymerized).	Transverse strength and maximum deflection – assay machine in three points (Instron 125) 5mm/m. Impact – Charpy assay.	There were differences among tested materials that can be resulted from composition and polymerization methods.

Table 3. Studies on hardness, transverse strength, flexural strength, shear bond strength, tensile bond strength, impact strength, roughness, and modulus of elasticity of microwaved resin.

5. Base adaptation, dimensional alteration, artificial tooth movement, and teeth occlusion

Considering dimensional alteration of denture or bases resin, when conventional water bath and microwave energy were compared, some authors (Reitz et al., 1985; Levin et al., 1989; Uchida et al., 1989; Takamata et al., 1989; Al-Hanbali et al., 1991; Nelson et al., 1991; Wallace et al., 1991; Sanders et al., 1991; Barbosa et al., 2002; Keenan et al., 2003) found no difference between the two techniques; also, these results are not in agreement with others (Sanders et al., 1991; Nelson et al, 1991). Sanders et al. (1991) observed in their study that microwave polymerization provided a lower degree of artificial tooth movement, while Nelson et al (1991) reported a greater degree of tooth positional changes when microwave polymerization was employed. Although different investing mediums or polymerization techniques, have been compared (Reitzet al., 1985; Levin et al. 1989; Nelson et al., 1991; Turck & Richards, 1992) the authors could not identify studies published concerning the outcomes provided by the combination of different flasking methods and polymerization techniques. Dimensional changes and distortion of the denture due to the investing stone mold and the heating of acrylic resin can promote tooth movement and, consequently alterations in the occlusal contacts and occlusal vertical dimension (OVD). (Rizzatti-Barbosa et al., 2006; Rizzatti-Barbosa et al., 2005). Acrylic resin processing methods do not avoid displacement of artificial teeth during denture inclusion and processing, which might increase occlusal vertical dimension due to production of premature contacts (Barbosa et al., 2002). It is hence necessary to adjust the oclusal surface of artificial teeth, which alters the occlusal anatomy, especially of posterior teeth (Lai et al., 2004). In addition, these alterations may cause mucosal injuries and affect the functionality of the prostheses, thus causing damage to the stomatognathic system, temporomandibular disorders and discomfort to the patient (Yagi et al., 2006). Simultaneous polymerization of maxillary and mandibular complete dentures with the teeth in occlusion by means of a special double flask (DF), has been described as a more rapid and simple method for investing and polymerizing prostheses (Rizzatti-Barbosa et al., 2005). This inclusion technique was claimed to save time and decrease occlusal alteration during denture processing (Meloto et al., 2006). It may be an easier and faster method of investing and polymerizing prostheses. The first designed DF was a metal copper–aluminum flask (DMF) for simultaneous polymerization of both maxillary and mandibular prostheses in a warm water bath (Dental VIPI Ltd, Pirassununga, Brazil). The double polyvinyl chloride flask (DPVCF) (Dental VIPI Ltd, Pirassununga, Brazil) was developed following the same principles for simultaneous processing of both dentures in occlusion through microwave energy heating (Rizzatti-Barbosa et al., 2005). This new technique associating acrylic curing with microwave energy can be considered a clean method that saves time, reduces occlusal interferences, preserves the teeth occlusion, and maintains the OVD. (Rizzatti-Barbosa & Ribeiro-Dasilva, 2009). In the classical literature, some data on the acrylic resins morphology alteration can be found (Huggett et al., 1984; Polyzois et al., 1987; Chen et al., 1988; Jagger, 1996). Some controversial results about microwave processing resin are shown on table 4, where the authors compared microwaving technique with other and related the results of teeth positioning, dimensional alteration, vertical measurement of dentures, etc.

Authors and date	Materials and groups	Methods and measurements	Comments and conclusions
Negreiros et al. (2009)	Clássico (Conventional long cycle in water bath). Onda-Cryl (Microwave energy). QC-20 (Fast cycle in boiling water). Post pressing process.	The linear distances - STM microscope (right premolar to left premolar; right molar to left molar; right incisor to right molar; left incisor to left molar).	Microwave polymerization was similar to that of the conventional cycles in water bath post-pressing time had no relevant effect on tooth movement.

Authors and date	Materials and groups	Methods and measurements	Comments and conclusions
Schibayama et al. (2009)	Dentures processed with: Acrylic resin for water bath polymerization (QC20); Acrylic resin for microwave polymerization (Acron MC). Flasking: (1) adding a second investment layer of gypsum and conventional water bath polymerization, (2) adding a second investment layer of gypsum and polymerization with microwave energy, (3) adding a second investment layer of silicone and conventional polymerization, and (4) adding a second investment layer of silicone and polymerization with microwave energy.	Comparison of the artificial tooth position changes following flasking and polymerization of dentures – linear microscope.	The use of a silicone investment layer when flasking complete dentures resulted in the least positional changes of the artificial teeth regardless of the polymerization technique.
Faot et al. (2008)	Onda Cryl resin Microwave processing: (3 min at 360 W, 4-min. pause, and 3 min at 810 W; and 6 min at 630 W).	Accuracy of fit: 3 points at the right and left ridge crests and at the midline on the posterior palatal seal for each denture base.	Both polymerization cycles are adequate to polymerize the studied denture resin.
Pavan et al. (2005)	Dentures in acrylic resin (Clássico – water bath processing). Storage in water for 30 days. Microwave disinfection: 3 min at 500 W; and 10 min at 604 W.	Dimensional accuracy along the posterior palatal border of maxillary acrylic resin denture bases.	Treatment in microwave oven at 604 W for 10 min produced the greatest discrepancies in the adaptation of maxillary acrylic resin denture bases to the stone casts.
Rizzatti-Barbosa et al. (2005)	Pairs of dentures processed by microwave energy and flasked in occlusion in double flasks.	Occlusal inclination of mandibular and maxillary artificial denture molars.	Dentures double flasking and microwave curing save time, reduce occlusal alteration, and reduce time exposure of dentures.

Authors and date	Materials and groups	Methods and measurements	Comments and conclusions
Lai et al. (2004)	Microwaved resin blocks, processed at 80, 160, 240, and 560 W for 15, 10, 7, and 2 min / 70°C for 9 h.	The morphology of the specimens after staining with osmium tetroxide was examined by transmission electron microscope.	Highly statistical differences in morphology favor of the water-bath method.
Botega et al. (2004)	Retangular samples and half-discs of microwaved-cure acrylic resin (Acron MC). 1, 2, 4, and 6 flasks simultaneous processing. Polymerization time starting in 3min/450W – adjustmentsbased on monomer release.	Monomer release: spectrophotometer Beckman DU-70. Knoop hardness: Shimadzu HMV 2000 hardness tester. Porosity: black inch immersion and count pores in microscope.	The number of flasks does not interfere with polymerization, knoop hardness, and porosity.
Phoenix et al. (2004)	6 commonly used polymethyl methacrylate denture base resins. ADA Specification No. 12. Thermal assessments (differential scanning calorimetry and dynamic mechanical analysis).	Adaptation denture bases.	The microwaveable resins displayed better adaptation than other denture base resins included in this investigation.
Keenan et al. (2003)	Identical maxillary denture bases in: PMMA (Trevalon - Compression flask - hot air oven); microwave polymerized resin (AcronMC Injection flask – 600-W microwave oven/3m); PMMA (Trevalon - Injection flask - hot air oven); resin injection flasks (Microbase Injection flask - 600-W microwave oven/ 6m)	Pre and post treatment intermolar width - traveling microscope. Pre and post treatment vertical dimension of occlusion – points on the superior and inferior members of the articulator.	All injection molding methods produced dentures with a slightly smaller increase in vertical dimension of occlusion. Both microwave polymerization methods produced maxillary complete dentures with a greater reduction in intermolar width.
Ganzarolli et al. (2002)	Heat-cured acrylic resin (Classico); two microwave-cured acrylic resins (Acron MC and Onda Cryl). Water storage.	Adaptation - weight of silicone impression material between the base and the master die.	Interaction of type of material and cooling procedure has effect on the final adaptation. Water storage was not a source of variance on the final adaptation.

Authors and date	Materials and groups	Methods and measurements	Comments and conclusions
Barbosa et al. (2002)	Maxillary dentures polymerization with different cycles by microwave radiation.	Changes in occlusal vertical dimension: average in articulator pin opening	There was no difference between the groups polymerized by the microwave method and the control group
Del Bel Cury et al. (2001)	Two microwave-cured acrylic resins (Acron MC and Onda Cryl). Gypsum moulding technique or silicone gypsum moulding technique.	Residual monomer – 24 or 48 h over a period of 288 h. Knoop hardness – after 24, 48, 72 h and 30 days. Transverse strength – 48 h of water storage.	Storage periods and moulding technique did not influence Knoop hardness. The type of mould did not affect tranverse strength. The acrylic resins differed from each other for all properties, regardless of the type of mould.
Braun et al. (2000)	Samples of acrylic resins (Classico, Lucitone 550, Acron MC). Water bath (long cicle), water bath (short cicle), and microwave energy (500W/3 m). Water sorption – 30 days.	Linear dimensional alteration – pre and post water sorption period (linear microscopy).	All samples presented expansion after water sorption period.
Wong et al. (1999)	Dentures polymerization by 3 processing techniques (dry and wet heat; different rates of cooling).	Dimensional changes - traveling microscope. Water sorption - electronic balance (water uptake).	Water uptake after deflasking was low. The dentures did not reveal differences in shrinkage at water saturation. Oven-processed and water bath processed acrylic resin dentures showed similar dimensional shrinkage at water saturation.
Sadamori et al. (1997)	Acrylic resin dentures. Different thickness. Processing: conventional method, fluid resin technique, and microwave curing method.	Residual monomer – gas liquid chromatography. Dimensional accuracy, and resin stability.	Dimensional accuracy and stability of acrylic resin dentures could be influenced by the processing method. The thickness of the bases, and the shape and size of the dentures influenced on dimensional changes.

Authors and date	Materials and groups	Methods and measurements	Comments and conclusions
Rodrigues Garcia et al. (1996)	Denture bases of conventional acrylic resin (water bath 73°C/ 9 hours) and specific resin for microwave polymerization (500 W/3m). Dentures were relined by addition method.	Accuracy - weight of an impression material put between the denture base and cast die.	Conventional resin cured by water bath or microwave energy showed better adaptation.
Rizzatti-Barbosa et al. (1995)	Dentures in acrylic resin: Acron MC (microwaved – 500W/3m); Lucitone 150 (water bath - 72°C/9h). Water Storage for 30 days.	Posterior palatal fit weight and measurement of impression material between the denture base and master cast.	There were no difference among the resins, polymerization methods and water storage period.
Salim et al. (1992)	Rectangular acrylic resin specimens processed by three methods: a conventional method, the SR-Ivocap system, and a microwave curing method.	Dimensional accuracy - change of the distance vector V (calculated by means of measurements of the distances between fixed points on specimens).	SR-Ivocap system exhibited less dimensional change. SR- Ivocap system might produce more accurate denture base than conventional and microwave curing methods.

Table 4. Studies on base adaptation, dimensional alteration, artificial tooth movement, and teeth occlusion of microwaved resin

6. Studies about the effects of microwave disinfection

Bacterial and yeast plaque on dentures may lead to serious infections, such as systemic candidal infection, particularly in patients who have debilitating diseases (Montagner et al., 2009). The use of microwave energy to disinfect dentures has been suggested to overcome the problems associated with denture cleaning. Microwave energy was introduced in 1985 for sterilization of nonautoclavable dental materials. It was shown that exposed to microwave energy for 10 minutes can kill microorganisms if the denture is attached to a three-dimensional rotating device (Rohler & Bulard, 1985).

Lining materials have been found to be more prone to microbial adhesion than acrylic resin base materials and have been demonstrated to interact with oral microorganisms because of their surface texture and the physical/chemical affinity between the materials. Surface roughness of the resilient liners may differ among materials (Zissis et al., 2000; Jin et al., 2003), and rougher surfaces enhance the adhesion of microorganisms onto resilient lining materials (Bulad et al., 2004) that may allow fungal growth (Brosky et al., 2003). The microorganisms from the plaque on the denture surface may expose patients and dental personnel to infection (Witt et al., 1990).

Denture disinfection has been recommended as an essential procedure for preventing cross-contamination and the maintenance of a healthy oral mucosa. The use of microwave irradiation to disinfect dentures and reliners has been suggested (Burns et al., 1990; Polyzois et al., 1995; Webb et al., 1998; Baysan et al., 1998; Thomas et al., 1995; Baysan et al., 1998; Webb et al., 1998) and stimulated as a disinfection model (Fitzpatrick et al., 1978; Lamb et al., 1983; Rohler et al., 1985; Jeng et al., 1987; Friedrich et al., 1988; Najdovski et al., 1991; Arikan et al., 1995; Atmaca et al., 1996; Lin et al., 1999; Kedjarune et al., 1999; Yeo et al., 1999; Nikawa et al., 2000; Jin et al., 2003; Pavarina et al., 2003; Gonçalves et al., 2006; Setlow, 2006; Gonçalves et al., 2007). Since that, researches have been developed in order to ensure the safe use of microwaving disinfection (Table 5).

Authors and date	Materials and groups	Methods and measurements	Comments and conclusions
Sanita et al. (2008)	Dentures incubation (37 °C / 48 h): (*C.albicans, C.dubliniensis, C. krusei,C. glabrata and C.tropicalis*). Dentures microwaved (650 W/ 3 min).	Microbial growth on the plates.	Microwave irradiation for 3 min at 650 W resulted in sterilisation of all complete dentures.
Machado et al. (2009)	2 hard chairside reline resins (Kooliner, DuraLiner II); 1 heat-polymerizing denture base resin (Lucitone 550). Microwave and chemical disinfection of samples.	Vickers hardness. Roughness measurements – profilometer (diamond stylus tip radius of 2 µm).	Disinfection by microwave irradiation did not adversely affect the hardness of all materials evaluated. Roughness varied among materials and the effect seems to be material dependent.
Novais et al. (2009)	Auto polymerised denture reline materials (Kooliner). Conventional heat polymerized denture base resin (Lucitone 550).	Porosity – after polymerization; after two cycles of microwave disinfection; after seven cycles of microwave disinfection; after 7 days storage in water at 37°C. Number of pores - Scanning electron microscopy at magnification x 100.	Differences in the porosity amongst the materials and for different experimental conditions were observed following microwave disinfection.
Paranhos et al. (2009)	Microwave polymerized acrylic (Onda-Cryl). Immersed in 0.5% NaOCl, and 1% NaOCl.	Color stability – portable colorimeter. Surface roughness – Surftest SJ-201P surface analyzer (resolution of 0.01 µm). Flexural strength – universal testing machine (50 kgf load cell / crosshead speed of 1 mm/ min).	Microwave showed similar results after treatment in all groups.
Dovigo et al. (2009)	70 water bath polymerized complete dentures.	Cultures were interpreted as positive or negative growth after disinfection	Microwave irradiation for 3 minutes at 650 W produced sterilization

Authors and date	Materials and groups	Methods and measurements	Comments and conclusions
	Inoculated – *Staphylococcus aureus* , *Pseudomona aeruginosa, Bacillus subtilis*, and incubated for 24 hours at 37° C. Microwave irradiation at 650 W for 3 minutes.	treatment.	of complete dentures contaminated with *S. aureus* and *P. aeruginosa*. Dentures contaminated with *B. subtilis* can be disinfected by microwave irradiation after 3 and 5 minutes at 650 W.
Montagner et al. (2009)	Microwave-cured acrylic resin specimens contaminated with *Candida albicans* antifungal action of different disinfection agents (2% chlorhexidine solution / 10 min; 0.5% sodium hypochlorite / 10 min; modified sodium hypochlorite / 10 min; effervescent agent / 5 min; hydrogen peroxide 10 v / 30 min).	Culture media turbidity - spectrophotometrically according to the transmittance degree (the higher the transmittance the stronger the antimicrobial action).	Sodium hypochloritebased substances and hydrogen peroxide are more efficient disinfectants against *C. albicans* than 2% chlorhexidine solution and the effervescent agent.
Sartori et al. (2008)	Denture base resin disinfection procedures performed twice: Chemical disinfection (immersion in 100ppm chloride solution for 24h), and Microwave Disinfection (irradiation at 690 W).	Knoop microhardness dimensional stability - weighing a vinyl polysiloxane film reproducing the gap between resin base and master model.	Knoop microhardness was not modified by any disinfection procedure, but decreased over time. Microwaved denture resin bases had gradual increase of distortion over time.
Mima et al. (2008)	Hard chairside reline resin samples inoculation by *Pseudomonas aeruginosa, Staphylococcus aureus, Candida albicans,* and *Bacillus subtilis.* Samples microwaved at 650 W for 1, 2, 3, 4, or 5 minutes before serial dilutions and platings.	Colonies on plates were counted after 48 hours of Incubation.	3 minutes of microwave irradiation can be used for acrylic resin sterilization.

Authors and date	Materials and groups	Methods and measurements	Comments and conclusions
Ribeiro et al. (2008)	Three polymethyl methacrylate (PMMA) resins: a conventional water-bath, heat activated acrylic resin (Lucitone 550), rapid polymerizing acrylic resin (QC-20-QC)), and microwave activated acrylic resin (Acron MC-AC). Two cycles of microwave disinfection (650W for 6 min – once, twice and seven times).	Shear bond strength between denture teeth and acrylic resins having different polymerization cycles - knife-edge shear test in a universal test machine (MTS-810).	The shear bond strength between the denture teeth and the acrylic resins Acron MC and Lucitone 550 was not affected by microwave disinfection. After two cycles of microwave disinfection, the shear bond strength of teeth to QC-20 acrylic resin was increased. Seven cycles of microwave disinfection significantly decreased the shear bond strength between teeth and QC-20 acrylic resin.
Pero et al. (2007)	Two heat-activated denture base resins — conventional (Clássico - water bath); designed for microwave polymerization (Onda-Cryl – manufacturing microwave cycle, short microwave cycle and long microwave cycle). Thicknesses - 2.0, 3.5, and 5.0 mm. Immersion in water.	Porosity.	Microwave polymerization cycles and the specimen thickness of acrylic resin influenced porosity. Porosity differences were not observed in the polymerized resin bases in the water bath cycle for any thickness.
Consani et al. (2007)	Maxillary denture base adaptation (Clássico - 74° C for 9 hours). 2 different flasks: closure methods (traditional flask closure and Restriction System flask closure methods). Simulated disinfection in 150 mL distilled water in a microwave oven at 650 W for 3 minutes.	3 transverse cuts on denture bases: distal of canines, mesial of first molars, and posterior palatal region. Measurements of desadaptation: optical micrometer at 5 points.	Simulated disinfection by microwave energy improved denture base adaptation.

Authors and date	Materials and groups	Methods and measurements	Comments and conclusions
Seo et al. (2007)	Lucitone 550 denture bases (2- and 4-mmthick). Intact and autopolymerizing resin relined. 1 cycle of microwave disinfection (650W/6min); daily microwave disinfection for 7 days.	Dimensional stability – area between 5 removable pins on the standard brass cast measured with a Nikon optical comparator, before and after treatment.	Microwave disinfection produced increased shrinkage of intact specimens and those relined.
Pavan et al. (2007)	Four soft lining materials (Molloplast-B, Ufi Gel P, Eversoft, Mucopren soft). Microwave oven for 3 minutes at 500 W. Disinfectant solutions for 10 minutes.	Hardness - Shore A durometer.	Application disinfection cycles did not change the Shore A hardness values for the materials.
Sartori et al. (2006)	Denture resin bases. Chemical disinfection. Microwave disinfection. Twice with a 7-day interval between them.	Internal adaptation: weighing a vinyl polysiloxane film reproducing the gap between the resin base and the master model.	Microwave disinfection had gradual increase of misfit bases immersed in chloride solution did not differ from the control group.
Silva et al. (2006)	Complete resin dentures. Inoculated by *Candida albicans, Streptoccus aureus, Bacillus subtilis,* and *Pseudomonas Aeruginosa.* Disinfection by microwave irradiation at 650 W for 6 minutes.	Effectiveness of microwave disinfection: count of the number of colony-forming units.	Microwave irradiation for 6 minutes at 650 W produced sterilization and disinfection of complete dentures.
Moura et al. (2006)	Classico (water bath - 9 h at 74°C). Onda Cryl (Microwave energy 3 min at 360 W + 4 min pause + 3 min at 810 W). Infection - (*Candida albicans, Candida tropicalis,Candida dubliniensis, Candida glabrata*).	Roughness – profilometer. Surface free energy – contact angle of a sessile drop of water.	The polymerization method, heat versus microwave, did not influence *Candida* species adherence values. There is no correlation regarding surface free energy, surface roughness and the adhesion of *Candida* species. Heat polymerized acrylic resin showed highest surface free energy values.

Authors and date	Materials and groups	Methods and measurements	Comments and conclusions
Machado et al. (2006)	Cylindrical (30 x 3.9 mm) resin specimens (Luciitone 199). Reline materials packed in resin. Twice microwave irradiation (650 W for 6 minutes).	Torsional test (0.1 Nm/min). Torsional strengths (MPa). Mode of failure.	Microwave disinfection cycles do not decrease the torsional bond strengths between the hard reline resins. Disinfection cycles on reline material may be clinically significant and requires further study.
Machado et al. (2005)	Acrylic resin (Lucitone 199). Resilient lining materials (GC Reline Extra Soft and Dentusil). Samples irradiated twice, with 650 W/6 min; samples irradiated daily for 7 total cycles.	To peel the resilient adhesive - screw tensile tester (Sintec 2/G) (visual analyzed of resin interface). Hadness - Shore A durometer (before and after treatment).	Microwave disinfection did not compromise the hardness of either resilient liners or their adhesion to the denture base resin Lucitone 199. The hardness of the Lucitone 550 denture base resin specimens was not affected by either disinfection method evaluated.
Campanha et al. (2005)	6 brands of artificial teeth. Microwave sterilization at 650W for 6 minutes.	Hardness - Vickers diamond indentator.	Two cycles of microwave sterilization did not affect the hardness of most of the acrylic resin denture teeth tested. Microwave sterilization significantly decreased the hardness of acrylic resin artificial teeth.
Pavarina et al. (2005)	Denture base resin (Lucitone 550). Five hard chairside reline resins (Kooliner, Duraliner II, Tokuso Rebase Fast, Ufi Gel Hard, and New Truliner). Microwave disinfection (650W/6 min): two cycles and seven cycles.	Flexural strength (MPa) – 3 point bend fixture in a MTS machine and loaded until failure.	Microwave disinfection do not adversely affect the flexural strength of all tested materials.
Rodrigues Garcia et al. (2004)	Microwave-cured acrylic resin disc shaped samples. 3 cleansing treatments (polident, manipulation pharmacy cleanser, and water).	Hardness. Roughness.	Manipulated cleanser containing sodium perborate increased surface roughness and hardness.

Authors and date	Materials and groups	Methods and measurements	Comments and conclusions
Seo et al. (2007)	2- and 4-mm-thick denture bases (Lucitone 550). Reline with 2 mm of autopolymerizing resin (Tokuso Rebase Fast, Ufi Gel Hard, Kooliner, or New Truliner).	Dimensional stability – 5 removable pins on the standard brass cast – area (mm) formed by the distance between 5 pins (Nikon optical comparator).	Microwave disinfection produced increased shrinkage of intact specimens and those relined with New Truliner and Kooliner.
Neppelenbroek et al. (2003)	Treatment of 15 denture patients with *Candida*- related denture stomatitis. Upper denture microwaved (650 W/6 min) three times per week for 30 days; conjunction with topical application of miconazole three times per day for 30 days; antifungal therapy only.	Cytological smears and mycological cultures – after and before treatment (days 15 and 30 follow-up).	Microwaving dentures was effective for the treatment of denture stomatitis.
Banting et al. (2001)	Thirty-four subjects with a positive test for *C. albicans pseudohyphae*. Subjects in the microwave treatment group – maxillary denture microwaved (850W/1m). Procedure repeated three times. Standard denture soak treatment – liquid disinfect the dentures (.2% chlorhexidine digluconate solution overnight for 14 days).	Infestation of the tissue surface of the maxillary denture; cytological smears.	Patients whose dentures were microwaved have delayed dramatically reinfestation of the denture surface and infection of the adjacent soft tissue. Microwave treatment is not recommended for all dentures and should be used with caution.
Dixon et al. (1999)	Denture base soft liners and heatpolymerized Acrylic resin denture base material. Inoculation with C. *albicans*. Irradiation in a 60 Hz microwave oven for	Efficacy of microwave irradiation against C. *albicans*. Effect of irradiation on the materials hardness. C. *Albicans* growth assessed with streaked blood agar plates and thioglycollate broth.	Five-minute irradiation, while immersed in water, killed all C. *albicans* present on the materials tested; repeated 5-minute irradiation significantly affected the hardness of only the PermaSoft.

Authors and date	Materials and groups	Methods and measurements	Comments and conclusions
	5minutes (10 - and 15-minute irradiation; repeated 5-minute irradiation cycles).	Shore A hardness material.	

Table 5. Studies about the effects of microwave disinfection

7. Conclusion

The use of microwave energy in the processing of acrylic resin is based on both, classic and recent studies. The observed differences when using microwave or water-bath curing usually are not clinically significant where mechanical properties of microwave and water-bath cured resins are not significant in the resins properties. The frequency and size of porosity in thick specimens could be reduced to 30% by a longer polymerization time at a lower wattage. Microwave curing as a rule, has little effect on the properties of resins when the choice of a suitable power and polymerization time are adequate, reducing porosity or dimensional alteration to a minimum level.

Because it offers some important physical properties as good as conventional processing, along with the advantage of being a quicker and easier method, it should also be considered in processing removable partial dentures or complete dentures, and as a disinfection method of resin prosthesis.

8. References

Al Doori, D.; Huggett, R. & Bates, J.F. (1988). A comparison of denture base acrylic resins polymerized by microwave irradiation and by conventional water bath curing systems. *Dental Materials*, Vol.4, No.1, 25-32, ISSN 0109-5641

Al-Hanbali, E.; Kalleway, J.P. & Howlett, J.A. (1991). Acrylic denture distortion following double processing with microwave or heat. *Journal of Dentistry*, Vol.91, No.3, (Jun) 176-180, ISSN 0300-5712

Alkhatib, M.B.; Goodacre, C.J.; Swartz, M.L.; Munoz-Viveros, C.A. & Andres, C.J. (1990). Comparison of microwave-polymerized denture base resins. *The International Journal of Prosthodontics*, Vol.3, No.3, (May-Jun) 249-255, ISSN 0893-2174

Al-Mulla, M.A.S.; Huggett, R.; Brooks, S.C. & Murphyt, W.M. (1988). Some physical and mechanical properties of a visible light-activated material. *Dental Materials*, Vol.4, No.4, (Aug), 197-200, ISSN 0109-5641

Arikan, A.; Kulak, Y. & Kadir, T. (1995). Comparison of different treatment methods for localized and generalized simple denture stomatitis. *Journal of Oral Rehabilitation*, Vol.22, No.5, (May) 365-369, ISSN 0305-182X

Assunção, W.G.; Barão, V.A.R.; Pita, M.S. & Goiato, M.C. (2009). Effect of polymerization methods and thermal cycling on color stability of acrylic resin denture teeth. *Journal of Prosthetic Dentistry*, Vol.102, No.6, (Dec) 385-392, ISSN 0022-3913

Atmaca, S.; Akdag, Z.; Dasdag, S. & Celik, S. (1996). Effect of microwaves on survival of some bacterial strains. *Acta microbiologica et immunologica Hungarica*, Vol.43, No.4, 371–378, ISSN 1217-8950

Austin, A.T. & Basker, R.M. (1982). Residual monomer levels in denture bases. The effects of varying short curing cycles. *British Dental Journal*, Vol.153, No.12, (Dec) 424-426, ISSN 0007-0610

Austin, A.T. & Basker, R.M. (1980). The level of residual monomer in acrylic denture base materials. *British Dental Journal*, Vol.149, No.10, (Nov) 281-286, ISSN 0007-0610

Azzarri, M.J.; Cortizo, M.S. & Alessandrini, J.L. (2003). Effect of the curing conditions on the properties of an acrylic denture base resin microwave-polymerised. *Journal of Dentistry*, Vol.31, 463–468, ISSN 0300-5712

Bafile, M.; Graser, G.N.; Myers, M.L. & Li, E.K. (1991). Porosity in denture resin cured by microwave energy. *Journal of Prosthetic Dentistry*, Vol.66, 269-274, ISSN 0022-3913

Banting, D.W. & Hill, S.A. (2001). Microwave disinfection of dentures for the treatment of oral candidiasis. *Special Care in Dentistry*, Vol.21, No.1, 4-8, ISSN 0275-1879

Barbosa, D.B.; Compagnoni, M.A. & Leles, C.R. (2002). Changes in occlusal vertical dimension in microwave processing of complete dentures. *Brazilian Dental Journal*, Vol.13, No.3, 197-200, ISSN 0103-6440

Bartoloni, J.A.; Murchison, D.F.; Wofford, D.t. & Sarkar, N.K. (2000). Degree of conversion in denture base materials for varied polymerization techniques. *Journal of OralRehabilitation*, Vol.27, 488–493, ISSN 0305-182X

Baysan, A.; Whitley, R. & Wright, P.S. (1998). Use of microwave energy to disinfect a long-term soft lining material contaminated with Candida albicans or Staphylococcus aureus. *Journal of Prosthetic Dentistry*, Vol.79, 454-458, ISSN 0022-3913

Blagojevic, V. & Murphy, V.M. (1999). Microwave polymerization of denture base materials. A comparative study. *Journal of Oral Rehabilitation*, Vol.26, 804–808, ISSN 0305-182X

Botega, D.M.; Machado, T.S.; Mello, J.A.N.; Rodrigues Garcia, R.C.M. & Del Bel Cury, A.A. (2004). Polymerization time for a microwavecured acrylic resin with multiple flasks. *Brazilian Oral Research*, Vol.18, No.1, 23-28, ISSN 1806-8324

Braun, K.O.; Rodrigues Garcia, R.C.M.; Rizzatti-Barbosa, C.M. & Del Bel Cury, A.A. (2000). Linear dimensional change of denture base resins cured by microwave activation. *Brazilian Oral Research*, Vol.14, No.3, 278-282, ISSN 1806-8324

Braun, K.O.; Del Bel Cury, A.A. & Cury J.A. (1988). Use of microwave energy for processing acrylic near metal. *Brazilian Oral Research*, Vol.12, No.2, 173-180, ISSN 1806-8324

Brosky, M.E.; Pesun, I.J.; Morrison, B.; Hodges, J.S.; Lai, J.H. & Liljemark, W. (2003). Clinical evaluation of resilient denture liners. Part 2: Candida count and speciation. *Journal of Prosthodontics*, Vol.12, 162-167, ISSN 1059-941X

Bulad, K.; Taylor, R.L.; Verran, J. & McCord, J.F. (2004). Colonization and penetration of denture soft lining materials by Candida albicans. *Dental Materials*, Vol.20, 167-175, ISSN 0109-5641

Burns, D.R.; Kazanoglu, A.; Moon, P.C. & Gunsolley, J.C. (1990). Dimensional stability of acrylic resin materials after microwave sterilization. *International Journal of Prosthodontics*, Vol.3, 489-493, ISSN 0893-2174

Campanha, N.H.; Pavarina, A.C.; Vergani, C.E. & Machado, A.L. (2005). Effect of microwave sterilization and water storage on the Vickers hardness of acrylic resin denture teeth. *Journal of Prosthetic Dentistry*, Vol.93, 483-487, ISSN 0022-3913

Caycik, S. & Jagger, R.G. (1992). The effect of cross-linking chain length on mechanical properties of a dough-molded poly(methylmethacrylate) resin. *Dental Materials*, Vol.8, 153-157, ISSN 0109-5641

Chen, M.; Siochi, E.J.; Ward, T.C. & McGrath, J.E. (1993). Basic ideas of microwave processing of polymers. *Polymer Engineering & Science*, Vol.33, 1092-1109, ISSN 0032-3888

Chen, J.C.; Lacefield, W.R. & Castleberry, D.J. (1988) Effect of denture thickness and curing cycle on the dimensional stability of acrylic resin denture bases. *Dental Materials*, Vol.4, No.1, (Feb) 20-24, ISSN 0109-5641

Compagnoni, M.A.; Barbosa, D.B.; de Souza, R.F. & Pero, A.C. (2004). The effect of polymerization cycles on porosity of microwave-processed denture base resin. *Journal of Prosthetic Dentistry*, Vol.91, 281–285, ISSN 0022-3913

Consani, R.L.X.; Mesquita, M.F.; Nobilo, M.A.A. & Henriques, G.E. (2007). Influence of simulated microwave disinfection on complete denture base adaptation using different flask closure methods. *Journal of Prosthetic Dentistry*, Vol.97, 173-178, ISSN 0022-3913

Craig, R.G. (1993). *Restorative Dental Materials*, 9th ed., pp. 514–530, Mosby, ISBN 0801668727, Saint Louis.

Davenport, J.C. (1970. The oral distribution of candida in denture stomatitis. *British Dental Journal*, Vol.129, 151-156, ISSN 0007-0610

De Clerck, J.P. (1987). Microwave polymerization of acrylic resins used in dental prostheses. *Journal of Prosthetic Dentistry*, Vol.57, 650-658, ISSN 0022-3913

Del Bel Cury, A.A.; Rached, R.N. & Ganzarolli, S.M. (2001). Microwave-cured acrylic resins and silicone-gypsum moulding technique. *Journal of Oral Rehabilitation*, Vol.28, 433–438, ISSN 0305-182X

Del Bel Cury, A.A.; Rodrigues Jr, A.L. & Panzeri, H. (1994). Acylic resins polymerized by microwave energy, conventional water bath curing and self-curing system: physical properties. *Revista da Faculdade de Odontologia da Universidade de São Paulo* , Vol.8, No.4, 243-249, ISSN 0103-0663

Dixon, D.L.; Breeding, L.C. & Faler, T.A. (1999). Microwave disinfection of denture base materials colonized with Candida albicans. *Journal of Prosthetic Dentistry*, Vol.81, 207-214, ISSN 0022-3913

Dovigo, L.N.; Pavarina, A.C.; Ribeiro, D.G.; Oliveira, J.A.; Vergani, C.E. & Machado, A.L. (2009). Microwave Disinfection of Complete Dentures Contaminated In Vitro with Selected Bacteria. *Journal of Prosthodontics*, Vol.18, 611–617, ISSN 1059-941X

Faot, F.; Rodrigues Garcia, R.C.M. & Del Bel Cury, A.A. (2008). Fractographic analysis, accuracy of fit and impact strength of acrylic resin. *Brazilian Oral Research*, Vol.22, No.4, 334-339, ISSN 1806-8324

Faraj, S.A. & Ellis, B. (1979). The effect of processing temperatures on the exotherm, porosity and properties of acrylic denture base. *British Dental Journal*, Vol.147, 209–212, ISSN 0007-0610

Fitzpatrick, J.A.; Kwao-Paul, J. & Massey, J. (1978). Sterilization of bacteria by means of microwave heating. *Journal of Clinical Engineering*, Vol.3, No.1, 44-47, ISSN 0363-8855

Frangou, M.J. & Polyzois, G.L. (1993). Effect of microwave polymerisation on indentation creep, recovery and hardness of acrylic denture base materials. *The European Journal of Prosthodontics and Restorative Dentistry*, Vol.1, 111-115, ISSN 0965-7452

Friedrich Jr, E.G. & Phillips, L.E. (1988). Microwave sterilization of Candida on underwear fabric. A preliminary report. *The Journal of Reproductive Medicine*, Vol.33, 421–422, ISSN 0024-7758

Ganzarolli, S.M.; Mello, J.A.N.; Shinkai, R.S. & Del Bel Cury, A.A. (2007). Internal Adaptation and Some Physical Properties of Methacrylate-based Denture Base Resins Polymerized by Different Techniques. *Journal of Biomedical Materials Research. Part B, Applied biomaterials*, Vol.82B, 169–173, ISSN 1552-4973

Ganzarolli, S.M.; Rached, R.N.; Rodrigues Garcia, R.C. M. & Del Bel Cury, A.A. (2002). Effect of cooling procedure on final denture base adaptation. *Journal of Oral Rehabilitation*, Vol.29, 787–790, ISSN 0305-182X

Gonçalves, A.R.; Machado, A.L.; Giampaolo, E.T.; Pavarina, A.C. & Vergani, C.E. (2006). Linear dimensional changes of denture base and hard chair-side reline resins after disinfection. *Journal of Applied Polymer Science*, Vol.102, No.2, 1821-1825, ISSN 0271-9460

Gonçalves, A.R.; Machado, A.L.; Vergani, C.E.; Giampaolo, E.T. & Pavarina, A.C. (2007). Effect of disinfection on adhesion of reline polymers. *Journal of Adhesion*, Vol.83, 139–150, ISSN 0021-8464

Gourdinne, A.; Maassarani, A.; Monchaux, P.; Aussudre, S. & Thourel, L. (1979). Cross-linking of thermosetting resins by microwave heating: quantitative approach. *Polymer Preprints*, Vol.20, 471-474, ISSN 0032-3934

Harman, I.E. (1949). Effects of time and temperature on polymerization of a methacrylate resin denture base. *Journal of the American Dental Association*, Vol.38, 188, ISSN 0002-8177

Hayden, W.J. (1986). Transverse strength of microwave-cured denture baseplates. *General Dentistry*, Vol.34, 367–371, ISSN 0363-6771

Hogan, P.F. & Mori, T. (1990). Development of a method of continuous temperature measurement for microwave denture processing. *Dental Materials Journal*, Vol.9, 1-11, ISSN 0287-4547

Huggett, R.; Brooks, S.C. & Bates, J.F. (1984). The effect of different curing cycles on levels of residual monomer in acrylic resin denture base materials. *Quintessence of Dental Technology*, Vol.8, 365-371, ISSN 0362-0913

Ilbay, S.G.; Guöener, S. & Alkumru, H.N. (1994). Processing dentures using a microwave technique. *Journal of Oral Rehabilitation*, Vol.21, 103-109, ISSN 0305-182X

Jacob, J.; Chia, L.H.L. & Boey, F.Y.C. (1997). Microwave polymerization of poly(methylacrylate): conversion studies at variable power. *Journal of Applied Polymer Science*, Vol.63, 787-797, ISSN 0271-9460

Jagger, R.G. (1996). Dimensional accuracy of thermoformed polymethyl methacrylate. *Journal of Prosthetic Dentistry*, Vol.76, No.6, (Dec) 573-575, ISSN 0022-3913

Jeng, D.K.; Kaczmarek, K.A.; Woodworth, A.G. & Balasky G. (1987). Mechanism of microwave sterilization in the dry state. *Applied and Environmental Microbiology*, Vol.53, No.9, (Sept) 2133-2137, ISSN 0099-2240

Jerolimov, V.; Brooks, S.C.; Huggett, R. & Bates, J.F. (1989). Rapid curing of acrylic denture-base materials. *Dental Materials*, Vol.5, 18-22, ISSN 0109-5641

Jin, C.; Nikawa, H.; Makihira, S.; Hamada, T.; Furukawa, M. & Murata, H. (2003). Changes in surface roughness and colour stability of soft denture lining materials caused by denture cleansers. *Journal of Oral Rehabilitation*, Vol.30, 125-130, ISSN 0305-182X

Kedjarune, U.; Charoenworaluk, N. & Koontongkaew, S. (1999). Release of methyl methacrylate from heat-cured and autopolymerized resins: cytotoxicity testing related to residual monomer. *Australian Dental Journal*, Vol.44, 25-30, ISSN 0045-0421

Keenan, P.L.J.; Radford, D.R. & Clark, R.K.F. (2003). Dimensional change in complete dentures fabricated by injection molding and microwave processing. *Journal of Prosthetic Dentistry*, Vol.89, 37-44, ISSN 0022-3913

Keller, J.C. & Lautenschlager, E.P. (1985). Porosity reduction and its associated effect on the diametral tensile strength of activated acrylic resins. *Journal of Prosthetic Dentistry*, Vol.53, 374-379, ISSN 0022-3913

Kimura, H.; Teraoka, F.; Ohnishi, H.; Saito, T. & Yato, M. (1983). Applications of microwave for dental technique, Part 1: Dough-forming and curing of acrylic resins. *The Journal of Osaka University Dental School*, Vol.23, 43–49, ISSN 0473-4599

Kimura, H.; Teraoka, F. & Saito, T. (1984). Applications of microwave for dental technique (Part 2): adaptability of cured acrylic resins. *The Journal of Osaka University Dental School*, Vol.23, 43-49, ISSN 0473-4599

Kimura, H.; Teraoka, F. & Sugita, M. (1987). Application of microwave for dental technique, Part 3: Development of model materials for microwave polymerization. *The Journal of Osaka University Dental School*, Vol.27, 41–50, ISSN 0473-4599

Lai, C.P.; Tsai, M.H.; Chen, M.; Chang, H.S. & Tay, H.H. (2004). Morphology and properties of denture acrylic resins cured by microwave energy and conventional water bath. *Dental Materials*, Vol.20, No.2, (Feb) 133-141, ISSN 0109-5641

Lamb, D.J.; Ellis, B. & Priestley, D. (1983). The effects of process variables on levels of residual monomer in autopolymerizing dental acrylic resin. *Journal of Dentistry*, Vol.11, 80-88, ISSN 0300-5712

Lbay, S.G.; Gu Vener, S. & Alkumru, H.N. (1994). Processing dentures using a microwave technique. *Journal of Oral Rehabilitation*, Vol.21, 103-109, ISSN 0305-182X

Levin, B.; Sanders, J.L. & Reitz, P.V. (1989). The use of microwave for processing acrylic resins. *Journal of Prosthetic Dentistry*, Vol.61, 381-383, ISSN 0022-3913

Lin, J.J.; Cameron, S.M.; Runyan, D.A. & Craft, D.W. (1999). Disinfection of denture base acrylic resin. *Journal of Prosthetic Dentistry*, Vol.81, No.2, (Feb) 202-206, ISSN 0022-3913

Machado, A.L.; Breeding, L.C. & Puckett, A.D. (2005). Effect of microwave disinfection on the hardness and adhesion of two resilient liners. *Journal of Prosthetic Dentistry*, Vol.94, 183-189, ISSN 0022-3913

Machado, A.L.; Breeding, L.C. & Puckett, A.D. (2006). Effect of microwave disinfection procedures on torsional bondstrengths of two hard chairside denture reline materials. *Journal of Prosthodontics*, Vol.15, 337–344, ISSN 1059-941X

Machado, A.L.; Breeding, L.C.; Vergani, C.E. & Perez, L.E.C. (2009). Hardness and surface roughness of reline and denture base acrylic resins after repeated disinfection procedures. *Journal of Prosthetic Dentistry*, Vol.102, 115-122, ISSN 0022-3913

Machado, C.; Rizzatti-Barbosa, C.M.; Gabriotti, M.; Joia, F.; Ribeiro, M.C. & Sous, R.L.S. (2004). Influence of mechanical and chemical polishing in the solubility of acrylic resins polymerized by microwave irradiation and conventional water bath. *Dental Materials*, Vol.20, 565–569, ISSN 0109-5641

May, K.B.; Razzoog, M.E.; Koran, A. 3rd & Robinson, E. (1992). Denture base resins: comparison study of color stability. *Journal of Prosthetic Dentistry*, Vol.68, No.1, (Jul) 78-82, ISSN 0022-3913

May, K.B.; Shotwell, J.R.; Koran, A. & Wang, R.F. (1996). Color stability: denture base resins processed with the microwave method. *Journal of Prosthetic Dentistry*, Vol.76, 581-589, ISSN 0022-3913

McCabe, J.F.& Basker, R.M. (1976) Tissue sensitivity to acrylic resin. *British Dental Journal*, Vol.140, No.10, (May) 347-350, ISSN 0007-0610

Meloto, C.B.; Silva-Concílio, L.R.; Machado, C.; Ribeiro, M.C.; Joia, F.A. & Rizzatti-Barbosa, C.M. (2006). Water Sorption of Heat-Polymerized Acrylic Resins Processed in Mono and Bimaxillary Flasks. *Brazilian Dental Journal*, Vol.17, No.2, 122-125, ISSN 0103-6440

Memon, M.S.; Yunus, N. & Razak, A.A. (2001). Some mechanical properties of a highly cross-linked, microwave-polymerised, injection molded denture base polymer. *The International Journal of Prosthodontics*, Vol.14, 214–218, ISSN 0893-2174

Mima, E.G.O; Pavarina, A.C.; Neppelenbroek, K.H.; Vergani, C.E.; Spolidorio, D.M. & Machado, A.L. (2008). Effect of different exposure times on microwave irradiation on the disinfection of a hard chairside reline resin. *Journal of Prosthodontics*, Vol.17, No.4, (Jun) 312-317, ISSN 1059-941X

Montagner, H.; Montagner, F.; Braun, K.O.; Peres, P.E. & Gomes, B.P. (2009). In vitro antifungal action of different substances over microwaved-cured acrylic resins. *Journal of Applied Oral Science*, Vol.17, 432-435, ISSN 1678-7757

Moura, J.; Silva, W.J.; Pereira, T.; Del Bel Cury, A. & Rodrigues Garcia, R.C.M. (2006). Influence of acrylic resin polymerization methods and saliva on the adherence of four Candida species. *Journal of Prosthetic Dentistry*, Vol.96, 205-211, ISSN 0022-3913

Najdovski, L.; Dragas, A.Z. & Kotnik, V. (1991). The killing activity of microwaves on some nonsporogenic and sporogenic medically important bacterial strains. *The Journal of Hospital Infection*, Vol.19, 239-247, ISSN 0195-6701

Negreiros, W.A.; Consani, R.L.X.; Verde, M.A.R.L.; Silva, A.M. & Pinto, L.P. (2009). The role of polymerization cycle and post-pressing time on tooth movement in complete dentures. *Brazilian Oral Research*, Vol.23, No.4, (Oct-Dec) 467-472, ISSN 1806-8324

Nelson, M.W.; Kotwal, K.R. & Sevedge, S.R. (1991). Changes in vertical dimension of occlusion in conventional and microwave processing of complete dentures. *Journal of Prosthetic Dentistry*, Vol.65, 306-308, ISSN 0022-3913

Neppelenbroek, K.H.; Pavarina, A.C.; Spolidorio, D.M.; Vergani, C.E.; Mima, E.G. & Machado, A.L. (2003). Effectiveness of microwave sterilization on three hard chairside reline resins. *The International Journal of Prosthodontics*, Vol.16, 616-620, ISSN 0893-2174

Nikawa, H.; Jin, C.; Hamada, T.; Makihira, S.; Kumagai, H. & Murata, H. (2000). Interactions between thermal cycled resilient denture lining materials, salivary and serum pellicles and Candida albicans in vitro. Part II. Effects on fungal colonization. *Journal of Oral Rehabilitation*, Vol.27, 124-130, ISSN 0305-182X

Nishii, M. (1968). Studies on the curing of denture base resins with microwave irradiation: with particular reference to heat-curing resins. *The Journal of Osaka University Dental School*, Vol.2, 23-40, ISSN 0473-4599

Novais, P.M.M.R.; Giampaolo, E.T.; Vergani, C.E.; Machado, A.L.; Pavarina, A.C. & Jorge, J.H. (2009). The occurrence of porosity in reline acrylic resins. Effect of microwave disinfection. *Gerodontology*, Vol.26, 65–71, ISSN 0734-0664

Oliveira, V.M.; Leon, B.L.; Del Bel Cury, A.A. & Consani, S. (2003). Influence of number and position of flasks in the monomer release, Knoop hardness and porosity of a microwave-cured acrylic resin. *Journal of Oral Rehabilitation*, Vol.30, 1104–1108, ISSN 0305-182X

Paranhos, H.F.O.; Davi, L.R.; Peracini, A.; Soares, R.B.; Lovato, C.H.S. & Souza, R.F. (2009). Comparison of Physical and Mechanical Properties of Microwave-Polymerized Acrylic Resin after Disinfection in Sodium Hypochlorite Solutions. *Brazilian Dental Journal*, Vol.20, No.4, 331-335, ISSN 0103-6440

Pavan, S.; Arioli Filho, J.N.; Dos Santos, P.H. & Mollo, F.A. Jr (2005). Effect of microwave treatments on dimensional accuracy of maxillary acrylic resin denture base. *Brazilian Dental Journal*, Vol.16, 119- 123, ISSN 0103-6440

Pavan, S.; Arioli Filho, J.N.; Santos, P.H.; Nogueira, S.S. & Batista, A.U.D. (2007). Effect of Disinfection Treatments on the Hardness of Soft Denture Liner Materials. *Journal of Prosthodontics*, Vol.20, 101-106, ISSN 1059-941X

Pavarina, A.C.; Neppelenbroek, K.H.; Guinesi, A.S.; Vergani, C.E.; Machado, A.L. & Giampaolo, E.T. (2005). Effect of microwave disinfection on the flexural strength of hard chairside reline resins. *Journal of Dentistry*, Vol.33, 741-748, ISSN 0300-5712

Pavarina, A.C.; Pizzolitto, A.C.; Machado, A.L.; Vergani, C.E. & Giampaolo, E.T. (2003). An infection control protocol: effectiveness of immersion solutions to reduce the microbial growth on dental prostheses. *Journal of Oral Rehabilitation*, Vol.30, 532-536, ISSN 0305-182X

Pero, A.C.; Barbosa, D.B.; Marra, J.; Ruvolo-Filho, A.C. & Compagnoni, M.A. (2007). Influence of Microwave Polymerization Method and Thickness on Porosity of Acrylic Resin. *Journal of Prosthodontics*, Vol.17, No.2, 125-129, ISSN 1059-941X

Peyton, F.A. (1950). Packing and processing denture base resins. *Journal of the American Dental Association*, Vol.40, 521-526, ISSN 0002-8177

Phoenix, R.D.; Mansueto, M.A.; Ackerman, N.A. & Jones, R.E. (2004). Evaluation of mechanical and thermal properties of commonly used denture base resins. *Journal of Prosthodontics*, Vol.13, 17–27, ISSN 1059-941X

Polyzois, G.L.; Karkazis, H.C.; Zissis, A.J. & Demetriou, P.P. (1987). Dimensional stability of denture processed in boilable acrylic resins: a comparative study. *Journal of Prosthetic Dentistry*, Vol.57, No.5, 639-647, ISSN 0022-3913

Polyzois, G.L.; Tarantili, P.A.; Frangou, M.J. & Andreopoulos, A.G. (2001). Fracture force, deflection at fracture, and toughness of repaired denture resin subjected to microwave polymerization or reinforced with wire or glass fiber. *Journal of Prosthetic Dentistry*, Vol.86, No.6, (Dec) 613-619, ISSN 0022-3913

Polyzois, G.L.; Yannikakis, S.A. & Zissis, A.J. (1999). Color stability of visible light-cured, hard direct denture reliners: an in vitro investigation. *The International Journal of Prosthodontics*, Vol.12, 140-146, ISSN 0893-2174

Polyzois, G.L.; Zissis, A.J. & Yannikakis, S.A. (1995). The effect of glutaraldehyde and microwave disinfection on some properties of acrylic denture resin. *The International Journal of Prosthodontics*, Vol.8, 150-154, ISSN 0893-2174

Polyzois, G.L., Handley, R.W. & Stafford, G.D. (1995). Repair strength of denture base resins using various methods. *European Journal of Prosthodontics and Restorative Dentistry*, Vol.3, No.4, (Jun) 183-186, ISSN 0965-7452

Rached, R.N. & Del Bel Cury, A.A. (2001). Heat-cured acrylic resin repaired with microwave-cured one: Bond strength and surface texture. *Journal of Oral Rehabilitation*, Vol.28, 370–375, ISSN 0305-182X

Reitz, P.V.; Sanders, J.L. & Levin, B. (1985). The curing of denture acrylic resins by microwave energy. Physical properties. *Quintessence International*, Vol.16, 547-551, ISSN 0033-6572

Ribeiro, D.G.; Pavarina, A.C.; Machado, A.L.; Giampaolo, E.T. & Vergani, C.E. (2008). Flexural strength and hardness of reline and denture base acrylic resins after different exposure times of microwave disinfection. *Quintessence International*, Vol.39, No.10, (Nov) 833-840, ISSN 0305-182X

Rizzatti-Barbosa CM, Del Bel Cury AA & Panzeri H. (1995). Influence of water sorption and microwave polymerization in the in adaptability of dentures. *Revista deOdontologia da Universidade Sao Paulo* Vol.9, 197-206, ISSN 0103-0663

Rizzatti-Barbosa, C.M.; Machado, C.; Jóia, F.A. & Sousa, R.L.S. (2005). A method to reduce tooth movement of complete dentures during microwave irradiation processing. *Journal of Prosthetic Dentistry*, Vol.94, No.3, 301-302, ISSN 0022-3913

Rizzatti-Barbosa, C.M. & Ribeiro da Silva, M.C. (2009). Influence of Double Flask Investing and Microwave Heating on the Superficial Porosity, Surface Roughness, and Knoop Hardness of Acrylic Resin. *Journal of Prosthodontics*, Vol.18, 503–506, ISSN 1059-941X

Rodrigues Garcia, R.C.M; Souza Jr, J.A.; Rached, R.N. & Del Bel Cury, A.A. (2004). Effect of denture cleansers on the surface roughness and hardness of a microwave-cured acrylic resin and dental alloys. *Journal of Prosthodontics*, Vol.13, 173-178, ISSN 1059-941X

Rodrigues Garcia, R.C.M. & Del Bel Cury, A.A. (1996). Accuracy and porosity of denture bases submitted to two polymerization cycles. *Indian Journal of Dental Research*, Vol.7, 122–126, ISSN 0970-9290

Rohler, D.M. & Bulard, R.A. (1985). Microwave sterilization. *Journal of the American Dental Association*, Vol.110, 194-198, ISSN 0002-8177

Sadamori, S.; Ganefiyanti, T.; Hamada, T. & Arima, T. (1994). Influence of thickness and location on the residual monomer content of denture base cured by three processing methods. *Journal of Prosthetic Dentistry*, Vol.72, No.1, (Jul) 19-22, ISSN 0022-3913

Sadamori, S., Ishii,T. & Hamada, T. (1997). Influence of thickness on the linear dimensional change, warpage, and water uptake of a denture base resin. *The International Journal of Prosthodontics*, Vol.10, No.1, 35-43, ISSN 0893-2174

Salim, S.; Sadamori, S. & Hamada, T. (1992). The dimensional accuracy of rectangular acrylic resin specimens cured by three denture base processing methods. *Journal of Prosthetic Dentistry*, Vol.67, 879-881, ISSN 0022-3913

Sanders, J.L.; Levin, B. & Reitz, P.V. (1991). Comparison of the adaptation of acrylic resin cured by microwave energy and conventional water bath. *Quintessence International*, Vol.22, 181–186, ISSN 0305-182X

Sanders, J.L.; Levin, B. & Reitz, P.V.(1987). Porosity in denture acrylic resins cured by microwave energy. *Quintessence International*, Vol.18, 453-456, ISSN 0305-182X

Sanita, P.V.; Vergani, C.E.; Giampaolo, E.T.; Pavarina, A.C. & Machado, A.L. (2009). Growth of Candida species on complete dentures: effect of microwave disinfection. *Mycoses*, Vol.52, No.2, (Mar) 154-160, ISSN 0933-7407

Sarac, Y.S.; Sarac, D.; Kulunk, T. & Kulunk, S. (2005). The effect of chemical surface treatments of different denture base resins on the shear bond strength of denture repair. *Journal of Prosthetic Dentistry*, Vol.94, 259-266, ISSN 0022-3913

Sartori, E.A.; Schmidt, C.B.; Mota, E.G.; Hirakata, L.M. & Shinkai, R.S. (2008). Cumulative effect of disinfection procedures on the microhardness and tridimensional stability of a poly(methyl methacrylate) denture base resin. *Journal of Biomedical Materials Research. Part B, Applied Biomaterials*, Vol.86B, 360-364, ISSN 1552-4973

Sartori, E.A.; Schmidt, C.B.; Walber, L.F. & Shinkai, R.S. (2006). Effect of microwave disinfection on denture base adaptation and resin surface roughness. *Brazilian Dental Journal*, Vol.17, 195-200, ISSN 0103-6440

Schibayama, R.; Gennari Filho, H.; Mazaro, J.V.Q.; Vedovatto, E. & Assunção, W.G. (2009). Effect of flasking and polymerization techniques on tooth movement in complete denture processing. *Journal of Prosthodontics*, Vol.18, No.3, (Apr) 259–264, ISSN 1059-941X

Schneider, R.L.; Curtis, E.R.; Clancy, J.M.S. (2002). Tensile bond strength of acrylic resin denture teeth to a microwave- or heat-processed denture base. *Journal of Prosthetic Dentistry*, Vol.88, 145-150, ISSN 0022-3913

Seo, R.S.; Vergani, C.E.; Pavarina, A.C.; Compagnoni, M.A. & Machado, A.L. (2007). Influence of microwave disinfection on the dimensional stability of intact and relined acrylic resin denture bases. *Journal of Prosthetic Dentistry*, Vol.98, No.3, (Sept) 216-223, ISSN 0022-3913

Setlow, P. (2006). Spores of Bacillus subtilis: their resistance to and killing by radiation, heat and chemicals. *Journal of Applied Microbiology*, Vol.101, No.3, (Sept) 514-525, ISSN 1364-5072

Shlosberg, S.R.; Goodacre, C.J.; Munoz, C.A.; Moore, B.K. & Schnell, R.J. (1989). Microwave energy polymerization of poly(methyl methacrylate) denture base resin. *The International Journal of Prosthodontics*, Vol.2, 453–458, ISSN 0893-2174

Silva, M.M.; Vergani, C.E.; Giampaolo, E.T.; Neppelenbroek, K.H.; Spolidorio, D.M. & Machado, A.L. (2006). Effectiveness of microwave irradiation on the disinfection of complete dentures. *The International Journal of Prosthodontics*, Vol.19, 288-293, ISSN 0893-2174

Smith, L.T.; Powers, J.M. & Ladd, D. (1992). Mechanical properties of new denture resins polymerized by visible light, heat, and microwave energy. *The International Journal of Prosthodontics*, Vol.5, 315-320, ISSN 0893-2174

Souza Jr, J.A.; Rodrigues-Garcia, R.C.M.; Moura, J.S. & Del Bel Cury, A.A. (2006). Knoop hardness of visible light-polymerized acrylic resins. *Journal of Applied Oral Science*, Vol.14, No.3, 208-212, ISSN 1678-7757

Stafford, G.D. & Handley, R.W. (1975). Transverse bend testing of denture base polymers. *Journal of Dentistry*, Vol.3, No.6, 251-255, ISSN 0300-5712

Stafford, G.D. & Huggett, R. (1978). Creep and hardness testing of some denture base polymers. *Journal of Prosthetic Dentistry*, Vol.39, No.6, 682-687, ISSN 0022-3913

Sweeney, W.T.; Paffenbarger, G.C. & Beall, J.R. (1942). Acrylic resins for dentures. *Journal of the American Dental Association*, Vol.29, 7-33, ISSN 0002-8177

Tacir, I.H.; Kama, J.D.; Zortuk, M. & Eskimez, S. (2006). Flexural properties of glass fibre reinforced acrylic resin polymers. *Australian Dental Journal*, Vol.51, No.1, 52-56, ISSN 0045-0421

Tager, A. (1978). *Physical Chemistry of Polymers*, pp.593-627, Mir Publisher, Moscow

Takahashi, Y.; Chai, J. & Kawaguchi, M. (1998). Effect of water sorption on the resistance to plastic deformation of a denture base material relined with four different denture reline materials. *The International Journal of Prosthodontics*, Vol.11, 49-54, ISSN 0893-2174

Takamata, T.; Setcos, J.C.; Phillips, R.W. & Boone, M.E. (1989). Adaptation of acrylic resin dentures as influenced by the activation mode of polymerization. *Journal of the American Dental Association*, Vol.119, 271-276, ISSN 0002-8177

Taylor, P.B. (1941). Acrylic resins: their manipulation. *Journal of the American Dental Association*, Vol.28, 373-387, ISSN 0002-8177

Thomas, C.J. & Webb, B.C. (1995). Microwaving of acrylic resin dentures. *The European Journal of Prosthodontics and Restorative Dentistry*, Vol.3, No.4, 179-182, ISSN 0965-7452

Truong, V.T. & Thomaz, F.G. (1988). Comparison of denture acrylic resins cured by boiling water and microwave energy. *Australian Dental Journal*, Vol.33, 201-204, ISSN 0045-0421

Uchida, K., Okamoto, F., Ogata, K. & Sato, T. (1989). Dimensional accuracy of microwave-cured denture base resin. *Nippon Hotetsu Shika Gakkai Zasshi*, Vol.33, No.1, 114-118, ISSN 0389-5386

Urabe, H.; Nomura, Y.; Shirai, K.; Yoshioka, M. & Shintani, H. (1999). Influence of polymerization initiator for base monomer on microwave curing of composite resin inlays. *Journal of Oral Rehabilitation*, Vol.26, 442–446, ISSN 0305-182X

Urban, V.M.; Machado, A.L.; Oliveira, R.V.; Vergani, C.E.; Pavarina, A.C. & Cass, Q.B. (2007). Residual monomer of reline acrylic resins. Effect of water-bath and microwave postpolymerization treatments. *Dental Materials*, Vol.23, 363-368, ISSN 0109-5641

Vergani, C.E.; Seo, R.S.; Pavarina, A.C. & Reis, J.M.S.N. (2005). Flexural strength of autopolymerizing denture reline resins with microwave postpolymerization treatment. *Journal of Prosthetic Dentistry*, Vol.93, 577-583, ISSN 0022-3913

Wallace, P.W.; Graser, G.N.; Myers, M.L. & Proskin, H.M. (1991). Dimensional accuracy of denture resin cured by microwave energy. *Journal of Prosthetic Dentistry*, Vol.66, 403–408, ISSN 0022-3913

Webb, B.C.; Thomas, C.J.; Harty, D.W. & Willcox, M.D. (1998). Effectiveness of two methods of denture sterilization. *Journal of Oral Rehabilitation*, Vol.25, 416-423, ISSN 0305-182X

Witt S, H.P. (1990). Cross-infection hazards associated with the use of pumice in dental laboratories. *Journal of Dentistry*, Vol.18, 281-283, ISSN 0300-5712

Wolfaardt, J.F.; Cleaton-Jones, P. & Fatti, P. (1986). The occurrence of porosity in a heatcured poly(methyl methacrylate) denture base resin. *Journal of Prosthetic Dentistry*, Vol.55, 393-400, ISSN 0022-3913

Wong, D.M.S.; Cheng, L.Y.Y.; Chow, T.W. & Clark, R.K.F. (1999). Effect of processing method on the dimensional accuracy and water sorption of acrylic resin dentures. *Journal of Prosthetic Dentistry*, Vol.81, 300-304, ISSN 0022-3913

Yagi T, Marimoto T, Hidako O, Iwata K, Masuda Y, Kabayashi M, Takada K. Adjustment of the oclusal vertical dimension in bit raised guinea pig. *Journal of Dental Research*, Vol.82,123-130, ISSN 0022-0345

Yannikakis, S.; Zissis, A.; Polyzois, G. & Andreopoulos, A. (2002). Evaluation of porosity in microwave-processed acrylic resin using a photographic method. *Journal of Prosthetic Dentistry*, Vol.87, 613-619, ISSN 0022-3913

Yeo, C.B.; Watson, I.A.; Stewart-Tull, D.E. & Koh, V.H. (1999). Heat transfer analysis of Staphylococcus aureus on stainless steel with microwave radiation. *Journal of Applied Microbiology*, Vol.87, No.3, 396-401, ISSN 1364-5072

Yunus, N.; Rashid, A.A.; Azmi, L.L. & Abu-Hassan, M.I. (2005). Some flexural properties of a nylon denture base polymer. *Journal of Oral Rehabilitation*, Vol.32, 65–71, ISSN 0305-182X

Yunus, N.; Harrison, A. & Hugget, R. (1994). Effect of microwave irradiation on the flexural strength and residual monomer levels of an acrylic resin repair material. *Journal of Oral Rehabilitation*, Vol.21, 641-648. ISSN 0305-182X

Zissis, A.J.; Polyzois, G.L.; Yannikakis, S.A. & Harrison, A. (2000). Roughness of denture materials: a comparative study. *The International Journal of Prosthodontics*, Vol.13, 136-140, ISSN 0893-2174

20

Ultra-fast Microwave Heating for Large Bandgap Semiconductor Processing

Mulpuri V. Rao
Department of Electrical and Computer Engineering
George Mason University, Fairfax, Virginia 22030,
U.S.A.

1. Introduction

The concept of using microwaves for fast semiconductor processing is known for the last three decades (Spinter et al., 1981; Scovell, 1984; Amada, 1987; Fukano et al., 1985; Zhang et al., 1994; Thompson et al., 2007). Conventional heating methods are slow because in these methods the heat is applied to the surface of the object to be heated, which needs to be transferred from the surface to the interior of the object by thermal conduction, requiring time for heating the whole object. In addition, in conventional heating methods we not only heat the object, but also the surrounding ambient, which slows down both heating and cooling rates. In contrast, microwave heating is fast because microwaves can permeate into objects heating both interior and exterior simultaneously. In addition, the ambient surrounding the object is not heated resulting in ultra-fast heating and cooling rates (Y. Tian & M.Y. Tian, 2009). It is this feature, which makes microwave heating attractive in semiconductor processing. This chapter focuses on the use of microwaves for large bandgap semiconductor material and device processing. More specifically, heating of semiconductors in a microwave cavity resonator and coupling of microwaves to a semiconductor using a microwave head are discussed. There are many applications of using microwave energy for ultra-fast thermal processing of semiconductors, such as impurity diffusion, annealing of ion-implantation generated lattice damage, ohmic contact alloying, nanowire growth etc. However, the primary focus of this chapter is on using microwave heating for annealing lattice damage in ion-implanted SiC and GaN and for growing SiC nanowires. Though microwave annealing can be used for processing any semiconductors, it is much more attractive in case of SiC, where ultra-fast heating (>1000 °C/s) to temperatures as high as 2100 °C is needed for achieving desirable electrical characteristics in ion-implantation doped material. Devices made of hexagonal polytypes of SiC and GaN are highly attractive for high-power, high-frequency and high-temperature electronic and also for optoelectronic applications.

2. Mechanisms of microwave heating

Interaction between electromagnetic field in the microwaves and the target material results in heating of the material. Mechanism and degree of interaction depends on type of the material. Materials with low complex dielectric constant are poor absorbers of microwave energy and are not suitable for microwave heating, whereas materials with high complex dielectric constant (such as SiC) are excellent absorbers of microwave energy and hence are strong candidates for microwave heating. In case of conductors such as metals, microwaves are reflected with absorption taking place only within a skin depth of the metal (which is in the range of a fraction of a micrometer to few micro-meters, depending on the frequency of the microwaves) inducing eddy current and consequently heat therein. Hence, metals are not good for microwave heating in general, unless the thickness of the metal is less than or in the same range as its skin depth. Since, the thickness of metal layers used for contacts in semiconductor device fabrication are in the same order as the skin depth, microwave heating is suitable for contact alloying. Though semiconductors such as Si and GaN are not strong absorbers of microwave energy, the highly doped conductive regions at the surface of the substrates of these materials absorb microwaves reasonably well causing selective heating of the doped regions. Similarly, in case of semiconductor hetero-structures comprising layers of strong and poor absorbers of microwaves, microwave heating can be targeted to specific layers in the hetero-structures (Y. Tian & M.Y. Tian, 2009).

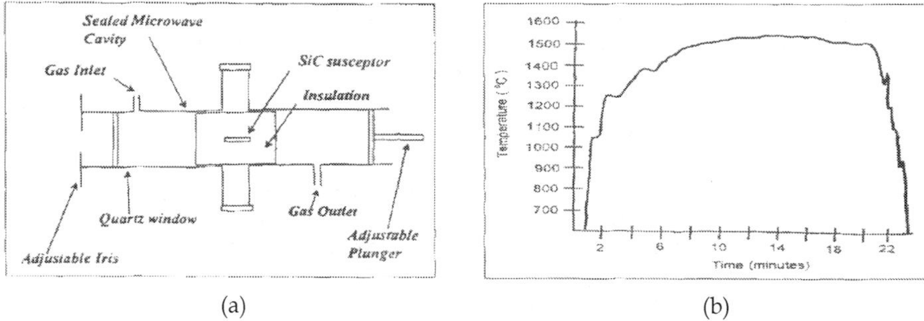

(a)   (b)

Fig. 1. Schematics of (a) microwave cavity annealing system, and (b) Typical temperature/time cycle for 1500 °C/10min microwave cavity annealing on implanted SiC.

The heating rate of microwave thermal process depends on the intensity of microwave electromagnetic (EM) field and the EM energy coupled to the material to be heated. Obviously a high intensity EM field is required for high heating rate. The EM energy can be coupled to the material in three different ways. In capacitive coupling, the energy is coupled to the material via electric (E) field and in inductive coupling, the energy is coupled to the material via magnetic (H) field; where as in cavity coupling the energy is coupled to the material via a combination of E and H fields. Cavity coupling is the most widely used method among microwave heating processes. In this method, the heating furnace is in the form of either a single-mode cavity or a multi-mode cavity operating at a fixed frequency. The single-mode cavity can generate a much higher intensity of EM field than the multi-mode cavity, therefore is more favorable for fast heating processes. Schematic of a typical single TE_{103} mode cavity resonator with a maximum power of 6 kW, used for heating SiC is shown in Fig. 1. Resonator was equipped with three different types of tuning mechanisms: (1) a three-stub tuner consisting of three movable screws positioned in a line on top of a waveguide just behind the iris, (2) a simple wheel device to increase or decrease the size of the iris, and (3) an adjustable back wall that can lengthen or shorten the resonator cavity; which allow the shape of the waveguide and resonator to be adjusted to maintain a resonance condition. This provides the intense electric field necessary to attain very high temperatures with short ramp times. Encapsulated ion-implanted SiC samples or ion-implanted SiC samples encased in an amorphous SiC crucible are then placed inside another enclosure made of an insulator such as MgO to minimize radiative loss during heating. This entire assembly is mounted inside the resonator for heating. The amorphous SiC crucible minimizes sublimation from the uncapped implanted SiC surface by maintaining Si overpressure in the vicinity of implanted SiC sample during high-temperature heating. Further details of the system shown in Fig. 1 can be obtained from (Gardner et al., 1997). For the cavity resonator system described so far the heating ramp rate is 200 °C/min, with the implanted SiC samples encased in an amorphous SiC crucible (see Fig. 1(b)). Heating rates as high as 10-100 °C/s are achievable in single-mode cavity resonators with a reduced thermal mass. But, these rates are not enough for the present day ultra-fast heating rate requirements of Si, GaN and SiC device processing. In addition, there are a number of technical barriers for achieving very high heating rate with a single mode cavity. First, the use of a fixed frequency magnetron as microwave source leads to a mismatch in resonant frequencies between the microwave source and the loaded cavity since the characteristic frequency of the loaded cavity shifts as the temperature changes during the thermal process. Second, the cavity is mechanically tuned, so its response to coupling change is slow, limiting the maximum achievable heating rate. Third problem with the cavity coupling technique is arcing and the breakdown of plasma inside the cavity as the microwave power reaches the threshold level to breakdown the ambient gas in the cavity. Especially in the presence of conductive material, such as a metal, inside the cavity, the electric field is significantly enhanced at the edge of the conductive material, resulting in arcing at a much lower power level than that of the threshold of the ambient gas. The limitation of input power due to the arcing problem also limits the ability of cavity technique in achieving high heating temperatures and ultra-fast heating rates. Fourth limitation with the cavity heating technique is the limitation on the size of the load (semiconductor wafer), which must be smaller than the size of the cavity. This problem gets worse in single mode cavity, where the size of the cavity decreases with increasing operating frequency. These limitations of cavity microwave heating demand an alternative way of coupling microwaves to the materials to be heated, especially for ultra-fast and high-temperature heating applications. In present day nano-dimensional Si integrated circuit applications ultra-fast rapid thermal heating is required and in case of large dimensional SiC and GaN devices, both ultra-fast and high-temperature heating are desired.

3. Need for ultra-fast and/or high-temperature thermal processing of semiconductor devices

Ion-implantation is an indispensable technique for achieving planar selective area doping in fabricating SiC, GaN and ZnO high-power electronic and opto-electronic devices. Other doping methods such as thermal diffusion are impractical for the SiC, GaN and ZnO technologies because the diffusion coefficients of the technologically relevant dopants in these materials is very small (Sadow & Agarwal, 2004; Pearton et al., 2006), for example even at temperatures in excess of 1800 °C for SiC. For the current nano-dimensional Si technology also ion-implantation is the only selective area doping method available due to its precise control over the doping concentration and doping depth. However, ion-implantation process causes damage to the semiconductor crystal lattice and also the as-implanted dopants do not reside in electrically active substitutional sites in the semiconductor lattice (Ghandhi, 1994). Therefore, the ion-implantation step always needs to be followed by a high-temperature annealing step for alleviating the implantation-induced lattice damage and for activating the implanted dopants (i.e. moving them from electrically inactive interstitial sites to the electrically active substitutional lattice sites).

For SiC, the implanted n-type dopants (nitrogen and phosphorus), depending on the implant dose, may require annealing temperatures up to 1900 °C, whereas implanted p-type dopants (aluminum and boron) may require temperatures in excess of 1900 °C (Seshadri et al.' 1998; Heera et al., 2004; Rambach et al., 2008). The higher annealing temperatures required for p-type dopants is a result of the higher activation energy required for forming the substitutional Al_{Si} acceptors compared to the P_{Si} donors. Also, the defects introduced by the Al implantation require higher annealing temperatures to be removed as opposed to the defects introduced by the P and N implantation (Seshadri et al. ,1998). Ultra-high temperature annealing not only helps to remove defects generated by the ion-implantation process but also the pre-existing defects in the material, which helps in obtaining a long ambipolar carrier lifetime in SiC p-i-n diodes (Jenny et al., 2006). Implanted p-type dopants (Mg) in GaN require annealing temperatures in excess of 1300 °C for satisfactorily removing implantation-induced defects, for activating the implanted dopants, and for recovering the luminescence properties (which are severely degraded by the ion-implantation) (Pearton et al., 2006; Feng 2006). A higher annealing temperature requirement for activating p-type implants compared to n-type implants in GaN is primarily due to the much larger formation energy of the substitutional Mg_{Ga} acceptors compared to the Si_{Ga} donors.

Traditionally, post-implantation annealing of SiC is performed in either resistively or inductively heated, high-temperature ceramic furnaces, since temperatures > 1600 °C are required. The furnaces used for annealing SiC have modest (few °C/s) heating and cooling rates, which makes annealing SiC at temperatures > 1500 °C using the traditional furnaces impractical because of an excessive SiC sublimation at such high temperatures when exposed for long durations. This problem can be alleviated to a certain extent by capping the SiC surface with a layer of graphite prior to annealing (Negoro et al., 2004; Vassilevski et al., 2005), but still the maximum conventional annealing temperatures are typically limited to about 1900 °C to preserve the surface integrity of SiC. For efficiently activating implanted p-type dopants in SiC and for removing the implantation-induced lattice defects, annealing temperatures > 1900 °C are required. In addition to the annealing temperature, the ramping rates also have an affect on the defect density and electrical characteristics of ion-implantation doped SiC (Poggi et al., 2006; Ottaviani, 2010). As for GaN, temperatures > 1300 °C are required for completely activating in-situ as well as ion-implanted p-type dopants. This is because of thermally stable defect complexes, which need temperatures as high as 1500 °C for breaking. However, when annealed at temperatures > 700 °C, GaN decomposes into Ga droplets due to nitrogen leaving the surface. This can be minimized by capping GaN with AlN. Annealing of GaN is performed in halogen lamp-based RTA systems, due to the rapid heating/cooling rates accorded by these RTA systems. However, due to their quartz hardware, these halogen lamp based RTA systems are limited to a maximum temperature of 1200 °C, which is not sufficient to effectively anneal p-type GaN. Laser annealing is known to result in high defect density and low implant activation for SiC (Ruksell and Ramirez, 2002; Tanaka et al., 2003), which is not suitable for achieving optimum device performance.

The conventional halogen lamp rapid thermal processing (RTP) technique, widely used in current Si device processing, is not suitable for future reduced size nano-dimensional devices due to its inability to reliably control annealing time to millisecond range (Tian, 2009), which is mandatory for future nm dimension Si devices for tight control of implant profiles. Millisecond range annealing method like laser annealing does not yield acceptable results for commercial production.

Based on what is presented above, raising the annealing temperature and reducing the temperature ramp-up and ramp-down rate are key elements in solving the most critical problems in the post-implantation annealing of both Si and large bandgap materials. Hence, for these two different device technologies, as shown in Fig. 2, the current annealing trend is either faster (for Si nano-devices) or both hotter and faster (for SiC, GaN and ZnO devices) compared to the existing annealing methods. Annealing temperatures as high as 2100 °C and temperature ramp-up and ramp-down rates as high as 2000 °C/s are desirable for these technologies.

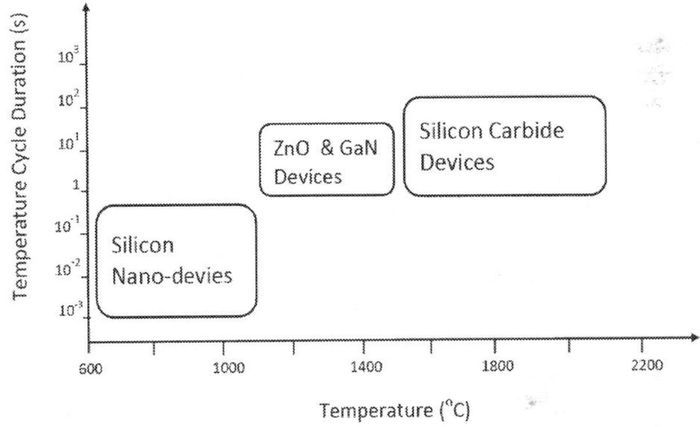

Fig. 2. Typical annealing temperature-time cycle requirements of different semiconductor device technologies.

One annealing method that satisfies the above requirements for both large bandgap and Si device applications is the solid-state microwave annealing (Sundaresan et al., 2007a, 2007b, 2008; Tian, 2009; Y. Tian & M.Y. Tian, 2009; Aluri et al., 2010). In this novel annealing method, microwaves coming from a microwave head are coupled to the sample directly. Microwave annealing provides ultra-fast heating and cooling rates and a good control over the annealing time when the semiconductor wafer is encased in microwave transparent materials. The heating rate is very high because the microwaves are absorbed only by the semiconductor wafer and not by the surroundings. The cooling rate is also high because the ambient surrounding the sample is not heated during the annealing process. Another attractive feature of microwave annealing is the selective heating of the doped region. Microwaves are absorbed more efficiently by the heavily doped region compared to the lightly doped regions in the same sample. Due to this reason, a heavily implantation doped region can be selectively annealed without adversely affecting the doping profile of lightly doped un-implanted region. This is possible because some of the implantation damage is recovered during ramp up transient, even before the sample reaches the annealing temperature, increasing its conductivity to a higher level than the background pre-implant conductivity. The solid-state microwave annealing is different from above explained microwave cavity method used for annealing Si (Spinter et al., 1981; Scovell, 1984; Amada, 1987; Zhang et al., 1994) and SiC (Gardner et al., 1997) , where microwaves in a cavity resonator are coupled to the sample for heating. The microwave cavity resonators require mechanical tuning and operate at a fixed resonant frequency, which compromise their efficiency and suitability for batch processing. The new solid-state microwave annealing method does not involve any cavity and has favorable features such as variable operating frequency, swift electronic tuning and other automation capabilities, which make it suitable for commercial mass semiconductor device production application.

4. Ultra-fast solid-state microwave annealing

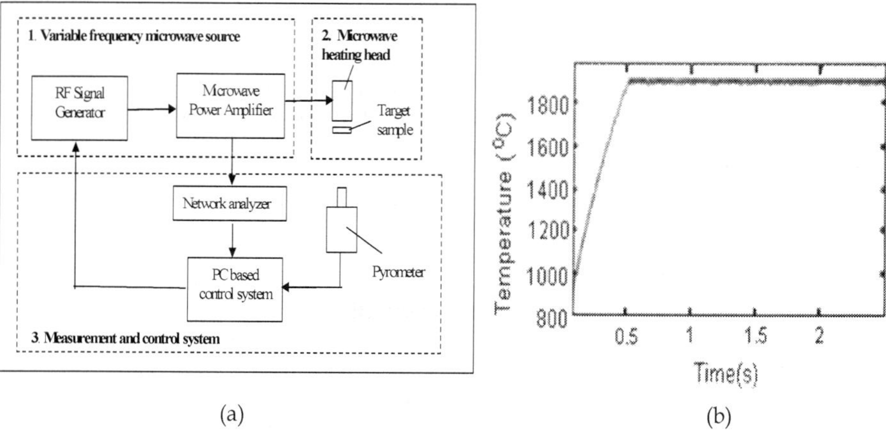

(a)   (b)

Fig. 3. (a) Block diagram (adapted from Tian, 2009) and (b) a typical temperature-time cycle profile, of a solid-state microwave annealing set-up.

Recently there is a rapid advancement of solid state microwave techniques for mobile phone and wireless communication application. Seizing this opportunity, LT Technologies (LTT) has developed a unique solid-state microwave Rapid Thermal Processing(RTP) system. Figure 3(a) shows the block diagram of the LTT's solid-state microwave RTP system. This microwave RTP unit can be divided into three main parts: (1) a variable frequency microwave power source which consists of a RF signal generator and a RF power amplifier, (2) a microwave heating head which is closely placed on the top of the target sample and couples microwave energy to the target, and (3) a measurement and control system which consists of a network analyzer, a PC based control system and an optical pyrometer.   The use of a variable frequency microwave source, instead of a fixed frequency magnetron, allows a broad bandwidth in the selection of a suitable operating frequency. It also provides the flexibility of sweeping the source frequency during the thermal process to compensate for the characteristic frequency shifting due to the temperature change of the specimen. In addition, the variable frequency source has the capability of generating pulsed or modulated microwave power for optimal energy dissipation. This solid state, cavity-less and variable frequency based microwave technology is capable of generating high-power, short-duration, spatially uniform, and material-specific microwave pulses. It offers many novel features and desirable characteristics for rapid thermal processing of compound semiconductors and other advanced materials at very high temperature. These novel features include: (1) very high heating temperature (> 2000 °C), (2)ultra-fast ramping rate (up to 2000 °C/s), (3) automation (a PC based computer control) to regulate the temperature uniformity and stability, and (4) scalability of the system to accommodate technology changes such as device feature size shrinkage and wafer size increase (Y. Tian and M.Y. Tian, 2009; Tian, 2009; Sundaresan et al., 2007b).

For SiC sample heating, the microwave power generated by the variable frequency power source is amplified and then coupled to the SiC sample through the heating head. The sample temperature is monitored by an infrared pyrometer. The SiC and GaN sample emissivities were both measured as 0.8 using a blackbody source and this emissivity value was keyed into the pyrometer for all temperature measurements. The microwave system can be tuned to efficiently heat semiconductor samples at variable frequencies. Since the samples are placed in an enclosure made of microwave transparent, high-temperature stable ceramics such as boron nitride and mullite, microwaves only heat the strong microwave absorbing (electrically conductive) semiconducting films, which present a very low thermal mass in comparison with a conventional furnace where the surroundings of the sample are also heated. Thus, heating rates > 1000 °C/s are possible. In fact, selective heating of thin, highly doped semiconductor layers is possible if the doped layers are formed on semi-insulating or insulating substrates. For efficient microwave annealing of implanted semi-insulating (SI) SiC substrates and GaN epilayers grown on (electrically insulating) sapphire substrates, placing a heavily doped 4H-SiC wafer as a susceptor in close proximity to the sample to be annealed is desired. It is possible to directly couple microwave power and heat GaN epilayers grown on sapphire, without using any susceptor. However, the spatial distribution of temperature across the sample in general is non-uniform. Placing a SiC susceptor sample underneath the GaN sample of interest solves this problem.

A typical temperature profile for microwave heating of a 4H-SiC sample is shown in Fig. 3(b). It can be seen that the high temperature of 1900 °C is reached within 0.5 second and the temperature oscillation is less than 10 °C. For microwave RTP of a large wafer, an array of microwave heating heads is placed on top of the wafer. The number of heating heads can be changed depending on the size of the wafer and the requirement of temperature uniformity across the wafer. Multiple microwave power sources and multiple temperature sensors are required to run and monitor the rapid thermal process effectively. Solid-state microwave heating is advantageous for high-temperature processing of wide bandgap semiconductors such as SiC, GaN and ZnO. The microwave heating system has a capability to reach sample temperatures > 2000 °C (for SiC wafers) with heating and cooling rates in excess of 1000 °C /s.

5. Microwave heating for synthesizing SiC nanowires

Over the past decade, one dimensional (1-D) semiconductor nanostructures, such as nanotubes and nanowires, have attracted special attention due to their high aspect and surface to volume ratios, small radius of curvature of their tips, absence of 3-D growth related defects such as threading dislocations, and fundamentally new electronic properties resulting from quantum confinement. These nanostructures are expected to play a crucial role as building blocks for future nanoscale electronic devices and nano-electro-mechanica-

systems (NEMS), designed using a bottom-up approach. The 1D and quasi-1D nanowires of Si, GaN, ZnO, SiC and other semiconductors have been synthesized (Huang and Lieber, 2004; Law et al., 2004; Andrievski, 2009; Fang et al., 2010). Silicon carbide, due to its wide bandgap, high electric breakdown field, mechanical hardness, and chemical inertness, offers exciting opportunities in fabricating nanoelectronic devices for chemical/biochemical sensing, for high-temperature, for high-frequency and for aggressive environment applications (Fan et al., 2006). Several techniques have been applied to synthesize SiC nanowires using physical evaporation, chemical vapor deposition, laser ablation, and various other techniques (Sundaresan et al., 2007c). A simple and cheaper method to grow SiC nanowires is highly desirable. A sublimation-sandwich method utilizing microwave heating satisfies this requirement.

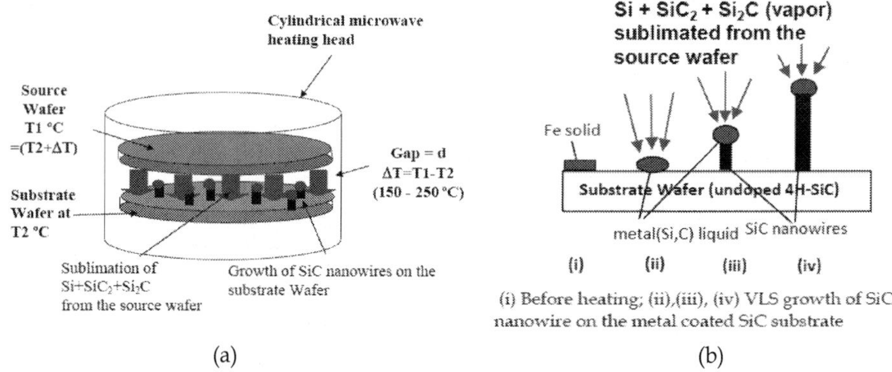

(a)   (b)

Fig. 4. Schematics of (a) sandwich cell used for SiC nanowire growth, and (b) SiC nanowire growth process.

Schematic of a typical "sandwich" cell used for SiC nanowire growth is shown in Fig. 4(a). The 'sandwich cell' in Fig. 4 consists of two parallel 4H-SiC wafers with a very small gap, 'd', between them. The bottom wafer in Fig. 4(a) is semi-insulating SiC, which will be referred to as the 'substrate wafer' hereafter. The inner surface of the substrate wafer is coated with a 5 nm layer of Fe, Ni, Pd, or Pt that acts as a catalyst for the vapor-liquid-solid (VLS) growth of SiC nanowires. The top wafer in Fig. 4(a) is a heavily n-type (nitrogen doped) in-situ doped SiC, which will be referred to as the 'source wafer'. As shown in Fig. 4(a), the microwave heating head is placed around the sandwich cell. Due to the difference in electrical conductivity of the source wafer and the substrate wafer, at a given microwave power, the source wafer temperature is higher than the substrate wafer temperature, resulting in a temperature gradient, ΔT, between the two wafers. When the Si- and C-containing species, such as Si, SiC_2, and Si_2C sublimate from the source wafer at temperatures > 1500 °C, the temperature gradient (ΔT) provides the driving force for transporting these species to the substrate wafer. On the substrate wafer surface, the metal film is either already molten at the growth temperature, or it melts after absorbing the Si species and forms spherical islands to minimize its surface free energy. The Si- and C-containing vapor species are absorbed by these metal islands, converting them into liquid droplets of metal-Si-C alloys. As shown in Fig. 4(b), once this alloy reaches a saturation point for SiC, a precipitation of SiC occurs at the liquid-substrate interface thereby leading to a VLS growth of the SiC nanowires. The nanowires always terminate in hemispherical metal-Si alloy end-caps. While group VIII metals facilitate growth of SiC nanowires, Au is not suitable as a catalyst in this VLS process due to its possible evaporation at the growth temperature.

The sublimation sandwich method described above can reliably grow SiC nanostructures with predictable morphologies, with very high growth rates. Since, the sublimation sandwich method is widely used in industry for growing SiC epilayers and substrates, there exists a vast body of information for controlling the polytype, doping, orientation, etc. of the SiC growth. The sandwich cell described above is a nearly closed system because of the small gap between the source and substrate wafers, which allows precise control of the composition of the vapor phase in the growth cell. At the same time, the system is open to the species exchange between the sandwich growth cell and the surrounding environment in the chamber. By appropriately adjusting the composition of the precursor species in the vapor, this approach has the potential to control the doping levels, or create heterostructures in the growing nanostructures. Another important feature of the sandwich growth cell is its compact size, which significantly reduces the volume of the surrounding chamber. The use of a small chamber not only saves the cost by utilization of small amount of expensive source materials, but also significantly reduces the vacuum pumping cycle time, which is needed for a high throughput fabrication. Yet another novel feature is the dynamic range of temperature ramping rates ($\geq$ 1000 °C/s) that are possible using the microwave heating system. This is another process parameter which can be tweaked to circumvent some thermodynamic restrictions.

6. Organization of rest of the chapter

In the rest of the chapter, results on the application of microwave annealing: to activate ion-implanted dopants in SiC, to activate in-situ doped and ion-implantated dopants in GaN, to grow SiC nanowires, and for other possible usages in device processing are presented. Unless otherwise specified, the phrase 'microwave annealing' hereafter refers to solid-state microwave head based annealing. The Polytype of SiC used for the solid-state microwave head based annealing was 4H. Wherever appropriate results of microwave cavity annealing on ion-implanted 6H-SiC are also presented. Implant activation behavior is similar for both 4H and 6H polytypes of SiC. The GaN results are for layers grown on sapphire substrate.

For brevity complete information on ion-implantation schedules are not provided, but the readers are referred to publications where this information is provided. In case of SiC, except for boron, for all other species, multiple-energy ion-implantations were performed to obtain a uniform implant concentration over a specific thickness from the surface of the wafer. Results are provided on the basis of volumetric concentration of the implant. Surface roughness, structural and electrical properties of: aluminum, boron, nitrogen and phosphorus implanted SiC, and magnesium ion-implantation (or in-situ) doped GaN, subjected to ultra-fast solid-state microwave annealing (or cavity microwave annealing) at temperatures in the range of 1300 °C – 2120 °C are presented.

7. Surface roughness of microwave annealed SiC and GaN

For high-yield device production the surface roughness of the heat-treated semiconductor should be close to the pre-annealed value. In case of conventional furnace annealed

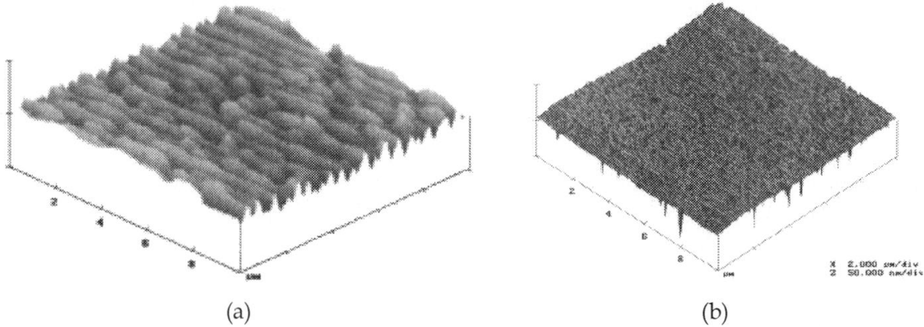

(a)   (b)

Fig. 5. Tapping mode AFM images of: (a) implanted SiC annealed by conventional furnace at 1600 °C for 10 min without cap and (b) implanted GaN annealed by microwaves with AlN cap at 1500 °C for15 s

implanted SiC the surface becomes rough due to continuous long furrows running in one direction across the sample surface (see Fig. 5(a)). These furrows are supposed to be caused by the thermal desorption of species such as Si, SiC_2, Si_2C, etc, a mechanism which is popularly known as 'step bunching' (Capano et al., 1999; Rao et al., 1999; Vathulya and White, 2000; Rambach et al., 2005). The step bunching puts fabrication constraints on SiC devices (Phelps et al., 2002). The step bunching is observed not only on uncapped furnace annealed SiC but also on epitaxial layers grown on off-axis cut SiC wafers. The step bunching related surface roughness can be decreased considerably by chemical mechanical polishing (Kato et al., 2010). For uncapped ion-implanted SiC samples microwave annealed in open air at temperatures $\leq$ 1850 °C, using another SiC piece proximity (in order to suppress Si sublimation during annealing), deep furrows are not formed and the surface roughness measured by tapping mode AFM measurements is $\leq$ 2 nm. The roughness increased quickly for higher annealing temperatures reaching a value of 8.5 nm for annealing at 1975 °C (Sundaresan et al., 2007a). Parabolic oxide growth on SiC sample surface was observed for open air microwave annealing. To prevent oxide growth, the microwave anneals need to be performed in a controlled inert atmosphere of N_2, Ar, or Xe. Arching due to breakdown of gas was observed for all inert gases except for the N_2. Due to this reason, N_2 is the preferred ambient for microwave annealing of SiC and GaN. In case of GaN, N_2 also provides nitrogen over-pressure to minimize nitrogen sublimation from the GaN surface during annealing, though it is not effective at normal pressures. High temperature microwave annealing in N_2 ambient not only prevents oxidation of the SiC surface during annealing but also results in a reduction of surface roughness. For SiC samples annealed in the N_2 ambient, without any deposited encapsulant (but with the SiC proximity), the RMS surface roughness after annealing at 2050 °C for 30 s is only about 2 nm (Sundaresan et al., 2007b). Microwave annealing without any deposited encapsulation resulted in severe sublimation of the implanted sample surface even though the surface roughness is low. The sublimated layer thickness for microwave annealing at 2050 °C for 30 s is about 120 nm, which is a substantial part of the implanted layer thickness. To prevent sublimation of this magnitude, a graphite cap formed by heat-treatment of a deposited photoresist layer is required (Vassilevski et al., 2005; Rao et al., 2010). Surface roughness of graphite cap protected microwave annealed samples is comparable to that of as-implanted/un-annealed samples (Sundaresan et al., 2008).

High-temperature microwave annealing of GaN results in a very rough ($\geq$ 10nm) surface with formation of hexagonal cavities, due to sublimation (Aluri et al., 2010; Sundaresan et al., 2007d). It is well known that nitrogen sublimates from GaN even at annealing temperatures as low as 700 °C, but temperatures as high as 1500 °C are required for activating ion-implanted Mg acceptors in GaN. Protecting GaN from sublimation even with deposited capping layers is not easy. It is quite tempting to use graphite cap, which successfully protected SiC surface during high-temperature microwave annealing, for the GaN also. Unfortunately, the graphite cap is ineffective in protecting GaN surface at annealing temperatures > 1000 °C, presumably because of the stress at the GaN/graphite interface. In general, capping layers having a good lattice matching with the semiconductor layer they are supposed to protect are more effective in withstanding high annealing temperatures. The MgO capping layer (200 nm thick), formed by e-beam evaporation, and having 6.5% lattice mismatch with GaN, could protect GaN only up to a microwave annealing temperature of 1300 °C (Sundaresan et al., 2007d). Pulsed-laser-deposited AlN(600 nm thick), having a 2.6% lattice mismatch with GaN, protects GaN layers reasonably well up to microwave annealing temperatures as high as 1500 °C (Sundaresan et al., 20007; Aluri et al., 2010). Surface roughness of AlN capped GaN is comparable to that of a virgin sample for 5s annealing but increases with increasing annealing time at 1500 °C (see Fig. 5(b) for 15 s anneal at 1500 °C). For additional protection, the AlN encapsulated GaN sample should be placed face down on the polished SiC susceptor substrate during annealing.

8. Thermal stability of implants in microwave annealed SiC and GaN

It is well known that dopants such as N, Al and P are thermally stable in SiC. No redistribution of these impurities was observed either in long duration conventional furnace annealed samples or in samples annealed by microwaves (for both microwave cavity and solid-state microwave head methods) even at temperatures as high as 2100 °C. On the other hand, the boron implant is known to redistribute in SiC even for low-temperature annealing, especially if the implant is a shallow implant. For the 50 keV boron implant, the 1670 °C / 10 s microwave annealing resulted in a substantial out-diffusion of boron from the SiC surface. The small atomic size of boron, resulting in a high transient enhanced diffusion (TED), is believed to be responsible for this behavior. However, for the completely buried 1 MeV boron implant, the 1670 °C / 10 s microwave annealing did not result in any significant boron redistribution (not shown). Boron atom density depth profiles obtained by SIMS measurements on 200 keV/1×10^{15} cm^{-2} B implanted SiC are shown in Fig. 6, before and after microwave and furnace annealing. The boron implant out- diffusion front in Fig. 6 probably is caused by the segregation of boron towards ~ 0.7 Rp, the depth where implant lattice damage is at its maximum. This is caused by the lattice strain at this location. A similar feature was observed in the B- depth profiles of 1 MeV/ 2 x 10^{15} cm^{-2} B for 1670 °C / 10 s microwave annealing. Out-diffusion of the boron is less for the microwave annealing compared to the furnace annealing even though the microwave annealing was performed at a temperature 270 °C higher than the furnace annealing. This again establishes the attractiveness of ultra-fast solid-state microwave annealing compared to the furnace annealing, which has much slower heating and cooling rates. Thus, the ultra-fast microwave annealing is reasonably effective in maintaining the integrity of buried boron implant profiles, but a substantial boron out-diffusion is still observed for 10 s microwave annealing at 1670 °C, for low-energy B ion implantation, where the implant profile is very close to the surface.

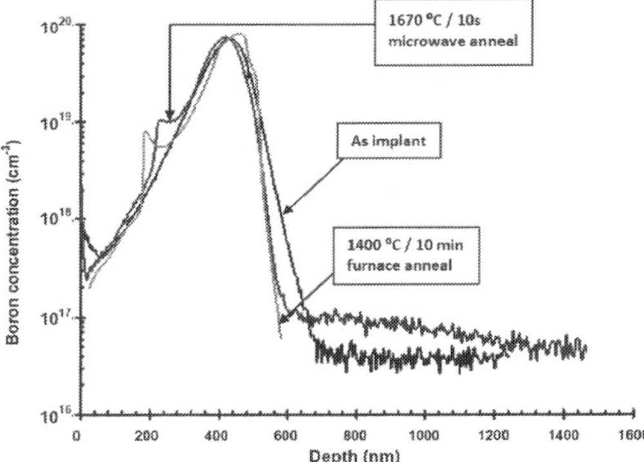

Fig. 6. SIMS depth profiles of 200 keV/1×10^{15} cm^{-2} B$^+$ implant in SiC before and after anneal.

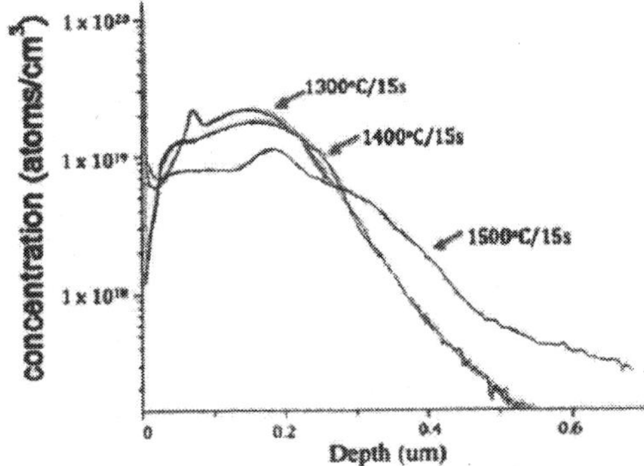

Fig. 7. SIMS depth profiles of 150 keV/5×10^{14} cm^{-2} Mg-implanted GaN after 15s microwave annealing at different temperatures.

The out-diffusion of the implant, as seen for the B-implant in SiC, is observed for Mg implant in GaN also. In addition, a significant in-diffusion of the Mg implant, as shown in Fig. 7, can be seen. The fast out-diffusion and in-diffusion of Mg in GaN at high annealing temperatures is due to a high diffusion coefficient of acceptors in III-V compounds. The extracted Mg implant doses in microwave annealed GaN samples are slightly lower compared to the extracted dose in the as-implanted GaN material due to out-diffusion of Mg into the deposited AlN cap during microwave annealing (Aluri et al, 2010). Despite of the out- and in-diffusion of the Mg implant during annealing, the implant profiles are reasonably stable at least up to a microwave annealing temperature of 1400 °C.

9. Lattice perfection of microwave annealed ion-implanted SiC and GaN

An effective annealing process removes all defects introduced in the material by the ion-implantation doping step. In case of SiC, Rutherford back-scattering (RBS) yield in ion-implanted microwave annealed samples is below that of the conventional furnace annealed samples indicating a better lattice quality for the ultra-fast microwave annealed samples. As

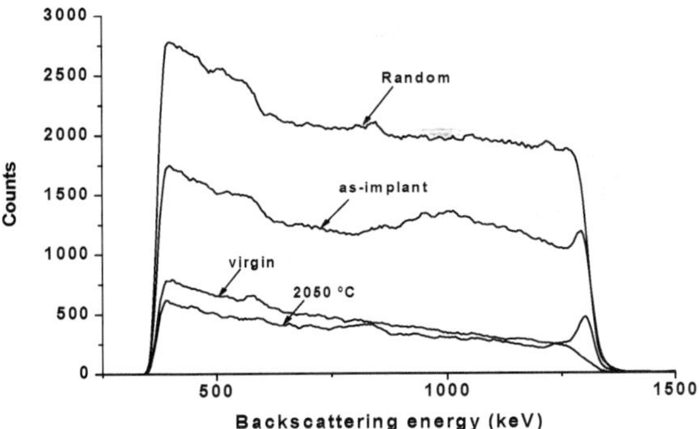

Fig. 8. RBS-C aligned spectra on multiple energy Al ion-implanted 4H-SiC, before and after 2050 °C/15 s microwave annealing. Random and aligned spectra on the virgin sample are also shown for comparison.

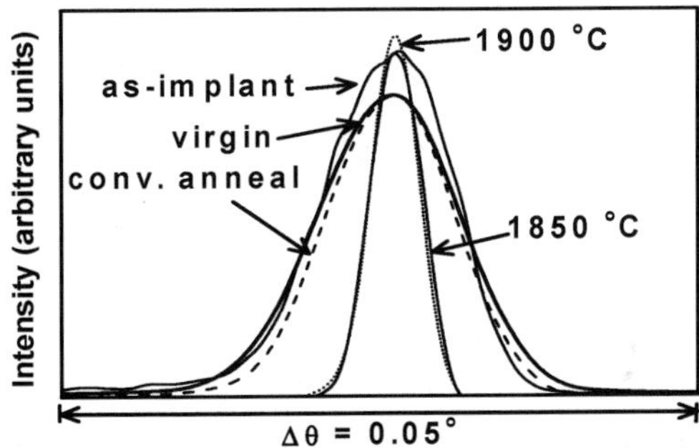

Fig. 9. X-ray rocking curves of the SiC (004) from the Al ion-implanted 4H-SiC, before and after 30 s microwave annealing. For comparison, curves on virgin and 1800 °C/5 min. conventional furnace annealed samples are also shown.

shown in Fig. 8 for aluminum ion-implanted 4H-SiC, the aligned RBS yield in the implanted ultra-fast microwave head annealed material is even below the aligned virgin level. The ultra high-temperature microwave annealing is very effective in removing not only the defects introduced by the ion-implantation step, but also the native defects present in the material before the ion-implantation step. This conclusion is based not only on the results of RBS via channeling (RBS-C) measurements but also the x-ray rocking curve measurements on the as-implanted, microwave annealed and virgin samples. The full-width-at-half-maximum (FWHM) of the high resolution x-ray rocking curves for the SiC (004) reflection of the microwave annealed Al-implanted SiC sample is below that of the virgin sample (see Fig. 9). The FWHM of the sample microwave annealed at 1900 °C for 30s is 15 arc-sec compared to the value of 22 arc-sec in the virgin sample. The x-rays penetrate much deeper (about 3 μm under dynamical diffraction condition) than the implanted region depth (0.3 μm) and hence probe the un-implanted region below the implanted surface region as well, which confirms the removal of defects in this region also (Mahadik et al., 2009). Microwave annealing on unimplanted SiC sample did not result in any reduction in defect density. It is speculated that the defects such as carbon interstitials (C_i) created by the ion-implantation process diffuse from the implanted surface region into the un-implanted bulk below the implanted region, annihilating native defects such as carbon vacancies (V_C) in the bulk region during the annealing process (Storasta and Tsuchida, 2007). It means preexistence of implantation process created defects in the surface region is required for reducing defect density in the bulk underneath during annealing. In case of microwave cavity annealing, for comparable annealing temperatures, the RBS yield of microwave annealed sample is similar to that of a conventional furnace annealed sample. So, the combination of ultra-fast temperature ramping rate (which is not high for microwave cavity annealing) and the very-high annealing temperature associated with the solid-state microwave head annealing are the contributing factors for reducing the defect density below the virgin level.

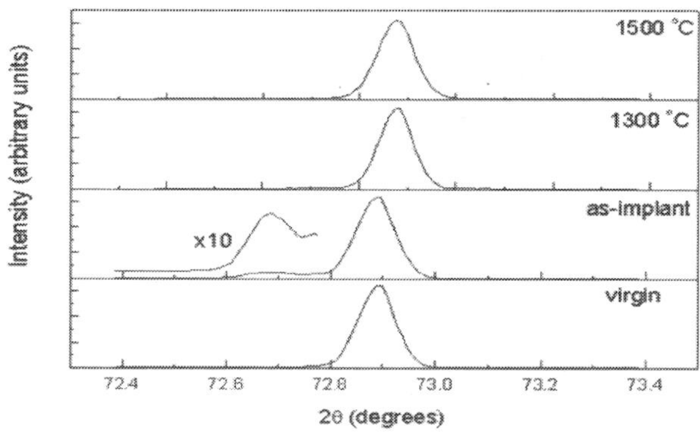

Fig. 10. X-ray diffraction scans of the GaN (004) for Mg-implanted GaN, before and after 15 s microwave annealing. For comparison a scan on the virgin sample is also shown.

Figure 10 shows the XRD scans of the GaN (004) reflections of the as-grown, as-implanted, and 15 s time duration microwave annealed 150 keV/5 $\times 10^{14}$ cm^{-2} Mg-implanted GaN samples. It can be seen that the as-implanted sample has the defect sublattice peak, which is due to the interference of the x-rays from the implanted impurity in the interstitial sites. The defect sublattice peak disappears after microwave annealing, confirming that the implanted Mg has moved to electrically and optically active substitutional lattice positions. In addition, the FWHM value of the GaN (004) peak also decreases upon annealing (from 310 arc-sec before annealing to 273 arc-sec after 1300 °C annealing), indicating that the implant induced damage (which is in the form of Ga and N interstitials) has been removed. It can be seen that upon 1300 °C/15 s annealing, the FWHM value decreased to below the virgin sample level (293 arc-sec). The FWHM value goes down even further for higher temperature anneals (252 arc-sec for 1500 °C annealing), reaching values much lower than that of the unimplanted (virgin) sample. This observation is consistent with the results of ion-implanted SiC. This suggests that the high-temperature microwave annealing process may remove some of the growth related defects in the unimplanted region of GaN films, improving their crystalline quality. This is not surprising, considering that the GaN films growth temperatures are way below the microwave annealing temperatures. However, this does not necessarily mean that there are no point defects left in microwave annealed Mg-implanted GaN. The electrical measurements on Mg-implanted microwave annealed material indicated the presence of residual defects with donor type behavior even after 1500 °C annealing for 15 s. In case of multiple-energy Mg-implanted GaN, the implantation damage created sublattice peak in XRD remained even after 1500 °C annealing (Aluri et al., 2010). In case of in-situ Mg doped GaN, microwave annealing at 1500 °C resulted in breaking of defect complexes, which have a deep donor behavior (Sundaresan et al., 2007d).

10. Electrical characteristics of microwave annealed ion-implanted SiC and GaN

Nitrogen and phosphorus are the most widely used donor (n-type) impurities in SiC. Phosphorus is a preferred donor dopant in SiC because of its higher solubility limit in SiC than that of nitrogen, which cannot be incorporated in excess of 3×10^{19} cm^{-3} due to precipitation during post-implantation annealing. For donor doping concentrations of $\leq 1\times10^{19}$ cm^{-3} either nitrogen or phosphorus implants can be used. But, for higher doping concentrations the phosphorus is the only viable donor dopant in SiC. Resistivity is an important material property that can be used to evaluate the electrical characteristics of an implanted layer, because a low resistivity is obtained only if both the implant electrical activation and the carrier mobility are high. Hence, room-temperature electrical resistivity is used as the prime figure of merit of the electrical characteristics of the microwave annealed implanted SiC layers. For a 170 °C higher temperature microwave annealing on nitrogen implanted SiC, the room-temperature electrical resistivity is 40% lower compared to the conventional furnace annealing at 1600 °C (Sundaresan et al., 2007a). Reduction of this magnitude in electrical resistivity is primarily obtained by a higher nitrogen implant substitutional activation (without sacrificing carrier mobility) in microwave annealed material compared to the conventional furnace annealed material.

The electrical resistivity of phosphorus implanted SiC material is more sensitive to the annealing temperature than the annealing duration (Sundaresan et al., 2007b). For 30 s time duration microwave anneals performed on 2×10^{20} cm^{-3} P$^+$-implanted 4H-SiC, with sample surface not protected by a graphite cap, the room-temperature electrical resistivity of the implanted layer dropped by 15 fold with an increase in annealing temperature from 1700°C to 1950°C. The electrical resistivity drop did not show any tendency towards minimum resistivity saturation even at 1950°C. While the initial decrease in electrical resistivity with increasing annealing temperature up to 1800°C is due to increasing substitutional electrical activation of the P donor, the later decrease at higher annealing temperatures is due to an increase in both P implant activation and carrier mobility, μ. The μ value tripled and the implant activation doubled with an increase in annealing temperature from 1800°C to 1950°C (Sundaresan et al., 2007b). High carrier mobilities are measured in microwave annealed samples due to not only the effective removal of implant lattice damage in the implanted region but also the removal of intrinsic lattice defects that existed in the pre-implanted material. Based on the results of this initial microwave annealing study on 2×10^{20} cm^{-3} P implanted 4H-SiC (only up to an annealing temperature of 1950 °C), during the later work on four other P concentrations; 5×10^{19} cm^{-3}, 1×10^{20} cm^{-3}, 4×10^{20} cm^{-3}, and 8×10^{20} cm^{-3}; the annealing temperature range was extended up to 2050°C, for the 30 s time duration anneals. The graphite cap enabled annealing at such high temperatures. The main reason for selecting higher annealing temperatures when compared to the initial study on 2×10^{20} cm^{-3} P implanted material is the non-saturation of the electrical resistivity in the 2×10^{20} cm^{-3} P implanted material, even at an annealing temperature of 1950°C. In addition, higher implant concentrations like 4×10^{20} cm^{-3} and 8×10^{20} cm^{-3} are expected to require higher annealing temperatures than the 2×10^{20} cm^{-3} P implant concentration, due to a higher degree of implant damage in the material for these implant concentrations. For the P implant concentrations: 1×10^{20} cm^{-3}, 4×10^{20} cm^{-3}, and 8×10^{20} cm^{-3}, there is an additional 25% - 30% reduction in the room-temperature electrical resistivity of the implanted region for higher annealing temperatures compared to the value for 1950°C annealing. For a given phosphorus doping concentration, saturation of the minimum obtainable room-temperature electrical resistivity for microwave annealing seems to occur at around an annealing temperature of 2050°C. Variation of implanted layer room-temperature electrical resistivity

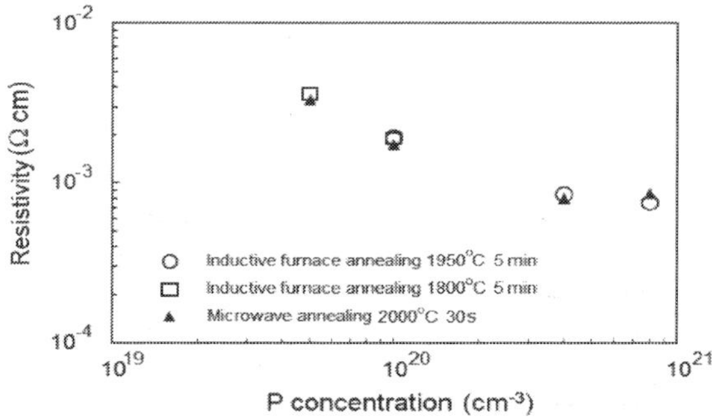

Fig. 11. Variation of measured room-temperature electrical resistivity with phosphorus implant concentration, in microwave annealed and conventional inductive heating furnace annealed 4H-SiC.

with the P-implant concentration for the 2000°C/30 s microwave annealing is shown in Fig. 11. For comparison, results of the conventional inductive heating furnace anneals are also shown in Fig. 11. The measured electrical resistivity data shows in general a lower resistivity values in microwave annealed material compared to the inductive furnace annealed material. A tendency towards saturation in the minimum obtainable resistivity is observed at around 6×10^{-4} Ω-cm, which is the lowest value reported to date on ion-implanted 4H-SiC. As far as resistivity is concerned, it seems, there is no appreciable benefit in going for P implant concentrations > 4×10^{20} cm^{-3}.

Aluminum has a lower acceptor ionization energy compared to the other acceptor impurities such as boron, gallium and indium (Handy et al., 2000). Due to this reason, Al is widely used for obtaining heavily doped p-type regions in SiC. Plots of room temperature sheet resistance (product of electrical resistivity and implant depth) and hole mobility, as a function of the microwave annealing temperature in the range 1750 °C – 1900 °C, for an anneal duration of 30 s, are shown in Fig. 12 for 1×10^{20} cm^{-3} Al implant to a depth of ~ 0.3 μm. It can be seen from Fig. 12 that increasing the microwave annealing temperature steadily lowers the sheet resistance, while increasing the hole mobility after an initial dip. Hole mobility in ion-implanted 4H-SiC is primarily controlled by scattering at ionized impurities and scattering at implantation-induced defects. The initial decrease in hole mobility (in Fig. 12) can be attributed to increased ionized Al impurity scattering. The later

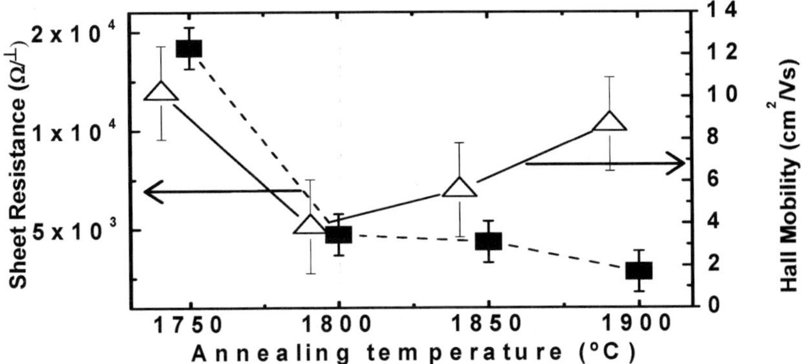

Fig. 12. Variation of measured room-temperature sheet resistance and hole mobility with annealing temperature for 30 s time duration microwave anneals.

increase in hole mobility with increasing temperature could be due to more effective defect removal by the microwave annealing. The measured room temperature sheet resistance (2.8 kΩ/$\square$) for the 1900 °C/30 s microwave annealed sample is a factor of 4.6 times lower compared to the value of 13 kΩ/$\square$ measured for the 1800 °C/5 min conventionally annealed sample, whereas the measured room temperature sheet carrier concentration (product of volumetric carrier concentration and implant depth, which is not shown) for the microwave annealed sample is almost one order of magnitude higher (3.2×10^{14} cm^{-2}) compared to the conventionally annealed sample (3.5×10^{13} cm^{-2}). But, the room temperature hole mobility for the microwave annealed sample is only about 40% lower than that for the conventionally annealed sample (9 cm^2/V.s vs. 14 cm^2/V.s), even though the holes in the microwave annealed material are subjected to a much higher ionized impurity scattering than in the conventionally annealed sample. This means that the decreased sheet resistance measured for the microwave annealed samples is due to a combination of a much more effective Al implant activation, as well as effective implant-induced damage annealing, compared to the conventional furnace annealing.

The electrical characteristics of the microwave cavity annealed material are comparable to the conventional furnace annealed material for similar annealing temperatures (Gardner et al., 1997), again confirming that it is not microwaves that contribute to the favorable electrical characteristics, but the annealing temperature and the temperature ramping rates associated with the microwave head annealing. The temperature ramping rate also has a profound effect on the electrical characteristics of the annealed material (Poggi et al., 2006). Higher ramping rates associated with the solid-state microwave head annealing compared to the microwave cavity annealing are helpful in achieving better electrical properties in implanted material.

Single-energy (150 keV) magnesium ion implanted GaN protected by the AlN cap did not show any electrical conduction for microwave annealing temperatures < 1300°C and also for annealing durations <10 s (Aluri et al., 2010). The material remained highly resistive, with no breakdown voltage seen in two-probe current–voltage (I-V) measurements. The two-probe I-V measurements indicated decreasing breakdown voltage with increasing annealing temperature for 15 s duration anneals performed at temperatures ≥ 1300°C. But, variable temperature van der Pauw–Hall measurements up to 200 °C on these samples did not show a consistent p-type behavior. While the sample showed a low p-type conductivity in one measurement, it showed an opposite type conductivity when the Hall measurements were repeated on the same sample at the same temperature. This could be due to close compensation between the Mg-acceptors which have a high ionization energy (≥ 200 meV) and the donor levels introduced in the material due to the implantation and annealing processes. Though the 15 s annealing is better than the 5 s annealing in decreasing the implant damage, a low residual implant damage in the form of point defects and defect complexes may still exist in the material, below the detectable level of the x-ray measurements. Some of the defect complexes have a deep donor behavior in GaN. In addition, it is well known that nitrogen vacancies have a donor behavior in GaN. Compensation of substitutional Mg acceptors by such donor levels limits the measurable net acceptor carrier concentration by the electrical methods; but, the photoluminescence (PL) measurements, which have no such limitation showed a Mg acceptor related donor-acceptor pair transition peak (Aluri et al., 2010). Hence, based on the XRD, PL, and I-V measurements it can be stated that the single-energy Mg-implanted films, after 15 s high temperature (≥ 1400 °C) microwave annealing have shown improved crystalline quality and acceptor activation of the Mg implant. But, the net Mg-acceptor activation seems to be compromised by compensating background donor defects left in the material even after 1500 °C/15 s annealing. The situation is much worse for multiple-energy Mg ion-implanted GaN. The microwave annealing couldn't remove the ion-implantation damage and hence the material remained highly resistive even after 1500 °C/15 s microwave annealing. Even in case of in-situ Mg doped GaN, microwave annealing at 1500 °C resulted in a decrease in compensating deep donor concentration and a consequent net Mg acceptor activation (Sundaresan et al., 2007d).

11. Results of differential microwave heating sublimation-sandwich method for growing SiC nano-structures

A schematic of the differential microwave heating sublimation sandwitch cell is already shown in Fig. 4. The substrate wafer temperature window for growing SiC nanostructures is 1550 °C to 1750 °C. In this growth method, the precursor Si and C containing species sublimate from the source wafer. Significant sublimation of Si and C species from a SiC wafer requires temperatures > 1400 °C (at 1 atm pressure). Therefore, the growth temperatures required for microwave heating sublimation sandwich method are higher than those typically employed for SiC nanowire growth (1000 °C – 1200 °C), since the traditional methods do not employ sublimated Si and C containing species from a SiC wafer as the source material. The growth in smicrowave heating sublimation sandwich method is performed for time durations of 15 s to 40 s in an atmosphere of UHP-grade nitrogen. Other inert gases such as Ar, He and Xe ionize due to the intense microwave field in the growth chamber. The ΔT between the source wafer and the substrate wafer can be varied from 150 °C to 250 °C by varying the spacing (d) from 300 μm to 600 μm. Growth of SiC nanowires was observed over a very narrow range of both substrate temperature 'T$_2$' (1650 °C -1750 °C) and ΔT (≈ 150 °C).

A plan-view FESEM image of the nanowires grown at 1700 °C for 40 s is shown in Fig. 13. The nanowires are 10 μm to 30 μm long with diameters in the range of 15 nm to 300 nm. The diameter distribution for the SiC nanowires grown at 1700 °C for 40 s revealed that 42 % of the nanowires have diameters in the range of 15 nm to 100 nm while 14% of nanowires have diameters in excess of 300 nm. EDAX analysis of the nanowires indicated that they mainly consist of Si and C with traces of nitrogen. The likely sources of this nitrogen are the ambient atmosphere and the nitrogen donor dopant in the source wafer. EDAX spectra taken on the droplets at the nanowire tips consist of the corresponding metal and Si.

Fig. 13. FESEM image of SiC nanowires grown at T_2 = 1700 °C and ΔT = 150 °C for 40 s.

(a)   (b)

Fig. 14. (a) Cone-shaped SiC nanostructures grown at T_2 = 1600 °C and ΔT=150 °C; (b) Needle-shaped SiC nanostructures grown at T_2 = 1700 °C and ΔT=250 °C.

In addition to SiC nanowires, growth of cone-shaped and needle-shaped SiC nanostructures was also observed under different growth conditions. For ΔT=150 °C, the substrate wafer temperature in the range of 1550 °C to 1650 °C (for 15 s to 1 min heating cycle duration) yielded mainly cone-shaped quasi 1-D SiC nanostructures (Fig. 14(a)) which are 2 μm – 5 μm long, whereas substrate wafer temperature > 1750 °C (for the same heating cycle durations) resulted in micron-sized SiC deposits (not shown). The "nanocones" shown in Fig. 14(a) taper off along their axis from thick catalytic metal tips. This suggests that the diameter of the droplets increased during the growth of the cones. The diameter of their thin ends are about 10 nm to 30 nm, while the diameter of the broad portion at the top (just under the catalytic metal tips), ranges from 100 nm to 200 nm. The fact that the diameter of the cones increases with growth duration must mean that there is an Oswald ripening effect, i.e. the metal is transferred from the smaller diameter droplets to the larger diameter ones, possibly via surface diffusion (Hannon et al., 2006). The short length of the cones results from a relatively low SiC growth rate for the experimental conditions under which the cones are grown. Thus, the surface diffusion length for the liquid metal to flow from the smaller diameter droplets to the larger diameter droplets is short.

Increasing the ΔT to 250 °C (by increasing 'd' from 300 µm to 600 µm) at a substrate wafer temperature (T_2) of 1700 °C resulted in mainly needle-shaped SiC nanostructures (Fig. 14(b)), which are 50 µm – 100 µm in length. These needles are narrow under the catalytic metal tips. It is obvious that the diameter of the metal droplets catalyzing the needle growth decreases with growth duration. Because the source wafer temperature for needle growth (1900 °C – 2000 °C) is high, it is possible that the metal droplets evaporate during crystal growth due to high temperatures in the vicinity of the droplets. The much longer needles (in comparison with the cones) also results in a greater surface diffusion length for the liquid metal to flow between droplets, which might have inhibited significant surface diffusion of the metal.

A much higher density of nanowires in comparison with other 2-D deposits are observed for the growth performed using Fe, Ni, and Pd catalysts. Selected area electron diffraction patterns (not shown) recorded from 10 nanowires were all consistent with a cubic 3C-SiC structure. The growth direction of the nanowire was identified as <112> which is in contrast to the <111> growth direction commonly observed for 3C-SiC nanowires. One of the reasons as to why the <112> growth direction is preferred for the SiC nanowires grown by microwave heating method over the commonly reported <111> direction could be the very high temperatures (1650 °C – 1750 °C) used for microwave heating method nanowire growth. The nanowire growth generally occurs along the direction, whose corresponding face has the highest surface energy, so that that particular face is not exposed. The {111} being a three cluster face must have a higher surface energy for SiC at lower temperatures, thereby driving the nanowire growth along the <111> direction. At higher temperatures, the nucleation rate along directions normal to lower atomic density planes such as {110} and {112} is known to be faster than {111}.

Factors affecting the SiC crystal polytype are the temperature (Knippenberg, 1963) and the pressure in the growth chamber, the polarity of the seed crystal, the presence of certain impurities and the Si/C ratio. Under more Si-rich (C-rich) conditions the formation of the cubic (hexagonal) polytype should be preferred (Omuri et al., 1989). Nucleation far from equilibrium conditions and a nitrogen atmosphere has been generally assumed to stabilize the cubic polytype (Limpijumnong and Lambrecht, 1998). This is supported by nucleation theory. Furthermore, 3C-SiC has the lowest surface energy among all polytypes. Since, in the microwave heating sublimation sandwich method, Si-rich precursor species are present (Si, Si_2C), and nucleation occurs far from equilibrium conditions in a nitrogen atmosphere, the growth of 3C-SiC is to be expected from the above considerations. Furthermore, since nanowires have a large surface to volume ratio, the low surface energy of the 3C-SiC polytype makes it much more favorable to grow 3C-SiC over other polytypes. However, it is worth trying to see if under C-rich precursor and Al vapor pressure conditions the hexagonal polytype SiC nanowires can be grown or not. Controlled diameter SiC nanowire growth is expected by pre-patterning the catalyst metal layer to desired dimensions, which controls the diameter of the catalyst liquid droplet during growth.

12. Other applications of microwave heating

Though microwave heating so far has been mainly used for activating ion-implanted and in-situ doped dopants in SiC and GaN, and for growing SiC nanowires, it can be used for other high-temperature device processing steps such as ohmic contact alloying. Since, the thickness of metal layers used for contacts in semiconductor device fabrication are in the same order as the skin depth, microwave heating is suitable for contact alloying and for formation of silicide films on substrates. Annealing temperature and temperature ramp rates have a profound effect on contact surface morphology and contact resistance, which are two important figures of merit of the ohmic contacts. Hence, microwave heating is desirable for forming ohmic contacts. Because ohmic contacts are always formed on heavily doped regions, differential heating of these regions helps in obtaining high-performance ohmic contacts without heating other regions of the devices.

Microwave heating is also attractive for achieving very high temperatures required for the activation of ion-implantation and in-situ doped impurities in other large bandgap materials such as diamond.

After processing, the semiconductor devices need to be bonded to the carrier and then hermetically sealed. Rapid and selective heating properties of microwaves are very useful for bonding and hermetic sealing of discrete semiconductor devices, and micro-electro-mechanical-system (MEMS) and integrated circuit (IC) devices to the carrier substrate. In some applications the bonding material used should be strong, chemically resistive and withstand high temperatures. Traditional Pb-Sn alloys are not good for these applications. Noble metals such as Au, Cr, Ni and Pt; glass; and ceramics are very good candidates as the bonding materials for these applications, but require high bonding temperatures due to their high melting point. Microwave heating is very attractive in reaching these high temperatures rapidly in the targeted area to be heated without adversely affecting the integrity of MEMS or IC devices to be bonded (Y. Tian and M.Y. Tian, 2009).

Microwave heating is also attractive to form stochiometric composition thin films of compounds such as $In_{2-x}Fe_xO_3$ on any substrate material. Since SiC is an excellent absorber of microwaves, it can be used as a susceptor to evaporate the compound. The compound in powder form can be placed on the SiC susceptor, which is coupled to the microwaves coming from a microwave head. As soon as the microwave power source is turned on, the compound is heated to a temperature greater than the evaporation temperature of all the constituent elements of the compound. This feature leads to the formation of the intended compound film on a nearby substrate, whose composition is similar to that of the powdered compound source material.

Microwave heating can be made as a part of multi-chamber semiconductor epitaxial growth systems for performing ultra-fast in-situ annealing of epitaxial layers in the intermittent stages of the growth process, without breaking the vacuum. Ultra-fast intermittent heat treatment helps to decrease dislocation density in epitaxial layers grown on lattice mismatched substrates (Chen et al., 2008).

13. Acknowledgment

Author thanks Dr. Y-L. Tian of LT Technologies for his collaboration with the author on microwave heating. He acknowledges the valuable contributions made by his past and current graduate students: Siddarth Sundaresan, Nadeemullah Mahadik, Madhu Gowda, Geetha Aluri and Anindya Nath, on this work. He also acknowledges the contributions made by Dr. Syed Qadri of the Naval Research Laboratory; Dr. Roberta Nipoti of IMM-CNR, Bologna, Italy; and Dr. Albert Davydov of NIST. He further acknowledges the financial support provided by the National Science Foundation, the Army Research Office

(ARO) and the DARPA (through the U.S. Naval Research Laboratory contract # N0017310-2-C006).

14. References

Aluri, G.S., Gowda, M., Mahadik, N.A., Sundaresan, S.G., Rao, M.V., Schreifels, J.A., Freitas, Jr., J.A., Qadri, S.B., and Tian, Y-L. (2010). Microwave annealing of Mg-implanted and *in situ* Be-doped GaN. *Journal of Applied Physics*, Vol. 108, 083103

Amada, H. (1987). Method and Apparatus for Microwave Heat-treatment of a Semiconductor Wafer, US Patent No. 4,667,076, May 19

Andrievski, R.A. 2009. Synthesis, Structure and Properties of Nanisized Silicon Carbide. *Review of Advanced Material Science*, Vol. 22, pp. 1-20

Capano, M.A., Ryu, S., Melloch, M.R., Cooper, Jr., J.A., and Buss, M.R. (1998). Dopant activation and surface morphology of ion implanted 4H- and 6H-silicon carbide. *Journal of Electronic Materials*, Vol. 27, No. 7, pp. 370-376

Chen, Y., Farrell, S., Brill, G., Wijewarnasuriya, P., and Dhar, N. (2008). Dislocation reduction in CdTe/Si by molecular beam epitaxy through in-situ annealing. *Journal of Crystal Growth*, Vol. 310, pp. 5303-5307

Fang, X., Hu, L., Ye, C., and Zhang, L. 2010. One-dimensional inorganic semiconductor nanostructures: A new carrier for nanosensors. *Pure and Applied Chemistry*, Vol. 82, No. 11, pp. 2185-2198

Fan, J.Y., Wu, X.L., Chu, P.K. 2006. Low-dimensional SiC nanostructures: Fabrication, luminescence, and electrical properties. *Progress in Material Science*, Vol. 51, No. 8, November, pp. 983-1031

Feng, Z.C. (ed.) 2006). *III-Nitride Semiconductor Materials*, Imperial College Press, London

Fukano, T., Ito, T., and Ishikawa, H. (1985). Microwave Annealing for Low Temperature VLSI Processing. *IEDM Technical Digest*, Vol. 31, pp.224-227

Gardner, J.A., Rao, M.V., Tian, Y.L., Holland, O.W., Roth, E.G., Chi, P.H., Ahmad, I. (1997). Rapid Thermal annealing of ion implanted 6H-SiC by microwave processing. *Journal of Electronic Materials*, Vol. 26, No. 3, pp. 144-150

Ghandhi, S.K. (1994). *VLSI Fabrication Principles: Silicon and Gallium Arsenide* (2nd edition), Wiley-Interscience, ISBN:0471580058, New York

Handy, E.M., Rao, M.V., Holland, O.W., Chi, P.H., Jones, K.A., Derenge, M.A., Vispute, R.D., and Venkatesan, T. (2000). Al, B, and Ga, ion-implantation doping of SiC.*Journal of Electronic Materials*, Vol. 29, pp. 1340-1345.

Hannon, J.B., Kodambaka, S., Ross, F.M., and Tromp, R.M. (2006). The influence of the surface migration of gold on the growth of silicon nanowires. *Nat ure*, Vol. 440, March, pp. 69-71

Heera, V., Mucklich, A., Dubois, C., Voelskow, M., and Skorupa, W. (2004). Layer morphology and Al implant profiles after annealing of supersaturated, single-crystalline, amorphous, and nanocrystalline SiC. *Journal of Applied Physics*, Vol. 96, No. 5, pp. 2841-2852

Huang, Y., Lieber, C.M. 2004. Integrated nanoscale electronics and optoelectronics : Exploring nanoscale science and technology through semiconductor nanowires. *Pure and Applied Chemistry*, Vol. 76, No. 12, pp. 2051-2068

Jenny, J. R., Malta, P., Tsvetkov, V.F., Das, M.K., Hobgood, McD, Carter, C.H., Kumar, R.J., Borrego, J.M., Gutman, R.J., Aavikko, R. (2006). Effects of annealing on carrier lifetime in 4H-SiC. *Journal of Applied Physics*, Vol. 100, No. 11, 113710

Kato, T., Kinoshita, A., Wada, K., Nishi, T., Hozomi, E., Taniguchi, H., Fukuda, K. and Okumura, H. (2010). Morphology improvement of step bunching on 4H-SiC wafers by polishing technique. *Material Science Forum*, Vol. 645-648, April, pp. 763-765

Knippenberg, W.F. (1963). Growth phenomenon in silicon carbide. *Phillips Research Reports*, Vol. 18, No. 3, pp. 161-274

Law, M., Goldberger, J., Yang, P. (2004). "Semiconductor nanowires and nanotubes. *Annual Review of Materials Research*, Vol. 34, August, pp. 83-122

Limpijumnong, S., and Lambrecht, W.R.L. (1998). Total energy differences of SiC polytypes revisited. *Physics Review* B, Vol. 57, pp. 12017-12022

Mahadik, N.A.Qadri, S.B., Sundaresan, S.G., Rao, M.V., Tian. Y., and Zhang, Q. (2009). Effects of microwave annealing on crystalline quality of ion-implanted SiC epitaxial layers. *Surface and Coatings Technology*, Vol. 203, pp. 2625-2627

Negoro, Y., Katsumoto, K., Kimoyo, T., and Matsunami, H. (2004). Electronic Behaviors of High-Dose Phosphorus-Ion Implanted 4H-SiC (0001). *Journal of Applied Physics*, Vol. 96, No. 1, July, pp. 224-228

Omuri, M., Takei, H., and Hukuda, T. (1989). *Japanese Journal of Applied Physics*, Vol. 28, pp. 1217-

Ottaviani, L., Blondo, S., Morata, S., Palais, O., Sauvage, T., and Torregrosa, F. (2010). Influence of Heating and Cooling Rates of Post-implantation Annealing Process on Al-implanted 4H-SiC Epitaxial Samples. *Materials Science Forum*, Vols. 645-648, pp. 717-720

Pearton, S.J., Abernathy, C.R., Ren, F. (2006). *Gallium Nitride Processing for Electronics,Sensors, and Spintronics*, Springer-Verlag, ISBN:1-85233-935-7, London

Phelps, G.J., Wright, N.G., Chester, E.G., Johnson, C.M., and O'Neill, A.G. (2002). Step bunching fabrication constraints in silicon carbide. *Semiconductor Science & Technology*, Vol. 17, No. 5, May, pp. L17-L21.

Poggi, A., Bergamini, F., Nipoti, R., Solmi, Canino, M., S., Carnera, A. (2006). Effects of heating ramp rates on the characteristics of Al implanted 4H-SiC junctions. *Applied Physics Letters*, Vol. 88, No. 16, April,162106.

Rambach, M., Schmid, F., Krieger, M., Frey, L., Bauer, A.J., Pensl, G., and Ryssel, H. (2005). Implantation and annealing of aluminum in 4H silicon carbide. *Nuclear Instruments and Methods in Physics Research* B, Vol. 237, No. 1-2, August, pp. 68-71

Rambach, M., Bauer, A.J., , and Ryssel, H. (2008). Electrical and topographical characterization of aluminum implanted layers in 4H silicon carbide. *Physica Status Solidi* (b),Vol. 245, No. 7, July, pp.1315-1326

Rao, M.V., Tucker, J.B., Ridgway, M.C., Holland, O.W., Capano, M.A., Papanicolaou, N., and Mittereder, J. (1999). Ion-implantation in bulk semi-insulating 4H-SiC. *Journal of Applied Physics*, Vol. 86, No. 2, pp. 752-758.

Rao, M.V., Nath, A., Nipoti, R., Qadri, S.B., and Tian, Y-L. (2010). Annealing of ion-implanted 4H-SiC, *AIP Proceedings : 18th International conference on Ion-implantation Technology (IIT 2010)*, kyoto, Japan, June 6-11, 2010.

Russell, S.D., and Ramirez, A.D. (1999). In situ boron incorporation and activation in silicon carbide using excimer laserrecrystallization. *Applied Physics Letters*, Vol. 74, No. 22, May, pp. 3368-3370

Saddow, S.E., Agarwal, A.K., (eds. (2004) *Advances in Silicon Carbide Processing and Applications*, Artech House, ISBN:1-58053-740-5, Norwood, MA, USA.

Seshadri, S., Eldridge, G.W., Agarwal, A.K. (1998). Comparison of the annealing behavior ofhigh-dose nitrogen-, aluminum-, and boron-implanted 4H–SiC. *Applied Physics Letters*, Vol. 72, No. 16, pp. 2026-2028.

Scovell, P.D. (1984). Method of Reactivating Implanted Dopants and Oxidation of Semiconductor Wafers by Microwaves, U.S. Patent No. 4,490,183, Dec. 25

Spinter, M.R., Palys, R.F., and Beguwala, M.M. (1981). *Low Temperature Microwave Annealing of Semiconductor Devices*, U.S. Patent No. 4,303,455, Dec. 1

Storasta, L. and Tsuchida, H. (2007). Reduction of traps and improvement of carrier lifetimein 4H-SiC epilayers by ion implantation. *Applied Physics Letters*, Vol. 90, No. 6, 062116

Sundaresan, S.G., Rao, M.V., Tian, Y., Schreifels, J.A., Wood, M.C., Jones, K.A., Davydov, A.V. (2007). Comparison of solid-state microwave annealing with conventional furnace annealing of ion-implanted SiC. *Journal of Electronic Materials*, Vol. 36, No. 4, pp. 324-331

Sundaresan, S.G., Rao, M.V., T-L. Tian, Ridgway, M.C., Schreifels, J.A., Kopanski, J.J. (2007). Ultrahigh-temperature microwave annealing of Al+- and P+-implanted 4H-SiC. *Journal of Applied Physics*, Vol. 101, 073708

Sundaresan, S.G., Davydov, A.V., Vaudin, M.D., Levin, I., Maslar, J.E., Tian, Y-L., and Rao, M.V. (2007). Growth of silicon carbide nanowires by a microwave heating-assisted physical vapor transport process using Group VIII metal catalysts. *Chemistry of Materials*, Vol. 19, pp. 5531-5537

Sundaresan, S.G., Murthy, M., Rao, M.V., Schreifels, J.A., Mastro, M.A., Eddy, Jr., C.R., Holm, R.T., Henry, R.L., Freitas, Jr., J.A., Gomar-Nidal, E., Vispute, R.D., and Tian, Y-L. (2007). Characteristics of in-situ Mg-doped GaN epi-layers subjected to ultra-high-temperature microwave annealing using protective caps. *Semiconductor Science and Technology*, Vol. 22, pp. 1151-1156

Sundaresan, S.G., Mahadik, N.A., Qadri, S.B., Schreifels, J.A., Tian, Y-L., Zhang, Q., Gomar-Nadal, E., Rao, M.V. (2008). Ultra-low resistivity Al+ implanted 4H-SiC obtained by microwave annealing and a protective graphite cap. *Solid-State Electronics*, Vol. 52, pp. 140-145

Tanaka, Y., Tanoue, H., Arai, K. (2003). Electrical activation of the ion implanted phosphorus in 4H-SiC by excimer laser annealing, *Journal of Applied Physics*, Vol. 93, pp. 5934-5936

Tian, Y., & Tian, M.Y. (2009). Method and Apparatus for Rapid Thermal Processing and Bonding of Materials Using RF and Microwaves. U.S. Patent No. US 7,569,800, Aug. 4

Tian, Y-L. (2009). Microwave based technique for ultra-fast and ultra-high temperature thermal processing of compound semiconductors and nano-scale Si semiconductors, *Proccedings of 17th IEEE International Conference on Advanced Thermal Processing of Semiconductors – RTP 2009*

Thompson, D.C., Decker, J., Alford, T.L., Mayer, J.W., Theodore, N.D. (2007). Microwave Activation of Dopants & Solid Phase Epitaxy in silicon, *MRS Proceedings*, Vol. 989, Title: Amorphous and Polycrystalline Thin-Film Silicon Science and Technology—2007, Spring 2007 Meeting, Symposium A

Vassilevski, K.V., Wright, N.G., Nikitina, I.P., Horshall, A.B., O'Neill, A.G., Uren, M.J., Hilton, K.P., Masterton, A.G., Hydes, A.J., and Johnson, C.M. (2005). Protection of selectively implanted and patterned silicon carbide surfaces with graphite capping layer during post-implantation annealing. *Semiconductor Science and Technology*, Vol. 20, pp. 271-278

Vathulya, V.R., and White, M.H. (2000). Characterization and performance comparison of the power DIMOS structure fabricated with a reduced thermal budget in 4H and 6H-SiC. *Solid-State Electronics*, Vol. 44, pp. 309-315.

Zhang, S.I., Buchta, R., and Sigurd, D. (1994). Rapid thermal processing with microwave heating. *Thin Solid Films*, Vol. 246,No. 1-2, 15 June, pp. 151-157

21

Magnetic Induction Heating of Nano-sized Ferrite Particles

Yi. Zhang[1] and Ya. Zhai[1,2]
[1]Department of Physics, Southeast University, Nanjing 211189,
[2]The center of Material Analysis, National Laboratory of Solid Microstructures,
Nanjing University, Nanjing 210093,
China

1. Introduction

Magnetic nano-structured materials have attracted much attention due to their unique properties and potential applications. Magnetic induction heating behavior in magnetic particles provides a benefit for biomedical applications, such as targeted drug delivery[1, 2], diagnostics[3], and magnetic separation[4, 5]. The magnetic nano-structured materials are also being explored as contrast agents in MRI[6, 7], thermo responsive drug carriers[8], and the hyperthermia[9, 10]. Hyperthermia is a promising approach for the thermal activation therapy of tumor[10, 11]. Inevitable technical problems for hyperthermia are the difficulty of heating only the local tumor to the intended temperature without damaging much of the surrounding healthy tissue and precise control of temperature. Induction heating of magnetic nanoparticles in an alternating field can solve the above problems with their unique characters of self-heating, self-temperature controlling due to the magnetic loss in AC magnetic field. In addition, nanoparticles can be conveniently transported to the local region of tumor tissue by the interventional techniques.

Magnetic particles heat inductively due to magnetic losses associated with the magnetization/demagnetization cycling. Ferrites (magnetic iron oxides) nanoparticles, are good candidates for Hyperthermia tumor therapy due to their excellent high-frequency response and high resistivity property, especially the possible large magnetic loss and high heating ability. Among them, Fe_3O_4 particles have attracted much interests because they are considered as a material with non-toxicity and biological compatibility due to their main composition of Fe ions. Besides, it is well known that Zn additives in magnetic oxides can change the final heated temperature and Zn is also an element in people's favor. In this chapter, we will introduce the magnetic induction heating behaviors of ferrite nanomaterials and fluid in following three aspects: (1) Basic knowledge of magnetic induction heating in ferrite nano-materials. (2) The effect of magnetic field and frequency on heating efficiency and speed. (3) Magnetic fluid hyperthermia behaviors.

2. Basic knowledge of magnetic induction heating of ferrite nano- materials

The magnetic induction heating of ferrite materials is originated from their power loss in alternating magnetic field. The total power loss (P_L) is composed of three parts, hysteresis loss (P_h), eddy current loss (P_e) and residual loss (P_r) [12, 13]. Hysteresis loss is due to the irreversible magnetization process in AC magnetic field. Eddy current loss is the Joule loss due to eddy current induced by the alternating magnetic field and hence depends much on the electrical resistivity of the material. The physical origin of residual loss is more complicated. The residual loss cannot be separated straight forwardly from eddy current loss, nor even from hysteresis loss easily. However, most ferrite materials have higher electrical resistivity, leading to very low eddy current loss. Thus the magnetic induction heating of ferrite materials is mainly caused by the hysteresis loss and residual loss in alternating magnetic field. The residual loss is originated from various relaxation effects of magnetization in magnetic field. Thus it is also called relaxation loss. In some low loss ferrites the relaxation effect shows resonance at certain high frequencies.

The hysteresis loss refers to the loss due to irreversible magnetization process in AC field. In DC field a hysteresis loop is observed during magnetization reversal for one cycle when the field is not very low. The hysteresis loss for one cycle can be easily estimated from the area of the hysteresis loop, W_h and the power loss is expressed as $P_h = W_h f$, where f is frequency. Note that the irreversible process and the hysteresis loop changes with the amplitude and frequency of the AC field. When the amplitude of the AC field increases, the H_c and B_r increase and hysteresis loss increases until the maximum value. However, when the frequency increases, the hysteresis loss due to irreversible process decreases and disappears at a high frequency when the irreversible process cannot catch the fast changing AC field. Nevertheless, an AC hysteresis loop can still be observed experimentally which is due to time lag of magnetization cause by after effect or relaxation effect relative to AC field. For bulk ferrites the hysteresis loss may disappear when the frequency is above MHz[14]. For induction heating, the frequency is generally not very high with relatively high field amplitude, thus the hyteresis loss often may still exist and makes the main contribution.

For low frequency(f <1kHz)and small amplitude of AC field ($H<<H_c$, $M<M_s/10$), the early studies by Jordan[15] and Legg[16] lead to the following well known expression for the losses of magnetic material in AC magnetic field:

$$W / B_m^{\ 2} = aB_m + ef + c \tag{1}$$

Here B_m is the amplitude of the magnetic induction in the material. The first term refers to magnetic hysteresis loss with a the coefficient of magnetic hysteresis loss. Second term stands for eddy loss which is proportional to the frequency of AC field with e the coefficient of eddy loss. The third term represents residual loss which is almost a constant in a low frequency for metals. However for ferrites c is found to be dependent on frequency so that it should be noted not to be confused with eddy current loss. .. Thus equation (1) should be changed to the following for ferrites:

$$W / B_m^{\ 2} = aB_m + ef + c(f) \tag{2}$$

When the amplitude of magnetic field is small, within so called "Rayleigh" region, the magnetic hysteresis loss per cycle can be expressed by

$$W_h = \frac{4}{3}\mu_0 bH_m^{\ 3} = \mu_0\pi\mu'' fH_m^{\ 2} \tag{3}$$

Here b is so called Rayleigh constant indicating a linear dependence of permeability on H, namely $\mu = \mu_i + bH$ in Rayleigh region. When the amplitude of magnetic field is highrt than that in Rayleigh region the hysteresis loss can only be determined empirically.

For nano-sized magnetic particles, the eddy currnt loss can be calculated without considering the effect of skin depth by assuming uniform distribution of magnetic field in the particle. The calculated eddy current power loss of the spherical particles is expressed by the following expression[14],

$$P_e = \frac{\pi}{20} B_m{}^2 d^2 \sigma f^2 [10^{-10} w / cm^2] \tag{4}$$

in CGS unit, where d is the diameter of the particle and σ is the conductivity. P_e is generally very small in ferrite materials. The mechanism of residual loss is complicated and the constant c can only be determined from experiment.

Earlier study for NiZn ferrite material reported that the magnetic hysteresis loss is dominant and for very soft MnZn ferrite material, in which the hysteresis loss is low and residual loss is dominant[17].

For medium and strong magnetic field and low frequency, The power loss was proved by experiment as blow[17]:

$$P = P_h + P_e = \eta B_m{}^n f + e B_m^2 f^2 \tag{5}$$

Here the constants η, and n for magnetic hysteresis loss are determined empirically. For Fe sheet, n was found to be 1.6. The coefficient e for eddy loss has been calculated theoretically in different cases.

For high frequency AC field, the magnetic loss is difficult to separate since the total loss P is not linear with f and the three components in equation (2) are all frequency dependent and have interaction among them.

As the size of the ferrite particle is larger than the critical size of single domain, the particle may contain several domains with domain walls between the adjacent domains. The magnetization reversal and hysteresis loss are realized by irreversible domain wall displacement. When the size of ferrite particle approaches the critical size of single domain, the magnetization reversal and hysteresis loss are realized by irreversible domain rotation in overcoming the energy barrier $E = KV$, where K is the anisotropy constant and V is the volume of the magnetic particle. Generally speaking, H_c and hysteresis loss due to irreversible domain wall displacement is smaller than that due to irreversible domain rotation. When the amplitude of AC field is large enough the hysteresis loss due to irreversible domain rotation dominates for single domain particles. When the particle size is smaller than the single domain critical value the superparamagnetic behavior appears, in which the thermal agitation helps the magnetization reversal to overcome the energy barrier. Then H_c is smaller and the hysteresis loss is smaller and down to zero for completes superparamagnetic tiny particles. Therefore the induction heating is optimized by selecting ferrite composition, particle size, AC field intensity and frequency.

For the superparamagnetic ferrite nano-materials in carrier liquid, induction heating can be due to either so-called Neel relaxation process or rotational Brownian motion within a carrier liquid, or to both generally. The Neel relaxation process refers to the heat assisted domain rotations in the particles by the AC magnetic field as mentioned above. The Browing relaxation refers to the rotation of the magnetic particle as a whole because of the torque exerted on the magnetic moment by the external AC magnetic field, and the energy barrier for reorientation of a particle is determined by rotational friction in the surrounding liquid. In general, the heating effects usually proceed via one of the two mechanisms: Neel and Brownian relaxation processes, or via both together. The total relaxation losses due to both processes could be calculated by the following equation[18]

$$P = (mH\omega\tau_{eff})^2 / [2\tau_{eff}\kappa TV(1 + \omega^2\tau^2_{eff})] \tag{6}$$

Here, m is the particle magnetic moment, ω is the AC field frequency, H is the ac field amplitude, V is the nanoparticle volume, and τ_{eff} is the effective relaxation time. When the ac magnetic field is applied to magnetic nanoparticles, their magnetic moments attempt to rotate following the magnetic field with time lag. The effective relaxation time (τ_{eff}) is given by [19]

$$\tau_{eff} = \frac{\tau_B\tau_N}{\tau_B + \tau_N} \tag{7}$$

in which the Brownian relaxation is

$$\tau_B = 3\eta V_H / kT \tag{8}$$

where η is the viscosity of the carrier fluid, k is the Boltzmann constant, T is the absolute temperature, and V_H is the hydrodynamic volume of the particle. Néel relaxation is

$$\tau_N = \tau_0 \exp(KV / kT) \tag{9}$$

where τ_0 is on the order of 10^{-9} S, and K is the anisotropy constant of the magnetic nanoparticle.

The Curie temperature (T_c) of the material provides a self-temperature- limiting behavior. Below T_c , the magnetic material exhibits the magnetic order and above T_c , it becomes paramagnetic (magnetic disorder) which does not show magnetic hysteresis. The magnetic material is heated up efficiently in an alternating magnetic field until it reaches its T_c, above which the material becomes paramagnetic and the heating effect almost stops. Then when temperature is cooled down below T_c, the nano-particles resume the strong magnetic state and heat up again and thus stay around T_c. Therefore the magnetic induction heating has a final equilibrium temperature, called dwell temperature, and we can control the temperature by changing T_c in terms of doping some impurities in magnetic particles. For Curie temperature-controlled heating, magnetic heat generation must greatly exceed other mechanisms of heat generation, thus the magnetic material will heat efficiently in an induction field until it reaches its T_c. Whereas T_c can be adjusted by changing the composition of ferrites. Examples are magnetite with a wide range of Zn substitution [20].

It is well-known that the Curie point T_c will drop as a result of the substitution of Zinc owing to the reduced A-B exchange interaction with increasing Zn content in the case of mixed Zn ferrites with spinel structure, $Me_{1-\delta}Zn_\delta Fe_2O_4$. Since the Curie temperature T_c decreases from 680°C to 0°C with increasing Zn content, the dwell temperature of various spinel structured ferrite particles under the AC field could be controlled by Zn substitution.

We have investigated the self-heating and self-temperature-limiting behavior tests of three series of $Zn_xFe_{3-x}O_4$, $Ni_{1-x}Zn_xFe_2O_4$ and $Co_{1-x}Zn_xFe_2O_4$ ferrite nano-particles with various Zn substitutions. The particles size of various Zn ferrites decreases from 23 nm to 6 nm with increasing Zn contents. By comparing the dwell temperature of various Zn ferrites nano-particles of the same weight of 2.0 g with their T_c, we obtained that the dwell temperature in AC field with 72kHz and 54.3mT is close to the Curie temperature T_c as shown in Fig.1. We see that the T_c does decrease with increasing Zn content in the spinel structured ferrite nano-materials. The final heating temperature is clearly determined by the Curie temperature of the material and is always slightly lower than its Curie point. This deviation is most likely due to heat losses to the environment, which are balanced by the small amount of heat which the material generates immediately below its Curie temperature. These environmental losses have been estimated based on the cooling rates of the samples after removal from the induction field, and are used to provide a minor correction in all subsequent quantitative calculations of induction heat generation.

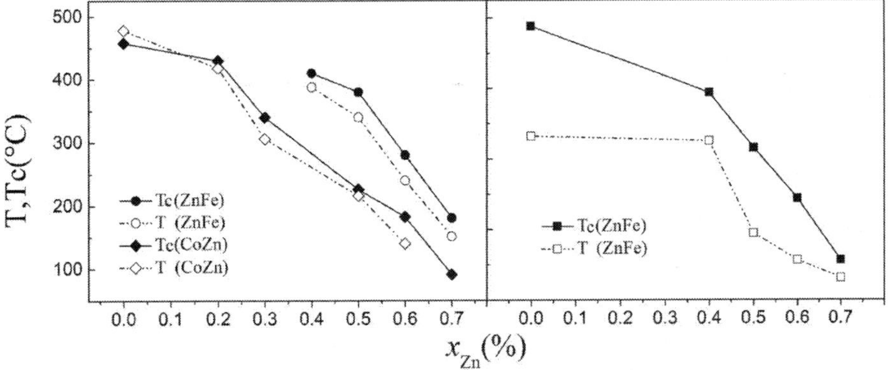

Fig. 1. The dwell temperature and T_c of Zn content of ferrite nano-particles

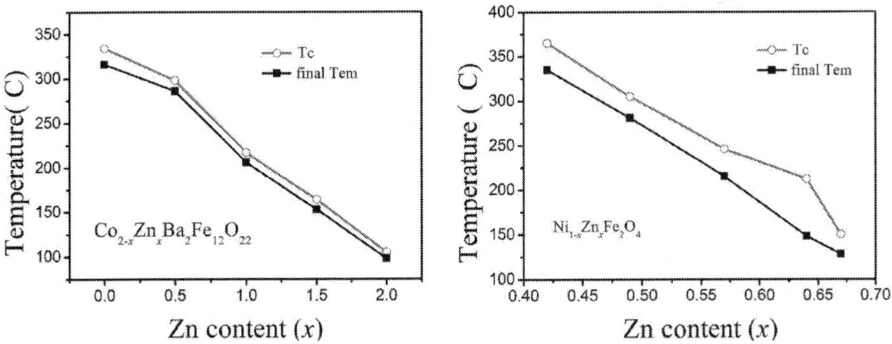

Fig. 2. The dwell temperature T and T_c in higher frequency(5.85 MHz)[21]

Similar results can be found in AC field with higher frequency[21]. John Xiao group investigated the Curie temperature-controlled induction heating behavior of $Co_2Ba_2Fe_{12}O_{22}$ semihard ferrites and $NiFe_2O_4$ soft ferrites in AC field with 5.85 MHz and 50–500 Oe, respectively. In scanning electron microscopy images, particle sizes of final products are in several microns range (average 2.52 μm for $Co_{1.5}Zn_{0.5}Ba_2Fe_{12}O_{22}$ and 2.85 μm for $Ni_{0.51}Zn_{0.49}Fe_2O_4$).

It is interesting to note that for Zn-ferrite particles, the change tendency of final dwell temperature with different Zn content in the ac field with both 72kHz and 5.85MHz is similar.

3. The effect of magnetic field intensity and frequency on heating efficiency and speed

With regard to the electromagnetic devices used for magnetic induction heating, the technology of AC magnetic field is still under development. Most of magnetic magnetic induction heating experiments are performed with laboratory-made generators in the frequency range of 50 kHz to 10MHz, with magnetic field amplitudes up to few tens of mT, using an induction coil (Fig.3) or in the air-gap of a magnetic inductor, which are also used in our experiments.

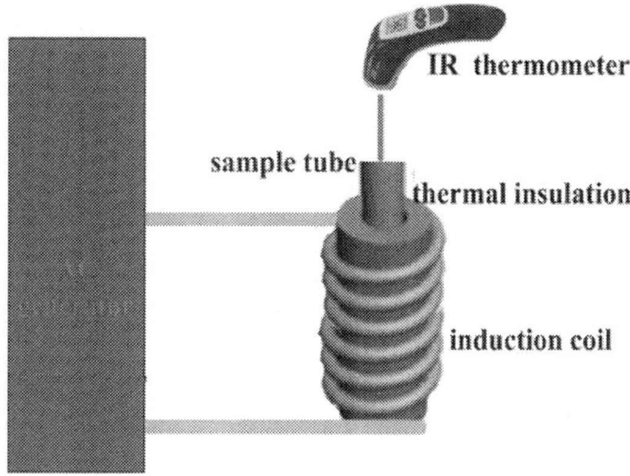

Fig. 3. Schematic diagram of a typical laboratory- made magnetic heating device

There are two kinds of sensor for detecting temperature including contact and non-contact type. The contact temperature sensor is usually thermistance. It is cheaper but poor reliability, and it required the circuit for signal treatment. We recommend DS18B20 digital temperature sensor made by DALLAS Company we have used, which is appropriate for detecting the heating temperature of magnetic liquid due to its wider temperature detecting range from -55°C to 125 °C and higher temperature resolution of 0.0625°C. In addition, DS18B20 digital temperature sensor has many advantages of smaller volume, stronger anti-jamming ability and easy connection with the microprocessor [22]. Such contact type sensor is suitable to measure the temperature of the ferrofluid. Two kinds of non-contact type temperature sensor were usually used, optical fiber and infrared ray sensor. The former should be more accurate than later, both are appropriate for detecting the heating temperature of naked materials.

In our experiment, a fundamental, quantitative understanding of the heating mechanisms in the presence of an induction field, whose frequency and intensity are typically 72kHz and 0-

60 mT, respectively, have been investigated on three series of spinel structured ferrites: ZnFe ferrite ($Zn_xFe_{3-x}O_4$, x = 0~0.8), NiZn ferrite ($Ni_{1-x}Zn_xFe_2O_4$, x=0~0.7) and CoZn ferrite ($Co_{1-x}Zn_xFe_2O_4$, x =0~0.8). Different level of Zn substitution allows a systematic variation in Curie temperature, from 100 to 480 °C. Both materials were synthesized using a chemical co-precipitation method. The chemical reaction equation is as follows:

$$xZn^{2+}+(1-x)M^{2+}+2Fe^{3+}+8OH^- \rightarrow Zn_xM_{1-x}Fe_2O_4+4H_2O, \quad M = Fe, Ni, Co$$

The compositions of various ferrite particles were determined by EDS, the structures were analyzed by XRD and the particle size were measured by TEM and XRD listed in Table 1.

x	size of particles (nm)		
	$Zn_xFe_{3-x}O_4$	$Ni_{1-x}Zn_xFe_2O_4$	$Co_{1-x}Zn_xFe_2O_4$
0	16	12	9
0.2	16	-	9
0.3	-	23	-
0.4	15	23	8
0.5	13	23	-
0.6	-	17	6
0.7	-	15	-
0.8	11	-	6

Table 1. The composition and the size of 3 kinds of ferrite nano-particles

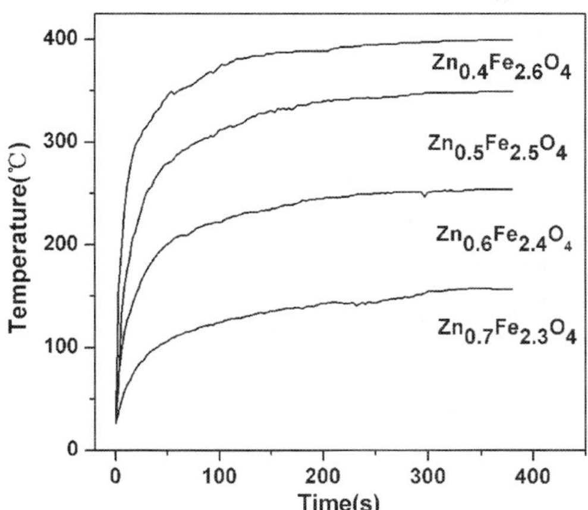

Fig. 4. The induction heating curves of $Zn_xFe_{3-x}O_4$ ferrite nanoparticles in ac field intensity with a frequency of 71 KHz and a intensity of 54.3mT.

Fig.4 shows the induction heating curves of $Zn_xFe_{3-x}O_4$ ferrite nanoparticles in ac field with a frequency of 71 KHz and an intensity of 54.3mT. It is seen that the heating temperature increases with increasing time and reaches equilibrium basically after 100-200 seconds. The dwell temperature can be obtained by extrapolating from linear part of the curve to 0 second. From Fig.4, we see the dwell temperature increases with decreasing Zn content. If we define a heating rate as $\Delta T/\Delta t$, the heating rate also increases with decreasing Zn additives. Same result of $Co_{1-x}Zn_xFe_2O_4$ is shown in Fig.5, the dwell temperature and heating rate increase with deceasing Zn content. However, for $Ni_{1-x}Zn_xFe_2O_4$, the dwell temperature and heating rate show different behaviors. The highest dwell temperature and heating rate appear at the Zn concentration of 0.5, which is the composition of largest magnetization. Also, the heating rate of $Co_{1-x}Zn_xFe_2O_4$ ferrets is obviously faster than that of $Ni_{1-x}Zn_xFe_2O_4$, which might be due the larger magnetic hysteresis loop area of the former than the latter.

The effect of AC magnetic field with fixed frequency of 72kHz and different amplitude on magnetic induction heat behavior of three Zn ferrite nano-materials has been studied. Fig.6 shows the heating curve of $Co_{0.8}Zn_{0.2}Fe_2O_4$ ferrite nanoparticles in different intensity of ac

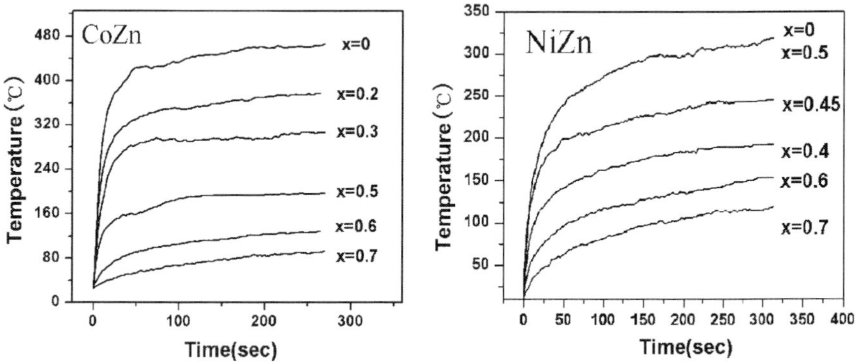

Fig. 5. The induction heating curves of $Ni_{1-x}Zn_xFe_2O_4$ and $Co_{1-x}Zn_xFe_2O_4$ ferrite nanoparticles in ac field with a frequency of 71 KHz and an intensity of 41.1mT.

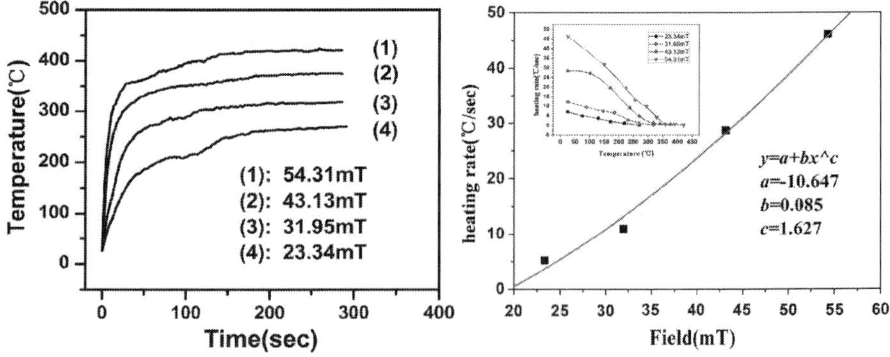

Fig. 6. The hearting curve of $Co_{0.8}Zn_{0.2}Fe_2O_4$ ferrite nanoparticles in different intensity of ac magnetic field (left) and the hearting rate as a function of amplitude of ac fields.

magnetic field and the heating rate as a function of amplitude of ac field. From Fig.6, we see that the dwell temperature and heating rate increase with increasing ac field (left) and the heating rate in different AC fields decreases as the temperature rises monotonously (inset). Fig.6 (right) shows the induction heating rate (when the temperature rise from room temperature to 100°C) as a function of field intensity, for $Co_{0.8}Zn_{0.2}Fe_2O_4$ ferrite. It is noted that the heating rate is approximately proportional to $H^{1.6}$.

For ferrites with the high resistivity eddy current has little contribution to heating. We assume that heat generation basically comes from hysteresis loss. The heat generation from hysteresis is given by [23]

$$\bar{C}_p m \, (dT/dt) = Q_{generation} = c_1 f W_h , \tag{9}$$

where in the ideal case, $c_1=1$. W_h refers to magnetic hysteresis loss for medium and strong magnetic field. Thus the expression of heating rate is

$$\frac{dT}{dt} = \frac{c_1}{C_p m} f W_h = \frac{c_1}{C_p m} f B_m^{1.6} \tag{10},$$

which just agrees well with our experimental results.

Comparing with higher frequency studied by John Xiao group[21], heating rates increase dramatically with increasing field intensity for Co2Y ferrite and Ni ferrite as shown in Fig.7, while dwell temperatures only increase slightly with increasing field intensity. The ac field intensity in their study is lower than ours. Furthermore, in their study, the heating rates for the Co2Y harder ferrite are at least twice as high as those of the softer Ni-ferrite, which is similar with ours.

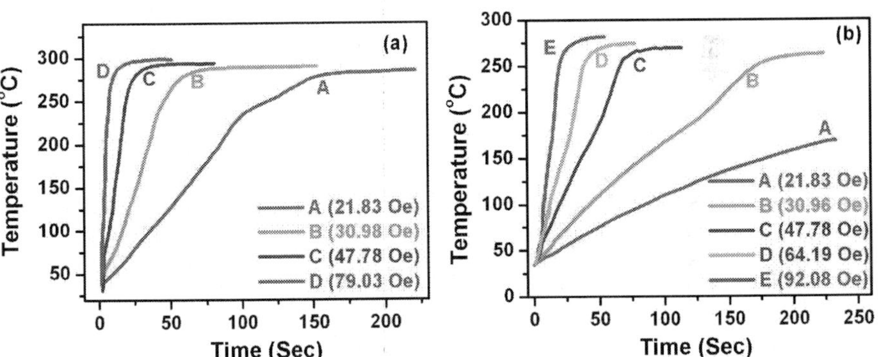

Fig. 7. Heating behaviors of (a) $Co_{1.5}Zn_{0.5}Ba_2Fe_{12}O_{22}$ and (b) $Ni_{0.51}Zn_{0.49}Fe_2O_4$ in a 5.85 MHz ac field[21].

Figure 8 shows the induction heating rate at a fixed temperature (90 °C) as a function of field intensity with 5.85 MHz[21], for one Co2Y ferrite. Note that the heating rate is approximately proportional to the square of the field intensity, as predicted theoretically in expression (4) of hysteresis loss in "Rayleigh" region. By using the correction factor of $c_1=0.418$, the theoretical volumetric heat generation rate based on the DC hysteresis loop and the induction field frequency, and the measured volumetric heat generation rate in the sample calculated from the measured heating rates as a function of field intensity are also shown in the Fig8(b).

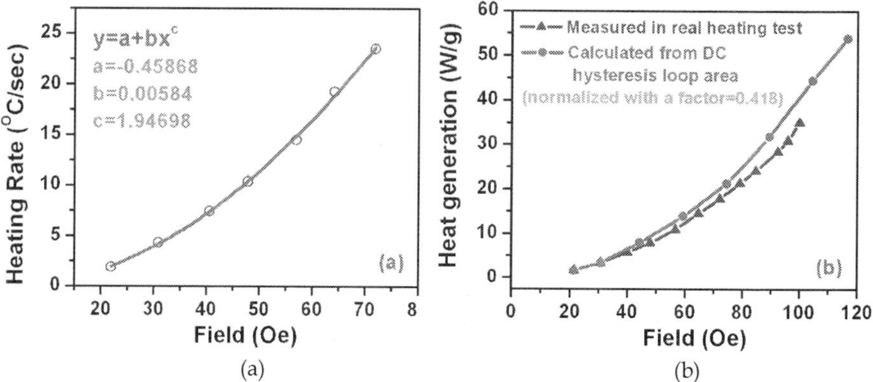

(a) (b)

Fig. 8. (a) Field dependence of heat generation for $Co_{1.5}Zn_{0.5}Ba_2Fe_{12}O_{22}$ (T=90 °C) in a 5.85 MHz ac field and (b) its comparison with a theoretical curve calculated from the dc hysteresis loop area. [21]

Figure 9 shows induction heating rate as a function of temperature at a fixed field intensity of 72 Oe and frequency of 5.85 MHz, for Co2Y ferrite and Ni ferrite studied by ref [21]. The heating rate shows a maximum at an intermediate temperature between room temperature and the final dwell temperature. We also observed the similar peak of hearting rate at lower frequency and higher field intensity.

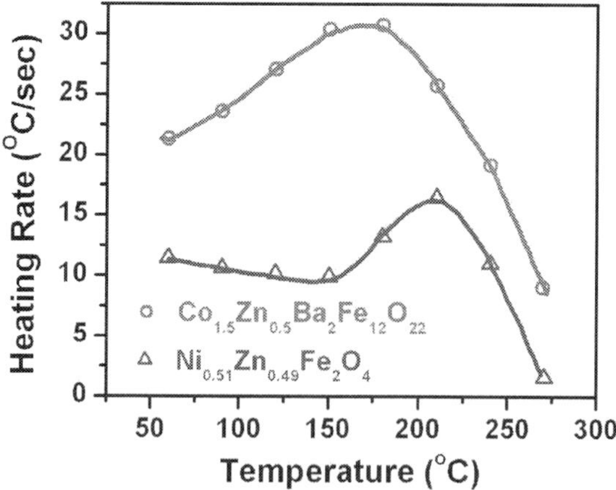

Fig. 9. Temperature dependence of heating rate for $Co_{1.5}Zn_{0.5}Ba_2Fe_{12}O_{22}$ and $Ni_{0.51}Zn_{0.49}Fe_2O_4$ in a 5.85 MHz ac field (72 Oe). [21]

In summary, we investigated the effect of AC magnetic field intensity on magnetic induction heating of three ferrits of $Zn_xFe_3O_4$, $Ni_{1-x}Zn_xFe_3O_4$ and $Co_{1-x}Zn_xFe_3O_4$ nanoparticles. The heating rate increases with increasing intensity of magnetic field and is approximately proportional to the $B_m^{1.6}$, which is consistent with expression (5) for loss at medium and strong magnetic field.

4. Magnetic fluid hyperthermia behavior

Magnetic fluid hyperthermia is based on nanoscale mediators in the form of intravenously injectable colloidal dispersion of magnetic particles. Magnetic nano-particles are dispersed in different liquid carriers to form different magnetic colloid, which can be used to simulate the hypermedia under various bionic or clinic conditions. Since the final temperature of magnetic colloid is limited by liquid boiling point, it is impossible to reach the dwell temperature of net particles (near T_c). Also, the heating rate is not only determined by magnetic particles, but also influenced by many factors such as liquid carrier and so on. Thus magnetic fluid induction heating behaviors is different from that of the net magnetic particles. Fig.10 shows the induction heating curves of the magnetic-water colloid (or suspension) containing $Ni_{0.55}Zn_{0.45}Fe_2O_4$ ferrite nanoparticles with a concentration of the 3mg/ml and the effect of ac field amplitude at a frequency of 71 kHz. It is seen that the heating rate ($\Delta T/\Delta t$) decreases when the liquid temperature increases and finally the temperature approaches a maximum value, dwell temperature. It increases with induction field intensity but the dwell temperature is far below the Curie temperature of the ferrite. This is because the ferrite particles are dispersed in water with boiling point of 100 °C which is the maximum dwell temperature. The dwell temperature geneally refers to the balance between the incoming energy of induction heating and outgoing energy due to evaporation

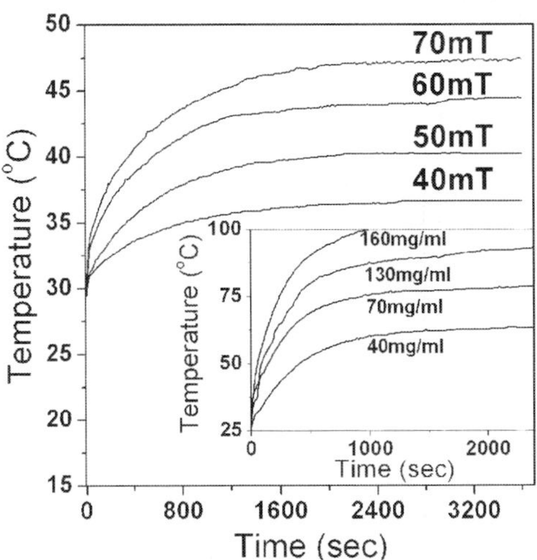

Fig. 10. The induction heating curves of $Ni_{0.55}Zn_{0.45}Fe_2O_4$ ferrite with different concentration at the AC magnetic field of 60 mT.

and heat leakage. The dwell temperature increases with the concentration of the ferrite particles in water and reaches boiling point when the concentration of liquid reaches 160 mg/ml as shown in the inset of Fig.10. The heating rate and dwell temperature increase with increasing field intensity. They are similar to and differ slightly from that in net particles above. The temperature increase of unit mass of magnetic materials due to induction heating in unit time, defined as Specific Absorption Rate (SAR), is determined from the linear temperature rise ΔT of the liquid measured in a time interval Δt after switching on the magnetic field, which is shown by following expression:

$$SAR = n\frac{\Delta T}{\Delta t}$$

where n is the heat capacity of the particle sample, and $\Delta T/\Delta t$ is the slope of the temperature rising data in a time interval. The value of SAR/c obtained from the data in Fig. 10 for Δt =500 sec is shown in Fig. 11. It is seen that the values of $\Delta T/\Delta t$ (SAR/c) are linearly proportional to the square of field strength H^2. The fitted proportional coefficient is SAR/n = $\Delta T/\Delta t$ = 2.8×10^{-4} H^2 (°C/Sec). It differs slightly from that in net particles in Fig.6, and agrees qualitatively with the power loss of magnetic materials in alternating magnetic field according to Rosensweig theory[24]:

$$P = \pi\mu_0\chi H_0^2 f\frac{2\pi f\tau}{1+2\pi f\tau}\infty H_0^2$$

τ is the relaxation time, μ_0 = $4\pi\times10^{-7}$ (T mA^{-1}) is the permeability of vacuum and χ is the magnetic susceptibility of the ferrite nanoparticles.

For $Zn_{0.3}Co_{0.7}Fe_2O_4$ nano-ferrite colloid, we also observed that the values of $\Delta T/\Delta t$ (SAR/c) are linearly proportional to the square of field strength as shown in Fig.12. However the different proportional coefficient is obtained by fitting of $\Delta T/\Delta t$ with respect of H^2 for different magnetic colloid.

According the equations of power loss, the theoretical values of SAR should be linearly proportional to the frequency. Dong-Hyun Kim has reported the tested results of

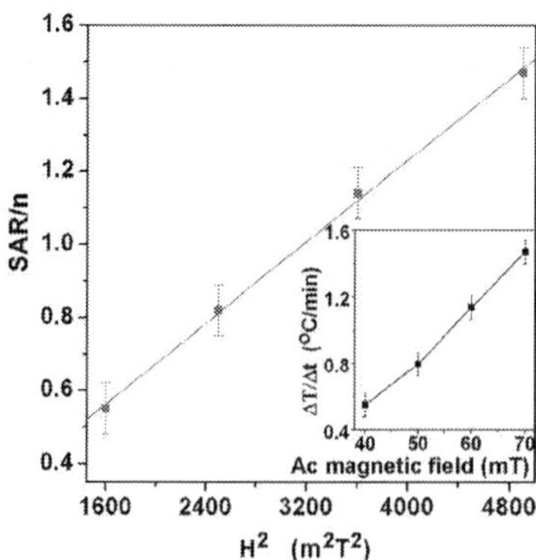

Fig. 11. The SAR/c ~ H^2 curve of the magnetic liquid containing $Ni_{0.55}Zn_{0.45}Fe_2O_4$ ferrite Nanoparticles with a concentration of the 3 mg/ml.

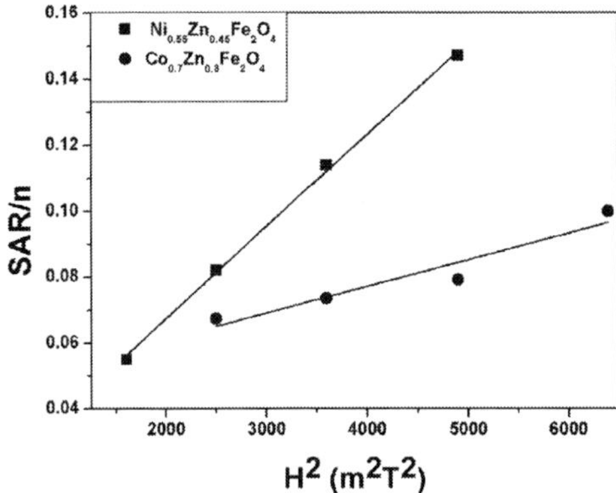

Fig. 12. The SAR/n ~ H^2 curve of the magnetic liquid containing $Ni_{0.55}Zn_{0.45}Fe_2O_4$ and $Zn_{0.3}Co_{0.7}Fe_2O_4$ ferrite nanoparticles with a concentration of the 3 mg/ml.

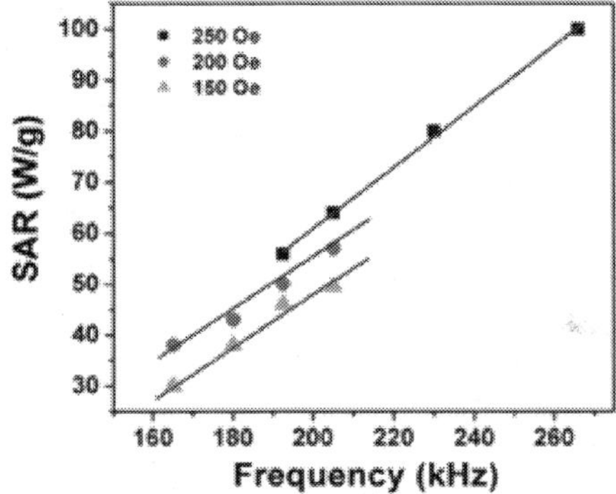

Fig. 13. Frequency of applied ac magnetic fields-dependency of the SAR in the $MnFe_2O_4$ nanoparticles with diameters 10.5 nm[25]

single-crystal $MnFe_2O_4$ nanoparticles with diameters 10.5 nm in three different magnetic field amplitudes. The values of SAR of $MnFe_2O_4$ nanoparticles were found to increase linearly with frequency as shown in Fig.13 [25]. Actually, as the frequency of applied ac magnetic field increases and reaches the resonant frequency of the material, the measured heating rates of net particles does not show a linearly change tendency with frequency, larger than the theoretical heating rates. John Q. Xiao[23] has reported the induction heating rate as a function of frequency at fixed temperature and field strength for a Co2Y ferrite as shown in Fig 14. The dashed line is the theoretical heating rate as a function of frequency and the square dot is the measured heating rate. Note that the measured and theoretical heating rates agree up to around 4.5 MHz, but above that frequency the measured heating rates are higher. This phenomenon is interpreted as due to domain wall resonance effects, and they confirmed the low-field permeability as a function of frequency showed a resonant peak for this material at around 10 MHz.

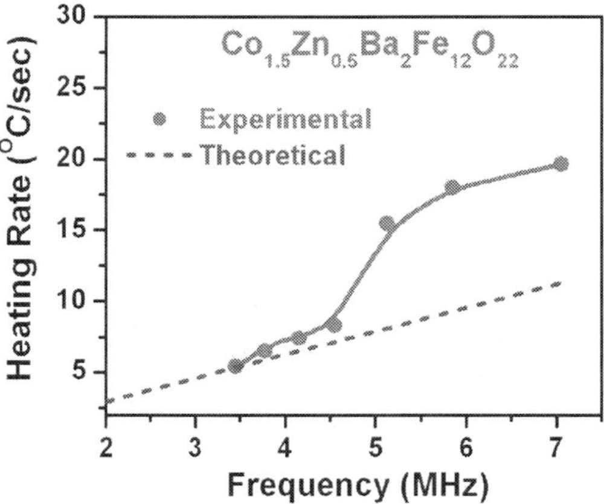

Fig. 14. The experimental and theoretical heating rate dependence of frequency for Co$_{1.5}$Zn$_{0.5}$Ba$_2$Fe$_{12}$O$_{22}$ at 150 °C and 60 Oe [21]

Fig.15 shows the heating curve and the heating rate of Ni$_{0.5}$Zn$_{0.5}$Fe$_2$O$_4$ fluid with different particle size. It is seen that the dwell temperature and heating rate increase with increasing particle size. The dwell temperature and heating rate reach the maximum (46.5°C and 1.65°C/min) when the particle size increases to critical size from single domain(45nm) to multidomain [26], then they decrease when the particles increase their size further. This tendency agrees with that of coecivity as a function of particle size. The results indicate that the SAR values of magnetite particles are strongly size dependent.

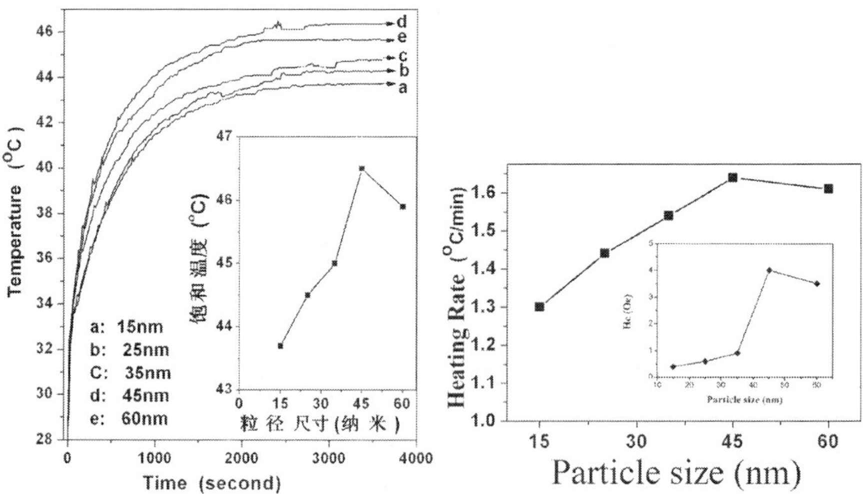

Fig. 15. The heating curve and the heating rate of Ni$_{0.5}$Zn$_{0.5}$Fe$_2$O$_4$ fluid with different particle size.

The power loss of two types of particles was studied by the group of Ning Gu[27]. They reported the specific absorption rate (SAR: specific loss power, defined as the heating power of the magnetic material per gram.) of aqueous suspensions of magnetite particles with different diameters varying from 7.5 to 416nm by measuring the time-dependent temperature curves in an external alternating magnetic field (80 kHz, 32.5 kA/m). For the magnetite particles larger than critical size of single domain, the SAR values increase as the particle size decreases and the coercivity Hc varies with the particle size which matches the variation of SAR values perfectly, indicating that hysteresis loss is the main contribution. For magnetite particles of 7.5 and 13nm which are superparamagnetic, hysteresis loss decreases to zero and, instead, relaxation losses (Neel loss and Brownian rotation loss) dominate, but Brown and Neel relaxation losses of the two samples are all relatively small in the applied frequency of 80 kHz.

Fe_3O_4 is a nery attractive nano-material for bio-applications because it contains only Fe ions which is suitable for body, thus it has been studied extensively [28, 29]. It is shown by studies that the liquid carrier is also an important factor for heating process. The heating rate and the dwell temperature exhibits different values and behavior for different stickiness in colloid. In our experiment, magnetic fluid with concentration of 2 mg/mL was prepared by combining the Fe_3O_4 particles of diameter of 20 nm and deioned water. With the fixed frequency of ac magnetic field at 72 kHz and the varied amplitude of ac fields of 43.1, 54.3, 65.5 mT, we obtained the heating curve of the colloid with Fe_3O_4 particles as shown in Fig. 16. From the left figures in Fig.16, it is seen that the SAR($\sim\Delta T/\Delta t$) also shows to be linearly proportional to the square of field amplitude H^2. While in the right figure of Fig.16, it is seen that the more the viscosity of liquid carrier, the higher the dwell temperature and heating rate.

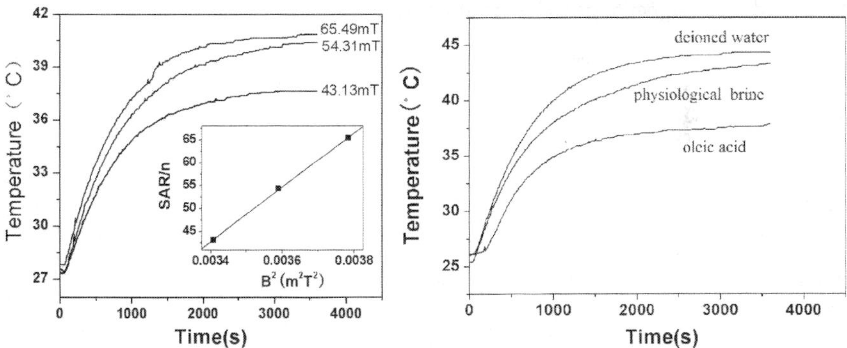

Fig. 16. The induction heating curves of Fe_3O_4 ferrite colloid with the concentration of 2 mg/ml at different amplitudes of AC magnetic field at 72 kHz (left) and in different liquid carrier(right)

On the basis of the physical and biological knowledge, when the magnetic therapy particles are applied for hyperthermia, how to decrease the toxicity and increase the biocompatibility of the particles must be considered. Dong-Lin Zhao et al reported the effect of Fe_3O_4 particles modified with chitosan[30]. the chitosan–Fe_3O_4 particles show multiple fine properties such as non-toxicity, well biocompatibility, biodegradable and anti-bacterial[31]. Babincova et al also reported dextran-magnetite has no measurable toxicity index LD50. [32, 33]

Fig.17 shows that the heating curves of Fe_3O_4 particles with and without SiO_2 shell in the ac magnetic field with 43.1 mT and 72kHz. The average particle size is 30 nm and the shell thickness is about 10 nm. The liquid carrier is deioned water and the concentration of the colloid is about 2 mg/ml. All the test samples have the same volume. From Fig.17 it is seen that the dwell temperature and the heating rate decrease for Fe_3O_4 @ SiO_2 core- shell structures in the colloid compared with Fe_3O_4 particles without shell. This may be partly due to the decrease of the amount of magnetic particle in the colloid with the same volume.

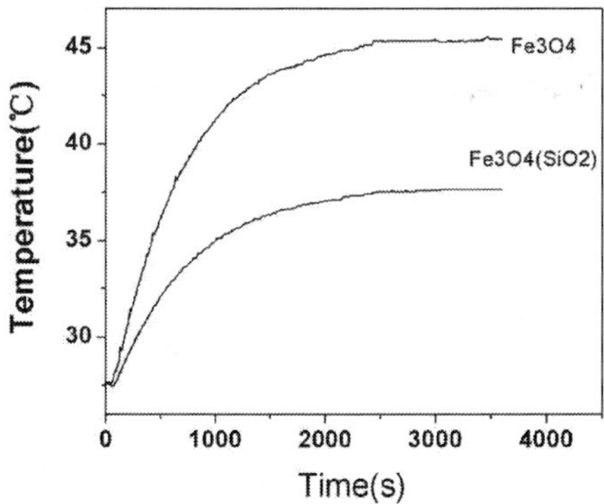

Fig. 17. The heating curves of magnetite with and without silicon oxide shell at ac field.

The heating for ferrofluid in hyperthermia is mainly due to hysteresis loss, Brownian and Neel relaxation loss. The ferrofluid contains the magnetic particles modified with polymer may enhance the Brownian and Neel relaxation loss, which may also increase the SAR value of the ferrofluid. Li-Ying Zhang reported magnetite particles with diameter of 50nm coated by dextran to form homogeneous ferrofluid. The ferrofluid presented a highest SAR value of 75 W/g, which is much higher than 4.5W/g for the 50nm uncoated particles at 55 kHz and 200 Oe. This results indicate that the magnetite particles modified with dextran not only can decrease the toxicity but also can increase the SAR of magnetite ferrofluid[34].

In summary, we report the magnetic induction heating behaviors of magnetic colloids with nanometer particles of ferrite containing Zn and magnetite particles. The SAR increases with increasing intensity of magnetic field and is approximately proportional to the square of the field intensity. They are different from that of the net particles.

5. The closing note

Since the pioneering work of Gilchrist et al. in 1957, magnetic hyperthermia has been the aim of numerous in vitro and in vivo investigations [35, 36], but most of the studies were unfortunately conducted with inadequate animal systems, inexact thermometry and poor AC magnetic field parameters, so that any clinical application was far behind the horizon.

6. Acknowledgment

This work is supported by NBRP (Grant Nos. 2010CB923401 and 2011CB707601), NSFC (Grant Nos. 50472049, 50871029 and 11074034), National Laboratory of Solid State

Microstructures at Nanjing University, and Jiangsu Key Laboratory for Design and Manufacture of Micro-Nano Biomedical Instruments, Southeast University.

7. Reference

[1] J. Q. Cao, Y. X. Wang, J. F. Yu, J. Y. Xia, C. F. Zhang, D. Z. Yin, and U. O. Urs O. Häfeli, J. Magn. Magn. Mater., 277, 165 (2004).

[2] D. K. Kim, Y. Zhang, J. Keher, T. Klason, B. Bjelke, and M. Muhammed, J. Magn. Magn. Mater., 225, 256 (2001).

[3] C. H. Dodd, H. C. Hsu, W. J. Chu, P. Yang, H. G. Zhang, Jr J. D. Mountz, K. Zinn, J. Forder, L. Josephoson, R. Weissleder, J. M. Mountz, and J. D. Mountz, J. Immunol. Methods, 256, 89 (2001).

[4] Q. A. Pankhurst, J. Connolly, S. K. Jones, and J. Dobson, J. Phy. D: Appl. Phys., 36, R167 (2003).

[5] P. Tartaj, M. P. Morales, S. Veintemillas-Verdaguer, T. González-Carreño, and C. J. Serna, J. Phy. D: Appl. Phys., 36, R182 (2003).

[6] P. S. Doyle, J. Bibette, A. Bancaud, and J.-L. Viovy, Science, 295, 2237 (2002).

[7] J.-M. Nam, S. I. Stoeva, and C. A. Mirkin, J. Am. Chem. Soc., 126, 5932 (2004).

[8] U. Häfeli, W. Schütt, J. Teller, M. Zborowski, Scientific and Clinical Applications of Magnetic Carriers. New York: Plenum Press, 57, 452 (1997).

[9] W. J. Parak, D. Gerion, T. Pellegrino, D. Zanchet, C. Micheel, S. C. Williams, R. Boudreau, M. A. L. Gros, C. A. Larabell, and A. P. Alivisatos, Nanotechnology, 14, R15 (2003).

[10] A. Jordan, R. Scholz, P. Wust, H. Fakhling, R. Felix, J. Magn. Magn. Mater., 201, 413 (1999)

[11] N. Kawai, A. Ito, Y. Nakahara, M. Futakuchi, T. Shirai, H. Honda, T. Kobayashi, and K. Kohri, The Prostate, 64, 373 (2005)

[12] D. Stoppels, J. Magn. Magn. Mater., 160, 323 (1996).

[13] O. Inoue, N. Matsutani, and K. Kugimiya, IEEE Trans. Magn., 29, 3532 (1993)

[14] J.Smit and H. P. J. Wijn, Ferrites ﹐

[15] H. Jordan, Elect. Nachr. Techn. 1, 7 (1924).

[16] V. E. Legg, Bell System Tech. J. 15, 39 (1936).

[17] R. M. Bozorth, Ferromagntism, Princeton, N.J. (1951).

[18] R. E. Rosensweig, J. Magn. Magn. Mater., 252, 370 (2002).

[19] M. I. Shliomis, Sov. Phys.-Uspech., 17, 153 (1974).

[20] E. D. Wetzel, B. K. Fink, Y. F. Li, and J. Q. Xiao, Army Sci. Conference, Baltimore, MD (2000).

[21] X. K. Zhang, Y. F. Li, J. Q. Xiao and E. D. Wetzel, J. Appl. Phys. 93, 7124 (2003).

[22] P. M. Levy, S. Zhang, Phys.Rev.Lett., 79 , 5110 (1997)

[23] S. B. Liao, Ferromagnetism (Science Publishing, Beijing, China, 1988), Vol. 3.

[24] R. E. Rosensweig, J. Magn. Magn. Mater., 252, 370 (2002).

[25] D. H. Kim, Y. T. Thai, D. E. Nikles, and C. S. Brazel, IEEE Trans. Magn. 45, 64 (2009)

[26] A. S. Albuquerque, J. D. Ardisson, and W. A. A. Macedo, J. Appl. Phys. 87, 4352 (2000).

[27] M. Ma, Y. Wu, J. Zhou, Y.K. Sun, Y. Zhang, N. Gu, J. Magn. Magn. Mater. 268, 33 (2004)

[28] D.L. Zhao, H. L. Zhang, X. W. Zeng, Q. S. Xia and J. T. Tang, Biomed. Mater., 1, 198 (2006)

[29] L. F. Gamarra, W. M. Pontuschka, J. B. Mamani, D. R. Cornejo, T. R. Oliveira, E. D. Vieira, A. J. Costa-Filho and E. Amaro Jr, J. Phys.: Condens. Matter 21, 115104 (2009)

[30] D.L. Zhao, X.X.Wang, X.W. Zeng, Q.S. Xia, J.T. Tang, J. Alloys Compd., 477, 739 (2009)

[31] J. Zhi, Y.J. Wang, Y.C. Lu, J.Y. Ma, G.S. Luo, React. Funct. Polym. 66, 1552 (2006).

[32] M. Babincova, D. Leszczynska, P. Sourivong and P. Babinec, Med. Hypoth. 54, 177 (2000).

[33] M. Babincova, P. Sourivong, D. Leszczynska and P. Babinec, Med. Hyptoth. 55, 459 (2000).

[34] Li-Ying Zhang, Hong-Chen Gu, Xu-Man Wang, J. Magn. Magn. Mater., 311, 228 (2007).

[35] S. Mornet, S. Vasseur, F. Grasset, E. Duguet. J. Mater. Chem., 14, 2161 (2004).

[36] P. Moroz P, S. K. Jones, B. N. Gray. Int. J. Hyperthermia, 18, 267 (2002)

Part 2

Microwave Heating of Organic Materials

22

Changes in Microwave-treated Wheat Grain Properties

Jerzy R. Warchalewski[1], Justyna Gralik[1], Stanisław Grundas[2],
Anna Pruska-Kędzior[3] and Zenon Kędzior[3]
*[1]Department of Biochemistry and Food Analysis,
Poznan University of Life Sciences, Poznan,
[2]Bohdan Dobrzanski Institute of Agrophysics, Polish Academy of Science, Lublin,
[3]Institute of Food Technology, Poznan University of Life Sciences, Poznan,
Poland*

1. Introduction

The world faces the challenge of feeding, housing and clothing an everincreasing population. On 12 October 1999, the United Nations marked the birth of the six billionth human currently living on the planet (Morris et al., 2000). Throughout our history there has been an overall increase in food production through agricultural innovations including the efforts of plant breeders. Cereals play such an integral part in global agriculture and diet. More than 50% of our food comes from three cereals: wheat, maize and rice (Morris et al., 2000). An important step forward in the feeding of the world was the green revolution. Advances in plant breeding and the adoption of highly efficient production systems bring about of fourfold increase in grain yield during the second half of twentieth century. But a continued, sustainable increase will be hard to realize without introducing modern cereals biotechnology and strong determination towards limitation of cereals grain losses during shipment and storage.

Pests are one of the most severe threat for food products, but often ignored by many manufactures. According to Codex Alimentarius monitoring as well as fighting pests in production process is essential to maintain safe food products. Wheat is the world's most important crop species providing about one-third of the global production of cereals. Stored grain is vulnerable to damage caused by internal and external insects. About half of annual grain production requires storage (Grundas et al., 2008). Cereal grain losses caused by pests during storage can reach 50% of the total harvest in some developing countries, which leads to a world-wide loss equivalent to thousands of millions of euros per year (Fornal et al., 2007).

Generally postharvest food losses are estimated to range from 8 to 10% in industrialized countries (Ciepielewska & Kordan, 2001; Brader et al., 2002). Insects are a problem in stored grain throughout the world because they reduce the quantity and quality of grain (Sinha & Watters, 1985; Madrid et al., 1990; Warchalewski & Gralik, 2010). Many chemical insecticides and fumigates are used as protectants against insect infestation in stored grains. But their indiscriminate use and residual toxicity effect the non-target animals and human beings

(Modgil & Samuels, 1998). Furthermore, fumigation is being increasingly restricted for environmental reasons.

Many countries such as Australia, Canada and France, have zero tolerance for live insects in grain (Fields, 2006). On the other hand, increasing concern by consumers over insecticide residues in processed cereal products call for new approaches to control stored-product insect pests. The use of low temperature, the higher temperatures, gamma radiation, microwave treatment, atmospheric gases, natural protectants, hormones and natural cereals grain resistance mechanisms are an alternative to chemical pesticides for the control of stored grain pests (Warchalewski & Nawrot, 1993; Prądzyńska, 1995; Shayesteh & Barthakur, 1996; Modgil & Samuels, 1998; Aldryhim & Adam, 1999; Warchalewski et al., 2000; Ciepielewska & Kordan, 2001; Cox & Collins, 2002; Nawrot et al., 2006; Nawrot et al., 2010). Among physical methods, a known nonchemical alternative for controlling hidden infestations inside kernels is the application of extreme temperatures, preferably heat because its lethal effects are usually faster than cold (Field & Muir, 1996).

The main advantage of microwave is fast and selective heating ability compared with the conventional heating methods. Because relative values of dielectric properties of insects and grain are extremely different, selective heating of insects can be achieved (Shayesteh & Barthakur, 1996).

Microwaves are a nonionising form of energy that interact with polar molecules and charged particles of penetrated medium to generate heat. Use of microwave heating has the advantage of saving time and energy, improving both nutritional quality and acceptability of some foods by consumers (Nijhuis et al., 1998; Sumnu, 2001). On the other hand, a significant reduction of stored-grain damage by insect pests can be achieved with the use of microwaves (Nelson, 1996; Plarre et al., 1999; Nawrot et al., 2006).

Microwave heating shortens processing time because microwaves penetrate solid matter and therefore work faster than systems which transfer heat by conduction from surface to centre. This improves product quality by minimizing changes in product texture, nutritional value and lowering moisture content.

In addition microwave treatment is a technique significantly attractive to be applied widely in food production (Venkatesh & Raghavan 2004). Today the food industry is a major user of microwave energy, especially in the drying of pasta and postbaking of biscuits (Vadivambal & Jayas, 2007). Microwave drying of grain is also conducted in pilot plants in the USA, Canada (Nijhuis et al, 1998; Jayas & White, 2003), and India (Walde et al., 2002). Microwave treatment can generally reduce drying time of biological products without quality degradation (Nijhuis et al., 1998). Controling heating uniformity is important to ensure to microbial safety and high quality of microwave-dried foods (Sundaram & Huai-Wen, 2007).

The industrial and the domestic use of microwaves has increased dramatically over the past few decades. While the use of large scale microwave process is increasing, recent improvements in the design of high-powered microwave ovens, reduced equipment manufacturing cost and trends in electrical energy costs offer a significant potential for developing new and improved industrial microwave process (Vadivambal & Jayas, 2007). The term food quality includes three principal areas: nutritional value, acceptability and safety. Almost any method of processing raw food will have adverse effect on some of its nutrients. Acceptability includes large array of attributes like visual appeal, aroma, flavour and texture (Warchalewski et al., 1998). The overall quality of a product is judged by the number of parameters. Good quality is judged by freshness, expected appearance, flavour and texture. The quality changes that can happen in any product during drying are changes in optical properties (colour, appearance), sensory properties (odour, taste, flavour), structural properties (density, porosity, specific volume), textural properties, rehydration properties (rehydration rate and capacity) and nutritional characteristics e.g. vitamins, proteins (Krokida & Maroulis, 2001). Food safety is protecting the food from physical (drying out, infestation), microbial, insects and chemical (rancidity, browning) hazardous or contamination that may occur during all stages of food production, growing, harvesting, processing, transporting, distribution and storage.

Nijhuis et al. (1998) presented advantages and disadvantages of alternative technologies of dry conservation process like: freeze drying, microwave and radio frequency. Microwave drying has positive ratings for: drying rate, flexibility, colour, flavour, nutritional value, microbial stability, enzyme inactivation, rehydration capacity, crispiness and fresh-like appearance. These qualities pointed out that microwave irradiation is a good technology for dry conservation process of fruits, vegetables and cereals grain on condition that the time of heating is properly chosen.

There is a lack of general models for predicting the heating pattern, moisture and vapour distribution during the drying procedure based on dielectric properties, water distribution, density, changes in microstructure and composition of food (Nijhijus et al., 1998). Reliable date on microwave-treated food products are still missing. Also more knowledge is needed about the influence of geometry, size of object being dried and phenomena like shrinkage of kernels, resistance to insect infestation or puffing and stress-cracking.

Currently, microwave heating is often combined with other techniques for example gamma radiation to kill or reduce insects or mites that attack wheat grain (Gralik & Warchalewski, 2006).

When applying microwave energy to reduce grain damage by insects it should always be considered how this treatment will affect wheat grain properties (Grundas et al., 2008). Therefore research on direct effects of microwave-induced changes remains an important interesting challenge. Despite the importance of microwave/heat-induced changes that directly affected wheat seeds may also induce some changes in the next generation crops which should be considered. When wheat grain is treated by microwave energy particular attention should be drawn to preserve nutritional and technological properties. In the case of grain treatment as seeds, the germination potency should be preserved with acceptance of some unintentionally mutation in next generation of wheat crops. The treatment of infested grain by microwave appears a reliable alternative to conventional post-harvest insect control in the near future, either with stationary or mobile applicators on the farm or quarantine purposes during the loading process (pre- or post-shipment) before grain storage and seed sowing (Plarre et al., 1999; Dolińska et al., 2004).

2. Direct effect of microwave heating of wheat grain

Microwave energy directly applied to wheat grain significantly changed the colour of grain as was reported earlier (Warchalewski et al., 1998). The analysis of variance and calculation of the smallest significant value proved that from 90 to 180 sec (64 to 98°C) of microwave treatment, the values of L* (lightness) were significantly higher in comparison to control sample. Also visual inspection of grain samples confirmed these results. The total colour difference (ΔH) between microwave-heated samples and control grain was increasing gradually with the rise of heating time. The yellowness (b*) and the redness (a*) values were statistically significantly higher in the case of 120 and 180 sec (79 and 98°C) of heating time in relation to the control sample. However, the increase of these values was not significant enough due to equate the L* value, which was visually confirmed. The significant increase of L* value could be associated with some thermal decomposition and modification of grain components which was correlated with the increase of the grain temperature from 64 to 98°C. The temperature at which damage of wheat starch begins is suggested between 53-64°C (Cornell & Hoveling, 1998).

Burnt-like flavour is usually observed in overheated food (Garcia et al., 1975; Baldwin et al., 1991). Appearance and sensoric properties of the microwave treated plain starch and blended with some compounds pointed to the reactions occurring after treatment (Sikora et al., 1997). The development of the secondary food aromas was observed (Warchalewski et al., 1998). In the case of plain, microwave treated starch 'intensive, caramel, smoked plums' aroma appeared, similar to 'brown' observed by Warchalewski et al. (1998) when wheat grain was microwave-heated to 98°C (180 sec). On the other hand, starch blended with glutamic acid was characterized, after microwave treatment, as 'pleasant, ginger bread' (Sikora et al., 1997).

Microwave treatment longer than 90 sec (64°C) caused marked changes in wheat grain endosperm structure (Błaszczak et al., 2002). There was a statistically significant increase of intragrain cracks within the wheat kernel treated with microwave energy. Increasing percentage of damages with prolonged time of treatment was observed. The highest level was found in 180 sec heated of grain samples to 98°C. The roentgenograms confirmed observations by light microscopy (LM) and scanning electron microscopy (SEM) techniques, showing the tendency of individual grains to form highly bursting regions within the grain, in some cases resulting in overall increase of grain volume (Fornal et al., 2001; Błaszczak et al., 2002). Those techniques showed denatured proteins creating filaments connecting starch granules at 120 sec (79°C) of microwave treatment and as a next stage dense protein matrix covering high gelatinized and deformed granules at 180 sec (98°C). Denaturation changes in proteins creating visible fibrils, as well as high swelling and deformation of starch granules (radial and tangential swelling), which were related to falling number (FN) increase, decrease of Zeleny test results, and lowering gluten content and extensogram values (Błaszczak et al., 2002). Lower quality of bread with increasing time of microwave heat-treatment of grain was also observed. Generally, marked changes in kernel microstructure and technological properties were started when the temperature of grain exceeded 64°C (90 sec). These marked changes in endosperm microstructure of starch granule and proteins well explain why the larval development time of confused flour beetle (*Tribolium confusum* Duv.) and Mediterranean flour moth (*Ephestia kuehniella* Zell.) reared on microwave treated wheat grain, which was secondary infested by these pests became statistically significantly short, when compared to non-heated grain (Warchalewski et al., 2000).

Changes in the physical characteristics of grain that may be induced directly by microwave rays are important in relation to milling properties, insect resistance, and protein content which affect the properties of flour and its suitability for various purposes. The test for milling involves physical qualities such as bulk density (BD), thousand kernel weight (TKW), kernel size, shape, hardness, vitreousness, colour, and kernel damage (Satumbaga et al., 1995) more than chemical qualities. In addition, hard, virteous wheat grain is usually associated with high protein content (Cornel & Hoveling, 1998).The vitreous or mealy character of grain is genetically controlled but it is also affected by environment. On the other hand, hardness is not materially affected by the environment but is determined almost entirely by genetics (Hoseney, 1987). The Single Kernel Characterization System (SKCS) developed for wheat classification has potential for determining wheat quality parameters (Martin et al., 1993; Satumbaga et al., 1995; Gaines et al., 1996; Osborne et al., 1997; Ohm et al., 1998).

Bulk density and thousand kernel weight are two of the physical criteria of wheat grain quality. Bulk density is one of the most common parameters of evaluating wheat quality. Kernel size has very little influence on bulk density, but shape and uniformity are significant factors, as is the moisture content. Average range of bulk density value of wheat grain was ≈77-84 kg/L; the range of thousand kernel weight was 20-40 g (Cornel & Hoveling, 1998). Grundas et al. (2008) reported that thousand kernel weight was well above range, which was significantly changed only in grain directly exposed to the microwave rays at 60, 120 and 180 sec where the grain temperature was 64, 79 and 98°C respectively.

Grain samples exposed directly to microwave heating showed decreased hardness index (HI), single kernel weight (SKW) and single kernel diameter (SKD) of all samples, with the exception of hardness index of M-120 (79°C) sample and single kernel diameter of M-180 (98°C) sample, where an increase of measured features was noted (Grundas et al., 2008).

Virtually biologically active wheat proteins were found in soluble proteins. The soluble proteins are therefore composed of what are known as albumins and globulins, together with glycoproteins, nucleoproteins, and many of the lipid-protein complexes in wheat flour (Kasarda et al., 1971). Microwave heating within the range of applied times of exposure caused a decrease in extractable protein content. However, a statistically significant drop in extractable protein contents were found only at 120 and 180 sec of microwave heating time, where grain reached 79 and 98°C, respectively (Warchalewski & Gralik, 2010). These temperatures were high enough to cause a decrease of >42% of extractable proteins.

Statistically significant changes in reducing sugars content was noted in all microwave-heated grain samples. However, the application of the lowest heating times of 15 sec (28°C), 45 sec (43°C), and 60 sec (48°C) caused significant increase in reducing sugars content. At the higher heating times of 90 sec (64°C), 120 sec (79°C), and 180 sec (98°C) a gradual decrease in reducing sugars content was observed. Grain temperatures ≤48°C probably enhance amylases activity within the grain during the heating process. Moreover, some changes in starch properties are possible due to differences reported by Dolińska et al. (2004) in falling number values, peak temperatures (T_p), and gelatinization enthalpy (ΔH). This could explain the increase in extractability of reducing sugars ≤60 sec of heating time, when grain temperature was not exceeded 48°C. Changes in reducing sugars content was well correlated with endogenous amylolytic activity determined in samples of grain at ≥64°C during microwave heating. Grain at 64, 79, and 98°C showed a statistically significant decrease in reducing sugars content and amylolytic activity (Warchalewski & Gralik, 2010). The gradual decrease in reducing sugars and amylolytic activity can be attributed to the secondary reactions induced by increasing grain temperature from 64 to 98°C. A distinct decrease of grain amylolytic activity to 10% of this activity in the control sample was noted when wheat grain reached 98°C. This suggests significant reduction of grain endogenous amylase by denaturation of enzyme protein. Also all endogenous biological activities were distinctly diminished in grain samples heated to 79°C. The decrease in wheat grain samples was 79% of amylolytic activity and inhibition activities against α-amylases from insects *Sitophilus granarius* L., *Tribolium confusum* Duv., *Ephestia kuehniella* Zell., also from human saliva and hog pancreas, as well as antitryptic activity by 50, 83, 31, 63, 39, and 20%, respectively. At the highest grain temperature of 98°C, the loss of all biological activities were even more pronounced due to denaturation of $\approx$45% of extractable proteins. Differences in the decrease of biological activities among wheat grain samples microwave-heated to 79 and 98ºC suggest that these activities are located in a number of soluble proteins with different thermal stabilities to the highest temperature. Substantial decrease in inhibitory activity against α-amylases and trypsin from mammalian sources should be considered beneficial from a nutritional point of view. The influence of microwave field on inactivation of antitryptic activity in soy (Boyes at al., 1997; Mitrus, 2000) and bean seeds (Bieżanowska et al., 2000) was also confirmed earlier.

Technological tests performed on microwave-heated grain proved changes in starch and proteins fractions which could be seen in deterioration of baking value (Błaszczak et al., 2002). Lowering of participation of vitreous grains especially in the case of two the longest heating times: 120 sec (79ºC) and 180 sec (98ºC) was correlated with deterioration in farinograph and extensograph tests. It can indicate some destruction of protein conformation within the grain. Also analysis of grain microstructure by scanning electron microscopy (SEM) proved some changes in protein structure (Błaszczak et al., 2002). In addition Sodium Dodecyl Sulphate (SDS) and Zeleny sedimentation tests confirmed this results. In the case of the two longest times of microwave exposure (120 and 180 sec) was observed decreasing volumes of precipitated proteins in sedimentation tests which indirectly indicate deterioration in gluten proteins hydration and in consequence in gluten quality. A distinct decrease of hand-washed wet gluten yield was noted when grain reached 79ºC after 120 seconds while at 98ºC (180 sec) wet gluten was not possible to wash at all (Gralik, 2003; Błaszczak et al., 2002).

Rheological properties of wheat gluten are very sensible to any physical or chemical modification of the gluten molecular and supramolecular structure (Shewry et al., 2002; Lefebvre et al., 2003; Belton, 2005). It could be therefore expected that even very fine changes in arrangement and physicochemical properties of the grain storage proteins induced due to microwave energy input to wheat grain would affect the gluten structure and its viscoelastic behaviour. Therefore wet gluten was isolated from the control and microwave-treated wheat grain. The grain microwave processing time ranged from 15 to 120 sec what corresponded to the final grain temperature of 28 to 79ºC. From the grain sample microwave-heated to 98ºC by 180 sec it was not possible to wash out wet gluten as was reported earlier (Błaszczak et al., 2002; Gralik, 2003). All hand washed out wet gluten samples were freeze dried and stored prior to analysis. Samples of freeze-dried gluten were rehydrated using a two-step procedure and next submitted to rheological measurement as previously described (Lefebvre et al., 2003; Pruska-Kedzior, 2006; Pruska-Kedzior et al., 2008). Mechanical spectra of studied gluten samples were plotted as functions of the storage and loss moduli (G', G'') (Fig. 1), or of the storage and loss compliances (J', J'') (Fig. 2), versus the angular frequency ω. The J', $J'' = f(\omega)$ mechanical spectra were fitted with phenomenological Cole-Cole functions yielding the viscoelastic plateau compliance J_N^0, viscoelastic plateau modulus $G_N^0 = 1/J_N^0$, loss peak characteristic frequency ω_0, and spread parameter n related to the peak broadness (Tschoegl, 1989). These parameters describe viscoelastic properties and provide information on the networking state of a polymer structure in the upper frequency limit of the viscoelastic plateau.

The storage protein system of wheat grain appeared very sensible to the grain microwaving as it reveal mechanical spectra of the gluten obtained from this material (Fig. 1). The absolute values of G' and G'' moduli increased slightly over entire frequency range following increasing microwave energy input to the grain for microwaving time 15 sec (28°C) to 90 sec (64°C). Very significant effect of the grain microwave heating on the gluten viscoelastic properties was observed at 120 sec (79°C) microwave treatment. The mechanical spectrum of the gluten obtained from the grain microwave-heated for 120 sec (79°C) is shifted up more than one logarithmic decade of G' and G'' values proving that the gluten structure has became much more networked.

In the studied frequency range (Fig. 1), the mechanical spectrum of gluten encompasses only a part of the viscoelastic plateau. In this part of the plateau, gluten shows a transient viscoelastic network structure (Hamer & Van Vliet, 2000; Dobraszczyk & Morgenstern, 2003; Lefebvre et al., 2003; Belton, 2005; Pruska- Kędzior, 2006). Viscoelastic properties of the material manifested in the upper frequency region of the plateau can be quantified using Cole–Cole functions. The Cole-Cole parameters fitted to the mechanical spectra of the studied gluten samples are shown in Table 1. Noticeable increase of modulus of viscoelastic plateau G_N^0 was observed for the material obtained from the grain submitted to at least 60 sec of microwave heating what corresponded to the final grain temperature of 48°C. Tremendous increase of G_N^0, above 4.2 times, appeared for gluten obtained from the grain microwave-heated for 120 sec (79°C). In Fig. 2 positioning of the compliance of viscoelastic plateau J_N^0 on the J' curve and the characteristic frequency ω_0 on the J'' curve of the mechanical spectra is shown for each studied gluten sample. The J_N^0 value of wheat gluten decreased as the grain microwaving time has been increased. The effect of the grain microwaving time on ω_0 values was more complex. Noticeable decrease of ω_0 comparing to the control gluten occurred for the grain microwaving time 15 sec to 90 sec and substantial increase of ω_0, more than one logarithmic decade, has been revealed when the microwaving time reached 120 sec (Table 1 and Fig. 2).

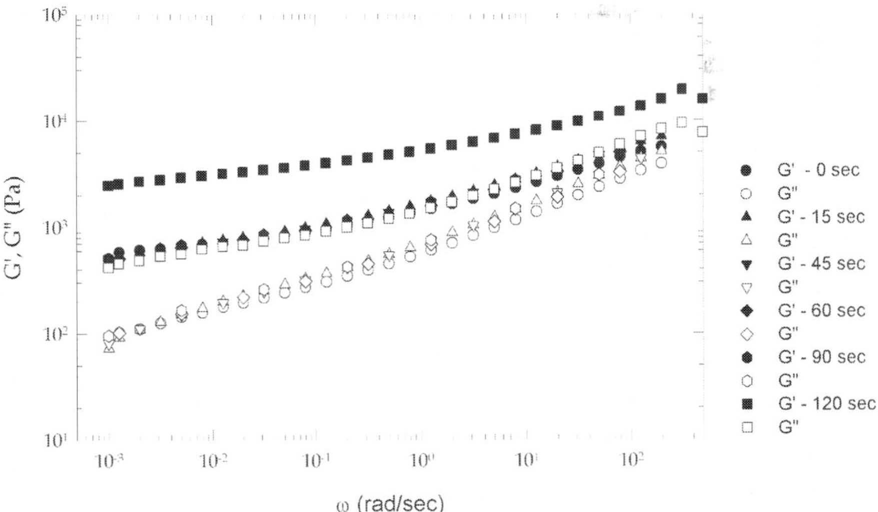

Fig. 1. The effect of microwave treatment of wheat grain on viscoelastic properties of wheat gluten. Mechanical spectra $G'(\omega)$, $G''(\omega)$

Exposure time [sec]	Grain temperature [°C]	G_N^0 [Pa]	J_N^0 [Pa^{-1}]	ω_0 [rad/s]	n	R^2
0	20	733	0.00136	0.19	0.409	0.996
15	28	751	0.00133	0.10	0.417	0.996
45	43	767	0.00130	0.12	0.418	0.996
60	48	838	0.00119	0.11	0.416	0.996
90	64	970	0.00103	0.17	0.437	0.997
120	79	3 795	0.00026	3.45	0.396	0.986

Table 1. The effect of microwave treatment of wheat grain on the Cole-Cole parameters calculated from the mechanical spectra of wheat gluten

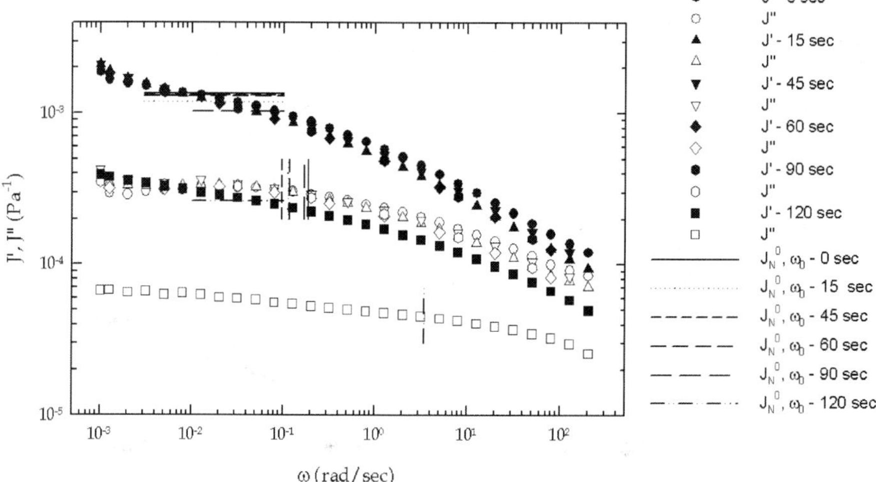

Fig. 2. The effect of microwave treatment of wheat grain on viscoelastic properties of wheat gluten. Positions of the Cole-Cole parameters J_N^0 (horizontal line crossing $J'(\omega)$ curve), and ω_0 (vertical line crossing $J''(\omega)$ curve) are marked

The effect of 120 sec microwave-treatment of wheat grain on the viscoelastic properties of gluten described in terms of J_N^0 and ω_0 values (Table 1) was intermediate to that observed for the control gluten submitted to the 5-hour contact preheating at 50°C and/or 70°C respectively, performed in the rheometer (Peltier effect). In this experiment J_N^0 value reached 0.00047 and 0.00024 Pa^{-1}, while ω_0 value varied from 3.4 for heating at 50°C to 12.6 rad/sec for heating at 70°C respectively (Pruska-Kedzior, 2006).

Studying viscoelastic properties of gluten appeared relevant to estimate degree of physicochemical denaturation of gluten protein system of wheat grain induced due to the controlled microwave treatment.

Viscoelastic effects are highly involved in such technological properties of wheat dough like dynamics of its structural development during kneading, dough stability, tolerance to mixing and extensibility, for instance. Wet vital gluten extensibility and spreadability are correlated with its overall viscoelasticity and the density of the structure forming networking bonds. General swelling capability of wheat flour which largely depends on water binding and swelling properties of the gluten proteins and therefore remains in some extent an indication of the potential structure-forming properties of the gluten proteins. Therefore, one could expect the existence of some correlation between the Cole-Cole parameters quantifying the wheat gluten viscoelastic properties and these technological qualitative parameters which depend significantly on the properties of gluten proteins. It was found that the viscoelastic plateau compliance J_N^0 is highly correlated with the technological baking quality parameters governed by the physicochemical properties of the gluten proteins. Correlation coefficient r varying from 0.87 for the dependency of J_N^0 versus farinographic valorimetric value to 0.99 for the dependencies of J_N^0 versus farinographic mixing tolerance index (MTI) and J_N^0 versus extensographic dough energy (Table 2). The technological parameters of this statistical analysis have been extracted from the study of Błaszczak et al. (2002).

Technological parameter	Terms of correlation equation		
	a	b	r
SDS SEDIMENTATION VALUE	75.85	9 774	0.916
Zeleny sedimentation value	40.36	12 613	0.911
Wet gluten	22.05	7 834	0.944
Gluten spreadability	-1.27	6 568	0.939
Dough stability	1.14	5 883	0.876
MTI	91.38	-45 990	0.992
Dough softening	86.89	-37 210	0.936
Valorimetric value	49.99	6 797	0.871
Dough energy	21.71	65 189	0.992
Dough extensibility	55.77	87 010	0.983
Ratio Number	5.58	-2 692	0.964

Table 2. Correlations between J_N^0 of wheat gluten and some technological parameters of wheat flour and dough obtained from the microwaved wheat grain

Particularly relevant correlation between J_N^0 and extensographic dough energy was also observed as presented in Figure 3.

Also farinograph and extensograph tests as well as baking test proved some changes in viscoelastic properties due to denaturation changes in gluten proteins (Błaszczak et al., 2002; Gralik, 2003). Bred volume was distinctly reduced in the case of all applied microwave heating times. All microwave doses also caused decreasing in bred score and shown some deterioration in bred freshness expressed by score points. On the basis of results obtained from technological tests it can be stated that longer time above 90 sec (64ºC) of microwave heating when the wheat grain moisture content was estimated on 11.7% disadvantage effect

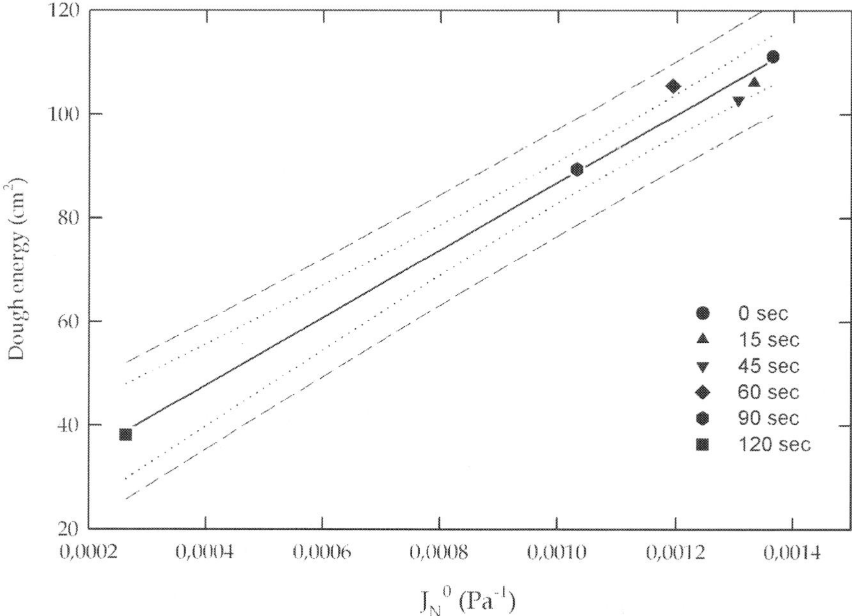

Fig. 3. Correlation between the viscoelastic plateau compliance J_N^0 of wheat gluten and dough energy. Legend: regression line – solid, dotted line – the 95% confidence interval, dashed line – the 95% prediction interval

on grain baking properties was observed (Błaszczak et al., 2002; Gralik, 2003). Similar results was reported earlier by MacArthur & D'Appolonia (1981) who analyzed of flour and bred and found that important qualities was adversely affected after 240 sec (61°C) of microwave exposure where the moisture content of flour was estimated on 14%. When exposed the flour to high levels of microwave radiation at 360 sec (75°C) and 480 sec (100°C) they found abnormal faringograph curve exhibiting two peaks whereas low levels up to 240 sec (61°C) produced bred with low volumes and overall bred characteristics equal to or better than those of the control flour. In general, storage improved all samples. The influence of microwave energy on certain chemical components was also investigated by MacArthur & D'Appolonia (1981). When starch pasting characteristics were examined, radiation appeared to have had no adverse effect on paste stability; however the rate of retrogradation decreased as microwave treatment increased. Decreasing starch intrinsic viscosity and total sugar values were observed with increased radiation levels after storage of flour for six months. A progressive decrease intrinsic viscosity was observed as irradiation increased except for 480 sec (100°C) sample. The decrease in intrinsinc viscosity suggests a possible alternation in starch structure as a result of the microwave treatment, even though peak viscosity as measured by amylograph did not differ. The reason while the starch sample microwave-heated for 480 sec (100°C) showed an increase in intrinsic viscosity is not clear; however, the increase might be related to thermal effects (MacArthur & D'Appolonia, 1981). As microwave treatment of flour increased, total sugar content at 0 and 30 days of storage showed a little change as the lengh of storage time was extended to 150 days, the sugar content for the control flour showed slight increase, which could be a result of flour maturation. However, with microwave-treated samples stored for 150 days, a trend toward reduction in sugar content with increased microwave energy was seen. The exact reason for such a result is unknown (MacArthur & D'Appolonia, 1981). Although the effects showed in MacArthur & D'Appolonia (1981) study can be explained in part as thermal and/or maturation-related, some alteration of the starch structure could be possibly occur during the microwave radiation of wheat flour.

The insect species that can attack stored grain and products are highly specialized and successfully exploit the storage environment. Internal insect infestation of wheat kernels not only degrades quality and technological value of wheat, but is one of the most difficult to detect and is generally considered to be the most damaging (Pederson, 1992, Fornal et al., 2007). Pest infestation causes grain loss by consumption, contaminates the grain with excrement and fragments, causes nutritional losses, and finally degrades end-use quality of flour (Nawrot et al., 2006). Additionally, during storage, grains can be contaminated by fungi and other microorganisms that will produce off odors and change the chemical composition of wheat and make it unsuitable for food products (Cornell & Hoveling, 1998). Many chemical insecticides are used as protectants against insect infestation in stored grains (Warchalewski et al., 2000). The high costs of developing new chemicals to replace those withdrawn from use because of pest resistance and the imposition of regulations to prevent environmental damage or human health risks have led to increased interest in nonchemical methods of pests control (Nelson, 1996; Watters, 1991). Many methods have been developed so far for the detection of pest-infested kernels, but none of them has been used in the monitoring of both storage and transport on a wide scale. Among the methods used in research laboratories it can be find: egg plug staining technique (Toews et al., 2006; Pearsons & Brabec, 2007), flotation and cracking (Brader et al., 2002; Haff & Slaughter,2004), acoustic method (Vick et al., 1988; Hagstrum et al., 1990; Nethirajan et al., 2007), immuno-assay method ELISA (Enzyme-Linked Immunosorbent Assay) (Piasecka-Kwiatkowska et al., 2005; Piasecka-Kwiatkowska et al., 2010), electrophoretic method (Piasecka-Kwiatkowska et al., 2006), near-infrared hyperspectral imaging (Mahesh et al.,2008; Singh et al., 2009), soft X-ray roentgenography (Schatzki & Fine, 1988; Grundas et al., 1999; Karunakaran et al., 2003, 2004; Fornal et al., 2007; Nawrocka et al., 2010,). Between physical methods, a known nonchemical alternative for controlling hidden infestations inside kernels is the application of extreme temperatures, preferably heat because its lethal effects are usually faster than cold (Fields & Muir, 1996). Higher temperatures increase locomotory activity in beetles generally (Cox & Collins, 2002). For example, *S. granarius* L. tend to move towards the top and sides of the bulk. Virtually all organisms, when they are exposed to temperatures 5-10°C above the optimal growth temperature, undergo a robust heat shock response. Within the cell, an increase of this magnitude can have many deleterious effects, including increase in membrane fluidity and the partial unfolding of proteins that can finally lead to death (Ng, 2004). In a search for effective, safe and cheap methods of control of insects in stored grains, the attention was drawn to temperatures employed in a process of grain drying (Prądzyńska, 1995). Temperatures from 45°C to 60°C at a proper exposure time, sterilize or kill insects in a mass of goods without any disadvantageous influence on product quality. Grain weevil beetles exposed to conventional heat-treatment, without any product, died much faster than those in grain at 45°C after 180 minutes while at 60°C after 5 minutes. At a high grain humidity over 20%, pests can stand heating for longer time (Prądzyńska, 1995).

Beetles of the granary weevil that survived a conventional heat-treatment gave far less progeny in a new F_1 generation than in the control tests. Lethal temperatures very considerably and depend on factors such as species, stage of development, acclimation and relative humidity (Cox & Collins, 2002). Some storage species will die eventually at temperatures a little above 35°C, while at 55°C death will occur within a few minutes (Fields, 1992).

Use of microwave heating has the advantage of saving time and energy, improving both nutritional quality and acceptability of some foods by consumers as well as significant reduction of stored-grain damage by insect pests (Nelson, 1996; Nijhuis et al., 1998; Plarre et al., 1999; Sumnu, 2001; Dolińska & Warchalewski, 2003; Nawrot et al., 2006). The interaction of electromagnetic wave with free water in insects is of interest because of the disparity of free water in relatively dry cereal grain and in insects. Free water of insects occurs in significant amount because it is the major constituent of the hemolymph (Wyatt, 1961). The application of high-power microwaves for stored products pest control have shown, that selective heating of insects and their resulting mortality is a non-linear function of frequency and increases at frequences above 2.45 GHz in the vicinity of increasing relaxation processes associated with free water (Plarre et al., 1999). The target temperature range of 40 to 60°C were successfully delivered to the infested products by *Sitophilus zeamais* using microwaves of different power levels at 28 GHz. The treatment of an infested product with 28 GHz showed good efficiency for the control of adults and older larvae of the mize weevil at rather low products temperatures. Complete kill of adult weevil achieved with product temperatures as low as approximately 40°C and minimum specific load energies of 20 Jg^{-1} to 30 Jg^{-1}. Older larvae did not survived product temperatures approximately 50°C or specific load energies above 45 Jg^{-1}. A mortality level of over 90% in younger larvae was achieved at temperatures slightly above 50°C and specific load energies of above 60 Jg^{-1}. Exposure times for this procedure were less than 1 second (Plarre et al., 1999). The reaction kinetics of heat-induced changes are governed mainly by time, temperature, and moisture content. Microwave treatment of insect-infested grain in a dynamic procedure with high through-put rates appears to be a reliable alternative to conventional postharvest pest control (Plarre et al., 1999). Nowadays, microwave heating is often combined with other techniques for killing or reducing insects and mites that attack wheat grain (Dolińska & Warchalewski, 2003).

From the ecological point of view disinfestations and microbial decontamination of cereal grain with the application of different physical methods seems to be attractive. (Gralik et al., 1999). However, the effect of microwave treatment of wheat grain on microflora reduction was comparatively negligible. Some reduction of microflora , but only in the case of bacterial cells was observed when wheat grain was treated with microwaves at 180 sec (98°). At this maximum time 180 sec of grain exposure to microwave energy the efficiency of bacterial reduction can be comparable to certain extend to wheat grain gamma irradiated. After six months of grain storage microwave treated wheat showed some decrease in bacterial count however, it was not significant (Gralik et al., 1999). In all microwave treated samples the number of fungi was very low and constant which indicated that microwave energy did not have any effect on fungi within the range of time 15 to 180 sec and temperature 28 to 98°C. This was supported by the results obtained after six months of grain storage when the final count was on the same level. Additionally the increase of temperature and microwave treated grain has some influence on gradual decrease of moisture content in wheat grain from 12.22 to 9.87 % (Warchalewski et al., 2000).

The major problem in the use of microwave heat-treatment for the eradication of pests and pathogens in seeds has been its adverse effect on seed quality (Stephenson et al., 1996). However, the limitation of seed quality loss can be minimized by rigorous control over the microwave treatment factors, like absorbed microwave power (AMP) and the duty cycle or pulsing (PUL). These variables can interact both with themselves and with other variables such as initial seed moisture content (SMC), time microwave power (TMP) and seeds vigour. In the case of *Ustilago nuda* complete control of this pathogen was not achieved at any level of absorbed microwave power and pulsing combinations of these parameters. Although significantly was reduced the ability of *U. nuda* to colonize the kernels and produce smutted heads, while germination and vigor were maintained at acceptable levels (Stephenson et al., 1996).

Head blight of wheat caused by *Fusarium graminearum* Schwabe is one of the most important diseases. Infected seeds may show reduced germination rates and seedling vigour (Shotwell et al., 1985). *F. graminearum* produces mycotoxins in cereal grains which affect human and livestock health (Goliński & Perkowski, 1998). Seed treatment with fungicides can create hazardous non-target residual effects. Therefore is an increasing demand for alternative methods of disease control. As an innovative approach to eradicate seedborne pathogens was the use of microwave energy (Seaman & Wallen, 1967; Lozano et al., 1986; James & Gilligan, 1988; Bhaskara Reddy et al., 1998). Microwave energy can be used to significantly reduce seedborne *F. graminearum* in wheat (Bhaskara Reddy et al., 1998). The percentage of seed infection was reduced to below 7% (from 36% for the controls), without reducing the seed quality below the commercial threshold of 85% seed germination. Such results were occurred under various combinations of seed moisture content, absorbed microwave power and time microwave power. However, the data were not sufficiently consistent to lead to an operational model due to the variability resulting from infection, as well as from incubation procedures and the natural variability of the seeds.

3. Indirect effect of microwave heating of wheat grain.

Among 650 million tones of world's annual production of wheat grain, about 46 million tones is used for seeding purposes (Grundas et al., 2008; Anonimus, 2010). Stored cereal's seeds are vulnerable to insects attack however, significant reduction of stored-grain damage by insect pests can be achieved with the use of microwaves (Nelson, 1996; Plarre et al., 1999; Dolińska and Warchalewski, 2003; Nawrot et al., 2006). The degree of changes is largely proportional to the microwave treatment dose, which may vary considerably depending upon the treatment objective. In the case of grain treatment as seeds, the germination potency should be preserved with acceptance of some unintentionally mutation in the next generation wheat grain crops.

When applying microwaves to reduce grain damage by stored-grain pests, it should be consider also the question of how this heat-treatment will affect wheat grain properties not only directly, but also indirectly, on seeds secondary infested as well used for the production of next-generation grain crops. So far only a few papers on indirect influence of microwaves on wheat seeds have been published (Dolińska & Warchalewski, 2002; Dolińska et al., 2004; Warchalewski et al., 2007; Grundas et al., 2008).

The major problem in the use of microwave energy for the eradication of pests and pathogens in seeds has been its adverse effect on seed quality (Stephenson et al., 1996). The lost of seed germination and deterioration technological quality are the major limitation of microwave heat treatment for the control of grain insect, so the growth of temperature during process must be under rigorous control (Nelson, 1996; Shayesteh & Barthakur, 1996; Stephenson et al., 1996; Warchalewski et al., 2000). Wheat grain microwave-treated for 120 sec and 180 sec before sowing caused a distinct decrease in the yield of the first generation crop due to the higher temperature of grain noted after treatment at 79°C and 98°C, respectively (Grundas et al., 2008). However, the yield of the crop in the second generation appeared to be higher, and in the third generation, the yield was restored independently to the level at microwave treatment. Possible genetic modification of microwave-treated wheat grain was also earlier reported (Dolińska et al., 2004; Grundas et al., 2008). Resistance mechanisms of cereal grain to attack by insect species are complex and depend on several chemical factors such a sterols and α-amylase inhibitors levels or to physical properties including thickness of the brain layer and hardness of the endosperm (Warchalewski & Nawrot, 1993; Cox & Collins, 2002; Warchalewski et al., 2002; Nawrot et al., 2006; Nawrot et al., 2010). Genetically modified cereals can contain proteins or other compounds detrimental to critical insect life functions (Gatehouse & Gatehouse, 1998). The indirect effect of microwave heating of wheat grain within temperature range from 28°C to 98°C, on seeds microstructure in two consecutive years was examined. The two highest temperatures applied during microwave heating process of wheat seeds with initial moisture content 12.2 % before sowing weakened the germination power in seed samples treated for 120 sec (79°C) and 180 sec (98°C), and in consequence lowered grain yield by 9 % and 57 % respectively (Warchalewski et al., 2007). No visible changes in microstructure of pericarp, aleurone layer, subaleurone and starchy endosperm of wheat kernels collected in two generations crops grown from microwave treated seeds were found. However, some small changes in external epidermal cells were noted, but only in the first generation crops examined by scanning electron microscopy (Dolińska, 2004).

Changes in grain colour were associated with some thermal decomposition and modification of grain components which was correlated with the increase of the grain temperature from 64 to 98°C (Warchalewski et al., 1998). The temperature at which damage of wheat starch begins is suggested at between 53 to 64°C (Cornell & Hoveling, 1998). However, it depends on the initial moisture content of the grain. On the other hand, alpha-amylases like lipooxygenases of wheat grain are fairly heat-stabile since no inactivation occurred below 60°C (Kruger & Reed, 1988). The colour of the flour, due to both its content of wheat and bran pigments, affects bred whiteness. But other factors influence the colour of bred crumb greatly as well. Foremost among these is the fineness of the grain. The brightness of grain and flour is probably only a minor factor in the formation in bred crust colour (Warchalewski et al., 1998). The statistically significant influence of microwave heating on changes of colour in the first and second generations grain was observed. The sample IM-90 (64°C) was characterized by higher value of lightness (L*) and more intensive red colour (b*) which was correlated with decreased length and increased width and thickness of external epidermal cell walls of wheat kernels (Dolińska, 2004). This grain sample shown also the highest hardness index (HI) as was determined by Grundas et al., (2008). In the case of grain samples IM-15 (28°C) and IM-45 (43°C) was observed more intensive redness (b*) while grain sample IM-180 (98°C) was characterized by slighter yellowness (a*) which was also correlated with the similar seeds diameter (Dolińska, 2004; Grundas et al., 2008). Grain samples collected in the second generation crop indicating some vanishing tendencies in differences of grain colour among grain samples. Flavour profile of the next generations crops of wheat grain obtained from microwave-treated seeds was similar to control grain samples (Dolińska, 2004).

All post-harvest grain insects infest new seed parts passively, rarely searching for food over long distances (Nawrot et al., 2010). Some wheat grain properties may constitute a source of resistance against pests, the amounts of proteins, enzyme inhibitors, starch, lipids and fiber, as well as selected physical and technological attributes, including water content, hardness, vitreosity, weight by volume and thousand kernel weight, are characteristic features of each variety that may be related to insects infestation levels (Warchalewski et al., 2002; Nawrot et al., 2006).

When microwave treated wheat seeds reached 98°C after 180 sec the total protein content (TPC) in the first generation crop significantly increased while in the third generation crop decreased (Dolińska, 2004; Grundas et al., 2008). All wheat samples of the second generation crop were characterized by similar total protein content (TPC). It is interesting, that the highest soluble protein content was found in all three generation crops breeded from wheat seeds microwave heated to 98°C by 180 sec. In the first and third generations, statistically significant tendencies in total protein content (TPC) were additionally described by quadratic and logistic equations (Grundas et al., 2008).

As a result of the indirect effect of microwave heat-treatment, thousand kernel weight (TKW) was statistically different in the three next-generation crops compared with the control samples (Grundas et al., 2008). Microwave heated wheat grain by 180 sec to 98°C before sowing caused increase in the total protein content to the highest level and the lowest thousand kernel weight (TKW) in the first generation crop, but this is not upheld in the next two generations. According to Slaughter et al. (1992) and Satumbaga et al. (1995), wheat kernel mass has a negative relationship to protein content. It is well known that protein content of wheat grain is influenced less by heredity than by edaphic factors (soil and climatic conditions) prevailing at the place of growth and by the fertilizer treatment. Since growing conditions were exactly the same each year, so the observed differences in protein content as an indirect effect of microwave treatment of wheat grain versus control grain samples should be rather attributed to some genetic modification of wheat seeds. Wheat albumins extracted from grain collected as the first and second generations crops which were obtained from microwave heat-treated seeds had statistically significant lower true and apparent digestibility of protein in comparison to untreated grain (Dolińska & Warchalewski, 2002). Indirect effect of microwaves caused statistically significant fluctuation in addition to total protein content (TPC) also thousand kernel weight (TKW), single kernel moisture content (SKM), single kernel weight (SKW), hardness index (HI) and single kernel diameter (SKD) in all three wheat grain crops in relation to their control samples (Grundas et al., 2008). As an indirect consequence of microwave heating, there was a statistically significant increase of hardness index (HI) in samples IM- 15 (28°C), IM-45 (43°C), IM-60 (48°C), IM-90 (64°C), IM-120 (79°C), IIM-60 (48°C) and IIM-180 (98°C) from the first and second generation crops, respectively. On the other hand, all samples from the third generation crop had lower kernel hardness in relation to the control sample. Also bulk density (BD) of the first generation crop showed a statistically significant tendency described by a quadratic equation but mainly due to seeds heated by 180 sec to the 98°C before sowing. The lowest bulk density value (BD) in this grain sample was well correlated with the lowest yield of grain harvested as well as the lowest statistically significant single kernel diameter (Grundas et al., 2008). The reducing sugars content in next generations of wheat grains was also affected by pre-sowing application of microwave heating of seeds from 28 to 180 sec. In general, the increasing tendency of reducing sugar content was noted in the first and second generation crops, while in the third grain crop the lower in relation to the control wheat grain samples (Dolińska, 2004). Statistically significant changes in falling number values and some changes in the peak temperatures (T_p) and enthalpy of gelatinization (ΔH) caused by indirect effect of microwaves were reported in all three generations of wheat grain crops (Dolińska et al., 2004). This support some minor changes in protein and starch granules causing them more susceptible to grain endogenous amylases which was documented by increasing reducing sugars content in the next generations crops.

Favorable changes in the inhibition activity against exogenous alpha-amylases of human saliva, *Sitophilus granarius* and *Tribolium confusum* were noted in all grain samples of three grain generations crops collected from microwave-treated seeds before sowing to the grain temperature 28ºC by 15 sec (Dolińska, 2004). From nutritional point of view substantial decrease of inhibition activity of human saliva alpha-amylase as well as an increase of inhibition activity against alpha-amylases of *Sitophilus granarius* and *Tribolium confusum* should be recognized as profitable. Cereal grains with the highest inhibition activities against alpha-amylases of insect pests was suggested to reduce to some extend insect populations in grain storage facilities (Warchalewski et al., 2002). This indicates that the studied physicochemical and biochemical properties of wheat grain were affected not only directly but also indirectly by microwave rays.

The technological properties of wheat grain collected in the three generations crops which were sown from microwave-heated seeds were presented in table 3, 4 and 5.

The highest temperature of microwave heating of wheat seeds before sowing weakened the germination power in most seeds treated by 180 sec (98ºC) and 120 sec (79ºC) as was reported by Grundas et al (2008). Statistically significant fluctuation in some parameters of farinograph (Table 3) and extensograph analysis (Table 4) as well as baking quality properties (Table 5) were noted in the next generation wheat grain collected in the consecutive year.

Generally indirect effect of microwave-heated seeds before sowing worsening technological properties of wheat grain collected in the next generations crop. In addition, as was earlier reported (Dolińska, 2004) in all next generations grain samples amount of hand washed wet gluten as well as wet gluten deliquescent were lower in relation to the wheat grain control samples. The most distinct changes are visible particularly in the case of Sodium Dodecyl Sulphate (SDS) sedimentation test in all three generations grain collected from wheat seeds microwave-heated by 60 sec (48ºC) and above. The results of sedimentation tests and gluten content proofed the negative effect of microwave heating applied before sowing on wheat grain collected in the consecutive years (Dolińska, 2004).

Cereals grain harvesting, drying and handling prior to storage will all affect the quality of the grain and its suitability for each potential pest (Cox & Collins, 2002). Grain damaged during harvesting and augering is more susceptible to attack by most storage beetles than are whole grains. Since, microwave may induce changes in a number of wheat grain properties not only directly but also indirectly in grain harvested in the next generation crops, it can be expected also some changes in susceptibility to insect infestation in wheat seeds as well as grain collected in the consecutive years. However, directly microwave-treated of wheat grain had no effect on secondary infested grain by the granary weevil (*Sitophilus granarius* L.) as was reported by Warchalewski et al. (2000). On the other hand, it

Farinograph analysis	Sample						
First generation	**IM-0**	**IM-15**	**IM-45**	**IM-60**	**IM-90**	**IM-120**	**IM-180***
Water absorption of flour [%]	61,3 a	61,7 a	61,7 a	62,6 a	62,7 a	62,8 a	-
Dough development [min]	2,0 a	2,0 a	**1,5 b**	2,0 a	2,0 a	2,0 a	-
Stability of dough [min]	6,0 a	6,0 a	6,0 a	**4,5 b**	5,5 a	**7,0 c**	-
MTI [FU]	45 a	45 a	**40 b**	**50 c**	45 a	**40 b**	-
Degree of dough softening [FU]	70 a	70 a	80 a	80 a	**85 b**	75 a	-
Valorimetric value [-]	48 a	48 a	47 a	46 a	47 a	48 a	-
Second generation	**IIM-0**	**IIM-15**	**IIM-45**	**IIM-60**	**IIM-90**	**IIM-120**	**IIM-180**
Water absorption of flour [%]	60,6 a	**61,2 b**	60,5 a	**60,0 c**	60,9 a	60,7 a	61,0 a
Dough development [min]	2,5 a	2,0 a	2,0 a	2,5 a	2,5 a	2,0 a	2,0 a
Stability of dough [min]	10,0 a	8,0 a	10,0 a	10,0 a	9,5 a	9,5 a	9,0 a
MTI [FU]	25 a	25 a	25 a	25 a	25 a	25 a	25 a
Degree of dough softening [FU]	50 a	**40 b**	30 a	30 a	30 a	30 a	30 a
Valorimetric value [-]	53 a	51 a	53 a	**55 b**	54 a	53 a	53 a
Third generation	**IIIM-0**	**IIIM-15**	**IIIM-45**	**IIIM-60**	**IIIM-90**	**IIIM-120**	**IIIM-180**
Water absorption of flour [%]	58,2 a	57,8 a	57,7 a	**57,2 b**	57,6 a	58,0 a	58,0 a
Dough development [min]	2,5 a	2,5 a	2,5 a	2,5 a	**2,0 b**	2,5 a	2,5 a
Stability of dough [min]	**12,0 a**	**5,5 b**	9,5 b	7,0 b	**6,0 b**	7,5 b	**11,5 a**
MTI [FU]	**40 a**	50 b	50 b	55 b	**60 c**	**40 a**	**40 a**
Degree of dough softening [FU]	40 a	60 a	50 a	60 a	**70 b**	55 a	40 a
Valorimetric value [-]	55 a	51 a	53a	52 a	**50 b**	52 a	55 a

* Due to the lowest yield of the first generation crop the technological analysis were not performed.

Table 3. Farinograph analysis of flour obtained from the three generation wheat crops (IM, IIM and IIIM) microwave-heated seeds before sowing. Mean values followed by the same letter in the column are not significantly different (P ≤0.05).

Extensograph analysis	Sample						
First generation	**IM-0**	**IM-15**	**IM-45**	**IM-60**	**IM-90**	**IM-120**	**IM-180***
Dough energy [cm²]	80,5 a	77,5 a	80,1 a	**72,6 b**	79,5 a	**86,6 c**	-
Resistance R_{50} [EU]	288 a	280 a	280 a	**265 b**	298 a	298 a	-
Extensibility-E [mm]	160 a	164 a	163 a	160 a	160 a	**168 b**	-
Ratio R_{50}/E [-]	1,80 a	1,71 a	1,72 a	**1,66 b**	**1,86 c**	1,77 a	-
Second generation	**IIM-0**	**IIM-15**	**IIM-45**	**IIM-60**	**IIM-90**	**IIM-120**	**IIM-180**
Dough energy [cm²]	95,2 a	**99,1 b**	94,9 a	94,3 a	87,8 a	88,9 a	81,0 a
Resistance R_{50} [EU]	355 a	**270 b**	302 a	322 a	290 a	312 a	**265 b**
Extensibility-E [mm]	158 a	**192 b**	178 a	170 a	171 a	164 a	168 a
Ratio R_{50}/E [-]	2,25 a	**1,41 b**	1,70 a	1,89 a	1,70 a	1,90 a	1,58 a
Third generation	**IIIM-0**	**IIIM-15**	**IIIM-45**	**IIIM-60**	**IIIM-90**	**IIIM-120**	**IIIM-180**
Dough energy [cm²]	123,6 a	**107,7 b**	**125,8 c**	**104,8 b**	112,8 a	120,5 a	120,5 a
Resistance R_{50} [EU]	470 a	462 a	**470 b**	**408 c**	435 a	440 a	440 a
Extensibility-E [mm]	158 a	**144 b**	160 a	154 a	156 a	162 a	161 a
Ratio R_{50}/E [-]	2,97 a	**3,21 b**	2,94 a	**2,65 c**	2,97 a	2,72 a	2,73 a

* Due to the lowest yield of the first generation crop the technological analysis were not performed.

Table 4. Extensograph analysis of flour obtained from the three generation wheat crops (IM, IIM and IIIM) microwave-heated seeds before sowing. Mean values followed by the same letter in the column are not significantly different (P ≤0.05).

Baking quality	Sample						
First generation	**IM-0**	**IM-15**	**IM-45**	**IM-60**	**IM-90**	**IM-120**	**IM-180***
Falling number in flour [sec]**	393 a	403 a	366 a	365 a	439b	391 a	-
Yield of dough [%]	166,8 a	167,0 a	167,1 a	167,3 a	167,6 a	167,8 a	-
Yield of bread [%]	147,7 a	147,8 a	145,8 a	147,3 a	148,3 b	148,1 a	-
Loaf volume [cm³]	587 a	581 a	**577 b**	580 a	579 a	**624 c**	-
Baking test [score]	127 a	124 a	122 a	123 a	1237 a	**158 b**	-
Estimation of loaf freshness [score]	69 a	61 a	63 a	**72 b**	56 a	**53 c**	-
Second generation	**IIM-0**	**IIM-15**	**IIM-45**	**IIM-60**	**IIM-90**	**IIM-120**	**IIM-180**
Falling number in flour [sec]**	427 a	**380 b**	413 a	427 a	439 a	452 a	392 a
Yield of dough [%]	165,7 a	**166,8 b**	166,1 a	165,8 a	166,3 a	166,6 a	166,4 a
Yield of bread [%]	147,8 a	147,1 a	**146,1 b**	146,9 a	1473 a	148,9 a	146,7 a
Loaf volume [cm³]	573 a	**551 b**	583 a	580 a	584 a	605 a	**603 c**
Baking test [score]	118 a	**109 b**	125 a	123 a	**117 b**	**139 c**	**139 c**
Estimation of loaf freshness [score]	89 a	**68 b**	79 a	78 a	69 a	71 a	72 a
Third generation	**IIIM-0**	**IIIM-15**	**IIIM-45**	**IIIM-60**	**IIIM-90**	**IIIM-120**	**IIIM-180**
Falling number in flour [sec]**	547 a	**511 b**	542 a	533 a	540 a	542 a	533 a
Yield of dough [%]	163,5 a	**163,9 b**	163,5 a	**163,2 c**	163,4 a	**163,9 b**	1634 a
Yield of bread [%]	145,8 a	147,0 a	145,8 a	**145,7 b**	146,0 a	**149,4 c**	147,2 a
Loaf volume [cm³]	687 a	**587 b**	633 a	604 a	**600 b**	**596 b**	635 a
Baking test [score]	118 a	**109 b**	125 a	123 a	**117 b**	**139 c**	**139 c**
Estimation of loaf freshness [score]	84 a	**57 b**	72 a	72 a	**57 b**	75 a	77a

* Due to the lowest yield of the first generation crop the technological analysis were not performed.
** Falling number value measured before standardization to 220 sec for the purpose of baking test.

Table 5. The baking quality of flour obtained from the three generation wheat crops (IM, IIM and IIIM) microwave-heated seeds before sowing. Mean values followed by the same letter in the column are not significantly different (P ≤0.05).

was observed some differences in progeny number, number of eggs, development time and weight of dust but they were not statistically significant. It was interesting that microwave-treated wheat grains which were secondary infested by the Mediterranean flour moth (*Ephestia kuehniella* Zell.) and confused flour beetle (*Tribolium confusum* Duv.) had statistically significantly shorter larval development time than on control seeds (Warchalewski et al., 2000). Although the range of damage depends on dosage and time of treatment. Table 6 presents the results of the granary weevil (*Sitophilus granarius* L.) development parameters on the next generation wheat grain crops bred from seeds microwave heated.

Sample	Survival of beetles after 14 days [%]	Beetle developed [%]	Progeny number [-]	Development time [days]	Weight of dust [mg]
IM-0	92 ab	100 a	9.0 a	58.4 a	46.5 a
IM-15	96 ab	86 a	7.8 a	59.4 a	46.8 a
IM-45	96 ab	89 a	8.0 a	58.2 a	40.8 a
IM-60	96 ab	**192 b**	**17.2 b**	58.6 a	**84.7 b**
IM-90	100 b	106 a	9.5 a	59.4 a	55.5 a
IM-120	86 a	136 ab	12.3 ab	57.0 a	62.6 ab
IM-180	98 b	**168 b**	**15.1 b**	56.8 a	**78.8 b**
IIM-0	88 ab	100 c	15.3 c	59.6 b	64.6 a
IIM-15	74 a	**63 ab**	**9.6 ab**	**54.2 a**	51.3 a
IIM-45	90 ab	**54 a**	**8.2 a**	61.2 b	39.7 a
IIM-60	100 b	**63 ab**	**9.6 ab**	60.4 ab	62.5 a
IIM-90	100 b	84 abc	12.8 abc	59.2 ab	65.5 a
IIM-120	98 ab	88 bc	13.5 bc	58.2 ab	62.8 a
IIM-180	98 ab	87 bc	13.4 bc	58.0 ab	60.9 a
IIIM-0	98 ab	100 a	10.5 a	58.8 a	65.8 ab
IIIM-15	100 b	126 a	13.2 a	59.2 a	65.9 ab
IIIM-45	100 b	138 a	14.2 a	59.2 a	103.6 b
IIIM-60	96 ab	137 a	14.4 a	61.0 a	89.3 ab
IIIM-90	78 a	130 a	13.6 a	56.8 a	40.1 a
IIIM-120	98 ab	133 a	13.9 a	58.8 a	60.5 ab
IIIM-180	98 ab	122 a	10.8 a	57.6 a	64.2 ab

Table 6. The effect of wheat seeds microwave treatment in the range of temperature 28 to 98ºC and time from 15 to 180 sec on *Sitophilus granarius* L. development parameters in three consecutive years. Mean values followed by the same letter in the column are not significantly different (P ≤0.05).

Statistically significant increase of developed beetle was found in the first generation crop in wheat grain samples coded IM-60 (48ºC) and IM-180 (98ºC) which resulted also in the higher progeny number and weight of dust. This was well correlated with the highest level of reducing sugars content (Dolińska, 2004). In the second generation crop were observed lower statistically significant beetle developed and progeny number in grain samples coded IIM-15 (28°C), IIM-45 (43°C) and IIM-60 (48°C) compared to the control grain. Although the development time of granary weevil was significantly shorter but only in grain coded IIM-15 (28°C). In the third generation crop no significant differences in development parameters of granary weevil was found which was well correlated with restoring tendencies of all physicochemical, biochemical and technological properties earlier described in relation to control grain samples. The development parameters of *Tribolium confusum* Duv. and *Ephesita kuehniella* Zell. fed on three generation wheat grain crops bred from seeds microwave heated shown some statistically significant differences as can be seen in Table 7. Extension statistically significant development time of *Tribolum confusum* Duv. in grain samples coded IIM-120 (79°C), IIM-180 (98°C) and IIIM-45 (43°C) by 7.2, 4.6 and 6.4 days respectively, as well as significantly less beetles developed in wheat grain IIIM-120 (79°C) and IIIM-180 (98°C) should be regarded as beneficial. Which means less population of *Tribolium confusum* Duv. will hatching during storage of wheat grain.

Changes in wheat grain physicochemical, biochemical, technological, nutritional and insect resistance properties may be ascribed not only to microwave heating but also to other

	Tribolium confusum Duv.		*Ephestia kuehniella* Zell.	
Sample	Beetles developed [%]	Development time [days]	Moths developed [%]	Development time [days]
IM-0	60 ab	23.6 a	47 ab	24.8 a
IM-15	80 b	22.6 a	66 ab	28.3 ab
IM-45	64 ab	24.0 a	40 ab	26.0 a
IM-60	72 ab	22.6 a	20 a	**33.0 b**
IM-90	72 ab	22.2a	40 ab	28.0 ab
IM-120	48 a	24.8 a	53 ab	27.0 ab
IM-180	60 ab	24.8 a	67 b	**33.0 b**
IIM-0	36 a	14.6 a	53 a	26.2 a
IIM-15	24 a	14.4 a	67 a	27.8 a
IIM-45	24 a	15.0 a	67 a	28.2 a
IIM-60	28 a	14.0 a	60 a	29.2 a
IIM-90	16 a	16.4 ab	67 a	25.4 a
IIM-120	24 a	**21.8 c**	73 a	27.8 a
IIM-180	28 a	**19.2 bc**	67 a	26.4 a
IIIM-0	84 c	20.0 bc	60 ab	25.4 a
IIIM-15	92 c	21.6 bcd	93 b	29.2 a
IIIM-45	72 bc	**26.4 d**	60 ab	26.2 a
IIIM-60	76 bc	24.8 cd	60 ab	24.4 a
IIIM-90	64 abc	23.2 bcd	80 ab	27.2 a
IIIM-120	**36 a**	**14.6 a**	53 a	26.8 a
IIIM-180	52 ab	18.2 ab	67 ab	30.4 a

Table 7. The effect of wheat seeds microwave treatment in the range of temperature 28 to 98°C and time from 15 to 180 sec on *Tribolium confusum* Duv. and *Ephestia kuehniella* Zell. development parameters in three consecutive years. Mean values followed by the same letter in the column are not significantly different (P ≤0.05).

factors such as heritability and possible genetic modification of the wheat crop. However, each year, the climate, soil and fertilization treatment of wheat seeds were exactly the same, the influence of different climate conditions during the three cultivation periods should not be excluded (Dolińska, 2004; Dolińska et al., 2004; Grundas et al., 2008).

4. Recapitulation

Microwave drying is most effective at product moisture content below 20 % like in the case in cereals grain. In addition drying operational cost is lower since microwave energy is consumed only by cereal grains not by surrounding environment. Direct application of microwave heating for drying and partial elimination of insect pests is possible with the use microwave heating time up to 90 sec where the grain temperature not exceeded 64°C. In addition, microwave heating of wheat grain to ≤ 64°C should be safe from a technological and baking point of view, as well as a reliable alternative to postharvest pest control during grain shipment and storage.

In general, it should be noted that microwave heating of wheat seeds did not have a negative effect on reproductive materials (seeds) in the second and third generation crops. Although, it should be pointed out that observed changes as direct effect of microwave heated wheat seeds showing some restoring tendency in all discussed wheat grain properties in the third generation crop.

Moreover, in the nearest future microwave heating combined with the other techniques like gamma radiation can be even more effective and attractive techniques for killing of insects and mites that infested stored cereals grain. However, in order to successfully using this techniques more knowledge is needed about the influence of microwave-treated different cereal varieties and theirs influence on physicochemical, biochemical, technological, nutritional and insect resistance properties, also shape and size of kernels as well as on seeds germination power

5. References

Aldryhim, Y.N. & Adam, E.E. (1999). Efficacy of gamma irradiation against *Sitophilus granarius* L. *(Coleoptera Curculionidae)*. *Journal of Stored Products Research*. Vol. 35, 225-232. ISSN 0022-474X.

Anonimus (2010). International Grain Council. *Grain Market Report*, No. 402.

Baldwin, A.J., Cooper, H.R. & Palmer, K.C. (1991). Effect of preheat treatment and storage on the properties of whole milk powder. Changes in sensory properties. *Netherlands Milk a nd Dairy Journal*. Vol. 45, 97-116, ISSN 0028-209X.

Belton, P. S. (2005). New approaches to study the molecular basis of the mechanical properties of gluten. *Journal of Cereal Science*. Vol. 41 , No. 2, 203-211, ISSN: 0733-5210.

Bhaskara Reddy, M.V., Raghavan, G.S.V., Kushalappa, A.C. & Paulitz, T.C. (1998). Effect of microwave treatment on quality of wheat seeds infected with *Fusarium graminearum*. *Journal of Agricultural Engineering Research*. Vol. 71, 113-117, ISSN 0021-8634.

Bieżanowska, R., Pysz, M., Kostogrys, R.B. & Pisulewski, P.M. (2000). The influence of microwave field on proteinase inhibitory activity in beans' seeds (in Polish), *Proceedings of XXXI Sesja Naukowa KTiChŻ PAN*. Komitet Chemii i Technologii Żywności PAN, Poznań Poland, September 2000, p. 247, ISBN 83- 86707-96-8.

Błaszczak, W., Gralik, J, Klockiewicz-Kamińska, E., Fornal, J. & Warchalewski, J.R. (2002). Effect of radiation and microwave heating on endosperm microstructure in relation to technological properties of wheat grain. *Nahrung/Food*. Vol. 46, No. 2, 122-129, ISSN 0027-769X.

Boyes, S., Chevis, P., Holden, J. & Perera, C. (1997). Microwave and water blanching of corn kernels: control of uniformity of heating during microwave blanching. *Journal of Food Preservation*. Vol. 21, 461-484, ISSN 0145-8892.

Brader, B, Lee, R. Plarre, R., Burkholder, W., Kitto, G.B., Kao, Ch., Polston, L., Dorneanu, E., Szabo, J, Mead, B., Rouse, B., Sullins, D. & Denning, R. (2002). A comparison of screening methods for insect contamination in wheat. *Journal of Stored Products Research*. Vol. 38, 75-86, ISSN 0022-474X.

Ciepielewska, D. & Kordan B. (2001). Natural peptide substances as a factor reducing the incidence of storage pests-short report. *Polish Journal of Food and Nutrition Sciences*. Vol. 3, No.3, 47-50, ISSN 1230-0322. Cornell, H.J. & Hoveling, A.W. (1998). *Wheat, Chemistry and Utilization*, Technomic Publishing, Basel, Switzerland, ISBN 1-56676-348-7.

Cox, P.D. & Collins, L.E. (2002). Factors affecting the behaviour of beetle pests in stored grain, with particular reference to the development of lures. *Journal of Stored Products Research*. Vol.38, 95-115. ISSN 0022-474X.

Dobraszczyk, B. J. &Morgenstern, M. P. (2003). Rheology and the breadmaking process. *Journal of Cereal Science*. Vol. 38, No. 3, 229-245, ISSN: 0733-5210.

Dolińska, R. (2004). The evaluation of gamma radiation and microwave heating on some changes in wheat grain properties of the next generations crops in relation to agro-environmental conditions. (in Polish). *PhDthesis*. University of Live Sciences, Poznan, Poland.

Dolińska, R. & Warchalewski, J.R. (2002). Effect of gamma radiation and microwave heating applied before swing the *in vitro* digestibility of albumin proteins from the first and second generations of wheat grain. (in Polish) *Food Science Technology Quality*. Vol. 33, No. 3, 102-116, ISSN 1425-6959.

Dolińska R., & Warchalewski, J.R. (2003). Future food technologies with microwave participation and their influence on food components. (in Polish) *Food Industry*. Vol. 57, No. 11, *pp.* 2,4,6-7,27, ISSN 0033-250X.

Dolińska, R., Warchalewski, J.R., Gralik, J. & Jankowski, T. (2004). Effect of γ-radiation and microwave heating of wheat grain on some starch properties in irradiated grain as well as in grain of the next generation crops. *Nahrung/Food*. Vol. 48, No. 3 195-200, ISSN 0027-769X.

Fields, P.G. (1992). The control of stored-product insects and mites with extreme temperatures. *Journal of Stored Products Research*. Vol. 28, 89-118. ISSN 0022-474X.

Fields, P.G. (2006). Effect of *Pisum sativum* fractions on the mortality and progeny production of nine stored-grain. *Journal of Stored Products Research*. Vol. 42, 86-96, ISSN 0022-474X.

Fields, P.G. & Muir, W.E. (1996). Physical control. In: *Integrated Management of Insects in Stored Products*, B. Subramanyam & D.W. Hagstrum, (Eds) 195-221, Marcel Dekker, New York. ISBN 978-0824795229.

Fornal, J., Grundas, S., Warchalewski, J.R., Błaszczak, W. & Gralik, J. (2001). Microdamages of wheat grain induced by microwaves and gamma irradiation. *Proceedings of Physical Method in Agriculture, International Conference*, August 27-30, University of Agriculture in Prague, Czech Republic, p. 84, ISBN 80-213-0787-0.

Fornal, J., Jeliński, T., Sadowska, J., Grundas, S., Nawrot, J., Niewiara, A., Warchalewski, J.R. & Błaszczak W.(2007). Detection of granary weevil *Sitophilus granarius* L. eggs and internal stage in wheat grain using soft X-ray and image analysis. *Journal of Stored Product Research*. Vol. 43, 142-148, ISSN 0022-474X.

Gaines, C.S., Finney, P.F., Fleege, L.M. & Andrews, L.C. (1996). Predicting a hardness measurement using the single-kernel characterization system. *Cereal Chemistry*. Vol. 73, No.2, 278-283, ISSN 0009-0352.

Garcia, R., Andrade, J. & Rolz, C. (1975). Effect of temperature and heating time on the detection of off-flavor in avocado paste. *Journal of Food Science*. Vol.40, No. 1, 200, ISSN 1750-3841.

Gatehouse, A.M.R. & Gatehouse, J.A. (1998). Identifying proteins with insecticidal; use of encoding genes to produce insect-resistant transgenic crops. *Pest Science*. Vol. 52, 165-175, ISSN 1612-4758.

Goliński, P. & Perkowski, J. (1998). Impurity of cereals and fodders by mycotoxins. In: *The Actual Status and Perspectives of Development Selected Branches of Food Production and Fodders* (in Polish). J.R. Warchalewski (Ed). 101-118, Publishing House PTTŻ – Oddział Wielkopolski, Poznań, Poland. ISSN 1429- 5946.

Gralik, J. (2003). *The influence of physical factors on wheat grain selected physicochemical, biochemical,technological properties and resistance against stored grain pests.* (in Polish) PhD thesis. University of Live Sciences, Poznań, Poland.

Gralik, J. & Warchalewski, J.R. (2006).The influence of γ-irradiation on some biological activities and electrophoresis pattern of albumin fraction. *Food Chemistry*, Vol. 99, 289-298, ISSN 0308-8146.

Gralik, J., Trojanowska, K., & Warchalewski (1999). Comparison of the effects of gamma and microwave irradiation of wheat grain on its microflora contamination. *Scientific Paper Agricultural University Poznan,Poland. Food Science and Technology*. Vol. 3, 59-66, ISSN 1505-4217.

Grundas, S., Velikanov, L., & Archipov, M. (1999). Importance of wheat grain orientation for the detection of internal mechanical damage by the X-ray method. *International Agrophysics*. Vol. 13, No. 3, 355-361, ISSN 0236-8722.

Grundas, S., Warchalewski, J.R., Dolińska, R. & Gralik, J. (2008). Influence of microwave heating on some physicochemical properties of wheat grain harvested in three consecutive years. *Cereal Chemistry*. Vol. 85, No. 2, 224-229, ISSN 0009-0352.

Haff, R.P. & Slaughter, D.C. (2004). Real-time X-ray inspection of wheat for infestation by the granary weevil, *Sitophilus granarius* L. *Transaction of the ASAE*. Vol. 47,531-537, ISSN 0001-2351.

Hagstrum, D.W.., Vick, K.W., & Webb, J.C. (1990). Acoustic monitoring of *Rhyzoperta dominica (Coleoptera: Bostrichidae)* populations in stored wheat. *Journal of Economic Entomology*. Vol. 83, No.2, 625-628, ISSN 0022-0493.

Hamer, R. J. &Van Vliet, T. (2000).Understanding the structure and properties of gluten: an overview. In: *Wheat gluten*, P. Shewry & A. S. Tatham (Eds), Royal Society of Chemistry, Cambridge. 125-131, ISBN: 0-85404-865-0.

Hoseney, R.C. (1987), Wheat hardness. *Cereal Foods World*. Vol. 32, No. 4, 320-322, ISSN 0146-6283.

James, R.L. & Gilligan, C.J. (1988). Microwave treatments to eradicate seed borne fungi on Douglas-fir seed. USDA Forest Service. Northern region. *Report 88-7.*

Jayas, D.S. & White, N.D.G. (2003). Storage and drying of grain in Canada: low cost approaches. *Food Control*. Vol. 14, 255-261, ISSN 0956-7135.

Karunakaran, C., Jayas, D.S., & White, N.D.G. (2003). X-ray image analysis to detect infestation caused by insect in grain. *Cereal Chemistry*. Vol. 80, No.5, 553-557, ISSN 0009-0352.

Karunakaran, C., Jayas, D.S., & White, N.D.G. (2004). Detection of internal wheat seed infestation by *Rhyzopertha dominica* F. using X-ray imaging. *Journal of Stored Product Research*. Vol. 40, 507-616, ISSN 0022-474X.

Kasarda, D.D., Nimmo, C.C. & Kohler, G.O. (1971). Proteins and the amino acid composition of wheat fraction. In: *Wheat Chemistry and Technology*. Y. Pomeranz (Ed.), American Association of Cereal Chemists Incorporated , St. Paul, Minnesota, USA. Vol. 3, 227-299.

Krokida, M.K. & Maroulis, Z.B. (2001). Quality changes during drying of food materials. In: *Drying technology in agriculture and food sciences*. A.S., Nujumdar (Ed.), Oxford IBH, Delhi, India. ISBN 978-1578081486.

Kruger, J.E. & Reed, G. (1988). Enzymes and color. In: *Wheat chemistry and technology*. Y. Pomeranz (Ed.), Vol. 1, 441-500. American Association of Cereal Chemists Inc., St. Paul, Minnesota, USA. ISBN 0-913250-65-1.

Lefebvre, J., Pruska-Kedzior, A., Kedzior, Z. &Lavenant, L. (2003). A phenomenological analysis of wheat gluten viscoelastic response in retardation and in dynamic experiments over a large time scale. *Journal of Cereal Science*. Vol. 38, No. 3, 257-267, ISSN 0733-5210.

Lozano, J.C, Laberry, R. & Bermudez, A. (1986). A microwave treatment to eradicate seed borne pathogens in cassava true seed. *Journal of Phytopathology*. Vol. 117, 1-8, ISSN 0931-1785.

MacArthur, L.A. & D'Appolonia, B.L. (1981). Effects of microwave radiation and storage on hard red spring wheat flour. *Cereal Chemistry*. Vol. 58, No.1, 53-56, ISSN 0009-0352.

Machesh, S., Manickawasagan, A., Jayas, D.S. Paliwal, J., & White, N.D.G. (2008). Feasibility of near-infrared   hyperspectral imaging to differentiate Canadian wheat classes. *Biosystems Engineering*. Vol. 101, No. 1, 50-57, ISSN 1537-5110.

Madrid, F.J., White, N.D.G.& Loschiavo, S.R. (1990). Insects in stored cereals, and their association with farming practices in southern Manitoba. *The Canadian Entomologist*. Vol. 122, 515-523, ISSN 0008-347X.

Martin, C.R., Rousser, R. & Brabec, D.L. (1993). Development of a single-kernel wheat characterization system. *Transactions of ASAE*. Vol. 36, 1399-1404, ISSN 0001-2351.

Mitrus, M. (2000). Applying of microwaves in food technology. (in Polish). *Postępy Nauk Rolniczych*. No. 4, 99-114, ISSN 0084-5477.

Modgil, E. & Samuels, R. (1998). Efficacy of mint and eucalyptus leaves on the physicochemical characteristics of stored wheat against insect infestation. *Nahrung/Food* . Vol. 5, No. 5, 304-308. ISSN 0027-769X.

Morris, P.C., Palmer, G.H. & Bryce, J.H. (2000). Summary and conclusions, In: *Cereal Biotechnology*. P.C. Morris & J.H. Bryce, (Eds) 237-242, Woodhead Publishing Limited, Cambridge, England. ISBN 1 85573 498 2.

Nawrocka, A., Grundas, S. & Grodek, J. (2010). Losses caused by granary weevil larva in wheat grain using digital analysis of X-ray images. *International Agrophysics*. Vol. 24, No.1, 63-68, 355-361, ISSN 0236- 8722.

Nawrot, J., Gawlak, M., Szafranek, J., Szafranek, B., Synak, E., Warchalewski, J.R., Piasecka-Kwiatkowska,D.,Błaszczak, W., Jeliński & T., Fornal, J., (2010). The effect of wheat grain composition, cuticular lipids and kernel surface microstructure on feeding, egg-laying, and development of the granary weevil,*Sitophilus granarius* (L.). *Journal of Stored Products Research.* Vol. 46, 133-141, ISSN 0022-474X.

Nawrot, J., Warchalewski, J.R., Piasecka-Kwiatkowska, D., Niewiadoma, A., Gawlak, M., Grundas, S.T., & Fornal, J. (2006). The effect of some biochemical and technological properties of wheat grain on granary weevil *(Sitophilus granaries L.) (Coleoptera: Crculionidae)* development. *Proceedings of 9th International Working Conference on Stored Product Protection.* Brazil, Brazilian Post-Harvest Association, Campinas. Pp. 400-407, ISBN 85-60234-00-4.

Nelson, S.O. (1996). Review and assessment of radio-frequency and microwave energy of stored-grain insect control. *Transactions of ASAE.* Vol. 39, 1475-1484, ISSN 0001-2351.

Neethirajan, S., Karunakaren, C., Jayas, D.S., & White, N.D.G. (2007). Detection techniques for stored-products insects in grain. *Food Control.* Vol. 18, No. 2, 157 -162, ISSN 0956-7135.

Ng, D.T.W. (2004). Heat/stress responses. In: *Encyclopedia of Biological Chemistry.* W.J. Lennarz & M.D. Lane, (Eds), Academic Press, , New York, USA. Pp. 343-347, ISBN 978-0124437104.

Nijhuis, H.H., Torringa, H.M., Muresan, S., Yuksel, D., Leguijt, C. & Kloek, W. 1998. Approaches to improving the quality of dried fruit and vegetables. *Trends in Food Scence and .Technology.* Vol. 9, 13-20, ISSN 0924-2244.

Ohm, J.B., Chung, O.K. & Deyoe, C.W. (1998). Single-kernel characteristics of hard winter wheat in relation to milling and baking quality. *Cereal Chemistry.* Vol. 75, No.1, 156-161, ISSN 0009-0352.

Osborne, B.G., Kotwal, Z., Blakeney, A.B., O'Brien, L.O., Shan, S. & Fearn, T. (1997). Application of the single-kernel characterization system to wheat receiving testing and quality predication. *Cereal Chemistry.* Vol. 74, No.4, 467-470, ISSN 0009-0352.

Pearson, T. & Brabec, D.I. (2007). Detection of wheat kernels with hidden insect infestation with an electrically conductive roller mill. *Applied Engineering in Agriculture.* Vol. 23, No. 5, 639-645, ISSN: 0883-8542.

Pederson, J. (1992). Insects: Identification, damage, and detection. In: *Storage of Cereal Grains.* D.B. Saur, (Ed.), 1- 615, American Association of Cereal Chemists, Incorporated, St. Paul, Minnesota, USA. ISBN 0913250740.

Piasecka-Kwiatkowska, D., Rosiński, M., Krawczyk, M., Warchalewski, J.R., Nawrot, J. & Gawlak, M. (2006). Possibilities of application SDS-PAGE electrophoresis for detection of cereal grains infestation caused by confused flour betele. (in Polish).*Progress in Plant Protection.* Vol. 46, No. 2, 453-456, ISSN 1427- 4337.

Piasecka-Kwiatkowska, D., Rosiński, M., & Warchalewski, J.R. (2005). Application of immunochemical method for detection od cereal grain infestation by granary weevil *(Sitopilus granarius L.)* (in Polish). *Progress in Plant Protection.* Vol. 45, No. 1, 369-374, ISSN 1427-4337.

Piasecka-Kwiatkowska, D., Zielińska-Dawidziak, M., Warchalewski, J.R. & Nawrot, J. (2010). ELISA test for detection of granary weevil in wheat grain (in Polish). 8th Polish Conference of Chemical Analysis "Analytics for community of 21 century", Poland, July 2010, Komitet Chemii Analitycznej & Polskie Towarzystwo Chemiczne, Kraków. ISBN 978-83-925269-7-1.

Plarre, R., Halverson, S.L., Burkholer, W.E., Bigelow, T.S., & Misenheimer, M.E. (1999). *Proceedings of 7th International Working Conference of Stored-Product Protection,* October,

1998, Sichuan Publishing House of Sciences and Technology, People Republic of China, Chengdu, Sichuan Province. Pp. 1152-1157, ISBN 7-5364-4098-7/TD15.

Prądzyńska, A. (1995). The role of higher temperatures to control of granary weevil (*Sitophilus granarius* L.). *Scientific Papers of Institute of Plant Protection*. Vol. 36, No. 1/2, 119-127, ISSN 0554-8004.

Pruska-Kedzior, A. (2006). Application of phenomenological rheology methods to quantification of wheat gluten viscoelastic properties (in Polish). *Scientific monographs*. Poznan University of Life Sciences, Poland. No. 373, ISSN 1896-1894.

Pruska-Kędzior, A., Kędzior, Z. & Klockiewicz-Kamińska, E. (2008). Comparison of viscoelastic properties of gluten from spelt and common wheat. *European Food Research and Technology*. Vol. 227, No. 1, 199-207, ISSN 1438-2377.

Satumbaga, R., Martin, C., Eustace, D. & Deyoe, C.W. (1995). Relationship of physical and milling properties of hard red winter wheat using the single kernel characterization system. *Association Operators of Millers Bulletin*. ISSN 6487- 6496.

Schatzky, T.F., & Fine, T.A. (1988). Analysis of radiograms of wheat kernel for quality control. *Cereal Chemistry*. Vol. 65, No. 3, 233-239, ISSN 0009-0352.

Seaman, W.L. & Wallen, V.R. (1967). Effect of exposure to radio frequency electric fields on seed borne microorganisms. *Canadian Journal of Plant Science*. Vol. 47, 39-49, ISSN 1918-1833.

Singh, C.B., Jayas, D.S., Paliwal, J., & White, N.D.G. (2009). Detection of insect-damaged wheat kernels using near-infrared hyperspectral imaging. *Journal of Stored Products Research*. Vol. 45, 151-154, ISSN 0022- 474X.

Sinha, R.N. & Watters, F.L. (1985). Insect Pests of Flour Mills, Grain Elevators and Feed Mills and Their Control. *Agriculture Canada*, Ottawa. Publication 1776.

Shayesteh, N. & Barthakur, N.N. (1996). Mortality and behaviour of two stored-product insects species during microwave irradiation. *Journal of Stored Products Research*. Vol. 32, 239-246, ISSN 0022-474X.

Shewry, P. R., Halford, N. G., Belton, P. S. &Tatham, A. S. (2002). The structure and properties of gluten: An elastic protein from wheat grain. *Philosophical Transactions of the Royal Society B: Biological Sciences*. Vol. 357, No.1418, 133-142, ISSN: 0962-8436.

Shotwell, O.L., Bennett, G.A., Stubbefield, R.D., Shannon, G.M., Kowolek, W.F. & Plattner, R.D. (1985). Deoxynivalenol in hard red winter wheat: relationship between toxin levels and factors that could be used in grading. *Journal of the Association of Analytical Chemists*. Vol. 68, 954-957, ISSN 0004-5756.

Sikora, M., Tomasik, P. & Pielichowski, K. (1997). Chemical starch modification in the field of microwaves. II. Reactions with alpha-amino and alpha-hydroxy acids. *Polish Journal of Food and Nutritional Sciences*. Vol. 6, No. 2, 23- 30, ISSN 1230-0322.

Slaughter, D.C., Norris, K.H. & Hruschka, W.R. (1992). Quality and classification of hard red wheat. *Cereal Chemistry*. Vol. 69, No. 4, 428-432, ISSN 0009-0352.

Stephenson, M.M.P., Kushalappa, A.C. & Raghavan, G.S.V. (1996). Effect of selected combinations of microwave treatment factors on inactivation of *Ustilago nuda* from barley seed. *Seed Science and Technology*. Vol. 24, 557-570, ISSN 0251-0952.

Sumnu, G. (2001). A review on microwave baking of foods. *International Journal of Food Science and Technology*. Vol. 36, 117-127, ISSN 0950-5423.

Sundaram, G. & Huai-Wen. Y. (2007). Effect of experimental parameters of temperature distribution during continous and pulsed microwave heating. *Journal of Food Engineering*. Vol. 78, 1452-1456, ISSN 0260-8774.

Toews, M.D., Pearson, M.C. & Campbel,l J.F. (2006). Imaging and automated detection of *Sitophilus oryzae (Coleoptera: Curculiobidae)* pupae in hard red winter wheat. *Journal of Economic Entomology.* Vol. 99, No. 2, 583-592, ISSN 0022-0493.

Tschoegl, N. W. (1989). The phenomenological theory of linear viscoelastic behavior : an introduction. Springer-Verlag, Berlin. ISBN 3-540-19173-9.

Vadivambal, R. & Jayas, D.S. (2007). Changes in quality of microwave-treated agricultural products – a review. *Biosystem Engineering.* Vol. 98, 1-16, ISSN 1537-5110.

Venkatesh, M.S. & Ragvan, G.S.V. (2004). An overview of microwave processing and dielectric properties of agri-food materials. *Biosystem Engineering.* Vol. 88, 1-18, ISSN 1537-5110.

Vick, K.W., Webb, J.C., Weaver, B.A., & Litzkow, C. (1988). Sound detection of stored-products insects that feed inside kernels of grain. *Journal of Economic Entomology.* Vol. 8, No. 4, 1489-1493, ISSN 0022-0493.

Walde, S.G., Balaswamy, V. & Rao, D.G. (2002). Microwave drying and grinding characteristics of wheat (*Triticum aestivum*). *Journal of Food Engineering.* Vol. 55, 271-276, ISSN 0260-8774.

Warchalewski, J.R., Dolińska, R. & Błaszczak, W. (2007). Microscope analysis of two generations of wheat grain crops grown from microwave heated seeds. In Polish. *Acta Agrophysica.* Vol. 10, No.3, 727-737, ISSN 1234-4125.

Warchalewski, J.R. & Gralik, J. (2010). Influence of microwave heating on biological activities and electrophoretic pattern of albumin fraction of wheat grain. *Cereal Chemistry.* Vol. 87, No. 1, 35-41, ISSN 0009-0352.

Warchalewski, J.R., Gralik, J. & Nawrot, J. (2000). Possibilities of reducing stored cereal-grain damage caused by insect pests. *Postępy Nauk Rolniczych.* No. 6, 85-96, ISSN 0032-5547.

Warchalewski, J.R., Gralik, J., Winiecki, Z., Nawrot, J. & Piasecka-Kwiatkowska, D. (2002). The effect of wheat α- amylase inhibitors incorporated into wheat-based artificial diets on development of *Sitophilus granarius* L., *Tribolium confusum* Duv., and *Ephestia kueniella* Zell. *Journal of Applied Entomology.* Vol. 126, No. 4, 161-168, ISSN 0931- 2048.

Warchalewski, J.R., Gralik, J., Zawirska-Wojtasiak, R., Zabielski, J. & Kuśnierz, R. (1998). The evaluation of wheat odor and colour after gamma and microwave irradiation. *Electronic Journal of Polish Agricultural Universities, Food Science and Technology.* 1, Available online: http://www.ejpau.media.pl/series/1998/food/art.-04.html. ISSN 1505-0297.

Warchalewski, J.R. & Nawrot, J. (1993). Insect infestation versus some properties of wheat grain. *Roczniki Nauk Rolniczych.* Seria E, Vol. 23,No. 1/2, 85-92, ISSN 1429-303X.

Warchalewski, J.R., Prądzyńska, A., Gralik, J. & Nawrot, J. (2000). The effect of gamma and microwave irradiation of wheat grain on development parameters of some stored grain pests. *Nahrung/Food.* Vol. 44, No. 6, 411-414, ISSN 0027-769X.

Watters, F.L. (1991). In: *Ecology and management of food-industry pests.* J.R. Gorham (Ed.), 399-443. FDA Technical Bulletin 4, AOAC, Arlington, VA , USA. ISBN 0-935584-45-5.

Wyatt, G.R. (1961). Biochemistry of insect hemolymph. *Annual Review of Entomology.* Vol. 6, 75-102, ISSN 066-4170.

23

Use of Microwave Radiation to Process Cereal-Based Products

Yoon Kil Chang[1], Caroline Joy Steel[1] and
Maria Teresa Pedrosa Silva Clerici[2]
[1]*University of Campinas (UNICAMP)*
[2]*Federal University of Alfenas (UNIFAL-MG)*
Brazil

1. Introduction

Within the electromagnetic spectrum, microwave radiation is characterized by being situated in the frequency interval between 300 MHz and 300 GHz, with those normally used for the industrial processing of foods being between 915 and 2450 MHz and, for domestic use, of 2450 MHz (Pei, 1982).

Microwave ovens can be built for industrial (continuous or batch processes), laboratory or domestic use. Industrial equipments present greater process control, but are costly and designed for large scale, which limits their use in research. Those developed for laboratory use are also costly, due to the need of heating, pressure and time control mechanisms to increase the reproducibility of the results. Therefore, many researchers end up adapting domestic microwave ovens, which results in lower cost equipments. Domestic microwave ovens do not have a uniform distribution of microwave radiation, as they were not designed for such. They produce interference between the microwaves and, thus, some parts of the oven receive a greater incidence than others. Therefore, it is necessary to map the distribution of microwave radiation for a more effective use of the energy generated (Pecoraro et al. 1997; Marsaioli Jr., 1991; Loewen & Gandolfi, 2004; Breslin, 1990).

A typical microwave apparatus consists of:

- Feed source: supplies tensions and currents for the functioning of the microwave generator;
- Microwave generator: is an oscillator that converts electric energy supplied by the feed source in microwave energy;
- Transmission section: propagates, irradiates or transfers the energy radiated by the oscillator to the applicator;
- Coupling: permits a more efficient transfer of microwave energy to the applicator;
- Microwave application cavity: is a volume limited by metallic walls in which the interaction material/microwaves occurs;
- Field agitator: rotating metallic reflector inside the cavity that modifies contour conditions periodically, with the objective of uniformizing energy distribution;
- Door or opening with electromagnetic blindage: permits an easy access to the interior of the applicator and must limit electromagnetic emissions to the surroundings to the maximum level permitted.

- Operation and safety control: permit the selection of heating conditions, the interruption of the potency flow and also indicate, through a sound and/or visual signal, the end of the processing cycle (Confort et al., 1991; Rocha, 2002).

Senise & Jermolovicius (2004) reported that the technically possible applications of microwaves, at the industrial level in the food industry are the drying of pasta products, dehydration, blanching, sterilization, pasteurization, cooking and thawing. However, nowadays these applications have increased with the use of microwaves for chemical reactions, insect and germination control, production of expanded cereals or tubers, etc.

The heating of foods using microwaves results of the coupling of electric energy from and electromagnetic field in a microwave cavity and of its dissipation within the product. This results in an instant temperature increase inside the product, in contrast to conventional heating processes that transfer energy from the surface, with high thermal time constants and a slow heat penetration (Goldblith, 1966).

The advantages of the use of microwaves, in comparison to conventional processing techniques, are a shorter processing time, higher yield and better quality of the final product (Decareau & Peterson, 1986), both sensory and nutritional (Sacharow & Schiffmann, 1992; Hoffman & Zabik, 1985).

The development of food processes and products with the use of microwaves should try to determine the interaction mechanisms between the food and the electromagnetic energy in the microwave frequencies. The main aspects that should be considered are:

- dielectric properties of foods;
- quantity of energy coupled by a food product and distribution within the product; microwave time and frequency necessary to heat the food product; and
- temperature, pressure and electric field parameters (Mudgett, 1982; Rocha, 2002).

The proposal of this chapter is to show the state-of-the-art of the use of microwaves in cereal products, with the aim of presenting the main results found in this promising area for research and industrial application.

2. Principles of the use of microwaves and main technological applications in cereal-based products

The basic phenomena involved in microwave heating are: the coupling of the energy from an electromagnetic field by the product and the attenuation of energy absorption within it. The most important intrinsic property of this form of energy for food technology is volumetric absorption by dielectric materials, in the form of heat (Engelder & Buffler, 1991).

In microwave processing, intermolecular friction caused mainly by bipolar rotation of polar molecules, generates heat internally in the material, with a lower heat gradient in the product and results in an instant temperature increase. Aqueous and polar ionic constituents in foods and their associated solid constituents have a direct influence on how the heating will be conducted (Buffler, 1992; Marsaioli Jr. 1991; Schiffmann, 1986).

Apart from the chemical composition of the product, its geometry, size and volume affect the distribution of microwave heating (Remmen et al., 1996; Khraisheh et al., 1997; Yang & Gunasekaran, 2004; Ohlsson & Risman, 1978).

In contrast to conventional methods, in microwave heating the production of heat is continuous and, therefore, there is a continuous and rapid temperature increase when the food is exposed to this radiation. During heating of a moist product, the heat will be used to evaporate water and the temperature will be maintained at 100°C. As soon as the free water is evaporated, product temperature can increase rapidly, with the risk of it burning (Sale, 1976).

2.1 Grain drying

In cereals, grain drying is directly related to conservation and time and quality maintenance. Keeping moisture content of cereal grains at low levels, of 10-12%, makes safe storage of approximately 5 years possible, as damaging effects, such as the plague attack, microorganism growth and grain respiration, are reduced. Grain drying can be natural and artificial (Frisvad, 1995; Marini et al., 2007).

In natural drying, the product is dried by sunlight, spread in open areas, being exposed to the environmental conditions of the region for a long period of time and occupying large areas, once the grains must be moved and spread around for a greater efficiency of the process.

Artificial or mechanical drying consists in submitting a layer of grains to a heated air current, which can be through electric energy or the use of fuels such as gas, wood or coal. At the beginning of drying, free water is evaporated rapidly, but after this the process becomes slow to evaporate the water from the interior of the product to the surface. The main controls are temperature, to avoid the loss of grains due to high temperatures, and aeration, with the aim of reducing the temperature and moisture of the grains during storage, avoiding the deleterious effects of moisture migration to the surface of the grain. This drying process takes long and can damage grain quality (Frisvad, 1995; Marini et al., 2007).

Microwave drying is an alternative drying method for a wide variety of grains, as it avoids the effects of conventional air drying that can cause serious damage to flavor, color and nutrients and can reduce bulk density and rehydration capacity of dried food product (Loewen & Gandolfi, 2004; Park, 2001).

During microwave drying there must be transportation of water from the interior of the solid to its surface, for evaporation to occur. The most important mechanisms involved in this process are:

- Liquid diffusion that occurs due to the existence of the concentration gradient;
- Vapor diffusion that occurs due to the vapor pressure gradient, caused by the temperature gradient;
- Liquid and vapor flow caused by the difference in external pressure, concentration, high temperature and capillarity;
- Initial moisture content of the material;
- Final moisture content that the material can reach, that is, its equilibrium moisture content;
- Thus, the water is related to the structure of the solid; and
- How water transportation is carried out from the interior to the surface of the solid (Park et al, 2001).

However, one of the problems of microwave drying is related to the non-uniform distribution of the electromagnetic field inside the applicator, which can induce a non-uniform heating of the product (Kudra, 1989).

Some researchers have developed equipments that aim to solve this problem:

Marsaioli Jr.(1991) developed a microwave equipment with a cylindrical rotatable applicator, where the movement of the product inside the cavity permits a more uniform distribution of heating.

Loewen & Gandolfi (2004) developed an equipment to dry cereals using microwaves, with elements that carry out the heating of water inside the cereal grains themselves, vaporizing only small quantities of water from the surface at the beginning of the operation. With the advance of heating, an internal pressure is created inside the grain that, in turn, causes an unbalance between internal and external pressure, thus favoring homogeneity and speed of drying. Once the desired temperatures are reached, adequate for each cereal, water migrates to the surface, being removed by a dry air current.

Breslin (1990) built a pilot scale microwave-convection dryer. The dryer had a working length of 4.25 m, with a maximum power of 20 kW. The residence time used was approximately 120 seconds and the drying time required to reduce the moisture content from 28 to 21% wet basis was reduced by 450% compared to the conventional convection system. He concluded that the reduction in residence time can be used to decrease the length of the dryer from 10.1 to 3.1 m, or to increase the capacity of the production unit by 235%.

Doty & Baker (1977) conditioned wheat in a closed system with up to 450 seconds of microwave application (625 watts), with a temperature variation from 22 (0 seconds) to 105°C (450 seconds). Flour analysis showed that the 90 second treatment maintained flour quality; however, quality was reduced after 270 seconds.

2.2 Enzyme inactivation

Cereals in the form of grains and wholegrain or refined flours, when raw, present variable enzymatic activity according to the type of enzyme, moisture content and storage temperature. Enzymatic inactivation of lipases and lipoxygenases in cereals and their flours increases stability during storage, by preventing lipid hydrolysis and oxidation that could lead to the formation of off-flavors (Saunders, 1990).

Brown rice has a higher nutritional value than polished (white) rice because it has more protein and B-complex vitamins, especially thiamin, riboflavin and niacin. Its production process results in greater yield with fewer broken grains (Juliano, 1985). However, brown rice is unstable during storage mainly due to the rapid decomposition of bran lipids, attributed to the action of the enzyme lipase also contained in rice bran (Takano, 1993). Brown rice lipids are readily hydrolyzed by lipase, releasing free fatty acids (FFAs), which contribute to the appearance of off-flavors.

Chang & El-Dash (1998) used microwave energy as a thermal treatment to inactivate lipase in brown rice, consequently improving its long term storage stability. The parameters studied were: (i) microwave energy level (maximum and medium); (ii) initial moisture content (13.4, 14.6 and 17.3%) and time of treatment (20 to 140 s). Microwave energy treatment was considered effective in the control of lipase activity in brown rice with the maximum inactivation rate at the maximum power/level of microwave energy, a treatment time of 80 s and initial moisture content of 14.6%. Moisture content of treated brown rice was reduced. During storage, a decrease in the production of free fatty acids (FFAs) was observed.

Vetrimani et al. (1992) used microwave energy to inactivate lipase and lipoxygenase in rice bran, germ and soybean. This treatment led to considerable inactivation of the lipase and complete inactivation of the lipoxygenase present in these materials.

Jiaxun-Tao et al. (1993) reported stabilization of rice bran by microwave heating at 2450 MHz for 3 minutes, for up to 4 weeks storage.

Wennermark (1993) verified that heat-inactivation of lipid-degrading enzymes with low-moisture processes, such as steam-flaking and microwave cooking, improved vitamin E retention during subsequent processing in the presence of water.

The feasibility of microwave energy for the inactivation of α-amylase in wheat and wheat flour has been successfully tested by Edwards (1964) and Aref et al. (1969). They showed that the levels of enzyme activity decreased without damaging the flour with respect to its capacity to form dough, maintaining its viscoelastic properties.

2.3 Insect control

When grain storage conditions are precarious, there can be damage to cereals and by-products through the action of biological agents such as insects, acarus, rodents and fungi. The factors that negatively affect grain storage are: moisture, temperature, storage period, initial contamination level, impurities, insects, intergranular CO_2 concentration, grain physical and sanitary conditions (Prado et al., 1991; Scudamore et al., 1999).

Grain damage causes economic losses by reducing quality and altering nutritional value. Sanitation measures must be taken in storage, production, maintenance and distribution practices, including steps such as cleaning and disinfection of the storage structures, grain expurgation, rodent control, among others.

Warchalewski & Gralik (2010) verified that microwave heating of wheat grains within the temperature range of 28-98°C, using microwaves, caused a decrease in water-extractable proteins, statistically significant when grain temperature reached 79 and 98°C. A statistically significant increase in reducing sugars content was noted in grain samples heated only to 48°C; a decrease was noted above this temperature. Above 48°C, inactivation of the enzyme α-amylase occurs. All biological activities studied (amylolytic and inhibition activities against α-amylases from insects (*Sithophilus granarius L., Tribolium confusum Duv., Ephestia kuehniella Zell.*)), human saliva, hog pancreas, antitryptic activity) were distinctly diminished in grain samples heated to 79°C. At the highest grain temperature of 98°C, the loss of all biological activities was even more pronounced due to denaturation of 45% of extractable proteins.

Microwave heating was evaluated to reduce the incidence of mycoflora on rice (Oryza sativa L.) grains without adversely affecting the aroma and cooking quality of the grains. The duration of exposure of microwave heat directly correlated with grain temperature and inversely correlated with grain moisture and mycoflora. The exposure of rice grains to microwave heat at 100% power level up to 2 min retained the moisture and cooking-quality attributes of rice grains but reduced the mycoflora incidence by > 50% (Gupta & Kumar, 2003).

2.4 Grain germination

Germination is an ample and complex phenomenon and can be characterized when, under appropriate conditions, the embryo axis of the seed continues its development, which was interrupted by physiological maturity (Carvalho & Nakagawa, 1988).

In this process, catabolic reactions, such as the degradation of storage substances, and anabolic reactions, in the production of new cells and organelles of the embryo, occur (Metivier, 1979).

Cereal grains are susceptible to post-harvest germination (Bushuk, 2001), especially when temperature and humidity conditions are favorable, which can lead to great losses during grain storage. Few studies have been carried out to evaluate the effect of microwave radiation on cereal grains.

Vadivambal et al. (2009) used a pilot-scale industrial microwave drier operating at 2450 MHz to heat grains and oilseed. An infra-red thermal camera was used to determine the temperature distribution in bulk rye and oats at 14, 16, and 18% moisture content (wet basis) and sunflower seeds at 8, 10, and 12% moisture content after being heated in the microwave drier. Fifty grams of grain were placed in a sample holder and allowed to pass on a conveyor belt under the applicator at 0, 200, 300, 400, and 500 W for 28 or 56 s. There were hot and cool regions in the samples. The temperature difference between hot and cool regions in a given sample varied between 23 and 62°C for rye, 7 and 25°C for oats and 7 and 29°C for sunflower seeds. The preliminary results of this study suggest that while using microwave as a heat treatment, the maximum temperature that can affect seed viability (potential to germinate under favorable conditions) should be taken into consideration, when developing microwave processing systems for grains and oilseeds.

2.5 Pre-cooking

The objective of pre-cooking operations is to reduce preparation time for the consumer. In the case of cereals, these operations consist basically of treating starch to reduce its gelatinization time during the final preparation of the food product.

By toasting cereal flours, Wang et al. (1993) obtained pre-cooked rice flour (with 13.4% moisture) after 11 min in a microwave oven (2450 MHz) and Caballero-Córdoba et al. (1994) obtained pre-cooked wheat flour (with 14% moisture) after 11 min in a microwave oven (2450 MHz), with good sensory and nutritional evaluations, which were used in combination with defatted soy flour with additional microwave treatment for 8 min, for porridge and soup products.

Martinez-Bustos et al. (2000) produced nixtamalized maize flours (NMF) for the production of masa and tortillas, with the use of microwaves, operating frequency 2450 MHz, high-power setting, heating for 10, 15 and 20 min during alkaline cooking with the objective of accelerating the conventional alkaline cooking process. The results showed that microwave heating reduced processing time (approximately 50%) and liquid waste discharges (cooking liquor) during NMF production. Samples of NMF from whole maize microwave heated for 20 min with 70 g.kg^{-1} water and 2.0 g.kg^{-1} Ca(OH)$_2$ were statistically similar to the control with respect to masa firmness.

Roberts (1977) obtained rice similar to conventional parboiled rice by processing at 30-35% moisture for 2 to 5 min with microwaves.

Velupillai et al. (1989) developed a process for parboiling rough rice, which includes: (1) soaking the rough rice; (2) subjecting the soaked rough rice to a first value of microwave energy to partially gelatinize the starch in the rice and raise its water content; (3) draining free water, if any, from the treated soaked rough rice; and (4) subjecting the drained rough rice from (3) to a second level of microwave energy to effect substantially complete gelatinization and to lower the rough rice water content.

2.6 Cooking

Cooking and thawing operations in microwave ovens are already well documented. Some authors reported that the cooked and/or thawed product presents sensory changes in flavor, color and texture. According to Li (1995), flavor loss due to steam distillation is a major problem when microwave foods are cooked or reheated and this results in a unbalanced flavor profile in the final product.

With respect to color, the promotion of browning in a microwave oven is a difficult problem. According to Domingues et al. (1992), the solutions to microwave browning can be divided into the following categories: packaging aided, cosmetic and reactive coating approaches, and they developed a browning system that includes Maillard browning reactants for developing the desired browning effect during microwave irradiation, and a carrier system which contains the Maillard browning reactants.

Regarding texture, Clarke & Farrell (2000) studied the impact of water-binding agents (potato starch, pea fibre, oat fibre, locust bean gum), emulsifiers, ingredient blends and a proteolytic enzyme (fungal protease) on the textural characteristics of frozen five-inch pizza bases topped with tomato and cheese, reheated for 120, 150 and 180 s in a domestic microwave oven. The textural characteristics of the microwave-reheated pizza base samples were influenced by the presence of the selected agents.

2.7 Baking

Baking using microwave energy has been limited due to poor product quality compared to products baked by using conventional energy sources, which can be a reflection of the differences in the mechanism of heat and mass transfer (Sakonidou et al., 2003).

In products such as breads, cakes and cookies, microwave baking can affect texture, moisture content and color of the final product, which represents a great challenge for research. Some researchers suggest adjustments in formulation and alterations in the baking process, while others study the interactions between microwave energy and the ingredients of the formulation.

2.7.1 Breads

The producton of bread with the requisite quality attributes presumes a carefully controlled baking process. Thus, the rate of heat application and the amount of heat supplied, the humidity level within the baking chamber, and the duration of the bake all exert a vital influence on the final quality of the bread (Pyler, 1988).

In the conventional baking process, heat is transmitted to the dough piece in three different ways: radiation, convection and conduction. In the transformation of dough to bread, different phases occur: oven-rise (rapid expansion of carbon dioxide), oven-spring (expansion of the dough by about one-third of its original volume), moisture evaporation, starch gelatinization and protein coagulation; finally, cell walls become firm and crust color is developed (Pyler, 1988; Young & Cauvain, 2007).

In studies with bread during microwave heating, there was a rapid loss of moisture and, after microwaving, the mechanical strength of bread was increased greatly (Jahnke, 2003; Chavan & Chavan, 2010). According to Yin & Walker (1995), the heat and mass transfer patterns, insufficient starch gelatinization due to very short microwave baking times, microwave-induced gluten changes, and rapidly generated gas and steam could be reasons for the poor quality of microwave-baked breads. The work of Umbach et al. (1990) reported differences in bagel structure by using microscopy following conventional or microwave baking.

Park (2000) studied microwave (MW) – water interaction in a bread system and verified that:

- there was a linear relationship between moisture losses and energy input within the range of experiment conditions.
- the creep compliance value of MW heated bread was significantly lower than those of non-MW heated bread although both had the same moisture content level, suggesting the collapse of the aerated structure during microwaving by losing water molecules.

- MW heated breads absorbed less water than non-MW heated breads (they were less hygroscopic).

Jahnke (2003) developed a microwave baking dough additive that has a gelling component, a gum component and an enzyme component, as a means of controlling moisture migration or starch recrystallization in yeast-leavened bakery products that are baked by microwave energy.

2.7.2 Cakes

During cake production, after the mixing process, the cake must be deposited into cake pans and rapidly conveyed to the oven. Baking time is inversely related to baking temperature and the optimum baking conditions for cake baking are determined by the sweetener level of the formula, amount of milk used in the batter, fluidity of the batter, pan size and others (Young & Cauvain, 2007; Bennion & Bamford, 1998).

Stinson (1986) verified that yellow cakes baked in microwave/convection (MW/C) and conventional (CON) ovens showed slight sensory differences, however were acceptable, indicating that high quality cakes can be baked 15–25% faster in a MW/C oven than in a CON oven.

The effect of modifications in cake ingredients was studied by Tsoubeci (1994). Sucrose substitution with a blend of WPI (whey protein isolate) and maltodextrin and fat substitution with a special fiber were evaluated in a model cake system during baking by conventional and microwave methods. Substitution of sucrose with WPI and maltodextrin affected the dielectric properties of the batters. Power absorption predictions indicated that the power absorbed by microwave heated cakes increased when sucrose was substituted. Removal of fat influenced the dielectric behavior of batters whereas substitution with the special fiber compensated for fat in both microwave and conventionally heated cakes.

Şummu et al. (2000) used the response surface methodology to optimize the formulation of microwave-baked cakes and found that cakes formulated with wheat starch, containing 0.3% polysorbate 60, 133.7% water and 45.2% shortening (flour substitute basis), baked for 6 min at 100% power yielded acceptable cakes that can compete with conventionally baked cakes.

Takashima (2005) patented a process to obtain a sponge cake free from bake shrinkage and good-looking voluminous appearance, through a batter prepared by adding a thermocoagulation protein to a sponge cake premix containing as main ingredient a cereal powder consisting of starch and a pregelatinized starch cooked under heat with a microwave oven.

2.7.3 Cookies

There is a wide variety of cookies and their formulations, weight, diameter and other factors directly influence the baking process. In general, when cookies are baked using conventional methods, in conveyor belt ovens of 80 to 110 m, with cyclothermal heating, and universal conveyor belts, after approximately 3.5 to 4 minutes, they exit the oven with 5 e 8% moisture content and are cooled so as to lose moisture, to a final content of 3.5 to 4%. If the moisture gradient between the border and the center of the cookie is too high (greater than 1.5%), it can present fissures and cracks (checking) during storage, due to expansion and contraction of the cookie. Other factors are involved in checking, such as formulation, oven temperature, air relative humidity, etc. (Bernussi et al., 1998; Manley, 2000).

Shiffmann (1992) used microwave heating for final baking of cookies and obtained a more uniform moisture distribution than with forced convection.

Bernussi et al. (1998) studied the effects of microwave heating after conventional baking on moisture gradient and product quality of cookies and showed that cookies prebaked in a conventional oven at 140°C for 4 min and subsequently baked with a microwave oven set at 2450 MHz and magnetron power output of 617.27 W for 29 s showed significant reductions in moisture gradient, from 2.16 to 0.88% (central disk-outer rim), and incidence of cracking, from 41.7 to 0%. So the association of conventional and microwave heating processes for baking cookies minimized cookie cracking by reducing the moisture gradient and preserved the normal characteristics of the product (color, texture, flavor and linear dimensions).

Lorenz et al. (1973) verified that the baking time of cookies with microwaves was 30-45 seconds, however without crust color formation, so they suggested small modifications in the formulation, such as the use of molasses, chocolate or cocoa.

Lou et al. (1990) patented the production of prebaked cookie dough pieces, containing gluten, pregelatinized starch and high fructose corn syrup, to be baked in a microwave oven.

2.8 Pasta products

In dry pasta products, drying is the most critical step, as product moisture content must be reduced from approximately 30% to 12-13%. The greatest quality defects in pasta products are due to problems in this step: if drying is too fast, fissures or ruptures can occur; if drying is too slow, microbiological and enzymatic deterioration of the product is accelerated (Banasik, 1981; Kruger et al., 1996).

Pasta products are difficult to dry because moisture slowly migrates to the surface. Hot air is, by itself, relatively efficient at removing free water at or near the surface, whereas the internal moisture takes time to move to the surface. Microwave energy can solve this problem by providing a positive moisture flow towards the surface, and studies show pasta drying through combined processes (hot air + microwave drying) in a unique way (Decareau, 1986).

Maurer et al. (1971) obtained pasta drying speeds 20 times greater using microwaves when compared to conventional drying, therefore pasta dried by microwaves required lower drying process times.

Altan and Maskan (2005) studied the effect of conventional and microwave drying on the quality parameters of cooked and uncooked macaroni. Macaroni samples were dried by conventional hot air, microwave alone and hot air followed by microwave drying methods. Drying only with microwave energy (70 and 210 W) or hot air–microwave energy (70 and 210 W) resulted in substantial shortening of the drying time, but starch was not completely gelatinized during drying.

Goksu et al. (2005) studied microwave assisted fluidized bed drying of macaroni beads using a household microwave oven. They found that the increase in microwave power and air temperature significantly reduced (at 50%) the drying time of the macaroni beads when compared with fluidized bed only.

Berteli & Marsaioli Jr. (2005) evaluated the efficiency of air drying of Penne-type short cut pasta with the assistance of microwave energy and observed that the average drying time was reduced by more than ten times when compared to conventional air drying, without negatively affecting the appearance of the final product. The greatest benefits of drying pasta products using microwaves are the reduction of drying time and the maintenance of product quality.

2.9 Expansion

Besides heating, cooking or baking, it is possible to use microwave energy for expansion of foods.

2.9.1 Popcorn

Freshly popped popcorn, with its characteristic aroma and taste, has long been a popular snack product, particularly when flavored (Chedid et al., 1998). Popcorn is a whole-grain product obtained from a special variety of corn and its consumption increased with the development of microwave popcorn, this being one of the most popular applications of microwave heating, especially in the United States (Lin & Anatheswaran, 1988; Zhuang, 1998; Dofing et al., 1990; Pordesimo et al., 1990, 1991; Mohamed et al., 1993; Singh & Singh, 1999a, 1999b; Allred-Coyle et al., 2000, 2001; Chedid et al., 1998; Moraru & Kolkini, 2003).

Schwartzberg et al. (1995) considered that starch is the major polymer responsible for popcorn expansion.

Singh & Singh (1999b) reported that >75% popped kernels can be obtained by using 10% hydrogenated oil, 2% butter, and 0.5% sodium chloride, and expanding it in a 2450 MHz, 660 W microwave oven, at 70% power.

Chedid et al. (1998) prepared flavored and/or colored unpopped popcorn by use of an amylase-treated low viscosity starch, to be popped in a microwave oven without the use of added fat or oil, thus having a decreased caloric content.

2.9.2 Pellets or third-generation snacks

Manufacturing of 3rd-generation snacks is one of the newest applications of microwave heating. Third-generation snacks bring tremendous extension to popcorn expansion as it becomes possible to use food polymers to form snacks of various biological origins, in any desirable shape. They are obtained from half-products (pellets) that can be further expanded by baking, deep-fat frying, or microwave heating (Boischot et al., 2003; Ernoult et al., 2002; Suknark et al., 1999; Van Lengerich et al., 1992; Van Hulle et al., 1981, 1983; Willoughby, 2001).

Microwave expandable snack food products have been developed using cereal-based formulations with additives prepared as pellets in extruders (van Lengerich et al., 1992; Boehmer et al., 1992).

The non-expanded pellets are formed by extruding cereal flours at high moisture contents (30 to 35%), moderate shear and temperature, and die temperatures below 100°C, followed by cooling and drying to 5 to 10% moisture (Suknark et al., 1999; Boischot et al., 2003; Ernoult et al., 2002).

Other processes were also developed, as that of Willoughby et al. (1999), who prepared pellets from starchy compositions such as cooked farinaceous dough or dehulled popcorn with an outer methylcellulose skin or casing of sufficient tensile strength to allow build-up of internally generated steam pressure upon microwave heating. Upon sufficient build-up of steam pressure, the skin fails suddenly, allowing the pellet to puff explosively, thereby simultaneously causing an audible popping sound.

During microwave heating, glassy cereal pellets simultaneously lose moisture and expand (Boischot et al., 2003; Ernoult et al., 2002). This was demonstrated by the study of Ernoult et al. (2002), who obtained no expansion for totally dehydrated pellets, and maximum expansion was observed at 10.8% moisture content (weight/weight), and collapse was visible at 11.9% moisture content (weight/weight).

During microwave expansion of cereal pellets, the microwave energy heats the product through the vibrational energy imparted on moisture. Upon heating, moisture generates the superheated steam necessary for expansion, which accumulates at nuclei in the glassy matrix, creating a locally high pressure (Boischot et al., 2002). As the cereal matrix undergoes a phase transition from the glassy to the rubbery state, it starts to yield under the high superheated steam pressure and expansion takes place. As moisture is lost from the matrix, and upon cessation of microwave heating, the matrix cools down and reverts to the glassy state and the final structure sets. When the matrix is too soft (that is, at high moisture), collapse occurs (Moraru & Kokini, 2003).

Chen & Yeh (2000) observed that the high moisture pellets resulted in expanded products with a coarse structure, a small number of relatively large cells, and thick cell walls. At a lower moisture content of the pellets, a much finer structure was obtained.

When used for the manufacturing of snacks from half-products (pellets), starch is usually pregelatinized, as the limited amount of moisture and heat available in the expansion stage as well as the short residence times in microwave expansion, would not allow it to gelatinize and form a network that can be expanded by superheated steam (Schiffmann, 1993; Wang, 1997).

The study of Lee et al. (2000) indicated a nonlinear correlation between the degree of gelatinization and the expansion bulk volume of extruded cornstarch pellets and an optimum level of starch gelatinization (50%) for maximum product expansion.

Ernoult et al. (2002) obtained maximum expansion ratio (of about 10) for amylopectin pellets containing 6% solid fat, while at 10% solid fat expansion decreased due to a slight collapse of the expanded structure.

Gimeno et al. (2003) added 1% xanthan gum or carboxymethylcellulose to glassy cereal pellets and significantly improved the uniformity of the expanded matrix. The authors assumed that the large hydrocolloid molecules disrupted the alignment of the starch chains and promoted a uniform distribution of moisture in the matrix.

Weisz (2007) developed a process to produce a snack product from an extruded dough to be prepared by expansion in a microwave oven before consumption, or through any cooking method that permits expansion, such as conventional or electric ovens or frying.

2.10 Nutrient and ingredient modifications in cereal products

The main modifications in cereals and their respective products during microwave applications have been related to protein and starch.

With respect to proteins, research indicates a reduction in available lysine, independent of the type of heating and, in the case of wheat, studies indicate damage to gluten (Campana et al., 1993). Monks et al. (2003) studied the effect of microwave drying of soft wheat on the rheological behavior of the flour and the alveographic analysis of wheat flour showed that the P/L relationship and the gluten strength (W) increased (p<0.05) with the increase of microwave conditioning time, producing doughs that were inadequate for bread making.

There are various studies of the behavior of starch submitted to heating by microwave radiation and they are based mainly on gelatinization properties. However, other studies of chemical and physical modifications have been carried out and will be described below.

2.10.1 Pre-gelatinized starch

The production of pre-gelatinized starches has been investigated and developed industrially, as the products can be easily used in various food preparations, once they present rapid swelling in cold water and an increase in viscosity. The gelatinization process promotes the rupture of the starch granule and, consequently, an increase in the linkage sites of starch hydroxyl groups with water and other nutrients in an aqueous medium. The most studied cooking processes are:

- Conventional cooking: a suspension of starch in water is heated above gelatinization temperature (>65°C), using a combination of time, agitation, water and temperature until complete cooking of the starch, which can be used after cooling or drying as pre-gelatinized starch;
- Drum-drier cooking: a suspension of starch in water is cooked and dried on heated rotating drums; and
- Thermoplastic extrusion cooking at high temperature and pressure: the starch, conditioned to moisture contents that vary from 20 to 30%, is transported through an extruder with heating and high pressure zones. When cooked starch exits the extruder, it must go through drying and pulverization processes (Leach, 1965; Chiang & Johnson, 1977; Harper, 1979).

Microwaves have been used to obtain pre-gelatinized starches, due to rapid cooking, however the published literature with respect to microwave heating of starch suspensions is limited and focuses primarily on the differences in the swelling of the starch granule during conduction and microwave heating (Palav & Seetharaman, 2007). A comprehensive investigation addressing the changes occurring in starch–water systems during microwave heating at molecular and macroscopic level is still necessary.

Various authors have shown that gels formed by heating starch slurries using microwave energy had significantly different properties than those heated by using conduction heat. The lack of granule swelling and the resulting soft gel are two key observations that highlight the differences in the two modes of heating. The significant differences in the other molecular properties, including enzyme susceptibility and amylopectin recrystallization, suggest a different mechanism of gelatinization during microwave heating (Goebel et al., 1984; Zylema et al., 1985, Palav & Seetharaman (2007).

Palav & Seetharaman (2007) proposed that during microwave heating starch granules lose their birefringence much earlier than gelatinization temperature due to the vibrational motion of the polar water molecules. The vibrational motion and the rapid increase in temperature also result in granule rupture and formation of film polymers coating the granule surface. This results in a soft gel even in the absence of a continuous network of amylose chains.

Goebel et al. (1984) and Zylema et al. (1985) compared the properties of starch–water systems at granular level following microwave heating. Zylema et al. (1985) found no differences in the swelling of granules when heated using microwave or conduction at the same heating rate. Goebel et al. (1984) worked with 1:1, 1:2, 1:4 and 5:95 starch/water ratios while Zylema et al. (1985) worked with 1:1, 1:2, 1:4 and 1:8 starch/water ratios. Despite the different sample sizes both studies found that their most dilute suspensions needed about 40% more time to heat-up compared to the most concentrated ones, an observation opposite to what might be intuitively expected (Buffler, 1993).

Ndife et al. (1998) reported the rates of gelatinization for different starches as influenced by microwave heating and also developed a quantitative model describing the relationship between water content and rate of gelatinization for corn, wheat and rice starches during microwave heating.

A relatively large volume of maize starch suspension (2.5–20% w/w solids) was heated to above its gelatinization temperature by two means: a microwave oven meant to provide a uniform global thermalization of the sample, and a conventional local electrical heater which, depending on agitation, yields different heating patterns on the sample. Contrary to what was observed in the conventionally heated samples, gelatinization was not completed in the microwave irradiated samples, although the temperatures reached were as high as with conventional heating. This was attributed to poor mass transfer of water molecules during microwave irradiation as a result of the short processing period and the absence of mixing of water with starch components (Sakonidou et al., 2003).

Niba (2003) tested autoclave, parboiling and microwave processes to gelatinize and enzymatically stabilize starches from various sources and, after storage under freezing conditions for 10 days, he verified that autoclaved and parboiled starch presented higher rapidly digestible starch (RDS) content that microwave processed starch.

Thed & Phillip (1995) studied domestic cooking methods comparatively to prepare potato products: microwave-heating and deep fat frying reduced an appreciable amount of in-vitro digestible starch and significantly increased both the resistant starch (RS) and water-insoluble dietary fiber (IDF), while boiling and baking had less effect. Water-soluble dietary content was not affected by any of the domestic cooking methods studied. The significant correlation between IDF and RS supported the idea that some of the starch in cooked potato had become indigestible by amylolytic enzymes, and this RS might contribute to the observed increment in the IDF fraction.

Roberts (1977) tested the effect of microwave energy on the gelatinization of hulled, whole or polished rice grains and verified that rice containing 30-35% moisture presented promising gelatinization results with microwave heating at atmospheric pressure.

Wadsworth & Koltun (1986) verified that there were no significant differences between rice dried in air (control) and rice dried under vacuum with microwaves, when peak viscosity and starch retrogradation were evaluated. The sensory evaluation team did not distinguish the samples with respect to differences in taste and texture.

2.10.2 Interactions between starch and other nutrients

Understanding the interactions between sugar, starch, protein and water, which are the main components of a baked product, will advance the development of high quality, microwaveable products (Sumnu et al. (1999); Chavan & Chavan, 2010), as less satisfactory results have been obtained for starch-based products.

Sumnu et al. (1999) evaluated the quantitative relationships between water, sugar and protein on the gelatinization of wheat starch following 20 s of microwave heat as determined by differential scanning calorimetry. They verified that the addition of sugar decreased the degree of gelatinization of starch due to microwave heating significantly. Water and protein were not found to be as significant as sugar in delaying gelatinization. The effects of sugar and protein on the gelatinization of starch were pronounced in water-limited systems.

The dielectric properties of hydrated whey protein isolate (WPI), Ca-caseinate, wheat starch and their mixtures were measured at ambient temperature and during heating to 90°C. WPI exhibited higher dielectric properties than starch at lower moisture contents and ambient temperature. At most moisture contents WPI showed increasing microwave absorption properties with increasing temperature. Addition of WPI in the starch:water system affected the dielectric loss and absorptivity of starch during heating (Tsoubeci, 1994).

2.10.3. Modified starches

Annealing and heat/moisture-treatment both cause a physical modification of starches without any gelatinization, or any other damage of the starch granules with respect to size, shape or birefringence only via a controlled application of heat/moisture (Stute, 1992).

Gupta et al (2003) investigated the effect of microwave heat-moisture and annealing processes on buckwheat starch that had been dried to three moisture levels: 30.3, 40.0, and 50.4 kg/100kg. DSC data indicated that moisture levels had a significant effect on onset melting temperature, peak melting temperature and enthalpy. In addition, heat treatment and interaction of moisture with heat treatment both had a significant effect on amylose leaching results. Resistance to amylose leaching and melting at higher temperatures for higher moisture level buckwheat starch samples was attributed to increased networking among the amylose and amylopectin components in the buckwheat starch.

Wheat, corn and waxy corn starches of intermediate moisture content (30%) were subjected to microwave processing and the effect of microwave radiation on physico-chemical properties and structure of cereal starches was studied. Microwave radiation was evidenced to cause a shift in the gelatinization range to higher temperatures, and a drop in solubility and crystallinity. The extent and type of these changes depended on the variety of starch. Normal corn and wheat starches suffered pronounced changes, whereas under the same conditions waxy corn starch was almost unchanged. It was concluded that susceptibility of different starches to changes due to microwave irradiation depended not only on their crystal structure, but also on their amylose content (Lewandowicz et al., 2000).

Shogren & Biswas (2006) prepared starch acetates of 0.1-1.5 substitution through microwave heating of corn starch, acetic acid and acetic anhydride in sealed, stirred, teflon vessels. They found that starch acetates of high cold water solubility with DS (degree of substitution) values of 0.3-1.1 were prepared quickly and efficiently by microwave heating of starch, acetic acid and acetic anhydride. Molecular weights of starch acetates were much lower than the native starch. This study contributed to increase the efficiency of the reaction and to economize reagents when compared to the conventional method to prepare starch acetates.

Resistant starches (RS), in other words, those that are not digested in the small intestine, have been used to substitute fiber in food products, because although both have the same physiological functions, RS does not interfere in the palatability of foods.

Oliveira et al. (2007) studied the cooking of starches in a microwave oven using different starch:water ratios (1:5 to 1:8) and cooking/cooling cycles (1 to 3). They evaluated the content of resistant starch (%RS) after each process, as well as water absorption index (WAI) and water solubility index (WSI) and verified that the %RS increased with the increase in the number of cooking/cooling cycles and the starch:water ratio, while WAI and WSI were higher with a reduced number of cycles and starch:water ratio.

3. Future perspectives

In applications involving cereal products, the use of microwaves is promising in pre-processing operations, such as the drying of cereal grains or flours, the production of starches and the control of moisture and germination during grain storage, once the results presented in the various studies show the efficiency of the process, which economizes time and energy.

One of the greatest successes of the use of microwaves is in the preparation of expanded products and research and development of new products in this area is growing, with an unlimited number of processes being patented. Future indicates the development of healthier expanded products, with the reduction of fat and the inclusion of nutrients that are more beneficial to health.

Cooking and baking processes to obtain cakes, breads, cookies and pasta show that when microwaves are use associated to the conventional process results are more satisfactory for final product quality, but the isolated use of microwaves to obtain these products requires changes in formulation and process conditions.

Effects on different food constituents have been studied. In cereal-based products, the most studied constituent is starch, as it is present in greatest proportion in grains, followed by protein. These two constituents directly influence the quality of the final product, both bakery products and pasta. There is a lack of information on the evaluation of the effect of microwaves on lipids, fibers, vitamins and antioxidant compounds in cereal grains. These constituents are present in smaller quantities, but due to the daily consumption of cereals, are of nutritional importance in the diet.

The use of microwaves as catalysts of chemical reactions is starting to be studied in the area of modified starches and tends to develop more, once the advantages shown are promising, such as speed of the reaction, lower quantity of reagents and final products with the desired physical-chemical characteristics.

4. Conclusions

The use of microwave radiation to process cereal-based products has progressed in a promising way, improving process efficiency with respect to economy of time and energy. While many processes are already considered of high efficiency, such as grain drying, moisture control in cookies, starch modifications and pop-corn and expanded snack production, others such as bread and cake baking still need greater investments in research and development to improve quality.

5. References

Allred-Coyle, T.A.; Toma, R.B.; Reiboldt, W. & Thakur, M. (2001). Effects of bag capacity, storage time and temperature, and salt on the expansion volume of microwave popcorn. *Journal of the Science of Food and Agriculture, Vol.* 81, No. 1, 121-5, *ISSN-1097-0010.*

Allred-Coyle, T.A.; Toma, R.B.; Reiboldt, W. & Thakur, M. (2000). Effects of moisture content, hybrid variety, kernel size, and microwave wattage on the expansion volume of microwave popcorn. *International Journal of Food Sciences and Nutrition, Vol.* 51, No. 5, 389-94, ISSN-1465-3478.

Altan, A. & Maskan, M. (2005). Microwave assisted drying of short-cut (ditalini) macaroni: Drying characteristics and effect of drying processes on starch properties. *Food Research International,* Vol. 38, No.7, 787–796, ISSN-0963-9969.

Aref, M. M.; Brach, E. I.; Tape, N. M. (1969). A pilot-plant continuous-process microwave oven. *Canadian Institute of Food Technology,* Vol. 2, No.1, 37-41, ISSN-0315-5463.

Banasik, O.J. (1981). Pasta processing. *Cereal Foods World,* Vol. 26, No. 4, 166-169, ISSN-0146-6283.

Bennion, E.B. & Bamford, G.S.T. (1997). *The Technology of Cake Making,* Chapman & Hall, 6ª ed., ISBN-10-0751403490, London.

Bernussi, A. L. M.; Chang, Y. K. & Martinez-Bustos, F. (1998). Effect of Production by Microwave Heating After Conventional Baking on Moisture Gradient and Product Quality of Biscuits (cookies). *Cereal Chemistry*, Vol. 75, No. 5, 606-611, ISSN- 0009-0352.

Berteli, M. N. & Marsaioli Jr., A. (2005). Evaluation of short cut pasta air dehydration assisted by microwaves as compared to the conventional drying process. *Journal of Food Engineering*, vol. 68, No. 2, 175–183, *ISSN – 0260-8774*.

Boehmer, E. W.; Bennet, W. L.; Guanella, T. J. & Levine, L. (1992). *Microwave expandable half product and process for its manufacture*. United States Patent 5,165,950.

Boischot C.; Moraru C. I.; Kokini J.L. (2003). Expansion of glassy amylopectin extrudates by microwave heating. *Cereal Chemistry*, Vol. 80, No.1, 56-61, ISSN- 0009-0352.

Breslin, J. C. (1990). *Microwave--convection drying of ready-to-eat cereals*. M.S. dissertation, Michigan State University, United States -- Michigan. Retrieved June 24, 2010, from Dissertations & Theses: A&I.(Publication No. AAT 1341402).

Buffler, C. R. (1992). *Microwave cooking and processing*. New York Van Nostrand Reinhold, ISBN-10- 0442008678, New York.

Bushuk, W.(2001). Rye production and uses worldwide. *Cereal Foods World*, Vol. 46, No.2, 70-73, ISSN- 0146-6283.

Caballero-Córdoba, G. M.; Wang, S. H. & Sgarbieri, V. C. (1994). Características nutricionais e sensoriais de sopa cremosa semi-instantânea à base de farinhas de trigo e soja desengordurada. *Pesquisa Agropecuária Brasileira*, Vol. 29, No. 7, 1137-1143, ISSN-0100-204X.

Campaña, L.E.; Sempé, M.E. & Filgueira, R. R. (1993). Physical, chemical, and baking properties of wheat dried with microwave energy. *Cereal Chemistry*, Vol. 70, No. 6, 760-762, ISSN- 0009-0352.

Carvalho, N.H. & Nakagawa, J. (1988). *Sementes: ciência, tecnologia e produção*. Fundação Cargill, 3 ed., *ISBN- 85-87632-01-9*, Campinas.

Chang, Y. K. & El-Dash, A. A. (1998). Effect of microwave energy on lipase inactivation and storage stability of brown rice. *Acta Alimentaria*, Vol. 27, No.2, 193-202, ISSN-0139-3006.

Chavan, R.S. & Chavan, S.R. (2010). Microwave baking in food industry: A review. *International Journal of Dairy Science*, Vol. 5, No. 3, 113-127, ISSN (on Line)-*1811-9751*.

Chedid, L.; Huang, D. P. & Baytan, P. (1998). *Flavored popping corn with low or no fat*. United States Patent 5,753,287.

Chen, C.M. & Yeh, A.L. (2000). Expansion of rice pellets: examination of glass transition and expansion temperature. *Journal of Cereal Science*, Vol.32, No.2, 137-45, ISSN- 0733-5210.

Chiang, B-Y. & Johnson, J.A. (1977). Gelatinization of starch in extruded products. *Cereal Chemistry*, Vol.54, No.3, 436-443, ISSN- 0009-0352.

Clarke, C.I. & Farrell, G.M. (2000). The effects of recipe formulation on the textural characteristics of microwave-reheated pizza bases. *Journal of the Science of Food and Agriculture*, Vol. 80, No.8, 1237-1244, ISSN-0022-5142.

Conforti, E.; Kieckbusch, T.G. & Marsaioli Jr., A. (1990). A prototype of a combined hot air and microwave rotary cylindrical oven for continuous drying of granular products. In: *Engineering and Food, Preservation Processes and Related Techniques*. Spiess, W.E.L. & Schubert, H. (ed). 679-685, Elsevier Sci. Publ., *ISBN-10-1851664661*, London & New York.

Decareau, R.V.; Peterson, R.A. (1986). *Microwave processing and engineering*. Ed. Wiley-VCH, 224 p.ISBN-10- 3527262105, Weinheim.

Dofing , S.M.; Thomas-Compton, M.A. & Buck, J.S. (1990). Genotype X popping method interaction for expansion volume in popcorn. *Crop Science* , Vol.30, No. 1, 62-5, *ISSN* 1435-0653.

Domingues, D.J.; Atwell, W. A.; Beckmann, P. J.; Panama, J. R.; Conn, R. E.; Matson, K. L.; Graf, E.; Feather, M. S.; Fahrenholtz, S. K. & Huang, V. T. (1992). *Process for microwave browning*. United States Patent 5,108,770.

Doty, N.C. & Baker, C.W. (1997). Microwave conditioning of hard red spring wheat. I. Effects of wide power range on flour and bread quality. *Cereal Chemistry*, Vol. 54, No. 4, 717- 727, ISSN- 0009-0352.

Edwards, G. H. (1964). Effects of microwave radiation on wheat and flour: the viscosity of the flour pastes. *Journal of the Science of Food and Agriculture*, Vol.15, No.1, 108-114, ISSN-0022-5142.

Engelder, D. S. & Buffler, C. R. (1991). Measuring dielectric properties of food products at microwave frequencies. *Microwave World*, Vol. 12, No.2, 6-15, ISSN-1759-0787.

Ernoult ,V.; Moraru, C.I. & Kokini, J.L. (2002). Influence of fat on the expansion of glassy amylopectin extrudates by microwave heating. *Cereal Chemistry*, Vol. 79, No.2, 265-273, ISSN- 0009-0352.

Frisvad, J.C. (1995). Mycotoxins and mycotoxigenic fungi in storage. In: *Stored –grain ecosystems*, Jayas, D.S. & White, W.E.M. M. Dekker, 251-288, ISBN - 0824789830, New York.

Gimeno, E.; Moraru, C.I.; Kokini, J.L. (2003). Effect of Xanthan Gum and CMC on the texture and microstructure of glassy corn flour extrudates expanded by microwave heating. *Cereal Chemistry*, Vol. 81, No.1, 100-107, ISSN- 0009-0352.

Goebel, N. K.; Grider, J.; Davis, E. A. & Gordon, J. (1984). The effects of microwave energy and convection heating on wheat starch granule transformations. *Food Microstructure*, Vol.3, No.1, 73–82, ISSN-1046-705X.

Goksu, E. I.; Summu, G. & Esin, A. (2005). Effect of microwave on fluidized bed drying of macaroni beads *Journal of Food Engineering*, Vol. 66, No. 4, 463 – 468, *ISSN – 0260-8774*.

Goldblith, S.A. (1966). Basic principles of microwaves and recent developments. *Advances Food Research*, Vol.15, p. 277-301. ISSN- 0065-2628

González Galán, A.; Wang, S. H.; Sgarbieri, V. C. & Moraes, M. A. C. (1991). Sensory and nutritional properties of cookies based on wheat-rice-soybean flours baked in a microwave oven. *Journal of Food Science*, Vol. 56, No. 6, 1699-1701, ISSN 0022-1147.

Gupta, A. & Kumar, P. (2003). Effect of microwave heating on rice (Oryza sativa) grain quality. *Indian Journal of Agricultural Sciences*, Vol. 73, No. 1, 41-43, ISSN- 0019-5022.

Gupta, M.; Gill, B. S.; Baw, A. S. (2008). Gelatinization and X-ray Crystallography of Buckwheat Starch: Effect of Microwave and Annealing Treatments. *International Journal of Food Properties*, Vol. 11, No. 1, 173 – 185, ISSN (electronic)- 1532-2386.

Harper, J.M. (1979). Food extrusion. *CRC - Critical Reviews in Food Science and Nutrition*, Vol.11, n.2, p.155-215, ISSN 1040-8398.

Hoffman, C.J. & Zabik, M.E. (1985). Effects of microwave cooking/reheating on nutrients and food systems: a review of recent studies. *Journal of the American Dietetic Association*, Vol. 85, No.8, 922-6, ISSN- 0002-8223.

Jahnke, M. (2003). *Method of making microwavable yeast-leavended bakery product containing dough additive*. United States Patent 6,579,546.

Jermolovicius, L. A.; Schneiderman, B. & Senise, J. T. (2006). Alteration of Esterification Kinetics Under Microwave Irradiation. In: *Advances in Microwave and Radio-Frequency Processing,*Willert-Porada, M. *(ed)*, Springer, Germany, ISBN978- 3-540-43252-4.

Jiaxun-Tao; Rao, R. & Liuuzo, J. (1993). Microwave heating for Rice bran stabilization. *Journal of Microwave, Power and Electromagnetic Energy*, Vol. 28, No.3, 156-164, ISSN-0832-7823.

Juliano, B. O. (1985). Criteria and tests for rice grain quality. In: *Rice Chemistry and Technology,* The American Association of Cereal Chemists (Ed.), pages 443-513. ISBN-10- 0913250414, St.Paul.

Khraisheh, M. A.; Cooper, T. J. R. & Magee, T. R. A. (1997). Microwave and air drying I. Fundamental considerations and assumptions for the simplified thermal calculations of volumetric power absorption. *Journal of Food Engineering*, Vol.33, No.1, 207–219. *ISSN − 0260-8774.*

Kruger, J.E.; Matsuo, R.B. & Dick, J.W. (1996). *Pasta and Noodle Technology.* American Association of Cereal Chemists, ISBN-10- 0913250899, St.Paul.

Kudra, T. (1989). Dielectric drying of particulate materials in a Fluidized in a fluidized state. *Drying Technology,* Vol. 1. No.7, 17-34. *ISSN − 0737-3937*

Leach, H.W. (1965). Gelatinization of starch. In: Starch Chemistry and Technology, vol I, ed. Whistler, R.L. & Paschalle.F. *Academic Press*, New York,. ISBN-10- 0127462759

Lee, E.Y.; Lim. K.I.; Lim, J.; Lim, S.T. (2000). Effects of gelatinization and moisture content of extruded starch pellets on morphology and physical properties of microwave expanded products. *Cereal Chemistry,* Vol.77, No.6, 769-73. ISSN-0009-0352

Lewandowicz, G.; Jankowski, T. & Fornal, J. (2000).Effect of microwave radiation on physico-chemical properties and structure of cereal starches. *Carbohydrate Polymers* vol. 42, No. 2, 193-199, *ISSN − 0144-8617.*

Li, H.-C. (1995). *Identification of aroma compounds and evaluation of volatile losses during the frying and microwave reheating of a flour-based batter system (pancake).* Ph.D. dissertation, University of Minnesota, United States -- Minnesota. Retrieved June 24, 2010, from Dissertations & Theses: A&I.(Publication No. AAT 9534129).

Lin ,Y.E. & Anantheswaran, R.C. (1988). Studies in popping of popcorn in a microwave oven. *Journal of Food Science*, Vol. 53, No.6,1746-9, ISSN 0022-1147.

Lin, Y. E.; Anantheswaran, R. C. & Puri, V. M. (1995). Finite element analysis of microwave heating of solid foods. *Journal of Food Engineering*, Vol.25, No.1, 85–112. ISSN − 0260-8774.

Loewen, R. H. & Gandolfi, O. (2004). *Equipamento para secagem de cereais por microondas.* PI0402177-0 A2, Brasil.

Lorenz, K.; Charman, E. & Dilsaver, W. (1973). Baking with microwave energy. *Food Technology,* Vol. 27, No.12, 28-36, ISSN- 0015-6639.

Lou, W. C. & Fazzolare, R. D. (1990). *Shelf-stable microwavable cookie dough.* United States Patent 4,911,939.

Manley, D. J. R. (2000). *Technology of biscuits, crackers and cookies* - Third Edition. CRC Woodhead Publishing Limited, ISBN- 1855735326, Abington.

Marini, L. J.; Gutkoski, L.C.; Elias, M. C. & Santin, J. A. (2007). Qualidade de grãos de aveia sob secagem intermitente em altas temperaturas. *Ciência Rural* [online], Vol.37, No.5, 1268-1273, ISSN (On line) 1678-4596.

Marotti Neto, A. S. (2007). *Secador por microondas modular com turbilhonamento em caracol por fluxo de ar sobre esteira.* MU8700836-0 U2, Brasil.

Marsaioli Jr., A. (1991). *Desenvolvimento de um protótipo de secador cilíndrico – rotativo a microondas e a ar quente para a secagem contínua de produtos sólidos granulados.* 197p. Tese (Doutor em Engenharia de Alimentos) – Faculdade de Engenharia de Alimentos, Universidade Estadual de Campinas.

Marsaioli Jr., A.; Berteli, M. N. & Pereira, N. R. (2009). Applications of microwave energy to postharvest technology of fruits and vegetables. *Stewart Postharvest Review*, Vol. 5, No.1, 1-5, ISSN- 1745-9656.

Martinez-Bustos, F. M.; Garcia, M.; Chang, Y. K.; Sanchezsinencio, F. & Figueroa, C. (2000). Characteristics of nixtamalised maize flours produced with the use of microwave heating during alkaline cooking. *Journal of the Science of Food and Agriculture*, vol. 80, No.6, 651-656, ISSN-0022-5142.

Maurer, R.L.; Trembly, M.R. & P.E. Chadwick, P.E. (1971). Microwave processing of pasta. *Food Technology* , Vol.25, 1240-1245, ISSN- 15574571.

Metivier, J.R. (1979).Dormência e germinação. In: *Fisiologia Vegetal.* FERRI, M.G. (Ed.).: EPU:EDUSP, 343-392 (Cap.12), ISBN-10: 8512119101, São Paulo.

Mohamed, A.A.; Ashman, R.B. & Kirleis, A.W. (1993). Pericarp thickness and other kernel physical characteristics relate to microwave popping quality of popcorn. *Journal of Food Science*, Vol. 58, No.2, 342-6, ISSN- 0022-1147.

Monks, L.F.; Costa, C.S. , & Soares, G.J.D. (2003). Microwave effects on wheat drying (T. aestivum, L.) and flour quality. *Alimentos e Nutrição*, Vol.14, No.2, 219-224, ISSN- 0103-4235.

Moraru, C.I. & Kokini, J.L. (2003). Nucleation and expansion during extrusion and microwave heating of cereal foods, *Comprehensive reviews in Food Science and Food Safety*, Vol. 2, No.3, 120-138, ISSN-1541-4337.

Mudgett, R. E. (1982). Electrical properties of foods in microwave processing. *Food Technology,* Vol.36. No. 2, 109-115, ISSN- 15574571.

Ndife, M.; Sumnu, G. & Bayındırlı, L. (1998). Differential scanning calorimetry determination of gelatinization rates in different starches due to microwave heating. *Lebensmittel-Wissenschaft Untersuchung-Technoology,* Vol. 31, No.5, 484–488. ISSN (Online) 1532-1738.

Niba, L. L. (2003). Processing effects on susceptibility of starch to digestion in some dietary starch sources. *International Journal of Food Sciences and Nutrition*, Vol.54, No.1, 97-109, ISSN (electronic) - 1465-3478.

Ohlsson, T., & Risman, P. O. (1978). Temperature distribution of microwave heating-spheres and cylinders. *Journal of Microwave Power*, Vol. 13, No.4, 303–310, ISSN- 0022-2739.

Oliveira, M.; Clerici, M. T.; Pompeu, F.; Chang, Y. K. & Martinez-Bustos, F. M. (2007). Physicochemical and technological properties of resistant starch obtained using microwave oven. In: 2007 IFT Annual Meeting, Chicago, 275-275.

Palav, T. & Seetharaman, K. (2007). Impact of microwave heating on the physico-chemical properties of a starch–water model system. *Carbohydrate Polymers*, Vol.67, No. 4, 596–604. ISSN – 0144-8617.

Park , J. K.; Bin, A. & Brod, F. P. R. (2001). Obtenção das isotermas de sorção e modelagem matemática para a pêra bartelett (*Pyrus sp.*) com e sem desidratação osmótica. *Ciência e Tecnologia de Alimentos*, Vol.21, No.1, 73-78, ISSN- 0101-2061.

Park, T. (2001). *A study of microwave-water interactions in bread system.* Ph.D. dissertation, Rutgers. The State University of New Jersey - New Brunswick, United States --

New Jersey. Retrieved June 24, 2010, from Dissertations & Theses: A&I.(Publication No. AAT 9973982).

Pecoraro, E. ; Davolos, M. R.; Jafelicci Jr., M. (1997). Adaptações em forno de microondas doméstico para utilização em laboratório. *Química Nova,* Vol. 20, No. 1, ISSN (On-Line) 1678-7064.

Pei, D.C.T. (1982). Microwave baking-new developments. *Baker's Digest,* Vol. 56, No.1, 8-10.

Pordesimo, L.O.; Anantheswaran, R.C.; Fleischmann, A.M.; Lin , Y.E.& Hanna, M.A. (1990). Physical properties as indicators of popping characteristics of microwave popcorn. *Journal of Food Science,* Vol. 55, No.5, 1352-5, ISSN- 0022-1147.

Pordesimo, L.O.; Anantheswaran, R.C. & Mattern, P.J. (1991). Quantification of horny and floury endosperm in popcorn and their effects on popping performance in a microwave oven. *Journal of Cereal Science,* Vol. 14, No.2, 189-98, ISSN- 0733-5210.

Prado, G.; Mattos, S. V. M. & Pereira, E. C. (1991). Efeito da umidade relativa na contaminação microbiana e produção de aflatoxinas em amendoim em grão. *Ciência e Tecnologia de Alimentos,* Vol.11, No. 2, 264-273, ISSN-0101-2061.

Pyler, E.J. (1988). *Baking Science and Technology. Sosland Publishing Company.* v.1 e 2. ISBN-0929005007, Chicago.

Remmen, H. H. J.; Ponne, C. T.; Nijhuis, H. H.; Bartels, P. V. & Kerkhof, P. J. A. M. (1996). Microwave heating distribution in slabs, spheres and cylinders with relation to food processing. *Journal of Food Science,* Vol. 61, n. 6, p. 1105-1113, ISSN- 0022-1147.

Roberts, R.L.(1977). Effect of microwave treatment of pre-soaked paddy, brown and white rice. *Journal of Food Science,* Vol. 42, No. 3, 804–806, ISSN- 0022-1147.

Rocha, C. R. (2002). *Efeitos do Tratamento por Microondas do Arroz Recém Colhido no Rendimento de Grãos Inteiros, na Qualidade de Cozimento e na Estabilização do Farelo.* 207 p. Tese (Doutorado em Engenharia de Alimentos) - Universidade Estadual de Campinas.

Sacharow, S. & Schiffmann, R. (1992). *Microwave Packaging.* Pira International, *Leatherhead* UK, 155p, ISBN 0-902799-70-3

Sakonidoua, E.P., Karapantsiosa, T.D. & Raphaelides, S.N. (2003). Mass transfer limitations during starch gelatinization. *Carbohydrate Polymers,* Vol. 53, No.1, 53–61, ISSN— 0144-8617.

Sale, A. J. H. (1976). A review of microwave for food processing. *International Journal of Food Science & Technology,* Vol. 11, No.4, 319–329, ISSN 0950-5423.

Saunders, R. M. (1990). The properties of rice bran as a foodstuff. *Cereal Foods World,* Vol.35, No. 7, 632-636, ISSN- 0146-6283.

Schiffmann, R.F. (1993). Understanding microwave reactions and interactions. *Food Product Design* (April), 72-88, ISSN - 1065-772X.

Schiffmann, R. F. (1986). Food product development for microwave processing. *Food Technology,* vol. 40, 94–98, ISSN- 15574571

Schiffmann, R. F. (1992). Microwave processing in the US food industry. *Food Technology,* Vol.56, No.1 50-52, ISSN- 15574571.

Scudamore, K. A.; Patel, S. & Breeze, V. (1999). Surveillance of stored grain from the 1997 harvest in the United Kingdom for ochratoxin A. *Food Additives and Contaminants,* Vol.16, No. 7, 81-290, ISSN- Online-1464-5122.

Senise, J. T. & Jermolovicius, L. A. (2004). Microwave Chemistry - A Fertile Field For Scientific Research And Industrial Applications. *Journal of Microwaves and Optoelectronics,* Vol.3, No.5, 97-112, ISSN- 1516-7399.

Shogren, R.L. & Biswas, A. (2006). Preparation of water-soluble and water-swellable starch acetates using microwave heating. *Carbohydrate Polymers*, Vol.64, No.1, 16-21, ISSN – 0144-8617.

Singh, J. & Singh, N. (1999a). *Effect of different ingredients and microwave power on snacks upon microwave heating*. U.S. Patent 4,409,250.

Singh, J. & Singh, N. (1999b). Effect of different ingredients and microwave power on popping characteristics of popcorn. *Journal of Food Engineering*, Vol. 42, No.3, 161-5. *ISSN – 0260-8774.*

Stinson, C. T. (1986). A Quality Comparison of Devil's Food and Yellow Cakes Baked in a Microwave/Convection Versus a Conventional Oven. *Journal of Food Science*, Vol. 51, No. 6, 1578–1579, ISSN- 0022-1147.

Stute, R. (1992). Hydrothermal Modification of Starches: The Difference between Annealing and Heat/Moisture –Treatment. *Starch – Stärke*, Vol. 44, No.6, 205-214, ISSN-0038-9056.

Suknark, K.; Phillips, R.D. & Huang, Y.W. (1999). Tapioca-fish and tapioca-peanut snacks by twin-screw extrusion and deep-fat frying. *Journal of Food Science*, Vol. 64, No.2, 303-8, ISSN- 0022-1147.

Şummu, G.; Ndife, M.K. & Bayindirli, L. (2000). Optimization of microwave baking of model layer cakes. *European Food Research and Technology*, Vol. 211, No. 3, 169-174, ISSN-(electronic)- 1438-2385.

Şummu, G. (2001). A review on microwave baking of foods. *International Journal of Food Science and Technology*, Vol. 36, No.2, 117-127, ISSN-0950-5423.

Şummu, G.; Ndife, M.K & Bayındırlı, L. (1999) . Effects of sugar, protein and water content on wheat starch gelatinization due to microwave heating. *European Food Research and Technology*, Vol. 209, No.1, 68–71, ISSN (electronic version) 1438-2385.

Takano, K.(1993).Mechanism of lipid hydrolysis in rice bran. *Cereal Foods World*, Vol.38, No.2, 95-698, ISSN- 0146-6283.

Takashima, H. (2005). Sponge cake premix and method of manufacturing sponge cake by using said premix. United States Patent 6,884,448.

Thed, S. T. & Phillip, R. D. (1995). Changes of dietary fiber and starch composition of processed potato products during domestic cooking. *Food Chemistry*, Vol. 52, No.3, 301-3.

Tsoubeci, M. N. (1994). Molecular interactions and ingredient substitution in cereal based model systems during conventional and microwave heating. Ph.D. dissertation, University of Minnesota, United States -- Minnesota. Retrieved June 24, 2010, from *Dissertations , & Theses*: A&I.(Publication No. AAT 9514682).

Umbach, S.L.; Davis, E.A. & Gordon, J. (1990). Effects of heat and water transport on the bagel-making process: Conventional and microwave baking. *Cereal Chemistry*, Vol. 67, 355-360, ISSN- 0009-0352.

Vadivambal, R.; Jayas, D.S.; Chelladurai, V. & White, N.D.G. (2009). Preliminary study of surface temperature distribution during microwave heating of cereals and oilseed. *Canadian Biosystems Engineering / Le Genie des biosystems au Canada*, Vol. 51, 3.45 - 3.52.

Van Hulle, G.J.; Anker, C.A. & Franssell, D.E. (1983). *Method for preparing sugar-coated, puffed snacks upon microwave heating*. U.S. Patent 4,409,250.

Van Lengerich, B. H. & Lou, W.C.(1992). *Filled, microwave expandable snack food product and method and apparatus for its production*. United States Patent 5,124,161.

Van Remmen, H.H.J.; Ponne, C.T.; Nijhuis, H.H.; Bartelis, P.V. & Kerkhof, P.J.A.M. (1996). Microwave heating distributions in slabs, spheres and cylinders with relation to food processing. *Journal of Food Science*, Vol. 61, No.6, 1105-13, *ISSN- 0022-1147*.

Velupillai, L.; Verma, L. R. & Tsangmuichung, M. (1989). *Process for parboiling rice*. United States Patent 4,810,511

Vetrimani, R.; Jyothirmayi, N.; Rao, H. P. & Ramadoss, C. S. (1992). Inactivation of lipase and lipoxygenase in cereal bran, germ and soybean by microwave treatment. *Lebensmittel Wissenshaft und Technologie*, Vol.6, No. 6, 532-535, ISSN (printed) - 0023-6438.

Wadsworth, J. I. & Koltun, S. P. (1986). Physicochemical properties and cooking quality of microwave dried rice. *Cereal Chemistry*, Vol.63, No. 4, 346-348, ISSN- 0009-0352.

Wang, S.W. (1997). Starches and starch derivatives in expanded snacks. *Cereal Foods World*, Vol.42, No.9, 743-5, ISSN- 0146-6283.

Wang, S. H.; Caballero-Córdoba, G. M. & Sgarbieri, V. C. (1992). Propriedades funcionais de misturas de farinhas de trigo e soja desengordurada, pré-tratadas por micRoondas. *Ciência e Tecnologia de Alimentos*, Vol. 12, No. 1, 14-25, ISSN- 0101-2061.

Wang, S. H.; Clerici, M. T. P. S. & Sgarbieri, V. C. (1993). Características sensoriais e nutricionais de mingau de preparo rápido à base de farinhas de arroz e soja desengordurada e leite em pó. *Alimentos e Nutrição*, Vol. 5, No.1, 77-86, ISSN- 0103-4235.

Warchalewski, J.R. & Gralik, J. (2010). Influence of microwave heating on biological activities and electrophoretic pattern of albumin fraction of wheat grain. *Cereal Chemistry*, Vol. 87, No. 1, 35-41, ISSN- 0009-0352.

Weisz, M. A. Co. 2007. *Processo de produção de salgadinho, snack ou pellet elaborado a partir de massa extrusada*. MU8701618-4 U2, Brasil.

Wennermark, B. (1993). *Vitamin E retention during processing of cereals*. Fil.Dr. dissertation, Lunds Universitet (Sweden), Sweden. Retrieved June 24, 2010, from *Dissertations & Theses*: A&I.(Publication No. AAT C299434).

Willoughby; C. L. & Engle, T. L. (2001). *Expandable food products and methods of preparing same* . United States Patent 6,171,631.

Yang, H. W. & Gunasekaran, S. (2004). Comparison of temperature distribution in model food cylinders based on Maxwell's equations and Lambert's law during pulsed microwave heating. *Journal of Food Engineering*, Vol.64, No.4, 445–453, *ISSN − 0260-8774*.

Yin, Y. & Walker, C.E. (1995). A quality comparison of breads baked by conventional versus non conventional ovens: a review. *Journal of the Science of Food and Agriculture*, Vol.67, No.3, 283-91, *ISSN* (Online) 1097-0010.

Young, L.; Cauvain, S.P. (2007). *Technology of Breadmaking*. Springer; 2 ed., ISBN-10-0751403458, Germany.

Zhuang, M. (1998). Microwave popcorn, waiting to explode. *Agexporter, Vol.* 10, No.3,10-1. ISSN- 1047-4781.

Zylema, B. J.; Grider, J. A.; Gordon, J. & Davis, E. A. (1985). Model wheat starch systems heated by microwave irradiation and conduction with equalized heating times. *Cereal Chemistry*, Vol.62, No.6 , 447–453, ISSN- 0009-0352.

24

Microwave Heating in Moist Materials

Graham Brodie

The University of Melbourne
Australia

1. Introduction

The definition of "microwaves" is somewhat arbitrary; however microwaves are usually considered to be electromagnetic waves in the frequency range from 300 MHz to 300 GHz. Before World War II, there is little evidence of work on radio frequency or microwave heating; however Kassner (1937b) mentions industrial applications of microwave energy in two of his patents on spark-gap microwave generators (Kassner, 1937a, 1937b, 1938). Unfortunately early studies in radio frequency heating concluded that microwave heating of food stuffs would be most unlikely because the calculated electric field strength required to heat biological materials would approach the breakdown voltage of air (Shaw & Galvin, 1949).

A fortuitous discover by Spencer that microwave energy could heat food (Murray, 1958) lead to a series of patents (Spencer, 1947, 1949, 1952) and the development of microwave cooking equipment. The major advantages of microwave heating are its short start-up, precise control and volumetric heating (Ayappa et al., 1991); however microwave heating suffers from: uneven temperature distributions (Van Remmen et al., 1996; Brodie, 2008); unstable temperatures (Vriezinga, 1996, 1998, 1999; Vriezinga et al., 2002); and rapid moisture movement (Brodie, 2007b). This chapter will explore the theory and practice of microwave heating in moist materials.

2. Theory

Microwave heating is governed by Maxwell's electromagnetic equations:

$$\nabla \cdot E = \frac{\rho_c}{\varepsilon^*}, \quad \nabla \times E = -\frac{\partial(\mu H)}{\partial t}, \quad and \quad \nabla \times H = J_c + J_s + \frac{\partial(\varepsilon^* E)}{\partial t} \tag{1}$$

From these, it is possible to derive a wave equation of the form:

$$\nabla^2 E = \mu \left[\frac{\partial^2(\varepsilon^* E)}{\partial t^2} + \frac{1}{2}\frac{\partial(\sigma_c E)}{\partial t} + \frac{\partial Js}{\partial t} \right] + \nabla\left(\frac{\rho_c}{\varepsilon^*}\right) \tag{2}$$

Equation (2) is a forced, damped wave equation made up of three components:

- $\nabla^2 E = \mu \left[\dfrac{\partial^2(\varepsilon^* E)}{\partial t^2} + \dfrac{1}{2}\dfrac{\partial(\sigma_c E)}{\partial t} \right]$ is the damped wave response;

- $\mu\dfrac{\partial Js}{\partial t}$ accounts for any time varying current source embedded in the space of interest that creates the damped wave; and

- $\nabla\left(\dfrac{\rho_c}{\varepsilon^*}\right)$ accounts for any static component to the field, associated with stationary charges embedded in the space of interest.

Assuming that all field sources lie outside the space occupied by the heated material and the incident microwave fields are monochromatic with an angular frequency ω, equation (2) simplifies to:

$$\nabla^2 E + f^2 E = 0 \tag{3}$$

Where:

$$f = \frac{\omega}{c}\sqrt{\kappa'\left[1 - j\frac{\left(\dfrac{2}{\omega}\dfrac{\partial\kappa'}{\partial t} + \kappa''\right)}{\kappa'}\right] - \left(\frac{1}{\omega^2}\frac{\partial^2\kappa'}{\partial t^2} + \frac{1}{\omega}\frac{\partial\kappa''}{\partial t}\right)} \tag{4}$$

Because the angular frequency ω is so large at microwave frequencies, equation (4) can be nicely approximated by:

$$f = \frac{\omega}{c}\sqrt{\kappa'\left[1 - j\frac{\kappa''}{\kappa'}\right]} \tag{5}$$

This can be rearranged to become:

$$f = \alpha + j\beta \tag{6}$$

Where:

$$\alpha = \frac{\omega}{c}\sqrt{\frac{\kappa'}{2}\left(\sqrt{1 + \left(\frac{\kappa''}{\kappa'}\right)^2} + 1\right)} \tag{7}$$

and

$$\beta = \frac{\omega}{c}\sqrt{\frac{\kappa'}{2}\left(\sqrt{1 + \left(\frac{\kappa''}{\kappa'}\right)^2} - 1\right)} \tag{8}$$

The form of the Laplacian operator (∇^2) in equation (3) depends entirely on the geometry of the problem.

2.1 Simultaneous heat and moisture movement

Any realistic analysis of microwave heating in moist materials must account for simultaneous heat and moisture diffusion through the material. The coupling between heat and moisture transfer is well known but not very well understood (Chu & Lee, 1993). Henry

(1948) first proposed the theory for diffusion of heat and moisture into a textile package. Crank (1979) presented a more thorough development of Henry's work. This theory has been rewritten and used by many authors (Chu & Lee, 1993; Vos et al., 1994; Fan et al., 2000; Wang et al., 2002; Fan, 2004; Brodie, 2007b).

The amount of water vapour moving into a small section of porous material is the sum of any net increase in moisture content in the air space and the net increase in moisture content of the material's fibres (Crank, 1979). Therefore:

$$a_v \tau_v D_a \cdot \nabla^2 M_v = a_v \frac{\partial M_v}{\partial t} + \left(1 - a_v\right)\rho_s \frac{\partial M_s}{\partial t} \tag{9}$$

Heat is evolved because of microwave heating and moisture absorption by the material; therefore the thermal diffusion equation, which includes a volumetric heat source, is:

$$C\rho \frac{\partial T}{\partial t} = k\nabla^2 T + L\rho \frac{\partial M_s}{\partial t} + q \tag{10}$$

If a linear relationship between the moisture content of a material, the moisture vapour concentration in the air spaces in the material and the temperature is assumed (Henry, 1948; Crank, 1979), then:

$$\frac{\partial M_s}{\partial t} = \sigma \frac{\partial M_v}{\partial t} - \omega \frac{\partial T}{\partial t} \tag{11}$$

where σ and ω are constants of association between heat transfer and moisture vapour concentration. Substituting from equation (11) into equations (9) and (10) and combining the two equations yields:

$$\nabla^2 \left(pM_v + nT\right) - \frac{\partial}{\partial t}\left\{ \begin{bmatrix} \left[\dfrac{1}{\tau_v D_a}\left(1 + \dfrac{\left(1 - a_v\right)\sigma\rho_s}{a_v}\right) - \dfrac{np\sigma L}{pk}\right]pM_v + \\ \left[\dfrac{C\rho}{k}\left(1 + \dfrac{\omega L}{C}\right) - \dfrac{p\left(1 - a_v\right)\omega\rho_s}{n\tau_v D_a a_v}\right]nT \end{bmatrix} \right\} + \frac{nq}{k} = 0 \tag{12}$$

This can be expressed in a simpler form if $\Omega = pM_v + nT$:

$$\nabla^2 \Omega - \frac{1}{\gamma}\frac{\partial \Omega}{\partial t} + \frac{nq}{k} = 0 \tag{13}$$

The constants of association, p and n, are calculated to satisfy:

$$\frac{1}{\gamma} = \left[\frac{1}{\tau_v D_a}\left(1 + \frac{\left(1 - a_v\right)\sigma\rho_s}{a_v}\right) - \frac{np\sigma L}{pk}\right] = \left[\frac{C\rho}{k}\left(1 + \frac{\omega L}{C}\right) - \frac{p\left(1 - a_v\right)\omega\rho_s}{n\tau_v D_a a_v}\right] \tag{14}$$

Equation (14) implies that the combined heat and moisture diffusion coefficient (γ) has two independent values. This is consistent with Henry's (1948) equation for simultaneous heat and moisture diffusion in textiles.

Henry (1948) showed that a change in external temperature or humidity (or both), "*results in a coupled diffusion of moisture and heat*". Henry (1948) also states that "*the diffusion constants appropriate to these two quantities are always such that one is greater and the other less than either of the diffusion constants which would be observed for the moisture or heat, were these not coupled by the interaction".*

The slower diffusion coefficient of the coupled system is less than either the isothermal diffusion constant for moisture or the constant vapour concentration coefficient for heat diffusion, whichever is less, but never by more than one half (Henry, 1948). The faster diffusion coefficient is always many times greater than either of these independent diffusion constants (Henry, 1948).

Considerable evidence exists in literature for rapid heating and drying during microwave processing (Rozsa, 1995; Zielonka & Dolowy, 1998; Torgovnikov & Vinden, 2003); therefore it is reasonable to assume that the faster diffusion wave dominates microwave heating in moist materials. A slow heat and moisture diffusion wave should also exist; however observing this slow wave may be difficult and no evidence of its influence on microwave heating has been seen in literature so far.

2.2 Solutions to the microwave heating equation

Approximate solutions to equation (13) are sought under assumptions of uniform material properties for rectangular, cylindrical and spherical geometries. The aim is to readily identify the influence of geometry over microwave fields and temperature distributions. It is recognised that the derived equations will only provide approximate solutions to the diffusion problem, however the solutions are sufficiently accurate to provide some insight into the influence of geometry over temperature distributions during microwave heating.

2.3 Microwave heating in rectangular coordinates

Beginning with Maxwell's equations, Ayappa et al. (1991) demonstrate that the average power density (W m^{-2}) at a distance z below the irradiated surface of a material block of thickness W is:

$$q(z) = \omega\varepsilon_o\kappa''(\tau E)^2 \frac{\left\{ e^{-2\beta z} + 2\Gamma e^{-2\beta W}\cos\left[\delta + 2\alpha(W-z)\right] + \Gamma^2 e^{-2\beta W}e^{-2\beta(W-z)} \right\}}{2\left(1 - 2\Gamma^2 e^{-2\beta W}\cos(2\delta + 2\alpha W) + \Gamma^4 e^{-4\beta W}\right)} \tag{15}$$

Where:

$$\delta = \tan^{-1}\left[\frac{2(\alpha_1\beta_2 - \alpha_2\beta_1)}{(\alpha_1^2 + \beta_1^2) - (\alpha_2^2 + \beta_2^2)}\right] \tag{16}$$

$$\Gamma = \sqrt{\frac{(\beta_1 - \beta_2)^2 + (\alpha_1 - \alpha_2)^2}{(\beta_1 + \beta_2)^2 + (\alpha_1 + \alpha_2)^2}} \tag{17}$$

$$\tau = \sqrt{\frac{4(\alpha_1^2 + \beta_1^2)}{(\beta_1 + \beta_2)^2 + (\alpha_1 + \alpha_2)^2}} \tag{18}$$

Provided the following boundary conditions are applied, solutions to equation (13) can be found using the Laplace transformation technique:

$$h\Omega\Big|_{surface} = -k\frac{\partial\Omega}{\partial z}\Big|_{surface} \tag{19a}$$

$$\frac{d\Omega}{dz}\Big|_{centre} = 0 \tag{19b}$$

$$\frac{d^2\Omega}{dz^2}\Big|_{centre} = 0 \tag{19c}$$

The Laplace transformation of equation (13) is:

$$\nabla^2\Omega(s) - \frac{s}{\gamma}\Omega(s) + n\omega\varepsilon_o\kappa''(\tau E)^2 \frac{\left\{e^{-2\beta z} + 2\Gamma e^{-2\beta W}\cos\left[\delta + 2\alpha(W-z)\right] + \Gamma^2 e^{-2\beta W}e^{-2\beta(W-z)}\right\}}{2sk\left(1 - 2\Gamma^2 e^{-2\beta W}\cos(2\delta + 2\alpha W) + \Gamma^4 e^{-4\beta W}\right)} = 0 \tag{20}$$

Solutions to equation (20) are found by combining the solution to the complementary function $\nabla^2\Omega_o(s) - \frac{s}{\gamma}\Omega_o(s) = 0$ with any particular solutions to the full differential equation (Bowman, 1931).

Adopting rectangular coordinates and applying the limit theorem, i.e., $\lim_{t\to 0} f(t) = \lim_{s\to\infty} F(s)$, to ensure that the solution is bounded at time $t = 0$ yields:

$$\Omega(s) = Ae^{-\sqrt{\frac{s}{\gamma}}z} + n\omega\varepsilon_o\kappa''(\tau E)^2 \frac{\left\{e^{-2\beta z} + 2\Gamma e^{-2\beta W}\cos\left[\delta + 2\alpha(W-z)\right] + \Gamma^2 e^{-2\beta W}e^{-2\beta(W-z)}\right\}}{2sk\left(\frac{s}{\gamma} - 4\beta^2\right)\left(1 - 2\Gamma^2 e^{-2\beta W}\cos(2\delta + 2\alpha W) + \Gamma^4 e^{-4\beta W}\right)} \tag{21}$$

This can now be solved using standard mathematical tables (Crank, 1979) to yield:

$$\Omega'(t) = \frac{Az}{2\sqrt{\pi\gamma t^3}}e^{-\frac{z^2}{4\gamma t}} + n\omega\varepsilon_o\kappa''(\tau E)^2\left(e^{4\gamma\beta^2 t} - 1\right)\frac{\left\{e^{-2\beta z} + 2\Gamma e^{-2\beta W}\cos\left[\delta + 2\alpha(W-z)\right] + \Gamma^2 e^{-2\beta W}e^{-2\beta(W-z)}\right\}}{8k\beta^2\left(1 - 2\Gamma^2 e^{-2\beta W}\cos(2\delta + 2\alpha W) + \Gamma^4 e^{-4\beta W}\right)} \tag{22}$$

Applying the first boundary condition to evaluate A yields:

$$\Omega'(t) = \frac{n\omega\varepsilon_o\kappa''(\tau E)^2\left(e^{4\gamma\beta^2 t} - 1\right)}{8k\beta^2\left(1 - 2\Gamma^2 e^{-2\beta W}\cos(2\delta + 2\alpha W) + \Gamma^4 e^{-4\beta W}\right)}(\Phi + \Psi + \Theta) + \Omega_o \tag{23}$$

where

$$\Phi = e^{-2\beta z} + \left(\frac{h}{k} + 2\beta\right)ze^{\frac{-z^2}{4\gamma t}} \tag{24}$$

$$\Psi = 2\Gamma e^{-2\beta W}\left\{\cos\left[\delta + 2\alpha(W-z)\right] + \left[\frac{h}{k}\cos(\delta + 2\alpha W) - 2\alpha\sin(\delta + 2\alpha W)\right]ze^{\frac{-z^2}{4\gamma t}}\right\} \tag{25}$$

And

$$\Theta = \Gamma^2 e^{-4\beta W} \left\{ e^{-2\beta z} + \left[\frac{h}{k} + 2\beta \right] z e^{\frac{-z^2}{4\gamma t}} \right\} \tag{26}$$

The analysis used to obtain equation (23) assumed that the incoming electromagnetic wave has a constant phase. In a multi-mode oven, this scenario is unlikely because the load is usually placed on a rotating turntable, or mode stirrers are employed to perturb the microwave fields. The net result is that the phase (δ) of the incoming wave varies considerably during the heating process. For simplicity, it has been assumed that δ regularly cycles through 2π radians during the heating process. Therefore:

$$\Omega(t) = \int_0^{2\pi} \Omega'(t) \cdot d\delta \tag{27}$$

This renders $\Psi = 0$, allowing equation (23) to simplify to:

$$\Omega(t) = \left\{ \frac{n\omega\varepsilon_o\kappa''(\tau E)^2 \left(e^{4\gamma\beta^2 t} - 1 \right)}{8k\beta^2 \left(1 - 2\Gamma^2 e^{-2\beta W} \cos(2\delta + 2\alpha W) + \Gamma^4 e^{-4\beta W} \right)} \right\} (\Phi + \Theta) + \Omega_o \tag{28}$$

If either β or W become sufficiently large or if there is no inner surface to reflect from (i.e., $\Gamma = 0$), equation (23) can be simplified to become:

$$\Omega(t) = \left\{ \frac{n\omega\varepsilon_o\kappa''(\tau E)^2 \left(e^{4\gamma\beta^2 t} - 1 \right)}{8k\beta^2 \left(1 - 2\Gamma^2 e^{-2\beta W} \cos(2\delta + 2\alpha W) + \Gamma^4 e^{-4\beta W} \right)} \right\} \Phi + \Omega_o \tag{29}$$

Equations (23) and (29) predict that when a slab or block is heated in a microwave oven, subsurface heating occurs with the maximum temperature located slightly below the material surface. This agrees with information presented by Van Remmen et al., (1996) and Zielonka and Gierlik (1999). When the thickness of the material is small, the maximum temperature occurs in the center of the heated slab because the "subsurface" temperature peaks coincide with the center of the slab. However when the thickness of the material increased, two temperature peaks appear. This is a direct result of field attenuation inside the material. When the slab or block is fairly small, very little wave attenuation occurs; however, when the slab or block becomes much larger, the fields are significantly attenuated near the center of the object.

2.4 Microwave heating in cylindrical coordinates

Temperature profiles in small-diameter cylinders usually exhibit pronounced core heating (Ohlsson & Risman, 1978; Van Remmen et al., 1996); on the other hand, temperature profiles in large cylinders exhibit subsurface heating, with the peak temperature occurring slightly below the surface (Van Remmen et al., 1996). The same is also true when the microwave fields can not penetrate very far into the cylinder (Van Remmen et al., 1996).

The refractive index, at microwave frequencies, for many moist materials is very high (refractive index > 8). Therefore, by Snell's law, it is reasonable to assume that any wave propagates radially inside the cylinder irrespective of the incident angle of the external wave. It is also assumed that the microwave oven's turntable or mode stirrer ensures that the long-term illumination of the cylinder is uniform over all exposed surfaces. Therefore, in cylindrical coordinates, equation (3) becomes:

$$\frac{d^2E}{dr^2} + \frac{1}{r}\frac{dE}{dr} + f^2E = 0 \tag{30}$$

The solution to equation (30) has the general form:

$$E = AJ_0(fr) + BY_0(fr) \tag{31}$$

For E to be bounded at the center of the cylinder, B must be equal to zero, and therefore $E = AJ_0(fr)$. Oliveira and Franca (2002) present the same solution for the electric field in a cylinder.

The Bessel function addition theorem is now applied to this complex field:

$$J_0(x+jy) = \sum_{m=-\infty}^{\infty} J_m(x)J_{o-m}(jy) = \sum_{m=-\infty}^{\infty} J_m(x)I_{o-m}(y) \tag{32}$$

When the complex argument is small, this simplifies to:

$$J_0(x+jy) \approx J_0(x)I_0(y) \tag{33}$$

Yousif and Melka (1997) further show that when the real part of the argument is small, as would be the case when heating relatively small cylindrical objects in domestic microwave ovens, $J_0(x+jy) \approx I_0(y) - jx[I_1(y)]$. Therefore, except near the central axis of the cylinder, the real component of the electric field has the form $E = AI_0(\beta r)$.

When the electric field in the cylinder is normalized to the field at the surface of the cylinder, equation (31) becomes:

$$E = \tau E_o \frac{I_0(\beta r)}{I_0(\beta r_o)} \tag{34}$$

Therefore, the heat and moisture diffusion equation in cylindrical coordinates is:

$$\frac{\partial^2\Omega}{\partial r^2} + \frac{1}{r}\frac{\partial\Omega}{\partial r} - \frac{1}{\gamma}\frac{\partial\Omega}{\partial t} + \frac{n\omega\varepsilon_o\kappa''\tau^2E_o^2}{k}\frac{I_0(2\beta r)}{I_0(2\beta r_o)} = 0 \tag{35}$$

Note: $[I_0(\beta r)]^2 \approx I_0(2\beta r)$.

Equation (35) can be solved using the Laplace transformation technique. It is customary to solve for the complementary (static) version of equations (35), by combining one or more solutions of $\dfrac{\partial^2\Omega(s)}{\partial r^2} + \dfrac{1}{r}\dfrac{\partial\Omega(s)}{\partial r} - \dfrac{s}{\gamma}\Omega(s) = 0$ with a specific solution to the full equation. In order to find a solution to the complementary function, a substitution of the form $z = \sqrt{\frac{s}{\gamma}}r$ is made. The complementary function then becomes a modified Bessel equation, which has the standard solution of the form:

$$\Omega_a(s) = AK_o(z) + BI_o(z) \tag{36}$$

Applying the limit theorem $\left[\lim_{t \to 0} f(t) = \lim_{s \to \infty} F(s)\right]$ to ensure that the solution is bounded at time $t = 0$ implies that $B = 0$.

In order to investigate what is happening near the surface of the heated cylinder, it is necessary to seek a particular solution to the complementary function. The most appropriate form of this solution would be: $\Omega_b(s) = e^{\lambda(r_o - r)}$. Substituting into the complementary function to evaluate λ, yields:

$$\Omega_b(s) = Ae^{\frac{1}{2r} + \sqrt{\frac{1}{4r^2} + \frac{s}{\gamma}}(r_o - r)} + Be^{\frac{1}{2r} - \sqrt{\frac{1}{4r^2} + \frac{s}{\gamma}}(r_o - r)} \tag{37}$$

Now γ is usually very small ($\approx 10^{-7}$), so except in the case when r is infinitesimally close to zero, equation (37) can be approximated by:

$$\Omega_b(s) = Ae^{\sqrt{\frac{s}{\gamma}}(r_o - r)} + Be^{-\sqrt{\frac{s}{\gamma}}(r_o - r)} \tag{38}$$

Again applying the limit theorem to keep the temperature bounded at time zero suggests that:

$$\Omega_b(s) = Be^{-\sqrt{\frac{s}{\gamma}}(r_o - r)} \tag{39}$$

A particular solution to equation (35) should be of the form $\Omega_c(s) = M \cdot I_o(2\beta r)$. Substituting into equation (35) and evaluating M yields:

$$\Omega_c(s) = \frac{n\omega\varepsilon_o\kappa''\tau^2 E_o^2}{ks\left(\frac{s}{\gamma} - 4\beta^2\right)} \frac{I_o(2\beta r)}{I_o(2\beta r_o)} \tag{40}$$

Therefore, the combined solution becomes:

$$\Omega(s) = \Omega_a(s) + \Omega_b(s) + \Omega_c(s) \tag{41}$$

Or

$$\Omega(s) = AK_0\left(\sqrt{\frac{s}{\gamma}}r\right) + Be^{-\sqrt{\frac{s}{\gamma}}(r_o - r)} + \frac{n\omega\varepsilon_o\kappa''\tau^2 E_o^2}{ks\left(\frac{s}{\gamma} - 4\beta^2\right)} \frac{I_o(2\beta r)}{I_o(2\beta r_o)} \tag{42}$$

This can be solved using standard mathematical tables (Crank, 1979) to yield:

$$\Omega(t) = \frac{A}{2t}e^{\frac{-r^2}{4\gamma t}} + \frac{B(r_o - r)}{2\sqrt{\pi\gamma t^3}}e^{\frac{-(r_o - r)^2}{4\gamma t}} + \frac{n\omega\varepsilon_o\kappa''\tau^2 E_o^2\left(e^{4\beta^2\gamma t} - 1\right)}{4k\beta^2}\frac{I_o(2\beta r)}{I_o(2\beta r_o)} + \Omega_o \tag{43}$$

To evaluate the full temperature/moisture distribution requires application of the following boundary conditions:

$$\left.\frac{d\Omega(t)}{dr}\right|_{r=0} = 0 \text{ and } -k\left.\frac{d\Omega(t)}{dr}\right|_{r=r_o} = h\Omega(t)\big|_{r=r_o} \tag{44}$$

For the sake of simplifying the mathematics somewhat, the second boundary condition is applied to:

$$\Omega(t) = \frac{B(r_o - r)}{2\sqrt{\pi\gamma t^3}}e^{\frac{-(r_o - r)^2}{4\gamma t}} + \frac{n\omega\varepsilon_o\kappa''\tau^2 E_o^2\left(e^{4\beta^2\gamma t} - 1\right)}{4k\beta^2}\frac{I_o(2\beta r)}{I_o(2\beta r_o)} \tag{45}$$

Because the approximation for the microwave power distribution used in equation (34) is not accurate near the center of the cylinder, the first boundary condition should be applied to:

$$\Omega(t) = \frac{A}{2t}e^{\frac{-r^2}{4\gamma t}} + \frac{n\omega\varepsilon_o\kappa''\tau^2 E_o^2\left(e^{4\beta^2\gamma t} - 1\right)}{4k\beta^2}\frac{\left[J_o(\alpha r)I_o(\beta r)\right]^2}{\left[J_o(\alpha r_o)I_o(\beta r_o)\right]^2} \tag{46}$$

The analysis eventually yields:

$$\Omega(t) = \frac{n\omega\varepsilon_o\kappa''\tau^2 E_o^2\left(e^{4\beta^2\gamma t} - 1\right)}{4k\beta^2 I_o(2\beta r_o)}(\Phi + \Psi + \Theta) + \Omega_o \tag{47}$$

In this case:

$$\Phi = \frac{4\alpha\gamma t}{\left[J_o(\alpha r_o)I_o(\beta r_o)\right]^2}e^{\frac{-r^2}{4\gamma t}} \tag{48}$$

$$\Psi = I_o(2\beta r) \tag{49}$$

And

$$\Theta = \left[2\beta I_1(2\beta r_o) + \frac{h}{k}I_o(2\beta r_o)\right](r_o - r)e^{\frac{-(r_o - r)^2}{4\gamma t}} \tag{50}$$

2.5 Microwave heating in spherical coordinates

Spherical objects respond in a similar way to cylindrical objects in that small-diameter spheres usually exhibit pronounced core heating (Kritikos et al., 1981; Van Remmen et al.,

1996), while larger-diameter spheres or spheres made from very absorbent materials exhibit subsurface heating (Van Remmen et al., 1996). When spherical coordinates are applied, equation (3) becomes:

$$\frac{d^2E}{dr^2} + \frac{2}{r}\frac{dE}{dr} + f^2E = 0 \tag{51}$$

Following the same process as with the cylindrical analysis described above, the solution to equation (51) becomes:

$$E = \tau E_o \frac{j_o(fr)}{j_o(fr_o)} \tag{52}$$

Vriezinga (1998) presents an expression for the radial component of the spherical field in terms of ordinary Bessel functions:

$$E \propto A\sqrt{\frac{\pi}{2fr}} J_{n+(1/2)}(fr) \tag{53}$$

Now $\sqrt{\dfrac{\pi}{2fr}} J_{n+(1/2)}(fr) = j_n(fr)$; therefore, equation (52) is similar to the equation presented

by Vriezinga. In the case of spherical coordinates, the heat/moisture diffusion equation becomes:

$$\frac{\partial^2\Omega}{\partial r^2} + \frac{2}{r}\frac{\partial\Omega}{\partial r} - \frac{1}{\gamma}\frac{\partial\Omega}{\partial t} + \frac{n\omega\varepsilon_o\kappa''\tau^2E_o^2}{k}\left[\frac{j_o(fr)}{j_o(fr_o)}\right]^2 = 0 \tag{54}$$

Following a similar derivation to the cylindrical case, the temperature/moisture distribution in spherical objects is eventually described by:

$$\Omega(t) = \frac{n\omega\varepsilon_o\kappa''\tau^2E_o^2\left(e^{4\beta^2\gamma t}-1\right)}{k\beta\cdot i_o(2\beta r_o)}(\Phi+\Psi+\Theta)+\Omega_o \tag{55}$$

In this case:

$$\Phi = \frac{\alpha\gamma t}{\beta\left[j_o(\alpha r_o)i_o(\beta r_o)\right]^2}e^{\frac{-r^2}{4\gamma t}} \tag{56}$$

$$\Psi = \frac{i_o(2\beta r)}{4\beta} \tag{57}$$

And

$$\Theta = \left[2\beta\cdot i_1(2\beta r_o) + \frac{h}{k}i_o(2\beta r_o)\right]\frac{(r_o-r)}{4\beta}e^{\frac{-(r_o-r)^2}{4\gamma t}} \tag{58}$$

Because cylinders and spheres tend to focus the microwave energy into the center of the workload, there is a stronger tendency towards core heating than in the rectangular slab. However, as the radius of the cylinder or sphere increases, the attenuation of the microwave fields becomes more pronounced, and there is a shift from predominantly core heating to predominantly subsurface heating. This phenomenon is well documented (Ohlsson & Risman, 1978; Kritikos et al., 1981; Van Remmen et al., 1996; Romano et al., 2005).

Even when the radius of the cylinder or sphere is fairly small, if the dielectric loss factor increases, there is a transition from predominantly core heating at lower loss factors to predominantly subsurface heating at higher loss factors. This was also observed by Van Remmen et al. (1996).

2.6 Multi-dimensional solutions

So far, this analysis has only considered single-dimensional heat transfer. Most real problems are multi-dimensional. The solution to multi-dimensional partial differential equations is a simple product of the single-dimensional solutions (Crank, 1979; Van Remmen et al., 1996). For example, the temperature distribution within a rectangular block, irradiated equally on all six faces, can be found by multiplying the temperature profiles associated with three slabs of thickness equal to the length, width, and height of the rectangular block.

Fig. 1 (a) shows the predicted temperature distribution in a rectangular block of agar gel heated in a multi-mode microwave oven. This distribution agrees with thermal images and other data found in the literature (Van Remmen et al., 1996; Zielonka & Gierlik, 1999). Agar gel is a useful medium for identifying hot spots during microwave heating because it melts at the hot spots, but remains solid where the temperature of the object is lower than its melting temperature. The theoretical temperature distributions in the agar gel block compare very well with the melting patterns in a block of agar gel (Fig. 1 (b)).

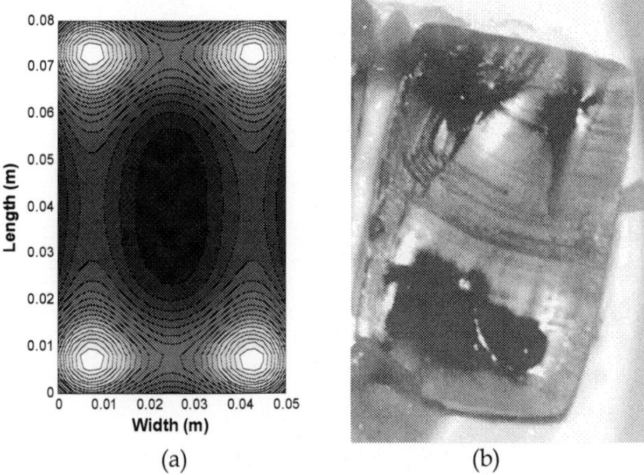

(a)   (b)

Fig. 1. (a) Theoretical multi-dimensional temperature distribution in a 50 × 50 × 80 mm rectangular block of agar gel heated for 90 seconds (lighter colours represent hottest places) compared to (b) the actual melting patterns in a rectangular block of agar gel heated for 90 seconds in a domestic microwave oven.

The temperature distribution within a cylinder, which is irradiated equally on all surfaces, can be found by multiplying the radial temperature profile for the cylinder by the temperature profile associated with a slab of thickness equal to the height of the cylinder. When the cylinder's radius and loss factor are small, the temperature distribution resembles a "dumbbell" with two temperature peaks along the longitudinal axis, as shown in fig. 2a.

As the loss factor of the cylinder increases, the temperature profile is transformed such that there is an annulus of high temperature below the upper and lower circular surfaces, as shown in Fig. 3b. The same is true when the outer radius of the cylinder increases, as shown in Fig. 3c. The predicted temperature profiles shown in Fig. 3a and 3b are directly comparable with thermal images for cylindrical objects presented by Van Remmen et al. (1996).

The temperature distribution within a sphere irradiated equally on all surfaces is purely radial. Fig. 4 shows the temperature distributions in spheres of various diameters and loss factors. Again, these predicted temperature profiles are directly comparable with thermal images for spherical objects presented by Van Remmen et al. (1996).

3. The effect of heating time on temperature and drying

As would be expected, increasing microwave heating time increases the temperature in moist objects (Fig. 5). However, the relationship between time and temperature at a particular location inside the heated material is often non-linear because of the complexity of synchronized heat and moisture diffusion, changes of the spatial distribution of heat with time, partial drying of the material and their combined effect on the absorption of microwave energy.

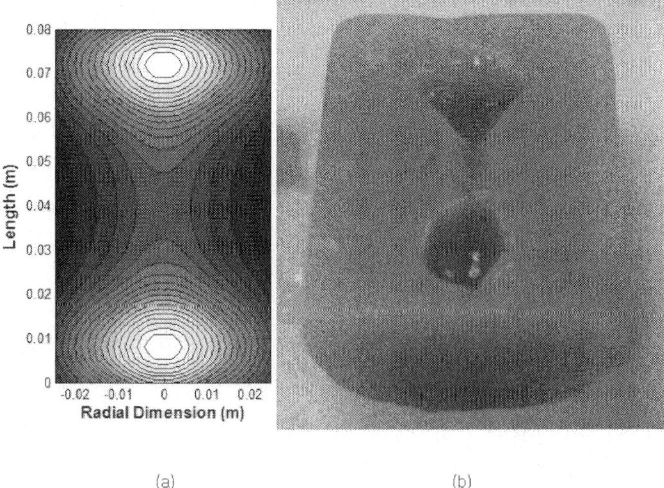

(a)   (b)

Fig. 2. (a) Theoretical multi-dimensional temperature distribution in a 50 mm diameter × 80 mm long cylinder of agar gel compared to (b) the actual melting patterns in a cylinder of agar gel heated for 90 seconds in a domestic microwave oven.

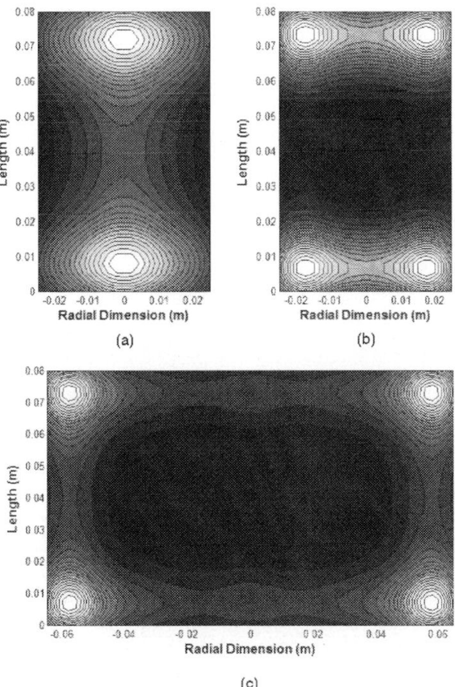

Fig. 3. Multi-dimensional temperature distribution along the longitudinal axis of 80 mm tall agar gel cylinders after 90 seconds of microwave heating for (a) radius = 25 mm, κ' = 76.0, κ'' =6.0; (b) radius = 25 mm, κ' = 76.0, κ'' = 15; and (c) radius = 65 mm, κ' = 76.0, κ'' = 6.0 (lighter colours represent hottest places).

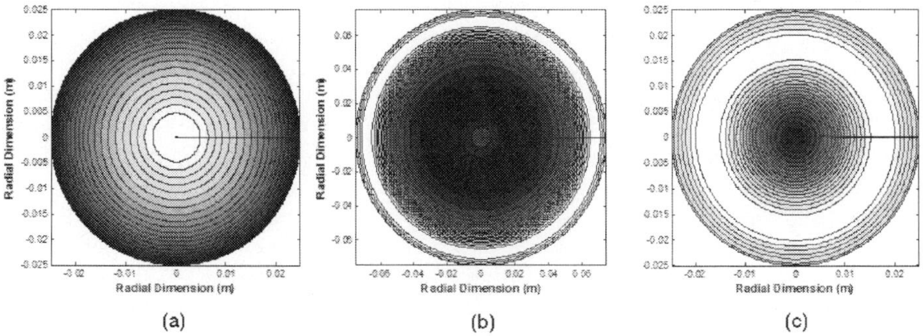

Fig. 4. Multi-dimensional temperature distribution in agar gel spheres after 90 seconds of microwave heating for (a) radius = 25 mm, κ' = 76.0, κ'' = 10.0; (b) radius = 75 mm, κ' = 76.0, κ'' = 10.0; and (c) radius = 25 mm, κ' = 76.0, κ'' = 20.0 (lighter colours represent hottest places).

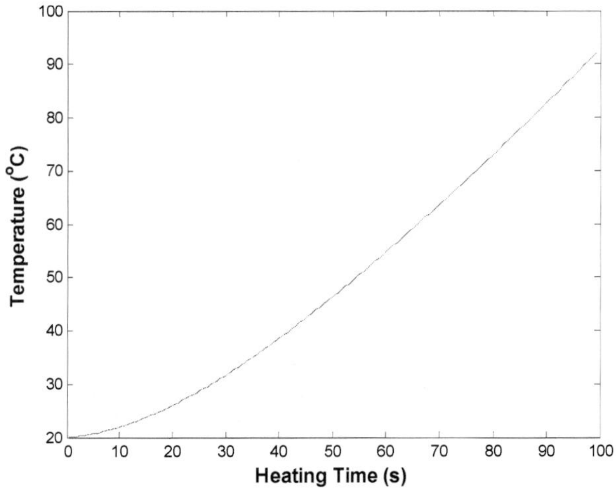

Fig. 5. Predicted temperature in the core of a 12 mm diameter cylinder of a generic plant material, initially having a moisture content of 80 % by weight, as a function of microwave heating time

Heating time not only affects the magnitude of the temperature but also the spatial distribution of heat. Fig. 6 shows a heating sequence for a rectangular block, while Fig. 7 shows the heating sequence for a cylinder. The change in temperature distribution as heating time increases is due to internal thermal diffusion from the zones where the microwave energy is being absorbed by the material to the rest of the object.

As may also be expected, microwave heating also dries moist materials. Conventional drying usually consists of a short constant rate drying period, where the rate of drying is controlled by surface evaporation, followed by a prolonged falling rate drying period (Fig. 8 a) during which the slow movement of moisture out of the object is controlled by the moisture diffusion properties of the material (Youngman et al., 1999). However, after a relatively short heat up period, microwave drying is usually quite linear with time (Fig. 8 b). This is because of the strong coupling between heat and moisture movement through the material. Therefore the rate of microwave drying is not limited by the internal diffusion properties of the material, but by the rate at which moisture can be evaporated from the surface of the material.

As the material dries, its dielectric properties will change significantly. These changes are too slow to affect the behaviour of individual cycles of the microwave fields; however they do affect microwave heating on the much longer thermal time scale.

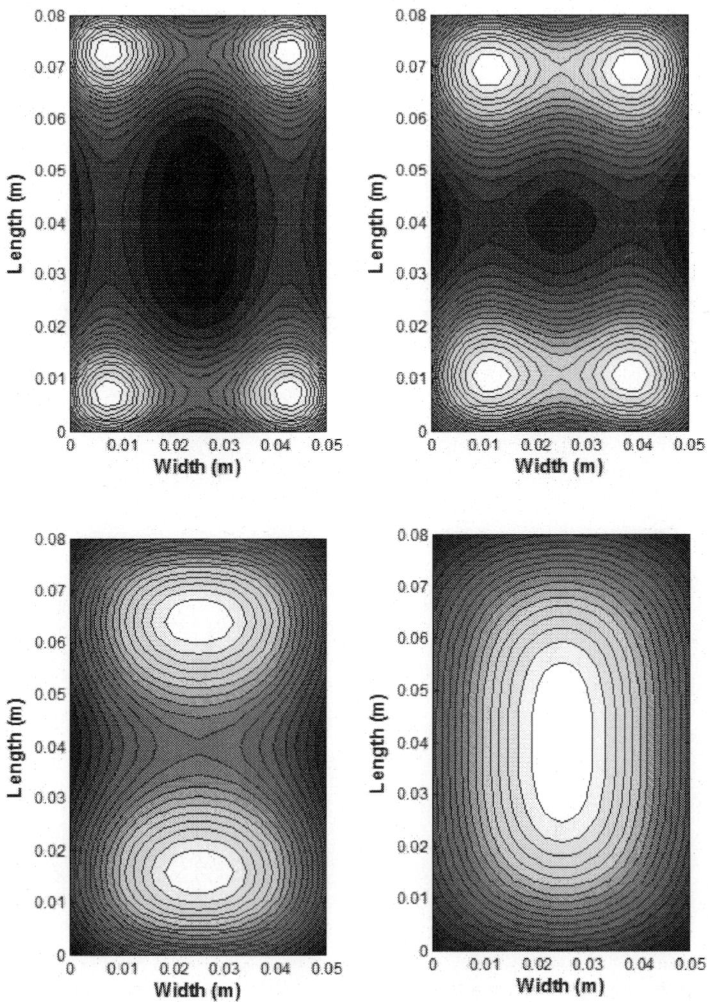

Fig. 6. Heating sequence for a 50 × 50 × 80 mm rectangular block of agar gel heated for (left to right, top to bottom) 90, 190, 400 and 800 seconds in a multi-mode microwave oven.

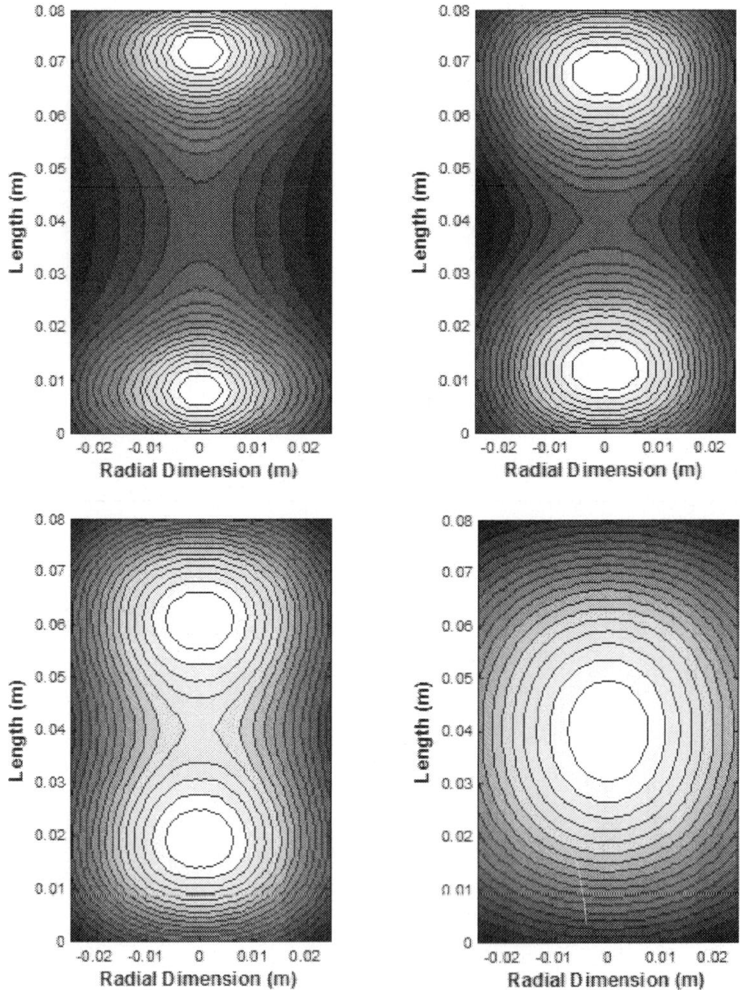

Fig. 7. Heating sequence for a 80 mm long × 50 mm diameter cylinder of agar gel heated for (left to right, top to bottom) 90, 190, 400 and 800 seconds in a multi-mode microwave oven.

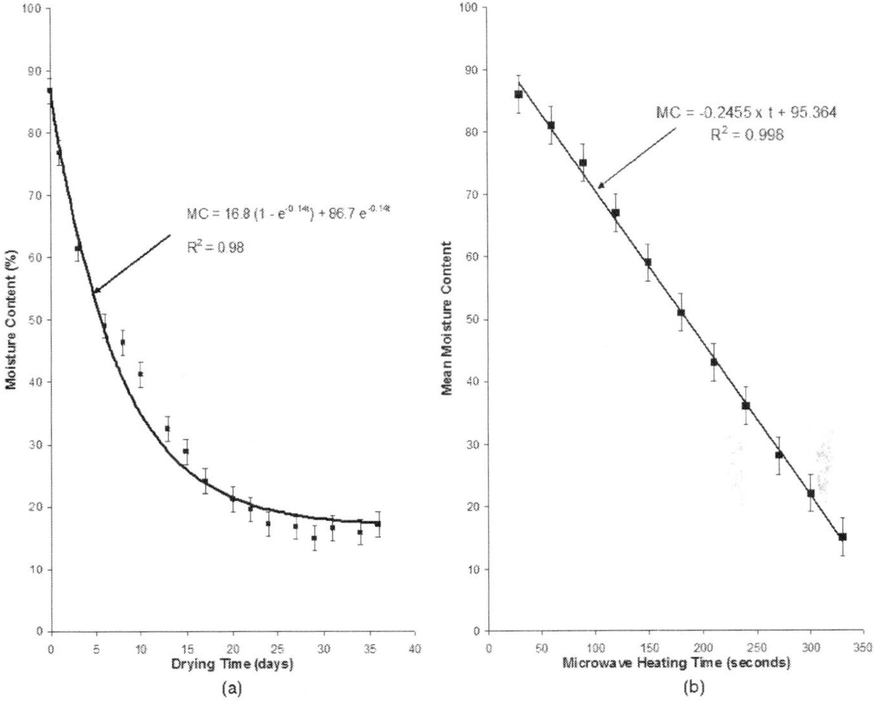

Fig. 8. Mean moisture content for 85 mm by 35 mm by 100 mm *Populus alba* wood samples (a) dried using conventional systems compared with (b) microwave drying.

4. The effect of changing dielectric properties on microwave heating

As an example of how the dielectric properties of moist materials change with moisture content, Ulaby and El-Rayes (1987) studied the dielectric properties of various plant materials at microwave frequencies. They determined that the dielectric behaviour of plant materials can be modeled using a Debye-Cole dual-dispersion dielectric model. Their model accounts for bound and free water in the plant material. Their equation for the dielectric constant of plant materials is:

$$\varepsilon = \varepsilon_s + v_{fw}\left(4.9 + \frac{73.5}{1+j\dfrac{\omega}{123.78\times10^9}} - j\frac{113.1\times10^9\times\sigma_c}{\omega}\right) + v_b\left(2.9 + \frac{55.0}{1+\sqrt{j\dfrac{\omega}{1.13\times10^9}}}\right) \quad (59)$$

where

$$\varepsilon_s = 1.7 - 0.74M_s + 6.16M_s^{\,2} \quad (60)$$

$$v_{fw} = M_s\left(0.55M_s - 0.076\right) \tag{61}$$

$$v_b = \frac{1.5306M_s - 2.5909M_s^{\,2} + 1.4355M_s^{\,3}}{1 - 0.60M_s} \tag{62}$$

and

$$\sigma_c = 1.27 \tag{63}$$

Plant material with high moisture content have higher dielectric constants (Fig. 9) and will therefore interact more with microwave fields.

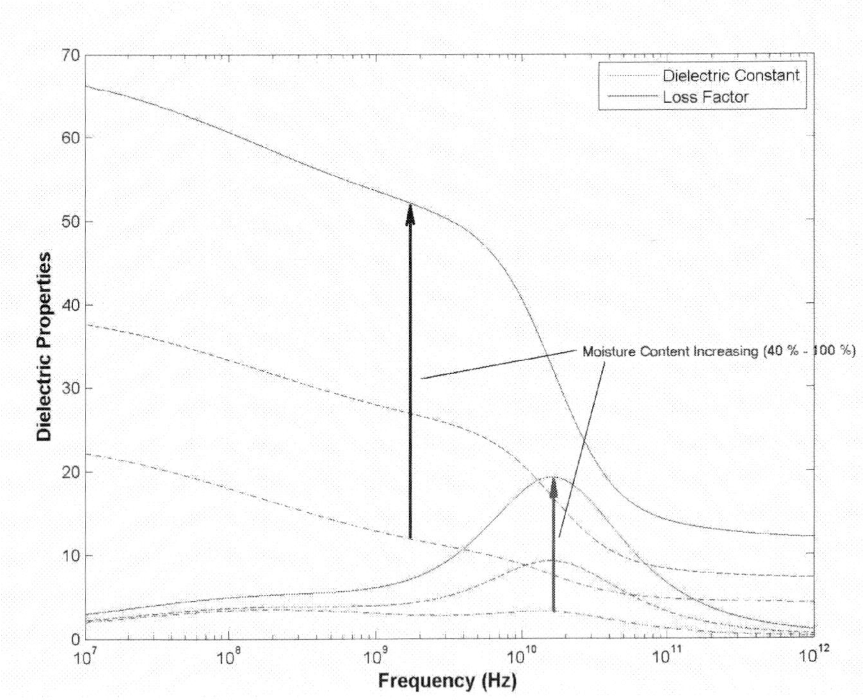

Fig. 9. Dielectric properties of plant materials as a function of frequency and moisture content.

It has also been well documented that the dielectric properties of most other materials are temperature and moisture dependent (Hill & Marchant, 1996; Vriezinga, 1998; Nelson et al., 2001). In order to explore what can happen when a material dries out during microwave heating, a semi-analytical technique can be used to analyse the microwave heating problem. As an example, the heating in a cylinder of plant material will be considered:

Equation (47) was used in an iterative calculation, where the temperature and moisture content of a plant based material was calculated after 0.5 second of microwave heating. This new temperature and moisture content was used to calculate the dielectric properties of the material using equation (59). These new dielectric properties were then used to calculate the next rise in temperature after another 0.5 seconds of microwave heating and so on. Moisture loss was assumed to have a linear relationship with microwave heating time as shown in Fig. 8 b. The results of this process are presented in Fig. 10.

The heating rate for 10 mm and 12 mm diameter cylinders is relatively constant with time; however there is a sudden temperature jump in the 15 mm diameter case when there is no change in the applied microwave power (Fig. 10). This sudden jump in temperature is the result of *"thermal runaway"*. Thermal runaway often destroys or chars the treated material, as seen in Fig. 11.

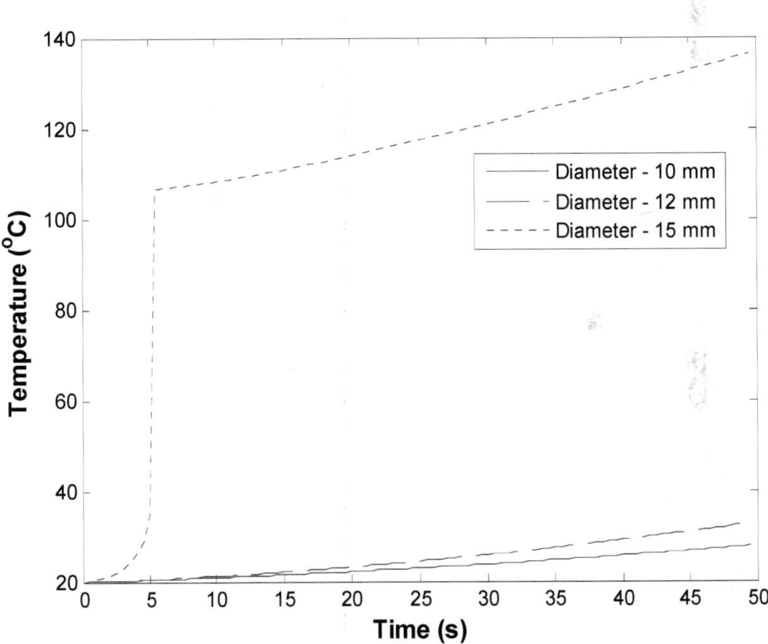

Fig. 10. Temperature response, at constant microwave power density, in the centre of a cylinder of plant based material, as a function of plant stem diameter, calculated using equation (47) and assuming constant moisture loss from a moisture content of 0.87 to 0.10 during microwave heating

Fig. 11. Charing, due to thermal runaway, in the core of a wood sample that was being dried in a microwave oven (Note: the location of the charring in comparison with the last image in Fig. 7)

4.1 Thermal runaway

Several authors (Vriezinga, 1998; Liu & Marchant, 2002) have suggested that the relationship between the applied microwave field strength and temperature follows a characteristic "S" shaped curve (Fig. 12).

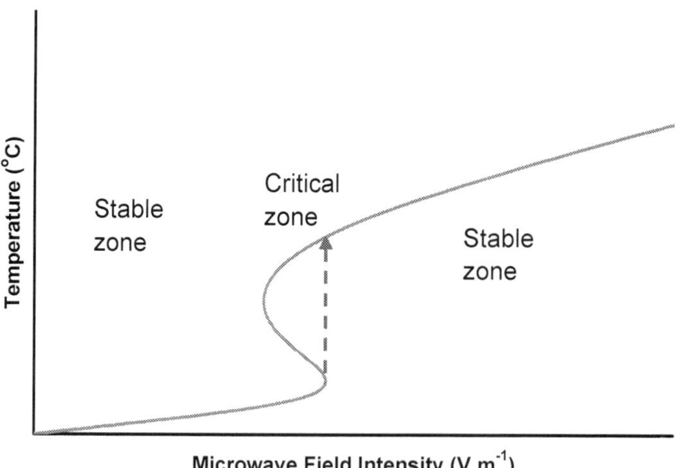

Fig. 12. The characteristic "S" shaped curve relating microwave field intensity and temperature

As microwave power is steadily increased, temperature rises steadily along the stable lower arm of the power curve shown in Fig. 12. In the critical power range, a small increase in electric field strength may shift the equilibrium temperature from the lower limb of the temperature curve to the higher limb, as indicated by the arrow in Fig. 12. The resulting change in temperature could be quite substantial and very rapid. This temperature jump gives rise to the potentially catastrophic phenomenon known as thermal runaway (Vriezinga, 1998).

Thermal runaway is predicted by equations (23), (48) and (56). In each case, under the right combination of physical dimensions and dielectric properties of the material, the denominators of these equations can become very small thus forcing the temperature to become very large.

Thermal runaway manifests itself as a sudden jump in temperature over a very short time. It is linked to the temperature and moisture dependence of the material's dielectric properties. In some cases this can result in the material absorbing more microwave energy as it is heated, which in turn leads to faster heating which in turn leads to more energy absorption, etc. The net result is a sudden and uncontrolled jump in temperature, depending on the strength of the electromagnetic fields and the heating time.

The transfer of microwave energy into a material depends on the transmission coefficient of the material's surface, which also depends on the dielectric properties of the material. In many moist materials, the dielectric constant decreases with temperature and moisture loss (Vriezinga, 1998). As the dielectric properties decrease with microwave heating, more energy enters the material due to increased transmission across the material's surface. This leads to faster heating, which leads to more energy transfer and so on.

Thermal runaway can also be caused by resonance (Vriezinga, 1999) of the electromagnetic waves inside the object. As dielectric properties decrease with increasing temperature and moisture loss, the wavelength inside the material will increase in length. At a certain temperature and moisture, the electromagnetic wave will fit exactly within the dimensions of the heated object, causing resonance (Vriezinga, 1999). Exactly at that moment the temperature will suddenly rise. Because heating and drying rates are dependent on the microwave field intensity, thermal runaway is strongly dependent on the intensity of the microwave's electric field strength (Fig. 13).

4.2 Examples of using thermal runaway in industrial microwave processes

In most cases, thermal runaway is a problem becasue it usually leads to destruction of the material (Zielonka et al., 1997); however it has been very effectively used in some applications:

Jerby et al. (2002) have developed a microwave drill that can drill holes through glass and ceramics by super-heating a very small section of the material using microwave induced thermal runaway. The system works by creating very intense microwave fields immediately in front of a needle like probe that extends into the material as the drilling process proceeds.

Vinden and Torgovnikov (2000) have also shown that the controlled application of intense microwave fields to green timber can rapidly boil free water inside the wood cells (Vinden & Torgovnikov, 2000; Ximing et al., 2002) causing microscopic fractures in the wood cells. When exposure to these intense fields is carefully managed, there is very little change in the vissible appearance or strength of the wood; however there is a substantial change in wood density.

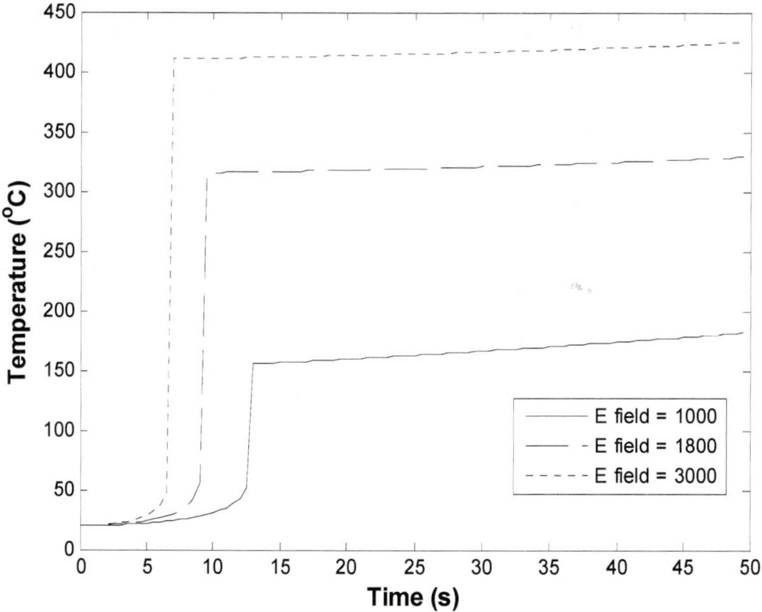

Fig. 13. Theoretical influence of microwave field strength over thermal runaway in a cylinder of plant based material of 16 mm diameter

Lawrence (2006) has demonstrated that this process can reduce the density of *Eucalyptus obliqua* wood by up to 12 %, as the wood cells are ruptured and the cross section of the wood is expanded due to internal steam explosions. This has been confirmed by other experimental work (Brodie, 2005, 2007a). This reduction in wood density leads to rapid drying in conventional systems with little loss in wood strength or quality. Microwave induced cell rupture has been linked to thermal runaway (Brodie, 2007a) in the timber.

5. Determining the electric field strength

The magnitude of the temperature/moisture vapour wave and the onset of phenomenon such as thermal runaway depend on the electric field strength at the surface of the heated object; therefore it is important to model the field strength in the immediate vicinity of the heated object. Analytical solutions to Maxwell's equations can be found for simple microwave applicators, especially when they are empty (Meredith, 1994); however modeling of the microwave field strength in a loaded applicator requires the solution of Maxwell's equations in three dimensions when complex boundary conditions are imposed onto the system. It is almost impossible to determine these fields using analytical techniques.

Several techniques can be employed to simulate microwave fields in complex systems, including numerical techniques such as the Finite Difference Time Domain (FDTD) technique, proposed by Yee (1966). The FDTD method is a simple and elegant way to transform the differential form of Maxwell's equations from equation (1) into difference equations. Yee used an electric field grid (Fig. 14), which was offset both spatially and temporally from a magnetic field grid to calculate the present field distribution throughout the computational domain in terms of the past field distribution.

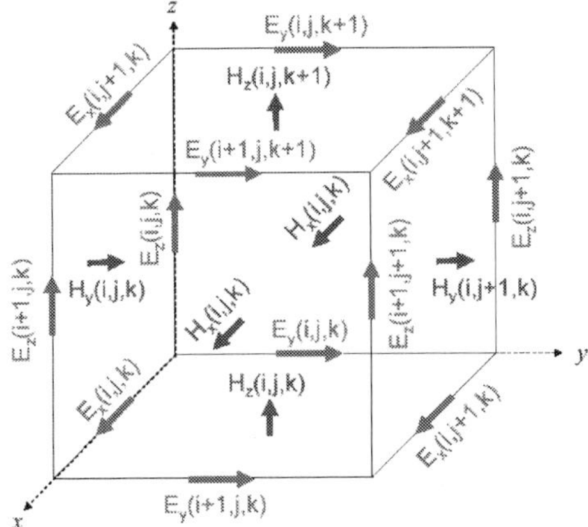

Fig. 14. Heating Single cell from the computational space used to compute Maxwell's equations using FDTD techniques

Two of Maxwell's equations can be transformed into difference equations for use in the FDTD algorithm:

$$E_x^{n+1}(i,j,k) = E_x^n(i,j,k) + \frac{\Delta t}{\varepsilon}\left[\frac{H_z^{n+\frac{1}{2}}(i,j+1,k) - H_z^{n+\frac{1}{2}}(i,j,k)}{\Delta y} - \frac{H_y^{n+\frac{1}{2}}(i,j,k+1) - H_y^{n+\frac{1}{2}}(i,j,k)}{\Delta z}\right] \tag{64}$$

And

$$H_x^{n+\frac{1}{2}}(i,j,k) = H_x^{n-\frac{1}{2}}(i,j,k) + \frac{\Delta t}{\mu}\left[\frac{E_y^n(i,j,k+1) - E_y^n(i,j,k)}{\Delta z} - \frac{E_z^n(i,j+1,k) - E_z^n(i,j,k)}{\Delta y}\right] \tag{65}$$

Equations (64) and (65) calculate the electric and magnetic fields in the x direction. Similar equations are required to calculate Ey, Ez, Hy and Hz. These equations are used in a leap-frog scheme to incrementally march the electric and magnetic fields forward in time; therefore this numerical technique is a simulation of the microwave field rather than a direct solution of the field equations in space and time.

The FDTD method was originally developed to solve electromagnetic transmission problems associated with the communications industry (Kopyt et al., 2003), but it is now commonly used to determine electromagnetic field distributions within industrial microwave applicators (Kopyt et al., 2003). Numerical simulations of the microwave fields can provide valuable insight into the behaviour of microwave heating systems.

5.1 Strengths of the FDTD method

Every modeling technique has some strengths and some weaknesses. FDTD is a very versatile modeling technique. It is fairly intuitive, so users can easily understand what to expect from a given model.

FDTD is a time domain technique, and when a time-domain pulse (such as a Gaussian pulse) is used as an input to the computational model, a wide frequency range can be explored in a single simulation. This is extremely useful in applications where resonant frequencies are not known or when broadband performance is desirable.

Since FDTD is a time-domain technique which finds the E/H fields everywhere in the computational domain, it lends itself to providing animation displays (movies) of the E/H field movement throughout the model. FDTD also allows the user to specify the material properties at all points within the computational domain, which provides useful insight into the field distributions inside dielectric materials as they are processed using microwave systems. For example, when the FDTD technique is applied to a domestic microwave oven cavity in which various sized samples of water are included in the simulation space, it becomes clear that the microwave fields are significantly attenuated as the dimensions of the water sample increase (Fig. 15). This is becasue the water sample reduces the quality factor

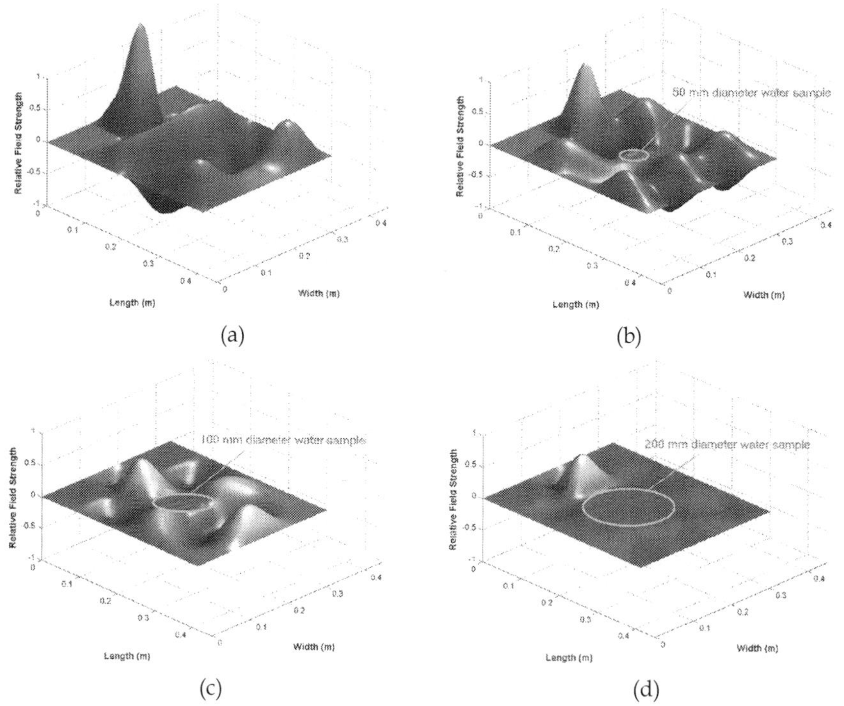

Fig. 15. FDTD simulations of the microwave fields inside a domestic microwave oven with various sized water loads (a) empty cavity, (b) a small water load, (c) a medium sized water load and (d) a large water load

(Q) of the field resonance inside the chamber and effectively dampens the wave resonance. This field damping will affect the microwave heating time needed for a sample to reach a desired temperature. Consequently, small samples will heat much faster per unit volume than larger samples.

Similarly, when the FDTD technique is applied to a domestic oven cavity in which a disk of wood with varying moisture content is placed, it becomes apparant that as the moisture content reduces, the microwave field amplitude inside the wood sample increases (Fig. 16). This supports the earlier discussion about wave resonance inside the heated material that may lead to thermal runaway.

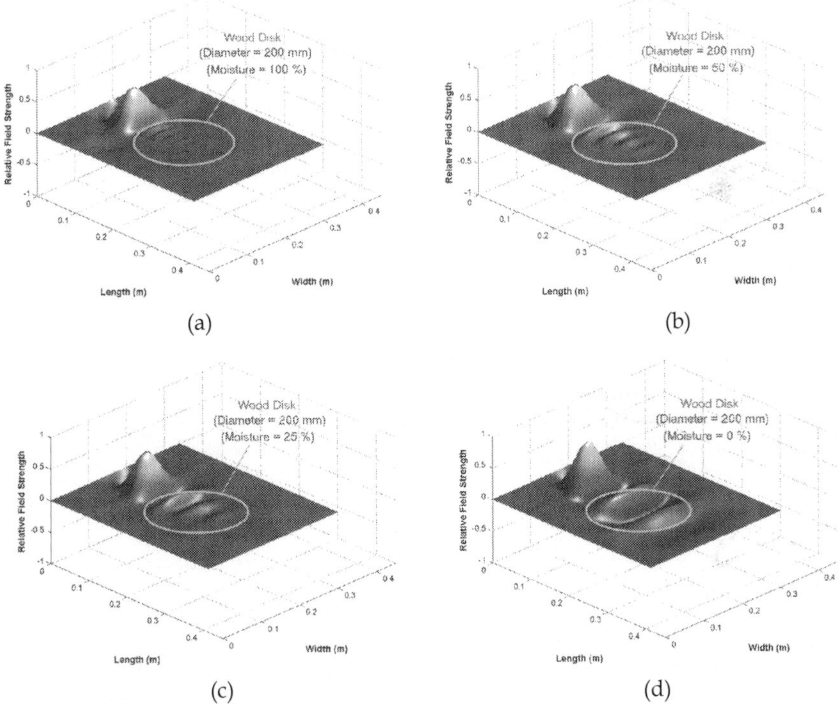

(a)   (b)

(c)   (d)

Fig. 16. FDTD simulations of the microwave fields inside a domestic microwave oven with a wooden disk of varying moisture content, measured as a percentage of oven dried wood.

5.2 Problems with the FDTD method

If aliasing of the final solution is to be avoided (Vandoren, 1982), the grid size must be small enough so that the electromagnetic field does not change significantly from one grid cell to the next. Similarly the time steps for each computational cycle must satisfy the Courant stability criterion (Yee, 1966; Taflove, 1988; Walker, 2001). This implies that the simulated time step for many problems may be no more than a few picoseconds. Therefore solving heating problems that may span many seconds or minutes of real time requires a significant number of computational iterations.

For computational stability to be satisfied (Yee, 1966; Walker, 2001), the time step used in the program must satisfy:

$$\Delta t \leq \frac{\sqrt{\left(\Delta x\right)^2 + \left(\Delta y\right)^2 + \left(\Delta z\right)^2}}{c} \tag{66}$$

When establishing an appropriate grid for analysis, it must be remembered that the wavelength of the microwaves inside a dielectric material is:

$$\lambda = \frac{\lambda_o}{\sqrt{\kappa'}} \tag{67}$$

The values of Δx, Δy and Δz must be small compared to λ to ensure accurate simulations.

Since FDTD requires that the entire computational domain be gridded, and these grids must be small compared to the smallest wavelength and smaller than the smallest feature in the model, very large computational domains can be developed. This implies that very long calculation times are needed to simulate real microwave heating systems.

To put this into perspective, simulating the first 20 nanoseconds of electromagnetic activity inside a 335 mm by 335 mm by 205 mm microwave oven that has a cup (cylinder) of water in it, using a 4 mm grid, requires approximately 30 minutes of uninterrupted computational time on a Pentium 4 computer. Trying to simulate 10 minutes of real time would take 950 years on the same machine using the same code.

FDTD finds the E/H fields everywhere in the computational domain. If the field values at some distance from the microwave source (like 10 meters away) are required, the computational domain will be excessively large. Far field extensions, which apply a Fast Fourier Transformation to the field distribution in a fixed plane of the FDTD computational domain are available, but this usually requires some post processing rather than direct calculations.

Most authors use FDTD and other numerical techniques to analyse "static systems". This implies that apart from the initial field excitation imposed onto the system, all other components are stationary. Commercial microwave applicators use mode stirrers (Metaxas & Meredith, 1983) or movement of the heated product relative to the microwave fields to deliberately perturb the fields and more evenly irradiate the load. Consequently, any attempt to evaluate the electric field, including numerical techniques, only provides an approximation of the field strength (Kopyt et al., 2003).

6. Applications of microwave heating

Throughout this discussion various applications of microwave heating have been mentioned. Microwave heating is now a standard method in food preparation and in some industrial processes (Metaxas & Meredith, 1983). In industry, microwave heating is used for drying of wood (Zielonka & Dolowy, 1998; Antti & Perre, 1999), paper and cardboard (Hasna et al., 2000), textiles, and leather (Ayappa et al., 1991). Other uses include oil extraction from tar sands, cross-linking of polymers, metal casting (Ayappa et al., 1991), medical applications (Bond et al., 2003), pest control (Nelson & Stetson, 1974; Nelson, 1996), enhancing seed germination (Nelson et al., 1976; Nelson & Stetson, 1985), and solvent-free chemistry (Arrieta et al., 2007).

More recently, microwave heating has been used to modify the structure of wood (Vinden & Torgovnikov, 2000; Torgovnikov & Vinden, 2005) allowing it to dry much faster (Brodie, 2007a) and be impregnated with resins to make it stronger (Przewloka et al., 2007). Microwave energy has also been used to manage weeds and other soil born pathogens (Nelson, 1987; Bansal, 2006) and to treat various animal feeds to improve their digestibility (Sadeghi & Shawrang, 2006, 2007; Brodie et al., 2010). Microwaves are commonly used to process ceramics (Takayama et al., 2005) and can be used to drill through ceramics and glass (Jerby et al., 2002).

7. Conclusion

The major advantages of microwave heating are its short startup, precise control, and volumetric heating (Ayappa et al., 1991); however non-uniform heating, rapid drying and unstable temperatures are commonly reported. This chapter has explored some of the reasons why these three features of microwave heating occur. Mathematical models and simulation techniques reveal that object geometry, size, moisture content, dielectric properties, and heating time all profoundly affect temperature/moisture distributions during microwave heating. Small slabs, cylinders, and spheres, with low dielectric loss, exhibit pronounced core heating, while increasing the size of the object or the dielectric loss of the material results in pronounced subsurface heating.

8. Nomenclature

Ω = combined temperature and moisture vapour parameter

Ω_o = Initial condition of the combined temperature and moisture vapour parameter

a_v = air space fraction in the material

c = speed on light (m s^{-1})

C = thermal capacity of the composite material (J kg^{-1} °C^{-1})

D_a = vapor diffusion coefficient of water vapor in air (m^2 s^{-1})

E = electric field associated with the microwave (V m^{-1})

E_o = magnitude of the electric field external to the work load (V m^{-1})

f = complex wave number of the form $f = \alpha + j\beta$

h = convective heat transfer at the surface of a heated object (W m^{-1} K^{-1})

H = magnetic field associated with the microwave

$i_o(x)$ = modified spherical Bessel function of the first kind of order zero

$I_o(z)$ = modified Bessel function of the first kind of order zero

j = complex operator $\sqrt{-1}$

J_c = current flux due to internal conduction

$j_o(x)$ = spherical Bessel function of the first kind of order zero

$J_o(z)$ = Bessel function of the first kind of order zero

J_s = current sources that may be embedded within the region of interest

k = thermal conductivity of the composite material (W m^{-1} °C^{-1})

$K_o(z)$ = modified Bessel function of the second kind of order zero

L = latent heat of vaporization of water (J kg^{-1})

M_s = moisture content of the solid material (kg kg^{-1})

M_v = moisture vapor concentration in the pores of the material (kg m^{-3})

n	=	constant of association relating water vapor concentration to internal temperature of a solid
p	=	constant of association relating internal temperature of a solid to water vapor concentration
q	=	volumetric heat generated by microwave fields (W m^{-3})
r	=	radial distance form the centre of a cylinder or sphere (m)
r_0	=	external radius of the cylinder or sphere (m)
s	=	Laplace transformation argument
t	=	heating time (s)
T	=	temperature (°C)
W	=	thickness of the slab (m)
$Y_o(z)$	=	Bessel function of the second kind of order zero
z	=	linear distance from the surface of a slab (m)
Δt	=	incrimental time step (s)
Δx	=	incrimental distance (m)
Δy	=	incrimental distance (m)
Δz	=	incrimental distance (m)
Γ	=	internal reflection coefficient
α	=	real part of the complex wave number f
β	=	imaginary part of the complex wave number f
χ	=	concentration of agar in agar gel (%)
δ	=	phase shift of microwave fields at the surface of a material
ε^*	=	complex electrical permittivity of the space through which the waves are propagating
ε_s	=	the dielectric properties of a material at very low frequencies
γ	=	combined diffusivity for simultaneous heat and moisture transfer
κ'	=	relative dielectric constant of the material
κ''	=	dielectric loss factor of the material
λ	=	wave length inside a material (m)
λo	=	wave length in free space (m)
μ	=	magnetic permeability of the space through which the waves are propagating
θ_A	=	phase angle associated with forward propagating wave in rectangular coordinate system
θ_B	=	phase angle associated with reverse propagating wave in rectangular coordinate system
ρ	=	composite material density (kg m^{-3})
ρ_c	=	charge density within the space through which the waves are propagating
ρ_s	=	density of the solid material (kg m^{-3})
σ	=	constant of association relating moisture vapor concentration to moisture content in a solid
σ_c	=	electrical conductivity of the material (S m^{-1})
τ	=	transmission coefficient for incoming microwave
τ_v	=	tortuosity factor
ω	=	angular frequency (rad s^{-1})
$\Re(z)$	=	real part of the complex number z

9. References

Antti, A. L. and Perre, P. 1999. A microwave applicator for on line wood drying: Temperature and moisture distribution in wood. *Wood Science and Technology*, vol. 33, no. 2, pp. 123-138, ISSN: 0043-7719

Arrieta, A., Otaegui, D., Zubia, A., Cossio, F. P., Diaz-Ortiz, A., delaHoz, A., Herrero, M. A., Prieto, P., Foces-Foces, C., Pizarro, J. L. and Arriortua, M. I. 2007. Solvent-Free Thermal and Microwave-Assisted [3 + 2] Cycloadditions between Stabilized Azomethine Ylides and Nitrostyrenes. An Experimental and Theoretical Study. *Journal of Organic Chemistry*, vol. 72, no. 12, pp. 4313-4322, ISSN: 0022-3263

Ayappa, K. G., Davis, H. T., Crapiste, G., Davis, E. J. and Gordon, J. 1991. Microwave heating: An evaluation of power formulations. *Chemical Engineering Science*, vol. 46, no. 4, pp. 1005-1016, ISSN: 0009-2509

Bansal, R. 2006. Microwave surfing: yard work [Agricultural machinery]. *Microwave Magazine, IEEE*, vol. 7, no. 4, pp. 26-28, ISSN: 1527-3342

Bond, E. J., Li, X., Hagness, S. C. and Van Veen, B. D. 2003. Microwave imaging via space-time beamforming for early detection of breast cancer. *IEEE Transaction on Antennas and Propagation*, vol. 51, no. 8, pp. 1690-1705, ISSN: 0018-926X

Bowman, F. 1931, *Elementary Calculus*, Longmans, Green and Co. Ltd., ISBN: 0582318076, London.

Brodie, G. 2005. Microwave preconditioning to accelerate solar drying of timber, In: *Microwave and Radio Frequency Applications: Proceedings of the Fourth World Congress on Microwave and Radio Frequency Applications*, Schulz, R. L. and Folz, D. C. (Eds), pp. 41-48, The Microwave Working Group, ISBN: 0 97862222 0 0, Arnold MD

Brodie, G. 2007a. Microwave treatment accelerates solar timber drying. *Transactions of the American Society of Agricultural and Biological Engineers*, vol. 50, no. 2, pp. 389-396, ISSN: 2151-0032

Brodie, G. 2007b. Simultaneous heat and moisture diffusion during microwave heating of moist wood. *Applied Engineering in Agriculture*, vol. 23, no. 2, pp. 179-187, ISSN: 0883-8542

Brodie, G. 2008. The Influence of Load Geometry on Temperature Distribution During Microwave Heating. *Transactions of the American Society of Agricultural and Biological Engineers*, vol. 51, no. 4, pp. 1401-1413, ISSN: 2151-0032

Brodie, G., Rath, C., Devanny, M., Reeve, J., Lancaster, C., Harris, G., Chaplin, S. and Laird, C. 2010. The effect of microwave treatment on lucerne fodder. *Animal Production Science*, vol. 50, no. 2, pp. 124–129, ISSN: 1836-0939

Chu, J. L. and Lee, S. 1993. Hygrothermal stresses in a solid: Constant surface stress. *Journal of Applied Physics*, vol. 74, no. 1, pp. 171-188, ISSN: 0021-8979

Crank, J. 1979, *The Mathematics of Diffusion*, J. W. Arrowsmith Ltd., ISBN: 0-19-853411-6, Bristol.

El-Rayes, M. A. and Ulaby, F. T. 1987. Microwave Dielectric Spectrum of Vegetation-Part I: Experimental Observations. *IEEE Transactions on Geoscience and Remote Sensing*, vol. GE-25, no. 5, pp. 541 - 549 ISSN: 0196-2892

Fan, J., Luo, Z. and Li, Y. 2000. Heat and moisture transfer with sorption and condensation in porous clothing assemblies and numerical simulation. *International Journal of Heat and Mass Transfer*, vol. 43, no. 16, pp. 2989-3000, ISSN: 0017-9310

Fan, J., Cheng, X., Wen, X., and Sun, W. 2004. An improved model of heat and moisture transfer with phase change and mobile condensates in fibrous insulation and comparison with experimental results. *International Journal of Heat and Mass Transfer*, vol. 47, no. 10-11, pp. 2343–2352, ISSN: 0017-9310

Hasna, A., Taube, A. and Siores, E. 2000. Moisture Monitoring of Corrugated Board During Microwave Processing. *Journal of Electromagnetic Waves and Applications*, vol. 14, no. 11, p. 1563, ISSN: 0920-5071

Henry, P. S. H. 1948. The diffusion of moisture and heat through textiles. *Discussions of the Faraday Society*, vol. 3, pp. 243-257, ISSN: 0366-9033

Hill, J. M. and Marchant, T. R. 1996. Modelling Microwave Heating. *Applied Mathematical Modeling*, vol. 20, no. 1, pp. 3-15, ISSN: 0307-904X

Jerby, E., Dikhtyar, V., Aktushev, O. and Grosglick, U. 2002. The microwave drill. *Science*, vol. 298, no. 5593, pp. 587-589, ISSN: 0193-4511

Kopyt, P., Celuch-Marcysiak, M. and Gwarek, W. K. 2003. Microwave processing of temperature-dependent and rotating objects: Development and experimental verification of FDTD algorithms, In: *Microwave and Radio Frequency Applications: Proceeding of the Third World Congress on Microwave and Radio Frequency Applications*, Folz, D. C., Booske, J. H., Clark, D. E. and Gerling, J. F. (Eds), pp. 7-16, The American Ceramic Society, ISBN: 1 57498 158 7, Westerville, Ohio

Kritikos, H. N., Foster, K. R. and Schwan, H. P. 1981. Temperature profiles in spheres due to electromagnetic heating. *Journal of Microwave Power and Electromagnetic Energy*, vol. 16, no. 3 and 4, pp. 327–340, ISSN: 0832-7823

Lawrence, A. 2006. Effect of microwave modification on the density of eucalyptus obliqua wood. *Journal of the Timber Development Association of India*, vol. 52, no. 1/2, pp. 26-31, ISSN: 0377-936X

Liu, B. and Marchant, T. R. 2002. The microwave heating of three-dimensional blocks: semi-analytical solutions. *IMA Journal of Applied Mathematics*, vol. 67, no. 2, pp. 145 -175, ISSN: 1110-757X

Meredith, R. J. 1994. A three axis model of the mode structure of multimode cavities. *The Journal of Microwave Power and Electromagnetic Energy*, vol. 29, no. 1, pp. 31-44, ISSN: 0832-7823

Metaxas, A. C. and Meredith, R. J. 1983, *Industrial Microwave Heating*, Peter Peregrinus, ISBN: 0 906048 89 3, London.

Murray, D. 1958. Percy Spencer and his itch to know. *Reader's Digest*, vol. August, ISSN: 0034-0413

Nelson, M. I., Wake, G. C., Chen, X. D. and Balakrishnan, E. 2001. The multiplicity of steady-state solutions arising from microwave heating. I. Infinite Biot number and small penetration depth. *The ANZIAM Journal*, vol. 43, no. 1, pp. 87-103, ISSN: 1446-1811

Nelson, S. O. 1987. Potential agricultural applications of RF and microwave energy. *Transactions of the ASAE*, vol. 30, no. 3, pp. 818-831, ISSN: 0001-2351

Nelson, S. O. 1996. Review and assessment of radio-frequency and microwave energy for stored-grain insect control. *Transactions of the ASAE*, vol. 39, no. 4, pp. 1475-1484, ISSN: 0001-2351

Nelson, S. O. and Stetson, L. E. 1974. Possibilities for Controlling Insects with Microwaves and Lower Frequency RF Energy. *IEEE Transactions on Microwave Theory and Techniques*, vol. December 1974, pp. 1303-1305, ISSN: 0018-9480

Nelson, S. O. and Stetson, L. E. 1985. Germination responses of selected plant species to RF electrical seed treatment. *Transactions of the ASAE*, vol. 28, no. 6, pp. 2051-2058, ISSN: 0001-2351

Nelson, S. O., Ballard, L. A. T., Stetson, L. E. and Buchwald, T. 1976. Increasing legume seed-germination by VHF and microwave dielectric heating. *Transactions of the ASAE*, vol. 19, no. 2, pp. 369-371, ISSN: 0001-2351

Ohlsson, T. and Risman, P. O. 1978. Temperature distributions of microwave heating - spheres and cylinders. *Journal of Microwave Power and Electromagnetic Energy*, vol. 13, no. 4, pp. 303–310, ISSN: 0832-7823

Oliveira, M. E. C. and Franca, A. S. 2002. Microwave heating of foodstuffs. *Journal of Food Engineering*, vol. 53, no. 4, pp. 347-359, ISSN: 0260-8774

Przewloka, S. R., Hann, J. A. and Vinden, P. 2007. Assessment of commercial low viscosity resins as binders in the wood composite material Vintorg. *Holz als Roh - und Werkstoff*, vol. 65, no. 3, pp. 209-214, ISSN: 0018-3768

Romano, V. R., Marra, F. and Tammaro, U. 2005. Modelling of microwave heating of foodstuff: study on the influence of sample dimensions with a FEM approach. *Journal of Food Engineering*, vol. 71, no. 3, pp. 233-241, ISSN: 0260-8774

Sadeghi, A. A. and Shawrang, P. 2006. Effects of microwave irradiation on ruminal protein and starch degradation of corn grain. *Animal Feed Science and Technology*, vol. 127, no. 1-2, pp. 113-123, ISSN: 0377-8401

Sadeghi, A. A. and Shawrang, P. 2007. Effects of microwave irradiation on ruminal protein degradation and intestinal digestibility of cottonseed meal. *Livestock Science*, vol. 106, no. 2-3, pp. 176-181, ISSN: 1871-1413

Shaw, T. M. and Galvin, J. A. 1949. High-Frequency-Heating Characteristics of Vegetable Tissues Determined from Electrical-Conductivity Measurements. *Proceedings of the IRE*, vol. 37, no. 1, pp. 83-86, ISSN: 0096-8390

Taflove, A. 1988. Review of the formulation and applications of the finite-difference time-domain method for numerical modeling of electromagnetic wave interactions with arbitrary structures. *Wave Motion*, vol. 10, no. 6, pp. 547-582, ISSN: 0165-2125

Takayama, S., Link, G., Sato, M. and Thumm, M. 2005. Microwave sintering of metal powder compacts, In: *Microwave and Radio Frequency Applications: Proceedings of the Fourth World Congress on Microwave and Radio Frequency Applications*, Schulz, R. L. and Folz, D. C. (Eds), pp. 311-318, The Microwave Working Group, ISBN: 0 97862222 0 0, Arnold MD

Torgovnikov, G. and Vinden, P. 2003. Innovative microwave technology for the timber industry, In: *Microwave and Radio Frequency Applications: Proceedings of the Third World Congress on Microwave and Radio Frequency Applications*, Folz, D. C., Booske, J. H., Clark, D. E. and Gerling, J. F. (Eds), pp. 349-356, The American Ceramic Society, ISBN: 1-57498-158-7, Westerville, Ohio

Torgovnikov, G. and Vinden, P. 2005. New microwave technology and equipment for wood modification, In: *Microwave and Radio Frequency Applications: Proceedings of the Fourth World Congress on Microwave and Radio Frequency Applications*, Schulz, R. L. and Folz, D. C. (Eds), pp. 91-98, The Microwave Working Group, ISBN: 0 97862222 0 0, Arnold MD

Van Remmen, H. H. J., Ponne, C. T., Nijhuis, H. H., Bartels, P. V. and Herkhof, P. J. A. M. 1996. Microwave Heating Distribution in Slabs, Spheres and Cylinders with

Relation to Food Processing. *Journal of Food Science*, vol. 61, no. 6, pp. 1105-1113, ISSN: 0022-1147

Vandoren, A. 1982, *Data Acquisition Systems*, Reston Publishing, ISBN: 0835912167, Reston, Virginia.

Vos, M., Ashton, G., Van Bogart, J. and Ensminger, R. 1994. Heat and Moisture Diffusion in Magnetic Tape Packs. *IEEE Transactions on Magnetics*, vol. 30, no. 2, pp. 237-242, ISSN: 0018-9464

Vriezinga, C. A. 1996. Thermal runaway and bistability in microwave heated isothermal slabs. *Journal of Applied Physics*, vol. 79, no. 3, pp. 1779 -1783, ISSN: 0021-8979

Vriezinga, C. A. 1998. Thermal runaway in microwave heated isothermal slabs, cylinders, and spheres. *Journal of Applied Physics*, vol. 83, no. 1, pp. 438 -442, ISSN: 0021-8979

Vriezinga, C. A. 1999. Thermal profiles and thermal runaway in microwave heated slabs. *Journal of Applied Physics*, vol. 85, no. 7, pp. 3774 -3779, ISSN: 0021-8979

Vriezinga, C. A., Sanchez-Pedreno, S. and Grasman, J. 2002. Thermal runaway in microwave heating: a mathematical analysis. *Applied Mathematical Modelling*, vol. 26, no. 11, pp. 1029 -1038, ISSN: 0307-904X

Wang, Z., Li, Y., Kowk, Y. L. and Yeung, C. Y. 2002. Mathematical simulation of the perception of fabric thermal and moisture sensation. *Textile Research Journal*, vol. 72, no. 4, pp. 327-334, ISSN: 0040-5175

XiMing, W., ZhengHua, X., LiHui, S. and WeiHua, Z. 2002. Preliminary study on microwave modified wood. *China Wood Industry Journal*, vol. 16, no. 4, pp. 16-19, ISSN: 1001-8654

Yee, K. S. 1966. Numerical solution of initial boundary value problems involving Maxwell's equations in isotropic media. *IEEE Transactions on Antennas and Propagation*, vol. 14, no. 3, pp. 302-307, ISSN: 0018-926X

Youngman, M. J., Kulasiri, D., Woodhead, I. M. and Buchan, G. D. 1999. Use of a combined constant rate and diffusion model to simulate kiln-drying of Pinus radiata Timber. *Silva Fennica*, vol. 33, no. 4, pp. 317-325, ISSN: 0037-5330

Yousif, H. A. and Melka, R. 1997. Bessel function of the first kind with complex argument. *Computer Physics Communications*, vol. 106, no. 3, pp. 199-206, ISSN: 0010-4655

Zielonka, P. and Dolowy, K. 1998. Microwave Drying of Spruce: Moisture Content, Temperature and Heat Energy Distribution. *Forest Products Journal*, vol. 48, no. 6, pp. 77-80, ISSN: 0015-7473

Zielonka, P. and Gierlik, E. 1999. Temperature distribution during conventional and microwave wood heating. *Holz als Roh- und Werkstoff*, vol. 57, no. 4, pp. 247-249, ISSN: 0018-3768

Zielonka, P., Gierlik, G., Matejak, M. and Dolowy, K. 1997. The comparison of experimental and theoretical temperature distribution during microwave wood heating. *Holz als Roh- und Werkstoff*, vol. 55, no. 6, pp. 395-398, ISSN: 0018-3768

25

Assessment of Microwave versus Conventional Heating Induced Degradation of Olive Oil by VIS Raman Spectroscopy and Classical Methods

Rasha M. El-Abassy, Patrice Donfack and Arnulf Materny
Jacobs University Bremen, Molecular Life Science Center,
Campus Ring 1, 28759 Bremen,
Germany

1. Introduction

Over the last few decades, the microwave heating process has experienced more common and routine use for both home and industrial applications. In industrial field, microwave heating has been used for many applications, including food processing and preservation, bleaching, pasteurization, and sterilization (Decareau, 1985; Farag et al., 2001; Knutson et al., 1987). Numerous advantages boosted the use of microwave heating making it in many cases a technique preferred to conventional heating. These advantages include precise timing, rapidity, and energy saving. The principle of microwave heating is based on the interaction of electromagnetic waves with the molecular constituents of food. Such interaction leads to heat generation in the entire volume at nearly the same rate due to internal thermal dissipation of the vibrational energy of the molecules in the food (Decareau, 1985; Kamel & Stauffer, 1993). On the contrary, conventional heating generates heat at the contact surface first, and then the heat diffuses inward. The effects of microwave heating and conventional heating on the food components are therefore expected to be completely different. Since processed foods by microwaves are heated as a result of molecular excitation (Stein, 1972), many researchers have been concerned with the evaluation of the effect of microwaves on food constituents, nutrient retention and the change of flavours and colours of heated food (Finot, 1995; Mudgett, 1982).

Microwave heating of roasted seeds and beans shows a better retention of flavour and antioxidant compounds without any significant chemical changes of the lipids (Behera et al., 2004; H. Yoshida & Kajimoto, 1989, 1994). With respect to lipid components, microwave heating was studied to verify eventual heat induced effects on different oils and fats (Farag, 1994; Hiromi Yoshida et al., 1990; H. Yoshida et al., 1992). For this purpose, peroxide value, carbonyl value and conjugated diene and triene levels were assessed.

Extra virgin olive oil that comes from the first pressing of the olive, without using heat or chemicals, contains natural antioxidants such as tocopherols, carotenoids, sterols, and phenolic compounds (Boskou, 1996). It should be mentioned that carotenoids play a significant role as antioxidants by scavenging free radicals, and as singlet oxygen quenchers (Burton & Ingold, 1984; Di Mascio et al., 1989). Since oil and fat have low specific heat constants and heat quickly (Jowitt, 1983), nowadays microwave frying of food has been introduced in order to improve the quality of the fried food (Oztop et al., 2007). Therefore, it is of paramount importance to consider the potential side effects of microwave and conventional heating on olive oil and their respective benefits. Owing to the above mentioned characteristics of olive oil, with both nutritional physiological benefits, this chapter is focused on the heat-induced degradation of extra virgin olive oil during microwave versus conventional heating. Emphasis is put on the repercussions in its natural antioxidant content, which is one of the most crucial factors for maintaining the quality and increasing the useful lifetime of frying oils in food manufacturing.

In our work, Raman spectroscopy was employed as a fast and non-destructive technique in order to reveal and compare the degradation of extra virgin olive oil induced by microwave and conventional heating processes. This spectroscopy technique is based on the inelastic scattering of laser light, giving rise to a frequency shift of the scattered light (usually to lower energies; Stokes shift) (Baeten & Dardenne, 2002). Since these energy losses reflect the internal vibrational energies of the scattering molecules, the Raman spectra have fingerprint properties making them very useful for analytical purposes. The major advantages of this technique lie in the fact that it requires little to no sample preparation, it is rapid and non-destructive, and it can be performed using miniaturized setups ideally suited for online industrial processing. It can provide content relevant information based on the energies of molecular vibrations yielding well defined and resolved spectral features in various sample categories including liquids, solids, and gases. Therefore, Raman spectroscopy has a considerable potential as simple, fast and reliable technique in the field of food analysis.

2. Methodology and design

2.1 Samples and heating procedures

In order to explore the effect of microwave and conventional heating processes on extra virgin olive oil degradation, olive oil was separately heated using a microwave oven and an electronic heater plate equipped with a magnetic stirrer.

In the microwave heating process, 200 ml of extra virgin olive oil was heated for a total of 15 min at 700 W. For conventional heating, 200 ml of extra virgin olive oil was heated using an electronic heater plate for 80 min. In order to avoid overheating of parts of the oil bath and to maintain average thermal equilibrium as far as possible, the oil bath was stirred with a magnetic stirrer either every 2 minutes or continuously, during microwave heating and conventional heating, respectively. 20 ml of oil was each time sampled from the heating oil bath at 50, 70, 100, 120, 140, 160, 180, 190, 215, and 225°C. These sampled aliquots were allowed to cool down to room temperature for free fatty acid and carotenoid content determination and Raman spectroscopy monitoring of the oil quality.

2.2 Chemical analysis
2.2.1 Determination of carotenoid content

The total carotenoid content was determined using the British standard method of analysis (British standard methods of analysis, 1977). This approach was used as a reference in order to construct a calibration model using Raman data. The sample was weighed and dissolved with hexane and diluted to the mark of the desired volume. Then it was filled into a quartz cuvette. UV-visible spectrophotometry was used to measure the sample absorbance at 445 nm against hexane contained in a separate identical quartz cuvette. The total carotenoid content was expressed as ppm of beta carotene. The calculation was done as follows:

$$Carotenoid\ content = \frac{[V \times 383 \times E]}{(100 \times W)} \tag{1}$$

Where, V is the volume used for analysis, 383 is the value diffusion coefficient of carotenoids, E is the observed difference in absorption between the sample solution and hexane, and W is the weight of the sample in g.

2.2.2 Determination of free fatty acid (FFA)

FFA was further determined as a reliability test using the American Oil Chemists' Society, (AOCS) Ca 5a-40 official method (Rukunudin et al., 1998). The principle of this method is based on dissolving a weighed sample of oil into a mixture of 1:1 ethanol and diethylether solvents, and then titrating the mixture under constant stirring against a 0.1 M KOH solution, using phenolphthalein as pH-indicator. The titration was run in triplicate for each sample as in the case of the Raman measurement. The results are presented as percentage oleic acid; the expression is given according to AOCS as follows:

$$\%FFA\ as\ oleic\ acid = \frac{alkali\ volume(ml) \times alkalinormality \times 28.2}{sample\ weight(g)} \tag{2}$$

Where 28.2 is the molecular weight of oleic acid divided by 10.

2.3 Experimental design

Tested samples were contained in 1800 µl quartz cuvettes (Starna) and illuminated by the 514.5 nm line of an Ar-ion laser (Coherent, Inova 308 Series) with an excitation power of 10 mW at the sample. The laser was focused within the sample using an inverted microscope setup equipped with a 10x ultra long working distance objective (10x ULWD, N.A 0.20; Olympus). The scattered signal was then recorded at a 180° backscattering geometry (Fig. 1) and dispersed by a single monochromator (TRIAX 550, Jobin Yvon) using a 1200 grooves/mm diffraction grating and an entrance slit width of 200 µm. The spectrometer was equipped with a liquid nitrogen cooled CCD detector with optimal sensitivity in the visible (blue/green) and a chip size of 2048 x 1024 pixels (Symphony 3500, Jobin Yvon). These unique features allowed for an absolute exposure time of only 3 s per spectral window (defined by detector area). Hence, the complete spectral range of interest [700 – 3100 cm^{-1}] was recorded in just 15 s within which 5 signal accumulations were averaged. Each sample aliquot was analyzed in triplicate to insure the reproducibility of the measurement. A spike filter was applied to the recorded spectra in order to remove cosmic ray peaks. Toluene was used under the same conditions as an external standard for calibration by recording the position and the intensity of its well known symmetry ring breathing Raman band at 1004 cm^{-1}. The baseline of each spectrum was approximated by a fourth-order polynomial fit in order to subtract the weak fluorescence background. Computer control of spectral recording and pre-processing was achieved using commercial software (NGSLabSpec, Jobin Ivon).

2.4 Statistical analysis

For chemometric analysis, partial least square (PLS) regression was performed using the robust commercial software package Unscrambler (v 9.7; CAMO A/S). Calibration models for the determination of the carotenoid content in the heated extra virgin olive oil were developed using PLS regression. This statistical process consists of two separate steps,

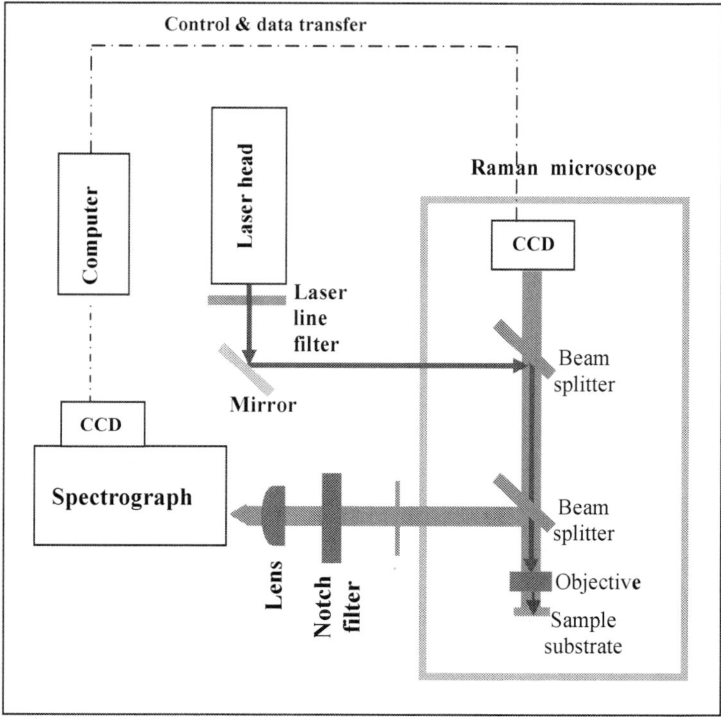

Fig. 1. Raman microscope setup

which are calibration and validation. First, calibration models between reference data (photospectrometry absorption) and Raman spectra were constructed. Here, information relevant for the prediction was extracted from the Raman data in a few components. Then, the validation step was performed to check whether the extracted components described the new data well enough. Cross validation was used in this approach, where one sample is left out from the calibration process, followed by the construction of a calibration model using the remaining samples, which is then tested on the sample left out; this procedure is repeated until each sample has been left out once. The model accuracy was assessed by calculating the root mean square error (RMSE), which measures the average difference between predicted and measured response values – it can be interpreted as the average modelling error, as expressed by the following equation:

$$RMSE = \sqrt{\frac{\sum_{i=1}^{N}(y_{ical} - y_{iref})^2}{N}} \qquad (3)$$

Here, y_{ical} is the calculated (predicted) value using the PLS model, y_{iref} the reference value and N the total number of samples used for calibration. To evaluate the degree of linearity of the model, the determination coefficients (R^2) (Næs et al., 2002) between the actual and predicted values were computed.

3. Results

3.1 Assignments of Raman spectra of olive oil

The change in the Raman spectra of olive oil during the microwave and conventional heating are shown in panels (a) and (b) of Fig. 2, respectively. Major bands in the Raman spectra of non-heated extra virgin olive oil are attributed to the main components in the oil, which are fatty acids. For example, the band at 1265 cm^{-1} can be assigned to δ(=C–H) of *cis* R-HC=CH-R and the band at 1300 cm^{-1} is characteristic of the C–H bending twist of the -CH$_2$ group, while the bands at 1440, 1650, and 1750 cm^{-1} correspond to δ(C–H) scissoring of -CH2, ν(C=C) of *cis* RHC=CHR, and ν(C=O) of RC=OOR, respectively. Furthermore, the bands at 2850, 2897, and 3005 cm^{-1} are attributed to the symmetric CH$_2$ stretch, ν_s(CH$_2$), the symmetric CH$_3$ stretch, ν_s(CH$_3$), and the *cis* RHC=CHR stretch, ν(=C-H), respectively (El-Abassy et al., 2009; Yang & Irudayaraj, 2001). The three bands, around 1008 cm^{-1} (C-CH$_3$ bend), 1150 cm^{-1} (C-C stretch), and 1525 cm^{-1} (C=C stretch), are attributed to carotenoids (Bernstein, 2002), which are responsible for the main characteristic variations in different brands of olive oil. The carotenoids in edible oils play a significant role as natural antioxidants and contribute for up to 50% of the vitamin A activity (Di Mascio et al., 1989). While these bands were not observed in previous studies using NIR excitation, they became prominently detectable under green excitation in our work and help to discriminate differently produced or processed olive oils. Table 1 summarizes the Raman bands of olive oil and their assignment.

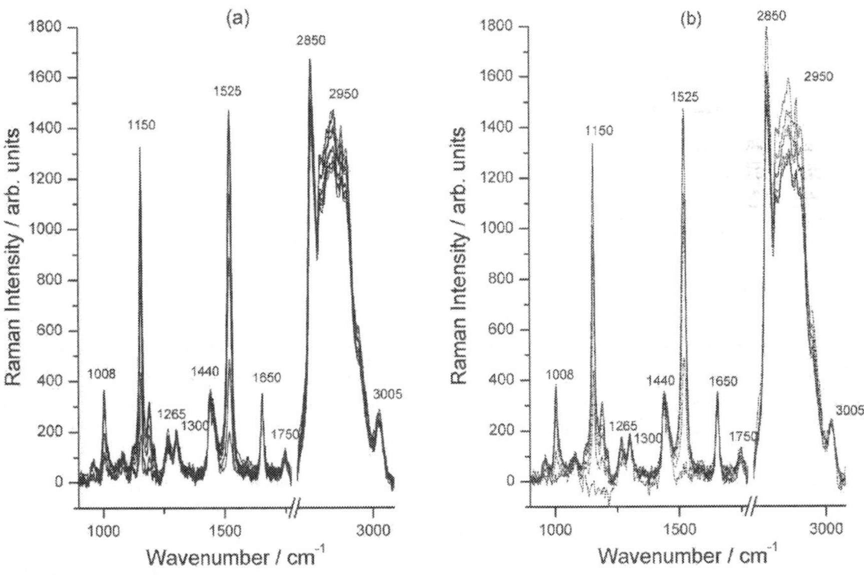

Fig. 2. Raman spectra of extra virgin olive oil, heated using (a) a microwave oven, and (b) a conventional heating setup

Wavenumber (cm⁻¹)	Molecule / Group	Vibrational mode
3005	*cis* RHC=CHR	=C–H symmetric stretching
2897	– C H$_3$	C–H symmetric stretching
2850	– C H$_2$	C–H symmetric stretching
1750	RC=OOR	C=O stretching
1650 1525	*cis* RHC=CHR RHC=CHR	C=C stretching C=C stretching
1440	– C H$_2$	C–H bending (scissoring)
1300	– C H$_2$	C–H bending (twisting)
1265 1150	*cis* RHC=CHR – (C H$_2$)n–	=C–H bending (scissoring) C–C stretching
1008	HC–CH$_3$	CH$_3$ bending

Table 1. Assignment of major Raman bands in olive oil.

3.2 Changes in Raman spectra of heated olive oil
3.2.1 Heat induced degradation after microwave heating

Significant changes in the Raman spectra during the microwave heating process have been observed to start after about 12 min of heating time and above 180°C, where the carotenoid bands at 1008, 1150, and 1525 cm⁻¹ show a gradual decrease in intensity and reach the lowest intensity at 225°C after about 15 min total heating time. Concerning the C=C vibrational mode of lipids at 1650 cm⁻¹, a gradual increase in the intensity can be noticed. This behaviour is certainly due to the conformational change of the methylene chains at higher temperatures, for which the C=C vibrational mode is a sensitive indicator (Wong, 1984). The unsaturation degree in the oil is highly correlated to the intensity ratio of the bands at 1265 and 1300 cm⁻¹ (Li-Chan, 1996). This ratio shows a linear decrease with increasing temperature reflecting a neat loss of the lipid chain unsaturation during the heating process. A slightly decreasing intensity of the C=O stretching vibration band at 1750 cm⁻¹ has also been observed indicating that hydrolysis is taking place leading to a slight increase of FFA content during the heating process (Innawong et al., 2004; Muik et al., 2005). This has been equally confirmed by titration, and is in agreement with the reported increase in FFA content due to thermal treatment.

In the high wavenumber region, a linear increase of the Raman peak at 2850 cm⁻¹ with temperature is observed. The relative intensities of the two Raman bands at 2850 and 2880 cm⁻¹ are related to the degree of disorder of the hydrocarbon chains (Li-Chan, 1996), and usually are employed to determine the lipid phase transition.

In the high wavenumber region, a linear increase of the Raman peak at 2850 cm⁻¹ with temperature is observed. The relative intensities of the two Raman bands at 2850 and 2880 cm⁻¹ are related to the degree of disorder of the hydrocarbon chains (Li-Chan, 1996), and are usually employed to determine the phase transition in lipids. In fact, the peak at 2850 cm⁻¹ is dominant in the liquid phase of lipids while the peak at 2897 cm⁻¹ is dominant in the solid state (Bergethon, 1998). Lipids liquefy as the temperature increases, which is reflected in our Raman results by the observed increase in the intensity of the 2850 cm⁻¹ band with increasing temperature (Larsson, 1973). The gradual decrease of the intensity of the band at

3005 cm⁻¹ which is used to estimate the degree of the total *cis* unsaturation (Wong, 1984), reveals the loss of unsaturation level during the heating process. This loss of the unsaturation degree could be due to the degradation of the natural antioxidants (carotenoids) of olive oil, since it has been reported that the oxidative degradation of the oil during microwave heating depends on its natural antioxidant content (Dostalova et al., 2005).

3.2.2 Heat induced degradation after conventional heating

Similar to the case of the microwave heating process, at the beginning of conventional heating, the intensities of the carotenoid bands at 1008, 1150, and 1525 cm⁻¹ follow the same trend. Especially, these bands temporary show an apparent stability. However, the intensity of these bands starts to decrease notably at 140°C. Moreover, unlike the situation observed in the microwave heating procedure, these carotenoid bands completely disappear at 203°C after 80 min of total heating time during the conventional heating process. The increase in the intensities of the band at 1650 cm⁻¹ has also been observed, which is due to the methylene chains disorder. As already mentioned above in the case of microwave heating, during conventional heating, the ratio of relative intensities of the bands at 1265 and 1300 cm⁻¹ is decreasing, revealing the general loss of *cis* double bonds during the heating process, and a gradual decrease in the intensity of the band at 1750 cm⁻¹ has been noticed as a result of hydrolysis reactions. Finally, in the high wavenumber region the Raman bands at 2850, 2897, and 3005 cm⁻¹ showed the same behaviour as in the microwave heated sample.

3.2.3 Effect of microwave versus conventional heating on carotenoid degradation

In comparison, the two different processes showed obvious differences. First, in the microwave heating process, carotenoid degradation starts at 180°C at 700 W, while it starts at 140°C in the conventional heating process as revealed by the decrease in the intensities of the carotenoid Raman bands and confirmed by quantitative analysis of carotenoid content. Moreover, the carotenoid bands completely vanish at 203°C with conventional heating, where the olive oil has been heated for about 60 min to reach this temperature. In contrast, during microwave heating, these bands can still be observed with much stronger intensities even at higher temperatures. In the last case, olive oil took only 15 min to reach the maximum temperature set at 225°C (panel (a) of Fig. 3). These observations indicate that the heating time is more effective on carotenoid degradation than the final temperature reached. In order to explore the role the heating time plays in the carotenoid degradation, additional experiments are carried out. When setting the power of the microwave oven to 240 W, the olive oil takes 40 min to heat up to a target temperature of 190°C; *i.e.*, the reduction of the power of the microwave oven results in an increase of the heating time. The differences in the Raman spectra of the heated olive oil using the conventional heating process and the microwave heating process with short and long heating times are shown in panel (b) of Fig. 3. It is obvious that the degradation of the carotenoid content is more pronounced with increasing heating time. In order to further confirm this behaviour and to obtain an accurate model that can explain the dependence of carotenoid degradation on temperature and heating time, these two independent parameters are tested using multiparametric regression. The multiparametric regression test reveals a significant determination between carotenoid degradation and heating time, as evidenced by the P values. The P value gives the probability that the result obtained in a statistical test is due to chance rather than true and it tells how strongly each independent variable is correlated with the observable. Small

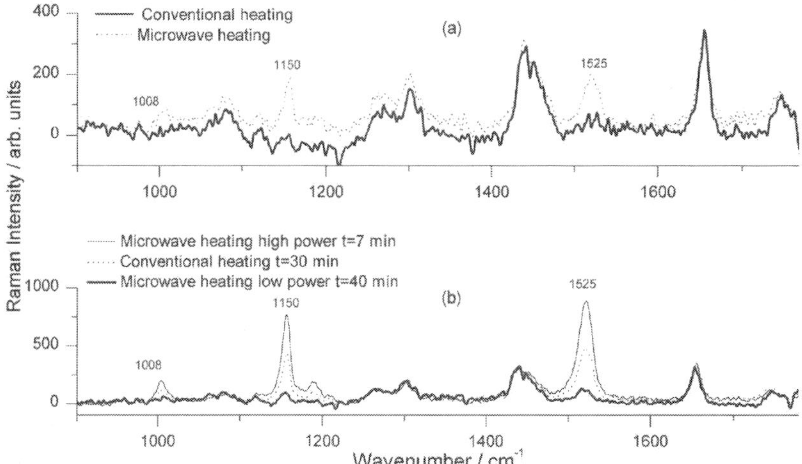

Fig. 3. Raman spectra of extra virgin olive oil heated in a microwave oven and in a conventional way at 225 and 203°C, respectively (a), and heated to 190°C conventionally and in a microwave oven set to high and low power (b).

P values indicate that there is a true relationship between the dependent and the independent variables. A *P* value < 0.05 is often considered statistically significant, and the smaller the P-value the more significant the relationship between the independent variable and the observable. From the evaluation of the experimental results, *P* values of less than 0.0001 for heating time and of approx. 0.04 for temperature in the microwave heating process were determined. On the other side, for the conventional heating process, a *P* value of 0.006 for heating time and a *P* value of 0.5 for temperature resulted. From this it becomes obvious that the heating time is the crucial factor.

Finally, slight changes in FFA content of the extra virgin olive oil have been observed in both heating processes. This is an expected result since it is well known that the FFA content increases in refined olive oil as a result of thermal treatment. The observed FFA content of olive oil analysis in terms of oleic acid percentage, as also indicated by the titration, ranges from 0.18 to 0.25% for conventional heating, while after the microwave heating process it ranges from 0.18 to 0.20%

3.3 Quantification of carotenoids degradation based on Raman data

Carotenoid values observed for heated olive oil in terms of beta carotene content range from 1.670 to 0.603 ppm as measured by the British standard method of analysis (1977) with the absorption photospectrometer. The calibration model of the carotenoid content based on Raman data has been constructed using PLS regression. The Raman spectral window [900-1570 cm-1], which includes the carotenoid bands, is used to construct a calibration model. Here, the multidimensional Raman data set is projected onto a reduced set of a few components describing the directions of the most important variations within the data. Figure 4 shows the predicted carotenoid content based on Raman spectra *vs.* reference values for the optimized model. The slope of the regression curve is close to 1, indicating a perfect linear relationship between the predicted carotenoid values based on the Raman spectra and the actual values determined using the standard method. The model errors (RMSE) for calibration and validation are calculated in order to assess the fitting of the model. This model shows a high determination coefficient $R^2 = 0.95$ and low RMSE of 0.0714 and 0.096 for calibration and validation respectively. This result proves that the selected region [900-1570 cm-1] yields an improved and optimized model with an explained spectral variation of 95% for only the first two components of the PLS model.

The equation used for the carotenoid content calibration based on the constructed PLS model of the Raman data is expressed as follows:

$$(Carotenoid)_p = 0.95 \ x \ (Carotenoid)_m + 0.06 \tag{4}$$

Where $(Carotenoid)_p$ is the predicted carotenoid content using PLS and $(Carotenoid)_m$ is the measured carotenoid content using absorption spectroscopy. These results indicate that the calibration model can be effectively used in online monitoring of frying oil antioxidant degradation.

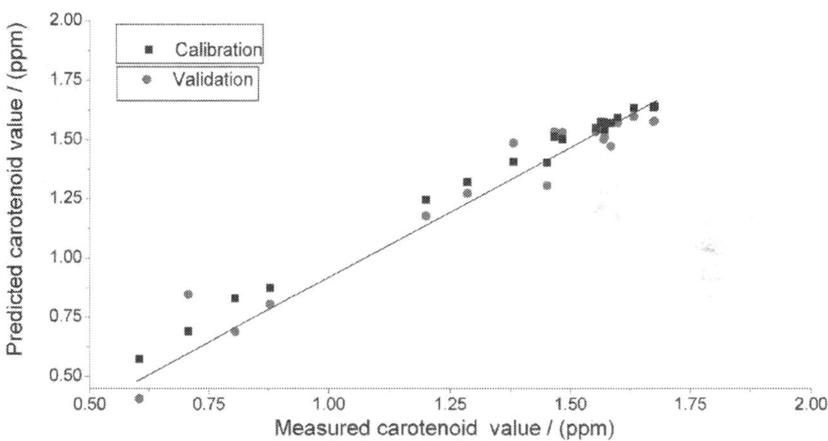

Fig. 4. Calibration curve for carotenoid content in extra virgin olive oil from PLS model

4. Conclusion

In this chapter, we have demonstrated the use of Raman spectroscopy for measuring the heat-induced degradation in extra virgin olive oil during microwave versus conventional heating processes. Particularly, the changes in the olive oils antioxidant content (carotenoids) during the microwave heating are clearly different from those observed during conventional heating. In the case of the conventional heating process, the degradation of carotenoids, occurring before the same given target heating temperature is reached, is much greater than in case of microwave heating. A progressive degradation in carotenoids is observed, starting at 180°C and 140°C during microwave and conventional heating, respectively. This is then followed by a rapid degradation at 180°C only with conventional heating. As the main difference, the Raman bands due to carotenoids completely disappear at 203°C with conventional heating, while these bands can still be observed even up to 225°C with microwave heating. Additionally, losses of *cis* double bonds and slight changes in the free fatty acid content have been observed for both heating processes. These differences have been found to be mainly due to the faster heating rates achievable with the microwave heating process. A slow and also less homogeneous heating process resulting from the use of conventional heating methods is considerably more aggressive than a fast homogeneous heating process in a microwave oven at higher powers. This means that in oil refinery the quality of the final product can be improved by reducing the heating time as much as possible. The observed changes of the molecular structures reflected by the changes of positions and intensities of Raman bands, which occur in the oils during both conventional and microwave heating, reflect the high sensitivity and specificity of the Raman spectroscopy technique. The degradation of the olive oil antioxidant content can be precisely monitored also in-line using Raman spectroscopy.

5. References

Baeten, V., & Dardenne, P. (2002). Spectroscopy: Developments in Instrumentation and Analysis, *Grasas y Aceites* Vol. 53: 45-63.

Behera, S., Nagarajan, S., & Rao, L. J. M. (2004). Microwave heating and conventional roasting of cumin seeds (Cuminum eyminum L.) and effect on chemical composition of volatiles, *Food Chemistry* Vol. 87(1): 25-29.

Bergethon, P. R. (1998). *The Physical Basis of Biochemistry: The Foundations of Molecular Biophysics* Springer.

Bernstein, P. S. (2002). New insights into the role of the macular carotenoids in age-related macular degeneration., *Pure and applied chemistry* Vol. 74(8): 1419-1426.

Boskou, D. (1996). *Olive Oil: Chemistry and Technology*, American Oil Chemists' Society.

British standard methods of analysis. (1977). Determination of carotene in vegetable oils, B.S 684, section 2.20.

Burton, G. W., & Ingold, K. U. (1984). beta-Carotene: an unusual type of lipid antioxidant, *Science* Vol. 224(4649): 569-573.

Decareau, R. V. (1985). *Microwaves in the food processing industry*, Academic Press.

Di Mascio, P., Murphy, M. E., & Sies, H. (1989, Oct 02-04). *Antioxidant defense systems: the role of carotenoids, tocopherols, and thiols.* Paper presented at the Conf on Antioxidant Vitamins and Beta-Carotene in Disease Prevention, London, England.

Dostalova, J., Hanzlik, P., Reblova, Z., & Pokorny, J. (2005). Oxidative changes of vegetable oils during microwave heating, *Czech Journal of Food Sciences* Vol. 23(6): 230-239.

El-Abassy, R. M., Donfack, P., & Materny, A. (2009). Rapid Determination of Free Fatty Acid in Extra Virgin Olive Oil by Raman Spectroscopy and Multivariate Analysis, *Journal of the American Oil Chemists Society* Vol. 86(6): 507-511.

Farag, R. S. (1994). Influence of Microwave and Conventional Heating on the Quality of Lipids in Model and Food Systems, *Fett Wissenschaft Technologie-Fat Science Technology* Vol. 96(6): 215-223.

Farag, R. S., Daw, Z. Y., Mahmoud, E. A., & El-Wahab, S. A. E. A. (2001). A comparison of g -irradiation and microwave treatments on the lipids and microbiological pattern of beef liver, *Grasas y Aceites* Vol. 52: 45-51.

Finot, P. A. (1995). Nutritional Values and Safety of Microwave-Heated Food, *Mitt. Gebiete Lebensm. Hyg.* Vol. 86: 128-139.

Innawong, B., Mallikarjunan, P., Irudayaraj, J., & Marcy, J. E. (2004). The determination of frying oil quality using Fourier transform infrared attenuated total reflectance, *Lebensmittel-Wissenschaft und-Technologie* Vol. 37(1): 23-28.

Jowitt, R. (1983). *Physical properties of foods,* Applied Science.

Kamel, B. S., & Stauffer, C. E. (1993). *Advances in Baking Technology,* Chapman & Hall

Knutson, K. M., Marth, E. H., & Wagner, M. K. (1987). Microwave Heating of Food, *Lebensm. Wiss. Technol.* Vol. 20: 101-110

Larsson, K. (1973). Conformation dependent features in the Raman spectra of simple lipids, *Chemistry and Physics of Lipids* Vol. 10(2): 165-176.

Li-Chan, E. C. Y. (1996). The applications of Raman spectroscopy in food science, *Trends in Food Science & Technology* Vol. 7(11): 361-370.

Mudgett, R. E. (1982). Electrical Properties of Food in Microwave Processing, *Food Technol.* Vol. 36 109-115

Muik, B., Lendl, B., Molina-Diaz, A., & Ayora-Canada, M. J. (2005). Direct monitoring of lipid oxidation in edible oils by Fourier transform Raman spectroscopy, *Chemistry and Physics of Lipids* Vol. 134(2): 173-182.

Næs, T., Isaksson, T., Fearn, T., & Davies, T. (2002). *A User-Friendly Guide to Multivariate Calibration and Classificationtion,* NIR Publications.

Oztop, M. H., Sahin, S., & Sumnu, G. (2007). Optimization of microwave frying of potato slices by using Taguchi technique, *Journal of Food Engineering* Vol. 79(1): 83-91.

Rukunudin, I. H., White, P. J., Bern, C. J., & Bailey, T. B. (1998). A modified method for determining free fatty acids from small soybean oil sample sizes, *Journal of the American Oil Chemists Society* Vol. 75(5): 563-568.

Stein, E. W. (1972). Application of microwave to bakery production, *Baker's Dig.* Vol. 46(21): 53-56.

Wong, P. T. T. (1984). Raman Spectroscopy of Thermotropic and High-Pressure Phases of Aqueous Phospholipid Dispersions, *Annual Review of Biophysics and Bioengineering* Vol. 13: 1-24.

Yang, H., & Irudayaraj, J. (2001). Comparison of near-infrared, fourier transform-infrared, and fourier transform-raman methods for determining olive pomace oil adulteration in extra virgin olive oil *Journal of the American Oil Chemists' Society* Vol. 78: 889-895.

Yoshida, H., Hirooka, N., & Kajimoto, G. (1990). Microwave Energy Effects on Quality of Some Seed Oils, *Journal of Food Science* Vol. 55(5): 1412-1416.

Yoshida, H., & Kajimoto, G. (1989). Effects of Microwave Energy on the Tocopherols of Soybean Seeds *Journal of Food Science* Vol. 54(6): 1596-1600.

Yoshida, H., & Kajimoto, G. (1994). Microwave Heating Affects Composition and Oxidative Stability of Sesame (*Sesamum indicum*) Oil, *Journal of Food Science* Vol. 59(3): 613-616.

Yoshida, H., Kondo, I., & Kajimoto, G. (1992). Effects of Microwave Energy on the Relative Stability of Vitamin E in Animal Fats, *Journal of the Science of Food and Agriculture* Vol. 58(4): 531-534.

26

Microwave Heating: a Time Saving Technology or a Way to Induce Vegetable Oils Oxidation?

Ricardo Malheiro[1], Susana Casal[2],
Elsa Ramalhosa[1] and José Alberto Pereira[1]
*[1]CIMO / School of Agriculture, Polytechnic Institute of Bragança,
Campus de Sta Apolónia, Apartado 1172, 5301-854 Bragança,
[2]REQUIMTE / Serviço de Bromatologia, Faculdade de Farmácia da Universidade do Porto,
Rua Aníbal Cunha, 164, 4099-030 Porto,
Portugal*

1. Introduction

The use of microwave radiation for food heating was discovered unintentionally in 1945 by Dr. Percy Spencer (Osepchuk, 1984). In the following years several experiments were conducted in order to improve the technology and food application fields, being exclusively used at the industrial level, mostly for drying, baking, and thawing. The popularization of domestic microwave ovens started in the 70's (Osepchuk, 1984) once the improvements allowed price and size reduction. Microwaves are now indispensible equipments in westerns modern kitchens.

Although there is no formal definition of the frequency range for microwave radiation, these electromagnetic waves occur in the 300MHz - 300GHz region. Nevertheless, and in accordance with the industrial, scientific and medical (ISM) frequency bands for non-communication purposes, only 915 MHz and 2.45GHz are used for food applications, especially the second due to its worldwide availability.

In domestic equipments, these microwaves (high frequency oscillating electric and magnetic fields) are produced inside the oven, when electrons resonate at high frequencies in an electron tube called magnetron. The electric field is created through an inner cathode and an outer anode presenting a large potential difference between both. Permanent magnets, which compose the magnetron, are responsible for the magnetic field. Once heated, the cathode releases the electrons, traveling from the cathode in an outward spiraling path, eventually making their way to the anode. As the electrons go by resonance chambers, energy is released and received as microwaves by the output antenna. The microwaves are applied to the cavity oven, where the food is heated, via waveguides (Mutyala et al., 2010). The microwaves are reflected and distributed by the stirrer fan and then reflected again by the metallic walls of the container, being absorbed by food and lead to its temperature rise (Fig. 1). This capability to absorb microwave energy is governed by food dielectric properties, and involves primarily two mechanisms: dipolar relaxation and ionic conduction. Food water is often the primary component responsible for dielectric heating. Because of its dipolar nature, water molecules tend to re-orientate with the high frequencies oscillating fields, producing heat from collisions with the near particles. The second mechanism occurs with the mineral ions present in food that migrate under the influence of the electric field, also generating heat. Once heat is generated within food, particularly in the regions with higher water content, the process is not limited by the rate of heat flow into the food, conditioned by its thermal characteristics, as specific heat, thermal conductivity, density and viscosity, while the uniformity of heat distribution is greatly improved and the whole food is heated at the same rate.

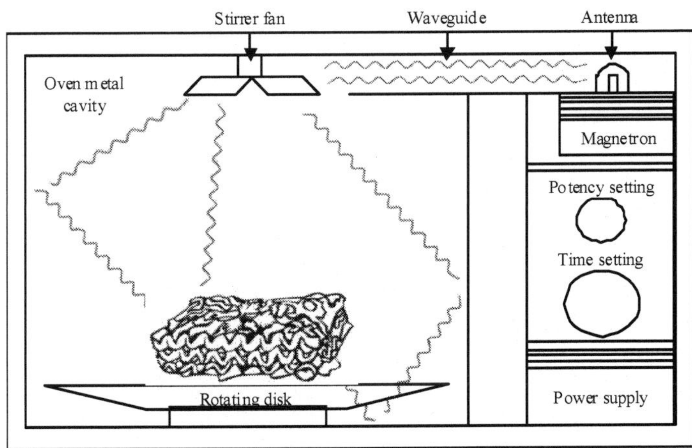

Fig. 1. Simplified scheme of microwave radiation path and absorption by food on a domestic microwave oven.

The heat generate within food is also dependent on the frequencies applied. Lower microwave frequencies (longer wavelengths) allow higher penetration of foods and consequently a more uniform and faster heating (Chavan & Chavan, 2010). For the occurrence of this mechanism, and to achieve convenient heating, the frequency range of the oscillating field should allow a proper inter-molecule interaction. If the frequency range is too high the inter-molecular forces will stop the movement of the polar molecules before trying to follow the field and to be in phase with it, resulting in an inadequate interaction. By other hand, if the frequency range is too low, polar molecules have sufficient time to align themselves in phase with the field and the interaction with near molecules is not observed.

Altogether, the main reasons for microwave efficiency for food heating, while compared to other conventional cooking methods, are related with the dielectric properties of foods and the microwave frequency, providing inner penetration and faster heat conduction, and the non-dielectric properties of the food containers and equipment, justifying less energy dissemination. Based on these properties, microwave ovens become very useful and convenient mainly at domestic level to defrost, heat and cook food, with reduced times, compatible with the demands of modern society. The food industry saw in this technology a way to follow the modern consumer, and actually this industry contributes largely to the use and dissemination of microwave ovens, once tendency is to create and produce convenient pre-prepared food products that only need to be defrosted, cooked or heated by a microwave oven.

Microwave food heating, like others technologies, presents also some disadvantages. Depending on the water content and its distribution in the food, different microwave absorption rates are created and the product can be unevenly cooked or with the so called "dead spots" and "hot spots". Some constraints regarding containers must not be disregarded. While metals and aluminum cannot be used, as they will reflect microwaves, only plastic containers specifically designed for this purpose must be used, avoiding migration of potentially dangerous chemicals into food. On the other hand, formation of health hazard compounds during microwave heating and loss of important minor compounds with biological properties is observed, a situation that will be further developed within this chapter, especially regarding food lipids.

2. Microwave heating effect on food lipids

The major food components: water, carbohydrates, lipids, protein and minerals interact differently with microwaves. The microwave heating effects on these food constituents, as well as color changes and flavor formation, have been extensively evaluated in several foods. A particularly focus has been devoted to lipids, recognized as being prone to thermo-oxidative degradation. These lipids include those naturally present in the raw food, but mostly the lipid ingredients added in pre-prepared meals or the lipids used for frying purposes.   Microwave frying became popular in recent years, being regarded as an alternative frying technology at the industrial level. While in conventional frying heating is transferred by conduction from the oil to the food, and then by conduction within the food, in microwave frying there is also heat generation directly within the food, due to the dipolar rotation induced in the water (Sahin & Sumnu, 2009). Among these applications, the main sources of lipids are vegetable oils, justifying their separate study.

Vegetable oils are constituted mainly by triglycerides, with reduced amounts of diglicerides, monoglycerides and free fatty acids. The unsaponifiable constituents, representing less than 10%, encompass important lipossoluble vitamins, as vitamins E and A, and other minor bioactive components as sterols, chlorophylls, or phenols. The vegetable oil composition will depend mostly on the lipid source (sunflower, soybean, maize, olives, palm, etc) and the technological treatment applied, with the refined seed oils being deprived of several constituents when compared with virgin oils as olive oil.

During microwave heating, as with other heating technologies, different chemical changes could occur that compromises the lipids quality and safety: some components are destroyed while other potentially hazard are formed. The main classical degradation patterns observed in the triacylglycerols include hydrolysis, oxidation, and thermal polymerization. The chemical nature of the fatty acids within the triacylglycerols, particularly their unsaturation degree, will determine its susceptibility towards oxidation and polymerization, while the presence of water will enforce hydrolysis. Other minor components, as vitamins and some sterols, can be regarded as protective factors, against oxidation and polymerization, respectively. The temperature achieved during food heating and the heating period will further determine the extent of the degradation effects, while comparing with other conventional heating technologies.

2.1 Effect of microwave heating on physical properties of edible oils
2.1.1 Viscosity and density
In opposition to other common liquids, an increase in viscosity and density values is observed when vegetable oils are heated, as shown by Albi et al. (1997a) for both

conventional (180°C, 120 min.) and microwave heating (500W, 120 min.) of olive oil, sunflower oil, and high oleic sunflower oil. However, microwave heating caused a higher increase in both parameters, as depicted in Fig. 2. Yet, when the oils were exposed to microwave energy without any increase of temperature, the results obtained were quite similar to those obtained in the unheated samples, concluding that microwave energy only by itself is not enough to cause viscosity and density increase (Albi et al. 1997a).

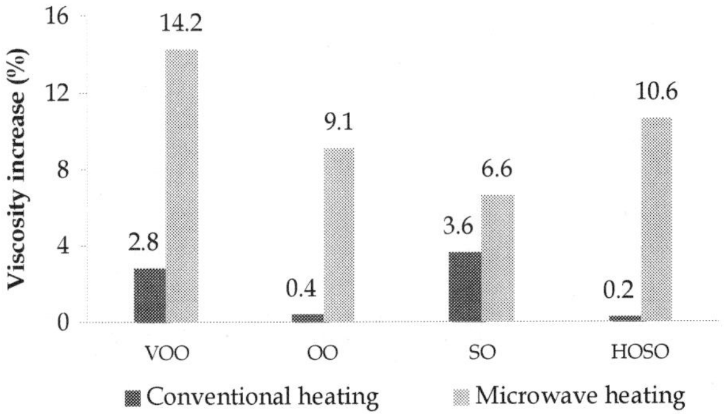

Fig. 2. Viscosity increase (%) in different vegetable oils subjected to microwave heating and conventional heating (VOO - Extra virgin olive oil; OO - Olive oil; SO - Sunflower oil; HOSO - High oleic sunflower oil). Adapted from Albi et al. (1997a).

Oomah et al. (1998), while studying the influence of microwave heating on grapeseed oil extractability and quality, compared conventional drying (50°C, 2h, fluid bed dryer) with microwave drying (950W, 9 min). Both methods provided increased oil viscosity, but a more pronounced effect (2-fold) was observed in the oils extracted from the microwave heated samples.

Viscosity and density increases are temperature-dependent and justified by chemical reactions that take place within the oils. The increased viscosity is related with the formation of cyclic monomers, dimmers and polymers in a non-radical mechanism. The double bonds migration within the unsaturated fatty acids enables the formation of conjugated dienes. These instable structures contribute to the formation of ring structures, allowing bonds between two triacylglycerols, therefore resulting in the formation of larger molecules. The increased proportion of saturated fatty acids, as a result of unsaturated fatty acids oxidation, will further contribute due to their higher melting points (Eskin et al., 1996). The increased density is related with the incorporation of oxygen in the fatty acids (oxidized triglycerides). These parameters will be detailed within the chemical alterations.

2.1.2 Color

Color is an important sensorial attribute perceived by the consumer, influencing significantly its acceptance towards vegetable oils, particularly for olive oil. Microwave heating influences the color indexes of seed and olive oils.

When seed oils (soybean, peanut, sunflower and mixture of soybean/peanut oils) were microwave heated from 2 to 18 minutes (120 to 227°C) their color changed gradually from yellow-brown to light brown. The color intensity at 420 nm in the four oils decreased in the following order: soybean oil > mixture of soybean/peanut oil (1:1) > peanut oil > sunflower oil. The darkening process of seed oils seems to be influenced by the presence of phospholipids, therefore justifying the soybean oil reduced performance regarding this parameter (Hassanein et al., 2003).

Olive oil submitted to microwave heating tends to lose its greenish-golden color becoming paler. In contrast, sunflower oil, as other seed oils, tends to get darker with the yellow light color turning into brown (Fig. 3).

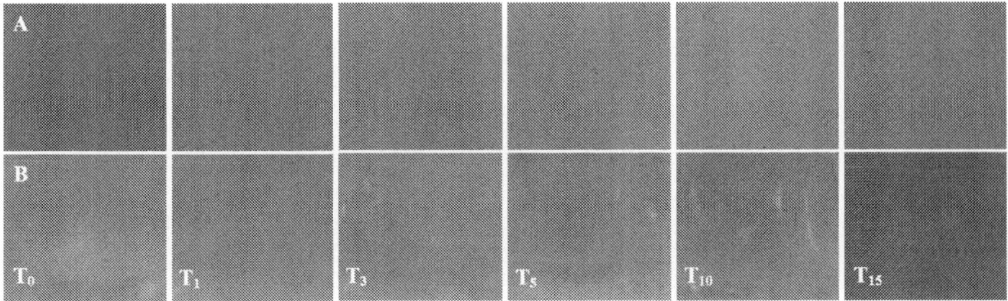

Fig. 3. Effect of microwave heating on color of "Trás-os-Montes" PDO olive oil (line **A**) and sunflower oil (line **B**) (T_0: unheated oils; T_1: 1 minute; T_3: 3 minutes; T_5: 5 minutes; T_{10}: 10 minutes and T_{15}: 15 minutes).

Malheiro et al. (2009) compared three PDO (Protected Designation of Origin) Portuguese olive oils submitted to increasing heating times, from 1 to 15 minutes (1000W), simulating general home practices with olive oil containing recipes, and evaluated different color parameters (x, y, λ_d, Y% and σ%) as shown in Fig. 4. The parameters x and y are the chromatic coordinates of the superficial point of the chromaticity diagram that corresponds to the light transmitted by the oil. Purity is the percentage of light transmitted by the oil in the prevailing wavelength (λ_d), while transparency corresponds to the light transmitted after passing through the oil layer.

 Color parameters were quite stable during the first 3 minutes of microwave heating, and began to decrease after 5 minutes, with the exception of transparency (Y%) where a gradual elevation was observed with the increased exposure time. The reported values and the reduction of green color could be related to the destruction of thermo-labile pigments like chlorophylls and carotenoids, naturally higher in cold-pressed virgin oils than in refined ones, as will be discussed ahead. The reported values are in accordance with eye-naked observations (see Fig. 3).

2.2 Effect of microwave heating on chemical quality parameters of edible oils
2.2.1 Free acidity

Triglycerides and diglycerides are prone to thermal hydrolysis, particularly in the presence of water, releasing fatty acids from their ester linkage, and increasing free acidity. This analytical parameter is therefore frequently used to evaluate hydrolysis extension, a very important quality issue. Hydrolysis is influenced by temperature and heat, interface between oil and water, and the existent quantity of water and steam.

Initial oil acidity is also an important parameter that influences oil thermal performance. Free fatty acids are eliminated during refining, providing commercial vegetable oils with low acidity. In opposition, virgin olive oils, being obtained only by physical processes at low

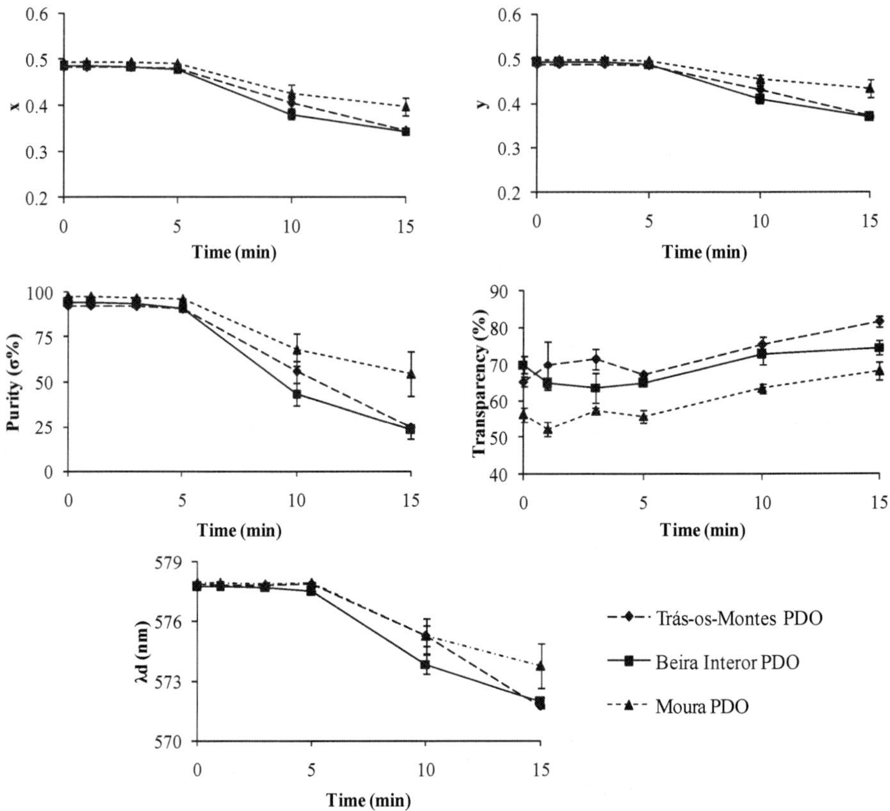

Fig. 4. Color characteristics of three Portuguese PDO olive oils subjected to increasing microwave heating exposure. Adapted from Malheiro et al. (2009).

temperature, could have naturally a higher acidity value, due to the lipases interaction with the triglycerides, and their prices reflect the care given to maintain the acidity as low as possible.

Several author's studied the effect of microwave heating on the free acidity of different vegetable oils: Portuguese virgin and extra virgin PDO olive oils (Malheiro et al., 2009); olive oil (refined + virgin olive oil) (Caponio et al., 2002); Spanish virgin olive oils (Brenes et al., 2002), and Italian extra-virgin olive oils (Cossignani et al., 1998), coconut, palm and safflower oils (Yoshida et al., 1991); peanut, high oleic sunflower and canola oils (Chiavaro et al., 2010); corn and soybean oils (Tan et al., 2001); virgin olive oil, olive oil, sunflower oil, high oleic sunflower oil, and lard (Albi et al., 1997 a,b). All the results provide clear evidence that microwave heating causes a slight increase in the free fatty acids amounts, higher than that obtained by conventional heating. Cerretani et al. (2009) reported a pronounced free acidity increase after 12 minutes of microwave heating (720 W) of extra virgin olive oil as compared to the other olive oil categories, but this occurred already at extremely high temperatures ($\geq$ 300 °C).

El-Abassy et al. (2010), using VIS Raman spectroscopy, were capable to compare free acidity increase during microwave and conventional heating in extra virgin olive oil, confirming the results obtained with the classic titration method.

2.2.2 Peroxide and ρ-Anisidine values

Oxidation is the most common chemical change in vegetable oils, promoted by several factors: oxygen/air, light, heat/temperature, trace metals, among others. It can occur through three distinct reaction pathways: 1) nonenzymatic chain autoxidation mediated by free radicals; 2) nonenzymatic and nonradical photooxidation; and 3) enzymatic oxidation. Among these, and under thermal influence, autoxidation via free radicals appears to be the key process in lipid oxidation. During the first oxidative stages hydroperoxydes are formed from unsaturated fatty acids, being the peroxide value (PV) a very sensitive indicator of the extension of primary oxidation of lipids. These unstable peroxides decompose swiftly into secondary oxidation products, particularly aldehydes, alcohols, ketones, acids, dimmers, trimers, polymers and cyclic compounds (Tan et al., 2001), including the volatile compounds responsible for off-flavors in seed oils and olive oils. Therefore, the peroxide value changes along the oxidative process, achieving a maximum due to the initial hydroperoxides formation, and then decreasing when the rate of conversion into secondary oxidation compounds surpasses the former. Therefore, the peroxide value is useful to evaluate the oxidation extent only during the inicial oxidation stages and should be complement with other evaluations. ρ-Anisidine value (ρ-AV) is an empirical determination used to assess advanced oxidative rancidity of oils. It is based on the reactiveness of the aldehyde carbonyl bond on de ρ-anisidine amine groups, leading to the formation of Schiff base that absorbs at 350 nm. It allows the estimation of secondary products of oxidation of unsaturated fatty acids, principally dienals and 2-alkenals (Labrinea et al., 2001). In opposition to peroxide value, ρ-anisidine value increases progressively during the oxidative process, which turns this determination useful to evaluate the oxidative history of the product, namely in abused oils of low peroxide value. The results however, must be interpreted carefully, as the response is more intense with di-unsaturated aldehydes than with mono-unsaturated aldehydes, which in turn are more sensitive than saturated aldehydes. Moreover, ρ-anisidine reacts with all aldehydes, irrespective of their origin. This is especially the case for some phenol compounds of virgin olive oil, such as decarboxymethyl oleuropeine dialdehyde, which could interfere in the assessment (Laguerre et al., 2007)

Tan et al. (2001) studied the effect of different microwave potencies on corn and soybean oils. For both the formation of hydroperoxides was higher at low-power setting than in the medium and high power settings. In opposition, the ρ-anisidine value increased faster at medium and high-power settings when compared with low-power setting. These results could be related to the temperatures achieved by oils at the different applied settings, indicating an effective increased oxidation at the higher power settings, despite the lower peroxide value. When the two oils are compared, soybean oil reported higher formation of secondary products of oxidation justified by its increased amount of unsaturated fatty acids, particularly linoleic and linolenic acids, both highly susceptible to oxidation. Cerretani et al. (2009) evaluated olive oils during several consecutive heating sessions and also observed that the peroxide value decreased while hydroperoxydes were transformed in secondary products of oxidation, increasing consequently the values of ρ-anisidine.

Several reports deal only with the peroxide value. Oomah et al. (1998) reported an increased peroxide value in the oil extracted from grapeseeds dried by microwave when compared with air dried ones. Caponio et al. (2003) compared olive oil, sunflower and peanut oils, and reported higher peroxide values for peanut and sunflower when compared with olive oil but the differences between the two heating systems, microwave and conventional, were reduced. While comparing different olive oil categories, Malheiro et al. (2009) observed higher peroxide values on virgin Portuguese PDO olive oils rather than in extra virgin Portuguese PDO olive oils. In this case, virgin olive oils were more vulnerable to oxidation processes than extra virgin olive oils, probably due to their different composition in antioxidant compounds such as phenolics and tocopherols which protects the oils against free radicals and oxidative pathways.

2.2.3 K_{232} and K_{270} extinction coefficients

UV spectrophotometrical analysis can provide a series of information about oil quality, its conservation status and possible changes occurred during the technological process. Over 90% of hydroperoxides formed by lipoperoxidation have a conjugated dienic system resulting from stabilization of the radical state by double bond rearrangement. These compounds absorb in the UV range (235 nm) forming a shoulder on the main absorption peak of nonconjugated double bonds (200–210 nm) (Laguerre et al., 2007). Several secondary oxidation products as well as triene conjugated double bounds absorb in the 270 nm region, and their presence is usually indicative of extensive oxidation or high thermal stress, as occurring during vegetable oil refining. For olive oil characterization the absorption at 232 nm and in the 270 nm region are of special importance, giving information on the oxidative status and enabling to distinguish virgin olive oils categories (EEC, 1991).

These extinction coefficients can provide additional information regarding oil oxidation and are influenced by the effect of microwave heating, as studied by several authors (Albi et al., 1997a; Vieira & Regitano-d'Arce, 1999; Caponio et al., 2003; Malheiro et al., 2009). Peanut oil and sunflower oil increase rapidly the K_{232}, K_{270} and ΔK values, showing that primary and secondary oxidation products are formed in greater extent after microwave heating than under conventional heating, independently of the vegetable oil (Caponio et al., 2003). Peanut oil shows to be more resistant to the formation of oxidation products then sunflower oil (Caponio et al., 2003) reporting lower values after heated for 15 minutes. The results obtained by Albi et al. (1997a) also clearly highlight that conventional heating inflicts lower damage to vegetable oils than microwave heating.

Concerning the microwave heating effect in olive oils, ΔK values increase considerably as well the other two parameters, increasing between 40.2-94% and 276-757%, respectively for K_{232} and K_{270} (Malheiro et al., 2009). Virgin PDO olive oil showed higher resistance to the temperature among the olive oils.

The addition of antioxidants to preserve edible oils, although forbidden for virgin olive oils, is a common practice in refined oils. However, no protective effect against microwave induced oxidation was observed in a supplemented corn oil, despite being more noticed when the same oil was heated in a conventional oven (Vieira & Regitano-d'Arce, 1999).

2.2.4 Oxidative stability

Oxidative stability is known as the resistance to oxidation processes established trough well defined conditions. Under normal heating conditions the oxidative oxidation process occurs slowly until a sudden increase in the oxidation rate is observed, the so called induction period, proportional to the unsaturated fatty acids type and amounts, as well as the presence of natural or added antioxidants. Accelerated experiments, by using high temperature in the presence of excess bubbling air, are usually performed to calculate and compare the induction period of different lipids.

Albi et al. (1997a) compared microwave and conventional heating (in an electric oven) in several vegetable oils (sunflower, high oleic sunflower and olive oil). A significantly reduced oxidative stability was observed in the microwave heated oils. Among the studied oils, the extra virgin olive oil exhibit better performance against oxidation with both heating methods, mainly due to the its composition, including minor compounds with antioxidant properties (phenolic compounds, carotenoids and tocopherols), lower percentage of linoleic acid (Albi et al., 1997a) and high oleic-linoleic fatty acids ratio. Similarly, the high oleic sunflower oil presented also higher oxidative stability than the sunflower oil.

Chiavaro et al. (2010) evaluated the oxidative resistance of peanut oil, high oleic sunflower oil and canola oil with increasing microwave heating times. Microwave heating during 15 minutes at (720W) inflicted an induction time reduction of nearly 26% and 23% comparing with initial values, respectively in peanut and canola oils. Concerning high oleic sunflower oil, a higher reduction of the induction time was observed (near 40%) but the final oxidative stability was still higher than that present by unheated peanut and canola oils, again a consequence of its chemical composition. Significant negative correlations between exposure time to microwave heating and the loss of oxidative stability were observed for all seed oils ($p < 0.05$ to canola oil and $p < 0.01$ to peanut and high oleic sunflower oil).

2.2.5 Polymeric compounds

Vegetable oils, mainly those rich in unsaturated fatty acids, tend to form larger molecules (polymers) when submitted to extreme conditions of temperature and time. During vegetable oils refining the development of polymers is insignificant but under consecutive frying conditions these compounds are gradually formed. Polymerization mechanism is not completely understood, but it is believed to be due to the formation of either carbon-to-carbon bonds or oxygen bridges between fatty acids, derived from terminating reactions within the oxidation pathway. When an appreciable amount of polymers is produced, an increase in viscosity is observed in the vegetable oils.

Polymeric compounds can be analyzed directly by liquid chromatography (HPSEC). When the polar compounds fraction is previously separated from the triacylglycerides, and then further separated by liquid chromatography, information on polymeric components as well as oxidized triacylglycerols, diglycerols and free fatty acid can be obtained simultaneously. However, the most frequent is to report as total polar compounds that includes all these chemical classes. This parameter is legally established with a maximum value of 25% for frying oils (Portaria No. 1135/95).

Despite the importance of these hazard compounds from a health point of view, information on total polar compounds and their constituents in microwave heated vegetable oils is scarce. No alterations in the total polar compounds were observed by Albi et al. (1997b) when vegetable oils were submitted to microwaves without being heated. Nevertheless, when microwave and conventional heating procedures were compared, a significantly higher formation of total polar compounds was observed in the microwave heated samples. Concerning to diglycerides and free fatty acids of the untreated and treated samples, no significant differences were observed as expected from the lower free acidity achieved. The amount of thermal induced alterations (polymers and dimmers) and oxidized triacylglicerols was higher in the microwave heated samples. Among the samples analyzed, sunflower and high oleic sunflower oils showed higher formation of polar compounds, probably derived from their fatty acid composition.

Dostálová et al. (2005) simulated microwave frying conditions with several vegetable oils heated during 20 minutes, at 500W, reaching nearly 200°C. The formation of polymers was observed for all vegetable oils with different rates depending on their fatty acid composition, as usual under conventional frying. Indeed, polymerization is known to arise faster in vegetable oils with higher linoleic acid content. As an adequate safety limit 10 to 12% of polymeric compounds is usually accepted, although the legal level of total polar compound being 25%. On microwave heating these values were not achieved, even in the samples submitted for 40 minutes. Therefore, if the oil is to be rejected after the microwave treatment, no safety issues are expected. When the oil is intended to be reserved for further use, then a more detailed study is still necessary in order to estimate the number of frying sessions or total heating time allowed, as usual in conventional frying.

2.3 Effect of microwave heating on chemical composition of edible oils
2.3.1 Fatty acids composition
Fatty acids are the main constitutes of the saponificable fraction of vegetable oils, being mostly conjugated with glycerol as triacylglycerides. Although, some individual fatty acids are of particular importance for nutritional purposes, they are frequently grouped in classes according with the unsaturation degree of the hydrocarbon chain, as saturated (SFA), monounsaturated (MUFA), and polyunsaturated fatty acids (PUFA). The reaction of oxygen with unsaturated fatty acids is the major cause of deterioration of lipids or lipid-containing foods leading to losses in quality and nutritional value, and to the development of off-flavors and hazard compounds.

Several works with vegetable oils investigated the effect of microwave heating in some of these fatty acid fractions, reporting a higher nutritional quality loss with microwave heating when compared to conventional heating (Caponio et al., 2003; Albi et al., 1997a).

Caponio et al. (2003) evaluated three vegetable oils (virgin olive oil, peanut and sunflower oils) by conventional electric oven and microwave heating and reported that while the saturated fraction did not suffer significant changes after heating, both unsaturated and polyunsaturated fractions were significantly decreased. Moreover, microwave heating gave rise to an increased oxidative degradation when compared to conventional electric oven, with significantly lower amounts of both mono- and polyunsaturated fatty acids (P < 0.05) than in conventionally heated oils, higher *trans*-isomers formation, and higher amounts of non-eluted materials (thermal-oxidized and polymerized materials). When the different oils were compared, the unsaturation degree was reported as influencing the non-eluted material proportion, with the vegetable oils with less polyunsaturated content, like olive oil, being less degraded (Caponio et al., 2003). Nevertheless, the antioxidant compounds present in virgin olive oil could also contribute to a lower formation of thermo-oxidized materials, by reducing the propagation of the oxidative processes. Studying the same oils, plus soybean oil, Hassanein et al. (2003) reported the decrease of oleic and linoleic acid in all samples.

The ratio between linoleic and stearic acids in corn and soybean oils continuously decreased during microwave heating exposure at different power settings (Tan et al., 2001). This is associated to the decrease of unsaturated fatty acids, like linoleic acid, and the increase in the percentage of the sum of saturated fatty acids (Tan et al., 2001) as also reported by Mahmoud et al. (2009). This ratio is often used to evaluate the extent of deterioration caused by oxidation once that linoleic acid is susceptible to oxidation whereas palmitic acid is quite stable to oxidation (Tan et al., 2001).

2.3.2 Sterols composition

Phytosterols are important component of vegetable oils. They are constant in botanic family and are also influenced by cultivar, oil extraction procedure refining procedures and storage conditions, as well as climatic and agronomic conditions (Cañabate-Díaz et al., 2007). In olive oil, phytosterols are the major constitutes of the unsaponifiable fraction (representing about 20%) and β-sitosterol is the most abundant phytosterol followed by campesterol, stigmasterol, Δ^5-avenasterol, Δ^7-avenasterol in low amounts. These compounds are very important once possess properties like anti-inflammatory, antibacterial, antifungical, antiulcerogenic and antitumoral activity (Li & Sinclair, 2002).

The effect of microwave heating in the sterols fraction of vegetable oils is not extensively studied. Albi et al. (1997a) by studying extra virgin olive oil, olive oil, sunflower oil and high oleic sunflower oil did not found significant differences after microwave heating in all the samples. Sterol compounds seem to be resistant to degradation through microwave radiation.

2.3.3 Tocopherols

Tocopherols and tocotrienols are natural lipophilic antioxidants, known to inhibit lipid oxidation in fats and oils by modifying the radical chain autoxidation process. These compounds present also an important vitaminic action (vitamin E), being essential from a nutritional point of view.

The effect of microwave heating on tocopherols content of seed oils (Yoshida et al., 1991a,b; Hassanein et al., 2003; Albi et al., 1997b) and different categories of olive oils (Albi et al., 1997b; Brenes, et al., 2002; Malheiro et al., 2009) was studied.

Yoshida et al. (1991) showed that tocopherols gradually decrease with microwave heating time, and that each tocopherol type exhibited different antioxidant activities. In particular, since sunflower oil contains mainly alpha-tocopherol, exhibiting the lowest antioxidant activity, the rate of tocopherol degradation was higher when compared with soybean oil, containing gamma- and delta-tocopherols. These authors also highlight that the free fatty acid amount, present in the unheated oil or formed during the microwave heating process, is an important parameter affecting tocopherols losses (Yoshida et al., 1992).

After microwave heating (18 min, 227 °C) of sunflower, soybean, peanut and a mixture of soybean and peanut oils, Hassanein et al. (2003) observed that tocopherol content and exposure time were inversely proportional but still detectable at the end of the process (Hassanein et al., 2003). These results are in accordance to those obtained by Albi et al. (1997b) that reported 72 and 85% of tocopherols losses after microwave heating (120 min, half-power) of sunflower oil and high oleic sunflower oil, respectively.

With the same exposure time and temperatures, extra virgin olive oil and olive oil (refined olive oil + virgin olive oil) α-tocopherol, the main tocopherol, was completely undetectable (Albi et al., 1997b). Malheiro et al. (2009) also reported the absence of α-tocopherol in Portuguese virgin and extra virgins PDO olive oils after 15 minutes (1000W) of exposure time, while Brenes et al. (2002) reported 60% to 80% losses after 10 minutes (500W). This high degradation rate of alpha-tocopherol in olive oil is also frequent in conventional frying mainly because olive oil contains smaller amounts of tocopherols than other vegetable oils, being its natural antioxidant activity attributed to other minor components.

When microwave heating was compared with conventional heating, higher degradation rates were observed in the former (Albi et al., 1997b, Brenes et al., 2002). Brenes et al. (2002) reported losses after 10 minutes of microwave heating (500W) equivalent to those obtained after almost 10h (180°C) under conventional oven heating.

2.3.4 Phenolic compounds

Virgin olive oil is a source of at least 30 different phenolic compounds in which oleuropein, tyrosol and hydroxytyrosol are present in higher amounts (Tuck & Hayball, 2002). Phenolic compounds contributes to the quality of virgin olive oils (Servili & Montedoro, 2002), and are responsible for a series of important characteristics and properties, such as color, texture and taste (Marsilio et al., 2001), antioxidant capacity (Ben Othman et al., 2009), antimicrobial activity (Sousa et al., 2006) and protection against micotoxins effects (Beekrum et al., 2003). As antioxidants, phenolic compounds contribute to a higher stability against oxidation by protecting the target lipids from oxidation initiators as well as stalling the propagation phase of the oxidative process (Laguerre et al., 2007). Refined oils are deprived of most of the natural phenolics during this technological process.

When exposed to 120 minutes of microwave heating, virgin olive oil and olive oil reported losses around 96 and 85%, respectively, in their total polyphenols content. Under conventional heating, only a 10% reduction was observed in the virgin olive oil, compared with a 64% reduction in olive oil (Albi et al., 1997a). Despite having used long heating periods in both experiments, the results highlight a significant reduction in total polyphenol content. Following these studied, Brenes et al. (2002) submitted virgin olive oils from two Spanish cultivars (Picual and Arbequina) to microwave heating during 5 and 10 minutes (500W) in order to simulate domestic conditions and evaluated the individual polyphenols. Under these conditions, the phenolic profile of both oils suffered minor changes due to microwave radiation. Hydroxytyrosol, its oleosidic forms (dialdehydic form of elenoic acid linked to hydroxytyrosol and oleuropein aglycon), and 4-(acetoxyethyl)-1,2-dihydroxybenzene suffered a 20-30% decrease when compared with the unheated samples (Brenes et al., 2002). A slight increase in ligustroside aglycon content and a rising in the formation of oxidized dialdehydic form of elenolic acid linked to tyrosol was observed in the samples heated for 10 minutes. Lignans like 1-acetoxypinoresinol and pinoresinol were not affected by microwave heating, confirming that these compounds are highly stable to microwave heating (Yoshida et al., 1995). The authors concluded that, when compared with frying, microwave heating is a less destructive culinary method regarding the phenol compounds (Brenes et al., 2002).

Pomace olive oil, as other vegetable oils submitted to refining, are naturally deprived or reduced in phenolic compounds, being unusual to report their phenolic content in heating experiments. Olive oil, a blend of refined olive oil with virgin olive oil, also reported lower phenolic compounds compared to extra virgin olive oil. Similarly, extra virgin olive oil, reported higher amounts of individual and total phenolic compounds. Cerretani et al. (2009) compared these olive oil categories under microwave heating and observed around 40% lost after 6 minutes of microwave, and total depletion after 15 minutes. Under the same heating conditions, a 28% loss was observed in the extra virgin olive oil at 6 minutes, and 83% after 15 minutes. The reported losses at 15 minutes are also related to the increasing temperatures observed (313 °C at 15 minutes of heating), contributing to the deterioration of phenolic compounds. Concerning the individual phenolic compounds that compose the extra virgin olive oil, lignans like (+)-pinoresinol and 1-acetoxypinoresinol were more stable among all the phenolic compounds detected after 15 minutes of microwave heating. Such results are in accordance with those obtained by Brenes et al. (2002). This is an important result once that these compounds are reported to exhibit beneficial effects on human health (Schouw et al., 2000).

In Fig. 5 the relative losses of some phenolic compounds with microwave heating are reported.

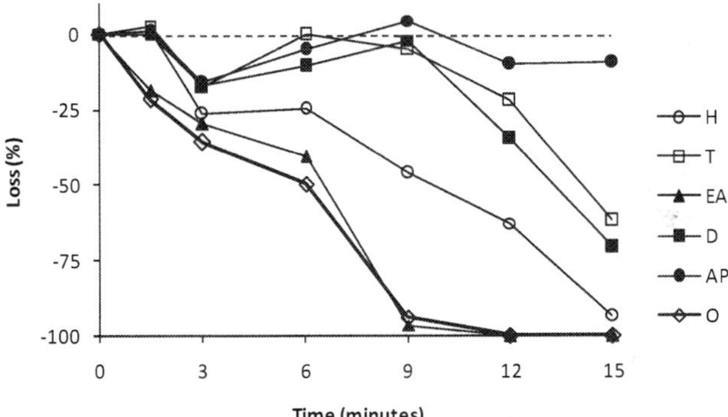

Fig. 5. Effect of microwave heating in some phenolic compounds of extra virgin olive oil.
H – Hydroxytyrosol; T – Tyrosol; EA – Elenolic acid; D – Decarboxymethyl-oleuropein aglycon; AP – 1-(+)-acetoxypinoresinol; and O – Oleuropein aglycone. Adapted from Cerretani et al. (2009).

At the first heating minutes, until 3 minutes, a constant decrease in the phenolic compounds of the extra virgin olive oil was observed, followed by abrupt losses mainly in hydroxytyrosol, oleuropein and elenolic acid. Amongst the phenolic compounds studied, hydroxytyrosol and oleuropein were those whith higher losses. Such fact could be attributed to their antioxidant capacity, once that *ortho*-diphenols present higher capacity to protect lipid targets from oxidation (Bester et al., 2008; Carrasco-Pancorbo et al., 2005). These compounds were the first to counterattack the oxidation process suffering higher losses than the remaining phenolic compounds.

2.3.5 Chlorophylls and carotenoids

Chlorophylls and carotenoids are the main responsible for the color of olive oils that changes from yellow-green to greenish gold (Criado et al., 2008). Green color is attributed to chlorophylls while the yellow color is mainly related with carotenoids content. These pigments play an important role in the oxidative stability of oils (β-caroten and others carotenoid compounds) but chlorophylls play a dualistic role, behaving as antioxidants in the dark, but becoming pro-oxidative agent when exposed to light (Sapino et al., 2005). These pigments are absent in most refined vegetable oils, being eliminated during the discoloration step (Albi et al., 1997a).

Concerning the effect of microwave heating in pigments of several categories of crude oils, the results are conclusive, microwave heating leads to the destruction of large quantities of chlorophylls, flavonoids and carotenoids, as reported in several works (Albi et al., 1997a; Oomah et al., 1998, Khan et al., 2010; Malheiro et al., 2009; El-Abassy et al., 2010). Grapeseed crude oil extracted from microwave dried seeds had reduced pigmentation at 410 nm and 670 nm, corresponding to total chlorophylls and carotenois, respectively, when compared with conventional drying (Oomah et al., 1998). Albi et al. (1997a) reported a similar behavior for both extra virgin olive oil and olive oil (refined + virgin olive oil) when exposed to microwave energy. No effects on chlorophylls and carotenoids contents were reported when the oils were put in contact with microwave energy without temperature increasing. When heated, higher generalized losses of carotenoids and chlorophyls were observed in the microwave heated samples when compared with conventional heated ones, more noticeable in the carotenoids (Albi et al., 1997a). A decrease in chlorophylls and carotenoids contents was also reported with extra virgins and virgin PDO Portuguese olive oils submitted to increasing microwave exposure (Malheiro et al., 2009), again higher for carotenoids than for chlorophylls: after 15 minutes of microwave heating, chlorophylls content decreased between 33.1 and 52.9% while carotenoids suffers considerable losses, between 77.3 and 89.3%. The color losses discussed previously can be chemically supported by these observations (see topic 2.1.2).

El-Abassy et al. (2010) reported that carotenoids degradation seems to be related with the exposure time and less with the final temperature reached. The authors monitored carotenoids degradation by VIS Raman spectroscopy and observed that the Raman bands corresponding to carotenoids (1008, 1150, and 1525 cm^{-1}) completely disappeared at 203 °C in conventional heating, being observed even up to 225 °C with microwave heating. In this case, microwave oven needed 12 minutes to reach the 225 °C while in the conventional oven the 203 °C were reached after 80 minutes of heating.

These results and facts show and corroborate that chlorophyll and carotenoids are thermo-labile pigments and that microwave heating induces a decrease in their content, causing higher damages in carotenoids content. Whether these observations are more dependent on the heat exposure time or the temperatures achieved within the process is an interesting issue that deserves more study.

3. Conclusions

Microwave heating features modern lifestyle. The use of this technology enables a significant reduction in the time dedicated to prepare meals and opens new perspectives at the industrial level. In this chapter we report the combined effect of time/applied potency of microwave heating in comparison to conventional electric oven on the quality of vegetable oils.

In general, the degradation pattern of vegetable oils under microwave heating was similar to that expected from other conventional heating methods, including oxidation, hydrolysis and polymerization. Nevertheless, microwave heating induced higher degradation extent, as confirmed by the physical and chemical evaluations reported, reducing the nutritional value of lipids. The combination of temperature and energy effects induced by microwaves, which could have strengthened the heating effect might lead to zonal overheating in the oils, a situation that deserves technological improvement. Nevertheless, most of the published studies compared high exposure times and temperatures, unusual on microwave domestic procedures, except for microwave frying. Therefore, the vegetable oils behavior under real cooking practices, power settings, and combined interaction of different ingredients, needs to be further exploit in order to provide concise information from nutritional, technological and health points of view. Based on these prospects and scientific findings the use of microwave is not discouraged, but vegetable oils heating should be reduced to the minimum, in order to reduce the degradation extent of important compounds, as lipossoluble vitamins, essential fatty acids, and phenolics, while reducing the formation of potentially hazard components, the oxidized lipids.

4. References

Albi, T.; Lanzón, A.; Guinda, A.; Pérez-Camino, M.C. & León, M. (1997a). Microwave and conventional heating effects on some physical and chemical parameters of edible fats. *Journal of Agricultural and Food Chemistry*, 45, 3000-3003.

Albi, T.; Lanzón, A.; Guinda, M.; Léon, M. & Pérez-Camino, M.C. (1997b). Microwave and conventional heating effects on thermoxidative degradation of edible oils. *Journal of Agricultural and Food Chemistry*, 45, 3795-3798.

Beekrum, S.; Govinden, R.; Padayachee, T. & Odhav, B. (2003). Naturally occurring phenols: a detoxification strategy for fumonisin B-1. *Food Additives & Contaminants: Part A – Chemistry*, 20, 490-493.

Ben Othman, N.; Roblain, D.; Chammen, N.; Thonart, P. & Hamdi, M. (2009). Antioxidant phenolic compounds loss during the fermentation of Chétoui olives. *Food Chemistry*, 116, 662-669.

Bešter, E.; Butinar, B.; Bučar-Miklavčič, M. & Golob, T. (2008). Chemical changes in extra virgin olive oil from Slovenian Istra after thermal treatment. Food Chemistry, 108, 446-454.

Brenes, M.; García, A.; Dobarganes, M.C.; Velasco, J. & Romero, C. (2002). Influence of thermal treatments simulating cooking processes on the polyphenol content in virgin olive oil. *Journal of Agricultural and Food Chemistry*, 50, 5962-5967.

Cañabate-Díaz, B.; Carratero, A.S.; Fernández-Gutiérrez, A.F.; Vega, A.B.; Frenich, A.G.; Vidal, J.L.M. & Martos, J.D. (2007). Separation and determination of sterols in olive oil by HPLC-MS. *Food Chemistry*, 102, 593-598.

Caponio, F.; Pasqualone, A. & Gomes, T. (2002). Effects of conventional and microwave heating on the degradation of olive oil. *European Food Research and Technology*, 215, 114-117.

Caponio, F.; Pasqualone, A. & Gomes, T. (2003). Changes in the fatty acids composition of vegetable oils in model doughs submitted to conventional or microwave heating. *International Journal of Food Science and Technology*, 38, 481-486.

Carrasco-Pancorbo, A.; Cerretani, L.; Bendini, A.; Segura-Carretero, A.; Del Carlo, M.; Gallina-Toschi, T.; Lercker, G.; Compagnone, D. & Fernández-Gutiérrez, A. (2005). Evaluation of the antioxidant capacity of individual phenolic compounds in virgin olive oil. *Journal of Agricultural and Food Chemistry*, 53, 8918-8925.

Cerretani, L.; Bendini, A.; Rodeiguez-Estrada, M.T.; Vittadini, E. & Chiavaro, E. (2009). Microwave heating of different commercial categories of olive oil: Part I. Effect on chemical oxidative stability indices and phenolic compounds. *Food Chemistry*, 115, 1381-1388.

Chavan, R.S. & Chavan, S.R. (2010). Microwave baking in food industry: a review. *International Journal of Dairy Science*, 5, 113-127.

Chiavaro, E.; Rodriguez-Estrada, M.T.; Vittadini, E. & Pellegrini, N. (2010). Microwave heating of different vegetable oils: Relation between chemical and thermal parameters. *LWT – Food Science and Tecnology*, 43, 1104-1112.

Cossignani, L.; Simonetti, M.S.; Neri, A. & Damiani, P. (1998). Changes in olive oil composition due to microwave heating. *Journal of the American Oil Chemists' Society*, 75, 931-937.

Criado, M. -N.; Romero, M. -P.; Casanovas, M. & Motilva, M.J. (2008). Pigment profile and colour of monovarietal virgin olive oils from Arbequina cultivar obtained during two consecutive crop seasons. *Food Chemistry*, 110, 873-880.

Dostálová J.; Hanzlík, P.; éblová Z. & Pokorný J. (2005). Oxidative changes of vegetable oils during microwave heating. *Czech Journal of Food Science*, 23, 230–239.

El-Abassy, R.M.; Donfack, P. & Materny, A. (2010). Assessment of conventional and microwave heating induced degradation of carotenoids in olive oil by VIS Raman spectroscopy and classical methods. *Food Research International*, 43, 694-700.

Eskin, N.A.M.; McDonald, B.E.; Przybylski, R.; Malcolmson, L.J.; Scarth, R.; Mag, T.; Ward, K.; & Adolph, D. (1996). Canola Oil. In *Bailey's Industrial Oil and Fat Products, Vol. 2, Edible Oil and Fat Products: Oil and Oilseeds*, 5th ed.; Y.H. Hui (Ed.), 1-95, Wiley, John & Sons, ISBN 9780471594246, New York, USA.

EEC, European Economic Community (1991). Commission Regulation 2568/91 of 11 July 1991 on the characteristics of olive oil and olive-residue oil and on the relevant methods of analysis. *Official Journal of European Communities*, L248.

Hassanein, M.M.; El-Shami, S.M. & El-Mallah, M.H. (2003). Changes occurring in vegetable oils composition due to microwave heating. *Grasas y Aceites*, 54, 343-349.

Khan, M.N.; Sarwar, A. & Wahab, M.F. (2010). Chemometric assessment of thermal oxidation of some edible oils. Effect of hot plate heating and microwave heating on physicochemical properties. *Journal of Thermal Analysis and Calorimetry*, in Press.

Labrinea, E.P.; Thomaidis, N.S.; & Georgiou, C.A., 2001. Direct olive oil anisidine value determination by flow injection. *Analytica Chimica Acta*, 448, 201-206.

Laguerre, M.; Lecomte, J. & Villeneuve, P. (2007). Evaluation of the ability of antioxidants to counteract lipid oxidation: Existing methods, new trends and challenges. *Progress in Lipid Research*, 46, 244-282.

Li, D. & Sinclair, A.J. (2002). Macronutrient innovations: The role of fats and sterols in human health. *Asia Pacific Journal of Clinical Nutrition*, 11, S155-S162.

Mahmoud, E.A.E-M.; Dostálová, J.; Pokorný, J.; Lukešová, D. & Doležal, M. (2009). Oxidation of olive oils during microwave and conventional heating for fast food preparation. *Czech Journal of Food Sciences*, 27, S173-S177.

Malheiro, R.; Oliveira, I.; Vilas-Boas, M.; Falcão, S.; Bento, A. & Pereira, J.A. (2009). Effect of microwave heating with different exposure times on physical and chemical parameters of olive oil. *Food and Chemical Toxicology*, 47, 92-97.

Marsilio, V.; Campestre, C. & Lanza, B. (2001). Phenolic compounds change during California-style ripe olive processing. *Food Chemistry*, 74, 55-60.

Mutyala, S.; Fairbridge, C.; Pare, J.R.J.; Bélanger, J.M.R.; Ng, S. & Hawkins R. (2010). Microwave applications to oil sands and petroleum: A review. *Fuel Processing Technology*, 91, 127-135.

Oomah, B.D.; Liang, J.; Godfrey, D. & Mazza, G. (1998). Microwave heating of grapeseed: effect on oil quality. *Journal of Agricultural and Food Chemistry*, 46, 4017-4021.

Osepchuk, J.M. (1984). A history of Microwave heating application. *IEEE Transactions on Microwave theory and Techniques*, MTT-32, 9, 1200-1224.

Portaria No. 1135/95, 1995. Diário da República. I Série-B, No. 214, p. 5836.

Sapino, S.; Carlotti, M.E.; Peira, E. & Gallarate, M. (2005). Hemp-seed and olive oils: their stability against oxidation and use in O/W emulsions. *Journal of Cosmetic Science*, 56, 227-251.

Sahin S. & Sumnu S.G (2009). Alternative Frying Technologies, In: *Advances in Deep-Fat Frying of Food*, Serpil Sahin and Servet Gülüm Sumnu (Ed.), 289-302, CRC Press, ISBN 9781420055580, Boca Raton.

Schouw, Y.T.; Kleijn, M.J.J.; Peeters, P.H.M. & Grobbee, D.E. (2000). Phytoestrogens and cardiovascular disease risk. *Nutrition, Metabolism and Cardiovascular Diseases*, 10, 154-167.

Servilli, M. & Montedoro, G. (2002). Contribution of phenolic compounds to virgin olive oil quality. *European Journal of Lipid Science and Technology*, 104, 602-613.

Sousa, A.; Ferreira, I.C.F.R.; Calhelha, R.; Andrade, P.B.; Valentão, P.; Seabra, R.; Estevinho, L.; Bento, A. & Pereira, J.A. (2006). Phenolics and antimicrobial activity of traditional stoned table olives "alcaparra". *Bioorganic & Medicinal Chemistry*, 14, 8533-8538.

Tan, C.P.; Che Man, Y.B.; Jinap, S. & Yusoff, M.S.A. (2001). Effects of microwave heating on changes in chemical and thermal properties of vegetable oils. *Journal of the American Oil Chemist's Society*, 78, 1227-1232.

Tuck, K.L. & Hayball, P.J. (2002). Major phenolic compounds in olive oil: metabolism and health effects. *Journal of Nutritional Biochemistry*, 13, 636-644.

Vieira, T.M.F.S. & Regitano-d'Arce, M.A.B. (1999). Ultraviolet spectrophotometric evaluation of corn oil oxidative stability during microwave heating and oven test. *Journal of Agricultural and Food Chemistry*, 47, 2203-2206.

Yoshida, H.; Hirooka, N. & Kajimoto, G. (1991a). Microwave heating effects on relative stabilities of tocopherols in oils. *Journal of Food Science*, 56, 1042-1046.

Yoshida, H.; Shigezaki, J.; Takagi, S. & Kajimoto, G. (1995). Variations in the composition of various acyl lipids, tocopherols and lignans in sesame seed oils roasted in a microwave oven. *Journal of the Science of Food and Agriculture*, 68, 407-415.

Yoshida, H.; Tatsumi, M. & Kajimoto, G. (1991b). Relationship between oxidative stability of vitamin E and production of fatty acids in oils during microwave heating. *Journal of the American Oil Chemists' Society*, 68, 566-570.

Yoshida, H.; Tatsumi, M. & Kajimoto, G. (1992). Influence of fatty acids on the tocopherol stability in vegetable oils during microwave heating. *Journal of the American Oil Chemists' Society*, 69, 119-125.

27

Experimental and Simulation Studies of the Primary and Secondary Vacuum Freeze Drying at Microwave Heating

Józef Nastaj and Konrad Witkiewicz
West Pomeranian University of Technology, Szczecin
Poland

1. Introduction

Freeze drying (FD) is a dehydration method used to obtain high-quality products, mainly foodstuffs, biomaterials, pharmaceuticals and other thermolabile materials. Compared with standard methods, the freeze drying technique ensures retaining original shape, color and texture of the product as well as the preservation of its flavor, nutritive content and biological activity. However, FD is an expensive and lengthy dehydration process carried out in three stages: pre-freezing, primary drying when ice sublimation takes place under vacuum, followed by desorption of residual, unfreezable bounded water during the secondary stage.

Material must be first cooled below its triple point to obtain frozen state. In the case of food and biomaterials freezing is done rapidly at temperature between -50°C and -80°C, below eutectic point to avoid destruction of cell walls by ice crystals. During the primary FD stage, the frozen water in dried material pores sublimates from the ice front, diffuses throughout the dried layer to the sample surface and next deposits on the condenser surface. The sublimation of water takes place in the range of temperature and pressure below the triple point (for water 273.16 K and 611.73 Pa, respectively). After primary drying, residual moisture content may be as high as 7%. Secondary drying is intended to reduce this to an optimum value for material stability – usually with moisture content between 0.5 and 2.0%. The typical freeze-dried products have a porous, nonshrunken structure resulted from structural rigidity achieved by frozen water and can be therefore easily rehydrated.

The use of conventional FD on industrial scale is restricted to rather high-value products. Recent research is being focused on reducing operating costs of FD by intensifying heat and mass transfer in dried material. Proposed various heating methods, cycled pressure strategies and formulated optimal control policies are limited by temperature constraints (Liapis & Bruttini, 2006). They must be set to avoid ice melting and scorching of exposed dried material layer. The major difficulty of process optimization results from the fact that imposed thermal gradient has direction opposite to vapor concentration gradient. Moreover, dried layer acts as a thermal insulation for heat fluxes being transferred toward frozen layer. New FD method that overcomes these disadvantages is microwave freeze drying (MFD). Microwaves penetrate very well into ice and supply energy for sublimation volumetrically and selectively, bypassing the problem of heat transport through the dried layer of the material. Microwaves can heat without the aid of thermal gradients what has a positive effect on product quality. With recent development in microwave hardware, the cost of MFD equipment is currently not too high what makes this method a promising alternative to conventional technologies.

Although MFD can greatly improve the drying rate, there are still many problems to be resolved in the practice. Application of microwave heating is limited because of difficulty to control the final product quality. Therefore constraints should be set to avoid specific problems of MFD such as corona discharge and non-uniform heating, which cause ice melting and overheating. To analyze and overcome these problems the experimental and theoretical studies should be performed to obtain optimal control policy of MFD in relation to particular material being dried.

Applying microwave energy in the MFD process may cause appearance of plasma discharge in the vacuum chamber. This phenomenon happens when the electric field intensity E is above the threshold value what may seriously damage the final product. Therefore upper limit on the electric field strength should be set and taken into account in mathematical modeling.

When microwaves, usually at standard frequency f=2.45 GHz, are applied in MFD, electric field strength E is the controllable process parameter which reflects heating intensity. The distribution of electric field strength in the microwave applicator and within the dried sample can be determined by solving Maxwell's equations. Such approach is mathematically difficult to apply and especially complicated for multimode applicators. Another difficulty results from the fact that the freeze-dried material consists of two layers: the frozen and the dried one, which are distinct dielectrics.

The exponential decay of the electric field intensity in a dielectric described by Lambert's law is usually omitted in MFD and simplified by the electric field strength value averaged for the whole material sample (Ma & Peltre, 1975a, 1975b; Péré et al., 2001; Schiffmann, 2006). Typical biomaterials and foodstuffs containing water in a frozen form as well as dried are characterized by extreme high penetration depth in order of few or even dozens of meters. For this reason the electric field strength distribution in the product formed as a thin layer or placed in vials is negligible.

For specific microwave system of constant frequency f, the efficiency of microwave energy absorption and dissipation into heat in the frozen and dried material layers depends on its dielectric properties: relative dielectric constant ε' and relative loss factor ε''. Both parameters vary with temperature therefore this dependency should be taken into consideration in the modelling for both sample regions.

Application of the microwave energy in MFD has been studied since the mid-20th century. The mathematical models of the MFD process are usually formulated with pseudo steady state assumption, and derived on the basis of heat and mass transfer analysis. Ma and Peltre in the 1975 described one of the first comprehensive mathematical models of the MFD of foodstuff. They obtained good agreement with experimental results of raw beef drying. Some researchers took into consideration the sublimation-condensation phenomena in MFD process (Liapis & Bruttini, 2006; Wang & Shi, 1998). Many mathematical analyses presented in papers concerning freeze drying at conventional heating can be adapted in the modeling of MFD after including the volumetric heat source term in model equations set (Liapis & Bruttini, 1996).

In this chapter the main ideas in MFD modeling are presented and key mechanisms of heat and mass transport governing the process are explained. The complex mathematical model of the primary and secondary FD stages at microwave heating is formulated to derive optimal control policy of the process depending on thermophysical properties and thickness of material being dried, total pressure and input microwave power. One-dimensional two-region model of the primary MFD was developed and then solved using various numerical approaches such as Landau's transformation (LT) and Variable Time-Step (VTS) methods. Varying during the process sublimation front temperature caused by varying with time vapor diffusional mass transfer resistances was taken into account. Some of the simulated drying curves were compared with experimental results giving fairly good agreement.

The mathematical model of the secondary MFD was also developed and solved using the Numerical Method of Lines (NMOL). Pressure drop in the material was taken into account and calculated using Ergun equation. As a result of the both models solution, the moisture contents and the temperature distributions in drying material were obtained.

The analysis performed basing on mathematical modeling enables computation of optimal ranges of significant material and process parameters, which ensure obtaining good product quality and optimal MFD drying times.

2. Experimental

2.1 MFD equipment

Experimental investigations of the primary freeze drying of random solids at microwave heating were performed by means of the set-up composed of the microwave circuit, the vacuum system, the refrigeration system, the temperature and weight measurements devices and the data acquisition system (Fig. 1).

A cylindrical Teflon container filled with material to be dried is hanged on the extensometer balance inside the vacuum chamber. Temperature of dried material is measured by the fluoroptic thermometer (FOT Lab Kit – Luxtron Corp.). A probe with a phosphorus sensor inserted to the sample is connected by optical fiber, via vacuum feedthrough with module m600 outside the chamber. The module emits photon pulses towards sensor and simultaneously records the decay of returning fluorescence signal which varies with material temperature. On this basis the module estimates measured temperature with a maximum absolute accuracy of 0.1°C. The fluoroptic thermometer essentially does not interfere with electromagnetic field in the applicator and is transparent for microwaves.

Temperature inside the chamber is controlled by the refrigeration system with a refrigerant circulating in the vacuum chamber's jacket. Sublimated vapors are removed from the chamber by the cold trap cooled by means of liquid nitrogen. The vacuum pump is operating constantly whilst self-regulated purge valve maintains pressure in the vacuum chamber at the level of 100 Pa.

Generated microwaves of 2450 MHz frequency are transmitted via the waveguide, the coaxial cable, the directional coupler and the vacuum feedthrough to the applicator inside the vacuum chamber. Microwave power reflected in the applicator returns to the directional coupler which directs it to the dummy load where is totally dissipated. The sample is inserted into the microwave applicator which is constructed as a section of rectangular brazen waveguide and acts as a mono-modal resonant cavity (Fig. 2). One of applicator walls is the movable tuner and its position can be adjusted remotely by stepping motor. Microwave is transmitted via the coaxial cable to the type N connector coupled with the antenna. Electromagnetic wave propagates inside the applicator in TE_{01n} mode where n is the number of half wavelengths.

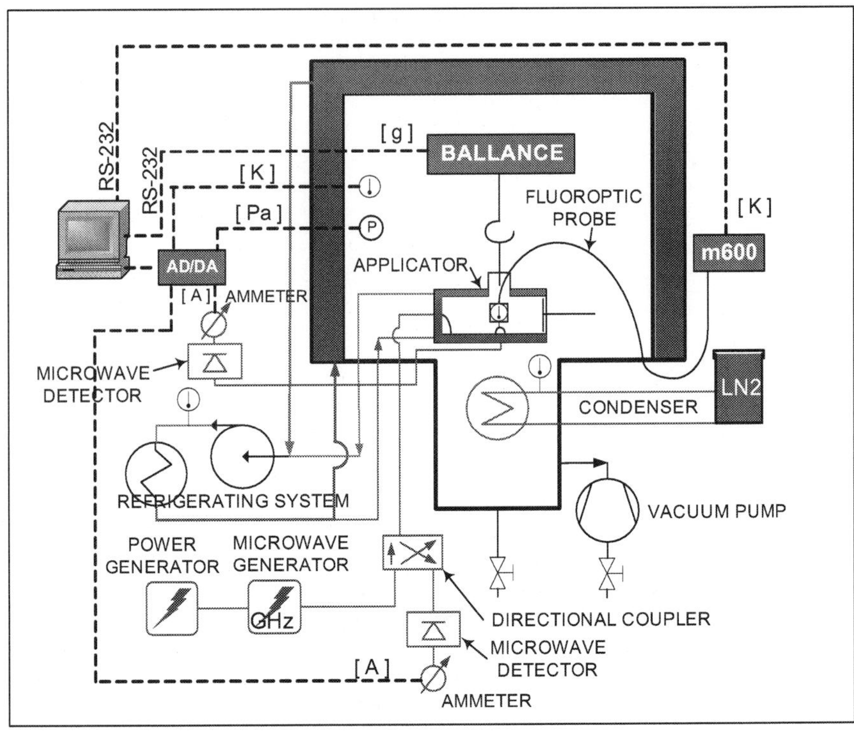

Fig. 1. The experimental set-up for investigations of the primary MFD

Theoretical wavelength in air for this construction amounts to about 15.5 cm, according to the following equation (Metaxas, 1996):

$$\lambda_w = \lambda_0 \bigg/ \sqrt{1 - \left(\frac{\lambda_0}{2a}\right)^2} \tag{1}$$

where: λ_0 – wavelength in air (12.24 cm for frequency f=2450 MHz); a – height of the waveguide (5 cm). A processed sample should be located at a distance of ¾ wavelength where electric field pattern achieves maximal values.

Tuning of the applicator is performed on the basis of readings of the microwave detector, which is coupled with a magnetic loop under the sample. The current intensity of the detector is directly proportional to electric field strength in the applicator. The applicator is considered to be tuned when current signal of the microwave detector coupled with the magnetic loop under the sample achieves maximum, which means maximal electric field strength in the sample.

2.2 Methodology

Experiments of the primary microwave freeze drying were performed it two stages: first with measurements of sample mass decrement and second with only sample temperature

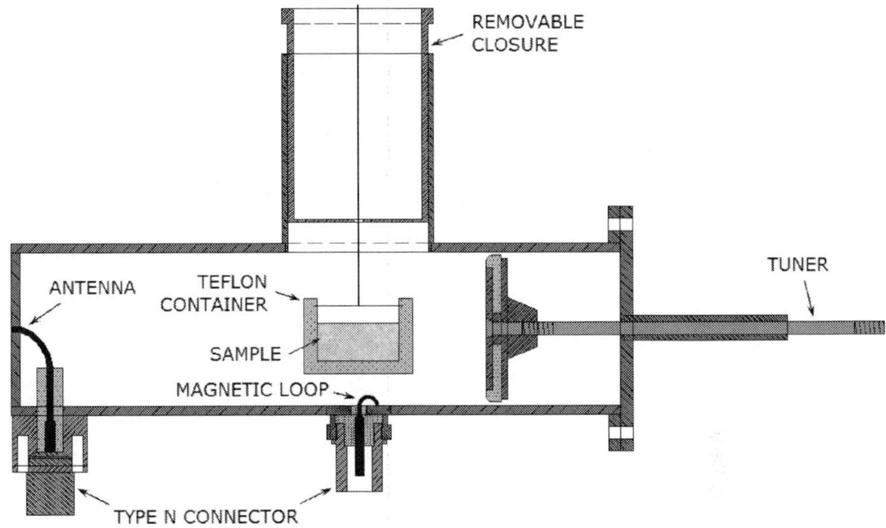

Fig. 2. Cross-section of a waveguide applicator

measurements. It was impossible to measure simultaneously both above mentioned values in the same experimental run. When temperature was measured, a fluoroptic probe was placed in the center of the sample.

After freezing of the wet material in the vacuum chamber at the temperature about -50 °C, the apparatus is sealed and the vacuum pump started. When demanded process pressure in the chamber is achieved, the microwave generator is turned on.

When a dried sample is inserted through the removable closure into the applicator, the waveguide is usually shortened and cavity needs to be tuned to a TE_{102} mode. However, as dielectric properties of a sample change as a result of moisture content or temperature fluctuations, the cavity length needs to be adjusted accordingly.

During the single experimental run, the sample temperature or sample weight decrement as well as temperature and pressure in the vacuum chamber are recorded by DASYLab data acquisition program. Additionally, continuous measurements of current signal of the microwave detector coupled with magnetic loop inside the applicator indicate when the resonant cavity needs to be tuned during the experiments. Varying during an experimental run current signal may indicate that electric field strength in a dried sample is not constant. These changes were not significant, but taking them into account in mathematical modeling would require derivation of electric field strength distribution in sample layers.

The average power absorbed volumetrically in the dried material (internal heat source capacity) can be calculated provided that $|E|$ - the magnitude of the electric field strength E inside the sample is known (Metaxas, 1996; Schiffmann, 2006):

$$Q_v = \frac{1}{2}\omega\varepsilon_0\varepsilon'\tan\,\delta|E|^2 = \pi f \varepsilon_0\varepsilon'\tan\,\delta|E|^2 \tag{2a}$$

or

$$Q_v = \frac{1}{2}\omega\varepsilon_0\varepsilon''|E|^2 = \pi f \varepsilon_0\varepsilon''|E|^2 \tag{2b}$$

where: $\tan\delta = \varepsilon''/\varepsilon'$. In the case of drying of small samples, we can assume uniform electric field inside the material, and then $|E| = E_0 = \sqrt{2}E_{rms.}$ where E_0 is the peak magnitude of the field and $E_{rms.}$ is the rms value. Applying this simplification to Eq. (2b) gives:

$$Q_v = 2\pi f \varepsilon_0\varepsilon''E^2 \tag{3}$$

In this chapter the symbol E is used instead of E_{rms} to define the electric field in the material.

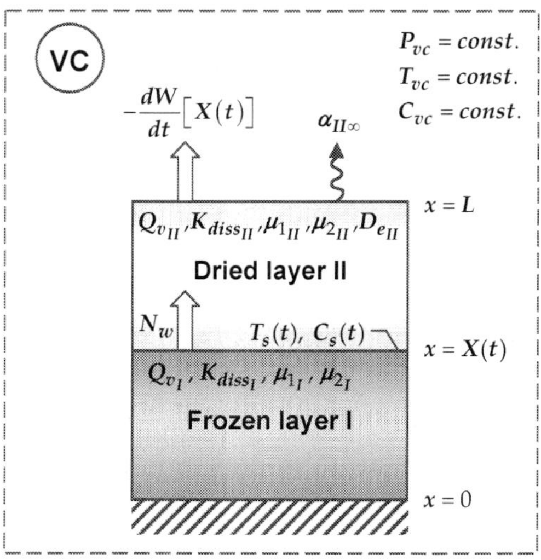

Fig. 3. The physical model of the primary MFD

3. Mathematical modeling of the primary MFD

3.1 Model formulation and assumptions

Consider a dried material which has geometry of a slab of infinite length. The sample bottom is insulated for heat and mass transfer and the upper surface is exposed to a vacuum at drying chamber temperature, as shown in Figure 3.

During the primary stage of the process, as sublimated vapors diffuse from the ice front towards exposed surface, moving boundary retreats until frozen free water is totally removed. The ice front (moving boundary) is assumed to be a plane of zero thickness and its initial position is arbitrary defined. In the frozen region I energy is transferred by conduction whereas conduction and convection are considered in the dried region II.

For dominant polarization of the electric field normal to the sample surface, it can be assumed that the electric field strength is constant throughout the dried material and equal to the value at the surface. Such simplification is justified when the penetration depth of microwaves is much greater than the size of the sample.

3.2 Governing equations

The physical system is described by the following differential equations.

Frozen Region

Heat transfer:

$$\frac{\partial T_I}{\partial t} = a_{e_I} \frac{\partial^2 T_I}{\partial x^2} + \frac{Q_{v_I}}{\rho_{bu_I} c_{p_I}} \tag{4}$$

Dried Region

Heat transfer:

$$\frac{\partial T_{II}}{\partial t} = a_{e_{II}} \frac{\partial^2 T_{II}}{\partial x^2} - \frac{c_{p_w}}{\rho_{bu_{II}} c_{p_{II}}} \frac{\partial(N_w T_{II})}{\partial x} + \frac{Q_{v_{II}}}{\rho_{bu_{II}} c_{p_{II}}} \,. \tag{5}$$

Steady capacities of internal heat sources in equations (4) and (5) are defined as follows:

$$Q_{v_i} = K_{diss_i}(T_i)E^2 \quad \text{for} \quad i = I, II \tag{6}$$

where dissipation coefficient:

$$K_{diss_i}(T_i) = 2\pi f \varepsilon_0 \varepsilon_i''(T_i) \quad \text{for} \quad i = I, II \tag{7}$$

Here, the dissipation coefficient in above equations is expressed as a linear function of material temperature:

$$K_{diss_i}(T) \approx \mu_{1_i} T + \mu_{2_i} \quad \text{for} \quad i = I, II \tag{8}$$

where parameters μ_1 and μ_2 are determined by linear regression of experimental data (Witkiewicz, 2006).

Mass transfer:

$$D_{e_{II}} \frac{\partial^2 C}{\partial x^2} = \frac{\partial C}{\partial t} \tag{9}$$

Effective diffusivity $D_{e_{II}}$:

$$\frac{1}{D_{e_{II}}} = \frac{1}{D_K} + \frac{1}{D_M} \tag{10}$$

is a combination of Knudsen diffusivity (Coulson et al., 1999):

$$D_K = 1.0638 \cdot r_p \sqrt{\frac{R T_{II_{avg}}}{M_w}} \tag{11}$$

and molecular diffusivity (Poling et al., 2001):

$$D_M = \frac{1.8829 \cdot T_{II_{avg}}{}^{3/2} \left(1/M_w + 1/M_{in}\right)^{1/2}}{P \sigma_{AB}^2 \Omega_{AB}} \tag{12}$$

where σ_{AB} and Ω_{AB} are determined for the system water vapor (A) – air (B) on the basis of tabulated constant of Lennard-Jones forces.

Sublimation front

The heat flux and mass flux at pseudo-steady-state conditions are related by:

$$q_s = N_w \Delta h_s \tag{13}$$

Then energy balance is:

$$-k_{e_I} \frac{\partial T_I}{\partial x} + k_{e_{II}} \frac{\partial T_{II}}{\partial x} = N_w \Delta h_s \tag{14}$$

The displacement of the moving boundary is related to the rate of sublimation:

$$N_w = -\left(W_{ini} - W_{eq}\right)\rho_{bu_{II}} \frac{dX(t)}{dt} \tag{15}$$

Initial Conditions

The initial moving boundary positions can be determined as follows:

$$X(t) = L - \delta, \quad \text{for} \quad t = 0 \tag{16}$$

where arbitrary initial thickness of the dried layer is chosen as $\delta = 3\%$ of sample thickness. This value of δ is used to start the numerical computations of the formulated model. In the frozen layer constant initial temperature is imposed:

$$T_I(x, 0) = T_{ini} \quad \text{for} \quad 0 \leq x \leq X \ \text{and}\ t = 0 \tag{17}$$

In the dried layer initial linear temperature and concentration profiles are assumed:

$$\frac{T_L(0) - T_{II}(x, 0)}{T_L(0) - T_{ini}(0)} = \frac{L - x}{\delta} \quad \text{for} \quad X \leq x \leq L \ \text{and}\ t = 0 \tag{18}$$

where:

$$T_L(0) = \frac{\alpha_{II\infty}(L - x)T_{vc} + k_{e_{II}} T_s}{\alpha_{II\infty}(L - x) + k_{e_{II}}} \quad \text{for} \quad x = L \ \text{and}\ t = 0 \tag{19}$$

$$C = C_{vc} - \left(\frac{\partial C}{\partial x}\right)_{x=X}(L - x) \quad \text{for} \quad X \leq x \leq L \ \text{and}\ t = 0 \tag{20}$$

Boundary Conditions

There is no heat transfer through the bottom boundary of the material, which is insulated by the wall of a Teflon container:

$$\left(-k_{e_I} \frac{\partial T_I}{\partial x}\right)_{x=0} = 0 \quad \text{for}\ t \geq 0 \tag{21a}$$

There is a heat transfer between the exposed material surface and surroundings:

$$k_{e_{II}}\left(\frac{\partial T_{II}}{\partial x}\right)_{x=L} = \alpha_{II\infty}(T_{vc} - T_L) \quad \text{for} \quad t \geq 0 \tag{21b}$$

At the ice front the thermodynamic equilibrium between water vapor and ice is assumed. Thus the vapor mass concentration at the moving front is related by Clausius-Clapeyron relation:

$$C_s(t) = f(T_s) = \exp\left(\frac{a}{T_s} + b\right) M_w / (RT_s) \quad \text{for} \quad x = X(t) \text{ and } t \geq 0 \tag{21c}$$

where: a=-6320.152 and b=29.558 (Wolff et al., 1989).
The condenser maintains low pressure of the vacuum chamber, therefore mass transfer resistance at the exposed surface is negligible:

$$C_L(t) = C_{vc} \quad \text{for} \quad x = L \text{ and } t \geq 0 \tag{21d}$$

3.3 Methods of model solution
The solution of the one-dimensional moving boundary problem in a planar medium described by Equations (4), (5), (9), (14)-(21) cannot be easily obtained unless some numerical techniques are used. Many approaches have been reported in the literature concerning general moving boundary problem with such phase change as melting, solidification or sublimation. Some researches use a fixed grid, fixed time-step formulation but it fails to give a reliable estimation of the material's temperature near the moving boundary (Yuen & Kleinman, 1980).
Another popular approach is the Landau's transformation (LT) method which mathematically immobilizes the moving boundary so that the number of spatial nodes in the frozen and dried region is constant (Fig. 4b). Such transformed mathematical model can be easily solved using commercial software.

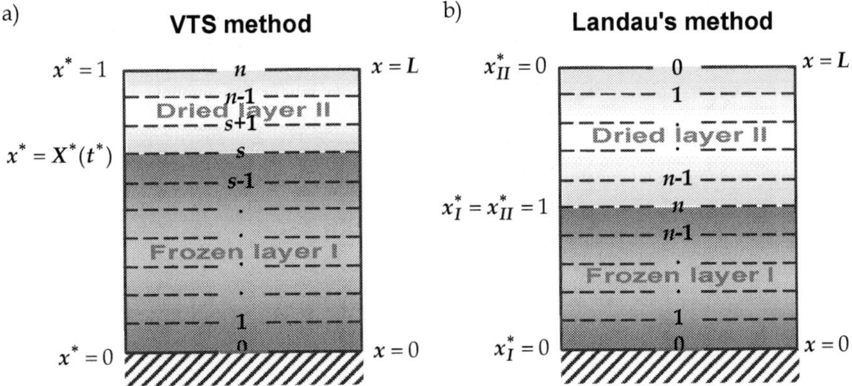

Fig. 4. Discretization grid of spatial variable

In this work, this method is compared with the variable time-step (VTS) approach, introduced as an alternative and effective numerical technique of moving boundary problem solution. The VTS method uses a fixed grid and the time step is adjusted to cause movement of the interface exactly one grid space (Fig. 4a). This approach can be also applied together with the Landau's transformation method.

The formulated mathematical model of the microwave freeze drying was solved numerically using two approaches: VTS and LT.

Equations (4), (5), (9), (14)-(15) make a set of mathematical model equations which are discretized by the implicit finite-difference Crank-Nicolson scheme and solved together with the adequate initial (16)-(20) and boundary conditions (21a)- (21d).

3.4 VTS method

The VTS method is a spatial fixed grid approach. The number of spatial nodes in computational grid resulting from discretization is constant. It means that in each step of computations the sublimation ice front should move exactly one grid space from the previous location. Instead of Eqs. (14) and (15) the following simultaneous heat and mass balance at the moving boundary must be fulfilled:

$$-k_{e_I}\frac{T_{s,j}-T_{I_{s-1,j}}}{\Delta x}-k_{e_{II}}\frac{T_{s,j}-T_{II_{s+1,j}}}{\Delta x}-(W_{ini}-W_{eq})\,\rho_{bu_{II}}\,\Delta h_s\frac{\Delta x}{\Delta t}-\rho_{bu_I}\,c_{p_I}\frac{\Delta x}{\Delta t}(T_{s,j}-T_{I_{s,j-1}})=0 \quad (22)$$

where in the above finite-difference scheme: the subscript s represents the space grid location of the ice sublimation interface (moving boundary) and the subscript j indicates time step.

The individual terms in above equation correspond to heat flux densities transferred from the frozen and dried material regions, heat flux utilized by sublimation and heat required to raise the temperature of interface node from previous time step to current one.

The following dimensionless variables are introduced for the sake of solution convenience:

$$x^* = \frac{x}{L} \quad (23)$$

$$X^* = \frac{X}{L} \qquad t^* = \frac{a_{el}\,t}{L^2} \qquad C^* = \frac{C-C_{vc}}{C_{s,3}-C_{vc}}$$

$$T_I^* = \frac{T_I-T_0}{T_{vc}-T_0} \qquad T_{II}^* = \frac{T_{II}-T_0}{T_{vc}-T_0} \qquad W^* = \frac{W-W_{eq}}{W_{ini}-W_{eq}} \tag{24}$$

where W denotes the average moisture content of the material bed.

The governing equations (4)-(5), (9) can be now rewritten as:

$$\frac{\partial^2 T_I^*}{\partial x^{*2}} - \frac{\partial T_I^*}{\partial t^*} = -\frac{Q_{V_I}\,L^2}{k_{e_I}(T_{vc}-T_0)} \quad (25)$$

$$\frac{\partial^2 T_{II}^*}{\partial x^{*2}} - \frac{c_{p_w}\rho_{bu_{II}}\,a_{e_I}(W_{ini}-W_{eq})}{k_{e_{II}}}\frac{dX^*}{dt^*}\frac{\partial T_{II}^*}{\partial x^*} - \frac{a_{e_I}}{a_{e_{II}}}\frac{\partial T_{II}^*}{\partial t^*} = -\frac{Q_{V_{II}}\,L^2}{k_{e_{II}}(T_{vc}-T_0)} \quad (26)$$

$$\frac{\partial^2 C^*}{\partial x^{*2}} - \frac{a_{e_I}}{D_{e_{II}}} \frac{\partial C^*}{\partial t^*} = 0 \tag{27}$$

Heat and mass balance at the sublimation front (22) can be rewritten as:

$$-k_{e_I} \frac{T^*_{s,j} - T^*_{I_{s-1,j}}}{\Delta x^*} - k_{e_{II}} \frac{T^*_{s,j} - T^*_{II_{s+1,j}}}{\Delta x^*} - \frac{W_{ini} - W_{eq}}{T_{vc} - T_0} \rho_{bu_{II}} \Delta h_s a_{e_I} \frac{\Delta x^*}{\Delta t^*} - k_{e_I} \frac{\Delta x^*}{\Delta t^*} (T^*_{s,j} - T^*_{I_{s,j-1}}) = 0 \tag{28}$$

In each calculation step, value of the actual time step corresponding to one spatial grid space movement is determined iteratively by bisection method as a root of Eq. (28).
The initial conditions (16)-(20) become:

$$X^* = \frac{\delta}{L} \quad \text{for} \quad t^* = 0 \tag{29a}$$

$$T^*_I = 0 \quad \text{for} \quad 0 \le x^* < X^* \quad \text{and} \quad t^* = 0 \tag{29b}$$

$$T^*_{II} = \left(1 - \frac{1-x^*}{1-X^*}\right) T^*_{II_L} \quad \text{for} \quad X^* \le x^* \le 1 \quad \text{and} \quad t^* = 0 \tag{29c}$$

$$C^* = \frac{k_{e_{II}} (T_{vc} - T_{ini}) T^*_{II_L}}{D_{e_{II}} (C_{s,3} - C_{vc}) \Delta h_s} x^*_{II} \quad \text{for} \quad X^* \le x^* \le 1 \quad \text{and} \quad t^* = 0 \tag{29d}$$

where

$$T^*_{II_L} = \frac{\alpha_{II\infty} L (1 - X^*)}{\alpha_{II\infty} L (1 - X^*) + k_{e_{II}}} \quad \text{for} \quad x^* = 1 \quad \text{and} \quad t^* = 0 \tag{29e}$$

The boundary conditions (21a)-(21d) become:

$$\frac{\partial T^*_I}{\partial x^*} = 0 \quad \text{for } x^* = 0 \text{ and } t^* > 0 \tag{30a}$$

$$T^*_{II} - \frac{k_{e_{II}}}{\alpha_{II\infty} L} \frac{\partial T^*_{II}}{\partial x^*} - 1 = 0 \quad \text{for} \quad x^* = 1 \text{ and } t^* > 0 \tag{30b}$$

$$C^* = f(T^*_{II}) \quad \text{for} \quad x^* = X^* \text{ and } t^* > 0 \tag{30c}$$

$$C^* = 0 \quad \text{for} \quad x^* = 0 \text{ and } t^* > 0 \tag{30d}$$

Finite difference schemes of Eqs. (25)-(28) together with the initial (29a)-(29e) and boundary conditions (30a)-(30d) constitute the algebraic linear equations set, and can be easily solved by the tridiagonal algorithm. The actual moving boundary position X^*_j versus time was computed as: $X^*_j = X^*_0 - \Delta x^* \cdot j$, where j means the actual number of time steps.

Analogically, the actual average moisture content W_j^* versus time was computed as: $W_j^* = W_0^* - \Delta W^* \cdot j$, where $\Delta W^* = \Delta x^*$.

3.5 LT method

The Landau's transformation converts the problem of a moving boundary to that of a fixed boundary (Ma & Peltre, 1973) by introducing the following definitions of dimensionless position coordinates in the frozen and dried region:

$$x_I^* = \frac{x}{X(t)} \tag{31a}$$

$$x_{II}^* = \frac{L - x}{L - X(t)} \tag{31b}$$

They are used in conjunction with the dimensionless variables (24) defined in the same way as in the VTS model. The mathematical model equations (4), (5), (9), (14)-(15) have now more complex dimensionless form.

Heat transfer in the frozen layer is:

$$\frac{\partial^2 T_I^*}{\partial x_I^{*2}} + x_I^* X^* \frac{dX^*}{dt^*} \frac{\partial T_I^*}{\partial x_I^*} - X^{*2} \frac{\partial T_I^*}{\partial t^*} = -\frac{Q_{V_I} L^2}{k_{e_I}(T_{vc} - T_0)} X^{*2} \tag{32}$$

Heat and mass transfer in the dried layer are as follows:

$$\frac{\partial^2 T_{II}^*}{\partial x_{II}^{*2}} - \left[\frac{c_{p_w} \rho_{bu_{II}} a_{e_I}(W_{ini} - W_{eq})}{k_{e_{II}}} + \frac{a_{e_I}}{a_{e_{II}}} x_{II}^* \right] (1 - X^*) \frac{dX^*}{dt^*} \frac{\partial T_{II}^*}{\partial x_{II}^*} +$$
$$- \frac{a_{e_I}}{a_{e_{II}}} (1 - X^*)^2 \frac{\partial T_{II}^*}{\partial t^*} = -\frac{Q_{V_{II}} L^2}{k_{e_{II}}(T_{vc} - T_{ini})} (1 - X^*)^2 \tag{33}$$

$$\frac{\partial^2 C^*}{\partial x_{II}^{*2}} - \frac{a_{e_I}}{D_{e_{II}}} x_{II}^* (1 - X^*) \frac{dX^*}{dt^*} \frac{\partial C^*}{\partial x_{II}^*} - \frac{a_{e_I}}{D_{e_{II}}} (1 - X^*)^2 \frac{\partial C^*}{\partial t^*} = 0 \tag{34}$$

Heat and mass balance at the sublimation front:

$$-\frac{T_{vc} - T_{ini}}{L} \left[\frac{k_{e_{II}}}{(1 - X^*)} \left(\frac{\partial T_{II}^*}{\partial x_{II}^*} \right)_{x_{II}^*=1} + \frac{k_{e_I}}{X^*} \left(\frac{\partial T_I^*}{\partial x_I^*} \right)_{x_I^*=1} \right] = N_w \Delta h_s \tag{35}$$

Displacement of the interface:

$$N_w = \frac{D_{e_{II}}(C_{s,3} - C_{vc})}{L(1 - X^*)} \left(\frac{\partial C^*}{\partial x_{II}^*} \right)_{x_{II}^*=1} = -\frac{\rho_{bu_{II}} a_{e_I}(W_{ini} - W_{eq})}{L} \frac{dX^*}{dt^*} \qquad \text{for } t^* \geq 0 \tag{36}$$

The initial conditions (16)-(20) become:

$$X^* = \frac{\delta}{L} \quad \text{for} \quad x_I^* = x_{II}^* = 1 \quad \text{and} \quad t^* = 0 \tag{37a}$$

$$T_I^* = 0 \quad \text{for} \quad 0 \le x_I^* \le 1 \quad \text{and} \quad t^* = 0 \tag{37b}$$

$$T_{II}^* = \left(1 - x_{II}^*\right)T_L^* \quad \text{for} \quad 0 \le x_{II}^* \le 1 \quad \text{and} \quad t^* = 0 \tag{37c}$$

$$C^* = \frac{k_{e_{II}}\left(T_{vc} - T_0\right)T_L^*}{D_{e_{II}}\left(C_{s,3} - C_{vc}\right)\Delta h_s} x_{II}^* \quad \text{for} \quad 0 \le x_{II}^* \le 1 \quad \text{and} \quad t^* = 0 \tag{37d}$$

where

$$T_L^* = \frac{\alpha_{II\infty}L\left(1 - X^*\right)}{\alpha_{II\infty}L\left(1 - X^*\right) + k_{e_{II}}} \quad \text{for} \quad x_{II}^* = 0 \quad \text{and} \quad t^* = 0 \tag{37e}$$

The boundary conditions (21a)-(21d) become:

$$\frac{\partial T_I^*}{\partial x_I^*} = 0 \quad \text{for} \quad x_I^* = 0 \text{ and } t^* \ge 0 \tag{38a}$$

$$T_{II}^* - \frac{k_{e_{II}}}{\alpha_{II\infty}(1 - X^*)}\frac{\partial T_{II}^*}{\partial x_{II}^*} - 1 = 0 \quad \text{for} \quad x_{II}^* = 0 \text{ and } t^* \ge 0 \tag{38b}$$

$$C^* = f(T_s^*) \quad \text{for} \quad x_I^* = x_{II}^* = 1 \text{ and } t^* \ge 0 \tag{38c}$$

$$C^* = 0 \quad \text{for} \quad x_{II}^* = 0 \text{ and } t^* \ge 0 \tag{38d}$$

Finite difference schemes of Eqs. (32)-(36) together with the initial (37a)-(37e) and boundary conditions (38a)-(38d) constitute the algebraic nonlinear equations set, and it must be solved by the adequate numerical algorithm.

Both the frozen and dried layer was divided into 20 equal space intervals. Resulting algebraic nonlinear equations set was solved in each time step by means of adequate numerical procedure in the Mathcad software.

In the LT method the actual moving boundary position X^* versus time was computed from solution of the set of discretized governing equations (32)-(36). The actual average moisture content of the drying bed W_j^* versus time was computed as: $W_j^* = W_0^* - \Delta W^* \cdot j$, where $\Delta W^* = \Delta x^*$.

3.6 Modeling results of the MFD primary stage

The simulations of the MFD were performed using two described above numerical methods for three selected materials: Sorbonorit 4, beef and mannitol. Thermophysical properties and parameters of the Sorbonorit 4 MFD system are presented in Table 1. The thermophysical parameters of the beef were taken from paper (Ma & Peltre, 1973, 1975a, b). While the analogical data for mannitol is included in the article (Wang & Chen, 2003). The exemplary simulations results of the MFD of selected materials using the VTS and LT approach are shown in Figs. 5 through 7.

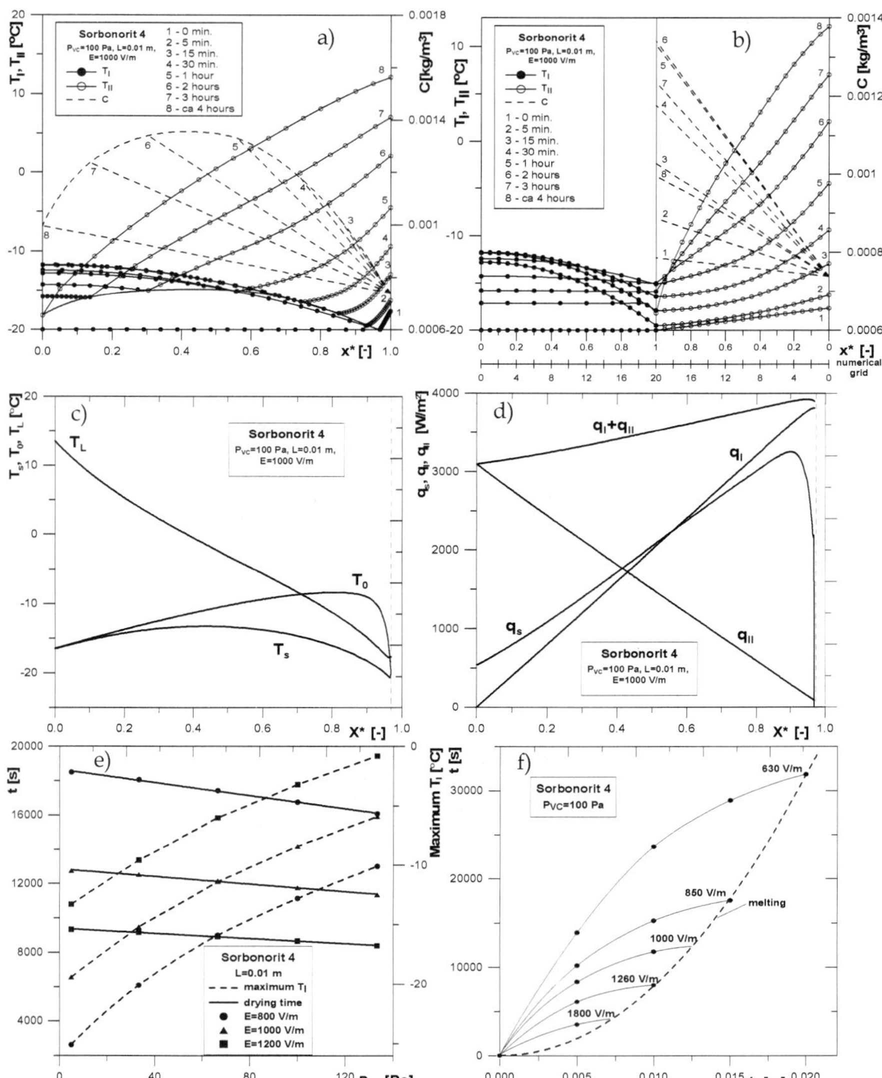

Fig. 5. Exemplary simulation results of Sorbonorit MFD: a) temperature and vapor concentration profiles for various process times – VTS method and b) LT method; c) temperature evolution on the sample surface T_L, sample bottom T_0 and sublimation front T_s versus dimensionless MB position; d) heat flux densities generated in the frozen layer q_I, in dried layer q_{II} and at MB q_s versus dimensionless MB position; e) effects of the total pressure on the drying time and maximum temperature of the frozen region; f) effects of the sample thickness on the total drying time for various electric field strengths.

Material thickness L	$0.005 \div 0.02$ m	Initial thickness δ	$0.03 \cdot L$ m
Elec. field strength E	$630 \div 1800$ V/m	VC pressure P_{vc}	$5 \div 133.33$ Pa
Ambient temp. T_{vc}	20 °C	Initial temperature T_{ini}	-20 °C
Heat trans. coeff. $\alpha_{II\infty}$	20 W/(m²K)	Bed porosity ε	0.71 m³/m³
Avg. moist. cont. W_{eq}	0.05 kg/kg	Eq. moisture content W_{ini}	1.63 kg/kg

Layer parameter	Frozen layer i=I	Dried layer i=II
Thermal diffusivity a_{e_i}	1.54 m²/s (100 Pa)	0.30 m²/s (100 Pa)
Thermal conductivity k_{e_i}	1.60 W/mK (100 Pa)	0.094 W/mK (100 Pa)
Parameter in Eq. (8) μ_{1_i}	0.00090 W/(KmV²)	0.00079 W/(KmV²)
Parameter in Eq. (8) μ_{2_i}	0.41148 W/(mV²)	0.30905 W/(mV²)
Density ρ_{e_i}	1048 kg/m³	400 kg/m³

Table 1. Thermophysical properties and parameters of the Sorbonorit 4 MFD system

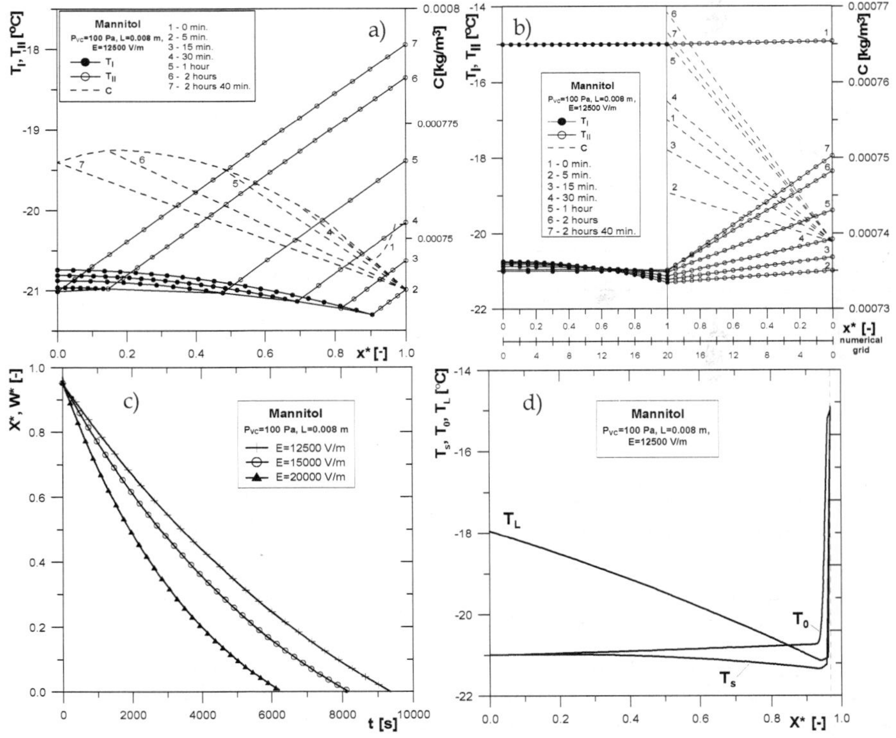

Fig. 6. Exemplary simulation results of mannitol MFD: a) temperature and vapor concentration profiles in the sample for various process times – VTS method and b) LT method; c) drying curves; d) temperature evolution on the sample surface T_L, sample bottom T_0 and sublimation front T_s versus dimensionless MB position.

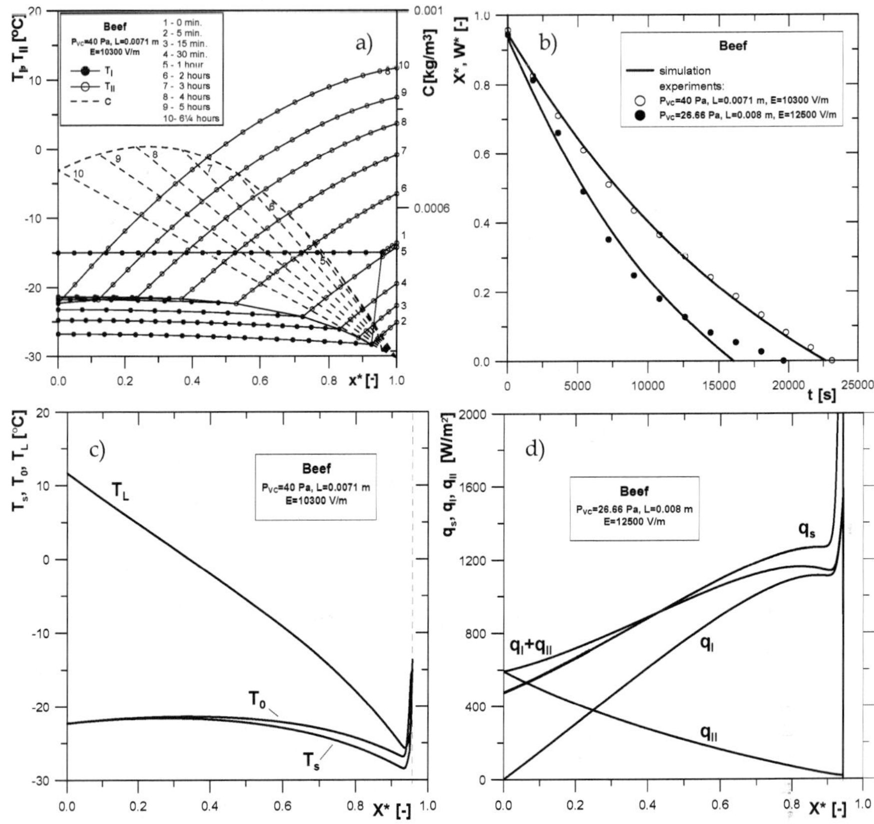

Fig. 7. Exemplary simulation and experimental results of beef MFD: a) temperature and vapor concentration profiles in the sample for various process times – VTS method and b) drying curve; c) temperature evolution on the sample surface T_L, sample bottom T_0 and sublimation front T_s versus dimensionless MB position; d) heat flux densities generated in the frozen layer q_I, in dried layer q_{II} and at the MB q_s versus dimensionless MB position.

The heat flux densities: q_I and q_{II} presented in the Figs. 5 through 7 were computed in the actual time step as follows: $q_i = \sum Q_{v_{i,j}} \Delta x$, for $i=1,2$, where j denotes number of the node in the difference scheme. The heat flux density at the sublimation front q_s was determined from the Eqs. (14) and (15) for the actual time step Δt and MB movement ΔX .

The objective of these simulations is to determine the optimal control policy of the MFD primary stage of the selected materials to improve drying rate and simultaneously achieve desired product quality. The vacuum chamber pressure P_{vc} and microwave power E are considered the controllable factors in the MFD system in order to avoid gas breakdown (corona discharge) and melting. The corona discharge occurs when the electric field intensity is above the threshold value. The ionization of the residual gases in the vacuum chamber leads to great energy losses and can seriously damage the final product. Since the electric field intensity E is proportional to the power applied by the microwave generator, its settings should be controlled during the process. The threshold value of the electric field is a function of chamber pressure and has the minimum in the typical pressure range used in freeze-drying (10 – 300 Pa). Therefore upper limit on the electric field strength should be set. Another limits need to be set on the temperature maxima in the dried product to keep its demanded quality. The temperature throughout the dried layer must be lower than the melting temperature and the temperature throughout the dried layer must not exceed the scorch temperature. The melting temperature of frozen foodstuffs is lower than the melting temperature of pure ice, due to existence of soluble chemical compounds in the material. This value amounts usually to $-3 \div -1.5\,°C$ for foodstuffs (Ma & Peltre, 1975; Ang et al., 1977). The scorch temperature of the dried layer is a temperature of material thermal degradation and typically amounts to $50 \div 60\,°C$. More restrictive temperature constraints are set in microwave freeze-drying of biological and pharmaceutical products, when biological activity is an important quality factor.

In Figure 5e the drying time and maximal temperatures reached in the frozen region are plotted as a function of the vacuum chamber pressure at various electric field strengths for Sorbonorit 4 bed (L=0.01 m). As can be seen, the drying time varies strongly with the electric field strength and is generally linearly depended on $1/E^2$. Increase of the vacuum chamber pressure decreases the drying time. It is caused by improved heat transfer in dried layer. Simultaneously, the temperatures in the material increase as a result of the greater diffusion mass transfer resistances in the dried layer. As dissipation coefficients are linearly temperature dependent, the internal heat source capacity is greater and implies intensification of the process. The maximal temperatures reached in the frozen region, shown in Figure 5e, indicate that increasing the vacuum chamber pressure and electric field strength has its limit and exceeding some constraints causes melting. The influence of the sample thickness on the drying time at various electric field strengths is shown in Figure 5f. Analysis of this relation leads to conclusion that smaller sample thickness would enable using higher electric field strengths. It should significantly shorten the drying time providing that all set constraints are not violated.

4. Mathematical modeling of the secondary MFD

4.1 Model formulation and assumptions

After primary drying, residual moisture content may be as high as 7% (Rowe & Snowman, 1978). Secondary drying is intended to reduce this to an optimum value for material stability – usually with moisture content between 0.5 and 2.0 %. In secondary stage of the MFD there is not any ice and bound water in the material bed (by means of surface forces) is removed by desorption (Fig. 8).

In formulating of the mathematical model of the secondary MFD the following assumptions are made:

- Initial average moisture content of material is equal to the equilibrium moisture content at average temperature of dried layer at the end of the primary freeze drying.
- Distribution of electric field in a sample is assumed to be uniform. Dissipation coefficient value in the whole material bed corresponds with that of dried region $K_{diss_{II}}$ in the primary freeze drying stage.

- There are two mechanisms of mass transport in the material: moisture desorption and diffusion of water vapor throughout the bed.
- Gas phase in dried material and in vacuum chamber consists of water vapor and air (inert).
- Adsorption equilibrium is described by multitemperature Langmuir isotherm.

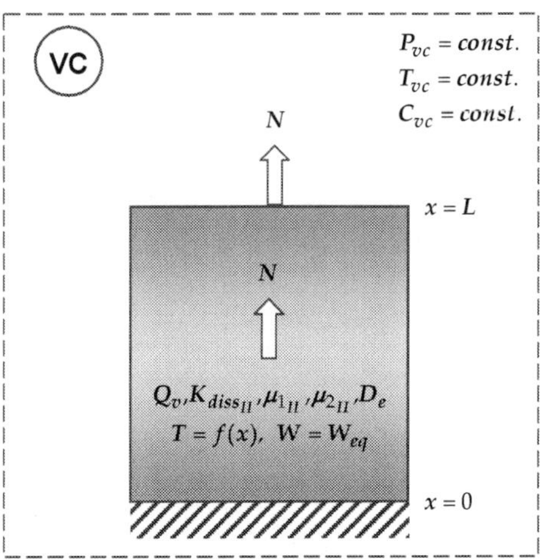

Fig. 8. The physical model of the secondary MFD

4.2 Governing equations

Mass conservation:

$$\frac{\partial(y_w \cdot C)}{\partial t} + \frac{\partial(y_w \cdot N)}{\partial x} + \frac{\rho_{bu_{II}}}{\varepsilon \cdot M_w}\frac{\partial W}{\partial t} = 0 \tag{39}$$

where $y_w + y_{in} = 1$ and $C = P/RT$ is a sum of molar concentrations of components in gas phase including inert (air).

Moisture mass balance in solid phase results from driving force between equilibrium moisture content and actual moisture content of a material:

$$\frac{\partial W}{\partial t} = K\left(W_{eq} - W\right) \tag{40}$$

Adsorption equilibrium is expressed by multitemperature extended Langmuir isotherm (Chahbani & Tondeur, 2001):

$$W_{eq} = W_L \exp\left(\frac{a}{T}\right)\left[\frac{b \cdot \exp\left(\dfrac{c}{T}\right)P \cdot y_w}{1 + b \cdot \exp\left(\dfrac{c}{T}\right)P \cdot y_w}\right] \tag{41}$$

where W_L, a, b and c are the Langmuir equation constants.

Mass transfer:
Kinetic coefficient K in Equation (40) is calculated applying the linear driving force conception (LDF) (Glueckauf, 1955):

$$K = \frac{60 \cdot D_{eff}}{d_e^2} \qquad (42)$$

The mechanism determining vacuum desorption process is the diffusion rate in material pores which is a combination of Knudsen and surface diffusion. Thus, effective diffusivity can be expressed as:

$$D_{eff} = D_S + D_K \frac{\varepsilon_p M_w}{\rho_p} \frac{\partial y_{w_{eq}}}{\partial W} \qquad (43)$$

where ε_p and ρ_p denote particle porosity and particle density, respectively.
Heat transfer:
Quasi-homogeneous heat balance equation can be expressed as:

$$-k_{e_{II}} \frac{\partial^2 T}{\partial x^2} + N \sum_{i=1}^{2} y_i c_{pg_i} \frac{\partial T}{\partial x} + c_\Sigma \frac{\partial T}{\partial t} + \Delta H_{ads} \frac{\rho_{bu_{II}}}{M_w} \frac{\partial W}{\partial t} - Q_v = 0 \qquad (44)$$

where c_Σ denotes total volumetric specific heat defined as::

$$c_\Sigma = \rho_{bu_{II}} \left(c_s + W c_w \right) + \varepsilon \cdot C \sum_{i=1}^{2} y_i c_{pg_i} \qquad (45)$$

Values c_w and c_s are the specific heat capacities of liquid water and dry material sample, respectively.
Isosteric adsorption heat of water vapor in dried material included in Equation (41) is estimated from Clausius-Clapeyron type equation (Do, 1998):

$$\Delta H_{ads} = -RT^2 \left(\frac{\partial \ln P}{\partial T} \right)_W \qquad (46)$$

Above equation after incorporating multitemperature Langmuir isotherm can be rewritten in analytical form:

$$\Delta H_{ads} = -\frac{R \cdot W_L \exp(a/T)}{W_L \exp(a/T) - W} \left(a + c - \frac{W \cdot c}{W_L \exp(a/T)} \right) \qquad (47)$$

Source term in Equation (44) resulting from dissipation of microwave energy in material volume is defined as:

$$Q_v = K_{diss_{II}} E^2 \qquad (48)$$

Dissipation coefficient $K_{diss_{II}}$ in Equation (48) is expressed as a linear function of material temperature (similarly as in equation (8)):

$$K_{diss_{II}}(T) \approx \mu_{1_{II}} T + \mu_{2_{II}} \tag{49}$$

where parameters μ_1 and μ_2 are determined by linear regression of experimental data (Witkiewicz, 2006).

Momentum balance

Pressure drop along the sample axis is described by Ergun relation (Ergun, 1952):

$$\frac{\partial P}{\partial x} = -\frac{\eta}{C \cdot k_D} N - \sum_{i=1}^{2} y_i M_i \frac{k_E}{C \, k_D} N^2 \tag{50}$$

where $i = 1$ denotes water vapor and $i = 2$ means inert gas (air).

Parameter k_D in above Equation defines permeability of dried bed and k_E parameter describes inertial effect:

$$k_D = \frac{d_e^2 \, \varepsilon^3}{150(1-\varepsilon)^2} ; \qquad k_E = \frac{1.75 \, d_e}{150(1-\varepsilon)} \tag{51}$$

In order to simplify calculation the following relation defining molar flux density of water vapor is derived (Sun et. al, 1995):

$$N = -\frac{2 C k_D \eta^{-1} \partial P / \partial x}{1 + \sqrt{1 + 4C \cdot (\sum_{i-1}^{2} y_i M_i) k_D k_E \eta^{-2} |\partial P / \partial x|}} \tag{52}$$

Boundary and initial conditions

Formulated mathematical model is solved together with the following boundary and initial conditions:

$$\frac{\partial y_i}{\partial x} = 0 ; \quad \frac{\partial P}{\partial x} = 0 ; \quad \frac{\partial T}{\partial x} = 0 \quad \text{for} \quad x = 0 \tag{53a}$$

$$\frac{\partial T}{\partial x} = 0 ; \quad P = P(t) \quad \text{for} \quad x = L \tag{53b}$$

$$W = W(x,0) ; \quad T = T(x,0) ; \quad P = P(x,0) \quad \text{for} \quad t = 0 \tag{54}$$

4.3 Solution of the mathematical model

Dimensionless variables

Model equations can be expressed in more convenient dimensionless form incorporating the following definitions of dimensionless variables:

$$t^* = \frac{a_{e_{II}} t}{L^2} ; \quad x^* = \frac{x}{L} ; \quad P^* = \frac{P}{\Delta P} ; \quad T^* = \frac{T - T_0}{\Delta T}$$

$$W^* = \frac{W - W_{eq}}{W_{ini} - W_{eq}} ; \quad C_w^* = \frac{y_w C}{C_{w0}} ; \quad Q_v^* = Q_v L^2 / (k_e \Delta T) ; \tag{55}$$

where pressure and temperature are normalized relatively to arbitrary chosen increments ΔP and ΔT respectively.

Governing equations in dimensionless form

Mathematical model after transformation consists of mass balance in gas phase:

$$a_{e_{II}} C_{w0} \frac{\partial C_w^*}{\partial t^*} + L \frac{\partial (y_w \cdot N)}{\partial x^*} + \frac{a_{e_{II}} \rho_{bu_{II}} \left(W_{ini} - W_{eq} \right)}{\varepsilon M_w} \frac{\partial W^*}{\partial t^*} = 0 \tag{56}$$

mass balance in solid phase:

$$\frac{\partial W^*}{\partial t^*} = -\frac{L^2}{a_{e_{II}}} K W^* \tag{57}$$

quasi-homogeneous heat balance:

$$-k_{e_{II}} \frac{\partial^2 T^*}{\partial x^{*2}} + LN \sum_{i=1}^{2} y_i c_{pg_i} \frac{\partial T^*}{\partial x^*} + Lc_\Sigma \frac{\partial T^*}{\partial t^*} + \frac{a_{e_{II}} \left(W_{ini} - W_{eq} \right) \rho_{bu_{II}}}{\Delta T} \frac{\Delta H_{ads}}{M_w} \frac{\partial W^*}{\partial t^*} - k_{e_{II}} Q_v^* = 0 \tag{58}$$

and Ergun equation:

$$\Delta P \frac{\partial \left(P^* \right)}{\partial x^*} = -\frac{L \, \eta}{C \, k_D} N - (\sum_{i=1}^{2} y_i M_i) \frac{L \, k_E}{C \, k_D} N^{\,2} \tag{59}$$

where W^* is expressed by multitemperature extended Langmuir isotherm.
Mathematical model is solved numerically by means of numerical method of lines (NMOL) which requires transformation of partial differential equations into set of ordinary differential equations for time derivatives, and approximation of space derivatives by adequate finite differences (Schiesser, 1991).

4.4 Modeling results of the MFD secondary stage

In theoretical analysis the vacuum desorption of water in Zeolite DAY-20F at microwave heating is considered. Parameters of multitemperature Langmuir isotherm (Eq. (41)) for water–Zeolite DAY-20F system listed in Table 2 were approximated by nonlinear Levenberg-Marquardt estimation of data within temperature range of 293÷373 K (Sun et al., 1995).

Component	Constants of multitemperature Langmuir isotherm equation			
	W_L, kg /kg	a, K	b, Pa^{-1}	c, K
Water	0.370300	1387.82	$0.170 \cdot 10^{-6}$	1511.02

Table 2. Constants of multitemperature Langmuir isotherm equation for water–Zeolite DAY- 20F system

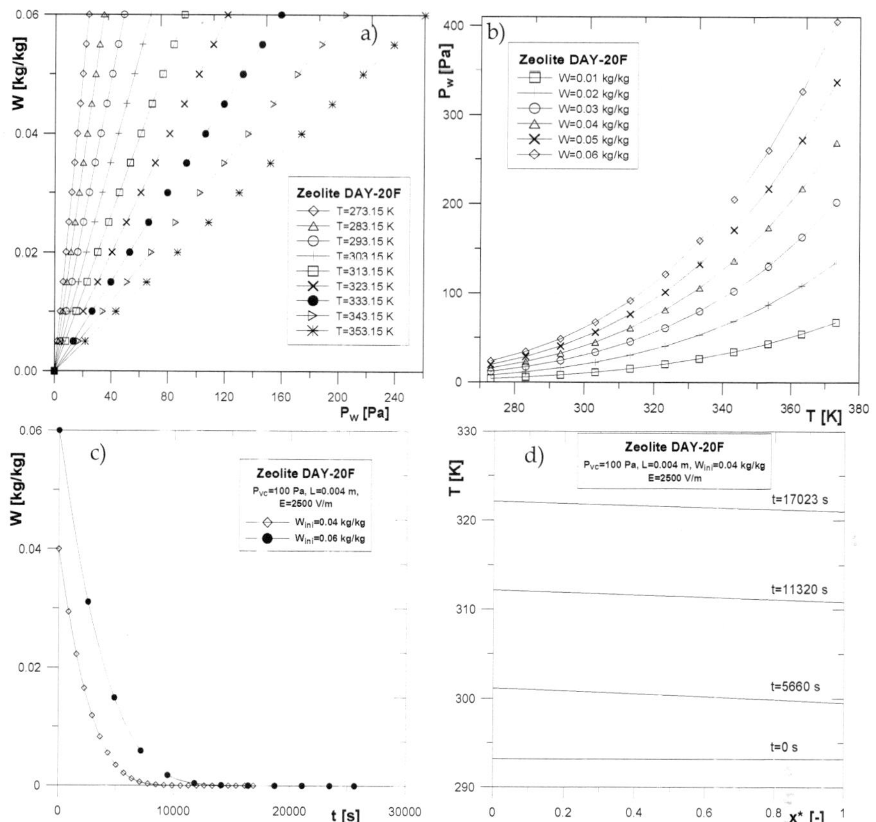

Fig. 9. Exemplary simulation of Zeolite DAY-20F secondary MFD: a) multitemperature adsorption isotherms and b) isosteres of water vapor; c) drying curves; d) temperature profiles in the sample for various process times.

Multitemperature Langmuir isotherm of water vapor on Zeolite DAY-20F in wide range of pressure is shown in Figure 9a. Additionally multitemperature Langmuir isostere for the same system is depicted in Figure 9b. Physical properties of Zeolite DAY-20F assumed at average bed temperature 303.15 K and process pressure 100 Pa are as follows: $k_{e_{II}} = 0.025$ W/(mK), $c_s = 900$ J/(kgK), $\rho_{bu_{II}} = 500$ kg/m³ ; $\varepsilon = 0.677$.

During the primary freeze drying stage equilibrium water vapor pressure in region II equals approximately vapor pressure at sublimation front, i.e. 100 Pa. Equilibrium material (Zeolite DAY-20F) moisture content relative to that pressure equals about 0.05 kg/kg for maximal material temperature: 323 K (Fig. 9b). Thus for dried layer temperatures range (region II), taking place during primary freeze drying (below 323 K) moisture desorption does not exist. Formulated here mathematical model of the secondary vacuum freeze drying at microwave heating consisting of equations (56)-(59) with adequate initial and boundary conditions (53,

54) was solved by 4th order Runge-Kutta method with the following process parameters: L=0.004 m, W_{ini}=0.04 or 0.06 kg/kg, $\mu_{1_{II}} = 1.107 \times 10^{-6}$ W/(mV^{2}K), $\mu_{2_{II}} = 1.021 \times 10^{-3}$ W/(mV^2), $E = 2500$ V/m and initial bed temperature $T(x,0) = 293.15$ K. Calculated dissipation coefficient at this conditions amounts to $K_{diss_{II}} = 1.054 \times 10^{-3}$ and average capacity of internal volumetric heat source throughout the material equals: Q_v= 6588 W/m^3. Obtained results of numerical simulation are presented in Figs. 9c and 9d. Moisture contents versus time for two initial bed moisture contents $W_{ini} = 0.04$ and $0.06\ kg/kg$ in Zeolite DAY-20F bed of thickness $L = 0.04\ m$ are shown in Figure 9c. Figure 9d depicts dried bed temperatures versus dimensionless positions in the material layer, for various process times at sample thickness $L = 0.04\ m$ and initial moisture content $W_{ini} = 0.04\ kg/kg$.

5. Conclusions

The mathematical model of both the primary and the secondary freeze drying stages of random solids at microwave heating was developed. Experimental and/or theoretical investigations of the primary freeze drying at microwave heating were presented for three selected materials: beef, Sorbonorit 4 activated carbon and mannitol. Furthemore, theoretical analysis of the secondary freeze drying at microwave heating was performed for Zeolite DAY-20F bed.

For dried granular materials such as: Sorbonorit 4, Zeolite DAY-20F, characterized by considerable internal porosity, primary freeze drying stage is not sufficient to remove entire moisture contained in the dried bed. During the primary freeze drying at microwave heating free, interstitial moisture is removed. The secondary freeze drying stage is then necessary to remove residual moisture which may be as high as 7%, after primary stage.

Kinetics of the microwave freeze drying process is enhanced in comparison with the freeze drying at conventional heating. It is caused mainly by the fact, that in conventional heating (e.g. contact or radiative) temperature and mass transfer gradients in the dried material have opposite directions. On the contrary, in the freeze drying process at microwave heating both temperature, and mass transfer gradients are cocurrent. It is very convenient phenomenon from point of view of the qualitative final dried material properties.

Assumption of constant electric field strength in a dried material is valid for sample dimensions up to half wavelength order in monomodal resonant cavities. Formulated mathematical model can be also applicable in the microwave heating of sample dimensions equaled multiple of wavelengths in multimodal applicators with uniform distribution of electric field strength.

The model for the secondary freeze drying at microwave heating takes into account mass transfer resistances which arise in vacuum desorption. The LDF mass transfer model with a variable, lumped-resistance coefficient K was used. This numerical model was applied to perform computer simulation of the vacuum desorption of pure water from Zeolite DAY-20F.

Experimental verification of the some model simulations of the primary freeze drying at microwave heating approved its fairly good usefulness for design applications. Furthemore, mathematical model of the vacuum desorption in the secondary freeze drying enables predictions of the drying kinetics for the granular solids.

6. Nomenclature

a_c	effective thermal diffusivity (m^2/s)
c_p, c_{p_w}	effective specific heat (material region, water vapor) (J/kg K)
c_{p_g}	effective specific heat of gas phase ((J/mol K))
C	gas phase molar concentration (mol/m^3)
C_w, $C_{s,3}$	water vapor mass concentration (general, triple point) (kg/m^3)
d_e	equivalent particle diameter (m)
$D_{e_{II}}$, D_K, D_M, D_S	water vapor diffusivity (effective, Knudsen, molecular, surface) in dry layer (m^2/s)
E, $\boldsymbol{E}$, E_0, E_{rms}	electric field strength (general, spatial vector, peak, r.m.s.) (V/m)
f	frequency (Hz)
K_{diss}	dissipation coefficient (W/mV^2)
L	sample thickness (m)
M_w, M_{in}	molar mass (water, inert) (kg/mol)
N	gas phase molar flux density in the secondary MFD (mol/m^{2}s)
N_w	water vapor mass flux density from sublimation front (kg/m^{2}s)
P	total pressure (Pa)
q	heat flux density (W/m^2)
Q_v	volumetric heat source capacity (W/m^3)
r_p	mean pore radius (m)
R	universal gas constant (J/mol K)
t	time (s)
T, $T_{II_{avg}}$, $T_{s,3}$, T_0	temperature (general, average in region II, triple point, starting value) (K, °C)
W, W_{eq}, W_{ini}, W_0	moisture content of dried bed (average, equilibrium, initial, starting value when microwave heating is turned on) (dry basis, kg/kg)
x, X, X_0	Cartesian position coordinate (general, moving boundary, starting) (m)
y_w, $y_{w_{eq}}$	water vapor mole fraction (general, equilibrium) (mol/mol)

Greek symbols

$\alpha_{II\infty}$	heat transfer coefficient at the surface of dried region II (W/m^{2}K)
δ	arbitrary initial thickness of the dried layer (m)
$\tan \delta$	dielectric loss tangent (-)
Δh_s	enthalpy of sublimation (J/kg)
Δh_{ads}	heat of adsorption (J/mol)
ε	bed porosity, (-)
ε_0	free space permittivity (F/m)
ε'	relative dielectric constant (–)
ε''	relative loss factor (–)
η	gas phase viscosity (Pa s)
λ_0, λ_w	wavelength (free space, waveguide) (m)
μ_1	constant in Eq. (8) (W/mV^{2}K)
μ_2	constant in Eq. (8) (W/mV^2)

ρ_{bu} bulk density (kg/m³)

ω angular frequency (rad/s)

Subscripts and superscripts

1, 2   gas phase (water vapor, air)

I, II   region number (frozen, dried)

L   at sample exposed surface

s   at moving boundary (sublimation front)

vc   vacuum chamber

*   dimensionless value

Abbreviations

FD   freeze drying

LDF   linear driving force

LT   Landau transformation

MB   moving boundary (sublimation front)

MFD   microwave freeze drying

NMOL   numerical method of lines

VTS   variable time-step

7. References

Ang, T.K.; Ford, J.D. & Pei, D.C.T. (1977). Microwave freeze-drying of food: a theoretical investigation, *International Journal of Heat and Mass Transfer*, 20, pp. 517-526.

Chahbani, M.H. & Tondeur, V. (2001). Pressure drop in fixed-bed adsorbers, *Chem. Eng. J.*, 81, pp. 23-34.

Coulson, J.M.; Richardson, J.F. & Sinnott, R.K. (1999). *Chemical Engineering*, Vol. 6, Butterworth-Heinemann, Oxford.

Do, D.D. (1998). *Adsorption Analysis: Equilibria and Kinetics*, Imperial College Press, London.

Ergun, S. (1952). Fluid flow through packed column, *Chem. Eng. Progress*, 48, pp. 89-94.

Glueckauf, E. (1955). Theory of chromatography, *Trans. Faraday Soc.*, 51, pp. 1540-1551.

Liapis, A.I. & Bruttini R. (1996). Mathematical models for the primary and secondary drying stages of the freeze drying of pharmaceuticals on trays and vials. In: *Mathematical Modeling and Numerical Techniques in Drying Technology*, Turner, I & Mujumdar, A.S., (Eds), pp. 481-536, CRC Press, ISBN: 9780824798185, Marcel Dekker, New York.

Liapis, A.I. & Bruttini, R. (2006). Freeze drying, In: *Handbook of Industrial Drying*, Mujumdar A.S., (Ed.), pp. 257-284, CRC Press, ISBN: 9781574446685, New York.

Ma, Y.H. & Peltre, P.R. (1973). Mathematical simulation of a freeze drying process using microwave energy, *AIChE Symp. Ser.*, 132, 69, 47-54.

Ma, Y.H. & Peltre, P.R. (1975a). Freeze dehydration by microwave energy: Part I. Theoretical investigation. *AIChE Journal*, 21, 2, pp. 335-344.

Ma, Y.H. & Peltre, P.R. (1975b). Freeze dehydration by microwave energy: Part II. Experimental Study. *AIChE Journal*, 21, 2, pp. 344-350.

Metaxas, A. C. (1996). *Foundations of Electroheat: A Unified Approach*, John Wiley & Sons, New York.

Péré, C.; Rodier, E. & Louisnard, O. (2001). Microwave vacuum drying of porous media: verification of a semi-empirical formulation of the total absorbed power. *Drying Technology*, 19, 6, pp. 1005-1022.

Poling, B.E.; Prausnitz, J.M. & O'Connell, J.P. (2001). *The properties of gases and liquids*, Fifth Edition, McGraw-Hill, New York.

Rowe, T.W. & Snowman, J.W. (1978). *Edwards Freeze drying Handbook*, Crawley, Cambridge.

Schiesser, W.E. (1991). *The Numerical Method of Lines: Integration of Partial Differential Equations*, Academic Press, New York.

Schiffmann, R.F. (2006). Microwave and dielectric drying. In.: *Handbook of Industrial Drying* Mujumdar A.S., (Ed.), pp. 286-306, CRC Press, ISBN: 9781574446685, New York.

Sun, L.M.; Amar, N.B. & Meunier F. (1995). Numerical study on coupled heat and mass transfer in an adsorber with external fluid heating, *Heat Recovery Systems & CHP*, 15, pp. 19-29.

Wang, Z.H. & Shi, M.H. (1998). Numerical study on sublimation-condensation phenomena during microwave freeze drying. *Chemical Engineering Science*, 53, 18, pp. 3189-3197.

Wang, W. & Chen, G. (2003). Numerical investigation on dielectric material assisted microwave freeze-drying of aqueous mannitol solution. *Drying Technology*, 21, 6, pp. 995-1017.

Witkiewicz, K. (2006). *The Numerical Modeling of Freeze Drying of Granular Solids at Microwave Heating*, Ph.D. Dissertation, Szczecin University of Technology, Szczecin, (in Polish).

Wolff, E.; Gilbert, H. & Rodolphe, F (1989). Vacuum freeze drying kinetics and modelling of a liquid in a vial, *Chemical Engineering and Processing*, 25, pp. 153-158.

Yuen, W.W. & Kleinman, A.M. (1980). Application of a variable time-step finite-difference method for the one-dimensional melting problem including the effect of subcooling, *AIChE Journal*, 26, 5, pp. 828-832.

Yun, J.-H.; Choi, D.-K. & Kim, S.-H. (1999). Equilibria and dynamics for mixed vapors of BTX in an activated carbon bed, *AIChE. Journal*, 45, 4, pp. 751- 760.

28

Application of Microwave Heating on the Facile Synthesis of Porous Molecular Sieve Membranes

Aisheng Huang, and Jürgen Caro

Institute of Physical Chemistry and Electrochemistry, Leibniz University of Hannover
Germany

1. Introduction

Microwaves are electromagnetic waves locating in the electromagnetic spectrum between infrared and radio waves, with wavelengths ranging from as long as one meter to as short as one millimetre, or equivalently, with a frequencies range between 0.3 and 30 GHz. Microwaves consist of an electric and magnetic field, and thus represent electromagnetic energy. This energy can act as a non-ionising radiation that causes molecular motions of charge carriers like ions or electrons and cause also molecules with a permanent dipole moment to follow the alternating field, but do not affect the molecular structure (Bougrin et al, 2005). It has been known for a long time that microwaves can be used to heat materials. In the case of heating solids, the kinetic energy of ions and electrons following the alternating electrical field is transformed into thermal energy. When heating liquids, the orientation polarization of systems of molecules with permanent dipoles like water, methanol or DMF is used. Actually, microwave heating technique has been widely used within the last 50 years not only in the domestic preparation and processing of food, but also in the field of materials synthesis. The classic work of Hippel and co-workers in the early 1950s provided a sound theoretical basis for these technological developments and his group provided an important database of dielectric properties on common substances, foodstuffs and materials (Von Hippel, 1954). Thereafter, Gabriel et al. have summarized the basic theory of underlying microwave dielectric heating and provided a database of dielectric parameters for organic solvents which are used in synthetic organic and inorganic chemistry (Gabriel et al., 1998). Nuchter et al. have given a critical technology overview on reaction engineering in microwave fields (Nuchter et al., 2004).

In the past two decades, microwave heating technique has attracted intense attention from chemists, material scientists and microwave engineers. Microwave heating has currently been employed in many chemical reaction studies, including organic and inorganic syntheses, selective sorption, oxidations/reductions, and polymerizations. Such review articles can be found elsewhere (de la Hoz et al., 2005; Langa et al., 1997; Nuchter, et al., 2003; Rao et al., 1999; Tompsett et al., 2006). The effect of microwave irradiation in chemical reactions is a combination of thermal and non-thermal effects (de la Hoz et al., 2005). Compared with the conventional heating for chemical synthesis, microwave heating has the following advantages (Gabriel et al., 1998): (i) The introduction of microwave energy into a chemical reaction system remarkably results in much higher heating rates than those which are achieved conventionally in air conditioned ovens. Therefore the synthesis times can be significantly reduced. (ii) The microwave energy is introduced into the chemical reactor remotely without direct contact between the energy source and the reacting chemicals, therefore, leading to a significantly different temperature profile for the reaction and, thus, to an alternative distribution of chemical products from a reaction. (iii) It is possible to achieve selective synthesis of desired materials and to change reaction kinetics and selectivity since chemicals and the containment materials for chemical reactions do not interact equally with microwaves. Specifically the containment materials for a chemical reaction may be chosen in such a way that the microwave energy passes through the walls of the vessel and heats only the reactants. This concept holds true also for addressing droplets of a dipolar solvent in an organic microwaves not-absorbing solvent (reverse microemulsions). (iv) "Hot spots" yielded on local boundaries by reflections and refractions of microwaves and may result in a "super-heating" effect, which can be described best as local overheating and which is comparable to the delayed boiling of overheated liquids under conventional conditions. The advantages outlined above have been exploited extensively during the last 20 years and to date more than 2000 articles have been published to apply microwave heating for chemical synthesis (Figure 1). A special application of microwaves is their use in the crystallization of porous molecular sieve membranes. In this chapter, microwave synthesis of porous molecular sieve membranes was summarized, including recent development and progress in microwave synthesis of porous molecular sieve membranes; combined microwave and conventional heating model; differences between microwave synthesis and conventional hydrothermal synthesis for membrane formation; formation mechanism in the case of microwave synthesis of porous molecular sieve membranes.

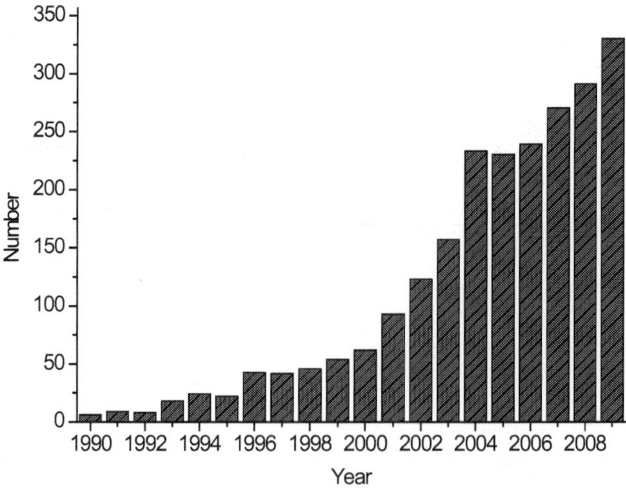

Fig. 1. Development of articles in open literatures with microwave heating synthesis.

2. Zeolite and zeolite-like porous molecular sieves

Zeolites are crystalline silicate and aluminosilicates microporous materials that possess three dimensional frameworks built from SiO_4 and AlO_4 tetrahedral units linked through oxygen atoms with periodic arrangements of cages and uniform channels of nanometer dimensions (Figure 2) (Baerlocher & McCusker, http://www.iza-structure.org/databases). Since the size of the pore aperture of zeolites is comparable to molecular dimensions (typically between 3 and 10 Å), zeolites can function as "molecular sieves", excluding molecules larger than their pore windows. So far, over 190 molecular sieve structures were registered to the Structure Commission of the International Zeolite Association (IZA) since the Swedish mineralogist Cronstedt discovered the first natural zeolite stilbite (STI) in 1756. Among those, only about 17 are of commercial interest and produced synthetically, i.e. AEL, AFY, BEA, CHA, EDI, FAU, FER, GIS, LTA, LTL, MER, MFI, MOR, MTT, MWW, TON, and RHO (Maesen, 2007). Zeolites have been widely used in the chemical industry as catalysts, ion exchangers, and adsorbents due to their uniform pore structure and high thermal stability. The total value of zeolites used in petroleum refining as catalysts and in detergents as water softeners is estimated at \$350 billion per year.

Based on the pioneering work of Barrer and Milton on the synthesis of zeolites (Barrer, 1948; Barrer et al., 1953; Barrer, 1981; Breeck et al., 1956), hydrothermal synthesis has become the basic route for zeolite synthesis. Zeolites are usually synthesized under hydrothermal conditions from reactive gels in alkaline media in an air-conditioned oven at temperature of 50 ~ 200 °C for days in a closed system. In the last ten years, microwave heating has been widely used in many chemical reactions, such as organic and inorganic synthesis, oxidation-reductions, and polymerizations, so in the fields of zeolite synthesis. The first report of the microwave synthesis of zeolites was given in an US patent by Mobil Oil Corp. in 1988 for the synthesis of zeolites LTA (NaA) and MFI (ZSM-5) (Chu et al., 1998). The crystallization was found to be significantly accelerated by the exposure of synthesis gels to microwave radiation. Thereafter, microwave heating synthesis has been applied for the synthesis of a number of zeolites, including zeolite A (LTA) (Panzarella et al., 2007; Slangan et al., 1997 (a)), faujasite (FAU) (Panzarella et al., 2007; Srinivasan & Grutzeck, 1999), analcime (ANA) and Gobbinsite (GIS) (Sathupunya et al., 2002), beta (BEA) (Panzarella et al., 2007), ZSM-5 (MFI) (Phiriyawirut et al., 2003; Somani et al., 2003), silicalite-1 (MFI) (Hu et al., 2009), $AlPO_4$-5 (AFI) (Girnus et al. 1995), VPI-5 (VFI) (Carmona et al., 1997), and cloverite (CLO) (Park & Komarneni, 2006). The number of publications of microwave heating synthesis of zeolite increases year by year. The microwave synthesis of zeolites and mesoporous materials was reviewed by Cundy (Cundy 1998) and by Tompsett et al. (Tompsett et al., 2006). Microwave heating synthesis turns out to be much faster, cleaner and more energy efficient than the conventional heating. The possible influences of microwave heating on the synthesis of zeolites were outlined in a previous review (Cundy 1998), mainly including: (i) Uniform heating of the synthetic mixture; (ii) Increase of the reaction rate; and (iii) Changing the association between species within the synthetic mixture. The specific reactions might be enhanced in a microwave field or by the absorption of microwave energy of reactants, intermediate species, and/or products.

For the crystallization of some zeolite structures like types LTA and FAU (zeolites X and Y), temperatures lower than 100 °C are usually applied. In these cases open glass vessels can be used in a microwave oven. However, most zeolites are crystallized at temperatures higher than 100 °C and usually stainless steel autoclaves with a Teflon insert are used. However, since steel is not transparent for microwaves, full-Teflon or full-ceramic autoclaves have to be used. Several companies offer research-microwave ovens operating at a frequency of 2.45 GHz. This is the world-wide standard frequency for the dipolar heating of water which also allows the efficient heating of other dipolar solvents like methanol or DMF.

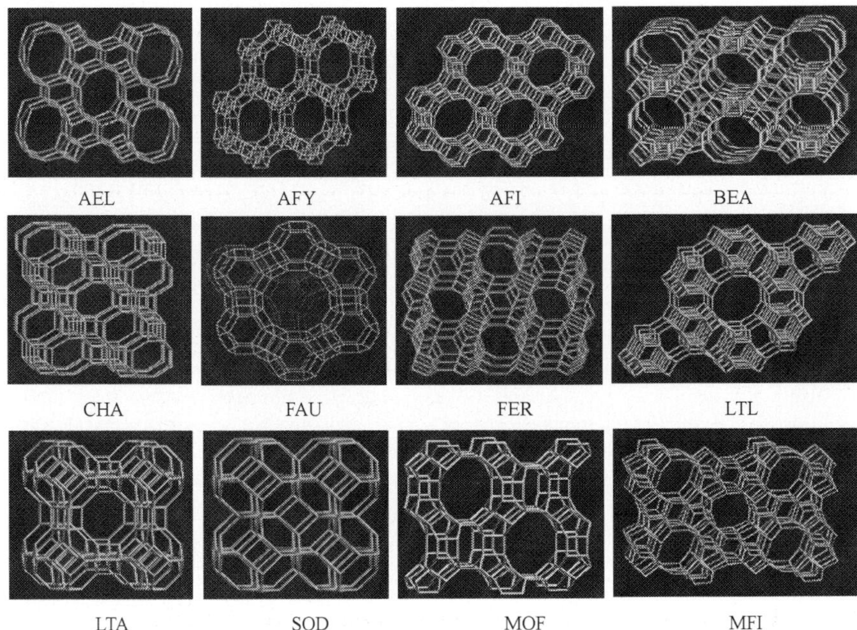

Fig. 2. Framework type of some commercial zeolite (Baerlocher & McCusker, http://www.iza-structure.org/databases).

3. Microwave synthesis of zeolite and zeolite-like porous molecular sieve membranes

Apart from the use of zeolites as powders, supported zeolite layers are of interest for many potential applications owing to their excellent properties such as uniform pore structure, high thermal stability and high mechanical strength. In the last 25 years, many research efforts have been focused on the preparation of supported zeolite layers due to their potential applications as separators (Lai et al., 2003), reactors (Casanave et al., 1995), sensors (Yu et al., 2007), and electrical insulators (Wang et al., 2001). A great deal of progress in the synthesis of zeolite membrane has been made during the last 25 years, over 2500 articles have been published (Figure 3). To date, MFI (van de Graaf et al., 1998), DDR (Tomita et al., 2004), LTA (Aoki et al., 1998; Huang et al., 2010; Huang & Caro, 2010), FAU (Kusakabe et al., 1997), SOD (Xu et al., 2004), CHA (Carreon et al., 2008), FER and MOR (Nishiyama et al., 1997), BEA (Sirikittikul et al., 2009), and AFI (Guan et al., 2003) membranes, have been prepared on different types of supports for liquid or gas separation. The first industrial application of zeolite membranes is the dewatering of bio-ethanol by using LTA zeolite membrane (Morigami et al., 2001; Sato et al., 2008). The significant progresses achieved in the rapidly growing field of supported zeolite membranes are outlined by a large number of review articles (Bein 1996; Caro et al., 2000; Caro & Noack, 2008; Lin, 2001; Tavolaro & Drioli, 1999).

Several synthesis strategies have been developed to prepare supported zeolite membranes/layers, including in-situ hydrothermal synthesis (Yan et al., 1995), secondary (seeded) growth synthesis (Lovallo et al., 1996; Lai & Gavalas, 1998), vapor phase transport synthesis (Nishiyama et al., 1996) and dry gel conversion (Ma et al., 2001). The in-situ hydrothermal synthesis is the best-studied method, in which the surface of the porous supports directly contact with the zeolite precursor solution and keep the system under controlled conditions in order to nucleate and grow a thin continuous zeolite film on the support surface. The process is relatively simple but the characteristics and quality of the obtained membrane are very sensitive to procedures and conditions. It is usually difficult to prepare high quality zeolite membrane by in-situ hydrothermal synthesis directly (Chau et al., 2000; Huang et al., 2007). The secondary growth method is now considered as one of the most attractive and flexible approaches for the formation of high quality zeolite membranes, exhibiting many advantages, such as better control over membrane microstructure (thickness, orientation) and enhancement of the membrane reproducibility (Xomeritakis et al., 1999).

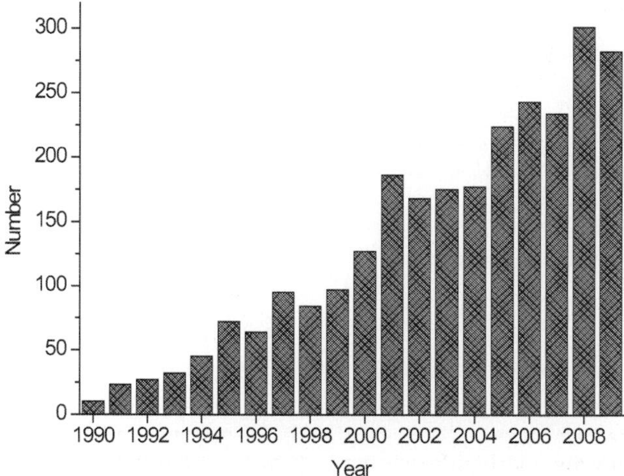

Fig. 3. Development of the articles of zeolite membrane/film in open literatures.

Microwave heating is a fast, simple and energy-efficient method, which could significantly reduce the synthesis time and cost. Microwave heating can also assist in the preparation of zeolite and zeolite-like porous molecular sieve membranes combining with the above synthesis strategies. Like in the case of the crystallization of zeolites powder, full-Teflon or full-ceramic autoclaves and standard microwave ovens operating at 2.45 GHz are used for membrane crystallization. A benefit of microwave heating in membrane preparation is that the autoclave material is transparent for microwaves, hence, the autoclave walls and the reaction system near to the wall remain cool. This effect suppresses heterogeneous nucleation at the autoclave wall. In the very beginning of microwave absorption by the liquid solution, steep temperature gradients occur. But usually these gradients are not harmful since the degradation of these temperature gradients with time is faster than the kinetically controlled nucleation. Therefore, volumetric heating produces a homogeneous nucleation and growth as well as fast supersaturation of the reaction mixture. To date, LTA, MFI, FAU, Zeolite T, SOD, AFI, CHA types of zeolite membranes as well as metal-organic frameworks (MOFs) membranes have been successfully synthesized by microwave heating. Usually these molecular sieve membranes are a few μm thick zeolite or MOF layer supported by a macroporous ceramic substrate like alumina, titania or zirconia. Whereas α-Al_2O_3 supports are transparent for microwaves, a technical TiO_2 as mixture of anatase and rutile absorbs microwaves thus influencing nucleation and crystal growth in a liquid layer directly attached to the support. Recently, Li et al. (Li & Yang, 2008) have given a comprehensive review on the microwave synthesis of zeolite membranes, summarizing the development of microwave synthesis of zeolite membranes. After this work, there have been increasing publications on the further developments in the field of microwave synthesis of zeolite membranes. During the last few years, microwave heating synthesis of molecular sieve membranes has extended from zeolite membrane to MOF membranes exploiting the shape-selectivity of membranes in gas separation. In this section, we try to make a comprehensive summary of the development of microwave synthesis of porous molecular sieve membranes.

3.1 LTA zeolite membranes

Zeolite LTA has a cubic unit cell of the composition $Na_{12}(Si_{12}O_{48})$. The LTA pore size can be fine-tuned by cation exchange, the pore diameters of the K^+(3A), Na^+ (4a) and Ca^{2+} (5A) cation exchanged forms are about 0.3 nm, 0.4 nm and 0.5 nm, respectively. More and more interest has been focused on synthesis and application of zeolite LTA (NaA) membranes due to their strong hydrophilicity in combination with a small pore size of about 0.4 nm for NaA (Kita et al., 1995; Kondo et al., 1997). Actually, among the molecular sieve membranes only the zeolite LTA membrane is produced so far on an industrial scale for the dewatering of bio-ethanol (Morigami et al., 2001; Sato et al., 2008). Probably due to the relative simple composition of the synthesis solution/gel (templates-free) and the relatively low crystallization temperature (usually <100 °C), LTA zeolite membranes have been widely studied by microwave synthesis.

Microwave heating was firstly applied by Han et al. to prepare LTA zeolite membranes in 1999 (Han et al., 1999). It is found that microwave heating markedly accelerated the crystallization rate and reduced the synthesis time from several hours to 20 min. However, no gas permeation or separation results were reported. An LTA zeolite membrane prepared by using microwave heating with separation performance was reported by Xu et al. in 2000 (Xu et al., 2000). It is proposed that the fast microwave synthesis can reduce the crystallization time to form thin and high-permeance LTA zeolite membrane, as illustrated in Figure 4. Results show that microwave heating is an effective method for the synthesis of high-permeance zeolite membranes. The synthesis time of the LTA zeolite membrane prepared by microwave heating was 8 ~ 12 times shorter than that of conventional heating. While the permeances of the zeolite membrane synthesized by microwave heating was four times higher than that of the zeolite membrane synthesized by conventional heating with comparable selectivity. However, the permeation of n-C_4H_{10} that has a larger kinetic diameter than LTA zeolite pore size indicates that the as-synthesized LTA zeolite membranes had some intergrowth defects. It is found that the permeance and selectivity of zeolite LTA zeolite membranes prepared by microwave heating could be improved through chemical modification of the macroporous alumina support surface with a thin mesoporous top layer to prevent the penetration of the reagent into the support (Chen et al., 2005). The H_2 permeance was about 5~16 times higher than that reported previously with a H_2/C_3H_8 selectivity of 14.7.

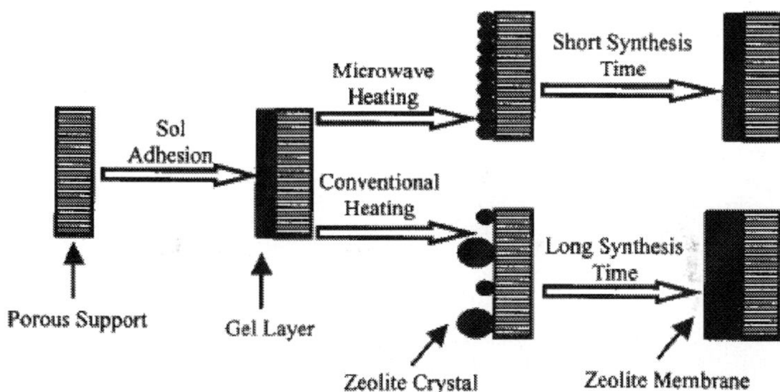

Fig. 4. Comparison of synthesis model of zeolite membrane prepare by microwave heating and conventional heating (Xu et al., 2000).

The previous studies indicated that in microwave heating synthesis the coating on the support surface with seed crystals was usually indispensable to promote the growth of LTA zeolite membranes to avoid the formation of foreign phases (Chen et al., 2003; Xu et al., 2001). A recent study in Yang's group in the utilization of microwave heating further supported the above results. They have developed a new strategy called "in-situ aging-microwave synthesis" method (Li et al., 2006). In the first-stage synthesis, the support is contacted with a clear synthesis solution and then in-situ aged to form germ nuclei on the support surface. In the second-stage synthesis, the consequential crystallization under fast and homogeneous microwave heating was carried out for the nucleation and crystals growth on the support surface. The "in-situ aging" stage in which germ nuclei produce was found to be necessary for the subsequent successful microwave synthesis. In this way, a much higher nucleation rate favored by microwave heating greatly enhanced the number of nuclei on the support surface. Therefore, LTA zeolite crystals could well intergrow in a short time to form high quality LTA zeolite membranes with a H_2/N_2 selectivity of 5.6.

Despite much progress in the preparation of LTA zeolite membranes, usually the separation factors for H_2 from such binary mixtures like H_2/N_2, H_2/O_2, H_2/CO_2 and H_2/CH_4 were found to be still close to the value of Knudsen diffusion (Aoki et al., 1998; Kondo et al., 1997). Therefore, the synthesis of LTA membranes with molecular sieving properties is still a challenge. Recently, we reported a novel synthesis strategy for the seeding-free preparation of dense LTA zeolite membranes by using 3-aminopropyltriethoxysilane (APTES) as covalent linker between zeolite membrane and porous Al_2O_3 support (Huang et al., 2010). As shown in Figure 5, APTES acts as a "bridge" between the zeolite layer and the support surface thus anchoring the LTA crystals, facilitating the formation of dense LTA zeolite membrane displaying good molecular sieving performance. Making use of the function of APTES, we prepare the LTA zeolite membrane on APTES modified alumina support via microwave heating. After crystallization for 30 min at 90 °C under microwave irradiation, a dense and thin LTA zeolite membrane with a thickness of about 2.5 µm can be formed (Figure 6). Comparing with our previous report on LTA zeolite membrane on APTES modified alumina support via conventional heating, the H_2 permeances can increase 2~3 times with a similar H_2/N_2 selectivity.

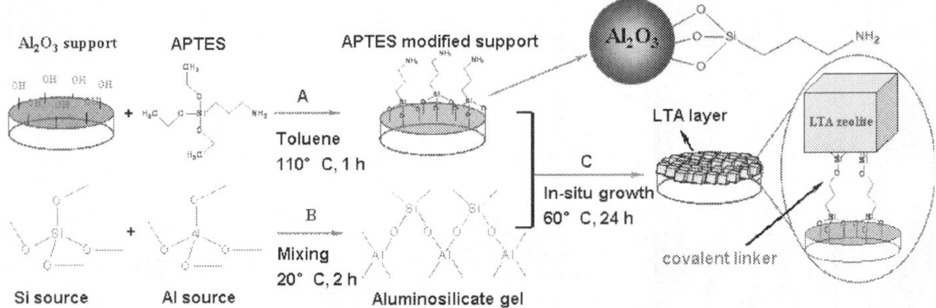

Fig. 5. Illustration of preparation of LTA membrane by using 3-aminopropyltriethoxysilane as covalent linker between the LTA zeolite layer and alumina support (Huang et al., 2010).

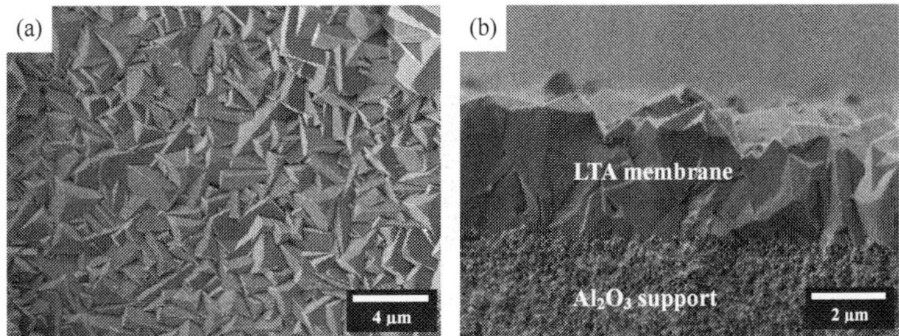

Fig. 6. (a) Top view and cross-section of SEM images of LTA zeolite membrane prepared with microwave heating on APTES-modified Al₂O₃ support.

3.2 MFI (Silicalite-1and ZSM-5) zeolite membrane

MFI type (including Al-free silicalite-1 and its Al-containing analogue ZSM-5) zeolite membranes have been widely studied in hydrocarbon separations due to its ease crystallization as well as high thermal and chemical stabilities (Arruebo et al., 2006; Coronas et al., 1998; Funke et al., 1996). However, the studies of microwave heating synthesis of MFI zeolite membranes are rather few. This is mainly due to the fact that MFI zeolite membranes are usually prepared at a relatively high temperature of 180 °C (Lin et al., 2001) and the organic structure-directing agent TPA$^+$ used to prepare MFI zeolite tends to degradate under microwave irradiation (Arafat et al., 1993).

The first study of microwave heating synthesis of silicalite-1 zeolite membranes was reported by Koegler et al. (Koegler et al., 1997). With microwave heating, a better control is possible over the temperature-time profile to obtain films containing 100 nm silicate-1 crystals on a silicon support. Madhusoodana et al. prepared MFI films on cordierite honeycomb substrates and alumina disks, involving the microwave heating to form a thin zeolite layer with nano-sized seeds and subsequent hydrothermal treatment to crystallize MFI zeolite films (Madhusoodana et al., 2006). It was found that the amount of zeolite layer formed on substrates by 10 min of microwave heating followed by 12 h of hydrothermal treatment was equivalent to that formed in 24 h by conventional hydrothermal treatment. The BET surface area of these samples was more than double that of samples made by conventional hydrothermal synthesis. In combination with secondary growth and microwave heating, Motuzas et al. prepared oriented silicalite-1 zeolite membranes on alumina supports (Motuzas et al., 2006), with the aim to investigate the morphology, thickness, homogeneity, crystal preferential orientation (CPO) and single gas permeation properties of the silicalite-1 membranes in correlation with the synthesis parameters. Thereafter, they have developed an ultra-rapid and reproducible synthesis method for preparation of thin and high quality (101) preferred orientation MFI membranes by coupling the microwave heating synthesis with a rapid template removal method in ozone (Motuzas et al., 2007). Tang et al. recently reported microwave synthesis of MFI-type zeolite membranes by secondary growth of nanocrystalline silicalite seed layers in pure inorganic precursors (Tang et al., 2009). The as-synthesized MFI membranes showed a rather high H_2/SF_6 selectivity of 1700 and H_2 permeance of 3×10^{-7} mol m^{-2} s^{-1} Pa^{-1} at 23 °C. Very recently, Sebastian et al. prepared MFI zeolite membranes on ceramic capillaries for pervaporation application under microwave heating with a high membrane surface area-to-volume ratio (Sebastian et al., 2010).

3.3 FAU zeolite membranes

The FAU zeolite (NaX and NaY) membrane, with a pore diameter of 0.74 nm, is suitable for separating large molecules, which cannot be handled effectively by MFI or LTA zeolite membrane (Jeong et al., 2003). Especially, FAU zeolite membrane display high adsorption affinity towards CO_2, showing high CO_2/N_2 permselectivity (Gu et al., 2005). In addition, FAU zeolite membrane was also reported for the dewatering of bio-ethanol due to its hydrophilic character (Li et al., 2002). Most of the FAU zeolite membranes were synthesized by secondary growth method using seeding technique with conventional heating (Lassinantti et al., 2000). Weh et al. made the first attempt to synthesize FAU composite membranes with either conventional heating or microwave heating (Weh et al., 2002). Results of gas permeation suggested that the separation of small gas molecules by the large and heteropolar FAU zeolite membrane was mainly based on differences of adsorption or diffusion rates, rather than on size exclusion like in LTA or MFI zeolite membranes.

Thereafter, by applying the so-called "in-situ aging-microwave synthesis" method in a clear solution, Zhu et al. prepared high-quality FAU zeolite membranes with highly preferable water-selectivity in the dehydration of ethanol and isopropanol through pervaporation (Zhu et al., 2009). The method coupled in situ aging at a low temperature under conventional heating and fast crystallization at a relatively high temperature under microwave heating. It was found that an amorphous gel layer containing plenty of pre-nuclei was first formed on the support surface after in-situ aging, which was indispensable to facilitate the nucleation of FAU zeolite and suppress the formation of LTA impure phase. The following rapid crystallization process under microwave heating promoted these pre-nuclei rapidly and simultaneously to develop into crystal nuclei, and then the crystals growth through propagation of gel network. Finally, a dense zeolite layer could be obtained without the aid of artificial seeding after the two-stage synthesis.

3.4 Zeolite T membranes

Pervaporation (PV) has gained widespread acceptance in the chemical industry as an effective process and energy-efficient technique for separation of azeotropic or close-boiling liquid mixtures (Feng & Huang, 1997). It has been applied to the dehydration of organic liquids mixtures by using zeolite membranes due to their better chemical and thermal stability than polymeric membranes (Shah et al., 2000). So far, most of the studies on dehydration of organic liquids mixtures by using zeolite membranes have been focused on LTA and FAU zeolite membrane (Kita et al., 1995; Li et al., 2002). LTA zeolite membranes with excellent separation performances for Pervaporation of water/organic liquids are currently commercially available and have been applied for the dehydration of organic solvents in a large scale (Morigami et al., 2001; Sato et al., 2008).

It is well known that, with an increase in Al content in the framework, the hydrophilicity of a zeolite increases, whereas the acid resistance simultaneously decreases because a strong acid leaches Al from the zeolite to decompose its framework structure (Muller et al., 2000). Therefore, LTA and FAU zeolite membranes cannot be applied for dehydration of organic liquids in acidic media. On the other hand, high silicate zeolite membranes such as MFI are stable toward acid, but have poor performance for the dehydration of organic liquids because of the hydrophobic character (Sano et al., 1997). Zeolite T with a middle Si/Al ratio of 3-4 is considered to be less hydrophilic but more stable to acid than zeolites LTA and FAU. Thus zeolite T membranes are preferable for the dehydration of organic liquids in acidic media. Zeolite T membranes were successfully prepared by the secondary growth synthesis with conventional heating and displayed the high PV performances for the dehydration of water/organic liquid mixtures even in acidic media (Cui et al., 2004). However, it was reported that there was no selectivity found for the membranes prepared on the unseeded tubes and dense and permselective crystalline layer was formed only on the seeded tubes.

As a simple preparation method, in situ synthesis is favored preparing zeolite membranes at a large scale. Recently, in situ synthesis of zeolite T membranes was developed by Zhou et al. through repeated microwave heating (Zhou et al. 2009 (a)). The obtained membranes displayed high separation performance with H_2O/alcohol selectivity higher than 10000 and rather high flux. Thereafter, Zhou et al. reported high-flux a&b-oriented zeolite T membranes by microwave-assisted hydrothermal synthesis method with high reproducibility (Zhou et al. 2009 (b)). When microwave heating was introduced in the synthesis process, the properties of the membranes were significantly improved and the synthesis time was markedly shortened to several hours.

3.5 SOD zeolite membranes

In the strictest sense of the word, sodalite (SOD) is not a zeolite, because it has only 6-ring windows with a small pore size of 2.8 $Å^3$ and thus has very limited sorption capacity. However, its framework density of 17.2 T-atoms per 1000 $Å^3$ is well within the zeolite range. Only small molecules, such as He (2.6 Å), NH_3 (2.5 Å) and water (2.65 Å), can enter the pore of SOD. Therefore, SOD zeolite membrane can be expected to display a better performance for the separation of small molecules from larger ones. SOD zeolite membrane can be prepared by both conventional heating and microwave heating. Khajavi et al. (Khajavi et al., 2007) have investigated the suitability of synthesis of a SOD membrane through conventional heating for water permeation. Julbe et al. (Julbe et al., 2003) have evaluated the potentialities of microwave heating for synthesis of SOD membranes on α-Al_2O_3 asymmetric tubes. It was found that the aluminium leaching from the support lead to a disturbing phenomenon for both the reproducibility and scaling-up of the direct membrane synthesis. This negative influence could be overcome by the pre-seeding/secondary growth method, which allowed synthesizing larger and homogeneous samples with a good reproducibility. For single gas permeation at 20~165 °C, gas through the SOD/α-Al_2O_3 membrane was essentially governed by adsorption and surface diffusion, yielding a maximum selectivity a He/N_2 of 6.2 at 115 °C.

Subsequently, Xu et al. have developed the microwave synthesis of high quality pure SOD zeolite membranes on α-Al_2O_3 supports (Xu et al., 2004). With detailed studied on the influence of microwave synthesis conditions, such as synthesis time and synthesis procedure, on the formation of SOD zeolite membrane, it was found that only 45 min was sufficient to form dense SOD zeolite membranes, which was eight times faster than the conventional heating hydrothermal synthesis. A pure SOD zeolite membrane was easily synthesized by microwave heating, while impure phases such as FAU zeolite, LTA zeolite, were usually found by conventional heating. The pure SOD zeolite membrane prepared by microwave heating displayed good gas separation performance with H_2/N_2 permselectivity larger than 1000.

3.6 AFI zeolite membranes

Aluminophosphate molecular sieves, known as series of $AlPO_4$-n and built from alternating AlO_4^- and PO_4^+ tetrahedra with a neutral framework, are useful in a number of applications as catalysts and adsorbent (Wilson et al., 1982). Among them, a well known $AlPO_4$-5 molecular sieve with an AFI type framework has attracted much attention due to its one-dimensional 12-membered-ring channels parallel to the c-axis with a pore diameter in micropore region of 0.73 nm (Gupta et al., 1995). The aluminophosphates are usually prepared under conventional hydrothermal heating conditions with a reaction time ranging from several hours to several days (Wan et al., 2000). For the first time, Girnus et al. (Girnus et al., 1995) reported the microwave synthesis of large $AlPO_4$-5 single crystals in 1995. After 60 s of microwave heating, well-shaped hexagonal column-like large $AlPO_4$-5 crystals up to 130 µm in length and 40 µm in thickness could be obtained and were found to be effective for one dimensional molecular sieving membrane. While Mintova et al. (Mintova et al., 1998) prepared nanosized $AlPO_4$-5 molecular sieves and submicron $AlPO_4$-5 films on gold coated quartz crystal microbalances by microwave treatment of aluminophosphate precursors. It was shown that the morphology, orientation, and size of the $AlPO_4$-5 crystals could be controlled by varying the chemical composition of the initial solution and the conditions of microwave treatment of aluminophosphate precursors. The application of microwave heating leads to improved preparation of nanosized molecular sieve crystals and thin $AlPO_4$-5 films in a very short crystallization times.

3.7 MOFs and ZIFs membranes

Since the discovery of metal-organic frameworks (MOFs), microporous MOFs have been attracted intense attention due to their potential applications in gas adsorption and storage, molecular separation, and catalysis (Yaghi et al., 1995; Rosi et al., 2003; Pan et al., 2006; Wu & Lin, 2007). MOFs generally consist of metal–oxygen polyhedra containing divalent (Zn^{2+}, Cd^{2+} and Cu^{2+}) or trivalent (Al^{3+}, Cr^{3+}) metal cations interconnected with a variety of organic linker molecules, resulting in tailored nanoporous materials. To date, mainly research efforts have been directed towards the synthesis of novel MOF structures through the assembly of molecular building blocks to construct tailored frameworks with tunable sizes, morphologies, structures, and desired properties (Li et al., 1999; Rowsell & Yagie, 2005). On the other hand, their highly diversified structures and pore size as well as high surface areas and specific adsorption affinities recommend MOFs as fascinating candidates for fabrication of superior molecular sieve membranes for gas separation. The potential of MOFs as membrane material has been well recognized both experimentally and computationally (Keskin & Sholl, 2007; Guo et al., 2009; Venna et al., 2010). A great deal of research effort has been focused on the preparation of supported MOF membranes. However, progress in MOF membranes is rather limited and synthesis of continuous MOF layers as membranes still remains a challenge. One major obstacle to grow MOF films on supports are the organic linkers in MOF materials, which usually do not provide additional linkage groups that can form bonds with OH groups on the surface of the supports. It was rather difficult to prepare continuous MOF membrane layers by a direct solvothermal synthesis route since the heterogeneous nucleation of MOF crystals on supports is very poor. Therefore, chemical modifications (Hermes et al., 2005) and seed coating (Ranjan & Tsapatsis, 2009) of the supports are usually indispensable to direct and control the nucleation and growth of the MOFs layers.

Recently, we have developed a novel synthesis strategy to prepare molecular sieve ZIF-22 membranes through covalent functionalization by using 3-aminopropyltriethoxysilane (APTES) as covalent linkers between the ZIF-22 layer and the titania support (Huang et al, 2010). As shown in Figure 7, in the first step, the porous alumina supports were treated with APTES for introducing the amino groups through a reaction between the ethoxy groups of the APTES and the surface hydroxyls of the Al_2O_3 support. In the second step, the amino groups of APTES coordinate to the free Zn^{2+} centres and bind the growing nanocrystals directly, leading to building of a "bridge" between the growing ZIF-22 membrane and the ceramic support to promote the nucleation and growth of the ZIF-22 on the support surface. Results strongly indicate that the covalent linkages between the ZIF-22 and the alumina support indeed facilitate the formation of a compact ZIF-22 layer on the APTES modified support. After solvothermal reaction, the polycrystalline ZIF-22 has grown as a compact layer on the APTES-modified TiO_2. The scanning electron microscopy (SEM) top view (Fig. 8a) and cross-section (Fig. 8b) show that the ZIF-22 membrane has a thickness of about 40 μm and is well intergrown, no cracks, pinholes or other defects are visible. On the contrary, separate ZIF-22 crystals and crystal islands rather than a continuous layer are formed if the support surface was not treated with APTES before ZIF-22 crystallization.

Microwave heating synthesis has been proved to be an effective way for preparation of high quality zeolite membrane in a rather short synthesis time. Rapid production of MOF crystals and films via microwave-assisted solvothermal synthesis has been recently reported (Ni & Masel 2006; Jhung et al., 2007). Yoo et al. made the first attempt to prepare MOF-5 films on carbon-coated anodic aluminium oxide (AAO) support by microwave heating synthesis (Yoo & Jeong, 2008). It was found that there was little MOF-5 deposition on bare AAO, but a coating with well-developed cubic crystallites was formed on amorphous carbon/AAO and a rather continuous film of smaller crystallites on graphite/AAO. The enhancement of

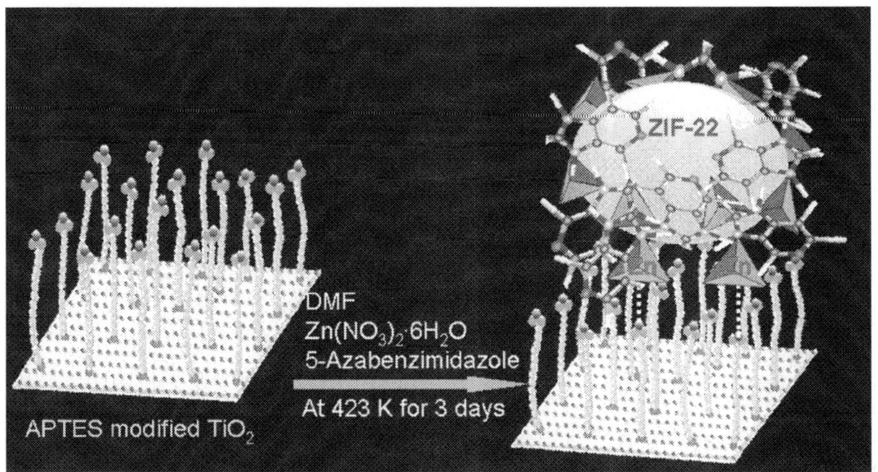

Fig. 7. Scheme of the preparation of a ZIF-22 membrane by using APTES as covalent linker between the ZIF-22 membrane and the titania support (Huang et al, 2010).

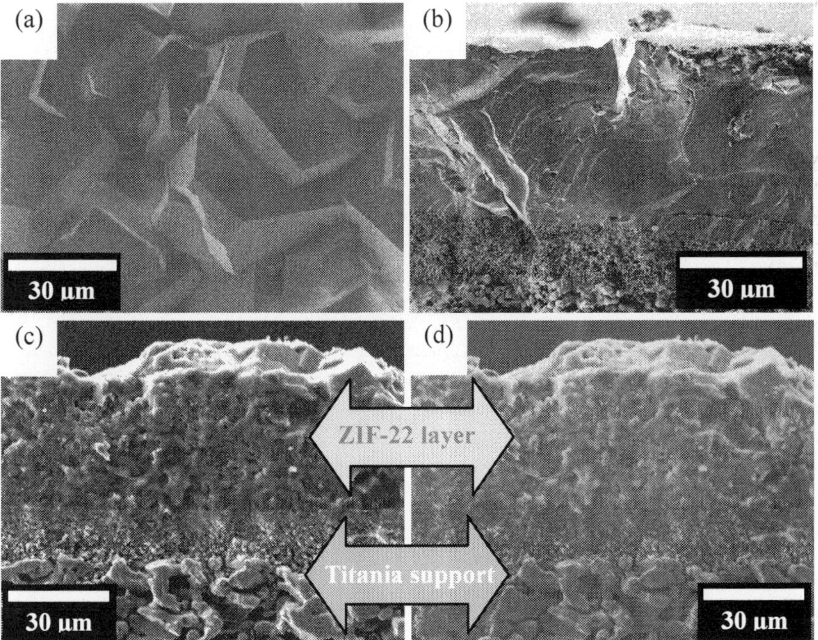

Fig. 8. (a) Top view and (b) cross-section SEM of the ZIF-22 membrane, (c, d) EDXS mapping of the sawn and polished ZIF-22 membrane (colour code: green, Zn; blue, Ti; brown is C from the polymer used for sample preparation) (Huang et al, 2010).

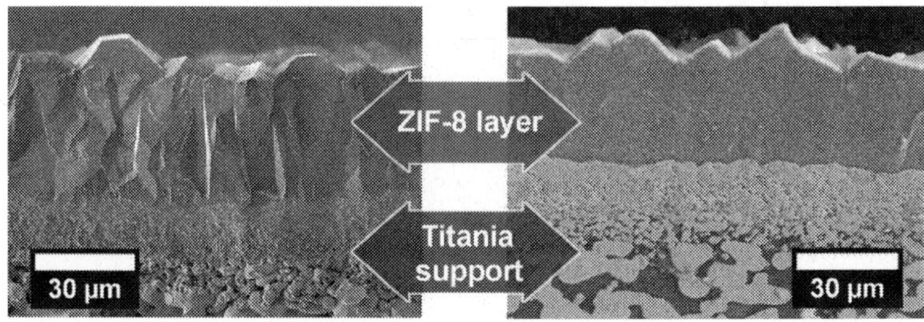

Fig. 9. (left) SEM image of cross-section of a ZIF-8 membrane, and (right) EDXS mapping of the sawn and polished ZIF-8 membrane (color code: orange, Zn; cyan, Ti) (Bux et al., 2009).

coating coverage on graphite/AAO was attributed to the better absorption of the microwaves. Thereafter, Yoo et al. prepared a MOF-5 seed layers on porous supports under microwaves, and then secondary growth was carried out to form a continuous MOF-5 membrane (Yoo et al., 2009). However, gas permeation results showed that the diffusion of small gas molecules (H_2, CH_4, N_2, and CO_2) through the MOF-5 membrane mainly followed the Knudsen diffusion behavior.

For MOF membranes synthesis, not only the problems with growing a dense polycrystalline layer on porous ceramic or metal supports but also the thermal and chemical stability of the MOFs have to be considered. Among the reported MOFs, zeolitic imidazolate frameworks (ZIFs), a subfamily of MOFs based on transition metals (Zn, Co) and imidazolates as linkers, have emerged as a novel type of crystalline porous material for the fabrication of molecular sieve membranes due to their zeolite-like properties such as permanent porosity, uniform pore size, exceptional thermal and chemical stability (Park et al., 2006; Phan et al., 2010). Bux et al. reported ZIF-8 membranes with molecular sieving performance by microwave-assisted synthesis method (Bux et al., 2009). With microwave heating synthesis, a crack-free, dense polycrystalline layer of ZIF-8 with thickness of about 30 µm could be formed on a porous titania support (Figure 9). Energy-dispersive X-ray spectroscopy (EDXS) reveals that there is a sharp transition between the ZIF-8 layer (Zn signal) and the titania support (Ti signal). The ZIF-8 membrane showed a good H_2/CH_4 selectivity higher than 10 (Figure 10), while a H_2/CO_2 selectivity close to Knudsen diffusion due to the fact that ZIF-8 is very flexible with pore size of 0.34 nm larger than the kinetic diameter of CO_2 (0.33 nm). For separation of H_2 from CO_2, ZIF-7 membrane (Li et al., 2010), which prepared on a seeded alumina support via microwave heating synthesis, displayed a better separation performance due to a smaller pore size of 0.30 nm, which is just in between the molecular size of H_2 (0.29 nm) and CO_2 (0.33 nm). After secondary growth with microwave irradiation, a well intergrown polycrystalline ZIF-7 layer without any pinholes or cracks was formed on the seeded alumina support (Figure 11a). Owing to the very thin seed layer and rather short secondary growth time by microwave heating, the thickness of ZIF-7 layer is only about 1.5 µm (Figure 11b). EDXS reveals that there is a sharp transition between the ZIF-7 layer (Zn signal) and the alumina support (Al signal) (Figure 11c). As shown in Figure 12, for both single and mixed gas permeation, there is a clear cut-off between H_2 and CO_2. The H_2/CO_2 ideal selectivity (calculated as the ratio of single gas permeances) and separation factor are 6.7 and 6.5, respectively, which exceed the Knudsen separation factor (~4.7). For the 1:1 binary mixtures, the H_2/N_2 and H_2/CH_4 separation factors are 7.7 and 5.9, respectively (at 200 °C and 1 bar), both are higher than the corresponding Knudsen separation factors (3.7 and 2.8, respectively) (Li et al., 2010).

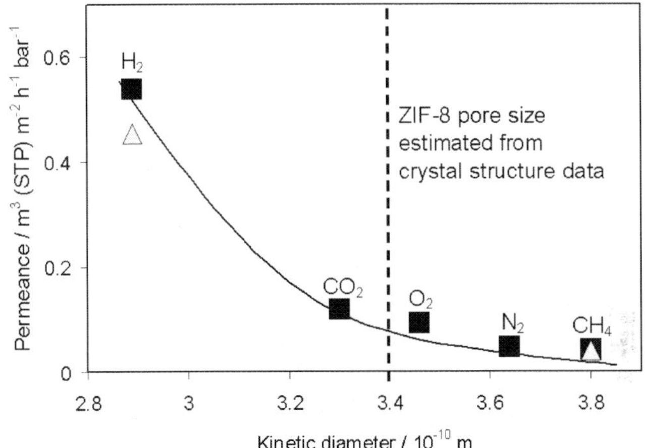

Fig. 10. Single (squares) and mixed (triangles) gas permeances for a ZIF-8 membrane as function of kinetic diameters (Bux et al., 2009).

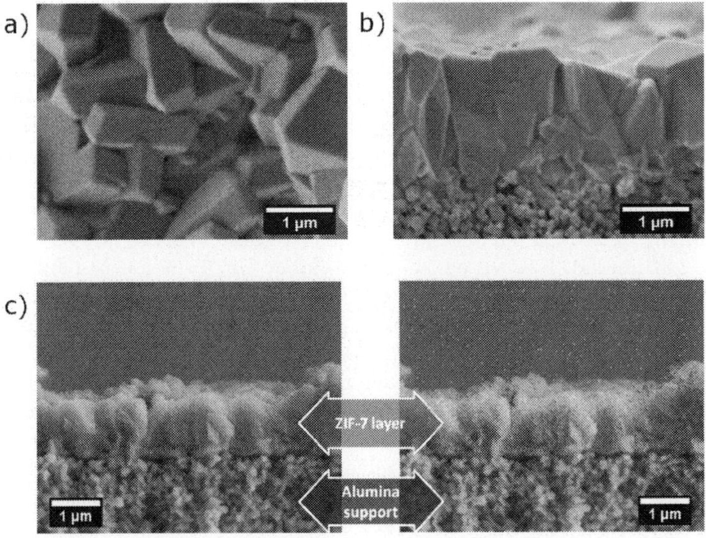

Fig. 11. a) Top view and b) cross-section SEM images of the ZIF-7 membrane; c) EDXS-mapping of the ZIF-7 membrane, color code: orange, Zn; cyan, Al (Li et al., 2010).

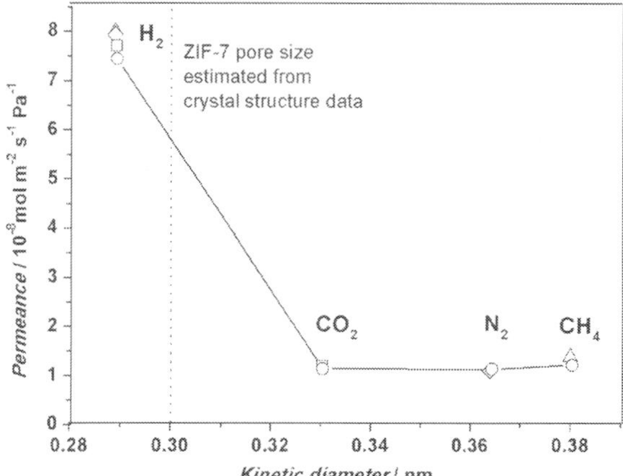

Fig. 12. Permeances of single gas (circles) and from 1:1 mixtures (squares: H_2/CO_2 mixture, rhombus: H_2/N_2 mixture, triangles: H_2/CH_4 mixture) of the ZIF-7 membrane at 200 °C as a function of molecular kinetic diameters (Li et al., 2010).

4. Combined microwave and conventional heating modes

In the past ten years, the fast synthesis of porous molecular sieving membranes by microwave heating is relatively well established. However, progress in understanding the mechanism and engineering for the rate enhancement of the syntheses is rather limited and still remains a challenge. The crystallisation of zeotype materials is frequently constrained by limitations at the nucleation stage. Therefore, it is common practice to age reaction mixtures or to add seeds (Cundy et al., 1998), thus leading to a reduction of the induction period and to a promotion of a dominant crystalline phase. In a previous report (Slangen et al., 1997 (b)), the synthesis time of microwave heating of LTA zeolite was reduced as short as 1 min by aging the reaction mixtures for various periods, and it was suggested that the rearrangement of the synthesis mixture to yield nuclei was the bottleneck in the microwave synthesis. Jhuang et al. quantitatively investigated the selectively accelerated stage between nucleation and crystal growth by various heating modes together with microwave irradiation (MW) and conventional electric heating (CE) (Jhuang et al., 2007). It was demonstrated that the microwave irradiation accelerates not only the nucleation but also crystal growth.

For microwave synthesis of molecular sieving membranes, multiple heating modes together with microwave heating (MH) and conventional heating (CH) are usually applied since it is difficult to prepare high performance zeolite membranes by use of microwave synthesis only. Recently, we prepared uniform and dense LTA zeolite membrane by hydrothermal synthesis method together with microwave heating and conventional heating (Huang & Yang, 2007). Two heating models were cooperated to prepare uniform and dense LTA zeolite membrane, as show in Figure 13. Before conventional heating, the pretreated support was quickly heated to 363 k for 25 min in a microwave oven. After microwave synthesis, the support surface was covered with uniform and small zeolite particles, with crystals size of

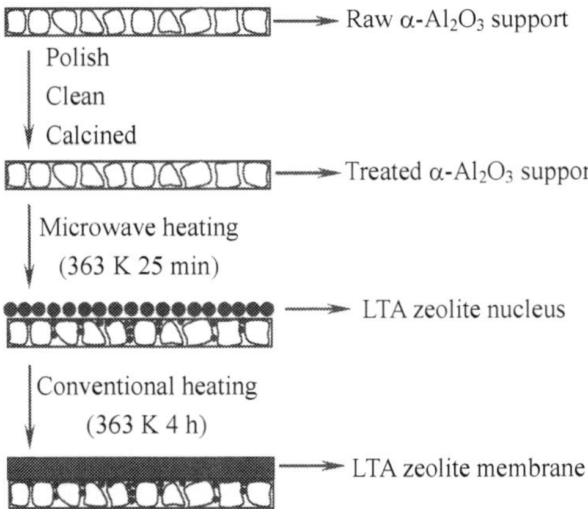

Fig. 13. Schematic diagram for synthesis of LTA zeolite membrane together with microwave heating and conventional heating method (Huang & Yang, 2007).

about 2~3 µm, which were used as seeds for further hydrothermal growth. The following hydrothermal growth in an air oven at 363 K for 4 h helps to increase the membrane density, facilitating to form well inter-grown and pure LTA zeolite membrane. In the case of microwave-free hydrothermal synthesis, no dense LTA zeolite membrane could be formed, and impure phases such as cabbage-like spherical crystals as well as octahedral crystals were usually formed among the cubic LTA zeolite crystals (Huang & Yang, 2007).

A different heating model containing conventional heating (CH) and microwave heating (MH) called "in-situ aging-microwave synthesis" was developed to prepare high quality LTA zeolite membrane as shown in Figure 14 (Li et al., 2006). Firstly the support is contacted with a clear synthesis solution and then in-situ aged at 50 °C for 7 h to form germ nuclei on the support surface. And then, the consequential crystallization under fast and homogeneous microwave heating at 90 °C for 25 min was carried out for the nucleation and

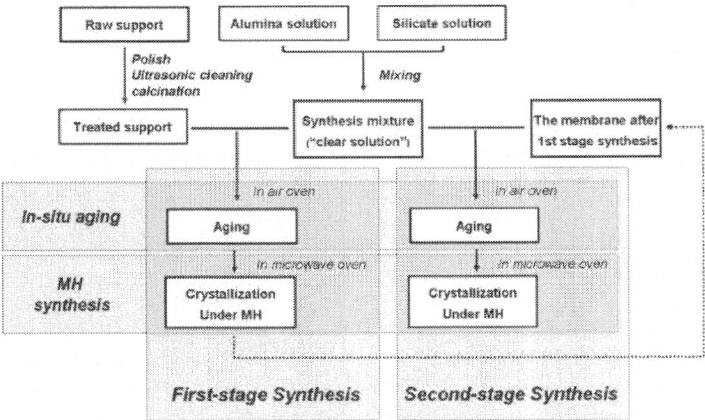

Fig. 14. The flow diagram of AM synthesis of zeolite membranes (Li et al., 2006).

growth of LTA zeolite membrane on the support surface. The "in-situ aging" stage to produce germ nuclei was found to be necessary for the subsequent successful microwave synthesis. In this way, a much higher nucleation rate favored by microwave heating greatly enhanced the number of nuclei on the support surface. Therefore, LTA zeolite crystals could well intergrow in a short time to form high quality LTA zeolite membranes. Applying this heating mode, FAU (Zhu et al., 2009) and T (Zhou et al. 2009 (b)) zeolite membrane were successfully synthesized. It should be noted that for both the above heating model, the first stage to produce nuclei as essential seeds for further growth is necessary to form dense membranes.

5. Differences between microwave and conventional hydrothermal synthesis

As shown above, microwave heating can remarkably reduce the synthesis time from days (hours) to hours (minutes). This is probably one of the best advantages of microwave synthesis compared with conventional heating synthesis. In the field of microwave synthesis of porous materials such as zeolites as well as its analogues, it has been well recognized that the microwave heating can effectively control particle size as well as size distribution (Hu et al., 2009), phases (Yoon et al., 2005), and morphology (Phiriyawirut et al., 2003). In the case of zeolite membranes, microwave synthesis also has great effect on the membrane characteristics such as crystals size, morphology and density, which essentially determine the membrane separation performances. According to the fast and homogeneous microwave heating leading to high nucleation rate, uniform and thinner zeolite membranes with higher fluxes/permeances can be obtained (Xu et al., 2000). Differences in crystals morphology of the zeolite membrane between microwave heating and conventional heating are usually observed. As shown in Figures 15 and 16, different to well-shaped cubic crystals of the LTA AlPO$_4$ and LTA aluminiumsilcate membrane prepared by conventional heating, the crystals of the two membranes prepared by microwave heating display a cone-shaped morphology with more uniform crystals of a smaller size. Different crystals morphology of the LTA zeolite membrane prepared by microwave heating and conventional heating were also reported by Li et al. (Li et al., 2006). The LTA zeolite membrane prepared by conventional heating was composed of well-shaped cubic LTA crystals, while the LTA zeolite membrane prepared by microwave heating consists of spherical grains without well-developed crystal faces. It was also reported that microwave heating synthesis leaded to differences in chemical composition of the zeolite membranes. Weh et al. reported the as-synthesized FAU

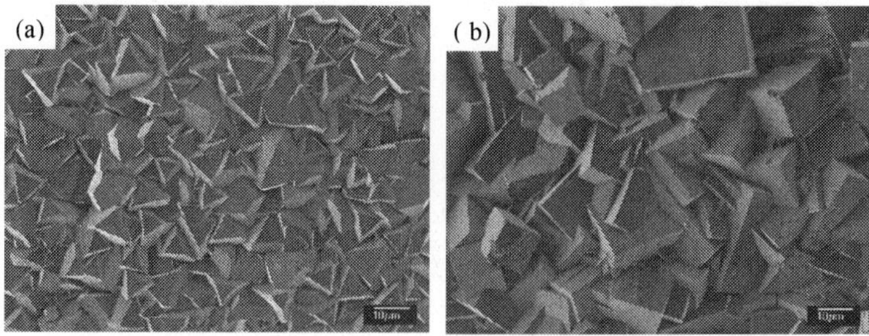

Fig. 15. SEM images of LTA AlPO$_4$ zeolite membrane prepared with (a) microwave heating synthesis (b) conventional heating synthesis.

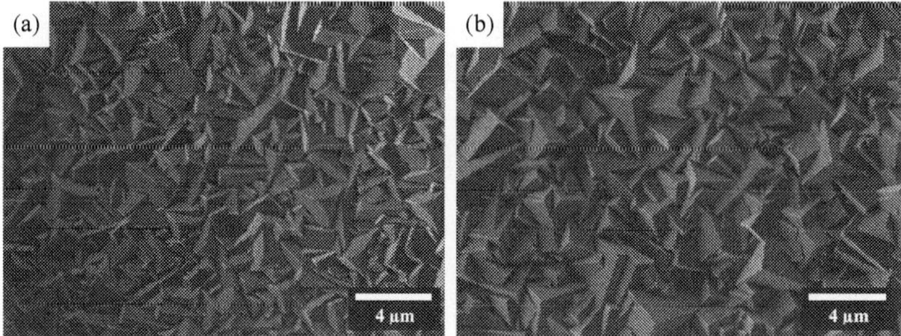

Fig. 16. SEM images of LTA aluminiumsilicate membrane prepared with (a) microwave heating synthesis (b) conventional heating synthesis.

zeolite membrane prepared by microwave heating had a slightly increased Si/Al ratio (from 1.4 to 1.8) (Weh et al., 2002). For microwave synthesis of LTA zeolite membranes, it was also found that the as-synthesized LTA zeolite membrane had a higher Si/Al ratio of 1.43 (usually 1) (Li et al., 2006).

Probably the most important difference between microwave heating and conventional heating is that of separation performances. Li et al. made a detailed comparison of separation performances of LTA zeolite membrane prepared by microwave heating and conventional heating synthesis (Li et al., 2006). LTA zeolite membrane prepared by in situ aging microwave synthesis displayed an improved permselectivity for the separation of H_2/N_2 with a selectivity of 5.6 with a high permeance. It was proposed that LTA zeolite membranes prepared by microwave heating containing spherical grains with undefined crystal facets could decrease the inter-crystalline defects, thus leading to the formation of compact defect-free zeolite membranes. The difference in separation performances of LTA zeolite membranes prepared by microwave heating and conventional heating synthesis was also confirmed by pervaporation separation (Li et al., 2007). For the membrane prepared by microwave heating (Figure 17), when the water concentration in the feed (WF) decreased from 9.6 to 0.2 wt.%, the water concentration in the permeate (WP) varied from 100.0 to 90.0

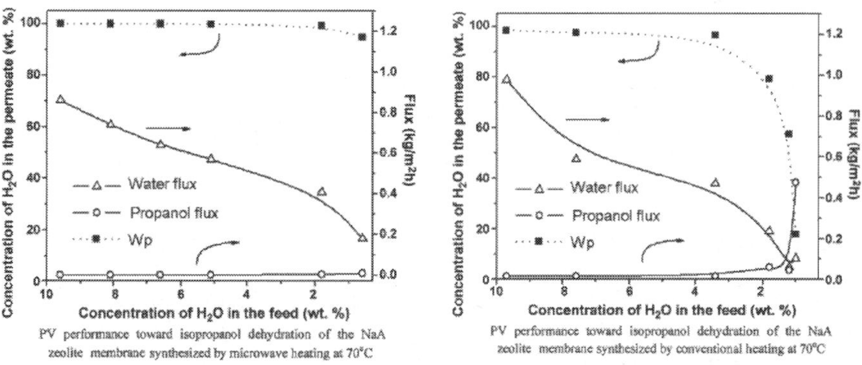

Fig. 17. Comparison of pervaporation performances between LTA zeolite membranes synthesized by microwave heating and conventional heating. Wp: the water concentration in permeate (Li et al., 2007).

wt.%, with a permeation flux decreasing approximately linearly from 0.86 to 0.08 kgm^{-2}h^{-1}. However, for the membrane prepared by conventional heating, the WP decreased rapidly when the WF was smaller 2.0 wt.%. When the WF was 1.0 wt.%, the WP was only 18.0 wt.%, while the permeate flux abnormally increased dramatically. This indicated that the membrane synthesized by conventional heating might have quite a few non-zeolitic defects. When the WF was low, the adsorbed water was not sufficient to block the defects to restrain isopropanol permeation, thus more isopropanol began to pass through the defects, which leaded to a decrease in the WP and to an increase of the permeate flux.

6. Formation mechanism of microwave synthesis of zeolite membranes

A full understanding of the formation mechanisms of zeolite membranes is of great significance to control and optimize membrane growth. Formation of a membrane could outlined in four different ways (Myatt et al., 1992), as shown in Figure 18: (1) production of nuclei and growth of crystals in the bulk solution followed by their attraction to and association with the substrate; (2) production of nuclei in the bulk solution, but diffusion to and accumulation on the substrate before significant growth has occurred; (3) diffusion of amorphous aluminosilicate material to and concentration on the substrate, providing more favorable conditions for nucleation and growth in the vicinity of the surface; (4) production of nuclei on the substrate surface, followed by growth. It is proposed that option (4) in conjunction with either option (2) or (3) was responsible for membrane formation.

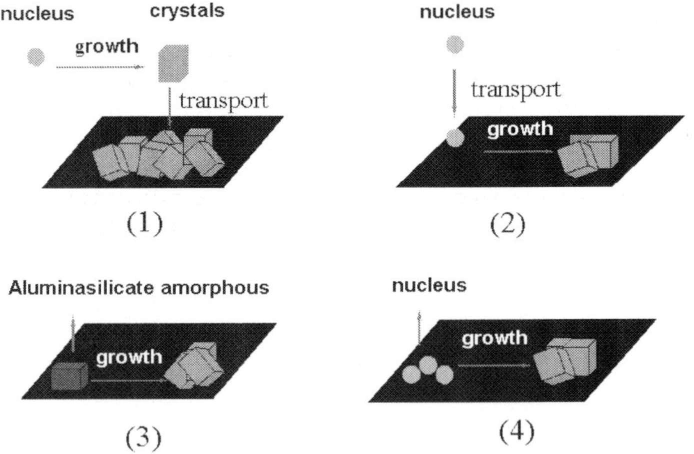

Fig. 18. Four possible growth mechanisms of zeolite membrane (Myatt et al., 1992).

Up to now, the molecular understanding of the formation mechanism of zeolite membranes by microwave heating synthesis is still rather limited. In the beginning of the 1990s, good experience in the crystallization of large and phase-pure zeolite crystals was made by using microwave heating synthesis. Pioneering papers reported the successful synthesis of MFI (ZSM-5) and FAU(Y) (Arafat et al., 1993), LTA and FAU(X) (Jansen et al., 1992), AFI (CoAPO-5 (Girnus et al., 1994) and AlPO$_4$-5 (Girnus et al., 1995 (a) and (b)). However, there was and is still some mystery about the molecular understanding of microwave heating.

- Homogeneous vs. heterogeneous heating: There can be a very high energy input by microwave absorption compared with the classical heating of autoclaves in air conditioned ovens. Microwave absorption takes place exponentially from outside to inside of an autoclave following Lambert Beer Law and leads – at least in a short time scale - to immense temperature gradients. Therefore, the synthesis mixture is heated heterogeneously by the microwaves. However, since the walls of the autoclave must be transparent for microwaves, these walls and a liquid layer near to the wall remain cold and heterogeneous nucleation is avoided.

- Freely rotating water molecules: The energy of one OH-bridging bond in water is about 20 kJ/mol (Greenwood & Earnshaw, 1997), i.e. 11.3×10^{-20} J per one water molecule. Since statistically every water molecule has about 3.4 OH-bridges, the energy of one OH-bridge is close to 3.3×10^{-20} J which is much larger than as the energy of one microwave quantum (1.6×10^{-24} J). Therefore, the existence of freely rotating water molecules as super-solvent seems not to be realistic (Caro & Noack, 2006). The dipolar heating of liquids like water, methanol or DMF is more a flip-flop orientation than a free rotation.

- Microwave effect: Probably there is no intrinsic microwave effect leading to a reduction of the nucleation and crystallization time compared to conventional heating if the latter can be done quick enough, e.g. by using induction heating which is based on the mobility of ions in an alternating field and results in a really homogeneous heating (Jansen, 2006).

However, there are other microwave effects which originate from the quick energy input and the resulting fast heating rate which brings the zeolite batch quickly to the crystallization temperature and suppresses kinetically the formation of nuclei. Hence, microwave heating can shorten the nucleation period. Furthermore, because of the accelerated heating of the synthesis mixture, the silicate species, due to a kinetic effect, are not in their thermal equilibrium since the heating rate is faster than decomposition rate of the macromolecular silica species. It is assumed, therefore, that the transport and the reactivity of alumina and silica species to precursor building units of a zeolite or MOF crystallization are influenced by microwave heating due to the kinetics of precursor formation, kinetics of nucleation and kinetics of crystal growth.

A remarkable progress in the utilization of microwave heating synthesis of zeolite membranes was achieved by Yang et al. in the last few years who developed the "in situ aging – microwave synthesis" method (AM method) (Li et al., 2006; Zhu et al., 2009; Zhou et al. 2009 (b)). Recently, they proposed the formation mechanism of LTA zeolite membranes by using gravimetric analysis, XRD, SEM, XPS, ATR/FTIR, and gas permeation to characterize the whole formation process of LTA zeolite membranes (Li et al., 2006), as illustrated in Figure 19. In this way, a gel layer containing plenty of pre-nuclei is firstly formed on the support after in situ aging. During the following microwave heating synthesis, these pre-nuclei rapidly and simultaneously develop into crystal nuclei, and then crystal growth, and finally formation of compact LTA zeolite membranes consisting of spherical grains with undefined crystal facets. It is believed that the essential mechanism of zeolite membrane formation in microwave heating is similar to that in conventional heating. Therefore, it can be concluded that the function of microwave irradiation during zeolite membrane formation is mainly derived from its thermal effect. The acceleration effect of microwave synthesis is mainly a result of fast and simultaneous nucleation.

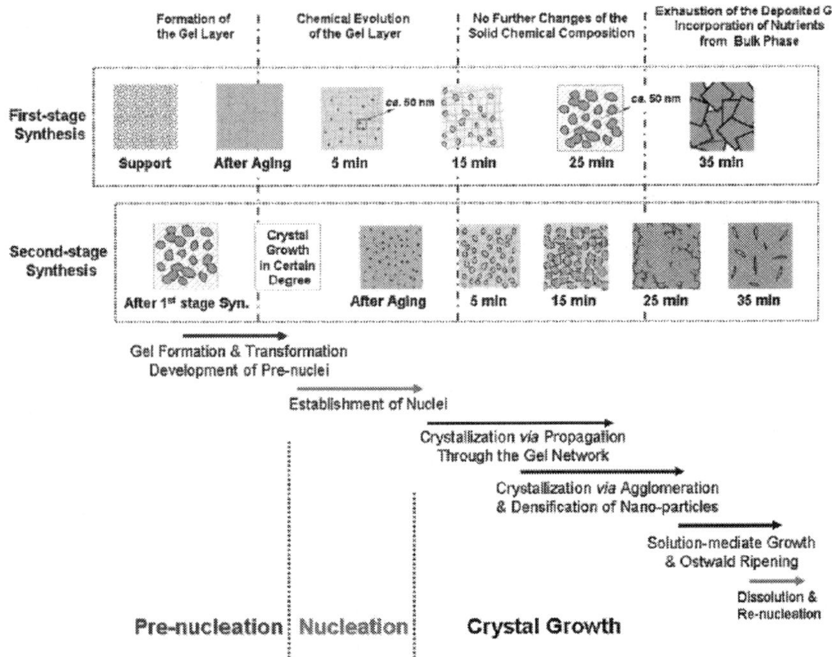

Fig. 19. Schematic illustration of the proposed formation mechanism of LTA type zeolite membrane synthesized by an "in-situ aging-microwave heating" method (Li et al., 2006).

7. Conclusions and outlook

Microwave heating synthesis is a promising method for the facile synthesis of porous molecular sieve membranes. There is an impressive progress in the development of porous molecular sieve membranes by microwave heating synthesis technique during the past decade. The microwave technology is now available to prepare porous molecular sieve membranes with sufficient quality and reliability. More and more molecular sieve membranes with good separation performances have been successfully synthesized under microwave irradiation. The combination of microwave and conventional heating is helpful to promote the nucleation and growth of zeolite membranes thus improving the membrane separation performance. Microwave heating synthesis can remarkably reduce the synthesis time, which will open a door for the production of zeolite membranes at industrial scale. Besides markedly accelerating the rate of membrane preparation, the microwave heating can effectively control the membrane morphology, orientation, composition, and thickness, thus leading to great improvement of separation performances of the zeolite membranes. Often only a few nuclei are formed since the nucleation step of crystallization is usually kinetically controlled. The high energy input causes a quick heating up of the crystallization system. Therefore, microwave heating is especially fruitful for secondary growth of membranes, i.e. in a first step seed crystals are attached to the support surface and in a second step these seeds grow to continuous layers at low supersaturation by microwave heating since a new homogeneous or heterogeneous nucleation is suppressed. Much effort has been made towards the understanding of the formation mechanism of zeolite membranes under microwave irradiation. Thermal effect and some specific microwave effects are proposed to be responsible for the influences of microwave heating on the synthesis of zeolites membranes. However, further research work is still needed to clarify the formation mechanism of zeolite membranes under microwave irradiation. This needs the interdisciplinary co-operation between material science, physical chemistry and microwave engineering. On the other hand, in situ characterization of the chemical and physical changes during microwave synthesis is required, such as neutron and X-ray scattering, and vibrational (Raman) spectroscopy.

8. References

Aoki, K.; Kusakabe, K. & Morooka, S. (1998). Gas permeation properties of A-type zeolite membrane formed on porous substrate by hydrothermal synthesis. *Journal of Membrane Science*, 141, 2, (April 1998) 197-205, 0376-7388

Arafat, A.; Jansen, J. C.; Ebaid, A. R. & Vanbekkum, H. (1993). Microwave preparation of zeolite-Y and ZSM-5. *Zeolites*, 13, 3, (March 1993) 162-165, 0144-2449

Arruebo, M.; Falconer, J. L. & Noble, R. D. (2006). Separation of binary C-5 and C-6 hydrocarbon mixtures through MFI zeolite membranes. *Journal of Membrane Science*, 269, 1-2, (February 2006) 171-176, 0376-7388

Baerlocher, Ch. & McCusker, L. B. Database of Zeolite Structures, http://www.iza-structure.org/databases.

Barrer, R. M. (1948). Synthesis of a zeolitic mineral with chabazite-like sorptive properties. *Journal of the Chemical Society*, (February 1948) 127-132, 03009246

Barrer, R. M.; Hinds, L. & White, E. A. (1953). The hydrothermal chemistry of silicates. Part III. Reactions of analcite and leucite. *Journal of the Chemical Society*, (May 1953) 1466-1475, 03009246

Barrer, R. M. (1981). Zeolite and their synthesis. *Zeolites*, 1, 3, (October 1981) 130-140, 0144-2449

Bein, T. (1996). Synthesis and applications of molecular sieve layers and membranes. *Chemistry of Materials*, 8, 8, (October 1996) 1636-1653, 0897-4756

Bougrin, K.; Loupy, A. & Soufiaoui, M. (2005). Microwave-assisted solvent-free heterocyclic synthesis. *Journal of Photochemistry and Photobiology C: Photochemistry Reviews*, 6, 2-3, (October 2005) 139-167, 1389-5567

Breck, D. W.; Eversole, W. G. & Milton, R. M. (1956). New synthetic crystalline zeolites. *Journal of the American Chemical Society*, 78, 10, (May 1956) 2838-2839, 0002-7863

Bux, H.; Liang, F. Y.; Li, Y. S.; Cravillon, J.; Wiebcke, M. & Caro, J. (2009). Zeolitic imidazolate framework membrane with molecular sieving properties by microwave-assisted solvothermal synthesis. *Journal of the American Chemical Society*, 131, 44, (November 2009) 16000-16001, 0002-7863

Carmona, J. G.; Clemente, R. R. & Morales, J. G. (1997). Comparative preparation of microporous VPI-5 using conventional and microwave heating techniques. *Zeolite*, 18, 5-6 (May 1997) 340-346, 0144-2449

Caro, J.; Noack, M.; Kolsch, P. & Schafer, R. (2000). Zeolite membranes-state of their development and perspective. *Microporous and Mesoporous Materials*, 38, 1, (July 2000) 3-24, 1387-1811

Caro, J. & Noack, M. (2006). Proceedings of the workshop on zeolite membranes, 9th International Conference on Inorganic Membranes, 13:978-82-14-04026-5, Lillehammer, Norway, June 2006

Caro, J. & Noack, M. (2008). Zeolite membranes - Recent developments and progress. *Microporous and Mesoporous Materials*, 115, 3, (November 2008) 215-233, 1387-1811

Carreon, M. A.; Li, S. G.; Falconer, J. L. & Noble, R. D. (2008). Alumina-supported SAPO-34 membranes for CO_2/CH_4 separation. *Journal of the American Chemical Society*, 130, 16, (April 2008) 5412–5413, 0002-7863

Casanave, D.; Giroir-Fendler, A.; Sanchez, J.; Loutaty, R. & Dalmon, J. A. (1995). Control of transport properties with a microporous membrane reactor to enhance yields in dehydrogenation reactions. *Catalysis Today*, 25, 3-4, (August 1995) 309-314, 0920-5861

Chau, J. L. H.; Tellez, C.; Yeung, K. L. & Ho, K. (2000). The role of surface chemistry in zeolite membrane formation. *Journal of Membrane Science*, 164, 1-2, (January 2000) 257-275, 0376-7388

Chen, X. B.; Yang, W. S.; Liu, J. & Lin, L. W. (2005). Synthesis of zeolite NaA membranes with high permeance under microwave radiation on mesoporous-layer-modified macroporous substrates for gas separation. *Journal of Membrane Science*, 255, 1-2, (September 2005) 201-211, 0376-7388

Cheng, Z, L.; Chao, Z. S. & Wan, H. L. (2003). Synthesis of compact NaA zeolite membrane by microwave heating method, *Chinese Chemical Letters*, 14, 8, (August 2003) 874-876, 1001-8417

Chu, P.; Dwyer, F. G. & Vartuli, J. C. (1998). Crystallization method employing microwave radiation, US Patent, 4778666, (October 1988).

Coronas, J.; Noble, R. D. & Falconer, J. L. (1998). Separations of C-4 and C-6 isomers in ZSM-5 tubular membranes. *Industrial & Engineering Chemistry Research*, 37, 1, (January 1998) 166-176, 0888-5885

Cui, Y.; Kita, H. & Okamoto, K. I. (2004). Zeolite T membrane: preparation, characterization, pervaporation of water/organic liquid mixtures and acid stability. *Journal of Membrane Science*, 236, 1, (June 2004) 17-27, 0376-7388

Cundy, C. S. (1998). Microwave techniques in the synthesis and modification of zeolite catalysts. A review. *Collection of Czechoslovak Chemical Communications*, 63, 11, (November 1998) 1699-1723, 0010-0765

Cundy, C. S.; Plaisted, R. J. & Zhao, J. P. (1998). Remarkable synergy between microwave heating and the addition of seed crystals in zeolite synthesis-a suggestion verified. *Chemical Communications*, 14, (June 1998) 1465-1466, 1359-7345

de la Hoz, A.; Diaz-Ortiz, A. & Moreno, A. (2005). Microwaves in organic synthesis. Thermal and non-thermal microwave effects. *Chemical Society Reviews*, 34, 2, (2005) 164–178, 0306-0012

Feng, X. S. & Huang, R. Y. M. (1997). Liquid separation by membrane pervaporation: A review. *Industrial & Engineering Chemistry Research*, 36, 4, (April 1997) 1048-1066, 0888-5885

Funke, H. H.; Kovalchick, M. G.; Falconer, J. L. & Noble, R. D. (1996). Separation of hydrocarbon isomer vapors with silicalite zeolite membranes. *Industrial & Engineering Chemistry Research*, 35, 5, (May 1996) 1575-1582, 0888-5885

Gabriel, C.; Gabriel, S.; Grant, E. H.; Halstead, B. S. J. & Mingos, D. M. P. (1998). Dielectric parameters relevant to microwave dielectric heating. *Chemical Society Reviews*, 27 (May 1998) 213-223, 0306-0012

Girnus, I.; Hoffmann, K.; Marlow, F.; Caro, J. & Döring, G. (1994). Large CoAPO-5 single crystals: Microwave synthesis and anisotropic optical absorption. *Microporous Materials*, 2, 6, (July 1994) 543-555, 0927-6513

Girnus, I.; Jancke, K.; Vetter, R.; Richter-Mendau, J. & Caro, J. (1995). Large $AlPO_4$-5 crystals by microwave heating. *Zeolites*, 15, 1, (January 1995) 33-39, 0144-2449

Girnus, I.; Pohl, M. M.; Richtermendau, J.; Schneider, M.; Noack, M.; Venzke, D. & Caro, J. (1995). Synthesis of $AlPO_4$-5 aluminumphosphate molecular sieve crystals for membrane applications by microwave heating. *Advanced Materials*, 7, 8, (August 1995) 711-714, 0935-9648

Greenwood, N. N. & Earnshaw, A. (1997). *Chemistry of the Elements*, Butterworth-Heinemann, 0-7506-33654, Oxford

Gu, X. H.; Dong, J. H. & Nenoff, T. M. (2005). Synthesis of defect-free FAU-type zeolite membranes and separation for dry and moist CO_2/N_2 mixtures. *Industrial & Engineering Chemistry Research*, 44, 4, (February 2002) 937-944, 0888-5885

Guan, G.; Tanaka, T.; Kusakabe, K.; Sotowa, K. & Morooka, S. (2003). Characterization of $AlPO_4$-type molecular sieving membranes formed on a porous α-alumina tube. *Journal of Membrane Science*, 214, 2, (April 2003) 191-198, 0376-7388

Guo, H. L.; Zhu, G. S.; Hewitt, I. J. & Qiu, S. L. (2009). "Twin copper source" growth of metal-organic framework membrane: Cu3(BTC)2 with high permeability and selectivity for recycling H_2. *Journal of the American Chemical Society*, 131, 5, (February 2009) 1646-1647, 0002-7863

Gupta, V.; Nivarthi, S. S.; McCormick, A. V. & Davis, H. T. (1995). Evidence for single file diffusion of ethane in the molecular sieve $AlPO_4$-5. *Chemical Physics Letters*, 247, 4-6, (December 1995) 596-600, 0009-2614

Han, Y.; Ma, H.; Qui, S. L. & Xiao, F. S. (1999). Preparation of zeolite A membranes by microwave heating. *Microporous and Mesoporous Materials*, 30, 2-3, (September 1999) 321-326, 1387-1811

Hermes, S.; Schroder, F.; Chelmowski, R.; Woll, C. & Fischer R. A. (2005). Selective nucleation and growth of metal-organic open framework thin films on patterned COOH/CF3-terminated self-assembled monolayers on Au (111). *Journal of the American Chemical Society*, 127, 40, (October 2005) 13744-13745, 0002-7863

Hu, Y. Y.; Liu, C.; Zhang, Y. H.; Ren. N. & Tang. Y. (2009). Microwave-assisted hydrothermal synthesis of nanozeolites with controllable size. *Microporous and Mesoporous Materials*, 119, 1-3, (March 2009) 306-314, 1387-1811

Huang, A. S. & Yang, W. S. (2007). Hydrothermal synthesis of NaA zeolite membrane together with microwave heating and conventional heating. *Materials Letters*, 61, 29, (December 2007) 5129-5132, 0167-577X

Huang, A. S.; Yang, W. & Liu, J. (2007). Synthesis and pervaporation properties of NaA zeolite membranes prepared with vacuum-assisted method. *Separation and Purification Technology*, 56, 2, (August 2007) 158-167, 1383-5866

Huang, A. S.; Liang, F. Y.; Steinbach, F.; Gesing, T. M. & Caro, J. (2010). Neutral and cation-free LTA-type aluminophosphate (AlPO₄) molecular sieve membrane with high hydrogen permselectivity. *Journal of the American Chemical Society*, 132, 7, (February 2010) 2140-2141, 0002-7863

Huang, A. S.; Liang, F. Y.; Steinbach F. & Caro, J. (2010). Preparation and separation properties of LTA membranes by using 3-aminopropyltriethoxysilane as covalent linker. *Journal of Membrane Science*, 350, 1-2, (March 2010) 5-9, 0376-7388

Huang, A. S. & Caro, J. (2010). Cationic polymer used to capture zeolite precursor particles for the facile synthesis of oriented zeolite LTA molecular sieve membrane. *Chemistry of Materials*, 22, 15, (August 2010) 4353-4355, 0897-4756

Huang, A. S. Bux, H.; Steinbach, F. & Caro, J. (2010). Molecular-sieve membrane with hydrogen permselectivity: ZIF-22 in LTA topology prepared with 3-aminopropyltriethoxysilane as covalent linker. *Angewandte Chemie International Edition*, 49, 29, (2010) 4958-4961, 1433-7851

Jansen, J. C.; Arafat, A.; Barakat, A. K. & Van Bekkum, H. (1992). Microwave techniques in zeolite synthesis, In: *Synthesis of Microporous Materials*, Occelli, M. L. & Robson, H. (Ed.), 507-521, Van Nostrand Reinhold, 0-442-00661-6, New York

Jansen, J. C. (2006). 9th International Conference on Inorganic Membranes. *Proceedings of the workshop on zeolite membranes*, 13:978-82-14-04026-5, Lillehammer, Norway, June 2006

Jeong, B. H.; Hasegawa, Y.; Sotowa, K, I.; Kusakabe, K. & Morooka, S. (2003). Permeation of binary mixtures of benzene and saturated C-4-C-7 hydrocarbons through an FAU-type zeolite membrane. *Journal of Membrane Science*, 213, 1-2, (March 2003) 115-124, 0376-7388

Jhung, S. H.; Lee, J. H.; Yoon, J. W.; Serre, C.; Ferey, G. & Chang, J. S. (2007). Microwave synthesis of chromium terephthalate MIL-101 and its benzene sorption ability. *Advanced Materials*, 19, 1, (January 2007) 121-124, 0935-9648

Jhung, S. H.; Jin, T. H.; Hwang, Y. K. & Chang, J. S. (2007). Microwave effect in the fast synthesis of microporous materials: which stage between nucleation and crystal growth is accelerated by microwave irradiation? *Chemistry - A European Journal*, 13, 16, (2007) 4410-4417, 0947-6539

Julbe, A.; Motuzas, J.; Cazevielle, F.; Volle, G. & Guizard, C. (2003). Synthesis of sodalite/alpha Al₂O₃ composite membranes by microwave heating. *Separation and Purification Technology*, 32, 1-3, (July 2003) 139-149, 1383-5866

Keskin, S. & Sholl, D. S. (2007). Screening metal-organic framework materials for membrane-based methane/carbon dioxide separations. *The Journal of Physical Chemistry C*, 111, 38, (September 2007) 14055-14059, 1932-7447

Khajavi, S.; Kapteijn, F. & Jansen, J. C. (2007). Synthesis of thin defect-free hydroxy sodalite membranes: New candidate for activated water permeation. *Journal of Membrane Science*, 299, 1-2, (August 2007) 63-72, 0376-7388

Kita, H.; Horiik, K.; Ohtoshi, Y.; Tanaka, K. & Okamoto, K. I. (1995), Synthesis of a zeolite NaA membrane for pervaporation of water/organic liquid mixtures. *Journal of Materials Science Letters*, 14, 3, (February 1995) 206-208, 0261-8028

Koegler, J. H.; Arafat, A.; van Bekkum, H. & Jansen, J. C. (1997). Synthesis of films of oriented silicalite-1 crystals using microwave heating. *Studies in Surface Science and Catalysis*, 105, (1997) 2163-2170, 0-444-82344-1

Kondo, M.; Komori, M.; Kita, H. & Okamoto, K. (1997). Tubular-type pervaporation module with zeolite NaA membrane. *Journal of Membrane Science*, 133, 1, (September 2000) 133-141, 0376-7388

Kusakabe, K.; Kuroda, T.; Murata, A. & Morooka, S. (1997). Formation of a Y-Type zeolite membrane on a porous α-alumina tube for gas separation. *Industrial & Engineering Chemistry Research*, 36, 3, (March 1997) 649-655, 0888-5885

Lai, R. & Gavalas, G. R. (1998). Surface seeding in ZSM-5 membrane preparation. *Industrial & Engineering Chemistry Research*, 37, 11, (November 1998) 4275-4283, 0888-5885

Lai, Z.; Bonilla, G.; Diaz, I.; Nery, J.; Sujaoti, K.; Amat, M. A.; Kokkoli, E.; Terasaki, O.; Thompson, R. W.; Tsapatsis, M. & Vlachos, D. G. (2003). Microstructural optimization of a zeolite membrane for organic vapor separation. *Science*, 300, 5618, (April 2003) 456-460, 0036-8075

Langa, F.; de la Cruz, P.; de la Hoz, A., Diaz-Ortiz, A. & Diez-Barra, E. (1997). Microwave irradiation: more than just a method for accelerating reactions. *Contemporary organic synthesis*, 4, 5, (October 1997) 373–386, 1350-4894

Lassinantti, M.; Hedlund, J. & Sterte, J. (2000). Faujasite-type films synthesized by seeding. *Microporous and Mesoporous Materials*, 38, 1, (July 2000) 25-34, 1387-1811

Li, H.; Eddaoudi, M.; O'Keeffe, M. & Yaghi, O. M. (1999). Design and synthesis of an exceptionally stable and highly porous metal-organic framework. *Nature*, 402, 6759, (November 1999) 276-279, 0028-0836

Li, S.; Tuan, V. A.; Falconer, J. L. & Noble, R. D. (2002). X-type zeolite membranes: preparation, characterization, and pervaporation performance. *Microporous and Mesoporous Materials*, 53, 1-3, (June 2002) 59-70, 1387-1811

Li, Y. S.; Chen, H. L.; Liu, J. & Yang, W. S. (2006). Microwave synthesis of LTA zeolite membranes without seeding. *Journal of Membrane Science*, 277, 1-2, (June 2006) 230-239, 0376-7388

Li, Y. S.; Liu, J. & Yang, W. S. (2006). Formation mechanism of microwave synthesized LTA zeolite membranes, *Journal of Membrane Science*, 281, 1-2, (September 2006) 646-657, 0376-7388

Li, Y. S.; Chen, H. L.; Liu, J.; Li, H. B. Yang, W. S. (2007). Pervaporation and vapor permeation dehydration of Fischer-Tropsch mixed-alcohols by LTA zeolite membranes. *Separation and Purification Technology*, 57, 1, (October 2007) 140-146, 1383-5866

Li, Y. S. & Yang, W. S. (2008). Microwave synthesis of zeolite membranes: A review. *Journal of Membrane Science*, 316, 1-2, (May 2008) 3-17, 0376-7388

Li, Y. S.; Liang, F. Y.; Bux, H.; Feldhoff, A.; Yang, W. S. & Caro, J. (2010). Molecular sieve membrane: supported metal-organic framework with high hydrogen selectivity. *Angewandte Chemie International Edition*, 49, 3, (2010) 548-551, 1433-7851

Lidstrom, P.; Tierney, J.; Wathey, B. & Westman, J. (2001). Microwave assisted organic synthesis-a review. *Tetrahedron*, 57, 51, (December 2001) 9225–9283, 0040-4020.

Lin, X.; Kita, H. & Okamoto K. (2001). Silicalite membrane preparation, characterization, and separation performance. *Industrial & Engineering Chemistry Research*, 40, 19, (September 2001) 4069-4078, 0888-5885

Lin, Y. S. (2001). Microporous and dense inorganic membranes: current status and prospective. *Separation and Purification Technology*, 25, 1-3, (October 2001) 39-55, 1383-5866

Lovallo, M. C. & Tsapatsis, M. (1996). Preferentially oriented submicron silicalite membranes. *AIChE Journal*, 42, 11, (1996) 3020-3029, 0001-1541.

Ma, Y. H.; Zhou, Y. J.; Poladi, R. & Engwall, E. (2001). The synthesis and characterization of zeolite A membranes. *Separation and Purification Technology*, 25, 1-3, (October 2001) 235-240, 1383-5866

Madhusoodana, C. D.; Das, R. N.; Kameshima, Y. & Okada, K. (2006). Microwave-assisted hydrothermal synthesis of zeolite films on ceramic supports. *Journal of Materials Science*, 41, 5, (March 2006) 1481-1487, 0022-2461

Maesen, T. (2007). The zeolite scene–An Overview, in: *Introduction to zeolite science and practice*, Jiří Čejka, Herman van Bekkum, Avelino Corma and Ferdi Schüth (Ed.), (1-9), Elsevier, 978-0-444-53063-9, Hungary

Mintova, S.; Mo, S. & Bein, T. (1998). Nanosized $AlPO_4$-5 molecular sieves and ultrathin films prepared by microwave synthesis. *Chemistry of Materials*, 10, 12, (December 1998) 4030–4036, 0897-4756

Morigami, Y.; Kondoa, M.; Abe, J., Kita, H. & Okamoto, K. (2001). The first large-scale pervaporation plant using tubular-type module with zeolite NaA membrane. *Separation and Purification Technology*, 25, 1-3, (October 2001) 251-260, 1383-5866

Motuzas, J.; Julbe, A.; Noble, R. D.; van der Lee, A. & Beresnevicius, Z. J. (2006). Rapid synthesis of oriented silicalite-1 membranes by microwave-assisted hydrothermal treatment. *Microporous and Mesoporous Materials*, 92, 1-3, (June 2006) 259–269, 1387-1811

Motuzas, J.; Heng, S.; Lau, P. P. S. Z.; Yeung, K. L.; Beresnevicius, Z. J. & Julbe, A. (2007). Ultra-rapid production of MFI membranes by coupling microwave assisted synthesis with either ozone or calcination treatment. *Microporous and Mesoporous Materials*, 99, 1-2, (February 2007) 197–205, 1387-1811

Muller, M.; Harvey, G. & Prins, R. (2000). Comparison of the dealumination of zeolites beta, mordenite, ZSM-5 and ferrierite by thermal treatment, leaching with oxalic acid and treatment with $SiCl_4$ by H-1, Si-29 and Al-27 MAS NMR. *Microporous and Mesoporous Materials*, 34, 2, (February 2000) 135-147, 1387-1811

Myatt, G. J.; Budd, P. M.; Price, C. & Carr, S. W. (1992). Synthesis of a zeolite NaA membrane. *Journal of Materials Chemistry*, 2, 10, (October 1992) 1103-1104, 0959-9428

Ni, Z. & Masel, R. I. (2006). Rapid production of metal-organic frameworks via microwave-assisted solvothermal synthesis. *Journal of the American Chemical Society*, 128, 38, (September 2006) 12394-12395, 0002-7863

Nikolakis, V.; Xomeritakis, G.; Abibi, A.; Dickson, M.; Tsapatsis, M. & Vlachos, D. G. (2001). Growth of a faujasite-type zeolite membrane and its application in the separation of saturated/unsaturated hydrocarbon mixtures. *Journal of Membrane Science*, 184, 2, (March 2001) 209-219, 0376-7388

Nishiyama, N.; Ueyama, K. & Matsukata, M. (1996). Synthesis of defect-free zeolite-alumina composite membranes by a vapor-phase transport method. *Microporous Materials*, 7, 6, (December 1996) 299-308, 0927-6513

Nishiyama, N.; Ueyama, K. & Matsukata, M. (1997). Gas permeation through zeolite-alumina composite membranes. *AIChE Journal*, 43, 11, (1997) 2724-2730, 0001-1541

Nuchter, M.; Muller, U.; Ondruschka, B.; Tied, A. & Lautenschlager, W. (2003). Microwave-assisted chemical reactions. *Chemical Engineering & Technology*, 26, 12, (December 2003) 1207–1216, 0930-7516

Nuchter, M.; Ondruschka, B.; Bonrath, W. & Gum, A. (2004). Microwave assisted synthesis-a critical technology overview. *Green Chemistry*, 6, 2, (2004) 128–141, 1463-9262

Pan, L.; Olson, D. H.; Ciemnolonski, L. R.; Heddy, R. & Li, J. (2006). Separation of hydrocarbons with a microporous metal-organic framework. *Angewandte Chemie International Edition*, 45, 4, (2006) 616-619, 1433-7851

Panzarella, B.; Tompsett, G.; Conner, W. C. & Jones, K. (2007). In Situ SAXS/WAXS of zeolite microwave synthesis: NaY, NaA, and Beta zeolites. *Chemphyschem*, 8, 3, (February 2007) 357-369, 1439-4235

Park, K. S.; Ni, Z.; Cote, A. P.; Choi, J. Y.; Huang, R. D.; Uribe-Romo, F. J.; Chae, H. K.; O'Keeffe, M. & Yaghi, O. M. (2006). Exceptional chemical and thermal stability of zeolitic imidazolate frameworks. *Proceedings of the National Academy of Sciences of the United States of America*, 103, 27, (July 2006) 10186-10191, 0027-8424

Park, M. & Komarneni, S. (2006). Rapid synthesis of $AlPO_4$-11 and cloverite by microwave-hydrothermal processing. *Microporous and Mesoporous Materials*, 20, 1-3, (February 2006) 39-44, 1387-1811

Phan, A.; Doonan, C. J.; Uribe-Romo, F. J.; Knobler, C. B.; O'Keeffe, M. & Yaghi, O. M. (2010). Synthesis, structure, and carbon dioxide capture properties of zeolitic imidazolate frameworks. *Accounts of Chemical Research*, 43, 1, (January 2010) 58-67, 0001-4842

Phiriyawirut, P.; Magaraphan, R.; Jamieson, A. M. & Wongkasemjit, S. (2003). Morphology study of MFI zeolite synthesized directly from silatrane and alumatrane via the sol-gel process and microwave heating. *Microporous and Mesoporous Materials*, 64, 1-3, (October 2003) 83-93, 1387-1811

Ranjan, R. & Tsapatsis, M. (2009). Microporous Metal organic framework membrane on porous support using the seeded growth method. *Chemistry of Materials*, 21, 20, (October 2009) 4920-4924, 0897-4756

Rao, K. J.; Vaidhyanathan, B.; Ganguli, M. & Ramakrishnan, P. A. (1999). Synthesis of inorganic solids using microwaves. *Chemistry of Materials*, 11, 4, (April 1999) 882–895, 0897-4756

Rosi, N. L.; Eckert, J.; Eddaoudi, M.; Vodak, D. T.; Kim, J.; O'Keeffe, M. & Yaghi, O. M. (2003). Hydrogen storage in microporous metal-organic frameworks. *Science*, 300, 5622, (May 2003) 1127-1129, 0036-8075

Rowsell, J. L. C. & Yaghi, O. M. (2005). Strategies for hydrogen storage in metal-organic frameworks. *Angewandte Chemie International Edition*, 44, 30, (2005) 4670-4679, 1433-7851

Sano, T.; Ejiri, S.; Yamada, K.; Kawakami, Y. & Yanagishita, H. (1997). Separation of acetic acid-water mixtures by pervaporation through silicalite membrane. *Journal of Membrane Science*, 123, 2, (January 1997) 225-233, 0376-7388

Sathupunya, M, Gulari, E, & Wongkasemjit S. (2002). ANA and GIS zeolite synthesis directly from alumatrane and silatrane by sol-gel process and microwave technique. *Journal of the European Ceramic Society*, 22, 13, (December 2002) 2305-2314, 0955-2219

Sato, K.; Aoki, K.; Sugimoto, K.; Izumi, K.; Inoue, S.; Saito, J.; Ikeda, S. & Nakane, T. (2008). Dehydrating performance of commercial LTA zeolite membranes and application to fuel grade bio-ethanol production by hybrid distillation/vapor permeation process. *Microporous and Mesoporous Materials*, 115, 1-2, (October 2008) 184-188, 1387-1811

Sebastian, V.; Mallada, R.; Coronas, J.; Julbe, A.; Terpstra, R. A. & Dirrix, R. W. J. (2010). Microwave-assisted hydrothermal rapid synthesis of capillary MFI-type zeolite-ceramic membranes for pervaporation application. *Journal of Membrane Science*, 355, 1-2, (June 2010) 28-35, 0376-7388

Shah, D.; Kissick, K.; Ghorpade, A.; Hannah, R. & Bhattacharyya, D. (2000). Pervaporation of alcohol-water and dimethylformamide-water mixtures using hydrophilic zeolite NaA membranes: mechanisms and experimental results. *Journal of Membrane Science*, 179, 1-2, (November 2000) 185-205, 0376-7388

Sirikittikul, D.; Fuongfuchat, A. & Booncharoen, W. (2009). Chemical modification of zeolite beta surface and its effect on gas permeation of mixed matrix membrane. *Polymers for Advanced Technologies*, 20, 10, (October 2009) 802-810, 1042-7147

Slangan, P. M.; Jansen, J. C. & van Bekkum, H. (1997). (a) Induction heating: A novel tool for zeolite synthesis. *Zeolites*, 18, 1, (January 1997) 63-66, 0144-2449; (b) The effect of ageing on the microwave synthesis of zeolite NaA. *Microporous Materials*, 9, 5-6, (May 1997) 259-265, 0927-6513

Somani, O. G.; Choudhari, A. L.; Rao B. S. & Mirajkar, S. P. (2003). Enhancement of crystallization rate by microwave radiation: synthesis of ZSM-5. *Materials Chemistry and Physics*, 82, 3, (December 2003) 538-545, 0254-0584

Srinivasan, A. & Grutzeck, M. W. (1999). The adsorption of SO_2 by zeolites synthesized from fly ash. *Environmental Science & Technology*, 33, 9, (May 1999) 1464-1469, 0013-936X

Tang, Z.; Kim, S. J.; Gu, X. H. & Dong, J. H. (2009). Microwave synthesis of MFI-type zeolite membranes by seeded secondary growth without the use of organic structure directing agents. *Microporous and Mesoporous Materials*, 118, 1-3, (February 2009) 224-231, 1387-1811

Tavolaro, A. & Drioli, E. (1999). Zeolite membranes. *Advanced Materials*, 11, 12, (August 1999) 975-996, 0935-9648

Tomita, T.; Nakayama, K. & Sakai, H. (2004). Gas separation characteristics of DDR type zeolite membrane. *Microporous and Mesoporous Materials*, 68, 1-3, (March 2004) 71-75, 1387-1811

Tompsett, G. A.; Conner, W. C. & Yngvesson, K. S. (2006). Microwave synthesis of nanoporous materials. *Chemphyschem*, 7, 2, (February 2006) 296-319, 1439-4235

van de Graaf, J. M.; van der Bijl, E.; Stol, A.; Kapteijn, F. & Moulijn, J. A. (1998). Effect of operating conditions and membrane quality on the separation performance of composite silicalite-1 membranes. *Industrial & Engineering Chemistry Research*, 37, 10, (October 1998) 4071-4083, 0888-5885

Venna, S. R. & Carreon, M. A. (2010). Highly permeable zeolite imidazolate framework-8 membranes for CO_2/CH_4 separation. *Journal of the American Chemical Society*, 132, 1, (January 2010) 76-78, 0002-7863

Von Hippel, A. R. (1954). *Dielectric and Waves*, Technology Press of MIT and John Wiley, New York

Wan, Y.; Williams, C. D.; Duke, C. V. A. & Cox, J. J. (2000). Systematic studies on the effect of water content on the synthesis, crystallisation, conversion and morphology of $AlPO_4$-5 molecular sieve. *Journal of Materials Chemistry*, 10, 12, (2000) 2857-2862, 0959-9428

Wang, Z. B.; Mitra, A. P.; Wang, H. T.; Huang, L. M. & Yan, Y. S. (2001). Pure silica zeolite films as low-k dielectrics by spin-on of nanoparticle suspensions. *Advanced Materials*, 13, 19, (October 2001) 1463-1466, 0935-9648

Weh, K.; Noack, M.; Sieber, I. & Caro, J. (2002). Permeation of single gases and gas mixtures through faujasite-type molecular sieve membranes. *Microporous and Mesoporous Materials*, 54, 1-2, (July 2002) 27-36, 1387-1811

Wilson, S. T.; Lok, B. M.; Messina, C. A.; Cannan, T. R. & Flanigen, E. M. (1982). Aluminophosphate molecular sieves: a new class of microporous crystalline inorganic solids. *Journal of the American Chemical Society*, 104, 4, (1982) 1146-1147, 0002-7863

Wu, C. D. & Lin, W. B. (2007). Heterogeneous asymmetric catalysis with homochiral metal-organic frameworks: Network-structure-dependent catalytic activity. *Angewandte Chemie International Edition*, 46, 7, (2007) 1075-1078, 1433-7851

Xomeritakis, G.; Gouzinis, A.; Nair, S.; Okubo, T.; He, M. Y.; Overney, R. M. & Tsapatsis, M. (1999). Growth, microstructure, and permeation properties of supported zeolite (MFI) films and membranes prepared by secondary growth. *Chemical Engineering Science*, 54, 15-16, (August 1999) 3521-3531, 0009-2509

Xu, X. C.; Yang, W. S.; Liu, J. & Lin, L.W. (2000). Synthesis of a high-permeance NaA zeolite membrane by microwave heating. *Advanced Materials*, 12, 3, (February 2000) 195-198, 0935-9648

Xu, X. C.; Yang, W. S.; Liu, J. & Lin, L.W. (2001). Synthesis of NaA zeolite membrane by microwave heating. *Separation and Purification Technology*, 25, 1-3, (October 2001) 241-249, 1383-5866

Xu, X. C.; Bao, Y.; Song, C. S.; Yang, W. S.; Liu, J. & Lin, L. W. (2004). Microwave-assisted hydrothermal synthesis of hydroxy-sodalite zeolite membrane. *Microporous and Mesoporous Materials*, 75, 3, (November 2004) 173-181, 1387-1811

Yaghi, O. M.; Li, G. & Li, H. (1995). Selective binding and removal of guests in a microporous metal-organic framework. *Nature*, 378, 6558, (December 1995) 703–706, 0028-0836

Yan, Y. S.; Davis, M. E. & Gavalas, G. R. (1995). Preparation of zeolite ZSM-5 membranes by in situ crystallization on porous α-Al₂O₃. *Industrial & Engineering Chemistry Research*, 34, 5, (May1995) 1152-1631, 0888-5885

Yoo, Y. & Jeong, H. K. (2008). Rapid fabrication of metal organic framework thin films using microwave-induced thermal deposition. *Chemical Communications*, 21, (June 2008) 2441-2443, 1359-7345

Yoo, Y.; Lai, Z. P. & Jeong, H. K. (2009). Fabrication of MOF-5 membranes using microwave-induced rapid seeding and solvothermal secondary growth. *Microporous and Mesoporous Materials*, 123, 1-3, (June 2009) 100-106, 1387-1811

Yoon, J. W.; Jhung, S. H.; Kim, Y. H.; Park, S. E. & Chang, J. S. (2005). Selective crystallization of SAPO-5 and SAPO-34 molecular sieves in alkaline condition: Effect of heating method. *Bulletin of the Korean Chemical Society*, 26, 4, (April 2005) 558-562, 0253-2964

Yu, M.; Falconer, J. L.; Amundsen, T. J.; Hong, M. & Noble, R. D. (2007). A controllable nanometer-sized valve. *Advanced Materials*, 19, 19, (October 2007) 3032-3036, 0935-9648

Zhou, H.; Li, Y. S.; Zhu, G. Q.; Liu, J. & Yang, W. S. (2009). (a) Preparation of zeolite T membranes by microwave-assisted in situ nucleation and secondary growth.

Materials Letters, 63, 2, (January 2009) 255-257, 0167-577X; (b) Microwave-assisted hydrothermal synthesis of a&b-oriented zeolite T membranes and their pervaporation properties. *Separation and Purification Technology*, 65, 2, (February 2009) 164-172, 1383-5866

Zhu, G. Q.; Li, Y. S.; Zhou, H.; Liu, J. & Yang, W. S. (2009). Microwave synthesis of high performance FAU-type zeolite membranes: Optimization, characterization and pervaporation dehydration of alcohols. *Journal of Membrane Science*, 337, 1-2, (July 2009) 47-54, 0376-7388

Microwave-assisted Domino Reaction in Organic Synthesis

Shu-Jiang Tu and Bo Jiang
Xuzhou Normal University
China

1. Introduction

The fields of combinatorial and automated medicinal chemistry have emerged to meet the increasing requirement of new compounds for drug discovery, where speed is of the essence (Loupy, 2002). In this regard, domino (or tandem, or cascade) reactions where "two or more bond-forming transformations take place under the same reaction conditions without adding additional reagents and catalysts, and in which the subsequent reactions result as a consequence of the functionality formed in the previous step" are especially suitable for the generation of libraries of bioactive small molecules (Tietze, 1996, 2000, 2006; Domling, 2006). When performed intermolecularly, they are used to couple small fragments to larger units. In intramolecular reactions, they can bring about cyclizations or bicyclizations, and thus astonishing changes of molecular structures and increases in molecular complexity. This effect can even be enhanced by repeating the same reaction type several times or combining it with a different transformation in a domino fashion. Such sequential processes offer a wide range of possibilities for the efficient construction of high structural diversity and molecular complexity in the desired scaffolds in a single synthetic step simply by proper variation of precursors, thus avoiding time-consuming and costly processes for purification of various precursors and tedious steps of protection and deprotection of functional groups (de Meijere et al., 2005; Tietze et al., 2009). Additionally, they frequently occur with enhanced regio-, diastereo-, and even enantioselectivity for the overall transformation (Ikeda, 2000; Domling & Ugi, 2000; D'Souza & Mueller, 2007).

Simultaneously, the emergence of sustainable microwave chemistry has further impacted synthetic chemistry significantly since the introduction of precision controlled microwave reactors. From the pioneering experiments of Gedye (Gedye et al., 1986) and Giguere (Giguere et al., 1986), the use of microwave irradiation as an energy-efficient heat source for accelerating chemical reactions including heterocycle-forming, condensation, and cycloaddition reactions has seen widespread application (Kappe, 2000; Kappe & Stadler, 2005, 2009; Loupy, 2006; Kappe et al., 2009). Therefore, the high density microwave irradiation has matured into a reliable and useful methodology for accelerating reaction processes for the collection of heterocycles. Not only is direct microwave heating able to reduce chemical reaction times from hours to minutes and seconds, but it is also known to reduce side reactions, increase yields, and improve reproducibility. As a result, many academic and industrial research groups are already using microwave-assisted organic synthesis (MAOS) as a forefront technology for rapid optimization of reactions, for the efficient synthesis of new chemical entities, and for discovering and probing new chemical reactivity. Thus, it has become clear that the combined approach of microwave superheating as an energy-efficient heat source with domino reaction, in particular those carried out using precision-controlled microwave reactors, offers an efficient synergistic strategy, which can be rapidly introduced or broadened structural diversity in heterocyclic chemistry. In this chapter, we have summarized our recent activity in the area of microwave chemistry for domino synthesis of bioactive small molecules.

2. MW-assisted domino cyclization

2.1 The construction of four-, five-membered ring

A three-component domino reaction of certain alkylhydroxylamine hydrochlorides (alkyl = benzyl, p-methoxybenzyl, benzhydryl, tert-butyl) with formaldehyde or an alkyl glyoxylate and bicyclopropylidene to furnish 3-spirocyclopropanated 2-azetidinones **1** has been described by Meijere and co-workers (Eq. 1) (Zanobini et al., 2006). Microwave heating of mixtures of the three components in the presence of sodium acetate in ethanol for 15–120 min furnished the β-lactams **1** in 49%–78% yield. In any case, the reaction time required under microwave heating is less than 2 h, whereas with traditional heating at 45 °C the 1,3-dipolar cycloaddition of the most reactive N-methyl-C-(ethoxycarbonyl)nitrone onto bicyclopropylidene requires 16 d, and only spirocyclopropanated piperidones was generated when the reaction was performed at higher.

$$R_1{-}NHOH \cdot HX \quad + \quad R_2{-}CHO \quad + \quad \text{bicyclopropylidene} \xrightarrow[\text{80-100 °C (MW)}]{\text{NaOAc, EtOH}} \quad \mathbf{1} \tag{1}$$

Yield, 49-78%

$$\tag{2}$$

Microwave-assisted unimolecular isomerizatio-Claisen domino reactions of 1,3-di(hetero)aryl propargyl trityl ethers selectively lead to the formation of tricyclo[3.2.1.0^{2,7}]oct-3-enes **2** depennding on the basicity of the amine (Eq. 2) (D'Souza et al., 2008). When triethylamine was as a base, the reaction generated structurally complex tricyclo[3.2.1.0^{2,7}]oct-3-enes in 82%-93% yield, but limited scope of this reaction, whereas the reaction gave indanes **3** in 62%-94% yields using DBU as a base at 100-130 °C for 5-20 min.

Based upon product analyses and computations, this base dependent dichotomy can be rationalized as a sequel of pericyclic reactions with intermediate protonation and deprotonation.

The high temperature microwave-accelerated ruthenium-catalysed domino ring-closing metathesis (RCM) reactions for the construction of a bicyclic and a tricyclic annulated ring system were described (Efskind & Undheim, 2003). The direct conversion of the dienyne to bicyclic annulated product 4 in the presence of standard Grubbs' catalyst $(PCy_3)_2Cl_2Ru=CHPh$ was observed in 76% yield by microwave heating in toluene at 160 °C for 45 min. For comparison, in the conventional preparative work very little RCM product was obtained, even after prolonged reaction times. Under the same reaction condition, triyne was highly converted to tricyclic annulated compound 5 in 100% yield for 10 min (Eq. 3).

$$(3)$$

The cyclic β-amino carbonyl derivatives 6 were synthesized via microwave-assisted tandem cross metathesis intramolecular aza-Michael reaction between vinyl ketone and amine (Fustero et al., 2007). During the optimization process, the best result was obtained in dichloromethane in the presence of Hoveyda-Grubbs catalyst using $BF_3 \cdot OEt_2$ as additive at 100 ºC for 20 min (Eq. 4). The same reaction was performed under conventional heating at 45 ºC for 4 days. Under optimized condition, a series of cyclic β-amino carbonyl derivatives were yielded. The domino process is independent of the ketone substitution affording good to excellent chemical yields in the formation of both five- and six-membered rings.

$$(4)$$

A new microwave-assisted rearrangement of 1,3-oxazolidines scaffolds 7 is the basis for a metal-free, direct, and modular construction of tetrasubstituted pyrroles 8 from terminal-conjugated alkynes, aldehydes, and primary amines (Tejedor et al., 2004). The synthetic protocol embodyed two coupled domino processes in a one-pot manner with both atom- and bond-efficiency and under very simple and environment-friendly experimental conditions. Microwave-assisted rearrangement of 1,3-oxazolidines absorbed on silica gel for 5 min cleanly afforded the 1,2,3,4-tetrasubstituted pyrrole derivative 8 with good yield (Eq. 5). The conjugated alkynoate (1 mmol) and the amine (1.3 mmol) were absorbed on 1 g of silica gel and irradiated at 900 W for 8 min to the corresponding 1,2,3,4-tetrasubstituted pyrroles in 38%-77% yields. The process is general for the amine and tolerates a range of functionalities in the aldehyde. The authors proposed plausible mechanism for this new MW-assisted rearrangement of 1,3-oxazolidines and further investigated reaction mechanism.

$$(5)$$

The group of Lindsay reported the synthesis of the acylpyrrolidines (**9**-*trans* and **9**-*cis*) *via* the tandem aza-Cope rearrangement–Mannich cyclization between the amino alcohol and aldehydes with varied functionalities (Eq. 6) (Johnson et al., 2007). Under microwave heating the reaction proceed smoothly at 60-90 ºC for 5-150 min to provide the acylpyrrolidines **9** with 22%-84% yields. Performing the reactions at lower temperatures led to the higher observed diastereoselectivities in some cases. This sequence provides in a single synthetic step while significantly reducing reaction times as compared to analogous reactions using conventional heating.

$$(6)$$

The high-temperature rearrangement cyclization of *o*-ethynylbenzyl-aminophosphonates to yield isoindoles **10** was achieved by microwave heating in a 1:1 mixture of benzene/CH_3CN (Dieltiens & Stevens, 2007). A set of structurally complex isoindoles **10** were generated in 40%-98% yield at 165 ºC for 60-180 min under microwave heating (Eq. 7). The reaction mechanism was proposed including addition, [1,3]-alkyl shift and aromatization based on [1,5]-H shift.

$$(7)$$

The group of Tu developed the cascade reaction of 4-(arylmethylene)-2-phenyloxazol-5(4*H*)-one with pyridin-2-amine to generate 13 new imidazo[1,2-*a*]pyridin-2-one derivatives **11** in 62%-78% yield in ethylene glycol at 120 ºC under microwave irradiation (Eq. 8) (Tu et al., 2007, a). The aromatic amine, instead of pyridin-2-amine, reacted with 4-(arylmethylene)-2-phenyloxazol-5(4*H*)-one to give open-ring prop-2-enamides **12** with good yield (82%-90%). The starting materials were easily obtained and the operation was convenient.

$$(8)$$

Recently, Tu and Li's groups established a concise and efficient four-component domino approach to highly substituted 2-(2'-azaaryl)imidazoles **13** under solvent-free and microwave-irradiation conditions (Jiang et al., 2009, a). The four-component reaction offers several advantages including broad scope of substrates where a wide range of common commercial aromatic aldehydes and heteroaryl nitriles can be used with shorter reaction times (15-34 min). The desired products were obtained in good to excellent chemical yields without the tedious workup isolations. A new mechanism involving an umpolung has been proposed for this reaction (Eq. 9).

$$(9)$$

The sequential domino annulation approaches for benzoxazole synthesis have been reported by the group of Batey (Viirre et al., 2008). Benzoxazoles **14** were formed in good yields for the reaction of 1,2-dibromobenzene, but the reaction has no regioselectivity using 3,4-dibromotoluene as inputs. Furthermore, the method is limited by the availability of 1,2-dihaloarenes. As a result of these limitations, an alternative more versatile one-pot domino annulation strategy was described involving reaction of 2-bromoanilines with acyl chlorides in the presence of Cs_2CO_3, catalytic CuI, and the non-acylatable ligand 1,10-phenanthroline. Under these conditions initial acylation of the aniline is followed by copper-catalyzed intramolecular cyclization of the resultant 2-haloanilide to form the Ar-O bond of the benzoxazole ring. Optimized conditions using microwave irradiation achieved much shorter reaction times (15 min) than conventional heating (24 h) and were applied to the synthesis of a small library of benzoxazoles in 21%-97% yields. These copper-catalyzed approaches using 2-haloanilines compliments existing approaches to benzoxazoles that instead rely upon the use of 2-aminophenols (Eq. 10).

$$\text{(10)}$$

Microwave irradiation promotes the rapid O,N-acylation–cyclodehydration cascade reaction of oximes and acid chlorides to give oxazoles **15** (Eq. 11) (Wipf et al., 2005). The microwave irradiation allowed considerable acceleration of the rate of oxazole formation and increased the yield of this synthetically attractive heterocycle formation process. The starting oximes were readily obtained from commercially available ketones in yields exceeding 90%, and the yields of isolated oxazoles ranged from 23% to 62%.

$$\text{(11)}$$

A tandem method for the synthesis of 2-hydrazolyl-4-thiazolidinones **16** from commercially available materials in a 3-component reaction was reported by the Mahler group(Eq. 12) (Saiz et al., 2009). The reaction connects aldehydes, thiosemicarbazides, and maleic anhydride, effectively assisted by microwave irradiation. The best yields were obtained using a co-solvent mixture of PhMe/DMF (1:1) at 120 °C. Under optimized microwave conditions, a range of aromatic and some aliphatic aldehydes were converted to the desired heterocycles. Aromatic aldehydes provided good yields from 45% to 82%, at 120 °C for 6–12 min reaction time, except for 2-thiophene-carboxaldehyde.

$$\text{(12)}$$

$$\text{(13)}$$

Mongin and co-workers presented the [3+2] cycloaddition reaction between carbonyl ylides generated from epoxides and ketones (α-ketone ester and isatin derivatives) to give substituted spirocyclic dioxolane indolinones **17** and dioxolanes **18** with a low stereoselectivity and a large regio- and chemoselectivity (Eq. 13) (Bentabed-Ababsa et al., 2008). This reaction is a domino process that comprises two steps: the first is the thermal ring opening of the epoxide to yield a carbonyl ylide intermediate, whereas the second step is a polar [3+2] cycloaddition to provide the final spiro cycloadducts. The more favorable channels are associated with the nucleophilic attack of the isatin carbonyl oxygen atom to the phenyl substituted carbon atom of the carbonyl ylide. Using microwave irradiation significantly reduced reaction times (30-55 min) in comparison to reaction in toluene at reflux (14-29 h).

Moses and co-workers reported a modification of the methodology, using microwave radiation to significantly enhance the rate of formation of a variety of 1,4-triazole products **19** with good to excellent yield (80%-99%) from a selection of readily available anilines and acetylenes (Moorhouse & Moses, 2008). The procedure is particularly amenable to electron-deficient anilines, and works well with a wide variety of alkynes including aromatic, conjugated, aliphatic, electron-rich and electron-deficient varieties. The practical and efficient one-pot azidation of anilines with the reagent combination *t*-BuONO and TMSN$_3$ has become a useful addition to the click-chemistry toolbox. The products were rapidly isolated by precipitation and filtration directly from the reaction mixture with no further purification (Eq. 14).

$$(14)$$

The Fang group described the direct conversion of an aldehyde with iodine in ammonia water to a nitrile intermediate, which without isolation was heated with dicyandiamide, using microwave heating at 80 °C, to furnish the [2+3] cycloaddition product 2,6-diamino-1,3,5-triazine **20** in a one-pot operation (Shie & Fang, 2007). The reaction time was shortened to 15-30 min. The aldehydes were subjected to oxidation with I$_2$ in ammonia water and in situ cycloadditions with NaN$_3$/ZnBr$_2$ by microwave irradiation at 80 °C for 10 min to give the 5-aryl-1,2,3,4-tetrazoles **21** in 70%-83% overall yields. In comparison with the conventional heating method using prolonged reflux (17-48 h) at a high temperature (>100 °C), the microwave-accelerated reaction in aqueous media is safer and more efficient (Eq. 15).

$$(15)$$

2.2 The construction of six- and seven-membered ring

Heo's group presented a one-pot cascade reaction of Suzuki-Miyaura coupling/aldol condensation as an efficient method for the construction of phenanthrene derivatives **22** (Kim et al., 2008). Coupling of aryl bromide with lboronic acid in the presence of a Pd catalyst and Cs_2CO_3 provided phenanthrenes **22** with 56%-90% yield using co-solvent of toluene and ethanol (v/v:1:4) under microwave heating. Also, the cascade reaction of pinacol boronate ester with 2-bromobenzaldehyde under the same conditions afforded the phenanthrene **22** in slightly lower yield (56%-71% yield) than that of aryl bromide with boronic acid (Eq. 16).

$$\tag{16}$$

Moreno and co-workers showed that under microwave irradiation the cycloaddition reactions of nitropyrroles in solvent-free conditions give 27%-71% yields of the aromatic indoles **23**, through elimination of the nitro group and subsequent aromatization (Eq. 17) (Victoria Gomez et al., 2009).

$$\tag{17}$$

The reaction of α-hydroxyketones with a double equivalent of t-butyl acetylacetonate in the presence of KF-alumina under microwave irradiation afforded isobenzofuran-1(3H)-ones **24** in moderate yield (42%-53%) (Villemin et al., 2006). This work involving one-pot three-component cascade process is significant, although only four products were synthesized (Eq. 18).

$$\tag{18}$$

Tu's group reported the synthesis of the N-hydroxylacridine derivatives **25** in 83%-91% yields through reaction of the substituted aryl aldeoxime with dimedone in glycol under microwave irradiation for 4-7 min (Eq. 19) (Tu et al., 2004). When ammonium acetate was added to this reaction system under the same conditions, the good to excellent yields of acridine derivatives **26** were obtained. In addition, these reactions could not take place when ketoximes were employed as starting materials. The same authors presented another highly efficient synthesis of a series of new pyrido[2,3-d]pyrimidine-4,7-dione derivatives **27** and **28** via a novel cascade reaction of 4-arylidene-3-methylisoxazol-5(4H)-ones (or 4-arylidene-2-

phenyl-5(4*H*)-oxazolones) with 2,6-diaminopyrimidin-4(3*H*)-one, or naphthalen-2-amine,respectively (Tu et al., 2007, b). The microwave-assisted reaction conditions was optimized, and co-solvent of DMF and HOAc (V/V = 2:1) as reaction media at 140 ºC gave best results. 31 desired pyrido[2,3-*d*]pyrimidine-4,7-dione derivatives with high diastereoselectivity were synthesized in 76%-91% yield (Eq. 20). This method has the advantages of shorter reaction time and convenient operation.

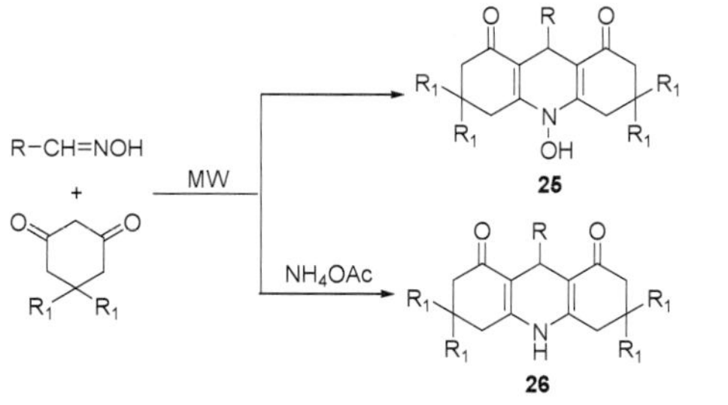

$$(19)$$

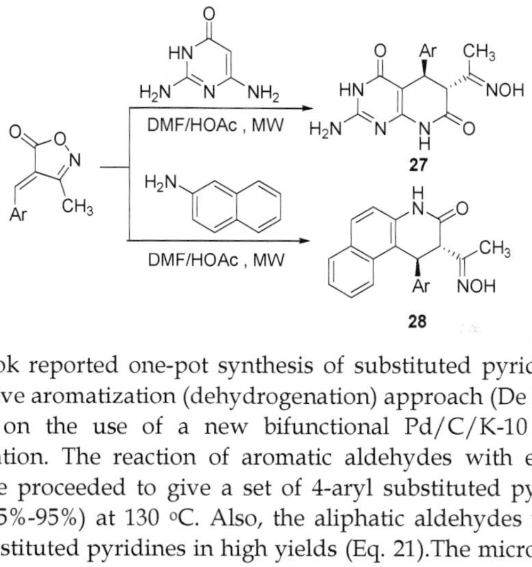

$$(20)$$

The group of Torok reported one-pot synthesis of substituted pyridines **29** via a domino cyclization–oxidative aromatization (dehydrogenation) approach (De Paolis et al., 2008). The process is based on the use of a new bifunctional Pd/C/K-10 montmorillonite and microwave irradiation. The reaction of aromatic aldehydes with ethyl acetoacetate and ammonium acetate proceeded to give a set of 4-aryl substituted pyridines **28** in good to excellent yields (45%-95%) at 130 ºC. Also, the aliphatic aldehydes were converted to the corresponding substituted pyridines in high yields (Eq. 21).The microwave-assisted process appears to be widely applicable and affords the products in good to excellent yields, high selectivities and in short reaction times compared to traditional heating.

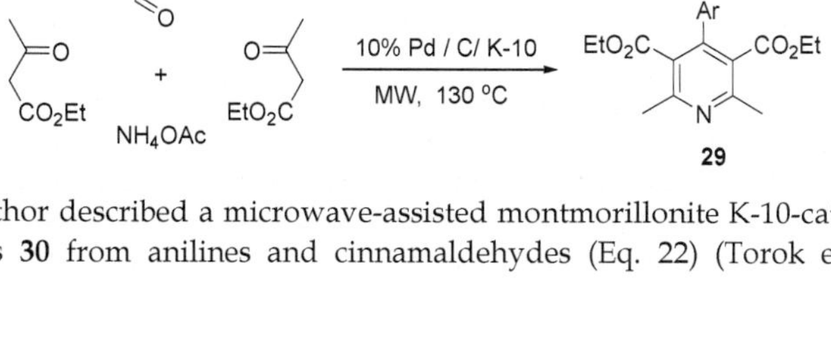

$$(21)$$

The same author described a microwave-assisted montmorillonite K-10-catalyzed synthesis of quinolines **30** from anilines and cinnamaldehydes (Eq. 22) (Torok et al., 2009). The cyclization and oxidation steps readily take place in a domino approach. The reaction works well at 90 ºC under free-solvent condition to give the substituted quinolines in 55%-95% yields, completed in a matter of minutes. As compared to conventional heating, microwave irradiation benefit significantly in regard to more efficient reaction time, yield and product purity. The efficient and ecofriendly catalyst and the convenience of the product isolation make this process an attractive alternative for the synthesis of these important heterocycles.

$$\text{(22)}$$

30

Tu's group reported a concise four-component reaction of aldehydes, Meldrum's acid, dimedone, and different amines in 95% ethanol without catalyst under microwave conditions, and a series of new N-substituted 4-aryl-tetrahydroquinoline-2,5-dione derivatives **31** with 82%-95% yields were afforded (Eq. 23) (Tu et al., 2006). The experimental results showed that structurally diverse aldehydes including electron-drawing, electron-donating, and hetero-aromatic aldehydes were tolerated in the reaction. It is also observed that the protocol can be applied to different amines (including aliphatic and alicyclic amines) but also to aromatic amines which highlight the wide scope of this four-component condensation.

$$\text{(23)}$$

31

$$\text{(24)}$$

32

In next work, four-component domino condensation for the synthesis of indenopyridine derivatives **32** has been described (Eq. 24) (Tu et al., 2007, c). Investigation of reaction condition was found to be N,N-dimethylformamide (DMF) and 120 ºC under microwave heating. Structurally diverse arone substrates were employed to react with different aldehydes and 1,3-indanedione in the present of ammonium acetate, providing the corresponding indeno[1,2-b]pyridin-5-ones in 62%–89% yields. The 2-(pyridin-2-yl)-indeno[1,2-b]pyridines was first synthesized in the MCRs. To realize green synthesis of 2-(pyridin-2-yl)-indeno[1,2-b]pyridines with important property, the same reaction were preformed using water as solvent in a sealed vial under microwave heating at 150 ºC (Tu et al., 2007, d). To broaden the diversity of indeno[1,2-b]pyridines, Tu's group developed a new domino three-component reaction of arylidenemalononitrile with 1,3-indanedione and mercaptoacetic acid (or 4-methylbenzenethiol, or aromatic amine) under microwave (MW) irradiation to afford 27 libraries of indeno[1,2-*b*]pyridines **33** with good to excellent yields (78%-91%) (Eq. 25) (Tu et al., 2007, e). The best reaction condition was found to be in DMF at 120 ºC for 4-12 min.

$$\text{(25)}$$

X = S, NH; R = alkyl, aryl

Huang and Torok reported rapid, microwave-assisted synthesis of β-carbolines **33** via a successive condensation / cyclization / dehydrogenation approach (Eq. 26) (Kulkarni et al., 2009). The use of the bifunctional catalyst Pd/C/K-10 combined with microwave irradiation enabled the synthesis of β-carbolines in short reaction times and in good to excellent yields. The reactions between tryptamine and aromatic glyoxals provide excellent yield (79%-96%) in 2–12 min, whereas tryptamine reacted with aromatic aldehydes give similar yield (71%-90%), demanding longer reaction time (20-45 min). The product imine undergoes a Pictet–Spengler cyclization followed by a final dehydrogenation to yield β-carbolines in a three-step domino reaction.

$$\text{(26)}$$

A method to efficiently prepare substituted 1,2-dihydroquinolines **35** and quinolines **36** by Au(I)-catalyzed tandem hydroamination-hydroarylation under microwave irradiation was developed (Eq. 27) (Liu et al., 2007). The best result was obtained when the reaction of primary arylamines with alkynes at 150 °C in the presence of additive NH_4PF_6 using Au(I) / AgOTf as catalysts with CH_3CN as solvent, and a set of quinoline derivatives **34** and **35** were obtained in 42-94% yields. This method requires short reaction time 10-70 min and has a broad substrate scope.

$$\text{(26)}$$

28 example
Yield, 42-94%

The Tu's group reported a new and highly stereoselective four-component protocol for the domino reactions of 2,6-diaminopyrimidine-4-one with structurally diverse aryl aldehydes and various barbituric acids, resulting in 19 examples of 6-spirosubstituted pyrido[2,3-d]pyrimidines **37** with high diastereoselectivities (up to 99 : 1) in which the major diastereomer bears a *cis* relationship between substituents at the 5- and 7-positions (Eq. 28) (Jiang et al., 2009, b). These reactions employ microwave heating and water as an environmentally benign reaction medium at 100 °C for 7-9 min. Furthermore, the mechanism for diastereoselectivity was confirmed by DFT (B3LYP) calculations. Subsqently, new multicomponent domino reactions of Meldrum's acid, aromatic aldehydes and electron-rich heteroaryl-amines have been established for the synthesis of spiro{pyrazolo[1,3]dioxanes-pyridine}-4,6-diones and spiro{isoxazolo [1,3]dioxanes-pyridine}-4,6-diones **38** in aqueous solution under microwave irradiation (Eq. 29) (Ma et al., 2010). A total of 26 examples were examined to show a broad substrate scope and high overall yields (76%-93%). A new mechanism has been proposed to explain the reaction process and the resulting chemo-, regio- and stereoselectivity.

$$ \text{(28)} $$

$$ \text{(29)} $$

Recently, Tu's group described an efficient reagent-controlled regiospecific synthesis of 2,2'-bipyridine and unsymmetrical 2,4,6-triarylpyridine derivatives **39** at high-temperature via microwave-assisted aqueous multicomponent domino reactions of aromatic aldehydes, 3-aryl-3-oxopropanenitrile, 2-acetylpyridine or aromatic ketone and ammonium acetate (Jiang et al., 2009, c). The reaction of aromatic aldehydes with 1,2-diphenylethanone resulted in structurally complex penta-arylpyridines **40**. This serves as an efficient general approach to diversity-oriented poly aryl pyridine skeletons (Eq. 30). Later on, the same group reported a new iodine-promoted domino reaction of 2-aminochromene-3-carbonitriles with various isocyanates, and 19 examples of polyfunctionalized N-substituted 2-aminoquinoline-3-carbonitriles with high regioselectivity were successfully synthesized under microwave heating (Jiang et al., 2010). The syntheses were finished within short periods (20-36 min) with good to excellent chemical yields (62%-85%) that avoided tedious work-up isolations (Eq. 31).

$$(30)$$

$$(31)$$

The same group also developed the multicomponent reaction of arylaldehyde, 4-hydroxy-2*H*-chromen-2-one and 2,6-diaminopyrimidin-4(3*H*)-one for the construction of pyrido[2,3-*d*]pyrimidine and chromeno[3′,4′:5,6]pyrido[2,3-*d*]pyrimidine skeletons **42** and **43** containing a pyridine unit controlled by the nature of solvent (Eq. 32) (Tu et al., 2008). The 9 examples of polysubstituted chromeno[3′,4′:5,6]pyrido[2,3-*d*]pyrimidines **42** in 83%-90% yields was generated using DMF as a solvent, whereas the reaction was found to undergo along another pathway leading to the formation of pyrido[2,3-*d*]pyrimidines **43** in 82%-92% chemical yields when the reaction was conducted in mixed solvent of HOAc and DMF.

$$(32)$$

Ma and co-workers presented the synthesis of dihydrocoumarin derivatives **44** through the reaction between phenols and cinnamoyl chloride using montmorillonite K-10 as eco-friendly solid-acid catalyst via a tandem esterification–Friedel–Crafts alkylation process under microwave irradiation (Zhang et al., 2008). 10 examples with 25%-92% yield were generated at 160 ºC in chlorobenzene (Eq. 33).

$$(33)$$

Beifuss et al. described a domino process for the preparation of libraries of substituted pyrano[4,3-*b*]pyran derivatives **45** in a single step by reaction of an α,β-unsaturated aldehyde with 6-substituted-pyran-2-one (Eq. 34) (Leutbecher et al., 2004). Microwave-assisted reaction was run at 110 ºC in the presence of calcium sulphate to give the heterocycles 3 with 18%-88% yields for 10-90 min. Even if the yields obtained are still lower than under conventional condition, the reactions under microwave heating condition benefit from considerable time savings and a substantially more rapid access to pyrano[4,3-b]pyran collections.

$$\text{(34)}$$

Estevez-Brau & Ravelo described the synthesis of novel pyrano embelin derivatives **46** through domino Knoevenagel hetero Diels–Alder reactions of embelin with paraformaldehyde and electron rich alkenes (Eq. 35) (Jimenez-Alonso et al., 2008). The author found that this synthetic approach is highly efficient when microwave irradiation is used, and the EtOH as a solvent at 120 ºC for 20 min gave the best result using the ratio of embelin, ethyl vinyl ether and paraformaldehyde in 1:3:8. A et of dihydropyran–embelin derivatives with good to excellent yield (50%-100%) were obtained employing several dienophiles.

$$\text{(35)}$$

Jimenez and co-workers introduced a new domino strategy, involving the coupling of enaminoesters with chlorosulfonyl isocyanate enables the straightforward preparation of the biologically important isoorotate bases **47** with moderate to good yields (Eq. 36) (Avalos et al., 2006). As compared to the refluxing in toluene within 48 h, the reaction was completed for 1 h under microwave irradiation.

$$\text{(36)}$$

$$\text{(37)}$$

The group of Tu described a highly efficient and chemoselective synthetic route to the benzothiazepinones **48** and thiazolidinones **49** *via* a microwave-assisted three-component reaction of an aromatic aldehyde with aniline and mercaptoacetic acid (Eq. 37) (Tu et al., 2009). The reaction gave best results in water at 110 ºC, and 25 benzothiazepinones were synthesized in 90%-96% yield under the optimum reaction conditions. The influences of electronic effect on the chemoselectivity were investigated in these reactions. The aromatic aldehydes bearing electron-donating groups (EDG) as well as heteroaromatic aldehydes resulted in benzothiazepinones whereas electron-withdrawing groups (EDG) on phenyl ring generated thiazolidinones **49**. When 3,4-(methylenedioxy)aniline was used as the inputs, aromatic aldehydes bearing both electron-withdrawing and electron-donating groups led to the benzothiazepinones **48** in all the case.

A number of substituted cyclohept-4-enone derivatives **50** in 45%-89% yield were synthesized via a microwave-assisted tandem oxyanionic 5-exo cyclization/Claisen rearrangement sequence using appropriately substituted 1-alkenyl-4-pentyn-1-ol systems served as useful precursors (Eq. 38) (Li, et al., 2007). All reactions were conducted at 210 ºC for 45 or 60 min in the presence of 10 mol % MeLi and using phenetole as the solvent and microwave heating. The reactions involving terminally substituted 4-pentyn-1-ols were found to be highly stereoselective (up to 93:7), with the α and β groups in the final product showing a strong preference for the *trans* orientation.

$$(38)$$

2.3 The construction of bicyclic ring

Indole-fused benzo-1,4-diazepines **51** were synthesized by copper-catalyzed domino three-component coupling-indole formation-N-arylation under microwave irradiation from a simple N-mesyl-2-ethynylaniline (Eq. 39) (Ohta et al., 2008). Investigation of the reaction solvent and loading of the catalyst revealed that 2.5 mol% of CuI in dioxane most effectively produced **51** in 88% yield within 40 min. this method was also applicable to the formation of heterocycle-fused 1,4-diazepines.

$$(39)$$

$$(40)$$

The group of Raghunathan described 4-hydroxy coumarin and its benzo-analogues undergo intramolecular domino Knoevenagel heter Diels-Alder reactions with O-prenylated aromatic aldehydes, the aliphatic aldehydes, and citronellal to afford pyrano[3-2c]coumarin derivatives **52** and **52'** (Eq. 40) (Shanmugasundaram et al., 2002, a). Compared with conventional condition (yield, 40%-75%), the higher yield (74%-92%) and chemoselectivity (up to 95: 5) of products were achieved by the application of microwave irradiation within a short period of time (10-180 s). The same authors reported the similar versions to pyranoquinolinone **53** and **53'** (Eq. 41) (Jayashankaran et al., 2002), thiopyrano coumarin/chromone **54** and **54'** (Eq. 42) (Raghunathan et al., 2006), and pyrrolo[2,3-d]pyrimidine annulated pyrano[5,6-c]coumarin/[6,5-c]chromone derivatives **55** and **55'** (Eq. 43) (Ramesh & Raghunathan, 2008) under microwave heating.

$$(41)$$

$$(42)$$

$$(43)$$

$$(44)$$

The group of Tu developed a high-speed and one-pot combinatorial strategy for the diverse quinoline collections, including imidazo[1,2-*a*]quinoline **56**, pyrimido[1,2-*a*]quinoline **57**, and quinolino[1,2-*a*]quinazoline **58** moiety, from readily available starting materials (Eq. 40) (Tu et al., 2007, f). The best yields (80%-89%) of products were obtained in ethylene glycol at 120 °C for 4-8 min. The authors combined the advantages of microwave heating with combinatorial chemistry to facilitate the rapid construction of biheterocyclic system such as imidazo[1,2-*a*]quinoline, pyrimido[1,2-*a*]-quinoline, and quinolino[1,2-*a*]quinazoline keletons, and 42 desired products were synthesized via multicomponent reaction of aromatic aldehyde, malononitrile, and *N*-substituted enaminones. The green chemoselective synthesis of thiazolo[3,2-*a*]pyridine derivatives **59** and **60** was achieved by the same group recently (Eq. 45) (Shi et al., 2009). The MCR of malononitrile, aromatic aldehydes and 2-mercaptoacetic acid was preformed to produce two different products by controlling the molar ratios of the starting materials. These products have been screened for their antioxidant activity and cytotoxicity in carcinoma HCT-116 cells and mice lymphocytes. Nearly all of the tested compounds possessed potent antioxidant activity.

$$(45)$$

The group of Tu and Li recently discovered a new four-component domino reaction between simple aldehydes, cycloketones and cyanoamides, and a diverse set of quinazoline derivatives **61** in 74%-90% yields were obtained, with the remarkable chemo-, regio- and stereoselectivity (Eq. 46) (Jiang et al., 2009, d). The reaction is easy to perform simply by mixing four common reactants and K_2CO_3 in ethylene glycol under microwave irradiation. The reaction is very fast and can be finished within 10-24 min with water as the major byproduct, which makes work-up convenient. Four stereogenic centers with one quaternary carbon-amino function have been controlled very well. Later on, the same group found that when aromatic aldehydes were replaced by their aliphatic counterparts, the quinazoline derivatives were not generated. Interestingly, the reaction resulted in the formation of multi-functionalized tricyclo[6.2.2.0^{1,6}] dodecanes **62**. (Eq. 47) (Jiang et al., 2010).

$$(46)$$

n = 0, 1, 2, X = CH_2, n = 1, X = N-Me, N-Bn **61**

$$(47)$$

$$n = 0, 1, X = CH_2;$$
$$n = 1, X = O, S, N\text{-}Boc, N\text{-}Bn$$

Kwong et al. reported microwave-assisted Rh–diphosphane-complex-catalyzed [2+2+1] cycloadducts to provide cyclopentenones **63** by sequential decarbonylation of aldehyde or formate and carbonylation of enynes within a short period of time (45 min) (Eq. 48) (Lee et al., 2008). Various O-, N-, and C-tethered enynes were transformed into the corresponding products in 34%-83% yields. The author also realized the enantioselective version of this microwave-accelerated cascade cyclization, and obtained the cyclopentenone products with *ee* values up to 90% in the presence of chiral Rh-(*S*)-bisbenzodioxanPhos complex.

$$X = O, NTs$$

$$(48)$$

Li and co-workers synthesized a set of fused tricyclic thiochromeno[2,3-*b*]pyridines by tandem [3+3] annulation and SNAr reaction (Wen et al., 2008). Under microwave irradiation, α-(2-chloroaryl)thioacetanilides reacted with activated 4-arylidene-2-phenyloxazol-5(4*H*)-ones to give a small library of 24 thiochromeno[2,3-*b*]pyridines **64** with 54%-85% yield in THF catalyzed by triethylamine; whereas the three-component reactions of α-(2-chloroaryl)thioacetanilides and aromatic aldehydes with ethyl 2-cyanoacetate resulted in the corresponding thiochromeno[2,3-*b*]pyridine derivatives **65** with 52%-76% yield in EtOH (Eq. 49). In the domino processes, at least seven reactive distinct chemical sites were involved and up to three new covalent bonds and one tricycle with only *cis* configuration were generated.

$$(49)$$

$$(50)$$

34-71% yields
16 examples

The microwave-assisted Wolff rearrangement of cyclic 2-diazo-1,3-diketones in the presence of aldehydes and primary amines was described by the Coquerel and Rodriguez et al. (Eq. 50) (Presset et al., 2009). The reaction provides a straightforward access to functionalized bi- and pentacyclic oxazinones **66** following an unprecedented three-component domino reaction. The above approach to 1,3-oxazin-4-ones is the first example of the exploitation of the Wolff rearrangement in a multicomponent reaction.

3. Conclusion

In conclusion, microwave-assisted domino reactions have quickly become a powerful and efficient tool in organic chemistry. The combination of different activation modes allows the design of innovative domino sequences to afford high molecular complexity (Enders et al., 2007). In addition to the bond-forming economy (multiple formations of carbon-carbon or - heteroatom bonds), the microwave-assisted domino reaction intrinsically has the following advantages: the rapid optimization of procedures, high reaction selectivity and efficiency, consecutive reaction pattern, structure economy (structural complexity), and environmentally benign. Therefore, the combination of microwave heating and domino reaction opens numerous options for devising new sustainable synthetic methodologies in organic chemistry that are diversity oriented and increase complexity of molecular scaffolds. Recent advances have witnessed many new microwave-assisted domino reaction being developed, and provided more convenient and rapid procedures for heterocycles synthesis. The further extension to domino processed and the combination of three and more steps for introduction of small bioactive molecules libraries is still an interesting field in the very near future.

4. References

Avalos, M.; Babiano, R.; Cintas, P.; Hursthouse, M. B.; Jimenez, J. L.; Lerma, E.; Light, M. E. & Palacios, J. C. (2006). A one-pot domino reaction in constructing isoorotate bases and their nucleosides. *Tetrahedron Letters*, Vol.47, No.12, 1989-1992, ISSN: 0040-4039.

Bentabed-Ababsa, G.; Derdour, A.; Roisnel, T.; Saez, J. A.; Domingo, L. R. & Mongin, F. (2008). Polar [3+2] cycloaddition of ketones with electrophilically activated carbonyl ylides. Synthesis of spirocyclic dioxolane indolinones. *Organic & Biomolecular Chemistry*, Vol.6, No.17, 3144-3157, ISSN: 1477-0520.

De Paolis, O.; Baffoe, J.; Landge, S. M. & Torok, B. (2008). Multicomponent domino cyclization-oxidative aromatization on a bifunctional Pd/C/K-10 catalyst: an environmentally benign approach toward the synthesis of pyridines. *Synthesis*, No.21, 3423-3428, ISSN: 0039-7881.

De Paolis, O.; Teixeira, L. & Torok, B. (2009). Synthesis of quinolines by a solid acid-catalyzed microwave-assisted domino cyclization-aromatization approach. *Tetrahedron Letters*, Vol.50, No.24, 2939-2942, ISSN: 0040-4039.

Domling, A. & Ugi, I. (2000). Multicomponent reactions with isocyanides. *Angewandte Chemie, International Edition*, Vol.39, No.18, 3168-3210, ISSN: 1433-7851.

de Meijere, A.; von Zezschwitz, P. & Brase, S. (2005). The virtue of palladium-catalyzed domino reactions - diverse oligocyclizations of acyclic 2-bromoenynes and 2-bromoenediynes. *Accounts of Chemical Research*, Vol.38, No.5, 413-422, ISSN: 0001-4842.

Dieltiens, N. & Stevens, C. V. (2007). Metal-free entry to phosphonylated isoindoles by a cascade of 5-exo-dig cyclization, a [1,3]-alkyl shift, and aromatization under microwave heating. *Organic Letters*, Vol.9, No.3, 465-468, ISSN: 1523-7060.

D'Souza, D. M.; Liao, W.-W.; Rominger, F.; Mueller, T. J. J. (2008). Dichotomies in microwave-assisted propargyl-isomerization-Claisen domino sequences dependent on base strengths. *Organic & Biomolecular Chemistry*, Vol.6, No.3, 532-539, ISSN: 1477-0520.

D'Souza, D. M. & Mueller, T. J. J. (2007). Multi-component syntheses of heterocycles by transition-metal catalysis. *Chemical Society Reviews*, Vol.36, No.7, 1095-1108, ISSN: 0306-0012.

Domling, A. (2006). Recent developments in isocyanide based multicomponent reactions in applied chemistry. *Chemical Reviews*, Vol. 106, No. 1, 17-89. ISSN:0009-2665.

Efskind, J. & Undheim, K. (2003). High temperature microwave-accelerated ruthenium-catalyzed domino RCM reactions. *Tetrahedron Letters*, Vol.44, No.14, 2837-2839, ISSN: 0040-4039.

Enders, D.; Grondal, C. & Huettl, M., R. M. (2007). Asymmetric organocatalytic domino reactions. *Angewandte Chemie, International Edition*, Vol.46, No.10, 1570-1581, ISSN: 1433-7851.

Fustero, S.; Jimenez, D.; Sanchez-Rosello, M. & del Pozo, C. (2007). Microwave-assisted tandem cross metathesis intramolecular aza-Michael reaction: An easy entry to cyclic β-amino carbonyl derivatives. *Journal of the American Chemical Society*, Vol.129, No.21, 6700-6701, ISSN: 0002-7863.Loupy, A. (2002). *Microwaves in Organic Synthesis*, 499pp, Wiley-VCH Verlag GmbH & Co. KGaA, ISBN: 3-527-31452-0, Weinheim, Germany.

Gedye, R.; Smith, F.; Westaway, K.; Ali, H.; Baldisera, L.; Laberge, L. & Rousell, J. (1986). The use of microwave ovens for rapid organic synthesis. *Tetrahedron Letters*, Vol.27, No.3, 279-282, ISSN: 0040-4039.

Giguere, R. J.; Bray, T. L.; Duncan, S. M. & Majetich, G. (1986). Application of commercial microwave ovens to organic synthesis. *Tetrahedron Letters*, Vol. 27, No. 41, 4945-4948, ISSN: 0040-4039.

Ikeda, S.-i. (2000). Nickel-Catalyzed Intermolecular Domino Reactions. *Accounts of Chemical Research*, Vol.33, No.8, 511-519, ISSN: 0001-4842.

Jayashankaran, J.; Manian, R. D. R. S. & Raghunathan, R. (2006). An efficient synthesis of thiopyrano[5,6-*c*]coumarin/[6,5-*c*]chromones through intramolecular domino Knoevenagel hetero-Diels-Alder reactions. *Tetrahedron Letters*, Vol.47, No.13, 2265-2270, ISSN: 0040-4039.

Jiang, B.; Cao, L.-J.; Tu, S.-J.; Zheng, W.-R. & Yu, H.-Z. (2009). Highly diastereoselective domino synthesis of 6-spirosubstituted pyrido[2,3-*d*]pyrimidine derivatives in water. *Journal of Combinatorial Chemistry*, Vol.11, No.4, 612-616, ISSN: 1520-4766, (b).

Jiang, B.; Hao, W.-J.; Wang, X.; Shi, F. & Tu, S.-J. (2009). Diversity-oriented synthesis of Krohnke pyridines. *Journal of Combinatorial Chemistry*, Vol.11, No.5, 846–850, ISSN: 1520-4766, (c).

Jiang, B.; Li, C.; Shi, F.; Tu, S.-J.; Kaur, P.; Wever, W. & Li, G. (2010). Four-component domino reaction providing an easy access to multifunctionalized tricyclo[6.2.2.0^{1,6}]dodecane derivatives. *Journal of Organic Chemistry*, Vol.75, No.9, 2962-2965, ISSN:0022-3263.

Jiang, B.; Li, C.; Tu, S.-J. & Shi, F. (2010) Iodine-promoted domino reaction leading to N-substituted 2-aminoquinoline-3-carbonitriles under microwave irradiation. *Journal of Combinatorial Chemistry*, Vol.12, No.4, 482-487, ISSN: 1520-4766.

Jiang, B.; Tu, S.-J.; Kaur, P.; Wever, W. & Li, G. (2009). Four-component domino reaction leading to multifunctionalized quinazolines. *Journal of the American Chemical Society*, Vol.131, No.33, 11660-11661, ISSN: 0002-7863, (d).

Jiang, B.; Wang, X.; Shi, F.; Tu, S.-J.; Ai, T.; Ballew, A. & Li, G. (2009). Microwave enabled umpulong mechanism based rapid and efficient four- and six-component domino formations of 2-(2'-azaaryl)imidazoles and anti-1,2-diarylethylbenzamides. *Journal of Organic Chemistry*, Vol.74, No.24, 9486–9489, ISSN:0022-3263, (a).

Jimenez-Alonso, S.; Chavez, H.; Estevez-Braun, A.; Ravelo, A. G.; Feresin, G. & Tapia, A. (2008). An efficient synthesis of embelin derivatives through domino Knoevenagel hetero Diels-Alder reactions under microwave irradiation. *Tetrahedron*, Vol.64, No.37, 8938-8942, ISSN: 0040-4020.

Johnson, B. F.; Marrero, E. L.; Turley, W. A. & Lindsay, H. A. (2007). Controlling diastereoselectivity in the tandem microwave-assisted aza-Cope rearrangement-Mannich cyclization. *Synlett*, No.6, 893-896, ISSN: 0936-5214.

Kappe, C. O. (2004). Synthetic methods. Controlled microwave heating in modern organic synthesis. *Angewandte Chemie, International Edition*, Vol.43, No.46, 6250-6284, ISSN: 1433-7851.

Kappe, C. O. & Dallinger, D. (2009). Controlled microwave heating in modern organic synthesis: highlights from the 2004-2008 literature. *Molecular Diversity*, Vol.13, No.2, 71-193, ISSN: 1381-1991.

Kappe, C. O.; Dallinger, D. & Kim, Y. H.; Lee, H.; Kim, Y. J.; Kim, B. T. & Heo, J.-N. (2008). Direct one-pot synthesis of phenanthrenes via Suzuki-Miyaura coupling/aldol condensation cascade reaction. *Journal of Organic Chemistry*, Vol.73, No.2, 495-501, ISSN: 0022-3263. Kappe, C. O. & Stadler, A. (2005), *Microwaves in Organic and Medicinal Chemistry*, 1-410. Wiley-VCH, Weinheim, Germany, ISBN: 3-527-31210-2.

Kulkarni, A.; Abid, M.; Torok, B. & Huang, X. (2009). A direct synthesis of β -carbolines via a three-step one-pot domino approach with a bifunctional Pd/C/K-10 catalyst. *Tetrahedron Letters*, Vol.50, No.16, 1791-1794, ISSN: 0040-4039.

Lee, H. W.; Lee, L. N.; Chan, A.t S. C. & Kwong, F. Y. (2008). Microwave-assisted rhodium-complex-catalyzed cascade decarbonylation and asymmetric Pauson-Khand-type cyclizations. *European Journal of Organic Chemistry*, No.19, 3403-3406, ISSN: 1434-193X.

Leutbecher, H.; Conrad, J.; Klaiber, I.; Beifuss, U. (2004). Microwave assisted domino Knoevenagel condensation/6π -electron electrocyclization reactions for the rapid and efficient synthesis of substituted 2*H*,5*H*-pyrano[4,3-*b*]pyran-5-ones and related heterocycles. *QSAR & Combinatorial Science*, Vol.23, No.10, 895-898, ISSN:1611-020X.

Li, X.; Kyne, R. E.; Ovaska, T. V. (2007). Synthesis of seven-membered carbocyclic rings via a microwave-assisted tandem oxyanionic 5-exo dig cyclization-claisen rearrangement process. *Journal of Organic Chemistry*, Vol.72, No.17, 6624-6627, ISSN: 0022-3263.

Liu, X.-Y.; Ding, P.; Huang, J.-S. & Che, C.-M. (2007). Synthesis of substituted 1,2-dihydroquinolines and quinolines from aromatic amines and alkynes by Gold(I)-catalyzed tandem hydroamination-hydroarylation under microwave-assisted conditions. *Organic Letters*, Vol.9, No.14, 2645-2648, ISSN:1523-7060.

Loupy, A. (2006). *Microwaves in Organic Synthesis: Second, Completely Revised and Enlarged Edition, Volume 2.* 1-482. Wiley-VCH, Weinheim, Germany.

Ma, N.; Jiang, B.; Zhang, G.; Tu, S.-J.; Wever, W. & Li, G. (2010). New multicomponent domino reactions (MDRs) in water: highly chemo-, regio- and stereoselective synthesis of spiro{[1,3]dioxanopyridine}-4,6-diones and pyrazolo[3,4-*b*]pyridines, Green Chemistry, Vol. 12, No.8, 1357-1361, ISSN:1463-9262.

Manikandan, S.; Shanmugasundaram, M. & Raghunathan, R. (2002). Competition between two intramolecular domino Knoevenagel hetero Diels-Alder reactions: a new entry into novel pyranoquinolinone derivatives. *Tetrahedron*, Vol.58, No.44, 8957-8962, ISSN: 0040-4020.

Moorhouse, A. D. & Moses, J. E. (2008). Microwave enhancement of a "one-pot" tandem azidation-"click" cycloaddition of anilines. *Synlett*, No.14, 2089-2092, ISSN: 0936-5214.

Murphree, S. S. (2009). *Practical Microwave Synthesis for Organic Chemists: Strategies, Instruments, and Protocols.* 1-299. Wiley-VCH Weinheim, Germany, ISBN: 978-3-527-32097-4.

Ohta, Y.; Chiba, H.; Oishi, S.; Fujii, N. & Ohno, H. (2008). Concise synthesis of indole-fused 1,4-diazepines through copper(I)-catalyzed domino three-component coupling-cyclization-N-arylation under microwave irradiation. *Organic Letters*, Vol.10, No.16, 3535-3538, ISSN: 1523-7060.

Presset, M.; Coquerel, Y. & Rodriguez, J. (2009). Microwave-assisted domino and multicomponent reactions with cyclic acylketenes: Expeditious syntheses of oxazinones and oxazindiones. *Organic Letters*, Vol.11, No.24, 5706-5709, ISSN:1523-7060.

Ramesh, E. & Raghunathan, R. (2008). An expedient microwave-assisted, solvent-free, solid-supported synthesis of pyrrolo[2,3-*d*]pyrimidine-pyrano[5,6-c]coumarin/[6,5-c]chromone derivatives by intramolecular hetero Diels-Alder reaction. *Tetrahedron Letters*, Vol.49, No.11, 1812-1817. ISSN: 0040-4039.

Saiz, C.; Pizzo, C.; Manta, E.; Wipf, P. & Mahler, S. G. (2009). Microwave-assisted tandem reactions for the synthesis of 2-hydrazolyl-4-thiazolidinones. *Tetrahedron Letters*, Vol. 50, No.8, 901-904, ISSN: 0040-4039.

Shanmugasundaram, M.; Manikandan, S. & Raghunathan, R. (2002). High chemoselectivity in microwave accelerated intramolecular domino Knoevenagel hetero Diels-Alder reactions - an efficient synthesis of pyrano[3-2*c*]coumarin frameworks. *Tetrahedron*, Vol.58, No.5, 997-1003, ISSN: 0040-4020.

Shi, F.; Li, C.; Xia, M.; Miao, K.; Zhao, Y.; Tu, S.; Zheng, W.; Zhang G. & Ma N. (2009). Green chemoselective synthesis of thiazolo[3,2-*a*]pyridine derivatives and evaluation of their antioxidant and cytotoxic activities. *Bioorganic & medicinal chemistry letters*, Vol.19, No.19, 5565-8, ISSN: 0960-894X.

Shie, J.-J. & Fang, J.-M.. (2007). Microwave-assisted one-pot tandem reactions for direct conversion of primary alcohols and aldehydes to triazines and tetrazoles in aqueous media. *Journal of Organic Chemistry*, Vol.72, No.8, 3141-3144, ISSN: 0022-3263.

Tejedor, D.; Gonzalez-Cruz, D.; Garcia-Tellado, F.; Marrero-Tellado, J. J. & Rodriguez, M. L. (2004). A diversity-oriented strategy for the construction of tetrasubstituted pyrroles via coupled domino processes. *Journal of the American Chemical Society*, Vol.126, No.27, 8390-8391, ISSN: 0002-7863.

Tietze, L. F. (1996). Domino Reactions in Organic Synthesis. *Chem. Rev.*, Vol. 96, No. 1, 115-136, ISSN:0009-2665.

Tietze, L. F.; Brasche, G. & Gericke K. (2006). *Domino Reactions in Organic Synthesis*, 1-617, Wiley-VCH, ISBN: 3-527-29060-5, Weinheim.

Tietze, L. F.; Kinzel, T. & Brazel, C. C. (2009). The Domino multicomponent allylation reaction for the stereoselective synthesis of homoallylic alcohols. *Accounts of Chemical Research*, Vol.42, No.2, 367-378, ISSN: 0001-4842.

Tietze, L. F. & Haunert, F. (2000). Domino reaction in organic synthesis. an approach to efficiency, elegance, ecological benefit, economic advantage and preservation of our resources in chemical transformations. in *Stimulating Concepts in Chemistry*, Vogtle, F.; Stoddart, J. F. & M. Shibasaki (Ed.), 39-64, Wiley-VCH, ISBN: 3527299785, Weinheim.

Tu, S.-J.; Cao, X.-D.; Hao, W.-J.; Zhang, X.-H.; Yan, S.; Wu, S.-S.; Han, Z.-G.; Shi, F. (2009). An efficient and chemoselective synthesis of benzo[e][1,4]thiazepin-2(1*H*,3*H*,5*H*)-ones via a microwave-assisted multi-component reaction in water. *Organic & Biomolecular Chemistry*, Vol.7, No.3, 557-563, ISSN: 1477-0520.

Tu, S.; Jiang, B.; Jia, R.; Zhang, J. & Zhang, Y. (2007). An efficient and expeditious microwave-assisted synthesis of 4-azafluorenones via a multi-component reaction. *Tetrahedron Letters*, Vol.48, No.8, 1369-1374, ISSN: 0040-4039, (c).

Tu, S.; Jiang, B.; Yao, C.; Jiang, H.; Zhang, J.; Jia, R. & Zhang, Y. (2007). An efficient and regiospecific synthesis of new 2,2'-bipyridine derivatives in water. *Synthesis*, No.9, 1366-1372, ISSN: 0039-7881, (d).

Tu, S.; Jiang, B.; Jiang, H.; Zhang, Y.; Jia, R.; Zhang, J.; Shao, Q.; Li, C.; Zhou, D. & Cao, L. (2007). A novel three-component reaction for the synthesis of new 4-azafluorenone derivatives. *Tetrahedron*, Vol.63, No.25, 5406-5414, ISSN: 0040-4020, (e).

Tu, S.; Li, C.; Li, G.; Cao, L.; Shao, Q.; Zhou, Di.; Jiang, B.; Zhou, J. & Xia, M. (2007). Microwave-assisted combinatorial synthesis of polysubstituent imidazo[1,2-*a*]quinoline, pyrimido[1,2-*a*]quinoline and quinolino[1,2-*a*]quinazoline derivatives. *Journal of Combinatorial Chemistry*, Vol.9, No.6, 1144-1148, ISSN: 1520-4766, (f).

Tu, S.; Li, C.; Shi, F.; Zhou, D.; Shao, Q.; Cao, L. & Jiang, B. (2008). An efficient chemoselective synthesis of pyrido[2,3-*d*]pyrimidine derivatives under microwave irradiation. *Synthesis*, No.3, 369–376, ISSN:0039-7881.

Tu, S.; Miao, C.; Gao, Y.; Fang, F.; Zhuang, Q.; Feng, Y. & Shi, D. (2004). A novel cascade reaction of aryl aldoxime with dimedone under microwave irradiation: The synthesis of N-hydroxyacridine. *Synlett*, No.2, 255-258, ISSN: 0936-5214.

Tu, S.; Zhang, J.; Jia, R.; Jiang, B.; Zhang, Y. & Jiang, H. (2007). An efficient route for the synthesis of a new class of pyrido[2,3-*d*]pyrimidine derivatives. *Organic & Biomolecular Chemistry*, Vol.5, No.9, 1450-1453, ISSN: 1477-0520, (b).

Tu, S.; Zhang, J.; Jia, R.; Zhang, Y.; Jiang, B. & Shi, F. (2007). A novel reaction of 4-(arylmethylene)-2-phenyloxazol-5(4*H*)-ones with pyridin-2-amine: formation of 3-(arylmethyl)-3-(benzoylamino)imidazo[1,2-*a*]pyridin-2(3*H*)-ones. *Synthesis*, No.4, 558-564, ISSN: 0039-7881, (a).

Tu, S.; Zhu, X.; Zhang, J.; Xu, J.; Zhang, Y.; Wang, Q.; Jia, R.; Jiang, B.; Zhang, J. & Yao, C. (2006). New potential biologically active compounds: Design and an efficient synthesis of *N*-substituted 4-aryl-4,6,7,8-tetrahydroquinoline-2,5(1*H*,3*H*)-diones under microwave irradiation. *Bioorganic & Medicinal Chemistry Letters*, Vol.16, No.11, 2925-2928, ISSN: 0960-894X.

Victoria Gomez, M.; Aranda, A. I.; Moreno, A.; Cossio, F. P.; de Cozar, A.; Diaz-Ortiz, A.; de la Hoz, A. & Prieto, P. (2009). Microwave-assisted reactions of nitroheterocycles with dienes. Diels-Alder and tandem hetero Diels-Alder/[3,3] sigmatropic shift. *Tetrahedron*, Vol.65, No.27, 5328-5336, ISSN: 0040-4020.

Viirre, R. D.; Evindar, G. & Batey, R. A. (2008). Copper-catalyzed domino annulation approaches to the synthesis of benzoxazoles under microwave-accelerated and conventional thermal conditions. *Journal of Organic Chemistry*, Vol.73, No.9, 3452-3459, ISSN: 0022-3263.

Villemin, D.; Cheikh, N.; Mostefa-Kara, B.; Bar, N.; Choukchou-Braham, N. & Didi, M. A. (2006). Solvent-free reaction on KF-alumina under microwave: Serendipitous one-pot domino synthesis of new isobenzofuran-1(3*H*)-ones from alpha-hydroxy ketones. *Tetrahedron Letters*, Vol.47, No.31, 5519-5521, ISSN: 0040-4039.

Wen, L.-R.; Sun, J.-H.; Li, M.; Sun, E.-T. & Zhang, S.-S. (2008). Application of β -(2-chloroaroyl) thioacetanilides in synthesis: An unusual and highly efficient access to thiochromeno[2,3-*b*]pyridine derivatives. *Journal of Organic Chemistry*, Vol.73, No.5, 1852-1863, ISSN: 0022-3263.

Wipf, P.; Fletcher, J. M. & Scarone, L. (2005). Microwave promoted oxazole synthesis: cyclocondensation cascade of oximes and acyl chlorides. *Tetrahedron Letters*, Vol.46, No.33, 5463-5466, ISSN: 0040-4039.

Zanobini, A.; Brandi, A. & de Meijere, A. (2006). A new three-component cascade reaction to yield 3-spirocyclopropanated β-lactams. *European Journal of Organic Chemistry*, No.5, 1251-1255, ISSN: 1434-193X.

Zhang, Z.; Ma, Y.; Zhao, Y. (2008). Microwave-assisted one-pot synthesis of dihydrocoumarins from phenols and cinnamoyl chloride. *Synlett*, No. 7, 1091-1095, ISSN: 0936-5214.

Application of Microwave Technology for Utilization of Recalcitrant Biomass

Shuntaro Tsubaki and Jun-ichi Azuma
Division of Environmental Science, Graduate School of Agriculture, Kyoto University
Japan

1. Introduction

Biomass is the sole organic material which is usable as an alternative to fossil resources on earth. Separation of constituent components (biorefinery) from biomass and their degradation are major steps for its practical use. Because of the diversity and recalcitrance of biomass, special techniques are required for these purposes. Microwave irradiation has been attracted much attention as a tool for degradation of biomass for its environment friendliness, known as green technology. Microwave heating is based on non-contacted irradiation of electromagnetic waves in a frequency range from 300 MHz to 300 GHz (wavelength about 1 m to 1 mm), which activate dipole molecules depending upon their dielectric constants (dielectric heating) and microwave sensitizers which induce electro-conductive heating by using modernized equipment with a computerized regulating system. In this Chapter authors introduce applicability of microwave technology for utilization of recalcitrant biomass as pioneers in this field.

2. Microwave apparatus for biomass utilization

Microwave irradiation method is a kind of autohydrolysis to separate hemicellulose and lignin from lignocellulose and it is utilizable as pretreatment before enzymatic saccharification to produce fermentable carbohydrates as well as an extraction method for biomass components such as polysaccharides. Fig. 1 shows the main concept of microwave irradiation technology for separation and applications of biomass components. In the simplest method, the heating medium is water since the water is the most environmental friendly and readily available solvent. The high temperature and pressurized water in a closed vessel provides decrease in dielectric constant and increase in ion products which enable solubilization of less polar substances and hydrolysis of biomass without catalyst. Moreover, the effects of microwave irradiation can be further enhanced by addition of organic acids, sulfuric acid, hydrogen peroxide, inorganic ion and aqueous alcohols to the medium depending on the targeting products. Recent sophisticated microwave ovens equipped with TFM vessel which can withstand <250°C, 55 bar satisfy use of these additives safely. Automatic temperature feedback system by PID is essential for precise control of reaction temperature, because the autohydrolysis of the biomass component is greatly dependent on the temperature. The feedback system also contributes to the safety of the reaction by quick shut down of microwave output in case of reaction overruns. Explosion protection system is further required especially when running severe oxidation-reduction reactions.

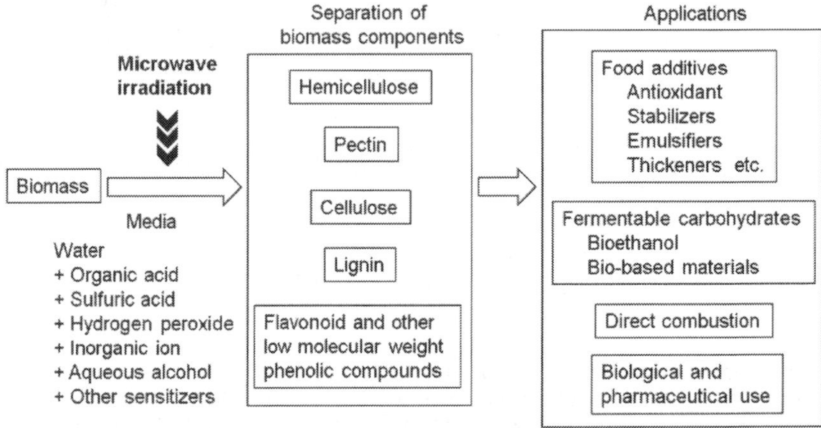

Fig. 1. Concept of microwave irradiation technology for refinery of biomass components.

Continuous flow system is crucial for scaling up of the microwave irradiation system. A continuous flow microwave irradiation apparatus was developed as shown in Fig. 2 (Tsumia & Azuma, 1994). The system provides 2.45 GHz of microwave with maximum output, temperature and flow rate of 4.9 kW, 240°C and 7-20 L/h, respectively. Generally, the powered materials are suspended in medium to give homogeneous slurry. The slurry is fed into the reactor at 20–30 kg/cm². The reactor is made of alumina fine ceramics (purity 99.7%, 2 m in length × 4 cm in diameter) to endure the high temperature and pressure reaction. The content inside the reactor was homogenized by rotating a stainless steel rod attached with blades. Finally the products are come out from a blow-down valve to atmospheric pressure. This final process gave a kind of defibrating effects like explosion.

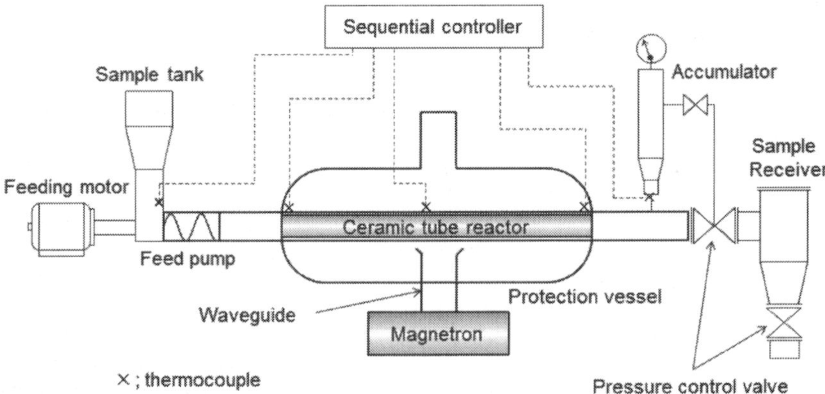

Fig. 2. Schematic illustration of continuous flow microwave irradiation apparatus.

3. Chemical composition of biomass

Biomass especially phytomass includes diverse group of plants which differed in chemical compositions, leading to difficulties in their utilization. Because about 70% (w/w) of phytomass was composed of carbohydrates, its utilization was usually directed to refinery (fractionate), decomposition and/or transformation of their carbohydrate moieties. For utilization of carbohydrates as biomass, polysaccharides were frequently targeted. In the phytomass two groups of polysaccharides were present, storage polysaccharides and constitutive polysaccharides present in cell walls consisting of primary cell wall and secondary cell wall in vascular plants. Primary cell wall is formed outside plasma membrane when daughter cell is separated from its mother cell and developed until the growth of cell cease. After finish of synthesis of primary cell wall, thick secondary cell wall is formed inside the primary cell wall. Lignification starts at cell corners and distributed throughout cell walls. Polysaccharides consisting of the primary and the secondary cell walls are different.

3.1 Storage polysaccharide

Starchy polysaccharides are typical storage polysaccharides present widely in plant tissues as an end-product of photosynthesis and accumulated in grains and tubers as crystallized granules (~1-100 μm). The predominant constituents in starch granules (98-99%, w/w) are amylose and amylopectin with a distribution of ~20-25% and ~80-75%, respectively. Amylose consists of a long linear chain having around 99% of α-(1→4) and a few α-(1→6) linkages (molecular mass in a range of 1.0×10^5 - 1.0×10^6) with around 9-20 branch points equivalent to 3-11 chains per molecule (Tester et al., 2004). Each chain contains 200-700 glucose residues having one reducing and one non-reducing ends. Amylopectin has a branched structure built from α-(1→4) linkages with 4.2-5.9% of α-(1→6) branch linkages (Tester et al., 2004; Robyt, 2008) and has 100-1,000 fold higher molecular mass (1.0×10^7 - 1.0×10^9) than amylose. Unit chain of amylopectin is much shorter than that of amylose and contains about 18-25 glucopyranose (Glc$p$) residues. Amylopectin molecule has thus one reducing and many non-reducing ends. It should be noted that the numerical values presented above differ in locations and species of plants. Exterior α-(1→4) linked chains of amylopectin form double helices to form crystalline region and branching point portions formed amorphous region leading to formation of radially oriented lamellae structure in 9 nm size (Fig. 3). Precise interactions between amylose and amylopectin are not known, but these polysaccharides attached together by intermolecular hydrogen and hydrophobic bonds in addition to intramolecular interactions to form water-insoluble granules. Three distinct X-ray patterns have been observed, A-, B- and C-types. Cereal grain starches showed A-type diffraction, and tuber, fruit and stem starches exhibit typical B-type diffraction. Root, bean and pea starches give intermediate diffraction profile between A- and B-types called as C-type. Although the double helical structures mostly due to exterior chains of amylopectin are common to all types, packing profiles of double helices are different between A- and B-types; A-type has a compact structure with a low water content, while B-type has a more open structure with a hydrated helical center.

Industrial enzymatic decomposition of starch proceeds through gelatinization (30-40% starch slurry, heating for 5 min at 105°C), liquefaction (1-2 h at 95-100°C) and saccharification (72 h at 60°C) to glucose whose concentration was 97% in the saccharified products (Van Der Maarel, 2010). Production of glucose from starch has economically superiority but has a defect in competition with supply of food from starchy polysaccharides. In addition starch industries produce a huge amount of residues after extraction of starch. Although the residues contain residual starch, presence of recalcitrant lignocellulosic materials hampers their utilization.

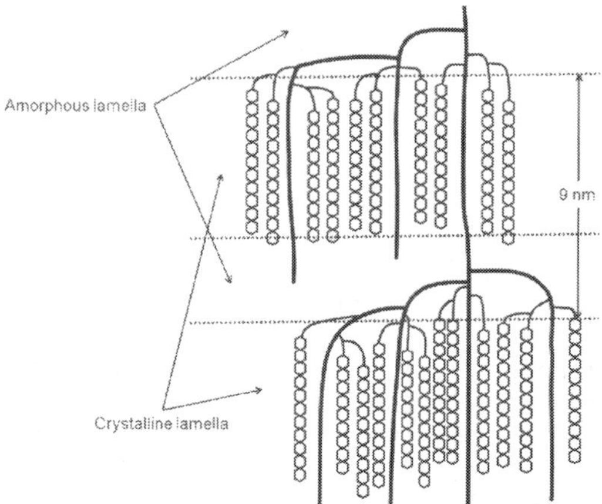

Fig. 3. Schematic illustration of lamellar structure of a starch granule.

3.2 Primary walls

Polysaccharides in the primary cell wall are grouped into three categories; skeletal polysaccharide such as cellulose, matrix polysaccharide mixture extractable with water and hot water, pectin fraction extractable with chelating agents, and hemicellulose fraction extractable with 1-24% sodium hydroxide or potassium hydroxide, and glycoconjugate rich in hydroxyproline such as extensin and arabinogalactan protein (AGP). Its composition was remarkably different in dicotyledonous and monocotyledonous plants. Typical constituent polymers are listed in Table 1. Plants grouped in gymnosperm have a similar composition as dicotyledonous plants. The primary cell wall is not lignified in the tissue cultured cells but is the most lignified of the whole cell wall layers to form middle lamella. Because of difficulty of separation of the middle lamellar structure from the other primary cell wall material, a whole region of primary cell wall is referred as compound middle lamella (CML).

Type	Monocotyledonous plant		Dicotyledonous plant	
Skeletal polysaccharide	Cellulose	(20-40)	Cellulose	(20-40)
Matrix polysaccharide	Pectin	(0.8-8)	Pectin	(20-35)
	(1→3,1→4)-β- Glucan	(12-16)	Arabinogalactan	(~10)
	Glucuronoarabino-xylan	(~30)	Gulucuronoarabino-xylan	(~5)
	Xyloglucan	(2-4)	Xyloglucan	(~20)
Glycoconjugate	Hydroxyproline-rich glycoprotein	(~10)	Hydroxyproline-rich glycoprotein	(~10)

Table 1. Polymers consisting primary cell wall of vascular plants (%, w/w)

Cellulose is a linear fibrous homopolysaccharide composed of β-(1→4) linked D-Glc$p$ residues with cellobiose as a unit of fiber axis (1.03 nm) (Glasser, W.G., 2008; Wertz et al., 2010). Degree of polymerization (DP) of primary cell wall cellulose is ~2,000-6,000, varied depending upon origin, age and location of plant tissues. Cellulose molecules packed together to form microfibrils (width 2-2.5 nm in the case of cotton primary cell wall) by two intramolecular hydrogen bonds, $O2'$-H•••$O6$ and $O3$-H•••$O5'$, and one intermolecular hydrogen bond, $O6$-H•••$O3$. Crystalline form of the native plant cellulose (cellulose I) is a mixture of two crystalline allomorphs, Iα (one-chain triclinic unit cell) and Iβ (two-chain monoclinic unit cell), rich in the latter. Orientation of microfibrils inside the primary cell wall is random but more perpendicularly oriented to the cell axis close to the secondary cell wall. Tensile strength of fibrous structure of plant cells is largely depending upon the structure of the cellulose.

Pectins are composed of complex networks of three categories of acidic polysaccharides, homogalacturonan (HG), rhamnogalacturonan I (RG-I) and rhamnogalacturonan II (RG-II) (Willats et al., 2001). The HG has a fundamentally linear homopolysaccharide (Fig. 4) composed of α-(1→4) linked D-galacturonic acid (Gal$p$A) residues with partial acetylation and methyl-esterification (DP about 70). Insertion of L-rhamnosyl units sometimes occurred. The HG is covalently linked to RG-I and RG-II.

The RG-I is a branched acidic polysaccharide composed of backbone disaccharide repeating units of [→2)-α-L-Rha$p$-(1→4)-α-D-Gal$p$A-(1→]$_n$ with n of 100-300 (Fig. 4). About a half rhamnopyranose (Rha$p$) residue has substitution at O-4 with at least about 30 kinds of mono- and oligo-saccharides consisting of D-galactose and/or L-arabinose with trace amount of L-fucose. In addition the Gal$p$A residues are acetylated on O-2 and O-3. The RG-I molecules are usually obtained by pectinase treatment. In plants belong to *Chenopodiaceae* some of the side chains of RG-I are esterified with ferulic and coumaric acids and cross-linked by dimerization due to oxidative coupling (Harris & Stone, 2008).

HG: →4)-α-D-Gal$p$A-(1→4)-α-D-Gal$p$A-(1→4)-α-D-Gal$p$A-(1→4)-α-D-Gal$p$A-(1→

RG-I: →4)-α-D-Gal$p$A-(1→2)-α-L-Rha$p$-(1→4)-α-D-Gal$p$A-(1→2)-α-L-Rha$p$-(1→ (Backbone)

Fig. 4. Structures of homogalacturonan (HG) and rhamnogalacturonan I (RG-I)

The RG-II is a kind of substituted galacturonans having a backbone structure similar to HG. The RG-II molecules are usually obtained by pectinase treatment as RG-I and thus thought to be covalently linked to HG. Substitutions occur at both O-2 and O-3 positions of D-Gal$p$A residues with 4 types of side chains consisting of more than 12 kinds of glycosyl residues including rarely observed sugars such as D-apiofuranose (Api$f$). The two RG-II molecules form a dimer with boric acid as a tetravalent 1:2 borate-diol ester at O-2 and O-3 of the two Api$f$ residues (Fig. 5) (Kobayashi et al, 1996; O'Neill et al., 2004). Acidic pectins are also known to form hydrogels by formation of egg-box structure with divalent cations such as Ca^{2+}.

β-Glucan (BG) is a linear polysaccharide mostly present in the primary cell wall of monocotyledonous plants especially in *Poaceae*. Its chemical structure is fundamentally composed of a mixture of cellotriose (~70%) and cellotetraose (~30%) units which are linked together by β-(1→3) linkages (Fig. 6). The ratio of β-(1→4) and β-(1→3) linkages vary depending upon species and location of plants. The BG has therefore a folded structure and shows viscous property in solution resulting in grouping into gum (Ebringerová et al., 2005).

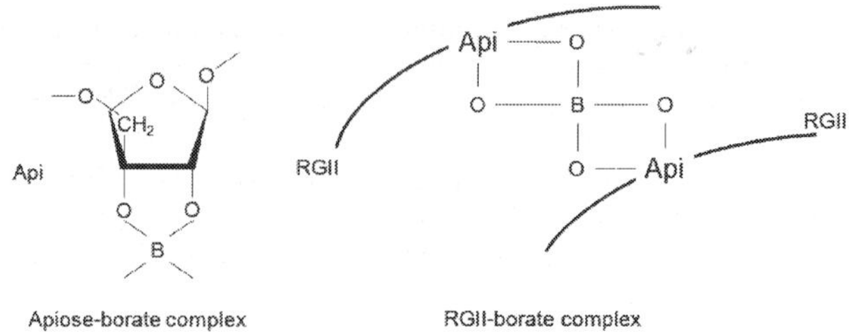

Fig. 5. Cross-linkage of two rhamnogalacturonan II (RG-II) chains by formation of a tetravalent borate ester at two apiofuranose residues

$$\beta\text{-D-Glc}p\text{-}(1\to4)\text{-}\beta\text{-D-Glc}p\text{-}(1\to4)\text{-}\beta\text{-D-Glc}p\text{-}(1\to4)\text{-}\beta\text{-D-Glc}p$$

Fig. 6. Structure of $(1\to3,1\to4)$-β glucan (BG)

Arabinogalactans (AG) are grouped into two types; β-$(1\to4)$-linked D-galactan having β-$(1\to5)$-linked L-arabinofuranosyl *(Araf)* short side chains at *O*-3 (Type I) (Fig. 7), and a highly branched arabino-3,6-galactan (Type II) (Ebringerová et al., 2005). The Type I AGs are pectinic and frequently considered to be originated from substituents in RG-I. The Type-II AG has an inner chain comprising β-$(1\to3)$-linked D-Gal*p* residues which are substituted with β-$(1\to6)$-linked D-Gal*p* residues. Other substituents such as D-Gal*p*, L-Ara*f*, L-Ara*p* and β-L-Ara*p*-$(1\to3)$-L-Ara*f* are also attached at *O*-3/*O*-6 of 3,6-galactan moiety (Fig. 7). The Type-II AG which is linked to hydroxyproline is known as arabinogalactan protein (AGP).

Type I AG: $\to4)$-β-D-Gal*p*-$(1\to4)$-β-D-Gal*p*-$(1\to4)$-β-D-Gal*p*-$(1\to4)$-β-D-Gal*p*-$(1\to$
β-L-Ara*p*-$(1\to5)$-α-L-Ara*f*

Type II AG: $\to3$-β-D-Gal*p*-$(1\to3)$-β-D-Gal*p*-$(1\to3)$-β-D-Gal*p*-$(1\to3)$-β-D-Gal*p*-$(1\to$
β-D-Gal*p*   R-3)-β-D-Gal*p* β-D-Gal*p*   R
R β-D-Gal*p*

Fig. 7. Structures of Type-I and Type-II agabinogalactans (AGs) (R: α-L-Ara*f/p*, β-L-Ara*p*-$(1\to3)$-L-Ara*f*, β-D-Gal*p*)

The primary cell walls of both dicotyledonous and monocotyledonous plants including gymnosperm contain up to 10% protein. Hydroxyproline-rich glycoproteins (HRGPs) are the well-known glycoproteins present in the primary cell wall. Extensin belongs to this type of glycoprotein and acts as a cross-linker between cell wall polysaccharides and lignin by formation of oxidative couplings through its tyrosine residues to provide mechanical strength. Carbohydrate portion of extensin is composed of arabinotriose and arabinotetraose linked α-glycosidic to hydroxyproline and monomeric β-D-Gal*p* residues linked to serine. In addition proline-rich and glycine-rich glycoproteins were also present in lignified cell walls. Xyloglucan is a kind of substituted cellulose and has a backbone structure composed of linear β-(1→4)-linked D-Glc*p* residues. Substitutions occurred at *O*-6 by α-D-Xy*p* residues which are further connected to 1,2-linked Ara*f*, Gal*p* and α-L-Fuc*p*-(1→2)-linked-Gal*p* residues (Fig. 8). Substitution profile is different in species and location of plants. Xyloglucans are firmly attached to cellulose by hydrogen bonds and also covalently bonded to pectin (Hayashi, 1989; Harris & Stone, 2008).

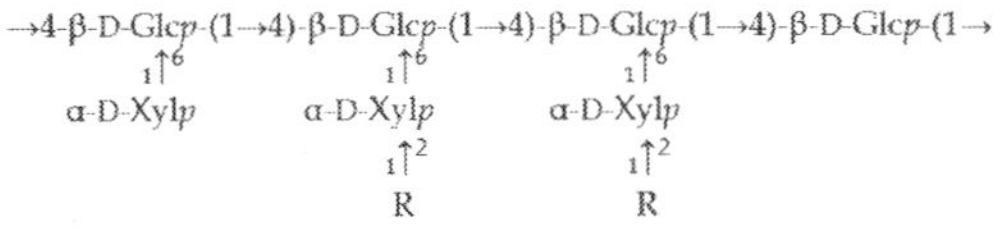

$$\rightarrow 4\text{-}\beta\text{-D-Glc}p\text{-}(1 \rightarrow 4)\text{-}\beta\text{-D-Glc}p\text{-}(1 \rightarrow 4)\text{-}\beta\text{-D-Glc}p\text{-}(1 \rightarrow 4)\text{-}\beta\text{-D-Glc}p\text{-}(1 \rightarrow$$

$_1\uparrow^6$ $_1\uparrow^6$ $_1\uparrow^6$

α-D-Xyl*p*   α-D-Xyl*p*   α-D-Xyl*p*

$_1\uparrow^2$ $_1\uparrow^2$

R   R

Fig. 8. Structure of xyloglucan (XG) (R: Ara*f*, Gal*p* and α-L-Fuc*p*-(1→2)-linked-Gal*p* residues)

3.3 Secondary walls

Secondary cell wall in the vascular plants is piled up by accumulation of layered structures in a concentric way inside the primary cell wall. In woods thickening of the secondary cell wall progressed in three steps mainly on account of cellulose synthesis to form three layers differing in orientation and thickness of cellulose microfibrils, a thin outer secondary wall (S_1) (0.1-0.2 µm, microfibril angle 50-70°), a thick secondary wall (S_2) (1-5 µm, microfibril angle 10-30° in earlywood and 0-10° in latewood) and a thin inner secondary wall (S_3) (~0.1 µm, microfibril angle 50-90°). DP of the cellulose in the secondary cell wall is higher than that in the primary cell wall (13,000~14,000 in cotton). Cellulose in wood has DP of 8,200-8,500 with width of microfibrils of 3-5 nm. Hemicelluloses are synthesized and deposited inside the secondary cell wall. Finally lignification starts from cell corner, then progressed into middle lamella, primary cell wall and secondary cell wall. Because of difficulty of separation of these cell walls, summative chemical compositions of softwoods and hardwoods in a temperate zone are listed in Table 2. Chemical composition of monocotyledonous plants is similar to that of hardwoods with peculiarly high content (1-3%) in hydroxycinnamic acids. Cellulose content in both woods is similar but softwoods are rich in lignin, in turn hardwoods are rich in hemicelluloses. Hemicelluloses are a group of polysaccharides extractable with alkali (Ebringerová et al., 2005). The most remarkable difference between both woods is in composition of hemicelluloses.

Xylans are ubiquitous in vascular plants having a common backbone structure of β-(1→4) linked D-xylopyranose (Xyl*p*) residues. Hardwood xylans have single 4-*O*-methyl-α-D-glucuronic acid (MeGlcA, α-D-4-*O*-Me-Glc*p*A) residues at *O*-2 with an average ratio of Xyl : MeGlcA being about 10 : 1 (distribution 4-16 : 1) and called as glucuronoxylan (4-*O*-methylglucuronoxylan, MeGX). In the native state Xyl*p* residues are partially acetylated at *O*-2 or *O*-3 and both positions (degree of substitution 0.3-0.7) and the native xylan is called as *O*-acetyl-4-*O*-methylglucuronoxylan (AcMeGX, Fig. 9). Some D-Xyl*p* residues bearing MeGlcA are further acetylated at *O*-3. In softwood xylans, no acetyl substitutions have been found and α-L-Ara*f* residues are further substituted at *O*-3 of Xyl*p* residues. Average ratios of Xyl : MeGlcA and Xyl : Ara*f* are 4-6 : 1 and 8-9 : 1, respectively. Thus softwood xylans are called as arabino-4-*O*-methylglucuronoxylan (AMeGX) or just arabinoglucurnoxylan (AGX) (Fig. 9). Presence of uronic acid substitutions on two contiguous D-Xyl*p* residues is noted in the softwood xylans. Before the reducing end →4)-β-D-Xyl$p$ residue of glucuronoxylan insertion of a peculiar disaccharide, →3)-α-L-Rha$p$-(1→2)-α-D-Gal$p$A-(1→, has been reported.

Component	Softwood	Hardwood
Cellulose	39-41	40-45
Lignin	25-36	18-25
Hemicellulose	15-26	23-38
Pectin	1	1
Extractives	1-5	1-5
Ash	0.1-1.0	0.1-1.0
Component of hemicellulose		
Glucuronoxylan	1-2	20-30
Glucomannan[a]	15-18	3-5[c]
Arabinoglucuronoxylan	8-10	-
Galactoglucomannan	1-4	Trace
Arabinogalactan	2-3	+

[a]Lower in Gal$p$ content around a molar ratio of Gal$p$: Glc$p$: Man$p$ = 0.1 : 1 : 3-4. [b]Higher in Gal$p$ content around a molar ratio of Gal$p$: Glc$p$: Man$p$ = 1 : 1 : 3-4. [c]A molar ratio of Glc$p$: Man$p$ = 1 : 1-2

Table 2. Summative chemical composition of woods grown in a temperate zone (%, w/w)

Average DP_n of wood xylans is 84-108. Non woody dicotyledonous plants also bear a lot of xylans similar to the native hardwood xylans with frequent presence of glucuronic acid (GlcA). Monocotyledonous plants, however, contain partially acetylated arabinoglucuronoxylans (AcAGX) and neutral xylans containing α-L-Ara$f$ residues at O-2 or O-3 and both positions of β-(1→4) linked D-Xyl$p$ residues (AX). These xylans carry additional esterified hydroxycinnamic acids which are further linked to lignin (Harris & Stone, 2008). Substitution profiles in these xylans varied very much depending upon species and location of plants.

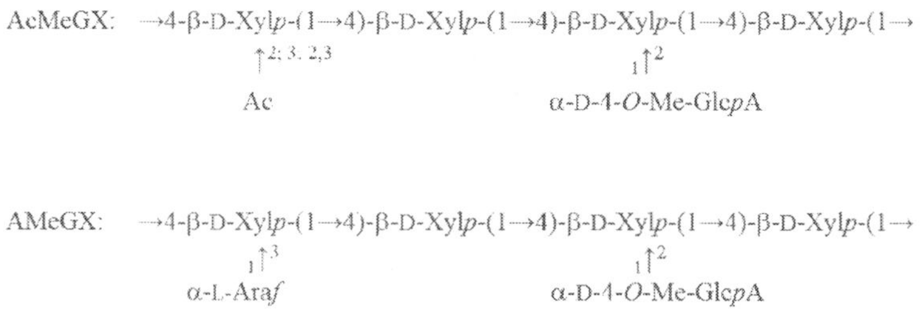

AcMeGX: →4-β-D-Xyl$p$-(1→4)-β-D-Xyl$p$-(1→4)-β-D-Xyl$p$-(1→4)-β-D-Xyl$p$-(1→

       ↑2; 3, 2,3   1↑2

        Ac   α-D-4-O-Me-Glc$p$A

AMeGX: →4-β-D-Xyl$p$-(1→4)-β-D-Xyl$p$-(1→4)-β-D-Xyl$p$-(1→4)-β-D-Xyl$p$-(1→

       1↑3   1↑2

     α-L-Ara$f$ α-D-4-O-Me-Glc$p$A

Fig. 9. Structures of hardwood native O-acetyl-4-O-methylglucuronoxylan (AcMeGX) and softwood arabino-4-O-methylglucuronoxylan (AMeGX)

Galactoglucomannans (GGM) showed a common structure consisting of a linear backbone of randomly distributed β-(1→4) linked D-Glc$p$ and D-Man$p$ residues with single substitutions of α-(1→6) linked D-Gal$p$ residues at O-6 of D-Man$p$ residues (Ebringerová et al., 2005). Softwoods contain two groups of galactoglucomannans differing in galactose contents, one has a low Gal$p$ (Gal$p$: Glc$p$: Man$p$ = 0.1 : 1 : 3-4), while the other has a high

Gal$p$ (Gal$p$: Glc$p$: Man$p$ = 1 : 1 : 3-4) (Table 2). In Table 2 content of the former is listed as glucomannan (GM). In the native state Man$p$ residues are partially acetylated at O-2 or O-3 (degree of substitution 0.2-0.4) and the native galactoglucomannan is called as O-acetyl-galactoglucomannan (AcGGM, Fig. 10). The average DP_n of softwood galactoglucomannans is 90-102. Hardwoods contains glucomannans having a Glc$p$: Man$p$ ratio of 1 : 3-4. Partial acetylation at D-Man$p$ residues is also observed similar to the softwood galactoglucomannans. The average DP_n of softwood galactoglucomannans is 60-70. O-Acetyl-glucomannans having similar structure are also present in many tissues of monocotyledonous plants, such as bulbs, tubers, roots, seeds and leaves. Konjac mannan is a typical glucomannan used as a food.

$$\rightarrow 4\text{-}\beta\text{-D-Man}p\text{-}(1\rightarrow 4)\text{-}\beta\text{-D-Glc}p\text{-}(1\rightarrow 4)\text{-}\beta\text{-D-Man}p\text{-}(1\rightarrow 4)\text{-}\beta\text{-D-Man}p\text{-}(1\rightarrow$$

$$\uparrow^{2;\,3} \qquad\qquad\qquad\qquad\qquad {}_1\uparrow^{6}$$

$$\text{Ac} \qquad\qquad\qquad\qquad\qquad \alpha\text{-D-Gal}p$$

Fig. 10. Structure of softwood native O-acetyl-galactoglucomannan (AcGGM)

Linear β-(1$\rightarrow$4) linked D-Mannans free of D-Glc$p$ residues and highly substituted by α-(1$\rightarrow$6) linked D-Gal$p$ residues at O-6 of D-Man$p$ residues are well known as galactomannans or gums. They are present in various kinds of plant storage tissues. Because of their high water-solubility with formation of very viscous transparent hydrogels, their industrial importance is high. The structure of arabinogalactan present in the secondary cell wall is similar to those belong to Type II AG. This kind of polysaccharide is abundantly present in the heartwood of the genus *Larix* and is called as larch gum. Because of its high water-solubility with low viscosity, its industrial importance is high.

3.4 Lignin and other components

Plant biomass contains polysaccharides, proteins and several groups of biopolymers which lack regular and ordered chemical structures like polysaccharides and proteins. Chemical properties of polysaccharides and glycoproteins in plant cell walls are reviewed above. Hydrophobic lignin, suberin and cuticular components such as cutin and cutan belong to the category of unordered polymers. These components wrap polysaccharides and glycoproteins inside the cell wall by direct covalent bonds and physical entanglements and indirect covering as hydrophobic layers, leading to give biomass recalcitrance. However, glass transition temperature of lignin (130-200°C) is usually lower than that of cellulose (230-250°C) and close to hemicellulose (160-200°C) under dry state. Glass transition temperatures of lignin and hemicelluloses unlike cellulose lowered with increase in moisture content. Therefore heating biomass in water close to the glass transition temperature of cellulose and extraction with solvents having various range of polarity is a strategy for refinery of biomass.

Lignins are radically coupled amorphous polymers having three fundamental monolignols, p-coumaryl alcohol, coniferyl alcohol and sinapyl alcohol, and a number of minor monolignols such as coniferaldehyde, acetylated coniferyl alcohol, ferulic acid, etc., and have a peculiar optically inactive characteristic. Lignins are present in vascular plant cell walls. Based on degrees of contribution of monolignols lignins are usually grouped into three types, softwood lignin or guaiacyl (G) lignin, hardwood lignin or syringyl-guaiacyl (SG) lignin, and grass or hydroxyphenyl-syringyl-guaiacyl (HSG) lignin. Monolignol ratios are related with evolution of plants and varied depending upon location and species of vascular plants. Polymerization of monolignols was previously proposed to progress by non-regulated couplings of peroxidase- or laccase-mediated monolignol radicals but recently a different theory was proposed by taking hypothetical low-molecular weight redox shuttle into consideration to regulate formation of monolignol radicals (Henriksson, 2009). Anyway inside lignins at least three kinds of ether bonds (β-aryl ether, diaryl ether and glyceraldehyde aryl ether types) and four kinds of carbon-carbon condensed bonds (dihydroxybiphenyl, phenyl coumarane, scoisolariciresinol and spirodienon types) connect the monolignols. In addition, hemicelluloses are covalently linked to lignins by ether and ester bonds at both benzyl and γ positions of the lignin moieties and also phenyl glycosides are formed to produce well-known lignin-carbohydrate complexes (LCCs) (Fig. 11) (Azuma, 1989; Harris & Stone, 2008). Presence of linkages between cellulose and lignin is also noted in woods. In monocotyledonous plant cell walls both ferulic and p-coumaric acids esterified to substituent sugars of hemicelluloses are further connected with lignins as described in 3.3.

Suberins are aliphatic-aromatic cross-linked polymers present in root, bark and fruit surface (Gandini et al., 2006). The contents of suberin in extractive-free outer bark of hardwoods of industrial relevance amount to 20-50%. Cork tissue contains suberin as the main constituent (Pinto et al. 2009). It serves as a protective barrier between the plants and environment. Their main aliphatic moieties are composed of C_{16}-C_{28} ω-hydroxy-fatty acids, C_{16}-C_{26} α,ω-dicarboxylic acids, C_{16}-C_{30} dihydroxy- or epoxy-fatty acids, and C_{16}-C_{26} aliphatic alcohols. Aromatic moiety is close to lignin but contains aromatic acids such as quinic acid, ferulic acid and 3,4-dihydroxybenzoic acid.

Cutin is an aliphatic polyester embedded with wax and cutan in the cuticular membranes of plant leaves, young stems and fruits, consisting of C_{16} and C_{18} hydroxy-fatty acids and hydroxyepoxy-fatty acids (Fig. 12) (Stark and Tian, 2006). Proportion of cutin and cutan varied based on maturity, location and species of plants. Its major functions are protection against microbial attack and intrusion of water and air. Supplying a polymer matrix for binding wax in the cuticle is also the primary role of cutin. Cutan is a common aliphatic biopolymer isolated as non-hydrolyzable and non-saponifiable residue in cuticular membrane (Boom et al., 2005). Presence of ether linkages was noted. Upon flash pyrolysis, cutan yields a series of C_7-C_{33} n-alkenes and n-alkanes with lack of cutin-derived C_{16} and C_{18} fatty acid monomers. Its function is in relation to drought resistance.

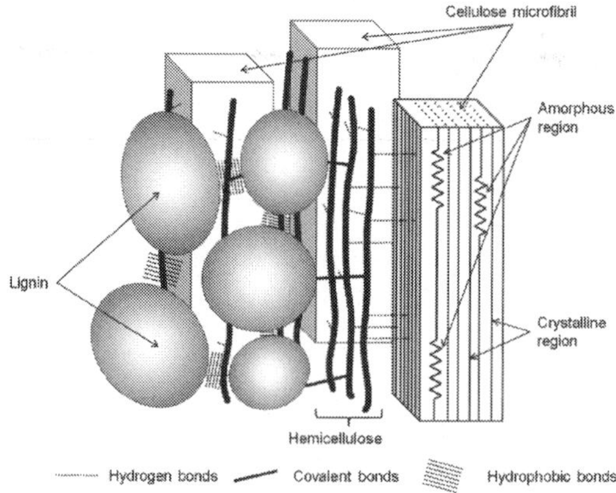

Fig. 11. Schematic illustration of lignin-polysaccharide network

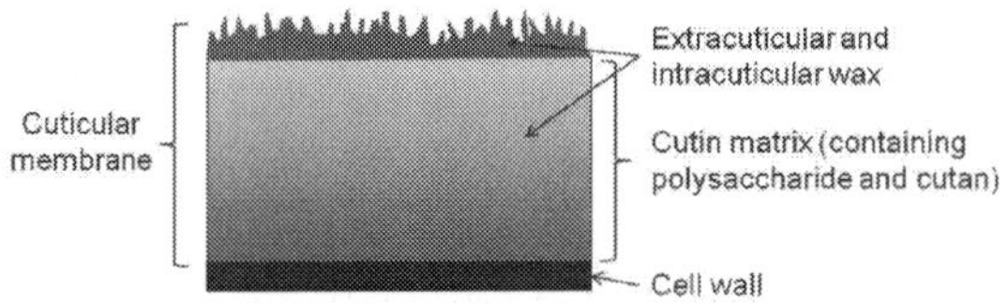

Fig. 12. Schematic illustration of cuticular membrane

4. Practical applications of microwave technology on biomass

This section summarizes microwave technology of three categories of biomass. The first category is a woody biomass including highly lignified softwoods such as cedar (Sugi, *Criptomeria japonica*), pine (Akamatsu, *Pinus densiflora*), etc., and hardwoods such as beech (Buna, *Fagus crenata*), birch (Shirakanba, *Betula platyphylla* var. *japonica*) etc. Treatments were also conducted in the presence of organic acids for saccharification and hydrogen peroxide for deliginification due to recalcitrant nature of woody biomass. The second category is monocotyledonous lignified biomass such as bamboo (Moso bamboo, *Phyllostachys pubescence*), rice straw, husk and hull, sugarcane bagasse, etc. The third category includes various kinds of non-utilized residues produced from food processing industries, agriculture and fisheries including *Okara* (soy bean residue), soy sauce residue, barley malt feed, tea residues, stones of fruits, waste of corn starch production, sea algae, fruiting bodies of mushrooms and peels of thinned fruits. Finally the microwave irradiation was applied to starch processing.

Although various kinds of biomass are present on earth, we are going to review the results we got on some typical woody biomass and agricultural and food residues. The chemical compositions of typical biomass we used are listed in Tables 3 and 4.

Component (%, w/w)	Pine (Akamatsu, *Pinus densiflora*)	Outer bark of Pine (Akamatsu, *Pinus densiflora*)	Beech (Buna, *Fagus crenata*)	Bamboo (Moso bamboo, *Phyllostachys pubescens*)
Ash	0.2	2.2	0.5	1.3
Alcohol benzene extract	3.2	5.8	2.2	3.3
1% NaOH soluble component	11.0	37.1	17.8	28.4
Water-soluble component	1.2	5.2	1.8	5.3
Hot-water soluble component	2.0	6.8	2.6	7.3
Lignin	26.6	47.4	21.3	22.6
Pentosan	11.3	8.5	23.8	23.7
Holocellulose	56.2	40.2	56.6	67.0
α-Cellulose	42.0	24.9	43.3	41.9
Relative monosaccharide composition (%, w/w)				
Rhamnose	0.4	10.3	0.7	Trace
Arabinose	3.7	1.4	1.5	2.2
Xylose	11.6	9.1	27.8	32.2
Glucose	59.8	57.1	64.5	65.6
Galactose	7.1	16.5	4.0	Trace
Mannose	17.4	5.6	1.5	Trace

Table 3. Chemical compositions of typical woody biomass used in this review

4.1 Lignified woody plants

As shown in section 3, lignified woody plants are highly resistant materials. Azuma et al. have utilized microwave irradiation technology for improvement of enzymatic susceptibility of these materials to produce fermentable sugars which can be converted into biofuels and bio-based materials. Several kinds of woody plants including softwoods, hardwoods and monocotyledonous plants were investigated for their reactivity by using batch-type microwave reactor.

Component (%, w/w)	Bagasse[a]	Coconut fiber[b]	Coconut coir dust[b]	Barley malt feed[c]	Green tea residue[d]	Stone of Japanese apricot[e]	Corn pericarp[f]	Soybean residue[g]
Ash	2.4	2.0	6.0	3.6	3.1	0.5	0.6	3.5-6.4
Extractive	2.9	0.9	0.6	10.7	10.4	1.4	7.7	6.9-22.2
1% NaOH soluble component	37.2	17.2	35.0	69.7	52.6	14.0	65.2	43.9-52.5
Water-soluble component	2.9	5.8	8.3	18.6	10.1	8.0	9.2	5.2-6.3
Hot-water soluble component	4.8	5.9	12.9	20.2	14.6	9.0	12.5	9.7-11.6
Lignin	19.5	30.9	51.8	24.8	30.8	25.3	4.0	Trace
Holocellulose	77.9	65.2	36.8	45.9	43.0	83.8	78.6	52.8-58.1
α-Cellulose	41.0	36.7	18.2	20.8	23.6	43.4	25.1	-
Protein	2.0	-	-	25.8	25.4	1.3	9.5	19.2-32.2
Relative monosaccharide composition (%, w/w)								
Rhamnose	0.2	0.9	1.4	-	0.8	0.3	Trace	0.5
Arabinose	5.8	9.1	24.9	14.1	12.7	0.6	20.0	10.2
Xylose	34.0	41.8	33.5	28.7	13.5	42.0	33.4	2.7
Glucose	57.4	41.3	29.8	53.2	54.3	56.0	36.8	52.0
Galactose	2.1	5.0	6.9	2.2	15.7	1.2	5.9	30.6
Mannose	0.5	2.0	3.6	1.7	3.0	Trace	4.1	4.1

[a]Including data listed in Kato et al., 1984; [b]Indrati & Azuma, 2000 [c]Matsui & Azuma, 2007; [d]Tsubaki et al., 2008; [e]Tsubaki et al., 2010a; [f]Yoshida et al., 2010; [g]O'Toole, 1999.

Table 4. Typical chemical compositions of food and agricultural biomass used in this review.

4.1.1 Hardwoods

For hardwoods, three kinds of sapwoods (Buna, Eucalypt and Shirakanba) and two kinds of green wood and dried chips (Poplar and Buna) were investigated for production of fermentable sugars by microwave irradiation (Azuma et al., 1984, 1985a, 1985b) (Fig. 13). Microwave irradiation was performed by using batch type glass tube sealed with stainless-steel stoppers connected together by screws and the microwave oven was equipped with 2 magnetrons (TMB-3210, Toshiba Co. Frequency; 2.45 GHz. Max output; 2.4 kW.). Heating above 160°C with solid to liquid ratio at 2.0 : 16-20 (g : mL) and 5-11 min of irradiation time was effective for solubilization of components in wood. The solubilization rate and the carbohydrate yield from sapwoods attained 30-38% and 15-20% (weight basis) by microwave irradiation at around 220-230°C, respectively. The pH of the water soluble fraction decreased to around 3.0-3.3 at 230°C, due to production of acids during microwave irradiation. The predominant neutral monosaccharide in the solubilized materials was only Xyl$p$ indicating release of xylan by microwave irradiation. The molecular weight distribution analysis by GPC revealed production of oligosaccharides having DP 2-8 and monosaccharide. This technology was further applied for production of xylo-oligosaccharides which has various biological effects (Azuma et al., 1994a,b). The microwave pretreated residues were further digested by a mixture of cellulose and hemicellulose degrading enzymes. The solubilization rate and carbohydrate production was widely improved to 70-80% at temperatures around 220-230°C. The carbohydrate recovery attained ≥90% of total hydrolysable carbohydrate content.

For practical application of hardwood refinery, green wood chips and dried chips were investigated for their reactivity under microwave irradiation (Azuma et al., 1985c). The saccharification rate increased above 160°C. The poplar green wood chips and Buna dried chips showed resistance to enzymatic saccharification, however, their enzymatic susceptibility could be improved by increasing the heating temperature up to 240-260°C.

4.1.2 Softwoods

For softwoods, ten kinds of sapwoods (Akamatsu, Ezomatsu, Hinoki, Karamatsu, Sugi, Todomatsu, Loblloy pine, Metasequioa, Bald cypress and Slash pine) and 5 kinds of barks (Akamatsu, Ezomatsu, Hinoki, Karamatsu and Sugi) were treated with microwave energy (Azuma et al., 1984b, 1985c, 1986a) (Fig.13). Sapwoods showed 24-33% (weight basis) of solubilization at temperatures around 230°C with microwave irradiation only. At the same condition, the production of carbohydrates attained 12-28% (weight basis). The major organic acid was acetic acid which was derived from acetyl ester of the native acetylated-hemicellulose. The most predominant carbohydrate released was mannose followed by glucose and galactose reflecting the native softwoods containing GM and GGM. Furfural was also produced 0.2-1.4% due to secondary degradation of released pentoses. These results showed that the hemicelluloses were released by autohydrolysis similar to steaming or steam explosion. The weight loss after enzymatic hydrolysis attained around 30-50%. The carbohydrate yield also attained around 30-50% (weight basis) and around 35-65% (total carbohydrate basis).

Softwoods are more resistant to microwave irradiation than hardwoods. Magara et al. has investigated the reason of low enzymatic susceptibility of softwoods by measuring pore size produced along microwave irradiation (1988, 1990). They have showed that hardwoods produced larger number of pores which are accessible by enzymes than softwoods leading to increase in the surface area of the treated cell walls. Proportion of condensed form in softwood lignins increased by microwave irradiation and limited the increase in surface areas accessible by enzymes.

In the case of softwood barks, microwave irradiation was also effective for production of carbohydrates (Azuma et al., 1986a). Microwave irradiation at 234-235°C produced 0.4-0.9 meq of acids and pH value attained to 3.6-4.0. The production of carbohydrates became predominant above 160°C, and reached 2.9% (Karamatsu outer bark) -8.6% (Akamatsu) at temperatures around 210-215°C. After successive enzymatic treatment, the maximum saccharification rate attained 35.5% (Karamatsu outer bark)-73.5% (Ezomatsu). The decrease in pH and carbohydrates yield was slightly lower than the sapwoods mentioned above since barks contain suberin.

4.1.3 Microwave irradiation of wood components

Reactivity of individual components of woody materials was also investigated for elucidation of the mechanism of chemical reaction of biomass under microwave irradiation. Avicel SF cellulose and Cellulose Powder D was subjected to microwave irradiation (Azuma et al., 1985a Elongated irradiation time (10-12 min) was necessary to induce degradation at >230°C. Increase in heating temperature increased the production of acids (up to 0.09 meq) and decreased pH down to 3.0-3.2. Production of carbohydrates were only 10.5% (Avicel SF Cellulose) and 6.5% (Cellulose Powder D), however the enzymatic susceptibility was improved above 220°C and the saccharification rate attained 81% (Avicel SF Cellulose) and 60% (Cellulose Powder D) at 245°C. Azuma et al. further investigated the effect of acetic acid, lignin and monomeric lignin model compounds on the extent of saccharification of crystalline cellulose (Whatman CF11) (Azuma et al., 1985d). Presence of acetic acid increased production of carbohydrate from 43.2% to 69.2% at 240°C. Although lignin did not induce degradation of cellulose by microwave irradiation alone, it facilitated the enzymatic susceptibility below 200°C. Moreover, addition of both acetic acid and lignin or monomeric lignin compounds showed synergistic improving effects on enzyme accessibility.
Effect of microwave irradiation on neutral LCC from Pine (C-I-M) was studied to investigate the correlation of polysaccharide and lignin (Azuma et al., 1986b). Heating at 237°C produced 45.7% of carbohydrates accompanied with production of oligosaccharides having DP of 2-5 and monosaccharides. Substantial amount of β-O-4 linkages and lignin-carbohydrate bonds were split by microwave irradiation.

4.1.4 Microwave irradiation of woody plants by continuous system

Based on the batch-scale studies mentioned above, microwave irradiation technology was scaled up by employing continuous system as introduced in section 2 (Magara et al., 1988) (Fig. 13). One kg of ground wood of Akamatsu and Buna was soaked in 10 L of water and microwave irradiation was performed by continuous irradiation plant (Japan Chemical Engineering and Machinery Co., Ltd. Frequency; 2.45 GHz. Max output; 4.9 kW) at a flow rate of 10-15 L/h. Reaction media were water or 0.5% (w/v) acetic acid solution. The maximum saccharification rate attained 55.7% (Akamatsu) and 61.0% (Buna) (total carbohydrate basis). Addition of acetic acid reduced the threshold temperature for removal of hemicelluloses by 10°C. The results showed that continuous system was also effective as a pretreatment of woody biomass for saccharification.

4.1.5 Catalyst assisted conversion of woody plants

Beyond the environmental-friendliness of microwave-irradiation technology based on water catalyzed reaction, Okahara et al. has developed hydrogen peroxide assisted liquefaction of woody materials to reduce severity of the process (2008) (Fig. 13). Hydrogen peroxide degrades into non-toxic water and oxygen easily by manganese dioxide or catalase, therefore it is recognized as an environmental-friendly activator of microwave irradiation. Akamatsu and Buna was loaded in TFM vessel with 20 mL of 10% hydrogen peroxide. Microwave irradiation was performed at 120-160°C and <7 min by microwave oven

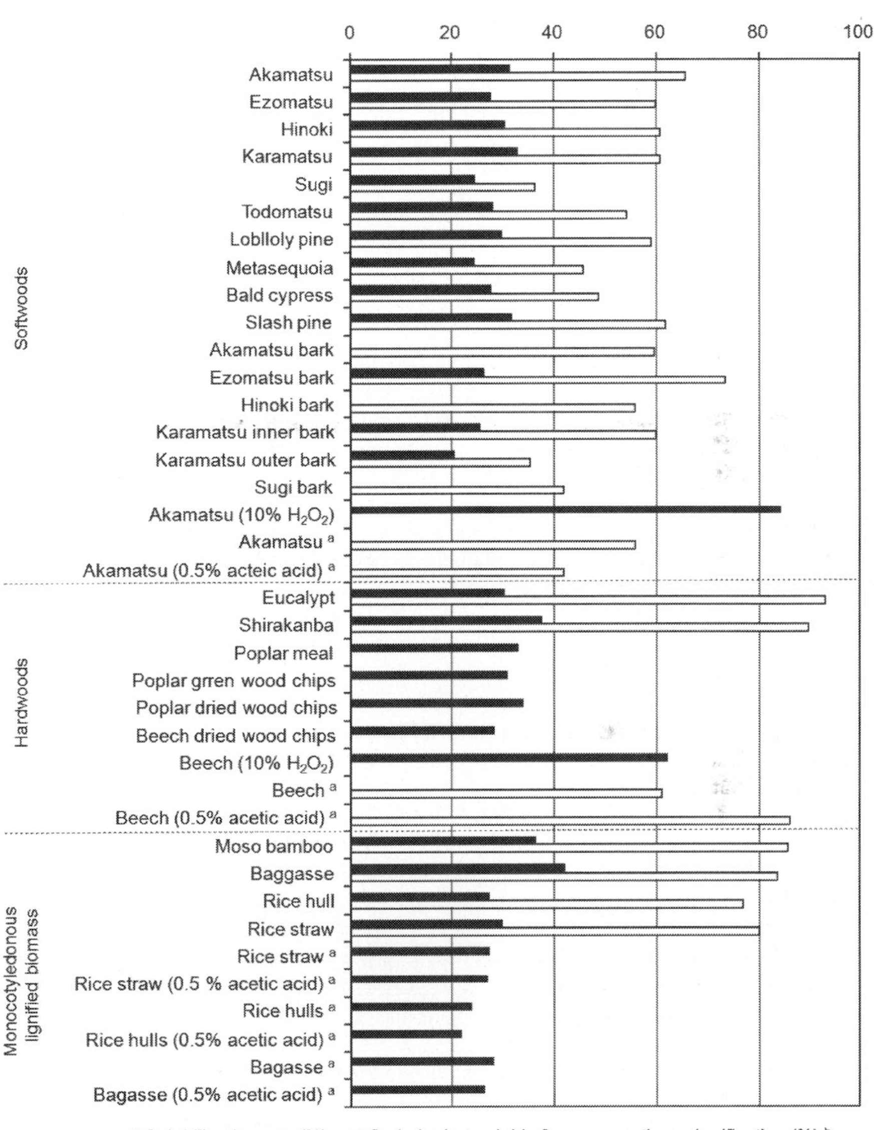

Fig. 13. Weight loss and carbohydrate yield after enzymatic saccharification (%, w/w) of softwoods, hardwoods and monocotyledonous lignified biomass. [a]Results obtained by using continuous flow system. [b]Results on the total carbohydrate basis.

MicroSYNTH (Milestone Inc., Shelton, CT, USA. Frequency; 2.45 GHz, Maximum output 1 kW. equipped with thermocouple thermometer and stirrer). Temperature was controlled by PID. The maximum solubilization rate attained 84.4% (Akamatsu) and 62.3% (Buna) at 160°C which were remarkably higher than reaction only with water or dilute organic acid solution. Neutral carbohydrate composition of extracted materials and residues showed that hemicelluloses were appreciably separated. The Klason lignin content decreased above 140°C and the lignin removal attained 61.1% (Akamatsu) and 91.7% (Buna) at 160°C. This technique was further applied to agricultural waste such as bagasse, rice straw, bamboo and lignified stone of Japanese apricot fruits (Azuma, Sakamoto & Onishi, 2009).

4.2 Monocotyledonous lignified biomass

Monocotyledonous plants such as rice, wheat, barley sugar cane and corn are important agricultural crops. Their productions reach 6.6×10^{14}, 6.1×10^{14}, 1.3×10^{14}, 1.6×10^{15} and 7.9×10^{14} tons in 2007, respectively (FAO, 2010). Agricultural residues of these crops such as husks, hulls, bagasse are important cellulosic resources to produce fermentable sugars. Furthermore, bamboo is one of the fast growing monocotyledonous plants. Therefore, it is valuable biomass feedstock. Although these plants are not classified to woods, they require pretreatment prior to enzymatic saccharification due to their lignified structures. Lab-scale microwave irradiation was demonstrated for Moso bamboo and three representative lignocellulosic wastes such as sugarcane bagasse, rice straw and rice hulls (Azuma et al., 1984a, 1984b) (Fig. 13). Moso bamboo shows 30-40% of weight loss at around 230°C. The released neutral carbohydrate fraction was mainly composed of Ara and Xyl reflecting the original carbohydrate composition (Table 3) and extraction of AX and/or AGX from these biomass. The maximal extent of enzymatic saccharification rate attained 85.7% (total carbohydrate basis), showing that the native material was comparably reactive to autohydrolysis by microwave irradiation as hardwoods. Sugarcane bagasse and rice straw showed almost the same saccharification rate (83.5 and 83.1%, respectively) as well as Moso bamboo (85.7%). In the case of rice hulls, the solubilization rate did not exceed 30%, since the recalcitrant nature of this biomass due to high content of Si.

Reactivity of lignin was further studied in detail for sugarcane bagasse, rice straw and rice hulls (Azuma et al., 1984a). Microwaved products were delignified by extraction with 90% aqueous dioxane or methanol. Bagasse lignin was the most susceptible to the pretreatment and approximately 70% of lignin became extractable in 90% aqueous dioxane at 226°C. In the case of rice straw and hulls, the maximal extents of delignification attained 58.8% and 54.4% at 226°C and 228°C, respectively. Delignification was accelerated above 180°C corresponding to release of hemicelluloses, supporting splitting of bonds between lignin and carbohydrates. The number average molecular weights of the solvent soluble lignin were measured by GPC. The extracted lignin showed smaller molecular weight (3,400-10,000) than the values of milled grass lignin isolated from the lignocellulosic wastes (5,000-15,000), showing the occurrence of sprit of bonds in lignin.

Three kinds of agricultural wastes (sugarcane bagasse, rice straw and rice hulls) were, further, pretreated by continuous microwave irradiation to scale-up the system (Magara et al., 1989). One kg of air-dried and ground material was suspended in 15 L of water and the slurry was fed to microwave irradiation at flow rates (8-20 L/h) and 160-225°C. Almost all of the hemicelluloses were removed by microwave irradiation at 210-220°C, and addition of acetic acid decreased the heating temperature by 10-30°C. Enzymatic saccharification was satisfactory progressed after microwave irradiation and the produced glucose was fully converted to ethanol by alcohol fermentation (rice straw; 378 mL/kg and bagasse; 285 mL/kg).

4.3 Unutilized agricultural biomass from industries

Various kinds and enormous amounts of residues are discharged from food processing industries, agriculture and fisheries. They are unutilized biomass capable of producing fermentable sugars, chemicals and materials. However, in many cases, they are burnt or dumped into landfills. Since chemical components of this kind of biomass are widely different depending upon species and parts of the mother plants, characterization of these components and their chemical reactivity under microwave irradiation should be elucidated for each biomass. The chemical compositions we got on typical food and agricultural biomass were already summarized in Table 4. Now the results of application of microwave irradiation for refinery of them including the results of sea algae are summarized in Fig. 14.

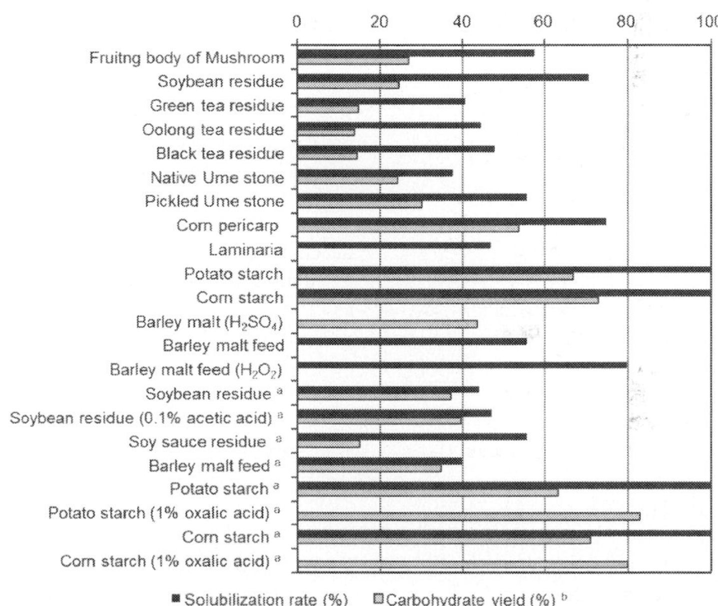

Fig. 14. Weight loss and carbohydrate yield after microwave irradiation (%, w/w) of unutilized biomass from food, agricultural and fishery industries. [a]Results obtained by using continuous flow system. [b]Data on weight basis.

4.3.1 Soybean residue

Soybean residue is a by-product from soybean curd (Tofu) production and its production reaches 700 thousand tons a year in Japan. Soybean residue is mainly composed of polysaccharide followed by protein and lipids. Microwave irradiation of soybean residue by using batch type oven (MicroSYNTH) at 200°C for 7 min (including 2 min of come-up time) with 1 : 20 of solid : liquid ratio resulted in approximately 70% of solubilization (Tsubaki et al., 2009) (Fig.14). Due to high content of pectic arabinogalactan in soybean residue, the maximum carbohydrate production (24.6% at initial weight basis) was achieved at 160°C and 7 min. Pectic polysaccharides suffer autohydrolysis greatly easier than hemicellulose which is autohydrolyzed above 180°C. Moreover, pentoses in extracted arabinogalactan suffer further degradation into furfural and other secondary decomposed materials, therefore, 2-step microwave irradiation was investigated at 170°C, 2 min for first step and 180°C, 2 min with 4% of citric acid for second step. The first step mainly solubilized arabinogalactan and the solubilized material was removed to prevent secondary degradation. At the second step, addition of citric acid was effective for separation of protein and the extent of solubilization attained 85% at the maximum with producing minimum browning.

Scaling up of this system was successfully done by employing continuous microwave irradiation. Solubilization rate and carbohydrate yield attained 45% (240°C) and 37% (220°C) at the maximum by loading slurry of 5 kg/20 L in water. The continuous system required higher temperature to initiate autohydrolysis, however addition of acetic acid slightly decreased the severity.

4.3.2 Soy sauce residue

Soy sauce is a brown colored traditional Japanese flavoring which is essential to most of the Japanese foods. Soy sauce is a fermented food of soybean and wheat with addition of salts. The fermentation liquor becomes soy sauce and the residual material is discarded as soy sauce residue. Solubilization of soy sauce residue was investigated by continuous microwave irradiation as soybean residue with addition of 0.1% (w/v) acetic acid. The highest solubilization rate and carbohydrate yield were 56% (200°C) and 22% (220°C), respectively (Fig.14). Heating at 170-220°C also produced oligosaccharides. The predominant monosaccharide contained in the extracted polysaccharide at 230°C was glucose (45.9%) followed by xylose (22.3%).

4.3.3 Barley malt feed

Barley malt feed (730 thousand tons/year) is a by-product from beer production. Fifty six percent of the barley malt feed was solubilized in water by the same microwave system as soybean residue (Azuma et al., 2008). The predominant extracted polysaccharide was AX and/or AGX. Addition of hydrogen peroxide greatly facilitated the solubilization of this biomass and the almost all the carbohydrates were solubilized by heating at 140°C under presence of 10% hydrogen peroxide with the maximal solubilization rate of 80% (Fig.14).

Barley malt feed was also saccharified by using two-step sulfuric acid hydrolysis (Matsui et al., 2007). At the optimized condition, 1 g of barley malt feed was suspended in 70% (w/w) sulfuric acid solution and microwaved for 2 min as the first step, and then the solution was diluted to 30% (w/w) by addition of water and microwaved for 3 min as the second step. The maximum saccharification rate attained 95.1% (total carbohydrate basis).

Additionally, barley malt feed was also liquefied by continuous microwave irradiation with addition of 1% (w/v) acetic acid. Heating at higher temperature than threshold temperature (200°C) was necessary to initiate solubilization of barley malt feed by continuous system. The highest solubilization rate was 41% at 240°C. Carbohydrate yield gave the maximum value of 36% at 220°C. The production of xylo-oligosaccharide was observed at 200-220°C.

4.3.4 Tea residues

Tea residues are herbal biomass discarded from tea drink production. The total emission amounts for 100 thousand tons a year in Japan. Three kinds of tea residues (green, oolong and black tea) were subjected to microwave irradiation (Tsubaki et al., 2008). Green tea residue is a non-fermented tea, while oolong and black teas are half and fully fermented by oxidase naturally exist in the native leaves, producing polymerized polyphenols and brown color. Tea residues are composed of carbohydrate (29.6-36.7%), protein (20.3-25.4%), acid-insoluble component (30.8-40.1%) and phenolic compounds (10.4-12.4%) (Fig.14). The specific aspect of this kind of biomass can be summarized in the abundance in phenolic compounds.

Microwave irradiation of tea residues solubilized 40.8-48.1% of solids and produced 13.9-14.7% of carbohydrates (initial weight basis) at 230°C, 2-min of microwave irradiation with 1.0 : 20 (g : mL) of solid to liquid ratio. The production of polyphenols attained 8.7-14.4% (initial weight basis) which was comparable to the amount of extracted carbohydrates. The extracted liquor showed antioxidant activity against hydroxyl radicals suggesting its applicability as biologically functional materials. Although catechins were fragile to microwave heating, heating above 180°C interestingly improved the polyphenol extraction, therefore, the composition was determined by using GC/MS (Tsubaki et al., 2010). The predominant phenolic compounds extracted were pyrogallol followed by catechol and dihydroconiferyl alcohol, indicating that autohydrolysis was responsible for the degradation of catechins and lignin and the production of these phenolic compounds.

4.3.5 Stones of Japanese apricot

Japanese apricot (*Ume*) has been appreciated for medicinal plant. The fruits are processed into Japanese traditional foods such as pickles (*Ume-boshi*) and liqueur (*Ume-shu*). Japanese apricot is a kind of stone fruits since it contains lignified stone in the center of the fruits. Lignified stone is a by-product of Pickles which amounts for 500 tons a year in Wakayama Prefecture, a leading producer in Japan. The stones are rich in ash (9.1%) since the pickling process includes immersion of fruits in fruit juice with addition of sodium chloride.

Heating above 200°C was necessary to solubilize components in stones and the solubilization rate and carbohydrate yield attained 55.9% and 30.3% (initial weight basis) at 230°C and 2 min of microwave irradiation (Tsubaki et al., 2010a) (Fig.14). Predominant extracted polysaccharide was xylan. Lignin also suffered partial degradation at syringyl moiety by splitting of β-O-4 linkages releasing appreciable amount of syringaldehyde, sinapaldehyde and syringic acid in the extracted liquor. Pickling process improved the solubilization of the solid material and the extraction of carbohydrates and phenolic compounds by 1.5, 1.3 and 1.4 fold, respectively. Additionally, organic acids mainly citric acid and salts contained in the stones improved the microwave absorption of the reaction medium, showing pickling process of lignocellulose in weak acidic and saline solution is capable as simple pretreatment before microwave irradiation for synergistic effect of pre-hydrolysis and microwave absorption.

4.3.6 Corn pericarp

Corn pericarp is a by-product from corn starch production. AX present in the corn pericarp could be extracted by microwave irradiation (Yoshida et al., 2010). The extraction condition was optimized by using response surface methodology and the highest carbohydrate yield was obtained as 70.8% (total carbohydrate basis) at heating temperature 176.5°C, come-up time 2min, heating time 16 min and solid to liquid ratio 1 : 20 (g : mL) (Fig.14). At the optimized condition, the solubilization rate attained 75% demonstrating high applicability of this method for utilization of corn pericarp.

4.3.7 Sea algae (*Makombu*)

Makombu (*Laminaria japonica*) is a kind of brown sea algae. Sea algae are regarded as potential renewable biomass widely distributed in the world. Makombu is capable of producing foods and polysaccharides such as alginate, fucoidan, etc. The solubilization rate increased with increase in heating temperature above 110°C. The maximum solubilization was achieved at 220°C attaining the value of 47% (Fig.14). Value of pH decreased from 6.2 to 3.8 which might accelerate autohydrolysis of the components.

4.3.8 Mushrooms

Mushrooms have been appreciated for pharmacological activities from ancient times. β-Glucan (BG) is one of the physiologically active polysaccharides. Ookushi et al. have investigated microwave-assisted extraction for BG from fruiting body of mushrooms (*Hericium erinaceum*) (2006, 2008a, b, 2009). The extraction of BG was improved by microwave irradiation for 7 min above 140°C, and the highest yield was achieved at 210°C (Fig.14). Comparing with the traditional conventional extraction method, microwave irradiation for 5 min at 140°C was equivalent to the heat conductive extraction for 6 h at 100°C. The difference in detailed structure was also investigated by methylation and ^{13}C-NMR analyses. Microwave-assisted extraction produced (1→3) rich (1→3;1→6)-β-D-glucan whereas conventional heating produced (1→6) rich (1→3;1→6)-β-D-glucan. Extraction of BG could be enhanced by treating with protease and chitin degrading enzymes.

4.3.9 Peels of thinned *Citrus unshiu*

Citrus unshiu is one of the most consumed fruits in Japan and 1 million tons are produced in 2009 in Japan (Statistics of Agriculture, Forestry and Fisheries, 2010). A large amount of green immature fruits are thinned to control the amount and quality of the fruits. Since *Citrus* peels are attracted as resource to produce pectin and flavonoids, Inoue et al. has employed microwave-assisted extraction for producing hesperidin from peels of thinned *C. unshiu* by using 70% aqueous ethanol as an extracting and crystallization solvent (2010). Hesperidin has been appreciated for biological and pharmacological activities such as antioxidant, hypocholesterolemic, hypoglycemic, protection against bone loss and antitumor activities. Fifty eight point six mg of hesperidin was extracted from 2 g of wet *C. unshiu* peels by microwave irradiation for 7 min at 140°C (Fig.15). The yield was comparable to that obtained by mixed solution of DMSO : methanol (1 : 1, v/v) for 30 min at room temperature, showing the effectiveness and environmental friendliness of this system. Storage of the extracted liquor at 5°C for 24 h produced significant amount of crystalline hesperidin as white precipitate. The maximum yield of the crystallized hesperidin attained 47.7 mg/g (86.6% of the total hesperidin).

4.3.10 Starch processing

Microwave irradiation of polysaccharides in water medium provides generation of oligosaccharide and glucose through autohydrolysis. As starches from potato, corn, wheat and tapioca are industrially important products, the effective processing of these materials could be done by employing microwave technology. The results indicate that both potato

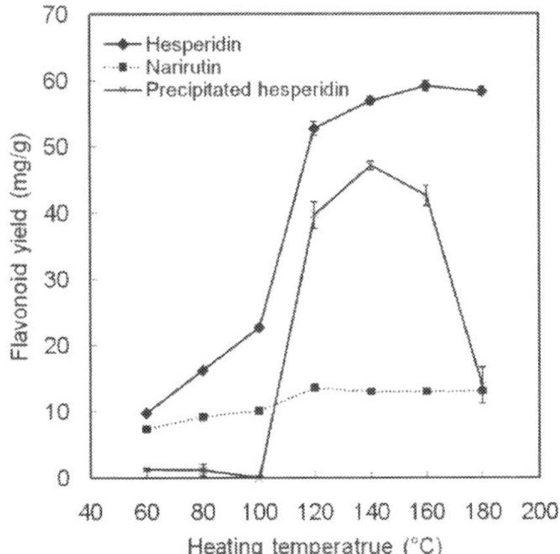

Fig. 15. Effect of microwave heating on extraction of flavonoids from thinned fruits peels of *Citrus unshiu*.

and corn starches have been saccharified by single microwave irradiation by using both batch and continuous type reactors. Gelatinization, liquefaction and saccharification processes could be done within 10 min. In the case of batch type reactor, efficient solubilization started above 120°C. Saccharification was started above 200°C accompanied with decrease in pH values. Maximal rate of saccharification of corn starch (73%) was given by batchwise microwave irradiation for 10 min at 220°C and solid to water ratio of 1 g : 20 mL. Corn starch was more susceptible to autohydrolysis than potato starch. In the case of potato starch maximal rate of saccharification (67%) was given at 230°C (Fig.14). Production of malto-oligosaccharides became prominent with increase in temperature and harder severity facilitated production of glucose. Scaling up of this method has been done by continuous system at 5-50% (w/w) and flow rate of 2.4-20.0 L/h. The results of saccharification rates of corn starch (71.0% at 220°C) and potato starch (63.6% at 225°C) were comparable to the values obtained by using the batch-type reactor. Addition of oxalic acid to 1% (w/v) was effective for lowering the severity of the process with increase in the saccharification rate.

Although microwave irradiation is a simple method for autohydrolysis of starch, secondary degradation of monosaccharides are unavoidable because the reaction was taken under elevated temperatures around 200°C. Addition of activated carbon is proved to be effective for removal of secondary degraded materials. Matsumoto et al. (2008) has applied such kind of activated carbons as sensitizers for reduction of secondary degradation and increase in microwave energy absorption. As a result, malto-oligosaccharides produced by heating for 10 min at 200-220°C were effectively adsorbed on the activated carbons. The malto-oligosaccharides adsorbed on the activated carbon could be recovered by elution with 50% aqueous ethanol. Additionally, the malto-oligosaccharides adsorbed on the activated carbon were stable and protected against microwave-assisted degradation.

5. The advantages of microwave irradiation

Microwave irradiation is different from the conventional heating in which thermal energy is delivered via conduction. Microwave energy is directly delivered to the materials and heat can be generated throughout their volumes, resulting in rapid and uniform heating. Although it is difficult to compare effects of microwave irradiation with conventional heating, some special effects notified by our results are summarized below.

5.1 Advantage of microwave irradiation over steam-explosion

Microwave irradiation in water is thought to be a kind of hydrothermal treatment. Is this microwave irradiation superior to usual hydrothermal treatment? Batch results on comparative analysis with steam-explosion shown in Table 5 indicate excellence of microwave treatment on account of enzymatic susceptibility in all three kinds of woody materials, hardwood (Japanese beech), softwood (Japanese red pine) and monocotyledon (Moso bamboo) (Azuma et al., 1984). The results on beech shown in Fig. 16A indicate further that continuous microwave irradiation is also superior to steam-explosion under the same duration of heating time of 3 min, enhancing enzymatic susceptibility at lower temperature after microwave irradiation (Yamashita & Azuma, 2010). The extents of methanol-extractable lignin are, however, similar in both treatments (Fig. 16B). These results indicate that microwave irradiation has more power to activate polysaccharides against enzymatic attacks than steam-explosion although the effects on lignin are similar.

Plant species		Temperature	Maximal extent of saccharification (%)
Pine (Akamatsu, *Pinus densiflora*)	Microwave	229 °C	65.6
	Steam explosion	227 °C	49.6
Beech (Buna, *Fagus crenata*)	Microwave	223 °C	93.0
	Steam explosion	227 °C	87.9
Bamboo (Moso bamboo, *Phyllostachys pabessence*)	Microwave	227 °C	85.7
	Steam explosion	227 °C	78.5

[a]Including data listed in Azuma et al. 1984.

Table 5. Comparison of microwave irradiation and steam explosion

5.2 Advantage of microwave irradiation over conventional heating

As described in 4.3.8 microwave irradiation shortens time necessary to extract valuable materials from biomass. Superiority of activating polysaccharides by microwave irradiation may be due to direct delivery of microwave energy to polysaccharides through molecular interactions with electromagnetic field. Similar to BG from mushroom microwave extraction was applied for extraction of hemicelluloses from plants (Ebringerová et al., 2005).

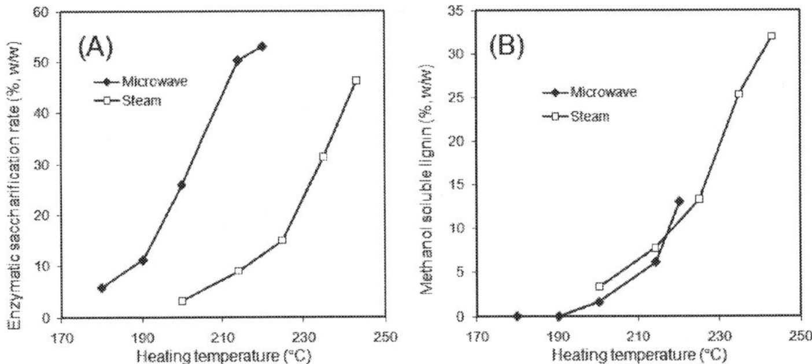

Fig. 16. Comparison of the values of saccharification rate (A) and methanol-soluble lignin content (B) got on continuous microwave irradiation and steam explosion (Parameter was set as duration of heating time 3 min, concentration 1 kg/L, flow rate of continuous microwave irradiation 8-15 L/h)

5.3 Advantage of microwave irradiation induced by electro-conductive reagents

As microwave energy is absorbed in electrically conductive materials, addition of any kind of sensitizers promotes heating. As described in 4.3.5 extraction efficiency of carbohydrates and phenolic compounds from fruits of Japanese apricot was improved for 1.4 fold by addition of sodium chloride for pickling. Furthermore addition of activated carbon was found to be useful for removal of secondary degradation products (4.3.10) with holding oligosaccharides inertly.

6. Conclusion

Effects of microwave irradiation on various kinds of recalcitrant biomass including woody and agricultural materials in aqueous media were reviewed. As a summary, microwave irradiation is demonstrated to have an efficient power for separation of biomass components. Its environment-friendliness, rapidity and easy to use safely opens unknown better ways for future welfare of human being. Finding new kind of sensitizers or catalysts which promote the power of microwave remarkably seems to be important for future innovation of this field of research. Development of scale up system for industrial use is also necessary.

7. Acknowledgements

This work was partly supported by The Towa Foundation for Food Research in 2010 in Japan.

8. References

Azuma, J. (1989). Analysis of Lignin-Cabohydrate Complexes of Plant Cell Walls, In: *Plant Fibers Modern Methods of Plant Analysis New Series Vol. 10*, Linskens, H.F. & Jackson, J.F. (Eds.), 100-126, Springer-Verlag, ISBN 3-540-18822-3, Berlin, Germany

Azuma, J.; Tanaka, F. & Koshijima, T. (1984a). Enhancement of enzymatic susceptibility of lignocellulosic wastes by microwave irradiation, *Journal of Fermentation Technology*, 62, 4, 377-384

Azuma, J.; Tanaka, F. & Koshijima, T. (1984b). Microwave irradiation of lignocellulosic materials I. Enzymatic susceptibility of microwave-irradiated woody plants. *Mokuzai Gakkaishi*, 30, 6, 501-509

Azuma, J.; Ozaki, A. & Koshijima, T. (1985a). Microwave irradiation of lignocellulosic materials VII. Microwave irradiation and enzymatic saccharification of celluloses. *Wood Research and Technical Notes*, 21, 78-86

Azuma, J. & Koshijima, T. (1985b). Microwave Irradiation of lignocellulosic materials III. Enzymatic susceptibility of microwave-irradiated green and dried wood chips. *Wood Research and Technical Notes*, 20, 22-30

Azuma, J.; Higashino, J.; Isaka, M. & Koshijima, T. (1985c). Microwave irradiation of lignocellulosic materials IV. Enhancement of enzymatic susceptibility of microwave-irradiated softwoods, *Wood Research*, 71, 13-24

Azuma, J.; Asai, T.; Isaka, M. & Koshijima, T. (1985d). Effects of microwave irradiation on enzymatic susceptibility of crystalline cellulose. *Journal of Fermentation Technology*, 63, 6, 529-536

Azuma, J.; Higashino, J. & Koshijima, T. (1986a). Microwave irradiation of lignocellulosic materials VI. Enhancement of enzymatic susceptibility of soft wood barks, *Mokuzai Gakkaishi*, 32, 5, 351-357

Azuma, J.; Katayama, T. & Koshijima, T. (1986b), Microwave irradiation of lignocellulosic materials VIII. Microwave irradiation of the neutral fraction (C-I-M) of pine Bjorkman LCC, *Wood Research*, 72, 1-11

Azuma, J.; Tsumiya, T. & Katata, T. (1994a) Jul 27. Jpn. Patent JP 1858751

Azuma, J.; Hosobuchi, T. & Katata, T. (1994b) Dec 7. Jpn Patent JP 1888744

Azuma, J.; Hosobuchi, T. & Katata, T. (1994c) Dec 7. Jpn Patent JP 1888745

Azuma, J.; Sakamoto, T.; Onishi, K. (2009) Production methods for solubilized lignin, saccharide raw material and monosaccharide raw material, and solubilized lignin. WO 2009/050882 A1

Azuma, J.; Okahara, K.; Sakamoto, M.; Kono, T. & Nomoto, H. (2008). Solubilization of barley malt feed by microwave heating in water. *Proceedings of Global Congress on Microwave Energy Applications (GCMEA 2008 MAJIC 1 st)*, pp. 785-788, ISBN978-4-904068-04-5, Otsu, Japan, August 2008. Japan Society of Electromagnetic Wave Energy Applications, Tokyo, Japan

Boom, A., Sinninge Damste, J.S. & de Leeuw, J.W. (2005). Cutan, a common aliphatic biopolymer in cuticles of drought-adapted plants, *Organic Geochemistry*, 36, 4, 595-601

Ebringerová, A., Hromádková, Z. & Heinze, T. (2005). Hemicellulose, *Advances in Polymer Science*, 186, 1-67

Gandini, A., Neto, C.P. & Silvestre, A.J.D. (2006). Suberin: A promising renewable resource for novel macromolecular materials, *Progress in Polymer Science*, 31, 4, 878-892

Gáspár, M., Juhasz, T., Szengyel, Z. & Reczey, K. (2004). Fractionation and utilization of corn fibre carbohydrates, *Process Biochemistry*, 40, 3-4, 1183-1188

Glasser, W.G. (2008). Cellulose and associated heteropolysaccharides, In: *Glycoscience Chemistry and Chemical Biology*, Fraser-Reid, B.O., Tatsuta, K. & Thiem, J. (Eds.), 1473-1512, Springer-Verlag, ISBN 978-3-540-36154-1, Berlin, Germany

Harris, P.J. & Stone, B.A. (2008). Chemistry and Molecular Organization of Plant Cell Walls, In: *Biomass Recalcitrance, Deconstructing the Plant Cell Wall for Bioenergy*, Himmel, M.E. (Ed.), 61-93, Blackwell Publishing Ltd., ISBN 978-1-4051-6360-6, West Sussex, UK

Hayashi, T. (1989). Measuring β–glucan deposition in plant cell walls, In: *Plant Fibers Modern Methods of Plant Analysis New Series Vol. 10*, Linskens, H.F. & Jackson, J.F. (Eds.), 138-160, Springer-Verlag, ISBN 3-540-18822-3, Berlin, Germany

Henriksson, G. (2009). Lignin In: *Pulp and Paper Chemistry and Technology Volume 1 Wood Chemistry and Wood Biotechnology*, Ek, M., Gellerstedt, G. & Henriksson, G. (Eds.) Walter de Gruyter GmBH & Co., ISBN 978-3-11-021339-3, Berlin, Germany

Hermiati, E. & Azuma, J. (2010). Saccharification of cassava pulp by microwave irradiation, will be presented at the 4th JEMEA Symposium, Fukuoka, Nov. 17-19, 2010

Inoue, T., Tsubaki, S., Ogawa, K., Onishi, K. & Azuma, J. (2010). Isolation of hesperidin from peels of thinned Citrus unshiu fruits by microwave-assisted extraction, *Food Chemistry*, 123, 2, 542-547

Kennedy, M., List, D., Lu, Y., Foo, Y., Newman, R.H., Sims, I.M., Bain, P.J.S., Hamilton, B. & Fenton, G. (1999). Apple pomace and products derived from apple pomace: Uses, composition and analysis. In: *Analysis of Plant Waste Materials Vol.20*, Linskens, H.F. & Jackson, J.F. (Eds.), 75-119, Springer-Verlag, ISBN 3-540-64669-8, Heidelberg, Germany

Kobayashi, M., Matoh, T. and Azuma, J. (1996). Two chains of rhamnogalacturonan II are cross-linked by borate-diol ester bonds in higher plant cell walls, *Plant Physiology*, 110, 3, 1017–1020

Magara, K.; Ueki, S.; Azuma, J. & Koshijima, T. (1988). Microwave irradiation of lignocellulosic materials IX. Conversion of microwave-irradiated lignocellulose into ethanol, *Mokuzai Gakkaishi*, 34, 5, 462-468

Maraga, K.; Tsubouchi, m. & Koshijima, T. (1988). Change of pore-size distribution of lignocellulose by microwave irradiation. *Mokuzai Gakkaishi*, 34, 10, 858-862

Magara, K.; Azuma, J. & Koshijima. T. (1989). Microwave-irradiation of lignocellulosic materials X. Conversion of microwave-irradiated agricultural wastes into ethanol, *Wood Research*, 76, 1-9

Magara, K. & Koshijima, T. (1990). Low level of enzymatic susceptibility of microwave-pretreated softwood, *Mokuzai Gakkaishi*, 36, 2, 159-164

Matsui, N. and Azuma, J. (2007). Unpublished results

Matsui, N.; Tsubaki, S.; Sakamoto, M. & Azuma, J. (2008). Saccharification of bamboo waste by concentrated sulfuric acid with irradiation of microwave energy. ASEAN COST+3: New Energy Forum for Sustainable Environment (NEFSE), Kyoto, Japan, May 2008

Matsumoto, A.; Tsubaki, S.; Sakamoto, M. & Azuma, J. (2008). Oligosaccharides adsorbed on activated charcoal powder escaped from hydrolysis by microwave heating in water, *Proceedings of Global Congress on Microwave Energy Applications (GCMEA 2008 MAJIC 1 st)*, pp. 785-788, ISBN 978-4-904068-04-5, Otsu, Japan, August 2008. Japan Society of Electromagnetic Wave Energy Applications, Tokyo, Japan

Okahara, K.; Tsubaki, S.; Sakamoto, M. & Azuma, J. (2008). Refinery of woody components by microwave irradiation in the presence of hydrogen peroxide, ASEAN COST+3: New Energy Forum for Sustainable Environment (NEFSE), Kyoto, Japan, May 2008.

O'Neill, M.A., Ishii,T., Albersheim, P. & Darvill, A.G. (2004). Rhamnogalacturonan II: Structure and function of a borate cross-linked cell wall pectic polysaccharide, *Annual Review of Plant Biology*, 55 109-139

Ookushi, Y.; Sakamoto, M. & Azuma, J. (2006). Optimization of microwave-assisted extraction of polysaccharides from the fruiting body of mushrooms, *Journal of Applied Glycoscience*, 53, 4, 267-272

Ookushi, Y.; Sakamoto, M. & Azuma, J. (2008). Extraction of β-glucan from the water-insoluble residue of *Hericium erinaceum, Journal of Applied Glycoscience*, 55, 4, 225-229

Ookushi, Y.; Sakamoto, M. & Azuma, J. (2008). β-Glucan in the water-insoluble residue of *Hericium erinaceum, Journal of Applied Glycoscience*, 55, 4, 231-234

Ookushi, Y.; Sakamoto, M. & Azuma, J. (2009). Effect of microwave irradiation on water-soluble polysacharides of the fruiting body of *Hericium erinaceum. Journal of Applied Glycoscience*, 56, 3, 153-157

O'Toole, D. K. (1999). Characteristics and use of Okara, the soybean residue from soy milk production-A review, *Journal of Agricultural and Food Chemistry*, 47, 2, 363-371

Pinto, P.C.R.O., Sousa, A.F. Silvestre, A.J.D., Neto, C.P., Gandini, A., Eckerman, C. & Holmbom, B. (2009). *Quercus suber* and *Betula pendula* outer barks as renewable sources of oleochemicals: A comparative study. *Industrial Crops and Products*, 29, 1, 126-132

Robyt, J.F. (2008). Starch: Structure, Properties, Chemistry, and Enzymology, In: *Glycoscience Chemistry and Chemical Biology*, Coté, G., Flitsch, S., Ito, Y., Kondo, H., & Nishimura, S.-I. (Eds.), 1437-1472, Springer-Verlag, ISBN 978-3-540-46154-1, Heidelberg, Germany

Stark, R.E. & Tian, S. (2006). The cutin biopolymer matrix, In: *Biology of the Plant Cuticle*, Riederer, M. & Muller, C. (Eds.) 126-144, Blackwell Publishing Co., ISBN 978-1405132688, Oxford, UK.

Tester, R.F., Karkalas, J. & Qi, X. (2004). Starch-composition, fine structure and architecture, *Journal of Cereal Science*, 39, 2, 151-165

Tsubaki, S., Iida, H., Sakamoto, M. & Azuma, J. (2008). Microwave heating of tea residue yields polysaccharides, polyphenols, and plant biopolyester. *Journal of Agricultural and Food Chemistry*, 56, 23, 11293-11299

Tsubaki, S., Nakauchi, M., Ozaki, Y. & Azuma, J. (2009). Microwave heating for solubilization of polysaccharide and polyphenol from soybean residue (Okara), *Food Science and Technology Research*, 15, 3, 307-314

Tsubaki, S., Ozaki, Y. & Azuma, J. (2010a). Microwave-assisted autohydrolysis of Prunus mume stone for extraction of polysaccharides and phenolic compounds. *Journal of Food Science*, 75, 2, C152-C159

Tsubaki, S., Sakamoto, M. & Azuma, J. (2010b). Microwave-assisted extraction of phenolic compounds from tea residues under autohydrolytic conditions, *Food Chemistry*, 123, 4, 1255-1258

Tsumiya, T. & Azuma, J. (1994). Continuous microwave hydrolyser of woods (A case study) Translation from Japanese, In: *Maikuroha Kanetsu Shusei*, Koshijima, T. (Ed.) 252-259, NTS Co., ISBN 4-86043-070-0, Tokyo, Japan

Van Der Maarel, M.J.E.C. (2010). Starch-processing enzymes, In: *Enzymes in Food Technology Second edition*, Whitehurst R. J. & Van Oort M. (Eds.), 320-331, Wiley-Blackwell, ISBN 978-1-4051-8366-6, West Sussex, UK

Wertz, J.-L., Bedue, O. & Pierre, J. (2010). *Cellulose and Technology*, The EPFL Press, Lausanne, ISBN 978-2-940222-41-4, Switzerland

Willats, W.G.T., McCartney, L., Mackie, W. & Knox, J.P. (2001). Pectin: Cell biology and prospects for functional analysis, *Plant Molecular Biology*, 47, 1-2, 9-27.

Yamashita, T. & Azuma, J. (2010). Unpublished results

Yoshida, T., Tsubaki, S., Teramoto, Y. & Azuma, J. (2010). Optimization of microwave-assisted extraction of carbohydrates from industrial waste of corn starch production using response surface methodology, *Bioresource Technology*, 101, 20, 7820-7826

31

Microwave Heating Applied to Pyrolysis

Yolanda Fernández, Ana Arenillas and J. Ángel Menéndez
Instituto Nacional del Carbón (CSIC)
Apartado 73, 33080 Oviedo
Spain

1. Introduction

The use of microwaves for heating is well established in society, being used in domestic and some industrial processes. However, there is potential for this technology to be introduced and applied to many other industrial heating processes, which offers unique advantages not attained with conventional heating. In this sense, microwave technology is being explored as one method to assist in waste management.

Currently, significant quantities of hazardous wastes are generated from a multitude of products and processes. The increase in both the quantity and of the diversity of waste production is now posing significant problems for their effective management. New technologies are being investigated to develop systems which shall support the safe handling, transportation, storage, disposal and destruction of the hazardous constituents of this waste. The recent interest in microwave technologies appears to offer the best solution to waste management, whereby a variety of microwave systems can be designed, developed and tailored to process many waste products. It is possible that microwave technologies shall provide for: (i) a reduction in waste volume, (ii) rapid heating, (iii) selective heating, (iv) enhanced chemical reactivity, (v) the ability to treat waste in-situ, (vi) rapid and flexible processes that can also be controlled remotely, (vii) ease of control, (viii) energy savings, (ix) overall cost effectiveness, (x) portability of equipment and processes, (xi) cleaner energy source compared to some more conventional systems, etc.

From existing processes for the harnessing of energy and raw materials from waste, thermochemical conversion routes are suitable candidates for the application of microwave technology. One of the thermochemical processes which is rapidly gaining in importance in this field is pyrolysis. This process not only allows for higher energy recovery from the waste, but it also generates a wide spectrum of products. Hitherto, most published work on the pyrolysis process has dealt with conventional heating systems, although recent interest in microwave-assisted pyrolysis (MP) has highlighted its unique advantages, not within the scope of traditional methods.

The aim of this chapter is to emphasize the principles of MP and to show recent research on the application of this technology to waste treatment. As an introduction to the topic, a brief background on the pyrolysis process and the fundamentals of microwave irradiation as an energy source are presented.

2. General overview of the pyrolysis process

It is imperative that we make use of appropriate technologies for the recovery of resources from non-conventional sources such as waste, in order to ease the energy crisis and to slow environmental degradation which shall, in turn, reduce the percentage of landfilled wastes. The choice of conversion process depends upon the type and quantity of waste feedstock, the desired form of energy (i.e., end use requirements), environmental standards, economic conditions and project specific factors. Many biochemical and thermochemical processes have been researched for the purpose of waste upgrading (Faaij, 2006). While both methods of processing can be used to produce fuels and chemicals, thermochemical processing can be seen as being the easiest to adapt to current energy infrastructures, and to deal with the inherent diversity in some wastes.

Three different thermochemical conversion routes are found according to the oxygen content in the process: combustion (complete oxidation), gasification (partial oxidation) and pyrolysis (thermal degradation without oxygen). Among them, combustion (also called incineration) is the most established route in industry but this is also associated with the generation of carbon oxides, sulphur, nitrogen, chlorine products (dioxins and furans), volatile organic compounds, polycyclic aromatic hydrocarbons, dust, etc. On the contrary, gasification and pyrolysis offer the potential for greater efficiencies in energy production and less pollution. Although pyrolysis is still under development in the waste industry, this process has received special attention, not only as a primary process of combustion and gasification, but also as independent process leading to the production of energy-dense products with numerous uses. This makes the pyrolysis treatment process self-sufficient in terms of energy use, and also significantly reduces operating costs.

2.1 Principles of the pyrolysis process

The term "pyrolysis" is defined as a thermal degradation in the absence of oxygen, which converts a raw material into different reactive intermediate products: solid (char), liquid (heavy molecular weight compounds that condense when cooled down) and gaseous products (light molecular weight gases). The understanding of the pyrolysis process is a complicated one since many factors have to be considered, such as raw material composition (see Section 2.2) and experimental conditions (see Section 2.3).

It is generally accepted that there are two possible steps in any pyrolysis process (Conesa et al., 1998): (i) *primary pyrolysis*, which comprises the devolatilization of the material where different reaction zones can appear corresponding to the thermal decomposition of the main constituents; and (ii) *secondary pyrolysis*, which covers the secondary decomposition reactions in the solid matrix, as well as secondary reactions between the volatiles release (homogeneous reactions), or between the volatiles and the carbonaceous residue (heterogeneous reactions). The first stage mainly involves dehydration, dehydrogenation, decarboxylation or decarbonilation reactions. The second comprises of processes such as cracking (thermal or catalytic), where heavy compounds further break into gases, or char is also converted into gases such as CO, CO_2, CH_4 and H_2 by reactions with gasifying agents, as well as partial oxidation, polymerization and condensation reactions.

2.2 Pyrolysis profiles

The diversity in chemical and physical properties of waste materials may imply significant differences between the corresponding pyrolysis profiles, since different levels of interactions between the components may occur. In this sense, an initial characterization of the material is of crucial importance to understand the pyrolysis dynamics, such as initial degradation temperature, conversion time, maximum volatile releasing rate and its corresponding temperature. Along with effective design and operation, each of the aforementioned represent the basic information required for full optimization of the process. Most of the studies on pyrolysis behaviour have been established for lignocellulosic materials, which comprise of a mixture of hemicellulose, cellulose, lignin and minor amounts of other organics. It is known that each of these components pyrolyze or degrade at different rates and by different mechanisms and pathways (Antal, 1983; Caballero et al., 1996; Branca & Di Blasi, 2003). While cellulose and hemicelluloses form mainly volatile products during pyrolysis due to the thermal cleavage of the sugar units, the lignin mainly forms char since it is not readily cleaved to lower molecular weight fragments. Wood, crops, agricultural and forestry residues, and sewage sludges are some of the main renewable energy resources available and subjected to pyrolysis processes, as are the biodegradable components of municipal solid waste (MSW) and commercial and industrial wastes.

There are different approaches that attempt to establish a correlation between the characteristics of the lignocellulosic materials and its final pyrolysis products. The first considers the biomass as a complex mixture of polymers consisting of carbon, hydrogen and oxygen (Couhert et al., 2009a). The second takes into account the functional groups presented (Savova et al., 2001), whilst the third is based on the biomass formed from cellulose, hemicelluloses and lignin (Couhert et al., 2009b). However, to date there is no model that predicts yield and composition of the final pyrolysis products, due largely to component interaction and the influence of mineral matter (Raveendran et al., 1995).

2.3 Pyrolysis technologies

Not only can raw material composition influence the yield and characteristics of the pyrolysis products, but the pyrolysis conditions can also modify the course of reactions and, therefore, strongly affect the yield and properties of products. Consideration should be taken of temperature, heating rate and residence time of vapours present in the reactor. Depending on these variables, the pyrolysis process can be divided into three subclasses: *slow, fast* and *flash pyrolysis* (see Table 1).

Pyrolysis technology	Residence time (s)	Heating rate (K / s)	Temperature (K)
Slow	450-550	0.1-1	550-950
Fast	0.5-10	10-200	850-1250
Flash	<0.5	>1000	1050-1300

Table 1. Range of the main operating parameters for pyrolysis processes (Elías, 2005)

Earlier literature generally equates pyrolysis to carbonization (*slow pyrolysis*), in which the principal product is a solid char (Antal & Gronli, 2003). Today, the term pyrolysis often describes processes in which oils are preferred products (Mohan et al., 2006). Hence *fast* and *flash pyrolysis* technologies have been considered as a good solution to convert materials to liquid fuel (Bridgwater & Peacoke, 2000). Nevertheless, pyrolysis as a means to convert a diversity of waste materials to combustible gas or syngas is receiving the increased attention which it deserves. In addition, microwave technology when applied to fast or flash pyrolysis is suitable for producing higher gas yield with more syngas content (Domínguez et al., 2007; 2008; Dufour et al. 2009).

Some interventions in the operating parameters may induce and/or alter particular chemical reactions, resulting in different chemical profiles of the volatiles. Generally, increasing the pyrolysis temperature reduces the char yield and increases the gas yield (Domínguez et al., 2007; 2008). The liquid yield reaches a maximum value at intermediate temperatures and decreases at higher temperatures due to thermal cracking of heavy compounds. Long residence times of volatiles in reactor and high temperatures decrease tar production but increase char formation as a result of the extension of secondary reactions (Fernández et al., 2009). Higher heating rates favour a quick release of volatiles, modifying the solid residue structure with an increased yield of the liquid and gaseous fractions (Menéndez et al., 2007). Other variables that have to be considered in a pyrolysis process such as the reactor type (Bridgwater, 2003), sample size (Tsai et al., 2007), pressure (Cetin et al., 2005), etc., might also alter the final product distribution.

The optimization of each of the final pyrolysis products can also be done by catalytic means. The use of catalysts or additives to improve the yield or quality of pyrolysis gas or liquid fuels is still in its infancy. While there is fundamental work underway (Bridgwater, 1996; Williams & Nugranad, 2000; Wang et at., 2006), more research is necessary to explore the wide range of conventional and unconventional catalysts. Catalytic pyrolysis has been reported to be a productive means to increase gas yield by decreasing the amount of liquid, as well as positively affecting the quality of the organic composition of the oils – *in situ* upgrading (Domínguez et al., 2003).

2.4 Pyrolysis products

Pyrolysis process has the ability to provide three end products: gas, oil and char, which all have the potential to be refined further if required. The concentrations and characteristics of each product can vary considerably according to the feed characteristics and the operating conditions of the pyrolysis process. The main properties and applications of each pyrolysis fraction are presented below:

Solid fraction

Pyrolysis char is a carbonaceous residue mainly composed of elemental carbon originating from thermal decomposition of the organic components, unconverted organic compounds, e.g. solid additives, and even carbon nanoparticles produced in the gas phase secondary reactions. This carbonaceous residue plays an important role in the pyrolysis process since it contains the mineral content of the original feed material, relevant to specific catalytic processes (Raveendran et al., 1995). The importance of the char cannot be understated as it may be involved as a reactive in heterogeneous or catalytic heterogeneous reactions (Menéndez et al., 2007).

The utilization of the char can vary considerably according to its characteristics. The main industrial uses of char can be summarized as follows: (i) as solid fuel for boilers which can be directly converted to pellets or mixed with other materials such as biomass, carbon, etc., to form the same, (ii) as feedstock for the production of activated carbon, (iii) as feedstock for making carbon nanofilaments, (iv) as feedstock for the gasification process to obtain hydrogen rich gas, (v) as feedstock for producing high surface area catalysts to be used in electrochemical capacitors, etc.

Liquid fraction

Pyrolysis oil is a complex mixture of several organic compounds which may be accompanied by inorganic species. In the case of biomass, the liquid or oil fraction (bio-oils) is found to be highly oxygenated and complex, chemically unstable and less miscible in conventional fuels (Demirbas, 2002). Thus, the liquid products still need to be upgraded by lowering the oxygen content and removing residues.

The oil obtained from pyrolysis can have the following industrial uses: (i) combustion fuel, (ii) used for power generation, (iii) production of chemicals and resins, (iv) transportation fuel, (v) production of anhydro-sugars like levoglucosan, (vi) as binder for pelletizing and briquetting combustible organic waste materials, (vii) bio-oil can be used as preservatives, e.g., wood preservatives, (viii) a suitable blend of a pyrolysis liquid with the diesel oil may be used as diesel engine fuels, (ix) bio-oils can be used in making adhesives, etc. Moreover, the oil may be stored and transported, and hence need not be used at the production site.

Gas fraction

The gas produced in a pyrolysis process is mainly composed of combustible gases, such as H_2, CO, C_2H_2, CH_4, C_2H_4, C_2H_6, etc. Other gases, such as CO_2 and pollutants (SO_2, NO_x), can also appear, although in lower concentrations.

The gas produced from pyrolysis can be used directly as a fuel in various energy applications, such as: (i) direct firing in boilers without the need for flue gas treatment, and (ii) in gas turbines/engines associated with electricity generation. Pyrolysis gas containing significant amounts of hydrogen and carbon monoxide might be utilized in syngas applications. It is known that synthesis gas (H_2 + CO) having different H_2/CO molar ratios is suitable for different applications. For example, synthesis gas having a high H_2/CO molar ratio is desirable for producing hydrogen for ammonia synthesis. This ratio is increased further during the water-gas shift reaction for the removal of CO.

3. Background information on microwave heating

Since Percy Spencer accidentally discovered the possibility of cooking food with microwaves in the 1940s, research on microwave heating has continued unabated. Current investigations into the application of microwave technology across a number of fields may lead to significant savings in both energy consumption and process time. Moreover, the unique internal heating phenomenon associated with microwave energy can enhance overall production quality, allowing for the development of new end products and processes that cannot be realised using conventional methods.

3.1 Fundamentals of microwave heating

Microwaves lie between infrared radiation and radio waves in the region of the electromagnetic spectrum. More specifically, they are defined as those waves with wavelengths between 0.001 and 1 m, which correspond to frequencies between 300 and 0.3 GHz. Within this portion of the electromagnetic spectrum there are frequencies that are used for cellular phones, radar, and television satellite communications (van Loock, 2006). For microwave heating, two frequencies reserved by the Federal Communications Commission (FCC) for industrial, scientific and medical (ISM) purposes are commonly used, which are 0.915 and 2.45 GHz.

As all electromagnetic waves, microwaves consist of electric and magnetic field components, both perpendicular to each other. Generally, there are three qualitative ways in which a material may be categorized with respect to its interaction with the electric field component of the microwave field: (i) insulators, where microwaves pass through without any losses (transparent), (ii) conductors, where microwaves are reflected and cannot penetrate, and (iii) absorbers. Materials that absorb microwave radiation are called dielectrics, thus, microwave heating is also referred to as dielectric heating.

There exist a number of mechanisms that contribute to the dielectric response of materials (Thostenson & Chou, 1999). These include electronic polarization, atomic polarization, ionic conduction, dipole (orientation) polarization, and interfacial or Maxwell-Wagner polarization mechanisms. At microwave frequencies, only dipole and Maxwell-Wagner polarizations result in the transfer of electromagnetic energy to thermal energy (Mijovic & Wijaya, 1990). In polar organic-solvent systems at non-extreme temperatures, the dipolar polarization mechanism accounts for the majority of the microwave heating effect, while in some carbon-based materials is the interfacial polarization the principal head. With respect to the former, dipoles may be a natural feature of the dielectric or they may be induced (Kelly & Rowson, 1995). Distortion of the electron cloud around non-polar molecules or atoms through the presence of an external electric field can induce temporary dipoles. The dipoles in the material exposed to the alternating electromagnetic field realign themselves approximately 2.5 billion times per second (for a microwave frequency of 2.45 GHz). This movement results in rotation of the dipoles, and energy is dissipated as heat from internal resistance to the rotation. On the other hand, the Maxwell-Wagner polarization occurs at the boundary of two materials with different dielectric properties, or in dielectric solid materials with charged particles which are free to move in a delimited region of the material, such as π-electrons in carbon materials (Zlotorzynski, 1995). When the charged particles cannot couple to the changes of phase of the electric field, the accumulation of charge at the material interface is produced and energy is dissipated in the form of heat due to the so-called Maxell-Wagner effect (Zlotorzynski, 1995). The interaction of microwaves with metals or metal powder (Marken et al., 2006) may further contribute to the energy absorption effect. The response to an applied electric field is dependent on the dielectric properties of the material (Thostenson & Chou, 1999). The polarization takes place when the effective current in the irradiated sample is out of phase with that of the applied field by a difference (termed δ). This difference defines the tangent loss factor, tan δ , often named the dissipation factor or the dielectric loss tangent. The word "loss" refers to the input microwave energy that is lost to the sample by being dissipated as heat. Thus, microwave energy is not transferred primarily by convection or by conduction, as with conventional heating, but by dielectric loss. The tangent loss factor is expressed as the quotient, tan $\delta = \varepsilon''/\varepsilon'$, where ε'' is the dielectric loss factor, indicative of the efficiency with which electromagnetic radiation is converted to heat, and ε' is the dielectric constant describing the ability of molecules to be polarized by the electric field. A high value for tan δ indicates a high susceptibility to microwave energy. Because the dielectric properties govern the ability of materials to heat in microwave fields, the measurement of these properties as a function of temperature frequency, or other relevant parameters (moisture content, density, material geometry, etc.) is important (Meredith, 1998). Many authors (Carpenter, 1991; Tinga, 1992; Guillon, 1994) have reviewed different techniques for dielectric property measurements at microwave frecuencies. Unfortunately, tan δ values are not easily found in the literature, and for most common solvents have only been determined at room temperature (Gabriel et al., 1998). Furthermore, the loss tangent of the material depends on the relaxation times of the molecules in the material, which, in turn, depends on the nature of the functional groups and the volume of the molecule (Gabriel et al., 1998). But not only microwave heating is a function of tan δ, but also other parameters related to the material, such as ionic strength, specific heat capacity, thermal conductivity and emissivity, and related to the applied field and the operating conditions have to be taken into account as well (Mingos & Baghurst, 1991).

3.2 Microwave monitoring conditions

When monitoring microwaves in heating applications several aspects should be considered. Amongst the most relevant are the design of the microwave device, temperature monitoring and temperature-feedback control.

The design of microwave equipment is particularly important since it determines the electromagnetic field applied to the material. Three major components are found in a microwave oven (Thostenson & Chou, 1999): (i) the source which generates the electromagnetic radiation, (ii) the transmission lines which deliver the electromagnetic energy from the source to the applicator (propagation), and (iii) the applicator where microwave energy is transferred to materials. Hence, the design of the applicator is critical because it is the area where the interaction between microwaves and the material takes place. The temperature fields within the material undergoing microwave heating are inherently linked to the distribution of the electric fields within the applicator. Common microwave applicators include waveguides, travelling wave applicators, single mode cavities, and multimode cavities. For processing materials, resonant applicators, such as single mode (which support one resonant mode) and multimode applicators (which sustain a number of high order modes at the same time) are most common because of their high field strengths (Chan & Reader, 2000). Examples of microwave ovens with both single- and multimode cavities appear in Figure 1.

While single mode cavities allow very well-defined electric fields in a relatively small volume, multimode cavities permit the electric fields to encompass a much larger volume, albeit with a compromise in the field definition. The geometry of multimode cavities is such that constructive and destructive interference of the microwaves occurs due to reflection off the cavity walls and the conductor elements of the sample. Hence there is no well-defined electric field but the field acts over a much larger volume, allowing larger samples to be treated. Although higher power densities are achievable in single mode cavities, multimode systems allow for much greater operational flexibility and are routinely used for initial assessment of the suitability of microwave treatment for a particular application.

In general, an efficient application of microwave irradiation when used as an energy source requires reliable temperature monitoring and temperature-feedback control during the irradiation. These control features enable a material or reaction mixture to be conveniently heated to a desired temperature without knowing the dielectric or conductive properties in detail. Different temperature sensors can be used attending to their non-contact or contact with the material or substance to be heated. In the first case, infrared optical pyrometers are usually employed. These are required for high temperature measurements, which are determined by the intensity of infrared radiation emitted by the materials and comparing it with an internal source. The emissivity of the material needs to be determined in the calibration procedure of the optical pyrometer (Menéndez et al., 1999). With respect to temperature sensor in contact with the sample, thermocouple probes consisting of two dissimilar metals joined together at one end are used. When the junction of the two metals is heated or cooled a voltage is produced that can be correlated back to the temperature. The general classification of thermocouples is made according the metal combination, which determines the working temperature range, the sheath material and the junction type. Sheathed thermocouple probes are available with one of three junction types: grounded, ungrounded or exposed. At the tip of a grounded junction probe, the thermocouple wires are physically attached to the inside of the probe wall. This results in good heat transfer from the outside, through the probe wall to the thermocouple junction. In an ungrounded

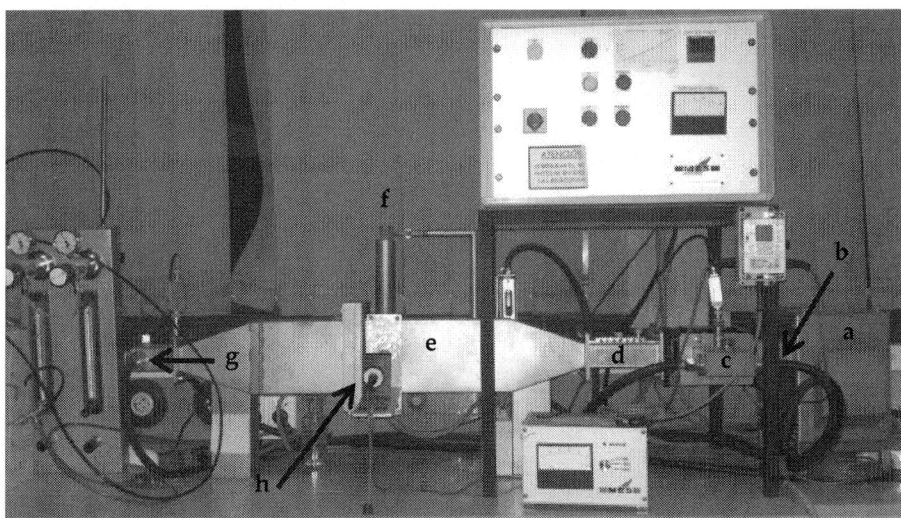

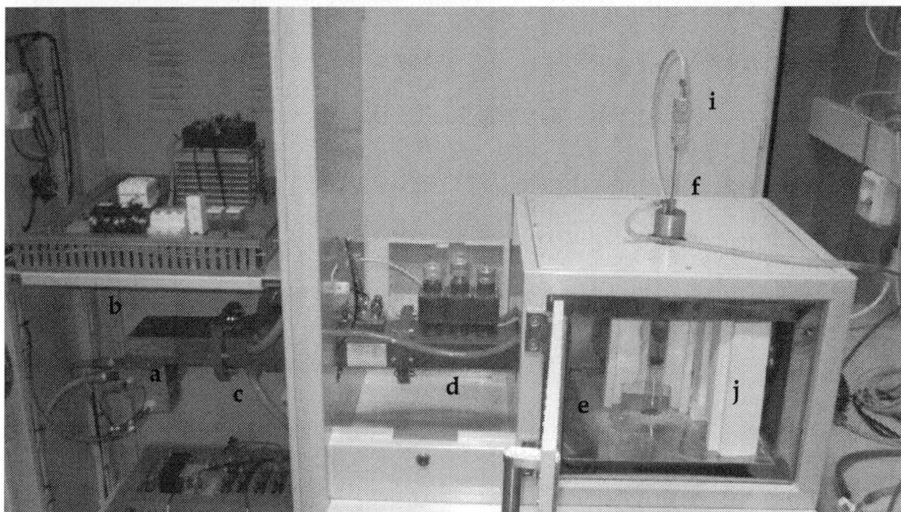

Fig. 1. Photographs of singlemode (upper) and multimode (lower) microwave ovens at INCAR (www.incar.csic.es) showing: a) magnetron; b) waveguide; c) circulator with reflected power meter; d) tuning screws; e) resonant cavity; f) reactor; g) terminator with transmitted power meter; h) infrared optical pyrometer; i) thermocouple; and j) insulating material

probe, the thermocouple junction is detached from the probe wall. Response time is slower than the grounded style, but the ungrounded offers electrical isolation, and therefore, no interaction with the electromagnetic field. Therefore, ungrounded probes are usually employed under microwave radiation.

However, accurate measurement of the evolution of temperature is difficult due to inherent difficulties involved in measuring this variable in microwave devices (Menéndez et al., 1999). On one hand, the way in which temperature is monitored is dependent upon the type of temperature sensor. While infrared optical pyrometers give a temperature measurement by default, the thermocouple probes provide an over-temperature (Cloete et al, 2001). The combination of optical pyrometers together with thermocouple probes may alleviate the intrinsic error in both instruments. On the other hand, because not all materials or substances are similar microwave absorbers, it is accepted that temperature measurement corresponds to the average temperature of the bulk. When heating non-polar substances, the temperature registered on the probe will in fact be that of the probe itself and not of the specimen. In this respect it is particularly difficult to determine the temperature in nonwater-soluble substances. Moreover, it is not always easy to determine the correct temperature for a substance during microwaving, even if it is soluble in water. As long as the load of water is large and the specimen not too small, the temperature of the water coincides with that of the substance. However, when specimens take the form of small droplets, these can be evaporated easily giving sudden unexpected temperature rises. Examples of accurate temperature measurements may be seen in the articles of Feirabend et al. (1991; 1992).

On the other hand, the temperature-feedback control should be capable of a fast reduction of power in cases where either the exothermic reaction energy becomes pronounced, or when the dissipation factor increases rapidly, to avoid thermal runaways. Most of the microwave ovens work at full power, and the power levels commonly fluctuate as a result of the patterns-of-switching of on-off cycles (Mingos & Baghurst, 1991). However, other models can set the power required, emitting only part of the full power, and thus, reaching a better control of temperature.

4. Approach to the microwave-assisted pyrolysis (MP)

Microwave heating is emerging as one of the most attractive alternative technologies in the pyrolysis process. MP not only overcomes the disadvantages of conventional pyrolysis (CP) methods such as slow heating and necessity of feedstock shredding, but also improves the quality of final pyrolysis products. At the same time it significantly saves processing time and energy. However, there are several major limitations which are preventing this technology from being widely employed in the waste manufacturing industry. These include the absence of sufficient data to quantify the dielectric properties of the treated waste streams, the need of a multi-disciplinary approach to design and develop the related conversion units, and the uncertainty about the actual costs.

Material properties and operating conditions are the main factors affecting any pyrolysis process, since they determine the characteristics and yields of the final products. However, when microwaves are used as heating method in the pyrolysis process, one must consider additional complications which may affect the heat, mass transportation mechanism and chemical reactions. In order to examine further the novel conditions achieved through MP, in this section the more traditional approach to the pyrolysis process is used as a reference point.

4.1 Role of microwave heating on the pyrolysis conditions

When microwave heating is applied to a pyrolysis process, novel conditions in the temperature distribution, the heating rate and the residence time of volatiles are observed when compared with those achieved by conventional heating. Therefore, different chemical profiles of the volatiles in both heating systems are obtained allowing for modification of final pyrolysis products.

Conventional heating means that heat is transferred from the surface towards the center of the material by convection, conduction and radiation. On the contrary, microwave heating represents the transfer of electromagnetic energy to thermal energy. Because microwaves can penetrate materials and deposit energy, heat can be generated throughout the volume of the material, rather than from an external source (volumetric heating). Therefore, microwave heating is energy conversion rather than heat transfer. As a result, opposite thermal gradients, i.e. temperature distribution, are found in both heating systems (see Fig.2). In microwave heating the material is at higher temperature than the surrounding area, unlike conventional heating where it is necessary that the conventional furnace cavity reach the operating temperature, to begin heating the material. Consequently, microwave heating favours the reactions involving the solid material, e.g. devolatilization or heterogeneous reactions (Zhang & Hayward, 2006), and conventional heating improves the reactions that take place in surroundings, such as homogeneous reactions in the gas-phase. Additionally, the lower temperatures in the microwave cavity can be useful to avoid undesirable reactions and condense the final pyrolysis vapours in this area.

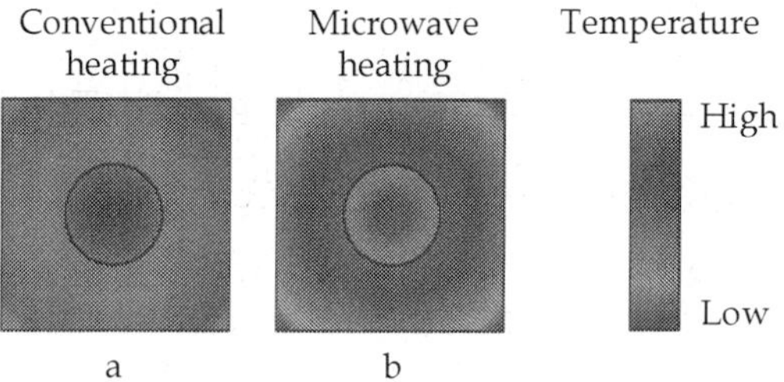

Fig. 2. Quality comparison of temperature gradient within samples heated by (a) conventional heating and (b) microwave dielectric heating. While the circles represent the sample, the squares correspond to the cavity used in both heating systems

From the different thermal gradients, it can be assumed that there are differences in final yields and composition of the pyrolysis products under both heating systems. Several authors (Menéndez et al., 2004; Domínguez et al., 2007) have observed greater gas yield and lesser carbonaceous residue in the MP experiments of different wastes, which demonstrate the efficiency of microwaves in carrying out heterogeneous reactions. Some of the heterogeneous reactions observed in pyrolysis processes, such as gasification reactions with CO_2 or H_2O, have been executed individually at different temperatures under both heating systems (Menéndez et al., 2007). The conversions in MW are always higher than those observed in conventional heating at any temperature. Additionally, the differences between both heating methods seem to be reduced with the temperature increase, which points to the higher efficiency of microwave heating at lower temperatures. Other heterogeneous reactions that can be found in a pyrolysis process are those in which the solid char not only act as a reactive, but also as a catalyst owing to the carbonaceous or metallic active centres located in the surface. These provide a catalytic effect for specific reactions, such as the methane decomposition reaction. This reaction has also been proved separately to give better conversion under microwave heating (Domínguez et al., 2008).

The differing performance between conventional and microwave heating is also translated into differential heating rates of the material. Microwave heating reveals higher heating rates due to the fact that microwave energy is delivered directly into the material through molecular interaction with the electromagnetic field, and no time is wasted in heating the surrounding area. Therefore, significant savings of time and energy are achieved in MP, although other effects can also be deduced affecting the volatiles profiles. Generally, high heating rates improve the devolatilization of the material reducing the conversion times. The heating rate also has an influence on the residence time of volatiles, whose flow occurs from the more internal hot zones towards the external cold regions of the sample. The higher heating rate, the shorter residence time, and the faster the volatiles arrive to the external cold regions which, in turn, reduces the activity of secondary reactions of vapour-phase products. This results in high yields of liquid and reduced deposition of refractory condensable material on the char's internal surface (Allan et al., 1980).

In the case of materials with significant moisture content, like most biomass feedstocks, MP offers a different paradigm in particle heating where the electromagnetic field penetrates the solid, and interacts directly with dipoles in the chemical structure. Due to the high affinity of water molecules with microwaves, moisture content within a given biomass particle is selectively targeted by incidental microwaves. Microwaves vaporize moisture in the depth of the particle, prior to volatilizing organic content. The steam generated is rapidly released into the surrounding area, not only sweeping volatiles, but also creating preferential channels in the carbonaceous solid that increase its porosity. This, in turn, favours the release of volatiles at low temperatures, and hence, its reaction with the steam produced (Minkova et al., 2001) which leads to partial oxidation and formation of permanent gases (H_2, CO, CH_4, CO_2).

4.2 Role of microwave heating on the material

In terms of its physical form (structural arrangements, thermal conductivity and specific heat) and chemical characteristics (organic and inorganic composition), the efficiency of MP processes depends greatly on the nature of the material being processed. Hence, not all materials present the same dielectric behaviour, and therefore, not all materials are similarly heated by microwaves. In fact, there are generally three qualitative ways in which a material may be categorized with respect to its interaction with the microwave field (insulators, conductors and absorbers), as mentioned in Section 3.1. However, a fourth type of interaction is that of a mixed absorber. This type of interaction is observed in composite or multi-phase materials where one of the phases is a high-loss material while the other is a low-loss material. Mixed absorbers take advantage of one of the most significant characteristics of microwave processing - that of selective heating. The microwaves are absorbed by the component that has high dielectric loss while passing through the low loss material with little drop in energy. In some processes and products, the heating of a specific component (whilst leaving the surrounding material relatively unaffected) would be of great advantage (Dernovsek et al., 2001). Microwave may also be able to initiate chemical reactions not possible in conventional processing through selective heating of reactants. Thus, new materials may be created.

Normally, waste materials have poor dielectric properties, being unable to absorb enough microwave energy to achieve the temperature necessary for pyrolysis. The amount of water present contributes to heating at insufficient temperatures at which it is merely possible to dry the material. Consequently, the use of a "microwave receptor", also called microwave-absorbing dopant (Robinson et al., 2010) or antenna (Hussain et al., 2010), is necessary. Different types of microwave receptors can be used, such as inorganic matter (Monsef-Mirzai et al., 1995) or substances with conduction electrons, e.g. activated carbon, char, graphite, etc (Chanaa et al., 1994; El harfi et al., 2000).

Influence of the microwave receptor on the pyrolysis conditions

The microwave receptor is perhaps the primary factor in the MP, directly affecting pyrolysis conditions, which mainly include temperature profiles and heating rate. Maximum temperature is primarily determined by the dielectric properties of the receptor, while heating rate is likely to be dependant upon the chemical composition of the material and, to a much lesser extent, by its physical structure. Numerous studies have shown that the temperature evolution of a material during MP could be divided in four stages (Zhao et al., 2010): stage I, the heating of the sample could be explained by the dielectric relaxation phenomenon of water molecules, which is responsible for the initial heating of the sample; stage II, the temperature reaches a plateau before attaining the pyrolysis temperature, which significantly depends upon the use of receptor; stage III, the temperature rises rapidly corresponding to an acute loss of mass, and normally a flex point appears; and stage IV, the thermal equilibrium is achieved, which depends on the receptor. Changes in the microwave receptor can alter the conductivity and permittivity of the sample, and hence the strength of the electric fields in the sample and the power dissipated in it. Meanwhile, a phenomenon known as "thermal runaway" can also occur, where the microwave energy is concentrated in the receptor, whose rate of absorption (referred to as the thermal absorptivity) increases with temperature, leading to an exponential increase in heating rate.

During the course of the MP, it is important to emphasize that not only the receptor will absorb microwave energy to produce heat; the solid residue produced in the process also contributes to the final pyrolysis temperature. In Fig. 3, a simple sketch shows the different stages during the MP of a mixture of materials/microwave receptors.

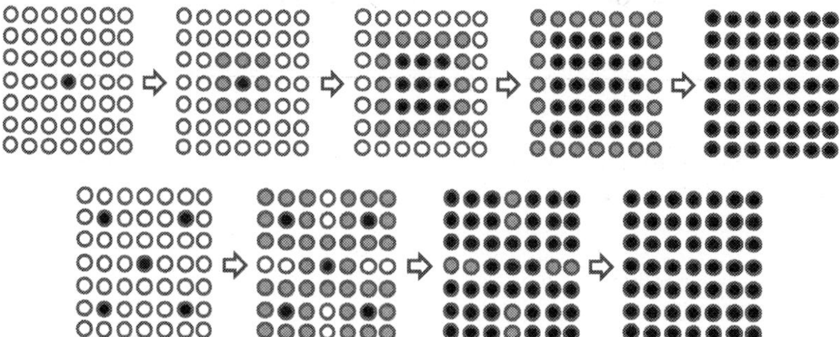

Fig. 3. Phases during MP of mixtures material particles (white circles)/microwave receptors (black circles), where it is shown the creation of first, second, third, etc. generations of microwave receptors (gray circles)

In the initial phase, only the added microwave receptor is able to absorb microwaves and produce heat, which allows the nearby particles of the materials to be heated by conventional methods (convection, conduction and radiation). Then, the removal of volatiles produces char, which will act as a microwave absorber, so that the pyrolysis process can then be sustained. Depending on the dispersal of the initial microwave receptor, first, second, third, etc. generations of microwave receptors may be found. The lower the number of generations, the better the dispersal of the added microwave receptor. This sustains higher temperature uniformity and faster heating rates, improving the pyrolysis rate of the material.

Although the majority of studies into MP consider microwave receptors to be an essential part of the process, more recent investigations have proved that there is a MP of biomass which does not require the addition of microwave absorbers. Robinson et al. (2010) have confirmed that MP can be achieved without the use of carbon-rich dopants, and that the heating of water alone can be used to induce the pyrolysis of wood.

Influence of the microwave receptor on the pyrolysis products

The incorporation of receptors in the material can also be used to alter or tailor the products of pyrolysis. Apart from its chemical composition, its dielectric properties, which determine the heating mechanism, have to be considered, because all these factors will establish the interactions with volatiles.

Previous studies have shown the advantageous effects of microwave receptors on pyrolysis products. Domínguez et al. (2003) compared the effects of char and graphite as microwave receptor on yield and property of oil products. The results showed that the oil yields obtained from MP with char as receptor were higher than those obtained with graphite. Moreover, it was proved that the use of graphite instead of char as a microwave receptor, favored the cracking of large aliphatic chains to lighter species in oil products. Another example is the influence of an iron mesh as antenna in the MP of polystyrene (Hussain et al., 2010). The rate of reaction was found to depend upon the size and shape of the mesh, in such a way that when the mesh was cylindrical, the conversion was greater and the rate of reaction was much faster. On the other hand, Wan et al. (2009) studied the influence of different catalysts to improve product selectivity during the MP of corn stover and aspen wood. They found that the catalysts, apart from working as microwave absorbents to speed up heating, participate in the so-called *in situ* upgrading of pyrolytic vapors.

4.3 Other special characteristics of microwave-assisted pyrolysis

The appearance of unexpected physical behaviors such as "hot spots" and "waiting time" phenomena may occur in a material irradiated by microwaves due to the non-linear dependence of the electromagnetic and thermal properties of the material on temperature (Reimbert et al., 1996). With respect to the former, the existence of hot spots, or large temperature gradients, within the pyrolysis bed subjected to microwave heating has been suggested by a number of researchers. These high temperature sites can arise either from microwave selective heating of materials with variable dielectric properties (higher thermal absorptivity or simply a geometrical feature such as a corner or edge), or by an uneven distribution of electromagnetic field strength (Zhang et al., 2001), which is related with the cavity design. The formation of standing waves within the microwave cavity results in some regions being exposed to higher energy than others. Regarding the second phenomenon, the waiting time behavior is exhibited by materials that either respond to microwave heating only after a finite amount of time has elapsed, or after heating by conventional means, they then respond to microwave energy.

One of the main characteristics that arise from the hot spot formation is the non-uniform heating of materials, which means that different interactions may appear. There have been numerous reports that the rate of some chemical reactions can be increased and the product selectivity changed when microwave heating is used (Zhang & Hayward, 2006, Huang et al., 2010). There has also been some debate about whether this effect is simply a thermal one, due to the higher temperatures attained at hot spots, or whether they are caused by nonthermal interactions between the material and the microwave radiation (Dreyfuss & Chipley, 1980; Kozempel et al., 1998).

Interestingly, the microwave heating of carbon-based material can give rise to hot spots that appear as small sparks or electric arcs that, in some occasions, could be attributed to "microplasmas" since these plasmas are confined to a tiny space and last for a fraction of a second (Menéndez et al., 2010). Clear evidence that plasmas are indeed formed, is presented in Fig.4. The formed microplasmas can be divided in two different types according to their shape and nature (Tendero et al., 2006). On one hand, the formation of quasi-spherical plasmas (ball lightning) can be observed, such as that shown in Figure 4a. These are relatively abundant during all the experiments no matter what the temperature of the carbon bed is. On the other hand, electric arcs (or arc discharges), like that shown in the Figure 4b are more abundant at higher temperatures.

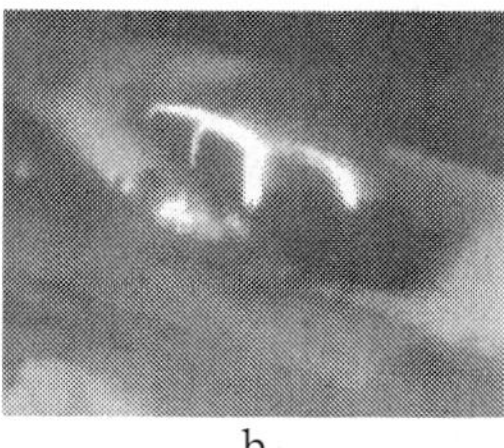

a   b

Fig. 4. Plasma formation during microwave heating of a biomass char: (a) ball lighting and (b) arc discharge

5. The state of the art in microwave-assisted pyrolysis

This section examines the state-of-the-art in the use of microwave technology as an energy-efficient alternative to current heating technologies employed in the pyrolysis process. A significant number of articles have concentrated on exposing the limitations of CP and suggesting where MP can offer both technical and financial benefits. A summary of the main advantages of the MP is presented as follows:

- Fast and uniform heating
- Selective heating
- Direct utilization of large-sized feedstocks
- Treatment of non-homogeneous wastes
- Waste reduction and material recovery
- Better production quality
- New materials and products
- Enhanced chemical reactivity

- Energy efficiency
- Overall cost effectiveness/savings
- Improved process control
- Ability to operate from an electrical source
- Ability to treat wastes in-situ
- Flexible process that can also be made remote
- Portability of equipment and process
- Reduce hazards to humans and the environment

5.1 Biomass wastes

Several pyrolysis studies have been conducted using microwaves as a heating source, with lignocellulosic feedstocks that included wood (Miura et al., 2004; Chen et al., 2008), coffee hulls (Domínguez et al., 2007), rice straw (Huang et al., 2008), waste tea (Yagmur et al., 2008) and wheat straw (Budarin et al., 2009). As a result of the aforementioned research, various products were generated, including gas, liquid and solid phase products. The primary factors affecting the product distribution and components include reaction temperature, reaction time, microwave power, particle size, additives, and the original characteristics of feedstock.

Microwave heating has also been considered as an alternative for carrying out the pyrolysis of sewage sludges (Menéndez et al., 2004). The use of microwave-assisted processes in sewage sludge handling or treatment is not entirely new. In fact, a variety of processes involving microwaves have already been investigated. These have focused mainly on the pretreatment, including enhancement of anaerobic digestion (Hong et al., 2006; Climent et al., 2007), desaturation (Wojciechowska, 2005), cell destruction (Hong et al., 2004) and solubilization of organic solids (Eskicioglu et al., 2006), but also on the vitrification of the solid residue using microwave heating for the purpose of decreasing the leaching of heavy metals (Menéndez et al., 2005). Various patent applications describing procedures or apparatus for carrying out the drying and/or pyrolysis of waste materials, including sewage sludge, have also been presented during the last 20 years (Sullivan III, 1986; Marks & Fluchel, 1991).

Several researchers have indicated that MP produces more gas and less oil than CP (Domínguez et al., 2007; 2008; Fernández et al., 2009; 2010). In addition, the major constituents in the gas pyrolysis products that can be seen as source of energy and fuel include H_2, CO, CH_4, and other light hydrocarbons. The MP of sewage sludge (Menéndez et al., 2004) produced H_2, CO and CH_4 with percentages of 22-43%, 23-30% and 2-7%, respectively. For the MP of coffee hulls (Domínguez et al., 2007), the percentages of H_2, CO and CH_4 in the gas product were 36-40%, 26-33% and 7-8%, respectively. Compared with CP, which is carried out using an electric furnace, MP produces more H_2 and CO content (Menéndez et al., 2004; Domínguez et al., 2007; Fernández et al., 2009; 2010), which is so-called syngas. The work of Huang et al. (2008) over the MP of rice straw shows that about half of rice straw sample is transformed into H_2-rich fuel gas, whose H_2, CO_2, CO, CH_4 percentages are 55, 17, 13 and 10 vol.%, respectively. These authors suggest that the difference in the production of H_2 obtained through MP, and through TA-MS analysis could be due to a hot spot generated by the focused heating of microwaves.

When considering bio-oils, MP satisfies the technical challenge directed to bring the biomass rapidly at the reaction temperature and to avoid exposition of vapor-phase pyrolysis products in a hot environment. In addition, MP is a novel approach to improve the quality and consistency of bio-oil through the *in situ* upgrading of the biomass pyrolysis vapors (evolved volatiles from thermal decomposition of organics). Three approaches of such *in situ* upgrading have been reported in the literature; (i) catalytic pyrolysis (Williams & Nugranad, 2000), where the volatiles react directly and immediately on catalysts pre-mixed with the biomass feedstock, (ii) microwave-assisted pyrolysis (Domínguez et al., 2007; 2008; Huang et al., 2008), where the microwaves are not only useful as alternative method of heating, but also cause specific molecular activations (De la Hoz et al., 2005), and recently (iii) microwave-assisted catalytic pyrolysis (Chen et al., 2008; Wan et al., 2009), which involves the combination of the former. Both the use of catalysts or microwaves during the pyrolysis process improves the yield or quality of the final products. In fact, both aspects together may be assigned to a synergistic effect in such a way that, the result of using catalysts is improved under microwave heating, and vice versa (Fernández et al., 2009). For example, Wan et al. (2009) ascribe catalysts an important role in solid-to-vapour conversion and in reforming of vapors during the pyrolysis of biomass. The use of microwaves is known to create a different temperature gradient which permits the catalyst to be at higher temperature, promoting its catalytic activity. Moreover, the appearance of hot spots is also an important factor to take into account. Their nature is not completely understood, but in a sense they can be put on a level with a catalytic effect.

Significant differences appear in the bio-oils generated with MP. Domínguez et al. (2003) carried out MP of different sewage sludges and found that the pyrolyzing oils have a high calorific value and a low proportion of compounds of considerable environmental concern, such as polycyclic aromatic hydrocarbons (PAHs). Apart from providing less hazardous compounds, MP generates some chemicals of interest in industry in higher proportions than CP. For their part, Wan et al. (2009) found that microwave-assisted catalytic pyrolysis usually produced additional water and coke-solid residue, and thus, reduces the yield of the organic phase of the bio-oil. Besides, that it is possible to control the product profiles by varying the catalysts and the dosages used. Much literature exists which have examined the use of a range of catalysts. Research has also been directed towards the design of selective catalysts for either increasing the production of specific compounds (e.g. phenols) or minimizing the formation of undesirable compounds (e.g. acids, carbonyls). However, it is still necessary to study how catalysts change the chemical profiles of bio-oils under MP conditions.

Apart from biomass wastes which have already been examined, studies have also concentrated on the main component of biomass: cellulose. Budarin et al. (2010) espouse a mechanism for the specific microwave effect on cellulose to explain the lower temperature at which carbonization occurs, the higher calorific value of the final char and the improvement in the properties of the oils produced, in comparison with CP. There are also studies that have investigated the pyrolysis or thermal cracking of glycerin to produce synthesis gas under microwave heating (Fernández et al., 2009; 2010). From those studies, it was concluded that the use of microwave heating in glycerol pyrolysis ensures a higher gas fraction with an elevated content of syngas compared to conventional heating, even at low temperatures (400 °C). It is accepted that vegetable oil residues offer little hope as a renewable fuel for compression ignition engines, due to its high viscosity, low volatility and the polyunsaturated character of triglycerides which leads to incomplete combustion and a decrease in the performance of the engine. The microwave treatment of these oil residues could present some interest as the transethylation can be significantly accelerated (Fukuda et al., 2001). The biodiesel preparation time is considerably reduced (i.e. more than 10 times in comparison with conventional heating), and its properties are similar to those of diesel. Interestingly, MP offers an alternative to most CP methods which generally require a reduction in the size of the material, largely necessary when materials have low thermal conductivity (Miura et al., 2004; Ciacci et al., 2010). Large-sized feedstocks can be heated with microwave irradiation, and thus it is possible to reduce pretreatment processing costs. A few studies of MP of wood block (Krieger-Brockett, 1994; Miura et al., 2004) and corn stalk bale (Zhao et al., 2010) demonstrate that it is possible to pyrolyze thick blocks given that the heat exchanged from the external surface is lower than the heat produced by dielectric loss. The radiation energy is dissipated within the sample more or less uniformly, giving rise to much higher heating rates, although significant temperature gradients may be established between the hot internal regions and the surface of the sample (see Fig. 5). Conversion times are also shorter than those of conventional heating. Consequently, this could result in huge reductions in the electricity consumption normally used in the grinding and shredding processes.

Fig. 5. Cross-section of wood after 3 minutes of microwave radiation (Reprinted from Miura et al., 2004, with permission from Elsevier)

5.2 Automotive industry wastes

Automotive technology consists of not only the industrial production of cars but also the aspects associated with their use, maintenance, and partial recycling. In this sense, significant amounts of automotive wastes are generated which are capable of being subjected to MP.

Currently, only about 75% of an end-of-life vehicles total weight is recycled in EU countries. The remaining 25%, which is called auto shredder residues (ASR), is mainly disposed of as landfill. ASR is a solid and extremely heterogeneous material containing metals, plastic, glass, textiles, fibre materials, wood, etc., and also hazardous compounds (PCB, cadmium, lead, etc.). Therefore, due to the complexity and diversification in chemical composition and material structure of ASR most authors (Nourreddine, 2007) would consider handling ASR via thermal treatment methods and by minimizing mechanical recycling. Donaj et al. (2010) have developed a novel and holistic approach to ASR waste management and other similar multi-component waste mixtures (electronic and electric wastes, cable residues, household appliances wastes, vehicle tires, construction and demolition waste, residues from oil refineries and petro-chemicals) based on combining both MP and the high temperature agent gasification (HTAG) process. This concept has been designed to maximize the conversion of organic contents of ASR into valuable products (carbon black, fuel gas and liquid), while preserving metals during each step of the process. Aside from the fact that the metals could be easier recovered from pyrolysed fractions than from raw ASR, the solid and liquid products after MP of ASR could be further subjected to HTSG for fuel gas production. Waste engine oil (WAO) is another environmentally hazardous and high-volume waste that is difficult to treat and dispose of due to the presence of undesirable elements such as soot, polycyclic aromatic hydrocarbons (PAHs), and impurities from additives such as chlorinated paraffins and polychlorinated biphenyls (PCBs). Existing treatment processes for WAO, such as incineration and hydro-treating, are becoming increasingly impracticable as concerns over environmental pollution and additional sludge disposal grow. Several studies have revealed the potential of pyrolysis as a disposal method for waste oil (Kim et al., 2003; Fuentes et al., 2007), and Lam et al. (2010) have conducted MP to treat and recycle WAO. They found that MP has huge potential as a means of recovering commercially valuable products from WAO. More specifically they realized the potential for recovering gaseous hydrocarbons with light olefins, and liquid hydrocarbon oils containing BTX and benzene derivatives. Apart from the significant influence of temperature on the overall yield and formation of the recovered pyrolysis gases and liquid oils, the MP process showed improved cracking reactions compared with CP processes.

Finally, much research has been undertaken to convert scrap tyres into value-added products such as olefins, chemicals and surface-activated carbons using CP. The technique permits volumetric reduction, material recovery and can render harmless many organic substances. Furthermore, the absence of oxygen means NO_x and SO_x emissions are significantly less compared to incineration. However, there are problems with this technique which are that insufficient temperatures result in incomplete pyrolysis of the forming char, making the application of microwave energy a viable alternative in heating scrap tyres more uniformly and quickly to the high temperatures required (2000 °C) (Lee, 2003).

5.3 Plastic wastes

Other non-biodegradable organic materials that represent problematic wastes on one hand, and provide valuable potential as secondary raw materials on the other hand are plastic solid wastes (PSW). These are polymers with specific chemicals and additives. Over the past seventy years, the plastic industry has witnessed a drastic growth, namely in the production of synthesis polymers represented by polyethylene (PE), polypropylene (PP), polystyrene (PS), polyethylene terephthalate (PET), polyvinyl alcohol (PVA) and polyvinyl chloride (PVC). Plastics also contribute many aspects of our daily life. Nowadays, household goods are mainly comprised of plastic or plastic reinforced materials from packaging, clothing, appliances and electrical and vehicle equipments, to insulations, industrial applications, greenhouses, automotive parts, aerospace and mulches.

Due to their use by huge populations the associated wastes are also on the rise, and the disposal of these wastes in landfill sites is not attractive, due to their non-biodegradable nature. Their incineration is costly in terms of the cost and wastage of valuable carbon resources, since the chemical value of the polymer is completely lost. When compared to existing plastic recycling processes, MP is an environmentally friendly technique whereby the hydrocarbon content of the polymers is recycled by thermal degradation to produce gaseous and liquid fuels. In addition, MP allows for the treatment of plastics that are often comprised of more than one polymer or alternatively fibres may be added to form a composite which provide extra strength and is difficult to separate, and hence recycle. On the other hand, the use of microwave heating overcomes the limitations of conventional treatments, such as the non-selective nature of conventional heating, as well as providing the high temperatures which are normally required.

MP techniques for treating plastic waste were initially developed by Tech-En Ltd in Hainault, UK (Ludlow-Palafox & Chase, 2001a). The process involves mixing plastic-containing wastes, which are known to have very high transparencies to microwaves, with a highly microwave-absorbent material such as particulate carbon (Ludlow-Palafox & Chase, 2001b). The carbon reaches temperatures around 1000 °C within a few minutes in the microwave field, and energy is transferred to the shredded plastic by conduction, providing the efficient energy-transfer associated with microwave-heating processes. Hussain et al. (2010) carried out the MP of PS using metals like iron as antenna to produce useful hydrocarbons. From their results it was concluded that MP provides a more even distribution of heat and better control over the heating process than conventional heating techniques. Ludlow-Palafox & Chase (2001b) performed a novel microwave-induced pyrolysis process to evaluate the degradation of high-density PE and aluminium/polymer laminates (toothpaste tube) in a semibatch bench-scale apparatus. The results showed that the new process has the same general features as other, more traditional, pyrolytic processes (Williams & Williams, 1997), but with the advantage that is able to deal with problematic wastes such as laminates. In the case of the toothpaste tube, the aluminum foil which was recovered was very high quality aluminium, and the organic content cracked to produce liquid and gas hydrocarbons. This trial proved that the process has excellent potential for the treatment of plastic wastes on a commercial scale.

5.4 Electronic wastes

The disposal of wastes of electrical and electronic equipments (WEEE) or electronic waste (e-waste) is an important issue worldwide. Many of the high-tech systems that are commonly used, containing electronic circuitry, are retired yearly. This increasing volume of WEEE must be disposed of in a cost effective and environmentally safe manner. Their disposal in landfills presents many drawbacks that must be avoided. These include the large and continuously increasing volume of these materials, the filling up of the available landfills, and the presence of hazardous materials contained in the electronic components which threatens the possibility of leaching. In addition, many useful materials, including precious metals, are contained within the WEEE that can be potentially re-used, providing a return on investment.

Several authors have investigated the use of microwave energy for the treatment of WEEE, and subsequently, reclamation of the important metals contained within it. Wicks et al. (1995) developed a new technology consisting of a tandem microwave system with two interconnected microwave chambers. While the first chamber treats primary waste, reducing its volume by up to 50% and effectively transforming it into glassy and metal products which are easily separated, the second chamber simultaneously treats the off-gases, significantly reducing the amount of hazardous components. Sung et al. (2010) studied the pyrolysis characteristics of WEEE under microwave irradiation, as well as the effects of microwave output power and adsorption additive. Their experiments proved that over 70% of WEEE would be decomposed through pyrolysis and the gas yield contains about 80% fuel gas such as CO, CH_4, H_2, etc. Moreover, after the pyrolysis stage, the waste glass product was found to immobilize electronic hazardous components, and the metal product formed allowed precious and other metals to be conveniently reclaimed for recycling.

A promising application of microwave technology has been reported in the processing of sediment sludge obtained from a high quality printed circuit boards (PCB) manufacturer in Northern Ireland (Gan, 2000). The process demands that the boards are washed after each individual process to avoid any cross-contamination. Consequently, a large quantity of wash water is required. Almost ninety-five percent of the metal ions are subsequently removed by means of flocculation, pH adjustment and alkaline precipitation processes. The resulting sludge is landfilled, and there is great potential for groundwater contamination. The microwave treatment involves the immobilization of the metal ions in the sludge. Volume reduction by evaporation of moisture is an important fringe benefit of this method and is shown to be particularly efficient when combined with standard convection drying.

5.5 Other microwave-assisted pyrolysis processes

Experimental investigations about MP date back to the 1980s, after the discovery of substances which heat rapidly in a microwave field (microwave receptors) inducing the pyrolysis of other materials. From that moment, several researchers deal with MP of coals of different ranks. Monsef-Mirzai et al. (1995) observed evidence that volatile material undergoes significant secondary cracking, which may be enhanced by the presence of the receptor, to still condensable products. The tar yields depend on the type of microwave receptor, e.g. up to 49 wt% with CuO, 27 wt% with Fe_3O_4, or 20 wt% with coke. Jie & Jiankang (1994) studied the behavior of coal pyrolysis desulfurization by microwave radiation. The microwave method decreases the volatile matter loss and slightly improves the desulfurization efficiency, as compared with external heating pyrolysis. Lester et al. (2006) have indicated a potential new method for coke production from high volatile bituminous coal using MP. The time taken to generate the graphitized coke under microwave irradiation is surprisingly short compared with conventional methods. They suggest that the sample has either reached temperatures well in excess of 1000 °C or that microwaves have a non-thermal effect on coal structure that increases the rate of the structural ordering of the carbon atoms. There are also studies that have evaluated the change in electric and dielectric properties of some Australian coals during the process of pyrolysis (Zubkova & Prezhdo, 2006).

Microwave irradiation has also been applied by El harfi et al. (2000) and Bilali et al. (2005) in the pyrolysis of oil shales and rock phosphate, respectively. The former found that the oils produced by microwave heating are more maltenic, less polar and contains less sulphur and nitrogen than those obtained by CP and the latter found to contain more paraffinic compounds.

Finally, some wastes are subjected to frequent variations and/or consisted of mixed materials which make the monitoring of the pyrolysis process difficult, such as municipal solid wastes (MSW). These ones are highly non-homogeneous in terms of material composition and size, and contain a large amount of ash, heavy metals and chlorine which increases the process cost and lower the quality of the products. However, studies on pyrolysis of MSW demonstrate that it is possible to produce valuable final products. To date, only conventional heating systems have been use in the pyrolysis of MSW (Buah et al., 2007; He et al., 2010), although the use of microwaves has a good potential for improving final pyrolysis products by reducing processing time and making cost savings. On the other hand, some segregated waste materials from MSW (e.g. paper, cardboard, waste wood, textiles or plastic) may also be suitable candidates for feedstock production by MP on small-scales.

5.6 Additional microwave heating applications in waste management

Microwave technology has also been employed in the sterilization of hospital wastes (Drake, 1993; Blenkharn, 1995). Hospital wastes (HW), comprising pathological, microbiological, sharps and associated contaminated waste, produce an environmental problem due to the very large volumes produced each year. The rapid and in-situ treatment of HW with microwave irradiation sterilizes and reduces it to inert ash prior to final disposal, as well as minimizes the contact with hazardous components.

In the case of nuclear wastes, there are certain components that are very good insulators and, therefore, conventional heating is not effective in their treatment. However, microwaves can selectively heat these waste mixtures, and thus, can be used for the *in situ* desaturation of low level nuclear waste, in the same waste storage vessel that the waste was deposited. This minimizes the handling of the radioactive waste and reduces the waste volume by about 5% (Oda, 1993).

Studies of thermal remediation using microwave heating enhanced by strong microwave absorbers have also been performed on soil contaminated with organic pollutants or toxic metal ions. George et al. (1992) investigated soil decontamination via microwave heating enhanced by carbon particles. The decontamination process was performed under reduced pressure and an inert gas presence to prevent the combustion of the contaminant. Their results showed that removal efficiencies of near 10% could be achieved for phenanthrene using microwave heating enhanced by 40 wt.% carbon particles. The decomposition of polycyclic aromatics and polychlorobiphenyl in soil using microwave energy, with the addition of NaOH in company with Cu_2O (powdered Al, metal wire, etc.) serving as both reaction catalysts and microwave absorbers was also investigated (Abramovitch & Huang, 1994; Abramovitch et al., 1998; 1999). The results show that the decomposition products were probably either mineralized or very highly bonded to the bed. Furthermore, Abramovitch et al., (2003) reported that, with the addition of pencil lead or iron wire, the contaminated soil could be remediated safely to preset depths without the toxic metal ions leaching out for a long time. Yuan et al. (2006) investigated the microwave remediation of the oil contaminated with hexachlorobenzene (HCB) using powdered MnO_2 as microwave receptor. Their results revealed that a complete removal of HCB was obtained with 10 min microwave treatment by the addition of 10 wt.% powdered MnO_2 and about 30 wt.% H_2SO_4 (50%). A successful remediation of crude oil-contaminated soil (99% oil removal) using carbon fibers has been recently reported by Li et al. (2009). They concluded that microwave thermal processes can not only clear rapidly up the contaminated soil, but also recover the usable oil contaminant efficiently without causing any secondary pollution.

6. Conclusions

The aim of this chapter has been to provide the reader with a fundamental understanding of microwave-assisted pyrolysis processes. The unique internal phenomenon associated with microwave energy not only leads to significant savings in both energy consumption and process time during the pyrolysis process, but also allows for the development of new chemical profiles of the volatiles, which enhances the overall quality of production. Both the influence of microwave heating on pyrolysis conditions (*distribution of temperature, heating rate and residence time of volatiles*) and raw materials have been discussed. Furthermore, special attention has been given to the microwave receptor, since it is an essential factor in any microwave heating process.

A significant number of articles have been reviewed for the purpose of demonstrating the limitations of conventional pyrolysis and suggesting where microwave-assisted pyrolysis can offer advantages. Essentially, microwave-assisted pyrolysis is a novel approach in the in-situ upgrading of pyrolysis vapors. The oils produced under this technique, offer less hazardous compounds and produce a greater number of chemicals which are of interest to industry, than those obtained under conventional pyrolysis. As equally desirable and advantageous, is the fact that the gas fraction presents higher yields of hydrogen or syngas. In essence, microwave-assisted pyrolysis offers a superb opportunity to divert waste from the present eco-damaging disposal techniques such as from landfill and incineration, and it also presents a worthy means of recovering commercially valuable products from wastes.

7. References

Abramovitch, R.A. & Huang, B. (1994). Decomposition of 4-bromobiphenyl in soil remediated by microwave energy. *Chemosphere*, Vol. 29, No. 12, (December 1994) 5 (2517-2521), 0045-6535.

Abramovitch, R.A.; Huang, B.; Mark, D. & Luke, P. (1998). Decomposition of PCBs and other polychlorinated aromatics in soil using microwave energy. *Chemosphere*, Vol. 37, No. 8, (October 1998) 10 (1427-1436), 0045-6535.

Abramovitch, R.A.; Huang, B.; Abramovitch, D.A. & Song, J. (1999). In situ decomposition of PAHs in soil desorption of organic solvents using microwave energy. *Chemosphere*, Vol. 39, No. 1, (1999) 7 (81-87), 0045-6535.

Abramovitch, R.A.; Lu, C.; Hicks, E. & Sinard, J. (2003). In situ remediation of soils contaminated with toxic metal ions using microwave energy. *Chemosphere*, Vol. 53, No. 9, (December 2003) 9 (1077-1085), 0045-6535.

Allan, G.G.; Krieger-Brockett, B. & Work, D.W. (1980). Dielectric loss microwave degradation of polymers: cellulose. *Journal of Applied Polymer Science*, Vol. 25, No. 9, (September 1980) 21 (1839-1859), 1097-4628.

Antal, M.J. (1983). Effect of reactor severity on the gas-phase pyrolysis of cellulose and kraft lignin-derived volatile matter. *Industrial & Engineering Chemistry Product Research and Development*, Vol. 22, No. 2, (June 1983) 10 (366-375), 0196-4321.

Antal, M.J. & Gronli, M.G. (2003). The art, science and technology of charcoal production. *Industrial and Engineering Chemistry Research*, Vol. 42, No. 8, (April 2003) 22 (1619-1640), 0888-5885.

Bilali, L.; Benchanaa, M.; El harfi, K.; Mokhlosse, A. & Outzourhit, A. (2005). A detailed study of the microwave pyrolysis of the Moroccan (Youssoufia) rock phosphate. *Journal of Analytical and Applied Pyrolysis*, Vol. 73, No. 1, (March 2005) 15 (1-15), 0165-2370.

Blenkharn, J.I. (1995). The disposal of clinical wastes. *Journal of Hospital Infection*, Vol. 30, No. 1, (June 1995) 7 (514-520), 0195-6701.

Branca, C. & Di Blasi, C. (2003). Kinetics of the isothermal degradation of wood in the temperature range 528-708 K. *Journal of Analytical and Applied Pyrolysis*, Vol. 67, No. 2, (May 2003) 13 (207-219), 0165-2370.

Bridgwater, A.V. (1996). Production of high grade fuels and chemicals from catalytic pyrolysis of biomass. *Catalysis Today*, Vol. 29, No. 1-4, (May 1996) 11 (285-295), 0920-5861.

Bridgwater, A.V. & Peacocke, G.V.C. (2000). Fast pyrolysis processes for biomass. *Renewable and Sustainable Energy Reviews*, Vol. 4, No. 1, (March 2010) 73 (1-73), 1364-0321.

Bridgwater, A.V. (2003). Renewable fuels and chemicals by thermal processing of biomass. *Chemical Engineering Journal*, Vol. 91, No. 2-3, (March 2003) 16 (87-102), 1385-8947.

Bua, W.K.; Cunliffe, A.M. & Williams, P.T. (2007). Characterization of products from the pyrolysis of municipal solid waste. *Process Safety and Environmental Protection*, Vol. 85, No. 5, (September 2007) 8 (450-457), 0957-5820.

Budarin, V.L.; James, H.C.; Lanigan, B.A.; Shuttleworth, P.; Breeden, S.W.; Wilson, A.J.; Macquarrie, D.J.; Milkowski, K. & Jones, J. (2009). The preparation of high-grade bio-oils through the controlled low temperature microwave activation of wheat straw. *Bioresource Technology*, Vol. 100, No. 23, (December 2009) 5 (6064-6068), 0960-8524.

Budarin, V.L.; Clark, J.H.; Lanigan, B.A.; Shuttleworth, P. & Macquarrie, D.J. (2010). Microwave assisted decomposition of cellulose: A new thermochemical route for biomass exploitation. *Bioresource Technology*, Vol. 101, No. 10, (May 2010) 4 (3776-3779), 0960-8524.

Caballero, J.A.; Font, R. & Marcilla, A. (1996). Comparative study of the pyrolysis of almond shells and their fractions, holocellulose and lignin. Product yields and kinetics. *Thermochimica Acta*, Vol. 276, (April 1996) 12 (57-77), 0040-6031.

Carpenter, J.A.Jr. (1991). Dielectric properties measurements and data. In: *Microwave Processing of Materials II. Materials Research Society Proceedings*, Snyder Jr., W.B.; Sutton, W.H.; Johnson, D.L. & Iskander, M.F. (Ed.), 11 (477-487), Materials Research Society, 1-55899-078-X, San Francisco, USA.

Cetin, E.; Gupta, R. & Moghtaderi, B. (2005). Effect of pyrolysis pressure and heating rate on radiate pine char structure and apparent gasification reactivity. *Fuel*, Vol. 84, No. 10, (July 2005) 9 (1328-1334), 0016-2361.

Chan, T.V.C.T. & Reader, H.C. (2000). Resonant Applicators. In: *Understanding Microwave Heating Cavities*, Chan, T.V.C.T. & Reader, H.C. (Ed.), 3 (12-14), Artech House, 1-58053-094-X, London, UK.

Chanaa, M.B.; Lallemant, M. & Mokhlisse, A. (1994). Pyrolysis of Timahdit, Morocco, oil shales under microwave field. *Fuel*, Vol. 73, No. 10, (October 1994) 7 (1643-1649), 0016-2361.

Chen, M.Q.; Wang, J.; Zhang, M.X.; Chen. M.G.; Zhu, X.F.; Min, F.F. & Tan, Z.C. (2008). Catalytic effects of eight inorganic additives on pyrolysis of pine wood sawdust by microwave heating. *Journal of Analytical and Applied Pyrolysis*, Vol. 82, No. 1, (May 2008) 6 (145-150), 0165-2370.

Ciacci, T.; Galgano, A. & Di Blasi, C. (2010). Numerical simulation of the electromagnetic field and the heat and mass transfer processes during microwave-induced pyrolysis of a wood block. *Chemical Engineering Science*, Vol. 65, No. 14, (July 2010) 17 (4117-4133), 0009-2509.

Climent, M.; Ferrer, I.; Baeza, M.M.; Artola, A.; Vazquez, F. & Font, X. (2007). Effects of thermal and mechanical pretreatments of secondary sludge on biogas production under thermophilic conditions. *Chemical Engineering Journal*, Vol. 133, No. 1-3, (September 2007) 8 (335-342), 1385-8947.

Cloete, J.H.; Bingle, M. & Davidson, D.B. (2001). The role of chirality and resonance in synthetic microwave absorbers. *AEU-International Journal of Electronics and Communications*, Vol. 55, No. 4, (March 2001) 7 (233-239), 1434-8411.

Conesa, J.A.; Marcilla, A.; Moral, R.; Moreno-Caselles, J. & Pérez-Espinosa, A. (1998). Evolution of the gases in the primary pyrolysis of different sewage sludges. *Thermochimica Acta*, Vol. 313, No. 1, (March 1998) 11 (63-73), 0040-6031.

Couhert, C.; Commandre, J.M. & Salvador, S. (2009a). Failure of the component additivity rule to predict gas yields of biomass in flash pyrolysis at 950°C. *Biomass & Bioenergy*, Vol. 33, No. 2, (February 2009) 11 (316-326), 0961-9534.

Couhert, C.; Commandre, J.M. & Salvador, S. (2009b). Is it possible to predict gas yields of any biomass after rapid pyrolysis at high temperature from its composition in cellulose, hemicelluloses and lignin? *Fuel*, Vol. 88, No. 3, (March 2009) 12 (408-417), 0016-2361.

De la Hoz, A.; Díaz-Ortiz, A. & Moreno, A. (2005). Microwaves in organic synthesis. Thermal and non-thermal microwave effect. *Chemical Society Reviews*, Vol. 34, No. 2, (February 2005) 15 (164-178), 0306-0012.

Demirbas, A. (2002). Partly chemical analysis of liquid fraction of flash pyrolysis products from biomass in the presence of sodium carbonate. *Energy Conversion and Management*, Vol. 43, No. 14, (September 2002) 8 (1801-1809), 0196-8904.

Dernovsek, O.; Naeini, A.; Preu, G.; Wersing, W.; Eberstein, M. & Schiller, W.A. (2001). LTCC glass-ceramic composites for microwave application. *Journal of the European Ceramic Society*, Vol. 21, No. 10-11, (February 2001) 5 (1693-1697), 0955-2219.

Dreyfuss, M.S. & Chipley, J.R. (1980). Comparison of effects of sublethal microwave radiation and conventional heating on the metabolic activity of Staphylococcus aureus. *Applied and Environmental Microbiology*, Vol. 39, No. 1, (January 1980) 4 (13-16), 0099-2240.

Domínguez, A.; Menéndez, J.A.; Inguanzo, M.; Bernad, P.L. & Pis, J.J. (2003). Gas chromatographic-mass spectrometric study of the oil fractions produced by microwave-assisted pyrolysis of different sewage sludges. *Journal of Chromatography A*, Vol. 1012, No. 2, (September 2003) 14 (193-206), 0021-9673.

Domínguez, A.; Menéndez, J.A.; Fernández, Y.; Pis, J.J.; Valente Nabais, J.M.; Carrott, P.J.M. & Ribeiro Carrott, M.M.L. (2007). Conventional and microwave induced pyrolysis of coffee hulls for the production of a hydrogen rich fuel gas. *Journal of Analytical and Applied Pyrolysis*, Vol. 79, No. 1-2, (May 2007) 8 (128-135), 0165-2370.

Domínguez, A.; Fernández, Y.; Fidalgo, B.; Pis, J.J. & Menéndez, J.A. (2008). Bio-gas production with low concentration of CO_2 and CH_4 from microwave-induced pyrolysis of wet and dried sewage sludge. *Chemosphere,* Vol. 70, No. 3, (January 2008) 7 (397-403), 0045-6535.

Donaj, P.; Yang, W.; Blasiak, W. & Forsgren, C. (2010). Recycling of automobile shredder residue with a microwave pyrolysis combined with high temperature steam gasification. *Journal of Hazardous Materials,* Vol. 182, No 1-3, (October 2010) 10 (80-89), 0304-3894.

Drake, C. (1993). Apparatus for sterilizing medical waste by microwave autoclaving, US Patent 5223231.

Dufour, A.; Girods, P.; Masson, E.; Rogaume, Y. & Zoulalian, A. (2009) Synthesis gas production by biomass pyrolysis: Effect of reactor temperature on product distribution. *International Journal of Hydrogen Energy,* Vol. 34, No. 4, (February 2009) 9 (1726-1734), 0360-3199.

El harfi, K.; Mohklisse, A.; Chanâa, M.B. & Outzourhit, A. (2000). Pyrolysis of the Moroccan (Tarfaya) oil shales under microwave irradiation. *Fuel,* Vol. 79, No. 7, (May 2000) 11 (733-742), 0016-2361.

Elías, X. (2005). La pirólisis, In: *Tratamiento y Valorización Energética de Residuos,* 63 (477-539), Díaz de Santos & Fundación Universitaria Iberoamericana, 84-7978-694-6, Barcelona, Spain.

Eskicioglu, C.; Kennedy, K.J. & Droste, R.L. (2006) Characterization of soluble organic matter of waste activated sludge before and after thermal pretreatment. *Water Research,* Vol. 40, No. 20, (December 2006) 11 (3725-3736), 0043-1354.

Faaij. A.P.C. (2006). Bio-energy in Europe: changing technology choices. *Energy Policy,* Vol. 34, No. 3, (February 2006) 21 (322-343), 0301-4215.

Feirabend, H.K.P. & Ploeger, S. (1991). Microwave applications in classical staining methods in formalin-fixed human brain tissue: a comparison between heating with microwave and conventional ovens. *European Journal of Morphology.,* Vol. 29, No. 3, (August 1991) 15 (183-197), 0924-3860.

Feirabend, H.K.P.; Ploeger, S.; Kok, P. & Choufoer, H. (1992). Does microwave irradiation have other than thermal effects on histological staining of the mammalian CNS?, *European Journal of Morphology.,* Vol. 30, No. 4, (November 1992) 16 (312-327), 0924-3860.

Fernández, Y.; Arenillas, A.; Díez, M.A.; Pis, J.J. & Menéndez, J.A. (2009). Pyrolysis of glycerol over activated carbons for syngas production. *Journal of Analytical and Applied Pyrolysis,* Vol. 84, No. 2, (March 2009) 6 (145-150), 0165-2370.

Fernández, Y.; Arenillas, A.; Bermúdez, J.M. & Menéndez, J.A. (2010). Comparative study of conventional and microwave-assisted pyrolysis, steam and dry reforming of glycerol for syngas production, using a carbonaceous catalyst. *Journal of Analytical and Applied Pyrolysis,* Vol. 88, No. 2, (July 2010) 5 (155-159), 0165-2370.

Fuentes, M.J.; Font, R.; Gómez-Rico, M.F. & Martín-Gullón, I. (2007). Pyrolysis and combustion of waste lubricant oil from diesel cars: decomposition and pollutants. *Journal of Analytical and Applied Pyrolysis,* Vol. 79, No. 1-2, (May 2007) 12 (215-226), 0165-2370.

Fukada, H.; Kondo, A. & Noda, H. (2001). Biodiesel fuel production by transesterification of oils. *Journal of Bioscience and Bioengineering*, Vol. 92, No. 5, (May 2001) 12 (405-416), 1389-1723.

Gabriel, C.; Gabriel, S.; Grant, E.H.; Halstead, B.S.J. & Mingos, D.M.P. (1998). Dielectric parameters relevant to microwave dielectric heating. *Chemical Society Reviews*, Vol. 27, No. 3, (May 1998) 11 (213-223), 0306-0012.

Gan, A. (2000). A case study of microwave processing of metal hydroxide sediment sludge from printed circuit board manufacturing wash water. *Waste Management*, Vol. 20, No. 8, (December 2000) 7 (695-701), 0956-053X.

George, C.E.; Lightsey, G.R.; Jun, I. & Fan, J. (1992). Soil decontamination via microwave and radio frequency co-volatilization. *Environmental Progress*, Vol. 11, No. 2, (March 1992) 4 (216-219), 0278-4491.

Guillon, P. (1994). Microwave techniques for measuring complex permittivity and permeability of materials, In: *Microwave processing of materials IV. Material Research Society Proceedings*, Iskander, M.F.; Lauf, R.J. & Sutton, W.H. (Ed.), 11 (79-89), Materials Research Society, 1-55899-247-2, San Francisco, USA.

Kelly, R.M. & Rowson, N.A. (1995). Microwave reduction of oxidised ilmenite concentrates. *Minerals Engineering*, Vol. 8, No. 11, (November 1995) 11 (1427-1438), 0892-6875.

Kim, S.S.; Chun, B.H. & Kim, S.H. (2003). Non-isothermal pyrolysis of waste automobile lubricating oil in a stirred batch reactor. *Chemical Engineering Journal*, Vol. 93, No. 3, (July 2003) 7 (225-231), 1385-8947.

Kozempel, M.F.; Annous, B.A.; Cook, R.D.; Scullen, O.J. & Whitting, R.C. (1998). Inactivation of microorganisms with microwaves at reduced temperatures. *Journal of Food Protection*, Vol. 61, No. 5, (May 1998) 4 (582-585), 0362-028X.

Krieger-Brockett, B. (1994). Microwave pyrolysis of biomass. *Research on Chemical Intermediates*, Vol. 20, No. 1, (January 1994) 10 (39-49), 1568-5675.

He, M.; Xiao, B.; Liu, S.; Hu, Z.; Guo, X.; Luo, S. & Yang, F. (2010). Syngas production from pyrolysis of municipal solid waste (MSW) with dolomite as downstream catalysts. *Journal of Analytical and Applied Pyrolysis*, Vol. 87, No. 2, (March 2010) 7 (181-187), 0165-2370.

Hong, S.M.; Park, J.K. & Lee, Y.O. (2004). Mechanisms of microwave irradiation involved in the destruction of fecal coliforms from biosolids. *Water Research*, Vol. 38, No. 6, (March 2006) 11 (1615-1625), 0043-1354.

Hong, S.M.; Park, J.K.; Teeradej, N.; Lee, Y.O.; Cho, Y.K. & Park, C.H. (2006). Pretreatment of sludge with microwaves for pathogen destruction and improved anaerobic digestion performance. *Water Environment Research*, Vol. 78, No. 1, (January 2006) 8 (76-83), 1061-4303.

Huang, Y.F.; Kuan, W.H.; Lo, S.L. & Lin, C.F. (2008). Total recovery of resources and energy from rice straw using microwave-induced pyrolysis. *Bioresource Technology*, Vol. 99, No. 17, (November 2008) 7 (8252-8258), 0960-8524.

Hussain, Z.; Khan, K.M. & Hussain, K. (2010). Microwave-metal interaction pyrolysis of polystyrene. *Journal of Analytical and Applied Pyrolysis*, Vol. 89, No. 1, (September 2010) 5 (39-43), 0165-2370.

Jie, W. & Jiankang, Y. (1994). Behaviour of coal pyrolysis desulfurization with microwave energy. *Fuel,* Vol. 73, No. 2, (February 1994) 5 (155-159), 0016-2361.

Kelly, R.M. & Rowson, N.A. (1995). Microwave reduction of oxidised ilmenite concentrates. *Minerals Engineering,* Vol. 8, No. 1, (November 1995) 10 (1427-1438), 0892-6875.

Lam, S.S.; Russell, A.D. & Chase, H.A. (2010). Microwave pyrolysis, a novel process for recycling waste automotive engine oil. *Energy,* Vol. 35, No. 7, (July 2010) 7 (2985-2991), 0360-5442.

Lee, A. (2003). Waving aside tyre incineration. *The Engineer,* Vol. 291, (March 2003) 2 (12-13).

Lester, E.; Kingman, S.; Dodds, C. & Patrick, J. (2006). The potential for rapid coke making using microwave energy. *Fuel,* Vol. 85, No. 14-15, (October 2006) 7 (2057-2063), 0016-2361.

Li, D.; Zhang, Y.; Quan, X. & Zhao, Y. (2009). Microwave thermal remediation of crude oil contaminated soil enhanced by carbon fiber. *Journal of Environmental Sciences,* Vol. 21, No. 9, (February 2009) 6 (1290-1295), 1001-0742.

Lulow-Palafox, C. & Chase, H.A. (2001a). Wave goodbye to plastic wastes. *Chemical Engineering,* Vol. 717, No. 1, (January 2001) 2 (28-29), 1385-8947.

Lulow-Palafox, C. & Chase, H.A. (2001b). Microwave induced pyrolysis of plastic wastes. *Industrial & Engineering Chemical Research,* Vol. 40, No. 22, (October 2001) 8 (4749-4756), 0888-5885.

Marken, F.; Sur, U.K.; Coles, B.A. & Compton, R.G. (2006) Focused microwaves in electrochemical processes. *Electrochimica Acta,* Vol. 51, No. 11, (February 2006) 9 (2195-2203), 0013-4686.

Marks, D.B. & Fluchel, D.G. (1991). Microwave sludge drying apparatus and method. US Patent 5003143.

Menéndez, J.A.; Menéndez, E.M.; García, A.; Parra, J.B. & Pis, J.J. (1999). Thermal treatment of active carbons: A comparison between microwave and electrical heating. *International Microwave Power Institute,* Vol. 34, No. 3, (March 1999) 7 (137-143), 0832-7823.

Menéndez, J.A.; Domínguez, A.; Inguanzo, M. & Pis, J.J. (2004). Microwave pyrolysis of sewage sludge: analysis of the gas fraction. *Journal of Analytical and Applied Pyrolysis,* Vol. 71, No. 2, (June 2004) 11 (657-667), 0165-2370.

Menéndez, J.A.; Domínguez, A.; Inguanzo, M. & Pis,J.J. (2005). Microwave-induced drying, pyrolysis and gasification (MWDPG) of sewage sludge: Vitrification of the solid residue. *Journal of Analytical and Applied Pyrolysis,* Vol. 74, No. 1-2, (August 2005) 7 (406-412), 0165-2370.

Menéndez, J.A.; Domínguez, A.; Fernández, Y. & Pis, J.J. (2007). Evidence of self-gasification during the microwave-induced pyrolysis of coffee hulls. *Energy & Fuels,* Vol. 21, No. 1, (January 2007) 6 (373-378), 0887-0624.

Menéndez, J.A.; Arenillas, A.; Fidalgo, B.; Fernández, Y.; Zubizarreta, L.; Calvo, E.G. & Bermúdez, J.M. (2010). Microwave heating processes involving carbon materials. *Fuel Processing Technology,* Vol. 91, No. 1, (January 2010) 8 (1-8), 0378-3820.

Meredith, R.J. (1998). *Engineers' Handbook of Industrial Microwave Heating*, The Institution of Electrical Engineers, 0852969163, London, UK.

Mijovic, J. & Wijaya, J. (1990). Review of cure of polymers and composites by microwave energy. *Polymer Composites*, Vol. 11, No. 3, (June 1990) 7 (184-190), 1548-0569.

Mingos, D.M.P. & Baghurst, D.R. (1991). Applications of microwave dielectric heating effects to synthetic problems in chemistry. *Chemical Society Reviews*, Vol. 20, No. 1, (January 1991) 47 (1-47), 0306-0012.

Minkova, V.; Razvigorova, M.; Bjornbom, E.; Zanzi, R.; Budinova, T & Petrov, N. (2001). Effect of water vapour and biomass nature on the yield and quality of the pyrolysis products from biomass. *Fuel Processing and Technology*, Vol. 70, No. 1, (April 2001) 9 (53-61), 0378-3820.

Miura, M.; Kaga, H.; Sakurai, A.; Kakuchi, T. & Takahashi, K. (2004). Rapid pyrolysis of wood block by microwave heating. *Journal of Analytical and Applied Pyrolysis*, Vol. 71, No. 1, (March 2004) 13 (187-199), 0165-2370.

Mohan, D.; Pittman, C.U. & Steele, P.H. (2006). Pyrolysis of wood/biomass for bio-oil: a critical review. *Energy Fuels*, Vol. 20, No. 3, (May 2006) 42 (848-889), 0887-0624.

Monsef-Mirzai, P.; Ravindran, M.; McWhinnie, W.R. & Burchill, P. (1995). Rapid microwave pyrolysis of coal: Methodology and examination of the residual and volatile phases. *Fuel*, Vol. 74, No. 1, (January 1995) 8 (20-27), 0016-2361.

Nourreddine, M. (2007). Recycling of auto shredder residue. *Journal of Hazardous Materials*, Vol. 139, No. 3, (January 2007) 10 (481-490), 0304-3894.

Oda, S.J. (1993). Microwave applications to process nuclear and hazardous waste, In: *Microwaves: Theory and Applications in Materials Processing II (Ceramic transactions)*, Clark, D.E.; Tinga, W.R. & Laia, J.R. (Ed.), 7 (73-79), American Ceramic Society, 1-55899-247-2, San Francisco, USA.

Raveendran, K.; Ganesh, A. & Khilar, K.C. (1995). Influence of mineral matter on biomass pyrolysis characteristics. *Fuel*, Vol. 74, No. 12, (December 1995) 11 (1812-1822), 0016-2361.

Reimbert, C.G.; Minzoni, A.A. & Smyth, N.F. (1996). Effect of radiation losses on hotspot formation and propagation in microwave heating. *IMA-Journal of Applied Mathematics*, Vol. 57, No. 2, (June 1996) 15 (165-179), 0111-757X.

Robinson, J.P.; Kingman, S.W.; Barranco, R.; Snape, C.E. & Al-Sayegh, H. (2010). Microwave pyrolysis of wood pellets. *Industrial & Engineering Chemical Research*, Vol. 49, No. 2, (January 2010) 5 (459-463), 0888-5885.

Savova, D.; Apak, E.; Ekinci, E.; Yardin, F.; Petrov, N.; Budinova, T.; Razvigorova, M. & Minkova, V. (2001). Biomass conversion to carbon adsorbents and gas. *Biomass & Bioenergy*, Vol. 21, No. 2, (August 2001) 10 (133-142), 0961-9534.

Sullivan III, D.W. (1986). Sewage treatment method and apparatus. US Patent 4592291.

Sung, J.; Wang, W.; Ma, C. & Dong, Y. (2010). Study on pyrolysis characteristics of electronic waste. *Chemical, Biological and Environmental Engineering*, doi: 10.1142/9789814295048_0003.

Tendero, C.; Tixier, C.; Tristant, P.; Desmaison, J. & Leprince, P. (2006) Atmospheric presure plasmas: A review. *Spectrochimica Acta Part B*, Vol. 61, No. 1, (January 2006) 28 (2-30), 0584-8547.

Thostenson, E.T. & Chou, T.W. (1999). Microwave processing: fundamentals and applications. *Composites: Part A*, Vol. 30, No. 9, (September 1999) 17 (1055-1071), 1359-835X.

Tinga, W.R. (1992). Rapid high temperature measurement of microwave dielectric properties, In: *Microwave Processing of materials III (Ceramic Transactions)*, Beatty, R.L.; Iskander, M.F. & Sutton, W.H. (Ed.), 12 (505-516), Materials Research Society, 1-55899-247-2, San Francisco, USA.

Tsai, W.T.; Lee, M.K. & Chang, Y.M. (2007). Fast pyrolysis of rice husk: Product yields and compositions. *Bioresource Technology*, Vol. 98, No. 1, (January 2007) 7 (22-28), 0960-8524.

van Loock, W. (2006). European regulations, safety issues in RF and microwaves power. In: *Advances in Microwave and Radio Frecuency Processing*, Willert-Porada, M. (Ed.), 8 (85-92), Springer, 3-540-43252-3, Netherlands.

Wan, Y.; Chen, P.; Zhang, B.; Yang, C.; Liu, Y.; Lin, X. & Ruan, R. (2009). Microwave-assisted pyrolysis of biomass: Catalysts to improve product selectivity. *Journal of Analytical and Applied Pyrolysis*, Vol. 86, No. 1, (September 2009) 7 (161-167), 0165-2370.

Wang, J.; Zhang, M.; Chen, M.; Min, F.; Zhang, S.; Ren, Z. & Yan, Y. (2006). Catalytic effects of six inorganic compounds on pyrolysis of three kinds of biomass. *Thermochimica Acta*, Vol. 444, No. 1, (May 2006) 5 (110-114), 0040-6031.

Wicks, G.G.; Clark, D.E.; Schulz, R.L. & Folz, D.C. (1995). Microwave technology for waste management applications including disposition of electronic circuitry, In: *Microwave: Theory &Applications in Materials Processing III (Ceramics Transactions)*, Clark, D.E.; Folz, D.C.; Oda, S.J. & Silberglitt, R. (Ed.), 11 (79-89), Materials Research Society, 1-55899-247-2, San Francisco, USA.

Williams, E.A. & Williams, P.T. (1997). Analysis of products derived from the fast pyrolysis of plastic waste. *Journal of Analytical and Applied Pyrolysis*, Vol. 40-41, (May 2007) 17 (347-363), 0165-2370.

Williams, P.T. & Nugranad, N. (2000). Comparison of products from the pyrolysis and catalytic pyrolysis of rice husks. *Energy*, Vol. 25, No. 6, (June 2000) 11 (493-513), 0360-5442.

Wojciechowska, E. (2005). Application of microwaves for sewage sludge conditioning. *Water Research*, Vol. 39, No. 19, (November 2005) 6 (4749-4754), 0043-1354.

Yagmur, E.; Ozmak, M. & Aktas, Z. (2008). A novel method for production of activated carbon from waste tea by chemical activation with microwave energy. *Fuel*, Vol. 87, No. 15-16, (November 2008) 8 (3278-3285), 0016-2361.

Yuan, S.H.; Tian, M. & Lu, X.H. (2006). Microwave remediation of soil contaminated with hexachlorobenzene. *Journal of Hazardous Materials*, Vol. 137, No. 2, (September 2006) 8 (878-885), 0304-3894.

Zhang, X.L.; Hayward, D.O. & Mingos, D.M.P. (2001). Microwave dielectric heating behaviour of supported MoS_2 and Pt catalyst. *Industrial Engineering Chemistry Research*, Vol. 40, No. 13, (June 2001) 8 (2810-2817), 0888-5885.

Zhang, X. & Hayward, D.O. (2006). Applications of microwave dielectric heating in environment-related heterogeneous gas-phase catalytic systems. *Inorganica Chimica Acta*, Vol. 359, No. 11, (August 2006) 13 (3421-3433), 0020-1693.

Zhao, X.; Song, Z. & Liu, H. (2010). Microwave pyrolysis of corn stalk bale: A promising method for direct utilization of large-sized biomass and syngas production. *Journal of Analytical and Applied Pyrolysis*, Vol. 89, No. 1, (September 2010) 8 (87-94), 0165-2370.

Zlotorzynski, A. (1995). The application of microwave radiation to analytical and environmental chemistry. *Critical Reviews in Analytical Chemistry*, Vol. 25, No. 1, (January 1995) 34 (43-76), 1040-8347.

Zubkova, V. & Prezhdo, V. (2006). Change in electric and dielectric properties of some Australian coals during the processes of pyrolysis. *Journal of Analytical and Applied Pyrolysis*, Vol. 75, No. 2, (March 2006) 10 (140-149), 0165-2370.